USEFUL CONSTANTS

Avogadro's number 6.02×10^{23} molecules $\cdot$ mol^{-1}

Gas constant (R)   8.314 J $\cdot$ K^{-1} $\cdot$ mol^{-1}

Faraday ($\mathcal{F}$)   96,485 J $\cdot$ V^{-1} $\cdot$ mol^{-1}

Kelvin (K) $^{\circ}$C + 273

KEY EQUATIONS

Henderson–Hasselbalch equation

$$pH = pK + \log \frac{[A-]}{[HA]}$$

Michaelis–Menten equation

$$v_0 = \frac{V_{max}[S]}{K_M + [S]}$$

Lineweaver–Burk equation

$$\frac{1}{v_0} = \left(\frac{K_M}{V_{max}}\right)\frac{1}{[S]} + \frac{1}{V_{max}}$$

Nernst equation

$$\mathcal{E} = \mathcal{E}^{\circ\prime} - \frac{RT}{n\mathcal{F}} \ln \frac{[A_{reduced}]}{[A_{oxidized}]} \quad \text{or} \quad \mathcal{E} = \mathcal{E}^{\circ\prime} - \frac{0.026\text{ V}}{n} \ln \frac{[A_{reduced}]}{[A_{oxidized}]}$$

Thermodynamics equations

$$\Delta G = \Delta H - T\Delta S$$

$$\Delta G^{\circ\prime} = -RT \ln K_{eq}$$

$$\Delta G = \Delta G^{\circ\prime} + RT \ln \frac{[C][D]}{[A][B]}$$

$$\Delta G^{\circ\prime} = -n\mathcal{F}\Delta\mathcal{E}^{\circ\prime}$$

Pratt's Essential Biochemistry

Global Edition

CHARLOTTE W. PRATT
Seattle Pacific University

KATHLEEN CORNELY
Providence College

WILEY

ISBN 978-1-119-45112-9

Printed in Singapore by Markono Print Media Pte Ltd

10   9   8   7   6   5   4   3   2

CHARLOTTE PRATT received a B.S. in biology from the University of Notre Dame and a Ph.D. in biochemistry from Duke University. She is a protein chemist who has conducted research in blood coagulation and inflammation at the University of North Carolina at Chapel Hill. She is currently Associate Professor in the Biology Department at Seattle Pacific University. Her interests include molecular evolution, enzyme action, and the relationship between metabolic processes and disease. She has written numerous research and review articles, has worked as a textbook editor, and is a co-author, with Donald Voet and Judith G. Voet, of *Fundamentals of Biochemistry*, published by John Wiley & Sons, Inc.

KATHLEEN CORNELY holds a B.S. in chemistry from Bowling Green (Ohio) State University, an M.S. in biochemistry from Indiana University, and a Ph.D. in nutritional biochemistry from Cornell University. She currently serves as a Professor in the Department of Chemistry and Biochemistry at Providence College where she has focused on expanding the use of case studies and guided inquiry across a broad spectrum of classes. Her interest in active pedagogy has led to her involvement in national programs including Project Kaleidoscope, the POGIL Project, and the Howard Hughes Medical Institute SEA PHAGES program, which has also fueled her current experimental research in phage genomics. She has been a member of the editorial board of *Biochemistry and Molecular Biology Education* and has served for several years as coordinator of the undergraduate poster competition at the annual meeting of the American Society for Biochemistry and Molecular Biology.

Brief Contents

Contents

Several years ago, we set out to write a short biochemistry textbook that combined succinct, clear chapters with extensive problem sets. We believed that students would benefit from a modern approach involving broad but not overwhelming coverage of biochemical facts, focusing on the chemistry behind biology, and providing students with practical knowledge and problem-solving opportunities. Our experience in the classroom continues to remind us that effective learning also requires students to become as fully engaged with the material as possible. To that end, we have embraced a strategy of posing questions and suggesting study activities throughout each chapter, so that students will not simply read and memorize but will explore and discover on their own—a truer reflection of how biochemists approach their work in the laboratory or clinic.

As always, we view our textbook as a guidebook for students, providing a solid foundation in biochemistry, presenting complete, up-to-date information, and showing the practical aspects of biochemistry as it applies to human health, nutrition, and disease. We hope that students will develop a sense of familiarity and comfort as they encounter new material, explore it, and test their understanding through problem solving.

New to Global Edition

Many details in the text and illustration program have been updated, with virtually no section left untouched. Some significant changes are worth mentioning: Chapter 3 includes an updated discussion of genomics and a completely new presentation of DNA sequencing technologies and the use of CRISPR-Cas to edit genes. Other new items include a discussion of archaeal lipids, details on the GLUT membrane transport protein, a box on exosomes, new illustrations of respiratory cilia and bacterial peptidoglycan, new molecular graphics of mitochondrial respiratory complexes, an updated presentation of the ribonucleotide reductase mechanism, and more information on the microbiome, cancer, and obesity. Descriptions of DNA replication and transcription have been extensively modified, with numerous new diagrams to present a more realistic picture of these processes. The histone code and readers, writers, and erasers are explained. New details on RNA splicing and protein translocation round out the revised text.

Eight health-related topics that were previously confined to short boxes have been updated and expanded to **Clinical Connection** sections to give them the appropriate attention: 2.5 Acid–Base Balance in Humans, 4.5 Protein Misfolding and Disease, 5.2 Hemoglobin Variants, 6.5 Blood Coagulation, 7.4 Drug Development, 13.5 Disorders of Carbohydrate Metabolism, 19.4 Cancer Metabolism, and 20.4 Cancer as a Genetic Disease.

With the same goal of making it easy for students to navigate complex topics, some material within sections has been reorganized, and several new sections of text now focus on key content areas: 14.3 Thermodynamics of the Citric Acid Cycle, 17.1 Lipid Transport, 18.5 Nucleotide Metabolism, 20.5 DNA Packaging, 21.1 Initiating Transcription, and 22.1 tRNA and the Genetic Code.

Above all, the focus of the global edition is ease of use, particularly for students and instructors taking advantage of new ways to assess student understanding. The **end-of-chapter problem sets** have been refreshed, with a total of 1,489 problems (averaging 68 per chapter). Problems are grouped by section and offered in pairs, with the answers to odd-numbered problems provided in an appendix.

Traditional Pedagogical Strengths

- "Do You Remember?" **review questions** start each chapter, to help students tie new topics to what they have already studied.
- **Key Concepts** at the beginning of each section prompt students to recognize the important takeaways or concepts in each section, providing the scaffolding for understanding by better defining these important points.
- **Concept Check** questions appear at the end of every section and are designed for students to check their mastery of the section's key concepts.
- **Figure Questions** that accompany key tables and figures prompt students to inspect information more closely.
- **Key Sentences** summarizing main points are printed in italics to assist with quick visual identification.
- **Tools and Techniques Sections** appear at the end of Chapters 2, 3, and 4, to showcase practical aspects of biochemistry and provide an overview of experimental techniques that students will encounter in their reading or laboratory experience.
- **Metabolism Overview Figures** introduced in Chapter 12 and revisited in subsequent chapters help students place individual metabolic pathways into a broader context.
- Chapter **Summaries**, organized by major section headings, highlight important concepts to guide students to the most important points within each section.
- **Key Terms** are in boldface. Their definitions are also included in the **Glossary**.
- An annotated list of **Selected Readings** for each chapter includes recent short papers, mostly reviews, that students are likely to find useful as sources of additional information.

Organization

We have chosen to focus on aspects of biochemistry that tend to receive little coverage in other courses or present a challenge to many students. Thus, in this textbook, we devote proportionately more space to topics such as acid–base chemistry, enzyme mechanisms, enzyme kinetics, oxidation–reduction reactions, oxidative phosphorylation, photosynthesis, and the enzymology of DNA replication, transcription, and translation. At the same time, we appreciate that students can become overwhelmed with information. To counteract this tendency, we have intentionally left out some details, particularly in the chapters on metabolic pathways, in order to emphasize some general themes, such as the stepwise nature of pathways, their evolution, and their regulation.

The 22 chapters of *Essential Biochemistry* are relatively short, so that students can spend less time reading and more time extending their learning through active problem-solving. Most of the problems require some analysis rather than simple recall of facts. Many problems based on research data provide students a glimpse of the "real world" of science and medicine.

Although each chapter of *Essential Biochemistry, Global Edition* is designed to be self-contained so that it can be covered at any point in the syllabus, the 22 chapters are organized into four parts that span the major themes of biochemistry, including some chemistry background, structure–function relationships, the transformation of matter and energy, and how genetic information is stored and made accessible.

Part 1 of the textbook includes an introductory chapter and a chapter on water. Students with extensive exposure to chemistry can use this material for review. For students with little previous experience, these two chapters provide the chemistry background they will need to appreciate the molecular structures and metabolic reactions they will encounter later.

Part 2 begins with a chapter on the genetic basis of macromolecular structure and function (Chapter 3). This is followed by chapters on structure of proteins (Chapter 4) and function of proteins (Chapter 5), with coverage of myoglobin and hemoglobin, and cytoskeletal and motor proteins. An explanation of how enzymes work (Chapter 6) precedes a discussion of enzyme kinetics (Chapter 7), an arrangement that allows students to grasp the importance of enzymes and to focus on the chemistry of enzyme-catalyzed reactions before delving into the more quantitative aspects of enzyme kinetics. A chapter on lipid chemistry (Chapter 8, Lipids and Biological Membranes) is followed by two chapters that discuss critical biological functions of membranes (Chapter 9, Transport through Membrane, and Chapter 10, Biochemical Signaling). The section ends with a chapter on carbohydrate chemistry (Chapter 11), completing the survey of molecular structure and function.

Part 3 begins with an introduction to metabolism that provides an overview of fuel acquisition, storage, and mobilization as well as the thermodynamics of metabolic reactions (Chapter 12). This is followed, in traditional fashion, by chapters on glucose and glycogen metabolism (Chapter 13); the citric acid cycle (Chapter 14); electron transport and oxidative phosphorylation (Chapter 15); the light and dark reactions of photosynthesis (Chapter 16); lipid catabolism and biosynthesis (Chapter 17); and pathways involving nitrogen-containing compounds, including the synthesis and degradation of amino acids, the synthesis and degradation of nucleotides, and the nitrogen cycle (Chapter 18). The final chapter of Part 2 explores the integration of mammalian metabolism, with extensive discussions of hormonal control of metabolic pathways, disorders of fuel metabolism, and cancer (Chapter 19).

Part 4, the management of genetic information, includes three chapters, covering DNA replication and repair (Chapter 20), transcription and RNA (Chapter 21), and synthesis of proteins (Chapter 22). Because these topics are typically also covered in other courses, Chapters 20–22 emphasize the relevant biochemical details, such as topoisomerase action, nucleosome structure, mechanisms of polymerases and other enzymes, structures of accessory proteins, proofreading strategies, and chaperone-assisted protein folding.

Flow of Genetic Information

The book's companion site (http://www.wiley.com/college/pratt) contains an extensive set of resources for enhancing student understanding of biochemistry.

Student Resources

Bioinformatics Projects   A set of newly updated exercises by Paul Craig, Rochester Institute of Technology, covering the contents and uses of databases related to nucleic acids, protein sequences, protein structures, enzyme inhibition, and other topics. The exercises use real data sets, pose specific questions, and prompt students to obtain information from online databases and to access the software tools for analyzing such data.

Instructor Resources

Image PowerPoint Slides   contain all images and tables in the text, optimized for viewing onscreen.

Digital Image Library   with images from the text, also available in JPEG format, can be used to customize presentations and provide additional visual support for quizzes and exams.

Lecture Power Points   by Mary Peek, Georgia Tech University, contain lecture outlines, Key Concepts, and relevant figures from the text.

Test Bank   by Scott Lefler, Arizona State University, contains over 1100 questions, with each multiple choice question keyed to the relevant section in the text and rated by difficulty level. A computerized version of the Test Bank is also available.

Classroom Response Questions ("Clicker Questions") by Gail Grabner, University of Texas at Austin, consist of interactive questions for classroom response systems to facilitate classroom participation and discussion. These questions can also be used by instructors as prelecture questions that help gauge students' knowledge of overall concepts, while addressing common misconceptions.

Solutions to all problems are available. Instructor resources are password protected.

Acknowledgments

We would like to thank everyone who helped develop *Essential Biochemistry, Global Edition,* including Biochemistry Editor Joan Kalkut, Product Designer Sean Hickey, Associate Development Editor Laura Rama, Senior Production Editor Elizabeth Swain, Senior Designer Tom Nery, and Senior Photo Editor Billy Ray.

We also thank all the reviewers who provided essential feedback on manuscript and media, corrected errors, and made valuable suggestions for improvements that have been so important in the writing and development of *Essential Biochemistry, Global Edition.*

Global Edition Reviews

Arkansas
Anne Grippo, *Arkansas State University*

California
Rakesh Mogul, *California State Polytechnic University, Pomona*
Brian Sato, *University of California, Irvine*

Colorado
Andrew Bonham, *Metropolitan State University of Denver*

Hawaii
Jon-Paul Bingham, *University of Hawaii-Manoa, College of Tropical Agriculture and Human Resources*

Florida
David Brown, *Florida Gulf Coast University*

Georgia
Chavonda Mills, *Georgia College*
Rich Singiser, *Clayton State University*

Illinois
Jon Friesen, *Illinois State University*
Stanley Lo, *Northwestern University*
Kristi McQuade, *Bradley University*

Indiana
Mohammad Qasim, *Indiana University Purdue, Fort Wayne*

Kansas
Peter Gegenheimer, *The University of Kansas*

Louisiana
Jeffrey Temple, *Southeastern Louisiana University*

Michigan
Kathleen Foley, *Michigan State University*
Deborah Heyl-Clegg, *Eastern Michigan University*

Mississippi
Arthur Chu, *Delta State University*

Missouri
Nuran Ercal, *Missouri University of Science and Technology*

Nebraska
Jodi Kreiling, *University of Nebraska, Omaha*

New Jersey
Bryan Speigelberg, *Rider University*
Yufeng Wei, *Seton Hall University*

New York
Sergio Abreu, *Fordham University*
Susan Rotenberg, *Queens College of CUNY*

Oregon
Jeannine Chan, *Pacific University*

Pennsylvania
Mahrukh Azam, *West Chester University of Pennsylvania*
Robin Ertl, *Marywood University*
Amy Hark, *Muhlengerg College*
Sandra Turchi-Dooley, *Millersville University*
Laura Zapanta, *University of Pittsburgh*

Rhode Island
Erica Oduaran, *Roger Williams University*

South Carolina
Weiguo Cao, *Clemson University*
Kerry Smith, *Clemson University*

Tennessee
Meagan Mann, *Austin Peay State University*

Texas
Johannes Bauer, *Southern Methodist University*

Utah
Craig Thulin, *Utah Valley University*

Previous Edition Reviews

Arkansas
Anne Grippo, *Arkansas State University*

Arizona
Allan Bieber, *Arizona State University*
Matthew Gage, *Northern Arizona University*
Scott Lefler, *Arizona State University, Tempe*
Allan Scruggs, *Arizona State University, Tempe*
Richard Posner, *Northern Arizona State University*

California
Elaine Carter, *Los Angeles City College*
Daniel Edwards, *California State University, Chico*
Gregg Jongeward, *University of the Pacific*
Pavan Kadandale, *University of California, Irvine*
Paul Larsen, *University of California, Riverside*
Rakesh Mogul, *California State Polytechnic University, Pomona*

Colorado
Paul Azari, *Colorado State University*
Andrew Bonham, *Metropolitan State University of Denver*
Johannes Rudolph, *University of Colorado*

Connecticut
Matthew Fisher, *Saint Vincent's College*

Florida
David Brown, *Florida Gulf Coast University*

Georgia
Chavonda Mills, *Georgia College*
Mary E. Peek, *Georgia Tech University*
Rich Singiser, *Clayton State University*

Hawaii
Jon-Paul Bingham, *University of Hawaii-Manoa, College of Tropical Agriculture and Human Resources*

Illinois
Lisa Wen, *Western Illinois University*
Gary Roby, *College of DuPage*
Jon Friesen, *Illinois State University*
Constance Jeffrey, *University of Illinois, Chicago*

Indiana
Brenda Blacklock, *Indiana University-Purdue University Indianapolis*
Todd Hrubey, *Butler University*
Christine Hrycyna, *Purdue University*
Mohammad Qasim, *Indiana University-Purdue University*

Iowa
Don Heck, *Iowa State University*

Kansas
Peter Gegenheimer, *The University of Kansas*
Ramaswamy Krishnamoorthi, *Kansas State University*

Louisiana
James Moroney, *Louisiana State University*
Jeffrey Temple, *Southeastern Louisiana University*

Maine
Robert Gundersen, *University of Maine, Orono*

Massachusetts
Jeffry Nichols, *Worcester State University*

Michigan
Marilee Benore, *University of Michigan*
Kim Colvert, *Ferris State University*
Kathleen Foley, *Michigan State University*
Deborah Heyl-Clegg, *Eastern Michigan University*
Melvin Schindler, *Michigan State University*
Jon Stoltzfus, *Michigan State University*
Mark Thomson, *Ferris State University*

Minnesota
Sandra Olmsted, *Augsburg College*
Tammy Stobb, *St. Cloud State University*

Mississippi
Jeffrey Evans, *University of Southern Mississippi*
James R. Heitz, *Mississippi State University*

Missouri
Karen Bame, *University of Missouri, Kansas City*
Nuran Ercal, *Missouri University of Science & Technology*

Nebraska
Jodi Kreiling, *University of Nebraska, Omaha*
Madhavan Soundararajan, *University of Nebraska*
Russell Rasmussen, *Wayne State College*

New Jersey
Yufeng Wei, *Seton Hall University*
Bryan Spiegelberg, *Rider University*

New Mexico
Beulah Woodfin, *University of New Mexico*

New York
Wendy Pogozelski, *SUNY Geneseo*
Susan Rotenberg, *Queens College of CUNY*

Ohio
Edward Merino, *University of Cincinnati*
Heeyoung Tai, *Miami University*
Lai-Chu Wu, *The Ohio State University*

Oklahoma
Charles Crittell, *East Central University*

Oregon
Jeannine Chan, *Pacific University*
Steven Sylvester, *Oregon State University*

Pennsylvania
Mahrukh Azam, *West Chester University of Pennsylvania*
Jeffrey Brodsky, *University of Pittsburgh*
David Edwards, *University of Pittsburgh School of Pharmacy*
Robin Ertl, *Marywood University*
Amy Hark, *Muhlenberg College*
Justin Huffman, *Pennsylvania State University, Altoona*
Michael Sypes, *Pennsylvania State University*
Sandra Turchi-Dooley, *Millersville University*

Rhode Island
Lenore Martin, *University of Rhode Island*
Erica Oduaran, *Roger Williams University*

South Carolina
Carolyn S. Brown, *Clemson University*
Weiguo Cao, *Clemson University*
Ramin Radfar, *Wofford College*
Paul Richardson, *Coastal Carolina University*
Kerry Smith, *Clemson University*

Tennessee
Meagan Mann, *Austin Peay State University*

Texas
Johannes Bauer, *Southern Methodist University*
David W. Eldridge, *Baylor University*
Edward Funkhouser, *Texas A&M University*
Gail Grabner, *University of Texas, Austin*
Barrie Kitto, *University of Texas at Austin*
Marcos Oliveira, *Feik School of Pharmacy, University of the Incarnate Word*
Richard Sheardy, *Texas Woman's University*
Linette Watkins, *Southwest Texas State University*

Utah
Craig Thulin, *Utah Valley University*

Wisconsin
Sandy Grunwald, *University of Wisconsin La Crosse*

Canada
Isabelle Barrette-Ng, *University of Calgary*

C H A P T E R 1

THE CHEMICAL BASIS OF LIFE

Astrid & Hanns-Frieder Michler/
Science Source Images

While no one has yet succeeded in reproducing all of a cell's chemical reactions in a test tube, it is possible to identify and quantify the thousands of molecules present in a cell, such as this amoeba. Understanding the structures and functions of those molecules is key to understanding how cells live, move, grow, and reproduce.

This first chapter offers a preview of the study of biochemistry, broken down into three sections that reflect how topics in this book are organized. First come brief descriptions of the four major types of small biological molecules and their polymeric forms. Next is a summary of the thermodynamics that apply to metabolic reactions. Finally, there is a discussion of the origin of self-replicating life-forms and their evolution into modern cells. These short discussions introduce some of the key players and major themes of biochemistry and provide a foundation for the topics that will be encountered in subsequent chapters.

WHAT IS BIOCHEMISTRY?

Biochemistry is the scientific discipline that seeks to explain life at the molecular level. It uses the tools and terminology of chemistry to describe the various attributes of living organisms. Biochemistry offers answers to such fundamental questions as "What are we made of?" and "How do we work?" Biochemistry is also a practical science: It generates powerful techniques that underlie advances in other fields, such as genetics, cell biology, and immunology; it offers insights into the treatment of diseases such as cancer and diabetes; and it improves the efficiency of industries such as wastewater treatment, food production, and drug manufacturing.

Some aspects of biochemistry can be approached by studying individual molecules isolated from cells. A thorough understanding of each molecule's physical structure and chemical reactivity helps lead to an understanding of how molecules cooperate and combine to form larger functional units and, ultimately, the intact organism (**Fig. 1.1**). But just as a clock completely disassembled no longer resembles a clock, information about a multitude of biological molecules does not necessarily reveal how an organism lives. Biochemists therefore investigate how organisms behave under different conditions or when a particular molecule is modified or absent. In addition, they collect vast amounts of information about molecular structures and functions—information that is stored and analyzed by computer, a field of study known as **bioinformatics.** A biochemist's laboratory is as likely to hold racks of test tubes as flasks of bacteria or computers.

FIGURE 1.1

Levels of organization in a living organism. Biochemistry focuses on the structures and functions of molecules. Interactions between molecules give rise to higher-order structures (for example, organelles), which may themselves be components of larger entities, leading ultimately to the entire organism. [*Photodisc/Rubberball/Getty Images*]

Chapters 3 through 22 of this book are divided into three groups that roughly correspond to three major themes of biochemistry:

1. *Living organisms are made of macromolecules.* Some molecules are responsible for the physical shapes of cells. Others carry out various activities in the cell. (For convenience, we often use cell interchangeably with organism since the simplest living entity is a single cell.) In all cases, the structure of a molecule is intimately linked to its function. Understanding a molecule's structural characteristics is therefore an important key to understanding its functional significance.

2. *Organisms acquire, transform, store, and use energy.* The ability of a cell to carry out metabolic reactions—to synthesize its constituents and to move, grow, and reproduce—requires the input of energy. A cell must extract this energy from the environment and spend it or store it in a manageable form.

3. *Biological information is transmitted from generation to generation.* Modern human beings look much like they did 100,000 years ago. Certain bacteria have persisted for millions, if not billions, of years. In all organisms, the genetic information that specifies a cell's structural composition and functional capacity must be safely maintained and transmitted each time the cell divides.

Several other themes run throughout biochemistry, and we will highlight these where appropriate.

4. *Cells maintain a state of homeostasis.* Even within its own lifetime, a cell may dramatically alter its shape or metabolic activities, but it does so within certain limits. And in order to remain in a steady, non-equilibrium state—**homeostasis**—the cell must recognize changing internal and external conditions and regulate its activities.

5. *Organisms evolve.* Over long periods of time, the genetic composition of a population of organisms changes. Examining the molecular makeup of living organisms allows biochemists to identify the genetic features that distinguish groups of organisms and to trace their evolutionary history.

6. *Diseases can be explained at the biochemical level.* Identifying the molecular defects that underlie human diseases, or investigating the pathways that allow one organism to infect another, is the first step in diagnosing, treating, preventing, or curing a host of ailments.

BIOLOGICAL MOLECULES

Even the simplest organisms contain a staggering number of different molecules, yet this number represents only an infinitesimal portion of all the molecules that are chemically possible. For one thing, *only a small subset of the known elements are found in living systems* (**Fig. 1.2**). The most abundant of these are C, N, O, and H, followed by Ca, P, K, S, Cl, Na, and Mg. Certain **trace elements** are also present in very small quantities.

Virtually all the molecules in a living organism contain carbon, so biochemistry can be considered to be a branch of organic chemistry. In addition, biological molecules are constructed from H, N, O, P, and S. Most of these molecules belong to one of a few structural classes, which are described below.

Similarly, *the chemical reactivity of biomolecules is limited relative to the reactivity of all chemical compounds.* A few of the functional groups and intramolecular linkages that are common in biochemistry are listed in **Table 1.1**. Familiarity with these functional groups is essential for understanding the behavior of the different types of biological molecules we will encounter throughout this book.

KEY CONCEPTS
- Biological molecules are composed of a subset of all possible elements and functional groups.
- Cells contain four major types of biological molecules and three major types of polymers.

CELLS CONTAIN FOUR MAJOR TYPES OF BIOMOLECULES

Most of the cell's small molecules can be divided into four classes. Although each class contains many members, *they are united under a single structural or functional definition.* Identifying a particular molecule's class may help predict its chemical properties and possibly its role in the cell.

1. Amino Acids Among the simplest compounds are the **amino acids,** so named because they contain an amino group ($-NH_2$) and a carboxylic acid group ($-COOH$). Under physiological conditions, these groups are actually ionized to $-NH_3^+$ and $-COO^-$. The common amino acid alanine—like other small molecules—can be depicted in different ways, for example, by a structural formula, a ball-and-stick model, or a space-filling model (**Fig. 1.3**). Other amino acids resemble alanine in basic structure, but instead of a methyl group ($-CH_3$), they have another group—called a side chain or R group—that may also contain N, O, or S; for example,

Glutamine   Cysteine

FIGURE 1.2

Elements found in biological systems. The most abundant elements are most darkly shaded; trace elements are most lightly shaded. Not every organism contains every trace element. Biological molecules primarily contain H, C, N, O, P, and S.

TABLE 1.1 Common Functional Groups and Linkages in Biochemistry

COMPOUND NAME	STRUCTURE[a]	FUNCTIONAL GROUP
Amine[b]	RNH_2 or RNH_3^+ R_2NH or $R_2NH_2^+$ R_3N or R_3NH^+	$-N\begin{smallmatrix}\\ \end{smallmatrix}$ or $\overset{+}{-}N-$ (amino group)
Alcohol	ROH	$-OH$ (hydroxyl group)
Thiol	RSH	$-SH$ (sulfhydryl group)
Ether	ROR	$-O-$ (ether linkage)
Aldehyde	$R-\overset{\displaystyle O}{\overset{\|}{C}}-H$	$-\overset{\displaystyle O}{\overset{\|}{C}}-$ (carbonyl group), $R-\overset{\displaystyle O}{\overset{\|}{C}}-$ (acyl group)
Ketone	$R-\overset{\displaystyle O}{\overset{\|}{C}}-R$	$-\overset{\displaystyle O}{\overset{\|}{C}}-$ (carbonyl group), $R-\overset{\displaystyle O}{\overset{\|}{C}}-$ (acyl group)
Carboxylic acid[b] (Carboxylate)	$R-\overset{\displaystyle O}{\overset{\|}{C}}-OH$ or $R-\overset{\displaystyle O}{\overset{\|}{C}}-O^-$	$-\overset{\displaystyle O}{\overset{\|}{C}}-OH$ (carboxyl group) or $-\overset{\displaystyle O}{\overset{\|}{C}}-O^-$ (carboxylate group)
Ester	$R-\overset{\displaystyle O}{\overset{\|}{C}}-OR$	$-\overset{\displaystyle O}{\overset{\|}{C}}-O-$ (ester linkage)
Amide	$R-\overset{\displaystyle O}{\overset{\|}{C}}-NH_2$ $R-\overset{\displaystyle O}{\overset{\|}{C}}-NHR$ $R-\overset{\displaystyle O}{\overset{\|}{C}}-NR_2$	$-\overset{\displaystyle O}{\overset{\|}{C}}-N\begin{smallmatrix}\\ \end{smallmatrix}$ (amido group)
Imine[b]	$R{=}NH$ or $R{=}NH_2^+$ $R{=}NR$ or $R{=}NHR^+$	$>C{=}N-$ or $>C{=}\overset{+}{N}\overset{H}{\diagdown}$ (imino group)
Phosphoric acid ester[b]	$R-O-\overset{\displaystyle O}{\underset{\displaystyle OH}{\overset{\|}{P}}}-OH$ or $R-O-\overset{\displaystyle O}{\underset{\displaystyle O^-}{\overset{\|}{P}}}-O^-$	$-O-\overset{\displaystyle O}{\underset{\displaystyle OH}{\overset{\|}{P}}}-O-$ (phosphoester linkage) $-\overset{\displaystyle O}{\underset{\displaystyle OH}{\overset{\|}{P}}}-OH$ or $-\overset{\displaystyle O}{\underset{\displaystyle O^-}{\overset{\|}{P}}}-O^-$ (phosphoryl group, P_i)
Diphosphoric acid ester[b]	$R-O-\overset{\displaystyle O}{\underset{\displaystyle OH}{\overset{\|}{P}}}-O-\overset{\displaystyle O}{\underset{\displaystyle OH}{\overset{\|}{P}}}-OH$ or $R-O-\overset{\displaystyle O}{\underset{\displaystyle O^-}{\overset{\|}{P}}}-O-\overset{\displaystyle O}{\underset{\displaystyle O^-}{\overset{\|}{P}}}-O^-$	$-O-\overset{\displaystyle O}{\underset{\displaystyle OH}{\overset{\|}{P}}}-O-\overset{\displaystyle O}{\underset{\displaystyle OH}{\overset{\|}{P}}}-O-$ (phosphoanhydride linkage) $-\overset{\displaystyle O}{\underset{\displaystyle OH}{\overset{\|}{P}}}-O-\overset{\displaystyle O}{\underset{\displaystyle OH}{\overset{\|}{P}}}-OH$ or $-\overset{\displaystyle O}{\underset{\displaystyle O^-}{\overset{\|}{P}}}-O-\overset{\displaystyle O}{\underset{\displaystyle O^-}{\overset{\|}{P}}}-O^-$ (diphosphoryl group, pyrophosphoryl group, PP_i)

[a] R represents any carbon-containing group. In a molecule with more than one R group, the groups may be the same or different.

[b] Under physiological conditions, these groups are ionized and hence bear a positive or negative charge.

❓ Cover the structure column and draw the structure for each compound listed on the left. Do the same for each functional group.

$$COO^-$$
$$H-\overset{\displaystyle |}{\underset{\displaystyle |}{C}}-CH_3$$
$$NH_3^+$$

(a)

In a structural formula, some bonds, such as the C—O and N—H bonds, are implied. Around the central carbon, the horizontal bonds extend slightly above the plane of the page, and the vertical bonds extend slightly behind it.

(b)

The atoms are color-coded by convention: C gray, N blue, O red, and H white. A ball-and-stick representation reveals the identities of the atoms and their positions in space.

(c)

In a space-filling model, each atom is presented as a sphere whose radius (the van der Waals radius) corresponds to the distance of closest approach by another atom.

FIGURE 1.3

Representations of alanine. The structural formula (a) indicates all the atoms and the major bonds. Because the central carbon atom has tetrahedral geometry, its four bonds do not lie flat in the plane of the paper. This tetrahedral arrangement is more accurately depicted in the ball-and-stick model (b), although the relative sizes and electrical charges of atoms are not shown. A space-filling model (c) best represents the actual shape of the molecule but may obscure some of its atoms and linkages.

2. Carbohydrates

Simple **carbohydrates** (also called **monosaccharides** or just sugars) have the formula $(CH_2O)_n$, where n is ≥ 3. Glucose, a monosaccharide with six carbon atoms, has the formula $C_6H_{12}O_6$. It is sometimes convenient to draw it as a ladder-like chain (*left*); however, glucose forms a cyclic structure in solution (*right*):

Glucose

In the representation of the cyclic structure, the darker bonds project in front of the page and the lighter bonds project behind it. In many monosaccharides, one or more hydroxyl groups are replaced by other groups, but the ring structure and multiple —OH groups of these molecules allow them to be easily recognized as carbohydrates.

3. Nucleotides

A five-carbon sugar, a nitrogen-containing ring, and one or more phosphate groups are the components of **nucleotides**. For example, adenosine triphosphate (ATP) contains the nitrogenous group adenine linked to the monosaccharide ribose, to which a triphosphate group is also attached:

Adenosine triphosphate (ATP)

The most common nucleotides are mono-, di-, and triphosphates containing the nitrogenous ring compounds (or "bases") adenine, cytosine, guanine, thymine, or uracil (abbreviated A, C, G, T, and U).

4. Lipids The fourth major group of biomolecules consists of the **lipids.** These compounds cannot be described by a single structural formula since they are a diverse collection of molecules. However, they all tend to be poorly soluble in water because the bulk of their structure is hydrocarbon-like. For example, palmitic acid consists of a highly insoluble chain of 15 carbons attached to a carboxylic acid group, which is ionized under physiological conditions. The anionic lipid is therefore called palmitate.

Palmitate

Cholesterol, although it differs significantly in structure from palmitate, is also poorly soluble in water because of its hydrocarbon-like composition.

Cholesterol

Cells also contain a few other small molecules that cannot be easily classified into the groups above or that are constructed from molecules belonging to more than one group.

THERE ARE THREE MAJOR KINDS OF BIOLOGICAL POLYMERS

In addition to small molecules consisting of relatively few atoms, organisms contain macromolecules that may consist of thousands of atoms. Such huge molecules are not synthesized in one piece but are built from smaller units. This is a universal feature of nature: *A few kinds of building blocks can be combined in different ways to produce a wide variety of larger structures.* This is advantageous for a cell, which can get by with a limited array of raw materials. In addition, the very act of chemically linking individual units (**monomers**) into longer strings (**polymers**) is a way of encoding information (the sequence of the monomeric units) in a stable form. Biochemists use certain units of measure to describe both large and small molecules (**Box 1.A**).

Amino acids, monosaccharides, and nucleotides each form polymeric structures with widely varying properties. In most cases, the individual monomers become covalently linked in head-to-tail fashion:

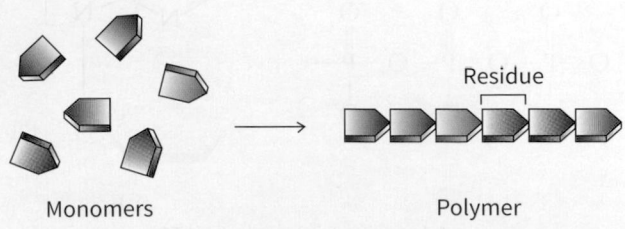

Monomers   Polymer

A CLOSER LOOK

Box 1.A Units Used in Biochemistry

Biochemists follow certain conventions when quantifying objects on a molecular scale. For example, the mass of a molecule can be expressed in atomic mass units; however, the masses of biological molecules—especially very large ones—are typically given without units. Here it is understood that the mass is expressed relative to one-twelfth the mass of an atom of the common carbon isotope ^{12}C (12.011 atomic mass units). Occasionally, units of daltons (D) are used (1 dalton = 1 atomic mass unit), often with the prefix kilo, k (kD). This is useful for macromolecules such as proteins, many of which have masses in the range from 20,000 (20 kD) to over 1,000,000 (1000 kD).

The standard metric prefixes are also necessary for expressing the minute concentrations of biomolecules in living cells. Concentrations are usually given as moles per liter ($mol \cdot L^{-1}$ or M), with the appropriate prefix such as m, μ, or n:

mega (M)	10^6		nano (n)	10^{-9}
kilo (k)	10^3		pico (p)	10^{-12}
milli (m)	10^{-3}		femto (f)	10^{-15}
micro (μ)	10^{-6}			

For example, the concentration of the sugar glucose in human blood is about 5 mM, but many intracellular molecules are present at concentrations of μM or less.

Distances are customarily expressed in angstroms, Å (1 Å = 10^{-10} m) or in nanometers, nm (1 nm = 10^{-9} m). For example, the distance between the centers of carbon atoms in a C—C bond is about 1.5 Å, and the diameter of a DNA molecule is about 20 Å.

? **The diameter of a typical spherical bacterial cell is about 1 μm. What is the cell's volume?**

The linkage between monomeric units is characteristic of each type of polymer. The monomers are called **residues** after they have been incorporated into the polymer. Strictly speaking, lipids do not form polymers, although they do tend to aggregate to form larger structures such as cell membranes.

1. Proteins

Polymers of amino acids are called **polypeptides** or **proteins.** Twenty different amino acids serve as building blocks for proteins, which may contain many hundreds of amino acid residues. The amino acid residues are linked to each other by amide bonds called **peptide bonds.** A peptide bond (*arrow*) links the two residues in a dipeptide (the side chains of the amino acids are represented by R_1 and R_2).

$$H_3\overset{+}{N}-\underset{\underset{H}{|}}{\overset{\overset{R_1}{|}}{C}}-\overset{O}{\overset{\|}{C}}-\underset{\underset{H}{|}}{\overset{\overset{}{|}}{N}}-\underset{\underset{H}{|}}{\overset{\overset{R_2}{|}}{C}}-\overset{O}{\overset{\|}{C}}-O^-$$

Because the side chains of the 20 amino acids have different sizes, shapes, and chemical properties, the exact **conformation** (three-dimensional shape) of the polypeptide chain depends on its amino acid composition and sequence. For example, the small polypeptide endothelin, with 21 residues, assumes a compact shape in which the polymer bends and folds to accommodate the functional groups of its amino acid residues (**Fig. 1.4**).

The 20 different amino acids can be combined in almost any order and in almost any proportion to produce myriad polypeptides, all of which have unique three-dimensional shapes. This property makes proteins as a class the most structurally variable and therefore the most functionally versatile of all the biopolymers. Accordingly, *proteins perform a wide variety of tasks in the cell, such as mediating chemical reactions and providing structural support.*

2. Nucleic Acids

Polymers of nucleotides are termed **polynucleotides** or **nucleic acids,** better known as DNA and RNA. Unlike polypeptides, with 20 different amino acids available for polymerization, each nucleic acid is made from just four different nucleotides. For example, the residues in RNA contain the bases adenine, cytosine, guanine, and uracil, whereas the residues in DNA contain adenine, cytosine, guanine, and thymine. Polymerization involves the phosphate and sugar groups of the nucleotides, which become linked by **phosphodiester bonds.**

(a)

(b)

FIGURE 1.4

Structure of human endothelin. The 21 amino acid residues of this polypeptide, shaded from blue to red, form a compact structure. In (a), each amino acid residue is represented by a sphere. The ball-and-stick model (b) shows all the atoms except hydrogen. [*Structure (pdb 1EDN) determined by B. A. Wallace and R. W. Jones.*]

Phosphodiester bond

CGUACG
(a)

(b)

FIGURE 1.5

Structure of a nucleic acid.
(a) Sequence of nucleotide residues, using one-letter abbreviations. (b) Ball-and-stick model of the polynucleotide, showing all atoms except hydrogen (this structure is a six-residue segment of RNA).
[*Structure (pdb ARF0108) determined by R. Biswas, S. N. Mitra, and M. Sundaralingam.*]

In part because nucleotides are much less variable in structure and chemistry than amino acids, nucleic acids tend to have more regular structures than proteins. *This is in keeping with their primary role as carriers of genetic information, which is contained in their sequence of nucleotide residues rather than in their three-dimensional shape* (**Fig. 1.5**). Nevertheless, many nucleic acids do bend and fold into compact globular shapes, as proteins do.

3. Polysaccharides

Polysaccharides usually contain only one or a few different types of monosaccharide residues, so even though a cell may synthesize dozens of different kinds of monosaccharides, most of its polysaccharides are homogeneous polymers. This tends to limit their potential for carrying genetic information in the sequence of their residues (as nucleic acids do) or for adopting a large variety of shapes and mediating chemical reactions (as proteins do). On the other hand, *polysaccharides perform essential cell functions by serving as fuel-storage molecules and by providing structural support*. For example, plants link the monosaccharide glucose, which is a fuel for virtually all cells, into the polysaccharide starch for long-term storage. The glucose residues are linked by **glycosidic bonds** (the bond is shown in red in this disaccharide):

Glucose monomers are also the building blocks for cellulose, the extended polymer that helps make plant cell walls rigid (**Fig. 1.6**). The starch and cellulose polymers differ in the arrangement of the glycosidic bonds between glucose residues.

The brief descriptions of biological polymers given above are generalizations, meant to convey some appreciation for the possible structures and functions of these macromolecules. *Exceptions to the generalizations abound*. For example, some small polysaccharides encode information that allows cells bearing the molecules on their surfaces to recognize each other. Likewise, some nucleic acids perform structural roles, for example, by serving as scaffolding in ribosomes, the small particles where protein synthesis takes place. Under certain conditions,

Glucose

Starch

Cellulose

FIGURE 1.6

Glucose and its polymers. Both starch and cellulose are polysaccharides containing glucose residues. They differ in the type of chemical linkage between the monosaccharide units. Starch molecules have a loose helical conformation, whereas cellulose molecules are extended and relatively stiff.

proteins are called on as fuel-storage molecules. A summary of the major and minor functions of proteins, polysaccharides, and nucleic acids is presented in **Table 1.2**.

 CONCEPT CHECK

1. List the six most abundant elements in biological molecules.

2. Name the common functional groups and linkages shown in Table 1.1.

3. Give the structural or functional definitions for amino acids, monosaccharides, nucleotides, and lipids.

4. Why is it efficient for macromolecules to be polymers?

5. What is the relationship between a monomer and a residue?

6. Summarize the structural definitions and major functions of proteins, polysaccharides, and nucleic acids.

TABLE 1.2 Functions of Biopolymers

BIOPOLYMER	ENCODE INFORMATION	CARRY OUT METABOLIC REACTIONS	STORE ENERGY	SUPPORT CELLULAR STRUCTURES
Proteins	—	✔	✔	✔
Nucleic acids	✔	✔	—	✓
Polysaccharides	✓	—	✔	✔

✔ major function
✓ minor function

ENERGY AND METABOLISM

Assembling small molecules into polymeric macromolecules requires energy. And unless the monomeric units are readily available, a cell must synthesize the monomers, which also requires energy. In fact, *cells require energy for all the functions of living, growing, and reproducing.*

It is useful to describe the energy in biological systems using the terminology of thermodynamics (the study of heat and power). An organism, like any chemical system, is subject to the laws of thermodynamics. According to the first law of thermodynamics, energy cannot be created or destroyed. However, it can be transformed. For example, the energy of a river flowing over a dam can be harnessed as electricity, which can then be used to produce heat or perform mechanical work. Cells can be considered to be very small machines that use chemical energy to drive metabolic reactions, which may also produce heat or carry out mechanical work.

ENTHALPY AND ENTROPY ARE COMPONENTS OF FREE ENERGY

The energy relevant to biochemical systems is called the Gibbs free energy (after the scientist who defined it) or just **free energy.** It is abbreviated G and has units of joules per mol $(J \cdot mol^{-1})$. Free energy has two components: enthalpy and entropy. *Enthalpy* (abbreviated H, with units of $J \cdot mol^{-1}$) *is taken to be equivalent to the heat content of the system. Entropy* (abbreviated S, with units of $J \cdot K^{-1} \cdot mol^{-1}$) *is a measure of how the energy is dispersed within that system.* Entropy can therefore be considered to be a measure of the system's disorder or randomness, because the more ways a system's components can be arranged, the more dispersed its energy. For example, consider a pool table at the start of a game when all 15 balls are arranged in one neat triangle (a state of high order or low entropy). After play has begun, the balls are scattered across the table, which is now in a state of disorder and high entropy (**Fig. 1.7**).

Free energy, enthalpy, and entropy are related by the equation

$$G = H - TS \qquad [1.1]$$

where T represents temperature in Kelvin (equivalent to degrees Celsius plus 273). Temperature is a coefficient of the entropy term because entropy varies with temperature; the entropy of a substance increases when it is warmed because more thermal energy has been dispersed

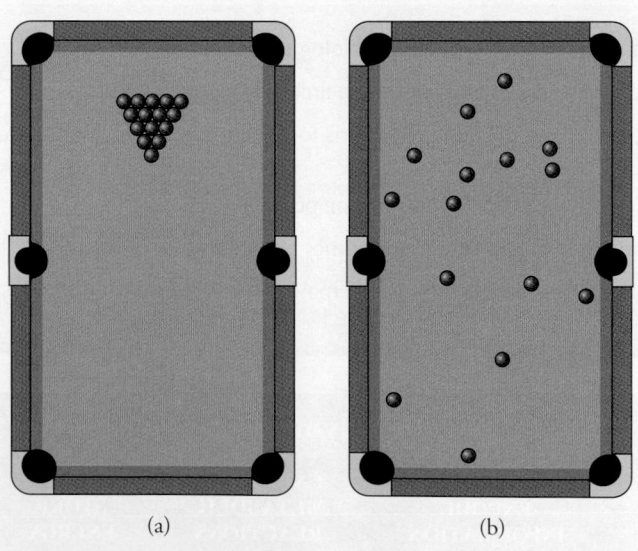

(a)   (b)

FIGURE 1.7

Illustration of entropy. Entropy is a measure of the dispersal of energy in a system, so it reflects the system's randomness or disorder. (a) Entropy is low when all the balls are arranged in a single area of the pool table. (b) Entropy is high after the balls have been scattered, because there are now a large number of different possible arrangements of the balls on the table.

? Compare the entropy of a ball of yarn before and after a cat has played with it.

within it. The enthalpy of a chemical system can be measured, although with some difficulty, but it is next to impossible to measure a system's entropy because this would require counting all the possible arrangements of its components or all the ways its energy could be spread out among them. Therefore, it is more practical to deal with *changes* in these quantities (change is indicated by the Greek letter delta, Δ) so that

$$\Delta G = \Delta H - T\Delta S \qquad [1.2]$$

Biochemists can measure how the free energy, enthalpy, and entropy of a system differ before and after a chemical reaction. For example, **exothermic reactions** are accompanied by the release of heat to the surroundings ($H_{final} - H_{initial} = \Delta H < 0$), whereas **endothermic reactions** absorb heat from the surroundings ($\Delta H > 0$). Similarly, the entropy change, $S_{final} - S_{initial} = \Delta S$, can be positive or negative. When ΔH and ΔS for a process are known, Equation 1.2 can be used to calculate the value of ΔG at a given temperature (see Sample Calculation 1.1).

SAMPLE CALCULATION 1.1

PROBLEM

Use the information below to calculate the change in enthalpy and the change in entropy for the reaction A → B.

	Enthalpy ($kJ \cdot mol^{-1}$)	Entropy ($J \cdot K^{-1} \cdot mol^{-1}$)
A	60	22
B	75	97

SOLUTION

$$\Delta H = H_B - H_A$$
$$= 75 \text{ kJ} \cdot \text{mol}^{-1} - 60 \text{ kJ} \cdot \text{mol}^{-1}$$
$$= 15 \text{ kJ} \cdot \text{mol}^{-1}$$
$$= 15,000 \text{ J} \cdot \text{mol}^{-1}$$

$$\Delta S = S_B - S_A$$
$$= 97 \text{ J} \cdot \text{K}^{-1} \cdot \text{mol}^{-1}$$
$$- 22 \text{ J} \cdot \text{K}^{-1} \cdot \text{mol}^{-1}$$
$$= 75 \text{ J} \cdot \text{K}^{-1} \cdot \text{mol}^{-1}$$

PRACTICE PROBLEMS

1. The change in enthalpy for a reaction is $-10 \text{ kJ} \cdot \text{mol}^{-1}$. Is heat absorbed or given off during the reaction?

2. A reacting system undergoes a change from disordered to ordered. Is the entropy change positive or negative?

ΔG IS LESS THAN ZERO FOR A SPONTANEOUS PROCESS

A china cup dropped from a great height will break, but the pieces will never reassemble themselves to restore the cup. The thermodynamic explanation is that the broken pieces have less free energy than the intact cup. *In order for a process to occur, the overall change in free energy (ΔG) must be negative.* For a chemical reaction, this means that the free energy of the products must be less than the free energy of the reactants:

$$\Delta G = G_{products} - G_{reactants} < 0 \qquad [1.3]$$

When ΔG is less than zero, the reaction is said to be **spontaneous** or **exergonic**. A **nonspontaneous** or **endergonic** reaction has a free energy change greater than zero; in this case, the reverse reaction is spontaneous.

A → B	B → A
$\Delta G > 0$	$\Delta G < 0$
Nonspontaneous	Spontaneous

Note that thermodynamic spontaneity does not indicate how *fast* a reaction occurs, only whether it will occur as written. (The rate of a reaction depends on other factors, such as the concentrations of the reacting molecules, the temperature, and the presence of a catalyst.) When a reaction, such as A → B, is at equilibrium, the rate of the forward reaction is equal to the rate of the reverse reaction, so there is no net change in the system. In this situation, $\Delta G = 0$.

A quick examination of Equation 1.2 reveals that *a reaction that occurs with a decrease in enthalpy and an increase in entropy is spontaneous at all temperatures because* ΔG *is always less than zero.* These results are consistent with everyday experience. For example, heat moves spontaneously from a hot object to a cool object, and items that are neatly arranged tend to become disordered, never the other way around. (This is a manifestation of the second law of thermodynamics, which states that energy tends to spread out.) Accordingly, reactions in which the enthalpy increases and entropy decreases do not occur. If enthalpy and entropy both increase or both decrease during a reaction, the value of ΔG then depends on the temperature, which governs whether the $T\Delta S$ term of Equation 1.2 is greater than or less than the ΔH term. This means that a large increase in entropy can offset an unfavorable (positive) change in enthalpy. Conversely, the release of a large amount of heat ($\Delta H < 0$) during a reaction can offset an unfavorable decrease in entropy (see Sample Calculation 1.2).

SAMPLE CALCULATION 1.2

PROBLEM   Use the information given in Sample Calculation 1.1 to determine whether the reaction A → B is spontaneous at 25°C.

SOLUTION   Substitute the values for ΔH and ΔS, calculated in Sample Calculation 1.1, into Equation 1.2. To express the temperature in Kelvin, add 273 to the temperature in degrees Celsius: $273 + 25 = 298$ K.

$$\Delta G = \Delta H - T\Delta S$$
$$= 15{,}000 \text{ J} \cdot \text{mol}^{-1} - 298 \text{ K} \ (75 \text{ J} \cdot \text{K}^{-1} \cdot \text{mol}^{-1})$$
$$= 15{,}000 - 22{,}400 \text{ J} \cdot \text{mol}^{-1}$$
$$= -7400 \text{ J} \cdot \text{mol}^{-1}$$
$$= -7.4 \text{ kJ} \cdot \text{mol}^{-1}$$

Because ΔG is less than zero, the reaction is spontaneous. Even though the change in enthalpy is unfavorable, the large increase in entropy makes ΔG favorable.

PRACTICE PROBLEMS   3. If $\Delta H = -20{,}000 \text{ J} \cdot \text{mol}^{-1}$ and $\Delta S = -50 \text{ J} \cdot \text{K}^{-1} \cdot \text{mol}^{-1}$, would the reaction be spontaneous at 25°C?

4. Use the information given in Sample Calculation 1.1 to determine the temperature at which $\Delta G = 0$.

LIFE IS THERMODYNAMICALLY POSSIBLE

In order to exist, life must be thermodynamically spontaneous. Does this hold at the molecular level? When analyzed in a test tube (*in vitro,* literally "in glass"), many of a cell's metabolic reactions have free energy changes that are less than zero, but some reactions do not. Nevertheless, the nonspontaneous reactions are able to proceed *in vivo* (in a living organism) because they occur in concert with other reactions that are thermodynamically favorable. Consider two reactions *in vitro,* one nonspontaneous ($\Delta G > 0$) and one spontaneous ($\Delta G < 0$):

$$\text{A} \rightarrow \text{B} \qquad \Delta G = +15 \text{ kJ} \cdot \text{mol}^{-1} \quad \text{(nonspontaneous)}$$
$$\text{B} \rightarrow \text{C} \qquad \Delta G = -20 \text{ kJ} \cdot \text{mol}^{-1} \quad \text{(spontaneous)}$$

When the reactions are combined, their ΔG values are added, so the overall process has a negative change in free energy:

$$\text{A} + \text{B} \rightarrow \text{B} + \text{C} \qquad \Delta G = (15 \text{ kJ} \cdot \text{mol}^{-1}) + (-20 \text{ kJ} \cdot \text{mol}^{-1})$$
$$\text{A} \rightarrow \text{C} \qquad \Delta G = -5 \text{ kJ} \cdot \text{mol}^{-1}$$

This phenomenon is shown graphically in **Figure 1.8**. In effect, the unfavorable "uphill" reaction A → B is pulled along by the more favorable "downhill" reaction B → C.

Cells couple unfavorable metabolic processes with favorable ones so that the net change in free energy is negative. Note that it is permissible to add ΔG values because the free energy, G,

FIGURE 1.8

Free energy changes in coupled reactions. A nonspontaneous reaction, such as A → B, which has a positive value of ΔG, can be coupled to another reaction, B → C, which has a negative value of ΔG and is therefore spontaneous. The reactions are coupled because the product of the first reaction, B, is a reactant for the second reaction.

> ❓ Which reaction occurs spontaneously in reverse: C → B, B → A, or C → A?

depends only on the initial and final states of the system, without regard to the specific chemical or mechanical work that occurred in going from one state to the other.

Most macroscopic life on earth today is sustained by the energy of the sun (this was not always the case, nor is it true of all organisms). In photosynthetic organisms, such as green plants, light energy excites certain molecules so that their subsequent chemical reactions occur with a net negative change in free energy. These thermodynamically favorable (spontaneous) reactions are coupled to the unfavorable synthesis of monosaccharides from atmospheric CO_2 (**Fig. 1.9**). In this process, the carbon is **reduced.** Reduction, the gain of electrons, is accomplished by the addition of hydrogen or the removal of oxygen (the oxidation states of carbon are reviewed in **Table 1.3**). The plant—or an animal that eats the plant—can then break down the monosaccharide to use it as a fuel to power other metabolic activities. In the process, the carbon is **oxidized**—it loses electrons through the addition of oxygen or the removal of hydrogen—and ultimately becomes CO_2. The oxidation of carbon is thermodynamically favorable, so it can be coupled to energy-requiring processes such as the synthesis of building blocks and their polymerization to form macromolecules.

Virtually all metabolic processes occur with the aid of catalysts called **enzymes,** most of which are proteins (a catalyst greatly increases the rate of a reaction without itself undergoing any net change). For example, specific enzymes catalyze the formation of peptide, phosphodiester, and glycosidic linkages during polymer synthesis. Other enzymes catalyze cleavage of these bonds to break the polymers into their monomeric units.

FIGURE 1.9

Reduction and reoxidation of carbon compounds. The sun provides the free energy to convert CO_2 to reduced compounds such as monosaccharides. The reoxidation of these compounds to CO_2 is thermodynamically spontaneous, so free energy can be made available for other metabolic processes. Note that free energy is not actually a substance that is physically released from a molecule.

TABLE 1.3 Oxidation States of Carbon	
COMPOUND[a]	FORMULA
Carbon dioxide *most oxidized* (*least reduced*)	O=C=O
Acetic acid	H—C—C with H, O, OH
Carbon monoxide	C≡O
Formic acid	H—C with O, OH
Acetone	H—C—C—C—H
Acetaldehyde	H—C—C with H, O, H
Formaldehyde	H,H C=O
Acetylene	H—C≡C—H
Ethanol	H—C—C—OH
Ethene	H,H C=C H,H
Ethane	H—C—C—H
Methane *least oxidized* (*most reduced*)	H—C—H

[a]Compounds are listed in order of decreasing oxidation state of the red carbon atom.

A living organism—with its high level of organization of atoms, molecules, and larger structures—represents a state of low entropy relative to its surroundings. Yet the organism can maintain this thermodynamically unfavorable state as long as it continually obtains free energy from its food. Thus, living organisms do indeed obey the laws of thermodynamics. When the organism ceases to obtain a source of free energy from its surroundings or exhausts its stored food, the chemical reactions in its cells reach equilibrium ($\Delta G = 0$), which results in death.

CONCEPT CHECK

1. What do enthalpy and entropy mean and how are they related to free energy?

2. What does the value of ΔG reveal about a biochemical process?

3. How do thermodynamically unfavorable reactions proceed *in vivo*?

4. Explain why an organism must have a steady supply of food.

5. Describe the cycle of carbon reduction and oxidation in photosynthesis and in the breakdown of a compound such as a monosaccharide.

THE ORIGIN AND EVOLUTION OF LIFE

KEY CONCEPTS
- Modern prokaryotic and eukaryotic cells apparently evolved from simpler nonliving systems.
- The three domains of life are bacteria, archaea, and eukarya.

Every living cell originates from the division of a parental cell. Thus, the ability to **replicate** (make a replica or copy of itself) is one of the universal characteristics of living organisms. *In order to leave descendants that closely resemble itself, a cell must contain a set of instructions—and the means for carrying them out—that can be transmitted from generation to generation.* Over time, the instructions change gradually, so that species also change, or **evolve.** By carefully examining an organism's genetic information and the cellular machinery that supports it, biochemists can draw some conclusions about the organism's relationship to more ancient life-forms. The history of evolution is therefore contained not just within the fossil record but also in the molecular makeup of all living cells. For example, nucleic acids participate in the storage and transmission of genetic information in all organisms, and the oxidation of glucose is an almost universal means for generating metabolic free energy. Consequently, DNA, RNA, and glucose must have been present in the ancestor of all cells.

THE PREBIOTIC WORLD

A combination of theory and experimental data leads to several plausible scenarios for the emergence of life from nonbiological (prebiotic) materials on the early earth. In one scenario, inorganic compounds such as H_2, H_2O, NH_3, and CH_4—which may have been present in the early atmosphere—could have given rise to simple biomolecules, such as amino acids, when sparked by lightning. Laboratory experiments with the same raw materials and electrical discharges to simulate lightning do in fact yield these molecules (**Fig. 1.10**). Other experiments suggest that hydrogen cyanide (HCN), formaldehyde (HCOH), and phosphate could have been converted to nucleotides with a similarly modest input of energy.

Over time, simple molecular building blocks could have accumulated and formed larger structures, particularly in shallow waters where evaporation would have had a concentrating effect. Eventually, conditions would have been ripe for the assembly of functional, living cells. Charles Darwin proposed that life might have arisen in some "warm little pond"; however, the early earth was probably a much more violent place, with frequent meteorite impacts and volcanic activity.

FIGURE 1.10

Laboratory synthesis of biological molecules. A mixture of gases—H_2, H_2O, NH_3, and CH_4—is subject to an electrical discharge. Newly formed compounds, such as amino acids, accumulate in the aqueous phase as water vapor condenses. Samples of the reaction products can be removed via the stopcock.

In an alternative scenario, supported by studies of the metabolism of some modern bacteria, the first cells could have developed at deep-sea hydrothermal vents, some of which are characterized by temperatures as high as 350°C and clouds of gaseous H_2S and metal sulfides (giving them the name "black smokers"; **Fig. 1.11**). In the laboratory, incubating a few small molecules in the presence of iron sulfide and nickel sulfide at 100°C yields acetic acid, an organic compound with a newly formed C—C bond:

$$CH_3SH \quad + \quad CO \quad + \quad H_2O \xrightarrow{FeS, NiS} CH_3COOH \quad + \quad H_2S$$

Methyl   Carbon   Acetic acid

thiol   monoxide

Under similar conditions, amino acids spontaneously form short polypeptides. Although the high temperatures that are necessary for their synthesis also tend to break them down, these compounds would have been stable in the cooler water next to the hydrothermal vent.

Regardless of how they formed, *the first biological building blocks would have had to polymerize*. This process might have been stimulated when the organic molecules—often bearing anionic (negatively charged) groups—aligned themselves on a cationic (positively charged) mineral surface.

FIGURE 1.11

A hydrothermal vent. Life may have originated at these "black smokers," where high temperatures, H_2S, and metal sulfides might have stimulated the formation of biological molecules. [*B. Murton/Southhampton Oceanography Centre/Science Photo Library/Photo Researchers.*]

In fact, in the laboratory, common clay promotes the polymerization of nucleotides into RNA. *Primitive polymers would have had to gain the capacity for self-replication.* Otherwise, no matter how stable or chemically versatile, such molecules would never have given rise to anything larger or more complicated: The probability of assembling a fully functional cell from

Original

1. *The polyA molecule serves as a template for the synthesis of a polymer containing uracil nucleotides, U, which are complementary to adenine nucleotides (in modern RNA, A pairs with U).*

2. *The two polymer chains separate.*

Original Complement

3. *The polyU molecule serves as a template for the synthesis of a new complementary polyA chain.*

4. *The chains again separate and the polyU polymer is discarded, leaving the original polyA molecule and its exact copy.*

Original Copy

FIGURE 1.12

Possible mechanism for the self-replication of a primitive RNA molecule. For simplicity, the RNA molecule is shown as a polymer of adenine nucleotides, A.

? Draw a diagram showing how polyU would be replicated.

a solution of thousands of separate small molecules is practically nil. Because RNA in modern cells represents a form of genetic information and participates in all aspects of expressing that information, it may be similar to the first self-replicating biopolymer. It might have made a copy of itself by first making a **complement,** a sort of mirror image, that could then make a complement of *itself,* which would be identical to the original molecule (**Fig. 1.12**).

ORIGINS OF MODERN CELLS

A replicating molecule's chances of increasing in number depend on **natural selection,** the phenomenon whereby the entities best suited to the prevailing conditions are the likeliest to survive and multiply (**Box 1.B**). This would have favored a replicator that was chemically stable and had a ready supply of building blocks and free energy for making copies of itself. Accordingly, it would have been advantageous to become enclosed in some sort of membrane that could prevent valuable small molecules from diffusing away. Natural selection would also have favored replicating systems that developed the means for synthesizing their own building blocks and for more efficiently harnessing sources of free energy.

The first cells were probably able to "fix" CO_2—that is, convert it to reduced organic compounds—using the free energy released in the oxidation of readily available inorganic compounds such as H_2S or Fe^{2+}. Vestiges of these processes can be seen in modern metabolic reactions that involve sulfur and iron.

Later, photosynthetic organisms similar to present-day cyanobacteria (also called blue-green algae) used the sun's energy to fix CO_2:

$$CO_2 + H_2O \rightarrow (CH_2O) + O_2$$

The concomitant oxidation of H_2O to O_2 dramatically increased the concentration of atmospheric O_2, about 2.4 billion years ago, and made it possible for **aerobic** (oxygen-using) organisms to take advantage of this powerful oxidizing agent. The **anaerobic** origins of life are still visible in the most basic metabolic reactions of modern organisms; these reactions proceed in the absence of oxygen. Now that the earth's atmosphere contains about 18% oxygen, anaerobic organisms have not disappeared, but they have been restricted to microenvironments where O_2 is scarce, such as the digestive systems of animals or underwater sediments.

The earth's present-day life-forms are of two types, which are distinguished by their cellular architecture:

1. *Prokaryotes* are small unicellular organisms that lack a discrete nucleus and usually contain no internal membrane systems. This group comprises two subgroups that are remarkably different metabolically, although they are similar in appearance: the eubacteria (usually just called **bacteria**), exemplified by *E. coli,* and the **archaea** (or archaebacteria), best known as organisms that inhabit extreme environments, although they are actually found almost everywhere (**Fig. 1.13**).

2. *Eukaryotic* cells are usually larger than prokaryotic cells and contain a nucleus and other membrane-bounded cellular compartments (such as mitochondria, chloroplasts, and endoplasmic reticulum). Eukaryotes may be unicellular or multicellular. This group (also called the **eukarya**) includes microscopic organisms as well as familiar macroscopic plants and animals (**Fig. 1.14**).

By analyzing the sequences of nucleotides in certain genes that are present in all species, it is possible to construct a diagram that indicates how the bacteria, archaea, and eukarya are

A CLOSER LOOK

Box 1.B How Does Evolution Work?

Documenting evolutionary change is relatively straightforward, but the mechanisms whereby evolution occurs are prone to misunderstanding. Populations change over time, and new species arise as a result of natural selection. Selection operates on individuals, but its effects can be seen in a population only over a period of time. Most populations are collections of individuals that share an overall genetic makeup but also exhibit small variations due to random alterations (mutations) in their genetic material as it is passed from parent to offspring. In general, the survival of an individual depends on how well suited it is to the particular conditions under which it lives.

Individuals whose genetic makeup grants them the greatest rate of survival have more opportunities to leave offspring with the same genetic makeup. Consequently, their characteristics become widespread in a population, and, over time, the population appears to adapt to its environment. A species that is well suited to its environment tends to persist; a poorly adapted species fails to reproduce and therefore dies out.

Because evolution is the result of random variations and changing probabilities for successful reproduction, it is inherently random and unpredictable. Furthermore, natural selection acts on the raw materials at hand. It cannot create something out of nothing but must operate in increments. For example, the insect wing did not suddenly appear in the offspring of a wingless parent but most likely developed bit by bit, over many generations, by modification of a gill or heat-exchange appendage. Each step of the wing's development would have been subject to natural selection, eventually making an individual that bore the appendage more likely to survive, perhaps by being able to first glide and then actually fly in pursuit of food or to evade predators.

Although we tend to think of evolution as an imperceptibly slow process, occurring on a geological time scale, it is ongoing and accessible to observation in the laboratory. For example, under optimal conditions, the bacterium *Escherichia coli* requires only about 20 minutes to produce a new generation. In the laboratory, a culture of *E. coli* cells can progress through about 2500 generations in a year (in contrast, 2500 human generations would require about 60,000 years). Hence, it is possible to subject a population of cultured bacterial cells to some "artificial" selection—for example, by making an essential nutrient scarce—and observe how the genetic composition of the population changes over time as it adapts to the new conditions.

? **Why can't acquired (rather than genetic) characteristics serve as the raw material for evolution?**

related. *The number of sequence differences between two groups of organisms indicates how long ago they diverged from a common ancestor:* Species with similar sequences have a longer shared evolutionary history than species with dissimilar sequences. This sort of analysis has produced the evolutionary tree shown in **Figure 1.15**.

The evolutionary history of eukaryotes is complicated by the fact that eukaryotic cells exhibit characteristics of both bacteria and archaea. Eukaryotic cells also contain organelles that are almost certainly the descendants of free-living prokaryotic cells. Specifically, the chloroplasts of plant cells, which carry out photosynthesis, closely resemble the photosynthetic cyanobacteria. The mitochondria of plant and animal cells, which are the site of

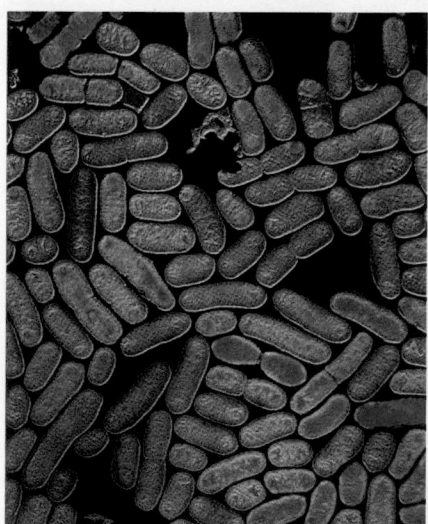

FIGURE 1.13

Prokaryotic cells. These single-celled *Escherichia coli* bacteria lack a nucleus and internal membrane systems. [*E. Gray/Science Photo Library/Photo Researchers.*]

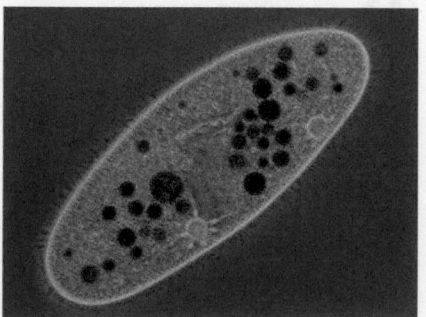

FIGURE 1.14

A eukaryotic cell. The paramecium, a one-celled organism, contains a nucleus and other membrane-bounded compartments. [*Dr. David Patterson/Science Photo Library/Photo Researchers.*]

  Describe the visible differences between prokaryotic and eukaryotic cells (Figures 1.13 and 1.14).

FIGURE 1.15

Evolutionary tree based on nucleotide sequences. This diagram reveals that the ancestors of archaea and bacteria separated before the eukarya emerged from an archaea-like ancestor. Note that the closely spaced fungi, plants, and animals are actually more similar to each other than are many groups of prokaryotes. [*After Wheelis, M. L., Kandler, O., and Woese, C. R., Proc. Natl. Acad. Sci. USA,* **89,** *2930–2934 (1992).*]

FIGURE 1.16

Possible origin of eukaryotic cells. The close association of different kinds of free-living cells gradually led to the modern eukaryotic cell, which appears to be a mosaic of bacterial and archaeal features and contains organelles that resemble whole bacterial cells.

much of the eukaryotic cell's aerobic metabolism, resemble certain bacteria. In fact, both chloroplasts and mitochondria contain their own genetic material and protein-synthesizing machinery.

It is likely that an early eukaryotic cell developed gradually from a mixed population of prokaryotic cells. Over many generations of living in close proximity and sharing each other's metabolic products, some of these cells became incorporated within a single larger cell. This arrangement would account for the mosaic-like character of modern eukaryotic cells (**Fig. 1.16**).

At some point, cells in dense populations might have traded their individual existence for a colonial lifestyle. This would have allowed for a division of labor as cells became specialized and would have eventually produced multicellular organisms.

The earth currently sustains about 9 million different species (although estimates vary widely). Perhaps some 500 million species have appeared and vanished over the course of evolutionary history. It is unlikely that the earth harbors more than a few mammals that have yet to be discovered, but new microbial species are routinely described. And although the number of known prokaryotes (about 10,000) is much less than the number of known eukaryotes (for example, there are about 900,000 known species of insects), prokaryotic metabolic strategies are amazingly varied. Nevertheless, by documenting characteristics that are common to all species, we can derive far-reaching conclusions about what life is made of, what sustains it, and how it has developed over the eons.

 CONCEPT CHECK

1. Describe how simple prebiotic compounds could give rise to biological monomers and polymers.

2. Discuss why anaerobic organisms came before aerobic organisms, and why are they relatively scarce now.

3. Differentiate between prokaryotes and eukaryotes.

4. Explain why eukaryotic cells appear to be mosaics.

5. Why is replication a requirement for life?

SUMMARY

1.2   Biological Molecules

• The most abundant elements in biological molecules are H, C, N, O, P, and S, but a variety of other elements are also present in living systems.

• The major classes of small molecules in cells are amino acids, monosaccharides, nucleotides, and lipids. The major types of biological polymers are proteins, nucleic acids, and polysaccharides.

1.3 Energy and Metabolism

• Free energy has two components: enthalpy (heat content) and entropy (disorder). Free energy decreases in a spontaneous process.

• Life is thermodynamically possible because unfavorable endergonic processes are coupled to favorable exergonic processes.

1.4 The Origin and Evolution of Life

• The earliest cells may have evolved in concentrated solutions of molecules or near hydrothermal vents.

• Eukaryotic cells contain membrane-bounded organelles. Prokaryotic cells, which are smaller and simpler, include the bacteria and the archaea.

KEY TERMS

bioinformatics	protein	endothermic reaction	evolution
homeostasis	peptide bond	ΔG	complement
trace element	conformation	spontaneous process	natural selection
amino acid	polynucleotide	exergonic reaction	aerobic
carbohydrate	nucleic acid	nonspontaneous process	anaerobic
monosaccharide	phosphodiester bond	endergonic reaction	prokaryote
nucleotide	polysaccharide	in vitro	bacteria
lipid	glycosidic bond	in vivo	archaea
monomer	free energy (G)	reduction	eukaryote
polymer	enthalpy (H)	oxidation	eukarya
residue	entropy (S)	enzyme	
polypeptide	exothermic reaction	replication	

BIOINFORMATICS

Brief Bioinformatics Exercises

1.1 The Periodic Table of the Elements and Domains of Life

1.2 Organic Functional Groups and the Three-Dimensional Structure of Vitamin C

PROBLEMS

1.2 Biological Molecules

1. Use Table 1.1 to assign the appropriate compound name to each molecule.

(a) $CH_3-CH_2-\overset{\displaystyle O}{\overset{\|}{C}}-CH_3$

(b) $H_3C-O-CH_2-CH_3$

(c) $H_3C-O-\overset{\displaystyle OH}{\underset{\displaystyle O}{\overset{|}{\underset{\|}{P}}}}-O-CH_3-CH_2$

(d) $H_3C-CH_2-CH_2-SH$

2. Use Table 1.1 to assign the appropriate compound name to each molecule.

(a) $H_3C-\overset{\displaystyle }{\underset{\displaystyle CH_3}{\overset{|}{C}H}}-OH$

(b) $H_3C-CH_2-CH_2-NH_2$

(c) $H_3C-\overset{\displaystyle O}{\overset{\|}{C}}-O-CH_2-CH_3$

(d) $H_3C-(CH_2)_{16}-\overset{\displaystyle O}{\overset{\|}{C}}-OH$

3. The structures of several molecules are shown below. Identify the functional groups in each structure.

Vitamin C

Vitamin K₃

Coenzyme Q

4. Investigators synthesized a series of compounds that showed promise as drugs for the treatment of Alzheimer's disease. The structure of one of the compounds is shown below. Identify the functional groups in this compound.

5. Name the four types of small biological molecules. Which three are capable of forming polymeric structures? What are the names of the polymeric structures that are formed?

6. To which of the four classes of biomolecules do the following compounds belong?

(a)

(b)

(c) $H_2N-CH_2-CH_2-CH_2-CH-COO^-$
                                                    $|$
                                                NH_3^+

(d)

7. The structures of the amino acids glutamine (Gln) and cysteine (Cys) are shown in Section 1.2. What functional group does Gln have that Cys does not? What functional group does Cys have that Gln does not?

8. The "straight-chain" structure of glucose is shown in Section 1.2. What functional groups are present in the glucose molecule?

9. Consider the monosaccharide fructose.

    (a) How does its molecular formula differ from that of glucose?

    (b) How does its structure differ from the structure of glucose?

Fructose

10. List two major biological roles of polysaccharides.

11. What are the structural components of the biological molecules called nucleotides?

12. The structures of the nitrogenous bases uracil and cytosine are shown below. How do their functional groups differ?

Uracil   Cytosine

13. A compound present in many foods has the formula $C_{44}H_{86}O_8NP$. To which class of molecules does this compound belong? Explain your answer.

14. The nutritive quality of food can be analyzed by measuring the amounts of the chemical elements it contains. Most foods are mixtures of the three major types of molecules: **(a)** fats (lipids), **(b)** carbohydrates, and **(c)** proteins. What elements are present in each of these types of molecules?

15. The structure of the compound urea is shown. Urea is a waste product of metabolism excreted by the kidneys into the urine. Why do doctors tell patients with kidney damage that they should consume a low-protein diet?

Urea

16. A healthy diet must include some protein. Assuming you had a way to measure the amount of each element in a sample of food, which element would you measure in order to tell whether the food contained protein?

17. The structures of three compounds are shown as follows. Based on your answer to Problem 16, which of the three compounds would you add to a food sample so that it would appear to contain more protein? Which of the three compounds would already be present in a food sample that actually did contain protein? Explain.

A   B   C

18. Compare the solubilities in water of serine, fructose, stearate, and cholesterol, and explain your reasoning.

19. What polymeric molecule forms a more regular structure, DNA or protein? Explain this observation in terms of the cellular roles of the two different molecules.

20. Plasma membranes have largely hydrophobic structures. Explain which compound will it allow to pass through more readily, galactose or picric acid?

Picric acid

21. Pancreatic amylase digests the glycosidic bonds that link glucose residues together in starch. Would you expect this enzyme to digest the glycosidic bonds in cellulose as well? Explain why or why not.

22. The complete digestion of starch in mammals yields 4 kilocalories per gram (see Problem 21). What is the energy yield for cellulose?

1.3 Energy and Metabolism

23. In which of the following reactions in aqueous solution does the entropy increase?

24. Which has the lesser entropy, a polymeric molecule or a mixture of its constituent monomers? Explain.

25. What is the sign of the entropy change for each of the following processes?

(a) Water freezes.
(b) Water evaporates.
(c) Dry ice sublimes.
(d) Sodium chloride dissolves in water.
(e) Several different types of lipid molecules assemble to form a membrane.

26. How does the entropy change when glucose is formed during photosynthesis?

$$6\ CO_2 + 6\ H_2O \rightarrow C_6H_{12}O_6 + 6\ O_2$$

27. Urea (NH_2CONH_2) dissolves readily in water; *i.e.*, this is a spontaneous process. The beaker containing the dissolved compound is cold to the touch. What conclusions can you make about the sign of the (a) enthalpy change and (b) entropy change for this process?

28. A soccer coach keeps a couple of instant cold packs in her bag in case one of her players suffers a muscle injury. Instant cold packs are composed of a plastic bag containing a smaller water bag and solid ammonium nitrate. In order to activate the cold pack, the bag is kneaded until the smaller water bag breaks, which allows the released water to dissolve the ammonium nitrate. The equation for the dissolution of ammonium nitrate in water is shown below. How does the cold pack work?

$$NH_4NO_3\,(s) \xrightarrow{H_2O} NH_4^+(aq) + NO_3^-(aq)$$
$$\Delta H = 26.4\ kJ \cdot mol^{-1}$$

29. Campers carry hot packs with them, especially when camping during the winter months or at high altitudes. The design is similar to that described in Problem 28, except that calcium chloride is used in place of the ammonium nitrate. The equation for the dissolution of calcium chloride in water is shown below. How does the hot pack work?

$$CaCl_2\,(s) \xrightarrow{H_2O} Ca^{2+}(aq) + 2\ Cl^-(aq) \qquad \Delta H = -81\ kJ \cdot mol^{-1}$$

30. For a given reaction, the value of ΔH is 25 kJ $\cdot$ mol^{-1} and the value of ΔS is 85 J $\cdot$ K^{-1} $\cdot$ mol^{-1}. Above what temperature will this reaction be spontaneous?

31. For the reaction in which reactant A is converted to product B, tell whether this process is favorable at (a) 4°C and (b) 37°C.

	H (kJ $\cdot$ mol^{-1})	S (J $\cdot$ K^{-1} $\cdot$ mol^{-1})
A	43	11
B	50	37

32. Phosphoenolpyruvate donates a phosphate group to ADP to produce pyruvate and ATP. The ΔG value for this reaction at 25°C is -63 kJ $\cdot$ mol^{-1} and the value of ΔS is 190 J $\cdot$ K^{-1} $\cdot$ mol^{-1}. What is the value of ΔH? Is heat absorbed from or released to the surroundings?

33. Which of the following processes are spontaneous?

(a) A reaction that occurs with any size decrease in enthalpy and any size increase in entropy.
(b) A reaction that occurs with a small increase in enthalpy and a large increase in entropy.
(c) A reaction that occurs with a large decrease in enthalpy and a small decrease in entropy.
(d) A reaction that occurs with any size increase in enthalpy and any size decrease in entropy.

34. When a stretched rubber band is allowed to relax, it feels cooler.

(a) Is the enthalpy change for this process positive or negative?
(b) Since a stretched rubber band spontaneously relaxes, what can you conclude about the entropy change during relaxation?

35. A monoclonal antibody binds to the protein cytochrome *c*. The ΔH value for binding at 25°C is -87.9 kJ $\cdot$ mol^{-1} and the ΔS is -118 J $\cdot$ K^{-1} $\cdot$ mol^{-1}.

(a) Does entropy increase or decrease when the antibody binds to the protein?
(b) Calculate ΔG for the formation of the antibody–protein complex. Does the complex form spontaneously?

(c) The ΔG value for the binding of a second monoclonal antibody to cytochrome c is -58.2 kJ $\cdot$ mol^{-1}. Which antibody would bind more readily to the protein?

36. Glucose can be converted to glucose-6-phosphate:

glucose + phosphate $\rightarrow$ glucose-6-phosphate + H_2O
$\Delta G = 13.8$ kJ $\cdot$ mol^{-1}

(a) Is this reaction favorable? Explain.

(b) Suppose the synthesis of glucose-6-phosphate is coupled with the hydrolysis of ATP shown below.

ATP + H_2O $\rightarrow$ ADP + phosphate $\Delta G = -30.5$ kJ $\cdot$ mol^{-1}

Write the overall equation for the coupled process and calculate the ΔG for the coupled reaction. Is the conversion of glucose to glucose-6-phosphate favorable under these conditions? Explain.

37. Glyceraldehyde-3-phosphate (GAP) is converted to 1,3-bisphosphoglycerate (1,3BPG) as shown.

GAP + P$_i$ + NAD$^+$ $\rightarrow$ 1,3BPG + NADH $\Delta G = +6.7$ kJ $\cdot$ mol^{-1}

(a) Is this reaction spontaneous?

(b) The reaction shown above is coupled to the following reaction in which 1,3BPG is converted to 3-phosphoglycerate (3PG):

1,3BPG + ADP $\rightarrow$ 3PG + ATP $\Delta G = -18.8$ kJ $\cdot$ mol^{-1}

Write the overall equation for the conversion of GAP to 3PG. Is the coupled reaction favorable? Explain.

38. The ΔG value at 37°C of the arginine kinase reaction was recently determined. The ΔH for the reaction is -8.19 kJ $\cdot$ mol^{-1} and the value of ΔS is 2.20 J $\cdot$ K^{-1} $\cdot$ mol^{-1}.

(a) Is the reaction exothermic or endothermic?

(b) What is the value of ΔG for the reaction? Is the reaction spontaneous?

(c) Which component makes a greater contribution to the free energy value—the ΔH or ΔS value? Comment on the significance of this observation.

39. Place these molecules in order from the most oxidized to the most reduced.

$$
\begin{array}{ccc}
\overset{\displaystyle H}{\underset{\displaystyle H}{H-C-H}} & O=C=O & \overset{\displaystyle O}{H-C-OH} \\
A & B & C
\end{array}
$$

40. Identify the process described in the following statements as an oxidation or reduction process.

(a) A tree synthesizing monosaccharides from carbon dioxide by using sunlight during photosynthesis.

(b) An animal eating the plant and breaking down the monosaccharide to obtain energy for cellular processes.

41. Given the following reactions, tell whether the reactant is being oxidized or reduced. Reactions may not be balanced.

(a)
$$
\underset{COO^-}{\overset{COO^-}{\underset{CH-OH}{\underset{CH_2}{|}}}} \longrightarrow \underset{COO^-}{\overset{COO^-}{\underset{C=O}{\underset{CH_2}{|}}}}
$$

(b) $CH_3{-}(CH_2)_{14}{-}\overset{\displaystyle O}{\overset{\|}{C}}{-}O^- \longrightarrow 8\ CH_3{-}\overset{\displaystyle O}{\overset{\|}{C}}{-}S{-}CoA$

(c)
$$
^+H_3N{-}\underset{\underset{\underset{\underset{\underset{\underset{-O{-}C{-}CH{-}NH_3^+}{\underset{\|}{O}}}{CH_2}}{S}}{S}}{CH_2}}{\overset{\displaystyle O}{\overset{\|}{CH}{-}C{-}O^-}} \ + H_2 \longrightarrow 2\ ^+H_3N{-}\underset{\underset{SH}{CH_2}}{\overset{\displaystyle O}{\overset{\|}{CH}{-}C{-}O^-}}
$$

(d)
$$
\underset{COO^-}{\overset{COO^-}{\underset{CH}{\underset{CH}{\overset{\|}{|}}}}} \longrightarrow \underset{COO^-}{\overset{COO^-}{\underset{CH-OH}{\underset{CH_2}{|}}}}
$$

42. For each of the reactions in Problem 41, tell whether an oxidizing agent or a reducing agent is needed to accomplish the reaction.

43. Which yields more free energy when completely oxidized, stearate or α-linolenate?

$$H_3C{-}(CH_2)_{16}{-}COO^-$$
Stearate

$$H_3C{-}CH_2{-}(CH{=}CHCH_2)_3{-}(CH_2)_6{-}COO^-$$
α-Linolenate

44. In some cells, lipids such as palmitate (shown in Section 1.2), rather than monosaccharides, serve as the primary metabolic fuel.

(a) Consider the oxidation state of palmitate's carbon atoms and explain how it fits into a scheme such as the one shown in Fig. 1.9.

(b) On a per-carbon basis, which would make more free energy available for metabolic reactions: palmitate or glucose?

1.4   The Origin and Evolution of Life

45. In order to give rise to more highly complex structures, what capabilities did the first biological molecules have to have?

46. Why is molecular information so important for classifying and tracing the evolutionary relatedness of bacterial species but less important for vertebrate species?

47. The first theories to explain the similarities between bacteria and mitochondria or chloroplasts suggested that an early eukaryotic cell actually engulfed but failed to fully digest a free-living prokaryotic cell. Why is such an event unlikely to account for the origin of mitochondria or chloroplasts?

48. In the 1920s, Oparin and Haldane independently suggested that the energy of lightning storms in the prebiotic world could have transformed gases in the early atmosphere to small organic molecules. In 1953, Miller and Urey conducted an experiment demonstrating that this scenario is plausible. They subjected a mixture of water, methane, ammonia, and hydrogen gas to electric discharge and refluxed the mixture so that any compounds formed would dissolve and accumulate in the water. At the end of a week, they analyzed the solution and found glycine, alanine, lactic acid, urea, and other amino acids and small organic acids. What was the significance of this experiment?

49. A portion of the evolutionary tree for a flu virus is shown here. Different strains are identified by an H followed by a number.

(a) Identify two pairs of closely related flu strains.

(b) Which strain(s) is(are) most closely related to strain H3?

50. Draw a simple evolutionary tree that shows the relationships between species A, B, and C based on the DNA sequences given here.

Species A	T G G A C C A G C C
Species B	T G G A C T A G C C
Species C	T C G T C G A G T C

SELECTED READINGS

Koonin, E. V., The origin and early evolution of eukaryotes in the light of phylogenomics, *Genome Biol.* **11,** 209 (2010). [Discusses several models for the origin of eukaryotic cells.]

Koshland, D. E., Jr., The seven pillars of life, *Science* **295,** 2215–2216 (2002). [Describes some of the essential attributes of all organisms, including a DNA program, ability to mutate, compartmentalization, need for energy, ability to regenerate, adaptability, and seclusion.]

Mora, C., Tittensor, D. P., Adl, S., Simpson, A. G. B., and Worm, B., How many species are there on earth and in the ocean? *PLoS Biol* **9(8):** e1001127. doi:10.1371/journal.pbio.1001127 (2011). [Shows how statistical analysis of databases can be used to estimate the total number of known and not-yet-described species.]

Nee, S., More than meets the eye, *Nature* **429,** 804–805 (2004). [A brief commentary about appreciating the metabolic diversity of microbial life.]

Nisbet, E. G., and Sleep, N. H., The habitat and nature of early life, *Nature* **409,** 1083–1091 (2001). [Explains some of the hypotheses regarding the early earth and the origin of life, including the possibility that life originated at hydrothermal vents.]

Tinoco, I., Jr., Sauer, K., Wang, J. C., Puglisi, J. C., Harbison, G., and Rovnyak, D., *Physical Chemistry. Principles and Applications in Biological Sciences* (5th ed.), Chapters 2–4, Prentice-Hall (2014). [This and other physical chemistry textbooks present the basic equations of thermodynamics.]

CHAPTER 2

AQUEOUS CHEMISTRY

Lisa Collins/robertharding/Getty Images, Inc.

Plotosus japonicus, a species of catfish, locates food by detecting the subtle change in pH caused by the release of carbon dioxide from hidden prey organisms.

DO YOU REMEMBER?

- Organisms maintain a state of homeostasis (Section 1.1)
- Biological molecules are composed of a subset of all possible elements and functional groups (Section 1.2).
- The free energy of a system is determined by its enthalpy and entropy (Section 1.3).

Water is a fundamental requirement for life, so it is important to understand the structural and chemical properties of water. Not only are most biological molecules surrounded by water, but their molecular structure is in part governed by how their component groups interact with water. And water plays a role in how these molecules assemble to form larger structures or undergo chemical transformation. In fact, water itself—or its H^+ and OH^- constituents—participates directly in many biochemical processes. Therefore, an examination of water is a logical prelude to exploring the structures and functions of biomolecules in the following chapters.

2.1 WATER MOLECULES AND HYDROGEN BONDS

KEY CONCEPTS
- The polar water molecule forms hydrogen bonds with other molecules.
- Noncovalent forces, including hydrogen bonds, ionic interactions, and van der Waals forces, act on biological molecules.
- Water dissolves both ionic and polar substances.

What is the nature of the substance that accounts for about 70% of the mass of most organisms? The human body, for example, is about 60% by weight water, most of it in the extracellular fluid (the fluid surrounding cells) and inside cells:

Intracellular water (40%)

Non-water (40%)

Extracellular water (15%)

Water in circulatory system (5%)

In an individual H_2O molecule, the central oxygen atom forms covalent bonds with two hydrogen atoms, leaving two unshared pairs of electrons. The molecule therefore has approximately tetrahedral geometry, with the oxygen atom at the center of the tetrahedron, the hydrogen atoms at two of the four corners, and electrons at the other two corners (**Fig. 2.1**).

As a result of this electronic arrangement, *the water molecule is **polar***; that is, it has an uneven distribution of charge. The oxygen atom bears a partial negative charge (indicated by the symbol δ^-), and each hydrogen atom bears a partial positive charge (indicated by the symbol δ^+):

This polarity is the key to many of water's unique physical properties.

Neighboring water molecules tend to orient themselves so that each partially positive hydrogen is aligned with a partially negative oxygen:

This interaction, shaded yellow here, is known as a **hydrogen bond.** This weak electrostatic attraction between oppositely charged particles actually has some covalent character. In addition, the bond has directionality, or a preferred orientation.

Each water molecule can potentially participate in four hydrogen bonds, since it has two hydrogen atoms to "donate" to a hydrogen bond and two pairs of unshared electrons that can "accept" a hydrogen bond. In ice, a crystalline form of water, each water molecule does indeed form hydrogen bonds with four other water molecules (**Fig. 2.2**). This regular, lattice-like structure breaks down when the ice melts.

In liquid water, each molecule can potentially form hydrogen bonds with up to four other water molecules, but each bond has a lifetime of only about 10^{-12} s. As a result, *the structure of water is continually flickering as water molecules rotate, bend, and reorient themselves.* Theoretical calculations and spectroscopic data suggest that water molecules participate in only two strong hydrogen bonds, one as a donor and one as an acceptor, generating transient hydrogen-bonded clusters such as the six-membered ring shown here:

Because of its ability to form hydrogen bonds, water is highly cohesive. This accounts for its high surface tension, which allows certain insects to walk on water (**Fig. 2.3**). The cohesiveness of water molecules also explains why water remains a liquid, whereas molecules of similar size, such as CH_4 and H_2S, are gases at room temperature (25°C). At the same time,

FIGURE 2.1

Electronic structure of the water molecule. Four electron orbitals, in an approximately tetrahedral arrangement, surround the central oxygen. Two orbitals participate in bonding to hydrogen (gray), and two contain unshared electron pairs.

FIGURE 2.2

Structure of ice. Each water molecule acts as a donor for two hydrogen bonds and an acceptor for two hydrogen bonds, thereby interacting with four other water molecules in the crystal. (Only two layers of water molecules are shown here.)

? Identify a hydrogen bond donor and acceptor in this structure.

water is less dense than other liquids because hydrogen bonding demands that individual molecules not just approach each other but interact with a certain orientation. This geometrical constraint also explains why ice floats; for other materials, the solid is denser than the liquid.

HYDROGEN BONDS ARE ONE TYPE OF ELECTROSTATIC FORCE

Powerful covalent bonds define basic molecular constitutions, but much weaker noncovalent bonds, including hydrogen bonds, govern the final three-dimensional shapes of molecules and how they interact with each other. For example, about $460 \text{ kJ} \cdot \text{mol}^{-1}$ ($110 \text{ kcal} \cdot \text{mol}^{-1}$) of energy is required to break a covalent O—H bond. But a hydrogen bond in water has a strength of only about $20 \text{ kJ} \cdot \text{mol}^{-1}$ ($4.8 \text{ kcal} \cdot \text{mol}^{-1}$). Other noncovalent interactions are weaker still.

Among the noncovalent interactions that occur in biological molecules are electrostatic interactions between charged groups such as carboxylate ($—COO^-$) and amino ($—NH_3^+$) groups. These **ionic interactions** are intermediate in strength to covalent bonds and hydrogen bonds (**Fig. 2.4**).

Hydrogen bonds, despite their partial covalent nature, are classified as a type of electrostatic interaction. At about 1.8 Å, they are longer and hence weaker than a covalent O—H bond (which is about 1 Å long). However, a completely noninteracting O and H would approach no closer than about 2.7 Å, which is the sum of their **van der Waals radii** (the van der Waals radius of an isolated atom is the distance from its nucleus to its effective electronic surface).

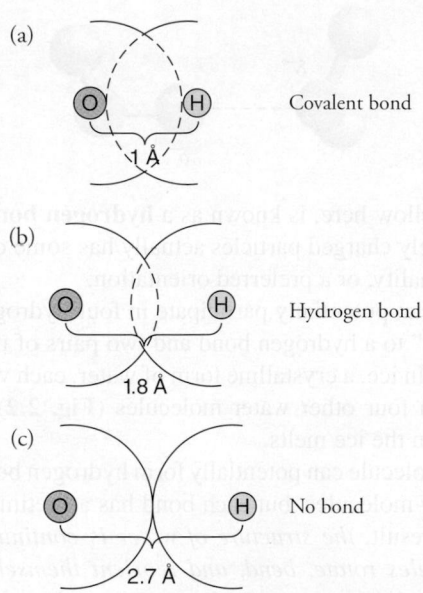

Hydrogen bonds usually involve N—H and O—H groups as hydrogen donors and the electronegative N and O and occasionally S atoms as hydrogen acceptors (**electronegativity** is a measure of an atom's affinity for electrons; **Table 2.1**). *Water, therefore, can form hydrogen bonds not just with other water molecules but with a wide variety of other compounds that bear N- and O-containing functional groups.*

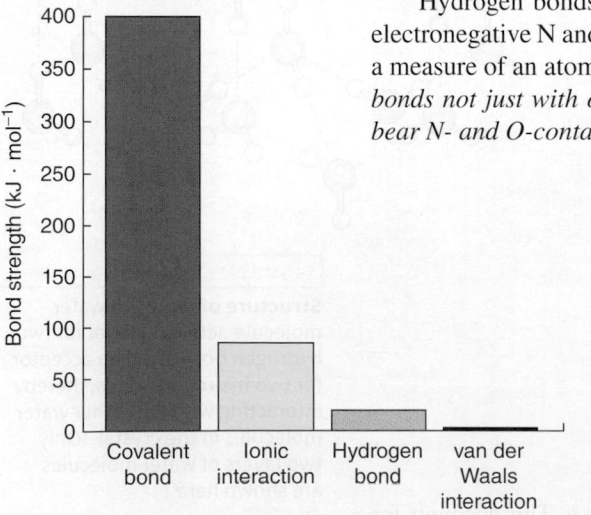

FIGURE 2.4

Relative strengths of bonds in biological molecules.

TABLE 2.1 Electronegativities of Some Elements	
ELEMENT	**ELECTRONEGATIVITY**
C	2.55
F	3.98
H	2.20
N	3.04
O	3.44
S	2.58

Likewise, these functional groups can form hydrogen bonds among themselves. For example, the complementarity of bases in DNA and RNA is determined by their ability to form hydrogen bonds with each other. Here, three N—H groups are hydrogen bond donors, and N and O atoms are acceptors:

Guanine

Cytosine

Other electrostatic interactions occur between particles that are polar but not actually charged, for example, two carbonyl groups:

$$\overset{\delta^-}{C=O}\text{-----}\overset{\delta^+}{C}=O$$

These forces, called **van der Waals interactions,** are usually weaker than hydrogen bonds. The interaction between two strongly polar groups is known as a **dipole–dipole interaction** and has a strength of about 9 kJ · mol^{-1}. Very weak van der Waals interactions, called **London dispersion forces,** occur between nonpolar molecules as a result of small fluctuations in their distribution of electrons that create a temporary separation of charge. Nonpolar groups such as methyl groups can therefore experience a small-attractive force, in this case about 0.3 kJ · mol^{-1}:

$$\overset{\delta^-}{-C}-\overset{\delta^+}{H}\text{-----}\overset{\delta^-}{H}-\overset{\delta^+}{C}-$$

Not surprisingly, these forces act only when the groups are very close, and their strength quickly falls off as the groups draw apart. If the groups approach too closely, however, their van der Waals radii collide and a strong repulsive force overcomes the attractive force.

Although hydrogen bonds and van der Waals interactions are individually weak, biological molecules usually contain multiple groups capable of participating in these intermolecular interactions, so their cumulative effect can be significant (**Fig. 2.5**). These weak forces also determine how biological molecules can "recognize" or bind noncovalently to each other. Drug molecules are typically designed to optimize the weak interactions that govern their therapeutic activity (**Box 2.A**).

FIGURE 2.5

The cumulative effect of small forces. Just as the fictional giant Gulliver was restrained by many small tethers at the hands of the tiny Lilliputians, the structures of macromolecules are constrained by the effects of many weak noncovalent interactions. [*Hulton Archive/Getty Images.*]

Box 2.A Why Do Some Drugs Contain Fluorine?

As mentioned in Section 1.2, the most abundant elements in biological molecules are H, C, N, O, P, and S. Fluorine only rarely appears in naturally occurring organic compounds. Why, then, do about one-quarter of all drug molecules, including the widely prescribed Prozac (fluoxetine, an antidepressant; Box 9.B), fluorouracil (an anticancer agent; Section 7.3), and Ciprofloxacin (an antibacterial agent; Section 20.5), contain F?

biological properties without significantly altering its shape. The small fluorine can take the place of hydrogen in a chemical structure, but with its high electronegativity (see Table 2.1), F behaves much more like O than H. Consequently, transforming a relatively inert C—H group into an electron-withdrawing C—F group can decrease the basicity of nearby amino groups (see Section 2.3). Fewer positive charges in a drug allow it to more easily pass through membranes to enter cells and exert its biological effect.

In addition, the polar C—F bond can participate in hydrogen bonding (C—F · · · H—C) or other dipole–dipole interactions (such as C—F · · · C=O), potentially augmenting the intermolecular attraction between a drug and its target molecule in the body. Better binding usually means that the drug will be effective at lower concentrations and will have fewer side effects.

Prozac (Fluoxetine)

In designing an effective drug, pharmaceutical scientists often intentionally introduce F in order to alter the drug's chemical or

? **Identify the hydrogen-bonding groups in Prozac.**

WATER DISSOLVES MANY COMPOUNDS

Unlike most other solvent molecules, water molecules are able to form hydrogen bonds and participate in other electrostatic interactions with a wide variety of compounds. Water has a relatively high **dielectric constant,** which is a measure of a solvent's ability to diminish the electrostatic attractions between dissolved ions (**Table 2.2**). The higher the dielectric constant of the solvent, the less able the ions are to associate with each other. The polar water molecules surround ions (for example, the Na^+ and Cl^- ions from the salt NaCl) by aligning their partial charges with the oppositely charged ions. Because the interactions between the polar water molecules and the ions are stronger than the attractive forces between the Na^+ and Cl^- ions, the salt dissolves (the dissolved particle is called a **solute**). Each solute ion surrounded by water molecules (shown here) is said to be **solvated** (or **hydrated,** to indicate that the solvent is water).

Biological molecules that bear polar or ionic functional groups are also readily solubilized, in this case because the groups can form hydrogen bonds with the solvent water molecules. Glucose, for example, with its six hydrogen-bonding oxygens, is highly soluble in water:

Note that when we describe the behavior of a single molecule, such as glucose in this example, we are really describing the average behavior of a huge number of molecules. (Most biochemical techniques cannot assess the activity of an individual molecule.)

The concentration of glucose in human blood is about 5 mM. In a solution of 5 mM glucose in water, there are about 10,000 water molecules for every glucose molecule (the water molecules are present at a concentration of about 55.5 M). However, biological molecules are never found alone in such dilute conditions *in vivo*, because a large number of small molecules,

TABLE 2.2 Dielectric Constants for Some Solvents at Room Temperature

SOLVENT	DIELECTRIC CONSTANT
Formamide ($HCONH_2$)	109
Water	80
Methanol (CH_3OH)	33
Ethanol (CH_3CH_2OH)	25
1-Propanol ($CH_3CH_2CH_2OH$)	20
1-Butanol ($CH_3CH_2CH_2CH_2OH$)	18
Benzene (C_6H_6)	2

? Compare the hydrogen-bonding ability of these solvents.

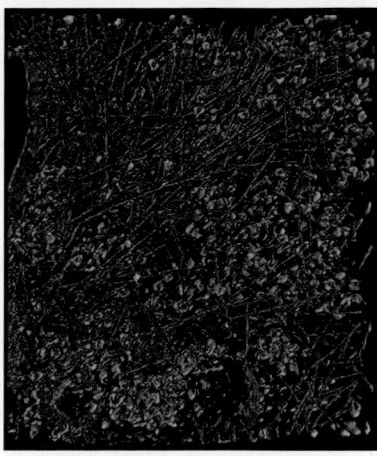

FIGURE 2.6

Portion of a *Dictyostelium* cell visualized by cryoelectron tomography. In this technique, the cells are rapidly frozen so that they retain their fine structure, and two-dimensional electron micrographs taken from different angles are merged to re-create a three-dimensional image. The red structures are filaments of the protein actin, ribosomes and other macromolecular complexes are colored green, and membranes are blue. Small molecules (not visible) fill the spaces between these larger cell components. [*Courtesy Wolfgang Baumeister, Max Planck Institute for Biochemistry.*]

large polymers, and macromolecular aggregates collectively form a solution that is more like a hearty stew than a thin, watery soup (**Fig. 2.6**).

Inside a cell, the spaces between molecules may be only a few Å wide, enough room for only two water molecules to fit. This allows solute molecules, each with a coating of properly oriented water molecules, to slide past each other. This thin coating, or shell, of water may be enough to keep molecules from coming into van der Waals contact (van der Waals interactions are weak but attractive), thereby helping maintain the cell's contents in a crowded but fluid state.

 CONCEPT CHECK

1. Why is a water molecule polar?

2. Draw three hydrogen-bonded water molecules.

3. Describe the structure of liquid water.

4. Compare the nature and strengths of covalent bonds, hydrogen bonds, ionic interactions, and van der Waals interactions.

5. Describe what happens when an ionic substance dissolves in water.

6. Explain why water is a more effective solvent than ammonia or methanol.

THE HYDROPHOBIC EFFECT

Glucose and other readily hydrated substances are said to be **hydrophilic** (water-loving). In contrast, a compound such as dodecane (a C_{12} alkane),

which lacks polar groups, is relatively insoluble in water and is said to be **hydrophobic** (water-fearing). Although pure hydrocarbons are rare in biological systems, many biological molecules contain hydrocarbon-like portions that are insoluble in water.

When a nonpolar substance such as vegetable oil (which consists of hydrocarbon-like molecules) is added to water, it does not dissolve but forms a separate phase. In order for the water and oil to mix, free energy must be added to the system (for example, by stirring

KEY CONCEPTS
- The hydrophobic effect, which is driven by entropy, excludes nonpolar substances from water.
- Amphiphilic molecules form micelles or bilayers.

Nonpolar molecule

Water

Layer of constrained water molecules

FIGURE 2.7

Hydration of a nonpolar molecule. When a nonpolar substance (green) is added to water, the system loses entropy because the water molecules surrounding the nonpolar solute (orange) lose their freedom to form hydrogen bonds. The loss of entropy is a property of the entire system, not just the water molecules nearest the solute, because these molecules are continually changing places with water molecules from the rest of the solution. The loss of entropy presents a thermodynamic barrier to the hydration of a nonpolar solute.

vigorously or applying heat). Why is it thermodynamically unfavorable to dissolve a hydrophobic substance in water? One possibility is that enthalpy is required to break the hydrogen bonds among solvent water molecules in order to create a "hole" into which a nonpolar molecule can fit. Experimental measurements, however, show that the free energy barrier (ΔG) to the solvation process depends much more on the entropy term (ΔS) than on the enthalpy term (ΔH; recall from Chapter 1 that $\Delta G = \Delta H - T\Delta S$; Equation 1.2). This is because when a hydrophobic molecule is hydrated, it becomes surrounded by a layer of water molecules that cannot participate in normal hydrogen bonding with each other but instead must align themselves so that their polar ends are not oriented toward the nonpolar solute. *This constraint on the structure of water represents a loss of entropy in the system,* because now the highly mobile water molecules have lost some of their freedom to rapidly form, break, and re-form hydrogen bonds with other water molecules (**Fig. 2.7**). But note that the loss of entropy is not due to the formation of a frozen "cage" of water molecules around the nonpolar solute, as commonly pictured, because in liquid water, the solvent molecules are in constant motion.

When a large number of nonpolar molecules are introduced into a sample of water, they do not disperse and become individually hydrated, each surrounded by a layer of water molecules. Instead, the nonpolar molecules tend to clump together, removing themselves from contact with water molecules. (This explains why small oil droplets coalesce into one large oily phase.) Although the entropy of the nonpolar molecules is thereby reduced, this thermodynamically unfavorable event is more than offset by the increase in the entropy of the water molecules, which regain their ability to interact freely with other water molecules (**Fig. 2.8**).

(a)   (b)

FIGURE 2.8

Aggregation of nonpolar molecules in water. (a) The individual hydration of dispersed nonpolar molecules (green) decreases the entropy of the system because the hydrating water molecules (orange) are not as free to form hydrogen bonds. (b) Aggregation of the nonpolar molecules increases the entropy of the system, since the number of water molecules required to hydrate the aggregated solutes is less than the number of water molecules required to hydrate the dispersed solute molecules. This increase in entropy accounts for the spontaneous aggregation of nonpolar substances in water.

? **Explain why it is incorrect to describe the behavior shown in part (b) in terms of "hydrophobic bonds."**

*The exclusion of nonpolar substances from an aqueous solution is known as the **hydrophobic effect.*** It is a powerful force in biochemical systems, even though it is not a bond or an attractive interaction in the conventional sense. The nonpolar molecules do not experience any additional attractive force among themselves; they aggregate only because they are driven out of the aqueous phase by the unfavorable entropy cost of individually hydrating them. The hydrophobic effect governs the structures and functions of many biological molecules. For example, each polypeptide chain of a protein folds into a globular mass so that its hydrophobic groups are in the interior, away from the solvent, and its polar groups are on the exterior, where they can interact with water. Similarly, the structure of the lipid membrane that surrounds all cells is maintained by the hydrophobic effect acting on the lipids.

AMPHIPHILIC MOLECULES EXPERIENCE BOTH HYDROPHILIC INTERACTIONS AND THE HYDROPHOBIC EFFECT

Consider a molecule such as the fatty acid palmitate:

The hydrocarbon "tail" of the molecule (on the right) is nonpolar, while its carboxylate "head" (on the left) is strongly polar. Molecules such as this one, which have both hydrophobic and hydrophilic portions, are said to be **amphiphilic** or **amphipathic.** What happens when amphiphilic molecules are added to water? In general, *the polar groups of amphiphiles orient themselves toward the solvent molecules and are therefore hydrated, while the nonpolar groups tend to aggregate due to the hydrophobic effect.* As a result, the amphiphiles may form a spherical **micelle,** a particle with a solvated surface and a hydrophobic core (**Fig. 2.9**).

Depending in part on the relative sizes of the hydrophilic and hydrophobic portions of the amphiphiles, the molecules may form a sheet rather than a spherical micelle. The amphiphilic lipids that provide the structural basis of biological membranes form two-layered sheets called **bilayers,** in which a hydrophobic layer is sandwiched between hydrated polar surfaces (**Fig. 2.10**). The structures of biological membranes are discussed in more detail in Chapter 8. The formation of micelles or bilayers is thermodynamically favored because the hydrogen-bonding capacity of the polar head groups is satisfied through interactions with solvent water molecules, and the nonpolar tails are sequestered from the solvent.

THE HYDROPHOBIC CORE OF A LIPID BILAYER IS A BARRIER TO DIFFUSION

To eliminate its solvent-exposed edges, a lipid bilayer tends to close up to form a **vesicle,** shown cut in half:

Many of the subcellular compartments (organelles) in eukaryotic cells have a similar structure.

When the vesicle forms, it traps a volume of the aqueous solution. Polar solutes in the enclosed compartment tend to remain there because they cannot easily pass through the hydrophobic interior of the bilayer. The energetic cost of transferring a hydrated polar group through the nonpolar lipid tails is too great. (In contrast, small nonpolar molecules such as O_2 can pass through the bilayer relatively easily.)

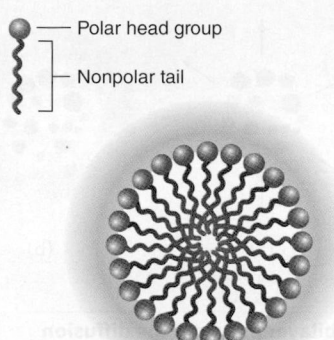
— Polar head group
— Nonpolar tail

FIGURE 2.9

Cross section of a micelle formed by amphiphilic molecules. The hydrophobic tails of the molecules aggregate, out of contact with water, due to the hydrophobic effect. The polar head groups are exposed to and can interact with the solvent water molecules.

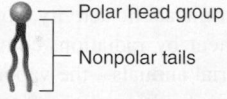
— Polar head group
— Nonpolar tails

FIGURE 2.10

A lipid bilayer. The amphiphilic lipid molecules form two layers so that their polar head groups are exposed to the solvent while their hydrophobic tails are sequestered in the interior of the bilayer, away from water. The likelihood of amphiphilic molecules forming a bilayer rather than a micelle depends in part on the sizes and nature of the hydrophobic and hydrophilic groups. One-tailed lipids tend to form micelles (see Fig. 2.9), and two-tailed lipids tend to form bilayers.

? Indicate where a sodium ion and a benzene molecule would be located.

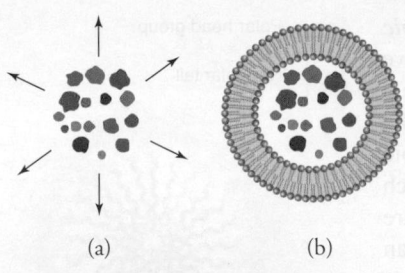

(a)   (b)

FIGURE 2.11

A bilayer prevents the diffusion of polar substances. (a) Solutes spontaneously diffuse from a region of high concentration to a region of low concentration. (b) A lipid barrier, which presents a thermodynamic barrier to the passage of polar substances, prevents the diffusion of polar substances out of the inner compartment (it also prevents the inward diffusion of polar substances from the external solution).

FIGURE 2.12

Ionic composition of intracellular and extracellular fluid. Human cells contain much higher concentrations of potassium than of sodium or chloride; the opposite is true of the fluid outside the cell. The cell membrane helps maintain the concentration differences.

Normally, substances that are present at high concentrations tend to diffuse to regions of lower concentration. (This movement "down" a concentration gradient is a spontaneous process driven by the increase in entropy of the solute molecules.) *A barrier such as a bilayer can prevent this diffusion* (**Fig. 2.11**). This helps explain why cells, which are universally enclosed by a membrane, can maintain their specific concentrations of ions, small molecules, and biopolymers even when the external concentrations of these substances are quite different (**Fig. 2.12**). The solute composition of intracellular compartments and other biological fluids is carefully regulated. Not surprisingly, organisms spend a considerable amount of metabolic energy to maintain the proper concentrations of water and salts, and losses of one or the other must be compensated (**Box 2.B**).

A CLOSER LOOK

Box 2.B Sweat, Exercise, and Sports Drinks

Animals, including humans, generate heat, even at rest, due to their metabolic activity. Some of this heat is lost to the environment by radiation, convection, conduction, and—in terrestrial animals—the vaporization of water. Evaporation has a significant cooling effect because about 2.5 kJ of heat is given up for every gram (mL) of water lost. In humans and certain other animals, an increase in skin temperature triggers the activity of sweat glands, which secrete a solution containing (in humans) about 50 mM Na^+, 5 mM K^+, and 45 mM Cl^-. The body is cooled as the sweat evaporates from its surface.

The evaporation of water accounts for a small portion of a resting body's heat loss, but sweating is the main mechanism for dissipating heat generated when the body is highly active. During vigorous exercise or exertion at high ambient temperatures, the body may experience a fluid loss of up to 2 L per hour. Athletic training not only improves the performance of the muscles and cardiopulmonary system, it also increases the capacity for sweating so that the athlete begins to sweat at a lower skin temperature and loses less salt in the secretions of the sweat glands. But regardless of training, a fluid loss representing more than 2% of the body's weight may impair cardiovascular function. In fact, "heat

exhaustion" in humans is usually due to dehydration rather than an actual increase in body temperature.

Numerous studies have concluded that athletes seldom drink enough before or during exercise. Ideally, fluid intake should match the losses due to sweat, and the rate of intake should keep pace with the rate of sweating. So what should the conscientious athlete drink? For activities lasting less than about 90 minutes, especially when periods of high intensity alternate with brief periods of rest, water alone is sufficient. Commercial sports drinks containing carbohydrates can replace the water lost as sweat and also provide a source of energy. However, this carbohydrate boost may be an advantage only during prolonged sustained activity, such as during a marathon, when the body's own carbohydrate stores are depleted. A marathon runner or a manual laborer in the hot sun might benefit from the salt found in sports drinks, but most athletes don't need the supplemental salt (although it does make the carbohydrate solution more palatable). A normal diet usually contains enough Na^+ and Cl^- to offset the losses in sweat.

 Compare the ion concentrations of sweat and extracellular fluid.

 CONCEPT CHECK

1. What does entropy have to do with the solubility of nonpolar substances in water?
2. Distinguish between hydrophobic and hydrophilic substances.
3. Explain why polar molecules dissolve more easily than nonpolar substances in water.
4. How can a molecule be both hydrophilic and hydrophobic? Give an example.
5. Why is lipid bilayer a barrier to the diffusion of polar molecules.

2.3 ACID–BASE CHEMISTRY

Water is not merely an inert medium for biochemical processes; it is an active participant. Its chemical reactivity in biological systems is in part a result of its ability to ionize. This can be expressed in terms of a chemical equilibrium:

$$H_2O \rightleftharpoons H^+ + OH^-$$

The products of water's dissociation are a hydrogen ion or proton (H^+) and a hydroxide ion (OH^-).

Aqueous solutions do not actually contain lone protons. Instead, the H^+ can be visualized as combining with a water molecule to produce a **hydronium ion** (H_3O^+):

However, the H^+ is somewhat delocalized, so it probably exists as part of a larger, fleeting structure such as

Because a proton does not remain associated with a single water molecule, it appears to be relayed through a hydrogen-bonded network of water molecules (**Fig. 2.13**). This rapid **proton jumping** means that the effective mobility of H^+ in water is much greater than the mobility of other ions that must physically diffuse among water molecules. Consequently, acid–base reactions are among the fastest biochemical reactions.

[H⁺] AND [OH⁻] ARE INVERSELY RELATED

Pure water exhibits only a slight tendency to ionize, so the resulting concentrations of H^+ and OH^- are actually quite small. According to the law of mass action, the ionization of water can be described by a dissociation constant, K, which is equivalent to the concentrations of the reaction products divided by the concentration of un-ionized water:

$$K = \frac{[H^+][OH^-]}{[H_2O]} \qquad [2.1]$$

The square brackets represent the molar concentrations of the indicated species.

FIGURE 2.13

Proton jumping. A proton associated with one water molecule (as a hydronium ion, at left) appears to jump rapidly through a network of hydrogen-bonded water molecules.

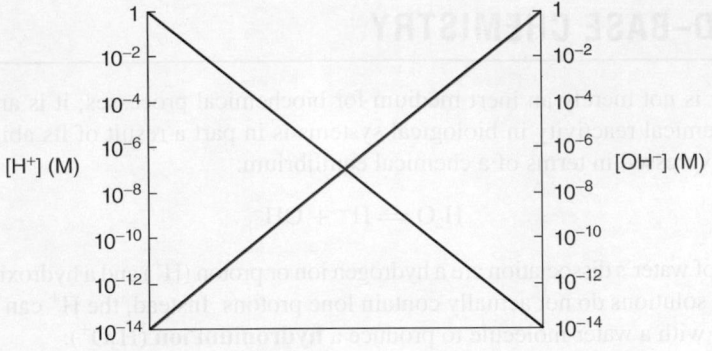

FIGURE 2.14

Relationship between [H⁺] and [OH⁻]. The product of [H⁺] and [OH⁻] is K_w, which is equal to 10^{-14}. Consequently, when [H⁺] is greater than 10^{-7} M, [OH⁻] is less than 10^{-7} M, and vice versa.

Because the concentration of H_2O (55.5 M) is so much greater than [H⁺] or [OH⁻], it is considered to be constant, and K is redefined as K_w, the **ionization constant of water:**

$$K_w = K[H_2O] = [H^+][OH^-] \qquad \text{[2.2]}$$

K_w is 10^{-14} at 25°C. In a sample of pure water, [H⁺] = [OH⁻], so [H⁺] and [OH⁻] must both be equal to 10^{-7} M:

$$K_w = 10^{-14} = [H^+][OH^-] = (10^{-7}M)(10^{-7}M) \qquad \text{[2.3]}$$

Since *the product of [H⁺] and [OH⁻] in any solution must be equal to 10^{-14}*, a hydrogen ion concentration greater than 10^{-7} M is balanced by a hydroxide ion concentration less than 10^{-7} M (**Fig. 2.14**).

A solution in which [H⁺] = [OH⁻] = 10^{-7} M is said to be **neutral;** a solution with [H⁺] > 10^{-7} M ([OH⁻] < 10^{-7} M) is **acidic;** and a solution with [H⁺] < 10^{-7} M ([OH⁻] > 10^{-7} M) is **basic.** To more easily describe such solutions, the hydrogen ion concentration is expressed as a **pH:**

$$pH = -\log[H^+] \qquad \text{[2.4]}$$

Accordingly, a neutral solution has a pH of 7, an acidic solution has a pH < 7, and a basic solution has a pH > 7 (**Fig. 2.15**). Note that because the pH scale is logarithmic, *a difference of one pH unit is equivalent to a 10-fold difference in [H⁺].* The so-called physiological pH, the normal pH of human blood, is a near-neutral 7.4. The pH values of some other body fluids are listed in **Table 2.3**. The pH of the environment is also a concern (**Box 2.C**).

THE pH OF A SOLUTION CAN BE ALTERED

The pH of a sample of water can be changed by adding a substance that affects the existing balance between [H⁺] and [OH⁻]. Adding an acid increases the concentration of [H⁺] and decreases the pH; adding a base has the opposite effect. *Biochemists define an acid as a substance that can donate a proton and a base as a substance that can accept a proton.* For

FIGURE 2.15

Relationship between pH and [H⁺]. Because pH is equal to –log[H⁺], the greater the [H⁺], the lower the pH. A solution with a pH of 7 is neutral, a solution with a pH < 7 is acidic, and a solution with a pH > 7 is basic.

? **What is the difference in H⁺ concentration between a solution at pH 4 and a solution at pH 8?**

TABLE 2.3 pH Values of Some Biological Fluids	
FLUID	**pH**
Pancreatic juice	7.8–8.0
Blood	7.4
Saliva	6.4–7.0
Urine	5.0–8.0
Gastric juice	1.5–3.0

Box 2.C Atmospheric CO_2 and Ocean Acidification

The human-generated increase in atmospheric carbon dioxide that is contributing to global warming is also impacting the chemistry of the world's oceans. Atmospheric CO_2 dissolves in water and reacts with it to generate carbonic acid. The acid immediately dissociates to form protons (H^+) and bicarbonate (HCO_3^-):

$$CO_2 + H_2O \rightleftharpoons H_2CO_3 \rightleftharpoons H^+ + HCO_3^-$$

The addition of hydrogen ions from CO_2-derived carbonic acid therefore leads to a decrease in the pH. Currently, the earth's oceans are slightly basic, with a pH of approximately 8.0. It has been estimated that over the next 100 years, the ocean pH will drop to about 7.8. Although the oceans act as a CO_2 "sink" that helps mitigate the increase in atmospheric CO_2, the increase in acidity in the marine environment represents an enormous challenge to organisms that must adapt to the new conditions.

Many marine organisms, including mollusks, many corals, and some plankton, use dissolved carbonate ions (CO_3^{2-}) to construct protective shells of calcium carbonate ($CaCO_3$). However, carbonate ions can combine with H^+ to form bicarbonate:

$$CO_3^{2-} + H^+ \rightleftharpoons HCO_3^-$$

Consequently, the increase in ocean acidity could decrease the availability of carbonate and thereby slow the growth of shell-building organisms. This not only would affect the availability of shellfish for human consumption but also would impact huge numbers of unicellular organisms at the base of the marine food chain. It is possible that acidification of the oceans could also dissolve existing calcium carbonate–based materials, such as coral reefs:

$$CaCO_3 + H^+ \rightleftharpoons HCO_3^- + Ca^{2+}$$

This could have disastrous consequences for these species-rich ecosystems.

? **Paradoxically, some marine organisms appear to benefit from increased atmospheric CO_2. Write an equation that describes how increased bicarbonate concentrations in seawater could promote shell growth.**

example, adding hydrochloric acid (HCl) to a sample of water increases the hydrogen ion concentration ($[H^+]$ or $[H_3O^+]$) because the HCl donates a proton to water:

$$HCl + H_2O \rightarrow H_3O^+ + Cl^-$$

Note that in this reaction, H_2O acts as a base that accepts a proton from the added acid.

Similarly, adding the base sodium hydroxide (NaOH) increases the pH (decreases $[H^+]$) by introducing hydroxide ions that can recombine with existing hydrogen ions:

$$NaOH + H_3O^+ \rightarrow Na^+ + 2\,H_2O$$

In this reaction, H_3O^+ is the acid that donates a proton to the added base. Note that acids and bases must operate in pairs: An acid can function as an acid (proton donor) only if a base (proton acceptor) is present, and vice versa. Water molecules can serve as both acid and base.

The final pH of the solution depends on how much H^+ (for example, from HCl) has been introduced or how much H^+ has been removed from the solution by its reaction with a base (for example, the OH^- ion of NaOH). Substances such as HCl and NaOH are known as "strong" acids and bases because they ionize completely in water. The Na^+ and Cl^- ions are called spectator ions and do not affect the pH. Calculating the pH of a solution of strong acid or base is straightforward (see Sample Calculation 2.1).

SAMPLE CALCULATION 2.1

Calculate the pH of 1 L of water to which is added (a) 10 mL of 5.0 M HCl or (b) 10 mL of 5.0 M NaOH.

PROBLEM

(a) The final concentration of HCl is $\dfrac{(0.01\ L)(5.0\ M)}{1.01\ L} = 0.050\ M$

SOLUTION

Since HCl dissociates completely, the added $[H^+]$ is equal to [HCl], or 0.050 M (the existing hydrogen ion concentration, 10^{-7} M, can be ignored because it is much smaller).

$$pH = -\log[H^+]$$
$$= -\log 0.050$$
$$= 1.3$$

(b) The final concentration of NaOH is 0.050 M. Since NaOH dissociates completely, the added $[OH^-]$ is 0.050 M. Use Equation 2.2 to calculate $[H^+]$.

$$K_w = 10^{-14} = [H^+][OH^-]$$
$$[H^+] = 10^{-14}/[OH^-]$$
$$= 10^{-14}/(0.050 \text{ M})$$
$$= 2.0 \times 10^{-13} \text{ M}$$
$$pH = -\log[H^+]$$
$$= -\log(2.0 \times 10^{-13})$$
$$= 12.7$$

PRACTICE PROBLEMS

1. Calculate the pH of 250 mL of water after the addition of 50 mL of 25 mM HCl.
2. Calculate the pH of 500 mL of water after mixing with 250 mL of 15 mM NaOH.

A p*K* VALUE DESCRIBES AN ACID'S TENDENCY TO IONIZE

Most biologically relevant acids and bases, unlike HCl and NaOH, do not dissociate completely when added to water. In other words, proton transfer to or from water is not complete. Therefore, the final concentrations of the acidic and basic species (including water itself) must be expressed in terms of an equilibrium. For example, acetic acid partially ionizes, or donates only some of its protons to water:

$$CH_3COOH + H_2O \rightleftharpoons CH_3COO^- + H_3O^+$$

The equilibrium constant for this reaction takes the form

$$K = \frac{[CH_3COO^-][H_3O^+]}{[CH_3COOH][H_2O]} \tag{2.5}$$

Because the concentration of H_2O is much higher than the other concentrations, it is considered constant and is incorporated into the value of K, which is then formally known as K_a, the **acid dissociation constant:**

$$K_a = K[H_2O] = \frac{[CH_3COO^-][H^+]}{[CH_3COOH]} \tag{2.6}$$

The acid dissociation constant for acetic acid is 1.74×10^{-5}. The larger the value of K_a, the more likely the acid is to ionize; that is, the greater its tendency to donate a proton to water. The smaller the value of K_a, the less likely the compound is to donate a proton.

Acid dissociation constants, like hydrogen ion concentrations, are often very small numbers. Therefore, it is convenient to transform the K_a to a **p*K*** value as follows:

$$pK = -\log K_a \tag{2.7}$$

The term pK_a is also used, but for simplicity, we will use p*K*. For acetic acid,

$$pK = -\log(1.74 \times 10^{-5}) = 4.76 \tag{2.8}$$

The larger an acid's K_a, the smaller its pK and the greater its "strength" as an acid.

Consider an acid such as the ammonium ion, NH_4^+:

$$NH_4^+ \rightleftharpoons NH_3 + H^+$$

Its K_a is 5.62×10^{-10}, which corresponds to a p*K* of 9.25. This indicates that the ammonium ion is a relatively weak acid, a compound that tends not to donate a proton. On the other hand, ammonia (NH_3), which is the **conjugate base** of the acid NH_4^+, readily accepts a proton. The p*K* values of some compounds are listed in **Table 2.4**. A **polyprotic acid,** a compound with

TABLE 2.4 pK Values of Some Acids

NAME	FORMULA[a]	pK
Trifluoroacetic acid	CF_3COOH	0.18
Phosphoric acid	H_3PO_4	2.15[b]
Formic acid	$HCOOH$	3.75
Succinic acid	$HOOCCH_2CH_2COOH$	4.21[b]
Acetic acid	CH_3COOH	4.76
Succinate	$HOOCCH_2CH_2COO^-$	5.64[c]
Thiophenol	C_6H_5SH	6.60
Phosphate	$H_2PO_4^-$	6.82[c]
N-(2-acetamido)-2-amino-ethanesulfonic acid (ACES)	$H_2NCOCH_2\overset{+}{N}H_2CH_2CH_2SO_3^-$	6.90
Imidazolium ion		7.00
p-Nitrophenol	$HO-\langle\rangle-NO_2$	7.24
N-2-hydroxyethylpiperazine-N'-2-ethanesulfonic acid (HEPES)	$HOCH_2CH_2H\overset{+}{N}\langle\rangle NCH_2CH_2SO_3^-$	7.55
Glycinamide	$^+H_3NCH_2CONH_2$	8.20
Tris(hydroxymethyl)-aminomethane (Tris)	$(HOCH_2)_3C\overset{+}{N}H_3$	8.30
Boric acid	H_3BO_3	9.24
Ammonium ion	NH_4^+	9.25
Phenol	C_6H_5OH	9.90
Methylammonium ion	$CH_3NH_3^+$	10.60
Phosphate	HPO_4^{2-}	12.38[d]

[a] The acidic hydrogen is highlighted in red; [b] pK_1; [c] pK_2; [d] pK_3.

more than one acidic hydrogen, has a pK value for each dissociation (called pK_1, pK_2, etc.). The first proton dissociates with the lowest pK value. Subsequent protons are less likely to dissociate and so have higher pK values.

THE pH OF A SOLUTION OF ACID IS RELATED TO THE pK

When an acid (represented as the proton donor HA) is added to water, the final hydrogen ion concentration of the solution depends on the acid's tendency to ionize:

$$HA \rightleftharpoons A^- + H^+$$

In other words, the final pH depends on the equilibrium between HA and A$^-$,

$$K_a = \frac{[A^-][H^+]}{HA}$$

[2.9]

so that

$$[H^+] = K_a \frac{[HA]}{[A^-]} \qquad [2.10]$$

We can express $[H^+]$ as a pH, and K_a as a pK, which yields

$$-\log[H^+] = -\log K_a - \log\frac{[HA]}{[A^-]} \qquad [2.11]$$

or

$$pH = pK + \log\frac{[A^-]}{[HA]} \qquad [2.12]$$

Equation 2.12 is known as the **Henderson–Hasselbalch equation.** It relates the pH of a solution to the pK of an acid and the concentration of the acid (HA) and its conjugate base (A^-). This equation makes it possible to perform practical calculations to predict the pH of a solution (see Sample Calculation 2.2) or the concentrations of an acid and its conjugate base at a given pH (see Sample Calculation 2.3).

SAMPLE CALCULATION 2.2

PROBLEM

Calculate the pH of a 1-L solution to which has been added 6.0 mL of 1.5 M acetic acid and 5.0 mL of 0.4 M sodium acetate.

SOLUTION

First, calculate the final concentrations of acetic acid (HA) and acetate (A^-). The final volume of the solution is 1 L + 6 mL + 5 mL = 1.011 L.

$$[HA] = \frac{(0.006\ L)(1.5\ M)}{1.011\ L} = 0.0089\ M$$

$$[A^-] = \frac{(0.005\ L)(0.4\ M)}{1.011\ L} = 0.0020\ M$$

Next, substitute these values into the Henderson–Hasselbalch equation using the pK for acetic acid given in Table 2.4:

$$pH = pK + \log\frac{[A^-]}{[HA]}$$

$$pH = 4.76 + \log\frac{0.0020}{0.0089}$$

$$= 4.76 - 0.65$$

$$= 4.11$$

PRACTICE PROBLEMS

3. Calculate the pH of a 1-L solution to which has been added 50 mL of 20 mM acetic acid and 50 mL of 30 mM sodium acetate.
4. Calculate the pH of a 500-mL solution to which has been added 10 mL of 50 mM boric acid and 20 mL of 20 mM sodium borate.

SAMPLE CALCULATION 2.3

PROBLEM

Calculate the concentration of formate in a 10-mM solution of formic acid at pH 4.15.

SOLUTION

The solution of formic acid contains both the acid species (formic acid) and its conjugate base (formate). Use the Henderson–Hasselbalch equation to determine the ratio of formate (A^-) to formic acid (HA) at pH 4.15, using the pK value given in Table 2.4.

$$pH = pK + \log \frac{[A^-]}{[HA]}$$

$$\log \frac{[A^-]}{[HA]} = pH - pK = 4.15 - 3.75 = 0.40$$

$$\frac{[A^-]}{[HA]} = 2.51 \text{ or } [A^-] = 2.51[HA]$$

Since the total concentration of formate and formic acid is 0.01 M, $[A^-] + [HA] = 0.01$ M, and $[HA] = 0.01$ M $- [A^-]$. Therefore,

$$[A^-] = 2.51[HA]$$
$$[A^-] = 2.51(0.01 \text{ M} - [A^-])$$
$$[A^-] = 0.0251 \text{ M} - 2.51[A^-]$$
$$3.51[A^-] = 0.0251 \text{ M}$$
$$[A^-] = 0.0072 \text{ M or } 7.2 \text{ mM}$$

5. Calculate the concentration of acetate in a 25 mM solution of acetic acid at pH 5.0.
6. Calculate the concentration of phosphoric acid in a 25-mM solution of phosphate at pH 3.0.

The Henderson–Hasselbalch equation indicates that *when the pH of a solution of acid is equal to the pK of that acid, then the acid is half dissociated; that is, exactly half of the molecules are in the protonated HA form and half are in the unprotonated A⁻ form.* You can prove to yourself that when $[A^-] = [HA]$, the log term of the Henderson–Hasselbalch equation becomes zero (log 1 = 0), and $pH = pK$. When the pH is far below the pK, the acid exists mostly in the HA form; when the pH is far above the pK, the acid exists mostly in the A⁻ form. Note that A⁻ signifies a deprotonated acid; if the acid (HA) bears a positive charge to begin with, the dissociation of a proton yields a neutral species, still designated A⁻.

Knowing the ionization state of an acidic substance at a given pH can be critical. For example, a drug that has no net charge at pH 7.4 may readily enter cells, whereas a drug that bears a net positive or negative charge at that pH may remain in the bloodstream and be therapeutically useless (see Sample Calculation 2.4).

Determine which molecular species of phosphoric acid predominates at pH values of **(a)** 1.5, **(b)** 4, **(c)** 9, and **(d)** 13.

From the pK values in Table 2.4, we know that:

Below pH 2.15, the fully protonated H_3PO_4 species predominates.

At pH 2.15, $[H_3PO_4] = [H_2PO_4^-]$.

Between pH 2.15 and 6.82, the $H_2PO_4^-$ species predominates.

At pH 6.82, $[H_2PO_4^-] = [HPO_4^{2-}]$.

Between pH 6.82 and 12.38, the HPO_4^{2-} species predominates.

At pH 12.38, $[HPO_4^{2-}] = [PO_4^{3-}]$.

Above pH 12.38, the fully deprotonated PO_4^{3-} species predominates.

Therefore, the predominant species at the indicated pH values are **(a)** H_3PO_4, **(b)** $H_2PO_4^-$, **(c)** HPO_4^{2-}, and **(d)** PO_4^{3-}.

7. Which species of phosphoric acid predominates at pH (a) 5.5 and (b) 8.5?
8. At what pH will $[NH_3] = [NH_4^+]$?
9. Over what pH range will the $HOOCCH_2CH_2COO^-$ form of succinic acid predominate?

The pH-dependent ionization of biological molecules is also key to understanding their structures and functions. *Many of the functional groups on biological molecules act as acids and bases.* Their ionization states depend on their respective pK values and on the pH ([H$^+$]) of their environment. For example, at physiological pH, a polypeptide bears multiple ionic charges because its carboxylic acid (—COOH) groups are ionized to carboxylate (—COO$^-$) groups and its amino (—NH$_2$) groups are protonated (—NH$_3^+$). This is because the pK values for the carboxylic acid groups are about 4, and the pK values for the amino groups are above 10. Consequently, below pH 4, both the carboxylic acid and amino groups are mostly protonated; above pH 10, both groups are mostly deprotonated.

pH < 4	4 < pH < 10	pH > 10
—COOH	—COO$^-$	—COO$^-$
—NH$_3^+$	—NH$_3^+$	—NH$_2$

Note that a compound containing a —COO$^-$ group is sometimes still called an "acid," even though it has already given up its proton. Similarly, a "basic" compound may already have accepted a proton.

CONCEPT CHECK

1. Draw the structure of a water molecule, including all its electrons, before and after ionization.

2. What is pH and how is it related to [H$^+$] and [OH$^-$]?

3. For each fluid in Table 2.3, determine whether the addition of acid or base would bring the pH to neutral.

4. Rearrange the Henderson–Hasselbalch equation to isolate each term.

5. What is the relationship between pH and the pK of an acid?

TOOLS AND TECHNIQUES: BUFFERS

2.4

When a strong acid such as HCl is added to pure water, all the added acid contributes directly to a decrease in pH. But when HCl is added to a solution containing a weak acid in equilibrium with its conjugate base (A$^-$), the pH does not change so dramatically, because some of the added protons combine with the conjugate base to re-form the acid and therefore do not contribute to an increase in [H$^+$].

$$\text{HCl} \rightarrow \text{H}^+ + \text{Cl}^- \quad \text{large increase in [H}^+\text{]}$$
$$\text{HCl} + \text{A}^- \rightarrow \text{HA} + \text{Cl}^- \quad \text{small increase in [H}^+\text{]}$$

Conversely, when a strong base (such as NaOH) is added to the solution of weak acid/conjugate base, some of the added hydroxide ions accept protons from the acid to form H$_2$O and therefore do not contribute to a decrease in [H$^+$].

$$\text{NaOH} \rightarrow \text{Na}^+ + \text{OH}^- \quad \text{large decrease in [H}^+\text{]}$$
$$\text{NaOH} + \text{HA} \rightarrow \text{Na}^+ + \text{A}^- + \text{H}_2\text{O} \quad \text{small decrease in [H}^+\text{]}$$

The weak acid/conjugate base system (HA/A$^-$) acts as a **buffer** against the added acid or base by preventing the dramatic changes in pH that would otherwise occur.

The buffering activity of a weak acid, such as acetic acid, can be traced by titrating the acid with a strong base (**Fig. 2.16**). At the start of the titration, all the acid is present in its protonated (HA) form. As base (for example, NaOH) is added, protons begin to dissociate from the acid, producing A$^-$. The continued addition of base eventually causes all the protons to dissociate, leaving all the acid in its conjugate base (A$^-$) form. At the midpoint of the titration,

FIGURE 2.16

Titration of acetic acid. At the start point (before base is added), the acid is present mainly in its CH₃COOH form. As small amounts of base are added, protons dissociate until, at the midpoint of the titration (where pH = pK), [CH₃COOH] = [CH₃COO⁻]. The addition of more base causes more protons to dissociate until nearly all the acid is in the CH₃COO⁻ form (the end point). The shaded area indicates the effective buffering range of acetic acid. Within one pH unit of the pK, additions of acid or base do not greatly perturb the pH of the solution.

....

? **Sketch the titration curve for ammonia.**

exactly half the protons have dissociated, so [HA] = [A⁻] and pH = pK (Equation 2.12). The broad, flat shape of the titration curve shown in Figure 2.16 indicates that the pH does not change drastically with added acid or base when the pH is near the pK. *The effective buffering capacity of an acid is generally taken to be within one pH unit of its pK.* For acetic acid (pK = 4.76), this would be pH 3.76–5.76.

Biochemists nearly always perform experiments in buffered solutions in order to maintain a constant pH when acidic or basic substances are added or when chemical reactions produce or consume protons. Without buffering, fluctuations in pH would alter the ionization state of the molecules under study, which might then behave differently. Before biochemists appreciated the importance of pH, experimental results were often poorly reproducible, even within the same laboratory.

A buffer solution is typically prepared from a weak acid and the salt of its conjugate base (see Sample Calculation 2.5). The two are mixed together in the appropriate ratio, according to the Henderson–Hasselbalch equation, and the final pH is adjusted if necessary by adding a small amount of concentrated HCl or NaOH. In addition to choosing a buffering compound with a pK value near the desired pH, a biochemist must consider other factors, including the compound's solubility, stability, toxicity to cells, reactivity with other molecules, and cost.

SAMPLE CALCULATION 2.5

How many mL of a 2.0-M solution of boric acid must be added to 600 mL of a solution of 10 mM sodium borate in order for the pH to be 9.45?   **PROBLEM**

SOLUTION

Rearrange the Henderson–Hasselbalch equation to isolate the $[A^-]/[HA]$ term:

$$pH = pK + \log \frac{[A^-]}{[HA]}$$

$$\log \frac{[A^-]}{[HA]} = pH - pK$$

$$\frac{[A^-]}{[HA]} = 10^{(pH-pK)}$$

Substitute the known pK (from Table 2.4) and the desired pH:

$$\frac{[A^-]}{[HA]} = 10^{(9.45-9.24)} = 10^{0.21} = 1.62$$

The starting solution contains $(0.6\ L)(0.01\ mol \cdot L^{-1}) = 0.006$ moles of borate (A^-). The amount of boric acid (HA) needed is $0.006\ mol/1.62 = 0.0037\ mol$.

Since the stock boric acid is 2.0 M, the volume of boric acid to be added is $(0.0037\ mol)/(2.0\ mol \cdot L^{-1}) = 0.0019\ L$ or 1.9 mL.

PRACTICE PROBLEMS

10. How many mL of a 5.0-M solution of sodium borate must be added to a 500-mL solution of 50 mM boric acid in order for the pH to be 9.4?

11. How many mL of a 1-M solution of imidazolium chloride must be added to a 500-mL solution of 10 mM imidazole in order for the pH to be 6.5?

One commonly used laboratory buffer system that mimics physiological conditions contains a mixture of dissolved NaH_2PO_4 and Na_2HPO_4 for a total phosphate concentration of 10 mM. The Na^+ ions are spectator ions and are usually not significant because the buffer solution usually also contains about 150 mM NaCl (see Fig. 2.12). In this "phosphate-buffered saline," the equilibrium between the two species of phosphate ions can "soak up" added acid (producing more $H_2PO_4^-$) or added base (producing more HPO_4^{2-}).

$$pK = 6.82$$

$$H_2PO_4^- \rightleftharpoons H^+ + HPO_4^{2-}$$

Acid added $H_2PO_4^- \longleftarrow H^+ + HPO_4^{2-}$

Base added $H_2PO_4^- \longrightarrow H^+ + HPO_4^{2-}$

This phenomenon illustrates **Le Châtelier's principle,** which states that a change in concentration of one reactant will shift the concentrations of other reactants in order to restore equilibrium. In the human body, the major buffering systems involve bicarbonate, phosphate, and other ions.

 CONCEPT CHECK

1. Write an equation that describes the equilibrium in a solution containing an acetate/acetic acid buffer and describe what happens when HCl or NaOH is added.

2. What is the useful pH range of a buffer?

CLINICAL CONNECTION: ACID–BASE BALANCE IN HUMANS

KEY CONCEPT
• The human body maintains a constant pH.

The cells of the human body typically maintain an internal pH of 6.9–7.4. The body does not normally have to defend itself against strong inorganic acids, but many metabolic processes generate acids, which must be buffered so that they do not cause the pH of blood to

Normal conditions

$$H^+ + HCO_3^- \rightleftharpoons H_2CO_3 \rightleftharpoons H_2O + CO_2$$

Excess acid

$$\mathbf{H^+} + HCO_3^- \longrightarrow H_2CO_3 \rightleftharpoons H_2O + CO_2$$
$$H^+ + HCO_3^- \rightleftharpoons \mathbf{H_2CO_3} \longrightarrow H_2O + CO_2$$
$$H^+ + HCO_3^- \rightleftharpoons H_2CO_3 \rightleftharpoons H_2O + \mathbf{CO_2}$$

Insufficient acid

$$H^+ + HCO_3^- \rightleftharpoons H_2CO_3 \longleftarrow H_2O + \mathbf{CO_2}$$
$$H^+ + HCO_3^- \longleftarrow \mathbf{H_2CO_3} \rightleftharpoons H_2O + CO_2$$
$$\mathbf{H^+} + HCO_3^- \rightleftharpoons H_2CO_3 \rightleftharpoons H_2O + CO_2$$

FIGURE 2.17

The bicarbonate buffer system. Elimination or retention of CO_2 can shift the equilibrium in order to promote or prevent the loss of H^+ from the body.

drop below its normal value of 7.4. The functional groups of proteins and phosphate groups can serve as biological buffers; however, the most important buffering system involves CO_2 (itself a product of metabolism) in the blood plasma (plasma is the fluid component of blood).

CO_2 reacts with water to form carbonic acid, H_2CO_3

$$CO_2 + H_2O \rightleftharpoons H_2CO_3$$

This freely reversible reaction is accelerated *in vivo* by the enzyme carbonic anhydrase, which is present in most tissues and is particularly abundant in red blood cells. Carbonic acid ionizes to bicarbonate, HCO_3^- (see Box 2.C):

$$H_2CO_3 \rightleftharpoons H^+ + HCO_3^-$$

so that the overall reaction is

$$CO_2 + H_2O \rightleftharpoons H^+ + HCO_3^-$$

The pK for this process is 6.1 (the ionization of HCO_3^- to CO_3^{2-} occurs with a pK of 10.3 and is therefore not significant at physiological pH).

Although a pK of 6.1 appears to be just outside the range of a useful physiological buffer (which would be within one pH unit of 7.4), the effectiveness of the bicarbonate buffer system is augmented by the fact that excess hydrogen ions can not only be buffered but can also be eliminated from the body. This is possible because after the H^+ combines with HCO_3^- to re-form H_2CO_3, which rapidly equilibrates with $CO_2 + H_2O$, some of the CO_2 can be given off as a gas in the lungs. If it becomes necessary to retain more H^+ to maintain a constant pH, breathing can be adjusted so that less gaseous CO_2 is lost during exhalation (**Fig. 2.17**).

Changes in pulmonary function can adequately adjust blood pH on the order of minutes to hours; however, longer-term adjustments of hours to days are made by the kidneys, which use a variety of mechanisms to excrete or retain H^+, bicarbonate, and other ions. In fact, the kidneys play a major role in the buffering of metabolic acids. Normal metabolic activity generates acids as the result of the degradation of amino acids, the incomplete oxidation of glucose and fatty acids, and the ingestion of acidic groups in the form of phosphoproteins and phospholipids. The HCO_3^- required to buffer these acids is initially filtered out of the bloodstream in the kidneys, but the kidneys actively reclaim most of this bicarbonate before it is lost in the urine (**Fig. 2.18**).

In addition to reabsorbing filtered HCO_3^-, the kidneys also generate additional HCO_3^- to offset losses due to the buffering of metabolic acids and the exhalation of CO_2. Metabolic activity

FIGURE 2.18

Bicarbonate reabsorption. H^+ leaves the kidney cells in exchange for Na^+ (step 1). The expelled H^+ combines with HCO_3^- in the filtrate, forming CO_2 (step 2). Because it is nonpolar, the CO_2 can diffuse into the kidney cell, where it is converted back to $H^+ + HCO_3^-$ (step 3).

FIGURE 2.19

Bicarbonate production. Carbonic acid–derived protons are pumped out of the kidney cell (step 1), and the bicarbonate remaining in the cell is returned to the bloodstream in exchange for Cl^- (step 2).

in the kidney cells produces CO_2, which is converted to $H^+ + HCO_3^-$. The cells actively secrete the H^+, which is lost via the urine, accounting for the mildly acidic pH of normal urine. The bicarbonate remaining in the cell is returned to the bloodstream in exchange for Cl^- (**Fig. 2.19**).

Some HCO_3^- also accumulates as a result of the metabolism of the amino acid glutamine in the kidneys.

$$\underset{\overset{|}{\overset{+}{N}H_3}}{H-\underset{|}{C}-CH_2-CH_2-C}\overset{O}{\underset{NH_2}{\diagdown}}$$

Glutamine

The two amino groups are removed as ammonia (NH_3), which is ultimately excreted in the urine. Because the ammonium ion (NH_4^+) has a pK value of 9.25, nearly all the ammonia molecules become protonated at physiological pH. The consumption of protons from carbonic acid (ultimately, CO_2) leaves an excess of HCO_3^-.

Certain medical conditions can disrupt normal acid–base balance, leading to **acidosis** (blood pH less than 7.35) or **alkalosis** (blood pH greater than 7.45). The activities of the lungs and kidneys, as well as other organs, may contribute to the imbalance or respond to help correct the imbalance.

The most common disorder of acid–base chemistry is **metabolic acidosis,** which is caused by the accumulation of the acidic products of metabolism and can develop during shock, starvation, severe diarrhea (in which bicarbonate-rich digestive fluid is lost), certain genetic diseases, and renal failure (in which damaged kidneys eliminate too little acid). Metabolic acidosis can interfere with cardiac function and oxygen delivery and contribute to central nervous system depression. Despite its many different causes, one common symptom of metabolic acidosis is rapid, deep breathing. The increased ventilation helps compensate for the acidosis by "blowing off" more acid in the form of CO_2 derived from H_2CO_3. However, this mechanism also impairs O_2 uptake by the lungs.

Metabolic acidosis can be treated by administering sodium bicarbonate ($NaHCO_3$). In chronic metabolic acidosis, the mineral component of bone serves as a buffer, which leads to the loss of calcium, magnesium, and phosphate and ultimately to osteoporosis and fractures.

Metabolic alkalosis, a less common condition, can result from prolonged vomiting, which represents a loss of HCl in gastric fluid. This specific disorder can be treated by infusing NaCl. Metabolic alkalosis is also caused by overproduction of mineralocorticoids (hormones produced by the adrenal glands), which leads to abnormally high levels of H^+ excretion and Na^+ retention. In this case, the disorder does not respond to saline infusion.

Individuals with metabolic alkalosis may experience apnea (cessation of breathing) for 15 s or more and cyanosis (blue coloration) due to inadequate oxygen uptake. Both these symptoms reflect the body's effort to compensate for the high blood pH by minimizing the loss of CO_2 from the lungs.

What happens when abnormal lung function is the *cause* of an acid–base imbalance? **Respiratory acidosis** can result from impaired pulmonary function caused by airway blockage, asthma (constriction of the airways), and emphysema (loss of alveolar tissue). In all cases, the kidneys respond by adjusting their activity, mainly by increasing the synthesis of the enzymes that break down glutamine in order to produce NH_3. Excretion of NH_4^+ helps correct the acidosis. However, renal compensation of respiratory acidosis takes hours to days (the time course for adjusting enzyme levels), so this condition is best treated by restoring pulmonary function through bronchodilation, supplemental oxygen, or mechanical ventilation (assisted breathing).

Some of the diseases that impair lung function (for example, asthma) can also contribute to **respiratory alkalosis,** but this relatively rare condition is more often caused by hyperventilation brought on by fear or anxiety. Unlike the other acid–base disorders, this form of respiratory alkalosis is seldom life-threatening.

 CONCEPT CHECK

1. Review the equations that describe the interconversion of carbon dioxide and bicarbonate.
2. Compare the metabolic adjustments made by the lungs and kidneys during acid–base homeostasis.
3. Discuss how impaired lung or kidney function could lead to acidosis or alkalosis.

SUMMARY

2.1 Water Molecules and Hydrogen Bonds

• Water molecules are polar; they form hydrogen bonds with each other and with other polar molecules bearing hydrogen bond donor or acceptor groups.

• The electrostatic forces acting on biological molecules also include ionic interactions and van der Waals interactions.

• Water dissolves polar and ionic substances.

2.2 The Hydrophobic Effect

• Nonpolar (hydrophobic) substances tend to aggregate rather than disperse in water in order to minimize the decrease in entropy that would be required for water molecules to surround each nonpolar molecule. This is the hydrophobic effect.

• Amphiphilic molecules, which contain both polar and nonpolar groups, may aggregate to form micelles or bilayers.

2.3 Acid–Base Chemistry

• The dissociation of water produces hydroxide ions (OH^-) and protons (H^+) whose concentration can be expressed as a pH value. The pH of

a solution can be altered by adding an acid (which donates protons) or a base (which accepts protons).

• The tendency for a proton to dissociate from an acid is expressed as a pK value.

• The Henderson–Hasselbalch equation relates the pH of a solution of a weak acid and its conjugate base to the pK and the concentrations of the acid and base.

2.4 Tools and Techniques: Buffers

• A buffered solution, which contains an acid and its conjugate base, resists changes in pH when more acid or base is added.

2.5 Clinical Connection: Acid–Base Balance in Humans

• The body uses the bicarbonate buffer system to maintain a constant internal pH. Homeostatic adjustments are made by the lungs, where CO_2 is released, and by the kidneys, which excrete H^+ and ammonia.

KEY TERMS

polarity	van der Waals radius	dipole–dipole interaction	solute
hydrogen bond	electronegativity	London dispersion forces	solvation
ionic interaction	van der Waals interaction	dielectric constant	hydration

hydrophilic
hydrophobic
hydrophobic effect
amphiphilic
amphipathic
micelle
bilayer
vesicle

hydronium ion
proton jumping
ionization constant of water (K_w)
neutral solution
acidic solution
basic solution
pH
acid

base
acid dissociation constant (K_a)
pK
conjugate base
polyprotic acid
Henderson–Hasselbalch
  equation
buffer

Le Châtelier's principle
acidosis
alkalosis
metabolic acidosis
metabolic alkalosis
respiratory acidosis
respiratory alkalosis

BIOINFORMATICS

Brief Bioinformatics Exercises

2.1 Structure and Solubility

2.2 Amino Acids, Ionization, and pK Values

PROBLEMS

2.1   Water Molecules and Hydrogen Bonds

1. Each C=O bond in CO_2 is polar, yet the whole molecule is nonpolar. Explain.

2. The H—C—H bond angle in the perfectly tetrahedral CH_4 molecule is 109°. Explain why the H—O—H bond angle in water is only about 104.5°.

3. Which compound has a higher boiling point, H_2O or H_2S? Explain.

4. Consider the following molecules and their melting points listed below. How can you account for the differences in melting points among these molecules of similar size?

	Molecular weight ($g \cdot mol^{-1}$)	Melting point (°C)
Water, H_2O	18.0	0
Ammonia, NH_3	17.0	−77
Methane, CH_4	16.0	−182

5. Identify the hydrogen bond acceptor and donor groups in the following molecules. Use an arrow to point toward each acceptor and away from each donor.

Aspartame

Sulfanilamide

Uric acid

6. The enzyme dihydrofolate reductase is required for DNA synthesis, and as such, is an attractive drug target in cancer therapy. The drug methotrexate (MTX) competes with the substrate DHF for binding to the enzyme. Identify the hydrogen bond acceptor and donor groups in MTX and DHF. Use an arrow to point toward each acceptor and away from each donor.

MTX

DHF

7. Rank the following compounds in order of melting points. Classify the compounds as soluble, slightly soluble, or insoluble in water and explain your answer.

(a) $H_3C-CH_2-O-CH_3$

(b) $H_3C-\overset{\overset{\displaystyle O}{\|}}{C}-NH_2$

(c) $H_2N-\overset{\overset{\displaystyle O}{\|}}{C}-NH_2$

(d) $H_3C-(CH_2)_3-CH_3$

(e) $H_3C-CH_2-\overset{\overset{\displaystyle O}{\|}}{CH}$

8. In 2007, pets consuming a brand of pet food containing melamine (Problem 1.10) and cyanuric acid (both nontoxic when consumed alone) suffered from renal failure when the two compounds combined to form crystalline melamine cyanurate in the kidney. Draw the structure of melamine cyanurate, which forms when melamine and cyanuric acid form hydrogen bonds with each other.

Melamine   Cyanuric acid

9. Examine Table 2.1.

   (a) Rank the six atoms listed in order of increasing electronegativity.
   (b) What is the relationship between an atom's electronegativity and its ability to participate in hydrogen bonding?

10. Do intermolecular hydrogen bonds form in the compounds below? Draw the hydrogen bonds where appropriate.

A $H_3C-\overset{\overset{\displaystyle O}{\|}}{C}-H$

B   (imidazole structure)

C H_3C-CH_2-OH

D H_3C-CH_2-Cl and $H-\overset{\displaystyle O}{\;}-H$

E $H_3C-CH_2-O-CH_2-CH_3$ and $H-\overset{\overset{\displaystyle N}{|}}{\;}-H$

11. What intermolecular interactions are likely to form between the two amino acid side chains shown in the protein diagrams below?

(a)

Asn / Ser

(b)

Lys / Asp

(c)

Phe / Ala

12. Practitioners of homeopathic medicine believe that good health can be restored by administering a small amount of a harmful substance that stimulates the body's natural healing processes. The remedy is typically prepared by serially diluting an animal or plant extract. In a so-called 30× dilution, the active substance is diluted 10-fold, 30 times in succession.

   (a) If the substance is initially present at a concentration of 1 M, what is its concentration after a 30× dilution?
   (b) A typical homeopathic dose is a few drops. How many molecules of the active substance would be present in 1 mL?
   (c) When confronted with the results of an exercise such as the one in part (b), proponents of homeopathy claim that a molecule can leave an imprint, or memory of itself, on the surrounding water molecules. What information presented in this chapter might be used to support this claim? To refute it?

13. Which of the four primary alcohols listed in Table 2.2 would be the best solvent for ammonium ions? What can you conclude about the polarity of these solvents, which all contain an —OH group that can form hydrogen bonds?

14. Ammonium sulfate, $(NH_4)_2SO_4$, is a water-soluble salt. Draw the structures of the hydrated ions that form when ammonium sulfate dissolves in water.

15. Water is unusual in that its solid form is less dense than its liquid form. This means that when a pond freezes in the winter, ice is found as a layer on top of the pond, not on the bottom. What are the biological advantages of this?

16. Ice skaters skate not on the solid ice surface itself but on a thin film of water that forms between the skater's blade and the ice. What unique property of water makes ice skating possible?

17. The amino acid glycine is sometimes drawn as structure A below. However, the structure of glycine is more accurately represented by structure B. Glycine has the following properties: white crystalline solid, high melting point, and high water solubility. Why does structure B more accurately represent the structure of glycine than structure A?

$$H_2N-CH_2-COOH \qquad\qquad {}^+H_3N-CH_2-COO^-$$
Structure A   Structure B

18. A typical spherical bacterial cell with a diameter of 1 μm contains 1000 molecules of a certain protein.

   (a) What is the protein's concentration in units of mM? (*Hint:* The volume of a sphere is $4\pi r^3/3$.)
   (b) How many molecules of glucose does the cell contain if the internal glucose concentration is 5 mM?

19. **(a)** Water has a surface tension that is nearly three times greater than that of ethanol. Explain.
(b) The surface tension of water decreases with increasing temperature. Explain.

2.2 The Hydrophobic Effect

20. The structures of **(a)** hexadecyltrimethylammonium (a cationic detergent) **(b)** cholate (an anionic detergent) **(c)** dimethyldecylphosphine oxide (a nonionic detergent), are shown here. Identify the polar and the nonpolar regions of these amphipathic molecules.

$$H_3C-(H_2C)_{15}-\overset{\overset{\displaystyle CH_3}{|}}{\underset{\underset{\displaystyle CH_3}{|}}{N^+}}-CH_3$$

Hexadecyltrimethylammonium

Cholate

$$H_3C-(CH_2)_9-\overset{\overset{\displaystyle O}{\|}}{\underset{\underset{\displaystyle CH_3}{|}}{P}}-CH_3$$

Dimethyldecylphosphine oxide

21. Consider the structures of the molecules below. Are these molecules polar, nonpolar, or amphiphilic?

(a) $H_3C-(CH_2)_{13}-\overset{\overset{\displaystyle CH_3}{|}}{\underset{\underset{\displaystyle CH_3}{|}}{N^+}}-CH_2COO^-$

(b) $H_3C-(CH_2)_{14}-CH_3$

(c) $H_3C-\overset{\overset{\displaystyle CH_3}{|}}{\underset{\underset{\displaystyle C_2H_5}{|}}{N^+}}-CH_3 \quad Cl^-$

(d) $H_3C-\overset{}{\underset{\underset{\displaystyle OH}{|}}{CH}}-COO^-$

(e) $\begin{array}{l} CH_2-O-\overset{\overset{\displaystyle O}{\|}}{C}-(CH_2)_{11}-CH_3 \\ HC-O-\overset{\overset{\displaystyle O}{\|}}{C}-(CH_2)_{11}-CH_3 \\ HO-CH_2 \end{array}$

Which of these compounds are capable of forming micelles and which are capable of forming bilayers?

22. Many household soaps are amphiphilic substances, often the salts of long-chain fatty acids, that form water-soluble micelles. An example is sodium dodecyl sulfate (SDS), an anionic detergent.

(a) Identify the polar and nonpolar portions of the SDS molecule.
(b) Draw the structure of the micelle formed by SDS.
(c) Explain how the SDS micelles "wash away" water-insoluble substances such as cooking grease.

$$H_3C-(CH_2)_{11}-O-\overset{\overset{\displaystyle O}{\|}}{\underset{\underset{\displaystyle O}{\|}}{S}}-O^-Na^+$$

Sodium dodecyl sulfate (SDS)

23. The compound bis-(2-ethylhexyl)sulfosuccinate (abbreviated AOT) is capable of forming "reverse" micelles in the hydrocarbon solvent isooctane (2,2,4-trimethylpentane). Scientists have investigated the use of reverse micelles for extracting water-soluble proteins. A two-phase system is formed: the hydrocarbon phase containing the reverse micelles and the water phase containing the protein. After a certain period of time, the protein is transferred to the reverse micelle.

(a) Draw the structure of the reverse micelle that AOT would form in isooctane.
(b) Where would the protein be located in the reverse micelle?

$$H_3C-(CH_2)_3-\overset{\overset{\displaystyle CH_2CH_3}{|}}{CH}-CH_2-O-\overset{\overset{\displaystyle O}{\|}}{C}-CH_2 \quad O$$
$$H_3C-(CH_2)_3-\overset{\overset{\displaystyle}{}}{\underset{\underset{\displaystyle CH_2CH_3}{|}}{CH}}-CH_2-O-\overset{\overset{\displaystyle O}{\|}}{C}-\overset{}{CH}-\overset{\overset{\displaystyle O}{\|}}{S}-O^-$$

Bis-(2-ethylhexyl)sulfosuccinate (AOT)

24. Which of the following substances might be able to cross a bilayer? Which substances could not? Explain your answer.

(a) CO_2

(b) Glucose

$$\begin{array}{c} CHO \\ H-\!\!-\!\!-OH \\ HO-\!\!-\!\!-H \\ H-\!\!-\!\!-OH \\ H-\!\!-\!\!-OH \\ CH_2OH \end{array}$$

Glucose

(c) 2,4-Dinitrophenol (DNP)

(d) Na^+

25. Just as a dissolved substance tends to move spontaneously down its concentration gradient, water also tends to move from an area of high concentration (low solute concentration) to an area of low concentration (high solute concentration), a process known as osmosis.

(a) Explain why a lipid bilayer would be a barrier to osmosis.
(b) Why are isolated human cells placed in a solution that typically contains about 150 mM NaCl? What would happen if the cells were placed in pure water?

26. Fresh water is obtained from seawater in a desalination process using reverse osmosis. In reverse osmosis, seawater is forced through a membrane; the salt remains on one side of the membrane and fresh water is produced on the other side. In what ways does reverse osmosis differ from osmosis described in Problem 25?

27. Drug delivery is a challenging problem because the drug molecules must be sufficiently water-soluble to dissolve in the blood, but also sufficiently nonpolar to pass through the cell membrane for delivery. A medicinal chemist proposes to encapsulate a water-soluble drug into a vesicle (see Section 2.2). How does this strategy facilitate the delivery of the drug to its target cell?

28. A specialized protein pump in the red blood cell membrane exports Na^+ ions and imports K^+ ions in order to maintain the sodium and potassium ion concentrations shown in Figure 2.12. Does the movement of these ions occur spontaneously, or is this an energy-requiring process? Explain.

29. Estimate the amount of Na^+ lost in sweat during 18 minutes of vigorous exercise. What is the mass of potato chips (200 mg Na^+ per ounce) you would have to consume in order to replace the lost sodium?

30. The bacterium *E. coli* can adapt to changes in the solute concentration of its growth medium. The cell consists of a cytoplasmic compartment bounded by a cell membrane surrounded by a porous cell wall; both the membrane and the cell wall allow the passage of water and ions.

(a) Under nongrowing conditions, only cytoplasmic water content is regulated. What happens to the cytoplasmic volume if *E. coli* is grown in a growth medium in a high osmolarity medium? In a low osmolarity medium?

(b) Under growth conditions, *E. coli* regulates cytoplasmic K^+ content (in addition to water) in order to prevent large changes in cell volume. How might *E. coli* regulate both the cytoplasmic concentrations of K^+ and water when grown in a low osmolarity medium? What happens when *E. coli* is grown in a high osmolarity medium?

2.3   Acid–Base Chemistry

31. Compare the concentrations of H_2O and H^+ in a sample of pure water at pH 7.0 at 25°C.

32. Determine the pH of a solution of

(a) 1.0×10^{-10} M HCl

(b) 1.0×10^{-10} M NaOH

33. Explain why H_3O^+ is the strongest acid that can exist in a biological system.

34. Draw a diagram similar to Fig. 2.13 showing how a hydroxide ion appears to jump through an aqueous solution.

35. Fill in the blanks of the following table:

	Acid, base, or neutral?	pH	$[H^+]$ (M)	$[OH^-]$ (M)
Blood		7.42		
Saliva	neutral			
Urine				6.3×10^{-8}
Gastric juice			7.9×10^{-3}	

36. Like all equilibrium constants, the value of K_w is temperature dependent. K_w is 1.47×10^{-14} at 30°C. What is "neutral" pH at this temperature?

37. Give the conjugate acid and base of the each of the following species: (a) $HC_2O_4^-$; (b) HSO_3^-; (c) $H_2PO_4^-$; (d) HCO_3^-; (e) $HAsO_4^{2-}$; (f) HPO_4^{2-}; (g) HO_2^-.

38. Identify the acidic hydrogens in the following compounds.

CH₂—COOH
|
HO—C—COOH
|
CH₂—COOH
Citric acid

Piperidine

H₃N⁺—CH—C—OH ... Lysine

Oxalic acid

Barbituric acid

4-Morphine ethanesulfonic acid (MES)

39. Rank the following according to their strength as acids:

	Acid	K_a	pK
A	citrate		4.76
B	succinic acid	6.17×10^{-5}	
C	succinate	2.29×10^{-6}	
D	formic acid	1.78×10^{-4}	
E	citric acid		3.13

40. Several hours after a meal, partially digested food leaves the stomach and enters the small intestine, where pancreatic juice is added. How does the pH of the partially digested mixture change as it passes from the stomach into the intestine (see Table 2.3)?

41. The pH of urine has been found to be correlated with diet. Acidic urine results when meat and dairy products are consumed, because the oxidation of sulfur-containing amino acids in the proteins produces protons. Consumption of fruits and vegetables leads to alkaline urine, because these foods contain plentiful amounts of potassium and magnesium carbonate salts. Why would the presence of these salts result in alkaline urine?

42. Calculate the pH of 200 mL of water to which (a) 10 mL of 1.0 M HNO_3 or (b) 5 mL of 1.0 M KOH has been added.

43. What is the pH of 2.0 L of water to which (a) 1.5 mL of 2.0 M HCl or (b) 1.5 mL of 2.0 M NaOH has been added?

44. A solution is made by mixing 100 mL of a stock solution of 2.0 M K_2HPO_4 and 50 mL of a stock solution of 2.0 M KH_2PO_4. The solution is diluted to a final volume of 500 mL. What is the final pH of the solution?

45. Calculate the ratio of imidazole to the imidazolium ion in a solution at pH 7.8.

46. Calculate the pH of a 200-mL solution to which has been added 15 mL of 50 mM boric acid and 25 mL of 20 mM sodium borate.

47. Calculate the concentration of acetate in a 50-mM solution of acetic acid buffer at pH 5.0.

48. What is the volume (in mL) of glacial acetic acid (17.4 M) that would have to be added to 500 mL of a solution of 0.20 M sodium acetate in order to achieve a pH of 5.0?

49. What is the mass of NaOH that would have to be added to 500 mL of a solution of 0.20 M acetic acid in order to achieve a pH of 5.0?

50. The pH of blood is maintained within a narrow range (7.35–7.45). Carbonic acid, H_2CO_3, participates in blood buffering.

(a) Write the equations for the dissociation of the two ionizable protons.

(b) The pK for the first ionizable proton is 6.35; the pK for the second ionizable proton is 10.33. Use this information to identify the weak acid and the conjugate base present in the blood.

(c) Calculate the concentration of carbonic acid in a sample of blood with a bicarbonate concentration of 24 mM and a pH of 7.40.

51. The pK of $CH_3CH_2NH_3^+$ is 10.7. Would the pK of $FCH_2CH_2NH_3^+$ be higher or lower?

52. The structure of pyruvic acid is shown.

(a) Draw the structure of pyruvate.

(b) Using what you have learned about acidic functional groups, which form of this compound is likely to predominate in the cell at pH 7.4? Explain.

CH₃
|
C=O
|
COOH
Pyruvic acid

53. The amino acid glycine (H_2N—CH_2—$COOH$) has pK values of 2.35 and 9.78. Indicate the structure and net charge of the molecular species that predominate at pH 2, 7, and 10.

2.4 Tools and Techniques: Buffers

54. The structure of acetylsalicylic acid (aspirin) is shown. Is aspirin more likely to be absorbed (pass through a lipid membrane) in the stomach (pH ~2) or in the small intestine (pH ~8)? Explain.

Acetylsalicylic acid (aspirin)

55. An experiment requires the buffer HEPES, pH = 8.0 (see Table 2.4).

(a) Write an equation for the dissociation of HEPES in water. Identify the weak acid and the conjugate base.
(b) What is the effective buffering range for HEPES?
(c) The buffer will be prepared by making 1.0 L of a 0.10 M solution of HEPES. Hydrochloric acid will be added until the desired pH is achieved. Describe how you will make 1.0 L of 0.10 M HEPES. (HEPES is supplied by the chemical company as a sodium salt with a molecular weight of 260.3 g · mol⁻¹.)
(d) What is the volume (in mL) of a stock solution of 6.0 M HCl that must be added to the 0.1 M HEPES to achieve the desired pH of 8.0? Describe how you will make the buffer.

56. Which would be a more effective buffer at pH 8.0 (see Table 2.4)?

(a) 10 mM HEPES buffer or 10 mM glycinamide buffer;
(b) 10 mM Tris buffer or 20 mM Tris buffer;
(c) 5 mM boric acid or 5 mM sodium borate.

57. Phosphoric acid, H_3PO_4, has three ionizable protons.

(a) Sketch the titration curve. Indicate the pK values and the species that predominate in each area of the curve.
(b) Write the equations for the dissociation of the three ionizable protons.
(c) Which two phosphate species are present in the blood at pH 7.4?
(d) Which two phosphate species would be used to prepare a buffer solution of pH 11?

58. What volume (in mL) of a 1.0 M solution of imidazolium chloride must be added to a 500 mL solution of 10 mM imidazole in order to obtain a pH of 6.5?

59. One liter of a 0.10 M Tris buffer (see Table 2.4) is prepared and adjusted to a pH of 8.2.

Tris(hydroxymethyl)aminomethane

(a) Write the equation for the dissociation of Tris in water. Identify the weak acid and the conjugate base.
(b) What is the effective buffering range for Tris?
(c) What are the concentrations of the conjugate acid and weak base at pH 8.2?
(d) What is the ratio of conjugate base to weak acid if 1.5 mL of 3.0 M HCl is added to 1.0 L of the buffer? What is the new pH? Has the buffer functioned effectively? Compare the pH change to that of Problem 43(a) in which the same amount of acid was added to the same volume of pure water.

(e) What is the ratio of conjugate base to weak acid if 1.5 mL of 3.0 M NaOH is added to 1.0 L of the buffer? What is the new pH? Has the buffer functioned effectively? Compare the pH change to that of Problem 43(b) in which the same amount of base was added to the same volume of pure water.

60. One liter of a 0.1 M Tris buffer (see Table 2.4) is prepared and adjusted to a pH of 2.0.

(a) What are the concentrations of the conjugate base and weak acid at this pH?
(b) What is the pH when 1.5 mL of 3.0 M HCl is added to 1.0 L of the buffer? Has the buffer functioned effectively? Explain.
(c) What is the pH when 1.5 mL of 3.0 M NaOH is added to 1.0 L of the buffer? Has the buffer functioned effectively? Explain.

2.5 Clinical Connection: Acid–Base Balance in Humans

61. Impaired pulmonary function can contribute to respiratory acidosis. Using the appropriate equations, explain how the failure to eliminate sufficient CO_2 through the lungs leads to acidosis.

62. Metabolic acidosis often occurs in patients with impaired circulation from cardiac arrest. Mechanical hyperventilation (a standard treatment for acidosis) cannot be used with these patients because they often have acute lung injury (ALI). A group of physicians at San Francisco General Hospital advocated using tris(hydroxymethyl)aminomethane (Tris) to treat the metabolic acidosis of these patients.

(a) How would mechanical hyperventilation help alleviate acidosis in patients who do not have ALI?
(b) Why would treatment with sodium bicarbonate be effective in treating metabolic acidosis? Why would this treatment also be unacceptable for patients with ALI?
(c) Explain how Tris (see Problem 59) works to treat metabolic acidosis. Why is this treatment acceptable for patients with ALI?

63. An individual who develops alkalosis by hyperventilating is encouraged to breathe into a paper bag for several minutes. Why does this treatment correct the alkalosis?

64. A patient who has taken an overdose of aspirin is brought into the emergency room for treatment. She suffers from respiratory alkalosis, and the pH of her blood is 7.5. Determine the ratio of HCO_3^- to H_2CO_3 in the patient's blood at this pH. How does this compare to the ratio of HCO_3^- to H_2CO_3 in normal blood? Can the H_2CO_3/HCO_3^- system work effectively as a buffer in this patient under these conditions?

65. Metabolic acidosis is a general term that describes a number of disorders in metabolism in the body that result in a lowering of the blood pH from 7.4 to 7.35 or below. The kidney plays a vital role in regulating blood pH. The kidney can either excrete or reabsorb various ions, including phosphate, $H_2PO_4^-$; ammonium, NH_4^+; or bicarbonate, HCO_3^-. Which ions are excreted and which ions are reabsorbed in metabolic acidosis? Explain, using relevant chemical equations.

66. Metabolic alkalosis occurs when the blood pH rises to 7.45 or greater. Which ions are excreted and which ions are reabsorbed in metabolic alkalosis (see Problem 65)?

67. Kidney cells excrete H^+ and reabsorb HCO_3^- from the filtrate. The movement of each of these ions is thermodynamically unfavorable. However, the movement of each becomes possible when it is coupled to another, thermodynamically favorable ion transport process. Explain how the movement of other ions drives the transport of H^+ and HCO_3^-.

68. In uncontrolled diabetes, the body converts fats to the so-called ketone bodies acetoacetate and 3-hydroxybutyrate, which accumulate in the bloodstream.

$$CH_3-\overset{\overset{\displaystyle O}{\|}}{C}-CH_2-COO^-\qquad CH_3-\overset{\overset{\displaystyle OH}{|}}{\underset{\underset{\displaystyle H}{|}}{C}}-CH_2-COO^-$$

Acetoacetate   3-Hydroxybutyrate

Do the ketone bodies contribute to acidosis or alkalosis? How might the body compensate for this acid–base imbalance?

69. Explain why metabolically generated CO_2 does not accumulate in tissues but is quickly converted to carbonic acid by the action of carbonic anhydrase in red blood cells.

70. Ammonia produced in kidney cells can diffuse out of the cell, but only in its unprotonated (NH_3) form.

(a) Explain.
(b) The NH_4^+ ion, however, can exit the cell in exchange for another ion. What ion is most likely involved?

SELECTED READINGS

Ball, P., Water, water, everywhere? *Nature* **427,** 19–20 (2004). [A brief discussion of water as the matrix of life.]

Ellis, R. J., Macromolecular crowding: obvious but underappreciated, *Trends Biochem. Sci.* **26,** 597–603 (2001). [Discusses how the large number of macromolecules in cells could affect reaction equilibria and reaction rates.]

Gerstein, M. and Levitt, M., Simulating water and the molecules of life, *Sci. Am.* **279(11),** 101–105 (1998). [Describes the structure of water and how water interacts with other molecules.]

Kropman, M. F. and Bakker, H. J., Dynamics of water molecules, *Science* **291,** 2118–2120 (2001). [Describes how water molecules in solvation shells move more slowly than bulk water molecules.]

Yucha, C., Renal regulation of acid–base balance, *Nephrol. Nursing J.* **31,** 201–206 (2004). [A short review of physiological buffer systems and the role of the kidney.]

Kletr/Shutterstock

The Anopheles mosquito is not deadly on its own, but as the carrier of the malaria parasite, it is linked to the deaths of several hundred thousand people every year. Researchers are investigating whether the latest genetic engineering techniques can produce a mosquito population that is unable to transmit the parasite.

DO YOU REMEMBER?

- Cells contain four major types of biological molecules and three major types of polymers (Section 1.2).
- Modern prokaryotic and eukaryotic cells apparently evolved from simpler nonliving systems (Section 1.4).
- Noncovalent forces, including hydrogen bonds, ionic interactions, and van der Waals forces, act on biological molecules (Section 2.1).

All the structural components of cells and the machinery that carries out the cell's activities are ultimately specified by the cell's genetic material—DNA. Therefore, before examining other types of biological molecules and their metabolic transformations, we must consider the nature of DNA, including its chemical structure and how its biological information is organized and expressed. The Tools and Techniques section of this chapter includes some of the methods used to study and manipulate DNA in the laboratory.

3.1 NUCLEOTIDES

KEY CONCEPT
- DNA and RNA are polymers of nucleotides, each of which consists of a purine or pyrimidine base, deoxyribose or ribose, and phosphate.

Gregor Mendel was certainly not the first to notice that an organism's characteristics (for example, flower color or seed shape in pea plants) were passed to its progeny, but in 1865 he was the first to describe their predictable patterns of inheritance. By 1903, Mendel's inherited factors (now called genes) were recognized as belonging to **chromosomes** (a word that means "colored bodies"), which are visible by light microscopy (**Fig. 3.1**).

Eventually, chromosomes were shown to be composed of proteins, which had first been described in 1838 by Gerardus Johannes Mulder, and **nucleic acids,** which had been discovered in 1869 by Friedrich Miescher. Proteins, with their 20 different types of amino acids and great diversity in size and shape, were the obvious candidates to be carriers of genetic information in chromosomes. Nucleic acids, in contrast, seemed uninteresting and contained only four different types of structural units, called **nucleotides.** In **DNA (deoxyribonucleic**

acid), these components—abbreviated A, C, G, and T—were thought to occur as simple repeating tetranucleotides, for example,

—ACGT-ACGT-ACGT-ACGT—

In 1950, when Erwin Chargaff showed that the nucleotides in DNA were not all present in equal numbers and that the nucleotide composition varied among species, it became apparent that DNA might be complex enough to be the genetic material after all. Several other lines of research also pointed to the importance of DNA, and the race was on to decipher its molecular structure.

NUCLEIC ACIDS ARE POLYMERS OF NUCLEOTIDES

Each nucleotide of DNA includes a nitrogen-containing **base.** The bases adenine (A) and guanine (G) are known as **purines** because they resemble the organic compound purine:

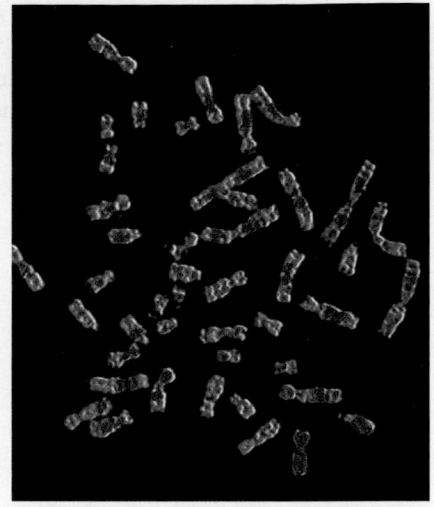

FIGURE 3.1

Human chromosomes from amniocentesis. In this image, the chromosomes have been stained with fluorescent dyes. [*Dr. P. Boyer/ Photo Researchers, Inc.*]

Adenine   Guanine   Purine

The bases cytosine (C) and thymine (T) are known as **pyrimidines** because they resemble the organic compound pyrimidine:

Cytosine   Thymine   Pyrimidine

Ribonucleic acid (RNA) contains the pyrimidine uracil (U) rather than thymine:

Uracil

so that DNA contains the bases A, C, G, and T, whereas RNA contains A, C, G, and U. The purines and pyrimidines are known as *bases* because they can participate in acid–base reactions. However, they donate or accept protons only at extremely low or high pH, so this behavior is not relevant to their function inside cells.

Linking atom N9 in a purine or atom N1 in a pyrimidine to a five-carbon sugar forms a **nucleoside.** *In DNA, the sugar is 2′-deoxyribose; in RNA, the sugar is ribose* (the sugar atoms are numbered with primes to distinguish them from the atoms of the attached bases).

Ribose   2′-Deoxyribose

A nucleotide is a nucleoside to which one or more phosphate groups are linked, usually at C5′ of the sugar. Depending on whether there are one, two, or three phosphate groups, the nucleotide is known as a nucleoside monophosphate, nucleoside diphosphate, or nucleoside triphosphate and is represented by a three-letter abbreviation, for example,

Adenosine monophosphate
(AMP)

Guanosine diphosphate
(GDP)

Cytidine triphosphate
(CTP)

Deoxynucleotides are named in a similar fashion, and their abbreviations are preceded by "d." The deoxy counterparts of the compounds shown above would therefore be deoxyadenosine monophosphate (dAMP), deoxyguanosine diphosphate (dGDP), and deoxycytidine triphosphate (dCTP). The names and abbreviations of the common bases, nucleosides, and nucleotides are summarized in **Table 3.1**.

SOME NUCLEOTIDES HAVE OTHER FUNCTIONS

In addition to serving as the building blocks for DNA and RNA, nucleotides perform a variety of functions in the cell. They are involved in energy transduction, intracellular signaling, and regulation of enzyme activity. Some nucleotide derivatives are essential players in the metabolic pathways that synthesize biomolecules or degrade them in order to "capture" free energy. For example, coenzyme A (CoA; **Fig. 3.2a**) is a carrier of other molecules during their synthesis and degradation. Two nucleotides are linked in the compounds nicotinamide adenine dinucleotide (NAD; Fig. 3.2b) and flavin adenine dinucleotide (FAD; Fig. 3.2c), which undergo reversible oxidation and reduction during a number of metabolic reactions. Interestingly, a portion of the structures of each of these molecules is derived from a **vitamin,** a compound that must be obtained from the diet.

TABLE 3.1 Nucleic Acid Bases, Nucleosides, and Nucleotides

BASE	NUCLEOSIDE[a]	NUCLEOTIDES[a]
Adenine (A)	Adenosine	Adenylate; adenosine monophosphate (AMP) adenosine diphosphate (ADP) adenosine triphosphate (ATP)
Cytosine (C)	Cytidine	Cytidylate; cytidine monophosphate (CMP) cytidine diphosphate (CDP) cytidine triphosphate (CTP)
Guanine (G)	Guanosine	Guanylate; guanosine monophosphate (GMP) guanosine diphosphate (GDP) guanosine triphosphate (GTP)
Thymine (T)[b]	Thymidine	Thymidylate; thymidine monophosphate (TMP) thymidine diphosphate (TDP) thymidine triphosphate (TTP)
Uracil (U)[c]	Uridine	Uridylate; uridine monophosphate (UMP) uridine diphosphate (UDP) uridine triphosphate (UTP)

[a]Nucleosides and nucleotides containing 2′-deoxyribose rather than ribose may be called deoxynucleosides and deoxynucleotides. The nucleotide abbreviation is then preceded by "d."

[b]Thymine is found in DNA but not in RNA.

[c]Uracil is found in RNA but not in DNA.

Coenzyme A (CoA)
(a)

Coenzyme A (CoA) contains a residue of pantothenic acid (pantothenate), also known as vitamin B₅. The sulfhydryl group is the site of attachment of other groups.

Nicotinamide adenine dinucleotide (NAD)
(b)

The nicotinamide group of nicotinamide adenine dinucleotide (NAD) is a derivative of the vitamin niacin (also called nicotinic acid or vitamin B₃; see inset) and undergoes oxidation and reduction. The related compound nicotinamide adenine dinucleotide phosphate (NADP) contains a phosphoryl group at the adenosine C2′ position.

Flavin adenine dinucleotide (FAD)
(c)

Oxidation and reduction of flavin adenine dinucleotide (FAD) occurs at the riboflavin group (also known as vitamin B₂).

FIGURE 3.2

Some nucleotide derivatives. The adenosine group of each of these compounds is shown in red. Note that each also contains a vitamin derivative.

? Locate the nitrogenous base(s) and sugar(s) in each structure.

CONCEPT CHECK

1. Draw the structures of the two purine bases, the three pyrimidine bases, ribose, and deoxyribose.

2. What are the relationships among purines, pyrimidines, nucleosides, nucleotides, and nucleic acids?

3.2 NUCLEIC ACID STRUCTURE

KEY CONCEPTS
- A DNA molecule contains two antiparallel strands that wind around each other to form a double helix in which A and T bases in opposite strands, and C and G bases in opposite strands, pair through hydrogen bonding.
- Double-stranded nucleic acids are denatured at high temperatures; at lower temperatures, complementary polynucleotides anneal.

In a nucleic acid, the linkage between nucleotides is called a **phosphodiester bond** because a single phosphate group forms ester bonds to both C5′ and C3′. During DNA synthesis in a cell, when a nucleoside triphosphate is added to the **polynucleotide** chain, a diphosphate group is eliminated. Once incorporated into a polynucleotide, the nucleotide is formally known as a nucleotide **residue.** Nucleotides consecutively linked by phosphodiester bonds form a polymer in which the bases project out from a backbone of repeating sugar–phosphate groups.

The end of the polymer that bears a phosphate group attached to C5′ is known as the **5′ end,** and the end that bears a free OH group at C3′ is the **3′ end.** By convention, the base sequence in a polynucleotide is read from the 5′ end (on the left) to the 3′ end (on the right).

DNA IS A DOUBLE HELIX

A DNA molecule consists of two polynucleotide strands linked by hydrogen bonds (hydrogen bonding is discussed in Section 2.1). The structure of this molecule, elucidated by James Watson and Francis Crick in 1953, incorporated Chargaff's earlier observations about DNA's base composition. Specifically, Chargaff noted that the amount of A is equal to the amount of T, the amount of C is equal to the amount of G, and the total amount of A + G is equal to the total amount of C + T. *Chargaff's "rules" could be satisfied by a molecule with two polynucleotide strands in which A and C in one strand pair with T and G in the other.* Two hydrogen bonds link adenine and thymine, and three hydrogen bonds link guanine and cytosine:

Adenine

Thymine

Guanine

Cytosine

10.85 Å

*All the **base pairs**, which consist of a purine and a pyrimidine, have the same molecular dimensions* (about 11 Å wide). Consequently, the **sugar–phosphate backbones** of the two strands of DNA are separated by a constant distance, regardless of whether the base pair is A:T, G:C, T:A, or C:G.

Although DNA can be shown as a ladder-like structure (*left*), with the two sugar–phosphate backbones as the vertical supports and the base pairs as the rungs, the two strands of DNA twist around each other to generate the familiar double helix (*right*).

Sugar–phosphate backbones

A	T
G	C
T	A
A	T
C	G
C	G
T	A
G	C
T	A
G	C
A	T
T	A

This conformation allows successive base pairs, which are essentially planar, to stack on top of each other with a center-to-center distance of only 3.4 Å. In fact, Watson and Crick derived this model for DNA not just from Chargaff's rules but also from Rosalind Franklin's studies of the diffraction (scattering) of an X-ray beam by a DNA fiber, which suggested a helix with a repeating spacing of 3.4 Å.

The major features of the DNA molecule include the following (**Fig. 3.3**):

1. The two polynucleotide strands are **antiparallel;** that is, their phosphodiester bonds run in opposite directions. One strand has a 5′ → 3′ orientation, and the other has a 3′ → 5′ orientation.

2. The DNA "ladder" is twisted in a right-handed fashion. (If you climbed the DNA helix as if it were a spiral staircase, you would hold the outer railing—the sugar–phosphate backbone—with your right hand.)

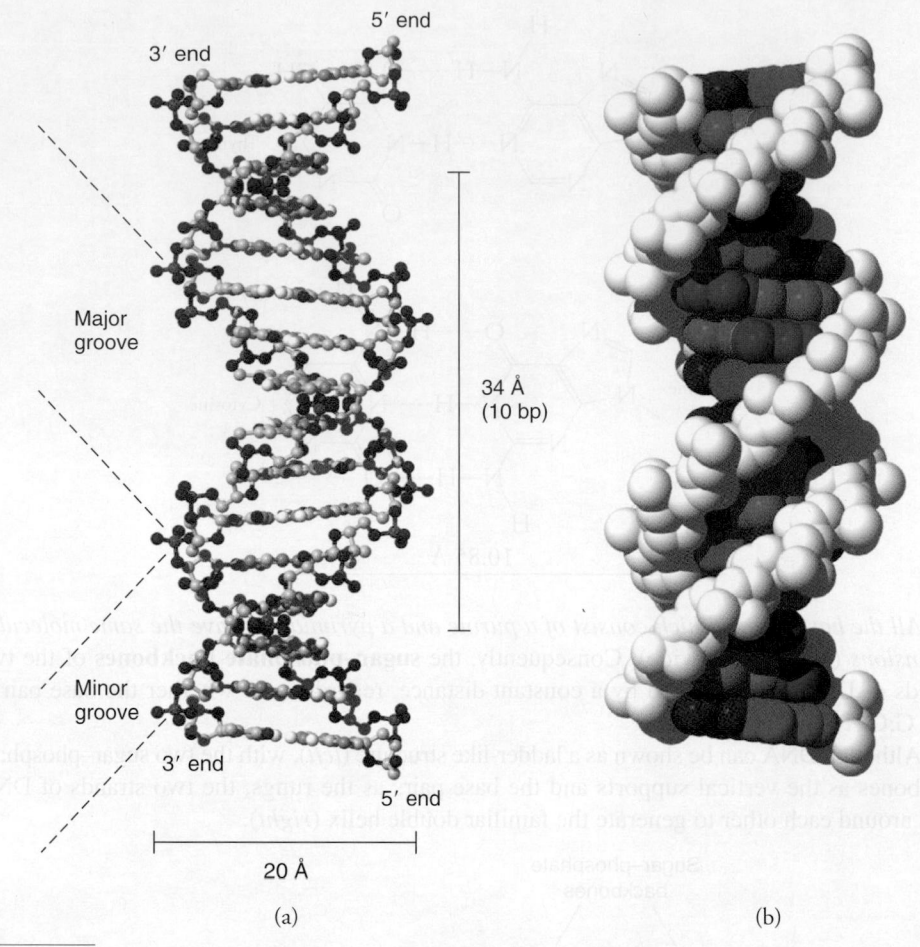

5′ end

3′ end

Major groove

34 Å
(10 bp)

Minor groove

3′ end

5′ end

20 Å

(a)

(b)

FIGURE 3.3

Model of DNA. (a) Ball-and-stick model with atoms colored: C gray, O red, N blue, and P gold (H atoms are not shown). (b) Space-filling model with the sugar–phosphate backbone in gray and the bases color-coded: A green, C blue, G yellow, and T red.

? **How many nucleotides are shown in this double helix?**

3. The diameter of the helix is about 20 Å, and it completes a turn about every 10 base pairs, which corresponds to an axial distance of about 34 Å.

4. The twisting of the DNA "ladder" into a helix creates two grooves of unequal width, the **major** and **minor grooves.**

5. The sugar–phosphate backbones define the exterior of the helix and are exposed to the solvent. The negatively charged phosphate groups bind Mg^{2+} cations *in vivo*, which helps minimize electrostatic repulsion between these groups.

6. The base pairs are located in the center of the helix, approximately perpendicular to the helix axis.

7. The base pairs stack on top of each other, so the core of the helix is solid (see Fig. 3.3b). Although the planar faces of the base pairs are not accessible to the solvent, their edges are exposed in the major and minor grooves (this allows certain DNA-binding proteins to recognize specific bases).

In nature, DNA seldom assumes a perfectly regular conformation because of small sequence-dependent irregularities. For example, base pairs can roll or twist like propeller blades, and the helix may wind more tightly or loosely at certain nucleotide sequences. DNA-binding proteins may take advantage of these small variations to locate their specific binding sites, and they in turn may further distort the DNA helix by causing it to bend or partially unwind.

The size of a DNA segment is expressed in units of base pairs (**bp**) or kilobase pairs (1000 bp, abbreviated **kb**). Most naturally occurring DNA molecules comprise thousands to millions of base pairs. A short single-stranded polymer of nucleotides is usually called an **oligonucleotide** (*oligo* is Greek for "few"). In a cell, nucleotides are polymerized by the action of enzymes known as **polymerases.** The phosphodiester bonds linking nucleotide residues can be hydrolyzed by the action of **nucleases.** An **exonuclease** removes a residue from the end of a polynucleotide chain, whereas an **endonuclease** cleaves at some other point along the chain. Polymerases and nucleases are usually specific for either DNA or RNA. In the absence of these enzymes, the structures of nucleic acids are remarkably stable. The hydrogen bonds between polynucleotide strands, however, are relatively weak and break to allow the strands to separate during replication and transcription, described below.

RNA IS SINGLE-STRANDED

RNA, which is a single-stranded polynucleotide, has greater conformational freedom than DNA, whose structure is constrained by the requirements of regular base-pairing between its two strands. *An RNA strand can fold back on itself so that base pairs form between complementary segments of the same strand.* **Complementarity** refers to the ability of bases to form hydrogen bonds with their standard partners: A is complementary to T and U, and G is complementary to C. Consequently, RNA molecules tend to assume intricate three-dimensional shapes (**Fig. 3.4**). Unlike DNA, whose regular structure is suited for the long-term storage of genetic information, RNA can assume more active roles in expressing that information. For example, the molecule shown in Figure 3.3, which carries the amino acid phenylalanine, interacts with a number of proteins and other RNA molecules during protein synthesis.

The residues of RNA are also capable of base-pairing with a complementary single strand of DNA to produce an RNA–DNA hybrid double helix (**Fig. 3.5**). A double helix involving RNA is wider and flatter than the standard DNA helix (its diameter is about 26 Å, and it makes one helical turn every 11 residues). In addition, its base pairs are inclined to the helix axis by about 20°. These structural differences relative to the standard DNA helix primarily reflect the presence of the 2′ OH groups in RNA.

A double-stranded DNA helix can adopt this same helical conformation; it is known as **A-DNA.** The standard DNA helix shown in Figure 3.3 is known as **B-DNA.** Other conformations of DNA have been described, and there is evidence that they exist *in vivo*, at least for certain nucleotide sequences, but their functional significance is not completely understood.

NUCLEIC ACIDS CAN BE DENATURED AND RENATURED

The pairing of polynucleotide strands in a double-stranded nucleic acid is possible because bases in each strand form hydrogen bonds with complementary bases in the other strand: A is the complement of T (or U), and G is the complement of C. However, the structural stability of a double helix does not depend significantly on hydrogen bonding between complementary bases. (If the strands were separated, the bases could still satisfy their hydrogen-bonding requirements by forming hydrogen bonds with solvent water molecules.) Instead, *stability depends mostly on* **stacking interactions**, *which are a form of van der Waals interaction, between adjacent base pairs.* A view down the helix axis shows that stacked base pairs do not overlap exactly, due to the winding of the helix (**Fig. 3.6**). Although individual stacking interactions are weak, they are additive along the length of a DNA molecule.

The stacking interactions between neighboring G:C base pairs are stronger than those of A:T base pairs (this is not related to the fact that G:C base pairs have one more hydrogen bond than A:T base pairs). Consequently, *a DNA helix that is rich in G and C is harder to disrupt than DNA with a high proportion of A and T.* These differences can be quantified in the **melting temperature** (T_m) of the DNA.

To determine the melting point of a sample of DNA, the temperature is slowly increased. At a sufficiently high temperature, the base pairs begin to unstack, hydrogen bonds break, and the two strands begin to separate. This process continues as the temperature rises, until the two strands come completely apart. The melting, or **denaturation,** of the DNA can be recorded as

FIGURE 3.4

A transfer RNA molecule. This 76-nucleotide single-stranded RNA molecule folds back on itself so that base pairs form between complementary segments. [*Structure (pdb 4TRA) determined by E. Westhoff, P. Dumas, and D. Moras.*]

FIGURE 3.5

An RNA–DNA hybrid helix. In a double helix formed by one strand of RNA (red) and one strand of DNA (blue), the planar base pairs are tilted and the helix does not wind as steeply as in a standard DNA double helix (compare with Fig. 3.3). [*Structure (pdb 1FIX) determined by N. C. Horton and B. C. Finzel.*]

FIGURE 3.6

Axial view of DNA base pairs. A view down the central axis of the DNA helix shows the overlap of neighboring base pairs (only the first two nucleotide pairs are highlighted).

? **Locate the base and sugar in the nucleotides with the blue bases.**

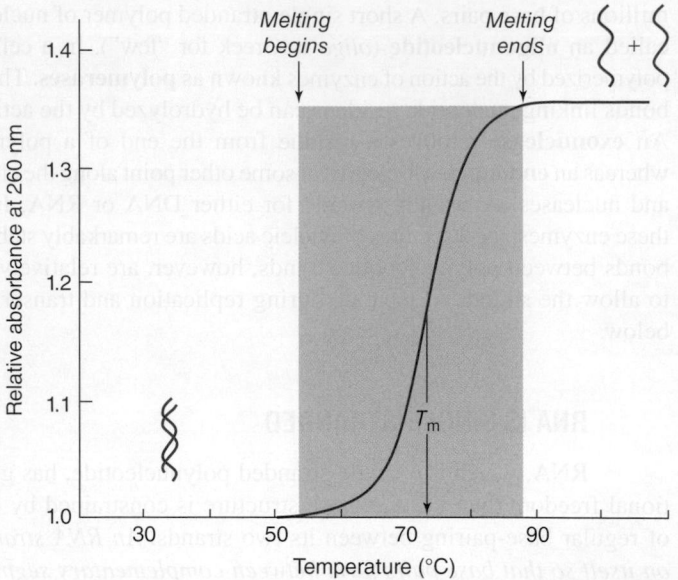

FIGURE 3.7

A DNA melting curve. Thermal denaturation (melting, or strand separation) of DNA results in an increase in ultraviolet absorbance relative to the absorbance at 25°C. The melting point, T_m, of the DNA sample is defined as the midpoint of the melting curve.

a melting curve (**Fig. 3.7**) by monitoring an increase in the absorbance of ultraviolet (260-nm) light (the aromatic bases absorb more light when unstacked). The midpoint of the melting curve (that is, the temperature at which half the DNA has separated into single strands) is the T_m. **Table 3.2** lists the GC content and the melting point of the DNA from different species. Since manipulating DNA in the laboratory frequently requires the thermal separation of paired DNA strands, it is sometimes helpful to know the DNA's GC content.

When the temperature is lowered slowly, denatured DNA can **renature;** that is, *the separated strands can re-form a double helix by reestablishing hydrogen bonds between the complementary strands and by restacking the base pairs.* The maximum rate of renaturation occurs at about 20–25°C below the melting temperature. If the DNA is cooled too rapidly, it may not fully renature because base pairs may form randomly between short complementary segments. At low temperatures, the improperly paired segments are frozen in place since they do not have enough thermal energy to melt apart and find their correct complements (**Fig. 3.8**). The rate of renaturation of denatured DNA depends on the length of the double-stranded molecule: Short segments come together (**anneal**) faster than longer segments because the bases in each strand must locate their partners along the length of the complementary strand.

The ability of short single-stranded nucleic acids (either DNA or RNA) to hybridize with longer polynucleotide chains is the basis for a number of useful laboratory techniques (described in detail in Section 3.5). For example, an oligonucleotide **probe** that has been tagged with a fluorescent group can be used to detect the presence of a complementary nucleic acid sequence in a complex mixture.

TABLE 3.2 GC Content and Melting Points of DNA

SOURCE OF DNA	GC CONTENT (%)	T_m (°C)
Dictyostelium discoideum (fungus)	23.0	79.5
Clostridium butyricum (bacterium)	37.4	82.1
Homo sapiens	40.3	86.5
Streptomyces albus (bacterium)	72.3	100.5

[Data from Brown, T. A. (ed.), *Molecular Biology LabFax*, vol. I., Academic Press (1998), pp. 233–237.]

High heat
(melting)

Cooling to
20–25°C below T_m
(renaturation)

Rapid cooling to
temperature much
lower than T_m
(improper base
pairing)

Rewarming
to 20–25°C
below T_m
(renaturation)

FIGURE 3.8

Renaturation of DNA. DNA strands that have been melted apart can renature at a temperature of 20–25°C below the T_m. At much lower temperatures, base pairs may form between short complementary segments within and between the single strands. Correct renaturation is possible only if the sample is rewarmed so that the improperly paired strands can separate and reanneal.

 CONCEPT CHECK

1. What did Chargaff's rules reveal about the structure of DNA?

2. Describe the arrangement of the base pairs and sugar–phosphate backbones in DNA.

3. Differentiate between RNA and DNA.

4. Describe the molecular events in DNA denaturation and renaturation.

5. How can a single-stranded oligonucleotide be used to reveal the presence of a particular segment of DNA?

THE CENTRAL DOGMA

The complementarity of the two strands of DNA is essential for its function as the storehouse of genetic information, since this information must be **replicated** (copied) for each new generation. As first suggested by Watson and Crick, the separated strands of DNA direct the synthesis of complementary strands, thereby generating two identical double-stranded molecules (**Fig. 3.9**). The parental strands are said to act as templates for the assembly of the new strands because their sequence of nucleotides determines the sequence of nucleotides in the new strands. Thus, genetic information—in the form of a sequence of nucleotide residues—is transmitted each time a cell divides.

A similar phenomenon is responsible for the **expression** of that genetic information, a process in which the information is used to direct the synthesis of proteins that carry out the cell's activities. First, *a portion of the DNA, a **gene**, is **transcribed** to produce a complementary strand of RNA; then the RNA is **translated** into protein.* This paradigm, known as the **central dogma of molecular biology,** was formulated by Francis Crick. It can be shown schematically as

replication

transcription

translation

DNA → RNA → Protein

KEY CONCEPT

- The biological information encoded by a sequence of DNA is transcribed to RNA and then translated into the amino acid sequence of a protein.

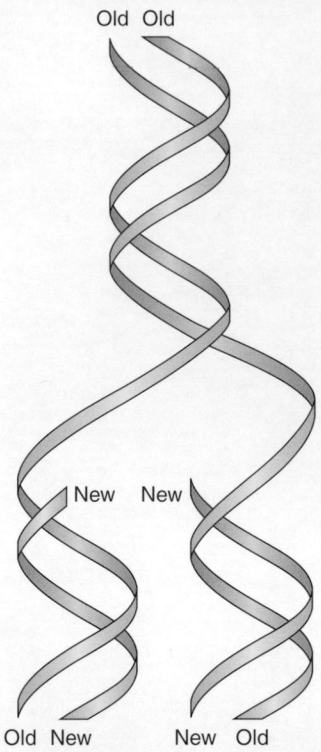

Old New   New Old

FIGURE 3.9

DNA replication. The double helix unwinds so that each parental strand can serve as a template for the synthesis of a new complementary strand. The result is two identical double-helical DNA molecules.

? Label the 5′ and 3′ ends of each strand.

Although it is tempting to think of a cell's DNA as its "brain," the DNA does not issue commands to the rest of the cell. Instead, DNA simply holds genetic information—the instructions for synthesizing proteins.

DNA MUST BE DECODED

Even in the simplest organisms, DNA is an enormous molecule, and many organisms contain several different DNA molecules (for example, the chromosomes of eukaryotes). An organism's complete set of genetic information is called its **genome.** A genome may comprise several hundred to perhaps 35,000 genes.

To transcribe a gene, one of the two strands of DNA serves as a template for an RNA polymerase to synthesize a complementary strand of RNA. The RNA therefore has the same sequence (except for the substitution of U for T) and the same 5′ → 3′ orientation as the nontemplate strand of DNA. This strand of DNA is often called the **coding strand** (the template strand is called the **noncoding strand**).

The transcribed RNA is known as **messenger RNA (mRNA)** because it carries the same genetic message as the gene.

The mRNA is translated by a **ribosome,** a cellular particle consisting of protein and **ribosomal RNA (rRNA).** At the ribosome, small molecules called **transfer RNA (tRNA),** which carry amino acids, recognize sequential sets of three bases (known as **codons**) in the mRNA through complementary base-pairing (a tRNA molecule is shown in Fig. 3.4). The ribosome covalently links the amino acids carried by successive tRNAs to form a protein. *The protein's amino acid sequence therefore ultimately depends on the nucleotide sequence of the DNA.*

The correspondence between amino acids and mRNA codons is known as the **genetic code.** There are a total of 64 codons: 3 of these are "stop" signals that terminate translation, and the remaining 61 represent, with some redundancy, the 20 standard amino acids found

TABLE 3.3 The Standard Genetic Code[a]

FIRST POSITION (5′ END)	SECOND POSITION				THIRD POSITION (3′ END)
	U	**C**	**A**	**G**	
U	UUU Phe	UCU Ser	UAU Tyr	UGU Cys	U
	UUC Phe	UCC Ser	UAC Tyr	UGC Cys	C
	UUA Leu	UCA Ser	UAA Stop	UGA Stop	A
	UUG Leu	UCG Ser	UAG Stop	UGG Trp	G
C	CUU Leu	CCU Pro	CAU His	CGU Arg	U
	CUC Leu	CCC Pro	CAC His	CGC Arg	C
	CUA Leu	CCA Pro	CAA Gln	CGA Arg	A
	CUG Leu	CCG Pro	CAG Gln	CGG Arg	G
A	AUU Ile	ACU Thr	AAU Asn	AGU Ser	U
	AUC Ile	ACC Thr	AAC Asn	AGC Ser	C
	AUA Ile	ACA Thr	AAA Lys	AGA Arg	A
	AUG Met	ACG Thr	AAG Lys	AGG Arg	G
G	GUU Vae	GCU Ala	GAU Asp	GGU Gly	U
	GUC Val	GCC Ala	GAC Asp	GGC Gly	C
	GUA Val	GCA Ala	GAA Glu	GGA Gly	A
	GUG Val	GCG Ala	GAG Glu	GGG Gly	G

[a]The 20 amino acids are abbreviated; Ala, alanine; Arg, arginine; Asn, asparagine; Asp, aspartate; Cys, cysteine; Gly, glycine; Gln, glutamine; Glu, glutamate; His, histidine; Ile, isoleucine; Leu, leucine; Lys, lysine; Met, methionine; Phe, phenylalanine; Pro, proline; Ser, serine; Thr, threonine; Trp, tryptophan; Tyr, tyrosine; and Val, valine.

? How many amino acids would be uniquely specified by a genetic code that consisted of just the first two nucleotides in each codon?

in proteins. **Table 3.3** shows which codons specify which amino acids. In theory, knowing a gene's nucleotide sequence should be equivalent to knowing the amino acid sequence of the protein encoded by the gene. However, as we will see, genetic information is often "processed" at several points before the protein reaches its mature form. Keep in mind that the rRNA and tRNA required for protein synthesis, as well as other types of RNA, are also encoded by genes. The "products" of these genes are the result of transcription without translation.

A MUTATED GENE CAN CAUSE DISEASE

Because an organism's genetic material influences the organism's entire repertoire of activities, it is vitally important to unravel the sequence of nucleotides in that organism's DNA, even by examining one gene at a time. Thousands of genes have been identified through studies of the genes' protein products, and millions more have been catalogued through genome-sequencing projects (discussed below in Section 3.4). Although the functions of many genes are not yet understood, some genes have come to light through the study of inherited diseases. In a traditional approach, researchers have used the defective protein associated with a particular disease to track down the relevant genetic defect. For example, the variant hemoglobin protein that causes sickle cell disease results from the substitution of the amino acid glutamate (Glu) by valine (Val). In the gene for that protein chain, the normal GAG codon has been **mutated** (altered) to GTG.

Normal gene ··· ACT CCT GAG GAG AAG ···

Protein ··· Thr – Pro – Glu – Glu – Lys ···

Mutated gene ··· ACT CCT GTG GAG AAG ···

Protein ··· Thr – Pro – Val – Glu – Lys ···

Many human diseases are **polygenic,** so sorting out all the genetic variations that contribute to the disease is difficult. But for almost 5000 **monogenic diseases,** such as sickle cell disease and cystic fibrosis, a defect in a specific gene explains the molecular basis of the disease.

In many cases, a variety of different mutations have been catalogued for a particular disease gene, which explains in part why symptoms of the disease vary between individuals. The database known as OMIM (Online Mendelian Inheritance in Man, omim.org) contains information on thousands of genetic variants, including the clinical features of the resulting disorder and its biochemical basis. The Genetic Testing Registry (www.ncbi.nlm.nih.gov/gtr/) is a database of the diseases that can be detected through analysis of DNA, carried out by either clinical or research laboratories.

 CONCEPT CHECK

1. Draw a diagram to illustrate each step of the central dogma.
2. How does DNA encode genetic information and how is this information expressed?
3. What is the relationship between the nucleotide sequence in a gene and the amino acid sequence of a protein?
4. List some reasons why knowing a gene's sequence might be useful.

3.4 GENOMICS

KEY CONCEPTS
- The genomes of different species vary in size and number of genes.
- Genes can be identified by their nucleotide sequences.
- Analysis of genetic data can provide information about gene function and risk of disease.

The ability to sequence large tracts of DNA has made it possible to study entire genomes, from the small DNA molecules of parasitic bacteria to the enormous multichromosome genomes of plants and mammals. Sequence data are customarily deposited in a public database such as GenBank. The data can be accessed electronically in order to compare a given sequence to sequences from other genes.

Some of the thousands of organisms whose genomes have been partially or fully sequenced are listed in **Table 3.4**. This list includes species that are widely used as model organisms for different types of biochemical studies (**Fig. 3.10**).

TABLE 3.4 Genome Size and Gene Number of Some Organisms		
ORGANISM	**GENOME SIZE (kb)**	**NUMBER OF GENES**
Bacteria		
Mycoplasma genitalium	580	525
Haemophilus influenzae	1,830	1,740
Synechocystis PCC6803	3,947	3,618
Escherichia coli	4,639	4,289
Archaea		
Methanocaldococcus jannaschii	1,740	1,830
Archaeoglobus fulgidus	2,178	2,486
Fungi		
Saccharomyces cerevisiae (yeast)	12,070	6,034
Plants		
Arabidopsis thaliana	119,200	~26,000
Oryza sativa (rice)	389,000	~35,000
Zea mays (corn)	~2,400,000	~32,000
Animals		
Caenorhabditis elegans (nematode)	97,000	19,099
Drosophila melanogaster (fruit fly)	139,500	13,061
Homo sapiens	3,038,000	~21,000

[Data from NCBI Genome Project.]

? What is the relationship between genome size and gene number in prokaryotes? How does this differ in eukaryotes?

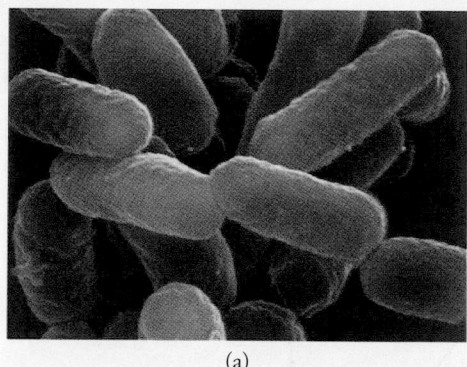

(a)

Escherichia coli, a normal inhabitant of the mammalian digestive tract, is a metabolically versatile bacterium that tolerates both aerobic and anaerobic conditions.

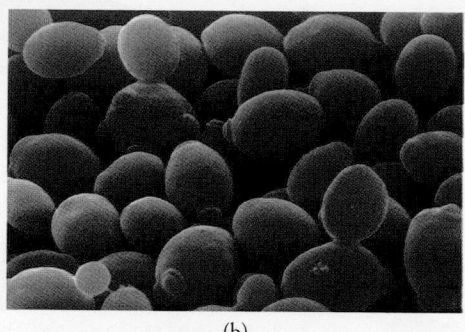

(b)

Baker's yeast, *Saccharomyces cerevisiae,* is one of the simplest eukaryotic organisms, with just over 6000 genes.

(c)

Caenorhabditis elegans is a small (1-mm) and transparent roundworm. As a multicellular organism, it bears genes not found in unicellular organisms.

(d)

The plant kingdom is represented by *Arabidopsis thaliana,* which has a short generation time and readily takes up foreign DNA.

FIGURE 3.10

Some model organisms. [*Dr. Kari Lounatmaa/Science Photo Library/Photo Researchers; Andrew Syred/Science Photo Library/Photo Researchers; Sinclair Stammers/Science Photo Library/Photo Researchers; Dr. Jeremy Burgess/Science Photo Library/Photo Researchers.*]

GENE NUMBER IS ROUGHLY CORRELATED WITH ORGANISMAL COMPLEXITY

Not surprisingly, *organisms with the simplest lifestyles tend to have the least amount of DNA and the fewest genes.* For example, *M. genitalium* and *H. influenzae* (see Table 3.4) are human parasites that depend on their host to provide nutrients; these organisms do not contain as many genes as free-living bacteria such as *Synechocystis* (a photosynthetic bacterium). Multicellular organisms generally have even more DNA and more genes, presumably to support the activities of their many specialized cell types. Interestingly, humans contain about as many genes as nematodes, suggesting that organismal complexity results not just from the raw number of genes but from how the genes are transcribed and translated into protein. Note that humans and many other organisms are **diploid** (having two sets of genetic information, one from each parent), so that each human cell contains roughly 6 billion base pairs of DNA. For simplicity, genetic information usually refers to the **haploid** state, equivalent to one set of genetic instructions.

FIGURE 3.11

Coding and noncoding portions of the human genome. Approximately 1.4% of the genome codes for proteins. Moderately repetitive sequences make up 45% of the genome and highly repetitive sequences about 3%, so that roughly half of the human genome consists of unique DNA sequences of unknown function. Up to 80% of the genome may be transcribed, however.

In prokaryotic genomes, all but a few percent of the DNA represents genes for proteins and RNA. The proportion of noncoding DNA generally increases with the complexity of the organism. For example, about 30% of the yeast genome, about half of the *Arabidopsis* genome, and over 98% of the human genome is noncoding DNA. Although up to 80% of the human genome may actually be transcribed to RNA, *the protein-coding segments account for only about 1.4% of the total* (**Fig. 3.11**).

Much of the noncoding DNA consists of repeating sequences with no known function. The presence of repetitive DNA helps explain why certain very large genomes actually include only a modest number of genes. For example, the maize (corn) and rice genomes contain about the same number of genes, but the maize genome is at least 6 times larger than the rice genome. Over half of the maize genome appears to be composed of **transposable elements,** short segments of DNA that are copied many times and inserted randomly into the chromosomes.

The human genome contains several types of repetitive DNA sequences, including the inactive remnants of transposable elements. About 45% of human DNA consists of **moderately repetitive sequences,** which are blocks of hundreds or thousands of nucleotides scattered throughout the genome. The most numerous of these are present in hundreds of thousands of copies. **Highly repetitive sequences** account for another 3% of the human genome. These segments of 2 to 10 bases are present in millions of copies. They are repeated tandemly (side by side), sometimes thousands of times. The number of repeats of a given sequence often varies between individuals, even in the same family, so this information can be analyzed to produce a DNA "fingerprint" (see Section 3.5).

GENES ARE IDENTIFIED BY COMPARING SEQUENCES

For many genomes, the exact number of genes has not yet been determined, and different methods for identifying genes yield different estimates. For example, a computer can scan a DNA sequence for an **open reading frame (ORF),** that is, a stretch of nucleotides that can potentially be transcribed and translated. For a protein-coding gene, the ORF begins with a "start" codon: ATG in the coding strand of DNA, which corresponds to AUG in RNA (see Table 3.3). This codon specifies methionine, the initial residue of all newly synthesized proteins. The ORF ends with one of the three "stop" codons: DNA coding sequences of TAA, TAG, or TGA, which correspond to the three mRNA stop codons (see Table 3.3). Other so-called *ab initio* ("from the beginning") gene-identifying methods scan the DNA for other features that characterize the beginnings and endings of genes.

Another method for identifying genes in a genome relies on sequence comparisons with known genes. Such genome-to-genome comparisons are possible because of the universal nature of the genetic code and the relatedness of all organisms through evolution (Section 1.4). *Genes with similar functions in different species tend to have similar sequences;* such genes

Portion of the human genome. The black line represents a 500,000-bp segment of human chromosome 6 that includes six protein-coding genes (purple boxes) distributed unevenly and separated by stretches of noncoding DNA. [*Based on information from the NCBI Map Viewer, www.ncbi.nlm.nih.gov/mapview.*]

? Why are the genes pointing in two directions?

are said to be **homologous.** Even an inexact match can still indicate a protein's functional category, such as enzyme or hormone receptor, although its exact role in the cell may not be obvious. Genes that appear to lack counterparts in other species are known as **orphan genes.** The RNA or protein products of many genes have not yet been isolated or otherwise identified. For example, about 20% of the genes in the well-studied organism *E. coli* have not yet been assigned functions.

Genome maps, such as the one shown in **Figure 3.12,** indicate the arrangement of genes on each strand of DNA in a chromosome. Human protein-coding genes are typically much longer than bacterial genes (27,000 bp versus 1000 bp on average), since they contain sequences removed from the mRNA transcripts before translation. In addition, the spaces between genes are much larger in the human genome.

In addition to the approximately 20,000 protein-coding genes in the human genome are a comparable number of genes that correspond to **noncoding RNA (ncRNA)** molecules. Many of these transcripts, which come in a variety of sizes, appear to be involved in regulating the expression of other genes.

Gene-mapping projects have uncovered some interesting aspects of evolution, including **horizontal gene transfer.** This occurs when a gene is transferred between species rather than from parent to offspring of the same species (vertical gene transfer). Horizontal gene transfer may be mediated by viruses, which can pick up extra DNA as they insert and excise themselves from the host's chromosomes. This activity can generate, for example, what appears to be a mammalian gene inside a bacterial genome. The ease with which many bacterial organisms trade their genes has given rise to the idea that groups of bacteria should be viewed as a continuum of genomic variations instead of separate species with discrete genomes.

GENOMIC DATA REVEAL BIOLOGICAL FUNCTIONS

Genomics, the study of genomes, has a number of practical applications. For one thing, *the number of genes and their putative functions provide a rough snapshot of the metabolic capabilities of a given organism.* For example, humans and fruit flies differ in the number of genes that code for developmental pathways and immune system functions (**Fig. 3.13**). An unusual number of genes belonging to one category might indicate some unusual biological property in an organism. This sort of knowledge can be useful for developing drugs to inhibit the growth of a pathogenic organism according to its unique metabolism.

Studies in which protein-coding genes are inactivated, one at a time, suggest that relatively simple organisms, such as *Saccharomyces* (yeast), have about 1000 essential genes. Humans likewise seem to absolutely require only about 10%—roughly 2000—of their genes. This core set of genes encodes proteins that are produced in abundance and carry out the most basic cellular activities. But clearly, the other 90% of genes are not useless. An analysis of 17,000 genes in 44 different human tissues has revealed that about half are expressed in all locations.

Genomic analysis also reveals variations among individuals, some of which can be linked to an individual's chance of developing a particular disease. In addition to genetic changes that are clearly associated with a single-gene disorder, millions more sequence variations have been catalogued. On average, *the DNA of any two humans differs at 3 million sites, or about once every thousand base pairs.* These **single-nucleotide polymorphisms** (**SNPs,** instances where the DNA sequence differs among individuals) are compiled in databases. Some of the factors that can alter DNA are discussed in Section 20.3. A person begins life with an average of 60 new genetic changes that were not present in either parent.

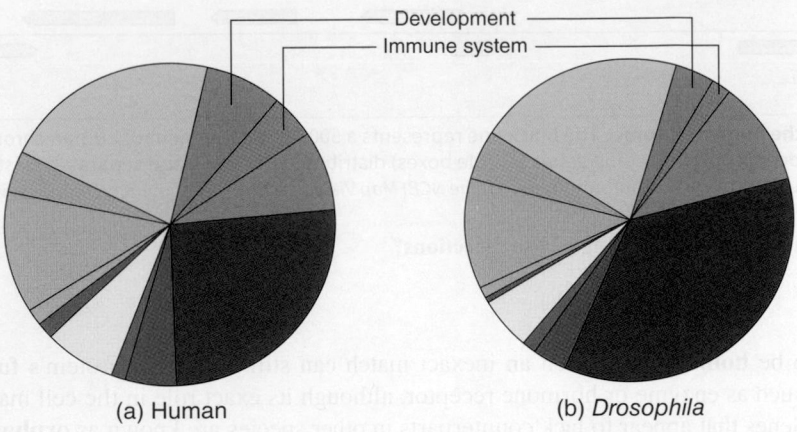

FIGURE 3.13

Functional classification of genes. This diagram is based on 19,184 human genes (a) and 10,787 *Drosophila* genes (b), grouped according to the biochemical function of the gene product. Humans devote a larger proportion of genes to development (7.7% versus 3.9% in *Drosophila*) and to the immune system (4.3% versus 2.0% in *Drosophila*). [*Data from the Protein Analysis through Evolutionary Relationships classification system, www.pantherdb.org/.*]

Researchers have attempted to correlate SNPs with disorders, such as cardiovascular disease or cancer, that likely depend on the contributions of many genes. **Genome-wide association studies (GWAS)** have identified, for example, 39 sites that are associated with type 2 diabetes and 71 that are associated with Crohn's disease, an autoimmune disorder. The risk tied to any particular genetic variant is low, but the entire set of variations can explain up to 50% of the heritability of the disease. Although the SNPs are only proxies for disease genes, these data should provide a starting point for researchers to explore the DNA near the SNPs to discover the genes that are directly involved in the disease. Several commercial enterprises offer individual genome-sequencing services, but until genetic information can be reliably translated into effective disease-prevention or treatment regimens, the practical value of "personal genomics" is somewhat limited.

 CONCEPT CHECK

1. Describe the rough correlation between gene number and organismal lifestyle.

2. Why is it useful to identify homologs of human genes?

3. List the approaches used to identify genes.

4. Explain the value and limitations of genome-wide association studies.

3.5   TOOLS AND TECHNIQUES: MANIPULATING DNA

Molecular biologists have devised clever procedures for manipulating DNA in the laboratory and in living organisms. *Many of the techniques take advantage of naturally occurring enzymes that cut, copy, and link nucleic acids.* These techniques also exploit the ability of nucleic acids to interact with complementary molecules. In this section we focus on currently used methods for generating recombinant DNA, amplifying specific DNA segments, determining the sequence of DNA, and altering the DNA.

KEY CONCEPTS

- A DNA molecule can be sequenced or amplified by using DNA polymerase to make a copy of a template strand.
- Linking together DNA fragments produces recombinant DNA molecules that can be used to study gene function, to express genes in host organisms, and to engineer genes for commercial and therapeutic purposes.

CUTTING AND PASTING GENERATES RECOMBINANT DNA

Long DNA molecules tend to break from mechanical stress during laboratory manipulations. However, such randomly fragmented DNA is not always useful, so researchers often employ enzymes that cut DNA in defined ways. Bacteria produce DNA-cleaving enzymes known as **restriction endonucleases** (or **restriction enzymes**) that catalyze the breakage of phosphodiester bonds at specific nucleotide sequences. These enzymes can thereby destroy foreign DNA that enters the cell, such as **bacteriophage** (viral) DNA. In this way, the bacterial cell "restricts" the growth of the phage. The bacterial cell protects its own DNA from endonucleolytic digestion by methylating it (adding a —CH$_3$ group) at the same sites recognized by its restriction endonucleases. Hundreds of these enzymes have been discovered; some are listed in **Table 3.5** along with their recognition sequences and cleavage sites.

Restriction enzymes typically recognize a 4- to 8-base sequence that is identical, when read in the same 5′ → 3′ direction, on both strands. DNA with this form of symmetry is said to be **palindromic** (words such as *madam* and *noon* are palindromes). One restriction enzyme isolated from *E. coli* is known as EcoRI (the first three letters are derived from the genus and species names). Its recognition sequence is

$$5' — GAATTC — 3'$$
$$3' — CTTAAG — 5'$$

The arrows indicate the phosphodiester bonds that are cleaved. Note that the sequence reads the same on both strands.

Because the EcoRI cleavage sites are symmetrical but staggered, the enzyme generates DNA fragments with single-stranded extensions known as **sticky ends:**

$$— G \qquad AATTC—$$
$$— CTTAA \qquad G—$$

In contrast, the *E. coli* restriction enzyme known as EcoRV cleaves both strands of DNA at the center of its 6-bp recognition sequence so that the resulting DNA fragments have **blunt ends:**

$$5' — GATATC — 3' \qquad — GAT \quad ATC—$$
$$3' — CTATAG — 5' \qquad — CTA \quad TAG—$$

TABLE 3.5 Recognition and Cleavage Sites of Some Restriction Endonucleases

ENZYME	RECOGNITION/CLEAVAGE SITE[a]
AluI	AG l CT
MspI	C l CGG
AsuI	G l GNCC[b]
EcoRI	G l AATTC
EcoRV	GAT l ATC
PstI	CTGCA l G
SauI	CC l TNAGG
NotI	GC l GGCCGC

[a]The sequence of one of the two DNA strands is shown.

The vertical bar indicates the cleavage site.

[b]N represents any nucleotide.

[An exhaustive source of information on restriction enzymes is available through the Restriction Enzyme Database: rebase.neb.com/rebase/rebase.html.]

23130
9416
6557
4361

2322
2027

FIGURE 3.14

Digestion of bacteriophage λ DNA by the restriction enzyme HindIII. The restriction enzyme cleaves the DNA to produce eight fragments of defined size, six of which are large enough to be separated by electrophoresis in an agarose gel. The DNA was applied to the top of the gel, and the negatively charged DNA fragments moved downward through the gel toward the positive electrode. The fragments were visualized by binding a fluorescent dye. The numbers indicate the number of base pairs in each fragment. [*Reprinted from www.neb.com, http://www.neb.com; © 2012 with permission from New England Biolabs.*]

❓ **Explain why the bands at the top appear brighter than the bands at the bottom.**

Vector

Foreign DNA

Recombinant DNA

FIGURE 3.15

Production of a recombinant DNA molecule. A circular vector and a sample of DNA are cut with the same restriction enzyme, generating complementary sticky ends, so that the fragment of foreign DNA can be ligated into the vector.

Restriction enzymes have many uses in the laboratory. For example, they are indispensable for reproducibly breaking large pieces of DNA into smaller pieces of manageable size. **Restriction digests** of well-characterized DNA molecules, such as the 48,502-bp *E. coli* bacteriophage λ, yield fragments of predictable size that can be separated by **electrophoresis,** a procedure in which molecules move through a gel-like matrix such as agarose or polyacrylamide under the influence of an electric field. Because all the DNA segments have a uniform charge density, they are separated on the basis of their size (the smallest molecules move fastest; **Fig. 3.14**).

A segment of DNA cut by a restriction enzyme or obtained by a method such as chemical synthesis or PCR (see below) can be joined to another DNA molecule. When different samples of DNA are digested with the same sticky end–generating restriction endonuclease, all the fragments have identical sticky ends. If the fragments are mixed together, the sticky ends can find their complements and re-form base pairs. The discontinuities in the sugar–phosphate backbone can then be mended by a **DNA ligase** (an enzyme that forms new phosphodiester bonds between adjacent nucleotide residues). DNA ligases that act on blunt-ended DNA segments are also available. *The cutting-and-pasting process allows a desired DNA segment to be incorporated into a carrier DNA molecule called a* vector, *leaving an unbroken, fully base-paired* **recombinant DNA** *molecule* (**Fig. 3.15**).

Some commonly used vectors are derived from **plasmids,** which are small, circular DNA molecules present in many bacterial cells. A single cell may contain multiple copies of a plasmid, which replicates independently of the bacterial chromosome and usually does not contain genes essential for the host's normal activities. However, plasmids often do carry genes for specialized functions, such as resistance to certain antibiotics (these genes often encode proteins that inactivate the antibiotics). An antibiotic resistance gene allows the **selection** of cells that harbor the plasmid: Only cells that contain the plasmid can survive in the presence of the antibiotic.

A recombinant plasmid—one that contains a foreign DNA sequence—can be introduced into a bacterial host, where it multiplies as the bacteria multiply. This is one way to produce large amounts of a desired DNA segment. (The segment can be harvested by recovering the plasmids and treating them with the same restriction enzyme used to insert the foreign DNA.) DNA that is copied in this manner is said to be cloned. Note that a **clone** is simply an identical copy of an original. The term is used to refer either to a gene that has been amplified, as described here, or to a cell or organism that is genetically identical to its parent.

An example of a plasmid used as a cloning vector is shown in **Figure 3.16**. This plasmid contains a gene (called *amp*^R) for resistance to the antibiotic ampicillin and a gene (called *lacZ*) encoding the enzyme β-galactosidase, which catalyzes the hydrolysis of certain galactose derivatives. The *lacZ* gene has been engineered to contain several restriction sites, any one of which can be used as an insertion point for a piece of foreign DNA with compatible sticky ends. Interrupting the *lacZ* gene with foreign DNA prevents the synthesis of the β-galactosidase protein.

Colonies of bacterial cells harboring the intact plasmid can be detected when their β-galactosidase cleaves a galactose derivative that generates a blue dye. Colonies of bacterial cells in which a foreign DNA insert has interrupted the *lacZ* gene are unable to cleave the galactose derivative and therefore do not turn blue (**Fig. 3.17**). A single white colony can then be removed from the culture plate and grown in order to harvest its recombinant DNA. Other screening techniques rely on signals such as fluorescence.

EcoRI
SacI
KpnI
AvaI
XmaI
SmaI
BamHI
XbaI
SalI
AccI
HincII
PstI
SphI
HindIII

amp^R

lacZ

FIGURE 3.16

Map of a cloning vector. This 2743-bp circular DNA molecule, called pGEM-3Z, has a gene for resistance to ampicillin so that bacterial cells containing the plasmid can be selected by their ability to grow in the presence of the antibiotic. The plasmid also has a site comprising recognition sequences for 14 different restriction enzymes. Inserting a foreign DNA segment at this site interrupts the *lacZ* gene, which encodes the enzyme β-galactosidase.

FIGURE 3.17

Culture dish with recombinant bacteria. Blue colonies arise from cells whose plasmids have an intact β-galactosidase gene. White colonies arise from cells whose plasmids contain an inserted DNA that interrupts the β-galactosidase gene. Colonies that lack the plasmid (and its amp^R gene) would also be white, but including ampicillin in the culture medium prevents their growth. [*Courtesy S. Kopczak and D. P. Snustad, University of Minnesota.*]

A wide variety of vectors have been developed to accommodate different sizes of DNA inserts and to deliver that DNA to different types of hosts, including human cells. If a gene that has been isolated and cloned in a host cell is also expressed (transcribed and translated into protein), it may affect the metabolism of that cell. The functions of some gene products have been assessed in this way. Sometimes a specific combination of vector and host cell are chosen so that large quantities of the gene product can be isolated from the cultured cells or from the medium in which they grow. This is an economical method for producing certain proteins that are difficult to obtain directly from human tissues (**Table 3.6**). Recombinant DNA can also be introduced into a different species to create a transgenic organism (**Box 3.A**).

THE POLYMERASE CHAIN REACTION AMPLIFIES DNA

Cloning DNA in cultured cells is a laborious process. A much more efficient technique for amplifying a particular DNA sequence was developed by Kary Mullis in 1985: the **polymerase chain reaction (PCR).** *One of the advantages of PCR over traditional cloning techniques is that the starting material need not be pure* (this makes the technique ideal for analyzing complex mixtures such as tissues or biological fluids).

PCR takes advantage of the enzyme DNA polymerase, which catalyzes the polymerization of nucleotides in the order determined by their base-pairing with a single-stranded DNA template. But because DNA polymerase cannot begin a new nucleotide strand (it can only extend

TABLE 3.6 Some Recombinant Protein Products

PROTEIN	PURPOSE
Insulin	Treat insulin-dependent diabetes
Growth hormone	Treat certain growth disorders in children
Erythropoietin	Stimulate production of red blood cells; useful in kidney dialysis
Coagulation factors IX and X	Treat hemophilia b and other bleeding disorders
Tissue plasminogen activator	Promote clot lysis following myocardial infarction or stroke
Colony stimulating factor	Promote white blood cell production after bone marrow transplant

Box 3.A Genetically Modified Organisms

Introducing a foreign gene into a single host cell via a vector alters the genetic makeup of that cell and all its descendants. But if the cell is part of a multicellular organism, such as an animal or plant, more work is required to generate a **transgenic organism** whose cells all contain the foreign gene. In mammals, the modified DNA must be injected into fertilized eggs, which are then implanted in a foster mother. Some of the resulting embryos' cells (possibly including their reproductive cells) will contain the foreign gene. When the animals mature, they must be bred in order to yield offspring whose cells are all transgenic.

Transgenic plants are produced by introducing recombinant DNA into a few cells, which can often develop into an entire plant whose cells all contain the foreign DNA. Desirable traits, such as resistance to insect pests and resistance to weed-killing herbicides, have been introduced into a number of important crop species. Approximately 80 to 90% of the U.S. corn (maize), soybean, and cotton harvest is genetically modified.

Concerns about the safety of foods containing foreign genes have limited their acceptance by consumers. Transgenic organisms also present some biological risks. For example, genes that code for insecticides (such as the bacterial toxin intended to kill the insect larvae that would otherwise feast on corn plants) can make their way into wild plants that support beneficial insects. Similarly, herbicide-resistance genes can jump to weed species, making them even more difficult to control.

Transgenic plants have also been engineered for better nutrition. Golden Rice, for example, contains foreign plant genes that encode the enzymes necessary to synthesize β-carotene (an orange pigment that is the precursor of vitamin A) and a gene for the iron-storage protein ferritin. This genetically modified rice could help alleviate vitamin A deficiencies (which afflict some 400 million people) and iron deficiencies (an estimated 30% of the world's population suffers from iron deficiency).

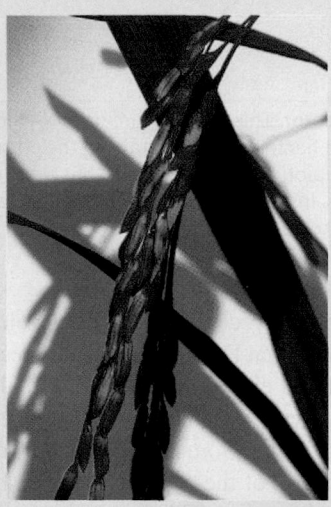

Golden Rice, whose high β-carotene content gives the normally white rice grains a yellow hue. [*phloen/Alamy Limited*]

a preexisting chain), a short single-stranded **primer** that base pairs with the template strand is added to the mixture. This means that the polymerase copies only what it is "told" to copy.

The PCR reaction mixture contains a DNA sample, DNA polymerase, all four deoxynucleotide substrates for the polymerase, and two synthetic oligonucleotide primers that are complementary to the 3′ ends of the two strands of the target DNA sequence. Keep in mind that the reaction mixture actually contains millions of molecules of each of these substances.

In the first step of PCR, the sample is heated to ~95°C to separate the DNA strands. Next, the temperature is lowered to about 55°C, cool enough for the primers to hybridize with the DNA strands. The temperature is then increased to about 75°C, and the DNA polymerase synthesizes new DNA strands by extending the primers (**Fig. 3.18**). The three steps—strand separation, primer binding, and primer extension—are repeated as many as 40 times. Because the primers represent the two ends of the target DNA, this sequence is preferentially amplified so that *it doubles in concentration with each reaction cycle*. For example, 20 cycles of PCR can theoretically yield $2^{20} = 1,048,576$ copies of the target sequence in a matter of hours.

PCR relies on bacterial DNA polymerases that can withstand the high temperatures required for strand separation (these temperatures inactivate most enzymes). Commercial PCR kits usually contain DNA polymerase from *Thermus aquaticus* ("Taq," which lives in hot springs) or *Pyrococcus furiosus* ("Pfu," which inhabits geothermally heated marine sediments), since their enzymes perform optimally at high temperatures.

One of the limitations of PCR is that choosing appropriate primers requires some knowledge of the DNA sequence to be amplified; otherwise the primers will not anneal with complementary sequences in the DNA and the polymerase will not be able to make any new DNA strands. However, since no new DNA is synthesized *unless* the primers can bind to the DNA sequence, *PCR can be used to verify the presence of that sequence*. Practical applications of PCR include identifying human genetic defects and diagnosing infections. PCR is the most efficient way to detect the presence of bacteria and viruses that are difficult or dangerous to work with, such as Ebola virus. Blood banks use PCR to test for the human immunodeficiency

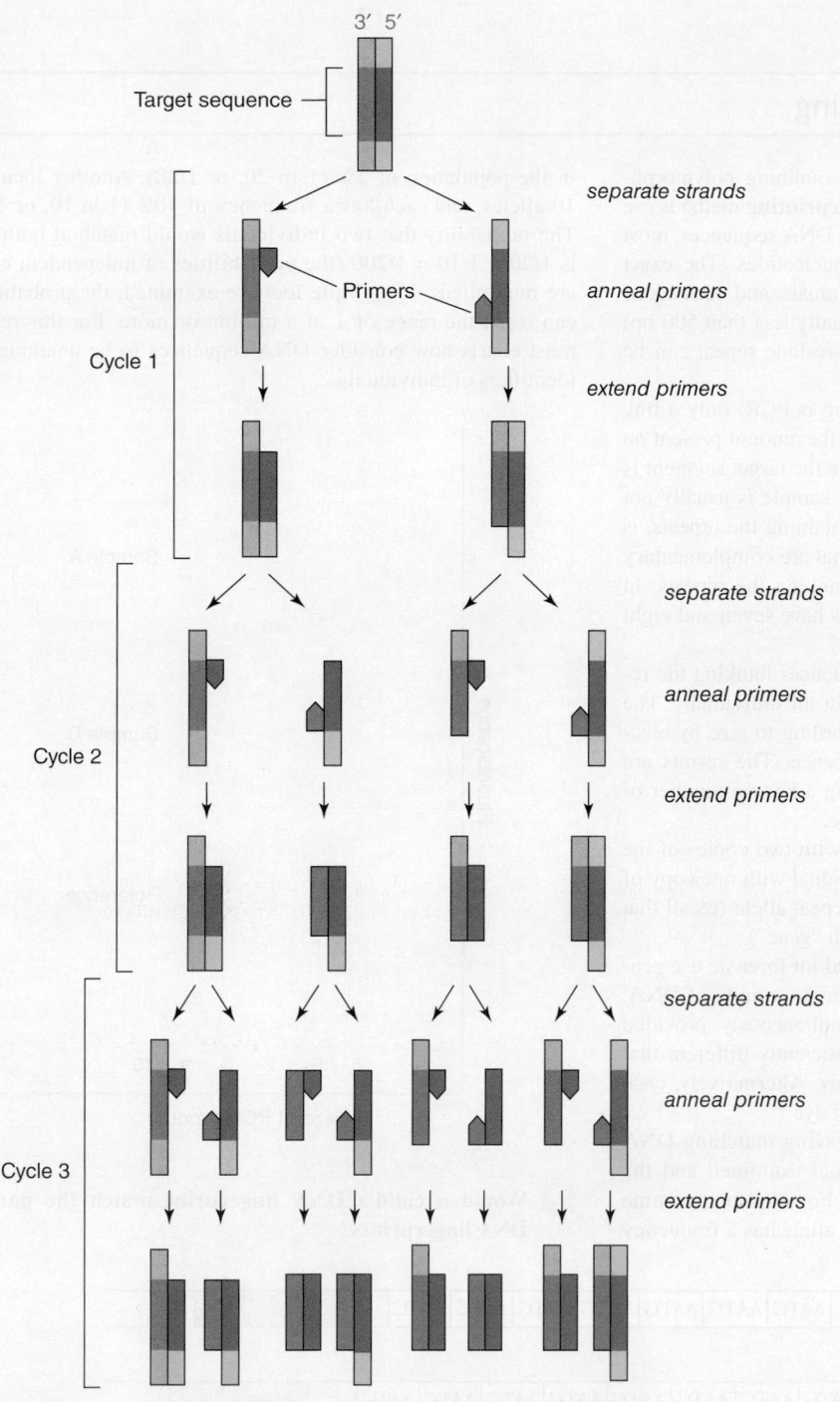

FIGURE 3.18

The polymerase chain reaction. Each cycle consists of separation of DNA strands, binding of primers to the 3′ ends of the target sequence, and extension of the primers by DNA polymerase. The target DNA doubles in concentration with each cycle.

? Indicate the temperature at which each step takes place.

virus (HIV), hepatitis viruses, and West Nile virus. Scientists have used PCR to amplify DNA from ancient bones, and, in the forensics laboratory, to analyze DNA of more recent origin (**Box 3.B**).

If merely detecting a specific DNA sequence does not provide enough information, **real-time PCR** (also known as quantitative PCR, qPCR) can be employed. In this technique, the polymerase chain reaction continually generates new DNA sequences that bind to fluorescent probes, so that the amount of the DNA sequence can be monitored over time (rather than at the end of many reaction cycles, as in standard PCR). This approach is useful for quantifying

A CLOSER LOOK

Box 3.B   DNA Fingerprinting

Individuals can be distinguished by examining polymorphisms in their DNA. Current **DNA fingerprinting** methods use PCR to examine segments of repetitive DNA sequences, most often short tandem repeats of four nucleotides. The exact number of repeats varies among individuals, and each set of repeats, or **allele,** is small enough (usually less than 500 bp) that alleles differing by just one four-residue repeat can be easily differentiated.

Because the first step of fingerprinting is PCR, only a tiny amount of DNA is needed—about 1 μg, or the amount present on a coffee cup or a licked envelope. And since the target segment is short, the purity and integrity of the DNA sample is usually not an issue. The **locus,** or region of DNA containing the repeats, is PCR-amplified using fluorescent primers that are complementary to the unique (nonrepeating) sequences flanking the repeats. In the example below, the two DNA segments have seven and eight tandem repeats of the AATG sequence.

The PCR primers hybridize with sequences flanking the repeats (shaded blue), which are the same in all individuals. The amplified products are then separated according to size by electrophoresis and detected by their fluorescence. The results are compared to reference standards containing a known number of AATG repeats, from 5 to 10 in this example.

Sample A comes from an individual with two copies of the 7-repeat allele and sample B from an individual with one copy of the 7-repeat allele and one copy of the 8-repeat allele (recall that humans are diploid, with two copies of each "gene").

Each of the loci that have been selected for forensic use generally have 7 to 30 different alleles. In a single sample of DNA, multiple loci can be amplified by PCR simultaneously, provided that the sizes of the PCR products are sufficiently different that they will not overlap during electrophoresis. Alternatively, each PCR primer can bear a different fluorescent dye.

The probability of two individuals having matching DNA fingerprints depends on the number of loci examined and the number of possible alleles at each locus. For example, assume that one locus has 20 alleles and that each allele has a frequency

in the population of 5% (1 in 20, or 1/20). Another locus has 10 alleles, and each has a frequency of 10% (1 in 10, or 1/10). The probability that two individuals would match at both sites is 1/20 × 1/10 = 1/200 (the probabilities of independent events are multiplied). If multiple loci are examined, the probabilities can reach the range of 1 in a million or more. For this reason, most courts now consider DNA sequences to be unambiguous identifiers of individuals.

? **Would a child's DNA fingerprint match the parents' DNA fingerprints?**

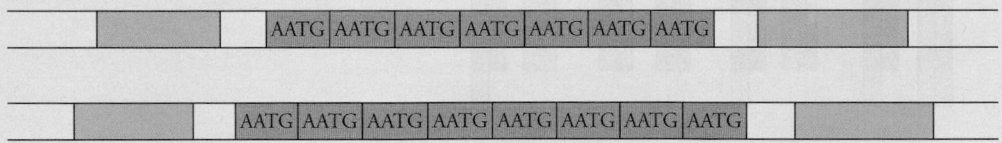

infectious agents such as bacteria and viruses. Real-time PCR methods are also used to assess the level of gene expression in cells: Cellular mRNA is first reverse-transcribed to DNA, then a specific DNA sequence is amplified by PCR. A gene's level of expression is sometimes reported relative to that of a gene that encodes a protein such as actin (Section 5.3), which is typically produced at constant levels in cells and can therefore serve as a sort of benchmark.

DNA SEQUENCING USES DNA POLYMERASE TO MAKE A COMPLEMENTARY STRAND

Determining the sequence of nucleotides in a DNA molecule involves using DNA polymerase to make copies. Various methods have been devised to detect the newly made DNA, such as labeling it with radioactive or fluorescent tags. The key, of course, *is to identify*

FIGURE 3.19

Pyrosequencing. PCR-amplified copies of a single-stranded template DNA are attached to a plastic bead, and a primer, DNA polymerase, and other components are added. Nucleotide solutions are passed, one at a time, over the beads. Polymerization of a nucleotide leads to a flash of light.

the added nucleotides one at a time in order to reconstruct the sequence of the original (template) DNA. The newest, most efficient DNA sequencing methods require specialized instruments, but they operate extremely quickly, sequence thousands of DNA segments simultaneously, and can handle huge tasks, such as sequencing an organism's entire genome within hours or days.

In **pyrosequencing** (also known as 454 sequencing), DNA segments to be sequenced are immobilized on tiny plastic beads, one segment per bead, and then a variation of PCR is used to make single-stranded copies that are attached to the same bead. DNA polymerase, a primer, and other reaction components are added to this template DNA. A solution of one of the four deoxynucleotides is then introduced to each bead, and if the nucleotide is a match—that is, if it can pair with the nucleotide in the immobilized template strand—it is incorporated into a new DNA chain. This polymerization reaction releases pyrophosphate (the diphosphate portion of the nucleotide, represented as PP_i). An enzyme called sulfurylase converts the PP_i product to ATP, which is used by another enzyme, firefly luciferase, in a reaction that produces a flash of light (**Fig. 3.19**).

A detector records the light generated by each bead as solutions of each of the four nucleotides are successively washed over the beads. If an incoming nucleotide does not pair with the template DNA, no polymerization reaction occurs and no light is produced. If an incoming nucleotide is incorporated into the new DNA strand more than once, the flash of light is proportionally brighter. The record of light produced by each bead reveals the sequence of nucleotides corresponding to the template DNA strand attached to that bead (**Fig. 3.20**).

The **Illumina sequencing** method uses a slightly different approach. First, segments of DNA are attached to a glass surface and amplified in place. After the addition of DNA polymerase and a primer, a solution containing all four deoxynucleotides, each with a different fluorescent group, is introduced. Unreacted nucleotides are washed away, and the nucleotides that remain—those that DNA polymerase incorporated into a new strand complementary to the template DNA—can be identified by their fluorescence. Before the next batch of nucleotides is delivered, the fluorescent groups on the growing DNA strands are detached so that new fluorescent signals can be detected in the next round of polymerization. The order of

FIGURE 3.20

Results of pyrosequencing. Nucleotides were added to the reaction system in the order shown at the bottom. A flash of light (peak) indicates that the nucleotide was polymerized. A double- or triple-height peak indicates multiple incorporations of a nucleotide. The absence of a peak means that no reaction occurred, because the complementary nucleotide did not occur in the template DNA. The deduced sequence of the newly made DNA strand is shown at the top.

 What is the sequence of the template DNA in this example?

FIGURE 3.21

Illumina sequencing. DNA polymerase uses immobilized single strands of DNA as templates to add a fluorescent nucleotide to a new strand. A detector records the fluorescence to identify the nucleotide added in that round of polymerization.

appearance of fluorescent signals corresponds to the nucleotide sequence of the growing DNA chain (**Fig. 3.21**).

Both of the sequencing methods described here generate "reads" of no more than a few hundred nucleotides, so deducing the sequence of a longer segment of DNA requires that the original sample be randomly broken into multiple short segments that are individually sequenced (this is known as a "shotgun" approach). The original sequence can then be reconstructed by computer analysis of the overlapping sequences.

The ability to rapidly sequence a huge amount of DNA, such as the genome of one individual, makes it possible to compare the DNA of children and parents to identify a genetic defect present only in the children. This approach has likely already saved lives in cases where no other diagnostic tests were available. Genomic data also provide the raw material for exploring the vast amounts of DNA that do not encode proteins and for comparing the genomes of different species in order to elucidate their evolutionary relationships.

DNA CAN BE ALTERED

In addition to rearranging, copying, and sequencing DNA, genetic engineers can modify the DNA in order to study how genes work, to make new gene products, and even to change the genetic makeup of an entire organism. DNA is constantly and randomly changing as a result of natural processes, but alterations can be introduced much more quickly and in a targeted manner in the laboratory. For example, researchers can produce genes with specific mutations so that after the genes are transcribed and translated in a cell, the encoded proteins exhibit altered structures and functions. One method for such **site-directed mutagenesis** of a gene is to perform a variation of PCR using oligonucleotide primers that are synthesized with the desired nucleotide changes. Several rounds of PCR yield a population of newly synthesized DNA molecules that contain the mutations specified by the primers. The altered DNA can then be introduced into a host cell.

An even more efficient technique for modifying genes takes advantage of a bacterial system that evolved to destroy bacteriophages. A bacterial cell carries traces of its ancestors' encounters with various bacteriophages in the form of multiple short segments of phage DNA that have been inserted into the bacterial chromosome among a set of **clustered regularly interspersed short palindromic repeats,** or **CRISPRs.**

When this DNA is transcribed, the resulting RNA is cleaved into short segments, each corresponding to about 30 nucleotides of phage DNA. These RNAs then bind to a nuclease called Cas9 (for *CRISPR-associated*) and guide it to complementary DNA sequences, which presumably represent bacteriophages trying to infect the cell. The Cas9 nuclease cleaves both strands of this DNA, destroying the incoming bacteriophage.

Without the CRISPR RNA guide, Cas9 cannot cleave DNA, but *it will cleave whatever DNA is complementary to the guide RNA*. This creates the ability to use the **CRISPR-Cas9 system** as a gene-editing tool. Using recombinant DNA technology, the Cas9 gene and an engineered guide RNA "gene" are introduced into a host cell. The cell makes the encoded Cas9 protein and the guide RNA, which act to cleave the DNA of the gene that is complementary to the guide RNA (**Fig. 3.22**).

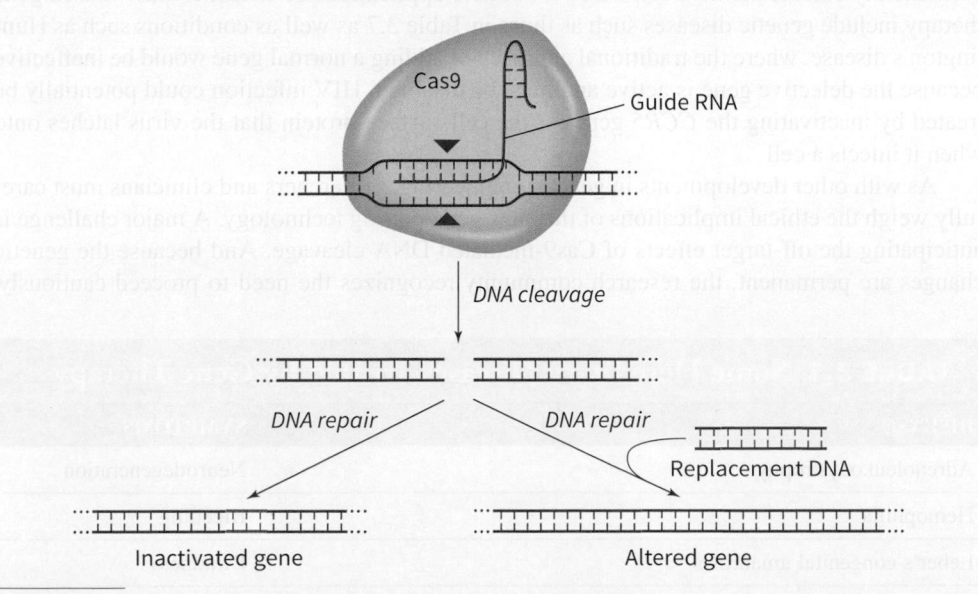

FIGURE 3.22

The CRISPR-Cas9 gene-editing system. A guide RNA containing a 22–23-bp segment that is complementary to the target DNA positions Cas9 to cleave the DNA at the points marked by arrowheads. This leads to permanent inactivation of the gene (left) or, if an altered DNA segment is provided, leads to replacement of the original gene sequence with the altered sequence (right).

Cells have various mechanisms for repairing broken DNA (described in more detail in Section 20.3). For example, the severed pieces are typically knit back together—but imperfectly, so the repair does not restore the original DNA sequence. When this occurs in the middle of a gene, the result is almost always an unreadable gene. The CRISPR-Cas9 system can therefore "knock out" a specific gene in an organism.

To "knock in" a gene, the target gene is replaced by a modified version of the gene. In this case, researchers deploy the CRISPR-Cas9 system to cleave the target gene but also introduce an altered segment of that same gene. The cell then undertakes a more elaborate repair process to rebuild the cleaved DNA using the intact—but altered—sequence as a blueprint. The result is a functional gene containing a modified DNA sequence.

Because the CRISPR-Cas9 gene-editing system can permanently disable a defective gene or replace a defective gene with a normal copy at the editing site, it has become the preferred approach for genetically modifying organisms. For example, CRISPR-Cas9 technology can generate animal models for human genetic diseases. *It is also an option for **gene therapy**, which introduces a functional gene into an individual in order to compensate for or correct a malfunctioning gene.*

In traditional gene therapy, an extra gene is delivered to a patient's cells, usually by a viral vector (viruses are already efficient at delivering DNA into mammalian cells). The first successful gene therapy trials treated children with severe combined immunodeficiency (SCID), a normally fatal condition caused by a single-gene defect. Bone marrow cells were removed from each patient and cultured in the presence of a viral vector containing a normal version of the defective gene. When the modified cells were infused into the patient, they differentiated into functional immune system cells.

Some of the diseases treated by gene therapy are listed in **Table 3.7**. Despite over 30 years of effort, the list of successes is still small. For one thing, the viral vectors used in gene therapy can behave unpredictably, sometimes triggering a fatal immune response or inserting themselves into the host cell's chromosomes at random, which may interrupt the functions of other genes and cause cancer. Another challenge is to deliver the therapeutic gene to a sufficient number of the appropriate type of host cells in order to "correct" the disease over the long term.

The CRISPR-Cas9 method shares some of the limitations of traditional gene therapy, such as the need for a suitable vector that can deliver its payload to the intended tissues. In addition, Cas9 sometimes cleaves DNA sequences that are not perfectly complementary to its guide RNA. Such "off target" gene inactivation could have disastrous unintended consequences. However, CRISPR-Cas9 offers some clear advantages. This gene-editing system can actually inactivate a misbehaving gene as well as introduce a replacement gene, and it can permanently correct the host cell's DNA. Possible applications of CRISPR-Cas9–based gene therapy include genetic diseases such as those in Table 3.7 as well as conditions such as Huntington's disease, where the traditional approach of adding a normal gene would be ineffective because the defective gene is active and must be disabled. HIV infection could potentially be treated by inactivating the *CCR5* gene for the cell-surface protein that the virus latches onto when it infects a cell.

As with other developments in genetic engineering, researchers and clinicians must carefully weigh the ethical implications of this new gene-editing technology. A major challenge is anticipating the off-target effects of Cas9-mediated DNA cleavage. And because the genetic changes are permanent, the research community recognizes the need to proceed cautiously.

TABLE 3.7 Some Hereditary Diseases Treated by Gene Therapy

DISEASE	SYMPTOMS
Adrenoleukodystrophy	Neurodegeneration
Hemophilia	Bleeding
Leber's congenital amaurosis	Blindness
Severe combined immunodeficiency (SCID)	Immunodeficiency
β-Thalassemia	Anemia
Wiskott-Aldrich syndrome	Immunodeficiency

Because a cell can transmit its modified DNA to its daughter cells when the cell divides, it is prudent to limit gene alterations to somatic (body) cells and avoid altering the DNA in reproductive cells that pass DNA to the next generation.

 CONCEPT CHECK

1. Summarize the steps of dideoxy DNA sequencing. How does pyrosequencing differ from the dideoxy approach?

2. How do researchers reconstruct long DNA sequences?

3. Explain why DNA sequencing and DNA fingerprinting rely on PCR.

4. Why are restriction endonucleases useful for constructing recombinant DNA?

5. How can cells containing recombinant DNA molecules be identified by selection or screening?

6. Describe how a cloned gene can be used to produce an altered protein, to generate a transgenic organism, or to cure a disease.

SUMMARY

3.1 Nucleotides

• The genetic material in virtually all organisms consists of DNA, a polymer of nucleotides. A nucleotide contains a purine or pyrimidine base linked to a ribose group (in RNA) or a deoxyribose group (in DNA) that also bears one or more phosphate groups.

3.2 Nucleic Acid Structure

• DNA contains two antiparallel helical strands of nucleotides linked by phosphodiester bonds. Each base pairs with a complementary base in the opposite strand: A with T and G with C. The structure of RNA, which is single-stranded and contains U rather than T, is more variable.

• Nucleic acid structures are stabilized primarily by stacking interactions between bases. The separated strands of DNA can reanneal.

3.3 The Central Dogma

• The central dogma summarizes how the sequence of nucleotides in DNA is transcribed into RNA, which is then translated into protein according to the genetic code.

• The sequence of nucleotides in a segment of DNA can reveal mutations that cause disease.

3.4 Genomics

• Genomes contain variable amounts of repetitive and other forms of noncoding DNA in addition to genes, which are identified by their sequence characteristics or similarity to other genes.

• Genetic variations can be linked to human diseases even when specific disease genes have not been identified.

3.5 Tools and Techniques: Manipulating DNA

• DNA molecules can be reproducibly fragmented by the action of restriction enzymes, which cleave DNA at specific sequences.

• DNA fragments can be joined to each other to generate recombinant DNA molecules that are then introduced into host cells.

• A DNA segment can be amplified many times by the polymerase chain reaction, in which a DNA polymerase makes complementary copies of the target DNA.

• Modern techniques for determining the sequence of nucleotides in a segment of DNA immobilize DNA segments, amplify them, use DNA polymerase to add nucleotides to a new complementary chain, and detect each added nucleotide by a flash of light or a fluorescent signal.

• The gene-editing CRISPR-Cas9 system uses the bacterial Cas9 nuclease to cut target DNA at a sequence complementary to a guide RNA. Subsequent DNA repair yields an inactivated gene or a gene with an altered sequence.

• In gene therapy, a normal gene is introduced or a gene is edited in order to cure a genetic disease.

KEY TERMS

chromosome	purine	vitamin	3′ end
nucleic acid	pyrimidine	phosphodiester bond	base pair
nucleotide	RNA (ribonucleic acid)	polynucleotide	sugar–phosphate backbone
DNA (deoxyribonucleic acid)	nucleoside	residue	antiparallel
base	deoxynucleotide	5′ end	major groove

BIOINFORMATICS

Brief Bioinformatics Exercises

3.1 Drawing and Visualizing Nucleotides

3.2 The DNA Double Helix

3.3 Melting Temperature and the GC Content of Duplex DNA

3.4 Analysis of Genomic DNA

3.5 Restriction Enzyme Mapping

Bioinformatics Project

Databases for the Storage and "Mining" of Genome Sequences

PROBLEMS

3.1 Nucleotides

1. The identification of DNA as the genetic material began with Griffith's "transformation" experiment conducted in 1928. Griffith worked with *Pneumococcus,* an encapsulated bacterium that forms smooth colonies when plated on agar and causes death when injected into mice. A mutant *Pneumococcus* lacking the enzymes needed to synthesize the polysaccharide capsule (required for virulence) forms rough colonies when plated on agar and does not cause death when injected into mice. Griffith found that heat-treated wild-type *Pneumococcus* did not cause death when injected into the mice because the heat treatment destroyed the polysaccharide capsule. However, if Griffith mixed heat-treated wild-type *Pneumococcus* and the mutant unencapsulated *Pneumococcus* together and injected this mixture, the mice died. Even more surprisingly, upon autopsy, Griffith found live, encapsulated *Pneumococcus* bacteria in the mouse tissue. Griffith concluded that the mutant *Pneumococcus* had been "transformed" into disease-causing *Pneumococcus,* but he could not explain how this occurred. Using your current knowledge of how DNA works, explain how the mutant *Pneumococcus* became transformed.

2. In 1944, Avery, MacLeod, and McCarty set out to identify the chemical agent capable of transforming mutant unencapsulated *Pneumococcus* to the deadly encapsulated form (see Problem 1). They isolated a viscous substance with the chemical and physical properties of DNA that was capable of transformation. If proteases (enzymes that degrade proteins) or ribonucleases (enzymes that

degrade RNA) were added prior to the experiment, transformation could still occur. What did these treatments tell the investigators about the molecular identity of the transforming factor?

3. In 1952, Alfred Hershey and Martha Chase carried out experiments using bacteriophages, which consist of nucleic acid enclosed by a protein capsid (coat). They first labeled the bacteriophages with the radioactive isotopes ^{35}S and ^{32}P. Because proteins contain sulfur but not phosphorus, and DNA contains phosphorus but not sulfur, each type of molecule was separately labeled. The radiolabeled bacteriophages were allowed to infect the bacteria, and then the preparation was treated to separate the empty capsids (ghosts) from the bacterial cells. The ghosts were found to contain most of the ^{35}S label, whereas 30% of the ^{32}P was found in the new bacteriophages produced by the infected cells. What does this experiment reveal about the roles of bacteriophage DNA and protein?

4. In February 1953 (two months before Watson and Crick published their paper describing DNA as a double helix), Linus Pauling and Robert Corey published a paper in which they proposed that DNA adopts a triple-helical structure. In their model, the three chains were tightly packed together, with the phosphates on the inside of the triple helix and the nitrogenous bases on the outside. They proposed that the triple helix was stabilized by hydrogen bonds between the interior phosphate groups. What are the flaws in this model?

5. In some organisms, DNA is modified by addition of methyl groups. Draw the structure of 3-methylguanine.

6. In certain pathogenic bacteria, the methylation of certain adenines in DNA is required in order for the bacteria to cause disease.

   (a) Draw the structure of N^6-methyladenine.

   (b) Why might scientists be interested in studying the bacterial enzyme N^6-DNA methyltransferase, which catalyzes the transfer of methyl groups to adenine?

7. Certain strains of *E. coli* incorporate the nitrogenous base shown here into nucleotides. For which base is this one a substitute?

8. An *E. coli* culture is grown in the presence of the base shown in Problem 7. A control culture is grown in the absence of this modified base. Compare the masses of the DNA isolated from *E. coli* in these two cultures.

9. Describe the chemical difference between uracil and thymine.

10. Many cellular signaling pathways involve the conversion of ATP to cyclic AMP, in which a single phosphate group is esterified to both C3′ and C5′. Draw the structure of cyclic AMP.

11. The simple synthesis of the antiviral compound 5-bromo-2′-deoxyuridine was recently reported. Draw the structure of this compound. For what base is this compound a substitute?

3.2 Nucleic Acid Structure

12. Draw a CA (ribo)dinucleotide and label the phosphodiester bond. How would the structure differ if it were DNA?

13. **(a)** Locate the base and monosaccharide in each residue of the dinucleotide NAD and the dinucleotide FAD (see Fig. 3.2).

   (b) What kind of linkage joins the two nucleotides in the dinucleotides NAD and FAD?

   (c) How do the adenosine groups in FAD and CoA differ?

14. The dinucleotide cyclic guanosine monophosphate–adenosine monophosphate (cGAMP) is an intracellular signaling molecule involved in antiviral defense. The molecule has two phosphodiester linkages, one between the 2′-OH of GMP and the 5′-phosphate of AMP and the other between the 3′-OH of AMP and the 5′-phosphate of GMP. Draw this dinucleotide.

15. Identify the base pair highlighted in blue in Figure 3.6.

16. The adenine derivative hypoxanthine can base pair with cytosine, adenine, and uracil. Show the structures of these base pairs.

Hypoxanthine

17. Explain whether the following statement is true or false: Because a G:C base pair is stabilized by three hydrogen bonds, whereas an A:T base pair is stabilized by only two hydrogen bonds, GC-rich DNA is harder to melt than AT-rich DNA.

18. Explain the structural basis for Chargaff's observation that the amount of A + G in DNA is equal to the total amount of C + T. Does this rule hold for RNA?

19. How can the hydrophobic effect (Section 2.2) explain why DNA adopts a helical structure?

20. A diploid organism with a 15,000-kb haploid genome contains 21% T residues. Calculate the number of A, C, G, and T residues in the DNA of each cell in this organism.

21. A well-studied bacteriophage has 98,534 base pairs in its complete genome.

   (a) There are 22,356 G residues in the genome. Calculate the number of C, A, and T residues.

   (b) Why does GenBank report a total of 49,267 bases for this bacteriophage genome?

22. The complete genome of a virus contains 1328 T residues, 1250 G residues, 1536 A residues, and 1212 C residues. What can you conclude about the structure of the viral genome, given this information?

23. **(a)** Would you expect proteins to bind to the major groove or the minor groove of DNA? Explain.

   (b) Eukaryotic DNA is packaged with histones, small proteins with a high lysine and arginine content. Why do histones have a high affinity for DNA?

24. **(a)** What is the T_m of the DNA sample whose melting curve is shown in Figure 3.7?

   (b) Draw melting curves that would be obtained from the DNA of *Dictyostelium discoideum* and *Streptomyces albus* (see Table 3.2).

25. **(a)** Using Table 3.2 as a guide, estimate the melting temperature of the DNA from an organism whose genome contains equal amounts of all four nucleotides.

   (b) To what temperature would you have to cool the DNA to allow it to reanneal?

26. What might you find in comparing the GC content of DNA from *Thermus aquaticus* or *Pyrococcus furiosus* and DNA from bacteria in a typical backyard pond?

27. Explain why the melting temperature of a sample of double-helical DNA increases when the Na^+ concentration increases.

28. **(a)** You have a short piece of synthetic RNA that you want to use as a probe to identify a gene in a sample of DNA. The RNA probe has a tendency to hybridize with sequences that are only weakly complementary. Should you increase or decrease the temperature to improve your chances of tagging the correct sequence?

   (b) You have a short piece of single-stranded DNA that you want to hybridize to another strand of DNA with one mismatched base pair between the two strands. Should you increase or decrease the temperature to improve your chances of annealing the two strands?

29. In the laboratory technique known as fluorescence *in situ* hybridization (FISH), a fluorescent oligonucleotide probe is allowed to hybridize with a cell's chromosomes, which are typically spread on a microscope slide. Explain why the chromosome preparation must be heated before the probe is added to it.

3.3 The Central Dogma

30. Discuss the shortcomings of the following definitions for *gene:*

   (a) A gene is the information that determines an inherited characteristic such as flower color.

(b) A gene is a segment of DNA that encodes a protein.

(c) A gene is a segment of DNA that is transcribed in all cells.

31. The semiconservative nature of DNA replication (as shown in Fig. 3.9) was proposed by Watson and Crick in 1953, but it wasn't experimentally verified until 1958. Meselson and Stahl grew bacteria on the "heavy" nitrogen isotope ^{15}N, producing DNA that was denser than normal. The food source was then abruptly switched to one containing only ^{14}N. Bacteria were harvested and the DNA isolated by density gradient centrifugation.

(a) What is the density of the DNA of the first-generation daughter DNA molecules? Explain.

(b) What is the density of the DNA isolated after two generations? Explain.

(c) What results would Meselson and Stahl have obtained had DNA replicated conservatively?

32. In the early 1960s, Marshall Nirenberg deciphered the genetic code by designing an experiment in which he synthesized a polynucleotide strand consisting solely of U residues, then added this strand to a test tube containing all of the components needed for protein synthesis.

(a) What polypeptide was produced by this "cell-free" system?

(b) What polypeptides were produced when poly A, poly C, and poly G were added to the cell-free system?

33. Har Gobind Korana extended Nirenburg's work by synthesizing polynucleotides with precisely defined sequences.

(a) What polypeptide(s) would be produced if a poly-GUGUGU··· were added to the cell-free system described in Problem 32?

(b) Do these results help to decipher the identities of the codons involved?

34. How many different codons are possible in nucleic acids containing four different nucleotides if a codon consisted of

(a) a single nucleotide, or consecutive sequences of

(b) two nucleotides,

(c) three nucleotides, or

(d) four nucleotides? Does your answer help explain why codons consist of three nucleotides?

35. A segment of the coding strand of a gene is shown below.

ATACGATGCTGAATTTGAGT

(a) Write the sequence of the complementary strand that DNA polymerase would make.

(b) Write the sequence of the mRNA that RNA polymerase would make from the gene segment.

36. A portion of a gene is shown below.

5′-ATCTAACCGGTCCGCACTCCTTAGTCGCTCCTGCG-
3′-TAGATTGGCCAGGCGTGAGGAATCAGCGAGGACGC-
GCTGACGCTGCTCGCCA-3′
CGACTGCGACGAGCGGT-5′

The sequence of the mRNA transcribed from this gene has the following sequence:

5′-AUCUAACCGGUCCGCACUCCUUAGUCGCUCCUGCG-
GUGACGCUGCUCGCCA-3′

(a) Identify the coding and noncoding strands of the DNA.

(b) Explain why only the coding strands of DNA are commonly published in databanks.

37. Synthetic biologists at the Scripps Institute expanded the genetic repertoire by adding two new bases into living bacterial cells. The two bases are named d5SICS and dNaM, and they base-pair with one another. How many different codons are possible in nucleic acids containing six different nucleotides if a codon consists of a consecutive sequence of three nucleotides?

38. A mutation occurs when there is a base change in the DNA sequence. Some base changes do not lead to changes in the amino acid sequence of the resulting protein. Explain why.

39. Is it possible for the same segment of DNA to encode two different proteins? Explain.

40. Examine the following nucleotide sequence from the coding strand of DNA. What is the amino acid sequence of the encoded protein?

GTAATTCAAAT ATG CCT TAC GCC
CCT GGA GAC GAA AAG AAG GGT
GCT ATT ACG TAT TTG AAG AAG
GCC ACC TCT GAG TAA ATGTGA

41. (a) One form of the disease adrenoleukodystrophy (ALD) is caused by the substitution of serine for asparagine in the ALD protein. List the possible single-nucleotide alterations in the DNA of the ALD gene that could cause this genetic disease.

(b) In another form of ALD, a CGA codon is converted to a UGA codon. Explain how this mutation affects the ALD protein.

42. An open reading frame (ORF) is a portion of the genome that potentially codes for a protein. A given mRNA nucleotide sequence potentially has three different reading frames, only one of which is correct (the selection of the correct ORF will be discussed more fully in Section 22.3). A portion of the gene for a type II human collagen is shown.

(a) What are the sequences of amino acids that can potentially be translated from each of the three possible reading frames?

(b) Collagen's amino acid sequence consists of repeating triplets in which every third amino acid is glycine. Does this information assist you in your identification of the correct reading frame?

AGGTCTTCAGGGAATGCCTGGCGAGAGGGGAGCAGCTG-
GTATCGCTGGGCCCAAAGGC

43. A type of gene therapy called RNA interference (RNAi) is being investigated to treat Huntington's disease. This disease is caused by a mutation in the DNA that results in the synthesis of an altered protein that leads to nervous system defects. To treat this disease, scientists synthesize short sequences of RNA (siRNA, or small interfering RNA) that form base pairs with the mRNA that codes for the mutated protein.

(a) Design an siRNA that will interfere with the synthesis of the protein shown in Problem 40.

(b) Explain how the addition of the siRNA will prevent the synthesis of the protein.

(c) What are the difficulties that must be overcome in order for RNA interference to be an effective technique for treating the disease?

44. The disease cystic fibrosis is the result of a mutation in the gene that encodes the cystic fibrosis transmembrane regulator (CFTR), a channel that allows chloride to exit the cell. A partial sequence of the CFTR gene is shown below, with the correct reading frame indicated. The most serious form of the disease results from the deletion of three consecutive nucleotides, as shown.

normal gene sequence 504 505 506 507 508 509 510 511 512
···GAA AAT ATC ATC TTT GGT GTT TCC TAT···

mutated gene sequence ···GAA AAT ATC AT– – –T GGT GTT TCC TAT···

(a) What are the sequences of the normal and mutated proteins?

(b) In another patient with a less severe form of the disease, the CF gene has the sequence ···AAT AGA TAC AG··· (the normal sequence of the CF gene in this region is ···AAT ATA GAT ACA G···). How has the DNA sequence changed and how does this affect the encoded protein?

3.4   Genomics

45. A portion of the nucleotide sequence from the DNA coding strand of the chick ovalbumin gene is shown. What is the partial amino acid sequence of the encoded protein?

CTCAGAGTTCACCATGGGCTCCATCGGTGCAGCAA-

GCATGGAA-(1104 bp)-TTCTTTGGCAGATGTGTTTCC-

CCTTAAAAAGAA

46. The genome of the bacterium *Carsonella ruddii* contains 159 kb of DNA with 182 ORFs. What can you conclude about the habitat or lifestyle of this bacterium?

47. For many years, biologists and others have claimed that humans and chimpanzees are 98% identical at the level of DNA. Both the human and chimp genomes, which are roughly the same size, have now been sequenced, and the data reveal approximately 35 million nucleotide differences between the two species. How does this number compare to the original claim?

48. When genomes of various organisms were sequenced, biologists expected that the DNA content (the C-value) would always be

positively correlated with organismal complexity. But no such correlation has been demonstrated. In fact, some plant and algae genomes are many times the size of the human genome. The *C-value paradox* is the term that refers to this puzzling lack of correlation between DNA content and organismal complexity. What questions do biologists need to ask as they attempt to solve the paradox?

49. A partial sequence of a newly discovered bacteriophage is shown below.

   (a) Identify the longest open reading frame (ORF).

   (b) Assuming that the ORF has been correctly identified, where is the most likely start site?

TATGGGATGGCTGAGTACAGCACGTTGAATGAGGCGAT-

GGCCGCTGGTGATG

50. The bacteriophage DNA described in Problem 49 contains 59 kb and 105 ORFs. None of the ORFs codes for tRNAs. How does the bacteriophage replicate its DNA and synthesize the structural proteins necessary to replicate itself?

51. Assuming that genes and SNPs are distributed evenly throughout the human genome, estimate how many protein-coding genes are likely to differ between two individuals.

52. If each person's genome contains a SNP every 300 nucleotides or so, how many SNPs are in that person's genome?

53. A genome-wide association study was carried out to identify the SNPs located on chromosome 1 that were correlated with an intestinal disease. The locations of three genes on chromosome 1 (between positions 6.3×10^7 and 6.6×10^7) are shown in the figure below. A $-\log_{10}$ P-value of 7 or greater is assumed to be associated with the disease.

   (a) What are the locations on the chromosome that show the strongest correlation with the disease?

   (b) Which genes contain SNPs associated with the disease and which do not?

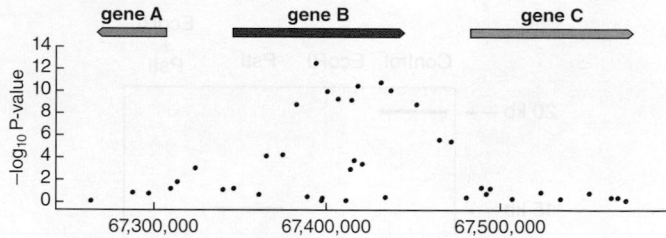

54. Use the information in the figure below to determine the chromosomal locations for the SNPs that are most closely associated with a colon disease. In this study, a $-\log_{10}$ P-value of 5 was used as the cutoff.

3.5   Tools and Techniques: Manipulating DNA

55. Which restriction enzymes in Table 3.5 generate sticky ends? Blunt ends?

56. Which is more likely to be called a "rare cutter": a restriction enzyme with a four-base recognition sequence or a restriction enzyme with an eight-base recognition sequence?

57. The sequence of a segment of the pET28 plasmid is shown below. Which of the restriction enzymes shown in Table 3.5 could cleave this DNA segment to insert a foreign gene into the plasmid?

CGCGGATCCGAATTCGAGCTCCGTCGACAAGCTTGCG-

GCCGCACTCGAG

58. A sample of the DNA segment shown below is treated with the restriction enzyme MspI. Next, the sample is incubated with an exonuclease that acts only on single-stranded polynucleotides. What mononucleotides will be present in the reaction mixture?

TGCTTAGCCGGAACGA

ACGAATCGGCCTTGCT

59. The plasmid pGEM-3Z shown in Figure 3.16 has three restriction sites for the enzyme HaeII, which cleaves at locations 323, 693 and 2564. Draw the result of agarose gel electrophoresis (see Figure 3.14) if the sample contains **(a)** linear or **(b)** circular plasmid digested with HaeII.

60. Restriction enzymes are used to construct restriction maps of DNA. These are diagrams of specific DNA molecules that show the sites where the restriction enzymes cleave the DNA. To construct a restriction map, purified samples of the DNA are treated with restriction enzymes, either alone or in combination, and then the reaction products are separated by agarose gel electrophoresis. Use the results of the agarose gel electrophoresis shown here to construct a restriction map for the sample of DNA.

61. The primer used in sequencing a cloned DNA segment often includes the recognition sequence for a restriction endonuclease. Explain.

62. Refer to the DNA sequence shown in Problem 36. Design two 15-bp primers that could be used to amplify this gene segment.

63. The "reverse" primer, which pairs with the coding strand, for the DNA segment shown in Problem 36 has a GC content of 67% (see Solution 62). Why might a primer with a high GC content be problematic in PCR?

64. Refer to the DNA sequence shown in Problem 40.

(a) Design two 18-bp PCR primers that could be used to amplify this gene.

(b) Suppose you wanted to add *EcoRI* restriction sites to each end of the gene. How would you modify the sequences of the PCR primers you designed in part a to amplify the gene?

65. Could you perform PCR with an ordinary DNA polymerase, that is, one that is destroyed by high temperatures? What modifications would you make in the PCR protocol?

66. Examine the sequence of the protein in Solution 40. Assume that the corresponding DNA sequence is *not* known. Using the amino acid sequence as a guide, design a pair of nine-base deoxynucleotide primers that could be used for PCR amplification of the protein-coding portion of the gene. (*Hint:* DNA polymerase can extend a primer only from its 3′ end.) How many different pairs of primers could you choose from?

67. A researcher trying to identify the gene for a known protein might begin by looking closely at the protein's sequence in order to design a single-stranded oligonucleotide probe that will hybridize with the DNA of the gene. Why would the researcher focus on a segment of the protein containing a methionine (Met) or tryptophan (Trp) residue (see Table 3.3)?

68. The following DNA fragments were produced by using the "shotgun" method. Show how the fragments should be aligned to determine the sequence of the original DNA.

ACCGTGTTTCCGACCG

ATTGTTCCCACAGACCG

CGGCGAAGCATTGTTCC

TTGTTCCCACAGACCGTG

69. The search for a gene sometimes starts with a DNA library, which is a set of cloned DNA fragments representing all the sequences in an organism's genome. If you were to construct a human DNA library to search for novel genes, would you choose to clone the DNA fragments in plasmids or in yeast artificial chromosomes? Explain.

70. In a CRISPR-Cas9 gene knock-in experiment, the replacement DNA sequence must differ from the original DNA sequence. Explain.

SELECTED READINGS

Dickerson, R. E., DNA structure from A to Z, *Methods Enzymol.* **211**, 67–111 (1992). [Describes the various crystallographic forms of DNA.]

Doudna, J. A. and Charpentier, E., The new frontier of genome engineering with CRISPR-Cas9, *Science* **346**, 1047 (2014) and DOI: 10.1126/science.1258096. [A review by two of the scientists who developed the CRISPR-Cas9 method; the full article is available at http://dx.doi.org/10.1126/science.1258096.]

International Human Genome Sequencing Consortium, Initial sequencing and analysis of the human genome, *Nature* **409**, 860–921 (2001) and Venter, J. C., et al., The sequence of the human genome, *Science* **291**, 1304–1351 (2001). [These and other papers in the same issues of *Nature* and *Science* describe the data that constitute the draft sequence of the human genome and discuss how this information can be used in understanding biological function, evolution, and human health.]

Lander, E. S., Initial impact of the sequencing of the human genome, *Nature* **470**, 187–197 (2011). [Reviews genome structure and provides insights into its variation and links to human diseases.]

Naldini, L. Gene therapy returns to centre stage, *Nature* **526**, 351–360 (2015). [Describes the various approaches for gene therapy and provides details on methods that have been successful in patients.]

Rigden, D. J., Fernández-Suárez, X. M., and Galperin, M. Y., The 2016 database issue of Nucleic Acids Research and an updated molecular biology database collection. *Nucl. Acids Res.* **44**, D1–D6 (2016). [Introduces the annual database issue, which includes a summary of bioinformatics resources and some additions to the list of 1685 molecular biology databases. The entire issue is available at http://nar.oxfordjournals.org/.]

Thieffry, D., Forty years under the central dogma, *Trends Biochem. Sci.* **23**, 312–316 (1998). [Traces the origins, acceptance, and shortcomings of the idea that nucleic acids contain biological information.]

STRUCTURE OF PROTEINS

The Asahi Shimbun/Getty Images

Many species of tree frog protect their newly laid eggs with the help of proteins called ranaspumins, which are usually secreted by the female and whipped into a frothy coating by the male. Foaming is an unusual characteristic for proteins, since most proteins lose their function when this occurs.

DO YOU REMEMBER?

- Cells contain four major types of biological molecules and three major types of polymers (Section 1.2).

- Noncovalent forces, including hydrogen bonds, ionic interactions, and van der Waals forces, act on biological molecules (Section 2.1).

- The hydrophobic effect, which is driven by entropy, excludes nonpolar substances from water (Section 2.2).

- An acid's pK value describes its tendency to ionize (Section 2.3).

- The biological information encoded by a sequence of DNA is transcribed to RNA and then translated into the amino acid sequence of a protein (Section 3.3).

Proteins are the workhorses of the cell. They provide structural stability and motors for movement; they form the molecular machinery for harvesting free energy and using it to carry out other metabolic activities; they participate in the expression of genetic information; and they mediate communication between the cell and its environment. In subsequent chapters, we will describe in more detail these protein-driven phenomena, but for now we will focus on protein structure.

We will look first at the amino acid components of proteins. Next comes a discussion of how the protein chain folds into a unique three-dimensional shape. Finally, the Tools and Techniques section of this chapter examines some of the procedures for purifying and sequencing proteins and determining their structures.

AMINO ACIDS, THE BUILDING BLOCKS OF PROTEINS

Proteins come in a huge variety of shapes and sizes (**Fig. 4.1**), but they are all built the same way. Each **protein** consists of one or more **polypeptides,** which are chains of polymerized amino acids. A cell may contain dozens of different amino acids, but only 20 of these—called the "standard" amino acids—are commonly found in proteins. As introduced in Section 1.2, an **amino acid** is a small molecule containing an amino group ($-NH_3^+$) and a carboxylate group ($-COO^-$) as well as a side chain of variable structure, called an **R group:**

$$
\begin{array}{c}
COO^- \\
| \\
H-C-R \\
| \\
NH_3^+
\end{array}
$$

Note that at physiological pH, the carboxyl group is unprotonated and the amino group is protonated, so an isolated amino acid bears both a negative and a positive charge.

DNA polymerase (*E. coli* Klenow fragment)
Synthesizes a new DNA chain using an existing DNA strand as a template (more in Section 20.1)

Plastocyanin (poplar)
Shuttles electrons as part of the apparatus for converting light energy to chemical energy (more in Section 16.2)

Insulin (pig)
Released from the pancreas to signal the availability of the metabolic fuel glucose (more in Section 19.2)

Maltoporin (*E. coli*)
Permits sugars to cross the bacterial cell membrane (more in Section 9.2)

Phosphoglycerate kinase (yeast)
Catalyzes one of the central reactions in metabolism (more in Section 13.1)

FIGURE 4.1

A gallery of protein structures. These space-filling models are all shown at approximately the same scale. In proteins that consist of more than one chain of amino acids, the chains are shaded differently. [*Structure of insulin (pdb 1ZNI) determined by M. G. W. Turkenburg, J. L. Whittingham, G. G. Dodson, E. J. Dodson, B. Xiao, and G. A. Bentley; structure of maltoporin (pdb 1MPM) determined by R. Dutzler and T. Schirmer; structure of phosphoglycerate kinase (pdb 3PGK) determined by P. J. Shaw, N. P. Walker, and H. C. Watson; structure of DNA polymerase (pdb 1KFS) determined by C. A. Brautigan and T. A. Steitz; and structure of plastocyanin (pdb 1PND) determined by B. A. Fields, J. M. Guss, and H. C. Freeman.*]

THE 20 AMINO ACIDS HAVE DIFFERENT CHEMICAL PROPERTIES

The identities of the R groups distinguish the 20 standard amino acids. The R groups can be classified by their overall chemical characteristics as hydrophobic, polar, or charged, as shown in **Figure 4.2**, which also includes the one- and three-letter codes for each amino acid. These compounds are formally called **α-amino acids** because the amino and carboxylate (acid) groups are both attached to a central carbon atom known as the α carbon (abbreviated **Cα**).

Figure 4.2 also includes the one- and three-letter codes for each amino acid. The three-letter code is usually the first three letters of the amino acid's name. The one-letter code is derived as follows: If only one amino acid begins with a particular letter, that letter is used: C = cysteine, H = histidine, I = isoleucine, M = methionine, S = serine, and V = valine. If more than one amino acid begins with a particular letter, the letter is assigned to the most abundant amino acid: A = alanine, G = glycine, L = leucine, P = proline, and T = threonine. Most of the others are phonetically suggestive: D = aspartate ("asparDate"), F = phenylalanine ("Fenylalanine"), N = asparagine ("asparagiNe"), R = arginine ("aRginine"), W = tryptophan ("tWyptophan"), and Y = tyrosine ("tYrosine"). The rest are assigned as follows: E = glutamate (near D, aspartate), K = lysine, and Q = glutamine (near N, asparagine). The carbon atoms of amino acids are sometimes assigned Greek letters, beginning with Cα, the carbon to which the R group is attached. Thus, glutamate has a γ-carboxylate group, and lysine has an ε-amino group.

Hydrophobic amino acids

Alanine (Ala, A)   Valine (Val, V)   Phenylalanine (Phe, F)   Tryptophan (Trp, W)

Leucine (Leu, L)   Isoleucine (Ile, I)   Methionine (Met, M)   Proline (Pro, P)

Polar amino acids

Serine (Ser, S)   Threonine (Thr, T)   Tyrosine (Tyr, Y)   Cysteine (Cys, C)

Asparagine (Asn, N)   Glutamine (Gln, Q)   Histidine (His, H)   Glycine (Gly, G)

Charged amino acids

Aspartate (Asp, D)   Glutamate (Glu, E)   Lysine (Lys, K)   Arginine (Arg, R)

FIGURE 4.2

Structures and abbreviations of the 20 standard amino acids. The amino acids can be classified according to the chemical properties of their R groups as hydrophobic, polar, or charged. The side chain (R group) of each amino acid is shaded.

? Identify the functional groups in each amino acid. Refer to Table 1.1.

Box 4.A Does Chirality Matter?

The importance of chirality in biological systems was brought home in the 1960s when pregnant women with morning sickness were given the sedative thalidomide, which was a mixture of right- and left-handed forms. The active form of the drug has the structure shown here.

Thalidomide

Tragically, its mirror image, which was also present, caused severe birth defects, including abnormally short or absent limbs.

Although the mechanisms of action of the two forms of thalidomide are not well understood, different responses to the two forms can be rationalized. An organism's ability to distinguish chiral molecules results from the handedness of its molecular constituents. For example, proteins contain all L amino acids, and polynucleotides coil in a right-handed helix (see Fig. 3.4). The lessons learned from thalidomide have made drugs more costly to develop and test but have also made them safer.

? Which of the 20 amino acids is not chiral?

Nineteen of the twenty standard amino acids are asymmetric, or chiral, molecules. Their **chirality,** or handedness (from the Greek *cheir,* "hand"), results from the asymmetry of the alpha carbon. The four different substituents of Cα can be arranged in two ways. For alanine, a small amino acid with a methyl R group, the possibilities are

$$H_3N^+ \text{—} C_\alpha \text{—} H \qquad H \text{—} C_\alpha \text{—} NH_3^+$$

You can use a simple model-building kit to satisfy yourself that the two structures are not identical. They are nonsuperimposable mirror images, like right and left hands.

The amino acids found in proteins all have the form on the left. For historical reasons, these are designated L amino acids (from the Greek *levo,* "left"). Their mirror images, which rarely occur in proteins, are the D amino acids (from *dextro,* "right").

Molecules related by mirror symmetry are physically indistinguishable and are usually present in equal amounts in synthetic preparations. However, the two forms behave differently in biological systems (**Box 4.A**).

It is advisable to become familiar with the structures of the standard amino acids, since their side chains ultimately help determine the three-dimensional shape of the protein, its solubility, its ability to interact with other molecules, and its chemical reactivity.

Amino Acids with Hydrophobic Side Chains *Some amino acids have nonpolar (hydrophobic) side chains that interact very weakly or not at all with water.* The aliphatic (hydrocarbonlike) side chains of alanine (Ala), valine (Val), leucine (Leu), isoleucine (Ile), and phenylalanine (Phe) obviously fit into this group. Although the side chains of methionine (Met) and tryptophan (Trp) include atoms with unshared electron pairs, the bulk of their side chains is nonpolar. Proline (Pro) is unique among the amino acids because its aliphatic side chain is also covalently linked to its amino group. Glycine (Gly) is sometimes included with the hydrophobic amino acids.

In proteins, the hydrophobic amino acids are almost always located in the interior of the molecule, among other hydrophobic groups, where they do not interact with water. And because they lack reactive functional groups, the hydrophobic side chains do not directly participate in mediating chemical reactions.

Amino Acids with Polar Side Chains *The side chains of the polar amino acids can interact with water because they contain hydrogen-bonding groups.* Serine (Ser), threonine (Thr), and tyrosine (Tyr) have hydroxyl groups; cysteine (Cys) has a thiol group; and asparagine (Asn) and glutamine (Gln) have amide groups. All these amino acids, along with histidine (His, which bears a polar imidazole ring), can be found on the solvent-exposed surface of a protein, although they also occur in the protein interior, provided that their hydrogen-bonding requirements are satisfied by their proximity to other hydrogen bond donor or acceptor groups. Glycine (Gly), whose side chain consists of only an H atom, cannot form hydrogen bonds but is included with the polar amino acids because it is neither hydrophobic nor charged.

Depending on the presence of nearby groups that increase their polarity, *some of the polar side chains can ionize at physiological pH values*. For example, the neutral (basic) form of histidine can accept a proton to form an imidazolium ion (an acid):

As we will see, the ability of histidine to act as an acid or a base gives it great versatility in catalyzing chemical reactions.

Similarly, the thiol group of cysteine can be deprotonated, yielding a thiolate anion:

Occasionally, cysteine's thiol group undergoes oxidation with another thiol group, such as another Cys side chain, to form a **disulfide bond:**

Disulfide bond

In certain situations, the hydroxyl groups of serine, threonine, and tyrosine undergo chemical reactions in which the O—H bond is cleaved.

Amino Acids with Charged Side Chains
Four amino acids have side chains that are virtually always charged under physiological conditions. Aspartate (Asp) and glutamate (Glu), which bear carboxylate groups, are negatively charged. Lysine (Lys) and arginine (Arg) are positively charged. Histidine, described above, can also bear a positive charge. Amino acids with charged side chains are usually located on the protein's surface, where the ionic groups can be surrounded by water molecules or interact with other polar or ionic substances. Note that the charges of these side chains depend on their ionization state, which is sensitive to the local pH. The amino acids with acidic or basic side chains can also participate in acid–base reactions.

Although it is convenient to view amino acids merely as the building blocks of proteins, many amino acids play key roles in regulating physiological processes (**Box 4.B**).

Box 4.B Monosodium Glutamate

A number of amino acids and compounds derived from them function as signaling molecules in the nervous system (we will look at some of these in more detail in Section 18.2). Among the amino acids with signaling activity is glutamate, which most often operates as an excitatory signal and is necessary for learning and memory. Because glutamate is abundant in dietary proteins and because the human body can manufacture it, glutamate deficiency is rare. But is there any danger in eating too much glutamate?

Glutamate binds to receptors on the tongue that register the taste of umami—one of the five human tastes, along with sweet, salty, sour, and bitter. By itself, the umami taste is not particularly pleasing, but when combined with other tastes, it imparts a sense of savoriness and induces salivation. For this reason, glutamate in the form of monosodium glutamate (MSG) is sometimes added to

processed foods as a flavor enhancer. For example, a low-salt food item can be made more appealing by adding MSG to it.

According to some popular accounts, "Chinese restaurant syndrome" can be attributed to the consumption of excess MSG added to prepared foods or present in soy sauce (MSG is also naturally present in many other foods, including cheese and tomatoes). The symptoms of the syndrome reportedly include muscle tingling, headache, and drowsiness—all of which could potentially reflect the role of glutamate in the nervous system. However, a definitive link between MSG intake and neurological symptoms has not been demonstrated in scientific studies and therefore remains mostly anecdotal.

? **Which group of glutamate is ionized to pair with a single sodium ion in MSG?**

PEPTIDE BONDS LINK AMINO ACIDS IN PROTEINS

The polymerization of amino acids to form a polypeptide chain involves the condensation of the carboxylate group of one amino acid with the amino group of another (a **condensation reaction** is one in which a water molecule is eliminated):

Peptide bond

The resulting amide bond linking the two amino acids is called a **peptide bond.** The remaining portions of the amino acids are called amino acid **residues.** In a cell, peptide bond formation is carried out in several steps involving the ribosome and additional RNA and protein factors (Section 22.3). Peptide bonds can be broken, or **hydrolyzed,** by the action of **exopeptidases** or **endopeptidases** (enzymes that act from the end or the middle of the chain, respectively).

By convention, a chain of amino acid residues linked by peptide bonds is written or drawn so that the residue with a free amino group is on the left (this end of the polypeptide is called the **N-terminus**) and the residue with a free carboxylate group is on the right (this end is called the **C-terminus**):

Note that, except for the two terminal groups, the charged amino and carboxylate groups of each amino acid are eliminated in forming peptide bonds. *The electrostatic properties of the*

TABLE 4.1 pK Values of Ionizable Groups in Amino Acids

GROUP[a]		pK
C-terminus	—COOH	3.5
Asp	—CH$_2$—C(=O)—OH	3.9
Glu	—CH$_2$—CH$_2$—C(=O)—OH	4.1
His	—CH$_2$— (imidazole, NH$^+$)	6.0
Cys	—CH$_2$—SH	8.4
N-terminus	—NH$_3^+$	9.0
Tyr	—CH$_2$—(phenyl)—OH	10.5
Lys	—CH$_2$—CH$_2$—CH$_2$—CH$_2$—NH$_3^+$	10.5
Arg	—CH$_2$—CH$_2$—CH$_2$—NH—C(NH$_2$)=NH$_2^+$	12.5

[a]The ionizable proton is indicated in red.

polypeptide therefore depend primarily on the identities of the side chains (R groups) that project out from the polypeptide **backbone.**

The pK values of all the charged and ionizable groups in amino acids are given in **Table 4.1** (recall from Section 2.3 that a pK value is a measure of a group's tendency to ionize). Thus, it is possible to calculate the net charge of a protein at a given pH (see Sample Calculation 4.1). At best, this value is only an estimate, since the side chains of polymerized amino acids do not behave as they do in free amino acids. This is because of the electronic effects of the peptide bond and other functional groups that may be brought into proximity when the polypeptide chain folds into a three-dimensional shape. The chemical properties of a side chain's immediate neighbors, its **microenvironment,** may alter its polarity, thereby altering its tendency to lose or accept a proton.

Nevertheless, *the chemical and physical properties of proteins depend on their constituent amino acids,* so proteins exhibit different behaviors under given laboratory conditions. These differences can be exploited to purify a protein, that is, to isolate it from a mixture containing other molecules (see Section 4.6). Most proteins contain all 20 amino acids, with some tending to appear more often than others (**Fig. 4.3**).

FIGURE 4.3

Percent occurrence of amino acids in proteins.

SAMPLE CALCULATION 4.1

PROBLEM   Estimate the net charge of the polypeptide chain below at physiological pH (7.4) and at pH 5.0.

Ala–Arg–Val–His–Asp–Gln

SOLUTION   The polypeptide contains the following ionizable groups, whose pK values are listed in Table 4.1: the N-terminus (pK = 9.0), Arg (pK = 12.5), His (pK = 6.0), Asp (pK = 3.9), and the C-terminus (pK = 3.5).

At pH 7.4, the groups whose pK values are less than 7.4 are mostly deprotonated, and the groups with pK values greater than 7.4 are mostly protonated. The polypeptide therefore has a net charge of 0:

Group	Charge
N-terminus	+1
Arg	+1
His	0
Asp	−1
C-terminus	−1
net charge	0

At pH 5.0, His is likely to be protonated, giving the polypeptide a net charge of +1:

Group	Charge
N-terminus	+1
Arg	+1
His	+1
Asp	−1
C-terminus	−1
net charge	+1

PRACTICE PROBLEMS   1. Estimate the net charge of Asp-His-Lys tripeptide at pH 8.0.

2. Estimate the net charge of Glu-Glu-Glu tripeptide at pH 7.0.

3. Estimate the net charge of Asp-Tyr dipeptide at pH 6.0.

Most polypeptides contain between 100 and 1000 amino acid residues, although some contain thousands of amino acids (**Table 4.2**). Short polypeptides are often called **oligopeptides** (*oligo* is Greek for "few") or just **peptides.** Since there are 20 different amino acids that can be polymerized to form polypeptides, even peptides of similar size can differ dramatically from each other, depending on their complement of amino acids.

The potential for sequence variation is enormous. For a modest-sized polypeptide of 100 residues, there are 20^{100} or 1.27×10^{130} possible amino acid sequences. This number is clearly unattainable in nature, since there are only about 10^{79} atoms in the universe, but it illustrates the tremendous structural variability of proteins.

Unraveling the amino acid sequence of a protein may be relatively straightforward if its gene has been sequenced (see Section 3.4). In this case, it is just a matter of reading successive sets of three nucleotides in the DNA as a sequence of amino acids in the protein. However, this exercise may not be accurate if the gene's mRNA is spliced before being translated or if the protein is hydrolyzed or otherwise covalently modified immediately after it is synthesized. And, of course, nucleic acid sequencing is of no use if the protein's gene has not been identified. The alternative is to use a technique such as mass spectrometry to directly determine the protein's amino acid sequence (Section 4.6).

TABLE 4.2 Composition of Some Proteins

PROTEIN	NUMBER OF AMINO ACID RESIDUES	NUMBER OF POLYPEPTIDE CHAINS	MOLECULAR MASS (D)
Insulin (bovine)	51	2	5733
Rubredoxin (*Pyrococcus*)	53	1	5878
Myoglobin (human)	153	1	17,053
Phosphorylase kinase (yeast)	416	1	44,552
Hemoglobin (human)	574	4	61,972
Reverse transcriptase (HIV)	986	2	114,097
Nitrite reductase (*Alcaligenes*)	1029	3	111,027
C-reactive protein (human)	1030	5	115,160
Pyruvate decarboxylase (yeast)	1112	2	121,600
Immunoglobulin (mouse)	1316	4	145,228
Ribulose bisphosphate carboxylase (spinach)	5048	16	567,960
Glutamine synthetase (*Salmonella*)	5628	12	621,600
Carbamoyl phosphate synthetase (*E. coli*)	5820	8	637,020

THE AMINO ACID SEQUENCE IS THE FIRST LEVEL OF PROTEIN STRUCTURE

The sequence of amino acids in a polypeptide is called the protein's **primary structure.** There are as many as four levels of structure in a protein (**Fig. 4.4**). Under physiological conditions, a polypeptide very seldom assumes a linear extended conformation but instead folds up to form a more compact shape, usually consisting of several layers. The conformation of the polypeptide backbone (exclusive of the side chains) is known as **secondary structure.** The complete three-dimensional conformation of the polypeptide, including its backbone atoms and all its side chains, is the polypeptide's **tertiary structure.** In a protein that consists of more than one polypeptide chain, the **quaternary structure** refers to the spatial arrangement of all the chains. In the following sections we will consider the second, third, and fourth levels of protein structure.

Primary structure	**Secondary structure**	**Tertiary structure**	**Quaternary structure**
The sequence of amino acid residues	The spatial arrangement of the polypeptide backbone	The three-dimensional structure of an entire polypeptide, including all its side chains	The spatial arrangement of polypeptide chains in a protein with multiple subunits

–Lys–Val–Asn–Val–Asp–

FIGURE 4.4

Levels of protein structure in hemoglobin. [*Structure of human hemoglobin (pdb 2HHB) determined by G. Fermi and M. F. Perutz.*]

CONCEPT CHECK

1. Draw the structures and give the one- and three-letter abbreviations for the 20 standard amino acids.

2. Divide the 20 amino acids into groups that are hydrophobic, polar, and charged.

3. Which polar amino acids are sometimes charged?

4. Draw a tripeptide and identify its peptide bonds, backbone, side chains, N-terminus, C-terminus, and net charge.

5. Describe the four levels of protein structure.

SECONDARY STRUCTURE: THE CONFORMATION OF THE PEPTIDE GROUP

KEY CONCEPTS

- The polypeptide backbone has limited conformational flexibility.
- The α helix and β sheet are common secondary structures characterized by hydrogen bonding between backbone groups.

In the peptide bond that links successive amino acids in a polypeptide chain, the electrons are somewhat delocalized so that the peptide bond has two resonance forms:

Due to this partial (about 40%) double-bond character, there is no rotation around the C—N bond. In a polypeptide backbone, the repeating N—Cα—C units of the amino acid residues can therefore be considered to be a series of planar peptide groups (where each plane contains the atoms involved in the peptide bond):

Here the H atom and R group attached to Cα are not shown.

The polypeptide backbone can still rotate around the N—Cα and Cα—C bonds, although rotation is somewhat limited. For example, a sharp bend at Cα could bring the carbonyl oxygens of neighboring residues too close:

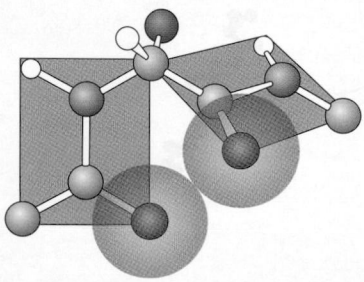

Here the atoms are color-coded: C gray, O red, N blue, and H white, and the van der Waals surfaces of the O atoms are shown.

As the resonance structures indicate, the groups involved in the peptide bond are strongly polar, with a tendency to form hydrogen bonds. The backbone amino groups are hydrogen bond donors, and the carbonyl oxygens are hydrogen bond acceptors. *Under physiological conditions, the polypeptide chain folds so that it can satisfy as many of these hydrogen-bonding requirements as possible.* At the same time, *the polypeptide backbone must adopt a conformation (a secondary structure) that minimizes steric strain.* In addition, side chains must be positioned in a way that minimizes their steric interference. To meet these criteria, the polypeptide backbone often assumes a repeating conformation, known as **regular secondary structure,** such as an α helix or a β sheet.

THE α HELIX EXHIBITS A TWISTED BACKBONE CONFORMATION

The **α helix** was first identified through model-building studies carried out by Linus Pauling. In this type of secondary structure, the polypeptide backbone twists in a right-handed helix (the DNA helix is also right-handed; see Section 3.2 for an explanation). There are 3.6 residues per turn of the helix, and for every turn, the helix rises 5.4 Å along its axis. In the α helix, the carbonyl oxygen of each residue forms a hydrogen bond with the backbone NH group four residues ahead. The backbone hydrogen-bonding tendencies are thereby met, except for the four residues at each end of the helix (**Fig. 4.5**). Most α helices are about 12 residues long.

Like the DNA helix, whose side chains fill the helix interior (see Figure 3.3b), the α helix is solid—the atoms of the polypeptide backbone are in van der Waals contact. However, in the α helix, the side chains extend outward from the helix (**Fig. 4.6**).

THE β SHEET CONTAINS MULTIPLE POLYPEPTIDE STRANDS

Pauling, along with Robert Corey, also built models of the **β sheet.** This type of secondary structure consists of aligned strands of polypeptide whose hydrogen-bonding requirements are met by bonding between neighboring strands. The strands of a β sheet can be arranged in two ways (**Fig. 4.7**): In a **parallel β sheet,** neighboring chains run in the same direction; in an **antiparallel β sheet,** neighboring chains run in opposite directions. Each residue forms two hydrogen bonds with a neighboring strand, so all hydrogen-bonding requirements are met. The first and last strands of the sheet must find other hydrogen-bonding partners for their outside edges.

FIGURE 4.5

The α helix. In this conformation, the polypeptide backbone twists in a right-handed fashion so that hydrogen bonds (dashed lines) form between C=O and N—H groups four residues farther along. Atoms are color-coded: Cα light gray, carbonyl C dark gray, O red, N blue, side chain purple, H white. [*Based on a drawing by Irving Geis.*]

? How many amino acid residues are shown here? How many hydrogen bonds?

(a)   (b)

FIGURE 4.6

An α helix from myoglobin. In (a) a ball-and-stick model and (b) a space-filling model of residues 100–118 of myoglobin, the backbone atoms are green and the side chains are gray. [*Structure of sperm whale myoglobin (pdb 1MBD) determined by S. E. V. Phillips.*]

FIGURE 4.7

β sheets. In a parallel
β sheet and an antiparallel
β sheet, the polypeptide
backbone is extended. In
both types of β sheet,
hydrogen bonds form
between the amino and
carbonyl groups of adjacent
strands. The H and R
attached to Cα are not shown.
Note that the strands are
not necessarily separate
polypeptides but may be
segments of a single chain
that loops back on itself.

❓ **How many amino
acid residues are
shown in each chain?**

Parallel   Antiparallel

A single β sheet may contain from 2 to more than 12 polypeptide strands, with an average
of 6 strands, and each strand has an average length of 6 residues. In a β sheet, the amino acid
side chains extend from both faces (**Fig. 4.8**).

PROTEINS ALSO CONTAIN IRREGULAR SECONDARY STRUCTURE

α Helices and β sheets are classified as regular secondary structures, because *their component residues exhibit backbone conformations that are the same from one residue to the next*.
In fact, these elements of secondary structure are easily recognized in the three-dimensional
structures of a huge variety of proteins, regardless of their amino acid composition. Of course,
depending on the identities of the side chains and other groups that might be present, α helices
and β sheets may be slightly distorted from their ideal conformations. For example, the final
turn of some α helices becomes "stretched out" (longer and thinner than the rest of the helix).

(a)   (b)

FIGURE 4.8

Side view of two parallel strands of a β sheet. In (a) a ball-and-stick model and (b) a space-filling model
of a β sheet from carboxypeptidase A, the backbone atoms are green. The amino acid side chains (gray)
point alternately to each side of the β sheet. [*Structure of carboxypeptidase (pdb 3CPA) determined by W. N.
Lipscomb.*]

In every protein, elements of secondary structure (individual α helices or strands in a β sheet) are linked together by polypeptide loops of various sizes. A loop may be a relatively simple hairpin turn, as in the connection of two antiparallel β strands (which are shown below as flat arrows; *left*). Or it may be quite long, especially if it joins successive strands in a parallel β sheet (*right*):

Usually, the loops that link β strands or α helices consist of residues with **irregular secondary structure;** that is, the polypeptide does not adopt a defined secondary structure in which successive residues have the same backbone conformation. Note that "irregular" does not mean "disordered": The peptide backbone almost always adopts a single, unique conformation. Most proteins contain a combination of regular and irregular secondary structure. On average, 31% of residues are in α helices, 28% are in β sheets, and most of the remainder are in irregular loops of different sizes. However, many proteins deviate from the average in dramatic ways.

 CONCEPT CHECK

1. Draw a polypeptide backbone and indicate which bonds can rotate freely.

2. What limits the rotation of the backbone?

3. Explain how an α helix and a β sheet satisfy a polypeptide's hydrogen-bonding needs.

4. Compare parallel and antiparallel β sheets.

5. Summarize the difference between regular and irregular secondary structure.

6. Why does virtually every protein contain loops?

TERTIARY STRUCTURE AND PROTEIN STABILITY

KEY CONCEPTS
- A folded polypeptide assumes a shape with a hydrophilic surface and a hydrophobic core.
- Protein folding and protein stabilization depend on noncovalent forces.
- Some proteins can adopt more than one stable conformation.

The three-dimensional shape of a protein, known as its tertiary structure, includes its regular and irregular secondary structure (that is, the overall folding of its peptide backbone) as well as the spatial arrangement of all its side chains. In a fully folded protein under physiological conditions, virtually every atom has a designated place.

One of the most powerful techniques for probing the atomic structures of macromolecules, including proteins, is X-ray crystallography (Section 4.6). The structure of myoglobin, the first protein structure to be determined by X-ray crystallography, came to light in 1958 through the efforts of John Kendrew, who painstakingly determined the conformation of every backbone and side-chain group. Kendrew's results—coming just a few years after Watson and Crick had published their elegant model of DNA—were a bit of a disappointment. The myoglobin structure lacked the simplicity and symmetry of a molecule such as DNA and was more irregular and complex than expected (myoglobin structure is examined in detail in Section 5.1).

The structure of another well-studied protein, the enzyme triose phosphate isomerase, is shown in **Figure 4.9a**. This enzyme is a **globular protein** (**fibrous proteins**, in contrast, are usually highly elongated; examples of these are presented in Section 5.3). The proteins shown in Figure 4.1 are all globular proteins. The tertiary structure of triose phosphate isomerase can be simplified by showing just the peptide backbone (Fig. 4.9b). Alternatively, the structure can be represented by a ribbon that passes through Cα of each residue (Fig. 4.9c). This rendering makes it easier to identify elements of secondary structure. However, it is important to keep in mind that proteins, like all molecules, are solid objects with no empty space inside.

Commonly used systems for analyzing protein structures are based on the presence of elements of regular secondary structure. For example, four classes are recognized by the CATH system (the name refers to a hierarchy of organizational levels: Class, Architecture, Topology, and Homology). Proteins may contain mostly α structure, mostly β structure, a combination of α and β, or very few regular secondary structural elements. Examples of each class are shown in **Figure 4.10**. Structural data—including the three-dimensional coordinates for each atom—for over a hundred thousand proteins and other macromolecules are available in online databases. The use of software for visualizing and manipulating such structures provides valuable insight into molecular structure and function.

PROTEINS HAVE HYDROPHOBIC CORES

Globular proteins typically contain at least two layers of secondary structure. This means that the protein has definite surface and core regions. On the protein's surface, some backbone and side-chain groups are exposed to the solvent; in the core, these groups are sequestered from the solvent. In other words, *the protein comprises a hydrophilic surface and a hydrophobic core.*

A polypeptide segment that has folded into a single structural unit with a hydrophobic core is often called a **domain.** Some small proteins consist of a single domain. Larger proteins may contain several domains, which may be structurally similar or dissimilar (**Fig. 4.11**).

The core of a domain or a small protein is typically rich in regular secondary structure. This is because the formation of α helices and β sheets, which are internally hydrogen-bonded, minimizes the hydrophilicity of the polar backbone groups. Irregular secondary structures (loops) are more often found on the solvent-exposed surface of the domain or protein, where the polar backbone groups can form hydrogen bonds to water molecules.

(a)

(b)

(c)

FIGURE 4.9

Triose phosphate isomerase.
(a) Space-filling model. All atoms (except H) are shown (C gray, O red, N blue). (b) Polypeptide backbone. The trace connects the α carbons of successive amino acid residues. (c) Ribbon diagram. The ribbon represents the overall conformation of the backbone. [*Structure (pdb 1YPI) determined by T. Alber, E. Lolis, and G. A. Petsko.*]

(a)   (b)

(c)   (d)

FIGURE 4.10

Classes of protein structure. In each protein, α helices are colored red, and β strands are colored yellow. (a) Growth hormone, an all-α protein. [*Structure (pdb 1HGU) determined by L. Chantalat, N. Jones, F. Korber, J. Navaza, and A. G. Pavlovsky.*] (b) γβ-Crystallin, an all-β protein. [*Structure (pdb 1GCS) determined by S. Najmudin, P. Lindley, C. Slingsby, D. Bateman, D. Myles, S. Kumaraswamy, and I. Glover.*] (c) Flavodoxin, an α/β protein. [*Structure (pdb 1CZN) determined by W. W. Smith, K. A. Pattridge, C. L. Luschinsky, and M. L. Ludwig.*] (d) Tachystatin, a protein with little secondary structure. [*Structure (pdb 1CIX) determined by N. Fujitani, S. Kawabata, T. Osaki, Y. Kumaki, M. Demura, K. Nitta, and K. Kawano.*]

The requirement for a hydrophobic core and a hydrophilic surface also places constraints on amino acid sequence. The location of a particular side chain in a protein's tertiary structure is related to its hydrophobicity: *The greater a residue's hydrophobicity, the more likely it is to be located in the protein interior.* In the protein interior, side chains pack together, leaving essentially no empty space or space that could be occupied by a water molecule.

Table 4.3 lists two scales for assessing the hydrophobicity of amino acid side chains. Such information is useful for predicting the locations of amino acid residues within proteins. For example, highly hydrophobic residues such as Phe and Met are almost always buried. Polar side chains, like hydrogen-bonding backbone groups, can participate in hydrogen bonding in the protein interior, which helps "neutralize" their polarity and allows them to be buried in a nonpolar environment. When a charged residue occurs in the protein interior, it is almost always located near another residue with the opposite charge, so the two groups can interact electrostatically to form an **ion pair.** By color-coding the amino acid residues of myoglobin according to their hydrophobicity, it is easy to see that hydrophobic side chains cluster in the interior while hydrophilic side chains predominate on the surface (**Fig. 4.12**).

PROTEIN STRUCTURES ARE STABILIZED MAINLY BY THE HYDROPHOBIC EFFECT

Surprisingly, the fully folded conformation of a protein is only marginally more stable than its unfolded form. The difference in thermodynamic stability amounts to about 0.4 kJ · mol⁻¹ per

TABLE 4.3 Hydrophobicity Scales

RESIDUE	SCALE A[a]	SCALE B[b]
Phe	2.8	3.7
Met	1.9	3.4
Ile	4.5	3.1
Leu	3.8	2.8
Val	4.2	2.6
Cys	2.5	2.0
Trp	−0.9	1.9
Ala	1.8	1.6
Thr	−0.7	1.2
Gly	−0.4	1.0
Ser	−0.8	0.6
Pro	−1.6	−0.2
Tyr	−1.3	−0.7
His	−3.2	−3.0
Gln	−3.5	−4.1
Asn	−3.5	−4.8
Glu	−3.5	−8.2
Lys	−3.9	−8.8
Asp	−3.5	−9.2
Arg	−4.5	−12.3

[a] Scale A is from Kyte, J., and Doolittle, R.F., *J. Mol. Biol.* **157,** 105–132 (1982).

[b] Scale B is from Engelman, D.M., Steitz, T.A., and Goldman, A., *Annu. Rev. Biophys. Chem.* **15,** 321–353 (1986).

FIGURE 4.11

A two-domain protein. In this model of glyceraldehyde-3-phosphate dehydrogenase, the small domain is red, and the large domain is green. [*Structure (pdb 1GPD) determined by D. Moras, K. W. Olsen, M. N. Sabesan, M. Buehner, G. C. Ford, and M. G. Rossmann.*]

 Which of the proteins in Figure 4.10 consists of two domains?

FIGURE 4.12

Hydrophobic and hydrophilic residues in myoglobin. Hydrophobic side chains belonging to Ala, Ile, Leu, Met, Phe, Pro, Trp, and Val (green) are located mostly in the protein interior, and polar and charged side chains (purple) are located primarily on the protein surface. Backbone atoms are gray.

Folded   Unfolded

Unfavorable solvation

FIGURE 4.13

The hydrophobic effect in protein folding. In a folded protein, hydrophobic regions (represented by blue segments of the polypeptide chain) are sequestered in the protein interior. Unfolding the protein exposes these segments to water. This arrangement is energetically unfavorable, because the presence of the hydrophobic groups interrupts the hydrogen-bonded network of water molecules.

amino acid, or about 40 kJ · mol^{-1} for a 100-residue polypeptide. This is equivalent to the amount of free energy required to break two hydrogen bonds (about 20 kJ · mol^{-1} each). This quantity seems incredibly small, considering the number of potential interactions among all of a protein's backbone and side-chain atoms. Nevertheless, most proteins do fold into a unique and stable three-dimensional arrangement of atoms.

The largest force governing protein structure is the hydrophobic effect (introduced in Section 2.2), which causes nonpolar groups to aggregate in order to minimize their contact with water. This partitioning is not due to any strong attractive force between hydrophobic groups. Rather, the hydrophobic effect is driven by the increase in entropy of the solvent water molecules, which would otherwise have to order themselves around each hydrophobic group. As we have seen, hydrophobic side chains are located predominantly in the interior of a protein. This arrangement stabilizes the folded polypeptide backbone, since unfolding it or extending it would expose the hydrophobic side chains to the solvent (**Fig. 4.13**). In the core of the protein, the hydrophobic groups experience only very weak attractive forces (van der Waals interactions); they stay in place mainly due to the hydrophobic effect.

Hydrogen bonding by itself is not a major determinant of protein stability, because in an unfolded protein, polar groups could just as easily form energetically equivalent hydrogen bonds with water molecules. Instead, hydrogen bonding—such as occurs in the formation of α helices or β sheets—may help the protein fine-tune a folded conformation that is already largely stabilized by the hydrophobic effect.

OTHER INTERACTIONS HELP STABILIZE PROTEINS

Once folded, many polypeptides appear to maintain their shapes through various other types of interactions, the most common being ion pairs, disulfide bonds, and interactions with zinc ions.

An ion pair can form between oppositely charged side chains or the N- and C-terminal groups of a polypeptide (**Fig. 4.14**). Although the resulting electrostatic interaction is strong,

Asp

Lys

Glu

Arg

(a)   (b)

FIGURE 4.14

Examples of ion pairs in myoglobin. (a) The ε-amino group of Lys 77 interacts with the carboxylate group of Glu 18. (b) The carboxylate group of Asp 60 interacts with Arg 45. The atoms are color-coded: C gray, N blue, and O red. Note that these intramolecular interactions occur between side chains that are near each other in the protein's tertiary structure but are far apart in the primary structure.

? **Identify residues that could form two other types of ion pairs.**

it does not contribute much to protein stability. This is because the favorable free energy of the electrostatic interaction is offset by the loss of entropy when the side chains become fixed in the ion pair. For a buried ion pair, there is the additional energetic cost of desolvating the charged groups in order for them to enter the hydrophobic core.

Disulfide bonds (a type of covalent bond, shown in Section 4.1) can form within and between polypeptide chains. Experiments show that even when the cysteine residues of certain proteins are chemically blocked, the proteins may still fold and function normally. This suggests that disulfide bonds are not essential for stabilizing these proteins. In fact, disulfides are rare in intracellular proteins, since the cytoplasm is a reducing environment. They are more plentiful in proteins that are secreted to an extracellular (oxidizing) environment (**Fig. 4.15**). Here, the bonds may help prevent protein unfolding under relatively harsh extracellular conditions.

Domains containing cross-links called **zinc fingers** are common in DNA-binding proteins. These structures consist of 20–60 residues with one or two Zn^{2+} ions. The Zn^{2+} ions are tetrahedrally coordinated by the side chains of cysteine and/or histidine and sometimes aspartate or glutamate (**Fig. 4.16**). Protein domains this size are too small to assume a stable tertiary structure without a metal ion cross-link. Zinc is an ideal ion for stabilizing proteins: It can interact with ligands (S, N, or O) provided by several amino acids, and it has only one oxidation state (unlike Cu or Fe ions, which readily undergo oxidation–reduction reactions under cellular conditions).

PROTEIN FOLDING BEGINS WITH THE FORMATION OF SECONDARY STRUCTURES

The crowded nature of the cell interior (see Fig. 2.6) demands that proteins and other macromolecules assume compact shapes. In the cell, a newly synthesized polypeptide begins to fold as soon as it emerges from the ribosome, so part of the polypeptide may adopt its mature tertiary structure before the entire chain has been synthesized. It is difficult to monitor this process in the cell, so studies of protein folding *in vitro* usually use full-length polypeptides that have been chemically unfolded (**denatured**) and then allowed to refold (**renature**). In the laboratory, proteins can be denatured by adding highly soluble substances such as salts or urea (NH_2—CO—NH_2). Large amounts of these solutes interfere with the structure of the solvent water, thereby attenuating the hydrophobic effect and causing the protein to unfold. When the solutes are removed, the proteins renature.

Protein renaturation experiments demonstrate that *protein folding is not a random process.* That is, the protein does not just happen upon its most stable tertiary structure (its **native structure**) by trial and error but approaches this conformation in a more coherent manner. A newly made polypeptide chain has high energy and high entropy and approaches a low-energy, low-entropy state as it folds. The progress of folding can be diagrammed as a funnel-shaped diagram where the wide mouth of the funnel represents high energy and high entropy and the

FIGURE 4.15

Disulfide bonds in lysozyme, an extracellular protein. This 129-residue enzyme from hen egg white contains eight Cys residues (yellow), which form four disulfide bonds that link different sites on the polypeptide backbone. [*Structure (pdb 1E8L) determined by H. Schwalbe, S. B. Grimshaw, A. Spencer, M. Buck, J. Boyd, C. M. Dobson, C. Redfield, and L. J. Smith.*]

(a)   (b)

FIGURE 4.16

Zinc fingers. (a) A zinc finger with one Zn^{2+} (purple sphere) coordinated by two Cys and two His residues, from *Xenopus* transcription factor IIIA. (b) A zinc finger with two Zn^{2+} coordinated by six Cys residues, from the yeast transcription factor GAL4. [*Structure of transcription factor IIIA (pdb 1TF6) determined by R. T. Nolte, R. M. Conlin, S. C. Harrison, and R. S. Brown; structure of GAL4 (pdb 1D66) determined by R. Marmorstein and S. Harrison.*]

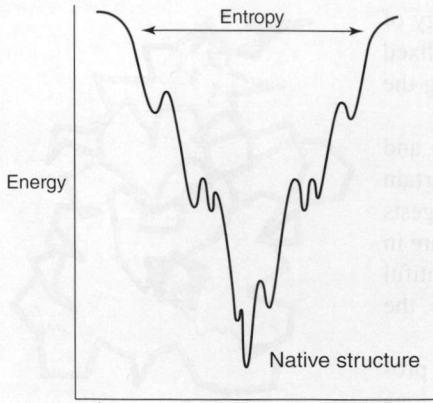

A protein folding funnel. An unfolded polypeptide has high energy and high entropy, but in its native conformation has low energy and low entropy.

narrow end respresents low energy and low entropy (**Fig. 4.17**). The folding process is not necessarily smooth and linear, since a partially folded polypeptide has many possible conformations in the folding "landscape" and sometimes gets stuck temporarily in a "valley" before reaching its most stable conformation.

Contrary to expectations, secondary structures such as α helices and β sheets do not form immediately. Instead, the first stages of protein folding can be described as a "hydrophobic collapse," when nopolar groups are forced out of contact with water. Once the hydrophobic core of the protein begins to take shape, the hydrogen bonding responsible for regular secondary structure can occur, and other structures and side chains jostle themselves into their final positions in the tertiary structure.

In the laboratory, certain small proteins can be repeatedly denatured and renatured, but in the cell, protein folding is more complicated and may require the assistance of other proteins. Some of these are known as **molecular chaperones** and are described in more detail in Section 22.4.

The pathway to full functionality may require additional steps beyond polypeptide folding. Some proteins contain several polypeptide chains, which must fold individually before assembling. In addition, many proteins undergo **posttranslational processing** before reaching their mature forms. Depending on the protein, this might mean removal of some amino acid residues or the covalent attachment of another group such as a lipid, carbohydrate, or phosphate group (**Fig. 4.18**). The attached groups usually have a discrete biological function and may also help stabilize the folded conformation of the protein. Finally, some proteins become functional only after associating with metal ions or a specific organic molecule.

SOME PROTEINS HAVE MORE THAN ONE CONFORMATION

All the information required for a protein to fold is contained in its amino acid sequence. Unfortunately, there are no completely reliable methods for predicting how a polypeptide chain will fold. In fact, it is difficult to determine whether a given amino acid sequence will form an α helix, β sheet, or irregular secondary structure. This presents a formidable obstacle to assigning three-dimensional structures—and possible functions—to the burgeoning number of proteins identified through genome sequencing (see Section 3.4).

In some cases, a protein can adopt more than one conformation. Nicknamed "transformer proteins," these molecules have two stable conformations that exist in dynamic equilibrium. A change in cellular conditions, such as pH or oxidation state, or the presence of a binding partner can tip the balance toward one conformation or the other (**Fig. 4.19**). Because structural biologists have traditionally examined proteins with fixed, stable shapes, the number of proteins known to have conformational flexibility is likely to increase.

Even more intriguingly, many proteins appear to include polypeptide segments that are relatively extended and have no fixed conformation at all (**Fig. 4.20**). Calculations suggest that up to 40% of the proteins in eukaryotes contain such segments, which comprise at least

(a)   (b)   (c)

$$H_3C-(CH_2)_{14}-\overset{\overset{\displaystyle O}{\|}}{C}-S-CH_2-$$

$$\cdots -O \quad NH-\overset{\overset{\displaystyle O}{\|}}{C}-CH_2-$$

$$-O-\overset{\overset{\displaystyle O^-}{|}}{\underset{\underset{\displaystyle O}{\|}}{P}}-O-CH_2-$$

Some covalent modifications of proteins. (a) A 16-carbon fatty acid (palmitate, in red) is linked by a thioester bond to a Cys residue. (b) A chain of several carbohydrate units (here only one sugar residue is shown, in red) is linked to the amide N of an Asn side chain. (c) A phosphoryl group (red) is esterified to a Ser side chain.

30 residues and tend to be rich in hydrophilic side chains. A few proteins may even be completely disordered from end to end. These **intrinsically unstructured proteins** or protein segments are particularly well suited for regulatory roles, in which one protein must interact with several others. The lack of a fixed structure makes it easier for a protein to inspect multiple potential binding partners and then, as it latches onto one, to adopt a defined tertiary structure that maximizes the strength of that interaction.

Of course, all proteins are inherently flexible, since they are built from polypeptide chains with some degree of rotational freedom and are stabilized by relatively weak noncovalent forces. Rather than existing as rigid blocks, proteins are surprisingly dynamic, undergoing minor movements as individual bonds rotate and stretch. In addition to these small random motions, other structural changes are necessary for many proteins to carry out their biological functions. For example, in order to catalyze a chemical reaction, an enzyme must closely contact the reacting substances while they are being transformed to reaction products. When an extracellular signaling molecule binds to a cell-surface receptor protein, the protein's conformation changes in a way that triggers additional events inside the cell. Although many dramatic protein conformational changes have been documented, most movements are more subtle adjustments that involve a few side chains or a loop of polypeptide, reposition an α helix or β strand, or allow one protein domain to tilt relative to another.

(a)   (b)

FIGURE 4.19

A protein with two stable conformations. One form of the protein lymphotactin consists of a three-stranded β sheet and an α helix (a). This form interconverts rapidly with an alternate form (b) with an all-β structure. [*Structures (a, pdb 1J9O and b, pdb 2JP1) determined by B. F. Volkman.*]

B-chain

Catalytic

Auto-inhibitory

CaM binding   Regulatory

FIGURE 4.20

An intrinsically unstructured protein. The structure of calcineurin includes domains with well-defined tertiary structure (orange and yellow) as well as a 95-residue disordered segment (blue and gold). [*Courtesy of David Weis, University of Kansas.*]

 CONCEPT CHECK

1. Explain why a folded polypeptide assumes a shape with a hydrophilic surface and a hydrophobic core.

2. Name some residues that are likely to be located on the protein surface and some residues that are likely to be located in its core.

3. List the covalent and noncovalent forces that can stabilize protein structures.

4. How does an unfolded polypeptide reach its native conformation?

5. Explain why proteins are not rigid.

QUATERNARY STRUCTURE

Most small proteins consist of a single polypeptide chain, but most proteins with masses greater than 100 kD contain multiple chains. The individual chains, called **subunits,** may all be identical, in which case the protein is known as a **homodimer, homotrimer, homotetramer,** and so on (*homo-* means "the same"). If the chains are not all identical, the prefix *hetero-* is used. The spatial arrangement of these polypeptides is known as the protein's quaternary structure.

The forces that hold subunits together are similar to those that determine the tertiary structures of the individual polypeptides. That is, the hydrophobic effect is mostly responsible for maintaining quaternary structure. Accordingly, the interface (the area of contact) between two subunits is mostly nonpolar, with closely packed side chains. Hydrogen bonds, ion pairs, and disulfide bonds contribute to a lesser extent but help dictate the exact geometry of the interacting subunits.

Among the most common quaternary structures in proteins are symmetrical arrangements of two or more identical subunits (**Fig. 4.21**). Even in proteins with nonidentical subunits, the symmetry is based on groups of subunits. For example, hemoglobin, a heterotetramer with two αβ units, can be considered to be a dimer of dimers (see Fig. 4.21c). A few multimeric proteins are known to adopt more than one possible quaternary structure and apparently shift among the alternatives by dissociating, adopting different tertiary structures, and reassembling.

The advantages of multisubunit protein structure are many. For starters, the cell can construct extremely large proteins through the incremental addition of small building blocks that are encoded by a single gene (we will see examples of this in the next chapter). This is clearly an asset for certain structural proteins that—due to their enormous size—cannot be synthesized all in one piece or must be assembled outside the cell. Moreover, the impact of the inevitable errors in transcription and translation can be minimized if the affected polypeptide is small and readily replaced.

Finally, the interaction between subunits in a multisubunit protein affords an opportunity for the subunits to influence each other's behavior or work cooperatively. The result is a way

FIGURE 4.21

Some proteins with quaternary structure. The alpha carbon backbone of each polypeptide is shown. (a) Nitrite reductase, an enzyme with three identical subunits, from *Alcaligenes.* [*Structure (pdb 1AS8) determined by M. E. P. Murphy, E. T. Adams, and S. Turley.*] (b) *E. coli* fumarase, a homotetrameric enzyme. [*Structure (pdb 1FUQ) determined by T. Weaver and L. Banaszak.*] (c) Human hemoglobin, a heterotetramer with two α subunits (blue) and two β subunits (red). [*Structure (pdb 2HHB) determined by G. Fermi and M. F. Perutz.*] (d) Bacterial methane hydroxylase, whose two halves (*right* and *left* in this image) each contain three kinds of subunits. [*Structure (pdb 1MMO) determined by A. C. Rosenzweig, C. A. Frederick, S. J. Lippard, and P. Nordlund.*]

(a)   (b)

(c)   (d)

? Which of the proteins shown in Fig. 4.1 has quaternary structure?

of regulating function that is not possible in single-subunit proteins or in multisubunit proteins whose subunits each operate independently. In Chapter 5, we will examine the cooperative behavior of hemoglobin, which has four interacting oxygen-binding sites.

CONCEPT CHECK

1. Explain how to tell whether a protein has quaternary structure.
2. What are the advantages of quaternary structure?

CLINICAL CONNECTION: PROTEIN MISFOLDING AND DISEASE

In a cell, a protein may occasionally fail to assume its proper tertiary or quaternary structure. Sometimes this is due to a genetic mutation, such as the substitution of one amino acid for another, that makes it difficult or impossible for the protein to fold. Regardless of the cause, the result may be disastrous for the cell. In fact, a variety of human diseases have been linked to faulty protein folding, more specifically, to the cell's failure to deal with the misfolded protein.

Normally, the chaperones that help new proteins to fold can also help misfolded proteins to refold. If the protein cannot be salvaged in this way, it is usually degraded to its component amino acids (**Fig. 4.22**). The operation of this quality control system could explain why some mutated proteins, which fold incorrectly, never reach their intended cellular destinations.

Other diseases result when misfolded proteins are not immediately refolded or degraded but instead aggregate, often as long insoluble fibers. Although the fibers can occur throughout the body, they appear to be deadliest when they occur in the brain. Among the human diseases characterized by such fibrous deposits are Alzheimer's disease, Parkinson's disease, and the transmissible spongiform encephalopathies, which lead to neurological abnormalities and loss of neurons (nerve cells). The aggregated proteins, a different type in each disease, are commonly called **amyloid deposits** (a name originally referring to their starchlike appearance).

Alzheimer's disease, the most common neurodegenerative disease, is accompanied by both intracellular "tangles" and extracellular "plaques" (**Fig. 4.23**). The fibrous material inside cells is made of a protein called tau, which is involved in the assembly of microtubules, a component of the cytoskeleton (Section 5.3). Tau deposits also appear in some other neurodegenerative diseases, and tau's role in Alzheimer's disease is not clear. The extracellular amyloid material in Alzheimer's disease consists primarily of a protein called amyloid-β, a 40- or 42-residue fragment of the much larger amyloid precursor protein, which is a membrane protein. Enzymes called β-secretase and γ-secretase cleave the precursor protein to generate amyloid-β. Normal brain tissue contains some extracellular amyloid-β, but neither its function

KEY CONCEPT
- Misfolded proteins accumulate as amyloid fibers and cause disease.

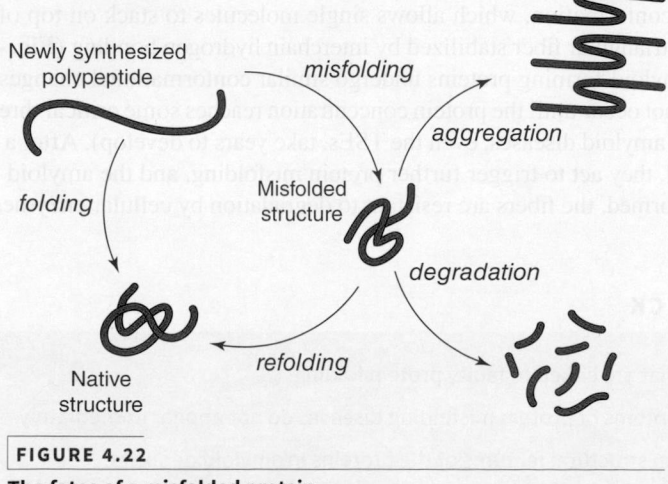

FIGURE 4.22

The fates of a misfolded protein.

FIGURE 4.23

Brain tissue from Alzheimer's disease. Amyloid deposits (large red areas) are surrounded by intracellular tangles (smaller dark shapes). [*Courtesy of Dennis Selkoe and Marcia Podlisny, Harvard University Medical School.*]

FIGURE 4.24

Model for amyloid formation. Five aggregated polypeptides are shown in different colors. The vertical arrow indicates the long axis of the amyloid fiber. [*From Meier, B., et al., Science,* **319,** *1523–1526 (2008). Reprinted with permission from AAAS.*]

nor the function of its precursor protein is completely understood. However, excess amyloid-β is clearly linked to Alzheimer's disease.

The neurodegeneration of Alzheimer's disease may begin many years before symptoms such as memory loss appear, and studies of the disease in animal models indicate that both of these signs can be detected before amyloid fibers accumulate in the brain. These and other experimental results support the hypothesis that the early stages of amyloid-β misfolding and aggregation are toxic to neurons and are the ultimate cause of Alzheimer's disease. The accumulation of extracellular fibers may actually be a protective mechanism to deal with excessive amyloid-β production.

In Parkinson's disease, neurons in a portion of the brain accumulate fragments of a protein known as α-synuclein. Like amyloid-β, α-synuclein's function is not fully known, but it appears to play a role in neurotransmission. α-Synuclein is a small soluble protein (140 amino acid residues) with an extended conformation, part of which appears to form α helices upon binding to other molecules. The intrinsic disorder of the protein may contribute to its propensity to form amyloid deposits, which are characterized by a high content of β secondary structure. Accumulation of this material is associated with the death of neurons, leading to the typical symptoms of Parkinson's disease, which include tremor, muscular rigidity, and slow movements. Mutations in the gene for α-synuclein that lead to increased expression of the protein or promote its self-aggregation are responsible for some hereditary forms of Parkinson's disease.

Mad cow disease is the best known of the **transmissible spongiform encephalopathies (TSEs),** a group of disorders that also includes scrapie in sheep and Creutzfeldt–Jakob disease in humans. These fatal diseases, which give the brain a spongy appearance, were once thought to be caused by a virus. However, extensive investigation has revealed that the infectious agent is a protein called a **prion.** Interestingly, normal human brain tissue contains the same protein, named PrP^C (C for cellular), which occurs on neural cell membranes and appears to play a role in normal brain function. The scrapie form of the prion protein, PrP^{Sc}, has the same 253–amino acid sequence as PrP^C but includes more β secondary structure. Introduction of PrP^{Sc} into cells apparently triggers PrP^C, which contains more α-helical structure, to assume the PrP^{Sc} conformation and thereby aggregate with it.

How is a prion disease transmitted? One route seems to be by ingestion, as illustrated by the incidence of bovine TSE in cows consuming feed prepared from scrapie-infected sheep. (This feeding practice has been discontinued, but not before a number of people—who presumably consumed the infected beef—developed a variant form of Creutzfeldt–Jakob disease.) PrP^{Sc} must be absorbed, without being digested, and transported to the central nervous system. Here, it causes neurodegeneration, possibly through the loss of PrP^C as it converts to PrP^{Sc} or through the toxic effects of the PrP^{Sc} as it accumulates.

Despite the lack of sequence or structural similarities among amyloid-β, α-synuclein, and PrP^{Sc}, their misfolded forms are all rich in β structure, and this seems to be the key to the formation of amyloid deposits. Studies of a fungal polypeptide similar to PrP^{Sc} show how amyloid formation might occur. In its original state, the protein is mostly α-helical. A segment of polypeptide shifts to an all-β conformation, which allows single molecules to stack on top of each other in parallel to form a triangular fiber stabilized by interchain hydrogen bonding (**Fig. 4.24**).

Presumably, other amyloid-forming proteins undergo similar conformational changes. Significant aggregation may not occur until the protein concentration reaches some critical threshold (this may explain why the amyloid diseases, even the TSEs, take years to develop). After a few β structures have assembled, they act to trigger further protein misfolding, and the amyloid fibers rapidly propagate. Once formed, the fibers are resistant to degradation by cellular enzymes.

 CONCEPT CHECK

1. List some diseases that are linked to faulty protein folding.

2. Explain why the symptoms of protein misfolding diseases do not appear immediately.

3. Describe the common structural features of the proteins in amyloid deposits.

TOOLS AND TECHNIQUES: ANALYZING PROTEIN STRUCTURE

Like nucleic acids (Section 3.5), proteins can be purified and analyzed in the laboratory. In this section, we will examine some commonly used approaches to isolating proteins and determining their sequence of amino acids and their three-dimensional structures.

CHROMATOGRAPHY TAKES ADVANTAGE OF A POLYPEPTIDE'S UNIQUE PROPERTIES

As described in Section 4.1, a protein's amino acid sequence determines its overall chemical characteristics, including its size, shape, charge, and ability to interact with other substances. A variety of laboratory techniques have been devised to exploit these features in order to separate proteins from other cellular components. Keep in mind that the exact charge and conformation of individual molecules vary slightly in a population of otherwise identical molecules, because acid–base groups exist in equilibrium between protonated and unprotonated forms and bonded atoms have some rotational freedom. Consequently, laboratory techniques assess the average chemical and physical properties of the population of molecules.

One of the most powerful techniques is **chromatography.** Originally performed with solvents moving across paper, chromatography now typically uses a column packed with a porous matrix (the stationary phase) and a buffered solution (the mobile phase) that percolates through the column. *Proteins or other solutes pass through the column at different rates, depending on how they interact with the stationary phase.*

In **size-exclusion chromatography** (also called gel filtration chromatography), the stationary phase consists of tiny beads with pores of a characteristic size. If a solution containing proteins of different sizes is applied to the top of the column, the proteins will move through the column as fluid drips out the bottom. Larger proteins will be excluded from the spaces inside the beads and will pass through the column faster than smaller proteins, which will spend time inside the beads. The proteins gradually become separated and can be recovered by collecting the solution that exits the column (**Fig. 4.25**).

A protein's net charge at a particular pH can be exploited for its purification by **ion exchange chromatography.** In this technique, the solid phase typically consists of beads derivatized

KEY CONCEPTS
- Chromatography is a technique for separating molecules on the basis of size, charge, or specific binding behavior.
- The sequence of amino acids in a polypeptide can be determined by mass spectrometry.
- The three-dimensional arrangement of atoms in a protein can be deduced by measuring the diffraction of X-rays or electrons or by analyzing nuclear magnetic resonance.

(a)   (b)

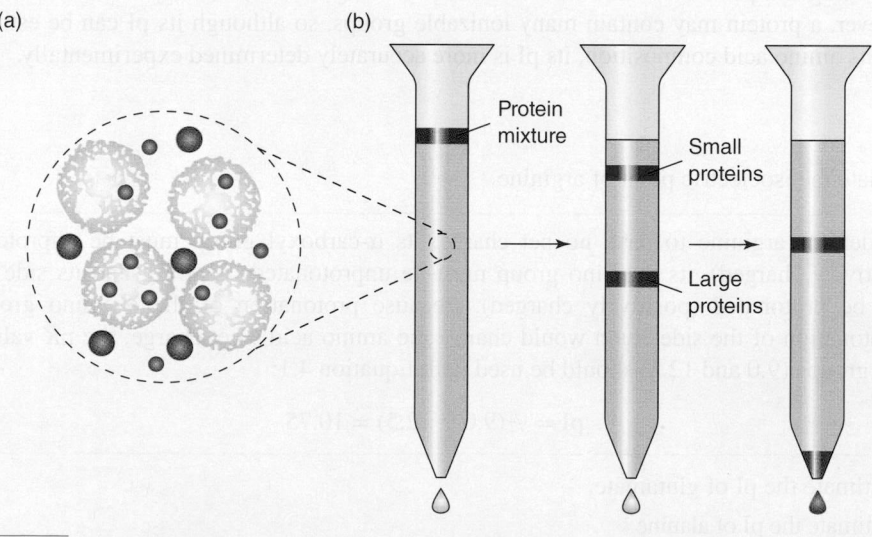

Protein mixture

Small proteins

Large proteins

FIGURE 4.25

Size-exclusion chromatography. (a) Small molecules (blue) can enter the spaces inside the porous beads of the stationary phase, while larger molecules (gold) are excluded. (b) When a mixture of proteins (green) is applied to the top of a size-exclusion column, the large proteins (gold) migrate more quickly than small proteins (blue) through the column and are recovered by collecting the material that flows out the bottom of the column. In this way, a mixture of proteins can be separated according to size.

Ion exchange chromatography.
When a mixture of proteins is applied to the top of a positively charged anion exchange column (e.g., a DEAE matrix), negatively charged proteins bind to the matrix, while uncharged and cationic proteins flow through the column. The desired protein can be dislodged by applying a high-salt solution (whose anions compete with the protein for binding to the DEAE groups).

with positively charged diethylaminoethyl (DEAE) groups or negatively charged carboxymethyl (CM) groups:

$$\text{DEAE} \quad \bullet\!-\!CH_2\!-\!CH_2\!-\!NH^+\!\!\begin{array}{c} CH_2\!-\!CH_3 \\ CH_2\!-\!CH_3 \end{array} \qquad\qquad \text{CM} \quad \bullet\!-\!CH_2\!-\!COO^-$$

Negatively charged proteins will bind tightly to the DEAE groups, while uncharged and positively charged proteins pass through the column. The bound proteins can then be dislodged by passing a high-salt solution through the column so that the dissolved ions can compete with the protein molecules for binding to DEAE groups (**Fig. 4.26**). Alternatively, the pH of the solvent can be decreased so that the bound protein's anionic groups become protonated, loosening their hold on the DEAE matrix. Similarly, positively charged proteins will bind to CM groups (while uncharged and anionic proteins flow through the column) and can subsequently be dislodged by solutions with a higher salt concentration or a higher pH.

The success of ion exchange can be enhanced by knowing something about the protein's net charge (see Sample Calculation 4.1) or its **isoelectric point, pI,** the pH at which it carries no net charge. For a molecule with two ionizable groups, the pI lies between the pK values of those two groups:

$$pI = \tfrac{1}{2}(pK_1 + pK_2) \tag{4.1}$$

Calculating the pI of an amino acid is relatively straightforward (Sample Calculation 4.2). However, a protein may contain many ionizable groups, so although its pI can be estimated from its amino acid composition, its pI is more accurately determined experimentally.

SAMPLE CALCULATION 4.2

PROBLEM   Estimate the isoelectric point of arginine.

SOLUTION   In order for arginine to have no net charge, its α-carboxyl group must be unprotonated (negatively charged), its α-amino group must be unprotonated (neutral), and its side chain must be protonated (positively charged). Because protonation of the α-amino group or deprotonation of the side chain would change the amino acid's net charge, the pK values of these groups (9.0 and 12.5) should be used with Equation 4.1:

$$pI = \tfrac{1}{2}(9.0 + 12.5) = 10.75$$

PRACTICE PROBLEMS   4. Estimate the pI of glutamate.

5. Estimate the pI of alanine.

Other binding behaviors can be adapted for chromatographic separations. For example, a small molecule can be immobilized on the chromatographic matrix, and proteins that can specifically bind to that molecule will stick to the column while other substances exit the column without binding. This technique, called **affinity chromatography,** is a particularly powerful separation method because it takes advantage of a protein's unique ability to interact with another molecule, rather than one of its general features such as size or charge. **High-performance liquid chromatography (HPLC)** is the name given to chromatographic separations, often analytical in nature rather than preparative, that are carried out in closed columns under high pressure, with precisely controlled flow rates and automated sample application.

Proteins are sometimes analyzed or isolated by electrophoresis, in which molecules move through a gel-like matrix such as polyacrylamide under the influence of an electric field (Section 3.5). In sodium dodecyl sulfate polyacrylamide gel electrophoresis **(SDS-PAGE),** both the sample and the gel contain the detergent SDS, which binds to proteins to give them a uniform density of negative charges. When the electric field is applied, the proteins all move toward the positive electrode at a rate depending on their size, with smaller proteins migrating faster than larger ones. After staining, the proteins are visible as bands in the gel (**Fig. 4.27**).

FIGURE 4.27

SDS-PAGE. Following electrophoresis, the gel was stained with Coomassie blue dye so that each protein is visible as a blue band. Lanes 1 and 8 contain proteins of known molecular mass that serve as standards for estimating the masses of other proteins. [*Courtesy of Bio-Rad Laboratories, ©2012.*]

MASS SPECTROMETRY REVEALS AMINO ACID SEQUENCES

A standard approach to sequencing a protein has several steps:

1. *The sample of protein to be sequenced is purified* (for example, by chromatographic or other methods) so that it is free of other proteins.

2. If the protein contains more than one kind of polypeptide chain, *the chains are separated* so that each can be individually sequenced. In some cases, this requires breaking (reducing) disulfide bonds.

3. *Large polypeptides must be broken into smaller pieces (<100 residues) that can be individually sequenced.* Cleavage can be accomplished chemically, for example, by treating the polypeptide with cyanogen bromide (CNBr), which cleaves the peptide bond on the C-terminal side of Met residues. Cleavage can also be accomplished with **proteases** (another term for peptidases) that hydrolyze specific peptide bonds. For example, the protease trypsin cleaves the peptide bond on the C-terminal side of the positively charged residues Arg and Lys:

$$
\begin{array}{c}
\text{NH}_3^+ \\
| \\
\text{CH}_2 \\
| \\
\text{CH}_2 \\
| \\
\text{CH}_2 \\
| \\
\text{CH}_2 \quad \text{O} \quad\quad \text{R} \quad\; \text{O} \\
| \quad\quad \| \quad\quad | \quad\; \| \\
\cdots\text{—NH—CH—C—NH—CH—C—}\cdots
\end{array}
\xrightarrow[\text{trypsin}]{\text{H}_2\text{O}}
\begin{array}{c}
\text{NH}_3^+ \\
| \\
\text{CH}_2 \\
| \\
\text{CH}_2 \\
| \\
\text{CH}_2 \\
| \\
\text{CH}_2 \quad \text{O} \quad\quad\quad\quad\quad \text{R} \quad\; \text{O} \\
| \quad\quad \| \quad\quad\quad\quad\quad | \quad\; \| \\
\cdots\text{—NH—CH—C—O}^- + \text{H}_3\overset{+}{\text{N}}\text{—CH—C—}\cdots
\end{array}
$$

Lys (or Arg)

Some commonly used proteases and their preferred cleavage sites are listed in **Table 4.4.**

4. The sequence of each peptide is determined. In a classic procedure known as **Edman degradation,** the N-terminal residue of a peptide is chemically derivatized, cleaved off, and identified; the process is then repeated over and over so that the peptide's sequence can be deduced, one residue at a time. Alternatively, each peptide can be sequenced by mass spectrometry.

5. To reconstruct the sequence of the intact polypeptide, *a different set of fragments that overlaps the first is generated* so that the two sets of sequenced fragments can be aligned.

TABLE 4.4 Specificities of Some Proteases

PROTEASE	RESIDUE PRECEDING CLEAVED PEPTIDE BOND[a]
Chymotrypsin	Leu, Met, Phe, Trp, Tyr
Elastase	Ala, Gly, Ser, Val
Thermolysin	Ile, Met, Phe, Trp, Tyr, Val
Trypsin	Arg, Lys

[a]Cleavage does not occur if the following residue is Pro.

Set 1
(cleaved with
trypsin)
Val—Leu—Lys   Ser—Phe—Gly—Arg   Tyr—Ala—Gln—Thr

Set 2
(cleaved with
chymotrypsin)
Val—Leu—Lys—Ser—Phe   Gly—Arg—Tyr   Ala—Gln—Thr

A more efficient approach for analyzing protein structure measures the sizes of peptide fragments by **mass spectrometry.** In standard mass spectrometry, a solution of the protein is sprayed from a tiny nozzle at high voltage. This yields droplets of positively charged molecular fragments, from which the solvent quickly evaporates. Each gas-phase ion then passes through an electric field. The ions are deflected, with smaller ions deflected more than larger ions, so that they are separated by their mass-to-charge ratios. In this way, the masses of the fragments can be measured and the mass of the intact molecule can be deduced.

Two mass spectrometers in series can be used to determine the sequence of amino acids in a polypeptide. The first instrument sorts the peptide ions so that only one emerges. This species is then allowed to collide with an inert gas, such as helium, which breaks the peptide, usually at a peptide bond. The second mass spectrometer then measures the mass-to-charge ratios of the peptide fragments (**Fig. 4.28**). *Because successively larger fragments all bear the same charge but differ by one amino acid, and because the mass of each of the 20 amino acids is known, the sequence of amino acids in a set of fragments can be deduced.* Although mass spectrometry is not practical for sequencing large polypeptides, even a partial sequence may be valuable (**Box 4.C**). Mass spectrometry can also reveal additional details of protein structure, such as covalently modified amino acid residues.

PROTEIN STRUCTURES ARE DETERMINED BY X-RAY CRYSTALLOGRAPHY, ELECTRON CRYSTALLOGRAPHY, AND NMR SPECTROSCOPY

Most proteins are too small to be directly visualized, even by electron microscopy, but their atomic structures are accessible to high-energy probes in the form of X-rays. **X-Ray crystallography** is performed on samples of protein that have been induced to form crystals.

A CLOSER LOOK

Box 4.C Mass Spectrometry Applications

Mass spectrometry has been used for several decades in clinical and forensics laboratories to identify normal metabolic compounds as well as toxins and drugs (both therapeutic and illicit). Instruments that detect traces of explosives at airport security checkpoints also use mass spectrometry, which is rapid, sensitive, and reliable. The analysis of small compounds by mass spectrometry is relatively straightforward. It is much more challenging, however, to analyze large molecules in complex mixtures for the purpose of diagnosing a disease such as cancer or tracking its effects on human tissues.

Fluids such as blood contain so many different proteins, with concentrations ranging over many orders of magnitude, that it is difficult to detect rare proteins against a backdrop of very abundant ones such as serum albumin and immunoglobulins, which together account for about 75% of the plasma proteins. Urinalysis by mass spectrometry holds more promise, as urine normally contains relatively few proteins and none with a mass greater than about 15,000 D. Even so, over 2000 different proteins are detectable in urine.

One common approach for identifying proteins in biological samples is to fractionate the mixture by electrophoresis, then extract the separated proteins from the gel, partially digest them with a protease, and subject them to mass spectrometry. The resulting pattern of peaks, a "fingerprint," can be compared to a database to identify the protein. Even when tandem mass spectrometers are used, it may not be necessary to completely sequence each polypeptide. A partial sequence of just a few amino acids is often sufficient to identify the parent protein. Of course, the availability of complete genomic sequences makes this approach possible, and a number of software programs have been developed to translate mass spectral data into query sequences for searching sequence databases.

? **Explain why it is possible that a 5-residue sequence can uniquely identify a protein whose total length is 200 residues. (Hint: compare the total number of possible pentapeptide sequences to the actual number of pentapeptide fragments in the polypeptide.)**

Protein

Ionization

MS1

Isolation

Selected peptide

Collision chamber

Fragmentation

MS2

Mass analysis

(a)

(b)

FIGURE 4.28

Peptide sequencing by mass spectrometry. (a) A solution of charged peptides is sprayed into the first mass spectrometer (MS1). One peptide ion is selected to enter the collision chamber to be fragmented. The second mass spectrometer (MS2) then measures the mass-to-charge ratio of the ionic peptide fragments. The peptide sequence is determined by comparing the masses of increasingly larger fragments. (b) An example of peptide sequencing by mass spectrometry. The difference in mass of each successive peak identifies each residue, allowing the amino acid sequence to be read from right to left. [*From Keough, T., et al., Proc. Nat. Acad. Sci.* **96,** *7131–7136 (1999). Reprinted with permission of PNAS.*]

A protein preparation must be exceptionally pure in order to crystallize without imperfections. A protein crystal, often no more than 0.5 mm in diameter, usually contains 40–70% water by volume and is therefore more gel-like than solid (**Fig. 4.29**).

When bombarded with a narrow beam of X-rays, the electrons of the atoms in the crystal scatter the X-rays, which reinforce and cancel each other to produce a **diffraction pattern** of light and dark spots that can be captured electronically or on a piece of X-ray film (**Fig. 4.30**).

Mathematical analysis of the intensities and positions of the diffracted X-rays yields a three-dimensional map of electron density in the crystallized molecule. The level of detail of the image depends in part on the quality of the crystal. Slight conformational variations among the crystallized protein molecules often limit resolution to about 2 Å. However, this is usually sufficient to trace the polypeptide backbone and discern the general shapes of the side chains (**Fig. 4.31**). When Kendrew determined the X-ray structure of myoglobin, its amino acid sequence was not known. Today, protein crystallographers usually take advantage of amino acid sequence information to simplify the task of elucidating protein structures.

Are X-ray structures accurate? One might expect a crystallized molecule to be utterly immobile, quite different from a protein in solution, which undergoes bending and stretching movements among its many bonds. But, in fact, crystallized proteins retain some of their ability to move. They can sometimes bind small molecules that diffuse into the crystal (which is

FIGURE 4.29

Crystals of the protein streptavidin. [*Courtesy I. Le Trong and R. E. Stenkamp, University of Washington.*]

5.0 Å 3.0 Å 1.5 Å

FIGURE 4.31

Protein structure at different resolutions. In this example, the green areas represent electron density, and the protein structure is superimposed in black. As the resolution improves, it becomes easier to trace the pattern of the protein backbone (a portion of an α helix). [*Based on a drawing by Wayne Hendrickson.*]

FIGURE 4.32

The NMR structure of glutaredoxin. Twenty structures for this 85-residue protein, all compatible with the NMR data, are shown as α-carbon traces, colored from the N-terminus (blue) to the C-terminus (red). [*Structure (pdb 1EGR) determined by P. Sodano, T. H. Xia, J. H. Bushweller, O. Bjornberg, A. Holmgren, M. Billeter, and K. Wuthrich.*]

about half water) and can sometimes mediate chemical reactions. These activities would not be possible if the structure of the crystallized protein did not closely approximate the structure of the protein in solution. Finally, nuclear magnetic resonance (NMR) methods (see below) for determining the structures of small proteins in solution appear to yield results consistent with X-ray crystallographic data.

Proteins that are difficult to crystallize, such as membrane proteins with extensive hydrophobic regions, can sometimes be analyzed by **electron crystallography,** a technique in which electron beams rather than X-rays probe the protein's structure. Protein samples, which do not have to be in a crystalline array, are placed in an electron microscope, and diffraction information is collected from many different angles. In this way, the three-dimensional structure of the protein can be reconstructed. Because electrons interact more strongly with atoms than X-rays do, the sample is susceptible to radiation damage. This can be minimized by rapidly freezing the sample and collecting data at the temperature of liquid nitrogen (−196°C), a method called **cryoelectron microscopy.** The structures of some macromolecular complexes, including the ribosome (Section 22.2), have been visualized using cryoelectron microscopy.

Proteins in solution can be analyzed by **nuclear magnetic resonance (NMR) spectroscopy,** which takes advantage of the ability of atomic nuclei (most commonly, hydrogen) to resonate in an applied magnetic field according to their interactions with nearby atoms. An NMR spectrum consists of numerous peaks that can be analyzed to reveal the distances between two H atoms that are close together in space or are covalently connected through one or two other atoms. These measurements, along with information about the protein's amino acid sequence, are used to construct a three-dimensional model of the protein. Due to the inherent imprecision of the measurements, NMR spectroscopy typically yields a set of closely related structures, which may convey a sense of the protein's natural conformational flexibility (**Fig. 4.32**).

 CONCEPT CHECK

1. What types of chromatography would separate proteins of different size? Of different charge?

2. Explain what happens in affinity chromatography.

3. Choose several amino acids with and without ionizable side chains and calculate their pI values.

4. Summarize the strategy used to determine the sequence of a large protein.

5. Compare the physical state of a protein to be analyzed by X-ray crystallography, electron crystallography, and NMR spectroscopy.

SUMMARY

4.1   Amino Acids, the Building Blocks of Proteins

• The 20 amino acid constituents of proteins are differentiated by the chemical properties of their side chains, which can be roughly classified as hydrophobic, polar, or charged.

• Amino acids are linked by peptide bonds to form a polypeptide.

4.2   Secondary Structure: The Conformation of the Peptide Group

• Protein secondary structure includes the α helix and β sheet, in which hydrogen bonds form between backbone carbonyl and amino groups. Irregular secondary structure has no regularly repeating conformation.

4.3   Tertiary Structure and Protein Stability

• The three-dimensional shape (tertiary structure) of a protein includes its backbone and all side chains. A protein may contain all α, all β, or a mix of α and β structures.

• A globular protein has a hydrophobic core and is stabilized primarily by the hydrophobic effect. Ion pairing, disulfide bonds, and other cross-links may also help stabilize proteins.

• A denatured protein may refold to achieve its native structure. In a cell, chaperones assist protein folding.

4.4   Quaternary Structure

• Proteins with quaternary structure have multiple subunits.

4.5   Clinical Connection: Protein Misfolding and Disease

• Proteins that do not fold properly and are not degraded may accumulate as amyloid fibers and damage cells, particularly neurons.

4.6   Tools and Techniques: Analyzing Protein Structure

• In the laboratory, proteins can be purified by chromatographic techniques that take advantage of the proteins' size, charge, and ability to bind other molecules.

• The sequence of amino acids in a polypeptide can be determined by chemically derivatizing and removing them in Edman degradation or by measuring the mass-to-charge ratio of peptide fragments in mass spectrometry.

• X-Ray crystallography, electron crystallography, and NMR spectroscopy provide information about the three-dimensional structures of proteins at the atomic level.

KEY TERMS

protein	microenvironment	zinc finger	prion
polypeptide	oligopeptide	denaturation	chromatography
amino acid	peptide	renaturation	size-exclusion chromatography
R group	primary structure	native structure	ion exchange chromatography
α-amino acid	secondary structure	molecular chaperone	pI
Cα	tertiary structure	post-translational processing	affinity chromatography
chirality	quaternary structure	intrinsically unstructured	HPLC
disulfide bond	regular secondary structure	protein	SDS-PAGE
condensation reaction	α helix	subunit	protease
peptide bond	β sheet	dimer	Edman degradation
residue	parallel β sheet	trimer	mass spectrometry
hydrolysis	antiparallel β sheet	tetramer	X-ray crystallography
exopeptidase	irregular secondary structure	homo-	diffraction pattern
endopeptidase	globular protein	hetero-	electron crystallography
N-terminus	fibrous protein	amyloid deposit	cryoelectron microscopy
C-terminus	domain	transmissible spongiform	NMR spectroscopy
backbone	ion pair	encephalopathy (TSE)	

BIOINFORMATICS

Brief Bioinformatics Exercises

4.1   Drawing and Analyzing Amino Acids and Peptides

4.2   Chemical Properties and Amino Acid Composition of Green Fluorescent Protein

4.3   Protein Properties and Purification Methods

4.4   Visualizing and Analyzing α Helices

4.5   Visualizing and Analyzing β Sheets

4.6   Visualizing and Analyzing Cytochrome *c*

Bioinformatics Projects

Visualizing Three-Dimensional Protein Structures Using the Molecular Visualization Programs JSmol and PyMOL

Structural Alignment and Protein Folding

PROBLEMS

4.1   Amino Acids, the Building Blocks of Proteins

1. Which of the 20 standard amino acids are **(a)** cyclic; **(b)** aromatic; **(c)** sometimes charged at physiological pH; **(d)** technically not hydrophobic, polar, or charged; **(e)** basic; **(f)** acidic; **(g)** sulfur-containing?

2. At what pH would an amino acid bear both a COOH and an NH_2 group? Is your answer consistent with the observation that most amino acids have high melting points ($\sim 300°C$) and are generally soluble in water?

3. Rank the solubility of the following amino acids in water at pH 6.5: His, Phe, Ser, Ile, Asp, Trp.

4. Amino acids (except for glycine) are chiral; that is, they contain a chiral carbon and thus can form mirror-image isomers, or enantiomers. Biochemists use D and L to distinguish enantiomers instead of the *RS* system used by chemists.

  (a) The structure of L-alanine is shown. Label the chiral carbon and draw the structure of D-alanine.

$$^+H_3N-\overset{\displaystyle COO^-}{\underset{\displaystyle CH_3}{\overset{|}{\underset{|}{C}}}}-H$$

L-Alanine

  (b) The vast majority of proteins in living systems consist of L-amino acids. However, D-amino acids are found in some short bacterial peptides that make up cell walls. Why might this be advantageous to the bacterium?

5. How many chiral carbons does threonine have? How many stereoisomers are possible? Draw the stereoisomers of threonine.

6. The side chains of asparagine and glutamine can undergo hydrolysis in aqueous acid. Draw the reactants and products for the deamination reaction. Name the products.

7. Draw the structure of the tripeptide His–Arg–Asp at pH 8.0. Label the following: **(a)** peptide bonds; **(b)** N-terminus; **(c)** C-terminus; **(d)** an α-amino group and an ε-amino group; **(e)** an α-carboxylate group and a γ-carboxylate group.

8. Estimate the net charge of the following peptides: **(a)** Glu–Tyr at pH 6.0; **(b)** Asp–Asp–Asp at pH 7.0; **(c)** His–Arg–Asp at pH 8.0 (see Problem 7).

9. The artificial non-nutritive sweetener aspartame is a dipeptide with the sequence Asp–Phe. The C-terminus is methylated. Draw the structure of aspartame at pH 7.0.

10. The zebrafish is a good animal model to test the actions of various drugs. The zebrafish produces a novel heptapeptide that binds to opiate receptors in the brain. The sequence of the peptide is Tyr–Gly–Gly–Phe–Met–Gly–Tyr. Draw the structure of the heptapeptide at pH 7.

11. Histones are basic proteins that bind to DNA. What amino acids are found in abundance in histones and why? What important intermolecular interactions form between histones and DNA?

12. An Asp residue in the RNase T1 enzyme is protonated. Does this make the side chain more or less likely to form hydrogen bonds with neighboring amino acid side chains in the enzyme?

13. Biochemists sometimes link a recombinant protein to a protein known as green fluorescent protein (GFP), which was first purified from bioluminescent jellyfish. The fluorophore in GFP (shown below) is a derivative of three consecutive amino acids that undergo cyclization of the polypeptide chain and an oxidation. Identify **(a)** the

three residues and the bonds that result from the **(b)** cyclization and **(c)** oxidation reactions.

14. Glutathione is a Cys-containing tripeptide found in red blood cells that functions to remove hydrogen peroxide and organic peroxides that can irreversibly damage hemoglobin and cell membranes.

  (a) The sequence of glutathione is γ-Glu–Cys–Gly. The γ-Glu indicates that the first peptide bond forms between the γ-carboxyl group of the Glu side chain and the α-amino group of Cys. Draw the structure of glutathione.

  (b) Glutathione is sometimes abbreviated as GSH to show the importance of the Cys side chain in reactions. For example, two molecules of GSH react with an organic peroxide as shown below. The glutathione is oxidized to GSSG, and the organic peroxide is reduced to a less harmful alcohol. Draw the structure of GSSG.

$$2\ GSH + R-O-O-H \xrightarrow[\text{peroxidase}]{\text{glutathione}} GSSG + ROH + H_2O$$

15. A fusion protein can be synthesized to carry a specific tag, which is useful in identifying the proteins in various experiments. Fusion proteins with the FLAG epitope bind to specific anti-FLAG antibodies and carry this extra sequence: Asp–Tyr–Lys–Asp–Asp–Asp–Asp–Lys. Draw the structure of the FLAG epitope at pH 7.0.

16. The biotech company that sells anti-FLAG antibodies states that fusion proteins carrying the FLAG epitope display the sequence on the surface of the protein (see Problem 15). Examine the sequence and explain why the company can make this claim.

17. Certain bacteria synthesize cyclic tetrapeptides. **(a)** Estimate the net charge at pH 7.0 of such a compound that consists of two Pro and two Tyr residues. **(b)** If the peptide were linear rather than cyclic, what would be its net charge?

18. Estimate the net charge of a His–His–His–His tetrapeptide at pH 6.0. Which is more soluble in water: a solution of the histidine tetrapeptide or a solution of free histidine?

19. You have isolated a tripeptide containing Arg, His, and Pro. How many different sequences are possible for this tripeptide?

20. You have isolated a tetrapeptide with an unknown sequence, and after hydrolyzing its peptide bonds, you recover Ala, Glu, Lys, and Thr. How many different sequences are possible for this tetrapeptide?

21. The pK values of the amino and carboxylate groups in free amino acids differ from the pK values of the N- and C-termini of polypeptides. Explain.

22. In structural studies of a staphylococcal nuclease enzyme, a valine residue buried in the protein's interior was experimentally changed, first to an Asp residue, and then, in a second experiment, to a Lys residue. How do the pK values of the **(a)** Asp and **(b)** Lys residues in the mutant nucleases compare to those listed for the free amino acids in Table 4.1?

23. Read the following sentences and identify each statement as describing the primary, secondary, tertiary, or quaternary structure of the protein:

  (a) The shape of myoglobin is roughly spherical.
  (b) Hemoglobin consists of four polypeptide chains.
  (c) About one-third of the amino acid residues in collagen are glycines.
  (d) The lysozyme molecule contains regions of helical structure.

4.2 Secondary Structure: The Conformation of the Peptide Group

24. The structure of the peptide bond is drawn in Section 4.1 in the *trans* configuration in which the carbonyl group and the amino hydrogen are on opposite sides of the peptide bond. Draw the structure of the peptide bond in the *cis* configuration. Which configuration is more likely to be found in proteins and why?

25. Compare and contrast the structures of the DNA helix and the α helix.

26. The α helix is stabilized by hydrogen bonds that are directed along the length of the helix. The peptide carbonyl group of the *n*th residue forms a hydrogen bond with the peptide —NH group of the (*n* + 4)th residue. Draw the structure of this hydrogen bond.

27. Proline is known as a helix disrupter; it sometimes appears at the beginning or the end of an α helix but never in the middle. Explain.

28. Interestingly, glycine, like proline (see Problem 27), is not usually found in an α helix, but for the opposite reason. Explain why glycine residues are not usually found in α helices.

29. In 1967, Schiffer and Edmonson developed a tool called a helical wheel that is still widely used today. A helical wheel is used to visualize an α helix in which the angle of rotation between two consecutive amino acid residues is 100°. Because the α helix consists of 3.6 residues per turn, the pattern repeats after five turns and 18 residues. In the final representation the view is down the helical axis.

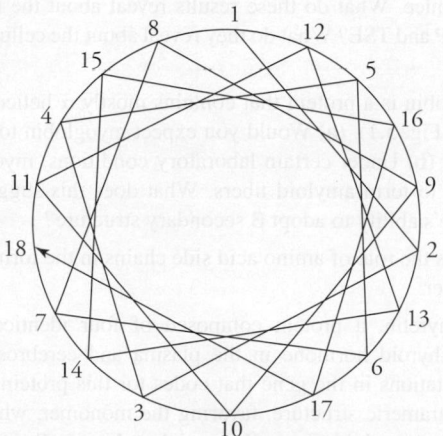

The sequence of a domain of the gp160 protein (found in the envelope of HIV) is shown below, using one-letter codes for the amino acids. Plot this sequence on the helical wheel. What do you notice about the types of amino acid residues on either side of the wheel?

MRVKEKYQHLWRWGWRWG

30. A 24-residue peptide called Pandinin 2, isolated from scorpion venom, was found to have both antimicrobial and hemolytic properties. The sequence of the first 18 residues of this peptide is shown below.

  (a) Use the helical wheel shown in Problem 29 to plot the sequence of this peptide.
  (b) Propose a hypothesis that explains why the peptide is able to lyse bacteria and red blood cells.

FWGALAKGALKLIPSLFS

31. The red blood cell protein glycophorin is a transmembrane protein that crosses the membrane once with an α-helical structure. A portion of the amino acid sequence is shown below. Identify the transmembrane domain of glycophorin.

PPEEETGERVQLAHHFSEPEITLIIFGVMAGVIGTILLI-
SYGIRRLIKKSPSDVKPLPSPDTD

4.3 Tertiary Structure and Protein Stability

32. Choose the amino acid in the following pairs that would be more likely to appear on the solvent-exposed surface of a protein.

  (a) His or Pro
  (b) Ser or Ala
  (c) Phe or Tyr
  (d) Met or Cys
  (e) Asn or Ile

33. Examine Figure 4.9 (triose phosphate isomerase). To which of the CATH classes does this protein belong?

34. **(a)** The guanidinium group on the Arg side chain is stabilized by resonance. Draw the contributing resonance structures for the Arg side chain. **(b)** A study of 60 proteins showed that Arg is about 50% more likely to be buried than Lys. Provide an explanation for this observation.

35. In site-directed mutagenesis experiments, Gly is often successfully substituted for Val, but Val cannot substitute for Gly. Explain.

36. In three separate site-directed mutagenesis experiments, scientists altered a Lys residue in corn PEP carboxylase to an Asn, Glu, or Arg. Which substitution would you expect to have the least effect on enzymatic activity? Which substitution would have the greatest effect? Explain.

37. Draw two amino acid side chains that can interact with each other via the following intermolecular interactions: **(a)** ion pair; **(b)** hydrogen bond; **(c)** van der Waals interaction (London dispersion forces).

38. A type of muscular dystrophy, called severe childhood autosomal recessive muscular dystrophy (SCARMD), results from a mutation in the gene for a 50-kD muscle protein. The defective protein leads to muscle necrosis. Detailed studies of this protein have revealed that an arginine residue at position 98 is mutated to a histidine. Why might replacing an arginine with histidine result in a defective protein?

39. Laboratory techniques for randomly linking together amino acids typically generate an insoluble polypeptide, yet a naturally occurring polypeptide of the same length is usually soluble. Explain.

40. Proteins can be unfolded, or denatured, by agents that alter the balance of weak noncovalent forces that maintain the native conformation. How would the following agents cause a protein to denature? Be specific about the type of intermolecular forces that would be affected. **(a)** Heat; **(b)** pH; **(c)** amphiphilic detergents; **(d)** reducing agents such as 2-dithiothreitol ($HSCH_2CHOHCHOHCH_2SH$).

41. In 1957, Christian Anfinsen carried out a denaturation experiment with ribonuclease (a pancreatic enzyme that catalyzes the digestion of RNA), which consists of a single chain of 124 amino acids cross-linked by four disulfide bonds. Urea and 2-mercaptoethanol were added to a solution of ribonuclease, which caused it to unfold, or denature. The loss of tertiary structure resulted in a loss of biological activity. When the denaturing agent (urea) and the reducing agent (mercaptoethanol) were simultaneously removed, the ribonuclease spontaneously folded back up to its native conformation and regained full enzymatic activity in a process called renaturation. What is the significance of this experiment?

42. Insulin consists of two chains, a shorter A chain and a longer B chain. The two chains are held together with disulfide bonds. *In vivo*, insulin is processed from proinsulin, a single polypeptide chain. The C chain is removed from proinsulin to form insulin.

(a) A denaturation/renaturation experiment similar to the one carried out by Anfinsen with ribonuclease (see Problem 41) was carried out using insulin. However, only 2–4% of the activity of the native protein was recovered when the urea and 2-mercaptoethanol were removed by dialysis (this is the level of activity to be expected if the disulfide bridges formed randomly). When the experiment was repeated with proinsulin, 60% of the activity was restored upon renaturation. Explain these observations.

(b) The investigators noted that the refolding of proinsulin depended on pH. For example, when the proinsulin was incubated in pH 7.5 buffer, only 10% of the proinsulin was renatured. But at pH 10.5, 60% of the proinsulin was renatured. Explain these observations. (*Hint:* The pK value of the cysteine side chain in proinsulin is similar to the pK value of the sulfhydryl group in free cysteine.)

43. In the mid-1980s, scientists noted that if cells were incubated at 42°C instead of the normal 37°C, the synthesis of a group of proteins dramatically increased. For lack of a better name, the scientists called these heat-shock proteins. It was later determined that the heat-shock proteins were chaperones. Why do you think that the cell would increase its synthesis of chaperones when the temperature increases?

44. A protein engineering laboratory studying monoclonal antibody proteins characterized the thermal stability of these proteins by measuring their melting temperature (T_m), defined as the temperature at which the proteins are half unfolded. The investigators found a positive correlation between T_m and the proteins' —SH content. In other words, proteins with more —SH groups were more thermally stable. Explain this observation.

45. Spectrin is a protein found in the cytoskeleton of the red blood cell. The cytoskeleton consists of proteins anchored to the cytosolic surface of the membrane and gives the cell the strength and flexibility to squeeze through capillaries. The spectrin protein chains consist of repeating segments that form α-helical bundles. Recently, a mutant spectrin was isolated in which a glutamine was replaced with proline. What would be the effect of this mutation on the spectrin protein? What are the consequences for the red blood cell?

46. During evolution, why do insertions, deletions, and substitutions of amino acids occur more often in loops than in elements of regular secondary structure?

4.4   Quaternary Structure

47. The restriction endonucleases *Eco*RI and *Eco*RV are dimeric (two-subunit) enzymes (see Section 3.5). Based on how these proteins interact with DNA, do you expect them to be homo- or heterodimeric?

48. A protein with two identical subunits can often be rotated 180° (halfway) around its axis so as to generate an identical structure; such a protein is said to have rotational symmetry. Why is it not possible for a protein to have mirror symmetry (that is, its halves would be related as if reflected in a mirror)?

49. Glutathione transferase consists of a homodimer in equilibrium with its constituent monomers. In site-directed mutagenesis studies, two arginine residues were mutated to glutamines and two aspartates were mutated to asparagines. These substitutions caused the equilibrium to shift in favor of the monomeric form of the enzyme. Where were the arginines and aspartates likely to be found on the protein and what is their role in stabilizing the dimeric form of the enzyme?

50. A tetrameric protein dissociates into dimers when the detergent sodium dodecyl sulfate (SDS) is added to a solution of the protein. But the dimers are termed SDS-resistant because they do not further dissociate into monomers in the presence of the detergent. What intermolecular forces might be acting at the dimer–dimer interface? Are the intermolecular forces acting at the monomer–monomer interface different? Explain.

4.5   Clinical Connection: Protein Misfolding and Disease

51. The gene for the amyloid precursor protein is located on chromosome 21. Individuals with Down syndrome (trisomy 21) have three rather than two copies of chromosome 21. Explain why individuals with Down syndrome tend to exhibit Alzheimer's-like symptoms by middle age.

52. Mice that have been genetically engineered to lack the PrP gene (but are normal in all other ways) do not develop spongiform encephalopathy after being inoculated with PrP^{Sc}, which causes disease in normal mice. What do these results reveal about the relationship between PrP and TSE? What do they reveal about the cellular function of PrP?

53. Myoglobin is a protein that contains mostly α helices and no β sheets (see Fig. 5.1). **(a)** Would you expect myoglobin to form amyloid fibers? **(b)** Under certain laboratory conditions, myoglobin can be induced to form amyloid fibers. What does this suggest about a polypeptide's ability to adopt β secondary structure?

54. Discuss the role of amino acid side chains in the formation of an amyloid fiber.

55. Transthyretin, a protein composed of four identical subunits, transports thyroid hormone in the plasma and cerebrospinal fluid. Several mutations in the gene that codes for this protein disrupt the protein's tetrameric structure, favoring the monomer, which is more likely to form amyloid fibers. Why might a Leu → Pro transthyretin mutant be amyloidogenic?

4.6   Tools and Techniques: Analyzing Protein Structure

56. Estimate the pI values of **(a)** valine, **(b)** aspartate, and **(c)** lysine.

57. Estimate the pI values of the following peptides: **(a)** Ala–Leu and **(b)** Gly–Tyr–Val.

58. What types of amino acids are likely to be relatively abundant in a protein with a pI of **(a)** 4.3 or **(b)** 11.0?

59. What can you conclude about the net charge of any protein at a pH **(a)** less than 3.5 or **(b)** greater than 11.0?

60. In your laboratory, you plan to use ion exchange chromatography to separate the peptide shown below from a mixture of different peptides at pH 7.0. Should you choose a matrix containing DEAE or CM groups?

Peptide: GLEDSLVELGDVQPSLGKESRADDFQRQ

61. The protein allergen from peanuts, a protein called Ara h8, was recently purified and characterized. The investigators initially had difficulty separating Ara h8 from a similar protein in peanuts called Ara h6 because the two proteins were of similar size and had nearly identical pI values. Separation of the two proteins was finally achieved when it was noted that the Ara h6 protein contained 10 cysteine residues involved in disulfide bridges, whereas Ara h8 contained no cysteines. The protein mixture was treated with a reducing agent, dithiothreitol (DTT), and then treated with iodoacetic acid (ICH_2COOH, a reagent that adds to, or alkylates, an —SH group and releases free iodine). The mixture was then loaded onto an anion exchange column and the two proteins were successfully separated.

(a) Show the structural changes that occur when Cys residues are exposed to DTT followed by iodoacetic acid.
(b) Draw a plausible elution profile (protein concentration versus solvent volume) for the separation of the proteins by anion exchange chromatography.
(c) Explain why this treatment resulted in successful separation of the two proteins.

62. Transforming growth factor-β (TGF-β) from human platelets has a molecular weight of 25 kD and consists of two identical subunits joined by disulfide bonds. Draw the gel electrophoresis pattern for an SDS-PAGE gel with a lane that shows the TGF-β with 2-mercaptoethanol (which reduces disulfide bonds) added to the sample and a lane in which the TGF-β was not treated with 2-mercaptoethanol. Be sure to include a third lane (similar to lanes 1 and 8 in Figure 4.27) in your drawing that contains molecular weight markers.

63. A protein isolated from a thermophilic bacterium shows a molecular weight of 160 kD when eluted from a size-exclusion chromatography column, but when the purified protein is run on an SDS-PAGE gel, a single band at 80 kD is observed. What do these observations tell you about the structure of the protein?

64. The sequence of crinia-angiotensin, an angiotensin II-like undecapeptide from the skin of the Australian frog, is determined. A single round of Edman degradation releases DNP-Ala. A second sample of the peptide is then treated with chymotrypsin. Two fragments are released with the following amino acid compositions: Fragment I (His, Pro, Phe, Val) and Fragment II (Ala, Asp, Arg, Gly, Pro, Ile, Tyr). Next, a third sample of peptide is treated with trypsin, which results in two fragments with the following amino acid compositions: Fragment III (Ala, Asp, Arg, Gly, Pro) and Fragment IV (His, Ile, Pro, Phe, Tyr, Val). Treatment of another sample with elastase yields three fragments, two of which are sequenced: Fragment V (His–Pro–Phe) and Fragment VI (Ala–Pro–Gly). What is the sequence of the undecapeptide?

65. Before sequencing, a protein whose two identical polypeptide chains are linked by a disulfide bond must be reduced and alkylated (to chemically block them). Why should reduction and alkylation also be performed for a single polypeptide chain that includes an intramolecular disulfide bond?

66. The peptide hormone glucagon from the *Nile tilapia* was treated with chymotrypsin and the resulting fragments were sequenced. A second sample of the polypeptide was treated with trypsin and the fragments were sequenced. What is the sequence of the polypeptide?

Chymotrypsin fragments	Trypsin fragments
LMNNKRSGAAE	AQDFVRWLMNNK
SNDY	HSEGTFSNDYSK
HSEGTF	RSGAAE
LEDRKAQDF	YLEDRK
VRW	
SKY	

67. Explain why sequencing using traditional methods (e.g., enzymatic cleavage) would be difficult to accomplish with this peptide:

METDTLLLWVLLLWVPGSTG

68. In prokaryotes, the error rate in protein synthesis may be as high as 5×10^{-4} per codon. How many polypeptides containing (a) 500 residues or (b) 2000 residues would you expect to contain an amino acid substitution?

69. The mass of each amino acid residue is shown below. Explain why mass spectrometry, which is highly accurate, cannot distinguish Leu and Ile.

Residue	Mass (D)	Residue	Mass (D)
Ala	71.0	Leu	113.1
Arg	156.1	Lys	128.1
Asn	114.0	Met	131.0
Asp	115.0	Phe	147.1
Cys	103.0	Pro	97.1
Gln	128.1	Ser	87.0
Glu	129.0	Thr	101.0
Gly	57.0	Trp	186.1
His	137.1	Tyr	163.1
Ile	113.1	Val	99.1

70. In sequencing by mass spectrometry, not every peptide bond may break. (a) If cleavage between two Gly residues does not occur, which amino acid would be identified in place of the two glycines? (b) What amino acid would be identified if a bond between Ser and Val did not break (see Problem 69 for amino acid masses)?

71. The peptide shown here is subjected to mass spectrometry to determine its sequence.

(a) If all its peptide bonds (but no other bonds) are broken, what is the mass of the smallest fragment? Assume that the only charged group is the N-terminus.
(b) What is the difference in mass between the smallest and the next-smallest fragment?

72. X-Ray crystallographic analysis of a protein crystal sometimes fails to reveal the positions of the first few residues of a polypeptide chain. Explain.

SELECTED READINGS

Branden, C. and Tooze, J., *Introduction to Protein Structure* (2nd ed.), Garland Publishing (1999). [A well-illustrated book with chapters introducing amino acids and protein structure, plus chapters on specific proteins categorized by their structure and function.]

Dill, K. A. and MacCallum, J. L., The protein folding problem, 50 years on, *Science* **338**, 1042–1046 (2012). [Describes the challenges of understanding how proteins fold and how to predict a protein's structure from its sequence.]

Goodsell, D. S. Visual methods from atoms to cells, *Structure* **13**, 347–354 (2005). [Discusses the challenges of presenting different features of molecular structures.]

Jucker, M. and Walker, L. C., Self-propagation of pathogenic protein aggregates in neurodegenerative diseases, *Nature* **501**, 45–51 (2013). [Summarizes evidence that protein misfolding and aggregation lie behind a variety of human diseases.]

Kaltashov, I. A., Bobst, C. E., and Abzalimov, R. R., Mass spectrometry-based methods to study protein architecture and dynamics, *Protein Science* **22**, 530–544 (2013). [Describes the use of mass spectrometry for sequencing proteins and analyzing their conformational dynamics.]

Oldfield, C. J. and Dunker, A. K., Intrinsically disordered proteins and intrinsically disordered protein regions, *Annu. Rev. Biochem.* **83**, 553–584 (2014). [Explains how proteins without stable structures can interact with other molecules to carry out a variety of biological functions.]

Proteopedia, www.proteopedia.org [This interactive resource showcases tutorials on the three-dimensional structures of proteins, nucleic acids, and other molecules.]

Richardson, J. S., Richardson, D. C., Tweedy, N. B., Gernert, K. M., Quinn, T. P., Hecht, M. H., Erickson, B. W., Yan, Y., McClain, R. D., Donlan, M. E., and Surles, M. C., Looking at proteins: representations, folding, packing, and design, *Biophys. J.* **63**, 1186–1209 (1992). [A highly readable review showing a variety of ways of visualizing protein folding patterns.]

Sillitoe, I., Lewis, T. E., Cuff, A., Das, S., Ashford, P., Dawson, N. L., Furnham, N., Laskowski, R. A., Lee, D., Lees, J. G., Lehtinen, S., Studer, R. A., Thornton, J., and Orengo, C. A., CATH: comprehensive structural and functional annotations for genome sequences, *Nuc. Acids Res.* **43**, D376–381 (2015). [Summarizes a widely used approach to classifying protein structures.]

CHAPTER 5

FUNCTIONS OF PROTEINS

Morales/Getty Images, Inc.

When a humpback whale feeds, its jaw opens wide and the underlying tissue expands to accommodate a huge mouthful of water and tiny prey. The nerves in the whale's mouth can double in length, thanks to arrays of folded collagen, which unfold and prevent the nerves from extending too far. In other organisms, collagen—and nerves—cannot stretch in this way.

DO YOU REMEMBER?

- Noncovalent forces, including hydrogen bonds, ionic interactions, and van der Waals forces, act on biological molecules (Section 2.1).
- A protein's structure may be described at four levels, from primary to quaternary (Section 4.1).
- Some proteins can adopt more than one stable conformation (Section 4.3).
- Proteins containing more than one polypeptide chain have quaternary structure (Section 4.4).

Every protein, with its unique three-dimensional structure, can perform some unique function for the organism that produces it. This chapter begins by examining myoglobin, an oxygen-binding protein that gives vertebrate muscle a reddish color, and hemoglobin, a major protein of red blood cells, which transports O_2 from the lungs to other tissues. These two proteins have been studied for many decades and provide a wealth of information about protein function.

Unlike myoglobin and hemoglobin, which are globular proteins, many of the most abundant proteins are fibrous proteins that form extended structures that determine the shape and other physical attributes of cells and entire organisms. These structural proteins include the extracellular matrix protein collagen and a variety of intracellular proteins. Other than forming fibrous networks, these proteins have little in common, exhibiting a variety of secondary, tertiary, and quaternary structures related to their distinct physiological functions.

The supportive role of fibrous proteins in cellular architecture may seem obvious, but it turns out that many of the dynamic functions of cells are also intricately tied to these proteins. The movements of cells and the movements of organelles within cells reflect the action of motor proteins that operate along fibrous protein tracks; they provide some additional lessons in protein structure and function.

MYOGLOBIN AND HEMOGLOBIN: OXYGEN-BINDING PROTEINS

Myoglobin is a relatively small protein with a compact shape about $44 \times 44 \times 25$ Å (**Fig. 5.1a**). Myoglobin lacks β structure entirely, and all but 32 of its 153 amino acids are part of eight α helices, which range in length from 7 to 26 residues and are labeled A through H (Fig. 5.1b). Hemoglobin is a tetrameric protein whose four subunits each resemble myoglobin.

The fully functional myoglobin molecule contains a polypeptide chain plus the iron-containing porphyrin derivative known as **heme** (shown below). The heme is a type of **prosthetic group,** an organic compound that allows a protein to carry out some function that the polypeptide alone cannot perform—in this case, binding oxygen.

Heme

(a)

(b)

FIGURE 5.1

Models of myoglobin structure.
(a) Space-filling model. All atoms (except H) are shown (C gray, O red, N blue). The heme group, where oxygen binds, is purple.
(b) Ribbon diagram with the eight α helices labeled A–H. [*Structure of myoglobin (pdb 1MBO) determined by S. E. V. Phillips.*]

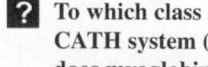

? To which class of the CATH system (Section 4.3) does myoglobin belong?

The planar heme is tightly wedged into a hydrophobic pocket between helices E and F of myoglobin. It is oriented so that its two nonpolar vinyl ($-CH=CH_2$) groups are buried and its two polar propionate ($-CH_2-CH_2-COO^-$) groups are exposed to the solvent. The central iron atom, with six possible coordination bonds, is liganded by four N atoms of the porphyrin ring system. A fifth ligand is provided by a histidine residue of myoglobin known as His F8 (the eighth residue of helix F). Molecular oxygen (O_2) can bind reversibly to form the sixth coordination bond. (This is what allows certain heme-containing proteins, such as myoglobin and hemoglobin, to function physiologically as oxygen carriers.) Residue His E7 (the seventh residue of helix E) forms a hydrogen bond to the O_2 molecule (**Fig. 5.2**). By itself, heme is not an effective oxygen carrier because the central Fe(II) (or Fe^{2+}) atom is easily oxidized to the ferric Fe(III) (or Fe^{3+}) state, which cannot bind O_2. Oxidation does not readily take place when the heme is part of a protein such as myoglobin or hemoglobin.

OXYGEN BINDING TO MYOGLOBIN DEPENDS ON THE OXYGEN CONCENTRATION

The muscles of diving mammals are especially rich in myoglobin. At one time, myoglobin was believed to be an oxygen-storage protein—which would be advantageous during a long dive—but it most likely just facilitates oxygen diffusion through muscle cells or binds other small molecules such as nitric oxide (NO).

Myoglobin's O_2-binding behavior can be quantified. To begin, the reversible binding of O_2 to myoglobin (Mb) is described by a simple equilibrium

$$Mb + O_2 \rightleftharpoons Mb \cdot O_2$$

with a dissociation constant, K:

$$K = \frac{[Mb][O_2]}{[Mb \cdot O_2]} \quad [5.1]$$

where the square brackets indicate molar concentrations. (Note that biochemists tend to describe binding phenomena in terms of dissociation constants, sometimes given as K_d, which are the reciprocals of the association constants, K_a, used by chemists.) The proportion of the total myoglobin molecules that have bound O_2 is called the **fractional saturation** and is abbreviated Y:

$$Y = \frac{[Mb \cdot O_2]}{[Mb] + [Mb \cdot O_2]} \quad [5.2]$$

Since $[Mb \cdot O_2]$ is equal to $[Mb][O_2]/K$ (Equation 5.1, rearranged),

$$Y = \frac{[O_2]}{K + [O_2]} \quad [5.3]$$

O_2 is a gas, so its concentration is expressed as pO_2, the **partial pressure of oxygen,** in units of torr (where 760 torr = 1 atm). Thus,

$$Y = \frac{pO_2}{K + pO_2} \quad [5.4]$$

In other words, *the amount of O_2 bound to myoglobin (Y) is a function of the oxygen concentration (pO_2) and the affinity of myoglobin for O_2 (K).*

A plot of fractional saturation (Y) versus pO_2 yields a hyperbola (**Fig. 5.3**). As the O_2 concentration increases, more and more O_2 molecules bind to the heme groups of myoglobin molecules until, at very high O_2 concentrations, virtually all the myoglobin molecules have bound O_2. Myoglobin is then said to be **saturated** with oxygen. The oxygen concentration at which myoglobin is half-saturated—that is, the concentration of O_2 at which Y is half-maximal—is equivalent to K. For convenience, K is usually called p_{50}, the oxygen pressure at 50% saturation. For human myoglobin, p_{50} is 2.8 torr (see Sample Calculation 5.1).

FIGURE 5.2

Oxygen binding to the heme group of myoglobin. The central Fe(II) atom of the heme group (purple) is liganded to four porphyrin N atoms and to the N of His F8 below the porphyrin plane. O_2 (red) binds reversibly to the sixth coordination site, above the porphyrin plane. Residue His E7 forms a hydrogen bond to O_2.

SAMPLE CALCULATION 5.1

Calculate the fractional saturation of myoglobin when $pO_2 = 1$ torr, 10 torr, and 100 torr.   **PROBLEM**

Use Equation 5.4 and let $K = 2.8$ torr.   **SOLUTION**

When $pO_2 = 1$ torr, $Y = \dfrac{1}{2.8 + 1} = 0.26$

When $pO_2 = 10$ torr, $Y = \dfrac{10}{2.8 + 10} = 0.78$

When $pO_2 = 100$ torr, $Y = \dfrac{100}{2.8 + 100} = 0.97$

PRACTICE PROBLEMS

1. Calculate the fractional saturation of myoglobin when $pO_2 = 6.8$ torr.
2. How would the value of Y in Practice Problem 1 change if $K = 1.4$ torr?
3. At what pO_2 value would myoglobin be 60% saturated with O_2?

MYOGLOBIN AND HEMOGLOBIN ARE RELATED BY EVOLUTION

Hemoglobin is a heterotetramer containing two α chains and two β chains. Each of these subunits, called a **globin,** looks a lot like myoglobin. The hemoglobin α chain, the hemoglobin β chain, and myoglobin have remarkably similar tertiary structures (**Fig. 5.4**). All have a heme group in a hydrophobic pocket, a His F8 that ligands the Fe(II) ion, and a His E7 that forms a hydrogen bond to O_2.

FIGURE 5.3

Myoglobin oxygen-binding curve. The relationship between the fractional saturation of myoglobin (Y) and the oxygen concentration (pO_2) is hyperbolic. When $pO_2 = K = 2.8$ torr, myoglobin is half-saturated ($Y = 0.5$). Note that this curve describes a population of myoglobin molecules, not the behavior of a single protein.

FIGURE 5.4

Tertiary structures of myoglobin and the α and β chains of hemoglobin. Backbone traces of α globin (blue) and β globin (red) are aligned with myoglobin (green) to show their structural similarity. The heme group of myoglobin is shown in gray. [*Structure of hemoglobin (pdb 2HHB) determined by G. Fermi and M. F. Perutz.*]

? **Which portion of globin structure shows the most variability? The least? Explain.**

Somewhat surprisingly, the amino acid sequences of the three globin polypeptides are only 18% identical. **Figure 5.5** shows the aligned sequences, with the necessary gaps (for example, the hemoglobin α chain has no D helix). The lack of striking sequence similarities among these proteins highlights an important principle of protein three-dimensional structure: *Certain tertiary structures—for example, the backbone folding pattern of a globin polypeptide—can accommodate a variety of amino acid sequences. In fact, many proteins with completely unrelated sequences adopt similar structures.*

```
Helix              A                    B              C          D          E
Mb     G-LSDGEWQLVLNVWGKVEADIPGHGQEVLIRLFKGHPETLEKFDKFKHLKSEDEMKASEDLKKHGATVLTALG
Hb α   V-LSPADKTNVKAAWGKVGAHAGEYGAEALERMFLSFPTTKTYFPHF-DLSH-----GSAQVKGHGKKVADALT
Hb β   VHLTPEEKSAVTALWGKV--NVDEVGGEALGRLLVVYPWTQRFFESFGDLSTPDAVMGNPKVKAHGKKVLGAFS
```

```
Helix              F             G                    H
Mb     GILKKKGHHEAEIKPLAQSHATKHKIPVKYLEFISECIIQVLQSKHPGDFGADAQGAMNKALELFRKDMASNYKELGFQG
Hb α   NAVAHVDDMPNALSALSDLHAHKLRVDPVNFKLLSHCLLVTLAAHLPAEFTPAVHASLDKFLASVSTVLTSKYR
Hb β   DGLAHLDNLKGTFATLSELHCDKLHVDPENFRLLGNVLVCVLAHHFGKEFTPPVQAAYQKVVAGVANALAHKYH
```

FIGURE 5.5

The amino acid sequences of myoglobin and the hemoglobin α and β chains. The sequence of human myoglobin (Mb) and the human hemoglobin (Hb) chains are written so that their helical segments (bars labeled A through H) are aligned. Residues that are identical in the α and β globins are shaded yellow; residues identical in myoglobin and the α and β globins are shaded blue, and residues that are invariant in all vertebrate myoglobin and hemoglobin chains are shaded purple. The one-letter abbreviations for amino acids are given in Figure 4.2. [*After Dickerson, R. E., and Geis, I.,* Hemoglobin, *pp. 68–69, Benjamin/Cummings (1983).*]

? **Can you identify positions occupied by structurally similar amino acids in all three globins?**

Clearly, the globins are **homologous proteins** that have evolved from a common ancestor through genetic mutation (see Section 3.4). The α and β chains of human hemoglobin share a number of residues; some of these are also identical in human myoglobin. A few residues are found in all vertebrate hemoglobin and myoglobin chains. The **invariant residues,** those that are identical in all the globins, are essential for the structure and/or function of the proteins and cannot be replaced by other residues. Some positions are under less selective pressure to maintain a particular amino acid match and can be **conservatively substituted** by a similar amino acid (for example, isoleucine for leucine or serine for threonine). Still other positions are **variable,** meaning that they can accommodate a variety of residues, none of which is critical for the protein's structure or function. *By looking at the similarities and differences in sequences among evolutionarily related proteins such as the globins, it is possible to deduce considerable information about elements of protein structure that are central to protein function.*

Sequence analysis also provides a window on the course of globin evolution, since *the number of sequence differences roughly corresponds to the time since the genes diverged.* An estimated 1.1 billion years ago, a single globin gene was duplicated, possibly by aberrant genetic recombination, leaving two globin genes that then could evolve independently (**Fig. 5.6**). Over time, the gene sequences diverged by mutation. One gene became the myoglobin gene. The other coded for a monomeric hemoglobin, which is still found in some primitive vertebrates such as the lamprey (an organism that originated about 425 million years ago). Subsequent duplication of the hemoglobin gene and additional sequence changes yielded the α and β globins, which made possible the evolution of a tetrameric hemoglobin (whose structure is abbreviated α₂β₂). Additional gene duplications and mutations produced the ζ chain (from the α chain) and the γ and ε chains (from the β chain). In fetal mammals, hemoglobin has the composition α₂γ₂, and early human embryos synthesize a ζ₂ε₂ hemoglobin. In primates, a recent duplication of the β chain gene has yielded the δ chain. An α₂δ₂ hemoglobin occurs as a minor component (about 2%) of adult human hemoglobin. At present, the δ chain appears to have no unique biological function, but it may eventually evolve one.

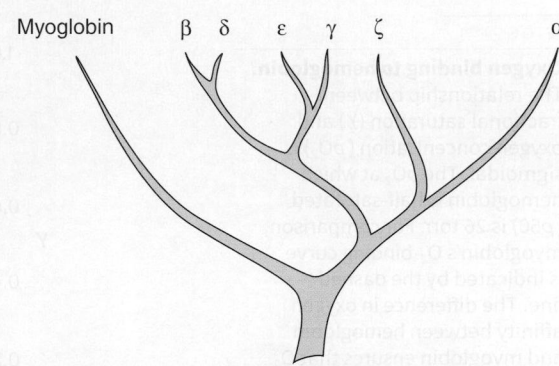

FIGURE 5.6

Evolution of the globins. Duplication of a primordial globin gene allowed the separate evolution of myoglobin and a monomeric hemoglobin. Additional duplications among the hemoglobin genes gave rise to six different globin chains that combine to form tetrameric hemoglobin variants at various times during development.

OXYGEN BINDS COOPERATIVELY TO HEMOGLOBIN

A milliliter of human blood contains about 5 billion red blood cells, each of which is packed with about 300 million hemoglobin molecules. Consequently, blood can carry far more oxygen than a comparable volume of pure water. The oxygen-carrying capacity of the blood can be quickly assessed by measuring the hematocrit (the percentage of the blood volume occupied by red blood cells, which ranges from about 40% (in women) to 45% (in men). Individuals with **anemia,** too few red blood cells, can sometimes be treated with the hormone erythropoietin to increase red blood cell (erythrocyte) production by bone marrow.

The hemoglobin in red blood cells, like myoglobin, binds O_2 reversibly, but it does not exhibit the simple behavior of myoglobin. A plot of fractional saturation (Y) versus pO_2 for hemoglobin is sigmoidal (S-shaped) rather than hyperbolic (**Fig. 5.7**). Furthermore, hemoglobin's overall oxygen affinity is lower than that of myoglobin: Hemoglobin is half-saturated at an oxygen pressure of 26 torr ($p_{50} = 26$ torr), whereas myoglobin is half-saturated at 2.8 torr.

Why is hemoglobin's binding curve sigmoidal? At low O_2 concentrations, hemoglobin appears to be reluctant to bind the first O_2, but as the pO_2 increases, O_2 binding increases sharply, until hemoglobin is almost fully saturated. A look at the binding curve in reverse shows that at high O_2 concentrations, oxygenated hemoglobin is reluctant to give up its first O_2, but as the pO_2 decreases, all the O_2 molecules are easily given up. This behavior suggests that the binding of the first O_2 increases the affinity of the remaining O_2-binding sites. Apparently, *hemoglobin's four heme groups are not independent but communicate with each other in order to work in a unified fashion.* This is known as **cooperative binding** behavior. In fact, the fourth O_2 taken up by hemoglobin binds with about 100 times greater affinity than the first.

Hemoglobin's relatively low oxygen affinity and its cooperative binding behavior are the keys to its physiological function (see Fig. 5.7). In the lungs, where the pO_2 is about 100 torr, hemoglobin is about 95% saturated with O_2. In the tissues, where the pO_2 is only about 20 to 40 torr, hemoglobin's oxygen affinity drops off rapidly (it is only about 55% saturated when

FIGURE 5.7

Oxygen binding to hemoglobin.
The relationship between fractional saturation (Y) and oxygen concentration (pO_2) is sigmoidal. The pO_2 at which hemoglobin is half-saturated ($p50$) is 26 torr. For comparison, myoglobin's O_2-binding curve is indicated by the dashed line. The difference in oxygen affinity between hemoglobin and myoglobin ensures that O_2 bound to hemoglobin in the lungs is released to myoglobin in the muscles. This oxygen-delivery system is efficient because the tissue pO_2 corresponds to the part of the hemoglobin binding curve where the O_2 affinity falls off most sharply.

? **Could this O_2-delivery system still operate if the $p50$ for myoglobin was twice as high?**

the pO_2 is 30 torr). Under these conditions, *the O_2 released from hemoglobin is readily taken up by myoglobin in muscle cells, since myoglobin's affinity for O_2 is much higher,* even at the lower oxygen pressure. Myoglobin can therefore relay O_2 from red blood cells in the capillaries to the muscle cells' mitochondria, where it is consumed in the oxidative reactions that sustain muscle activity. Agents such as carbon monoxide, which interferes with O_2 binding to hemoglobin, prevent the efficient delivery of O_2 to cells (**Box 5.A**).

A CONFORMATIONAL SHIFT EXPLAINS HEMOGLOBIN'S COOPERATIVE BEHAVIOR

The four heme groups of hemoglobin must be able to sense one another's oxygen-binding status so that they can bind or release their O_2 in concert. But the four heme groups are 25 to 37 Å apart, too far for them to communicate via an electronic signal. Therefore, the signal must be mechanical. In a mechanism worked out by Max Perutz, the four globin subunits undergo conformational changes when they bind O_2.

In **deoxyhemoglobin** (hemoglobin without any bound O_2), the heme Fe ion has five ligands, so the porphyrin ring is somewhat dome-shaped and the Fe lies about 0.6 Å out of the

Box 5.A   Carbon Monoxide Poisoning

The affinity of hemoglobin for carbon monoxide is about 250 times higher than its affinity for oxygen. However, the concentration of CO in the atmosphere is only about 0.1 ppm (parts per million by volume), compared to an O_2 concentration of about 200,000 ppm. Normally, only about 1% of the hemoglobin molecules in an individual are in the carboxyhemoglobin (Hb·CO) form, probably as a result of endogenous production of CO in the body (CO acts as a signaling molecule, although its physiological role is not well understood).

Danger arises when the fraction of carboxyhemoglobin rises, which can occur when individuals are exposed to high levels of environmental CO. For example, the incomplete combustion of fuels, as occurs in gas-burning appliances and vehicle engines, releases CO. The concentration of CO can rise to about 10 ppm in these situations and to as high as 100 ppm in highly polluted urban areas. The concentration of carboxyhemoglobin may reach 15% in some heavy smokers, although the symptoms of CO poisoning are usually not apparent.

CO toxicity, which occurs when the concentration of carboxyhemoglobin rises above about 25%, causes neurological impairment, usually dizziness and confusion. High doses of CO, which cause carboxyhemoglobin levels to rise above 50%, can trigger coma and death. When CO is bound to some of the heme groups of hemoglobin, O_2 is not able to bind to those sites because its low affinity means that it cannot displace the bound CO. In addition, the carboxyhemoglobin molecule remains in a high-affinity conformation, so that even if O_2 does bind to some of the hemoglobin heme groups in the lungs, O_2 release to the tissues is impaired. The effects of mild CO poisoning are largely reversible through the administration of O_2. But because the CO remains bound to hemoglobin with a half-life of several hours, recovery is slow.

? **Sketch the oxygen-binding curves of hemoglobin and carboxyhemoglobin [Hb·(CO)$_2$].**

plane of the porphyrin ring. As a result, the heme group is bowed slightly toward His F8 (**Fig. 5.8**). When O_2 binds to produce **oxyhemoglobin,** the Fe—now with six ligands—moves into the center of the porphyrin plane. This movement of the Fe ion pulls His F8 farther toward the heme group, and this in turn drags the entire F helix so that it moves as much as 1 Å. The F helix cannot move in this manner unless the entire protein alters its conformation, culminating in the rotation of one αβ unit relative to the other. Consequently, *hemoglobin has two quaternary structures, corresponding to the oxy and deoxy states.*

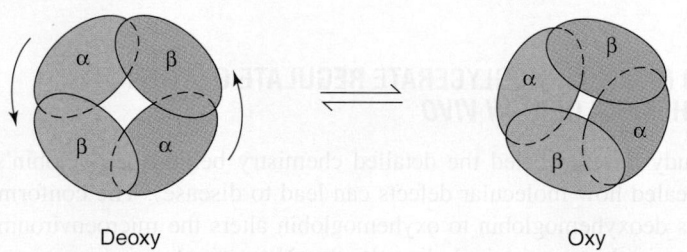

The shift in conformation between the oxy and deoxy states primarily involves rotation of one αβ unit relative to the other. Oxygen binding decreases the size of the central cavity between the four subunits and alters some of the contacts between subunits. The two conformational states of hemoglobin are formally known as **T** (for "tense") and **R** (for "relaxed"). The T state corresponds to deoxyhemoglobin, and the R state corresponds to oxyhemoglobin.

Deoxyhemoglobin is reluctant to bind the first O_2 molecule because the protein is in the deoxy (T) conformation, which is unfavorable for O_2 binding (the Fe atom lies out of the heme plane). However, once O_2 has bound, probably to the α chain in each αβ pair, the entire tetramer switches to the oxy (R) conformation as the Fe atom and the F helix move. An intermediate conformation is not possible, because the contacts between the αβ units do not allow it (**Fig. 5.9**). Molecular dynamics studies suggest that hemoglobin does not instantaneously snap from one conformation to the other but instead undergoes fluctuations in tertiary structure that precede the shift in quaternary structure.

Subsequent O_2 molecules bind with higher affinity because the protein is already in the oxy (R) conformation, which is favorable for O_2 binding. Similarly, oxyhemoglobin tends to retain its bound O_2 molecules until the oxygen pressure drops significantly. Then some O_2 is released, triggering the change to the deoxy (T) conformation. This decreases the affinity

FIGURE 5.8

Conformational changes in hemoglobin upon O_2 binding. In deoxyhemoglobin (blue), the porphyrin ring is slightly bowed down toward His F8 (shown in ball-and-stick form). The remainder of the F helix is represented by its alpha carbon atoms. In oxyhemoglobin (purple), the heme group becomes planar, pulling His F8 and its attached F helix upward. The bound O_2 is shown in red.

(a)   (b)

FIGURE 5.9

Some of the subunit interactions in hemoglobin. The interactions between the αβ units of hemoglobin include contacts between side chains. The relevant residues are shown in space-filling form. (a) In deoxyhemoglobin, a histidine residue on the β chain (blue, left) fits between a proline and a threonine residue on the α chain (green, right). (b) Upon oxygenation, the His residue moves between two Thr residues. An intermediate conformation (between the deoxy and oxy conformations) is disallowed in part because the highlighted side chains would experience strain. [*Structure of human deoxyhemoglobin (pdb 2HHB) determined by G. Fermi and M. F. Perutz; structure of human oxyhemoglobin (pdb 1HHO) determined by B. Shaanan.*]

of the remaining bound O_2 molecules, making it easier for hemoglobin to unload its bound oxygen. Because measurements of O_2 binding reflect the average behavior of many individual hemoglobin molecules, the result is a smooth curve (as shown in Fig. 5.7).

Hemoglobin and many other proteins with multiple binding sites are known as **allosteric proteins** (from the Greek *allos,* meaning "other," and *stereos,* meaning "space"). In these proteins, *the binding of a small molecule (called a **ligand**) to one site alters the ligand-binding affinity of the other sites.* In principle, the ligands need not be identical, and their binding may either increase or decrease the binding activity of the other sites. In hemoglobin, the ligands are all oxygen molecules, and O_2 binding to one subunit of the protein increases the O_2 affinity of the other subunits.

H⁺ IONS AND BISPHOSPHOGLYCERATE REGULATE OXYGEN BINDING TO HEMOGLOBIN *IN VIVO*

Decades of study have revealed the detailed chemistry behind hemoglobin's activity (and have also revealed how molecular defects can lead to disease). The conformational change that transforms deoxyhemoglobin to oxyhemoglobin alters the microenvironments of several ionizable groups in the protein, including the two N-terminal amino groups of the α subunits and the two histidine residues near the C-terminus of the β subunits. As a result, these groups become more acidic and release H⁺ when O_2 binds to the protein:

$$Hb \cdot H^+ + O_2 \rightleftharpoons Hb \cdot O_2 + H^+$$

Therefore, increasing the pH of a solution of hemoglobin (decreasing [H⁺]) favors O_2 binding by "pushing" the reaction to the right, as written above. Decreasing the pH (increasing [H⁺]) favors O_2 dissociation by "pushing" the reaction to the left. *The reduction of hemoglobin's oxygen-binding affinity when the pH decreases is known as the **Bohr effect.***

The Bohr effect plays an important role in O_2 transport *in vivo.* Tissues release CO_2 as they consume O_2 in respiration. The dissolved CO_2 enters red blood cells, where it is rapidly converted to bicarbonate (HCO_3^-) by the action of the enzyme carbonic anhydrase (see Section 2.5):

$$CO_2 + H_2O \rightleftharpoons HCO_3^- + H^+$$

The H⁺ released in this reaction induces hemoglobin to unload its O_2 (**Fig. 5.10**). In the lungs, the high concentration of oxygen promotes O_2 binding to hemoglobin. This causes the release of protons that can then combine with bicarbonate to re-form CO_2, which is breathed out.

FIGURE 5.10

Oxygen transport and the Bohr effect. Hemoglobin picks up O_2 in the lungs. In the tissues, H⁺ derived from the metabolic production of CO_2 decreases hemoglobin's affinity for O_2, thereby promoting O_2 release to the tissues. Back in the lungs, hemoglobin binds more O_2, releasing the protons, which recombine with bicarbonate to re-form CO_2.

? Write the net equation for the process shown in the diagram.

Red blood cells use one additional mechanism to fine-tune hemoglobin function. These cells contain a three-carbon compound, 2,3-bisphosphoglycerate (BPG):

FIGURE 5.11

Effect of BPG on hemoglobin. BPG binds to deoxyhemoglobin but not to oxyhemoglobin. It therefore reduces hemoglobin's O_2 affinity by stabilizing the deoxy conformation.

2,3-Bisphosphoglycerate (BPG)

BPG binds in the central cavity of hemoglobin, but only in the T (deoxy) state. The five negative charges in BPG interact with positively charged groups in deoxyhemoglobin; in oxyhemoglobin, these cationic groups have moved and the central cavity is too narrow to accommodate BPG. Thus, *the presence of BPG stabilizes the deoxy conformation of hemoglobin.* Without BPG, hemoglobin would bind O_2 too tightly to release it to cells. In fact, hemoglobin stripped of its BPG *in vitro* exhibits very strong O_2 affinity, even at low pO_2 (**Fig. 5.11**).

The fetus takes advantage of this chemistry to obtain O_2 from its mother's hemoglobin. Fetal hemoglobin has the subunit composition $\alpha_2\gamma_2$. In the γ chains, position H21 is not histidine (as it is in the mother's β chain) but serine. His H21 bears one of the positive charges important for binding BPG in adult hemoglobin. The absence of this interaction in fetal hemoglobin reduces BPG binding. Consequently, *hemoglobin in fetal red blood cells has a higher O_2 affinity than adult hemoglobin, which helps transfer O_2 from the maternal circulation across the placenta to the fetus.*

 CONCEPT CHECK

1. Prove that $pO_2 = p_{50}$ when $Y = 0.5$.

2. Why does myoglobin require a prosthetic group?

3. Explain how the sequences of homologous proteins provide information about residues that are essential and nonessential for a protein's function.

4. What is the relationship between myoglobin's fractional saturation and the oxygen concentration?

5. Link hemoglobin's structural features to its ability to bind O_2 cooperatively.

6. Describe how myoglobin and hemoglobin together make an efficient O_2-delivery system.

7. Explain how the Bohr effect and BPG regulate O_2 transport *in vivo*.

CLINICAL CONNECTION: HEMOGLOBIN VARIANTS

In addition to illustrating many basic principles of protein structure and function, hemoglobin also supplies biochemists with a rich repertoire of natural experiments in the form of inherited disorders of hemoglobin synthesis and activity. Analyzing these variant proteins has provided further insights into protein chemistry and the nature of disease in humans.

As much as 7% of the world's population carries a variant of the genes that encode the α and β chain of hemoglobin. These mutations produce hemoglobin proteins with altered amino acid sequences. In most cases, the mutation is benign and the hemoglobin molecules function more or less normally. But in other cases, the mutation results in serious physical complications for the individual, as the ability of the mutant hemoglobin to deliver oxygen to cells is compromised. Mutant hemoglobins are often unstable, which may also result in destruction of the

KEY CONCEPTS

- Mutations in DNA sequence for the genes that encode protein produce proteins with altered amino acid sequence.
- Mutant hemoglobins are often unstable and may cause anaemia.

FIGURE 5.12

Polymerized hemoglobin S.
In this model, the heme groups
are red and the mutant valine
residues are blue. [*From W. Royer
and D. Harrington, J. Mol. Biol. **272**,
398–407 (1992).*]

red blood cell, leading to anemia. Over a thousand hemoglobin variants have been discovered, but one of the best-known is sickle cell hemoglobin (known as hemoglobin S or Hb S; the normal hemoglobin is called hemoglobin A). Individuals with two copies of the defective gene develop sickle cell disease, a debilitating condition that predominantly affects populations of African descent.

The discovery of the molecular defect that causes sickle cell disease was a groundbreaking event in biochemistry. The disease was first described in 1910, but for many years, there was no direct evidence that sickle cell disease—or any genetic disease—was the result of an alteration in the molecular structure of a protein. Then in 1949, Linus Pauling (who was already on his way to discovering the α helix) showed that hemoglobin from patients with sickle cell disease had a different electrical charge than hemoglobin from healthy individuals. Eight years later, in 1957, Vernon Ingram identified a single amino acid difference: glutamate at position A3 in the β chain (the sixth residue of the polypeptide) is replaced by valine in sickle cell hemoglobin. This was the first evidence that an alteration in a gene caused an alteration in the amino acid sequence of the corresponding polypeptide. The mutation is described in Section 3.3.

In normal hemoglobin, the switch from the oxy to the deoxy conformation exposes a hydrophobic patch on the protein surface between the E and F helices. The hydrophobic valine residues on hemoglobin S are optimally positioned to bind to this patch. This intermolecular association leads to the rapid aggregation of hemoglobin S molecules to form long, rigid fibers (**Fig. 5.12**).

These fibers physically distort the red blood cell, producing the familiar sickle shape. Because hemoglobin S aggregation occurs only among deoxyhemoglobin S molecules, sickling tends to occur when the red blood cells pass through oxygen-poor capillaries. The misshapen cells can obstruct blood flow and rupture, leading to the intense pain, organ damage, and loss of red blood cells that characterize the disease (**Fig. 5.13**).

The high frequency of the gene for sickle cell disease (that is, the mutated β globin gene) was at first puzzling: Genes that lead to disabling conditions tend to be rare because individuals with two copies of the gene usually die before they can pass the gene to their offspring. However, carriers of the sickle cell variant appear to have a selective advantage. They are protected against malaria, an illness caused primarily by the intracellular parasite *Plasmodium falciparum*. Malaria afflicts about 225 million people and kills about 1 million each year, mostly children. In fact, the sickle cell hemoglobin variant is common in regions of the world where malaria is endemic.

In heterozygotes (individuals with one normal and one defective β globin gene), only about 2% of red blood cells undergo sickling. The sickling and loss of a small proportion of parasite-infected red blood cells probably has no significant impact on the overall number of parasites circulating in the body, as was once thought. Instead, the lysis (breakage) of the sickled cells likely releases hemoglobin, whose free heme groups are toxic. In response to the heme groups, the body increases production of an enzyme, called heme oxygenase, to

FIGURE 5.13

Normal and sickled red blood cells. Normal cells (left) are rounded and can flex slightly when they pass through capillaries. A sickled cell (right) is more likely to rupture when passing through a small capillary. [*From Andrew Skred/Science Photo Library/Photo Researchers and Jacki Lewin, Royal Free Hospital/Science Photo Library/Photo Researchers.*]

TABLE 5.1 Some Hemoglobin Variants

VARIANT[a]	CHAIN	POSITION OF MUTATION	AMINO ACID CHANGE	ROLE OF NORMAL RESIDUE
Milledgeville	α	44	Pro → Leu	Participates in the formation of the α–β interface in the deoxy form but not the oxy form.
Chesapeake	α	92	Arg → Leu	Participates in α–β contacts.
Singapore	α	141 (C-terminus)	Arg → Pro	Its COO⁻ forms an ion pair with Lys 127 and its side chain forms an ion pair with Asp 126 in the deoxy form.
Providence	β	82	Lys → Asn	Forms an ion pair to BPG in the central cavity.
Kansas	β	102	Asn → Thr	Participates in subunit–subunit contacts.
Syracuse	β	146	His → Pro	The side-chain imidazole ring forms an ion pair with Asp 94. It also forms an ion pair with BPG in the central cavity.

[a]Hemoglobin variants are usually named for the place where they were first observed or characterized.

degrade the heme. A byproduct of this reaction is carbon monoxide, which in small amounts is a signal for cells to minimize their response to inflammatory signals. Consequently, the inflammation and tissue damage that would normally occur during a *Plasmodium* infection are less severe. Experiments in mice indicate that hemoglobin S does not interfere with the life cycle of the *Plasmodium* parasite, but it makes a typical bout of malaria less likely to lead to death.

Like hemoglobin S, hemoglobin C confers protection against malaria. Coincidentally, the same amino acid of the hemoglobin β chain is affected, but in this case the glutamate is replaced by lysine. The mutant hemoglobin C does not aggregate as hemoglobin S does, but it makes the red blood cells a bit more rigid than normal, leading to mild anemia. The antimalarial effect of hemoglobin C is not well understood; one possibility is that the variant hemoglobin interferes with the placement of a parasite protein on the surface of the red blood cell. As a result, parasite-infected cells are less able to stick to the walls of blood vessels, a feature of *Plasmodium* infection, and instead are more likely to be destroyed by the spleen, where worn-out red cells are normally taken out of circulation.

The **thalassemias** result from genetic defects that reduce the rate of synthesis of the α or β globin chains. These disorders are prevalent in the Mediterranean area (the name comes from the Greek word *thalassa,* meaning "sea") and in South Asia. Depending on the nature of the mutation, individuals with thalassemia may experience mild to severe anemia and their red blood cells may be smaller than normal. But like heterozygotes for hemoglobin S and hemoglobin C, individuals with thalassemia are apparently more likely to survive malaria.

Table 5.1 lists some hemoglobin residues that are critical for normal function; their mutation produces clinical symptoms. Because hemoglobin is a transport protein that must efficiently bind O_2 in the lungs but give it up in the tissues, variations that interfere with O_2 binding or allow it to bind too tightly can both lead to impaired O_2 delivery.

 CONCEPT CHECK

1. Predict the physiological effects of hemoglobin genetic changes that lead to higher O_2 affinity, lower O_2 affinity, loss of cooperativity, unstable protein structure, or red blood cell lysis.

Actin filaments   Intermediate filaments   Microtubules

FIGURE 5.14

Distribution of cytoskeletal fibers in a single cell. To make these micrographs, each type of fiber was labeled with a fluorescent probe that binds specifically to one type of cytoskeletal protein. Note how the distribution of actin filaments differs somewhat from that of intermediate filaments and microtubules. [*Courtesy J. Victor Small, Austrian Academy of Sciences, Vienna, Austria.*]

? What type of fiber best defines the cell's nucleus?

STRUCTURAL PROTEINS

KEY CONCEPTS
- Globular actin subunits associate in a double chain to form a microfilament.
- The growth and regression of actin filaments can change a cell's shape.
- Microtubules are hollow tubes built from tubulin dimers.
- Intermediate filaments are long-lasting fibrous proteins consisting of coiled α helices.
- Three left-handed Gly-rich helical polypeptides form the collagen triple helix.

The shapes of eukaryotic cells, particularly those without an external cell wall, are determined by an intracellular network of proteins known as the **cytoskeleton.** Typically, three types of cytoskeletal proteins form fibers that extend throughout the cell (**Fig. 5.14**). These are **actin filaments** (with a diameter of about 70 Å), **intermediate filaments** (with a diameter of about 100 Å), and **microtubules** (with a diameter of about 240 Å). In large multicellular organisms, fibers of the protein collagen provide structural support extracellularly. Bacterial cells also contain proteins that form structures similar to actin filaments and microtubules. In the following discussion, note how the structure of each protein influences the overall structure and flexibility of the fiber as well as the fiber's ability to disassemble and reassemble.

ACTIN FILAMENTS ARE MOST ABUNDANT

A major portion of the eukaryotic cytoskeleton consists of actin filaments, also known as microfilaments, which are polymers of the protein actin. In many cells, a network of actin filaments supports the plasma membrane and therefore determines cell shape (see Fig. 2.7 and Fig. 5.14). Certain proteins cross-link individual actin polymers to help form bundles of filaments, thereby increasing their strength.

Monomeric actin is a globular protein with about 375 amino acids (**Fig. 5.15**). On its surface is a cleft in which adenosine triphosphate (ATP) binds. The adenosine group slips

FIGURE 5.15

Actin monomer. This protein assumes a globular shape with a cleft where ATP (green) binds. [*Structure of rabbit actin (pdb 1J6Z) determined by L. R. Otterbein, P. Graceffa, and R. Dominguez.*]

into a pocket on the protein, and the ribose hydroxyl groups and the phosphate groups form hydrogen bonds with the protein.

Polymerized actin is sometimes referred to as **F-actin** (for filamentous actin, to distinguish it from **G-actin,** the globular monomeric form). *The actin polymer is actually a double chain of subunits in which each subunit contacts four neighboring subunits* (**Fig. 5.16**). Each actin subunit has the same orientation (for example, all the nucleotide-binding sites point up in Fig. 5.16), so the assembled fiber has a distinct polarity. The end with the ATP site is known as the **(–) end,** and the opposite end is the **(+) end.**

Initially, polymerization of actin monomers is slow because actin dimers and trimers are unstable. But once a longer polymer has formed, subunits add to both ends. Addition is usually much more rapid at the (+) end (hence its name) than at the (–) end (**Fig. 5.17**).

Actin polymerization is driven by the hydrolysis of ATP (splitting ATP by the addition of water) to produce ADP + inorganic phosphate (**P_i**):

FIGURE 5.16

Model of an actin filament. The structure of F-actin was determined from X-ray diffraction data and computer model-building. Fourteen actin subunits are shown (all are different colors except the central actin subunit, whose two halves are blue and gray). [*Courtesy Ken Holmes, Max Planck Institute for Medical Research.*]

Adenosine triphosphate (ATP)

$+ H_2O \longrightarrow$

Adenosine diphosphate (ADP)   Inorganic phosphate (P_i)

This reaction is catalyzed by F-actin but not by G-actin. Consequently, *most of the actin subunits in a filament contain bound ADP.* Only the most recently added subunits still contain ATP. Because ATP-actin and ADP-actin assume slightly different conformations, proteins that interact with actin filaments may be able to distinguish rapidly polymerizing (ATP-rich) actin filaments from longer-established (ADP-rich) actin filaments.

ACTIN FILAMENTS CONTINUOUSLY EXTEND AND RETRACT

Actin filaments are dynamic structures. *Polymerization of actin monomers is a reversible process, so the polymer undergoes constant shrinking and growing as subunits add to and dissociate from one or both ends of the microfilament* (see Fig. 5.17). When the net rate of addition of subunits to one end of an actin filament matches the net rate of removal of subunits at the other end, the polymer is said to be **treadmilling** (**Fig. 5.18**).

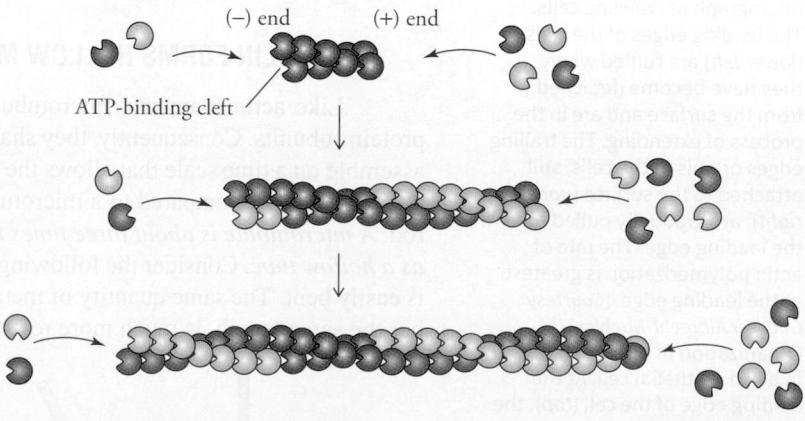

FIGURE 5.17

Actin filament assembly. An actin filament grows as subunits add to its ends. Subunits usually add more rapidly to the (+) end, which therefore grows faster than the (–) end. The original segment is shaded more darkly. Actual actin filaments are much longer than depicted here.

FIGURE 5.18

Actin filament treadmilling. Net assembly at one end balances net dissociation at the other end. The original segment (darker color) appears to travel along the filament during treadmilling.

? **Explain why the original segment usually moves toward the (–) end.**

Calculations suggest that under cellular conditions, the equilibrium between monomeric actin and polymeric actin favors the polymer. However, the growth of actin filaments *in vivo* is limited by capping proteins that bind to and block further polymerization at the (+) or (–) ends. A process that removes an actin filament cap will target growth to the uncapped end. New actin filament growth can also occur as branches form along existing microfilaments.

A supply of actin monomers to support actin filament growth in one area must come at the expense of actin filament disassembly elsewhere. In a cell, certain proteins sever *actin filaments* by binding to a polymerized actin subunit and inducing a small structural change that weakens actin–actin interactions and thereby increases the likelihood that the filament will break at that point. Actin subunits can then dissociate from the newly exposed ends unless they are subsequently capped.

Capping, branching, and severing proteins, along with other proteins whose activity is sensitive to extracellular signals, regulate the assembly and disassembly of actin filaments. A cell containing a network of actin filaments can therefore change its shape as the filaments lengthen in one area and regress in another. Certain cells use this system to move. When a cell crawls along a surface, actin polymerization extends its "leading" edge, while depolymerization helps retract its "trailing" edge (**Fig. 5.19a**). The high density of growing filament ends at the leading edge of the cell (Fig. 5.19b) illustrates how the rapid formation and outward extension of actin filaments can modulate cell shape and drive cell locomotion. Not only do actin filaments provide structural support and generate cell movement by assembly and disassembly, they also participate in generating tensile force. This system is well developed in muscle cells, where actin filaments are an essential part of the contractile apparatus (see Section 5.4).

TUBULIN FORMS HOLLOW MICROTUBULES

Like actin filaments, microtubules are cytoskeletal fibers built from small globular protein subunits. Consequently, they share with actin filaments the ability to assemble and disassemble on a time scale that allows the cell to rapidly change shape in response to external or internal stimuli. Compared to a microtubule, however, an actin filament is a thin and flexible rod. *A microtubule is about three times thicker and much more rigid because it is constructed as a hollow tube.* Consider the following analogy: A metal rod with the dimensions of a pencil is easily bent. The same quantity of metal, fashioned into a hollow tube with a larger diameter but the same length, is much more resistant to bending.

(a)

(b)

FIGURE 5.19

Actin filament dynamics in cell crawling. (a) Scanning electron micrograph of crawling cells. The leading edges of the cells (*lower left*) are ruffled where they have become detached from the surface and are in the process of extending. The trailing edges or tails of the cells, still attached to the surface (*upper right*), are gradually pulled toward the leading edge. The rate of actin polymerization is greatest at the leading edge. [*Courtesy Guenter Albrecht-Buehler.*] (b) Organization of actin filaments in a fish epithelial cell. At the leading edge of the cell (*top*), the filaments form a dense and highly branched network. Deeper within the cell (*bottom*), the filaments are sparser. [*Courtesy Tatyana Svitkina, Northwestern University Medical School.*]

FIGURE 5.20

Structure of β-tubulin. The strands of the two β sheets are shown in blue, and the 12 α helices that surround them are green. [*Structure of pig tubulin (pdb 1TUB) determined by E. Nogales and K. H. Downing.*]

FIGURE 5.21

The tubulin dimer. The guanine nucleotide (gold) in the α-tubulin subunit (*bottom*) is inaccessible in the dimer, whereas the nucleotide in the β-tubulin subunit (*top*) is more exposed to the solvent.

Bicycle frames, plant stems, and bones are built on this same principle. Cells use hollow microtubules to reinforce other elements of the cytoskeleton (see Fig. 5.14), to construct cilia and flagella, and to align and separate pairs of chromosomes during mitosis.

The basic structural unit of a microtubule is the protein tubulin. Two monomers, known as α-tubulin and β-tubulin, form a dimer, and a microtubule grows by the addition of tubulin dimers. Each tubulin monomer contains about 450 amino acids, 40% of them identical in α- and β-tubulin. The core of tubulin consists of a four-stranded and a six-stranded β sheet surrounded by 12 α helices (**Fig. 5.20**).

Each tubulin subunit includes a nucleotide-binding site. Unlike actin, tubulin binds a guanine nucleotide, either guanosine triphosphate (GTP) or its hydrolysis product, guanosine diphosphate (GDP). When the dimer forms, the α-tubulin GTP-binding site becomes buried in the interface between the monomers. The nucleotide-binding site in β-tubulin remains exposed to the solvent (**Fig. 5.21**). After the tubulin dimer is incorporated into a microtubule and another dimer binds on top of it, the β-tubulin nucleotide-binding site is also sequestered from solvent. The GTP is then hydrolyzed, but the resulting GDP remains bound to β-tubulin because it cannot diffuse away (the GTP in the α-tubulin subunit remains where it is and is not hydrolyzed).

Assembly of a microtubule begins with the end-to-end association of tubulin dimers to form a short linear **protofilament.** Protofilaments then align side-to-side in a curved sheet, which wraps around on itself to form a hollow tube of 13 protofilaments (**Fig. 5.22**). The microtubule extends as tubulin dimers add to both ends. Like an actin filament, the microtubule

FIGURE 5.22

Assembly of a microtubule. (a) αβ Dimers of tubulin initially form a linear protofilament. Protofilaments associate side by side, ultimately forming a tube. Tubulin dimers can add to either end of the microtubule, but growth is about twice as fast at the (+) end. (b) Cryoelectron microscopy view of a microtubule. [*Courtesy Kenneth Downing, Lawrence Berkeley National Laboratory.*]

? Compare a microtubule and an actin filament (Fig. 5.16) in terms of strength and speed of assembly.

is polar and one end grows more rapidly. *The (+) end, terminating in β-tubulin, grows about twice as fast as the (−) or α-tubulin end because tubulin dimers bind preferentially to the (+) end.*

Disassembly of a microtubule also takes place at both ends but occurs more rapidly at the (+) end. Under conditions that favor depolymerization, the ends of the microtubule appear to fray (**Fig. 5.23**). This suggests that tubulin dimers do not simply dissociate individually from the microtubule ends but that the interactions between protofilaments weaken before the tubulin dimers come loose.

Under certain conditions, microtubule treadmilling can occur when tubulin subunits add to the (+) end as fast as they leave the (−) end. *In vivo*, the (−) ends are often anchored to some sort of organizing center in the cell. This means that most microtubule growth and regression occur at the (+) end. Microtubule dynamics are also regulated by proteins that cross-link microtubules and promote or prevent depolymerization.

SOME DRUGS AFFECT MICROTUBULES

Compounds that interfere with microtubule dynamics can have drastic physiological effects. One reason is that during mitosis, chromosomes separate along a spindle made of microtubules (**Fig. 5.24**). The drug colchicine, a product of the meadow saffron plant, causes microtubules to depolymerize, thereby blocking cell division.

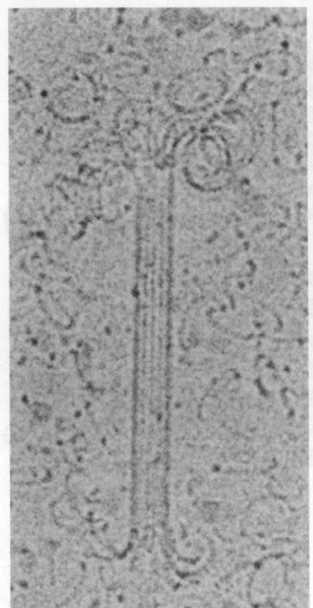

FIGURE 5.23

Electron micrograph of a depolymerizing microtubule. The ends of protofilaments apparently curve away from the microtubule and separate before tubulin dimers dissociate. [*Courtesy Ronald Milligan, The Scripps Research Institute.*]

Colchicine

Colchicine binds at the interface between α- and β-tubulin in a dimer, facing the inside of the microtubule cylinder. The bound drug may induce a slight conformational change that weakens the lateral contacts between protofilaments. If enough colchicine is present, microtubules shorten and eventually disappear. Colchicine was first used over 2000 years ago to treat **gout** (inflammation stemming from the precipitation of uric acid in the joints) because it inhibits the action of the white blood cells that mediate inflammation.

Paclitaxel binds to β-tubulin subunits in a microtubule, but not to free tubulin, so it stabilizes the microtubule, preventing its depolymerization.

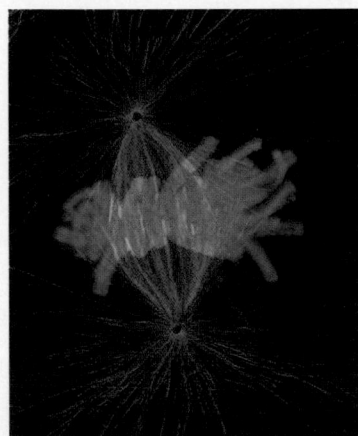

FIGURE 5.24

Microtubules in a dividing cell. During mitosis, microtubules (green fluorescence) connect replicated chromosomes (blue fluorescence) to two organizing centers (red dots) at opposite sides of the cell. The microtubules disassemble and shorten to pull the chromosomes apart before the cell splits in half. [*Micrograph by Alexey Khodjakov and Conly L. Rieder, Division of Translational Medicine, Wadsworth Center, Albany, New York 12201-0509.*]

Paclitaxel

FIGURE 5.25

Scanning electron micrograph of sectioned human skin.
The layers of dead epidermal cells at the top consist mostly of
keratin. [*Science Photo Library/Photo Researchers.*]

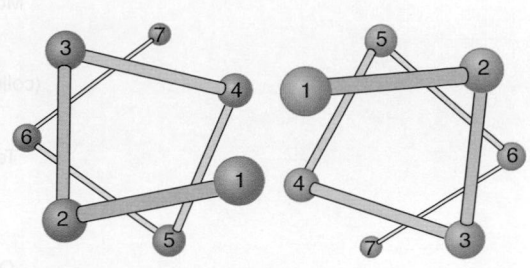

FIGURE 5.26

Arrangement of residues in a coiled coil. This view down the
axis of two seven-residue α helices shows that amino acids at
positions 1 and 4 line up on one side of each helix. Nonpolar
residues occupying these positions form a hydrophobic strip
along the sides of the helices.

The paclitaxel–tubulin interaction appears to include close contacts between paclitaxel's
phenyl groups and hydrophobic residues such as phenylalanine, valine, and leucine. Paclitaxel
was originally extracted from the slow-growing and endangered Pacific yew tree, but it can
also be purified from more renewable sources or chemically synthesized. *Paclitaxel is used as
an anticancer agent because it blocks cell division and is therefore toxic to rapidly dividing
cells such as tumor cells.*

KERATIN IS AN INTERMEDIATE FILAMENT

In addition to actin filaments and microtubules, eukaryotic cells—particularly those
in multicellular organisms—contain intermediate filaments. With a diameter of about 100 Å,
these fibers are intermediate in thickness to actin filaments and microtubules. *Intermediate filaments are exclusively structural proteins.* They play no part in cell motility, and unlike actin
filaments and microtubules, they have no associated motor proteins. However, they do interact
with actin filaments and microtubules via cross-linking proteins.

Intermediate filament proteins as a group are much more heterogeneous
than the highly conserved actin and tubulin. For example, humans have about
65 intermediate filament genes. The lamins are the intermediate filaments
that help form the nuclear lamina in animal cells, a 30–100-Å-thick network
inside the nuclear membrane that helps define the nuclear shape and may play
a role in DNA replication and transcription. In many cells, intermediate filaments are much more abundant than actin filaments or microtubules and are
most prominent in the dead remnants of epidermal cells—that is, in the hard
outer layers of the skin—where they may account for 85% of the total protein
(**Fig. 5.25**). The best-known intermediate filament proteins are the keratins,
a large group of proteins that include the "soft" keratins, which help define
internal body structures, and the "hard" keratins of skin, hair, and claws.

*The basic structural unit of an intermediate filament is a dimer of α helices
that wind around each other—that is, a **coiled coil.*** The amino acid sequence
in such a structure consists of seven-residue repeating units in which the first
and fourth residues are predominantly nonpolar. In an α helix, these nonpolar
residues line up along one side (**Fig. 5.26**). Because a nonpolar group appears
on average every 3.5 residues but there are 3.6 residues per α-helical turn, the
strip of nonpolar residues actually winds slightly around the surface of the
helix. Two helices whose nonpolar strips contact each other therefore adopt a
coiled structure with a left-handed twist (**Fig. 5.27**).

Each intermediate filament subunit contains a stretch of α helix flanked
by nonhelical regions at the N- and C-termini. Two of these polypeptides
interact in register (parallel and with ends aligned) to form a coiled coil.
The dimers then associate in a staggered antiparallel arrangement to form
higher-order fibrous structures (**Fig. 5.28**). The fully assembled intermediate
filament may consist of 16 to 32 polypeptides in cross-section. Note that no

(a)   (b)   (c)

FIGURE 5.27

Three views of a coiled coil. These models show
a segment of the coiled coil from the protein
tropomyosin. (a) Backbone model. (b) Stick model.
(c) Space-filling model. Each α-helical chain contains
100 residues. The nonpolar strips along each helix
contact each other, so the two helices wind around
each other in a gentle left-handed coil. [*Structure of
tropomyosin (pdb 1C1G) determined by F. G. Whitby and
G. N. Phillips, Jr.*]

? **Which nonpolar residues would be most likely
to appear at positions 1 and 4? Which would
be least likely?**

Monomer

Dimer
(coiled coil)

Tetramer

Octamer

Intermediate
filament

FIGURE 5.28

Model of an intermediate filament. Pairs of polypeptides form coiled coils. These dimers associate to form tetramers and so on, ultimately producing an intermediate filament composed of 16 to 32 polypeptides in cross-section. Although drawn as straight rods here, the intermediate filament and its component structures are probably all twisted around each other in some way, much like a man-made rope or cable.

nucleotides are required for intermediate filament assembly. The N- and C-terminal domains may help align subunits during polymerization and interact with proteins that cross-link intermediate filaments to other cell components. Keratin fibers themselves are cross-linked through disulfide bonds between cysteine residues on adjacent chains.

An animal hair—for example, sheep's wool or the hair on your head—consists almost entirely of keratin filaments (**Fig. 5.29**). Hair resists deformation but can be stretched. Tensile stress breaks the hydrogen bonds between carbonyl and amino groups four residues apart in the keratin α helix. The helices can then be pulled until the polypeptides are fully extended. Additional force causes the polypeptide chain to break. If unbroken, the protein can spring back—at least partially—to its original α-helical conformation when the force is removed. This is why a wool sweater stretched out of shape gradually reverts to its former style.

Perhaps due to their cable-like construction, intermediate filaments undergo less remodeling than cellular fibers constructed from globular subunits such as actin and tubulin. For example, nuclear lamins disassemble and reassemble only once during the cell cycle, when the cell divides. Keratins, as part of dead cells, remain intact for years. In the innermost layers of animal skin, epidermal cells synthesize large amounts of keratin. As layers of cells move outward and die, their keratin molecules are pushed together to form a strong waterproof coating. Because keratin is so important for the integrity of the epidermis, mutations in keratin genes are linked to certain skin disorders. In diseases such as epidermolysis bullosa simplex (EBS), cells rupture when subject to normal mechanical stress. The result is separation of epidermal layers, which leads to blistering. The most severe cases of the disease arise from mutations in the most highly conserved regions of the keratin molecules, near the ends of the α-helical regions.

COLLAGEN IS A TRIPLE HELIX

Unicellular organisms can get by with just a cytoskeleton, but multicellular animals must have a way to hold their cells together according to some characteristic body plan. Large animals—especially nonaquatic ones—must also support the body's weight. This support is provided by collagen, which is the most abundant animal protein. It plays a major structural role in the **extracellular matrix** (the

FIGURE 5.29

Scanning electron micrograph of a human hair. Bundles of intermediate filaments, not the individual proteins, are visible. [*Tony Brain/Science Photo Library/Photo Researchers.*]

material that helps hold cells together), in connective tissue within and between organs, and in bone. Its name was derived from the French word for *glue* (at a time when glue was derived from animal connective tissue).

There are at least 28 types of collagen with different three-dimensional structures and physiological functions. The most familiar is the collagen from animal bones and tendons, which forms thick, ropelike fibers (**Fig. 5.30**). This type of collagen is a trimeric molecule about 3000 Å long but only 15 Å wide. As in all forms of collagen, the polypeptide chains have an unusual amino acid composition and an unusual conformation. Except in the extreme N- and C-terminal regions of the polypeptides (which are cleaved off once the protein exits the cell), every third amino acid is glycine, and about 30% of the remaining residues are proline and hydroxyproline (Hyp). Hyp residues result from the hydroxylation of Pro residues after the polypeptide has been synthesized, in a reaction that requires ascorbate (vitamin C; **Box 5.B**).

FIGURE 5.30

Electron micrograph of collagen fibers. Thousands of aligned collagen proteins yield structures with a diameter of 500 to 2000 Å, which are visible here. [*J. Gross/ Science Photo Library/Photo Researchers.*]

Proline   Hydroxyproline (Hyp)

For a stretch of about 1000 residues, each collagen chain consists of repeating triplets, the most common of which is Gly–Pro–Hyp. Glycine residues, which have only a hydrogen atom for a side chain, can normally adopt a wide range of secondary structures. However, the **imino groups** of proline and hydroxyproline residues (that is, their connected side chains and amino groups) constrain the geometry of the peptide group. *The most stable conformation for a polypeptide sequence containing repeating units of Gly–Pro–Hyp is a narrow left-handed helix* (**Fig. 5.31a**).

A CLOSER LOOK

Box 5.B Vitamin C Deficiency Causes Scurvy

Ascorbate (vitamin C)

In the absence of ascorbate (vitamin C), collagen contains too few hydroxyproline residues and hydroxylated lysine residues, so the resulting collagen fibers are relatively weak. Ascorbate also participates in enzymatic reactions involved in fatty acid breakdown and production of certain hormones. The symptoms of ascorbate deficiency include poor wound healing, loss of teeth, and easy bleeding—all of which can be attributed to abnormal collagen synthesis—as well as lethargy and depression.

Historically, ascorbate deficiency, known as scurvy, was common in sailors on long voyages where fresh fruit was unavailable. A remedy, in the form of a daily ration of limes, was discovered in the mid-eighteenth century. Unfortunately, citrus juice,

which was also widely administered, proved much less effective in preventing scurvy because ascorbate is destroyed by heating and by prolonged exposure to air. For this reason, factors such as exercise and good hygiene were also believed to prevent scurvy.

Fruit is not the only source of ascorbate. Most animals— with the exception of bats, guinea pigs, and primates—produce ascorbate, so a diet containing fresh meat can supply sufficient ascorbate, which is vital in locations, such as the far North, where fruit is not available.

Given the presence of ascorbate in many foods, a true deficiency is rare in adults. Scurvy does still occur, however, as a side effect of general malnutrition or in individuals consuming odd diets. Fortunately, the symptoms of scurvy, which is otherwise fatal, can be easily reversed by administering ascorbate or consuming fresh food.

? **Refer to the structure of ascorbate to explain why it does not accumulate in the body but instead is readily lost via the kidneys.**

FIGURE 5.31

Collagen structure. (a) A sequence of repeating Gly–Pro–Hyp residues adopts a secondary structure in which the polypeptide forms a narrow left-handed helix. The residues in this stick model are color-coded: Gly gray, Pro orange, Hyp red. H atoms are not shown. (b) Space-filling model of a single collagen polypeptide. (c) Space-filling model of the triple helix. (d) Backbone trace showing the three polypeptides in different shades of gray. Each polypeptide has a left-handed twist, but the triple helix has a right-handed twist. [*Model of collagen (pdb 2CLG) constructed by J. M. Chen.*]

In collagen, three polypeptides wind around each other to form a right-handed **triple helix** (Fig. 5.31b–d). The chains are parallel but staggered by one residue so that glycine appears at every position along the axis of the triple helix. The glycine residues are all located in the center of the helix, whereas all other residues are on the periphery. A look down the axis of the triple helix shows why glycine—but no other residue—occurs in the center of the helix (**Fig. 5.32**). The side chain of any other residue would be too large to fit. In fact, replacing glycine with alanine, the next-smallest amino acid, greatly perturbs the structure of the triple helix.

FIGURE 5.32

Cross-section of the collagen triple helix. In this view, looking down the axis of the three-chain molecule, each ball represents an amino acid, and the bars represent peptide bonds. Glycine residues (which lack side chains and are marked by "G") are located in the center of the triple helix, whereas the side chains of other residues point outward from the triple helix.

❓ Compare this axial view of a collagen triple helix to the axial view of a coiled coil (Fig. 5.26).

The collagen triple helix is stabilized through hydrogen bonding. One set of interactions links the backbone N—H group of each glycine residue to a backbone C=O group in another chain. The geometry of the triple helix prevents the other backbone N—H and C=O groups from forming hydrogen bonds with each other, but they are able to interact with a highly ordered network of water molecules surrounding the triple helix like a sheath.

COLLAGEN MOLECULES ARE COVALENTLY CROSS-LINKED

Trimeric collagen molecules assemble in the endoplasmic reticulum. After they are secreted from the cell, they are trimmed by proteases and align side-to-side and end-to-end to form the enormous fibers visible by electron microscopy (see Fig. 5.30). The fibers are strengthened by several kinds of cross-links. Because collagen polypeptides contain almost no cysteine, these links are not disulfide bonds. Instead, *the cross-links are covalent bonds between side chains that have been chemically modified following polypeptide synthesis.* For example, one kind of cross-link requires the enzyme-catalyzed oxidation of two lysine side chains, which then react to form a covalent bond. The number of these and other types of cross-links tends to increase with age, which explains why meat from older animals is tougher than meat from younger animals.

Collagen fibers have tremendous tensile strength. On a per-weight basis, collagen is stronger than steel. Not all types of collagen form thick linear fibers, however. Many nonfibrillar collagens form sheetlike networks of fibers that support layers of cells in tissues. Often, several types of collagen are found together. Not surprisingly, defects in collagen affect a variety of organ systems (**Box 5.C**).

A CLOSER LOOK

Box 5.C Bone and Collagen Defects

Connective tissue, such as cartilage and bone, consists of cells embedded in a matrix containing proteins (mainly collagen) and a space-filling "ground substance" (mostly polysaccharides; see Section 11.3). The polysaccharides, which are highly hydrated, are resilient and return to their original shape after being compressed. The collagen fibers are strong and relatively rigid, resisting tensile (stretching) forces. Together, the polysaccharides and collagen give ligaments (which attach bone to bone) and tendons (which join muscles to bones) the appropriate degree of resistance and flexibility. The connective tissues that surround muscles and organs contain collagen fibers arranged in sheetlike networks with similar physical properties.

The regular arrays of collagen fibers (*horizontal bands*) in tendon allow the tissue to resist tension. Fibroblasts, collagen-producing cells, occupy the spaces between the fibers. Cartilage typically lacks blood vessels. [*Mark Nielsen.*]

In bone, the extracellular matrix is supplemented by minerals, mainly hydroxyapatite, $Ca(PO_4)_3(OH)$. This form of calcium phosphate forms extensive white crystals that account for up to 50% of the mass of bone. By itself, calcium phosphate is brittle. Yet bone is thousands of times stronger than the crystals because it is a composite structure in which the mineral is interspersed with collagen fibers. This layered arrangement dissipates stresses so that the bone remains strong but can "give" a little under pressure without shattering. Bone tissue also includes passageways for small blood vessels.

During development, most of the skeleton takes shape as cartilaginous tissue becomes mineralized, forming hard bone. The repair of a fractured bone follows a similar process in which fibroblasts moving to the injured site synthesize large amounts of collagen, chondroblasts produce cartilage, and osteoblasts gradually replace the cartilage with bony tissue. Mature bone continues to undergo remodeling, which begins with the release of enzymes and acid from cells known as osteoclasts. The enzymes digest collagen and other extracellular matrix components, while the low pH helps dissolve calcium phosphate. Osteoblasts then fill the void with new bone material. Some bones are remodeled faster than others, and the system responds to physical demands so that the bone becomes stronger and thicker when subjected to heavy loads. This is the basis for orthodontics: Teeth are realigned by placing stress on the bone in the tooth sockets for a period long enough for remodeling to take place. Bone remodeling also plays a role in cancer, as microscopic sites of bone destruction may provide access to metastatic (spreading) cancer cells; bone is a common site for metastatic tumor growth, especially in breast cancer.

The importance of collagen for the structure and function of connective tissues means that irregularities in the collagen protein itself or in the enzymes that process collagen molecules can lead to serious physical abnormalities. Hundreds of collagen-related mutations have been identified. Because most tissues contain more than one type of collagen, the physiological manifestations of collagen mutations are highly variable.

Defects in collagen type I (the major form in bones and tendons) cause the congenital disease **osteogenesis imperfecta.** The primary symptoms of the disease include bone fragility leading to easy fracture, long-bone deformation, and abnormalities of the skin and teeth.

Collagen type I, a trimeric molecule, contains two different types of polypeptide chains. Therefore, the severity of the disease depends in part on whether one or two chains in a collagen molecule are affected. Furthermore, *the location and nature of the mutation determine whether the abnormal collagen retains some normal function.* For example, in one severe form of osteogenesis imperfecta, a 599-base deletion in a collagen gene represents the loss of a large portion of triple helix. The resulting protein is unstable and is degraded intracellularly. Milder cases of osteogenesis imperfecta result from amino acid substitutions, for example, the replacement of glycine by a bulkier residue. Other amino acid changes may slow intracellular processing and excretion of collagen polypeptides, which affects the assembly of collagen fibers. Osteogenesis imperfecta affects about one in 10,000 people.

Mutations in collagen type II, a form found in cartilage, lead to osteoarthritis. This genetic disease, which becomes apparent in childhood, is distinct from the osteoarthritis that can develop later in life, often after years of wear and tear on the joints. Defects in the proteins that process collagen extracellularly and help assemble collagen fibers lead to disorders such as dermatosparaxis, which is characterized by extreme skin fragility.

Ehlers–Danlos syndrome results from abnormalities in collagen type III, a molecule that is abundant in most tissues but is scarce in skin and bone. Symptoms of this phenotypically variable disorder include easy bruising, thin or elastic skin, and joint hyperextensibility. In one form of the disease, which is accompanied by a high risk for arterial rupture, the molecular defect is a mutation in a collagen type III gene. In another form of the disease, in which individuals often suffer from scoliosis (curvature of the spine), the collagen genes are normal. In these cases, the disease results from a deficiency of lysyl oxidase, the enzyme that modifies lysine residues so that they can participate in collagen cross-links. Ehlers–Danlos syndrome is both rarer and less severe than osteogenesis imperfecta, with many affected individuals surviving to adulthood.

X-Ray of a child with a moderately severe case of osteogenesis imperfecta. [*ISM/Phototake.*]

 CONCEPT CHECK

1. Make a table to compare the subunit structure, overall size, dynamic character, and biological functions of actin filaments, microtubules, intermediate filaments, and collagen.
2. Why does normal physiological function require regulation of the assembly and disassembly of cytoskeletal elements?
3. What is the basis of actin filament polarity?
4. How do two α helices form a coiled coil?
5. List some ways that cells can increase the strength of structural fibers.
6. What is the structural importance of Gly residues in collagen?

MOTOR PROTEINS

5.4

In eukaryotic cells, **motor proteins** act on structural elements such as actin filaments and microtubules to generate the movements that allow the cells to reorganize their contents, change their shape, and even crawl or swim. For example, muscle contraction is accomplished by the motor protein myosin pulling on actin filaments. Eukaryotic cells that move via the wavelike motion of cilia or flagella rely on the motor protein dynein, which acts by bending a bundle of microtubules. Intracellular transport is carried out by motor proteins that move along actin filament and microtubule tracks. In all cases, the molecular machinery uses the chemical energy of ATP hydrolysis to carry out mechanical work. In this section, we focus on two well-characterized motor proteins, myosin and kinesin.

KEY CONCEPTS
- The motor protein myosin couples the steps of ATP hydrolysis to conformational changes, resulting in muscle contraction.
- Kinesin transports cargo by moving processively along a microtubule track.

MYOSIN HAS TWO HEADS AND A LONG TAIL

There are at least 20 different types of myosin, which is present in nearly all eukaryotic cells. Myosin works with actin to produce movement by transducing the free energy of the ATP hydrolysis reaction to mechanical energy. Muscle myosin, with a total molecular mass of about 540 kD, consists mostly of two large polypeptides that form two globular heads attached to a long tail (**Fig. 5.33**). Each head includes a binding site for actin and a binding site for an adenine nucleotide. In the tail region, the two polypeptides twist around each other to form a single rodlike coiled coil (the same structural motif that occurs in intermediate filaments). The "neck" that joins each myosin head to the tail region consists of an α helix about 100 Å long, around which are wrapped two small polypeptides (**Fig. 5.34**). These so-called light chains help stiffen the neck helix so that it can act as a lever.

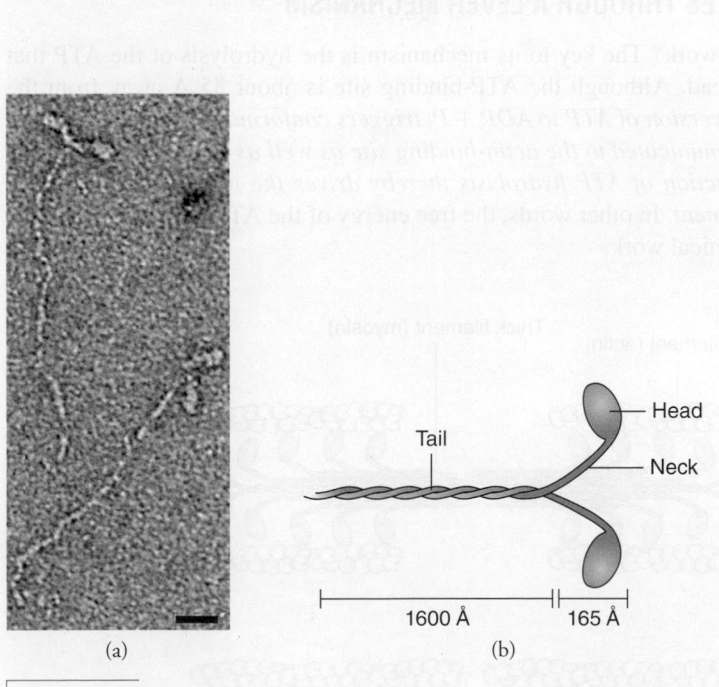

(a)

(b)

Tail

Head

Neck

1600 Å

165 Å

FIGURE 5.33

Structure of muscle myosin. (a) Electron micrograph. [*Courtesy John Trinick, University of Leeds.*] (b) Drawing of a myosin molecule. Myosin's two globular heads are connected via necks to myosin's tail, where the polypeptide chains form a coiled coil.

? **In what part of the protein are large hydrophobic residues most likely to occur?**

Neck

ATP-binding pocket

Head

Actin-binding site

FIGURE 5.34

Myosin head and neck region. The myosin neck forms a molecular lever between the head domain and the tail. Two light chains (lighter shades of purple) help stabilize the α-helical neck. The actin-binding site is at the far end of the myosin head. ATP binds in a cleft near the middle of the head. Only the alpha carbons of the light chains are visible in this model. [*Structure of chicken myosin (pdb 2MYS) determined by I. Rayment and H. M. Holden.*]

FIGURE 5.35

Electron micrograph of a thick filament. The heads of many myosin molecules project laterally from the thick rod formed by the aligned myosin tails. [*From Trinick, J., and Elliott, A., J. Mol. Biol. 131, 15 (1977).*]

Each myosin head can interact noncovalently with a subunit in an actin filament, but the two heads act independently, and only one head binds to the actin filament at a given time. *In a series of steps that include protein conformational changes and the hydrolysis of ATP, the myosin head releases its bound actin subunit and rebinds another subunit closer to the (+) end of the actin filament.* Repetition of this reaction cycle allows myosin to progressively walk along the length of the actin filament.

In a muscle cell, hundreds of myosin tails associate to form a **thick filament** with the head domains sticking out (**Fig. 5.35**). These heads act as cross-bridges to **thin filaments,** which each consist of an actin filament and actin-binding proteins that regulate the accessibility of the actin subunits to myosin heads. When a muscle contracts, the multitude of myosin heads individually bind and release actin, like rowers working asynchronously, which causes the thin and thick filaments to slide past each other (**Fig. 5.36**). Because of the arrangement of filaments in the muscle cell, the action of myosin on actin results in an overall shortening of the muscle. This phenomenon is commonly called contraction, but the muscle does not undergo any compression and its volume remains constant—it actually becomes thicker around the middle. A shortening on the order of 20% of a muscle's length is typical; 40% is extreme.

MYOSIN OPERATES THROUGH A LEVER MECHANISM

How does myosin work? The key to its mechanism is the hydrolysis of the ATP that is bound to the myosin head. Although the ATP-binding site is about 35 Å away from the actin-binding site, *the conversion of ATP to ADP + P_i triggers conformational changes in the myosin head that are communicated to the actin-binding site as well as to the lever (the neck region). The chemical reaction of ATP hydrolysis thereby drives the physical movement of myosin along an actin filament.* In other words, the free energy of the ATP hydrolysis reaction is transformed into mechanical work.

Thin filament (actin)   Thick filament (myosin)

Relaxed

Contracted

FIGURE 5.36

Movement of thin and thick filaments during muscle contraction.

1. *The reaction sequence begins with a myosin head bound to an actin subunit of the thin filament. ATP binding alters the conformation of the myosin head so that it releases actin.*

2. *The rapid hydrolysis of ATP to ADP + P_i triggers a conformational change that rotates the myosin lever and increases the affinity of myosin for actin.*

3. *Myosin binds to an actin subunit farther along the thin filament.*

4. *Binding to actin causes P_i and then ADP to be released. As these reaction products exit, the myosin lever returns to its original position. This causes the thin filament to move relative to the thick filament (the power stroke).*

ATP replaces the lost ADP to repeat the reaction cycle.

FIGURE 5.37

The myosin–actin reaction cycle. For simplicity, only one myosin head is shown.

 Write an equation that describes the reaction catalyzed by myosin.

The four steps of the myosin–actin reaction sequence are shown in **Figure 5.37**. Note how each event at the nucleotide-binding site correlates with a conformational change related to either actin binding or bending of the lever. An α helix makes an ideal lever because it can be quite long. It is also relatively incompressible, so it can pull the coiled-coil myosin tail along with it. Altogether, the lever swings by about 70° relative to the myosin head. The return of the lever to its original conformation (step 4 of the reaction cycle) is the force-generating step. The shortening and bulging of a contracting muscle actually augment its power by optimizing the angle at which the myosin heads tug on actin filaments. When adjusted for the difference in mass, the myosin–actin system has a power output comparable to a typical automobile.

Each cycle of ATP hydrolysis moves the myosin head by an estimated 50–100 Å. Since individual actin subunits are spaced about 55 Å apart along the thin filament, the myosin head advances by at least one actin subunit per reaction cycle. Because the reaction cycle involves several steps, some of which are essentially irreversible (such as ATP → ADP + P_i), the entire cycle is unidirectional.

Myosin operates in many cells, not just in muscles. For example, myosin works with actin during **cytokinesis** (the splitting of the cell into two halves following mitosis), and some myosin proteins use their motor activity to transport certain cell components along actin filament tracks. Myosin molecules may also act as tension rods to cross-link the actin filaments of the cytoskeleton. This is one reason why mutations in myosin in the sensory cells of the ear cause deafness and other abnormalities (**Box 5.D**).

KINESIN IS A MICROTUBULE-ASSOCIATED MOTOR PROTEIN

Many cells also contain motor proteins, such as kinesin, that move along microtubule tracks. There are several different types of kinesins; we will describe the prototypical one, also known as conventional kinesin.

A CLOSER LOOK

Box 5.D Myosin Mutations and Deafness

Inside the cochlea, the spiral-shaped organ of the inner ear, are thousands of hair cells, each of which is topped with a bundle of bristles known as **stereocilia.**

[P. Motta/Science Photo Library/Photo Researchers.]

Each stereocilium contains several hundred cross-linked actin filaments and is therefore extremely rigid, except at its base, where there are fewer actin filaments. Sound waves deflect the stereocilia at the base, initiating an electrical signal that is transmitted to the brain.

Myosin molecules probably help control the tension inside each stereocilium, so the ratcheting activity of the myosin motors along the actin filaments may adjust the sensitivity of the hair cells to different degrees of stimulus. Other myosin molecules whose tails bind certain cell constituents may use their motor activity to redistribute these substances along the length of the actin filaments. Abnormalities in any of these proteins could interfere with normal hearing.

About half of all cases of deafness have a genetic basis, and over 100 different genes have been linked to deafness. One of these codes for myosin type VIIa, which is considered to be an "unconventional" myosin because it differs somewhat from the "conventional" myosin of skeletal muscle (also known as myosin type II). Gene-sequencing studies indicate that myosin VIIa has 2215 residues and forms a dimer with two heads and a long tail. Its head domains probably operate by the same mechanism as muscle myosin, converting the chemical energy of ATP into mechanical energy for movement relative to an actin filament. Over a hundred mutations in the myosin VIIa gene have been identified, including premature stop codons, amino acid substitutions, and deletions—all of which compromise the protein's function. Such mutations are responsible for many cases of **Usher syndrome,** the most common form of deaf-blindness in the United States. Usher syndrome is characterized by profound hearing loss, retinitis pigmentosa (which leads to blindness), and sometimes vestibular (balance) problems.

The congenital deafness of Usher syndrome results from the failure of the cochlear hair cells to develop properly. The unresponsiveness of the stereocilia to sound waves probably also accounts for their inability to respond normally to the movement of fluid in the inner ear, which is necessary for maintaining balance. Abnormal myosin also plays a role in the blindness that often develops in individuals with Usher syndrome, usually by the second or third decade. The intracellular transport function of myosin VIIa is responsible for distributing bundles of pigment in the retina. In retinitis pigmentosa, retinal neurons gradually lose their ability to transmit signals in response to light; in advanced stages of the disease, pigment actually becomes clumped on the retina.

? **Not all mutations are associated with a loss of function. Explain how a myosin mutation that *increases* the rate of ADP release could lead to deafness.**

Kinesin, like myosin, is a relatively large protein (with a molecular mass of 380 kD) and has two large globular heads and a coiled-coil tail domain (**Fig. 5.38**). Each 100-Å-long head consists of an eight-stranded β sheet flanked by three α helices on each side and includes a tubulin-binding site and a nucleotide-binding site. The light chains, situated at the opposite end of the protein, bind to proteins in the membrane shell of a **vesicle.** The vesicle

(a)

Heads

Tail

Light chains

(b)

FIGURE 5.38

Structure of kinesin. (a) Diagram of the molecule. (b) Model of the head and neck region. Each globular head, which contains α helices and β sheets, connects to an α helix that winds around its counterpart to form a coiled coil. Two light chains at the end of the coiled-coil tail can interact with a membranous vesicle, the "cargo." [*Structure of rat kinesin (pdb 3KIN) determined by F. Kozielski, S. Sack, A. Marx, M. Thormahlen, E. Schonbrunn, V. Biou, A. Thompson, E.-M. Mandelkow, and E. Mandelkow.*]

? Compare the structure of kinesin to the structure of myosin (Fig. 5.33).

and its contents become kinesin's cargo. *Kinesin moves its cargo toward the (+) end of a microtubule by stepping along the length of a single protofilament.* Other microtubule-associated motor proteins appear to use a similar mechanism but move toward the (−) end of the microtubule.

The motor activity of kinesin requires the free energy of ATP hydrolysis. However, *kinesin cannot follow the myosin lever mechanism because its head domains are not rigidly fixed to its neck regions.* In kinesin, a relatively flexible polypeptide segment joins each head to an α helix that eventually becomes part of the coiled-coil tail (see Fig. 5.38b). (Recall that in myosin, the lever is a long α helix that extends from the head to the coiled-coil region and is stiffened by the two light chains; see Fig. 5.34.) Nevertheless, the relative flexibility of kinesin's neck is critical for its function.

Kinesin's two heads are not independent but work in a coordinated fashion so that the two heads alternately bind to successive β-tubulin subunits along a protofilament, as if walking. Conformational changes elicited by ATP binding and hydrolysis are relayed to other regions of the molecule (**Fig. 5.39**). *This transforms the free energy of ATP hydrolysis into*

1. *ATP binding to the leading head (orange) induces a conformational change in which the neck docks against the head. This movement swings the trailing head (yellow) forward by 180° toward the (+) end of the microtubule. This is the force-generating step.*

2. *The new leading head (yellow) quickly binds to a tubulin subunit and releases its ADP. This step moves kinesin's cargo forward along the protofilament.*

3. *In the trailing head (orange), ATP is hydrolyzed to ADP + P_i. The P_i diffuses away, and the trailing head begins to detach from the microtubule.*

4. *ATP binds to the leading head to repeat the reaction cycle.*

FIGURE 5.39

The kinesin reaction cycle. The cycle begins with one kinesin head bound to a tubulin subunit in a protofilament of a microtubule. The trailing head has ADP in its nucleotide-binding site. For clarity, the relative size of the neck region is exaggerated.

? **What prevents kinesin from moving backward?**

FIGURE 5.40

Electron micrograph of neurons. Microtubule-associated motor proteins move cargo between the cell body and the ends of the axon and other cell processes. [*CNRI/Science Photo Library/Photo Researchers.*]

the mechanical movement of kinesin. Each ATP-binding event yanks the trailing head forward by about 160 Å, so the net movement of the attached cargo is about 80 Å, or the length of a tubulin dimer.

KINESIN IS A PROCESSIVE MOTOR

As in the myosin–actin system (see Fig. 5.37), the kinesin–tubulin reaction cycle proceeds in only one direction. Although most molecular movement is associated with ATP binding, ATP hydrolysis is a necessary part of the reaction cycle. The slowest step of the reaction cycle shown in Figure 5.39 is the dissociation of the trailing kinesin head from the microtubule. ATP binding to the leading head may help promote trailing-head release as part of the forward-swing step. Because the free head quickly rebinds tubulin, the kinesin heads spend most of their time bound to the microtubule track.

One consequence of kinesin's almost constant hold on a microtubule is that many—perhaps 100 or more—cycles of ATP hydrolysis and kinesin advancement can occur before the motor dissociates from its microtubule track. *Kinesin is therefore said to have high* **processivity.** A motor protein such as myosin, which dissociates from an actin filament after a single stroke, is not processive.

In a muscle cell, low processivity is permitted because the many myosin–actin interactions occur more or less simultaneously to cause the thin and thick filaments to slide past each other (see Fig. 5.36). *High processivity is advantageous for a transport engine such as kinesin, because its cargo (which is relatively large and bulky) can be moved long distances without being lost.* Consider the need for efficient transport in a neuron. Neurotransmitters and membrane components are synthesized in the cell body, where ribosomes are located, but must be moved to the end of the axon, which may be several meters long in some cases (**Fig. 5.40**).

 CONCEPT CHECK

1. Compare and contrast the myosin–actin and kinesin–tubulin reaction cycles.
2. Describe the steps of myosin and kinesin reaction cycles.
3. Explain the role of nucleotide binding and hydrolysis in molecular movement.
4. List the cellular activities that require processive motors.

SUMMARY

5.1 Myoglobin and Hemoglobin: Oxygen-Binding Proteins

- Myoglobin contains a heme prosthetic group that reversibly binds oxygen. The amount of O_2 bound depends on the O_2 concentration and on myoglobin's affinity for oxygen.
- Hemoglobin's α and β chains are homologous to myoglobin, indicating a common evolutionary origin.
- O_2 binds cooperatively to hemoglobin with low affinity, so hemoglobin can efficiently bind O_2 in the lungs and deliver it to myoglobin in the tissues.
- Hemoglobin is an allosteric protein whose four subunits alternate between the T (deoxy) and R (oxy) conformations in response to O_2

binding to the heme groups. The deoxy conformation is favored by low pH (the Bohr effect) and by the presence of BPG.

5.2 Clinical Connection: Hemoglobin Variants

- Hemoglobin variants provide information about protein structure and function. The substitution of glutamate by valine in the β chain of hemoglobin S leads to polymerization that causes red blood cell sickling and anemia. Carriers of the genetic variant are more likely to survive malaria, as are carriers of the variant that produces hemoglobin C.
- Genetic variations that affect hemoglobin production, oxygen binding affinity, and intersubunit interactions may have mild to severe clinical effects.

5.3 Structural Proteins

• The microfilament elements of a cell's cytoskeleton are built from ATP-binding globular actin subunits that polymerize as a double chain. Polymerization is reversible, so actin filaments undergo growth and regression. Their dynamics may be modified by proteins that mediate filament capping, branching, and severing.

• GTP-binding tubulin dimers polymerize to form a hollow microtubule. Polymerization is more rapid at one end, and the microtubule can disassemble rapidly by fraying. Drugs that affect microtubule dynamics interfere with cell division.

• The intermediate filament keratin contains two long α-helical chains that coil around each other so that their hydrophobic residues are in contact. Keratin filaments associate and are cross-linked to form semipermanent structures.

• Collagen polypeptides contain a large amount of proline and hydroxyproline and include a glycine residue at every third position.

Each chain forms a narrow left-handed helix, and three chains coil around each other to form a right-handed triple helix with glycine residues at its center. Covalent cross-links strengthen collagen fibers.

5.4 Motor Proteins

• Myosin, a protein with a coiled-coil tail and two globular heads, interacts with actin filaments to perform mechanical work. ATP-driven conformational changes allow the myosin head to bind, release, and rebind actin. This mechanism, in which myosin acts as a lever, is the basis of muscle contraction.

• The motor protein kinesin has two globular heads connected by flexible necks to a coiled-coil tail. Kinesin transports vesicular cargo along the length of a microtubule by a processive stepping mechanism that is driven by conformational changes triggered by ATP binding and hydrolysis.

KEY TERMS

heme
prosthetic group
Y
pO_2
saturation
p_{50}
globin
homologous proteins
invariant residue
conservative substitution
variable residue
anemia

cooperative binding
deoxyhemoglobin
oxyhemoglobin
T state
R state
allosteric protein
ligand
Bohr effect
thalassemia
cytoskeleton
actin filament
intermediate filament

microtubule
F-actin
G-actin
(−) end
(+) end
P_i
treadmilling
protofilament
gout
coiled coil
extracellular matrix
imino group

triple helix
osteogenesis imperfecta
Ehlers–Danlos syndrome
motor protein
thick filament
thin filament
cytokinesis
stereocilia
Usher syndrome
vesicle
processivity

BIOINFORMATICS

Brief Bioinformatics Exercises

5.1 Visualizing and Analyzing Hemoglobin
5.2 Globin Sequence Alignment and Evolution
5.3 Deoxyhemoglobin and Oxyhemoglobin

Bioinformatics Projects

Using Databases to Compare and Identify Related Protein Sequences
The Pfam Database

PROBLEMS

5.1 Myoglobin and Hemoglobin: Oxygen-Binding Proteins

1. Explain why globin or heme alone is not effective as an oxygen carrier.

2. Myoglobin transfers oxygen obtained from hemoglobin to mitochondrial proteins involved in the catabolism of metabolic fuels to produce energy for the muscle cell. Using what you have learned in this chapter about myoglobin and hemoglobin, what can you conclude about the structures of these mitochondrial proteins?

3. Myoglobin most effectively facilitates oxygen diffusion through muscle cells when the intracellular oxygen partial pressure is comparable to the p_{50} value of myoglobin. Explain why.

4. Calculate the fractional saturation for myoglobin when p_{50} is (a) 15 torr and (b) 75 torr.

5. At what pO_2 value is myoglobin (a) 30% saturated and (b) 80% saturated?

6. Heart and muscle cells, where myoglobin resides, maintain an intracellular pO_2 of about 2.5 torr. Explain why a small change in oxygen partial pressure of 1 torr in either direction results in a dramatic change in myoglobin oxygen binding.

7. Oxygenated myoglobin is bright red, whereas deoxygenated myoglobin is a purplish color. Myoglobin in which the Fe^{2+} has been oxidized to Fe^{3+} is referred to as met-myoglobin and appears brown. The sixth ligand in met-myoglobin is water rather than oxygen.

   (a) When a fresh, uncooked beef roast is sliced in two, the surface of the meat first appears purplish but then turns red. Explain why.
   (b) Why does meat turn brown when it is cooked?
   (c) Why might a retailer choose to package meat using oxygen permeable plastic wrap instead of a vacuum-sealed package?

8. The oxygen binding site in myoglobin, shown in Figure 5.2 and in the figure below, resides in a heme pocket in which a valine residue provides one of the pocket boundaries. How is oxygen binding affected when the Val residue is mutated to (a) isoleucine or (b) serine? *Hint:* Is it more advantageous for oxygen binding for the pocket to be polar or nonpolar?

9. Myoglobins from three species of fish—tuna, bonito, and mackerel—were purified, and their oxygen binding affinities were measured at 20°C. The results are shown in the figure above right.

   (a) Use the graph to estimate the p_{50} values for each species.
   (b) Which species of fish has the highest oxygen-binding affinity at the experimental temperature? The least?

10. Below is a list of the first 10 residues of the B helix in myoglobin from different organisms. Based on this information, which positions (a) cannot tolerate substitution, (b) can tolerate conservative substitution, and (c) are highly variable?

Position	Human	Chicken	Alligator	Turtle	Tuna	Carp
1	D	D	K	D	D	D
2	I	I	L	L	Y	F
3	P	A	P	S	T	E
4	G	G	E	A	T	G
5	H	H	H	H	M	T
6	G	G	G	G	G	G
7	Q	H	H	Q	G	G
8	E	E	E	E	L	E
9	V	V	V	V	V	V
10	L	L	I	I	L	L

11. Hexacoordinate Fe(II) in heme is bright red. Pentacoordinate Fe(II) is blue. Explain how these electronic changes account for the different colors of arterial (scarlet) and venous (purple) blood.

12. Invariant residues are those that are essential for the structure and/or function of the protein and cannot be replaced by other residues. Which amino acid residue in the globin chain is most likely to be invariant? Explain.

13. Hemoglobin is 50% saturated with oxygen when $pO_2 = 26$ torr. If hemoglobin exhibited hyperbolic binding (as myoglobin does), what would be the fractional saturation when $pO_2 = 30$ torr and 100 torr? What does this tell you about the physiological importance of hemoglobin's sigmoidal oxygen-binding curve?

14. Hemoglobin does not exhibit hyperbolic binding, as assumed in Problem 13; its oxygen-binding curve has a sigmoidal shape. The equation for hemoglobin's oxygen binding curve is modified from Equation 5.4 and can be used to calculate the fractional saturation for hemoglobin:

$$Y = \frac{(pO_2)^n}{(p_{50})^n + (pO_2)^n}$$

The quantity n is the Hill coefficient and its value increases with the degree of cooperation of ligand binding (in this case, O_2). For hemoglobin, $n \approx 3$. (The value of n is 1 in the absence of cooperativity.) What is Y when $pO_2 = 25$ torr, a typical venous pO_2? What is the Y when $pO_2 = 120$ torr, a typical oxygen partial pressure in the lungs? (*Note:* At $pCO_2 = 5$ torr, $p_{50} = 15$ torr.)

15. Calculate the fractional saturation of hemoglobin for $pO_2 = 25$ torr and 120 torr (see Problem 14) when the CO_2 partial pressure is 80 torr. (*Note:* Under these conditions, the p_{50} value for hemoglobin is 40 torr.)

16. Compare your answers to Problems 14 and 15. How much oxygen is delivered to tissues when the pCO_2 is 5 torr? How does this compare to the amount of oxygen delivered to tissues when the pCO_2 is 80 torr? What role does CO_2 play in assisting oxygen delivery to tissues?

17. Carbon monoxide, CO, binds to hemoglobin about 250 times as tightly as oxygen (see Box 5.A). The resulting complex is a red color

that is much brighter than that observed when oxygen is bound. Why might a researcher studying the properties of oxygenated Hb decide to use Hb·CO instead?

18. Highly active muscle generates lactic acid by respiration so fast that the blood passing through the muscle actually experiences a drop in pH from about 7.4 to about 7.2. Under these conditions, hemoglobin releases about 10% more O_2 than it does at pH 7.4. Explain.

19. About two dozen histidine residues in hemoglobin are involved in binding the protons produced by cellular metabolism. In this manner, hemoglobin contributes to buffering in the blood, and the imidazole groups able to bind and release protons contribute to the Bohr effect. One important contributor to the Bohr effect is His 146 on the β chain of hemoglobin, whose side chain is in close proximity to the side chain of Asp 94 in the deoxy form of hemoglobin but not the oxy form.

 (a) What kind of interaction occurs between Asp 94 and His 146 in deoxyhemoglobin?
 (b) The proximity of Asp 94 alters the pK value of the imidazole ring of His. In what way?

20. While most humans are able to hold their breath for only a minute or two, crocodiles can stay submerged under water for an hour or longer. This adaptation allows the crocodiles to kill their prey by drowning. Investigators hypothesized that bicarbonate (HCO_3^-) might act as an allosteric effector in crocodiles in a manner similar to BPG in humans.

 (a) What is the source of HCO_3^- in crocodile tissues?
 (b) Draw oxygen-binding curves for crocodile hemoglobin in the presence and absence of HCO_3^-. Which conditions increase the p_{50} value for crocodile hemoglobin? What does this tell you about the oxygen-binding affinity of hemoglobin under those conditions?
 (c) The investigators found that the HCO_3^- binding site on crocodile hemoglobin was located at the α–β subunit interface, where the two subunits slide with respect to each other during the oxy ↔ deoxy transition. Based on their results, the researchers modeled a stereochemically plausible binding site that included the phenolate anions of Tyr 41β and Tyr 42α and the ε-amino group of Lys 38β. What interactions might form between these amino acid side chains and HCO_3^-?
 (d) Fish use ATP or GTP as allosteric effectors to encourage hemoglobin to release its bound O_2. What structural characteristics do these allosteric effectors have in common with the BPG allosteric effector used by mammals and the HCO_3^- effector used by crocodiles?

21. Lampreys are among the world's most primitive vertebrates and are therefore interesting organisms to study. It has been shown that lamprey hemoglobin forms a tetramer when deoxygenated, but when the hemoglobin becomes oxygenated, the tetramers dissociate into monomers. The binding of oxygen to lamprey hemoglobin is influenced by pH; as in human hemoglobin, the deoxygenated form is favored when the pH decreases. Glutamate residues on the surface of the monomers play an important role. How might glutamate residues influence the equilibrium between the monomer and tetramer form with changes in pH?

22. The grelag goose lives year-round in the Indian plains, while its close relative, the bar-headed goose, spends the summer months in the Tibetan lake region. The p_{50} value for bar-headed goose hemoglobin is less than the p_{50} for grelag goose hemoglobin. What is the significance of this adaptation?

23. In the developing fetus, fetal hemoglobin (Hb F) is synthesized beginning at the third month of gestation and continuing until

birth. After the baby is born, the concentration of Hb F declines and is replaced by age six months with adult hemoglobin (Hb A) as synthesis of the γ chain declines and synthesis of the β chain increases.

 (a) In the graph, which curve represents fetal hemoglobin?
 (b) Why does Hb F have a higher oxygen affinity than Hb A? Why does this provide an advantage to the developing fetus?

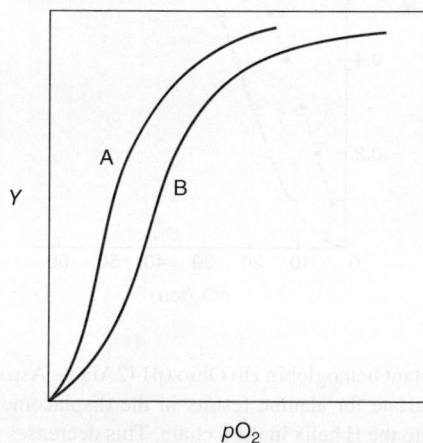

24. People who live at or near sea level undergo *high altitude acclimatization* when they go up to moderately high altitudes of about 5000 m.

 (a) The low pO_2 at these altitudes initially induces hypoxia, which results in hyperventilation. Explain why this occurs. What happens to blood pH during hyperventilation?
 (b) After a period of weeks, the alveolar pCO_2 decreases and the 2,3-BPG concentration increases. Explain these observations.

25. Animals indigenous to high altitudes, such as yaks, llamas, and alpacas, have adapted to the low pO_2 at high altitudes because they have high-affinity hemoglobins, achieved through a combination of globin chain mutations and the continued synthesis of fetal hemoglobin in the mature animal. This is a true evolutionary adaptation and not the high altitude acclimatization described in Problem 24.

 (a) Compare the p_{50} values of the high-altitude animals with the p_{50} values of their low-altitude counterparts.
 (b) How does the continued synthesis of fetal hemoglobin help these animals survive at high altitudes?

26. Drinking a few drops of a preparation called "vitamin O," which consists of oxygen and sodium chloride dissolved in water, purportedly increases the concentration of oxygen in the body.

 (a) Use your knowledge of oxygen transport to evaluate this claim.
 (a) Would vitamin O be more or less effective if it was infused directly into the bloodstream?

5.2   Clinical Connection: Hemoglobin Variants

27. The oxygen-binding curves for normal hemoglobin (Hb A) and a mutant hemoglobin (Hb Great Lakes) are shown in the figure.

 (a) Compare the shapes of the curves for the two hemoglobins and comment on their significance.
 (b) Which hemoglobin has a higher affinity for oxygen when $pO_2 = 20$ torr?
 (c) Which hemoglobin has a higher affinity for oxygen when $pO_2 = 75$ torr?
 (d) Which hemoglobin is most efficient at delivering oxygen from arterial blood ($pO_2 = 75$ torr) to active muscle ($pO_2 = 20$ torr)?

28. In the mutant hemoglobin Hb Ohio (β142Ala → Asp), the substitution of aspartate for alanine results in the displacement of the G helix relative to the H helix in the β chain. This decreases the stability of the β146His–β94Asp ion pair (see Problem 19). Draw an oxygen-binding curve that compares the relative p_{50} values of normal hemoglobin (Hb A) and Hb Ohio. What is the effect of the decreased stability of the His–Asp ion pair on Hb Ohio?

29. Hb Providence (β82Lys → Asn) results from a single point mutation of the DNA (see Table 5.1).

(a) Compare the oxygen affinities of Hb Providence and Hb A.
(b) There are actually two forms of Hb Providence in affected individuals. Hb Providence in which the β82Lys has been replaced with Asn is referred to as Hb Providence Asn. This hemoglobin variant can undergo deamidation to produce Hb Providence Asp. Draw the reaction that converts Hb Providence Asn to Hb Providence Asp.
(c) Compare the oxygen affinities of Hb Providence Asp and Hb Providence Asn.

30. Hb Milledgeville (α44Pro → Leu) results in a mutated hemoglobin with altered oxygen affinity. Explain how the oxygen affinity is altered (see Table 5.1).

31. *Plasmodium falciparum*, the protozoan that causes malaria, slightly decreases the pH of the red blood cells it infects. Invoke the Bohr effect to explain why *Plasmodium*-infected cells are more likely to undergo sickling in individuals with the Hb S variant.

32. The drug hydroxyurea can be used to treat sickle cell disease, although it is not used often because of undesirable side effects. Hydroxyurea is thought to function by stimulating the afflicted person's synthesis of fetal hemoglobin. In a clinical study, patients who took hydroxyurea showed a 50% reduction in frequency of hospital admissions for severe pain. Why would this drug alleviate the symptoms of sickle cell disease?

33. Investigators of mutant human hemoglobins often subject the proteins to cellulose acetate electrophoresis at pH 8.5, a pH at which most hemoglobins are negatively charged. The proteins are applied at the negative pole and migrate toward the positive pole when the current is turned on. The more negatively charged the hemoglobin, the faster it migrates to the positive pole. Results for normal (Hb A) and sickle-cell hemoglobin (Hb S) are shown. Draw a diagram that shows the results of electrophoresis of Hb A and Hb Ohio (β142Ala → Asp).

34. Because networks of collagen fibers lend resilience to the deep (living) layers of skin, some "anti-aging" cosmetics contain purified collagen. Explain why this exogenous (externally supplied) collagen is unlikely to attenuate the skin wrinkling that accompanies aging.

35. A high-affinity hemoglobin was isolated from a subject in which lysine at β144 was replaced with asparagine. (The normal C-terminal sequence of the β chain is –Lys^{144}–Tyr^{145}–His^{146}–COO⁻.) The C-terminal portion of the β chain, which resides in hemoglobin's central cavity, normally forms interactions that affect the position of the F helix that contains His F8. How might the Lys → Asn substitution result in a higher-affinity hemoglobin?

36. Patients with the mutation described in Problem 35 are able to undergo high altitude acclimatization (described in Problem 24) much more readily than "lowlanders" with normal hemoglobin. Explain why.

5.3 Structural Proteins

37. Compare and contrast globular and fibrous proteins.

38. Explain why actin filaments and microtubules are polar whereas intermediate filaments are not.

39. Of the various proteins highlighted in Sections 5.3 and 5.4 (actin, tubulin, keratin, collagen, myosin, and kinesin), (a) which are exclusively structural (not involved in cell shape changes); (b) which are considered to be motor proteins; (c) which are not motor proteins but can undergo structural changes; and (d) which contain nucleotide-binding sites?

40. In order to obtain crystals suitable for X-ray crystallography, G-actin was first mixed with another protein.

(a) Why was this step necessary?
(b) What additional information was required to solve the structure of G-actin?

41. Phalloidin, a peptide isolated from a poisonous mushroom, binds to F-actin but not G-actin. How does the addition of phalloidin affect cell motility?

42. Phalloidin (see Problem 41) is used as an imaging tool by covalently attaching a fluorescent tag to it. What structures can be visualized in cells treated with fluorescently labeled phalloidin? What is not visible?

43. How could a microtubule-binding protein distinguish a rapidly growing microtubule from one that was growing more slowly?

44. Why would fraying be a faster mechanism for microtubule disassembly than dissociation of tubulin dimers?

45. Microtubule ends with GTP bound tend to be straight, whereas microtubule ends with GDP bound are curved and tend to fray. An experiment was carried out in which monomers of β-tubulin were allowed to polymerize in the presence of a nonhydrolyzable analog of GTP called guanylyl-(αβ)-methylenediphosphonate (GMP · CPP). Compare the stability of this polymer with one polymerized in the presence of GTP.

GMP · CPP

46. Explain why colchicine (which promotes microtubule depolymerization) and paclitaxel (which prevents depolymerization) both prevent cell division.

47. A recent review listed dozens of compounds that are being investigated as possible agents for treating cancer. Tubulin was the target for all of the compounds. Why is tubulin a good target for anticancer drugs?

48. The three-dimensional structure of paclitaxel bound to the tubulin dimer has been solved. Paclitaxel binds to a pocket in β-tubulin on the inner surface of the microtubule and causes a conformational change that counteracts the effects of GTP hydrolysis. How does the binding of paclitaxel affect the stability of the microtubule?

49. Gout is a disease characterized by the deposition of sodium urate crystals in the synovial (joint) fluid. The crystals are ingested by neutrophils, which are white blood cells that circulate in the bloodstream before they crawl through tissues to reach sites of injury. Upon ingestion of the crystals, a series of biochemical reactions leads to a release of mediators from the neutrophils that cause inflammation, resulting in joint pain. Colchicine is used as a drug to treat gout because colchicine can inhibit neutrophil mobility. How does colchicine accomplish this?

50. The microtubules involved in cell division are less stable than microtubules found in axon extensions of nerve cells. Why is this the case?

51. Vinblastine, a compound in the periwinkle, has been shown to affect microtubule assembly by stabilizing the (+) ends and destabilizing the (−) ends of microtubules.

(a) How does vinblastine affect the formation of the mitotic spindle during mitosis?

(b) Why would you expect vinblastine to have a more dramatic effect on rapidly dividing cells such as cancer cells?

52. Hydrophobic residues usually appear at the first and fourth positions in the seven-residue repeats of polypeptides that form coiled coils.

(a) Why do polar or charged residues usually appear in the remaining five positions?

(b) Why is the sequence Ile–Gln–Glu–Val–Glu–Arg–Asp more likely than the sequence Trp–Gln–Glu–Tyr–Glu–Arg–Asp to appear in a coiled coil?

53. Globular proteins are typically constructed from several layers of secondary structure, with a hydrophobic core and a hydrophilic surface. Is this true for a fibrous protein such as keratin?

54. Straight hair can be curled and curly hair can be straightened by exposing wet hair to a reducing agent, repositioning the hair with

rollers, then exposing the hair to an oxidizing agent. Explain how this procedure alters the shape of the hair.

55. "Hard" keratins such as those found in hair, horn, and nails have a high sulfur content whereas "soft" keratins found in the skin have a lower sulfur content. Explain the structural basis for this observation.

56. Describe the primary, secondary, tertiary, and quaternary structures of (a) actin and (b) collagen. Why is it difficult to assign the four structural categories to these proteins?

57. The triple helix of collagen is stabilized through hydrogen bonding. Draw an example of one of these hydrogen bonds, as described in the text.

58. In a collagen polypeptide consisting mostly of repeating Gly–Pro–Hyp sequences, which residue(s) is(are) most likely to be substituted with 5-hydroxylysine?

59. A retrospective study conducted over 25 years examined a group of patients with similar symptoms: bruising, joint swelling, fatigue, and gum disease. Examination of patient records showed that some patients suffered from gastrointestinal diseases, poor dentition, or alcoholism, while others followed various fad diets.

(a) What disease afflicts this diverse group of patients?

(b) Explain why the patients suffer from the symptoms listed.

(c) Explain why your diagnosis is consistent with the patients' medical histories.

60. The effect of the drug minoxidil on collagen synthesis in skin fibroblasts was investigated by measuring the activities of the prolyl hydroxylase and lysyl hydroxylase enzymes in the cells.

(a) What is the effect of minoxidil on the cultured fibroblasts?

(b) Why would this drug be effective in treating fibrosis (a skin condition associated with an accumulation of collagen)?

(c) What are the potential dangers in the long-term use of minoxidil as a topical ointment to treat hair loss?

	Prolyl hydroxylase activity units	Lysyl hydroxylase activity units
Control cells	13,100	12,400
Minoxidil-treated cells	13,200	1,900

61. The T_m (melting temperature) is defined as that temperature at which collagen is half-denatured and is used as a criterion for collagen stability. T_m measurements for collagen from rats and sea urchins are presented in the table.

(a) Assign each collagen to the proper organism, and

(b) Explain the correlation between imino acid content (defined as the combined content of Pro and Hyp) and T_m.

Collagen	Number of Hyp per 1000 residues	Number of Pro per 1000 residues	T_m (°C)
A	48.5	81.3	27.0
B	68.5	111	38.5

62. A series of collagen-like peptides, each containing 30 amino acid residues, were synthesized in order to study the important interactions among the three chains in the triple helix. Peptide 1 is (Pro–Hyp–Gly)₁₀, peptide 2 is (Pro–Hyp–Gly)₄–Glu–Lys–Gly–(Pro–Hyp–Gly)₅, and peptide 3 is Gly–Lys–Hyp–Gly–Glu–Hyp–Gly–Pro–Lys–Gly–Asp–Ala–(Gly–Ala–Hyp)₂–(Gly–Pro–Hyp)₄. Their structural properties and melting temperatures (see Problem 57) are summarized in the table below. Note that "imino acid content" refers to the combined content of Pro and Hyp.

Peptide	Forms trimer	Imino acid content	T_m (°C)	
1	Yes	67%	pH = 1	61
			pH = 7	58
			pH = 11	60
2	Yes	60%	pH = 1	44
			pH = 7	46
			pH = 1	49
3	Yes	30%	pH = 7	18
			pH = 7	26.5
			pH = 13	19

(a) Rank the stability of the three collagen-like peptides. What is the reason for the observed stability?

(b) Compare the T_m values of peptide 3 at the various pH values. Why does peptide 3 have a maximum T_m value at pH = 7? What interactions are primarily responsible?

(c) Compare the T_m values of peptide 1 at the various pH values. Why is there less of a difference of T_m values at different pH values for peptide 1 than for peptide 3?

(d) Consider the answers to the above questions. Which factor is more important in stabilizing the structure of the collagen-like peptides: ion pairing or imino acid content?

63. The deep-sea hydrothermal vent worm has a thick collagen-containing cuticle that protects it from the drastic temperature changes and low oxygen content of its habitat. Sequence analyses indicated that the worm collagen has the customary Gly–X–Y triplet but Hyp occurs only in the X position and Y is often a glycosylated threonine (a threonine side chain with a covalently attached galactose sugar residue). The melting temperatures (see Problem 62) of a series of synthetic peptides were measured to assess their stability. The T_m values are shown in the table.

(a) Compare the melting temperatures of (Pro–Pro–Gly)$_{10}$ and (Pro–Hyp–Gly)$_{10}$. What is the structural basis for the difference?

(b) Compare the melting temperature of (Pro–Pro–Gly)$_{10}$ and (Gly–Pro–Thr(Gal))$_{10}$ and provide an explanation.

(c) Why was (Gly–Pro–Thr)$_{10}$ included by the investigators?

Gal-Thr

Synthetic peptide	Forms a triple helix?	T_m (°C)
(Pro–Pro–Gly)$_{10}$	Yes	41
(Pro–Hyp–Gly)$_{10}$	Yes	60
(Gly–Pro–Thr)$_{10}$	No	N/A
(Gly–Pro–Thr(Gal))$_{10}$	Yes	41

64. Why do Hyp-containing collagen molecules have greater stability? To investigate this question, a group of investigators synthesized a collagen containing 4-fluoroproline (Flp), which resembles hydroxyproline except that fluorine substitutes for the hydroxyl group. The T_m values (see Problem 62) of three synthetic collagen molecules are shown in the table. **(a)** Compare the structures of hydroxyproline and fluoroproline. Why did the investigators chose fluorine? **(b)** Compare the T_m values of the three synthetic collagens. What factor contributes the most to the stability of the molecule?

Synthetic peptide	T_m (°C)
(Pro–Pro–Gly)$_{10}$	41
(Pro–Hyp–Gly)$_{10}$	60
(Pro–Flp–Gly)$_{10}$	91

65. Because collagen molecules are difficult to isolate from the connective tissue of animals, investigators who study collagen structure often use synthetic peptides (see Problems 63–64). Would it be practical to try to purify collagen molecules from cultures of bacterial cells that have been engineered to express collagen genes?

66. The highly pathogenic bacterium *Clostridium perfringens* causes gangrene, a disease which results in the destruction of animal tissue. The bacterium secretes two types of collagenases that cleave the peptide bonds linking glycine and proline residues. A biochemical company sells a preparation of these collagenases to investigators interested in culturing cells derived from bone, cartilage, muscle, or endothelial tissue. How does this enzyme preparation assist the investigator in obtaining cells for culturing?

67. Gelatin is a food product obtained from the thermal degradation of collagen. Powdered gelatin is soluble in hot water and forms a gel upon cooling to room temperature. Why is gelatin nutritionally inferior to other types of protein?

68. Radioactively labeled [^{14}C]-proline is incorporated into collagen in cultured fibroblasts. The radioactivity is detected in the collagen protein. But collagen synthesized in the presence of [^{14}C]-hydroxyproline is not radiolabeled. Explain why.

69. A pharmaceutical company sells a collagenase ointment that is applied to the surface of skin wounds. How does the ointment promote the healing process? What precaution should patients take when using the ointment?

70. An enzyme found in tadpoles cleaves the bond indicated in the figure below. What kind of enzyme is this, and what is its role in the metamorphosis of the tadpole to the adult frog?

—Gly—Pro—Gln—Gly↓Ile—Ala

71. Papaya and pineapple fruits contain proteolytic enzymes that can degrade collagen. **(a)** Cooking meat with these fruits helps to tenderize the meat. How does this work? **(b)** A cook decides to flavor a gelatin dessert with fresh papaya and pineapple. The fruits are added to a solution of gelatin in hot water, then the mixture is allowed to cool. What is the result and how could the cook have avoided this?

72. Osteogenesis imperfecta is a genetic disease, but most cases result from new mutations rather than ones that are passed from parent to child. Explain.

73. A small portion of the middle of the gene for the α1(II) chain of type II collagen (the entire gene contains 3421 base pairs) is shown below as the sequence of the nontemplate, or coding, strand of the DNA. Also shown is the gene isolated from the skin fibroblasts of a patient with a mutant α1(II) gene. The patient was a neonate delivered at 38 weeks who died shortly after birth. The neonate had shortened limbs and a large head, with a flat nasal bridge and short bones and ribs. The symptoms are consistent with the disease hypochondrogenesis. Analysis of tissue taken from the neonate revealed a single point mutation in the α1(II) collagen gene.

Normal α1(II) collagen gene:
… TAACGGCGAGAAGGGAGAAGTTGGACCTCCT…

Mutant α1(II) collagen gene:
…TAACGGCGAGAAGGCAGAAGTTGGACCTCCT…

(a) What is the sequence of the protein encoded by both the normal and mutant genes? (Note that you must first determine the correct open reading frame.) What is the change in the protein sequence?

(b) How would the Tm values for normal collagen and mutant collagen compare?

(c) Why did the patient die shortly after birth?

5.4   Motor Proteins

74. Is myosin a fibrous protein or a globular protein? Explain.

75. Myosin type V is a two-headed myosin that operates as a transport motor to move its attached cargo along actin filaments. Its mechanism is similar to that of muscle myosin, but it acts processively, like kinesin. The reaction cycle diagrammed here begins with both myosin V heads bound to an actin filament. ADP is bound to the leading head, and the nucleotide-binding site in the trailing head is empty.

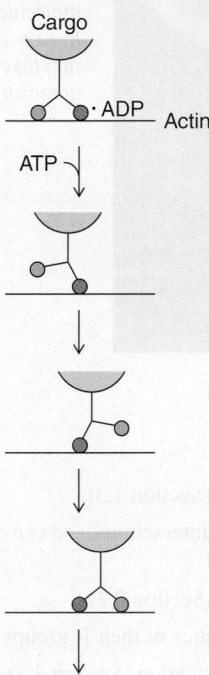

Based on your knowledge of the muscle myosin reaction cycle (Fig. 5.37), propose a reaction mechanism for myosin V, starting with the entry of ATP. How does each step of the ATP hydrolysis reaction correspond to a conformational change in myosin V?

76. Early cell biologists, examining living cells under the microscope, observed that the movement of certain cell constituents was rapid, linear, and targeted (that is, directed toward a particular point).

(a) Why are these qualities inconsistent with diffusion as a mechanism for redistributing cell components?

(b) List the minimum requirements for an intracellular transport system that is rapid, linear, and targeted.

77. Explain why the movement of myosin along an actin filament is "hopping," whereas the movement of kinesin along a microtubule is "walking."

78. Rigor mortis is the stiffening of a corpse that occurs in the hours following death.

(a) Using what you know about the mechanism of motor proteins, explain what causes this stiffness.

(b) Why is rigor mortis of concern to those in the meat industry?

(c) How could a forensic pathologist use the concept of rigor mortis to investigate the circumstances of an individual's death?

79. In individuals with muscular dystrophy, a mutation in a gene for a structural protein in muscle fibers leads to progressive muscle weakening beginning in early childhood. Explain why muscular dystrophy patients typically also exhibit abnormal bone development.

80. Explain why a one-headed kinesin motor would be ineffective.

SELECTED READINGS

Gardner, M. K., Hunt, A. J., Goodson, H. V., and Odde, D. J., Microtubule assembly dynamics: new insights at the nano-scale, *Curr. Opin. Cell Biol.* **20,** 64–70 (2008). [Describes some of the techniques used to characterize microtubule formation.]

Kull, J. F. and Endow, S. A., Force generation by kinesin and myosin cytoskeletal motor proteins, *J. Cell Sci.* **126,** 9–19 (2013). [Provides details about the protein conformational changes during force generation in motor proteins.]

Oshima, R. G., Intermediate filaments: a historical perspective, *Exp. Cell Res.* **313,** 1981–1994 (2007). [Describes what is known about various types of intermediate filaments, including keratin.]

Reisler, E. and Egelman, E. H., Actin structure and function: what we still do not understand, *J. Biol. Chem.* **282,** 36113–36137 (2007). [Summarizes some areas of research on actin structure and its assembly into filaments.]

Ricard-Blum, S., The collagen family, *Cold Spring Harb. Perspect. Biol.* 2011;3:a004978. doi: 10.1101/cshperspect.a004978 (2011). [Reviews the structure, assembly, and functions of vertebrate collagens.]

Schechter, A. N., Hemoglobin research and the origins of molecular medicine, *Blood* **112,** 3927–3938 (2008). [Reviews hemoglobin structure and function, including hemoglobin S.]

Thom, C. S., Dickson, C. F., Gell, D. A., and Weiss, M. J., Hemoglobin variants: Biochemical properties and clinical correlates, *Cold Spring Harb. Perspect. Med.* 2013;3:a011858. doi: 10.1101/cshperspect. a011858 (2013). [Summarizes hemoglobin structure and function and how these are disrupted by different types of genetic changes.]

Wilson, M. T. and Reeder, B. J., Oxygen-binding haem proteins, *Exp. Physiol.* **93,** 128–132 (2007). [Discusses the lesser-known functions of hemoglobin, myoglobin, and some of their relatives.]

HOW ENZYMES WORK

Jurgen Freund/Nature Picture Library

The ability of a firefly to produce light depends on the presence of the enzyme luciferase, which catalyzes a reaction involving O_2, ATP, and a nonluminescent molecule called luciferin. The free energy released in the reaction takes the form of a flash of greenish light.

DO YOU REMEMBER?

- Living organisms obey the laws of thermodynamics (Section 1.3).
- Noncovalent forces, including hydrogen bonds, ionic interactions, and van der Waals forces, act on biological molecules (Section 2.1).
- An acid's pK value describes its tendency to ionize (Section 2.3).
- The 20 amino acids differ in the chemical characteristics of their R groups (Section 4.1).
- Some proteins can adopt more than one stable conformation (Section 4.3).

We have already seen how protein structures relate to their physiological functions. We are now ready to examine enzymes, which directly participate in the chemical reactions by which matter and energy are transformed by living cells. In this chapter, we examine the fundamental features of enzymes, including the thermodynamic underpinnings of their activity. We describe various mechanisms by which enzymes accelerate chemical reactions, focusing primarily on the digestive enzyme chymotrypsin to illustrate how different structural features influence catalytic activity. The following chapter continues the discussion of enzymes by describing how enzymatic activity is quantified and how it can be regulated.

6.1 WHAT IS AN ENZYME?

KEY CONCEPTS
- Enzymes differ from simple chemical catalysts in their efficiency and specificity.
- An enzyme's name may reflect the reaction it catalyzes.

In vitro, under physiological conditions, the peptide bond that links amino acids in peptides and proteins has a half-life of about 20 years. That is, after 20 years, about half of the peptide bonds in a given sample of a peptide will have been broken down through **hydrolysis** (cleavage by water):

$$\cdots -\underset{\underset{H}{|}}{N}-\underset{\underset{}{|}}{\overset{\overset{R}{|}}{C}}H-\overset{\overset{O}{||}}{C}-\underset{\underset{H}{|}}{N}-\underset{\underset{}{|}}{\overset{\overset{R}{|}}{C}}H-\overset{\overset{O}{||}}{C}-\cdots \quad + \quad H_2O$$

$$\cdots -\underset{\underset{H}{|}}{N}-\underset{\underset{}{|}}{\overset{\overset{R}{|}}{C}}H-C\overset{\diagup O}{\underset{\diagdown O^-}{}} \quad + \quad H-\overset{\overset{H}{\overset{|}{+}}}{\underset{\underset{H}{|}}{N}}-\underset{\underset{}{|}}{\overset{\overset{R}{|}}{C}}H-\overset{\overset{O}{||}}{C}-\cdots$$

Obviously, the long half-life of the peptide bond is advantageous for living organisms, since many of their structural and functional characteristics depend on the integrity of proteins. On the other hand, many proteins—some hormones, for example—must be broken down very rapidly so that their biological effects can be limited. Clearly, an organism must be able to accelerate the rate of peptide bond hydrolysis.

In general, there are three ways to increase the rate of a chemical reaction, including hydrolysis:

1. *Increasing the temperature (adding energy in the form of heat).* Unfortunately, this is not very practical, since the vast majority of organisms cannot regulate their internal temperature and thrive only within relatively narrow temperature ranges. Furthermore, an increase in temperature accelerates all chemical reactions, not just the desired reaction.

2. *Increasing the concentrations of the reacting substances.* Higher concentrations of **reactants** increase the likelihood that they will encounter each other in order to react. But a cell may contain tens of thousands of different types of molecules, space is limited, and many essential reactants are scarce inside as well as outside the cell.

3. *Adding a **catalyst**.* A catalyst is a substance that participates in the reaction yet emerges at the end in its original form. A huge variety of chemical catalysts are known. For example, the catalytic converter in an automobile engine contains a mixture of platinum and palladium that accelerates the conversion of carbon monoxide and unburned hydrocarbons to the relatively harmless carbon dioxide. *Living systems use catalysts called **enzymes** to increase the rates of chemical reactions.*

Most enzymes are proteins, but a few are made of RNA (these are called **ribozymes** and are described more fully in Section 21.3). One of the best-studied enzymes is chymotrypsin, a digestive protein that is synthesized in the pancreas and secreted into the small intestine, where it helps break down dietary proteins. Perhaps because it can be purified in relatively large quantities from the pancreas of cows, chymotrypsin was one of the first enzymes to be crystallized (it is also widely used in the laboratory; for example, see Table 4.4). Chymotrypsin's 241 amino acid residues form a compact two-domain structure (**Fig. 6.1**). Hydrolysis of polypeptide substrates takes place in a cleft between the two domains, near the side chains of three residues (His 57, Asp 102, and Ser 195). This area of the enzyme is known as the **active site.** The active sites of nearly all known enzymes are located in similar crevices on the enzyme surface.

Chymotrypsin catalyzes the hydrolysis of peptide bonds at a rate of about 190 per second, which is about 1.7×10^{11} times faster than in the absence of a catalyst. This is also orders of magnitude faster than would be possible with a simple chemical catalyst. In addition, *chymotrypsin and other enzymes act under mild conditions* (atmospheric pressure and physiological temperature), whereas many chemical catalysts require extremely high temperatures and pressures for optimal performance.

Chymotrypsin's catalytic power is not unusual: *Rate enhancements of 10^8 to 10^{12} are typical of enzymes* (**Table 6.1** gives the rates of some enzyme-catalyzed reactions). Of course, the slower the rate of the uncatalyzed reaction, the greater the opportunity for rate enhancement by an enzyme (see, for example, orotidine-5′-monophosphate decarboxylase in Table 6.1).

FIGURE 6.1

Ribbon model of chymotrypsin. The polypeptide chain (gray) folds into two domains. Three residues essential for the enzyme's activity are shown in red. [*Structure (pdb 4CHA) determined by H. Tsukada and D. M. Blow.*]

TABLE 6.1 Rate Enhancements of Enzymes

ENZYME	HALF-TIME (UNCATALYZED)[a]	UNCATALYZED RATE (s⁻¹)	CATALYZED RATE (s⁻¹)	RATE ENHANCEMENT (CATALYZED RATE/ UNCATALYZED RATE)
Orotidine-5′-monophosphate decarboxylase	78,000,000 years	2.8×10^{-16}	39	1.4×10^{17}
Staphylococcal nuclease	130,000 years	1.7×10^{-13}	95	5.6×10^{14}
Adenosine deaminase	120 years	1.8×10^{-10}	370	2.1×10^{12}
Chymotrypsin	20 years	1.0×10^{-9}	190	1.7×10^{11}
Triose phosphate isomerase	1.9 years	4.3×10^{-6}	4,300	1.0×10^{9}
Chorismate mutase	7.4 hours	2.6×10^{-5}	50	1.9×10^{6}
Carbonic anhydrase	5 seconds	1.3×10^{-1}	1,000,000	7.7×10^{6}

[a]The half-times of very slow reactions were estimated by extrapolating from measurements made at very high temperatures.

[Data mostly from Radzicka, R., and Wolfenden, R., *Science* **267**, 90–93 (1995).]

Interestingly, even relatively fast reactions are subject to enzymatic catalysis in biological systems. For example, the conversion of CO_2 to carbonic acid in water

$$CO_2 + H_2O \rightleftharpoons H_2CO_3$$

has a half-time of 5 seconds (half the molecules will have reacted within 5 seconds). This reaction is accelerated over a millionfold by the enzyme carbonic anhydrase (see Table 6.1).

Another feature that sets enzymes apart from nonbiological catalysts is their **reaction specificity.** Most enzymes are highly specific for their reactants (called **substrates**) and products. The functional groups in the active site of an enzyme are so carefully arranged that the enzyme can distinguish its substrates from among many others that are similar in size and shape and can then mediate a single chemical reaction involving those substrates. This reaction specificity stands in marked contrast to the permissiveness of most organic catalysts, which can act on many different kinds of reactants and, for a given reactant, sometimes yield more than one product.

Chymotrypsin and some other digestive enzymes are somewhat unusual in acting on a relatively broad range of substrates and, at least in the laboratory, catalyzing several types of reactions. For instance, chymotrypsin catalyzes the hydrolysis of the peptide bond following almost any large nonpolar residue such as phenylalanine, tryptophan, or tyrosine. It can also catalyze the hydrolysis of other amide bonds and ester bonds. This behavior has proved to be convenient for quantifying the activity of purified chymotrypsin. An artificial substrate such as p-nitrophenylacetate (an ester) is readily hydrolyzed by the action of chymotrypsin (the name of the enzyme appears next to the reaction arrow to indicate that it participates as a catalyst):

p-Nitrophenylacetate (colorless)

Acetate   p-Nitrophenolate (yellow)

The *p*-nitrophenolate reaction product is bright yellow, so the progress of the reaction can be easily monitored by a spectrophotometer.

Finally, enzymes differ from nonbiological catalysts in that *the activities of many enzymes are regulated so that the organism can respond to changing conditions or follow genetically determined developmental programs.* For this reason, biochemists seek to understand *how* enzymes work as well as *when* and *why*. These aspects of enzyme behavior are fairly well understood for chymotrypsin, which makes it an ideal subject to showcase the fundamentals of enzyme activity.

ENZYMES ARE USUALLY NAMED AFTER THE REACTION THEY CATALYZE

The enzymes that catalyze biochemical reactions have been formally classified into six subgroups according to the type of reaction carried out (**Table 6.2**). Basically, *all biochemical reactions involve either the addition of some substance to another, or its removal, or the rearrangement of that substance.* Keep in mind that although the substrates of many biochemical reactions appear to be quite large (for example, proteins or nucleic acids), the action really involves just a few chemical bonds and a few small groups (sometimes H_2O or even just an electron).

The name of an enzyme frequently provides a clue to its function. In some cases, an enzyme is named by incorporating the suffix *-ase* into the name of its substrate. For example, fumarase is an enzyme that acts on fumarate (Reaction 7 in the citric acid cycle; see Section 14.2). Chymotrypsin can similarly be called a proteinase, a protease, or a peptidase. Most enzyme names contain more descriptive words (also ending in *-ase*) to indicate the nature of the reaction catalyzed by that enzyme. For example, pyruvate decarboxylase catalyzes the removal of a CO_2 group from pyruvate:

TABLE 6.2	Enzyme Classification
CLASS OF ENZYME	**TYPE OF REACTION CATALYZED**
1. Oxidoreductases	Oxidation–reduction reactions
2. Transferases	Transfer of functional groups
3. Hydrolases	Hydrolysis reactions
4. Lyases	Group elimination to form double bonds
5. Isomerases	Isomerization reactions
6. Ligases	Bond formation coupled with ATP hydrolysis

Alanine aminotransferase catalyzes the transfer of an amino group from alanine to an α-keto acid:

Such a descriptive naming system tends to break down in the face of the many thousands of known enzyme-catalyzed reactions, but it is adequate for the small number of well-known reactions that are included in this book. A more precise classification scheme systematically groups enzymes in a four-level hierarchy and assigns each enzyme a unique number. For example, chymotrypsin is known as EC 3.4.21.1 (*EC* stands for Enzyme Commission, part of the nomenclature committee of the International Union of Biochemistry and Molecular Biology; the EC database can be accessed at enzyme.expasy.org).

Keep in mind that even within an organism, more than one protein may catalyze a given chemical reaction. Multiple enzymes catalyzing the same reaction are called **isozymes.** Although they usually share a common evolutionary origin, isozymes differ in their catalytic properties. Consequently, the various isozymes that are expressed in different tissues or at different developmental stages can perform slightly different metabolic functions.

 CONCEPT CHECK

1. Make a list of things that enzymes do but purely chemical systems cannot do.

2. Explain why an enzyme's common name may not reveal its biological role.

CHEMICAL CATALYTIC MECHANISMS

KEY CONCEPTS
- The height of the activation energy barrier determines the rate of a reaction.
- An enzyme provides a lower-energy pathway from reactants to products.
- Enzymes accelerate chemical reactions using acid–base catalysis, covalent catalysis, and metal ion catalysis.
- Chymotrypsin's catalytic triad participates in acid–base and covalent catalysis.

In a biochemical reaction, the reacting species must come together and undergo electronic rearrangements that result in the formation of products. In other words, some old bonds break and new bonds form. Let us consider an idealized transfer reaction in which compound A—B reacts with compound C to form two new compounds, A and B—C:

In order for the first two compounds to react, they must approach closely enough for their constituent atoms to interact. Normally, atoms that approach too closely repel each other. But if the groups have sufficient free energy, they can pass this point and react with each other to form products. The progress of the reaction can be depicted on a diagram (**Fig. 6.2**) in which the horizontal axis represents the progress of the reaction (the **reaction coordinate**) and the vertical axis represents the free energy (G) of the system. The energy-requiring step of the reaction is shown as an energy barrier, called the **free energy of activation** or **activation energy** and symbolized $\Delta G^{\ddagger}$. The point of highest energy is known as the **transition state** and can be considered as something midway between the reactants and products.

The lifetimes of transition states are extremely short, on the order of 10^{-14} to 10^{-13} seconds. Because they are too short-lived to be accessible to most analytical techniques, the transition states of many reactions cannot be identified with absolute certainty. However, it is useful to visualize the transition state as a molecular species in the process of breaking old bonds and forming new bonds. For the reaction above, we can represent this as A····B····C. The reactants require free energy ($\Delta G^{\ddagger}$) to reach this point (for example, some energy is required to break existing bonds and bring other atoms together to begin forming new bonds). The analogy of going uphill in order to undergo a reaction is appropriate.

The height of the activation energy barrier determines the rate of a reaction (the amount of product formed per unit time). The higher the activation energy barrier, the less likely the

FIGURE 6.2

Reaction coordinate diagram for the reaction A—B + C → A + B—C. The progress of the reaction is shown on the horizontal axis, and free energy is shown on the vertical axis. The transition state of the reaction, represented as A···B···C, is the point of highest free energy. The free energy difference between the reactants and the transition state is the free energy of activation ($\Delta G^{\ddagger}$).

(a)   (b)

FIGURE 6.3

Reaction coordinate diagram for a reaction in which reactants and products have different free energies. The free energy change for the reaction (ΔG) is equivalent to $G_{products} - G_{reactants}$. (a) When the free energy of the reactants is greater than that of the products, the free energy change for the reaction is negative, so the reaction proceeds spontaneously. (b) When the free energy of the products is greater than that of the reactants, the free energy change for the reaction is positive, so the reaction does not proceed spontaneously (however, the reverse reaction does proceed).

❓ **In a cell, some enzyme-catalyzed reactions proceed in both the forward and reverse directions. Sketch a reaction coordinate diagram for such a reaction.**

reaction is to occur (the slower it is). Although the reactant molecules have varying free energies, very few of them have enough free energy to reach the transition state during a given time interval. But the lower the energy barrier, the more likely the reaction is to occur (the faster it is), because more reactant molecules happen to have enough free energy to achieve the transition state during the same time interval. Note that the transition state, at the peak, can potentially roll down either side of the free energy hill. Therefore, not all the reactants that get together to form a transition state actually proceed all the way to products; they may return to their original state. Similarly, the products (A and B—C in this case) can react, pass through the same transition state (A····B····C), and yield the original reactants (A—B and C).

In nature, the free energies of the reactants and products of a chemical reaction are seldom identical, so the reaction coordinate diagram looks more like **Figure 6.3a**. When the products have a lower free energy than the reactants, then the overall free energy change of the reaction ($\Delta G_{\text{reaction}}$, or $G_{products} - G_{reactants}$) is less than zero. A negative free energy change indicates that a reaction proceeds spontaneously as written. Note that "spontaneously" does not mean "quickly." A reaction with a negative free energy change is thermodynamically favorable, but the height of the activation energy barrier ($\Delta G^{\ddagger}$) determines how fast the reaction actually occurs. If the products have greater free energy than the reactants (Fig. 6.3b), then the overall free energy change for the reaction ($\Delta G_{\text{reaction}}$) is greater than zero. This reaction does not proceed as written (because it cannot go "uphill"), but it does proceed in the reverse direction.

A CATALYST PROVIDES A REACTION PATHWAY WITH A LOWER ACTIVATION ENERGY BARRIER

A catalyst, whether inorganic or enzymatic, allows a reaction to proceed with a lower activation energy barrier ($\Delta G^{\ddagger}$; Fig. 6.4). It does so by interacting with the reacting molecules such that they are more likely to assume the transition state. A catalyst speeds up the reaction because as more reacting molecules achieve the transition state per unit time, more molecules of product can form per unit time. (An increase in temperature increases the rate of a reaction for a similar reason: The input of thermal energy allows more molecules to achieve the transition state per unit time.) Thermodynamic calculations indicate that lowering $\Delta G^{\ddagger}$ by about 5.7 kJ·mol^{-1} accelerates the reaction 10-fold. A rate increase of 10^6 requires lowering $\Delta G^{\ddagger}$ by six times this amount, or about 34 kJ·mol^{-1}.

An enzyme catalyst does not alter the net free energy change for a reaction; it merely provides a pathway from reactants to products that passes through a transition state that has lower free energy than the transition state of the uncatalyzed reaction. *An enzyme, therefore,*

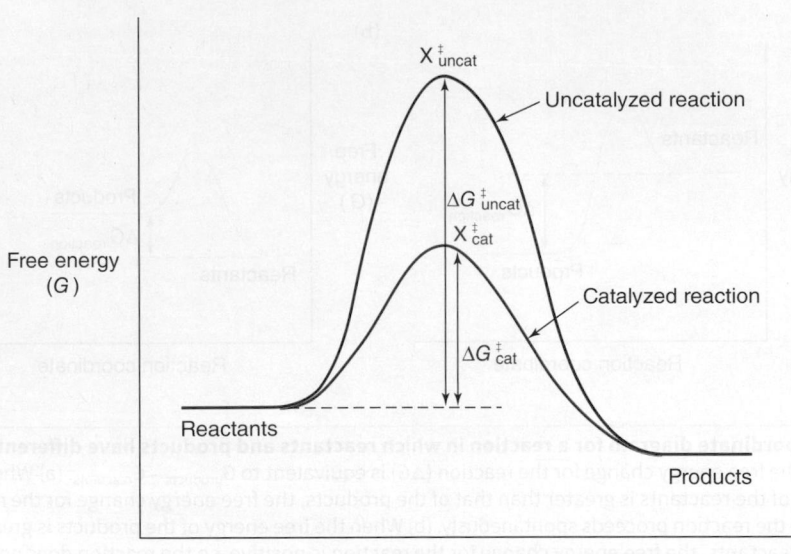

FIGURE 6.4

Effect of a catalyst on a chemical reaction. Here, the reactants proceed through a transition state symbolized $X^\ddagger$ during their conversion to products. In the presence of a catalyst, the free energy of activation ($\Delta G^\ddagger$) for the reaction is lower, so that $\Delta G^\ddagger_{cat} < \Delta G^\ddagger_{uncat}$. Lowering the free energy of the transition state ($X^\ddagger$) accelerates the reaction because more reactants are able to achieve the transition state per unit time.

lowers the height of the activation energy barrier ($\Delta G^\ddagger$) by lowering the energy of the transition state. The hydrolysis of a peptide bond is always thermodynamically favorable, but the reaction occurs quickly only when a catalyst (such as the protease chymotrypsin) is available to provide a lower-energy route to the transition state. Note that enzymes do not work at a distance; they must closely interact with their substrates to catalyze reactions.

ENZYMES USE CHEMICAL CATALYTIC MECHANISMS

The idea that living organisms contain agents that can promote the change of one substance into another has been around at least since the early nineteenth century, when scientists began to analyze the chemical transformations carried out by organisms such as yeast. However, it took some time to appreciate that these catalytic agents were not part of some "vital force" present only in intact, living organisms. In 1878, the word *enzyme* was coined to indicate that there was something *in* yeast (Greek *en* = "in," *zyme* = "yeast"), rather than the yeast itself, that was responsible for breaking down (fermenting) sugar. In fact, the action of enzymes can be explained in purely chemical terms. What we currently know about enzyme mechanisms rests solidly on a foundation of knowledge about simple chemical catalysts.

In an enzyme, certain functional groups in the enzyme's active site perform the same catalytic function as small chemical catalysts. In some cases, the amino acid side chains of an enzyme cannot provide the required catalytic groups, so a tightly bound **cofactor** participates in catalysis. For example, many oxidation–reduction reactions require a metal ion cofactor, since a metal ion can exist in multiple oxidation states, unlike an amino acid side chain. Some enzyme cofactors are organic molecules known as **coenzymes,** which may be derived from vitamins. Enzymatic activity still requires the protein portion of the enzyme, which helps position the cofactor and reactants for the reaction (this situation is reminiscent of myoglobin and hemoglobin, where the globin and heme group together function to bind oxygen; see Section 5.1). Some coenzymes, termed cosubstrates, enter and exit the active site as substrates do; a tightly bound coenzyme that remains in the active site between reactions is called a prosthetic group (**Fig. 6.5**).

There are three basic kinds of chemical catalytic mechanisms used by enzymes: acid–base catalysis, covalent catalysis, and metal ion catalysis. We will examine each of these, using model reactions to illustrate some of their fundamental features.

1. Acid–Base Catalysis

Many enzyme mechanisms include **acid–base catalysis,** in which a proton is transferred between the enzyme and the substrate. This mechanism of catalysis

FIGURE 6.5

Types of enzyme cofactors.

can be further divided into **acid catalysis** and **base catalysis.** Some enzymes use one or the other; many use both. Consider the following model reaction, the tautomerization of a ketone to an enol (**tautomers** are interconvertible isomers that differ in the placement of a hydrogen and a double bond):

$$
\begin{array}{ccccc}
\text{R} & & \left[\text{R}\right] & & \text{R} \\
| & & \quad\quad | & & | \\
\text{C}{=}\text{O} & \rightleftharpoons & \text{C}{\cdots}\text{O}^{-} & \rightleftharpoons & \text{C}{-}\text{O}{-}\text{H} \\
| & & \quad\quad \vdots & & | \\
\text{CH}_2 & & \text{CH}_2 & & \text{CH}_2 \\
| & & \quad\quad \vdots & & \\
\text{H} & & \text{H}^{+} & &
\end{array}
$$

$$\text{Ketone} \qquad\qquad \text{Transition state} \qquad\qquad \text{Enol}$$

Here the transition state is shown in square brackets to indicate that it is an unstable, transient species. The dotted lines represent bonds in the process of breaking or forming. The uncatalyzed reaction occurs slowly because formation of the carbanion-like transition state has a high activation energy barrier (a **carbanion** is a compound in which the carbon atom bears a negative charge).

If a catalyst (symbolized H—A) donates a proton to the ketone's oxygen atom, it reduces the unfavorable carbanion character of the transition state, thereby lowering its energy and hence lowering the activation energy barrier for the reaction:

$$
\begin{array}{ccccc}
\text{R} & & \left[\text{R}\right] & & \text{R} \\
| & & \quad\quad | & & | \\
\text{C}{=}\text{O} + \text{H}{-}\text{A} & \rightleftharpoons & \text{C}{\cdots}\text{O}^{-}{\cdots}\text{H}^{+}{\cdots}\text{A}^{-} & \rightleftharpoons & \text{C}{-}\text{O}{-}\text{H} + \text{H}{-}\text{A} \\
| & & \quad\quad \vdots & & \| \\
\text{CH}_2 & & \text{CH}_2 & & \text{CH}_2 \\
| & & \quad\quad \vdots & & \\
\text{H} & & \text{H}^{+} & &
\end{array}
$$

This is an example of acid catalysis, since the catalyst acts as an acid by donating a proton. Note that *the catalyst is returned to its original form at the end of the reaction.*

The same keto-enol tautomerization reaction shown above can be accelerated by a catalyst that can accept a proton, that is, by a base catalyst. Here, the catalyst is shown as :B, where the dots represent unpaired electrons:

$$
\begin{array}{cccccc}
\text{R} & & \left[\text{R}\right] & \text{H}^{+} & \text{R} & \\
| & & \quad\quad | & & | & \\
\text{C}{=}\text{O} + :\text{B} & \rightleftharpoons & \text{C}{\cdots}\text{O}^{-} & \rightleftharpoons & \text{C}{-}\text{O}{-}\text{H} + :\text{B} + \text{H}^{+} \\
| & & \quad\quad \vdots & & \| & \\
\text{CH}_2 & & \text{CH}_2 & & \text{CH}_2 & \\
| & & \quad\quad \vdots & & & \\
\text{H} & & \text{H}^{+} & & & \\
& & \quad\quad \vdots & & & \\
& & \text{B} & & &
\end{array}
$$

Base catalysis lowers the energy of the transition state and thereby accelerates the reaction.

In enzyme active sites, several amino acid side chains can potentially act as acid or base catalysts. These are the groups whose pK values are in or near the physiological pH range. The residues most commonly identified as acid–base catalysts are shown in **Figure 6.6**. Because the catalytic functions of these residues depend on their state of protonation or deprotonation, the catalytic activity of the enzyme may be sensitive to changes in pH.

2. Covalent Catalysis
In **covalent catalysis,** the second major chemical reaction mechanism used by enzymes, a covalent bond forms between the catalyst and the substrate during formation of the transition state. Consider as a model reaction the decarboxylation of acetoacetate. In this reaction, the movement of electron pairs among atoms is indicated by red curved arrows (**Box 6.A**).

$$
\text{H}_3\text{C}{-}\overset{\overset{\text{O}}{\|}}{\text{C}}{-}\text{CH}_2{-}\overset{\overset{\nearrow\text{O}}{}}{\text{C}}{\underset{\text{O}^-}{\diagdown}} \xrightarrow{\text{CO}_2} \left[\text{H}_3\text{C}{-}\overset{\overset{\text{O}^-}{\|}}{\text{C}}{=}\text{CH}_2\right] \xrightarrow{\text{H}^+} \text{H}_3\text{C}{-}\overset{\overset{\text{O}}{\|}}{\text{C}}{-}\text{CH}_3
$$

$$\text{Acetoacetate} \qquad\qquad\qquad \text{Enolate} \qquad\qquad\qquad \text{Acetone}$$

FIGURE 6.6

Amino acid side chains that can act as acid–base catalysts. These groups can act as acid or base catalysts, depending on their state of protonation in the enzyme's active site. The side chains are shown in their protonated forms, with the acidic proton highlighted.

? **At neutral pH, which of these side chains most likely function as acid catalysts? Which most likely function as base catalysts?** (*Hint:* See Table 4.1.)

The transition state, an enolate, has a high free energy of activation. This reaction can be catalyzed by a primary amine (RNH_2), which reacts with the carbonyl group of acetoacetate to form an **imine,** a compound containing a C=N bond (this adduct is known as a **Schiff base**):

Acetoacetate + RNH_2 → Schiff base (imine) + OH^-

A CLOSER LOOK

Box 6.A Depicting Reaction Mechanisms

While it is often sufficient to draw the structures of a reaction's substrates and products, a full understanding of the reaction mechanism requires knowing what the electrons are doing. Recall that a covalent bond forms when two atoms share a pair of electrons, and a vast number of biochemical reactions involve breaking and forming covalent bonds. Although single-electron reactions also occur in biochemistry, we will focus on the more common two-electron reactions.

The curved arrow convention shows how electrons are rearranged during a reaction. The arrow emanates from the original location of an electron pair. This can be either an unshared electron pair, on an atom such as N or O, or the electrons of a covalent bond. The curved arrow points to the final location of the electron pair. For example, to show a bond breaking:

$$X-Y \longrightarrow X^+ + Y^-$$

and to show a bond forming:

$$X^+ + :Y^- \longrightarrow X-Y$$

A familiarity with Lewis dot structures and an understanding of electronegativity (see Section 2.1) is helpful for identifying electron-rich groups (often the source of electrons during a reaction) and electron-poor groups (where the electrons often end up).

A typical biochemical reaction requires several curved arrows, for example,

? **Draw arrows to show the electron movements in the reaction $2 H_2O \rightarrow OH^- + H_3O^+$.**

FIGURE 6.7

Diagram for a reaction accelerated by covalent catalysis. Two transition states flank the covalent intermediate. The relative heights of the activation energy barriers (to achieve the two transition states, $X_1^{\ddagger}$ and $X_2^{\ddagger}$) vary depending on the reaction.

? **Explain why the free energy of the reaction intermediate must be greater than the free energy of either the reactants or products.**

In this covalent intermediate, the protonated nitrogen atom acts as an electron sink to reduce the enolate character of the transition state in the decarboxylation reaction:

$$
\underset{\substack{\\ \text{}}}{\overset{R}{\underset{H_3C-\overset{\displaystyle N^+ -H}{\underset{\displaystyle |}{C}}-CH_2-C{\overset{\displaystyle O}{\underset{\displaystyle O^-}{}}}}}} \xrightarrow{CO_2}
H_3C-\overset{R}{\underset{\displaystyle N\cdot}{\overset{\displaystyle H}{C}}}=CH_2
\xrightarrow{H^+}
H_3C-\overset{R}{\underset{\displaystyle N^+ -H}{C}}-CH_3
$$

Finally, the Schiff base decomposes, which regenerates the amine catalyst and releases the product, acetone:

$$
\underset{\text{Schiff base}}{H_3C-\overset{R\,\,N^+\! -H}{\underset{\displaystyle |}{C}}-CH_3} + OH^- \longrightarrow \underset{\text{Acetone}}{H_3C-\overset{\displaystyle O}{\underset{\displaystyle \|}{C}}-CH_3} + RNH_2
$$

In enzymes that use covalent catalysis, an electron-rich group in the enzyme active site forms a covalent adduct with a substrate. This covalent complex can sometimes be isolated; it is much more stable than a transition state. *Enzymes that use covalent catalysis undergo a two-part reaction process, so the reaction coordinate diagram contains two energy barriers with the reaction intermediate between them* (**Fig. 6.7**).

Many of the same groups that make good acid–base catalysts (see Fig. 6.6) also make good covalent catalysts because they contain unshared electron pairs (**Fig. 6.8**). Covalent catalysis is often called nucleophilic catalysis because the catalyst is a **nucleophile,** that is, an electron-rich group in search of an electron-poor center (a compound with an electron deficiency is known as an **electrophile**).

3. Metal Ion Catalysis

Metal ion catalysis occurs when metal ions participate in enzymatic reactions by mediating oxidation–reduction reactions, as mentioned earlier, or by promoting the reactivity of other groups in the enzyme's active site through electrostatic effects. A protein-bound metal ion can also interact directly with the reacting substrate. For example, during the conversion of acetaldehyde to ethanol as catalyzed by the liver enzyme alcohol

Ser, Tyr	—ÖH	—Ö: $^-$
Cys	—ṠH	—Ṡ: $^-$
Lys	—N$^+$H$_3$	—NH$_2$
His	⟨imidazole⟩ N$^+$H	⟨imidazole⟩ N:

FIGURE 6.8

Protein groups that can act as covalent catalysts. In their deprotonated forms (*right*), these groups act as nucleophiles. They attack electron-deficient centers to form covalent intermediates.

dehydrogenase, a zinc ion stabilizes the developing negative charge on the oxygen atom during formation of the transition state:

$$H_3C-C{\overset{O}{\underset{H}{}}} \rightleftharpoons \left[H_3C-C{\overset{O^{\cdots-}{\cdots}Zn^{2+}}{\underset{H}{}}} \right] \xrightarrow{2\,H\cdot} H_3C-\overset{OH}{\underset{H}{C}}-H$$

Acetaldehyde   Ethanol

THE CATALYTIC TRIAD OF CHYMOTRYPSIN PROMOTES PEPTIDE BOND HYDROLYSIS

Chymotrypsin uses both acid–base catalysis and covalent catalysis to accelerate peptide bond hydrolysis. These activities depend on three active-site residues whose identities and catalytic importance have been the object of intense study since the 1960s. Two of chymotrypsin's catalytic residues were identified using a technique called **chemical labeling.** When chymotrypsin is incubated with the compound diisopropylphosphofluoridate (DIPF), one of its 27 serine residues (Ser 195) becomes covalently tagged with the diisopropylphospho (DIP) group, and the enzyme loses activity.

$$-CH_2-OH \;+\; \underset{\underset{CH(CH_3)_2}{\overset{\overset{CH(CH_3)_2}{|}}{\overset{O}{|}}}}{F-P{=}O} \xrightarrow{\;HF\;} \underset{\underset{CH(CH_3)_2}{\overset{\overset{CH(CH_3)_2}{|}}{\overset{O}{|}}}}{-CH_2-O-P{=}O}$$

Ser 195   Diisopropylphosphofluoridate   DIP-Ser 195
(active)   (DIPF)   (inactive)

This observation provided strong evidence that Ser 195 is essential for catalysis. Chymotrypsin is therefore known as a **serine protease.** It is one of a large family of enzymes that use the same Ser-dependent catalytic mechanism. A similar labeling technique was used to identify the catalytic importance of His 57. The third residue involved in catalysis by chymotrypsin—Asp 102—was identified only after the fine structure of chymotrypsin was visualized through X-ray crystallography.

The hydrogen-bonded arrangement of the Asp, His, and Ser residues of chymotrypsin and other serine proteases is called the **catalytic triad (Fig. 6.9).** The substrate's **scissile bond** (the bond to be cleaved by hydrolysis) is positioned near Ser 195 when the substrate binds to the enzyme. The side chain of serine is not normally a strong enough nucleophile to attack an amide bond. However, His 57, acting as a base catalyst, abstracts a proton from Ser 195 so that the oxygen can act as a covalent catalyst. Asp 102 promotes catalysis by stabilizing the resulting positively charged imidazole group of His 57.

Chymotrypsin-catalyzed peptide bond hydrolysis actually occurs in two phases that correspond to the formation and breakdown of a covalent reaction intermediate. The steps of catalysis are detailed in **Figure 6.10**. Nucleophilic attack of Ser 195 on the substrate's carbonyl carbon leads to a transition state in which the carbonyl carbon assumes tetrahedral geometry. This structure then collapses to an intermediate in which the N-terminal portion of the substrate remains covalently attached to the enzyme. The second part of the reaction, during which the oxygen of a water molecule attacks the carbonyl carbon, also includes a tetrahedral transition state. Although the enzyme-catalyzed reaction requires multiple steps, the net reaction is the same as the uncatalyzed reaction shown in Section 6.1.

Asp 102   His 57   Ser 195

FIGURE 6.9

The catalytic triad of chymotrypsin. Asp 102, His 57, and Ser 195 are arrayed in a hydrogen-bonded network. Atoms are color-coded (C gray, N blue, O red), and the hydrogen bonds are shaded yellow.

? **Add hydrogen atoms to the three side chains.**

The peptide substrate enters the active site of chymotrypsin so that its scissile bond (red) is close to the oxygen of Ser 195 (the N-terminal portion of the substrate is represented by R_N, and the C-terminal portion by R_C).

1. Removal of the Ser hydroxyl proton by His 57 (a base catalyst) allows the resulting nucleophilic oxygen (a covalent catalyst) to attack the carbonyl carbon of the substrate.

Tetrahedral intermediate (transition state)

2. The transition state, known as the tetrahedral intermediate, decomposes when His 57, now acting as an acid catalyst, donates a proton to the nitrogen of the scissile peptide bond. This step cleaves the bond. Asp 102 promotes the reaction by stabilizing His 57 through hydrogen bonding.

Acyl–enzyme intermediate (covalent intermediate)

The departure of the C-terminal portion of the cleaved peptide, with a newly exposed N-terminus, leaves the N-terminal portion of the substrate (an acyl group) linked to the enzyme. This relatively stable covalent complex is known as the acyl–enzyme intermediate.

3. Water then enters the active site. It donates a proton to His 57 (again a base catalyst), leaving a hydroxyl group that attacks the carbonyl group of the remaining substrate. This step resembles Step 1 above.

4. In the second tetrahedral intermediate, His 57, now an acid catalyst, donates a proton to the Ser oxygen, leading to collapse of the intermediate. This step resembles Step 2 above.

Tetrahedral intermediate (transition state)

5. The N-terminal portion of the original substrate, now with a new C-terminus, diffuses away, regenerating the enzyme.

FIGURE 6.10

The catalytic mechanism of chymotrypsin and other serine proteases. Two tetrahedral transition states lead to and from the acyl–enzyme intermediate.

❓ Choose one step of the reaction and identify all the bonds that break and form in that step.

The roles of Asp, His, and Ser in peptide bond hydrolysis, as catalyzed by chymotrypsin and other members of the serine protease family, have been tested through site-directed mutagenesis (see Section 3.5). Replacing the catalytic aspartate with another residue decreases the rate of substrate hydrolysis about 5000-fold. Adding a methyl group to histidine by chemical labeling (so that it can't accept or donate a proton) has a similar effect. Replacing the catalytic serine with another residue decreases enzyme activity about a millionfold. Surprisingly, replacing all three catalytic residues—Asp, His, and Ser—through site-directed mutagenesis does not completely abolish protease activity: The modified enzyme still catalyzes peptide bond hydrolysis at a rate about 50,000 times greater than the rate of the uncatalyzed reaction. Clearly, chymotrypsin and its relatives rely on the acid–base catalysis and covalent catalysis carried out by the Asp–His–Ser catalytic triad, but these enzymes must have additional catalytic mechanisms that allow them to achieve reaction rates 10^{11} times greater than the rate of the uncatalyzed reaction.

 CONCEPT CHECK

1. Draw a free energy diagram for a reaction with and without a catalyst, and label reactants, products, transition state, activation energy, and free energy change for the reaction.

2. Why must a reaction that proceeds via covalent catalysis have two transition states?

3. Make a list of amino acid side chains that can function as acid–base catalysts or covalent catalysts.

4. Explain why some enzymes require a metal ion or organic group in the active site.

5. Draw the groups that make up chymotrypsin's catalytic triad and explain their roles.

6. How do these residues participate in peptide bond hydrolysis as catalyzed by chymotrypsin?

6.3 UNIQUE PROPERTIES OF ENZYME CATALYSTS

If only a few residues in an enzyme directly participate in catalysis (for example, Asp, His, and Ser in chymotrypsin), why are enzymes so large? One obvious answer is that the catalytic residues must be precisely aligned in the active site, so a certain amount of surrounding structure is required to hold them in place. In 1894, long before the first enzyme structure had been determined (and several decades before it had been shown that enzymes are proteins), Emil Fischer noted the exquisite substrate specificity of enzymes and proposed that the substrate fit the enzyme like a key in a lock. However, the **lock-and-key model** for enzyme action does not explain how an active site that perfectly accommodates substrates could also accommodate reaction products before they are released from the enzyme. Moreover, the lock-and-key model cannot explain how an enzyme inhibitor can bind tightly in the active site but not react. The answer is that enzymes, like other proteins (see Section 4.3), are not rigid molecules but instead can flex while binding substrates. In other words, the enzyme–substrate interaction must be more dynamic than a key in a lock, more like a hand in a glove. In some cases, the enzyme can physically distort the substrate as it binds, in effect pushing it toward a higher-energy conformation closer to the reaction's transition state. Current theories attribute much of the catalytic power of enzymes to these and other specific interactions between enzymes and their substrates.

ENZYMES STABILIZE THE TRANSITION STATE

The lock-and-key model does contain a grain of truth, a principle first formulated by Linus Pauling in 1946. He proposed that *an enzyme increases the reaction rate not by binding tightly to the substrates but by binding tightly to the reaction's transition state* (that is, substrates that have been strained toward the structures of the products). In other words, the tightly bound key of the lock-and-key model is the transition state, not the substrate. In an enzyme, tight binding (stabilization) of the transition state occurs in addition to acid–base, covalent, or metal ion catalysis. In general, transition state stabilization is accomplished through the close complementarity in shape and charge between the active site and the transition state. A nonreactive substance that mimics the transition state can therefore bind tightly to the enzyme and block its catalytic activity (enzyme inhibition is discussed further in Section 7.3).

Transition state stabilization appears to be an important part of the chymotrypsin reaction. In this case, the two tetrahedral transition states (see Fig. 6.10), are stabilized through interactions that do not occur at any other point in the reaction. Rate acceleration is believed to result from an increase in both the number and strength of the bonds that form between active site groups and the substrate in the transition state.

1. During formation of the tetrahedral intermediate, the planar peptide group of the substrate changes its geometry, and the carbonyl oxygen, now an anion, moves into a previously unoccupied cavity near the Ser 195 side chain. In this cavity, called the **oxyanion hole,** the substrate oxygen ion can form two new hydrogen bonds with the backbone NH groups of Ser 195 and Gly 193 (**Fig. 6.11**). The backbone NH group of the substrate residue preceding the scissile bond forms another hydrogen bond to Gly 193 (not shown in Fig. 6.11). Thus, the transition state is stabilized (its energy lowered) by three hydrogen bonds that cannot form when the enzyme first binds its substrate. The stabilizing effect of these three new hydrogen bonds could account for a significant portion of chymotrypsin's catalytic power, since the energy of a standard hydrogen bond is about 20 kJ · mol^{-1} and the reaction rate increases 10-fold for every decrease in $\Delta G^{\ddagger}$ of 5.7 kJ · mol^{-1}.

FIGURE 6.11

Transition state stabilization in the oxyanion hole. (a) The chymotrypsin active site is shown with the oxyanion hole shaded in pink. The carbonyl carbon of the peptide substrate has trigonal geometry, so the carbonyl oxygen cannot occupy the oxyanion hole. (b) Nucleophilic attack by the oxygen of Ser 195 on the substrate carbonyl group leads to a transition state, in which the carbonyl carbon assumes tetrahedral geometry. At this point, the substrate's anionic oxygen (the oxyanion) can move into the oxyanion hole, where it forms hydrogen bonds (shaded yellow) with two enzyme backbone groups.

His 57

CH₂

$$O \atop C\!=\!\bar{O}\cdots H\!-\!N$$

Asp 102 — CH₂

↓

His 57

CH₂

$$O \atop C\!=\!\bar{O}\cdots H\!-\!N \overset{+}{\cdots} N\!-\!H$$

Asp 102 — CH₂

FIGURE 6.12

Formation of a low-barrier hydrogen bond during catalysis in chymotrypsin. The Asp 102—His 57 hydrogen bond becomes shorter and stronger, so the imidazole proton comes to be shared equally between the O of aspartate and the N of histidine in a low-barrier hydrogen bond.

2. NMR studies, which can identify individual hydrogen-bonding interactions, suggest that the hydrogen bond between Asp 102 and His 57 becomes shorter during formation of the two transition states (**Fig. 6.12**). Such a bond is called a **low-barrier hydrogen bond** because the hydrogen is shared equally between the original donor and acceptor atoms (in a standard hydrogen bond, the hydrogen still "belongs" to the donor atom and there is an energy barrier for its full transfer to the acceptor atom). A decrease in bond length from ∼2.8 Å to ∼2.5 Å in forming the low-barrier hydrogen bond is accompanied by a three- to fourfold increase in bond strength. The low-barrier hydrogen bond that forms during catalysis in chymotrypsin helps stabilize the transition state and thereby accelerate the reaction.

EFFICIENT CATALYSIS DEPENDS ON PROXIMITY AND ORIENTATION EFFECTS

Enzymes increase reaction rates by bringing reacting groups into close proximity so as to increase the frequency of collisions that can lead to a reaction. Furthermore, when substrates bind to an enzyme, their translational and rotational motions are frozen out so that they can be oriented properly for reaction (**Fig. 6.13**). These **proximity and orientation effects** likely explain some of the residual activity of chymotrypsin whose catalytic residues have been altered. Nevertheless, *an enzyme must be more than a template for assembling and aligning reacting groups.*

THE ACTIVE-SITE MICROENVIRONMENT PROMOTES CATALYSIS

In nearly all cases, an enzyme's active site is somewhat removed from the solvent, with its catalytic residues in some sort of cleft or pocket on the enzyme surface. Upon binding substrates, some enzymes undergo a pronounced conformational change so that they almost fully enclose the substrates. Daniel Koshland has called this phenomenon **induced fit.** Many studies of enzyme activity, supplemented by X-ray crystallographic data, support this feature of enzyme action, further highlighting the shortcomings of the lock-and-key model. In fact, techniques that assess protein structure on millisecond time scales suggest that conformational fluctuations may be critical for all stages of an enzyme-catalyzed reaction, not just substrate binding.

(a)   (b)

FIGURE 6.13

Proximity and orientation effects in catalysis. In order to react, two groups must come together and collide with the correct orientation. (a) Reactants in solution are separated in space and have translational and rotational motions that must be overcome. (b) When the reactants bind to an enzyme, their motion is limited, and they are held in close proximity and with the correct alignment for a productive reaction.

A classic case of induced fit occurs in hexokinase, which catalyzes the phosphorylation of glucose by ATP (Reaction 1 of glycolysis; Section 13.1):

Glucose + ATP $\xrightarrow{\text{hexokinase}}$ Glucose-6-phosphate + ADP

(a)

(b)

The enzyme consists of two hinged lobes with the active site located between them (**Fig. 6.14a**). When glucose binds to hexokinase, the lobes swing together, engulfing the sugar (Fig. 6.14b). The result of hinge bending is that the substrate glucose is positioned near the substrate ATP such that a phosphoryl group can be easily transferred from the ATP to a hydroxyl group of the sugar. Not even a water molecule can enter the closed active site. This is beneficial, since water in the active site could lead to wasteful hydrolysis of ATP:

$$ATP + H_2O \rightarrow ADP + P_i$$

A glucose molecule in solution is surrounded by ordered water molecules in a hydration shell (see Section 2.1). These water molecules must be shed in order for glucose to fit into the active site of an enzyme, such as hexokinase. However, once the desolvated substrate is in the enzyme active site, the reaction can proceed quickly because there are no solvent molecules to interfere. In solution, rearranging the hydrogen bonds of surrounding water molecules as the reactants approach each other and pass through the transition state is energetically costly. *By sequestering substrates in the active site, an enzyme can eliminate the energy barrier imposed by the ordered solvent molecules, thereby accelerating the reaction.*

This phenomenon is sometimes described as **electrostatic catalysis** since the nonaqueous active site allows more powerful electrostatic interactions between the enzyme and substrate than could occur in aqueous solution (for example, a low-barrier hydrogen bond can form in an active site but not in the presence of solvent molecules that would form ordinary hydrogen bonds).

FIGURE 6.14

Conformational changes in hexokinase. (a) The enzyme consists of two lobes connected by a hinge region. The active site is located in a cleft between the lobes. (b) When glucose (not shown) binds to the active site, the enzyme lobes swing together, enclosing glucose and preventing the entry of water. [*Structure of "open" hexokinase (pdb 2YHX) determined by T. A. Steitz, C. M. Anderson, and R. E. Stenkamp; structure of "closed" hexokinase (pdb 1HKG) determined by W. S. Bennett Jr. and T. A. Steitz.*]

 CONCEPT CHECK

1. Discuss why the lock-and-key model does not fully describe enzyme action.
2. Why does exclusion of water from the active site promote catalysis?
3. Describe the roles of the oxyanion hole and low-barrier hydrogen bonds in chymotrypsin.
4. What are the roles of proximity and orientation effects in enzyme catalysis?

CHYMOTRYPSIN IN CONTEXT

Chymotrypsin serves as a model for the structures and functions of a large family of serine proteases. And as with myoglobin and hemoglobin (Section 5.1), a close look at chymotrypsin reveals some general features of enzyme function, including evolution, substrate specificity, and inhibition.

KEY CONCEPTS

- Evolution has produced a number of serine proteases that differ in overall structure and substrate specificity.
- Inactive zymogens are activated by proteolysis.
- Active enzymes may be regulated by inhibitors.

FIGURE 6.15

Structures of chymotrypsin, trypsin, and elastase. The superimposed backbone traces of bovine chymotrypsin (blue), bovine trypsin (green), and porcine elastase (red) are shown along with the side chains of the active site Asp, His, and Ser residues. [*Chymotrypsin structure (pdb 4CHA) determined by H. Tsukada and D. M. Blow; trypsin structure (pdb 3PTN) determined by J. Walker, W. Steigemann, T. P. Singh, H. Bartunik, W. Bode, and R. Huber; elastase structure (3EST) determined by E. F. Meyer, G. Cole, R. Radhakrishnan, and O. Epp.*]

NOT ALL SERINE PROTEASES ARE RELATED BY EVOLUTION

The first three proteases to be examined in detail were the digestive enzymes chymotrypsin, trypsin, and elastase, which have strikingly similar three-dimensional structures (**Fig. 6.15**). This was not expected on the basis of their limited sequence similarity (**Table 6.3**). However, careful examination revealed that most of the sequence variation is on the enzyme surface, and the positions of the catalytic residues in the three active sites are virtually identical. It is believed that *these proteins are the result of* **divergent evolution;** *that is, they evolved from a common ancestor and have retained their overall structure and catalytic mechanism.*

Some bacterial proteases with a catalytically essential serine are structurally related to the mammalian digestive serine proteases. However, the bacterial serine protease subtilisin (**Fig. 6.16**) shows no sequence similarity to chymotrypsin and no overall structural similarity, although it has the same Asp–His–Ser catalytic triad and an oxyanion hole in its active site. Subtilisin is an example of **convergent evolution,** a phenomenon whereby unrelated proteins evolve similar characteristics.

As many as five groups of serine proteases, each with a different overall backbone conformation, have undergone convergent evolution to arrive at the same Asp, His, and Ser catalytic groups. In some other hydrolases, the substrate is attacked by a nucleophilic serine or threonine residue that is located in a catalytic triad such as His–His–Ser or Asp–Lys–Thr. It would appear that natural selection favors this sort of arrangement of catalytic residues.

ENZYMES WITH SIMILAR MECHANISMS EXHIBIT DIFFERENT SUBSTRATE SPECIFICITY

Despite similarities in their catalytic mechanisms, chymotrypsin, trypsin, and elastase differ significantly from one another in their substrate specificity. Chymotrypsin preferentially cleaves peptide bonds following large hydrophobic residues. Trypsin prefers the basic residues arginine and lysine, and elastase cleaves the peptide bonds following small hydrophobic residues such as alanine, glycine, and valine (these residues predominate in elastin, an animal protein responsible for the elasticity of some tissues). *The varying specificities of these enzymes are largely explained by the chemical character of the so-called* **specificity pocket,** *a cavity on the enzyme surface at the active site that accommodates the residue on the N-terminal side of the scissile peptide bond* (**Fig. 6.17**).

In chymotrypsin, the specificity pocket is about 10 Å deep and 5 Å wide, which offers a snug fit for an aromatic ring (whose dimensions are 6 Å × 3.5 Å). The specificity pocket in trypsin is similarly sized but has an aspartate residue rather than serine at the bottom. Consequently, the trypsin specificity pocket readily binds the side chain of arginine or lysine, which has a diameter of about 4 Å and a cationic group at the end. In elastase, the specificity pocket is only a small depression due to the replacement of two glycine residues on the walls of the specificity pocket (residues 216 and 226 in chymotrypsin) with the bulkier valine and threonine. Elastase therefore preferentially binds small nonpolar side chains. Although these same side chains could easily enter the chymotrypsin and trypsin specificity pockets, they do not fit well enough to immobilize the substrate at the active site, as required for

FIGURE 6.16

Structure of subtilisin from *Bacillus amyloliquefaciens*. The residues of the catalytic triad are highlighted in red. [*Structure (pdb 1CSE) determined by W. Bode.*]

❓ **Compare the structure of this enzyme with those of the three serine proteases shown in Figure 6.15.**

TABLE 6.3 Percent Sequence Identity among Three Serine Proteases	
Bovine trypsin	100%
Bovine chymotrypsinogen	53%
Porcine elastase	48%

FIGURE 6.17

Specificity pockets of three serine proteases. The side chains of key residues that determine the size and nature of the specificity pocket are shown along with a representative substrate for each enzyme. Chymotrypsin prefers large hydrophobic side chains; trypsin prefers Lys or Arg; and elastase prefers Ala, Gly, or Val. For convenience, the residues of all three enzymes are numbered to correspond to the sequence of residues in chymotrypsin.

❓ **What would the specificity pocket look like in a protease that cleaved bonds following Asp or Glu residues?**

efficient catalysis. Other serine proteases, such as those that participate in blood coagulation, exhibit exquisite substrate specificity, in keeping with their highly specialized physiological functions.

CHYMOTRYPSIN IS ACTIVATED BY PROTEOLYSIS

Relatively nonspecific proteases could do considerable damage to the cells where they are synthesized unless their activity is carefully controlled. In many organisms, *the activity of proteases is limited by the action of protease inhibitors* (some of which are discussed further below) *and by synthesizing the proteases as inactive precursors (called* **zymogens***) that are later activated when and where they are needed.*

The inactive precursor of chymotrypsin is called chymotrypsinogen, and it is synthesized by the pancreas along with the zymogens of trypsin (trypsinogen), elastase (proelastase), and other hydrolytic enzymes. All these zymogens are activated by proteolysis after they are secreted into the small intestine. An intestinal protease called enteropeptidase activates trypsinogen by catalyzing the hydrolysis of its Lys 6—Ile 7 bond. Enteropeptidase catalyzes a highly specific reaction; it appears to recognize a string of aspartate residues near the N-terminus of its substrate:

$$H_3\overset{+}{N}-Val-Asp-Asp-Asp-Asp-Lys-Ile-\cdots$$

$$H_2O \searrow \text{enteropeptidase}$$

$$H_3\overset{+}{N}-Val-Asp-Asp-Asp-Asp-Lys-COO^- + H_3\overset{+}{N}-Ile-\cdots$$

Trypsin, now itself active, cleaves the N-terminal peptide of the other pancreatic zymogens, including trypsinogen. The activation of trypsinogen by trypsin is an example of **autoactivation.**

The Arg 15—Ile 16 bond of chymotrypsinogen is susceptible to trypsin-catalyzed hydrolysis. Cleavage of this bond generates a species of active chymotrypsin (called π chymotrypsin), which then undergoes two autoactivation steps to generate fully active chymotrypsin (also called α chymotrypsin; **Fig. 6.18**). A similar process, in which zymogens are sequentially activated through proteolysis, occurs during blood coagulation (Section 6.5).

Activation of chymotrypsinogen. Trypsin activates chymotrypsinogen by catalyzing hydrolysis of the Arg 15—Ile 16 bond of the zymogen. The resulting active chymotrypsin then excises the Ser 14–Arg 15 dipeptide (by cleaving the Leu 13—Ser 14 bond) and the Thr 147–Asn 148 dipeptide (by cleaving the Tyr 146—Thr 147 and Asn 148—Ala 149 bonds). All three species of chymotrypsin (π, δ, and α) have proteolytic activity.

Location of the dipeptides removed during the activation of chymotrypsinogen. The Ser 14–Arg 15 dipeptide (lower *right*, green) and the Thr 147–Asn 148 dipeptide (*right*, blue) are located at some distance from the active site residues (red) in chymotrypsinogen. [*Chymotrypsinogen structure (pdb 2CGA) determined by D. Wang, W. Bode, and R. Huber.*]

The two dipeptides that are excised during chymotrypsinogen activation are far removed from the active site (**Fig. 6.19**). How does their removal boost catalytic activity? A comparison of the X-ray structures of chymotrypsin and chymotrypsinogen reveals that the conformations of their active site Asp, His, and Ser residues are virtually identical (in fact, the zymogen can catalyze hydrolysis extremely slowly). However, the substrate specificity pocket and the oxyanion hole are incompletely formed in the zymogen. Proteolysis of the zymogen elicits small conformational changes that open up the substrate specificity pocket and oxyanion hole. *Thus, the enzyme becomes maximally active only when it can efficiently bind its substrates and stabilize the transition state.*

PROTEASE INHIBITORS LIMIT PROTEASE ACTIVITY

The pancreas, in addition to synthesizing the zymogens of digestive proteases, synthesizes small proteins that act as **protease inhibitors.** The liver also produces a variety of protease inhibitor proteins that circulate in the bloodstream. If the pancreatic enzymes were prematurely activated or escaped from the pancreas through trauma, they would be rapidly inactivated by protease inhibitors. *The inhibitors pose as protease substrates but are not completely hydrolyzed.* For example, when trypsin attacks a lysine residue of bovine pancreatic trypsin inhibitor, the reaction halts during formation of the first transition state. The inhibitor remains in the active site, preventing any further catalytic activity (**Fig. 6.20**). The noncovalent complex between trypsin

The complex of trypsin with bovine pancreatic trypsin inhibitor. Ser 195 of trypsin (gold) attacks the peptide bond of Lys 15 (green) in the inhibitor, but the reaction is arrested on the way to the tetrahedral intermediate. [*Structure (pdb 2PTC) determined by R. Huber and J. Deisenhofer.*]

and bovine pancreatic trypsin inhibitor is one of the strongest protein—protein interactions known, with a dissociation constant of 10^{-14} M. An imbalance between the activities of proteases and the activities of protease inhibitors may contribute to disease (Section 6.5).

 CONCEPT CHECK

1. How are chymotrypsin, trypsin, and elastase similar?

2. What is the chemical basis for their different substrate specificities?

3. Describe how to determine whether two proteins are related by convergent or divergent evolution.

4. How is chymotrypsin activated and why is this step necessary?

5. Why are protease inhibitors necessary?

CLINICAL CONNECTION: BLOOD COAGULATION

When a blood vessel is injured by mechanical force, infection, or some other pathological process, red and white blood cells and the plasma (fluid) that surrounds them can leak out. Except in the most severe trauma, the loss of blood can be halted through formation of a clot at the site of injury. The clot consists of aggregated platelets (tiny cell fragments that rapidly adhere to the damaged vessel wall and to each other) and a mesh of the protein fibrin, which reinforces the platelet plug and traps larger particles such as red blood cells (**Fig. 6.21**).

Fibrin polymers can form rapidly because they are generated at the site of injury from the soluble protein fibrinogen, which circulates in the blood plasma. Fibrinogen is an elongated molecule with a molecular mass of 340,000 and consists of three pairs of polypeptide chains. The proteolytic removal of short (14- or 16-residue) peptides from the N-termini of four of the six chains causes the protein to polymerize in end-to-end and side-to-side fashion to produce a thick fiber:

KEY CONCEPT
- Blood coagulation process as a cascade of proteases.

The conversion of fibrinogen to fibrin is the final step of **coagulation,** a series of proteolytic reactions involving a number of proteins and additional factors from platelets and damaged tissue. The enzyme responsible for cleaving fibrinogen to fibrin is known as thrombin (**Fig. 6.22**). It is similar to trypsin in its sequence (38% of their residues are identical) and in its structure and catalytic mechanism.

Like trypsin, thrombin cleaves peptide bonds following arginine residues, but it is highly specific for the two cleavage sites in the fibrinogen sequence. Thrombin, like fibrinogen, circulates as an inactive precursor. Its zymogen, called prothrombin, contains a serine protease domain along with several other structural motifs. These elements interact with other coagulation factors to help ensure that thrombin—and therefore fibrin—is produced only when needed.

FIGURE 6.21

A blood clot. Red blood cells are trapped in a mesh of fibrin and platelets. [*P. Motta/Dept. of Anatomy, University La Sapienza, Rome/Science Photo Library/Photo Researchers.*]

FIGURE 6.22

Thrombin. The catalytic residues (aspartate, histidine, and serine) are highlighted in red. [*Structure (pdb 1PPB) determined by W. Bode.*]

? **Compare the structures of thrombin and chymotrypsin (Fig. 6.1).**

A serine protease known as factor Xa catalyzes the specific hydrolysis of prothrombin to generate thrombin. Factor Xa (the *a* stands for *active*) is the protease form of the zymogen factor X. To initiate coagulation, factor X is activated by a protease known as factor VIIa, working in association with an accessory protein called tissue factor, which is exposed when a blood vessel is broken. During the later stages of coagulation, factor Xa is generated by the activity of factor IXa, which is generated from its zymogen by the activity of factor XIa or factor VIIa–tissue factor. Factor XIa is in turn generated by the proteolysis of its zymogen by trace amounts of thrombin produced earlier in the coagulation process. The cascade of activation reactions is depicted in **Figure 6.23**. Many of these enzymatic reactions require accessory factors that are not shown in this simplified diagram.

Note that the coagulation proteases are named according to their order of discovery, not their order of action (thrombin is also known as factor II). All the coagulation proteases appear to have evolved from a trypsin-like enzyme but have acquired narrow substrate specificity and a correspondingly narrow range of physiological activities.

The coagulation reactions have an amplifying effect because each protease is a catalyst for the activation of another catalyst. Thus, a very small amount of factor IXa can activate a larger amount of factor Xa, which can then activate an even larger amount of thrombin. This amplification effect is reflected in the plasma concentrations of the coagulation factors (**Table 6.4**).

A complex process such as coagulation is subject to regulation at a variety of points, including the activation and inhibition of the various proteases. Protease inhibitors account for about 10% of the protein that circulates in plasma. One inhibitor, known as antithrombin, blocks the proteolytic activities of factor IXa, factor Xa, and thrombin, thereby limiting the extent and duration of clot formation. An arginine residue serves as a "bait" for the arginine-specific coagulation proteases (**Fig. 6.24**). The protease recognizes the inhibitor as a substrate but is unable to complete the hydrolysis reaction. The protease and inhibitor form a stable acyl–enzyme intermediate that is removed from the circulation within a few minutes.

Heparin, a sulfated polysaccharide (Section 11.3) first purified from liver, enhances the activity of antithrombin by two mechanisms: A short segment (5 monosaccharide residues) acts as an allosteric activator of antithrombin, and a longer heparin polymer (containing at least 18 residues) can bind simultaneously to both antithrombin and its target protease such that their co-localization dramatically increases the rate of their reaction. Heparin or a synthetic version of it is used clinically as an anticoagulant following surgery.

Defects in many of the proteins involved in blood coagulation or its regulation have been linked to bleeding (or clotting) disorders. For example, one form of hemophilia, a tendency to bleed following minor trauma, results from a genetic deficiency

Thrombin

XI

    XIa
      or VIIa–tissue factor

IX

    IXa
      or VIIa–tissue factor

X   *

    Xa

Prothrombin

    Thrombin

Fibrinogen

    Fibrin

FIGURE 6.23

The coagulation cascade. The triggering step is marked with an asterisk.

TABLE 6.4 Plasma Concentrations of Some Human Coagulation Factors

FACTOR	CONCENTRATION (μM)[a]
XI	0.06
IX	0.09
VII	0.01
X	0.18
Prothrombin	1.39
Fibrinogen	8.82

[a]Concentrations calculated from data in High, K. A., and Roberts, H. R., eds., *Molecular Basis of Thrombosis and Hemostasis*, Marcel Dekker (1995).

FIGURE 6.24

Antithrombin. A loop of the protein (yellow, residues 377–400 of the 432-residue protein) offers arginine 393 (red) as a "substrate" for several proteases. [*Antithrombin structure (pdb 2ANT) determined by R. Skinner, J. P. Abrahams, J. C. Whisstock, A. M. Lesk, R. W. Carrell, and M. R. Wardell.*]

of factor IX. A deficiency of antithrombin results in an increased risk of clot formation in the veins. If dislodged, the clots may end up blocking an artery in the lungs or brain, with dire consequences.

 CONCEPT CHECK

1. Describe the coagulation process beginning with the step marked by an asterisk in Fig. 6.23.
2. Discuss the advantages of a cascade system involving zymogen activation.
3. Why can antithrombin inhibit more than one coagulation protease?

SUMMARY

6.1 What Is an Enzyme?

• Enzymes accelerate chemical reactions with high specificity under mild conditions.

6.2 Chemical Catalytic Mechanisms

• A reaction coordinate diagram illustrates the change in free energy between the reactants and products as well as the activation energy required to reach the transition state. The higher the activation energy, the fewer the reactant molecules that can reach the transition state and the slower the reaction.

• An enzyme provides a route from reactants (substrates) to products, which has a lower activation energy than the uncatalyzed reaction. Enzymes, sometimes with the assistance of a cofactor, use chemical catalytic mechanisms such as acid–base catalysis, covalent catalysis, and metal ion catalysis.

• In chymotrypsin, an Asp–His–Ser triad catalyzes peptide bond hydrolysis through acid–base and covalent catalysis and by stabilizing the transition state via the oxyanion hole and low-barrier hydrogen bonds.

6.3 Unique Properties of Enzyme Catalysts

• In addition to transition state stabilization, enzymes use proximity and orientation effects, induced fit, and electrostatic catalysis to facilitate reactions.

6.4 Chymotrypsin in Context

• Serine proteases that have evolved from a common ancestor share their overall structure and catalytic mechanism but differ in their substrate specificity.

• The activities of some proteases are limited by their synthesis as zymogens that are later activated and by their interaction with protease inhibitors.

6.5 Clinical Connection: Blood Coagulation

• Formation of a blood clot involves the sequential activation of serine protease zymogens, culminating in the generation of active thrombin, which converts fibrinogen to fibrin.

• Antithrombin inhibits several proteases to limit clot formation.

KEY TERMS

hydrolysis	transition state	Schiff base	proximity and orientation effects
reactant	$\Delta G_{reaction}$	nucleophile	induced fit
catalyst	cofactor	electrophile	electrostatic catalysis
enzyme	coenzyme	metal ion catalysis	divergent evolution
ribozyme	acid–base catalysis	chemical labeling	convergent evolution
active site	acid catalysis	serine protease	specificity pocket
reaction specificity	base catalysis	catalytic triad	zymogen
substrate	tautomer	scissile bond	autoactivation
isozymes	carbanion	lock-and-key model	protease inhibitor
reaction coordinate	covalent catalysis	oxyanion hole	coagulation
$\Delta G^{\ddagger}$	imine	low-barrier hydrogen bond	

BIOINFORMATICS

Brief Bioinformatics Exercises

6.1 Introduction to Enzyme Informatics

6.2 Interactions in the Serine Protease Active Site

Bioinformatics Projects

Enzyme Commission Classes and Catalytic Site Alignments with PyMOL

Enzyme Evolution

PROBLEMS

6.1 What Is an Enzyme?

1. Why are most enzymes globular rather than fibrous proteins?

2. Explain why the motor proteins myosin and kinesin (described in Section 5.4) are enzymes and write the reaction that each catalyzes.

3. The uncatalyzed rate of amide bond hydrolysis in hippuryl-phenylalanine is 1.3×10^{-10} s^{-1}, whereas the rate of hydrolysis of the amide bond in the same substrate catalyzed by carboxypeptidase is 61 s^{-1}. What is the rate enhancement for amide bond hydrolysis by this enzyme for this particular substrate?

4. The half-life for the hydrolysis of the glycosidic bond in the sugar trehalose is 6.6×10^6 years.

(a) What is the rate constant for the uncatalyzed hydrolysis of this bond? [*Hint*: For a first-order reaction, the rate constant, k, is equal to $0.693/t_{1/2}$.]

(b) What is the rate enhancement for glycosidic bond hydrolysis catalyzed by trehalase if the rate constant for the catalyzed reaction is 2.6×10^3 s^{-1}?

5. Compare the rate enhancements of adenosine deaminase and triose phosphate isomerase (see Table 6.1).

6. What is the relationship between the rate of an enzyme-catalyzed reaction and the rate of the corresponding uncatalyzed reaction? Do enzymes enhance the rates of slow uncatalyzed reactions as much as they enhance the rates of fast uncatalyzed reactions?

7. (a) Draw an arrow to identify the bonds in the peptide below that would be hydrolyzed in the presence of the appropriate peptidase. (b) To what class of enzymes does the peptidase belong?

8. The reactions catalyzed by the enzymes listed here are presented in this chapter. To which class does each enzyme belong? Explain your answers. (a) Pyruvate decarboxylase; (b) alanine aminotransferase; (c) alcohol dehydrogenase; (d) hexokinase; (e) chymotrypsin.

9. To which class do the enzymes that catalyze the following reactions belong?

(d)

10. (a) The molecule shown below is a substrate for trypsin. Draw the reaction products.

(b) Design an assay that would allow you to measure the rate of trypsin-catalyzed hydrolysis of the molecule.

11. (a) Draw the oxidized product of the reaction catalyzed by succinate dehydrogenase.

(b) Malate dehydrogenase catalyzes a reaction in which C2 of malate is oxidized. Draw the structure of the product.

(c) To which class of enzymes do lactate dehydrogenase and malate dehydrogenase belong?

Succinate   Malate

12. (a) Examine the reaction catalyzed by hexokinase in Section 6.3. Draw the product of the reaction catalyzed by creatine kinase, which acts on creatine in a similar manner. What would you predict to be the usual function of a kinase?

(b) The reaction catalyzed by hexokinase can be reversed by the enzyme glucose-6-phosphatase. To what class of enzymes do phosphatases belong? What would you predict to be the usual function of a phosphatase?

Creatine

13. The ALT and AST tests measure the plasma activities of the liver enzymes alanine aminotransferase and aspartate aminotransferase, respectively, which are released from the liver into the blood when the liver is damaged. Draw the reaction catalyzed by aspartate aminotransferase.

14. Propose a name for the enzymes that catalyze the following reactions (reactions may not be balanced):

(a)

Glucose-6-phosphate   6-Phosphoglucono-δ-lactone

(b)

Isocitrate   Succinate   Glyoxylate

(c)

1,3-Bisphosphoglycerate   3-Phosphoglycerate

(d)

Glyceraldehyde-3-phosphate   Dihydroxyacetone phosphate

15. The amino acid tryptophan is converted to the hormone melatonin as shown here. Indicate which reaction is catalyzed by each of the following types of enzymes. **(a)** methyltransferase; **(b)** hydroxylase; **(c)** acetyltransferase; **(d)** decarboxylase.

16. Use the enzyme nomenclature database at www.expasy.org/enzyme to determine what reaction is catalyzed by the enzyme **(a)** catalase and **(b)** glutathione peroxidase.

17. Given the following EC numbers, provide the common name of the enzyme **(a)** 2.7.1.1 and **(b)** 1.7.2.2.

6.2 Chemical Catalytic Mechanisms

18. Approximately how much do the enzymes **(a)** adenosine deaminase and **(b)** triose phosphate isomerase (see Table 6.1) decrease the activation energy ($\Delta G^\ddagger$) of the reactions they catalyze?

19. The rate of an enzyme-catalyzed reaction is measured at several temperatures, generating the curve shown below. Explain why the enzyme activity increases with temperature and then drops off sharply.

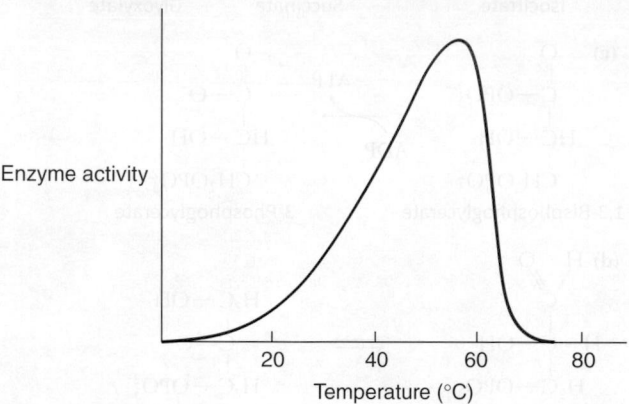

20. Draw pairs of free energy diagrams that show the differences between the following reactions:

(a) A fast reaction vs. a slow reaction.

(b) A one-step reaction vs. a two-step reaction.

(c) An endergonic reaction vs. an exergonic reaction.

(d) A two-step reaction with an initial slow step vs. a two-step reaction with an initial fast step.

21. Consult Figure 6.10 and draw a reaction coordinate diagram for the chymotrypsin-catalyzed hydrolysis of a peptide bond.

22. Under certain conditions, peptide bond formation is more thermodynamically favorable than peptide bond hydrolysis. Would you expect chymotrypsin to catalyze peptide bond formation?

23. Solutions of chymotrypsin prepared for experimental purposes are stored in a dilute solution of hydrochloric acid as a "preservative." Explain.

24. What is the relationship between the nucleophilicity and the acidity of an amino acid side chain?

25. Amino acids such as glycine, alanine, and valine are not known to participate directly in acid–base or covalent catalysis.

(a) Explain why this is the case.

(b) Mutating a glycine, alanine, or valine residue in an enzyme's active site can still have dramatic effects on catalysis. Why?

26. Using what you know about the structures of the amino acid side chains and the mechanisms presented in this chapter, choose an amino acid side chain to play the following roles in an enzymatic mechanism: **(a)** participate in proton transfer, **(b)** act as a nucleophile.

27. Ribozymes are RNA molecules that catalyze chemical reactions.

(a) What features of a nucleic acid would be important for it to act as an enzyme?

(b) What part of the RNA structure might fulfill the roles listed in Problem 26?

(c) Why can RNA but not DNA act as a catalyst?

28. RNA is susceptible to base catalysis in which the hydroxide ion abstracts a proton from the 2′ OH group and then the resulting 2′ O⁻ nucleophilically attacks the 5′ phosphate.

(a) Is DNA susceptible to base hydrolysis?

(b) Does this observation help you to explain why DNA rather than RNA evolved as the genetic material?

29. Use what you know about the mechanism of chymotrypsin to explain why DIPF and TPCK inactivates the enzyme.

30. Sarin is an organophosphorus compound similar to DIPF. In 1995, terrorists released it on a Japanese subway. Injuries resulted from the reaction of sarin with a serine esterase involved in nerve transmission, called acetylcholinesterase. Draw the structure of the enzyme's catalytic residue modified by sarin.

$$H_3C\text{---}CH\text{---}O\text{---}\underset{\underset{CH_3}{|}}{\overset{\overset{O}{||}}{P}}\text{---}F$$

Sarin

31. (a) Diagram the hydrogen-bonding interactions of the catalytic triad His–Lys–Ser during catalysis in a hypothetical hydrolytic enzyme. **(b)** In a chemical labeling study of the enzyme, a reagent that covalently modifies Lys residues also abolishes enzyme activity. Why doesn't this observation prove that the active site includes a Lys residue?

32. How would chymotrypsin's catalytic triad be affected by extremely low and extremely high pH values (assuming that the rest of the protein structure remained intact)?

33. Enzymes with Cys residues in their active sites can often be inactivated by iodoacetate (ICH_2COO^-). Show the chemical reaction for the modification of the Cys side chain by iodoacetate. Why would such a modification inactivate an enzyme that required a Cys side chain for activity?

34. Angiogenin, an enzyme involved in the formation of blood vessels, was treated with bromoacetate ($BrCH_2COO^-$). An essential histidine was modified, and the activity of the enzyme was reduced by 95%. Show the chemical reaction for the modification of the histidine side chain with bromoacetate. Why would such a modification inactivate an enzyme that requires a His side chain for activity?

35. Treatment of the enzyme D-amino acid oxidase with 1-fluoro-2,4-dinitrophenol (FDNP, below) inactivates the enzyme. Analysis of the derivatized enzyme revealed that one of the tyrosine side chains in the enzyme was unusually reactive. In a second experiment, when benzoate (a compound that structurally resembles the enzyme's substrate) was added prior to the addition of the FDNP, the tyrosine side chain did not react. **(a)** Show the chemical reaction between FDNP and the tyrosyl residue. **(b)** Why does benzoate prevent the reaction? What does this tell you about the role of the tyrosine side chain in this enzyme?

FDNP

36. Sulfhydryl groups can react with the alkylating reagent *N*-ethylmaleimide (NEM). When NEM is added to a solution of creatine kinase, Cys 278 is alkylated, but no other cysteine residues in the protein are modified. What can you infer about the role of the Cys 278 residue based on this information?

37. A bioinformatics study revealed that the enzymes called vasoinhibins have a Cys–His–Ser catalytic triad. These enzymes regulate angiogenesis (the formation of new blood vessels) and are thus potential drug targets for cancer treatment. If the cysteine residue serves as the nucleophile, what are the possible roles of histidine and serine in enzyme catalysis?

38. The crystal structure of a protease from an avian virus was recently solved. The enzyme has a Cys–His–Asp catalytic triad. Draw the structure of a plausible tetrahedral intermediate for this enzyme.

39. *E. coli* ribonuclease H1 is an enzyme that catalyzes the hydrolysis of phosphodiester bonds in RNA. Its proposed mechanism involves a "carboxylate relay."

(a) Using the structures below as a guide, draw arrows that indicate how the hydrolysis reaction might be initiated.
(b) The pK values for all of the histidines in ribonuclease H1 were determined and include values of 7.1, 5.5, and < 5.0. Which value is most likely to correspond to His 124? Explain.
(c) Substituting an alanine residue for His 124 resulted in a dramatic decrease in enzyme activity. Explain.

His 124   Asp 70   RNA Substrate

40. Lysozyme catalyzes the hydrolysis of a polysaccharide component of bacterial cell walls. The damaged bacteria subsequently lyse (rupture). Part of lysozyme's mechanism is shown below. The enzyme catalyzes cleavage of a bond between two sugar residues (represented by hexagons). Catalysis involves the side chains of Glu 35 and Asp 52. One of the residues has a pK of 4.5; the other has a pK of 5.9.

(a) Assign the given pK values to Glu 35 and Asp 52.
(b) Lysozyme is inactive at pH 2.0 and at pH 8.0. Explain.
(c) The lysozyme mechanism proceeds via a covalent intermediate. Use this information to complete the drawing of the lysozyme mechanism.

41. RNase A is a digestive enzyme secreted by the pancreas into the small intestine, where it hydrolyzes RNA into its component nucleotides. The optimum pH for RNAse A is about 6, and the pK values of the two histidines that serve as catalytic residues are 5.4 and 6.4. The first step of the mechanism is shown.

(a) Does ribonuclease proceed via acid–base catalysis, covalent catalysis, metal ion catalysis or some combination of these strategies? Explain.
(b) Assign the appropriate pK values to His 12 and His 119.

(c) Explain why the pH optimum of ribonuclease is pH 6.
(d) Ribonuclease catalyzes the hydrolysis of RNA but not DNA. Explain why.

42. Chymotrypsin hydrolyzes *p*-nitrophenylacetate using a similar mechanism to that of peptide bond hydrolysis. When *p*-nitrophenylacetate is mixed with chymotrypsin, *p*-nitrophenolate initially forms extremely rapidly. This is followed by a steady-state phase in which the product forms at a uniform rate.

(a) Using what you know about the mechanism of chymotrypsin, explain these observations.
(b) Draw a reaction coordinate diagram for the reaction.
(c) Would *p*-nitrophenylacetate be useful for monitoring the activity of trypsin *in vitro*?

43. Renin belongs to a class of enzymes called aspartyl proteases in which aspartate plays a catalytic role similar to serine in chymotrypsin. Renin inhibitors could potentially be used to treat high blood pressure. Two aspartate residues in the renin active site, Asp 32 and Asp 215, constitute a catalytic dyad, rather than the catalytic triad found in chymotrypsin.

(a) Begin with the figure provided and draw the reaction mechanism for peptide hydrolysis catalyzed by renin.
(b) What type of catalytic strategy is utilized by renin?
(c) Compare the pK values of the Asp 32 and Asp 215 residues.

44. The enzyme bromelain (found in pineapple) is a cysteine protease in which the cysteine residue plays a catalytic role similar to serine in chymotrypsin. The active site also contains a histidine residue but lacks aspartate. Bromelain relieves inflammation and may have antitumor activity.

(a) Using chymotrypsin as a model, draw a reaction mechanism for the bromelain protease.
(b) Does the mechanism you have drawn employ acid–base catalysis, covalent catalysis, or both?
(c) Draw a reaction coordinate diagram consistent with the mechanism of bromelain.
(d) Why did natives of tropical countries use pineapple as a meat tenderizer?

(e) The pK values for the active site residues in bromelain are ~3 and ~8. The pH optimum for the reaction is 6.0. Assign pK values to the appropriate amino acid side chains and explain your reasoning.

45. The botulinum neurotoxin secreted by *Clostridium botulinum* contains a protease whose active site includes a Zn^{2+} ion coordinated by two histidine residues and a glutamate residue. The active site contains a second glutamate residue (Glu 224). Use the figure provided as a starting point to draw the mechanism of peptide bond hydrolysis catalyzed by this enzyme.

46. The enzyme carbonic anhydrase catalyzes the hydration of carbon dioxide to form carbonic acid (see Section 6.1), which dissociates to form a bicarbonate ion and a hydrogen ion. The enzyme's active site contains a zinc ion coordinated to the imidazole rings of three histidine residues. (A fourth histidine residue is located nearby and participates in catalysis.) A fourth coordination position is occupied by a water molecule. Draw the reaction mechanism (the first step is shown) of the hydration of carbon dioxide and show the regeneration of the enzyme.

47. Asparagine and glutamine residues in proteins can be nonenzymatically hydrolytically deamidated to aspartate and glutamate, respectively, a process that might function as a molecular timer for protein turnover. Studies show that asparagine residues susceptible to deamidation are more likely to be preceded by serine, threonine, or lysine and to be followed by glycine, serine, or threonine.

(a) Write the balanced chemical equation for the hydrolytic deamidation of asparagine to aspartate.
(b) The first step of the mechanism of the deamidation of glutamine is shown below, along with a required acid catalyst (HA). Draw the rest of the reaction mechanism.

(c) The HA catalyst in nonenzymatic deamidation is believed to be provided by neighboring amino acids in the deamidated protein. Describe how amino acid side chains that either precede or follow the labile Asn could serve as catalytic groups in the deamidation process.
(d) Amino-terminal glutamine residues undergo deamidation much more rapidly than internal glutamine residues. Write a

mechanism for this deamidation process, which includes the formation of a five-membered pyrrolidone ring. Amino-terminal asparagine residues do not undergo deamidation. Explain why.
(e) It has been observed that Asn and Gln residues in the interior of a protein are deamidated at a much slower rate than Asn and Gln residues on the surface of the protein. Explain why.

48. Proteases in the amidase family have a Ser–Ser–Lys catalytic triad, as shown below. Propose a mechanism for the amidase enzyme.

6.3 Unique Properties of Enzyme Catalysts

49. When substrates bind to an enzyme, their free energy may be lowered. Why doesn't this binding defeat catalysis?

50. Substrates and reactive groups in an enzyme's active site must be precisely aligned in order for a productive reaction to occur. Why, then, is some conformational flexibility also a requirement for catalysis?

51. (a) Daniel Koshland has noted that hexokinase undergoes a large conformational change on substrate binding, which apparently prevents water from entering the active site and participating in hydrolysis. However, the serine proteases do not undergo large conformational changes on substrate binding. Explain.
(b) When ATP and the sugar xylose (which resembles glucose but has only 5 carbons) are added to hexokinase, the enzyme produces a small amount of xylose-5-phosphate along with a large amount of ADP. Does this observation support Koshland's induced fit hypothesis?

52. Refer to Problem 45. What is the role of the zinc ion in catalysis? In transition state stabilization?

53. Refer to Problem 48. How might the amidase stabilize the transition state?

54. The enzyme adenosine deaminase catalyzes the conversion of adenosine to inosine. The compound 1,6-dihydropurine ribonucleoside binds to the enzyme with a much greater affinity than the adenosine substrate. What does this tell you about the mechanism of adenosine deaminase?

1,6-Dihydropurine ribonucleoside

55. Flurofamide is a potent inhibitor of urease, an enzyme that catalyzes the conversion of urea to CO_2 and NH_3 (shown below). What does this tell you about the transition state for the reaction?

Urea

Flurofamide

56. Imidazole reacts with *p*-nitrophenylacetate to produce the *N*-acetylimidazolium and *p*-nitrophenolate ions.

p-Nitrophenylacetate

Imidazole

N-Acetylimidazolium + *p*-Nitrophenolate

(a) The compound shown below reacts to form *p*-nitrophenolate. Draw the reaction mechanism for this reaction.

(b) The compound described in part a reacts 24 times faster than imidazole. Explain why.

(c) What do the results of this experiment tell you about how enzymes speed up reaction rates?

6.4 Chymotrypsin in Context

57. Many genetic mutations prevent the synthesis of a protein or give rise to an enzyme with diminished catalytic activity. Is it possible for a mutation to increase the catalytic activity of an enzyme?

58. The protease from the human immunodeficiency virus (HIV) is a target of the drugs used to treat HIV/AIDS. The protease has a mechanism similar to that of renin, described in Problem 43, except that renin consists of a single protein with two catalytic Asp residues, whereas the HIV protease consists of two identical subunits, each of which contributes an Asp residue. Is this an example of convergent or divergent evolution? Explain.

59. The amino acid Asp 189 lies at the base of the substrate specificity pocket in the enzyme trypsin (see Fig. 6.17). **(a)** How is this related to trypsin's substrate specificity? What kinds of interactions take place between the Asp 189 and the amino acid side chain on the carboxyl side of the scissile bond? **(b)** In site-directed mutagenesis studies, Asp 189 was replaced with lysine. How do you think this would affect substrate specificity? **(c)** The investigators who carried out the experiment described in part b analyzed the three-dimensional structure of the mutant enzyme and found that Lys 189 is actually not located in the substrate specificity pocket. Instead, the lysine side chain reaches out of the base of the pocket, rendering the specificity pocket nonpolar. With this additional information, how would the substrate specificity differ in the Lys 189 mutant enzyme?

60. The aspartyl aminopeptidase enzyme (involved in regulation of blood pressure) hydrolyzes peptide bonds on the carboxyl side of aspartate and is involved in the regulation of blood pressure. In order to study the enzyme's substrate binding site, a number of tetrapeptides were synthesized and the ability of the enzyme to bind to these peptides was measured. It is likely that both the P1′ residue and the P2′ residue fit in adjacent "pockets" on the aspartyl aminopeptidase enzyme, with hydrolysis occurring between residues P1 and P1′. Describe the characteristics of these pockets on the enzyme, using the data in the table.

Peptide (P1–P1′–P2′–P3′)	Catalyzed rate (s^{-1})
Asp–Phe–Ala–Leu	9.9
Asp–Lys–Ala–Leu	2.8
Asp–Ala–Phe–Leu	17.2
Asp–Ala–Lys–Leu	5.0
Asp–Ala–Asp–Leu	2.3

61. Chymotrypsin is usually described as an enzyme that catalyzes hydrolysis of peptide bonds following phenylalanine, tryptophan, or tyrosine residues. Is this information consistent with the description of the bonds cleaved during chymotrypsin activation (Fig. 6.18)? What does this tell you about chymotrypsin's substrate specificity?

62. Describe how the activation of the digestive enzyme chymotrypsin follows a cascade mechanism. Draw a diagram of your cascade.

63. Would chymotrypsin catalyze hydrolysis of the substrate shown in Problem 8? Explain why or why not.

64. Does DIPF inactivate trypsin or elastase (see Problem 29)? Explain.

65. Hereditary pancreatitis is caused by a mutation in the autocatalytic domain of trypsinogen that results in persistent activity of the enzyme. **(a)** What are the physiological consequences of this disease? **(b)** Describe a strategy to treat this disease.

66. Some plants contain compounds that inhibit serine proteases. It has been hypothesized that these compounds protect the plant from proteolytic enzymes of insects and microorganisms that would damage the plant. Tofu, or bean curd, possesses these compounds. Manufacturers of tofu treat it to eliminate serine protease inhibitors. Why is this treatment necessary?

67. Not all proteases are synthesized as zymogens or have inhibitors that block their activity. What limits the potentially destructive power of these proteases?

68. A chymotrypsin inhibitor isolated from snake venom resembles the bovine pancreatic trypsin inhibitor (Fig. 6.20) but has an asparagine residue in place of Lys 15. Why is this finding unexpected?

6.5 Clinical Connection: Blood Coagulation

69. Hemophiliacs who are deficient in factor IX can be treated with injections of pure factor IX. Occasionally, a patient develops antibodies to this material and it becomes ineffective. In such cases, the patient can be given factor VII instead. Explain why factor VII injections would be a useful treatment for a factor IX deficiency.

70. Of all the proteases included in the cascade diagram of coagulation, which one is the oldest in evolutionary terms?

71. Factor IX deficiency has been successfully treated by gene therapy, in which a normal copy of the gene is introduced into the body (see Section 3.5). Explain why restoration of factor IX levels to just a few percent of normal is enough to cure the abnormal bleeding.

72. A genetic defect in factor IX causes hemophilia, a serious disease. However, a genetic defect in factor XI may have no clinical symptoms. Explain this discrepancy in terms of the cascade mechanism for activation of coagulation proteases.

73. A vitamin K–dependent enzyme catalyzes the modification of glutamate residues in the amino-terminal domain of prothrombin to produce γ-carboxyglutamate. **a.** Draw the structure of a γ-carboxyglutamate residue. **b.** The γ-carboxyglutamate side chain binds Ca^{2+}

(which is essential for the blood-clotting process) more effectively than a glutamate side chain. Explain why.

74. Following severe trauma or infection, a patient may develop disseminated intravascular coagulation (DIC) in which numerous small clots form throughout the circulatory system. Explain why patients with DIC subsequently exhibit excessive bleeding.

75. In one thrombin variant, a point mutation produces a protease with normal activity toward fibrinogen but decreased activity toward antithrombin. Would this genetic defect increase the risk of bleeding or of clotting?

76. Why can't heparin be administered orally?

77. Researchers have developed drugs based on hirudin, a thrombin inhibitor from the medicinal leech *Hirudo medicinalis*. Why would the leech be a good source for an anticoagulant?

SELECTED READINGS

Di Cera, E., Serine proteases, *IUBMB Life* **61**, 510–515 (2009). [Summarizes some of the biological roles and mechanistic features of proteases.]

Fersht, A., *Structure and Mechanism in Protein Science: A Guide to Enzyme Catalysis and Protein Folding,* W. H. Freeman (1999). [Includes detailed reaction mechanisms for chymotrypsin and other enzymes.]

Gutteridge, A. and Thornton, J. M., Understanding nature's catalytic toolkit, *Trends Biochem. Sci.* **30**, 622–629 (2005). [Describes how certain sets of amino acid side chains form catalytic units that appear in many different enzymes.]

Radisky, E. S., Lee, J. M., Lu, C.-J. K., and Koshland, D. E., Jr., Insights into the serine protease mechanism from atomic resolution structures of trypsin reaction intermediates, *Proc. Nat. Acad. Sci.* **103**, 6835–6840 (2006).

Ringe, D. and Petsko, G. A., How enzymes work, *Science* **320**, 1428–1429 (2008). [Briefly summarizes some general features of enzyme function.]

CHAPTER 7

ENZYME KINETICS AND INHIBITION

Enzymes complete their catalytic cycles in a fraction of a second. Yet these fleeting events can sustain an organism across much longer time scales— hundreds to thousands of years in the case of the giant barrel sponge.

DO YOU REMEMBER?

- O_2 binds to myoglobin such that binding is half-maximal when the oxygen concentration is equal to the dissociation constant (Section 5.1).
- O_2 can bind cooperatively to hemoglobin as the protein shifts conformation (Section 5.1).
- Enzymes differ from simple chemical catalysts in their efficiency and specificity (Section 6.1).
- The height of the activation energy barrier determines the rate of a reaction (Section 6.2).
- The catalytic activity of enzymes depends on transition state stabilization (Section 6.3).

In the preceding chapter, we examined the basic features of enzyme catalytic activity, primarily by exploring the various catalytic mechanisms used by chymotrypsin. This chapter extends the discussion of enzymes by introducing enzyme kinetics, the mathematical analysis of enzyme activity. Here we describe how an enzyme's reaction speed and specificity can be quantified and how this information can be used to evaluate the enzyme's physiological function. We also look at the regulation of enzyme activity by inhibitors, including drugs, that bind to the enzyme and alter its activity. The discussion also focuses on allosteric regulation, a mechanism for inhibiting as well as activating enzymes.

7.1 INTRODUCTION TO ENZYME KINETICS

The structure and chemical mechanism of an enzyme (for example, chymotrypsin, as discussed in Chapter 6) often reveal a great deal about how that enzyme functions *in vivo*. However, structural information alone does not provide a full accounting of an enzyme's physiological role. For example, one might need to know exactly how fast the enzyme catalyzes a reaction or how well it recognizes different substrates or how its activity is affected by other substances. These questions are not trivial—consider that a single cell contains thousands of different enzymes, all operating simultaneously and in the presence of one anothers' substrates and products. To fully describe enzyme activity, enzymologists apply mathematical tools to quantify an enzyme's catalytic power and its substrate affinity as well as its response to inhibitors.

KEY CONCEPT
- An enzyme's activity, measured as the rate of product formation, varies with the substrate concentration.

FIGURE 7.1

Progress of the triose phosphate isomerase reaction. Over time, the concentration of the substrate glyceraldehyde-3-phosphate decreases and the concentration of the product dihydroxyacetone phosphate increases.

? Extend the lines in the graph to show that the reaction eventually reaches equilibrium, when the ratio of product concentration to reactant concentration is about 20.

This analysis is part of the area of study known as enzyme **kinetics** (from the Greek *kinetos*, which means "moving").

Some of the earliest biochemical studies examined the reactions of crude preparations of yeast cells and other organisms. Even without isolating any enzymes, researchers could mathematically analyze their activity by measuring the concentrations of substrates and reaction products and observing how these quantities changed over time. For example, consider the simple reaction catalyzed by triose phosphate isomerase, which interconverts two three-carbon sugars (trioses):

Glyceraldehyde-3-phosphate — triose phosphate isomerase → Dihydroxyacetone phosphate

Over the course of the reaction, the concentration of the substrate glyceraldehyde-3-phosphate falls as the concentration of the product dihydroxyacetone phosphate rises (**Fig. 7.1**). *The progress of this or any reaction can be expressed as a **velocity (v)**, either the rate of disappearance of the substrate (S) or the rate of appearance of the product (P):*

$$v = -\frac{d[S]}{dt} = \frac{d[P]}{dt} \qquad [7.1]$$

where [S] and [P] represent the molar concentrations of the substrate and product, respectively. Not surprisingly, *the more catalyst (enzyme) present, the faster the reaction* (**Fig. 7.2**).

Enzymologists typically study the early stages of a reaction, when the substrate concentration is just starting to decrease and the product concentration is just starting to increase. It wouldn't make sense to wait until the reaction goes to completion (that is, reaches equilibrium), since the substrate and product concentrations don't change at that point. In fact, in a cell, many reactions never reach equilibrium, since new substrates are constantly arriving and products are used up as soon as they are made.

How does the rate of an enzyme-catalyzed reaction change when the substrate concentration changes? When the enzyme concentration is held constant, the reaction velocity varies with the substrate concentration, but in a nonlinear fashion (**Fig. 7.3**). The shape of this velocity versus substrate curve is an important key to understanding how enzymes interact with their substrates. *The hyperbolic, rather than linear, shape of the curve suggests that an enzyme physically combines with its substrate to form an **enzyme–substrate (ES) complex**.* Therefore, the enzyme-catalyzed conversion of S to P

$$S \xrightarrow{E} P$$

FIGURE 7.2

Progress of the triose phosphate isomerase reaction. The more enzyme present, the faster the reaction.

Reaction velocity, v

Substrate concentration, [S]

FIGURE 7.3

A plot of reaction velocity versus substrate concentration. Varying amounts of substrate are added to a fixed amount of enzyme. The reaction velocity is measured for each substrate concentration and plotted. The resulting curve takes the form of a hyperbola, a mathematical function in which the values first increase steeply but eventually approach a maximum.

? **Compare this diagram to Figure 5.3, which shows oxygen binding to myoglobin.**

can be more accurately written as

$$E + S \rightarrow ES \rightarrow E + P$$

As small amounts of substrate are added to the enzyme preparation, enzyme activity (measured as the reaction velocity) appears to increase almost linearly. However, the enzyme's activity increases less dramatically as more substrate is added. At very high substrate concentrations, enzyme activity appears to level off as it approaches a maximum value. This behavior shows that at low substrate concentrations, the enzyme quickly converts all the substrate to product, but as more substrate is added, the enzyme becomes **saturated** with substrate—that is, there are many more substrate molecules than enzyme molecules, so not all the substrate can be converted to product in a given time. These so-called saturation kinetics are a feature of many binding phenomena, including the binding of O_2 to myoglobin (see Section 5.1).

The curve shown in Figure 7.3 reveals considerable information about a given enzyme and substrate under a chosen set of reaction conditions. All simple enzyme-catalyzed reactions yield a hyperbolic velocity versus substrate curve, but the exact shape of the curve depends on the enzyme, its concentration, the concentrations of enzyme inhibitors, the pH, the temperature, and so on. By analyzing such curves, it is possible to address some basic questions, for example,

- How fast does the enzyme operate?
- How efficiently does the enzyme convert different substrates to products?
- How susceptible is the enzyme to various inhibitors and how do these inhibitors affect enzyme activity?

The answers to these questions, in turn, may reveal

- Whether an enzyme is likely to catalyze a particular reaction *in vivo*
- What substances are likely to serve as physiological regulators of the enzyme's activity
- Which enzyme inhibitors might be effective drugs

 CONCEPT CHECK

1. Describe two ways you could measure the rate of an enzyme-catalyzed reaction.
2. Sketch a velocity versus substrate curve like the one in Fig. 7.3.
3. Why does a plot of velocity versus substrate concentration yield a curve?
4. Make a list of the questions that can be answered by studying an enzyme's kinetics.

DERIVATION AND MEANING OF THE MICHAELIS–MENTEN EQUATION

The mathematical analysis of enzyme behavior centers on the equation that describes the hyperbolic shape of the velocity versus substrate plot (see Fig. 7.3). We can analyze an enzyme-catalyzed reaction, such as the one catalyzed by triose phosphate isomerase, by conceptually breaking it down into smaller steps and using the terms that apply to simple chemical processes.

RATE EQUATIONS DESCRIBE CHEMICAL PROCESSES

Consider a **unimolecular reaction** (one that involves a single reactant) such as the conversion of compound A to compound B:

$$A \rightarrow B$$

The progress of this reaction can be mathematically described by a **rate equation** in which the reaction rate (the velocity) is expressed in terms of a constant (the **rate constant**) and the reactant concentration [A]:

$$v = -\frac{d[A]}{dt} = k[A] \qquad [7.2]$$

Here, k is the rate constant and has units of reciprocal seconds (s^{-1}). This equation shows that the reaction velocity is directly proportional to the concentration of reactant A. Such a reaction is said to be **first-order** because its rate depends on the concentration of one substance.

A **bimolecular** or **second-order reaction**, which involves two reactants, can be written

$$A + B \rightarrow C$$

Its rate equation is

$$v = -\frac{d[A]}{dt} = -\frac{d[B]}{dt} = k[A][B] \qquad [7.3]$$

Here, k is a second-order rate constant and has units of $M^{-1} \cdot s^{-1}$. The velocity of a second-order reaction is therefore proportional to the product of the two reactant concentrations (see Sample Calculation 7.1).

SAMPLE CALCULATION 7.1

PROBLEM   Determine the velocity of the reaction $X + Y \rightarrow Z$ when the sample contains 3 µM X and 5 µM Y and k for the reaction is 400 $M^{-1} \cdot s^{-1}$.

SOLUTION   Use Equation 7.3 and make sure that all units are consistent:

$$v = k[X][Y]$$
$$= (400 \ M^{-1} \cdot s^{-1})(3 \ \mu M)(5 \ \mu M)$$
$$= (400 \ M^{-1} \cdot s^{-1})(3 \times 10^{-6} \ M)(5 \times 10^{-6} \ M)$$
$$= 6 \times 10^{-9} \ M \cdot s^{-1} = 6 \ nM \cdot s^{-1}$$

PRACTICE PROBLEMS
1. Calculate k for the reaction $X + Y \rightarrow Z$ when the sample contains 8 µM X and 8 µM Y and the velocity is 8 $\mu M \cdot s^{-1}$.
2. Determine the velocity of the reaction in Problem 1 when [X] = 20 µM and [Y] = 10 µM.
3. If k for the reaction $X + Y \rightarrow Z$ is 0.5 $mM^{-1} \cdot s^{-1}$ and [X] = [Y], determine the substrate concentration at which $v = 12.5 \ mM \cdot s^{-1}$.

THE MICHAELIS–MENTEN EQUATION IS A RATE EQUATION FOR AN ENZYME-CATALYZED REACTION

In the simplest case, an enzyme binds its substrate (in an enzyme–substrate complex) before converting it to product, so *the overall reaction actually consists of first-order and second-order processes, each with a characteristic rate constant:*

$$E + S \underset{k_{-1}}{\overset{k_1}{\rightleftharpoons}} ES \xrightarrow{k_2} E + P \qquad [7.4]$$

The initial collision of E and S is a bimolecular reaction with the second-order rate constant k_1. The ES complex can then undergo one of two possible unimolecular reactions: k_2 is the first-order rate constant for the conversion of ES to E and P, and k_{-1} is the first-order rate constant for the conversion of ES back to E and S. The bimolecular reaction that would represent the formation of ES from E and P is not shown because we assume that the formation of product from ES (the step described by k_2) does not occur in reverse.

The rate equation for product formation is

$$v = \frac{d[P]}{dt} = k_2[ES] \qquad [7.5]$$

To calculate the rate constant, k_2, for the reaction, we would need to know the reaction velocity and the concentration of ES. The velocity can be measured relatively easily, for example, by using a synthetic substrate that is converted to a light-absorbing or fluorescent product. (The velocity of the reaction is just the rate of appearance of the product as monitored by a spectrophotometer or fluorometer.) However, measuring [ES] is more difficult because the concentration of the enzyme–substrate complex depends on its rate of formation from E and S and its rate of decomposition to E + S and E + P:

$$\frac{d[ES]}{dt} = k_1[E][S] - k_{-1}[ES] - k_2[ES] \qquad [7.6]$$

To simplify our analysis, we choose experimental conditions such that the substrate concentration is much greater than the enzyme concentration ($[S] \gg [E]$). Under these conditions, after E and S have been mixed together, *the concentration of ES remains constant until nearly all the substrate molecules have been converted to product.* This is shown graphically in **Figure 7.4**. [ES] is said to maintain a **steady state** (it has a constant value) and

$$\frac{d[ES]}{dt} = 0 \qquad [7.7]$$

According to the steady-state assumption, the rate of ES formation must therefore balance the rate of ES consumption:

$$k_1[E][S] = k_{-1}[ES] + k_2[ES] \qquad [7.8]$$

At any point during the reaction, [E]—like [ES]—is difficult to determine, but the total enzyme concentration, $[E]_T$, is usually known:

$$[E]_T = [E] + [ES] \qquad [7.9]$$

FIGURE 7.4

Changes in concentration for a simple enzyme-catalyzed reaction. For most of the duration of the reaction, [ES] remains constant while S is converted to P. In this idealized reaction, all the substrate is converted to product.

Concentration

Time

Thus, $[E] = [E]_T - [ES]$. This expression for $[E]$ can be substituted into Equation 7.8 to give

$$k_1([E]_T - [ES])[S] = k_{-1}[ES] + k_2[ES] \qquad [7.10]$$

Rearranging (by dividing both sides by $[ES]$ and k_1) gives an expression in which all three rate constants are together:

$$\frac{([E]_T - [ES])[S]}{[ES]} = \frac{k_{-1} + k_2}{k_1} \qquad [7.11]$$

At this point, we can define the **Michaelis constant, K_M,** as a collection of rate constants:

$$K_M = \frac{k_{-1} + k_2}{k_1} \qquad [7.12]$$

Consequently, Equation 7.11 becomes

$$\frac{([E]_T - [ES])[S]}{[ES]} = K_M \qquad [7.13]$$

or

$$K_M[ES] = ([E]_T - [ES])[S] \qquad [7.14]$$

Dividing both sides by $[ES]$ gives

$$K_M = \frac{[E]_T[S]}{[ES]} - [S] \qquad [7.15]$$

or

$$\frac{[E]_T[S]}{[ES]} = K_M + [S] \qquad [7.16]$$

Solving for $[ES]$ yields

$$[ES] = \frac{[E]_T[S]}{K_M + [S]} \qquad [7.17]$$

The rate equation for the formation of product (Equation 7.5) is $v = k_2[ES]$, so we can express the reaction velocity as

$$v = k_2[ES] = \frac{k_2[E]_T[S]}{K_M + [S]} \qquad [7.18]$$

Now we have an equation containing known quantities: $[E]_T$ and $[S]$. Although some S is consumed in forming the ES complex, we can ignore it because $[S]_T \gg [E]_T$.

Typically, kinetic measurements are made soon after the enzyme and substrate are mixed together, before more than about 10% of the substrate molecules have been converted to product molecules (this is also the reason why we can ignore the reverse reaction, $E + P \rightarrow ES$). Therefore, the velocity at the start of the reaction (at time zero) is expressed as v_0 (the **initial velocity**):

$$v_0 = \frac{k_2[E]_T[S]}{K_M + [S]} \qquad [7.19]$$

We can make one additional simplification: When $[S]$ is very high, virtually all the enzyme is in its ES form (it is saturated with substrate) and therefore approaches its point of maximum activity (see Fig. 7.3). The maximum reaction velocity, designated V_{max}, can be expressed as

$$V_{max} = k_2[E]_T \qquad [7.20]$$

which is similar to Equation 7.5. By substituting Equation 7.20 into Equation 7.19, we obtain

$$v_0 = \frac{V_{max}[S]}{K_M + [S]} \qquad [7.21]$$

This relationship is called the **Michaelis–Menten equation** after Leonor Michaelis and Maude Menten, who derived it in 1913. *It is the rate equation for an enzyme-catalyzed reaction and is the mathematical description of the hyperbolic curve shown in Figure 7.3.* (See Sample Calculation 7.2.)

An enzyme-catalyzed reaction has a K_M of 1 mM and a V_{max} of 5 nM·s^{-1}. What is the reaction velocity when the substrate concentration is 0.25 mM?

PROBLEM

Use the Michaelis–Menten equation (Equation 7.21):

SOLUTION

$$v_0 = \frac{(5 \text{ nM} \cdot \text{s}^{-1})(0.25 \text{ mM})}{(1 \text{ mM}) + (0.25 \text{ mM})}$$

$$= \frac{1.25}{1.25} \text{ nM} \cdot \text{s}^{-1}$$

$$= 1 \text{ nM} \cdot \text{s}^{-1}$$

4. Using the above information, determine the reaction velocity when the substrate concentration is 10 mM.

PRACTICE PROBLEMS

5. If V_{max} for a reaction is 9.5 μM·s^{-1} and the reaction velocity is 5 μM·s^{-1} when the substrate concentration is 1 μM, what is the K_M?

6. Using the same information as in Problem 5, determine the substrate concentration at which $3.5 \text{ μM} \cdot \text{s}^{-1}$.

K_M IS THE SUBSTRATE CONCENTRATION AT WHICH VELOCITY IS HALF-MAXIMAL

We have seen that the Michaelis constant, K_M, is a combination of three rate constants (Equation 7.12), but it can be fairly easily determined from experimental data. Kinetic measurements are usually made over a range of substrate concentrations. When $[S] = K_M$, the reaction velocity (v_0) is equal to half its maximum value ($v_0 = V_{max}/2$), as shown in **Figure 7.5**. You can prove that this is true by substituting K_M for $[S]$ in the Michaelis–Menten equation (Equation 7.21). *Since K_M is the substrate concentration at which the reaction velocity is half-maximal, it indicates how efficiently an enzyme selects its substrate and converts it to product.* The lower the value of K_M, the more effective the enzyme is at low substrate concentrations; the higher the value of K_M, the less effective the enzyme is. The K_M is unique for each enzyme–substrate pair. Consequently, K_M values are useful for comparing the activities of two enzymes that act on the same substance or for assessing the ability of different substrates to be recognized by a single enzyme.

FIGURE 7.5

Graphical determination of K_M.
K_M corresponds to the substrate concentration at which the reaction velocity is half-maximal. It can be visually estimated from a plot of v_0 versus $[S]$.

? Explain why doubling the substrate concentration does not necessarily double the reaction rate.

In practice, K_M *is often used as a measure of an enzyme's affinity for a substrate.* In other words, it approximates the dissociation constant of the ES complex:

$$K_M \approx \frac{[E][S]}{[ES]} \qquad [7.22]$$

Note that this relationship is strictly true only when the rate of the ES $\rightarrow$ E + P reaction is much slower than the rate of the ES $\rightarrow$ E + S reaction (that is, when $k_2 \ll k_{-1}$).

THE CATALYTIC CONSTANT DESCRIBES HOW QUICKLY AN ENZYME CAN ACT

It is also useful to know how fast an enzyme operates after it has bound its substrate. In other words, how fast does the ES complex proceed to E + P? This parameter is termed the **catalytic constant** and is symbolized k_{cat}. For any enzyme-catalyzed reaction,

$$k_{cat} = \frac{V_{max}}{[E]_T} \qquad [7.23]$$

For a simple reaction, such as the one diagrammed in Equation 7.4,

$$k_{cat} = k_2 \qquad [7.24]$$

Thus, k_{cat} *is the rate constant of the reaction when the enzyme is saturated with substrate* (when $[ES] \approx [E]_T$ and $v_0 \approx V_{max}$). We have already seen this relationship in Equation 7.20. k_{cat} is also known as the enzyme's **turnover number** because it is the number of catalytic cycles that each active site undergoes per unit time, or the number of substrate molecules transformed to product molecules by a single enzyme in a given period of time. The turnover number is a first-order rate constant and therefore has units of s^{-1}. As shown in **Table 7.1**, the catalytic constants of enzymes vary over many orders of magnitude.

Keep in mind that the rate of an enzymatic reaction is a function of the number of reactant molecules that can achieve the high-energy transition state per unit time (as explained in Section 6.2). While an enzyme can accelerate a chemical reaction by providing a mechanistic pathway with a lower activation energy barrier, the enzyme cannot alter the free energies of the reactants and products. This means that the enzyme can make a reaction happen faster, but only when the overall change in free energy is less than zero (that is, the products have lower free energy than the reactants).

k_{cat}/K_M INDICATES CATALYTIC EFFICIENCY

An enzyme's effectiveness as a catalyst depends on how avidly it binds its substrates and how rapidly it converts them to products. Thus, a measure of catalytic efficiency must reflect both binding and catalytic events. The quantity k_{cat}/K_M satisfies this requirement. At low concentrations of substrate ($[S] < K_M$), very little ES forms and $[E] \approx [E]_T$. Equation 7.18 can then be simplified (the [S] term in the denominator becomes insignificant):

$$v_0 = \frac{k_2[E]_T[S]}{K_M + [S]} \qquad [7.25]$$

$$v_0 \approx \frac{k_2}{K_M}[E][S] \qquad [7.26]$$

Equation 7.26 is the rate equation for the second-order reaction of E and S. k_{cat}/K_M, which has units of $M^{-1} \cdot s^{-1}$, is the apparent second-order rate constant. As such, it indicates how the reaction velocity varies according to how often the enzyme and substrate combine with each other. *The value of k_{cat}/K_M, more than either K_M or k_{cat} alone, represents the enzyme's overall ability to convert substrate to product.*

What limits the catalytic power of enzymes? Electronic rearrangements during formation of the transition state occur on the

TABLE 7.1   Catalytic Constants of Some Enzymes

ENZYME	k_{cat} (s^{-1})
Staphylococcal nuclease	95
Cytidine deaminase	299
Triose phosphate isomerase	4300
Cyclophilin	13,000
Ketosteroid isomerase	66,000
Carbonic anhydrase	1,000,000

[Data from Radzicka, A., and Wolfenden, R., *Science* **267**, 90–93 (1995).]

order of 10^{-13} s, about the lifetime of a bond vibration. But enzyme turnover numbers are much slower than this (see Table 7.1). An enzyme's overall speed is further limited by how often it collides productively with its substrate. The upper limit for the rate of this second-order reaction (a bimolecular reaction) is about 10^8 to 10^9 $M^{-1} \cdot s^{-1}$, which is the maximum rate at which two freely diffusing molecules can collide with each other in aqueous solution.

This so-called **diffusion-controlled limit** for the second-order reaction between an enzyme and a substrate is achieved by several enzymes, including triose phosphate isomerase, whose value of k_{cat}/K_M is 2.4×10^8 $M^{-1} \cdot s^{-1}$. This enzyme is therefore said to have reached **catalytic perfection** because *its overall rate is diffusion-controlled: It catalyzes a reaction as rapidly as it encounters its substrate.* However, many enzymes perform their physiological roles with more modest k_{cat}/K_M values.

K_M AND V_{max} ARE EXPERIMENTALLY DETERMINED

Kinetic data are usually collected by adding a small amount of an enzyme to varying amounts of substrate and then monitoring the reaction mixture for the appearance of product over a period of time. In order to meet the assumptions of the Michaelis–Menten model, the concentration of the substrate must be much greater than the concentration of the enzyme (so that the concentration of the ES complex will be constant and the formation of the ES complex will be limited by the affinity of E for S, not the amount of S available), and measurements must be taken of the initial velocity, before product begins to accumulate and the reverse reaction becomes significant.

Velocity versus substrate plots such as Figure 7.5 can be useful for visually estimating the kinetic parameters K_M and V_{max} (from which k_{cat} can be derived by Equation 7.23). However, in practice, hyperbolic curves are prone to misinterpretation because it is difficult to estimate the upper limit of the curve (V_{max}). In order to more accurately determine V_{max} and K_M (the substrate concentration at $V_{max}/2$), it is necessary to perform one of the following steps:

1. Analyze the data by a curve-fitting computer program that mathematically calculates the upper limit for the reaction velocity.

2. Transform the data to a form that can be plotted as a line. The best-known linear transformation of the velocity versus substrate curve is known as a **Lineweaver–Burk plot,** whose equation is

$$\frac{1}{v_0} = \left(\frac{K_M}{V_{max}}\right)\frac{1}{[S]} + \frac{1}{V_{max}} \qquad \text{[7.27]}$$

Equation 7.27 has the familiar form $y = mx + b$. A plot of $1/v_0$ versus $1/[S]$ gives a straight line whose slope is K_M/V_{max} and whose intercept on the $1/v_0$ axis is $1/V_{max}$. The extrapolated intercept on the $1/[S]$ axis is $-1/K_M$ (**Fig. 7.6**). A comparison of Figures 7.5 and 7.6, made from the same data, illustrates how evenly spaced points on a velocity versus substrate plot become compressed in a Lineweaver–Burk plot (see Sample Calculation 7.3).

FIGURE 7.6

A Lineweaver–Burk plot. Plotting the reciprocals of [S] and v_0 yields a line whose slope and intercepts yield values of K_M and V_{max}. The plotted points correspond to the points in Figure 7.5.

? Sketch a line to represent a reaction with a larger value of K_M and the same value of V_{max}.

SAMPLE CALCULATION 7.3

PROBLEM The velocity of an enzyme-catalyzed reaction was measured at several substrate concentrations. Calculate K_M and V_{max} for the reaction.

[S] (μM)	v_0 (mM · s^{-1})
0.25	0.75
0.5	1.20
1.0	1.71
2.0	2.18
4.0	2.53

SOLUTION Calculate the reciprocals of the substrate concentration and velocity, then make a plot of $1/v_0$ versus $1/[S]$ (a Lineweaver–Burk plot).

1/[S] (μM^{-1})	$1/v_0$ (mM^{-1} · s)
4.0	1.33
2.0	0.83
1.0	0.58
0.5	0.46
0.25	0.40

The intercept on the $1/[S]$ axis (which is equal to $-1/K_M$) is -1.33 μM^{-1}. Therefore,

$$K_M = -\left(\frac{1}{-1.33\ \mu M^{-1}}\right) = 0.75\ \mu M$$

The intercept on the $1/v_0$ axis (which is equal to $1/V_{max}$) is 0.33 mM^{-1} · s. Therefore,

$$V_{max} = \frac{1}{0.33\ mM^{-1} \cdot s} = 3.0\ mM \cdot s^{-1}$$

PRACTICE PROBLEM 7. Calculate K_M and V_{max} from the following data.

[S] (mM)	v_0 (mM · s^{-1})
1	1.82
2	3.33
4	5.71
8	8.89
18	12.31

Ideally, experimental conditions are chosen so that velocity measurements can be made for substrate concentrations that are both higher and lower than K_M. This yields the most accurate values for K_M and V_{max}. A Lineweaver–Burk plot, whether constructed manually or by computer, offers the advantage that K_M and V_{max} can be quickly estimated by eye. Linear plots are also more convenient than curves for comparing multiple data sets, such as different enzyme preparations, or a single enzyme in the presence of different concentrations of an inhibitor.

NOT ALL ENZYMES FIT THE SIMPLE MICHAELIS–MENTEN MODEL

So far, the discussion has focused on the very simplest of enzyme-catalyzed reactions, namely, a reaction with one substrate and one product. Such reactions represent only a small portion of known enzymatic reactions, which often involve multiple substrates and products, proceed via multiple steps, or do not meet the assumptions of the Michaelis–Menten kinetic model for other reasons. Nevertheless, the kinetics of these reactions can still be evaluated.

1. Multisubstrate Reactions

More than half of all known biochemical reactions involve two substrates. Most of these **bisubstrate reactions** are either oxidation–reduction reactions or transfer reactions. In an oxidation–reduction reaction, electrons are transferred between substrates:

$$X_{oxidized} + Y_{reduced} \longrightarrow X_{reduced} + Y_{oxidized}$$

In a transfer reaction, such as the one catalyzed by transketolase, a group is transferred between two molecules:

| Fructose-6-phosphate | Glyceraldehyde-3-phosphate | | Erythrose-4-phosphate | Xylulose-5-phosphate |

The transketolase reaction is ubiquitous in nature; it functions in the synthesis and degradation of carbohydrates. As written here, it transforms a six-carbon sugar and a three-carbon sugar to a four-carbon sugar and a five-carbon sugar. *Each of the substrates interacts with the enzyme with a characteristic K_M.* To experimentally determine each K_M, the reaction velocity is measured at different concentrations of one substrate while the other substrate is present at a saturating concentration. *V_{max} is the maximum reaction velocity when both substrates are present at concentrations that saturate their binding sites on the enzyme.*

In some bisubstrate reactions, the substrates can bind in any order, as long as they both end up in the active site at the same time. These reactions are said to follow a **random mechanism.** Enzymes in which one substrate must bind before the other follow an **ordered mechanism.** In a **ping pong mechanism,** one substrate binds and one product is released before the other substrate binds and the second product is released. Transketolase catalyzes a ping pong reaction: Fructose-6-phosphate binds first and surrenders a two-carbon fragment to the enzyme, and the first product (erythrose-4-phosphate) leaves the active site before the second substrate (glyceraldehyde-3-phosphate) binds and receives the two-carbon fragment to yield the second product (xylulose-5-phosphate).

2. Multistep Reactions

As the transketolase reaction illustrates, an enzyme-catalyzed reaction may contain many steps. In this example, the reaction includes an intermediate in which the two-carbon fragment removed from fructose-6-phosphate remains bound to the enzyme while awaiting the arrival of the second substrate. (The chymotrypsin reaction mechanism outlined in Fig. 6.10 similarly requires several steps.) The multistep transketolase reaction can be broken down into a series of simple mechanistic steps and diagrammed as follows:

FIGURE 7.7

Effect of cooperative substrate binding. The velocity versus substrate curve is sigmoidal rather than hyperbolic when substrate binding to one active site in an oligomeric enzyme alters the catalytic activity of the other active sites. The maximum reaction velocity is V_{max}, and K_M is the substrate concentration when the velocity is half-maximal.

? Compare this diagram to Figure 5.7, which shows oxygen binding to hemoglobin.

KEY CONCEPTS

- Substances that bind irreversibly to an enzyme can inhibit its activity.
- A competitive inhibitor appears to increase K_M without affecting V_{max}.
- Transition state analogs can act as competitive inhibitors.
- Noncompetitive, mixed, and uncompetitive inhibitors decrease k_{cat}.
- Allosteric regulators can inhibit or activate enzymes.

Each step of this process has characteristic forward and reverse rate constants. Consequently, k_{cat} for the overall reaction

$$\text{fructose-6-phosphate} + \text{glyceraldehyde-3-phosphate} \rightleftharpoons$$
$$\text{erythrose-4-phosphate} + \text{xylulose-5-phosphate}$$

is a complicated function of many individual rate constants (only for a very simple reaction, for example, Equation 7.4, does $k_{cat} = k_2$). Nevertheless, *the meaning of k_{cat}—the enzyme's turnover number—is the same as for a single-step reaction.*

The rate constants of individual steps in a multistep reaction can sometimes be measured during the initial stages of the reaction, that is, *before* a steady state is established. This requires instruments that can rapidly mix the reactants and then monitor the mixture on a time scale from 1 s to 10^{-7} s.

3. Nonhyperbolic Reactions Many enzymes, particularly oligomeric enzymes with multiple active sites, do not obey the Michaelis–Menten rate equation and therefore do not yield hyperbolic velocity versus substrate curves. In these **allosteric enzymes,** *the presence of a substrate at one active site can affect the catalytic activity of the other active sites.* This **cooperative** behavior occurs when the enzyme subunits are structurally linked to each other so that a substrate-induced conformational change in one subunit elicits conformational changes in the remaining subunits. (Cooperative behavior also occurs in hemoglobin, when O_2 binding to the heme group in one subunit alters the O_2 affinity of the other subunits; see Section 5.1.) Like hemoglobin, an allosteric enzyme has two possible quaternary structures. The T (or "tense") state has lower catalytic activity, and the R (or "relaxed") state has higher activity. Because of the interactions between individual subunits, the entire enzyme can switch between the T and R conformations. The result of allosteric behavior is a sigmoidal (S-shaped) velocity versus substrate curve (**Fig. 7.7**). Although the standard Michaelis–Menten equation does not apply here, K_M and V_{max} can be estimated and used to characterize enzyme activity.

 CONCEPT CHECK

1. Write the rate equations for first-order and second-order reactions.
2. Describe the two possible fates of an ES complex.
3. Plot the concentrations of S, P, E_T, and ES during the course of an enzyme-catalyzed reaction.
4. Write the rate equation for an enzyme-catalyzed reaction.
5. Write definitions for K_M, v_0, V_{max}, k_{cat}, and k_{cat}/K_M.
6. Explain why k_{cat}/K_M is a better indicator of enzyme efficiency than either k_{cat} or K_M alone.
7. Why is triose phosphate isomerase considered to be a catalytically perfect enzyme?
8. Why is a Lineweaver–Burk plot, rather than a velocity versus substrate plot, often used to determine K_M and V_{max}?
9. Give the number of K_M values and V_{max} values that pertain to a bisubstrate reaction.
10. Describe the shape of the velocity versus substrate curve when an enzyme exhibits cooperative behavior.

ENZYME INHIBITION

Inside a cell, an enzyme is subject to a variety of factors that can influence its behavior. Substances that interact with the enzyme can interfere with substrate binding and/or catalysis. Many naturally occurring antibiotics, pesticides, and other poisons are substances that inhibit the activity of essential enzymes. From a strictly scientific point of view, these inhibitors are useful probes of an enzyme's active-site structure and catalytic mechanism. Enzyme inhibitors are also used therapeutically as drugs. The ongoing pursuit of more effective drugs requires knowledge of how enzyme inhibitors work and how they can be altered to better inhibit their target enzymes.

SOME INHIBITORS ACT IRREVERSIBLY

Certain compounds interact with enzymes so tightly that their effects are essentially irreversible. For example, diisopropylphosphofluoridate (DIPF), the reagent used to identify the active-site Ser residue of chymotrypsin (see Section 6.2), is an **irreversible inhibitor** of the enzyme. When DIPF reacts with chymotrypsin, leaving the DIP group covalently attached to the Ser hydroxyl group, the enzyme becomes catalytically inactive. In general, *any reagent that covalently modifies an amino acid side chain in a protein can potentially act as an irreversible enzyme inhibitor.*

Some irreversible enzyme inhibitors are called **suicide substrates** because they enter the enzyme's active site and begin to react, just as a normal substrate would. However, they are unable to undergo the complete reaction and hence become "stuck" in the active site. For example, thymidylate synthase is the enzyme that converts the nucleotide deoxyuridylate (dUMP) to deoxythymidylate (dTMP) by adding a methyl group to C5:

Deoxyribose-5-phosphate
dUMP

thymidylate synthase →

Deoxyribose-5-phosphate
dTMP

When the synthetic compound 5-fluorouracil is taken up by cells, it is readily converted to the nucleotide 5-fluorodeoxyuridylate. This compound, like dUMP, enters the active site of thymidylate synthase, where a Cys —SH group adds to C6. Normally, this enhances the nucleophilicity (electron richness) of C5 so that it can accept an electron-poor methyl group. However, the presence of the electron-withdrawing F atom prevents methylation. The inhibitor therefore remains in the active site, bound to the cysteine side chain, rendering thymidylate synthase inactive. For this reason, 5-fluorouracil is used to disrupt DNA synthesis in rapidly dividing cancer cells.

5-Fluorouracil

COMPETITIVE INHIBITION IS THE MOST COMMON FORM OF REVERSIBLE ENZYME INHIBITION

As its name implies, reversible enzyme inhibition results when a substance binds reversibly (that is, noncovalently) to an enzyme so as to alter its catalytic properties. A reversible inhibitor may affect the enzyme's K_M, k_{cat}, or both. The most common form of reversible enzyme inhibition is known as **competitive inhibition.** In this situation, *the inhibitor is a substance that directly competes with a substrate for binding to the enzyme's active site* (**Fig. 7.8**). As expected, the inhibitor usually resembles the substrate in overall size and chemical properties so that it can bind to the enzyme, but it lacks the exact electronic structure that allows it to react.

One well-known competitive inhibitor affects the activity of succinate dehydrogenase, which catalyzes the oxidation (dehydrogenation) of succinate to produce fumarate:

COO⁻
|
CH₂
|
CH₂   succinate dehydrogenase →
|
COO⁻
Succinate

COO⁻
|
CH
‖
HC
|
COO⁻
Fumarate

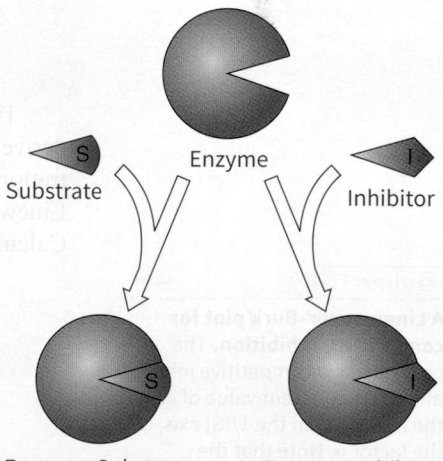

FIGURE 7.8

Competitive enzyme inhibition. In its simplest form, competitive inhibition of an enzyme occurs when the inhibitor and substrate compete for binding in the enzyme active site. A competitive inhibitor often resembles the substrate in size and shape but cannot undergo a reaction. In all cases, binding of the inhibitor and the substrate is mutually exclusive.

FIGURE 7.9

Effect of a competitive inhibitor on reaction velocity. In a plot of velocity versus substrate concentration, the inhibitor increases the apparent K_M because it competes with the substrate for binding to the enzyme. The inhibitor does not affect k_{cat}, so at high [S], v_0 approaches V_{max}.

The compound malonate

$$
\begin{array}{c}
COO^- \\
| \\
CH_2 \\
| \\
COO^-
\end{array}
$$
Malonate

inhibits the reaction because it binds to the dehydrogenase active site but cannot be dehydrogenated. Apparently, the enzyme active site can accommodate either the substrate succinate or the competitive inhibitor malonate, although they differ slightly in size.

A plot of an enzyme's reaction velocity in the presence of a competitive inhibitor, as a function of the substrate concentration, is shown in **Figure 7.9**. Because the inhibitor prevents some of the substrate from reaching the active site, the K_M appears to increase (the enzyme's affinity for the substrate appears to decrease). But because the inhibitor binds reversibly, it constantly dissociates from and reassociates with the enzyme, which allows a substrate molecule to occasionally enter the active site. High concentrations of substrate can overcome the effect of the inhibitor because when [S] $\gg$ [I], the enzyme is more likely to bind S than I. The presence of a competitive inhibitor does not affect the enzyme's k_{cat}, so as [S] approaches infinity, v_0 approaches V_{max}. To summarize, *a competitive inhibitor increases the apparent K_M of the enzyme but does not affect k_{cat} or V_{max}*.

The Michaelis–Menten equation for a competitively inhibited reaction has the form

$$v_0 = \frac{V_{max}[S]}{\alpha K_M + [S]} \qquad [7.28]$$

where α is a factor that makes K_M appear larger. The value of α—the degree of inhibition—depends on the inhibitor's concentration and its affinity for the enzyme:

$$\alpha = 1 + \frac{[I]}{K_I} \qquad [7.29]$$

K_I is the **inhibition constant**; it is the dissociation constant for the **enzyme–inhibitor (EI) complex:**

$$K_I = \frac{[E][I]}{[EI]} \qquad [7.30]$$

The lower the value of K_I, the tighter the inhibitor binds to the enzyme. It is possible to derive α (and therefore K_I) by plotting the reaction velocity as a function of substrate concentration in the presence of a known concentration of inhibitor. When the data are replotted in Lineweaver–Burk form, the intercept on the 1/[S] axis is $-1/\alpha K_M$ (**Fig. 7.10**; also see Sample Calculation 7.4).

FIGURE 7.10

A Lineweaver–Burk plot for competitive inhibition. The presence of a competitive inhibitor alters the apparent value of $-1/K_M$, the intercept on the 1/[S] axis, by the factor α. Note that the inhibitor does not affect the value of $1/V_{max}$, the intercept on the $1/v_0$ axis.

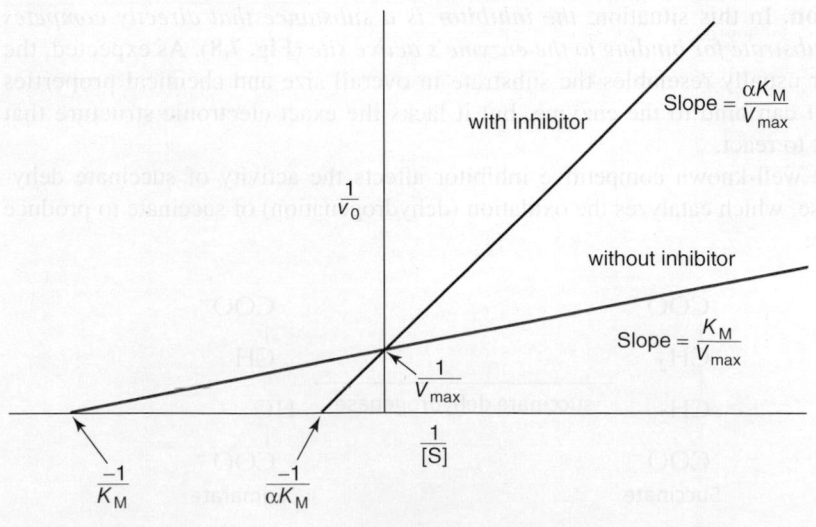

An enzyme has a K_M of 8 μM in the absence of a competitive inhibitor and an apparent K_M of 12 μM in the presence of 3 μM of the inhibitor. Calculate K_I. **PROBLEM**

The inhibitor increases K_M by a factor α (Equation 7.28). Since the value of K_M with the inhibitor is 1.5 times greater than the value of K_M without the inhibitor (12 μM ÷ 8 μM), α = 1.5. Equation 7.29, which gives the relationship between α, [I], and K_I, can be rearranged to solve for K_I: **SOLUTION**

$$K_I = \frac{[I]}{\alpha - 1}$$
$$= \frac{3 \text{ μM}}{1.5 - 1}$$
$$= \frac{3 \text{ μM}}{0.5} = 6 \text{ μM}$$

PRACTICE PROBLEMS

8. An enzyme has a K_M of 1 μM in the absence of a competitive inhibitor and an apparent K_M of 2 mM in the presence of 5 μM of the inhibitor. Calculate K_I.

9. The K_I value for a certain inhibitor is 2 μM. When no inhibitor is present, the K_M value is 8 μM. Calculate the apparent K_M when 6 μM inhibitor is present.

10. Inhibitor A at a concentration of 2 μM doubles the apparent K_M for an enzymatic reaction, whereas inhibitor B at a concentration of 9 μM quadruples the apparent K_M. What is the ratio of the K_I for inhibitor B to the K_I for inhibitor A?

K_I values are useful for assessing the inhibitory power of different substances, such as a series of compounds being tested for usefulness as drugs. For example, atorvastatin, a widely used cholesterol-lowering drug, binds to the enzyme HMG-CoA reductase with a K_I value of about 8 nM (the enzyme's substrate has a K_M of about 4 μM). Keep in mind that an effective drug is not necessarily the compound with the lowest K_I (the tightest binding), since other factors, such as the drug's solubility or its stability, must also be considered.

Product inhibition occurs when the product of a reaction occupies the enzyme's active site, thereby preventing the binding of additional substrate molecules. This is one reason why measurements of enzyme activity are made early in the reaction, before product has significantly accumulated.

TRANSITION STATE ANALOGS INHIBIT ENZYMES

Studies of enzyme inhibitors can reveal information about the chemistry of the reaction and the enzyme's active site. For example, the inhibition of succinate dehydrogenase by malonate, shown earlier, suggests that the dehydrogenase active site recognizes and binds substances with two carboxylate groups. Similarly, the ability of an inhibitor to bind to an enzyme's active site may confirm a proposed reaction mechanism. The triose phosphate isomerase reaction (introduced in Section 7.1) is believed to proceed through an enediolate transition state (shown inside brackets):

Glyceraldehyde-3-phosphate   Enediolate   Dihydroxyacetone phosphate

Recall from Section 6.2 that the transition state corresponds to a high-energy structure in which bonds are in the process of breaking and forming. The compound phosphoglycohydroxamate

$$
\begin{array}{c}
\text{OH} \\
\text{N} \\
\parallel \\
\text{C}-\text{O}^- \\
\mid \\
\text{CH}_2\text{OPO}_3^{2-}
\end{array}
$$

Phosphoglycohydroxamate

resembles the proposed transition state and, in fact, binds to triose phosphate isomerase about 300 times more tightly than glyceraldehyde-3-phosphate or dihydroxyacetone phosphate binds to the enzyme.

Numerous studies demonstrate that *whereas substrate analogs make good competitive inhibitors,* **transition state analogs** *make even better inhibitors.* This is because in order to catalyze a reaction, the enzyme must bind to (stabilize, or lower the energy of) the reaction's transition state. A compound that mimics the transition state can take advantage of features in the active site in a way that a substrate analog cannot. For example, the nucleoside adenosine is converted to inosine as follows:

Adenosine   Ribose   →   [Ribose]   →   Inosine
                                                                  Ribose

The K_M of the enzyme for the substrate adenosine is 3×10^{-5} M. The product inosine acts as an inhibitor of the reaction, with a K_I of 3×10^{-4} M. The transition state analog 1,6-dihydroinosine inhibits the reaction with a K_I of 1.5×10^{-13} M.

Not only do such inhibitors shed light on the probable structure of the reaction's transition state, they may provide a starting point for the design of even better inhibitors. Some of the drugs used to treat infection by HIV (the human immunodeficiency virus) were first designed by considering how transition state analogs inhibit viral enzymes (**Box 7.A**).

Ribose
1,6-Dihydroinosine

Box 7.A Inhibitors of HIV Protease

Human immunodeficiency virus (HIV), the causative agent of acquired immune deficiency syndrome (AIDS), has an RNA genome that codes for 15 different proteins. Six of these are structural proteins (green in the diagram below), three are enzymes (purple), and six are accessory proteins (blue) that are required for viral gene expression and assembly of new viral particles. Several genes overlap, and two genes are composed of noncontiguous segments of RNA.

HIV genome

HIV's structural proteins and its three enzymes are initially synthesized as **polyproteins** whose individual members are later separated by proteolysis (at the sites indicated by arrows). The enzyme responsible for this activity is HIV protease, one of the three viral enzymes. A small amount of the protease is present in the virus particle when it first infects a cell; more is generated

as the viral genome is transcribed and translated. HIV protease catalyzes the hydrolysis of Tyr—Pro or Phe—Pro peptide bonds in the viral polyproteins. Catalytic activity is centered on two Asp residues (shown in green in the model below), each contributed by one subunit of the homodimeric enzyme. The gold structure represents a peptide substrate analog. The side chains of peptide substrates bind in hydrophobic pockets near the active site.

HIV protease. [*Structure (pdb 1HXW) determined by C. H. Park, V. Nienaber, and X. P. Kong.*]

HIV protease inhibitors are the result of rational drug design (see Section 7.4) based on detailed knowledge of the enzyme's structure. For example, studies of inhibitor–protease complexes revealed that strong inhibitors must be at least the size of a tetrapeptide but need not be symmetrical (although the enzyme itself is symmetrical). Saquinavir (Invirase®) was the first widely used HIV protease inhibitor. It is a transition state analog with bulky side chains that mimics the protease's natural Phe—Pro substrates (the scissile bonds are shown in red).

Saquinavir acts as a competitive inhibitor with a K_I of 0.15 nM (for comparison, synthetic peptide substrates have K_M values of about 35 μM). Efforts to develop other drugs have focused on improving the solubility (and therefore the bioavailability) of protease inhibitors by adding polar groups without diminishing binding to the protease. Such efforts have yielded, for example, ritonavir (Norvir®), with a K_I of 0.17 nM.

One of the challenges of developing antiviral drugs is to *select a target that is unique to the virus so that the drugs will not disrupt the host's normal metabolic reactions.* The HIV protease inhibitors are effective antiviral agents because mammalian proteases do not recognize compounds containing amide bonds to proline or proline analogs. Nevertheless, the drugs that target HIV protease do have side effects. In addition, the high rate of mutation in HIV increases the risk of the virus developing resistance to a drug. For this reason, HIV infection is typically treated with a combination of several drugs, including protease inhibitors and inhibitors of reverse transcriptase and integrase, the virus's other two enzymes (see Box 20.A).

? Can you identify the amino acid residues in saquinavir and ritonavir?

—Phe—Pro—

Saquinavir

Ritonavir

OTHER TYPES OF INHIBITORS AFFECT V_{max}

Some reversible enzyme inhibitors diminish an enzyme's activity not only by interfering with substrate binding (as approximated by K_M) but also by directly affecting k_{cat}. This situation usually occurs when the inhibitor binds to a site on the enzyme other than the active site and triggers a conformational change that affects the structure or chemical properties of the active site. As a result, k_{cat} *and the apparent V_{max} decrease but K_M does not change.* This situation is called **noncompetitive inhibition** (Fig. 7.11).

Often, however, the inhibitor binding to the enzyme alters its conformation in such a way that both V_{max} and K_M are affected, although not necessarily in the same way. This phenomenon is called **mixed inhibition,** and the apparent K_M may increase or decrease. A Lineweaver–Burk plot for mixed inhibition is shown in **Figure 7.12**.

Metal ions may act as noncompetitive enzyme inhibitors. For example, trivalent ions such as aluminum (Al^{3+}) inhibit the activity of acetylcholinesterase, which catalyzes the hydrolysis of the neurotransmitter acetylcholine:

FIGURE 7.11

Noncompetitive enzyme inhibition. The inhibitor and substrate bind to separate sites. Inhibitor binding does not prevent substrate binding but alters the catalytic activity of the enzyme (thereby decreasing the apparent V_{max}). In mixed inhibition, the inhibitor also affects substrate binding so that the K_M appears to increase or decrease.

Enzyme   Enzyme–Substrate

Enzyme–Inhibitor   Enzyme–Substrate–Inhibitor

This reaction limits the duration of certain nerve impulses (see Section 9.4). Al^{3+} inhibits acetylcholinesterase noncompetitively by binding to the enzyme at a site distinct from the active site. Consequently, Al^{3+} can bind to the free enzyme or to the enzyme–substrate complex.

In a multisubstrate reaction, an inhibitor can bind to the enzyme *after* one substrate has bound, in a way that prevents the reaction from continuing and yielding product (**Fig. 7.13**). In the presence of such a substance, called an **uncompetitive inhibitor,** k_{cat} is affected so that the apparent V_{max} is lowered, and the apparent K_M is lowered to the same degree.

Competitive inhibition can be distinguished from mixed, noncompetitive, and uncompetitive inhibition because increasing the concentration of the substrate reverses the effects of a competitive inhibitor. Increasing the substrate concentration does not alleviate mixed, noncompetitive, and uncompetitive inhibition because the inhibitor does not prevent substrate binding to the active site.

ALLOSTERIC ENZYME REGULATION INCLUDES INHIBITION AND ACTIVATION

Oligomeric enzymes—those with multiple active sites in one multisubunit protein—are commonly subject to **allosteric regulation,** which includes inhibition as well as activation. Just as ligand binding to one subunit of an oligomeric enzyme may alter the activity of the other active sites (as occurs in hemoglobin; see Section 5.1), inhibitor (or activator) binding to one subunit of an enzyme may decrease (or increase) the catalytic activity of all the subunits.

FIGURE 7.12

Effect of a mixed inhibitor on reaction velocity. As shown here, the inhibitor affects both substrate binding (represented as K_M) and k_{cat} so that the apparent K_M increases and the apparent V_{max} decreases. In some cases, the apparent K_M may decrease or remain unchanged (noncompetitive inhibition).

| Enzyme | Enzyme–Substrate | Enzyme–Substrate–Inhibitor |

FIGURE 7.13

Uncompetitive enzyme inhibition. The inhibitor binds to the enzyme after the substrate binds. As a result, V_{max} and K_M appear to be reduced by the same amount.

> **?** Sketch a velocity versus substrate curve for an enzymatic reaction in the absence and presence of an uncompetitive inhibitor.

Allosteric effects are part of the physiological regulation of the enzyme phosphofructo-kinase, which catalyzes the reaction

$$^{-2}O_3POCH_2 \quad O \quad CH_2OH + ATP \xrightarrow{\text{phosphofructokinase}} \, ^{-2}O_3POCH_2 \quad O \quad CH_2OPO_3^{2-} + ADP$$

Fructose-6-phosphate → Fructose-1,6-bisphosphate

This phosphorylation reaction is step 3 of glycolysis, the glucose degradation pathway that is an important source of ATP in virtually all cells (see Section 13.1). The phosphofructokinase reaction is inhibited by phosphoenolpyruvate, the product of Reaction 9 of glycolysis:

$$\begin{array}{c} O \quad O^- \\ \backslash \!\! / \\ C \\ | \\ C-OPO_3^{2-} \\ \| \\ CH_2 \end{array}$$

Phosphoenolpyruvate

Phosphoenolpyruvate is an example of a **feedback inhibitor:** When its concentration in the cell is sufficiently high, *it shuts down its own synthesis by blocking an earlier step in its biosynthetic pathway:*

Glucose → → ———————→ → → → → → → Phosphoenolpyruvate
phosphofructokinase

Phosphofructokinase from the bacterium *Bacillus stearothermophilus* is a tetramer with four active sites. The subunits are arranged as a dimer of dimers (**Fig. 7.14**). Each of the four fructose-6-phosphate binding sites is made up of residues from both dimers. *B. stearothermophilus* phosphofructokinase binds fructose-6-phosphate with hyperbolic kinetics and a K_M of 23 µM. In the presence of 300 µM of the inhibitor phosphoenolpyruvate, fructose-6-phosphate binding becomes sigmoidal and the K_M increases to about 200 µM (**Fig. 7.15**). The inhibitor does not affect V_{max}, but phosphofructokinase becomes less active because its apparent affinity for fructose-6-phosphate decreases.

How does phosphoenolpyruvate exert its inhibitory effects? The sigmoidal velocity versus substrate curve (see Fig. 7.15) indicates that the phosphofructokinase active sites behave cooperatively in the presence of phosphoenolpyruvate. In each subunit, the inhibitor binds in a pocket that is separated from the fructose-6-phosphate binding site of the neighboring dimer by a loop of protein. When phosphoenolpyruvate occupies its binding site, the protein closes in around it. This causes a conformational change in which two residues in the loop switch positions: Arg 162 moves

FIGURE 7.14

Structure of phosphofructokinase from B. stearothermophilus. The four identical subunits are arranged as a dimer of dimers (one dimer is shown with blue subunits, the other with purple subunits). [*Structure (pdb 6PFK) determined by P. R. Evans, T. Schirmer, and M. Auer.*]

FIGURE 7.15

Effect of phosphoenolpyruvate on phosphofructokinase activity. In the absence of the inhibitor (green line), *B. stearothermophilus* phosphofructokinase binds the substrate fructose-6-phosphate with a K_M of 23 μM. In the presence of 300 μM phosphoenolpyruvate (red line), the K_M increase to about 200 μM. [*Data from Zhu, X., Byrnes, N., Nelson, J. W., and Chang, S. H., Biochemistry 34, 2560–2565 (1995).*]

away from the fructose-6-phosphate binding site of the neighboring subunit and is replaced by Glu 161 (**Fig. 7.16**). This conformational switch diminishes fructose-6-phosphate binding because the positively charged side chain of Arg 162, which helps stabilize the negatively charged phosphate group of fructose-6-phosphate, is replaced by the negatively charged side chain of Glu 161, which repels the phosphate group. *The effect of phosphoenolpyruvate is communicated to the entire protein (thereby explaining the cooperative effect) because phosphoenolpyruvate binding to one subunit of phosphofructokinase affects fructose-6-phosphate binding to the neighboring subunit in the other dimer.* Using the terminology for allosteric proteins, phosphoenolpyruvate

FIGURE 7.16

Conformational change upon phosphoenolpyruvate binding to phosphofructokinase. The green structure represents the conformation of the enzyme that readily binds the substrate fructose-6-phosphate (labeled F6P). The red structure represents the enzyme with a bound allosteric inhibitor (a phosphoenolpyruvate analog labeled PEP). Phosphoenolpyruvate binding to the enzyme causes a conformational change in which Arg 162 (which forms part of the fructose-6-phosphate binding site of the neighboring subunit) changes places with Glu 161. Because the subunits act cooperatively, the inhibitor diminishes substrate binding to the entire enzyme, causing the K_M to increase. [*After Schirmer, T., and Evans, P. R., Nature 343, 142 (1990).*]

binding causes the entire tetramer to switch to the T (low-activity) conformation, as measured by fructose-6-phosphate binding affinity. Phosphoenolpyruvate is therefore known as a **negative effector** of the enzyme.

Phosphofructokinase is allosterically inhibited by phosphoenolpyruvate, but it can also be allosterically activated by ADP, a **positive effector** of the enzyme. Although ADP is a product of the phosphofructokinase reaction, it is also a general signal of the cell's need for more ATP since the metabolic consumption of ATP yields ADP:

$$ATP + H_2O \longrightarrow ADP + P_i$$

Because phosphofructokinase catalyzes step 3 of the 10-step glycolytic pathway (one of whose ultimate products is ATP), increasing phosphofructokinase activity can increase the rate of ATP produced by the pathway as a whole.

Interestingly, the activator ADP binds not to the active site (which accommodates the substrate ATP and the reaction product ADP) but to the same site where the inhibitor phosphoenolpyruvate binds. But because ADP is much larger than phosphoenolpyruvate, the enzyme cannot close around it, and the conformational change to the low-activity T state cannot occur. Instead, ADP binding forces Arg 162 to remain where it can stabilize fructose-6-phosphate binding (in other words, it helps keep the enzyme in the high-activity R state). The overall result is that ADP counteracts the inhibitory effect of phosphoenolpyruvate and boosts phosphofructokinase activity.

SEVERAL FACTORS MAY INFLUENCE ENZYME ACTIVITY

So far, we have examined how small molecules that bind to an enzyme inhibit (or sometimes activate) that enzyme. These relatively simple phenomena are not the only means for regulating enzyme activity *in vivo*. Listed below and in **Figure 7.17** are some additional mechanisms. Keep in mind that several of these mechanisms, along with enzyme inhibition or activation, may operate in concert to precisely adjust the activity of a given enzyme.

1. A change in the rate of an enzyme's synthesis or degradation can alter the amount of enzyme available to catalyze a reaction (since $V_{max} = k_{cat}[E]_T$; Equation 7.23).

2. A change in subcellular location, for example, from an intracellular membrane to the cell surface, can bring an enzyme into proximity to its substrate and thereby increase reaction velocity. The opposite effect—sequestering an enzyme away from its substrate—dampens the reaction velocity.

3. An ionic "signal" such as a change in pH or the release of stored Ca^{2+} ions can activate or deactivate an enzyme by altering its conformation.

4. Covalent modification of an enzyme can affect the enzyme's K_M or k_{cat}, just as with an allosteric activator or inhibitor. Most commonly, a phosphoryl ($-PO_3^{2-}$) group or a fatty acyl (lipid) group is added to an enzyme so as to alter its catalytic activity. The effects of covalent modification are actually considered to be reversible, since cells contain enzymes that catalyze the removal of the modifying group as well as its addition. As we will see in Chapter 10 on signaling, phosphorylation and dephosphorylation can dramatically alter the activities of certain proteins.

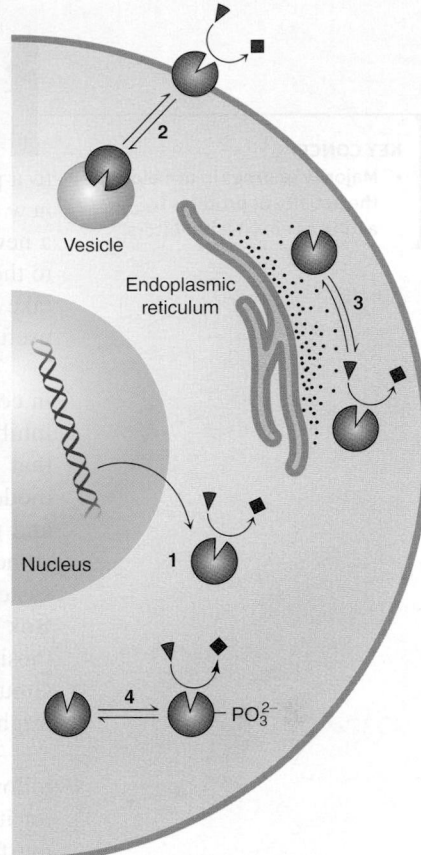

FIGURE 7.17

Some mechanisms for regulating enzyme activity. In this diagram, enzymes are shown as circular shapes, substrates as small triangles, and products as small squares. The amount of enzyme may depend on its rates of synthesis and degradation (1). The reaction velocity may depend on the enzyme's location (2). A signal such as a burst of Ca^{2+} ions released from the endoplasmic reticulum may affect the enzyme's activity (3). Covalent modification, such as phosphorylation, may activate an enzyme; the enzyme may be inactive when dephosphorylated (4).

 Which of the mechanisms shown in the diagram would be fastest (or slowest) to alter an enzyme's activity?

 CONCEPT CHECK

1. Why does an agent that chemically modifies an amino acid residue sometimes act as an irreversible enzyme inhibitor?

2. Why do some reaction products and transition state analogs act as competitive enzyme inhibitors?

3. How can noncompetitive, mixed, and uncompetitive inhibition be distinguished from competitive inhibition?

4. Explain why quaternary structure is required for allosteric regulation.

5. List all the ways that a cell could increase or decrease the activity of an enzyme.

CLINICAL CONNECTION: DRUG DEVELOPMENT

The development of a drug from an enzyme inhibitor in a biochemist's laboratory to a pill in a patient's medicine cabinet is typically long, arduous, and expensive. Whether a new drug results from a surprise discovery or dedicated efforts by pharmaceutical scientists, a newly identified drug candidate is almost never exactly the same compound that makes it to the pharmacist's shelves. Drug development is a process of refinement and testing that can take years and cost billions of dollars, but it usually yields a substance that is therapeutically useful and safe.

The majority of drugs currently in use block the activity of proteins that participate in cell-signaling pathways, but they have much in common with drugs that act as enzyme inhibitors. Every potential drug starts out as a synthetic compound or a natural product that is then altered and tested for the desired biological effect. For enzyme inhibitors, modern approaches take advantage of knowledge about an enzyme's active-site structure and mechanism to design a compound that will precisely block catalysis. This process is sometimes called **rational drug design.** A drug candidate, or lead compound, may be systematically altered by adding or deleting various chemical groups (including fluorine, Box 2.A) and then retested for inhibitory activity. Robotic procedures for chemical synthesis and analysis can handle thousands of substances at a time. Alternatively, computer simulations, based on the enzyme's structure, can predict whether a modified structure might be a better inhibitor.

The goal throughout the drug development process is to create a compound with the following properties: The drug must be a tight-binding inhibitor, it must be highly selective for its target enzyme (so that it doesn't interfere with the activity of other enzymes), and it must not be toxic. In addition, the drug's **pharmacokinetics,** or behavior in the body, must be assessed. For example, the drug must be water-soluble so that it can be transported via the bloodstream; at the same time, it must be lipid-soluble so that it can pass through the walls of the intestine to be absorbed and through the walls of blood vessels so that it can reach the tissues. In many cases, the drug is inactive until this point; once inside a cell, it is converted to its bioactive form by cellular enzymes.

Acyclovir (Zovirax®), which is used to treat herpes virus infections, provides an interesting example of this phenomenon. The drug mimics guanosine with an incomplete ribose group. Infected cells contain a viral kinase that converts the drug to a nucleotide analog that then interferes with DNA synthesis.

Acyclovir

Uninfected cells lack the kinase and are therefore unaffected by acyclovir.

A large part of the research relevant to drug development focuses on how the body metabolizes foreign compounds. The cytochrome P450 class of enzymes is responsible for detoxifying a variety of naturally occurring compounds that make their way into the body; drugs are also potential substrates. Cytochrome P450 contains a heme prosthetic group (see Section 5.1) that participates in an oxidation–reduction reaction that adds a hydroxyl group to the substrate to make it more water-soluble and therefore more easily excreted. Consequently, the activity of a cytochrome P450 enzyme can decrease the effectiveness of a drug. In some cases, the hydroxylation reaction can convert a drug to a toxin. This occurs when acetaminophen, a commonly used fever-reducer (also known as paracetamol), is consumed in large doses:

Acetaminophen → cytochrome P450 (O_2, H_2O) → (intermediate) → spontaneous (H_2O) → Acetimidoquinone (toxic)

Individuals vary in the amount and type of cytochrome P450 enzymes they express, so it is difficult to predict how a drug might be metabolized.

Consequently, the best laboratory-based predictions—or even testing in animals—cannot guarantee a successful outcome in the human body. Ultimately, a drug's effectiveness and safety must be assessed in **clinical trials.** This type of testing is organized into three consecutive phases, each involving a larger number of subjects. In Phase I clinical trials, the drug candidate is administered to a small group of healthy volunteers at levels up to the expected therapeutic dose. The goal of Phase I trials is to ensure that the drug is safe and is well tolerated by humans. Researchers can also examine the drug's pharmacokinetics and select a version of the drug and a dosing regimen suitable for larger-scale testing.

After the safety of the drug has been verified, its effectiveness is tested in Phase II trials, which typically involve several hundred subjects who have the disease targeted by the drug. Assessments of safety and optimum dosing continue during Phase II. Patients are usually randomly assigned to test or control groups. Since many clinical trials test whether a new drug works better than an existing drug, the control group receives the old drug rather than no drug at all. To avoid the placebo effect, in which a patient's or physician's expectations can alter the outcome, the trial is "blinded" so that the patients do not know which treatment they are receiving (a single-blind trial) or, better yet, neither the patients nor their physicians know (a double-blind trial). Drug trials are usually easy to blind, since the test and control substances can be made to appear identical, but it is difficult if not impossible to conduct blind trials in other situations, such as comparing a drug to a surgical intervention. In these cases, objectivity can be maximized by not telling the statisticians who analyze the results which patient group received which treatment.

Phase III clinical trials are conducted with large numbers of patients (hundreds to thousands). Large numbers are needed to provide robust statistical evidence that the new drug works as hoped and is safe for widespread use. Like Phase II trials, Phase III trials are randomized and blinded if possible. Successful completion of Phase III, the longest phase of testing, usually leads to regulatory approval of the drug by the Food and Drug Administration (in the United States), although the drug may be marketed to a limited extent before full approval has been granted.

Even after a drug has been approved, marketed, and widely adopted, the patient population is continually scrutinized for rare side effects or side effects that might not have developed during the short time period of Phase II or Phase III trials. This surveillance period is sometimes called Phase IV. Its importance is highlighted by the case of the widely used pain reliever rofecoxib (Vioxx®), which had been approved but was withdrawn when it was discovered that it increased the risk of heart attacks. Similarly, the use of the antidiabetic drug rosiglitazone (Avandia®) has been limited since analysis of large numbers of patients revealed an increase in heart attacks.

Ongoing challenges for the drug industry include the development of new drugs that are more effective than existing drugs, so that a patient can use less of a drug and therefore experience fewer side effects. However, bringing the new drug to market is expensive, and drug developers must recover their investment. This can be difficult to accomplish without pricing the drug out of patients' reach. A consequence of this trade-off is that some of the most widely used (and most profitable) drugs target common disorders that are not immediately life-threatening (**Table 7.2**), while rarer and deadlier diseases have fewer drug treatment options.

TABLE 7.2 Some Commonly Prescribed Drugs

DRUG[a]	DRUG ACTION	TARGET DISEASE OR CONDITION
Albuterol (Ventolin®)	binds to adrenergic receptors to trigger bronchodilation	asthma, chronic obstructive pulmonary disease
Amoxicillin	antibiotic, inhibits enzyme required for synthesis of bacterial cell walls	bacterial infections
Atorvastatin (Lipitor®)	inhibits HMG-CoA reductase, a key enzyme of cholesterol synthesis	hypercholesterolemia, atherosclerosis
Duloxetine (Cymbalta®)	inhibits reuptake of neurotransmitters serotonin and norepinephrine in central nervous system	depression, anxiety disorder
Esomeprazole (Nexium®)	inhibits a proton pump to block stomach acid production	peptic ulcers, gastroesophageal reflux disease
Hydrocodone	narcotic, binds to opioid receptors in central nervous system	moderate to severe pain
Levothyroxine (Synthroid®)	mimics thyroid hormone, binds to thyroid hormone receptor	hypothyroidism
Lisdexamfetamine (Vyvanse®)	activates catecholamine receptors in central nervous system	attention deficit hyperactivity disorder, narcolepsy
Lisinopril	inhibits angiotensin converting enzyme to block vasoconstriction	hypertension (high blood pressure), congestive heart failure

[a]The trade name is given in parentheses.

 CONCEPT CHECK

1. Compare rational drug design and design by trial and error.
2. Explain why developers study a drug's pharmacokinetics.
3. Summarize the purpose of each phase of a clinical trial.

SUMMARY

7.1 Introduction to Enzyme Kinetics

• Rate equations describe the velocity of simple unimolecular (first-order) or bimolecular (second-order) reactions in terms of a rate constant.

7.2 Derivation and Meaning of the Michaelis–Menten Equation

• An enzyme-catalyzed reaction can be described by the Michaelis–Menten equation. The overall rate of the reaction is a function of the rates of formation and breakdown of an enzyme–substrate (ES) complex.

• The Michaelis constant, K_M, is a combination of the three rate constants relevant to the ES complex. It is also equivalent to the substrate concentration at which the enzyme is operating at half-maximal velocity. The maximum velocity is achieved when the enzyme is fully saturated with substrate.

• The catalytic constant, k_{cat}, for a reaction is the first-order rate constant for the conversion of the enzyme–substrate complex to product. The quotient k_{cat}/K_M, a second-order rate constant for the overall conversion of substrate to product, indicates an enzyme's

catalytic efficiency because it accounts for both the binding and catalytic activities of the enzyme.

• Values for K_M and V_{max} (from which k_{cat} can be calculated) are often derived from Lineweaver–Burk, or double-reciprocal, plots. Not all enzymatic reactions obey the simple Michaelis–Menten model, but their kinetic parameters can still be estimated.

7.3 Enzyme Inhibition

• Some substances react irreversibly with enzymes to permanently block catalytic activity.

• The most common reversible enzyme inhibitors, which may be transition state analogs, compete with substrate for binding to the active site, thereby increasing the apparent K_M.

• Reversible enzyme inhibitors that affect an enzyme's V_{max} may be noncompetitive, mixed, or uncompetitive inhibitors.

• Oligomeric enzymes such as bacterial phosphofructokinase are regulated by allosteric inhibitors and activators.

• The activities of enzymes may also be regulated by changes in enzyme concentration, location, ion concentrations, and covalent modification.

7.4 Clinical Connection: Drug Development

- Some drugs act as enzyme inhibitors; these substances are modified to maximize their binding to target proteins and to optimize their pharmacokinetics.

- The effectiveness and safety of a prospective drug are tested in clinical trials, starting with a small group of subjects and including blinded assessments in larger groups.

KEY TERMS

kinetics	v_0	ping pong mechanism	mixed inhibition
v	V_{max}	allosteric enzyme	uncompetitive inhibition
ES complex	Michaelis–Menten equation	cooperative binding	allosteric regulation
saturation	catalytic constant (k_{cat})	irreversible inhibitor	feedback inhibitor
unimolecular reaction	turnover number	suicide substrate	negative effector
rate equation	k_{cat}/K_M	competitive inhibition	positive effector
rate constant (k)	diffusion-controlled limit	inhibition constant (K_I)	rational drug design
first-order reaction	catalytic perfection	EI complex	pharmacokinetics
bimolecular reaction	Lineweaver–Burk plot	product inhibition	clinical trial
second-order reaction	bisubstrate reaction	transition state analog	
steady state	random mechanism	polyprotein	
Michaelis constant (K_M)	ordered mechanism	noncompetitive inhibition	

BIOINFORMATICS

Brief Bioinformatics Exericses

7.1 Enzyme Cofactors and Kinetics

7.2 Survey of Enzyme–Inhibitor Complexes

Bioinformatics Project

Enzyme Inhibitors and Rational Drug Design

PROBLEMS

7.1 Introduction to Enzyme Kinetics

1. At about the time scientists began analyzing the hyperbolic velocity versus substrate curve (Fig. 7.3), Emil Fischer was formulating his lock-and-key hypothesis of enzyme action (Section 6.3), which described the enzyme as a lock and the substrate as a key. Later experiments showed that the enzyme–substrate relationship is more dynamic, but kinetic data are generally consistent with his model. Explain.

2. Explain why it is usually easier to calculate an enzyme's reaction velocity from the rate of appearance of the product rather than the rate of disappearance of the substrate.

3. The rate of hydrolysis of sucrose to glucose and fructose is quite slow in the absence of a catalyst.

$$\text{sucrose} + H_2O \rightarrow \text{glucose} + \text{fructose}$$

If the initial concentration of sucrose is 0.10 M, it takes 440 years for the concentration of the sucrose to decrease by half to 0.05 M. What is the rate of disappearance of sucrose in the absence of a catalyst?

4. If a catalyst is present, the hydrolysis of sucrose (see Problem 3) is much more rapid. If the initial concentration of sucrose is 0.10 M, it takes 6.9×10^{-5} s for the concentration to decrease by half to 0.05 M. What is the rate of disappearance of sucrose in the presence of a catalyst?

5. The rate of hydrolysis of trehalose to its constituent glucose monomers in the absence of a catalyst is even slower than hydrolysis of sucrose (see Problem 3) and may indicate that the two sugars are hydrolyzed by different mechanisms. If the initial concentration of trehalose is 0.10 M, it takes 6.6×10^6 years for the concentration to decrease by half. What is the rate of disappearance of trehalose in the absence of a catalyst?

6. If a catalyst is present, the hydrolysis of trehalose (see Problem 5) is much more rapid. It takes 2.7×10^{-4} s for the concentration of trehalose to decrease by half from 0.10 M to 0.05 M.

(a) What is the rate of disappearance of trehalose in the presence of a catalyst?

(b) Is the rate enhancement greater for sucrose (see Problem 4) or trehalose?

7. The uncatalyzed rate of amide bond hydrolysis in hippuryl-phenylalanine was measured by monitoring the formation of the product, phenylalanine. A 60 mM solution of hippurylphenylalanine was monitored for amide bond hydrolysis for 50 days. At the end of this time, 50 μM phenylalanine was detected in the solution.

(a) What is the rate of formation of phenylalanine (in units of $M \cdot s^{-1}$) in this reaction?

(b) Draw a graph of the reaction to show the velocity of the reaction (measured as the rate of the appearance of phenylalanine per unit time) as a function of the reactant concentration.

8. The amide bond in hippurylphenylalanine (see Problem 7(a)) is hydrolyzed 4.7×10^{11} times faster in the presence of an enzyme.

(a) What is the rate of formation of phenylalanine in the presence of an enzyme?

(b) Draw a graph of the reaction described to show the rate of appearance of phenylalanine as a function of substrate concentration.

(c) Compare this graph to the one you drew in Problem 7 (b). How is this graph different, and why?

9. A bacterial enzyme catalyzes hydrolysis of the disaccharide maltose to produce two glucose monosaccharides.

$$\text{maltose} + H_2O \rightarrow 2 \text{ glucose}$$

During an interval of one minute, the concentration of maltose decreases by 45 mM.

(a) What is the rate of disappearance of maltose in the enzyme-catalyzed reaction?

(b) For the reaction, what is the rate of appearance of glucose?

7.2 Derivation and Meaning of the Michaelis–Menten Equation

10. (a) Complete the table for the following one-step reactions:

Reaction	Molecularity	Rate equation	Units of k	Reaction velocity is proportional to ...	Order
$A \rightarrow B + C$					
$A + B \rightarrow C$					
$2\,A \rightarrow B$					
$2\,A \rightarrow B + C$					

(b) The formation of an enzyme–substrate complex for a one-substrate reaction can be expressed as $E + S \rightarrow ES$. For an enzymatic reaction with two substrates, A and B, the expression would be $E + A + B \rightarrow EAB$. Explain why the second process is approached as two bimolecular reactions rather than as a single termolecular (three-reactant) reaction.

11. The rates of the reactions described in Problems 3 and 5 double when the concentrations of sucrose and trehalose double. The reaction rates are not affected by the concentration of water. Write rate equations for the reactions that are consistent with these observations. What is the order of the reactions?

12. Refer to Sample Calculation 7.1 and to the rate equation you wrote for Problem 11 to

(a) Determine the velocity of the uncatalyzed reaction when the concentration of trehalose is 0.025 M. (The rate constant k for the uncatalyzed reaction is $3.3 \times 10^{-15}\ \mathrm{s}^{-1}$).

(b) Determine the velocity of the enzyme-catalyzed hydrolysis of trehalose at the same concentration. (The rate constant k for the catalyzed reaction is $2.6 \times 10^{3}\ \mathrm{s}^{-1}$.)

13. At what substrate concentration is $v = 5.0\ \mathrm{mM \cdot s}^{-1}$ for the enzyme-catalyzed hydrolysis of trehalose (see Problem 12b)?

14. Sugars entering bacteria are phosphorylated during transport via a complex series of reactions. In one of these reactions, an enzyme is phosphorylated; the phosphate group is subsequently transferred from the enzyme to the sugar.

(a) Write a rate equation for the one-step reaction: Enzyme + $P_i \rightarrow$ Enzyme–P_i

(b) What is the velocity of this reaction when the concentration of phosphate is 25 mM and the concentration of the enzyme is 12 pM? The rate constant k is $3.9 \times 10^{6}\ \mathrm{M}^{-1} \cdot \mathrm{s}^{-1}$.

15. An unprotonated primary amino group in a blood protein can react with carbon dioxide to form a carbamate as shown here:

$$R{-}NH_2 + CO_2 \longrightarrow R{-}NH{-}COO^- + H^+$$
$$\text{Carbamate}$$

The rate constant k for this reaction is $4950\ \mathrm{M}^{-1} \cdot \mathrm{s}^{-1}$.

(a) What is the order of this reaction?

(b) Calculate the velocity of the reaction of an α-amino group in a blood protein at 37°C if its concentration is 0.5 mM and the partial pressure of carbon dioxide is 45 torr. (*Hint*: Convert the units of partial pressure to molar concentration using the ideal gas law. The value of R is $0.0821\ \mathrm{L \cdot atm \cdot K}^{-1} \cdot \mathrm{mol}^{-1}$.)

(c) How would the rate constant for this reaction vary with pH? Explain.

(d) What CO_2 partial pressure is required to yield a velocity of $0.045\ \mathrm{M \cdot s}^{-1}$ for the reaction?

16. What portions of the velocity versus substrate curve (Fig. 7.3) correspond to zero-order and first-order processes?

17. Draw curves that show the appropriate relationship between the variables of each plot. (a) [P] *vs.* time; (b) [S] *vs.* time; (c) [ES] *vs.* time; (d) [E] *vs.* time; (e) v_0 *vs.* [E]; (f) v_0 *vs.* [S].

18. The enzyme-catalyzed reaction described in Problem 9 has a K_M of 0.135 µM and a V_{max} of 65 µmol · min^{-1}. What is the reaction velocity when the concentration of maltose is 1.0 µM?

19. You are attempting to determine K_M by measuring the reaction velocity at different concentrations, but you do not realize that the substrate tends to precipitate under the experimental conditions you have chosen. How would this affect your measurement of K_M?

20. You are constructing a velocity versus substrate concentration curve for an enzyme whose K_M is believed to be about 2 µM. The enzyme concentration is 200 nM and the substrate concentrations range from 0.1 µM to 10 µM. What is wrong with this experimental setup and how could you fix it?

21. (a) Is it necessary for measurements of reaction velocity to be expressed in units of concentration per unit time (M/s, for example) in order to calculate an enzyme's K_M?

(b) Is it necessary to know $[E]_T$ in order to determine K_M, V_{max}, or k_{cat}?

22. The enzyme phenylalanine hydroxylase (PAH) catalyzes the hydroxylation of phenylalanine to tyrosine and is deficient in patients with the disease phenylketonuria (PKU). The K_M for PAH is 0.5 mM and the V_{max} is 7.5 µmol · min^{-1} · mg^{-1}. What is the velocity of the reaction when the phenylalanine concentration is 0.25 mM?

23. The enzyme polyphenol oxidase acts on several substrates, one of which is dopamine. What is the K_M for the enzyme if the velocity for the reaction is 0.23 U · min^{-1} · mL^{-1} when the concentration of substrate is 10 mM? The V_{max} for the reaction is 0.36 U · min^{-1} · mL^{-1} (U stands for enzyme units).

24. In the absence of allosteric effectors, the enzyme phosphofructokinase displays Michaelis–Menten kinetics (see Fig. 7.15). The v_0/V_{max} ratio is 0.9 when the concentration of the substrate, fructose-6-phosphate, is 0.10 mM. Calculate the K_M for phosphofructokinase under these conditions.

25. Brain glutaminase has a V_{max} of 1.1 µmol · min^{-1} · mL^{-1} and a K_M of 0.6 mM. What is the substrate concentration when the velocity is 0.3 µmol · min^{-1} · mL^{-1}?

26. Use the plot to estimate values of K_M and V_{max} for an enzyme-catalyzed reaction.

27. Estimate the K_M and the V_{max} for each enzyme from the plot.

 (a) Which enzyme generates product more rapidly when [S] = 1 mM? What is the relationship between [S] and K_M for each enzyme at this substrate concentration?

 (b) Answer part (a) when [S] = 10 mM.

28. What relationship exists between K_M and [S] when an enzyme-catalyzed reaction proceeds at **(a)** 80% V_{max} and **(b)** 60% V_{max}?

29. When [S] = $5K_M$, how close is v_0 to V_{max}? When [S] = $20K_M$, how close is v_0 to V_{max}? What do these results tell you about the accuracy of estimating V_{max} from a plot of v_0 versus [S]?

30. Three enzymatic reactions carried out under different conditions are shown in the figure above right. Estimate the K_M and V_{max} values from the plot. Which reaction has the lowest K_M? Which reaction has the highest V_{max}?

31. The K_M values for the reaction of chymotrypsin with two different substrates are given. **(a)** Which substrate has the higher apparent affinity for the enzyme? Use what you learned about chymotrypsin in Chapter 6 to explain why. **(b)** Which substrate is likely to give a higher value for V_{max}?

Substrate	K_M (M)
N-Acetylvaline ethyl ester	8.8×10^{-2}
N-Acetyltyrosine ethyl ester	6.6×10^{-4}

32. The enzyme hexokinase acts on both glucose and fructose. Using the K_M and V_{max} values in the table, compare and contrast the interaction of hexokinase with each substrate.

Substrate	K_M (M)	V_{max} (relative)
Glucose	1.0×10^{-4}	1.0
Fructose	7.0×10^{-4}	1.8

33. (a) Calculate k_{cat} for the polyphenol oxidase described in Problem 23 for the dopamine substrate. The enzyme concentration is 2.0×10^{-2} U·mL^{-1}.

 (b) The polyphenol oxidase also acts on catechol as a substrate, with a V_{max} of 0.39 U·min^{-1}·mL^{-1}. Assuming the same enzyme concentration, what is the k_{cat} for the oxidase-catalyzed catechol reaction?

34. Using data from Problems 23 and 33, calculate and compare the catalytic efficiencies of the polyphenol oxidase enzyme using **(a)** dopamine or **(b)** catechol as a substrate. The K_M for catechol is 7.9×10^{-4} M.

35. (a) Use the data provided to determine whether the reactions catalyzed by enzymes A, B, and C are diffusion-controlled.

 (b) A reaction is carried out in which 5 nM substrate S is added to a reaction mixture containing equivalent amounts of enzymes A, B, and C. After 30 seconds, which product will be more abundant, P, Q or R?

Enzyme	Reaction	K_M	k_{cat}
A	S → P	0.3 mM	5000 s^{-1}
B	S → Q	1 nM	2 s^{-1}
C	S → R	2 µM	850 s^{-1}

36. The enzyme γ-glutamylcysteine synthetase (γ-GCS) from the protozoan *Trypanosoma brucei,* the parasite that causes African sleeping sickness, catalyzes the first step of the biosynthesis of trypanothione, a compound the parasite requires to maintain proper redox balance. The K_M values for each substrate have been measured.

 (a) Does this reaction obey Michaelis–Menten kinetics?

 (b) Describe how the K_M values for each of the three substrates were determined.

 (c) How would V_{max} be achieved for the γ-GCS reaction?

$$\text{aminobutyric acid} + \text{glutamate} + \text{ATP} \xrightarrow{\gamma\text{-GCS}} \rightarrow \rightarrow \text{trypanothione}$$

Substrate	K_M (mM)
Glutamate	5.9
Aminobutyric acid	6.1
ATP	1.4

37. A galactosyl transferase, β3-GalT, catalyzes the reaction shown below, an early step in the production of glycoproteins important in immune recognition. Use the Lineweaver–Burk plot provided to determine the K_M and the V_{max} for β3-GalT.

38. Subtilisin, an alkaline serine protease produced by *Bacillus* bacteria, is used in the detergent and food-processing industries. Protein engineers constructed a mutant subtilisin enzyme in an attempt to improve its catalytic activity. The isoleucine at position 31 (adjacent to the Asp residue in the Asp–His–Ser catalytic triad) was replaced with leucine. Both the mutant and wild-type enzymes were assayed for catalytic activity using the artificial substrate *N*-succinyl-Ala-Ala-Pro-Phe-*p*-nitroanilide (AAPF). The results are shown in the table.

Enzyme	K_M (mM)	k_{cat} (s^{-1})	k_{cat}/K_M (mM$^{-1} \cdot$ s^{-1})
Ile 31 wild-type subtilisin E	1.9 ± 0.2	21 ± 4	11
Leu 31 mutant subtilisin E	2.0 ± 0.3	120 ± 15	60

(a) Explain how reaction velocity is measured using the AAPF substrate.

(b) What effect did the replacement of Leu for Ile at position 31 have on the catalytic activity of subtilisin E for the hydrolysis of AAPF? Comment on the K_M and k_{cat} values.

(c) Both enzymes were tested for their ability to cleave peptide bonds in a "natural" substrate, milk casein. The results are shown in the table. Compare the activities of the two enzymes for the case in substrate.

Enzyme	Specific activity (units/mg)
Ile 31 wild-type subtilisin E	109 ± 9
Leu 31 mutant subtilisin E	297 ± 30

(d) Why did the Ile → Leu substitution alter the enzyme's catalytic efficiency?

(e) Would subtilisin as a detergent additive be effective in removing protein-based stains (such as milk or blood) from clothing?

39. Aspartate aminotransferase (AspAT) catalyzes the following reaction:

$$NH_3^+{-}CH{-}COO^- \xrightarrow[\alpha\text{-Keto acid}]{\substack{\alpha\text{-Amino acid}\\ \text{AspAT}}} \begin{array}{c} COO^- \\ | \\ C{=}O \\ | \\ CH_2 \\ | \\ COO^- \end{array}$$

Aspartate → Oxaloacetate

The AspAT enzyme has two active site arginines, Arg 386 and Arg 292, which interact with the α-carboxylate and β-carboxylate groups on the aspartate substrate, respectively. Mutant AspAT enzymes were constructed in which either or both of the essential arginines were replaced with a lysine residue. The kinetic parameters for the wild-type enzyme and mutant enzymes are shown in the table.

(a) Compare the ability of aspartate to bind to the wild-type and mutant enzymes.
(b) Why does replacing arginine with lysine affect substrate binding to AspAT?
(c) Evaluate the catalytic efficiency of the wild-type and mutant enzymes.
(d) Why does the replacement of arginine with lysine affect the catalytic activity of the enzyme?

Enzyme	K_M Aspartate (mM)	k_{cat} (s^{-1})
Wild-type AspAT (Arg 292, Arg 386)	4	530
Mutant AspAT (Lys 292, Arg 386)	326	4.5
Mutant AspAT (Arg 292, Lys 386)	72	9.6
Mutant AspAT (Lys 292, Lys 386)	300	0.055

40. The kinetics of a bacterial dehalogenase were investigated. Calculate K_M and V_{max} from the following data:

[S] (mM)	v_0 (nmol · min^{-1})
0.04	0.229
0.13	0.493
0.40	0.755
0.90	0.880
1.30	0.917

7.3 Enzyme Inhibition

41. Based on some preliminary measurements, you suspect that a sample of enzyme contains an irreversible inhibitor. You decide to dilute the sample 100-fold and remeasure the enzyme's activity. What would your results show if the inhibitor in the sample is (a) irreversible or (b) reversible?

42. How would diisopropylphosphofluoridate (DIPF) affect the apparent K_M and V_{max} of a sample of chymotrypsin?

43. Shown below are five Lineweaver–Burk plots. The reaction without inhibitor is shown in blue; the reaction with inhibitor is shown in red.

(a) Identify the type of inhibition in each plot.
(b) For each of the plots, describe the changes that occur (if any) in the values of K_M and V_{max} in the presence of the inhibitor.

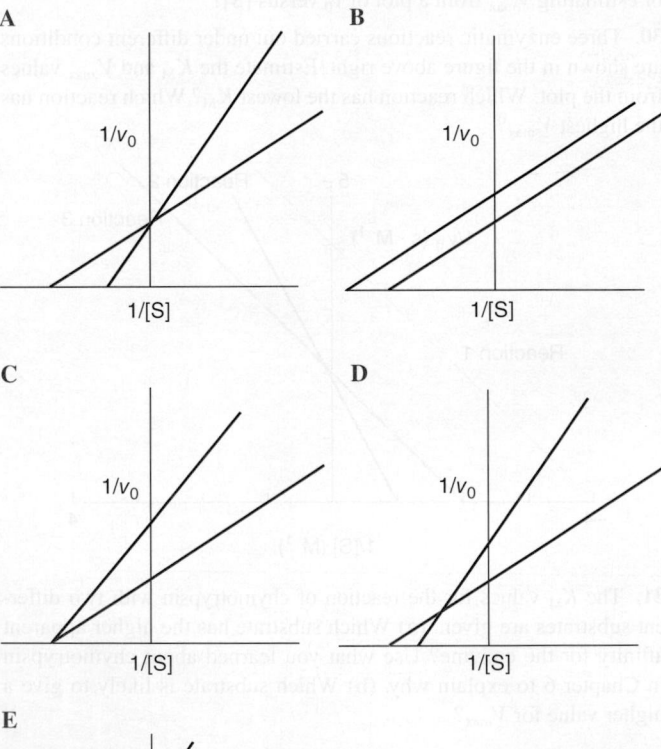

44. Decamethonium (a muscle relaxant) is an inhibitor of acetylcholinesterase. The normal substrate for acetylcholinesterase is acetylcholine. Structures of both molecules are shown.

Acetylcholine

Decamethonium

(a) What kind of inhibitor is decamethonium? Explain.
(b) Can inhibition by decamethonium be overcome *in vitro* by adding large amounts of substrate? Explain.
(c) Does this inhibitor bind reversibly or irreversibly to the enzyme? Explain.

45. Would indole be a more effective competitive inhibitor of chymotrypsin, trypsin, or elastase? Explain.

Indole

46. Explain why there are so few examples of inhibitors that decrease V_{max} but do not affect K_M.

47. Computer-modeling studies have shown that uncompetitive and noncompetitive inhibitors of enzymes are more effective than competitive inhibitors. These studies have important implications in drug design. Propose a hypothesis that explains these results.

48. Inhibitor A at a concentration of 3 μM doubles the apparent K_M for an enzymatic reaction, whereas inhibitor B at a concentration of 12 μM quadruples the apparent K_M. What is the ratio of the K_I for inhibitor B to the K_I for inhibitor A?

49. Glucose-6-phosphate dehydrogenase catalyzes the reaction shown below (see Section 13.4): Kinetic data for the enzyme isolated from the thermophilic bacterium *T. maritima* are presented in the table.

(a) NADPH inhibits glucose-6-phosphate dehydrogenase. What kind of inhibitor is NADPH?
(b) How are the values of K_M and V_{max} likely to differ from those listed in the table if NADPH is present?
(c) Does glucose-6-phosphate dehydrogenase prefer NAD$^+$ or NADP$^+$ as a cofactor?

Glucose-6-phosphate → 6-Phosphogluccono-δ-lactone

Substrate	K_M (mM)	V_{max} (U · mg^{-1})
Glucose-6-phosphate	0.15	20
NADP$^+$	0.03	20
NAD$^+$	12.0	6

50. Glucose-6-phosphate dehydrogenase catalyzes the same reaction shown in Problem 47 in yeast. The K_M for yeast glucose-6-phosphate is 2.0×10^{-5} M and the K_M for NADP$^+$ is 2.0×10^{-6} M. The yeast enzyme can be inhibited by a number of cellular agents whose K_I values are listed in the table.

(a) Which is the most effective inhibitor? Explain.
(b) Which inhibitor(s) is (are) likely to be completely ineffective under normal cellular conditions? Explain.

Inhibitor	K_I (M)
Inorganic phosphate	1.0×10^{-1}
Glucosamine-6-phosphate	7.2×10^{-4}
NADPH	2.7×10^{-5}

51. The bacterial enzyme proline racemase catalyzes the interconversion of two isomers of proline:

The compound shown below is an inhibitor of proline racemase. Explain why.

52. Cytidine deaminase catalyzes the following reaction:

Cytidine → Uridine

Both of the compounds shown below inhibit the reaction. The compounds have K_I values of 3×10^{-5} M and 1.2×10^{-12} M. Assign the appropriate K_I value to each inhibitor. Which compound is the more effective inhibitor? Give a structural basis for your answer.

A   B

53. The observation that adenosine deaminase is inhibited by 1,6-dihydroinosine allowed scientists to propose a structure for the transition state of this enzyme. The compound coformycin also inhibits adenosine deaminase; its K_I value is about 0.25 μM. Does this observation support or refute the proposed transition state for adenosine deaminase?

Coformycin

54. The flu virus enzyme neuraminidase hydrolyzes sialic acid residues from cell-surface glycoproteins, allowing newly made viruses to escape from the host cell. The drugs oseltamivir (Tamiflu®) and zanamivir (Relenza®) are transition state analogs that inhibit neuraminidase to block viral infection. Wild-type and mutant strains of the flu virus, in which neuraminidase residue 274 is changed from His to Tyr, exhibit the kinetic parameters shown below.

(a) For the wild-type virus, which drug would work better?
(b) Does the mutation appear to affect substrate binding or turnover?
(c) How does the mutation affect inhibition by oseltamivir and zanamivir? Which drug would be more effective against the mutant flu strain?

	K_M for sialic acid (μM)	V_{max} (relative units)
Wild-type enzyme	6.3	1.0
His 274 → Tyr mutant enzyme	27.0	0.8

	K_I for oseltamivir (nM)	K_I for zanamivir (nM)
Wild-type enzyme	0.32	0.1
His 274 → Tyr mutant enzyme	85	0.19

55. A phospholipase hydrolyzes its lipid substrate with a K_M of 10 μM and a V_{max} of 7 μmol $\cdot$ mg^{-1} $\cdot$ min^{-1}. In the presence of 30 μM palmitoylcarnitine inhibitor, the K_M increases to 40 μM and the V_{max} remains unchanged. What is the K_I of the inhibitor?

56. A compound named P56 was found to inhibit a malarial parasite cysteine protease and has the potential to be used as a drug to treat malaria. Data are plotted below.

(a) What kind of inhibitor is P56?

(b) Use the equations of the lines provided to calculate K_M and V_{max} for the enzyme with and without the inhibitor.

(c) Calculate the value of K_I in the presence of 0.22 mM inhibitor.

y = 4.93x + 2.35

y = 1.93x + 2.25

1/v_0 (μmol^{-1} $\cdot$ min)

1/[S] (mL $\cdot$ mg^{-1})

♦ Inhibitor P56 ■ No inhibitor

57. The tyrosinase enzyme catalyzes reactions that produce brown-colored products. In mammals, it is responsible for the production of melanin in the skin; in plants, it is responsible for the browning that occurs when foods such as apples or mushrooms are sliced open and exposed to air. Food scientists interested in preventing the browning process in plant-based foods have shown that dodecyl gallate inhibits tyrosinase. Using the Lineweaver–Burk plot provided, calculate the V_{max} and K_M for the enzyme in the presence and absence of the inhibitor. What type of inhibitor is dodecyl gallate?

y = 1.58x + 4.27

1/v (OD^{-1} $\cdot$ min)

y = 1.52x + 1.51

1/[S] (mM^{-1})

♦ No inhibitor ■ With dodecyl gallate

58. The HIV protease, an important enzyme in the life cycle of the human immunodeficiency virus, is a good drug target for the treatment of HIV and AIDS (see Box 7.A). A protein produced by the virus, p6*, inhibits the protease. The activity of the protease was measured in the presence (10 μM) and absence of p6* using an assay involving an artificial substrate, as shown in the figure (Nle is norleucine, a nonstandard amino acid). Data are shown in the table.

(a) Construct a Lineweaver–Burk plot and determine the K_M and V_{max} in the presence and in the absence of inhibitor.

(b) What type of inhibitor is p6*? Explain.

(c) Calculate the K_I for the inhibitor.

Lys
|
Ala
|
Arg
|
Val—Nle—NH—CH—C—NH—Glu
                    | ‖
                   CH$_2$ O
                    |

NH$_2$
|
Nle
|
Ala

NO$_2$

Lys
|
Ala
|
Arg
|
Val—Nle—NH—CH—COO$^-$ +   NH$_3^+$—Glu
                    |
                   CH$_2$
                    |

H$_2$O → HIV-1 protease

NH$_2$
|
Nle
|
Ala

Absorbs light at 295 nm

NO$_2$

[S] (μM)	v_0 without p6* (nmol $\cdot$ min^{-1})	v_0 with p6* (nmol $\cdot$ min^{-1})
10	4.63	2.70
15	5.88	3.46
20	6.94	4.74
25	9.26	6.06
30	10.78	6.49
40	12.14	8.06
50	14.93	9.71

59. The drug troglitazone was used to treat diabetes but was withdrawn from the market when patients who took the drug suffered from severe side effects. The activity of a hydroxylase enzyme, which acts on progesterone as a substrate in a steroid synthetic pathway, was measured in the presence and in the absence of 10 μM troglitazone.

(a) Use the data in the table to construct a Lineweaver–Burk plot. What type of inhibitor is troglitazone?

(b) Calculate the K_I for the inhibitor.

[Progesterone] (μM)	v_0 without I (pmol $\cdot$ min^{-1})	v_0 with I (pmol $\cdot$ min^{-1})
0.50	0.00272	0.0015
1.00	0.00457	0.0026
2.00	0.00672	0.0044
4.00	0.00769	0.0067

60. The phosphatase enzyme PTP1B catalyzes the removal of phosphate groups from specific proteins and is involved in insulin signaling. Phosphatase inhibitors such as vanadate may be useful in the treatment of diabetes. The activity of PTP1B was measured in the presence and absence of vanadate using an artificial substrate, fluorescein diphosphate (FDP), which produces a light-absorbing product. The reaction is shown below; data are shown in the table.

(a) Construct a Lineweaver–Burk plot using the data provided. Calculate K_M and V_{max} for PTP1B in the absence and in the presence of vanadate.

(b) What kind of inhibitor is vanadate? Explain.

fluorescein diphosphate $\xrightarrow{\text{PTP1B}}$ fluorescein monophosphate + HPO_4^{2-}

[FDP] (µM)	v_0 without vanadate (nM · s⁻¹)	v_0 with vanadate (nM · s⁻¹)
6.67	5.7	0.71
10	8.3	1.06
20	12.5	2.04
40	16.7	3.70
100	22.2	8.00
200	25.4	12.5

61. An alternative way to calculate K_I for the vanadate inhibitor in the PTP1B reaction (see Problem 60) is to measure the velocity of the enzyme-catalyzed reaction in the presence of increasing amounts of inhibitor and a constant amount of substrate. These data are shown in the table for a substrate concentration of 6.67 µM. To calculate K_I, rearrange Equation 7.28 and solve for α. Then construct a graph plotting α versus [I]. Since $\alpha = 1 + [I]/K_I$, the slope of the line is equal to $1/K_I$. Determine K_I for vanadate using this method.

[Vanadate] (µM)	v_0 (nM · s⁻¹)
0.0	5.70
0.2	3.83
0.4	3.07
0.7	2.35
1.0	2.04
2.0	1.18
4.0	0.71

62. Homoarginine has been shown to inhibit the activity of an alkaline phosphatase found in bone. Alkaline phosphatase enzymes are widely distributed in tissues, and serum levels of these enzymes are often used as a diagnostic tool for various diseases. The artificial substrate phenyl phosphate was incubated with bone alkaline phosphatase in the presence of 4 mM homoarginine. The data are shown in the table.

[Phenyl phosphate] (mM)	v_0, no inhibitor (nM · min⁻¹)	v_0, with inhibitor (nM · min⁻¹)
4.00	1.176	0.476
2.00	0.909	0.436
1.00	0.667	0.385
0.67	0.556	0.333
0.50	0.455	0.286
0.33	0.345	0.244

(a) Construct a Lineweaver–Burk plot using the data provided. Calculate K_M and V_{max} for the alkaline phosphatase enzyme in the absence and in the presence of homoarginine.

(b) What kind of inhibitor is homoarginine? Explain.

(c) Homoarginine does not inhibit intestinal alkaline phosphatase. Why might homoarginine inhibit bone alkaline phosphatase but not the enzyme found in the intestine?

63. Protein phosphatase 1 (PP1) helps regulate cell division and is a possible drug target to treat certain types of cancer. The enzyme catalyzes the hydrolysis of a phosphate group from myelin basic protein (MBP). The activity of PP1 was measured in the presence and absence of the inhibitor phosphatidic acid (PA).

(a) Use the data to construct a Lineweaver–Burk plot for the PP1 enzyme in the presence and absence of PA. What kind of inhibitor is PA?

(b) Report the K_M and V_{max} values for PP1 in the presence and absence of the inhibitor.

[MBP] (mg · mL⁻¹)	v_0 without PA (nmol · mL⁻¹ · min⁻¹)	v_0 with PA (nmol · mL⁻¹ · min⁻¹)
0.010	0.0209	0.00381
0.015	0.0355	0.00620
0.025	0.0419	0.00931
0.050	0.0838	0.01400

64. Aspartate transcarbamoylase (ATCase) catalyzes the formation of N-carbamoylaspartate from carbamoyl phosphate and aspartate, a step in the multienzyme process that synthesizes cytidine triphosphate (CTP). Measurements of ATCase activity as a function of aspartate concentration yield the results shown in the graph.

(a) Is ATCase an allosteric enzyme? How do you know?

(b) What kind of an effector is CTP? Explain. What is the biological significance of CTP's effect on ATCase?

(c) What kind of an effector is ATP? Explain. What is the biological significance of ATP's effect on ATCase?

carbamoyl phosphate + aspartate $\xrightarrow{\text{ATCase}}$
N-carbamoylaspartate $\rightarrow\rightarrow\rightarrow$UMP$\rightarrow\rightarrow$UTP$\rightarrow$CTP

65. The activity of ATCase (see Problem 64) was measured in the presence of several nucleotide combinations. The results are shown in the table. A value greater than 1 indicates that the enzyme was activated, a value less than 1 indicates inhibition.

(a) What combination gives the most effective inhibition?

(b) What is the physiological significance of this combination?

(c) Redraw the graph shown in Problem 64 to include a line representing ATCase activity in the presence of the most inhibitory nucleotide combination. How do the K_M and V_{max} values of this combination compare with the values for ATCase with ATP and CTP alone?

Nucleotide effectors	Activity in the presence of 5 mM aspartate
ATP	1.35
CTP	0.43
UTP	0.95
ATP + CTP	0.85
ATP + UTP	1.52
CTP + UTP	0.06

66. The redox state of the cell (the likelihood that certain groups will be oxidized or reduced) is believed to regulate the activities of some enzymes. Explain how the reversible formation of an intramolecular disulfide bond (—S—S—) from two Cys —SH groups could affect the activity of an enzyme.

67. Pyruvate kinase catalyzes the last reaction in the glycolytic pathway:

$$\text{Phosphoenolpyruvate (PEP) + ADP} \longrightarrow \text{pyruvate + ATP}$$

There are four different mammalian forms of pyruvate kinase. All catalyze the same reaction but differ in their response to the glycolytic metabolite fructose-1,6-bisphosphate (F16BP). The activity of the M_1 form of pyruvate kinase was measured at various concentrations of PEP, both in the presence (blue) and absence (red) of F16BP. The results are shown in the graph.

(a) Is the M_1 form of pyruvate kinase an allosteric enzyme? Does F16BP affect enzyme activity, and if so, how?

(b) The investigators carried out site-directed mutagenesis to replace Ala 398 with Arg. The mutated residue is located at one of the intersubunit contact positions. The activity of the mutant enzyme in the presence and absence of F16BP was measured as in part (a). The results are shown in the graph. How did the mutation affect the enzyme?

7.4 Clinical Connection: Drug Development

68. Medicinal chemists apply certain rules to assess the suitability of drug candidates. For example, they prefer to work with compounds that have a molecular mass less than 500 g·mol^{-1} and contain fewer than 5 hydrogen bond donor groups and fewer than 10 hydrogen bond acceptor groups. Explain why these properties would be desirable in a drug.

69. The structure of rosiglitazone is shown. Does it meet the "rules" described in Problem 68?

70. Explain why some drugs must be administered intravenously rather than swallowed in the form of a pill.

71. The toxicity of the acetaminophen derivative acetamidoquinone results from its ability to react with the Cys groups of proteins. Explain why the toxic effects of acetamidoquinone are localized to the liver.

72. The drug warfarin is widely prescribed as a "blood thinner" because it inhibits the post-translational modification of some of the proteins involved in blood coagulation (see Section 6.5). In order to choose an effective dose of warfarin, physicians may test patients to identify which cytochrome P450 enzymes they express. Explain the purpose of this genetic testing.

73. Enalapril (Vasotec®) inhibits an enzyme that converts the peptide angiotensin I to angiotensin II, a potent vasoconstrictor.

(a) What type of disorders might be treated by enalapril?

(b) Enalapril is inactive until acted upon by an esterase. Draw the structure of the resulting bioactive derivative.

Enalapril

SELECTED READINGS

Corey, E. J., Czakó, B., and Kürti, L., *Molecules and Medicine,* Wiley (2007). [Describes the discovery, structure, and action of over a hundred drugs affecting a variety of medical conditions.]

Cornish-Bowden, A., *Fundamentals of Enzyme Kinetics* (4th ed.), Wiley-Blackwell (2012). [Provides detailed coverage of enzyme kinetics.]

Lv, Z., Chu, Y., and Wang, Y., HIV protease inhibitors: a review of molecular selectivity and toxicity, *HIV AIDS (Auckl.)* **7,** 95–104

(2015). [A short review of the structures and action of HIV protease inhibitors.]

Schramm, V. L., Enzymatic transition state theory and transition state analogue design, *J. Biol. Chem.* **282,** 28297–28300 (2007). [Describes how transition state analog inhibitors work, with examples.]

Segel, I. H., *Enzyme Kinetics,* Wiley (1993). [Focuses on the equations that describe enzyme activity and various forms of inhibition.]

LIPIDS AND BIOLOGICAL MEMBRANES

The Toxoplasma gondii *parasite, which infects animals as well as humans, produces an unusual membrane lipid, phosphatidylthreonine. The synthesis of this nonstandard lipid is a potential target for antiparasitic drugs, and genetically modified* T. gondii *cells that cannot synthesize this lipid are much less infectious and could be used in a vaccine.*

DO YOU REMEMBER?

- Cells contain four major types of biological molecules (Section 1.2).
- Noncovalent forces, including hydrogen bonds, ionic interactions, and van der Waals forces, act on biological molecules (Section 2.1).
- The hydrophobic effect, which is driven by entropy, excludes nonpolar substances from water (Section 2.2).
- Amphipathic molecules form micelles or bilayers (Section 2.2).
- A folded polypeptide assumes a shape with a hydrophilic surface and a hydrophobic core (Section 4.3).

All cells—and the various compartments inside eukaryotic cells—are surrounded by membranes. In fact, the formation of a membrane is believed to be a defining event in the evolutionary history of the cell (Section 1.4); without membranes, a cell would be unable to retain essential resources. To begin to understand how membranes work, we will examine them as composite structures containing both lipids and proteins.

The lipids that occur in biological membranes aggregate to form sheets that are impermeable to ions and other solutes. The key to this behavior is the hydrophobicity of the lipid molecules. Hydrophobicity is also a useful feature of lipids that perform other roles, such as energy storage. Although lipids exhibit enormous variety in shape and size and carry out all sorts of biological tasks, they are united by their hydrophobicity.

8.1 LIPIDS

The molecules that fit the label of *lipid* do not follow a single structural template or share a common set of functional groups, as nucleotides and amino acids do. In fact, *lipids are defined primarily by the absence of functional groups.* Because they consist mainly of C and H atoms and have few if any N- or O-containing functional groups, they lack the ability to form hydrogen bonds and are therefore largely insoluble in water (most lipids are soluble in nonpolar organic solvents). Although some lipids do contain polar or charged groups, the bulk of their structure is hydrocarbon-like.

KEY CONCEPTS

- Lipids are predominantly hydrophobic molecules that can be esterified but cannot form polymers.
- Glycerophospholipids and sphingolipids are amphipathic molecules.
- Cholesterol and other lipids that do not form bilayers have a variety of other functions.

FATTY ACIDS CONTAIN LONG HYDROCARBON CHAINS

The simplest lipids are the **fatty acids,** which are long-chain carboxylic acids (at physiological pH, they are ionized to the carboxylate form). These molecules may contain up to 24 carbon atoms, but the most common fatty acids in plants and animals are the even-numbered C_{16} and C_{18} species such as palmitate and stearate:

Palmitate	Stearate	Oleate	Linoleate

Such molecules are called **saturated fatty acids** because all their tail carbons are "saturated" with hydrogen. **Unsaturated fatty acids** (which contain one or more double bonds) such as oleate and linoleate are also common in biological systems. In these molecules, the double bond usually has the *cis* configuration (in which the two hydrogens are on the same side). Some common saturated and unsaturated fatty acids are listed in **Table 8.1.** Although human cells can synthesize a variety of unsaturated fatty acids, they are unable to make any with double bonds past carbon 9 (counting from the carboxylate end). Some organisms can do so, however, producing what are known as omega-3 fatty acids (**Box 8.A**).

A CLOSER LOOK

Box 8.A Omega-3 Fatty Acids

An **omega-3 fatty acid** has a double bond starting three carbons from its methyl end (the last carbon in the fatty acid chain is called the omega carbon). Marine algae are notable producers of the long-chain omega-3 fatty acids EPA and DHA (see Table 8.1); these lipids tend to accumulate in the fatty tissues of cold-water fish. Consequently, fish oil has been identified as a convenient source of omega-3 fatty acids, which have purported health benefits. Somewhat shorter omega-3 fatty acids such as α-linoleic acid are manufactured by plants. Humans whose diets don't include fish oil acquire α-linoleic acid from plant sources and convert it to the longer varieties of omega-3 fatty acids by lengthening the fatty acid chain from the carboxyl end (Section 17.2).

Omega-3 fatty acids were identified as essential for normal human growth in the 1930s, but it was not until the 1970s that consumption of omega-3 fatty acids such as EPA was linked to decreased risk of cardiovascular disease. The correlation came to light from the observation that native Arctic populations, who ate fish and meat but few vegetables, had a surprisingly low incidence of heart disease. One possible biochemical explanation is that the omega-3 fatty acids compete with omega-6 fatty acids for the enzymes that convert the fatty acids to certain signaling molecules.

The omega-6 derivatives are stronger triggers of inflammation, which underlies conditions such as atherosclerosis. The relative amounts of omega-3 and omega-6 fatty acids might therefore matter more than the absolute amount of omega-3 fatty acids consumed.

The 22-carbon DHA is abundant in the brain and retina, and its concentration decreases with age. DHA is converted *in vivo* to substances that are believed to protect neural tissues from damage following a stroke. However, DHA supplements do not appear to reverse the cognitive decline associated with disorders such as Alzheimer's disease (see Section 4.5).

Numerous studies have attempted to demonstrate the ability of omega-3 fatty acids to prevent or treat other conditions such as arthritis and cancer, but the findings have been mixed, with the omega-3 fatty acid supplements in some cases exacerbating the disease. For this reason, the true role of omega-3 fatty acids in human health demands additional investigation.

? **Explain why vegans exhibit DHA levels only slightly lower than the levels in individuals who consume large amounts of fish. What does this information reveal about the usefulness of consuming DHA supplements?**

TABLE 8.1 Some Common Fatty Acids

NUMBER OF CARBON ATOMS	COMMON NAME	SYSTEMATIC NAME[a]	STRUCTURE
Saturated fatty acids			
12	Lauric acid	Dodecanoic acid	$CH_3(CH_2)_{10}COOH$
14	Myristic acid	Tetradecanoic acid	$CH_3(CH_2)_{12}COOH$
16	Palmitic acid	Hexadecanoic acid	$CH_3(CH_2)_{14}COOH$
18	Stearic acid	Octadecanoic acid	$CH_3(CH_2)_{16}COOH$
20	Arachidic acid	Eicosanoic acid	$CH_3(CH_2)_{18}COOH$
22	Behenic acid	Docosanoic acid	$CH_3(CH_2)_{20}COOH$
24	Lignoceric acid	Tetracosanoic acid	$CH_3(CH_2)_{22}COOH$
Unsaturated fatty acids			
16	Palmitoleic acid	9-Hexadecenoic acid	$CH_3(CH_2)_5CH{=}CH(CH_2)_7COOH$
18	Oleic acid	9-Octadecenoic acid	$CH_3(CH_2)_7CH{=}CH(CH_2)_7COOH$
18	Linoleic acid	9,12-Octadecadienoic acid	$CH_3(CH_2)_4(CH{=}CHCH_2)_2(CH_2)_6COOH$
18	α-Linolenic acid	9,12,15-Octadecatrienoic acid	$CH_3CH_2(CH{=}CHCH_2)_3(CH_2)_6COOH$
18	γ-Linolenic acid	6,9,12-Octadecatrienoic acid	$CH_3(CH_2)_4(CH{=}CHCH_2)_3(CH_2)_3COOH$
20	Arachidonic acid	5,8,11,14-Eicosatetraenoic acid	$CH_3(CH_2)_4(CH{=}CHCH_2)_4(CH_2)_2COOH$
20	EPA	5,8,11,14,17-Eicosapentaenoic acid	$CH_3CH_2(CH{=}CHCH_2)_5(CH_2)_2COOH$
22	DHA	4,7,10,13,16,19-Docosahexaenoic acid	$CH_3CH_2(CH{=}CHCH_2)_6CH_2COOH$

[a] Numbers indicate the starting position of the double bond; the carboxylate carbon is in position 1.

? **Which of the unsaturated fatty acids listed here are omega-3 fatty acids?**

Free fatty acids are relatively scarce in biological systems. Instead, they are usually esterified, for example, to glycerol. The fats and oils found in animals and plants are **triacylglycerols** (sometimes called **triglycerides**) in which the **acyl groups** (the R—CO— groups) of three fatty acids are esterified to the three hydroxyl groups of glycerol. The ester bond linking each acyl group is the result of a condensation reaction. This is as close as lipids come to forming polymers: They cannot be linked end-to-end to form long chains, as the other types of biological molecules can.

The three fatty acids of a given triacylglycerol may be the same or different. For reasons that are outlined below, triacylglycerols do not form bilayers and so are not important components of biological membranes. However, they do aggregate in large globules, acting as a storage depot for fatty acids that serve as metabolic fuels (these reactions are described in Section 17.1). The hydrophobic nature of triacylglycerols and their tendency to aggregate means that cells can store a large amount of this material without it interfering with other activities that take place in an aqueous environment.

SOME LIPIDS CONTAIN POLAR HEAD GROUPS

Among the major lipids of biological membranes in bacteria and eukaryotes are the **glycerophospholipids,** which contain a glycerol backbone with fatty acyl groups esterified at positions 1 and 2 and a phosphate derivative, called a head group, esterified at position 3. As in triacylglycerols, the fatty acyl components of glycerophospholipids vary from molecule to molecule. These lipids are usually named according to their head group, for example,

Phosphatidylcholine

$$
\begin{array}{c}
O^- \\
| \\
O=P-O-CH_2-CH_2-\overset{+}{N}(CH_3)_3 \\
| \\
O \\
| \\
CH_2-CH-CH_2 \\
|| \\
OO \\
| | \\
C=O\ \ C=O \\
| | \\
R R
\end{array}
$$

Phosphatidylethanolamine

$$
\begin{array}{c}
O^- \\
| \\
O=P-O-CH_2-CH_2-\overset{+}{N}H_3 \\
| \\
O \\
| \\
CH_2-CH-CH_2 \\
OO \\
C=O\ \ C=O \\
R R
\end{array}
$$

Phosphatidylglycerol

$$
\begin{array}{c}
O^- \\
| \\
O=P-O-CH_2-CH-CH_2OH \\
| | \\
O OH \\
| \\
CH_2-CH-CH_2 \\
OO \\
C=O\ \ C=O \\
R R
\end{array}
$$

Phosphatidylserine

$$
\begin{array}{c}
O^- \\
| \\
O=P-O-CH_2-CH-COO^- \\
| | \\
O \overset{+}{N}H_3 \\
| \\
CH_2-CH-CH_2 \\
OO \\
C=O\ \ C=O \\
R R
\end{array}
$$

Note that the glycerophospholipids are not completely hydrophobic: *They are **amphipathic**, with hydrophobic tails attached to polar or charged head groups.* As we will see, their structure is ideal for forming bilayers.

The bonds that link the various components of a glycerophospholipid can be hydrolyzed by **phospholipases** to release the acyl chains or portions of the head group (**Fig. 8.1**). These enzymatic reactions are not just for degrading lipids. Some products derived from membrane lipids act as signaling molecules inside cells or between neighboring cells.

Eukaryotic membranes also contain amphipathic lipids known as **sphingolipids.** The **sphingomyelins,** with phosphocholine or phosphoethanolamine head groups, are sterically similar to their glycerophospholipid counterparts. The major difference is that sphingomyelins are not built on a glycerol backbone. Instead, their basic component is sphingosine, a derivative of serine and the fatty acid palmitate. In a sphingolipid, a second fatty acyl group is attached via an amide bond to the serine nitrogen (**Fig. 8.2**). Some sphingolipids include head groups consisting of one or more carbohydrate groups. These **glycolipids** are known as cerebrosides and gangliosides.

Archaebacterial membranes contain a different sort of amphipathic lipid that has the same overall shape and function as a bacterial or eukaryotic glycerophospholipid. But in the archaeal version, the hydrocarbon tails are linked to the glycerol backbone by ether rather than ester linkages, and the glycerol group has the opposite chirality (handedness, see Box 4.A).

phospholipase D

phospholipase C →

phospholipase A$_1$ →

← phospholipase A$_2$

$$
\begin{array}{c}
O^- \\
| \\
O=P-O-X \\
| \\
O \\
| \\
H_2C-CH-CH_2 \\
OO \\
O=C\ \ C=O \\
R R
\end{array}
$$

FIGURE 8.1

Sites of action of phospholipases.

❓ Which phospholipase-catalyzed reactions release charged portions of the phospholipid?

FIGURE 8.2

Sphingolipids. (a) The sphingosine backbone is derived from serine and palmitate. (b) The attachment of a second acyl group and a phosphocholine (or phosphoethanolamine) head group yields a sphingomyelin. (c) A cerebroside has a monosaccharide as a head group rather than a phosphate derivative. (d) A ganglioside includes an oligosaccharide head group.

? Compare the structure of a sphingomyelin to the structure of the glycerophospholipids.

In addition, the tails are typically 20-carbon branched structures rather than straight chains (**Fig. 8.3**). The polar and charged head groups of the archaeal membrane lipids are similar to those in bacteria and eukaryotes.

LIPIDS PERFORM A VARIETY OF PHYSIOLOGICAL FUNCTIONS

In addition to glycerophospholipids and sphingolipids, many other types of lipids occur in membranes and elsewhere in the cell. One of these is cholesterol, a 27-carbon, four-ring molecule:

Cholesterol

$$H_3\overset{+}{N}-CH_2-CH_2-O-\overset{\overset{\displaystyle O^-}{|}}{\underset{\underset{\displaystyle O}{|}}{P}}=O$$

$$H_2C-CH-CH_2$$

FIGURE 8.3

An archaeal membrane lipid with an ethanolamine head group.

? **Compare this structure to that of the phosphatidylethanolamine lipid shown earlier.**

Cholesterol is an important component of membranes and is also a metabolic precursor of steroid hormones such as estrogen and testosterone.

Cholesterol is one of many types of terpenoids, or **isoprenoids,** lipids that are constructed from 5-carbon units with the same carbon skeleton as isoprene. For example, the isoprenoid ubiquinone is a compound that is reversibly reduced and oxidized in the mitochondrial membrane (it is described further in Section 12.2):

$$H_2C=CH-\overset{\overset{\displaystyle CH_3}{|}}{C}=CH_2$$

Isoprene

Isoprenoid units

Ubiquinone

The tails of archaeal membrane lipids (Fig. 8.3) are also isoprenoids, as are the molecules known as vitamins A, D, E, and K, which perform a variety of physiological roles not related to membrane structure (**Box 8.B**).

What are some other functions of lipids that are not used to construct bilayers? Due to their hydrophobicity, some lipids function as waterproofing agents. For example, waxes produced by plants protect the surfaces of leaves and fruits against water loss. Beeswax contains an ester of palmitate and a 30-carbon alcohol that makes this substance extremely water-insoluble.

$$H_3C-(CH_2)_{14}-\overset{\overset{\displaystyle O}{||}}{C}-O-(CH_2)_{29}-CH_3$$

In humans, derivatives of the C_{20} fatty acid arachidonate are signaling molecules that help regulate blood pressure, pain, and inflammation (Section 10.4). Many plant lipids function as attractants for pollinators or repellants for herbivores. For example, geraniol is produced by

Box 8.B The Lipid Vitamins A, D, E, and K

The plant kingdom is rich in isoprenoid compounds, which serve as pigments, molecular signals (hormones and pheromones), and defensive agents. During the course of evolution, vertebrate metabolism has co-opted several of these compounds for other purposes. The compounds have become **vitamins,** which are substances that an animal cannot synthesize but must obtain from its food. Vitamins A, D, E, and K are lipids, but many other vitamins are water-soluble. Other than being lipids, vitamins A, D, E, and K have little in common.

The first vitamin to be discovered was vitamin A, or retinol.

Retinol
(vitamin A)

It is derived mainly from plant pigments such as β-carotene (an orange pigment that is present in green vegetables as well as carrots and tomatoes). Retinol is oxidized to retinal, an aldehyde, which functions as a light receptor in the eye. Light causes the retinal to isomerize, triggering an impulse through the optic nerve. A severe deficiency of vitamin A can lead to blindness. The retinol derivative retinoic acid behaves like a hormone by stimulating tissue repair. It is sometimes used to treat severe acne and skin ulcers.

The steroid derivative vitamin D is actually two similar compounds—one (vitamin D_2) derived from plants and the other (vitamin D_3) from endogenously produced cholesterol.

Vitamin D_2

Vitamin D_3

Ultraviolet light is required for the formation of vitamins D_2 and D_3, giving rise to the saying that sunlight makes vitamin D. Two hydroxylation reactions carried out by enzymes in the liver and

kidney convert vitamin D to its active form, which stimulates calcium absorption in the intestine. The resulting high concentration of Ca^{2+} in the bloodstream promotes Ca^{2+} deposition in the bones and teeth. Rickets, a vitamin D–deficiency disease characterized by stunted growth and deformed bones, is easily prevented by good nutrition and exposure to sunlight.

α-Tocopherol, or vitamin E,

α-Tocopherol
(vitamin E)

is a highly hydrophobic molecule that is incorporated into cell membranes. The traditional view is that it reacts with free radicals generated during oxidative reactions. Vitamin E activity would thereby help prevent the peroxidation of polyunsaturated fatty acids in membrane lipids. However, compounds that are closely related to α-tocopherol in structure do not exhibit this free radical–scavenging activity, and it has been proposed that the observed antioxidant effect of vitamin E instead stems from its activity as a regulatory molecule that may suppress free radical formation by inhibiting the production or activation of oxidative enzymes. In this respect, vitamin E resembles other lipids that function as signaling molecules.

Vitamin K is named for the Danish word *koagulation*. It participates in the enzymatic carboxylation of glutamate residues in some of the proteins involved in blood coagulation (see Section 6.5). A vitamin K deficiency prevents Glu carboxylation, which inhibits the normal function of the proteins, leading to excessive bleeding. Vitamin K can be obtained from green plants, as phylloquinone,

Phylloquinone
(vitamin K)

However, about half the daily uptake of the vitamin is supplied by intestinal bacteria.

Because vitamins A, D, E, and K are water-insoluble, they can accumulate in fatty tissues over time. Excessive vitamin D accumulation can lead to kidney stones and abnormal calcification of soft tissues. High levels of vitamin K have few adverse effects, but extremely high levels of vitamin A can produce a host of non-specific symptoms as well as birth defects. In general, vitamin toxicities are rare and usually result from overconsumption of commercial vitamin supplements rather than from natural causes.

many flowering plants (it has a roselike smell). Limonene gives citrus fruits their characteristic odor.

Geraniol

Limonene

Capsaicin, the compound that gives chili peppers their "hot" taste, is an irritant to the digestive tracts of many animals, but it is consumed by humans worldwide.

Capsaicin

Its hydrophobicity explains why it cannot be washed away with water. Capsaicin has been used therapeutically as a pain reliever. It appears to activate receptors on neurons that sense both pain and heat; by overwhelming the receptors with a "hot" signal, capsaicin prevents the neurons from receiving pain signals.

 CONCEPT CHECK

1. What are the structural features that distinguish fatty acids, triacylglycerols, glycerophospholipids, sphingomyelins, and cholesterol?

2. Explain why lipids cannot form true polymers.

3. Describe how the hydrophobic character of a lipid determines its location or function.

KEY CONCEPTS
- Bilayer fluidity depends on the length and saturation of its lipids and on the presence of cholesterol.
- Lipid asymmetry is maintained by the slow rate of diffusion between leaflets.

8.2 THE LIPID BILAYER

The fundamental component of a biological membrane is the **lipid bilayer,** *a two-dimensional array of amphipathic molecules whose tails associate with each other, out of contact with water, and whose head groups interact with the aqueous solvent* (**Fig. 8.4**). The beauty of the bilayer as a barrier for biological systems is that it forms spontaneously (without the input of free energy) due to the influence of the hydrophobic effect, which favors the aggregation of nonpolar groups that are energetically costly to individually hydrate. In other words, the lipids do not aggregate because they are attracted to each other; instead they are forced together because this maximizes the entropy of the surrounding water molecules (Section 2.2). In addition, a bilayer is self-sealing and, despite its thinness, it can enclose a relatively vast compartment or an entire cell. Once it has formed, a bilayer is quite stable.

Glycerophospholipids and sphingolipids—with two tails and a large head group—have the appropriate geometry to form bilayers. Although they are amphipathic, fatty acids form spherical particles (micelles, Fig. 2.9) rather than bilayers. Triacylglycerols are almost completely nonpolar and therefore cannot form bilayers. Likewise, pure cholesterol, with only a single polar hydroxyl group, cannot form a bilayer on its own. Nevertheless, molecules of these non-bilayer-forming lipids can be found buried in the hydrophobic region of a membrane, among the acyl chains of other lipids.

FIGURE 8.4

A lipid bilayer.

FIGURE 8.5

Simulation of a lipid bilayer.
In this model of a dipalmitoyl phosphatidylcholine bilayer, C atoms are gray (except for the terminal carbon of each lipid tail, which is yellow), ester O atoms are red, phosphate groups are green, and choline head groups are magenta. Water molecules on each side of the bilayer are shown as blue spheres. [*Courtesy Richard Pastor and Richard Venable, National Institutes of Health.*]

? Indicate where the molecules shown in Box 8.B would be located.

THE BILAYER IS A FLUID STRUCTURE

Naturally occurring bilayers are mixtures of many different lipids. This is one reason why *a lipid bilayer has no clearly defined geometry.* Most lipid bilayers have a total thickness of between 30 and 40 Å, with a hydrophobic core about 25 to 30 Å thick. The exact thickness varies according to the lengths of the acyl chains and how they bend and interdigitate. In addition, the head groups of membrane lipids are not of uniform size, and their distance from the membrane center depends on how they nestle in with neighboring head groups. Finally, the lipid bilayer is impossible to describe in precise terms because it is not a static structure. Rather, it is a dynamic assembly: The head groups bob up and down, and the hydrocarbon tails of the lipids are in constant rapid motion. At its very center, the bilayer is as fluid as a sample of pure hydrocarbon (**Fig. 8.5**). The bilayer is flexible enough to accommodate changes in cell shape, but its hydrocarbon interior remains intact, as an oily layer between two aqueous compartments.

It is useful to describe the fluidity of a given membrane lipid in terms of its **melting point,** the temperature of transition from an ordered crystalline state to a more fluid state. Note that the melting point does not refer to the conformational shift of just one individual molecule but instead describes the collective behavior of a population of molecules. In the crystalline phase, all the acyl chains in the sample pack together tightly in van der Waals contact. In the fluid phase, the methylene (—CH$_2$—) groups of the acyl chains in the sample can rotate freely. The melting point of a particular acyl chain depends on its length and degree of saturation. For a saturated acyl chain, the melting point increases with increasing chain length. This is because more free energy (a higher temperature) is required to disrupt the more extensive van der Waals interactions between longer chains. A shorter acyl chain melts at a lower temperature because it has less surface area to make van der Waals contacts. A double bond introduces a kink into the acyl chain, so an unsaturated acyl chain is less able to pack efficiently against its neighbors:

Consequently, the melting point of an acyl chain decreases as the degree of unsaturation increases.

Of course, biological membranes contain more than one type of acyl chain and do not typically experience sudden changes in temperature. But in a mixed bilayer at constant temperature, *longer acyl chains tend to be less mobile (more crystalline) than shorter acyl chains, and saturated acyl chains are less mobile than unsaturated chains.* Because a fluid membrane is essential for many metabolic processes, organisms endeavor to maintain constant membrane fluidity by adjusting the lipid composition of the bilayer. For example, during adaptation to lower temperatures, an organism may increase its production of lipids with shorter and less saturated acyl chains.

The membranes of most organisms remain fluid over a range of temperatures. This is partly because biological membranes include a variety of different lipids (with different melting points) and do not undergo a sharp transition between liquid and crystalline phases, as a sample of pure lipid would. In eukaryotic membranes, the planar cholesterol slips in between the acyl tails of membrane lipids and helps maintain constant membrane fluidity over a range of temperatures through two opposing mechanisms:

1. In a bilayer of mixed lipids, cholesterol's rigid ring system restricts the movements of nearby acyl chains, which prevents the bilayer from becoming too fluid.

2. By inserting between membrane lipids, cholesterol prevents their close packing (that is, their crystallization), which prevents the bilayer from becoming too solid.

The isoprenoid tails of the lipids in archaeal membranes, which lack cholesterol, sometimes include rings of five or six carbons, which appear to modulate bilayer fluidity in the same manner as cholesterol.

Different areas of a naturally occurring membrane may be characterized by different degrees of fluidity. For example, regions known as membrane **rafts** are thought to contain tightly packed cholesterol and sphingolipids and have a near-crystalline consistency. Certain proteins appear to associate with rafts, so these structures may have functional importance for processes such as transport and signaling. However, the physical characteristics of lipid rafts have been difficult to pin down, and it is possible that such structures may have only a fleeting existence.

NATURAL BILAYERS ARE ASYMMETRIC

The two leaflets of the bilayer in a biological membrane seldom have identical compositions. For example, sphingolipids with carbohydrate head groups occur almost exclusively on the outer leaflet of the animal plasma membrane, facing the extracellular space. The polar head groups of phosphatidylcholine also usually face the cell exterior, whereas phosphatidylserine is usually found in the inner leaflet.

Lipid asymmetry in eukaryotic cells is mostly a consequence of the orientation of lipid-synthesizing enzymes in the endoplasmic reticulum, but the distinct compositions of the inner and outer leaflets are preserved by the extremely slow rate at which most membrane lipids undergo **transverse diffusion,** or **flip-flop** (the movement from one leaflet to the other):

Transverse diffusion (flip-flop)

very slow

This movement is thermodynamically unfavorable since it would require the passage of a solvated polar head group through the hydrophobic interior of the bilayer. However, cells can and do move certain lipids between leaflets with the assistance of enzymes called **translocases** or **flippases.** Lipid molecules undergo rapid **lateral diffusion,** that is, movement within one leaflet:

Lateral diffusion

rapid

In a membrane bilayer, a lipid changes places with its neighbors as often as 10^7 times per second. Thus, the image of the bilayer in Figure 8.5 represents a bilayer frozen for an instant in time.

CONCEPT CHECK

1. Explain why glycerophospholipids and sphingolipids form bilayers while fatty acids, triacylglycerols, and cholesterol do not.

2. Draw a diagram of a lipid bilayer and add arrows to indicate all the ways that membrane lipids can move.

3. How do chain length and saturation influence the melting point of a fatty acid?

4. Describe the effects of cholesterol on membrane fluidity.

5. What are the differences between transverse and lateral diffusion?

MEMBRANE PROTEINS

Biological membranes consist of proteins as well as lipids. On average, a membrane is about 50% protein by weight, but this value varies widely, depending on the source of the membrane. Some bacterial plasma membranes and organelle membranes are as much as three-quarters protein. By itself, a lipid bilayer serves mainly as a barrier to the diffusion of polar substances, and virtually all the additional functions of a biological membrane depend on membrane proteins. For example, some membrane proteins sense exterior conditions and communicate them to the cell interior. Other membrane proteins carry out specific metabolic reactions or function as transporters to move substances from one side of the membrane to the other. Membrane proteins fall into different groups, depending on how they are specialized for interaction with the hydrophobic interior of the lipid bilayer (**Fig. 8.6**).

KEY CONCEPTS

- Integral membrane proteins completely span the bilayer by forming one or more α helices or a β barrel.
- Covalently attached lipids anchor some proteins in the bilayer.

INTEGRAL MEMBRANE PROTEINS SPAN THE BILAYER

In the membrane proteins known as **integral** or **intrinsic membrane proteins,** a portion of the structure is fully buried in the lipid bilayer. These proteins are customarily contrasted

Integral membrane protein Peripheral membrane protein Lipid-linked protein

FIGURE 8.6

Types of membrane proteins. This schematic cross-section of a membrane shows an integral protein spanning the width of the membrane, a peripheral membrane protein associated with the membrane surface, and a lipid-linked protein whose attached hydrophobic tail is incorporated into the lipid bilayer.

? Which type of membrane protein would be easiest to separate from the lipid bilayer?

(a) Pro–Glu–Trp–Ile–Trp–Leu–Ala–
Leu–Gly–Thr–Ala–Leu–Met–
Gly–Leu–Gly–Thr–Leu–Tyr–
Phe–Leu–Val–Lys–Gly

(b)

FIGURE 8.7

A membrane-spanning α helix.
(a) A portion of the amino acid
sequence from the protein
bacteriorhodopsin. (b) Three-
dimensional structure of the same
sequence. Polar residues are
purple and nonpolar residues are
green.

with **peripheral** or **extrinsic membrane proteins,** which are more loosely associated with the membrane via interactions with lipid head groups or integral membrane proteins (see Fig. 8.6). Except for their weak affiliation with the membrane, peripheral proteins are not notably different from ordinary water-soluble proteins.

All but a few integral membrane proteins completely span the lipid bilayer, so they are exposed to the hydrophobic interior as well as the aqueous environment on each side of the membrane. The solvent-exposed portions of an integral membrane protein are typical of other proteins: a polar surface surrounding a hydrophobic core. However, the portion of the protein that penetrates the lipid bilayer must have a hydrophobic surface, since the energetic cost of burying a polar protein group (and its solvating water molecules) is too great.

AN α HELIX CAN CROSS THE BILAYER

One way for a polypeptide chain to cross a lipid bilayer is by forming an α helix whose side chains are all hydrophobic. The hydrogen-bonding tendencies of the functional groups of the backbone are satisfied through hydrogen bonding in the α helix (see Fig. 4.5). The hydrophobic side chains project outward from the helix to mingle with the acyl chains of the lipids.

To span a 30-Å hydrophobic bilayer core, an α helix must contain at least 20 amino acids. A transmembrane helix is often easy to spot by its sequence: It is rich in highly hydrophobic amino acids such as isoleucine, leucine, valine, and phenylalanine. Polar aromatic groups (tryptophan and tyrosine) and asparagine and glutamine often occur where the helix approaches the more polar lipid head groups. Charged residues such as aspartate, glutamate, lysine, and arginine often mark the point where the polypeptide leaves the membrane and is exposed to the solvent (**Fig. 8.7**).

Many integral membrane proteins contain several membrane-spanning α helices bundled together (**Fig. 8.8**). These α helices interact much like the left-handed coiled coils in keratin (Section 5.3). Some of the helix–helix interactions involve the electrostatic pairing of polar residues, but the surface of the helix bundle—where it contacts the lipid tails—is predominantly hydrophobic.

A TRANSMEMBRANE β SHEET FORMS A BARREL

A polypeptide that crossed the membrane as a β strand would leave its hydrogen-bonding backbone groups unsatisfied. However, if several β strands together form a fully

EXTRACELLULAR
SPACE

FIGURE 8.8

Bacteriorhodopsin. This integral membrane protein consists of a bundle of seven membrane-spanning α helices connected by loops that project into the solution on each side of the membrane. The helices are colored in rainbow order from blue (N-terminus) to red (C-terminus). The horizontal lines approximate the outer surfaces of the membrane. [*Structure (pdb 1QHJ) determined by H. Belrhali, P. Nollert, A. Royant, C. Menzel, J. P. Rosenbusch, E. M. Landau, and E. Pebay-Peyroula.*]

hydrogen-bonded β sheet, they can cross the membrane in an energetically favorable way. In order to maximize hydrogen bonding, the β sheet must close up on itself to form a **β barrel.**

The smallest possible β barrel contains eight strands. The exterior surface of the barrel includes a band, about 22 Å wide, of hydrophobic side chains. This band is flanked on each side by aromatic side chains, which are more polar and form an interface with the lipid head groups (**Fig. 8.9**). Larger β barrels, containing up to 22 strands, sometimes include a central water-filled passageway that allows small molecules to diffuse from one side of the membrane to the other (Section 9.2).

Because the side chains in a β sheet point alternately to each face, some side chains in a β barrel point into the barrel interior, and others face the lipid bilayer. The absence of a discrete stretch of hydrophobic residues, as in a membrane-spanning α helix, makes it difficult to detect membrane-spanning β strands by examining a protein's sequence.

LIPID-LINKED PROTEINS ARE ANCHORED IN THE MEMBRANE

A second group of membrane proteins consists of **lipid-linked proteins.** Many of these are otherwise soluble proteins that are anchored in the lipid bilayer by a covalently attached lipid group. A few lipid-linked proteins also contain membrane-spanning polypeptide segments. In some lipid-linked proteins, a fatty acyl group such as a myristoyl residue (from the 14-carbon saturated fatty acid myristate) is attached via an amide bond to the N-terminal glycine residue of a protein (**Fig. 8.10a**). Other proteins contain a palmitoyl group (from the 16-carbon palmitate) attached to the sulfur of a cysteine side chain via a thioester bond (Fig. 8.10b). Palmitoylation, in contrast to myristoylation, is reversible *in vivo.* Consequently, proteins with a myristoyl group are permanently anchored to the membrane, but proteins with a palmitoyl group may become soluble if the acyl group is removed.

FIGURE 8.9

A membrane-spanning β barrel. The eight strands of this *E. coli* protein, known as OmpX, are fully hydrogen-bonded where they span the width of the bilayer. Hydrophobic side chains (green) on the barrel exterior face the bilayer core. Aromatic residues (gold) are located mostly near the lipid head groups. [*Structure (pdb 1QJ9) determined by J. Vogt and G. E. Schulz.*]

FIGURE 8.10

Attachment of lipids in some lipid-linked proteins. Proteins are shown in blue, and the lipid anchors are in green. Other groups are purple. (a) Myristoylation. (b) Palmitoylation. (c) Prenylation. The lipid anchor is the 15-carbon farnesyl group. (d) Linkage to a glycosylphosphatidylinositol group. The hexagons represent different monosaccharide residues.

Other lipid-linked proteins in eukaryotes are prenylated; that is, they contain a 15- or 20-carbon isoprenoid group linked to a C-terminal cysteine residue via a thioether bond (Fig. 8.10c). The C-terminus is also usually carboxymethylated.

Finally, many eukaryotes, particularly protozoans, contain proteins linked to a lipid–carbohydrate group, known as glycosylphosphatidylinositol, at the C-terminus (Fig. 8.10d). These lipid-linked proteins almost always face the external surface of the cell and are often found in sphingolipid–cholesterol rafts.

CONCEPT CHECK

1. Describe the sequence requirements for a membrane-spanning α helix and β barrel.

2. List the similarities and differences between integral membrane proteins, peripheral membrane proteins, and lipid-linked proteins.

THE FLUID MOSAIC MODEL

KEY CONCEPT

• A membrane's structure can be described as proteins diffusing laterally within a lipid bilayer.

Biological membranes consist of both proteins and lipids, although the mixture may not be entirely random. For example, *certain lipids appear to associate specifically with certain proteins,* possibly to stabilize the protein's structure or modulate its function. A given membrane protein has a characteristic orientation; that is, it faces one side of the membrane or the other. After it has assumed its mature conformation and orientation, it does not undergo flip-flop because this would require the passage of large polar polypeptide regions through the hydrophobic bilayer core. However, lateral movement is still possible. An integral membrane protein or a lipid-linked protein can diffuse within the plane of the bilayer, albeit more slowly than a membrane lipid. This sort of movement is a key feature of the **fluid mosaic model** of membrane structure described in 1972 by S. Jonathan Singer and Garth Nicolson. According to their model, membrane proteins are like icebergs floating in a lipid sea (**Fig. 8.11**).

FIGURE 8.11

The fluid mosaic model of membrane structure. According to this model, integral membrane proteins (blue) float in a sea of lipid and can move laterally but cannot undergo transverse movement (flip-flop). The gold structures are the carbohydrate chains of glycolipids and glycoproteins.

Over the years, the fluid mosaic model has remained generally valid, although it has been refined. For example, *many membrane proteins do not diffuse as freely as first imagined.* Their movements are hindered to some degree according to whether they interact with other membrane proteins or with cytoskeletal elements that lie just beneath the membrane. Thus, a given membrane protein may be virtually immobile (if it is firmly attached to the cytoskeleton), mobile within a small area (if it is confined within a space defined by other membrane and cytoskeletal proteins), or fully free to diffuse (**Fig. 8.12**). The presence of lipid rafts, another feature not described in the original fluid mosaic model, may further define the boundaries for a membrane protein.

FIGURE 8.12

Limitations on the mobility of membrane proteins. A protein (labeled A) that interacts tightly with the underlying cytoskeleton appears to be immobile. Another protein (B) can diffuse within a space defined by cytoskeletal proteins. Some proteins appear to diffuse throughout the membrane with no constraints.

MEMBRANE GLYCOPROTEINS FACE THE CELL EXTERIOR

Like membrane lipids, membrane proteins are distributed asymmetrically between the two leaflets. For example, most lipid-linked proteins face the cell interior (glycosylphosphatidylinositol-linked proteins are an exception). The exterior face of the membrane in vertebrate cells is rich in carbohydrate-bearing glycolipids (such as cerebrosides and gangliosides) and **glycoproteins.** The oligosaccharide chains (polymers of monosaccharide residues) that are covalently attached to membrane lipids and proteins shroud the cell in a fuzzy coat (see Fig. 8.11). When fully solvated, the highly hydrophilic carbohydrates tend to occupy a large volume.

As we will see in Chapter 11, monosaccharide residues can be linked to each other in different ways and in potentially unlimited sequences. This diversity, present in both glycolipids and glycoproteins, is a form of biological information. For example, the well-known ABO blood group system is based on differences in the composition of the carbohydrate components of glycolipids and glycoproteins on red blood cells (discussed in Box 11.B). Many other cells appear to recognize each other through mutual interactions between membrane proteins.

 CONCEPT CHECK

1. Describe the fluid mosaic model of membrane structure.
2. List some factors that limit membrane protein mobility.
3. Why do glycoproteins and glycolipids face the cell exterior?

SUMMARY

8.1   Lipids

• Lipids are largely hydrophobic molecules. Fatty acids can be esterified to form triacylglycerols.

• Glycerophospholipids contain two fatty acyl groups attached to a glycerol backbone that bears a phosphate derivative head group. Sphingomyelins are functionally similar but lack a glycerol backbone. Cholesterol, archaeal membrane lipids, and some other lipids are isoprenoids.

8.2   The Lipid Bilayer

• Lipid bilayers are dynamic structures. Their fluidity depends on the length and degree of saturation of their fatty acyl groups: Shorter and less saturated chains are more fluid. Cholesterol helps maintain membrane fluidity over a range of temperatures.

• Membrane lipids can freely diffuse laterally but undergo transverse diffusion very slowly. Membranes may contain crystalline rafts composed of cholesterol and sphingolipids.

8.3   Membrane Proteins

• An integral membrane protein spans the lipid bilayer as one or a bundle of α helices or as a β barrel. Some membrane proteins are anchored in the bilayer by a covalently linked lipid group.

8.4   The Fluid Mosaic Model

• According to the fluid mosaic model, membrane proteins diffuse within the plane of the bilayer. The mobility of proteins may be limited by their interaction with cytoskeletal proteins. Membrane glycolipids and glycoproteins face the cell exterior.

KEY TERMS

lipid	glycerophospholipid	vitamin	integral (intrinsic) protein
fatty acid	amphipathic	lipid bilayer	peripheral (extrinsic) protein
saturated fatty acid	phospholipase	melting point	β barrel
unsaturated fatty acid	sphingolipid	raft	lipid-linked protein
omega-3 fatty acid	sphingomyelin	transverse diffusion (flip-flop)	fluid mosaic model
triacylglycerol (triglyceride)	glycolipid	translocase (flippase)	glycoprotein
acyl group	isoprenoid	lateral diffusion	

BIOINFORMATICS

Brief Bioinformatics Exercises

8.1 Drawing and Naming Fatty Acids

8.2 Properties of Membrane Proteins

PROBLEMS

8.1 Lipids

1. The structures of fatty acids can be represented in a shorthand form consisting of two numbers separated by a colon. The first number is the number of carbons; the second number is the number of double bonds. For example, palmitate would be represented by the shorthand 16:0. For unsaturated fatty acids, the quantity n-x is used, where n is the total number of carbons and x is the last double-bonded carbon counting from the methyl end. Unless indicated otherwise, it is assumed that the double bonds are *cis* and that one methylene group separates each double bond. For example, oleate would be represented by the shorthand 18:1n-9. Using the shorthand form as a guide, draw the structures of the following fatty acids: **(a)** palmitate, 16:0; **(b)** palmitoleate, (16:1n-7); **(c)** γ-linolenate, 18:3n-3; **(d)** nervonate, 24:1n-9.

2. Fish oils contain the fatty acids EPA and DHA. People who eat at least two fish meals a week have a lower incidence of cardiovascular disease because of the positive physiological effects of these lipids (see Box 8.A). Use the shorthand form described in Problem 1 to draw the structures of **(a)** EPA (eicosapentaenoate, 20:5n-3) and **(b)** DHA (docosahexaenoate 22:6n-3).

3. Some species of plants contain desaturase enzymes capable of producing fatty acids not found in animals. An example of an unusual fatty acid is sciadonate, which is designated as all-*cis*-$\Delta^{5,11,14}$-eicosatrienoate. In this shorthand nomenclature, the superscripts refer to the position of the double bonds beginning at the carboxyl end. This nomenclature is less common than the scheme described in Problem 1 but is sometimes used when the positions of the double bonds do not conform to the pattern described in Problem 1. Using this form of shorthand as a guide, draw the structure of sciadonate.

4. Several hundred unusual fatty acids have been found in plants. Some of these fatty acids are unusually short or long; some have double bonds in unexpected positions (see Problem 3); others have additional functional groups. These fatty acids are important as starting materials in the industrial synthesis of lubricants and polymers. Using the shorthand notation described in Problem 3, draw the structures of the following plant fatty acids:

(a) erucic acid (*cis*-Δ^{13}-docosenoic acid)
(b) calendic acid (*trans, trans, cis*-$\Delta^{8,10,12}$-octadecatrienoic acid)
(c) ricinoleic acid (12-hydroxy-*cis*-Δ^9-octadecenoic acid)

5. A class of fatty acids called demospongic fatty acids was so named because of their occurrence in *Demospongia* sponges; however it has since been discovered that these fatty acids have a wider distribution. Use the shorthand method described described in Problem 3 as a guide to draw the structures of the following lipids found in marine mollusks: **(a)** *cis*, *cis*-$\Delta^{5,9}$-tetracosodienoate; **(b)** all *cis*-$\Delta^{5,9,15,18}$-tetracosotetraenoate.

6. Minerval (2-hydroxyoleate) is a fatty acid that induces apoptosis (a form of programmed cell death) when added to leukemia cells in culture. Draw the structure of minerval.

7. *Trans* fatty acids occur naturally in beef and milk products but are also produced when oils undergo partial hydrogenation to convert the liquid oil into a semi-solid fat. Draw the structure of elaidic acid (18:1 $\Delta^{9\ trans}$). How does the melting point of elaidic acid compare with the melting point of oleic acid (18:1Δ^9) in which the double bond is in the *cis* configuration?

8. Marine organisms are good sources of unusual fatty acids that have potential as therapeutic agents. A monoacylglycerol isolated from a sponge contains 10-methyl-9-*cis*-octadecenoic acid esterified to C1 of glycerol. Draw its structure.

9. When certain nutrients are limiting, some marine phytoplankton can change their membrane lipid composition, producing substitute lipids such as sulfoquinovosyldiacylglycerol (SQDG).

(a) Is SQDG more likely to substitute for phosphatidylethanolamine or phosphatidylglycerol?
(b) What element must be in short supply to induce the organism to increase its synthesis of SQDG?

Sulfoquinovosyldiacylglycerol
(SQDG)

10. Draw the structure of tripalmitin, a triacylglycerol containing three palmitate fatty acyl chains (see Table 8.1).

11. During digestion, pancreatic lipase catalyzes hydrolysis of the fatty acids from the 1 and 3 positions of triacylglycerols.

(a) Draw the products of lipase-catalyzed digestion of tripalmitin (see Problem 10).
(b) What type of structure is formed by the products? (*Hint:* See Section 2.2.)

12. Soap (a sodium salt of a fatty acid) is made by treating fat (triacylglycerols) with lye (an aqueous solution of sodium hydroxide) in a process called saponification. Draw the products of the saponification of trilaurin.

13. Phosphatidylinositols (PI) are glycerophospholipids important in cell signaling.

(a) Draw the structure of a PI, given the structure of *myo*-inositol. The hydroxyl group involved in the bond between the inositol and the phosphate group is circled.
(b) Indicate the polar and nonpolar domains of the molecule.

myo-Inositol

14. Some signaling pathways generate signaling molecules derived from phosphatidylinositol that has been phosphorylated at multiple sites. How many additional phosphate groups can potentially be attached to phosphatidylinositol?

15. Dipalmitoylphosphatidylcholine (DPPC) is the major lipid of lung surfactant, a protein–lipid mixture essential for pulmonary function. Surfactant production in the developing fetus is low until just before birth, so infants may develop respiratory difficulties if born prematurely. Draw the structure of DPPC.

16. An unusual sphingosine variant has recently been isolated from the nerve of the squid *Loligo pealeii*. Its chemical name is 2-amino-9-methyl-4,8,10-octadecatriene-1,3-diol. Draw the structure of this sphingosine variant.

17. Complex lipids in mammalian skin serve as a waterproof layer. One of these lipids is a glucocerebroside in which the amide-linked

acyl group has 28 carbons and an omega hydroxyl group to which linoleate is esterified. Draw the structure of this lipid.

18. The lipid in Problem 17 undergoes hydrolysis to remove the glucose and linoleate groups, followed by linkage of the omega hydroxyl group to the side chain of a protein Glu residue. Draw the structure of the protein–lipid complex.

19. Which of the glycerophospholipids shown in Section 8.1 have hydrogen-bonding head groups?

20. Which of the glycerophospholipids shown in Section 8.1 are charged? Which are neutral?

21. Identify the polar heads and the nonpolar tails in the glycerophospholipids shown in Section 8.1.

22. In some autoimmune diseases, an individual develops antibodies that recognize cell constituents such as DNA and phospholipids. Some of the antibodies actually react with both DNA and phospholipids. What is the biochemical basis for this cross-reactivity?

23. The points of attack of several phospholipases are indicated in Figure 8.1. Draw the products of the following reactions:

(a) phosphatidylserine + phospholipase A_1;
(b) phosphatidylcholine + phospholipase C;
(c) phosphatidylglycerol + phospholipase D.

24. Archaeal lipids (see Fig. 8.3) are found in thermophiles, bacteria that thrive at extremely high temperatures. Why are archaeal lipids more stable than glycerophospholipids at high temperatures?

25. Spicy Indian dishes flavored with hot peppers are often served with a side dish made from whole-milk yogurt. Why is a spoonful of yogurt preferable to a drink of water after a mouthful of spicy food?

26. Olestra®, a synthetic lipid used to make "low-calorie" chips and snacks, passes through the intestinal tract undigested. Prior to approving Olestra®, the FDA required Procter & Gamble, which markets Olestra®, to add vitamins A, D, and K to products containing the synthetic lipid. Explain.

27. In a nutrition study, volunteers consumed salads with and without added avocado (a rich source of monounsaturated lipids). Blood samples drawn from the volunteers showed a dramatic increase in β-carotene (an orange pigment found in plants that is processed by cells to yield vitamin A) following the consumption of salad containing avocado. Explain these findings.

28. Why does the consumption of excessive amounts of vitamins D and A cause adverse health effects whereas consumption of vitamin C (see Box 5.B) well in excess of the recommended daily allowance generally does not lead to toxicity?

29. Bacterial infection stimulates nearby host cells to increase production of an enzyme that oxidizes the aldehyde group of retinal. Identify the product of the reaction.

30. Because it stimulates the activity of immune system cells, retinoic acid has been proposed as a treatment for infections. However, administering retinoic acid can also exacerbate conditions such as arthritis and other inflammatory diseases. Explain.

31. (a) Does vitamin D fit the definition of a vitamin as a substance that an organism requires but cannot synthesize?
(b) Vitamin D deficiency is hypothesized to contribute to the development of multiple sclerosis because the prevalence of this autoimmune disease increases with increasing latitude. Explain.

32. Explain the following:

(a) Long-term use of antibiotics may lead to a vitamin K deficiency.
(b) Obese individuals require larger amounts of vitamins A, D, E, and K in their diets.

8.2 The Lipid Bilayer

33. Identify the following molecules as polar, nonpolar, or amphipathic.

(a) $CH_3CH_2(CH=CHCH_2)_3(CH_2)_8COO^-$

(b)

$$CH_2-OH$$
$$HC-OH$$
$$CH_2-OH$$

(c)

(d)

(e)

34. Which of these molecules in Problem 33 can form bilayers? For the molecules that do not form bilayers, explain why not.

35. The lipid shown below is a plasmalogen.

(a) How does it differ from a glycerophospholipid?
(b) Would the presence of this lipid have a dramatic effect on a bilayer that contained only phosphatidylcholine?

A plasmalogen

36. Use a simple diagram to show why bilayer curvature would be affected by replacing glycerophospholipids bearing two saturated acyl chains with ones bearing two highly unsaturated acyl chains.

37. Why can't triacylglycerols form a lipid bilayer?

38. Why is fluidity greatest at the center of a lipid bilayer?

39. Red blood cells lyse (break apart) when treated with phospholipase A$_1$, an enzyme found in the venom of many poisonous insects (see Fig. 8.1). Why does treatment with the enzyme result in the destruction of the red blood cell membrane?

40. The melting points of some common saturated and unsaturated fatty acids are shown in the table. What important factors influence a fatty acid's melting point?

Fatty acid	Melting point (°C)
Laurate (12:0)	44.2
Linoleate (18:2)	–9
Linolenate (18:3)	–17
Myristate (14:0)	52
Oleate (18:1)	13.4
Palmitate (16:0)	63.1
Stearate (18:0)	69.1

41. How does the melting point of elaidic acid (see Problem 7) compare with the melting point of oleic acid (see Table 8.1)?

42. Rank the melting points of the following fatty acids: (a) *cis*-oleate (18:1); (b) *trans*-oleate (18:1); (c) linoleate (18:2).

43. The triacylglycerols of animals tend to be solids (fats), whereas the triacylglycerols of plants tend to be liquids (oils) at room temperature. What can you conclude about the nature of the fatty acyl chains in animal and plant triacylglycerols?

44. Peanut oil contains a high percentage of monounsaturated triacylglycerols (having acyl chains with only one double bond), whereas vegetable oil contains a higher percentage of polyunsaturated triacylglycerols (having acyl chains with more than one double bond). A bottle of peanut oil and a bottle of vegetable oil are stored in a pantry with an outside wall. During a cold spell, the peanut oil freezes but the vegetable oil remains liquid. Explain why.

45. Membrane lipids in tissue samples obtained from different parts of the leg of a reindeer show different fatty acid compositions. The proportion of unsaturated fatty acyl chains increases from the top of the leg to the hoof. Provide an explanation for this observation.

46. Phytol is an alcohol produced from chlorophyll that becomes part of the diet of mammals consuming plants. Phytol is converted to phytanic acid in a three-step process, then oxidized to obtain metabolic energy. In individuals with a defect in one of the enzymes of the oxidative pathway, phytanic acid accumulates in the membranes of nerve cells and impairs neurological functions. How does the presence of phytanic acid affect nerve cell membrane fluidity?

Phytol (3,7,11,15-Tetramethyl-2-hexadecen-1-ol)

Phytanic acid (3,7,11,15-Tetramethylhexadecanoic acid)

47. Bacteria of the genus *Lactobacillus* colonize the human digestive tract and are considered "friendly" bacteria that are often used to treat digestive disorders. These bacteria produce lactobacillic acid, a 19-carbon fatty acid containing a cyclopropane ring. Is the melting point of this fatty acid closer to the melting point of stearate (18:0) or oleate (18:1)? Rank the melting points of these three fatty acids.

48. Bacteria are typically grown at a temperature of 37°C. What happens to the membrane lipid composition if the temperature is increased to 42°C?

49. A membrane consisting only of phospholipids undergoes a sharp transition from the crystalline form to the fluid form as it is heated. However, a membrane containing 80% phospholipid and 20% cholesterol undergoes a more gradual change from crystalline to fluid form when heated over the same temperature range. Explain why.

50. Plants can synthesize trienoic acids (fatty acids with three double bonds) by introducing another double bond into a dienoic acid. Would you expect plants growing at higher temperatures to convert more of their dienoic acids into trienoic acids?

51. *Chorispora bungeana* is a plant that is well adapted to growth at freezing temperatures. Plants grown at −4°C have a greater percentage of 18:3 fatty acids compared to control plants grown at 25°C. The increase in 18:3 fatty acids was accompanied by a decrease in 18:0, 18:1, and 18:2 fatty acids. Propose a hypothesis consistent with these data.

52. The lipid distribution in membranes is asymmetric. Phosphatidylserine (PS) is exclusively found in the cytosol-facing leaflet of the membrane bilayer. Phosphatidylethanolamine (PE) is also more likely to be found in this leaflet. In contrast, phosphatidylcholine (PC) and sphingomyelin (SM) are more likely to be found in the extracellular leaflet of the membrane bilayer.

(a) What functional group do PS and PE have in common?

(b) What functional group do PC and SM have in common?

(c) Is one side of a membrane more likely to carry a charge than the other side, or do both sides of the membrane have the same charge?

53. A phospholipid flippase was studied in red blood cells. Experiments showed that the flippase translocated phospholipids from the extracellular leaflet to the cytosolic leaflet of the membrane. The flippase had a preference for phosphatidylserine and translocated phosphatidylethanolamine more slowly. Phosphatidylcholine was not translocated. Translocation did not occur if cells were deprived of ATP or magnesium ions. Translocation did not occur if red blood cells were treated with a reagent that alkylated sulfhydryl groups. Write a paragraph that describes the essential features of the flippase. Are these observations consistent with the data presented in Problem 52?

8.3 Membrane Proteins

54. The interleukin-1 receptor is an integral membrane protein that includes a single membrane-spanning segment. A portion of the protein sequence is shown below. Identify the portion of the protein that crosses the membrane.

DAAYIQLIYPVTNFQKHMIGICVTLTVIIVCSVFIYKIFKIDIV-
LWYRDS

55. Purification of transmembrane proteins requires the addition of detergents to the buffers in order to solubilize the proteins.

(a) Why would transmembrane proteins be insoluble in the absence of detergent?

(b) Draw a schematic diagram that shows how the detergent sodium dodecyl sulfate interacts with a transmembrane protein.

56. Cytochrome *c*, a protein of the electron transport chain in the inner mitochondrial membrane, can be removed by relatively mild means, such as extraction with salt solution. In contrast, cytochrome oxidase from the same source can be removed only by extraction into detergent solutions or organic solvents. What kind of membrane proteins are cytochrome *c* and cytochrome oxidase? Explain. Draw a schematic diagram of what each protein looks like in the membrane.

57. Draw the structures of the lipid-protein linkages described:

(a) A protein involved in the Wnt signaling pathway is modified by the attachment of palmitoleate (see Table 8.1) to a Ser residue. Blocking this modification may be an effective cancer treatment.

(b) Ghrelin, an appetite-stimulating factor, requires the attachment of an octanoate residue for full activity. Blocking this attachment may be an effective strategy for treating obesity.

58. Glycophorin A is a 131-residue integral membrane protein that includes one bilayer-spanning segment. Identify that segment in the glycophorin A amino acid sequence (which uses one-letter abbreviations):

LSTTEVAMHTTTSSSVSKSYISSQTNDTHKRDTYAAT-
PRAHEVSEISVRTVYPPEEETGERVQLAHHFSEPEITLII-
FGVMAGVIGTILLISYGIRRLIKKSPSDVKPLPSPDTDVPLSS-
VEIENPETSDQ

59. Proteins that form a transmembrane β barrel always have an even number of β strands.

(a) Explain why.

(b) Why are the strands antiparallel?

(c) Could some of them possibly be parallel?

60. Peptide hormones must bind to receptors on the extracellular surface of their target cells before their effects are communicated to the cell interior. In contrast, receptors for steroid hormones such as estrogen are intracellular proteins. Why is this possible?

61. Melittin, a 26-amino acid peptide, is known to associate with membranes. The presence of a tryptophan residue in the peptide allows fluorescence spectroscopy to be used to monitor the structure assumed by melittin when it is membrane-associated. Artificial membranes were prepared using phosphatidylcholine (PC) esterified with palmitate (16:0) at position 1. Lipids at position 2 were varied. Melittin was conformationally restricted when associated with PC containing oleate (18:1) at position 2. But when the lipids contained arachidonate (20:4) at position 2, the melittin peptide was less conformationally restricted. Propose a hypothesis consistent with the observed results.

62. There is some evidence that membranes contain structured domains called lipid rafts. These structures are thought to be composed of loosely packed glycosphingolipids, with the gaps filled in with cholesterol. The fatty acyl chains of the phospholipids associated with lipid rafts tend to be saturated.

(a) Why do glycosphingolipids pack together loosely?

(b) Given the description provided here, do you expect a lipid raft to be more or less fluid than the surrounding membrane?

63. Cholesterol is transported through the blood in association with phospholipids and proteins that form complexes called lipoproteins. The particle consists of an inner core of cholesteryl ester molecules (see Problem 33d) covered with a layer of phospholipids and cholesterol. A single molecule of apolipoprotein B (apoB) winds around the particle. High levels of LDL are associated with an increased risk of cardiovascular disease.

(a) Why is it necessary for cholesterol and cholesteryl esters to be packaged into LDL for transport through the blood?

(b) How is the structure of LDL similar to the structure of a membrane? How is it different?

(c) The protein apolipoprotein B was purified in the mid-1980s. Why was this protein so difficult to purify?

64. The glycosylphosphatidylinositol (GPI) group shown in Figure 8.10d is susceptible to cleavage by phospholipase C (see Fig. 8.1) *in vitro*. If this same reaction were to occur *in vivo*, what would be the result?

8.4 The Fluid Mosaic Model

65. Around the turn of the twentieth century, Charles Overton noted that low-molecular-weight aliphatic alcohols, ether, chloroform, and acetone could pass through membranes easily, while sugars, amino acids, and salts could not. This was a radical notion at the time, since most scientists believed that membranes were impermeable to all compounds but water.

(a) Using what you know about membrane structure, explain Charles Overton's results.

(b) Propose a hypothesis to explain how the polar water molecule could be transported across a membrane.

66. In 1935, Davson and Danielli described a "sandwich model" for membrane structure, which proposed that the membrane consisted of outer and inner layers of protein (the sandwich bread) with a "filling" of lipid. This model is no longer accepted because of inconsistencies between the model and experimental data. Using what we know now about membrane structure, explain some of the shortcomings of the sandwich model.

67. In a famous experiment, Michael Edidin labeled the proteins on the surface of mouse and human cells with green and red fluorescent markers, respectively. The two types of cells were induced to fuse, forming hybrid cells. Immediately after fusion, green markers could be seen on the surface of one half of a hybrid cell and red markers on the other half. After a 40-minute incubation at 37°C, the green and red markers became intermingled over the entire surface of the hybrid cell. If the hybrid cells were instead incubated at 15°C, this mixing did not occur. Explain these observations and why they support the fluid mosaic model of membrane structure.

68. In fluorescence photobleaching recovery studies, fluorescent groups are attached to membrane components in a cell. An intense laser beam pulse focused on a very small area destroys (bleaches) the fluorophores in that area. What happens to the fluorescence in that area over time?

SELECTED READINGS

Albers, S.-V. and Meyer, B. H., The archaeal cell envelope, *Nature Rev. Microbiol.* **9,** 414–426 (2011). [Includes a description of archaeal membrane lipids.]

Edidin, M., Lipids on the frontier: A century of cell-membrane bilayers, *Nat. Rev. Mol. Cell Biol.* **4,** 414–418 (2003). [Briefly reviews the history of the study of membranes.]

Engel, A. and Gaub, H. E., Structure and mechanics of membrane proteins, *Annu. Rev. Biochem.* **77,** 127–148 (2008). [Discusses new techniques for examining the structures of membrane proteins.]

Lingwood, D. and Simons, K., *Science* **327,** 46–50 (2010). [Discusses lipid rafts as well as some general features of membranes and membrane proteins.]

Nicolson, G. L., The fluid-mosaic model of membrane structure: Still relevant to understanding the structure, function and dynamics of biological membranes after more than 40 years. *Biochim. Biophys. Acta* **1838,** 1451–1466 (2014). [Describes the model and how it has been updated.]

van Meer, G., Voelker, D. R., and Feigenson, G. W., Membrane lipids: Where they are and how they behave, *Nat. Rev. Mol. Cell Biol.* **9,** 112–124 (2008). [Reviews the structures and functions of membrane lipids, including their asymmetry and their liquid and gel phases.]

Watson, H., Biological membranes, *Essays Biochem.* **59,** 43–69 (2015). [An overview of membrane structure, membrane proteins, and membrane functions.]

C H A P T E R 9

TRANSPORT THROUGH MEMBRANE

© Brandon Cole Marine Photography/
Alamy Stock Photo

Among the most dangerous substances are the conotoxins, small peptides found in the venom of Conus *snails. Many of the toxins block ion channels in nerve and muscle cell membranes, causing paralysis in the snails' fish and invertebrate prey.*

DO YOU REMEMBER?

- Living organisms obey the laws of thermodynamics (Section 1.3).
- Amphiphilic molecules form micelles or bilayers (Section 2.2).
- An enzyme provides a lower-energy pathway from reactants to products (Section 6.2).
- Integral membrane proteins completely span the bilayer by forming one or more α helices or a β barrel (Section 8.3).

Some of the best-understood membrane-related events occur during nerve signaling. The ability of neurons to transmit signals from cell to cell depends on electrical changes that result from the regulated flow of charged particles through the cells' plasma membranes. But because the plasma membrane presents a barrier to the free diffusion of ions and other substances, membrane proteins are needed to mediate their movement across membranes. In this chapter, we'll examine some of these proteins as well as the membrane shape changes involved in nerve signaling and other processes.

THE THERMODYNAMICS OF MEMBRANE TRANSPORT

9.1

All animal cells—including neurons—maintain intracellular ion concentrations that differ from those outside the cell (see Fig. 2.12). For example, intracellular sodium ion concentrations are much lower than extracellular sodium ion concentrations, and the opposite is true for potassium ions. Neither ion is at equilibrium. To reach equilibrium, Na^+ would have to enter the cell, by spontaneously moving down its concentration gradient. Likewise, K^+ would have to exit the cell, also moving down its concentration gradient (**Fig. 9.1**). This occurs extremely slowly, because animal membranes are largely impermeable to ions.

Even so, a small percentage of K^+ ions do leak out of the cell. The movement of K^+ and other ions places relatively more positive charges outside the cell and leaves relatively more

KEY CONCEPTS

- During a nerve impulse, ion movements alter membrane potential, producing an action potential that travels along the axon.

- Transporters obey the laws of thermodynamics, providing a way for solutes to move down their concentration gradients or using ATP to move substances against their gradients.

FIGURE 9.1

Distribution of Na⁺ and K⁺ ions in an animal cell. The extracellular Na⁺ concentration (about 150 mM) is much greater than the intracellular concentration (about 12 mM), whereas the extracellular K⁺ concentration (about 4 mM) is much less than the intracellular concentration (about 140 mM). If the plasma membrane were completely permeable to ions, Na⁺ would flow into the cell down its concentration gradient (blue arrow), and K⁺ would flow out of the cell down its concentration gradient (orange arrow).

? **Would ion movement ever stop? What would be the final ion concentrations?**

negative charges inside the cell. The resulting charge imbalance, though small, generates a voltage across the membrane, which is called the **membrane potential** and is symbolized $\Delta\psi$. In the simplest case, $\Delta\psi$ is a function of the ion concentration on each side of a membrane:

$$\Delta\psi = \frac{RT}{Z\mathcal{F}} \ln \frac{[\text{ion}]_{in}}{[\text{ion}]_{out}} \qquad [9.1]$$

where R is the **gas constant** ($8.3145 \text{ J} \cdot \text{K}^{-1} \cdot \text{mol}^{-1}$), T is temperature in Kelvin ($20°\text{C} = 293 \text{ K}$), Z is the net charge per ion, $\mathcal{F}$ is the **Faraday constant,** the charge of one mole of electrons ($96,485 \text{ coulombs} \cdot \text{mol}^{-1}$ or $96,485 \text{ J} \cdot \text{V}^{-1} \cdot \text{mol}^{-1}$), and the cytosol is *in*. $\Delta\psi$ is expressed in units of volts (V) or millivolts (mV). For a monovalent ion ($Z = 1$) at $20°\text{C}$, the equation reduces to

$$\Delta\psi = 0.058 \text{ V} \log_{10} \frac{[\text{ion}]_{in}}{[\text{ion}]_{out}} \qquad [9.2]$$

See Sample Calculation 9.1. In a neuron, the membrane potential is actually a more complicated function of the concentrations and membrane permeabilities of several different ions, although K⁺ is the most important.

SAMPLE CALCULATION 9.1

PROBLEM   Calculate the intracellular concentration of Na⁺ when the extracellular concentration is 160 mM. Assume that the membrane potential, –50 mV at 20°C, is due entirely to Na⁺.

SOLUTION   Use Equation 9.2 and solve for $[\text{Na}^+]_{in}$:

$$\Delta\psi = 0.058 \text{ V} \log \frac{[\text{Na}^+]_{in}}{[\text{Na}^+]_{out}}$$

$$\frac{\Delta\psi}{0.058 \text{ V}} = \log [\text{Na}^+]_{in} - \log [\text{Na}^+]_{out}$$

$$\log [\text{Na}^+]_{in} = \frac{\Delta\psi}{0.058 \text{ V}} + \log [\text{Na}^+]_{out}$$

$$\log [\text{Na}^+]_{in} = \frac{-0.050 \text{ V}}{0.058 \text{ V}} + \log (0.160)$$

$$\log [\text{Na}^+]_{in} = -0.862 - 0.796$$

$$\log [\text{Na}^+]_{in} = -1.66$$

$$[\text{Na}^+]_{in} = 0.022 \text{ M} = 22 \text{ mM}$$

PRACTICE PROBLEMS   1. Calculate the intracellular concentration of Na⁺ when the membrane potential is –50 mV at 20°C.

2. Calculate the membrane potential at 20°C when $[\text{Na}^+]_{in} = 10$ mM and $[\text{Na}^+]_{out} = 100$ mM.

ION MOVEMENTS ALTER MEMBRANE POTENTIAL

Most animal cells maintain a membrane potential of about −70 mV. The negative sign indicates that the inside (the cytosol) is more negative than the outside (the extracellular fluid). A sudden flux of ions across the cell membrane can dramatically alter the membrane potential, and this is exactly what happens when a neuron fires.

When a nerve is stimulated, either mechanically or by a signal ultimately derived from one of the sensory organs, Na^+ channels in the plasma membrane open. Sodium ions immediately move into the cell, since their concentration inside is much less than outside. The inward movement of Na^+ makes the membrane potential more positive, increasing it from its resting value of −70 mV to as much as +50 mV. This reversal of membrane potential, or depolarization, is called the **action potential.**

The Na^+ channels remain open for less than a millisecond. However, the action potential has already been generated, and it has two effects. First, *it triggers the opening of nearby voltage-gated K^+ channels* (these channels open only in response to the change in membrane potential). The open K^+ channels allow K^+ ions to diffuse out of the cell, following their concentration gradient. This action restores the membrane potential to about −70 mV (**Fig. 9.2**).

*The action potential also stimulates the opening of additional Na^+ channels farther along the **axon*** (the elongated portion of the cell). This induces another round of depolarization and repolarization, and then another. In this way, the action potential travels down the axon. The signal cannot travel backward because once the ion channels have shut, they remain closed for a few milliseconds. These events are summarized in **Figure 9.3**.

In mammals, action potentials propagate extremely rapidly because the axons are insulated by a so-called **myelin sheath.** This structure consists of several layers of membrane, derived from another cell, coiled around the axon (**Fig. 9.4**). The myelin sheath is rich in sphingomyelins and contains little protein (about 18%; a typical membrane contains about 50% protein). Because the myelin sheath prevents ion movements except at the points, or nodes, in between myelinated segments of the axon, the action potential appears to jump from node to node, propagating about 20 times faster than it would in an unwrapped axon. Deterioration of the myelin sheath in diseases such as multiple sclerosis results in the progressive loss of motor control.

FIGURE 9.2

An action potential. The neuron membrane undergoes depolarization as Na^+ channels are opened (green dashed line), then repolarizes as K^+ channels are opened (red dashed line). Following the action potential, the membrane may be hyperpolarized ($\Delta\psi < -70$ mV) but returns to normal within a few milliseconds.

MEMBRANE PROTEINS MEDIATE TRANSMEMBRANE ION MOVEMENT

The Na^+ and K^+ channels that participate in the propagation of an action potential are just two members of a large group of transport proteins that occur in the plasma membranes of all cells and in the internal membranes of eukaryotes. Transport proteins go by many different names, depending somewhat arbitrarily on their mode of action: transporters, translocases, permeases, pores, channels, and pumps, to list a few. These proteins can also be classified by the type of substance they transport across the membrane and by whether they are always open or gated (open only when stimulated). Over 10,000 transporters have been grouped into five mechanistic classes in the Transporter Classification Database (www.tcdb.org). The most important distinction among transport proteins, however, is whether they require a source of free energy to operate. The neuronal Na^+ and K^+ channels are considered **passive transporters** because they provide a means for ions to move down a concentration gradient, a thermodynamically favorable event.

For any transport protein operating independently of the effects of membrane potential, the free energy change for the transmembrane movement of a substance X from the outside to the inside is

$$\Delta G = RT \ \ln \frac{[X]_{in}}{[X]_{out}} \qquad \text{[9.3]}$$

Consequently, *the free energy change is negative (the process is spontaneous) only when X moves from an area of high concentration on the outer side of the membrane to an area of low concentration on the inner side of the membrane* (see Sample Calculation 9.2).

Cell body

Stimulus

Axon

−70 mV

1. For simplicity, the stimulus is assumed to originate where the axon leaves the neuronal cell body. The stimulus opens Na⁺ channels in the membrane.

● Na⁺

+50 mV

2. The influx of Na⁺ causes depolarization, which triggers the opening of voltage-sensitive K⁺ channels. The efflux of K⁺ restores the resting potential.

−70 mV

● K⁺

3. The initial depolarization also triggers the opening of additional Na⁺ channels, and then K⁺ channels, farther along the axon.

● Na⁺

● K⁺

4. The action potential moves down the length of the axon as a wave of depolarization and repolarization. The action potential travels in only one direction because previously open channels remain closed.

● Na⁺

Direction of impulse

● K⁺

Propagation of a nerve impulse.

Myelination of an axon. (a) Cross-sectional diagram showing how an accessory cell coils around the axon so that multiple layers of its plasma membrane coat the axon. (b) Electron micrograph of myelinated axons. The myelin sheath may be 10 to 15 layers thick. [*Courtesy Cedric S. Raine, Albert Einstein College of Medicine.*]

Axon

(a)

(b)

Show that $\Delta G < 0$ when glucose moves from outside the cell (where its concentration is 10 mM) to the cytosol (where its concentration is 0.1 mM).

The cytosol is *in* and the extracellular space is *out*.

$$\Delta G = RT \ln \frac{[\text{glucose}]_{in}}{[\text{glucose}]_{out}}$$

$$= RT \ln \frac{[10^{-4}]}{[10^{-2}]} = RT(-4.6)$$

Because the logarithm of $(10^{-4}/10^{-2})$ is a negative quantity, ΔG is also negative.

3. Calculate the value of ΔG for the process described above. Assume that $T = 25°C$.

4. Calculate the value of ΔG for the movement of glucose from outside to inside at 20°C when the extracellular concentration is 5 mM and the cytosolic concentration is 0.5 mM.

5. What is the free energy cost of moving glucose from the outside of the cell (where its concentration is 0.8 mM) to the cytosol (where its concentration is 8 mM) when $T = 20°C$?

If the transported substance is an ion, there will be a charge difference across the membrane, so a term containing the membrane potential must be added to Equation 9.3:

$$\Delta G = RT \ln \frac{[X]_{in}}{[X]_{out}} + Z\mathcal{F}\Delta\psi \qquad \textbf{[9.4]}$$

Equation 9.4 can be used to determine the free energy change for transporting an ion when *out* is the ion's initial location and *in* is the final location (see Sample Calculation 9.3). Note that for an anionic substance with charge Z, transport may not be thermodynamically favored, depending on the membrane potential $\Delta\psi$, even if the concentration gradient alone favors transport.

Calculate the free energy change for the movement of Na^+ into a cell when its concentration outside is 150 mM and its cytosolic concentration is 10 mM. Assume that $T = 20°C$ and $\Delta\psi = -50$ mV (inside negative).

Use Equation 9.4:

$$\Delta G = RT \ln \frac{[X]_{in}}{[X]_{out}} + Z\mathcal{F}\Delta\psi$$

$$= (8.3145 \text{ J} \cdot \text{K}^{-1} \cdot \text{mol}^{-1})(293 \text{ K}) \ln \frac{(0.010)}{(0.150)}$$

$$+ (1)(96,485 \text{ J} \cdot \text{V}^{-1} \cdot \text{mol}^{-1})(-0.05 \text{ V})$$

$$= -6600 \text{ J} \cdot \text{mol}^{-1} - 4820 \text{ J} \cdot \text{mol}^{-1}$$

$$= -11,600 \text{ J} \cdot \text{mol}^{-1} = -11.6 \text{ kJ} \cdot \text{mol}^{-1}$$

6. Calculate the free energy change for the movement of K^+ into a cell when the K^+ concentration outside is 10 mM and the cytosolic K^+ concentration is 50 mM. Assume that $T = 20°C$ and $\Delta\psi = -50$ mM (inside negative). Is this process spontaneous?

7. Calculate the free energy cost of moving Na^+ ions across a membrane from a compartment (*outside*) where $[Na^+] = 100$ mM to a compartment (*inside*) where $[Na^+] = 25$ mM. Assume that $T = 20°C$ and $\Delta\psi = +50$ mV. Is this process spontaneous?

In contrast to the passive ion channels in neurons, the protein that initially establishes and maintains the cell's Na^+ and K^+ gradients is an **active transporter** that needs the free energy of the ATP hydrolysis reaction to move ions against their concentration gradients. In the following sections we will examine various types of transport proteins. Keep in mind that small nonpolar substances can cross a membrane without the aid of any transport protein; they simply diffuse through the lipid bilayer.

 CONCEPT CHECK

1. What is the membrane potential?

2. Explain the role of cell membrane in maintaining membrane potential.

3. Describe how an action potential is generated and propagated.

4. Explain how concentration differences and membrane potential influence the free energy change for ion movements across a membrane.

5. Summarize the differences between active and passive transport.

PASSIVE TRANSPORT

KEY CONCEPTS

- Porins are β barrel channels with some solute selectivity.
- Ion channels include a selectivity filter and may be gated.
- Aquaporins allow only water molecules to pass through.
- Transport proteins alternate between conformations to expose binding sites on each side of the membrane.

Transporters of one kind or another have been described for virtually every substance that cannot easily diffuse through a bilayer on its own. Some transporters create a straightforward opening through the membrane, while more complicated transporters function much like enzymes. We begin our discussion with porins, the simplest transporters.

PORINS ARE β BARREL PROTEINS

Porins are located in the outer membranes of bacteria, mitochondria, and chloroplasts (some bacteria and the organelles descended from them have a second outer membrane in addition to the membrane that encloses the cytosol). *All known porins are trimers in which each subunit forms a 16- or 18-stranded membrane-spanning β barrel* (**Fig. 9.5**). A β barrel of this size has a water-filled core lined with hydrophilic side chains, which forms a passageway for the transmembrane movement of ions or molecules with a molecular mass up to about 1000 D. In the eight-stranded β barrel shown in Fig. 8.9, the protein core is too tightly packed with amino acid side chains for it to function as a pore.

(a)   (b)   (c)

FIGURE 9.5

The *E. coli* OmpF porin. Each subunit of the trimeric protein forms a transmembrane β barrel that permits the passage of ions or small molecules. (a) Ribbon model, viewed from the extracellular side of the membrane. (b) Stick model. (c) In each subunit, the 16 β strands are connected by loops, one of which (blue) constricts the barrel core and makes the porin specific for small cationic solutes. [*Structure (pdb 1OPF) determined by S. W. Cowan, T. Schirmer, R. A Pauptit, and J. N. Jansonius.*]

FIGURE 9.6

**Structure of the K⁺ channel from
S. lividans.** The four subunits are
shown in different colors. Each
subunit consists mostly of an
inner helix that forms part of the
central pore and an outer helix that
contacts the membrane interior.
[*Structure (pdb 1BL8) determined
by D. A. Doyle, J. M. Cabral, R. A.
Pfuetzner, A. Kuo, J. M. Gulbis, S. L.
Cohen, B. T. Chait, and R. MacKinnon.*]

EXTRACELLULAR
SPACE

CYTOSOL

In the 16-stranded OmpF barrel, long loops connect the β strands (Fig. 9.5c). One of these loops in each monomer folds down into the β barrel and constricts its diameter to about 7 Å at one point, thereby preventing the passage of substances larger than 600 D. The loop bears several carboxylate side chains, which make this porin weakly selective for cationic substances. Other porins exhibit a greater degree of solute selectivity, depending on the geometry of the barrel interior and the nature of the side chains that project into it. For example, some porins are specific for anions or small carbohydrates. A porin is considered to be always open, and a solute can travel through it in either direction, depending on which side of the membrane has a lower solute concentration.

ION CHANNELS ARE HIGHLY SELECTIVE

The ion channels in neurons and in other eukaryotic and prokaryotic cells are more complicated proteins than the porins. Many are multimers of identical or similar subunits with α-helical membrane-spanning segments. The ion passageway itself lies along the central axis of the protein, where the subunits meet. One of the best known of these proteins is the K⁺ channel from the bacterium *Streptomyces lividans*. Each subunit of this tetrameric protein includes two long α helices. One helix forms part of the wall of the transmembrane pore, and the other helix faces the hydrophobic membrane interior (**Fig. 9.6**). A third, smaller helix is located on the extracellular side of the protein.

The K⁺ channel is about 10,000 times more permeant to K⁺ than to Na⁺, even though Na⁺ is smaller and should easily pass through the central pore. *The high selectivity for K⁺ reflects the geometry of the selectivity filter,* an arrangement of protein groups that define the extracellular mouth of the pore. At one point, the pore narrows to ~3 Å, and the four polypeptide backbones fold so that their carbonyl groups project into the pore. The carbonyl oxygen atoms are arranged with a geometry suitable for coordinating desolvated K⁺ ions (diameter 2.67 Å) as they move through the pore. A desolvated Na⁺ ion (diameter 1.90 Å) is too small to coordinate with the carbonyl groups and is therefore excluded from the pore (**Fig. 9.7**).

The voltage-gated K⁺ channel in neurons is larger than the bacterial channel, with six helices in each of its four subunits, and it associates with other proteins to form a large complex. However, like most K⁺ channels, it contains the same type of selectivity filter. Other ion channels, such as those for Na⁺ and Ca²⁺, necessarily have different filtering mechanisms. Membrane channels that are specific for water molecules form an entirely different family of proteins (described below).

FIGURE 9.7

The K⁺ channel selectivity filter. This model shows a portion of the pore looking from the extracellular space into the selectivity filter. The pore is lined by backbone carbonyl groups with a geometry suitable for coordinating a K⁺ ion (purple sphere). A rigid protein network that includes tyrosine residues prevents the pore from contracting to accommodate the smaller Na⁺ ion. Atoms are color-coded: C green, N blue, and O red.

? **Explain why a small anion would not pass through the channel.**

Box 9.A Pores Can Kill

The antifungal agent amphotericin B kills a variety of pathogenic fungi. Amphotericin B is a relatively small cyclic compound with decidedly hydrophobic and hydrophilic faces. Atoms are color-coded: C green, N blue, O red, and H white.

An estimated four to six amphotericin molecules insert into the fungal cell membrane, where the hydrophobic portions interact strongly with ergosterol (a fungal counterpart of the cholesterol found in mammalian cell membranes), and the hydrophilic portions define a passageway from one side of the membrane to the other. The amphotericin molecule, with a length of only 23 Å, barely spans the hydrophobic core of the lipid bilayer, and the opening it forms is too small to allow the mass exit of the cell's contents. However, the pore is apparently sufficient to permit the flow of Na$^+$, K$^+$, and other ions. The resulting disruption of ion concentration gradients and the loss of membrane potential are lethal to the cell, which otherwise remains intact.

Amphotericin, produced by soil bacteria, is just one example of a wide array of microbial products that are intended to disrupt the membrane integrity of competitors or predators. Typically, these chemical weapons are excreted individually and, in pairs or small multimers, assemble in the target cell's membrane to form a passageway for ions. Some of these compounds, often in chemically modified form, have been adopted for use as antibiotics.

The mammalian immune system also relies on pore-forming mechanisms to combat bacterial and fungal infections. The presence of certain cell-wall components in these organisms triggers the activation of **complement,** a set of circulating proteins that sequentially activate each other and lead to the formation of a doughnut-shaped structure, the so-called membrane attack complex, that creates a pore in the target cell's membrane. The resulting loss of ions kills the cell. The inappropriate assembly of the membrane attack complex on the surface of human cells contributes to the pathology of some diseases that are caused by the immune system mistakenly responding to the body's own components.

? **What prevents pore formation in the membrane of the cell that produces amphotericin or a similar compound? Explain why bacteria with elaborate outer membranes are more resistant to pore-forming antibiotics.**

GATED CHANNELS UNDERGO CONFORMATIONAL CHANGES

If K$^+$ or Na$^+$ channels were always open, nerve cells would not experience action potentials, and the intracellular and extracellular concentrations of ions would quickly reach equilibrium, thereby killing the cell (**Box 9.A**). Consequently, these channels—and many others—are **gated;** that is, they open or close in response to a specific signal. Some ion channels respond to changes in pH or to the binding of a specific ligand such as Ca^{2+} or a small molecule. Some channel proteins open when phosphorylated (have a phosphate group covalently attached).

Analysis of gated channels reveals a variety of mechanisms for opening and closing off a pore. The neuronal K$^+$ channel is voltage-gated; it opens in response to depolarization. The gating mechanism involves the motion of helices near the intracellular side of the membrane, which move enough to expose the entrance of the pore without significantly disrupting the structure of the rest of the protein (**Fig. 9.8**).

In addition to voltage gating, the K$^+$ channel in neurons is subject to inactivation by a process in which an N-terminal segment of the protein (not shown in Fig. 9.8) is repositioned

FIGURE 9.8

Operation of the voltage-gated K$^+$ channel. In the closed conformation (right), the so-called S4-S5 linker helix (shown in red) pushes down on the S6 helix, pinching off the intracellular end of the pore. When depolarization occurs, the linker helix swings upward, and the S6 helix bends, opening up the pore (left). [*Courtesy Roderick MacKinnon, Rockefeller University and Howard Hughes Medical Institute.*]

closed

open

MscS   MscL

FIGURE 9.9

Closed and open conformations of mechanosensitive channels. In the bacterial proteins MscS and MscL, a set of α helices surround the pore (the rest of the proteins are not shown). The helices slide past each other, opening and closing the pore, much like the iris of the eye. Hydrophobic residues that block the pore in the closed state are shown in magenta. [*Courtesy Douglas C. Rees, California Institute of Technology.*]

to block the cytoplasmic opening of the pore. This inactivation occurs a few milliseconds after the K^+ channel first opens and explains why the channel cannot immediately reopen. As a result, the action potential can only travel forward.

In bacterial mechanosensitive channels, which open in response to membrane tension, a set of α helices slide past each other to alter their packing arrangement (**Fig. 9.9**). Interestingly, in the closed state, the pore is not 100% occluded. However, neither water nor ions can pass through because the opening is lined with bulky hydrophobic residues. Although a single water molecule or desolvated ion might fit geometrically, the high energetic cost of passing a polar solute past this hydrophobic barrier effectively closes off the pore.

AQUAPORINS ARE WATER-SPECIFIC PORES

For many years, water molecules were assumed to cross membranes by simple diffusion (technically **osmosis,** the movement of water from regions of low solute concentration to regions of high solute concentration). Because water is present in large amounts in biological systems, this premise seemed reasonable. However, certain cells, such as in the kidney, can sustain unexpectedly rapid rates of water transport, which suggested the existence of a previously unrecognized pore for water. The elusive protein was discovered in 1992 by Peter Agre, who coined the term **aquaporin.**

Aquaporins are widely distributed in nature; plants may have as many as 50 different aquaporins. The 13 mammalian aquaporins are expressed at high levels in tissues where fluid transport is important, including the kidney, salivary glands, and lacrimal glands (which produce tears). Most aquaporins are extremely specific for water molecules and do not permit the transmembrane passage of other small polar molecules such as glycerol or urea.

$$H_2N-\overset{\displaystyle O}{\overset{\displaystyle \|}{C}}-NH_2$$
Urea

The best-defined member of the aquaporin family (aquaporin 1, or AQP1) is a homotetramer with carbohydrate chains on its extracellular surface. Each subunit consists mostly of six membrane-spanning α helices plus two shorter helices that lie within the dimensions of the bilayer (**Fig. 9.10**).

(a)

(b)

FIGURE 9.10

Structure of an aquaporin subunit. (a) Ribbon model viewed from within the membrane. (b) Stick model viewed from one end. In the intact aquaporin, four of these subunits associate via hydrogen bonding between helices and through interactions among the loops outside the membrane. [*Structure (pdb 1FQY) determined by K. Murata, K. Mitsuoka, T. Hirai, T. Walz, P. Agre, J. B. Heymann, A. Engel, and Y. Fujiyoshi.*]

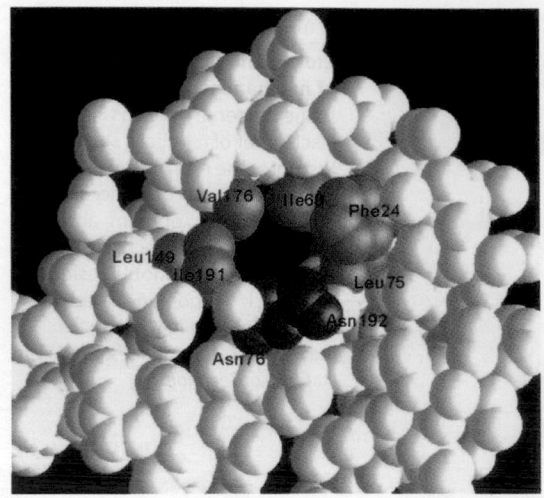

FIGURE 9.11
View of the aquaporin pore. Hydrophobic residues are colored yellow and the two asparagine residues are red. [*Courtesy Yoshinori Fujiyoshi, Kyoto University.*]

Unlike the K^+ channel, whose pore lies in the center of the four protein subunits, each aquaporin subunit contains a pore. At its narrowest, the pore is about 3 Å in diameter (the diameter of a water molecule is 2.8 Å). The dimensions of the pore clearly restrict the passage of larger molecules. The pore is lined with hydrophobic residues except for two asparagine side chains, which have an important function (**Fig. 9.11**).

If water were to pass through aquaporin as a chain of hydrogen-bonded molecules, then protons could also easily pass through (recall from Section 2.3 that a proton is equivalent to H_3O^+ and that a proton can appear to jump rapidly through a network of hydrogen-bonded water molecules). However, aquaporin does not transport protons (other proteins that do transport protons play important roles in energy metabolism). To prevent proton transport, aquaporin interrupts the hydrogen-bonded chain of water molecules in its pores, which occurs when the asparagine side chains transiently form hydrogen bonds to a water molecule passing by.

SOME TRANSPORT PROTEINS ALTERNATE BETWEEN CONFORMATIONS

Not all proteins that mediate transmembrane traffic have an obvious membrane-spanning pore, as in porins and ion channels. Many transport proteins undergo conformational changes in order to move a solute from one side of the membrane to the other. Some of the best known transporters are the glucose-transporting family of GLUT proteins. There are 14 forms of GLUT in humans, and similar proteins occur in all types of cells.

The GLUT protein has a glucose-binding site that alternately faces the cell exterior and interior. In a typical transport cycle, extracellular glucose binds to the outward-facing conformation of the protein. Ligand binding triggers conformational changes that involve a sort of rocking motion in the protein, which exposes the bound glucose to the intracellular side of the membrane (**Fig. 9.12**). Because the glucose concentration is typically lower inside the cell, the bound glucose dissociates. The GLUT protein is not perfectly symmetrical, however, so the inward-facing conformation switches back to the outward-facing conformation, ready to bind another glucose molecule. The two conformational states of GLUT are in equilibrium, so this passive transporter can move glucose in either direction across the cell membrane, depending on the relative concentrations of glucose inside and outside the cell.

The GLUT transporters consist of 12 membrane-spanning α helices arranged in two domains (**Fig. 9.13**). Modest shifts in the orientations of a few helices convert the proteins from outward-facing to inward-facing and back again. *Many other transport proteins structurally resemble the GLUT proteins and use the same alternating-access mechanism to bind and release a ligand on opposite sides of the membrane.* These transporters function like enzymes by accelerating the rate at which a substance crosses the membrane. And like enzymes, they can be saturated by high concentrations of their "substrate," and they are susceptible to competitive and other types of inhibition. For obvious reasons, transport proteins tend to be more solute-selective than porins or ion channels. Their great variety reflects the need to transport many different kinds of metabolic fuels and building blocks into and out of cells and organelles. An estimated 10% of the genes in microorganisms encode transport proteins.

EXTRACELLULAR SPACE

Transporter

CYTOSOL

Glucose

1. *Glucose (hexagon) binds to a site on the transporter that faces the cell exterior.*

2. *Glucose binding triggers a conformational change that exposes the glucose-binding site to the cell interior.*

3. *Glucose dissociates from the transporter.*

4. *The transporter reverts to its original conformation.*

FIGURE 9.12

Operation of the red blood cell glucose transporter.

 Would this transport protein allow water or ions to move across the membrane?

outward-facing   inward-facing

CYTOSOL

FIGURE 9.13

Conformations of GLUT proteins. Two GLUT proteins are shown as ribbon models. Analogs of glucose (red) occupy the binding sites of the proteins in the outward-facing (left) and inward-facing (right) conformations. [*Structure of GLUT3 (left, pdb 4ZWC) determined by D. Deng, P. C. Sun, C. Y. Yan, and N. Yan, and structure of GLUT1 (right, pdb 4PYP) determined by D. Deng, C. Y. Yan, C. Xu, J. P. Wu, P. C. Sun, M. X. Hu, and N. Yan.*]

Some transport proteins can bind more than one type of ligand, so it is useful to classify them according to how they operate (**Fig. 9.14**):

1. A **uniporter** such as the glucose transporter moves a single substance at a time.
2. A **symporter** transports two different substances across the membrane.
3. An **antiporter** moves two different substances in opposite directions across the membrane.

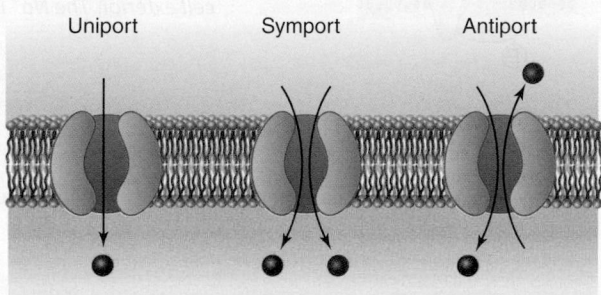

Uniport   Symport   Antiport

FIGURE 9.14

Some types of membrane transport systems.

 CONCEPT CHECK

1. Compare the overall structure, solute selectivity, and general mechanism of porins, ion channels, aquaporins, and the GLUT transporters.
2. Which of these transporters is continually open?
3. Explain why these transport systems allow solute movement in either direction.

ACTIVE TRANSPORT

The differing Na^+ and K^+ concentrations inside and outside of eukaryotic cells are maintained largely by an antiport protein known as the Na,K-ATPase. This active transporter pumps Na^+ out of and K^+ into the cell, working against the ion concentration gradients. As the name **ATPase** implies, ATP hydrolysis is its source of free energy. Other ATP-requiring transport proteins pump a variety of substances against their concentration gradients.

KEY CONCEPTS
- Conformational changes resulting from ATP hydrolysis drive Na^+ and K^+ transport in the Na,K-ATPase.
- Secondary active transport of a substance is driven indirectly by the ATP-dependent formation of a gradient of a second substance.

EXTRACELLULAR SPACE

Na⁺   CYTOSOL

1. Three intracellular Na⁺ ions bind.

ATP

2. ATP binds.

ATP

3. A phosphoryl group is transferred from ATP to an Asp side chain of the pump. ADP is released.

ADP

7. The protein conformation changes, exposing the K⁺-binding sites to the cell interior. The K⁺ ions dissociate.

K⁺

6. The aspartyl phosphate group is hydrolyzed. P_i is released.

P_i

K⁺

5. Two extracellular K⁺ ions bind.

Na⁺

4. The protein conformation changes, exposing the Na⁺-binding sites to the cell exterior. The Na⁺ ions dissociate.

FIGURE 9.15

The reaction cycle of the Na,K-ATPase.

? What does the ion transport mechanism have in common with the mechanism of motor proteins (see Figs. 5.37 and 5.39)?

THE Na,K-ATPase CHANGES CONFORMATION AS IT PUMPS IONS ACROSS THE MEMBRANE

With each reaction cycle, the Na,K-ATPase hydrolyzes 1 ATP, pumps 3 Na⁺ ions out, and pumps 2 K⁺ ions in:

$$3 \text{ Na}^+_{in} + 2 \text{ K}^+_{out} + \text{ATP} + \text{H}_2\text{O} \rightarrow 3 \text{ Na}^+_{out} + 2 \text{ K}^+_{in} + \text{ADP} + \text{P}_i$$

Like other membrane transport proteins, the Na,K-ATPase has two conformations that alternately expose the Na⁺ and K⁺ binding sites to each side of the membrane. As diagrammed in **Figure 9.15**, the protein pumps out 3 Na⁺ ions at a time, then transports in 2 K⁺ ions at a time as it hydrolyzes ATP. *The energetically favorable reaction of converting ATP to ADP + P_i drives the energetically unfavorable transport of Na⁺ and K⁺.* The ATP hydrolysis reaction is coupled to ion transport so that phosphoryl-group transfer from ATP to the protein triggers one conformational change (steps 3 and 4) and the subsequent release of the phosphoryl group as P_i triggers another conformational change (steps 5 and 6). This multistep process, which involves a phosphorylated protein intermediate, ensures that the transporter operates in only one direction and prevents Na⁺ and K⁺ from diffusing back down their concentration gradients. A similar mechanism operates in motor proteins (Section 5.4), where ADP and P_i are released in separate steps and so cannot recombine to re-form ATP and drive the reaction cycle in reverse.

The Na,K-ATPase consists of a large α subunit with 10 transmembrane helices plus smaller β and γ subunits containing one transmembrane helix each. The structure of the pump in its outward-facing form is shown in **Figure 9.16**. The ATP binding site and the aspartate residue

FIGURE 9.16

Structure of the Na,K-ATPase. The α subunit is green and the β and γ subunits are blue. Three Na$^+$ ions (orange) occupy the ion-binding sites, and an ATP analog (yellow) marks the site of phosphorylation. [*Structure (pdb 4HQJ) determined by M. Nyblom, L. Reinhard, P. Gourdon, and P. Nissen.*]

EXTRACELLULAR SPACE

Membrane

CYTOSOL

that becomes phosphorylated during the reaction cycle are located in cytoplasmic domains, indicating that ATP-binding and phosphate-transfer events must be communicated over a considerable distance to the membrane-spanning region where cations are bound and released.

The Na,K-ATPase is known as a P-type ATPase (P stands for phosphorylation). Other types of ATP-dependent pumps are the V-type ATPases, which operate in plant vacuoles and other organelles, and the F-type ATPases, which actually operate in reverse to *synthesize* ATP in mitochondria (Section 15.4) and chloroplasts (Section 16.2).

ABC TRANSPORTERS MEDIATE DRUG RESISTANCE

All cells have some ability to protect themselves from toxic substances that insert themselves into the lipid bilayer and alter membrane structure and function. This defense depends on the action of membrane proteins known as **ABC transporters** (ABC refers to ATP-binding cassette, a common structural motif in these proteins). Unfortunately, many antibiotics and other drugs are lipid-soluble and are therefore substrates for these same transporters. Drug resistance in cancer chemotherapy and antibiotic resistance in bacteria have been linked to the expression or overexpression of ABC transporters. In humans, this transporter is also known as P-glycoprotein or the multidrug-resistance transporter.

ABC transporters function much like other transport proteins and ATPase pumps: *ATP-dependent conformational changes in cytoplasmic portions of the protein are coupled to conformational changes in the membrane-embedded portion of the protein.* As expected, the transporter consists of two halves that presumably reorient relative to each other to expose the ligand-binding site to each side of the membrane in turn (as in the GLUT transporter mechanism shown in Fig. 9.12). Each half of the transporter includes a bundle of membrane-spanning α helices linked to a globular nucleotide-binding domain where the ATP reaction takes place (**Fig. 9.17**).

Some ABC transporters are specific for ions, sugars, amino acids, or other polar substances. P-glycoprotein and other drug-resistance transporters prefer nonpolar substrates. In this case, the transmembrane domain of the protein allows the entry of substances from within the lipid bilayer. The substance may then be entirely expelled from the cell, as occurs in drug resistance. Alternatively, the substance may simply move from one leaflet to the other. Some lipid flippases (Section 8.2) are ABC transporters that transport lipids between leaflets, generating nonequilibrium distributions of certain lipids in membranes.

SECONDARY ACTIVE TRANSPORT EXPLOITS EXISTING GRADIENTS

In some cases, the "uphill" transmembrane movement of a substance is not directly coupled to the conversion of ATP to ADP + P$_i$. Instead, *the transporter takes advantage of a gradient already established by another pump,* which is often an ATPase. This indirect use of the free energy of ATP is known as **secondary active transport.** For example, the high Na$^+$ concentration outside of intestinal cells (a gradient established by the Na,K-ATPase) helps drive glucose into the cells via a symport protein (**Fig. 9.18**). The free energy released by the movement of Na$^+$ into the cells (down its concentration gradient) drives the inward movement of glucose (against its gradient). This mechanism allows the intestine to collect glucose from digested food and then release it into the bloodstream. Other types of secondary active transporters take advantage of the free energy of a proton gradient, transporting solutes across a membrane along with a proton.

EXTRACELLULAR SPACE

CYTOSOL

FIGURE 9.17

Structure of mouse P-glycoprotein. The transporter, which is built from a single polypeptide chain, is shown as a ribbon model. Note that the internal cavity is open to both the cytoplasm and the inner leaflet of the membrane. [*Structure (pdb 3G5U) determined by S. G. Aller, J. Yu, A. Ward, Y. Weng, S. Chittaboina, R. Zhuo, P. M. Harrell, Y. T. Trinh, Q. Zhang, I. L. Urbatsch, and G. Chang.*]

FIGURE 9.18

Glucose transport into intestinal cells. The Na, K-ATPase establishes a concentration gradient in which $[Na^+]_{out} > [Na^+]_{in}$. Sodium ions move into the cell, down their concentration gradient, along with glucose molecules via a symport protein that transports Na^+ and glucose simultaneously. Glucose thereby becomes more concentrated inside the cell, which it then exits, down its concentration gradient, via a passive uniport GLUT transporter. Energetically favorable movements are indicated by green arrows; energy-requiring movements are indicated by red arrows.

 CONCEPT CHECK

1. Compare the mechanisms of active and passive transporters.

2. List the ways that the ATP reaction is used to move solutes across membranes.

3. How do secondary active transporters compare to other transporters in terms of mechanism and energy requirement?

9.4 MEMBRANE FUSION

KEY CONCEPTS
- Neurotransmitters are released by the process of exocytosis.
- Membrane fusion, driven by the action of SNARE proteins, requires changes in bilayer curvature.

The final steps in the transmission of a signal from one neuron to the next, or to a gland or muscle cell, culminate in the release of substances known as **neurotransmitters** from the end of the axon. Common neurotransmitters include amino acids and compounds derived from them. In the case of a synapse linking a motor neuron and its target muscle cell, the neurotransmitter is acetylcholine:

$$H_3C-\overset{\overset{\displaystyle O}{\|}}{C}-O-CH_2-CH_2-\overset{\overset{\displaystyle CH_3}{|}}{\underset{\underset{\displaystyle CH_3}{|}}{N^+}}-CH_3$$

Acetylcholine

Acetylcholine is stored in membrane-bounded compartments, called **synaptic vesicles,** about 40 nm in diameter. When the action potential reaches the axon terminus, voltage-gated Ca^{2+} channels open and allow the influx of extracellular Ca^{2+} ions. The increase in the local intracellular Ca^{2+} concentration, from less than 1 µM to as much as 100 µM, triggers **exocytosis** of the vesicles (exocytosis is the fusion of the vesicle with the plasma membrane such that the vesicle contents are released into the extracellular space). The acetylcholine diffuses across the synaptic cleft, the narrow space between the axon terminus and the muscle cell, and binds to integral membrane protein receptors on the muscle cell surface. This binding initiates a sequence of events that result in muscle contraction (**Fig. 9.19**). In general, the response of the cell that receives a neurotransmitter depends on the nature of the neurotransmitter and the cellular proteins that are activated when the neurotransmitter binds to its receptor. We will explore other receptor systems in Chapter 10.

Ca²⁺ channel

Action potential

SYNAPTIC CLEFT

MUSCLE CELL

AXON

Acetylcholine receptors

Synaptic vesicles

Acetylcholine

1. When an action potential reaches the axon terminus, it causes voltage-gated Ca²⁺ channels to open.

Ca²⁺

2. The increase in intracellular Ca²⁺ ion concentration triggers the fusion of synaptic vesicles with the plasma membrane so that the neurotransmitter acetylcholine is released into the synaptic cleft.

3. Acetylcholine binding to receptors on the surface of the muscle cell leads to muscle contraction. The signal is short-lived because acetylcholine remaining in the synaptic cleft is rapidly degraded.

Muscle contraction

FIGURE 9.19

Events at the nerve–muscle synapse.

 Describe the events that must occur to restore the neuron and muscle cell to their original states.

The events at the nerve–muscle synapse occur rapidly, within about one millisecond, but the effects of a single action potential are limited. First, the Ca^{2+} that triggers neurotransmitter release is quickly pumped back out of the cell by a Ca^{2+}-ATPase. Second, acetylcholine in the synaptic cleft is degraded within a few milliseconds by a lipid-linked or soluble acetylcholinesterase, which catalyzes the reaction

$$H_3C-\overset{\overset{\displaystyle O}{\|}}{C}-O-CH_2-CH_2-\overset{+}{N}(CH_3)_3 \xrightarrow[\text{acetylcholinesterase}]{H_2O \quad\quad H^+}$$

Acetylcholine

$$H_3C-\overset{\overset{\displaystyle O}{\|}}{C}-O^- + HO-CH_2-CH_2-\overset{+}{N}(CH_3)_3$$

Acetate   Choline

Other types of neurotransmitters are recycled rather than destroyed. They are transported back into the cell that released them by the action of secondary active transporters (**Box 9.B**). A neuron may contain hundreds of synaptic vesicles, only a small percentage of which undergo exocytosis at a time, so the cell can release neurotransmitters repeatedly (as often as 50 times per second).

A CLOSER LOOK

Box 9.B Antidepressants Block Serotonin Transport

The neurotransmitter serotonin, a derivative of tryptophan, is released by cells in the central nervous system.

Serotonin

Fluoxetine

Sertraline

Serotonin signaling leads to feelings of well-being, suppression of appetite, and wakefulness, among other things. Seven different families of receptor proteins respond to serotonin signals, sometimes in opposing ways, so the pathways by which this neurotransmitter affects mood and behavior have not been completely defined.

Unlike acetylcholine, serotonin is not broken down in the synapse, but instead about 90% of it is transported back into the cell that released it and is reused. Because the extracellular concentration of serotonin is relatively low, it reenters cells via symport with Na^+, whose extracellular concentration is greater than its intracellular concentration. The rate at which the serotonin transporter takes up the neurotransmitter helps regulate the extent of signaling. Research suggests that genetic variation in the transporter protein may explain an individual's susceptibility to conditions such as depression and post-traumatic stress disorder, but such correlations are difficult to prove, since the level of expression of the transporter appears to vary between individuals and even within an individual.

Drugs known as selective serotonin reuptake inhibitors (SSRIs) block the transporter and thereby enhance serotonin signaling. Some of the most widely prescribed drugs are SSRIs, including fluoxetine (Prozac®) and sertraline (Zoloft®).

These drugs are used primarily as antidepressants, although they are also prescribed for anxiety disorders and obsessive-compulsive disorder. Despite decades of research, the interactions between the serotonin transporter and the drugs are not completely understood at the molecular level, and some studies indicate that different inhibitors may bind to different sites on the transporter.

The results of rigorous clinical tests suggest that SSRIs are most effective at treating severe disorders, whereas in mild cases of depression, the SSRIs are about as effective as a placebo. One of the challenges in assessing the clinical effectiveness of drugs such as fluoxetine and sertraline is that depression is difficult to define biochemically. Furthermore, serotonin signaling pathways are complex, and the body responds to SSRIs with changes in gene expression and adjustments in other signaling pathways so that antidepressive effects may not be apparent for several weeks. The list of SSRI side effects is long and highly variable among individuals, and, disturbingly, includes a small increase in risk of suicidal behavior.

? **What types of food could contribute to serotonin production in the body?**

SNAREs LINK VESICLE AND PLASMA MEMBRANES

Membrane fusion is a multistep process that begins with the targeting of one membrane (for example, the vesicle) to another (for example, the plasma membrane). A number of proteins participate in tethering the two membranes and readying them for fusion. However, many of these proteins may be only accessory factors for the SNAREs, the proteins that physically pair the two membranes and induce them to fuse.

SNAREs are integral membrane proteins (their name is coined from the term "soluble N-ethylmaleimide-sensitive-factor attachment protein receptor"). Two SNAREs from the plasma membrane and one from the synaptic vesicle form a complex that includes a 120-Å-long coiled-coil structure containing four helices (two of the SNAREs contribute one helix each, and one SNARE contributes two helices). The four helices, each with about 70 residues, line up in parallel fashion (**Fig. 9.20**). Unlike other coiled-coil proteins such as keratin (Section 5.3), the four-helix bundle is not a perfectly geometric structure but varies irregularly in diameter.

The mutual interactions between SNAREs in the vesicle and plasma membranes serve as an addressing system so that the proper membranes fuse with each other. Initially, the individual SNARE proteins are unfolded, and they spontaneously zip up to form the four-helix complex. This action necessarily brings the two membranes close together (**Fig. 9.21**). *Because the formation of the SNARE complex is thermodynamically favorable, membrane fusion is also spontaneous.* The rapid rate of acetylcholine release *in vivo* indicates that at least some synaptic vesicles are already docked at the plasma membrane, awaiting the Ca^{2+} signal for fusion to proceed.

Experiments with pure lipid vesicles demonstrate that SNAREs are not essential for membrane fusion to occur *in vitro,* but the rate of fusion does depend on the membranes' lipid compositions. The explanation for this observation is that the lipid bilayers of the fusing membranes undergo rearrangement: The lipids in the contacting leaflets must mix before a pore forms (**Fig. 9.22**). Certain types of lipids appear to promote the required changes in membrane curvature.

In living cells, bilayer shape changes could be facilitated by the tension exerted by the SNARE complex. In addition, membrane lipids may undergo active remodeling. For example, the enzymatic removal of an acyl chain would convert a cylindrical lipid to a cone-shaped lipid. Clustering of such lipids would cause the bilayer to bow outward.

FIGURE 9.20

Structure of the four-helix bundle of the SNARE complex. The three proteins (one includes two helices) are in different colors. Portions of the SNAREs that do not form the helix bundle were cleaved off before X-ray crystallography. [*Structure (pdb 1SFC) determined by R. B. Sutton and A. T. Brunger.*]

 ← Removal of lipid tail Removal of head group →

Conversely, removing large lipid head groups would cause the bilayer to bow inward.

FIGURE 9.21

Model for SNARE-mediated membrane fusion. Formation of the complex of SNAREs from the vesicle and plasma membranes brings the membranes close together so that they can fuse.

? Why does disassembly of the SNARE complex require ATP?

FIGURE 9.22

Schematic view of membrane fusion. For simplicity, the vesicle and plasma membranes are depicted as bilayers.

ENDOCYTOSIS IS THE REVERSE OF EXOCYTOSIS

When neurotransmitter vesicles fuse with the plasma membrane during exocytosis, the plasma membrane gains membrane proteins and lipids. These materials must be removed and recycled in order to maintain the shape of the neuron and to generate new neurotransmitter vesicles for the next round of neurotransmission. One mechanism is simply the reverse of the process shown in Figure 9.22. Soon after the vesicle and plasma membranes fuse, the "fusion pore" closes and the vesicle re-forms. The empty vesicle can later be refilled with neurotransmitters.

Another mechanism involves **endocytosis,** in which a new vesicle forms by budding inward from the plasma membrane. This process also requires the membrane to pinch off (Figure 9.22 in reverse) but could occur anywhere on the cell surface. Cells are capable of several kinds of endocytosis (**Fig. 9.23**). **Pinocytosis** is the formation of small intracellular vesicles that contain extracellular fluid plus whatever solutes happen to be present. In **receptor-mediated endocytosis,** the materials brought into the cell must first bind specifically to a protein receptor on the cell surface. This binding event triggers the membrane shape changes that lead to formation of an intracellular vesicle. The vesicle may subsequently fuse with a lysosome (or lysosomal enzymes may be delivered to the vesicle) so that its contents can be digested by the lysosomal enzymes.

Formation of membranous vesicles—either at the cell surface or involving other membrane systems—often involves "coated" vesicles. So-called coat proteins form a lattice or cage on the cytosolic side of the membrane, forcing the membrane to bud into a spherical shape and then maintaining that structure as the vesicle makes its way to another site in the cell. Vesicle trafficking between the endoplasmic reticulum and Golgi apparatus, and between the Golgi apparatus and the plasma membrane, involves coated vesicles. One well known coat protein is clathrin, whose spindly subunits assemble into a regular geometric structure around a vesicle (**Fig. 9.24**). Clathrin and other coat

FIGURE 9.24

Clathrin assembly. Three clathrin subunits are shown in a model derived from electron micrographs. Multiple subunits assemble to form a lattice that surrounds and defines the shape of an intracellular vesicle. [*Structure (pdb 1XI4) determined by A. Fotin, Y. Cheng, P. Sliz, N. Grigorieff, S. C. Harrison, T. Kirchhausen, and T. Walz.*]

FIGURE 9.23

Endocytosis. Pinocytosis captures extracellular materials nonspecifically. In receptor-mediated endocytosis, a molecule binds to a specific receptor, which is then internalized along with its bound ligand. Coat proteins just beneath the membrane interact with the receptors and help mediate the membrane shape changes required to pinch off a portion of the cell membrane and form a vesicle.

Box 9.C Exosomes

Exosomes, also known as **microvesicles,** are small extracellular membrane-enclosed particles. Although they were once believed to function as a sort of microscopic garbage can containing unwanted cellular components, they are now believed to participate in informal cell–cell communication. Most animal cells release exosomes, which are typically 50 to 150 nm in diameter and contain an assortment of the cell's proteins, lipids, and several types of RNA.

Rather than budding directly from the plasma membrane, exosomes are formed within an intracellular compartment, which then fuses with the plasma membrane to release the exosomes (see diagram below).

Some cells produce exosomes constitutively (constantly), but others seem to release them in response to stimulation.

Neighboring cells take up the exosomes and their cargo, so this transport system may be a mechanism for sharing information about what's going on inside the cells. The RNA components may be particularly informative, since mRNAs reflect what genes are being expressed and small RNAs help regulate gene expression. Cancer cells release exosomes containing components that could make neighboring cells more accommodating of tumor expansion. Exosomes also appear to play a role in coordinating defenses during an immune response.

Because exosomes are present in all body fluids, including blood and urine, they may provide diagnostic information that previously required more invasive procedures such as tissue biopsies. Exosomes have also attracted attention as systems for delivering drugs or other materials that promote tissue healing after injury.

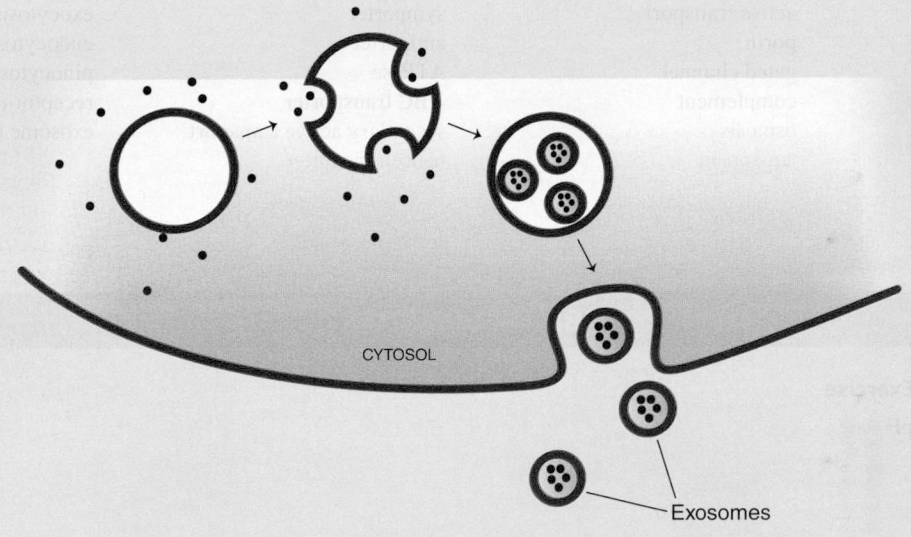

CYTOSOL

Exosomes

proteins must form strong yet flexible networks that change shape as the vesicle matures and separates from its parent membrane. An interesting type of vesicle formation occurs in the production of exosomes (**Box 9.C**).

CONCEPT CHECK

1. Explain how acetylcholine inside a nerve cell reaches a muscle cell.
2. How is the structure of the SNARE complex related to its function?
3. Why are changes in bilayer curvature necessary for endocytosis and exocytosis?

SUMMARY

9.1 The Thermodynamics of Membrane Transport

• The transmembrane movements of ions generate changes in membrane potential, $\Delta\psi$, during neuronal signaling.

• The free energy change for the transmembrane movement of a substance depends on the concentrations on each side of the membrane and, if the substance is charged, on the membrane potential.

9.2 Passive Transport

• Passive transport proteins such as porins allow the transmembrane movement of substances according to their concentration gradients. Aquaporins mediate the transport of water molecules.

• Ion channels have a selectivity filter that allows passage of one type of ion. Gated channels open and close in response to some other event.

• Membrane proteins such as the passive glucose transporter undergo conformational changes that alternately expose ligand-binding sites on each side of the membrane.

9.3   Active Transport

• Active transporters such as the Na,K-ATPase and ABC transporters require the free energy of ATP to drive the transmembrane movement of substances against their concentration gradients.

• Secondary active transport allows the favorable movement of one substance to drive the unfavorable transport of another substance.

9.4   Membrane Fusion

• During the release of neurotransmitters, intracellular vesicles fuse with the cell membrane. SNARE proteins in the vesicle and target membranes form a four-helix structure that brings the fusing membranes close together. Changes in bilayer curvature are also necessary for fusion to occur.

• In endocytosis, a vesicle buds off from an existing membrane to form an intracellular vesicle.

KEY TERMS

membrane potential ($\Delta\psi$)	passive transport	uniporter	synaptic vesicle
gas constant (R)	active transport	symporter	exocytosis
Z	porin	antiporter	endocytosis
Faraday constant ($\mathcal{F}$)	gated channel	ATPase	pinocytosis
action potential	complement	ABC transporter	receptor-mediated endocytosis
axon	osmosis	secondary active transport	exosome (microvesicle)
myelin sheath	aquaporin	neurotransmitter	

BIOINFORMATICS

Brief Bioinformatics Exercise

9.1 Maltoporin and OmpF

PROBLEMS

9.1   The Thermodynamics of Membrane Transport

1. Calculate the membrane potential at 20°C when **(a)** $[Na^+]_{in}$ = 15 mM and $[Na^+]_{out}$ = 150 mM and **(b)** $[Na^+]_{in}$ = 56 mM and $[Na^+]_{out}$ = 35 mM.

2. The resting membrane potential maintained in most nerve cells is about –70 mV. Use Equation 9.2 to calculate the ratio of $[Na^+]_{in}$/$[Na^+]_{out}$ at the resting potential.

3. When a nerve is stimulated, the membrane potential increases from –70 mV to +50 mV. Calculate the ratio of $[Na^+]_{in}$/$[Na^+]_{out}$ in the depolarized nerve cell. How does this ratio compare to the ratio you calculated in Problem 2 and what is the significance of the change?

4. Use Equation 9.4 to calculate the free energy change for the movement of Na^+ into **(a)** cell at the resting potential (described in Problem 2). **(b)** a depolarized nerve cell (described in Problem 3). Assume the temperature is 37°C. Which one of these processes are favourable?

5. In typical marine organisms, the intracellular concentrations of Na^+ and Ca^{2+} are 10 mM and 0.1 μM, respectively. Extracellular concentrations of Na^+ and Ca^{2+} are 450 mM and 4 mM, respectively. Use Equation 9.4 to calculate the free energy changes at 20°C for the transmembrane movement of these ions. In which direction do the ions move? Assume the membrane potential is –70 mV.

6. Calculate the free energy changes at 20°C for the transmembrane movement of Na^+ and K^+ ions using the conditions presented in Figure 9.1. Assume the membrane potential is –70 mV. In which direction do the ions move?

7. The concentration of Ca^{2+} in the endoplasmic reticulum (*outside*) is 1 mM, and the concentration of Ca^{2+} in the cytosol (*inside*) is 0.1 μM. Calculate ΔG at 37°C when the membrane potential is **(a)** –50 mV (cytosol negative) and **(b)** +50 mV. In which case is Ca^{2+} movement more thermodynamically favorable?

8. The concentration of Ca^{2+} in the cytosol (*inside*) is 0.1 μM and the concentration of Ca^{2+} in the extracellular medium (*outside*) is 2 mM. Calculate ΔG at 37°C, assuming the membrane potential is –50 mV. In which direction is Ca^{2+} movement thermodynamically favored?

9. Calculate the free energy cost of moving Na^+ across a membrane from a compartment (*outside*) where $[Na^+]$ = 150 mM to a compartment (*inside*) where $[Na^+]$ = 30 mM. Assume that T = 20°C and $\Delta\psi$ = +50 mV. Is this process spontaneous?

10. Calculate the free energy change for the movement of Cl^- from the extracellular medium, where $[Cl^-]$ = 100 mM, into the cytosol, where $[Cl^-]$ = 10 mM. Assume that T = 37°C and $\Delta\psi$ = –50 mV. Is this process spontaneous?

11. A high fever can interfere with normal neuronal activity. Since temperature is one of the terms in Equation 9.1, which defines membrane potential, a fever could potentially alter a neuron's resting membrane potential.

(a) Calculate the effect of a change in temperature from 98°F to 104°F (37°C to 40°C) on a neuron's membrane potential. Assume that the normal resting potential is −70 mV and that the distribution of ions does not change.

(b) How else might an elevated temperature affect neuronal activity?

12. During mitochondrial electron transport (Chapter 15), protons are transported across the inner mitochondrial membrane from inside the mitochondrial matrix to the intermembrane space. The pH of the mitochondrial matrix is 7.78. The pH of the intermembrane space is 6.88.

(a) What is the membrane potential under these conditions?

(b) Use Equation 9.4 to determine the free energy change for proton transport at 37°C.

13. (a) Calculate the value of ΔG for the movement of glucose from outside to inside at 20°C when the extracellular concentration is 5 mM and the cytosolic concentration is 0.5 mM.

(b) What is the free energy cost of moving glucose from the outside of the cell (where its concentration is 0.5 mM) to the cytosol (where its concentration is 5 mM) when $T = 20°C$?

14. Glucose absorbed by the epithelial cells lining the intestine leaves these cells and travels to the liver via the portal vein. After a high-carbohydrate meal, the concentration of glucose in the portal vein can reach 15 mM.

(a) What is the ΔG for transport of glucose from portal vein blood to the interior of the liver cell where the concentration of glucose is 0.5 mM?

(b) What is the ΔG for transport under fasting conditions when the blood glucose level falls to 4 mM? Assume the temperature is 37°C.

15. What is the ΔG for transport of glutamate from the outside of the cell, where the concentration is 0.1 mM, to the inside of the cell, where the concentration is 10 mM? Assume the membrane potential is −70 mV.

16. Rank the rate of transmembrane diffusion of the following compounds:

$$H_3C-\overset{\overset{\displaystyle O}{\|}}{C}-NH_2 \qquad H_3C-CH_2-CH_2-\overset{\overset{\displaystyle O}{\|}}{C}-NH_2$$

A. Acetamide   B. Butyramide

$$H_2N-\overset{\overset{\displaystyle O}{\|}}{C}-NH_2$$

C. Urea

17. The permeability coefficient indicates a solute's tendency to move from an aqueous solvent to a nonpolar lipid membrane.

(a) How do the permeability coefficients assist you in ranking the rate of the transmembrane diffusion of these compounds?

(b) Compare the permeability coefficients for the two solutes (structures shown below). Which solute moves more easily through the synthetic bilayer and why?

(c) Compare the permeability coefficients of each solute for the two types of membranes. For which solute is the difference more dramatic and why?

Permeability coefficient (cm·s⁻¹)

	Glucose	Mannitol
Synthetic lipid bilayer	2.4×10^{-10}	4.4×10^{-11}
Red blood cell membrane	2.0×10^{-4}	5.0×10^{-9}

Glucose   Mannitol

18. Explain why carbon dioxide can cross a cell membrane without the aid of a transport protein.

9.2 Passive Transport

19. The bacterium *Pseudomonas aeruginosa* expresses a phosphate-specific porin when phosphate in its growth medium is limiting. Noting that there are three lysines clustered in the surface-exposed amino terminal region of the protein, investigators constructed mutants in which the lysine residues were replaced with glutamate residues.

(a) Why did the investigators hypothesize that lysine residues might play an important role in phosphate transport in the bacterium?

(b) Predict the effect of the Lys → Glu substitution on the transport activity of the porin.

20. As noted in the text, the OmpF porin in *E. coli* has a constricted diameter that prevents the passage of large substances. The protein loops in this constricted site contain a D-E-K-A sequence, which makes the porin weakly selective for cations. If you wanted to construct a mutant porin that was highly selective for calcium ions, what changes would you make to the amino acid sequence at the constricted site?

21. In addition to neurons, muscle cells undergo depolarization, although smaller and slower than in the neuron, as a result of the activity of the acetylcholine receptor.

(a) The acetylcholine receptor is also a gated ion channel. What triggers the gate to open?

(b) The acetylcholine receptor/ion channel is specific for Na^+ ions. Do Na^+ ions flow in or out?

(c) How does the Na^+ flow through the ion channel change the membrane potential?

22. Explain why acid-gated channel proteins include a set of Asp or Glu residues in their acid sensors. How would these groups participate in gating?

23. When a bacterial cell is transferred from a solute-rich environment to pure water, mechanosensitive channels open and allow the efflux (outflow) of cytoplasmic contents. Use osmotic effects to explain how this prevents cell death.

24. Ammonia, like water, was long believed to diffuse across membranes without the aid of a channel protein. Researchers deleted the gene *Rhcg* from mice and examined the NH_3 permeability of their kidney cells. Cells from wild-type mice exhibited an NH_3 flux about three times higher than the mutant cells.

(a) What do these results suggest about the role of the protein encoded by *Rhcg*?

(b) In light of what you know about aquaporin, does this surprise you?

25. The selectivity filter in the bacterial chloride channel ClC is formed in part by the hydroxyl groups of Ser and Tyr side chains and main chain NH groups. Explain how these groups would allow Cl⁻ but not cations to pass through.

26. The selectivity filter of the KcsA channel has the sequence TVGYG. The corresponding sequences of some other ion channels, and their ion specificities, are given below. What do you expect would be the ion selectivity of channels with the following sequences: (a) SVGFG, (b) ITMEG?

Channel	Sequence	Ion specificity
KcsA	T V G Y G	K^+
NaK	T V G D G	Na^+ and K^+
Ca^{2+} II	L T G E D	Ca^{2+}
Na^+ IV	T T S A G	Na^+

27. A plot of the glucose transport rate versus glucose concentration for the passive glucose transporter of red blood cells is shown here.

(a) Explain why the plot yields a hyperbolic curve.
(b) Use what you have learned about Michaelis–Menten kinetics to estimate the V_{max} and the K_M values for this glucose transporter.

28. The compound 6-O-benzyl-D-galactose competes with glucose for binding to the glucose transporter. Sketch a curve similar to the one shown in Problem 27 that illustrates the kinetics of glucose transport in the presence of this inhibitor.

29. Experiments with erythrocyte "ghosts" were carried out to learn more about the glucose transporter in red blood cells. "Ghosts" are prepared by lysing red blood cells in a hypotonic medium and washing away the cytoplasmic contents. Suspension in an isotonic buffer allows the ghost membranes to reseal. If ghosts are prepared so that the enzyme trypsin is incorporated into the ghost interior, glucose transport does not occur. But glucose transport is not affected if trypsin is located in the extracellular ghost medium. What can you conclude about the glucose transporter, given these observations?

30. If a propyl group is added to the hydroxyl group on C1 of glucose, the modified glucose is unable to bind to its transporter on the extracellular surface. Another experiment showed that if a propyl group is added to the hydroxyl group on C6 of glucose, the modified glucose is unable to bind to its transporter on the cytosolic surface of the membrane. What do these observations tell you about the mechanism of passive glucose transport?

31. Glutamate acts as a neurotransmitter in the brain, and it is reabsorbed and recycled by glial cells associated with the neurons. The glutamate transporter also transports 3 Na^+ and 1 H^+ along with glutamate, and it transports 1 K^+ in the opposite direction. What is the net charge imbalance across the cell membrane for each glutamate transported?

32. The bacterium *Oxalobacter formigenes* is found in the intestine, where it plays a role in digesting the oxalic acid found in some fruits and vegetables (spinach is a particularly rich source). The bacterium takes up oxalic acid from the extracellular medium in the form of oxalate. Once inside the cell, the oxalate undergoes decarboxylation to produce formate, which leaves the cell in antiport with oxalate.

(a) What is the net charge imbalance generated by this system?
(b) Why do you suppose the investigators who elucidated this mechanism referred to the process as a "hydrogen pump"?

33. As discussed in Section 5.1, tissues produce CO_2 as a waste product during respiration. The CO_2 enters the red blood cell and combines with water to form carbonic acid in a reaction catalyzed by carbonic anhydrase. A red blood cell protein called Band 3 transports HCO_3^- ions in exchange for Cl^- ions.

(a) What role does Band 3 play in transporting CO_2 to the lungs where it can be exhaled?
(b) Band 3 is unusual in that it can transport HCO_3^- in exchange for Cl^- ions, in either direction, via passive transport. Why is it important that Band 3 operate in both directions?

34. Glucose uptake by cancer cells via the GLUT1 transport protein occurs far more rapidly than in normal cells. The cancer cell metabolizes the glucose to lactic acid, which exits the cell via the MCT4 transport protein. In response, the activity of carbonic anhydrase (see Section 2.5 and Problem 33a), whose catalytic domain is on the outside of the cancer cell, increases. The carbonic anhydrase is associated with a bicarbonate transporter, which assists in maintaining the slightly alkaline pH that cancer cells prefer.

(a) Why is it advantageous for the cancer cell to up-regulate the activity of carbonic anhydrase?
(b) List several drug targets that have the potential to interfere with this process and kill the cancer cell.

35. Liver cells use a choline transport protein to import choline from the portal circulation. Choline transport was measured in mouse cells transfected with the gene for the hepatic transporter. The uptake of radioactively labeled choline by the transfected cells was measured at increasing choline concentrations.

$$HO-(CH_2)_2-\overset{CH_3}{\underset{CH_3}{\overset{|}{\underset{|}{N^+}}}}-CH_3$$
Choline

(a) Use the language of Michaelis and Menten as described in Problem 27 to estimate K_M and V_{max} from the curve shown.

(b) Plasma concentrations of choline range from 10 to 80 μM, although the concentration may be higher in portal circulation after ingestion of choline. Does the transporter operate effectively at these concentrations?
(c) The choline transport protein is inhibited by low external pH and stimulated by high external pH. What role might protons play in the transport of choline?

(d) Propose a mechanism that explains how tetramethylammonium (TEA) ions inhibit choline transport.

Tetraethylammonium (TEA)

36. Unidirectional glucose transport into the brain was measured in the presence and absence of phlorizin. The velocity of transport was determined at various glucose concentrations and the data are displayed as a Lineweaver–Burk plot.

(a) Calculate the K_M and the V_{max} in the presence and absence of phlorizin.

(b) What kind of inhibitor is phlorizin? Explain.

With inhibitor
$y = 13.8x + 0.65$

No inhibitor
$y = 5.4x + 0.64$

9.3 Active Transport

37. The Na,K-ATPase first binds sodium ions, then reacts with ATP to form an aspartyl phosphate intermediate. Draw the structure of the phosphorylated Asp residue.

38. Ouabain, an extract of the East African ouabio tree, has been used as an arrow poison. Ouabain binds to an extracellular site on the Na,K-ATPase and prevents hydrolysis of the phosphorylated intermediate. Why is ouabain a lethal poison?

39. The parasite *Trypanosoma brucei*, the causative agent of sleeping sickness, uses proline as an energy source during one stage of its life cycle. The properties of its proline-specific transporter were investigated in a series of experiments. It was discovered that L-hydroxyproline (see Section 5.3) is a potent inhibitor of the transporter whereas D-proline is not, and that proline transport is not affected by pH, Na+, K+, or ouabain (see Problem 38). Incubation of the cells with iodoacetate (which decreases intracellular ATP) decreases the rate of proline transport into the parasite. What do these experiments tell you about the nature of the proline transporter in trypanosomes?

40. In eukaryotes, ribosomes (approximate mass 4×10^6 D) are synthesized inside the nucleus, which is enclosed by a double membrane. Protein synthesis occurs in the cytosol. (a) Would a glucose transporter–type protein be capable of transporting ribosomes into the cytoplasm? Would a porin-type transporter be able to do so? Explain why or why not. (b) Do you think that free energy would be required to move ribosomes from the nucleus to the cytoplasm? Why or why not?

41. The retina of the eye contains equal amounts of endothelial and pericyte cells. Basement membrane thickening in pericytes occurs during the early stages of diabetic retinopathy, eventually leading to blindness. Glucose uptake was measured in both types of cells in culture in the presence of increasing amounts of sodium. The results are shown in the graph.

(a) What is your interpretation of these results?

(b) What information is conveyed by the shapes of the curves?

(c) By what mechanism might the pericytes use sodium ions to assist with glucose import?

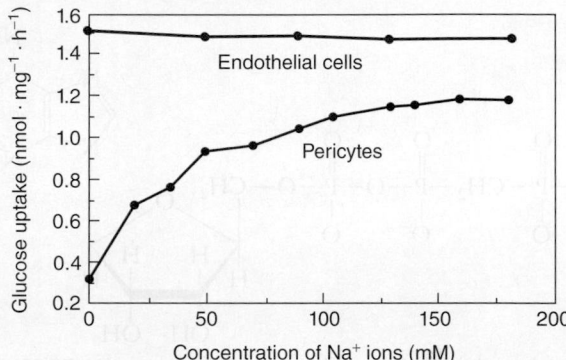

42. The kinetics of transport through protein transporters can be described using the language of Michaelis and Menten. The transported substance is analogous to the substrate, and the protein transporter is analogous to the enzyme. K_M and V_{max} values can be determined to describe the binding and transport process where K_M is a measure of the affinity of the transported substance for its transporter and V_{max} is a measure of the rate of the transport process. Use the information provided in the figure to estimate the K_M and V_{max} for glucose uptake by the pericyte transporter described in Problem 41?

43. Many ABC transporters are inhibited by vanadate, a phosphate analog. Why is vanadate an effective inhibitor of these transporters?

44. The ABC multidrug transporter LmrA from *Lactococcus* exports cytotoxic compounds like vinblastine. There are two binding sites for vinblastine on the transporter; one vinblastine binds with an association constant (equivalent to $1/K_d$) of 150 nM; the association constant of the second vinblastine is 30 nM. What do these values tell you about the mode of vinblastine binding to LmrA?

45. Kidney cells include two antiport proteins, an H+/Na+ exchanger and a Cl−/HCO3− exchanger (see Fig. 2.18). What is the source of free energy that drives the transmembrane movement of all these ions?

46. Many cells have a mechanism for exporting ammonium ions. Describe how this could occur through secondary active transport.

47. The PEPT1 transporter aids in digestion by transporting di- and tripeptides into the cells lining the small intestine. There are three components of this system: (a) a symport transporter that ferries di- and tripeptides across the membrane, along with an H^+ ion, (b) a Na^+–H^+ antiporter (H^+/Na^+ exchanger), and (c) a Na,K-ATPase. Draw a diagram that illustrates how these three transporters work together to transport peptides into the cell.

48. The X-ray structure of a Ca^{2+}-ATPase was determined by crystallizing the enzyme along with adenosine-5′-(βγ-methylene) triphosphate (AMPPCP), an ATP analog (structure shown below). How did the co-crystallization strategy assist the crystallographers in imaging this protein?

Adenosine-5′-(β,γ-methylene) triphosphate (AMPPCP)

9.4 Membrane Fusion

49. Myasthenia gravis is an autoimmune disorder characterized by muscle weakness and fatigue. Patients with the disease generate antibodies against the acetylcholine receptor in postsynaptic cells; this results in a decrease in the number of receptors. The disease can be treated by administering drugs that inhibit acetylcholinesterase. Why is this an effective strategy for treating the disease?

50. Lambert–Eaton syndrome is another autoimmune muscle disorder (see Problem 49), but in this disease, antibodies against the voltage-gated calcium channels in the presynaptic cell prevent the channels from opening. Patients with this disease suffer from muscle weakness. Explain why.

51. Like chymotrypsin, acetylcholinesterase is a member of the serine protease family because it contains an active site serine, and like chymotrypsin, it reacts with DIPF. Draw the structure of the enzyme's catalytic residue modified by DIPF.

52. Parathion and malathion are organophosphorus compounds similar to DIPF (see Problem 51). These compounds are sometimes used as insecticides. Why are these compounds deadly poisons?

53. The Na,K-ATPase is essential for establishing the ion gradients that make neuronal signaling possible. Explain why the pump is also required to recycle serotonin.

54. Serotonin and other so-called monoamine signaling molecules are degraded in the liver, beginning with a reaction catalyzed by a monoamine oxidase (MAO). Explain why doctors avoid prescribing an MAO inhibitor along with an SSRI.

55. The toxin produced by *Clostridium tetani*, which causes tetanus, is a protease that cleaves and destroys SNAREs. Explain why this activity would lead to muscle paralysis.

56. The drug known as Botox is a preparation of botulinum toxin, which is similar to the tetanus toxin (Problem 55). Describe the biochemical basis for its use by plastic surgeons, who inject small amounts of it to alleviate wrinkling in areas such as around the eyes.

57. Phosphatidylinositol is a membrane glycerophospholipid whose head group includes a monosaccharide (inositol) group. A certain kinase adds another phosphate group to a phosphorylated phosphatidylinositol:

Why might this activity be required during the production of a new vesicle, which forms by budding from an existing membrane?

58. Some studies show that prior to membrane fusion, the proportion of diacylglycerol in the bilayer increases. Explain how the presence of this lipid would aid the fusion process.

59. In autophagy (literally, "self-eating"), a damaged or unneeded cellular organelle becomes enclosed in a compartment called an autophagosome. Autophagosome formation begins with an assembly of lipids and proteins that grows by acquiring additional lipids until it forms a small bilayer-enclosed compartment called a phagophore. This structure continues to expand, encircling the damaged organelle. A membrane fusion event closes off the autophagosome. Complete the diagram of autophagosome formation shown here. How many membranes now separate the damaged organelle from the rest of the cell?

Developing phagophore

Damaged organelle

60. After an autophagosome has formed (Problem 59), a lysosome fuses with it to deliver hydrolytic enzymes that will eventually degrade the damaged organelle inside the autophagosome. Using the diagram you made for Problem 55 as a starting point, show that the lysosomal enzymes must degrade a lipid bilayer before they can begin to degrade the organelle.

61. Draw a graph (similar to the one shown in Problem 27) that compares the kinetics of substance delivery into the cell via pinocytosis and receptor-mediated endocytosis (see Fig. 9.23).

62. Low-density lipoprotein (LDL, see Problem 8.63 for a description) enters the cell via receptor-mediated endocytosis by binding to a specific cell-surface receptor that localizes to an invaginated area on the membrane surface called a clathrin-coated pit (see Fig. 9.23). Once internalized, the LDL particle is degraded; cholesterol is available to the cell for biosynthetic reactions, and the rate-limiting enzyme that catalyzes the cell's own synthesis of cholesterol is inactivated. The experimental data used to elucidate this process are shown below. Redraw these graphs to show how the normal process would be altered in (**a**) a patient with a defective receptor that could not bind LDL, and (**b**) a patient with an LDL receptor that does not localize to a clathrin-coated pit.

SELECTED READINGS

Gouaux, E. and MacKinnon, R., Principles of selective ion transport in channels and pumps, *Science* **310,** 1461–1465 (2005). [Compares several transport proteins of known structure and discusses the selectivity of Na^+, K^+, Ca^{2+}, and Cl^- transport.]

Higgins, C. F., Multiple molecular mechanisms for multidrug resistance transporters, *Nature* **446,** 749–757 (2007). [Describes some general features of ABC transporters and others that function in drug resistance.]

Jahn, R. and Fasshauer, D., Molecular machines governing exocytosis of synaptic vesicles, *Nature* **490,** 201–207 (2012). [Reviews many of the events involved in nerve cell signaling, including vesicle fusion.]

Kreida, S. and Tömroth-Horsefield, S., Structural insights into aquaporin selectivity and regulation, *Curr. Opin. Struct. Biol.* **33,** 126–134 (2015). [Provides up-to-date information on aquaporin structures and functions.]

Morth, J. P., Pederson, B. P., Buch-Pederson, M. J., Andersen, J. P., Vilsen, B., Palmgren, M. G., and Nissen, P., A structural overview of the plasma membrane Na^+, K^+-ATPase and H^+-ATPase ion pumps, *Nat. Rev. Mol. Cell Biol.* **12,** 60–70 (2011). [Summarizes the structures and physiological importance of two active transporters.]

C H A P T E R **10**

BIOCHEMICAL SIGNALING

Hysteria/Shutterstock

Tetrahydrocannabinol, the active ingredient in Cannabis sativa, *mimics endogenous (natural) endocannabionoid signaling molecules such as anandamide, which act on receptors in the central nervous system. Substances that bind to the endocannabinoid receptors are useful because they stimulate appetite and reduce anxiety, but they also interfere with neural development and short-term memory.*

DO YOU REMEMBER?

- Some proteins can adopt more than one stable conformation (Section 4.3).
- Allosteric regulators can inhibit or activate enzymes (Section 7.3).
- Cholesterol and other lipids that do not form bilayers have a variety of other functions (Section 8.1).
- Integral membrane proteins completely span the bilayer by forming one or more α helices or a β barrel (Section 8.3).
- Conformational changes resulting from ATP hydrolysis drive Na^+ and K^+ transport in the Na,K-ATPase (Section 9.3).

All cells, including prokaryotes, must have mechanisms for sensing external conditions and responding to them. Because the cell membrane creates a barrier between outside and inside, communication typically involves an extracellular molecule binding to a cell-surface receptor. The receptor then changes its conformation to transmit information to the cell interior. A signaling pathway requires many proteins, from the receptor itself to the intracellular proteins that ultimately respond to the signal by changing their behavior. We begin this chapter by describing some characteristics of signaling pathways and then examine some well-known systems that involve G proteins and receptor tyrosine kinases.

10.1 GENERAL FEATURES OF SIGNALING PATHWAYS

KEY CONCEPT
- Receptor–ligand binding is described in terms of a dissociation constant.
- G protein–coupled receptors and receptor tyrosine kinases are the two major types of receptors that transduce extracellular signals to the cell interior.
- Regulatory mechanisms limit the extent of signaling.

Every signaling pathway requires a **receptor,** most commonly an integral membrane protein, that specifically binds a small molecule called a **ligand.** A receptor does not merely bind its ligand in the way that hemoglobin binds oxygen; rather, *a receptor interacts with its ligand in such a way that some kind of response occurs inside the cell.* This is **signal transduction:** the signal itself does not enter the cell, but information is transmitted.

A LIGAND BINDS TO A RECEPTOR WITH A CHARACTERISTIC AFFINITY

Extracellular signals can take many forms, including amino acids and their derivatives, peptides, lipids, and other small molecules (**Table 10.1**). Some are formally called **hormones,**

TABLE 10.1 Examples of Extracellular Signals

HORMONE	CHEMICAL CLASS	SOURCE	PHYSIOLOGICAL FUNCTION
Auxin	Amino acid derivative	Most plant tissues	Promotes cell elongation and flowering in plants
Cortisol	Steroid	Adrenal gland	Suppresses inflammation
Epinephrine	Amino acid derivative	Adrenal gland	Prepares the body for action
Erythropoietin	Polypeptide (165 residues)	Kidneys	Stimulates red blood cell production
Growth hormone	Polypeptide (19 residues)	Pituitary gland	Stimulates growth and metabolism
Nitric oxide	Gas	Vascular endothelial cells	Triggers vasodilation
Thromboxane	Eicosanoid	Platelets	Activates platelets and triggers vasoconstriction

which are substances produced in one tissue that affect the functions of other tissues, but many signals go by other names. Keep in mind that other stimuli—such as light, mechanical stress, odorants, and tastants—also serve as signals for cells, although we will not discuss them here. Some bacterial signal molecules are described in **Box 10.A.**

Signaling molecules behave much like enzyme substrates: *They bind to their receptors with high affinity, reflecting the structural and electronic complementarity between each ligand and its binding site.* Receptor–ligand binding can be written as a reversible reaction, where R represents the receptor and L the ligand:

$$R + L \rightleftharpoons R \cdot L$$

Biochemists express the strength of receptor–ligand binding as a **dissociation constant, K_d,** which is the reciprocal of the association constant. For this reaction,

$$K_d = \frac{[R][L]}{[R \cdot L]} \qquad \textbf{[10.1]}$$

(See Sample Calculation 10.1.) In keeping with other binding phenomena, such as oxygen binding to myoglobin (Section 5.1) or substrate binding to an enzyme (Section 7.2), K_d *is the ligand concentration at which the receptor is half-saturated with ligand;* in other words, half the receptor molecules have bound ligand (**Fig. 10.1**).

A ligand that binds to a receptor and elicits a biological effect is known as an **agonist.** For example, adenosine is the natural agonist of the adenosine receptor. Adenosine signaling in cardiac muscle slows the heart, and adenosine signaling in the brain leads to a decrease in neurotransmitter release, producing a sedative effect.

Adenosine

FIGURE 10.1

Receptor–ligand binding. As the ligand concentration [L] increases, more receptor molecules bind ligand. Consequently, the fraction of receptors that have bound ligand [R·L] approaches 1.0. $[R]_T$ is the total concentration of receptors. The dissociation constant K_d is the ligand concentration at which half of the receptor molecules have bound ligand.

? Compare this graph to Figures 5.3 and 7.5.

Box 10.A Bacterial Quorum Sensing

Even free-living single-celled organisms need to communicate with others of their kind. In bacteria, one form of intercellular communication is known as **quorum sensing,** which allows the cells to monitor population density and adjust gene expression accordingly. As a result, groups of cells can undertake communal endeavors such as producing the polysaccharides and other materials that form a protective matrix called a biofilm (discussed in Section 11.2). Quorum sensing also prepares cells to take up or exchange DNA, an occasional necessity for asexually reproducing organisms. Pathogenic bacteria may use quorum sensing to wait until their numbers are sufficiently high before synthesizing toxins and other proteins needed to attack a host organism.

The essence of quorum sensing is that *cells respond to some signal molecule that increases in concentration as cell density increases.* Only a few types of these molecules have been identified. One type consists of acyl homoserine lactones. The acyl chains of these molecules, which may include from 4 to 18 carbons, are derived either from fatty acids within the cell or from exogenous lipids.

A number of different acyl chains can be attached to the homoserine lactone group, making this class of compounds highly diverse. These lipids are poorly soluble in water, so they may be released from cells in the form of lipid vesicles (Section 2.2) that also contain other types of molecules. Because they are hydrophobic, the molecules can diffuse across the membranes of recipient cells and combine with a receptor protein in the cytosol. The receptor–ligand complex then binds to DNA to turn on the expression of certain genes.

A given bacterial species may produce dozens of different molecules that could be used for quorum sensing. Some of these molecules appear to do double duty as toxins for other species of bacteria, thus coordinating the growth of one species at the expense of others. In retaliation, some bacteria have evolved mechanisms to degrade the signals produced by other species or to synthesize molecules that competitively inhibit receptor binding by the other signals.

An acyl homoserine lactone

? Microbiologists have proposed that drugs that interfere with quorum sensing would be useful as antibiotics, particularly since there would be little selective pressure for individual bacteria to evolve resistance to the drug. Explain.

SAMPLE CALCULATION 10.1

PROBLEM

A sample of cells has a total receptor concentration of 10 mM. Twenty-five percent of the receptors are occupied with ligand, and the concentration of free ligand is 15 mM. Calculate K_d for the receptor–ligand interaction.

SOLUTION

Because 25% of the receptors are occupied, $[R \cdot L]$ = 2.5 mM and $[R]$ = 7.5 mM. Use Equation 10.1 to calculate K_d:

$$K_d = \frac{[R][L]}{[R \cdot L]}$$
$$= \frac{(0.0075)(0.015)}{(0.0025)}$$
$$= 0.045 \text{ M} = 45 \text{ mM}$$

PRACTICE PROBLEMS

1. A sample of cells has a total receptor concentration of 28 μM, and 45% of the receptors are occupied with ligand. The concentration of free ligand is 10 μM. Calculate K_d.

2. K_d for a receptor–ligand interaction is 5 mM. When the concentration of free ligand is 20 mM and the concentration of free receptor is 8 mM, what is the concentration of receptor that is occupied by ligand?

3. The total concentration of receptors in a sample is 20 mM. The concentration of free ligand is 5 mM, and K_d is 10 mM. Calculate the percentage of receptors that are occupied by ligand.

Caffeine is an **antagonist** of the adenosine receptor because it binds to the receptor but does not trigger a response. It functions like a competitive enzyme inhibitor (Section 7.3). As a result, caffeine keeps the heart rate high and produces a sense of wakefulness. Like caffeine, the majority of drugs currently in clinical use act as agonists or antagonists for various receptors involved in regulating such things as blood pressure, reproduction, and inflammation.

Ligands typically bind to a receptor with high affinity and high specificity, but because the ligand–receptor interactions are noncovalent, binding is reversible. Consequently, a cell responds to a signal molecule—or a drug—only while that molecule is associated with its receptor.

Caffeine

MOST SIGNALING OCCURS THROUGH TWO TYPES OF RECEPTORS

When an agonist binds to the adenosine receptor, which is a transmembrane protein, the receptor undergoes a conformational change so that it can interact with an intracellular protein called a **G protein.** Such receptors are therefore called **G protein–coupled receptors (GPCRs).** G proteins are named for their ability to bind guanine nucleotides (GTP and GDP). In response to receptor–ligand binding, the G protein becomes activated and in turn interacts with and thereby activates additional intracellular proteins. Often, one of these is an enzyme that generates a small molecule product that diffuses throughout the cell. These small molecules are called **second messengers** because they represent the intracellular response to the extracellular, or first, message that binds to the GPCR. A variety of substances serve as second messengers in cells, including nucleotides, nucleotide derivatives, and the polar and nonpolar portions of membrane lipids. The presence of a second messenger can alter the activities of cellular proteins, leading ultimately to changes in metabolic activity and gene expression. These events are summarized in **Figure 10.2a**.

FIGURE 10.2

Overview of signal transduction pathways. Ligand binding to a cell-surface receptor causes a signal to be transduced to the cell interior, leading eventually to changes in the cell's behavior. (a) Ligand binding to a G protein–coupled receptor triggers the activation of a G protein, which then activates an enzyme that produces a second messenger. Second messenger molecules diffuse away to activate or inhibit the activity of target proteins in the cell. (b) Ligand binding to a receptor tyrosine kinase activates the kinase activity of the receptor so that intracellular proteins become phosphorylated. A series of kinase reactions activates or inhibits target proteins by adding phosphoryl groups to them.

? Which pathway components are enzymes?

A second type of receptor, also a transmembrane protein, becomes activated as a kinase as a result of ligand binding. A **kinase** is an enzyme that transfers a phosphoryl group from ATP to another molecule. In this case, the phosphoryl group is condensed with the hydroxyl group of a tyrosine side chain on a target protein, so these receptors are termed **receptor tyrosine kinases.** In some signal transduction pathways involving receptor tyrosine kinases, the target protein is also a kinase that becomes catalytically active when phosphorylated. The result may be a series of kinase-activation events that eventually lead to changes in metabolism and gene expression (Fig. 10.2b).

Some receptor systems include both G proteins and tyrosine kinases, and others operate by entirely different mechanisms. For example, the acetylcholine receptor on muscle cells (Fig. 9.19) is a ligand-gated ion channel. When acetylcholine is released into the neuromuscular synapse and binds to the receptor, Na^+ ions flow into the muscle cell, causing depolarization that leads to an influx of Ca^{2+} ions to trigger muscle contraction.

THE EFFECTS OF SIGNALING ARE LIMITED

The multistep nature of signaling pathways and the participation of enzyme catalysts ensure that the signal represented by an extracellular ligand will be amplified as it is transduced inside the cell (see Fig. 10.2). Consequently, *a relatively small extracellular signal can have a dramatic effect on a cell's behavior.* The cell's responses to the signal, however, are regulated in various ways.

The speed, strength, and duration of a signaling event may depend on the cellular location of the components of the signaling pathway. There is evidence that components for some pathways are preassembled in multiprotein complexes in or near the plasma membrane so that they can be quickly activated when the ligand docks with its receptor. Components that must diffuse long distances to reach their targets, or that move from the cytoplasm to the nucleus, may need more time to trigger cellular responses.

Because signaling pathways tend to be branched rather than completely linear, the same intracellular components may participate in more than one signal transduction pathway, so two different extracellular signals could ultimately achieve the same intracellular results. Alternatively, two signals could cancel each other's effects. The response of a given cell, which expresses many different types of receptors, therefore depends on how various signals are integrated. Similarly, different types of cells may include different intracellular components and therefore respond to the same ligand in different ways.

In a biological system that obeys the law of homeostasis, *any process that is turned on must eventually be turned off.* Such control applies to signaling pathways. For example, shortly after a G protein has been activated by its associated receptor, it becomes inactive again. The action of kinases is undone by the action of enzymes that remove phosphoryl groups from target proteins. These and other reactions restore the signaling components to their resting state so that they can be ready to respond again when another ligand binds to its receptor.

Finally, most people are aware that a strong odor loses its pungency after a few minutes. This occurs because olfactory receptors, like other types of receptors, become **desensitized.** In other words, the receptors become less able to transmit a signal even when continually exposed to ligand. Desensitization may allow the signaling machinery to reset itself at a certain level of stimulation so that it can better respond to subsequent changes in ligand concentration.

 CONCEPT CHECK

1. Compare hormone receptors to enzymes and simple binding proteins.
2. Why does an extracellular signal need a second messenger?
3. How are extracellular signals amplified inside a cell?
4. Draw simple diagrams of signaling pathways involving G proteins and receptor tyrosine kinases.
5. Explain why a receptor would need to be turned off or desensitized.

G PROTEIN SIGNALING PATHWAYS

Over 800 genes in the human genome encode G protein–coupled receptors, and these proteins are responsible for transducing the majority of extracellular signals. In this section we will describe some features of these receptors, their associated G proteins, and various second messengers and their intracellular targets.

G PROTEIN–COUPLED RECEPTORS INCLUDE SEVEN TRANSMEMBRANE HELICES

The GPCRs are known as 7-transmembrane (7TM) receptors because they include seven α helices, which are arranged much like those of the membrane protein bacteriorhodopsin (Fig. 8.8). Many G protein–coupled receptors are palmitoylated at a cysteine residue, so they are also lipid-linked proteins (Section 8.3). In the GPCR family, the helical segments are more strongly conserved than the loops that join them on the intracellular and extracellular sides of the membrane.

The structure of one of these proteins, the β_2-adrenergic receptor, is shown in **Figure 10.3**. The ligand-binding site of a GPCR is defined by a portion of the helical core of the protein as well as its extracellular loops. Aside from this general location, there are few similarities in the structures of the binding sites in different GPCRs, which is consistent with the observation that each receptor is specific for only one or a few of many possible ligands, which may be large or small, polar or nonpolar substances.

The physiological ligands for the β_2-adrenergic receptor are the hormones epinephrine and norepinephrine, which are synthesized by the adrenal glands from the amino acid tyrosine.

Epinephrine

Norepinephrine

These same substances, sometimes called adrenaline and noradrenaline, also function as neurotransmitters. They are responsible for the fight-or-flight response, which is characterized by fuel mobilization, dilation of blood vessels and bronchi (airways), and increased cardiac action. Antagonists that prevent signaling via the β_2-adrenergic receptor, known as β-blockers, are used to treat high blood pressure.

How does the receptor transmit the extracellular hormonal signal to the cell interior? *Signal transduction depends on conformational changes involving the receptor's transmembrane helices.* Two of the helices shift slightly to accommodate the ligand, which repositions one of the cytoplasmic protein loops. Studies with a variety of different ligands indicate that the receptor can actually adopt a range of conformations, suggesting that the receptor is not merely an on–off switch but can mediate the effects of strong as well as weak agonists.

THE RECEPTOR ACTIVATES A G PROTEIN

Ligand-induced conformational changes in a G protein–coupled receptor open up a pocket on its cytoplasmic side to create a binding site for a specific G protein (**Fig. 10.4a**). The G protein is presumably already close to the receptor, since it is lipid-linked. The trimeric G proteins associated with GPCRs consist of three subunits, designated α, β, and γ (Fig. 10.4b; other types of G proteins do not have this three-part structure). In the resting state, GDP is bound to the α subunit, but

KEY CONCEPTS
- Ligand binding to a G protein–coupled receptor alters its conformation so that an intracellular G protein becomes activated.
- The G protein stimulates adenylate cyclase to produce the second messenger cAMP, which activates protein kinase A.
- G protein–dependent signaling is limited by several mechanisms.
- The phosphoinositide signaling system activates a G protein, which leads to the production of lipid-derived second messengers and the activation of protein kinase C.
- Cross-talk results when signaling pathways share components.

FIGURE 10.3

The β_2-adrenergic receptor. The backbone structure of the protein is colored in rainbow order from the N-terminus (blue) to the C-terminus (red). A ligand is shown in space-filling form in blue. [*Structure (pdb 2RH1) determined by V. Cherezov, D. M. Rosenbaum, M. A. Hanson, S. G. F. Rasmussen, F. S. Thian, T. S. Kobilka, H. J. Choi, P. Kuhn, W. I. Weis, B. K. Kobilka, and R. C. Stevens.*]

? Explain why the receptor for a polar hormone must be a transmembrane protein.

FIGURE 10.4

A GPCR–G protein complex.
(a) Side view. The GPCR is purple, a bound agonist is red, and the G protein is yellow, green, and blue. [*Structure (pdb 3SN6) determined by S. G. F. Rasmussen et al.*] (b) The β subunit (green) has a propeller-like structure. The small γ subunit (yellow) associates tightly with the β subunit. The α subunit (blue) binds a guanine nucleotide (GDP, orange) in a cleft between two domains. The α and β subunits are covalently attached to lipids, so they are anchored to the cytosolic leaflet of the plasma membrane near the receptor. [*Structure (pdb 1GP2) determined by M. A. Wall and S. R. Sprang.*].

(a)   (b)

association with the hormone–receptor complex causes the G protein to release GDP and bind GTP in its place. The third phosphate group of GTP is not easily accommodated in the αβγ trimer, so the α subunit dissociates from the β and γ subunits, which remain together. Once they dissociate, the α subunit and the βγ dimer are both active; that is, *they interact with additional cellular components in the signal transduction pathway.* However, since both the α and β subunits include lipid anchors, the G protein subunits do not diffuse far from the receptor that activated them.

The signaling activity of the G protein is limited by the intrinsic GTPase activity of the α subunit, which slowly converts the bound GTP to GDP:

$$GTP + H_2O \rightarrow GDP + P_i$$

Hydrolysis of the GTP allows the α and βγ units to reassociate as an inactive trimer (**Fig. 10.5**). The cost for the cell to switch a G protein on and then off again is the free energy of the GTP hydrolysis reaction (GTP is energetically equivalent to ATP).

ADENYLATE CYCLASE GENERATES THE SECOND MESSENGER CYCLIC AMP

Cells contain a number of different types of G proteins that interact with various targets in the cell and activate or inhibit them. A single receptor may interact with more than one G protein, so the effects of ligand binding are amplified at this point. One of the major targets of the activated G protein is an integral membrane enzyme called adenylate cyclase. When the α subunit of the G protein binds, the enzyme's catalytic domains convert ATP to a molecule known as cyclic AMP (**cAMP**). cAMP is a second messenger that can freely diffuse in the cell.

ATP $\xrightarrow[\text{adenylate cyclase}]{PP_i}$ Cyclic AMP (cAMP)

CYCLIC AMP ACTIVATES PROTEIN KINASE A

Among the targets of cAMP is an enzyme called protein kinase A or PKA. In the absence of cAMP, this kinase is an inactive tetramer of two regulatory (R) and two catalytic (C) subunits (**Fig. 10.6**). A segment of each R subunit occupies the active site in a C subunit so that the kinase is unable to phosphorylate any substrates. cAMP binding to the regulatory subunits relieves the inhibition, causing the tetramer to release the two active catalytic subunits.

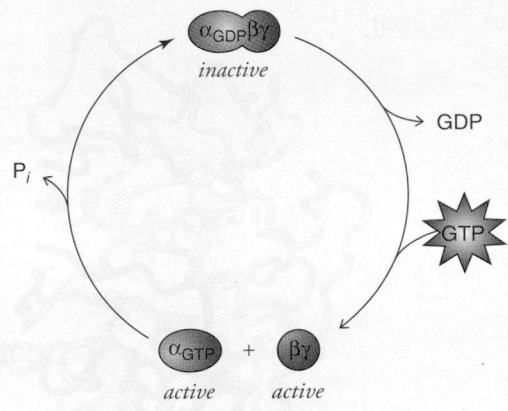

FIGURE 10.5

The G protein cycle. The αβγ trimer, with GDP bound to the α subunit, is inactive. Ligand binding to a receptor associated with the G protein triggers a conformational change that causes GTP to replace GDP and the α subunit to dissociate from the βγ dimer. Both portions of the G protein are active in the signaling pathway. The GTPase activity of the α subunit returns the G protein to its inactive trimeric state.

Consequently, cAMP functions as an allosteric activator of the kinase, and *the level of cAMP determines the level of activity of protein kinase A.*

Protein kinase A is known as a Ser/Thr kinase because it transfers a phosphoryl group from ATP to the serine or threonine side chain of a target protein.

$$-CH_2-O-PO_3^{2-} \qquad\qquad -CH-O-PO_3^{2-}$$
$$\phantom{-CH_2-O-PO_3^{2-} \qquad\qquad -CH-O-}|$$
$$\phantom{-CH_2-O-PO_3^{2-} \qquad\qquad -CH-O-}CH_3$$

Phospho-Ser   Phospho-Thr

The substrates for the reaction bind in a cleft defined by the two lobes of the protein (**Fig. 10.7a**). Other kinases share this same core structure but often have additional domains that determine their subcellular location or provide additional regulatory functions.

In addition to regulation by cAMP binding to the R subunit, *protein kinase A itself is regulated by phosphorylation.* The protein's so-called activation loop, a segment of polypeptide near the entrance to the active site, includes a phosphorylatable threonine residue. When the loop is not phosphorylated, the kinase's active site is blocked. When phosphorylated, the loop swings aside and the kinase's catalytic activity increases; for some protein kinases, activity increases by several orders of magnitude. This activation effect is not merely a matter of improving substrate access to the active site but also appears to involve conformational changes that affect catalysis. For example, the negatively charged phospho-Thr interacts with a positively charged arginine residue in the active site. Efficient catalysis requires that this arginine residue and an adjacent aspartate residue be re-positioned for phosphoryl-group transfer from ATP to a protein substrate (Fig 10.7b).

Among the many targets of protein kinase A are enzymes involved in glycogen metabolism (Section 19.2). One result of signaling via the β₂-adrenergic receptor, which leads to protein kinase A activation by cAMP, is the phosphorylation and activation of an enzyme called glycogen phosphorylase, which catalyzes the removal of glucose residues, the cell's primary metabolic fuel, from glycogen, the cell's glucose-storage depot. Consequently, a signal such as epinephrine can mobilize the metabolic fuel needed to power the body's fight-or-flight response.

The enzymes that phosphorylate the activation loops of protein kinase A and other cell-signaling kinases apparently operate when the kinase is first synthesized, so the kinase is already "primed" and needs only to be allosterically activated by the presence of a second messenger. However, this regulatory mechanism begs the question of what activates the kinase that phosphorylates the kinase. As we will see, kinases that act in series are common in biological signaling pathways, and many of these pathways are interconnected, making it difficult to trace simple cause-and-effect relationships.

FIGURE 10.6

Inactive protein kinase A. In the inactive complex, the two regulatory subunits (R) block the active sites of the two catalytic subunits (C). [*Structure (pdb 3TNP) determined by P. Zhang, E. V. Smith-Nguyen, M. M. Keshwani, M. S. Deal, A. P. Kornev, and S. S. Taylor.*]

SIGNALING PATHWAYS ARE ALSO SWITCHED OFF

What happens after a ligand binds to a receptor, a G protein responds, a second messenger is produced, an effector enzyme such as protein kinase A is activated, and target proteins are phosphorylated? *To restore the cell to a resting state, any or all of the events of the signal transduction pathway can be blocked or reversed.* We have already seen that the

(a)　　　　　　　　　(b)

FIGURE 10.7

Protein kinase A. (a) The backbone of the catalytic subunit is light green, with its activation loop dark green. The phospho-Thr residue (right side) and ATP (left side) are shown in stick form. A peptide that mimics a target protein is blue. (b) Close-up view of the active site region. When the activation loop is phosphorylated, the phospho-Thr residue interacts with an Arg residue, and the adjacent Asp residue is positioned near the third phosphate group of ATP and the peptide substrate. Atoms are color-coded: C gray or green, O red, N blue, and P gold. [*Structure (pdb 1ATP) determined by J. Zheng, E. A. Trafny, D. R. Knighton, N.-H. Xuong, S. S. Taylor, L. F. Teneyck, and J. M. Sowadski.*]

intrinsic GTPase activity of G proteins limits their activity. And second messengers often have short lifetimes due to their rapid degradation in the cell. For example, cAMP is hydrolyzed by the enzyme cAMP phosphodiesterase:

cAMP ⟶ (phosphodiesterase, H_2O) ⟶ AMP

Caffeine, in addition to being an adenosine receptor antagonist, can diffuse inside cells and inhibit cAMP phosphodiesterase. As a result, the cAMP concentration remains high, the action of protein kinase A is sustained, and stored metabolic fuels are mobilized, readying the body for action rather than sleep.

Some of the cell's G proteins may inhibit rather than activate adenylate cyclase and therefore decrease the cellular cAMP level. Some G proteins activate cAMP phosphodiesterase, with similar effects on cAMP-dependent processes. *A cell's response to a hormone signal depends in part on which G proteins respond.* Because a single type of hormone may stimulate several types of G proteins, the signaling system may be active for only a brief time before it is turned off.

The phosphorylations catalyzed by protein kinase A (and other kinases) can be reversed by the work of protein **phosphatases,** which catalyze a hydrolytic reaction to remove phosphoryl groups from protein side chains. Like kinases, phosphatases are generally specific for serine or threonine, or tyrosine, although some "dual specificity" phosphatases remove phosphoryl groups from all three side chains. The active-site pocket of the Tyr phosphatases is deeper than the pocket of Ser/Thr phosphatases in order to accommodate the larger phospho-Tyr side chain. Some protein phosphatases are transmembrane proteins; others are entirely intracellular. They tend to have multiple domains or multiple subunits, consistent with their ability to form numerous protein–protein interactions and participate in complex regulatory networks.

Ultimately, the dissociation of an extracellular ligand from its receptor can halt a signal transduction process, or the receptor can become desensitized. Desensitization of a G protein–coupled receptor begins with phosphorylation of the ligand-bound receptor by a GPCR kinase. The phosphorylated receptor is then recognized by a protein known as arrestin (**Fig. 10.8**), which includes lysine and arginine residues that bind the phosphoryl group. This binding halts signaling, by blocking interactions with a G protein (hence the name *arrestin*), and promotes the movement of the receptor from the cell surface to an intracellular compartment by the process of endocytosis. Interestingly, arrestins also serve as scaffold proteins for organizing components of other signaling pathways. Experimental evidence suggests that some GPCRs, including the β_2-adrenergic receptor, interact with arrestin so that they can initiate an intracellular response without ever recruiting a G protein.

EXTRACELLULAR SPACE

CYTOSOL

FIGURE 10.8

A receptor–arrestin complex. Rhodopsin, a model G protein–coupled receptor, is shown as a purple ribbon, and the intracellular arrestin protein as a pink ribbon. [*Structure (pdb 4ZWJ) determined by Y. Kang et al.*]

❓ **Compare this figure and Figure 10.4a.**

THE PHOSPHOINOSITIDE SIGNALING PATHWAY GENERATES TWO SECOND MESSENGERS

The diversity of G protein–coupled receptors, together with the diversity of G proteins, creates almost unlimited possibilities for adjusting the levels of second messengers and altering the activities of cellular enzymes. Epinephrine, the hormone that activates the β_2-adrenergic receptor, also binds to a receptor known as the α-adrenergic receptor, which is part of the **phosphoinositide signaling system.** The α- and β-adrenergic receptors are situated in different tissues and mediate different physiological effects, even though they bind the same hormone. The G protein associated with the α-adrenergic receptor activates the cellular enzyme phospholipase C, which acts on the membrane lipid phosphatidylinositol bisphosphate. Phosphatidylinositol is a minor component of the plasma membrane (4–5% of the total phospholipids), and the bisphosphorylated form (with a total of three phosphate groups) is rarer still. Phospholipase C converts this lipid to inositol trisphosphate and diacylglycerol.

Phosphatidylinositol bisphosphate

$\xrightarrow[\text{phospholipase C}]{H_2O}$

Inositol trisphosphate

Diacylglycerol

The highly polar inositol trisphosphate is a second messenger that triggers the opening of calcium channels in the endoplasmic reticulum membrane, allowing Ca^{2+} ions to flow into the cytosol. The flux of Ca^{2+} initiates numerous events in the cell, including the activation of a Ser/Thr kinase known as protein kinase B or Akt. Inositol trisphosphate also directly activates kinases and can undergo additional phosphorylations to generate a series of second messengers containing up to eight phosphoryl groups.

The hydrophobic diacylglycerol product of the phospholipase C reaction is also a second messenger. Although it remains in the cell membrane, it can diffuse laterally to activate protein kinase C, which phosphorylates its target proteins at serine or threonine residues. In its resting state, protein kinase C is a cytosolic protein with an activation loop blocking its active site. Noncovalent binding to diacylglycerol docks the enzyme at the membrane surface so that it changes its conformation, repositions the activation loop, and becomes catalytically active. As in protein kinase A (which shares about 40% sequence identity), phosphorylation of a threonine residue in the activation loop, a requirement for catalytic activity, has already occurred. Full activation of some forms of protein kinase C also requires Ca^{2+}, which is presumably available

(a)

(b)

FIGURE 10.9

Calmodulin. (a) Isolated calmodulin has an extended shape. The four bound Ca^{2+} ions are shown as blue spheres. (b) When bound to a target protein (blue helix), calmodulin's long central helix unwinds and bends so that the protein can wrap around its target. [*Structure of calmodulin (pdb 3CLN) determined by Y. S. Babu, C.E. Bugg, and W. J. Cook. Structure of calmodulin bound to a 26-residue target (pdb 2BBM) determined by G. M. Clore, A. Bax, M. Ikura, and A. M. Gronenborn.*]

as a result of the activity of the inositol trisphosphate second messenger. Among the targets of protein kinase C are proteins involved in the regulation of gene expression and cell division. Certain compounds that mimic diacylglycerol can activate protein kinase C, leading eventually to the uncontrolled growth characteristic of cancer.

Phospholipase C can be activated not only by a G protein–coupled receptor such as the α-adrenergic receptor but also by other signaling systems involving receptor tyrosine kinases. This is an example of **cross-talk,** the interconnections between signaling pathways that share some intracellular components. The phosphoinositide signaling pathway is regulated in part by the action of lipid phosphatases that remove phosphoryl groups from the phosphatidylinositol bisphosphate substrate that gives rise to the second messengers.

Another example of the overlap between signaling pathways involves sphingolipids such as sphingomyelin (Fig. 8.2), which is a normal component of membranes. Ligand binding to certain receptor tyrosine kinases leads to activation of sphingomyelinases that release sphingosine and ceramide (ceramide is sphingomyelin without its phosphocholine head group). Ceramide is a second messenger that activates kinases, phosphatases, and other cellular enzymes. Sphingosine, which undergoes phosphorylation (by a receptor tyrosine kinase–dependent mechanism) to sphingosine-1-phosphate, is both an intracellular and extracellular signaling molecule. It inhibits phospholipase C inside the cell, and it apparently exits the cell via an ABC transport protein (Section 9.3), then binds to a G protein–coupled receptor to trigger additional cellular responses.

CALMODULIN MEDIATES SOME Ca^{2+} SIGNALS

In some cases where Ca^{2+} ions elicit a change in an enzyme's activity, the change is mediated by a Ca^{2+}-binding protein known as calmodulin. This small (148-residue) protein binds two Ca^{2+} ions in each of its two globular domains, which are separated by a long α helix (**Fig. 10.9a**). Free calmodulin has an extended shape, but in the presence of Ca^{2+} and a target protein, the helix partially unwinds and calmodulin bends in half to grasp the target protein to activate or inhibit it (Fig. 10.9b).

 CONCEPT CHECK

1. Draw a simple diagram that includes all the components of the $β_2$-adrenergic receptor signaling pathway. Draw a similar diagram for the α-adrenergic receptor signaling pathway.

2. Describe the structure of a G protein–coupled receptor.

3. What is the relationship between kinases and phosphatases?

4. Draw a diagram to illustrate the phosphoinositide signaling pathway.

5. For each pathway, identify points where signaling activity can be switched off.

6. Explain why it is an advantage for receptors and G proteins to be lipid-linked.

7. What is the relationship between an extracellular hormone and a second messenger in terms of concentration and specificity?

8. Discuss the advantages and disadvantages of cross-talk.

KEY CONCEPTS
- Ligands such as insulin activate the tyrosine kinase activity of their receptors.
- Receptor tyrosine kinases trigger cellular responses by phosphorylating target proteins and by activating Ras.

10.3 RECEPTOR TYROSINE KINASES

A number of hormones and other signaling molecules that regulate cell growth, division, and immune responses bind to cell-surface receptors that operate as tyrosine kinases. These receptors are typically built from two protein subunits, each with a single membrane-spanning segment. At one time, ligand binding was believed to bring two monomeric receptors together to form a catalytically active dimer, but more recent evidence indicates that these receptors are already dimerized in the inactive state. When a ligand molecule binds, the receptor is activated by conformational changes that lead to a scissor-like rearrangement of the receptor's extracellular and intracellular domains.

FIGURE 10.10

The insulin receptor. (a) Schematic diagram. The receptor consists of two α and two β subunits joined by disulfide bonds (horizontal lines). The α subunits bind one molecule of insulin, and the β subunits each include a membrane-spanning segment (coil) and a cytoplasmic tyrosine kinase domain. (b) The extracellular portion of the insulin receptor. The cell surface is at the bottom. One αβ pair of subunits is shown in space-filling form, and the other is shown as a backbone trace. Insulin binds to one of the two binding sites indicated by arrows. [*Structure (pdb 2DTG) determined by M. C. Lawrence and V. A. Streltsov.*]

Although the receptor's two halves are identical and would be expected to have identical ligand-binding sites, only one ligand molecule appears to bind to the dimer. In some cases, binding to a site on one half of the dimer, and then to the other half, causes the dimer to become structurally and functionally asymmetric. However, in all cases, ligand binding to the extracellular portion of the receptor leads to tyrosine kinase activity in the intracellular portion. Due to extensive analysis, the insulin receptor serves as a useful model for the approximately 60 other receptor tyrosine kinases in humans.

THE INSULIN RECEPTOR DIMER BINDS ONE INSULIN

Insulin, a 51-residue polypeptide hormone that regulates many aspects of fuel metabolism in mammals, binds to receptors that are present in most of the body's tissues. The receptor is constructed from two long polypeptides that are cleaved after their synthesis, so the mature receptor has an $\alpha_2\beta_2$ structure in which all four polypeptide segments are held together by disulfide bonds (**Fig. 10.10a**).

The extracellular portion of the insulin receptor has an inverted V shape with multiple structural domains (Fig. 10.10b). Segments of the α subunits define the two insulin-binding sites, but the sites are too far apart (~65 Å) to simultaneously bind a single hormone molecule. Instead, biochemical evidence indicates that binding to just one site pulls the two α subunits together in such a way that the second binding site cannot bind insulin. Interdomain interactions, along with the disulfide linkages, suggest that the receptor is rigid, and this feature is believed to be important in transducing the message (insulin binding to the extracellular α subunits) to the intracellular signaling apparatus (the tyrosine kinase domains of the β subunits). As the ligand-binding domains come together, the intracellular tyrosine kinase domains separate (hence the analogy to scissors) but in a way that switches on their catalytic activity.

THE RECEPTOR UNDERGOES AUTOPHOSPHORYLATION

Inside the cell, *the two tyrosine kinase domains phosphorylate each other,* using ATP as a source of the phosphoryl groups. Because the receptors appear to phosphorylate themselves, this process is termed **autophosphorylation.** Each tyrosine kinase domain has the typical kinase

FIGURE 10.11

Activation of the insulin receptor tyrosine kinase. The backbone structure of the inactive tyrosine kinase domain of the insulin receptor is shown in light blue, with the activation loop in dark blue. The structure of the active tyrosine kinase domain is shown in light green, with the activation loop in dark green. Note that three Tyr side chains in the activation loop have been phosphorylated (a result of each tyrosine kinase domain phosphorylating the other) so that the activation loop has swung aside to better expose the active site. The phospho-Tyr side chains are shown in stick form with atoms color coded: C green, O red, and P orange. [*Structure of the inactive kinase domain (pdb 1IRK) determined by S. R. Hubbard, L. Wei, L. Ellis, and W. A. Hendrickson. Structure of the active kinase domain (pdb 1IR3) determined by S. R. Hubbard.*]

? **How would a phosphatase affect the activity of the insulin receptor?**

core structure, including an activation loop that lies across the active site to prevent substrate binding. In the insulin receptor, phosphorylation of three tyrosine residues in the kinase activation loop causes a conformational change that allows the enzyme to interact with additional protein substrates and to transfer a phosphoryl group from ATP to a tyrosine side chain on these target proteins (**Fig. 10.11**). However, some receptor tyrosine kinases do not phosphorylate any other proteins but activate their intracellular targets by other mechanisms.

For example, to initiate the responses that promote cell growth and division, many growth factor receptors phosphorylate various intracellular target proteins. They also switch on other pathways that involve small monomeric G proteins (which are also GTPases) such as Ras. The tyrosine kinase domain of the receptor does not directly interact with Ras but instead relies on one or more adapter proteins that form a bridge between Ras and the phospho-Tyr residues of the receptor (**Fig. 10.12**). These proteins also stimulate Ras to release GDP and bind GTP.

Ligand binding to the receptor triggers autophosphorylation of the tyrosine kinase domains.

Adapter proteins bridge the receptor and Ras and induce Ras to release GDP and bind GTP.

The active Ras can now initiate a kinase cascade.

Receptor tyrosine kinase

Ligand

GDP

Ras (inactive)

Adapter proteins

Kinase (inactive)

GTP

Ras (active)

ATP

ADP

GTP

Kinase (active)

FIGURE 10.12

The Ras pathway. Ras links receptor–hormone binding to an intracellular kinase cascade.

? Describe the role of protein conformational changes in each step of the pathway.

Box 10.B Cell Signaling and Cancer

The progress of a cell through the cell cycle, from DNA replication through the phases of mitosis, depends on the orderly activity of signaling pathways. Cancer, which is the uncontrolled growth of cells, can result from a variety of factors, including overactivation of the signaling pathways that stimulate cell growth. In fact, *the majority of cancers include mutations in the genes for proteins that participate in signaling via Ras and the phosphoinositide pathways.* These altered genes are termed **oncogenes,** from the Greek *onkos,* meaning "tumor."

Oncogenes were first discovered in certain cancer-causing viruses. The viruses presumably picked up the normal gene from a host cell, then mutated. In some cases, an oncogene encodes a growth factor receptor that has lost its ligand-binding domain but retains its tyrosine kinase domain. As a result, the kinase is constitutively (constantly) active, promoting cell growth and division even in the absence of the growth factor. Some *RAS* oncogenes generate mutant forms of Ras that hydrolyze GTP extremely slowly, thus maintaining the signaling pathway in the "on" state. Note that oncogenic mutations can strengthen an activating event or weaken an inhibitory event; in either case, the outcome is excessive signaling activity.

The importance of various kinases in triggering or sustaining tumor growth has made these enzymes attractive targets for anticancer drugs. Some forms of leukemia (a cancer of white blood cells) are triggered by a chromosomal rearrangement that generates a kinase with constitutive signaling activity, called Bcr-Abl. The drug imatinib (Gleevec®) specifically inhibits this kinase without affecting any of the cell's numerous other kinases. The result is an effective anticancer treatment with few side effects.

Gleevec® (imatinib)

The engineered antibody known as trastuzumab (Herceptin®) binds as an antagonist to a growth factor receptor that is overexpressed in many breast cancers. Other antibody-based drugs target similar receptors in other types of cancers. Clearly, understanding the operation of growth-signaling pathways—both normal and mutated—is essential for the ongoing development of effective anticancer treatments.

? Using Figures 10.2 and 10.12 as a guide, identify other types of signaling proteins that are potential anticancer drug targets.

Like other G proteins, *Ras is active as long as it has GTP bound to it.* The Ras·GTP complex allosterically activates a Ser/Thr kinase, which becomes active and phosphorylates another kinase, activating it, and so on. Cascades of several kinases can therefore amplify the initial growth factor signal.

The ultimate targets of Ras-dependent signaling cascades are nuclear proteins, which, when phosphorylated, bind to specific sequences of DNA to induce (turn on) or repress (turn off) gene expression. The altered activity of these **transcription factors** means that the original hormonal signal not only alters the activities of cellular enzymes on a short time scale (seconds to minutes) via phosphorylation but also affects protein synthesis, a process that may require several hours or more.

Ras signaling activity is shut down by the action of proteins that enhance the GTPase activity of Ras so that it returns to its inactive GDP-bound form. In addition, phosphatases reverse the effects of the various kinases. Like the other signaling pathways we have examined, the receptor tyrosine kinase pathways are not linear and they are capable of cross-talk. For example, some receptor tyrosine kinases directly or indirectly (via Ras) activate the kinase that phosphorylates phosphatidylinositol lipids, thereby promoting signaling through the phosphoinositide pathway. Abnormalities in these signaling pathways can promote tumor growth (**Box 10.B**).

CONCEPT CHECK

1. Draw a diagram to illustrate receptor tyrosine kinase signaling, including tyrosine phosphorylation and Ras activation.

2. Explain how the free energy of ATP and GTP hydrolysis is used to turn on cellular responses to extracellular signals.

3. Explain how kinases and transcription factors mediate cellular responses over different time scales.

4. Compare phosphorylation and proteolysis (Section 6.4) as mechanisms for activating an enzyme.

LIPID HORMONE SIGNALING

Some hormones do not need to bind to cell-surface receptors because they are lipids and can cross the membrane to interact with intracellular receptors. For example, retinoic acid and the thyroid hormones thyroxine (T_4) and triiodothyronine (T_3) belong to this class of hormones (**Fig. 10.13**). Retinoic acid (retinoate), a compound that regulates cell growth and differentiation, particularly in the immune system, is synthesized from retinol, a derivative of β-carotene (Box 8.B). The thyroid hormones, which generally stimulate metabolism, are derived from a large precursor protein called thyroglobulin: Tyrosine side chains are enzymatically iodinated, then two of these residues undergo condensation, and the hormones are liberated from thyroglobulin by proteolysis.

The 27-carbon cholesterol, introduced in Section 8.1, is the precursor of a large number of hormones that regulate metabolism, salt and water balance, and reproductive functions. Androgens (which are primarily male hormones) have 19 carbons, and estrogens (which are primarily female hormones) contain 18 carbons. Cortisol, a C_{21} glucocorticoid hormone, affects the metabolic activities of a wide variety of tissues.

Cortisol

Retinoic acid, thyroid hormones, and steroids are all hydrophobic molecules that are carried in the bloodstream either by specific carrier proteins or by albumin, a sort of all-purpose binding protein.

The receptors to which the lipid hormones bind are located inside the appropriate target cell, either in the cytoplasm or the nucleus. Ligand binding often—but not always—causes the receptors to form dimers. Each receptor subunit is constructed from several modules, which include a ligand-binding domain and a DNA-binding domain. The ligand-binding domains are as varied as their hormone ligands, but the DNA-binding domains exhibit a common structure that includes two zinc fingers, which are cross-links formed by the interaction of four cysteine side chains with a Zn^{2+} ion (Section 4.3). In the absence of a ligand, the receptor cannot bind to DNA.

Following ligand binding and dimerization, the receptor moves to the nucleus (if it is not already there) and binds to specific DNA sequences called **hormone response elements.** Although the hormone response elements vary for each receptor–ligand complex, they are all composed of two identical 6-bp sequences separated by a few base pairs. Simultaneous binding of the two hormone response element sequences explains why many of the lipid hormone receptors are dimers (**Fig. 10.14**).

The receptors function as transcription factors so that the genes near the hormone response elements may experience higher or lower levels of expression. For example, glucocorticoids such as cortisol stimulate the production of phosphatases, which dampen the stimulating effects of kinases. This property makes cortisol and its derivatives useful as drugs to treat conditions such as chronic inflammation or asthma. However, because so many tissues respond to glucocorticoids, the side effects of these drugs can be significant and tend to limit their long-term use.

The changes in gene expression triggered by steroids and other lipid hormones require many hours to take effect. However, cellular responses to some lipid hormones are evident within seconds or

Retinoate

Thyroxine (T_4)

Triiodothyronine (T_3)

FIGURE 10.13

Some lipid hormones.

minutes, suggesting that the hormones also participate in signaling pathways with shorter time courses, such as those centered on G proteins and/or kinases. In these instances, the receptors must be located on the cell surface.

EICOSANOIDS ARE SHORT-RANGE SIGNALS

Many of the hormones discussed in this chapter are synthesized and stockpiled to some extent before they are released, but some lipid hormones are synthesized as a response to other signaling events (sphingosine-1-phosphate is one example; Section 10.2). The lipid hormones called **eicosanoids** are produced when the enzyme phospholipase A_2 is activated by phosphorylation and by the presence of Ca^{2+}. One substrate of the phospholipase is the membrane lipid phosphatidylinositol. In this lipid, cleavage of the acyl chain attached to the second glycerol carbon often releases arachidonate, a C_{20} fatty acid (the term *eicosanoid* comes from the Greek *eikosi*, meaning "twenty").

Arachidonate, a polyunsaturated fatty acid with four double bonds, is further modified by the action of enzymes that catalyze cyclization and oxidation reactions (**Fig. 10.15**). A wide variety of eicosanoids can be produced in a tissue-dependent fashion, and their functions are similarly varied. Eicosanoids regulate such things as blood pressure, blood coagulation, inflammation, pain, and fever. The eicosanoid thromboxane, for example, helps activate platelets (cell fragments that participate in blood coagulation) and induces vasoconstriction. Other eicosanoids have the opposite effects: They prevent platelet activation and promote vasodilation. The use of aspirin as a "blood thinner" stems from its ability to inhibit the enzyme that initiates the conversion of arachidonate to thromboxane (see Fig. 10.15). A number of other drugs interfere with the production of eicosanoids by blocking the same enzymatic step (**Box 10.C**).

The receptors for eicosanoids are G protein–coupled receptors that trigger cAMP-dependent and phosphoinositide-dependent responses. However, eicosanoids are degraded relatively quickly. This instability, along with their hydrophobicity, means that *their effects are relatively limited in time and space.* Eicosanoids tend to elicit responses only in the cells that produce them and in nearby cells. In contrast, many other hormones travel throughout the body, eliciting effects in any tissue that exhibits the appropriate receptors. For this reason, eicosanoids are sometimes called local mediators rather than hormones.

Similar signaling molecules operate in plants. For example, the lipid hormone jasmonate is synthesized as a result of localized herbivore damage and quickly spreads to other parts of the plant to trigger defensive responses.

FIGURE 10.14

The glucocorticoid receptor–DNA complex. The two DNA-binding zinc finger domains of the glucocorticoid receptor are blue and green. The Zn^{2+} ions are shown as gray spheres. The hormone response element sequences of the DNA (bottom) are colored red. Two protein helices make sequence-specific contacts with nucleotides. [*Structure (pdb 1GLU) determined by B. F. Luisi, W. X. Xu, Z. Otwinowski, L. P. Freedman, K. R. Yamamoto, and P. B. Sigler.*]

? **What is the surface charge of the receptor's DNA-binding domain?**

Arachidonate

cyclooxygenase

Prostaglandin H_2

other biologically active lipids

Jasmonate

Thromboxane

FIGURE 10.15

Arachidonate conversion to eicosanoid signal molecules. The first step is catalyzed by cyclooxygenase. Only two of the many dozens of eicosanoids are shown here.

Volatile derivatives of jasmonate can travel through the air to spread the alarm to neighboring plants.

Box 10.C Aspirin and Other Inhibitors of Cyclooxygenase

The bark of the willow *Salix alba* has been used since ancient times to relieve pain and fever. The active ingredient is acetylsalicylate, or aspirin.

Acetylsalicylate (aspirin)

Aspirin was first prepared in 1853, but it was not used clinically for another 50 years or so. Effective promotion of aspirin by the Bayer chemical company at the start of the twentieth century marked the beginning of the modern pharmaceutical industry.

Despite its universal popularity, aspirin's mode of action was not discovered until 1971. It inhibits the production of prostaglandins (which induce pain and fever, among other things) by inhibiting the activity of cyclooxygenase (also known as COX), the enzyme that acts on arachidonate (see Fig. 10.15). COX inhibition results from acetylation of a serine residue located near the active site in a cavity that accommodates the arachidonate substrate. Other pain-relieving substances such as ibuprofen also bind to COX to prevent the synthesis of prostaglandins, although these drugs do not acetylate the enzyme.

Ibuprofen

One shortcoming of aspirin is that it inhibits more than one COX isozyme. COX-1 is a constitutively expressed enzyme that is responsible for generating various eicosanoids, including those that maintain the stomach's protective layer of mucus. COX-2 expression increases during tissue injury or infection and generates eicosanoids involved in inflammation. Long-term aspirin use suppresses the activity of both isozymes, which can lead to side effects such as gastric ulcers.

Rational drug design (Section 7.4) based on the slightly different structures of COX-1 and COX-2 led to the development of the drugs Celebrex and Vioxx. These compounds bind only to the active site of COX-2 (they are too large to fit into the COX-1 active site) and therefore can selectively block the production of proinflammatory eicosanoids without damaging gastric tissue. Unfortunately, the side effects of these drugs include an increased risk of heart attacks, through a mechanism that is not fully understood, and as a result Vioxx has been taken off the market and the use of Celebrex is limited. If nothing else, this story illustrates the complexity of biological signaling pathways and the difficulty of understanding how to manipulate them for therapeutic reasons.

A third COX isozyme, COX-3, is expressed at high levels in the central nervous system. It is the target of the widely used drug acetaminophen (Section 7.4), which reduces pain and fever and does not appear to incur the side effects of the COX-2–specific inhibitors.

Acetaminophen

? **Which of the drugs shown here is chiral (see Section 4.1), with two different configurations?**

CONCEPT CHECK

1. Explain how lipid hormones can bypass G protein–coupled receptors and receptor tyrosine kinases to alter gene expression.

2. Explain why lipid hormones have intracellular receptors.

3. Compare the timing of the responses to steroid hormones and eicosanoids.

4. Draw a model of a cell and add shapes to represent all the types of receptors, target enzymes, and other signal-transduction machinery mentioned in the chapter, placing each component in the membrane, cytosol, or nucleus as appropriate.

5. Make a list of the drugs mentioned in this chapter, and indicate how they interfere with signaling.

SUMMARY

10.1 General Features of Signaling Pathways

• Agonist or antagonist binding to a receptor can be quantified by a dissociation constant.

• G protein–coupled receptors and receptor tyrosine kinases are the most common types of receptors.

- While signaling systems amplify extracellular signals, they are also regulated so that signaling can be turned off, and the receptor may become desensitized.

10.2 G Protein Signaling Pathways

- A ligand such as epinephrine binds to a G protein–coupled receptor. A G protein responds to the receptor–ligand complex by releasing GDP, binding GTP, and splitting into an α subunit and a βγ dimer.
- The α subunit of the G protein activates adenylate cyclase, which converts ATP to cAMP. cAMP is a second messenger that triggers a conformational change in protein kinase A that repositions its activation loop to achieve full catalytic activity.
- cAMP-dependent signaling activity is limited by the reduction of second messenger production through the GTPase activity of G proteins and the action of phosphodiesterases and by the activity of phosphatases that reverse the effects of protein kinase A. Ligand dissociation and receptor desensitization through phosphorylation and arrestin binding also limit signaling via G protein–coupled receptors.
- G protein–coupled receptors that lead to activation of phospholipase C generate inositol trisphosphate and diacylglycerol second messengers, which activate protein kinase B and protein kinase C, respectively.

- Signaling pathways originating with different G protein–coupled receptors and receptor tyrosine kinases overlap through activation or inhibition of the same intracellular components, such as kinases, phosphatases, and phospholipases.

10.3 Receptor Tyrosine Kinases

- Many receptor tyrosine kinases are dimeric molecules with a single ligand-binding site. Extracellular ligand binding activates the cytoplasmic tyrosine kinase domains to phosphorylate each other.
- In addition to acting as kinases, the receptor tyrosine kinases initiate other kinase cascades by activating the small monomeric G protein Ras.

10.4 Lipid Hormone Signaling

- Steroids and other lipid hormones bind primarily to intracellular receptors that dimerize and bind to hormone response elements on DNA to induce or repress the expression of nearby genes.
- Eicosanoids, which are synthesized from membrane lipids, function as signals over short ranges and for a limited time.

KEY TERMS

receptor	agonist	receptor tyrosine kinase	cross-talk
ligand	antagonist	desensitization	autophosphorylation
signal transduction	G protein	cAMP	transcription factor
hormone	GPCR	phosphatase	oncogene
quorum sensing	second messenger	phosphoinositide signaling	hormone response element
dissociation constant (K_d)	kinase	system	eicosanoid

BIOINFORMATICS

Brief Bioinformatics Exercises

10.1. G Protein–Coupled Receptors and Receptor Tyrosine Kinases

10.2. Biosignaling and the KEGG Database

PROBLEMS

10.1 General Features of Signaling Pathways

1. Which of the signal molecules listed in Table 10.1 would not require a cell-surface receptor?

2. The structure of the drug prednisone is shown. What kind of molecule is this and by what pathway is it likely to exert its effects?

3. A sample of cells has a total receptor concentration of 40 mM. Fifty percent of the receptors have bound ligand and the concentration of free ligand is 8 mM. **(a)** What is the K_d for the receptor–ligand interaction? **(b)** What is the relationship between K_d and [L]?

4. The K_d for a receptor–ligand interaction is 4 mM. When the concentration of free ligand is 24 mM and the concentration of free receptor is 6 mM, what is the concentration of receptor that is occupied by ligand?

5. The total concentration of receptors in a sample is 20 mM. The concentration of free ligand is 5 mM. Calculate the percentage of receptors that are occupied by ligand when **(a)** K_d is 2.5 mM and **(b)** K_d is 0.6 mM.

Compare the answer in both the cases and explain the difference.

6. Use the plot below to estimate a value for K_d.

7. In an experiment, the ligand adenosine is added to heart cells in culture. The number of receptors with ligand bound is measured and the data yield a curve like the one shown in Figure 10.1. What would the results look like if the experiment was repeated in the presence of caffeine?

8. The definition of K_d as given in Equation 10.1, shows the relationship between the free receptor concentration [R], the ligand concentration [L], and the concentration of receptor–ligand complexes [R·L]. The value of [R], like [R·L], is difficult to evaluate, but various experimental techniques can be used to determine $[R]_T$, the total number of receptors. $[R]_T$ is the sum of [R] and [R·L]. Using this information, begin with Equation 10.1 and derive an expression for the $[R·L]/[R]_T$ ratio. (Note that your derived expression will be similar to the Michaelis–Menten equation and that Equations 7.9 through 7.17 may give you an idea how to proceed.)

9. The $[R·L]:[R]_T$ ratio gives the fraction of receptors that have bound ligand. Use the expression you derived in Problem 8 to express [R·L] as a fraction of $[R]_T$ for the following situations: (a) $K_d = 3[L]$, (b) $K_d = [L]$, and (c) $3K_d = [L]$.

10. If a cell has 1000 surface receptors for erythropoietin, and if only 15% of those receptors need to bind ligand to achieve a maximal response, what ligand concentration is required to achieve a maximal response? Use the equation you derived in Problem 8. The K_d for erythropoietin is 1.0×10^{-10} M.

11. Suppose the number of surface receptors on the cell described in Problem 10 decreases to 200. What ligand concentration is required to achieve a maximal response?

12. ADP binds to platelets in order to initiate the activation process. Two binding sites were identified on platelets, one with a K_d of 0.35 μM and one with a K_d of 7.9 μM.

   (a) Which site is a low-affinity binding site and which is a high-affinity binding site?
   (b) The ADP concentration required to activate a platelet is in the range of 0.1–0.5 μM. Which receptor will be more effective at activating the platelet?
   (c) Two ADP agonists were also found to bind to platelets: 2-methythio-ADP bound with a K_d of 7 μM and 2-(3-aminopropylthio)-ADP bound with a K_d of 200 μM. Can these agonists effectively compete with ADP for binding to platelets?
   (d) In the study described above, 160,000 high-affinity binding sites were identified on each platelet. What is the concentration of ADP required to achieve 85% binding? Use the equation you derived in Problem 8.

13. Like the Michaelis–Menten equation, the equation derived in Problem 8 can be converted to an equation for a straight line. A double-reciprocal plot for a ligand binding to its receptor is shown below. Use the information in the plot to estimate a value for K_d.

14. A Scatchard plot is another method of representing ligand binding data using a straight line. In a Scatchard plot, [R·L]/[L] is plotted versus [R·L]. The slope is equal to $-1/K_d$. Use the Scatchard plot provided to estimate a value for K_d for calmodulin binding to calcineurin.

$y = -0.067x + 6.890$

15. Why might it be difficult to purify cell-surface receptors using the techniques described in Section 4.6?

16. Affinity chromatography is often used as a technique to purify cell-surface receptors. Describe the steps you would take to purify a cell-surface receptor using this technique.

17. Epinephrine can bind to several different types of G protein–linked receptors. Each of these receptors triggers a different cellular response. Explain how this is possible.

18. In the liver, both glucagon and epinephrine bind to different members of the G protein–coupled receptor family, yet binding of each of these ligands results in the same response—glycogen breakdown. How is it possible that two different ligands can trigger the same cellular response?

19. Many receptors become desensitized in the presence of high concentrations of signaling ligand. This can occur in a variety of ways, such as removal of the receptors from the cell surface by endocytosis. Why is this an effective desensitization strategy?

20. What is the advantage of activating D using the strategy shown in the figure? Why is this strategy more effective than a simple activation of D that occurs in one step?

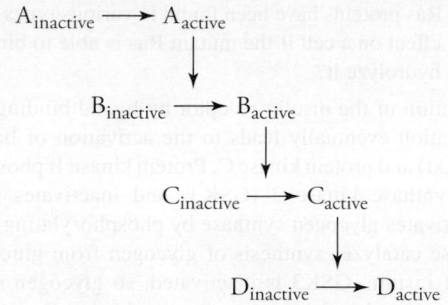

GTPγS

10.2   G Protein Signaling Pathways

21. As described in the text, G protein–coupled receptors are often palmitoylated at a Cys residue.

  (a) Draw the structure of the palmitate (16:0) residue covalently linked to a Cys residue.

  (b) What is the role of this fatty acyl chain?

  (c) What is the expected result if the Cys residue is mutated to a Gly?

22. The cell contains many types of guanine nucleotide exchange factors (GEFs) that regulate the activity of monomeric GTPases by promoting the exchange of GDP for GTP. Explain how G protein–coupled receptors also function as GEFs.

23. (a) Naturally occurring mutations in the genes that code for GPCRs have provided insights into GPCR function and the diseases that often result from these mutations. List the types of mutations in a GPCR that would result in loss of the receptor's function.

(b) Mutations in a GPCR gene occasionally result in increased receptor activity. List the types of mutations in a GPCR that would result in gain of function of the receptor.

24. An Asp residue in the third transmembrane helix of the epinephrine receptor (a GPCR) interacts with the ligand epinephrine.

  (a) What type of interaction is likely to form between the receptor and epinephrine?

  (b) How might an Asp → Glu mutation affect ligand binding?

  (c) How might an Asp → Asn mutation affect ligand binding?

25. How do epinephrine and norepinephrine differ from tyrosine, their parent amino acid?

26. Why are the antagonists known as β-blockers effective at treating high blood pressure?

27. A toxin secreted by the bacterium *Vibrio cholerae* catalyzes the covalent attachment of an ADP-ribose group to the α subunit of a G protein. This results in the inhibition of the intrinsic GTPase activity of the G protein. How does this affect the activity of adenylate cyclase? How are intracellular levels of cAMP affected?

28. Some G protein–linked receptors are associated with a protein called RGS (regulator of G protein signaling). RGS stimulates the GTPase activity of the G protein associated with the receptor. What effect does RGS have on the signaling process?

29. Some G protein–linked receptors are associated with G proteins that inhibit rather than stimulate the activity of adenylate cyclase. A toxin secreted by the bacterium *Bordetella pertussis* (the causative agent of whooping cough) catalyzes the covalent attachment of an ADP-ribose group to the α subunit of the inhibitory G protein, preventing it from carrying out its normal function. How does this affect the activity of adenylate cyclase? How are intracellular levels of cAMP affected?

30. Addition of the nonhydrolyzable analog GTPγS to cultured cells is a common practice in signal transduction experiments. What effect does GTPγS have on cellular cAMP levels?

31. How does the addition of a single phosphate group to a protein change that protein's activity?

32. (a) Draw the reaction that shows the protein kinase A–catalyzed phosphorylation of a threonine residue on a target protein.

(b) Draw the reaction that shows the phosphatase-catalyzed hydrolysis of the phosphorylated threonine.

(c) Some bacterial signaling systems involve kinases that transfer a phosphoryl group to a His side chain. Draw the structure of the phospho-His side chain.

33. Phorbol esters, which are compounds isolated from plants, are structurally similar to diacylglycerol. How does the addition of phorbol esters affect the cellular signaling pathways of cells in culture?

34. As described in the text, ligand binding to certain receptor tyrosine kinases results in the activation of a sphingomyelinase enzyme. Draw the reaction that shows the sphingomyelinase-catalyzed hydrolysis of sphingomyelin to ceramide.

35. In unstimulated T cells, a transcription factor called NFAT (nuclear factor of activated T cells) resides in the cytosol in a phosphorylated form. When the cell is stimulated, the cytosolic Ca^{2+} concentration increases and activates a phosphatase called calcineurin. The activated calcineurin catalyzes the hydrolysis of the phosphate group from NFAT, exposing a nuclear localization signal that allows the NFAT to enter the nucleus and stimulate the expression of genes essential for T cell activation. Describe the cell-signaling events that result in the activation of NFAT.

36. The immunosuppressive drug cyclosporine A is an inhibitor of calcineurin (see Problem 35). Why is cyclosporine A an effective immunosuppressant?

37. Pathways that lead to the activation of protein kinase B (Akt) are considered to be anti-apoptotic (apoptosis is programmed cell death). In other words, protein kinase B stimulates a cell to grow and proliferate. Like all biological events, signaling pathways that are turned on must also be turned off. A phosphatase called PTEN plays a role in removing phosphate groups from proteins, but it is highly specific for removing a phosphate group from inositol trisphosphate.

  (a) If PTEN is overexpressed in mammalian cells, do these cells grow or do they undergo apoptosis?

  (b) Would you expect to find mutations in the gene for PTEN in human cancers? Explain why or why not.

38. Nitric oxide (NO) is a naturally occurring signaling molecule (see Table 10.1) that is produced from the decomposition of arginine to NO and citrulline in endothelial cells. The enzyme that catalyzes this reaction, NO synthase, is stimulated by cytosolic Ca^{2+}, which increases when acetylcholine binds to endothelial cells.

  (a) What is the source of the acetylcholine ligand?

  (b) Propose a mechanism that describes how acetylcholine binding leads to the activation of NO synthase.

  (c) NO formed in endothelial cells quickly diffuses into neighboring smooth muscle cells and binds to a cytosolic protein that catalyzes the formation of the second messenger cyclic GMP.

Cyclic GMP then activates protein kinase G, resulting in smooth muscle cell relaxation. How might protein kinase G bring about smooth muscle relaxation?

39. As discussed in the text, any signal transduction event that is turned on must subsequently be turned off. Refer to your answer to Problem 38 and describe the events that would lead to the cessation of each step of the signaling pathway you described.

40. NO synthase knockout mice (animals missing the NO synthase enzyme) have elevated blood pressure, an increased heart rate, and enlarged left ventricle chambers. Explain the reasons for these symptoms.

41. Clotrimazole is a calmodulin antagonist (see Solution 38b). How does the addition of clotrimazole affect endothelial cells in culture?

42. Nitroglycerin placed under the tongue has been used since the late nineteenth century to treat angina pectoris (chest pains resulting from reduced blood flow to the heart). But only recently have scientists elucidated its mechanism of action. Propose a hypothesis that explains why nitroglycerin placed under the tongue relieves the pain of angina.

Nitroglycerin

43. Viagra, a drug used to treat erectile dysfunction, is a cGMP phosphodiesterase inhibitor. Propose a mechanism that explains why the drug is effective in treating this condition.

44. *Bacillus anthracis*, the cause of anthrax, produces a three-part toxin. One part facilitates the entry of the two other toxins into the cytoplasm of a mammalian cell. The toxin known as edema factor (EF) is an adenylate cyclase. **(a)** Explain how EF could disrupt normal cell signaling. **(b)** EF must first be activated by Ca^{2+}-calmodulin binding to it. Explain how this requirement could also disrupt cell signaling.

45. *Bacillus anthracis* also makes a toxin called lethal factor (LF; see Problem 44). LF is a protease that specifically cleaves and inactivates a protein kinase that is part of a pathway for stimulating cell proliferation. Explain why the entry of LF into white blood cells promotes the spread of *B. anthracis* in the body.

46. Ligand binding to some growth-factor receptors triggers kinase cascades and also leads to activation of enzymes that convert O_2 to hydrogen peroxide (H_2O_2), which acts as a second messenger. Describe the likely effect of H_2O_2 on the activity of cellular phosphatases.

47. Hydrogen peroxide acts as a second messenger, as described in Problem 46, and affects PTEN (see Problem 37) as well as other cellular phosphatases. Does H_2O_2 activate or inhibit PTEN?

10.3   Receptor Tyrosine Kinases

48. The activity of Ras is regulated in part by two proteins, a guanine nucleotide exchange factor (GEF) and a GTPase activating protein (GAP). The GEF protein binds to Ras·GDP and promotes dissociation of bound GDP. The GAP protein binds to Ras·GTP and stimulates the intrinsic GTPase activity of Ras. How is downstream activity of a signaling pathway affected by the presence of GEF? By the presence of GAP?

49. Mutant Ras proteins have been found in various types of cancers. What is the effect on a cell if the mutant Ras is able to bind GTP but is unable to hydrolyze it?

50. Stimulation of the insulin receptor by ligand binding and autophosphorylation eventually leads to the activation of both protein kinase B (Akt) and protein kinase C. Protein kinase B phosphorylates glycogen synthase kinase 3 (GSK3) and inactivates it. (Active GSK3 inactivates glycogen synthase by phosphorylating it.) Glycogen synthase catalyzes synthesis of glycogen from glucose. In the presence of insulin, GSK3 is inactivated, so glycogen synthase is not phosphorylated and is active. Protein kinase C stimulates the translocation of glucose transporters to the plasma membrane by a mechanism not currently understood. One strategy for treating diabetes is to develop drugs that act as inhibitors of the phosphatases that remove phosphate groups from the phosphorylated tyrosines on the insulin receptor. Why might this be an effective treatment for diabetes?

51. When insulin binds to its receptor, a conformational change occurs that results in autophosphorylation of the receptor on specific Tyr residues. In the next step of the signaling pathway, an adaptor protein called IRS-1 (insulin receptor substrate-1) docks with the phosphorylated receptor (the involvement of adaptor proteins in cell signaling is shown in Fig. 10.12). This step is essential for the downstream activation of protein kinases B and C (see Problem 50). If IRS-1 is overexpressed in muscle cells in culture, what effects, if any, would you expect to see on glucose transporter translocation and glycogen synthesis?

52. As shown in Figure 10.12, Ras can activate a kinase cascade. The most common cascade is the MAP kinase pathway, which is activated when growth factors bind to cell surface receptors and activate Ras. This leads to the activation of transcription factors and other gene regulatory proteins and results in growth, proliferation, and differentiation. Use this information to explain why phorbol esters (see Problem 33) promote tumor development.

53. How might a signaling molecule activate the MAP kinase cascade (see Problem 52) via a G protein–coupled receptor rather than a receptor tyrosine kinase?

54. Ras can activate a cascade of reactions (see Fig. 10.12) involving membrane phosphatidylinositol (PI) lipids (see Solutions 8.13 and 8.14). Ras activates PI 3-kinase (PI3K), which phosphorylates PI (4,5) bisphosphate (PIP_2) to PI (3,4,5) trisphosphate (PIP_3). PIP_3 then activates Akt (see Problem 37).
   (a) Why would PI3K inhibitors be effective at treating cancer?
   (b) PTEN (see Problem 37) can catalyze the hydrolysis of PIP_3 to PIP_2. Does PTEN promote cell survival or cell death?

55. PKR is a protein kinase that recognizes double-stranded RNA molecules such as those that form during the intracellular growth of certain viruses. The structure of PKR includes the standard kinase domains as well as an RNA-binding module. In the presence of viral RNA, PKR undergoes autophosphorylation and is then able to phosphorylate cellular target proteins that initiate antiviral responses. Short (<30 bp) RNAs inhibit activation of PKR, but RNAs longer than 33 bp are strong activators of PKR. Explain the role of RNA in PKR activation.

56. The bacterium *Yersinia pestis,* the pathogen responsible for bubonic plague, caused the deaths of about a third of the population of Europe in the fourteenth century. The bacterium produces a phosphatase called YopH, which hydrolyzes phosphorylated tyrosines and is much more catalytically active than mammalian phosphatases.

(a) What happens when the *Yersinia* bacterium injects YopH into a mammalian cell?

(b) Why is the bacterium itself not affected by YopH?

(c) Scientists are interested in developing YopH inhibitors in order to treat *Yersinia* infection, a re-emerging disease. What are some important considerations in the development of a YopH inhibitor?

10.4 Lipid Hormone Signaling

57. Steroid hormone receptors have different cellular locations. The progesterone receptor is located in the nucleus and interacts with DNA once progesterone has bound. But the glucocorticoid receptor is located in the cytosol and does not move into the nucleus until its ligand has bound. What structural feature must be different in these two receptor molecules?

58. Abnormal changes in steroid hormone levels in the breast, uterus, ovaries, prostate, and testes are observed in cancers of these steroid-responsive tissues.

(a) Using what you know about the mechanism of steroid hormone–induced stimulation in these tissues, what strategies could you use to design drugs to treat cancers in these tissues?

(b) Given what you know about cell signaling and cancer (see Box 10.B), is it reasonable to suspect that a "nonclassical" pathway may be involved in the development of cancers in these tissues?

59. The production of sphingosine-1-phosphate and ceramide-1-phosphate is shown in the diagram. These signaling molecules can act in the cell where they are produced (intracellularly) or they can exit the cell and act on neighboring cells, as described in Section 10.2. There is quite a bit of cross-talk between the pathway shown in the diagram and other pathways discussed in this chapter. Ceramide-1-phosphate (C1P) promotes the release of arachidonate from the membrane. Sphingosine-1-phosphate (S1P) stimulates the activity of COX-2. Are these observations consistent with the inflammatory properties attributed to C1P and S1P?

60. Sphingosine-1-phosphate (see Problem 59) stimulates the activation of protein kinase B (Akt) (see Problem 37). What effect would this have on the cell?

61. Sphingosine-1-phosphate's pro-survival effects probably result from the interaction of S1P with multiple cellular pathways. In addition to S1P's ability to activate Akt (see Problem 60), how else might S1P act to promote cell survival?

62. There is a strong association between inflammation and cancer. Using the information presented in Problem 59, identify possible targets for anticancer drugs.

63. Aspirin inhibits COX by acetylating a Ser residue on the enzyme (see Box 10.C).

(a) Draw structures to show how aspirin acetylates the serine side chain.

(b) Propose a hypothesis that explains why Ser acetylation inhibits the enzyme.

(c) What type of inhibitor is aspirin (see Section 7.3)?

64. Cyclooxygenase uses arachidonate as a substrate for the synthesis of prostaglandins (see Box 10.C). In platelets, a similar pathway yields thromboxanes, compounds that stimulate vasoconstriction and platelet aggregation (see Fig. 10.15). Why do some people take a daily aspirin as protection against heart attacks?

65. Analogs of cortisol, such as prednisone, are used as anti-inflammatory drugs, although their mechanism of action is not entirely understood. Explain how inhibitors of phospholipase A_2 by prednisone could decrease inflammation.

66. Tetrahydrocannabinol (THC), the active ingredient in marijuana, binds to a receptor in the brain. The natural ligand for the receptor is anandamide. Anandamide has a short half-life because it is rapidly broken down by a hydrolase. One product is ethanolamine. Name the other product of anandamide breakdown.

Anandamide

67. A complex signaling pathway in yeast allows the cells to accumulate high concentrations of glycerol if they are exposed to high extracellular concentration of salt or glucose. The increased osmolarity of the extracellular medium activates Ras, which in turn activates adenylate cyclase. A second pathway, the HOG (high osmolarity glycerol) pathway activates the MAP kinase pathway (see Problem 52). The target protein is the enzyme PFK2, which is activated by phosphorylation. (PFK2 produces an allosteric regulator that activates glycolysis, which ultimately produces glycerol.) Draw a diagram that shows how the Ras and MAP kinase pathway converge to result in the phosphorylation and activation of PFK2.

68. Yeast mutants lacking components of the HOG pathway (see Problem 67) were first exposed to high concentrations of glucose and then the PFK2 activity was measured. How does the PFK2 activity in the mutants compare with the PFK2 activity in mutants exposed to isotonic conditions?

apoptosis

Sphingomyelin

sphingomyelinase

Sphingosine ← Ceramide

sphingosine kinase   ceramide kinase

Sphingosine-1-phosphate   Ceramide-1-phosphate

cell survival and proliferation, inflammation

SELECTED READINGS

Eyster, K. M., The membrane and lipids as integral participants in signal transduction: Lipid signal transduction for the non-lipid biochemist, *Adv. Physiol. Educ.* **31**, 5–16 (2007). [Summarizes the roles of phospholipases, lipid kinases and phosphatases, and lipid hormones and second messengers.]

Lemmon, A. and Schlessinger, J., Cell signaling by receptor tyrosine kinases, *Cell* **141**, 1117–1134 (2010). [Describes the many events that occur following ligand binding to the receptor tyrosine kinases.]

Milligan, G. and Kostenis, E., Heterotrimeric G-proteins: A short history, *Br. J. Pharmacol.* **147**, S46–55 (2006).

Newton, A. C., Lipid activation of protein kinases, *J. Lipid Res.* **50**, S266–S271 (2009). [Describes how lipid second messengers affect kinases such as protein kinase C.]

Rasmussen, S. G. F., DeVree, B. T., Zou, Y., Kruse, A. C., Chung, K. Y., Kobilka, T. S., Thian, F. S., Chae, P. S., Pardon, E., Calinski,

D., Mathiesen, J. M., Shah, S. T. A., Lyons, J. A., Caffrey, M., Gellman, S. H., Steyaert, J., Skiniotis, G., Weis, W. I., Sunahara, R. K., and Kobilka, B. K., Crystal structure of the β_2 adrenergic receptor–Gs protein complex, *Nature* **477**, 549–555 (2011). [Provides detailed information about the interaction between a G protein and its associated receptor.]

Taylor, S. S. and Kornev, A. P., Protein kinases: Evolution of dynamic regulatory proteins, *Trends Biochem. Sci.* **36**, 65–77 (2011). [Reviews the general structure of protein kinases and their mechanism of activation by phosphorylation of an activation loop.]

Venkatakrishnan, A. J., Deupi X., Lebon, G., Tate, C. G., Schertler, G. F., and Babu, M., Molecular signatures of G-protein-coupled receptors, *Nature* **494**, 185–194 (2013). [Includes an overview of GPCR structures and some of their ligands.]

CHAPTER **11**

CARBOHYDRATES

Gerd Guenther/Science Source Images

Cultures of brown microalgae are the source of carbohydrates—primarily alginate, mannitol, and laminarin—that can be converted to biofuels as well as feed for farm animals and fish. Unlike land plants and some other algae, these organisms do not produce lignin, a structural polymer that is difficult to break down.

DO YOU REMEMBER?

- Cells contain four major types of biological molecules and three major types of polymers (Section 1.2).
- The polar water molecule forms hydrogen bonds with other molecules (Section 2.1).

Of all the classes of biological molecules, carbohydrates are the simplest, in terms of their atomic composition and molecular structure. They are typically constructed of C, H, and O atoms and adhere to the molecular formula $(CH_2O)_n$, where $n \geq 3$ (hence the name carbohydrate). Even carbohydrate derivatives—many of which include groups containing nitrogen, phosphorus, and other elements—are easy to recognize by their large number of hydroxyl (—OH) groups. Despite this regularity, carbohydrates participate in activities ranging from energy metabolism to cellular structure. This chapter surveys simple and complex carbohydrates and examines some of their biological functions.

11.1 MONOSACCHARIDES

Carbohydrates, also known as sugars or saccharides, occur as **monosaccharides** (simple sugars), small polymers (**disaccharides, trisaccharides,** and so on), and larger **polysaccharides** (sometimes called complex carbohydrates). The simplest sugars are the three-carbon compounds glyceraldehyde and dihydroxyacetone:

Glyceraldehyde

Dihydroxyacetone

KEY CONCEPTS
- Monosaccharides contain different numbers of carbon atoms and may be aldoses or ketoses with multiple enantiomeric and epimeric forms.
- α and β anomers can freely interconvert unless the anomeric carbon participates in a glycosidic bond.
- The functional groups of a monosaccharide can be modified to form a variety of derivatives.

A sugar such as glyceraldehyde, in which the carbonyl group is part of an aldehyde, is known as an **aldose,** and a carbohydrate such as dihydroxyacetone, in which the carbonyl group is part of a ketone, is known as a **ketose**. In most ketoses, the carbonyl group occurs at the second carbon (C2).

Monosaccharides can also be described according to the number of carbon atoms they contain; for example, the three-carbon compounds shown above are **trioses**. **Tetroses** contain four carbons, **pentoses** five, **hexoses** six, and so on. The aldopentose ribose is a component of ribonucleic acid (RNA; its derivative 2′-deoxyribose occurs in deoxyribonucleic acid, DNA). By far the most abundant monosaccharide is glucose, an aldohexose. It is the preferred metabolic fuel for most types of cells; it is stored in significant amounts in plants and animals, and it is a key component of plant cell walls. A common ketohexose is fructose, which is also a metabolic fuel.

Ribose   Glucose   Fructose

MOST CARBOHYDRATES ARE CHIRAL COMPOUNDS

Note that glucose is a **chiral** compound because several of its carbon atoms (all except C1 and C6) bear four different substituents (see Section 4.1 for a discussion of chirality). As a result, *glucose has a number of stereoisomers, as do nearly all monosaccharides* (the symmetric dihydroxyacetone is one exception). Several types of stereoisomerism apply to carbohydrates.

Like the amino acids (Section 4.1), glyceraldehyde has two different structures that exhibit mirror symmetry. *Such pairs of structures, known as **enantiomers,** cannot be superimposed by rotation.* By convention, these structures are given the designations L and D, derived from the Latin *laevus,* "left," and *dexter,* "right." The enantiomeric forms of larger monosaccharides are given the D or L designation by comparing their structures to D- and L-glyceraldehyde. In a **D sugar,** the asymmetric carbon farthest from the carbonyl group (that would be C5 in glucose) has the same spatial arrangement as the chiral carbon of D-glyceraldehyde. In an **L sugar,** that carbon has the same arrangement as in L-glyceraldehyde. Thus, every D sugar is the mirror image of an L sugar.

L-Glyceraldehyde   D-Glyceraldehyde

Although enantiomers behave identically in a strictly chemical sense, they are not biologically equivalent. This is because biological systems, which are built of other chiral compounds, such as L-amino acids, can distinguish D and L sugars. Most naturally occurring sugars have the D configuration, so the D and L prefixes are often omitted from their names.

Glucose, in addition to its enantiomeric carbon (C1), has four other asymmetric carbons, so there are stereoisomers for the configuration at each of these positions. *Carbohydrates that* *differ in configuration at one of these carbons are known as **epimers***. For example, the common monosaccharide galactose is an epimer of glucose, at position C4:

D-Glucose   D-Galactose

Both ketoses and aldoses have epimeric forms. And like enantiomers, epimers are not biologically interchangeable: An enzyme whose active site accommodates glucose may not recognize galactose at all.

CYCLIZATION GENERATES α AND β ANOMERS

The numerous hydroxyl groups that characterize carbohydrate structures also provide multiple points for chemical reactions to occur. One such reaction is an intramolecular rearrangement in which the sugar's carbonyl group reacts with one of its hydroxyl groups to form a cyclic structure (**Fig. 11.1**). The cyclic sugars are represented as **Haworth projections** in which the darker horizontal lines correspond to bonds above the plane of the paper, and the lighter lines correspond to bonds behind the plane of the paper. A simple rule makes it easy to convert a structure from its linear **Fischer projection** (in which horizontal bonds are above the plane of the paper and vertical bonds are behind it) to a Haworth projection: Groups projecting to the right in a Fischer projection will point down in a Haworth projection, and groups projecting to the left will point up.

As a result of the cyclization reaction, the hydroxyl group attached to what was the carbonyl carbon (C1 in the case of glucose) may point either up or down. In the **α anomer,** this hydroxyl group lies on the opposite side of the ring from the CH_2OH group of the chiral carbon that determines the D or L configuration (in the α anomer of glucose, the hydroxyl group points down; see Fig. 11.1). In the **β anomer,** the hydroxyl group lies on the same side of the ring as the CH_2OH group of the chiral carbon that determines the D or L configuration (in the β anomer of glucose, the hydroxyl group points up; see Fig. 11.1).

*Unlike enantiomers and epimers, which are not interchangeable, **anomers** in an aqueous solution freely interconvert between the α and β forms, unless the hydroxyl group attached to the anomeric carbon is linked to another molecule.* In fact, a solution of glucose molecules

Fischer projection   Haworth projections

FIGURE 11.1

Representation of glucose. In a linear Fischer projection, the horizontal bonds point out of the page and the vertical bonds point below the page. Glucose cyclizes to form a six-membered ring represented by a Haworth projection, in which the heaviest bonds point out from the plane of the page. The α and β anomers freely interconvert.

consists of about 64% β anomer, about 36% α anomer, and only trace amounts of the linear or open-chain form.

Hexoses and pentoses, which also undergo cyclization, do not form planar structures, as a Haworth projection might suggest. Instead, the sugar ring puckers so that each C atom can retain its tetrahedral bonding geometry. The substituents of each carbon may point either above the ring (axial positions) or outward (equatorial positions). Glucose can adopt a chair conformation in which all its bulky ring substituents (the —OH and —CH₂OH groups) occupy equatorial positions.

Glucose

In all other hexoses, some of these groups must occupy the more crowded—and therefore less stable—axial positions. The greater stability of glucose may be one reason for its abundance among monosaccharides.

MONOSACCHARIDES CAN BE DERIVATIZED IN MANY DIFFERENT WAYS

The anomeric carbon of a monosaccharide is easy to recognize: It is the carbonyl carbon in the straight-chain form of the sugar, and it is the carbon bonded to both the ring oxygen and a hydroxyl group in the cyclic form of the sugar. The anomeric carbon can undergo oxidation, so it can reduce substances such as Cu(II) to Cu(I). This chemical reactivity, often assayed using a copper-containing solution known as Benedict's reagent, can distinguish a free monosaccharide, called a **reducing sugar,** from a monosaccharide in which the anomeric carbon has already condensed with another molecule. For example, when a glucose molecule (a reducing sugar) reacts with methanol (CH₃OH), the result is a **nonreducing sugar (Fig. 11.2).** Because the anomeric carbon is involved in the reaction, the methyl group can end up in either the α or β position. The bond that links the anomeric carbon to the other group is called a **glycosidic bond,** and a molecule consisting of a sugar linked to another molecule is called a **glycoside.** Glycosidic bonds link the monomers in oligo- and polysaccharides (Section 11.2) and also link the ribose groups to the purine and pyrimidine bases of nucleotides (Section 3.1).

FIGURE 11.2

Reaction of glucose with methanol. The addition of methanol to the anomeric carbon blocks the ability of glucose to function as a reducing sugar. The glycosidic bond that forms between the anomeric carbon and the oxygen of methanol may have the α or β configuration.

Phosphorylated sugars, including glyceraldehyde-3-phosphate and fructose-6-phosphate, appear as intermediates in the metabolic pathways for breaking down glucose (glycolysis; Section 13.1) and synthesizing it (photosynthesis; Section 16.3).

Glyceraldehyde-3-phosphate

Fructose-6-phosphate

Other metabolic processes replace a hydroxyl group with an amino group to produce an amino sugar, such as glucosamine (**Fig. 11.3a**). Oxidation of a sugar's carbonyl and hydroxyl groups can yield uronic acids (sugars containing carboxylic acid groups; Fig. 11.3b), and reduction can yield molecules such as xylitol, a sweetener used in "sugarless" foods (Fig. 11.3c). One metabolically essential carbohydrate-modifying reaction is the one catalyzed by ribonucleotide reductase, which reduces the 2′-OH group of ribose to convert a ribonucleotide to a deoxyribonucleotide for DNA synthesis (Section 18.5):

Ribose   **2′-Deoxyribose**

(a)

Glucosamine

(b)

Glucuronate

(c)

Xylitol

FIGURE 11.3

Some monosaccharide derivatives. In an amino sugar (a), $-NH_3^+$ replaces an $-OH$ group. Oxidation and reduction reactions yield sugars with carboxylate groups (b) or additional hydroxyl groups (c).

? Identify the net charge of each sugar.

CONCEPT CHECK

1. Draw the straight-chain form of D-glucose, a ketose isomer of glucose, the L enantiomer of glucose, and one of its epimers.

2. Explain why the α and β anomers of a monosaccharide can interconvert.

3. List some types of monosaccharide derivatives.

11.2 POLYSACCHARIDES

Monosaccharides are the building blocks of polysaccharides, in which glycosidic bonds link successive residues. Unlike amino acids and nucleotides—the other polymer-forming biological molecules—which are linked in only one configuration, monosaccharides can be hooked together in a variety of ways to produce a dizzying array of chains. *Each monosaccharide contains several free —OH groups that can participate in a condensation reaction, which permits different bonding arrangements and allows for branching.* While this expands the structural repertoire of carbohydrates, it makes studying them difficult.

In the laboratory, carbohydrate chains, or **glycans,** can be sequenced using mass spectrometry (Section 4.6), although the results are sometimes ambiguous due to the inability to distinguish isomers, which have the same mass. Glycan three-dimensional structures are typically studied using NMR techniques (Section 4.6), since these yield an average conformation for the molecules, which tend to be highly flexible in solution. Due to the challenges of defining carbohydrate sequences and structures, **glycomics,** the systematic study of carbohydrates, is not as fully developed as genomics or proteomics.

The most complex glycans are the oligosaccharides that are commonly linked to other molecules, for example, in glycoproteins. Polysaccharides, some of which are truly enormous molecules, generally do not exhibit the heterogeneity and complexity of oligosaccharides.

KEY CONCEPTS

- Monosaccharides can be linked by glycosidic bonds in various arrangements.
- Lactose and sucrose are disaccharides that are used as metabolic fuel.
- Glucose polymers include the fuelstorage polysaccharides starch and glycogen and the structural polysaccharide cellulose.
- Other structural polysaccharides include chitin and components of bacterial biofilms.

Instead, they tend to consist of one or a pair of monosaccharides that are linked over and over in the same fashion. This sort of structural homogeneity is well suited to the function of polysaccharides as fuel-storage molecules and architectural elements. We begin our survey with the simplest of the polysaccharides, the disaccharides.

LACTOSE AND SUCROSE ARE THE MOST COMMON DISACCHARIDES

A glycosidic bond links two monosaccharides to generate a disaccharide. In nature, disaccharides occur as intermediates in the digestion of polysaccharides and as a source of metabolic fuel. For example, lactose, secreted into the milk of lactating mammals, consists of galactose and glucose:

Lactose

The curved lines in the diagram are ordinary covalent bonds; drawing the linkages in this fashion makes it easier to inspect the monosaccharides side by side.

Note that the anomeric carbon (C1) of galactose is linked to C4 of glucose via a β-glycosidic bond. If the two sugars were linked by an α-glycosidic bond, or if the galactose anomeric carbon were linked to a different glucose carbon, the result would be an entirely different disaccharide. Lactose serves as a major food for newborn mammals. Most adult mammals, including humans, produce very little lactase (also called β-galactosidase), the enzyme that breaks the glycosidic bond of lactose, and therefore cannot efficiently digest this disaccharide.

In addition to lactose, milk contains hundreds of different oligosaccharides. Many of these cannot be digested by mammalian enzymes and therefore cannot be a food source for the newborn. Instead, these carbohydrates appear to sustain certain types of bacteria that are essential for establishing a healthy microbial community in the newborn's intestine.

Sucrose, or table sugar, is the most abundant disaccharide in nature:

Sucrose

In this molecule, the anomeric carbon of glucose (in the α configuration) is linked to the anomeric carbon of fructose (in the β configuration). Sucrose is the major form in which newly synthesized carbohydrates are transported from a plant's leaves, where most photosynthesis occurs, to other plant tissues to be used as a fuel or stored as starch for later use.

STARCH AND GLYCOGEN ARE FUEL-STORAGE MOLECULES

Starch and glycogen are polymers of glucose residues linked by glycosidic bonds designated α(1 → 4); in other words, the anomeric carbon (carbon 1) of one residue is linked by an α-glycosidic bond to carbon 4 of the next residue:

Plants manufacture a linear form of starch, called amylose, which can consist of several thousand glucose residues. Amylopectin, an even larger molecule, includes $\alpha(1 \rightarrow 6)$ glycosidic linkages every 24 to 30 glucose residues to generate a branched polymer:

Compiling many monosaccharide residues in a single polysaccharide is an efficient way to store glucose, the plant's primary metabolic fuel. The α-linked chains curve into helices so that the entire molecule forms a relatively compact particle (**Fig. 11.4**).

Animals store glucose in the form of glycogen, a polymer that resembles amylopectin but with branches every 12 residues or so. Due to its highly branched structure, a glycogen molecule can be quickly assembled or disassembled according to the metabolic needs of the cell because the enzymes that add or remove glucose residues work from the ends of the branches.

CELLULOSE AND CHITIN PROVIDE STRUCTURAL SUPPORT

Cellulose, like amylose, is a linear polymer containing thousands of glucose residues. However, *the residues are linked by β(1 → 4) rather than α(1 → 4) glycosidic bonds:*

This simple difference in bonding has profound structural consequences: Whereas starch molecules form compact granules inside the cell, cellulose forms extended fibers that lend rigidity and strength to plant cell walls. Individual cellulose polymers form bundles with

FIGURE 11.4

Structure of amylose. This unbranched polysaccharide, consisting of α(1 → 4)-linked glucose residues (six are shown here), forms a large left-handed helix. C atoms are gray, O atoms red, and hydroxyl H atoms white (other H atoms are not shown).

? In what part of a cell would amylose be located?

FIGURE 11.5

Cellulose structure. Glucose residues are represented as hexagons, with C atoms gray and O atoms red. Not all H atoms (small circles) are shown. Hydrogen bonds (dashed lines) link residues in the same and adjacent chains so that a bundle of cellulose polymers forms an extended rigid fiber.

extensive hydrogen bonding within and between adjacent chains (**Fig. 11.5**). Plant cell walls include other polymers that, together with cellulose, yield a strong but resilient substance. Recovering the carbohydrates from materials such as wood remains a challenge for the biofuels industry (**Box 11.A**).

A CLOSER LOOK

Box 11.A Cellulosic Biofuel

Cellulose is by far the most abundant polysaccharide in nature, with chains containing thousands of monosaccharide residues. For this reason, cellulose-rich materials, including wood and agricultural waste, are sources of sugars that can be converted to biofuels such as ethanol, potentially replacing petroleum-derived fuels. Unfortunately, cellulose does not exist in pure form in nature; plant cell walls typically contain other polymers, including hemicellulose, pectin, and lignin.

Hemicellulose is the name given to a class of polysaccharides whose chains are shorter than cellulose (500–3000 residues) and may be branched. Hemicellulose is a heteropolymer, indicating that it contains a variety of monomeric units, in this case 5- and 6-carbon sugars. Xylose is the most abundant of these:

$$
\begin{array}{c}
\mathrm{CHO} \\
| \\
\mathrm{HCOH} \\
| \\
\mathrm{HOCH} \\
| \\
\mathrm{HCOH} \\
| \\
\mathrm{CH_2OH}
\end{array}
$$

Xylose

While cellulose forms rigid fibers and hemicellulose forms a network, the spaces in between these polymers are occupied by pectin, a heteropolymer containing galacturonate and rhamnose residues, among others:

Galacturonate   Rhamnose

The large number of hydroxyl groups makes pectin highly hydrophilic, so it "holds" a great deal of water and has the physical properties of a gel.

Lignin—in contrast to cellulose, hemicellulose, and pectin—is not a polysaccharide at all. It is a highly heterogeneous, difficult-to-characterize polymer built from aromatic (phenolic) compounds. With few hydroxyl groups, it is relatively hydrophobic. Lignin is covalently linked to hemicellulose chains, so it contributes to the mechanical strength of cell walls.

All of the components of wood, including lignin, represent a large amount of stored free energy, which can be released by combustion (for example, when wood burns). The industrial conversion of this stored energy into other types of fuels is known as bioconversion. The first step is the hardest: separating the polysaccharides from lignin. Physical methods such as grinding and pulverizing consume energy, but chemical methods—which may include strong acids or organic solvents—come with their own hazards. An additional drawback is the generation of reaction products that can inhibit subsequent steps in biofuel production, which depend on living organisms or enzymes derived from them.

Once freed from lignin, the carbohydrate polymers are accessible to hydrolytic enzymes, most of which originate in bacteria and fungi that are adept at degrading plant materials. The result is a mixture of monosaccharides. Fungi such as yeast efficiently ferment the glucose to ethanol (Section 13.1), which can be distilled and used as a fuel. Other organisms can convert monosaccharides such as xylose to ethanol, but these pathways are not always efficient, and they may yield other substances, such as lactate and acetate, as end products. The most promising approach appears to be bioengineering microorganisms to carry out polysaccharide hydrolysis and then convert the resulting monosaccharides to ethanol, which is relatively stable and easy to transport and store. Alternative strategies use microorganisms to convert the sugar mixture into hydrocarbons that can be used in place of diesel fuel.

? **How do xylose, galacturonate, and rhamnose differ from glucose?**

Animals do not synthesize cellulose, and most cannot digest it in order to use its glucose residues as an energy source. Organisms such as termites and ruminants (grazing mammals), who do derive energy from cellulose-rich foods, harbor microorganisms that produce cellulases capable of hydrolyzing the β(1 → 4) bonds between glucose residues. Humans lack these microorganisms, so although up to 80% of the dry weight of plants consists of glucose, much of this is not available for metabolism in humans. However, its bulk, called fiber, is required for the normal function of the digestive system.

The exoskeletons of insects and crustaceans and the cell walls of many fungi contain a cellulose-like polymer called chitin, in which the β(1 → 4)-linked residues are the glucose derivative *N*-acetylglucosamine (glucosamine with an acetyl group linked to its amino group):

FIGURE 11.6

A *Pseudomonas aeruginosa* biofilm. These pathogenic bacteria growing on the surface of an agar plate form a biofilm with a complex three-dimensional shape. [*Courtesy Roberto Kolter, Harvard Medical School.*]

BACTERIAL POLYSACCHARIDES FORM A BIOFILM

Prokaryotes do not synthesize cellulosic cell walls (see Section 11.3) or store fuel as starch or glycogen, but they do produce extracellular polysaccharides that provide a protective matrix for their growth. *A **biofilm** is attached to a surface and harbors a community of embedded bacteria that contribute to biofilm production and maintenance* (**Fig. 11.6**). The extracellular material of the biofilm includes an assortment of highly hydrated polysaccharides containing glucuronate and *N*-acetylglucosamine. A biofilm can be difficult to characterize because it typically houses a mixture of species and the proportions of its component polysaccharides depend on many environmental factors.

The gel-like consistency of a biofilm, such as the plaque that forms on teeth, prevents bacterial cells from being washed away and protects them from desiccation. Biofilms that develop on medical apparatus, such as catheters, are problematic because they offer a foothold for pathogenic organisms and create a barrier for antibiotics and cells of the immune system.

◆ CONCEPT CHECK

1. Explain why it is possible for two monosaccharides to form more than one type of disaccharide.
2. Summarize the physiological roles of lactose, sucrose, starch, glycogen, cellulose, and chitin.
3. How do the physical properties of polysaccharides—such as overall size, shape, branching, composition, and hydrophilicity—relate to their biological functions?

GLYCOPROTEINS

Because there are so many different monosaccharides and so many ways to link them, the number of possible structures for oligosaccharides with even just a few residues is enormous. Organisms take advantage of this complexity to mark various structures—mainly proteins and

KEY CONCEPTS
- Carbohydrates are attached to proteins as *N*-linked or *O*-linked oligosaccharides.
- The long glycosaminoglycan chains of proteoglycans are highly hydrated.
- Bacteria build cell walls from peptidoglycan, a three-dimensional network of carbohydrate chains and short peptides.

lipids—with unique oligosaccharides. Most of the proteins that are secreted from eukaryotic cells or remain on their surface are glycoproteins in which one or more **oligosaccharide** chains are covalently attached to the polypeptide chain shortly after its synthesis.

OLIGOSACCHARIDES ARE *N*-LINKED OR *O*-LINKED

In eukaryotes, the oligosaccharides attached to glycoproteins are usually linked either to an asparagine side chain (***N*-linked oligosaccharides**) or to a serine or threonine side chain (***O*-linked oligosaccharides).**

N-Linked oligosaccharide

O-Linked oligosaccharide

N-Glycosylation begins while a protein is being synthesized by a ribosome associated with the rough endoplasmic reticulum (ER). As the protein is translocated into the ER lumen (the internal space), an oligosaccharide chain of 14 residues is attached to an asparagine residue (**Fig. 11.7**). When the newly synthesized protein leaves the ER and traverses the Golgi apparatus (a series of membrane-bounded compartments), *enzymes known as* **glycosidases** *remove various monosaccharide residues, and other enzymes, called* **glycosyltransferases,** *add new monosaccharides.* These processing enzymes are highly specific for the identities of the monosaccharides and the positions of the glycosidic bonds.

Apparently, the amino acid sequence or local structure of the protein, as well as the set of processing enzymes present in the cell, roughly determines which sugars are added and deleted. The net result is a great deal of heterogeneity in the oligosaccharide chains

---

FIGURE 11.7

Processing of an *N*-linked oligosaccharide. Glycosidases and glycosyltransferases in the Golgi apparatus process the 14-residue oligosaccharide that is attached to the newly synthesized protein in the ER. The 5-residue core oligosaccharide (with 3 mannose and 2 *N*-acetylglucosamine residues, center) is common to all *N*-linked oligosaccharides. Only one of many possible mature oligosaccharides is shown.

? **How many different glycosyltransferases would be required to synthesize the oligosaccharide shown here?**

polypeptide

Newly synthesized polypeptide with 14-residue oligosaccharide

■ *N*-Acetylglucosamine
● Mannose
● Galactose
◆ *N*-Acetylneuraminic acid
● Glucose
▲ Fucose

The oligosaccharide initially attached to the protein is trimmed by glycosidases.

Partially processed glycoprotein containing core oligosaccharide

Additional monosaccharide residues are added by the action of glycosyltransferases.

Glycoprotein with mature oligosaccharide

attached to different glycoproteins or even to different molecules of the same glycoprotein. An example of an *N*-linked oligosaccharide is shown in **Figure 11.8**. Despite the variability, attaching oligosaccharides to proteins is essential, an idea supported by the documentation of at least 100 congenital disorders linked to faulty glycosylation.

O-Linked oligosaccharides are built, one residue at a time, primarily in the Golgi apparatus, through the action of glycosyltransferases. Unlike *N*-linked oligosaccharides, the *O*-linked oligosaccharides do not undergo processing by glycosidases. Glycoproteins with *O*-linked saccharide chains tend to have many such groups, and the glycan chains tend to be longer than those of *N*-linked oligosaccharides.

OLIGOSACCHARIDE GROUPS ARE BIOLOGICAL MARKERS

Because oligosaccharides are highly hydrophilic and are conformationally flexible, they occupy a large effective volume above the protein's surface. This may serve a protective function or help stabilize the protein's structure. In fact, certain chaperones recognize partially glycosylated proteins and help them fold to their native conformations. In some cases, oligosaccharide groups constitute a sort of intracellular addressing system so that newly synthesized proteins can be delivered to their proper cellular location, such as a lysosome. In other cases, the oligosaccharide groups act as recognition and attachment points for interactions between different types of cells. For example, the familiar A, B, and O blood types are determined by the presence of different oligosaccharides on the surface of red blood cells (**Box 11.B**). Circulating white blood cells latch onto glycoproteins on the cells lining the blood vessels in order to leave the bloodstream and migrate to sites of injury or infection.

FIGURE 11.8

Structure of an *N*-linked oligosaccharide. The 11 monosaccharide residues, linked by glycosidic bonds, experience considerable conformational flexibility, so the structure shown here is just one of many possible. The arrow indicates the N atom of the Asn side chain to which the glycan is attached. [*Structure of the carbohydrate chain of soybean agglutinin determined by A. Darvill and H. Halbeek.*]

A CLOSER LOOK

Box 11.B The ABO Blood Group System

The carbohydrates on the surface of red blood cells and other human cells form 15 different blood group systems. The best known and one of the clinically important carbohydrate-classification schemes is the ABO blood group system, which has been known for over a century. Biochemically, the ABO system involves the oligosaccharides attached to sphingolipids and proteins on red blood cells and other cells.

In individuals with type A blood, the oligosaccharide has a terminal *N*-acetylated galactose group. In type B individuals, the terminal sugar is galactose. Neither of these groups appears in the oligosaccharides of type O individuals.

Blood groups are genetically determined: Type A and B individuals have slightly different versions of the gene for a glycosyltransferase that adds the final monosaccharide residue to the oligosaccharide. Type O individuals have a mutation such that they lack the enzyme entirely and therefore produce an oligosaccharide without the final residue.

Type A individuals develop antibodies that recognize and cross-link red blood cells bearing the type B oligosaccharide. Type B individuals develop antibodies to the type A oligosaccharide. Therefore, a transfusion of type B blood cannot be given to a type A individual, and vice versa. Individuals with type AB blood bear both types of oligosaccharides and therefore do not develop antibodies to either type. They can receive transfusions of either type A or type B blood. Type O individuals develop both anti-A and anti-B antibodies. If they receive a transfusion of type A, type B, or type AB blood, their antibodies will react with the transfused cells, which causes them to lyse or to clump together and block blood vessels. On the other hand, type O individuals are universal donors: Type A, B, or AB individuals can safely receive type O blood (these individuals do not develop antibodies to the O-type oligosaccharide because it occurs naturally in these individuals as a precursor of the A-type and B-type oligosaccharides).

? In rare cases, individuals do not produce an A-, B-, or O-type oligosaccharide at all. Can these individuals receive type A, B, or O blood transfusions? Can they donate blood to others?

Unfortunately, many viruses and pathogenic bacteria also recognize specific carbohydrate groups on cell surfaces and attach themselves to these sites before invading the host cell.

Intracellular eukaryotic proteins are also glycosylated, most often by the addition of a single *N*-acetylglucosamine to serine or threonine side chains in cytosolic, mitochondrial, and nuclear proteins. This protein modification is much more common than previously thought and appears to influence a protein's localization, rate of degradation, or interactions with other molecules—all of which can in turn regulate metabolic activity and patterns of gene expression. In this respect, intracellular protein glycosylation is analogous to the protein phosphorylation and dephosphorylation that occur during signal transduction (Chapter 10). *The monosaccharide groups are derived directly from glucose (the cell's major fuel) and glutamine (a key amino acid and a cellular source of amino groups) and therefore serve as a sort of nutrient sensor to help coordinate the cell's activities with its food supply.* This regulatory scheme could become a target for drugs to treat conditions where fuel metabolism is abnormal, such as in diabetes and cancer.

PROTEOGLYCANS CONTAIN LONG GLYCOSAMINOGLYCAN CHAINS

Proteoglycans are glycoproteins in which the protein chain serves mainly as an attachment site for enormous linear *O*-linked polysaccharides called *glycosaminoglycans.* Most glycosaminoglycan chains consist of a repeating disaccharide of an amino sugar (often *N*-acetylated) and a uronic acid (a sugar with a carboxylate group). After synthesis, various hydroxyl groups may be enzymatically sulfated (an $-OSO_3^-$ group added). The repeating disaccharide of the proteoglycan known as chondroitin sulfate is shown in **Figure 11.9.** Proteoglycans may be transmembrane proteins or lipid-linked (Section 8.3), but *the glycosaminoglycan chains are invariably on the extracellular side of the plasma membrane.* Extracellular proteoglycans and glycosaminoglycan chains that are not attached to a protein scaffold play an important structural role in connective tissue.

The many hydrophilic groups on the glycosaminoglycans attract water molecules, so glycosaminoglycans are highly hydrated and occupy the spaces between cells and other components of the extracellular matrix, such as collagen fibrils (Section 5.3). Under mechanical pressure, some of the water can be squeezed out of the glycosaminoglycans, which allows connective tissue and other structures to accommodate the body's movements. Pressure also brings the negatively charged sulfate and carboxylate groups of the polysaccharides close together. When the pressure abates, the glycosaminoglycans quickly spring back to their original shape as the repulsion between anionic groups is relieved and water is drawn back into the molecule. This spongelike action of glycosaminoglycans in the spaces of the joints provides shock absorption.

The proteoglycans known as mucins form the protective mucus lining the respiratory, gastrointestinal, and reproductive tracts. These proteins, half of whose residues may be serine and threonine, serve as scaffolding for numerous glycosaminoglycan chains, so that the entire molecule is often enormous (up to 10,000,000 D, with carbohydrate accounting for as much as 80% of the total mass). The viscous, tangled mucin chains act as lubricants and also trap bacteria and other foreign particles to prevent their access to the body. In the respiratory tract, the cilia protruding from epithelial cells create a current that moves the

FIGURE 11.9

The repeating disaccharide of chondroitin sulfate. A chondroitin sulfate chain may include hundreds of these disaccharide units, and the degree of sulfation may vary along its length.

 What ions are likely to be associated with chondroitin sulfate?

Chondroitin sulfate

FIGURE 11.10

The airway surface. The outward-facing cell surface is ciliated. A mucus layer lies above the cilia so that ciliary movement sweeps away the mucus and trapped particles. [*Photomicrograph provided by Dr. Scott H. Randell (The University of North Carolina at Chapel Hill).*]

mucus
cilia
endothelial cells

overlying mucus layer and any captured particles out of the airways so that they can be swallowed (**Fig. 11.10**).

BACTERIAL CELL WALLS ARE MADE OF PEPTIDOGLYCAN

A network of cross-linked carbohydrate chains and peptides constitutes the cell walls of bacteria. This material, called **peptidoglycan,** *surrounds the plasma membrane of the cell and determines its overall shape*. The carbohydrate component in many species is a repeating $\beta(1 \rightarrow 4)$-linked disaccharide with a typical total length of about 20 to 40 disaccharides.

CH₂OH / H / OH / H / NH—C—CH₃ / O / CH₃CHCOO⁻ / n

Peptides of four or five amino acids covalently cross-link the saccharide chains in three dimensions to form a structure as thick as 250 Å in some species. Antibiotics of the penicillin family block the formation of the peptide cross-links, thereby killing bacteria without harming their eukaryotic hosts. Archaebacterial cell walls contain proteins and sometimes polysaccharides but not peptidoglycans.

The overall structure of the peptidoglycan cell wall is not known. However, a model of peptidoglycan from the bacterium *Escherichia coli* suggests that the glycan chains are parallel to the cell surface and are arranged with enough space in between to allow nutrients and wastes to easily diffuse to and from the cell surface (**Fig. 11.11**). Although the carbohydrate chains are relatively stiff, calculations indicate that the peptide cross-links are more elastic and could accommodate cell shape changes of 25 to 33%.

FIGURE 11.11

Model of *E. coli* peptidoglycan. In this model, looking down on the cell surface, a single layer of peptidoglycan is shown with carbohydrate strands in blue and cross-linking peptides in green. The scale bar is 10 nm. [*From Gumbart, J. C., Beeby, M., Jensen, G. J., and Roux, B.,* PLoS Comput. Biol. *10(2): e1003475. doi:10.1371/ journal.pcbi.1003475 (2014).*]

 CONCEPT CHECK

1. Summarize the processing of an *N*-linked oligosaccharide.
2. List some functions of the carbohydrate groups of glycoproteins.
3. Compare *N*- and *O*-linked oligosaccharide structures and functions.
4. How do proteoglycans function as shock absorbers?
5. Compare proteoglycans and peptidoglycans to polysaccharides such as starch and cellulose.

SUMMARY

11.1 Monosaccharides

• Carbohydrates, which have the general formula $(CH_2O)_n$, exist as monosaccharides and polysaccharides of different sizes. The monosaccharides may be aldoses or ketoses and exist as enantiomers, which are mirror images, and as epimers, which differ in configuration at individual carbon atoms.

• Cyclization of a monosaccharide produces α and β anomers. Formation of a glycosidic bond prevents the interconversion of the α and β forms.

• Monosaccharide derivatives include phosphorylated sugars; sugars with amino, carboxylate, and extra hydroxyl groups; and deoxy sugars.

11.2 Polysaccharides

• Lactose consists of galactose linked β(1 → 4) to glucose. Sucrose consists of glucose linked α(1 → 2) to fructose.

• Starch is a linear polymer of α(1 → 4)-linked glucose residues; glycogen also contains α(1 → 6) branch points. Cellulose consists of β(1 → 4)-linked glucose residues; in chitin, the residues are N-acetylglucosamine.

• A bacterial biofilm is a community of cells embedded in an extracellular polysaccharide matrix.

11.3 Glycoproteins

• Oligosaccharides are attached to proteins as N- and O-linked oligosaccharides. N-Linked oligosaccharides undergo processing by glycosidases and glycosyltransferases. The carbohydrate chains of glycoproteins function as protection and as recognition markers.

• Proteoglycans consist mainly of long glycosaminoglycan chains that can function as shock absorbers, lubricants, and protective coatings.

• Peptidoglycan in bacterial cell walls is constructed of cross-linked oligosaccharides and peptides.

KEY TERMS

carbohydrate	pentose	α anomer	biofilm
monosaccharide	hexose	β anomer	oligosaccharide
disaccharide	chirality	anomers	N-linked oligosaccharide
trisaccharide	enantiomers	reducing sugar	O-linked oligosaccharide
polysaccharide	D sugar	nonreducing sugar	glycosidase
aldose	L sugar	glycosidic bond	glycosyltransferase
ketose	epimers	glycoside	proteoglycan
triose	Haworth projection	glycan	glycosaminoglycan
tetrose	Fischer projection	glycomics	peptidoglycan

BIOINFORMATICS

Brief Bioinformatics Exercises

11.1 Drawing Monosaccharides

11.2 Introduction to Glycomics

11.3 Oligosaccharides, Polysaccharides, and Glycoproteins

Bioinformatics Projects

Glycomics and the H1N1 Flu

PROBLEMS

11.1 Monosaccharides

1. Classify the following sugars as adloses or ketoses:

(a)
```
       CHO
        |
  HO—C—H
        |
  HO—C—H
        |
   H—C—OH
        |
      CH₂OH
```

(b)
```
      CH₂OH
        |
       C=O
        |
  HO—C—H
        |
      CH₂OH
```

(c)
```
       CHO
        |
  HO—C—H
        |
  HO—C—H
        |
  HO—C—H
        |
   H—C—OH
        |
      CH₂OH
```

2. Glucose can be described as an aldohexose. Use similar terminology to describe the following sugars:

(a)
```
       CHO
        |
   H—C—OH
        |
   H—C—OH
        |
   H—C—OH
        |
      CH₂OH
```

(b)
```
      CH₂OH
        |
       C=O
        |
   H—C—OH
        |
      CH₂OH
```

(c)
```
        CHO
         |
  HO — C — H
         |
  HO — C — H
         |
  HO — C — H
         |
   H — C — OH
         |
       CH₂OH
```

(d)
```
       CH₂OH
         |
        C = O
         |
   H — C — OH
         |
   H — C — OH
         |
   H — C — OH
         |
       CH₂OH
```

3. Which of the sugars shown in Problem 2 has the greatest number of stereoisomers? (*Hint*: The number of stereoisomers is equal to 2^n where n is the number of chiral carbons.)

4. Identify the monosaccharide(s) present in coenzyme A, NAD, and FAD (see Fig. 3.2).

5. Identify the following sugars as D or L.

(a)
```
       CH₂OH
         |
        C = O
         |
  HO — C — H
         |
       CH₂OH
```

(b)
```
        CHO
         |
   H — C — OH
         |
   H — C — OH
         |
   H — C — OH
         |
       CH₂OH
```

(c)
```
       CH₂OH
         |
   H — C — OH
         |
   H — C — OH
         |
   H — C — OH
         |
   H — C — OH
         |
       CH₂OH
```

(d)
```
        CHO
         |
        C = O
         |
   H — C — OH
         |
  HO — C — H
         |
  HO — C — H
         |
       CH₂OH
```

6. Which type of isomer is represented by each pair of sugars?

(a) D-Gulose and D-idose

```
        CHO                      CHO
         |                        |
   H — C — OH              HO — C — H
         |                        |
   H — C — OH               H — C — OH
         |                        |
  HO — C — H               HO — C — H
         |                        |
   H — C — OH               H — C — OH
         |                        |
       CH₂OH                    CH₂OH

     D-Gulose                  D-Idose
```

(b) D-fructose and L-fructose
(c) D-ribose (structure shown in Problem 2b) and D-ribulose
(d) D-sorbose and D-psicose

7. Which of the following are isomers of glucose? **(a)** glucose-6-phosphate; **(b)** fructose; **(c)** galactose; **(d)** ribose.

8. Mannose is the C2 epimer of glucose. Draw its structure.

9. Tagatose is an "artificial" sweetener that is similar to sucrose in sweetness and replaces sucrose in Pepsi Slurpees®. Tagatose is a C4 epimer of fructose.
(a) Draw the structure of D-tagatose.

(b) L-Tagatose is also about as sweet as sugar, but it costs more to manufacture than its D isomer, thus an early decision to market L-tagatose was abandoned. Draw the structure of L-tagatose.
(c) Only about 30% of tagatose is absorbed in the small intestine; effectively, tagatose has 30% of the calories of sucrose. Why is D-tagatose absorbed less efficiently in the small intestine?

10. D-Psicose, the C3 epimer of fructose, has no calories and has also been found to have hypoglycemic effects, which might be useful in treating diabetes. Draw the structure of D-psicose.

11. Which is more stable: the α or the β anomer of galactose?

12. Carry out a cyclization reaction with mannose and draw the Haworth structures of the two possible reaction products.

13. Like glucose, ribose can undergo a cyclization reaction with its aldehyde group and the C5 hydroxyl group to form a six-membered ring. Draw the structures of the two possible reaction products.

14. Ribose can undergo a cyclization reaction different from the one presented in Problem 13. In this case, the aldehyde group reacts with the C4 hydroxyl group. Draw the structures of the two possible reaction products. What is the size of the ring?

15. (a) Like glucose, fructose can undergo a cyclization reaction. The most common reaction involves the ketone group and the C5 hydroxyl group. Draw the structures of the two possible reaction products. What is the size of the ring?
(b) Repeat the exercise described in part (a) but use the C6 hydroxyl group instead. What is the size of the resulting ring?

16. The Fischer structure of N-acetylneuraminic acid (NANA) is shown below. It is synthesized from N-acetylmannosamine and pyruvic acid. This monosaccharide can undergo a cyclization reaction to form a six-membered ring. Draw the structure of the α-anomer of NANA.

```
                COOH       ⎫
                 |         ⎬ Pyruvic
                C = O      ⎪ acid
                 |         ⎭ residue
                CH₂
                 |
   O        H — C — OH     ⎫
   ‖            |          ⎪
CH₃ — C — NH — C — H       ⎪
                |          ⎬ N-Acetyl-
           HO — C — H      ⎪ mannosamine
                |          ⎪
            H — C — OH     ⎪
                |          ⎪
            H — C — OH     ⎪
                |          ⎭
              CH₂OH
```

N-Acetylneuraminic acid
(Fischer structure)

17. An enzyme recognizes only the β anomer of glucose as a substrate and converts it to product. If the enzyme is added to a mixture of the α and β anomers, explain why all the sugar molecules in the sample will eventually be converted to product.

18. As described in the text, a solution of glucose molecules consists of about 64% β anomer and about 36% α anomer. Why isn't the mixture 50% β anomer and 50% α anomer? In other words, why is the β anomer favored?

19. When glucose from the blood enters a cell, intracellular enzymes convert it to glucose-6-phosphate. This strategy results in the "trapping" of glucose in the cell. Explain.

20. During baking, reducing sugars react with proteins in food to generate adducts with brown color and different flavors (this is why bread becomes darker and has a different taste when toasted). The first step in this process is a condensation between a carbonyl group and an amino group. For example, the carbonyl carbon of glucose can condense with the ε-amino group of a lysine side chain to form a Schiff base. Draw the product of this reaction.

21. (a) Oxidation of the aldehyde group of glucose yields gluconate. Draw the Fischer structure of gluconate.
(b) Gluconate cyclizes to form a cyclic ester called a lactone. Draw the Haworth projection of the lactone.
(c) Reduction of the aldehyde group of glucose yields sorbitol. Draw the Fischer structure of sorbitol.

22. Benedict's solution is an alkaline copper sulfate solution which is used to detect the presence of aldehyde groups. In the presence of Benedict's solution, the aldehyde group is oxidized and the aqueous blue Cu^{2+} ion is reduced to a red Cu_2O precipitate. Sugars such as glucose, which produce the red precipitate when Benedict's solution is added, are called reducing sugars because they can reduce Cu^{2+} to Cu^+. Which of the following carbohydrates would give a positive reaction with Benedict's reagent? **(a)** galactose; **(b)** sorbitol (see Solution 21c); **(c)** β-ethylglucoside; **(d)** gluconate (see Solution 21a).

23. (a) If methanol is added to a sugar in the presence of an acid catalyst, only the anomeric hydroxyl group becomes methylated, as shown in Figure 11.2. If a stronger methylating agent is used, such as methyl iodide, CH_3I, all hydroxyl groups become methylated. Draw the product that results when methyl iodide is added to a solution of α-D-glucose.
(b) If the product that formed in part a is treated with a strong aqueous acid solution, the glycosidic methyl group is readily hydrolyzed, but the methyl ethers are not. Draw the structure of the product that results when the compound formed in part a is treated with strong aqueous acid.

24. When an unknown monosaccharide is treated with methyl iodide followed by strong aqueous acid (see Problem 23), the product is 2,3,5,6-tetra-*O*-methyl-D-glucose. Draw a Haworth structure of the unmodified monosaccharide.

25. In algae, the breakdown of certain monosaccharides yields 2-keto-3-deoxygluconate (KDG). Draw the structure of KDG.

26. Use the examples in the textbook to deduce the structures of the following sugars: **(a)** fructose-1,6-bisphosphate; **(b)** galactosamine; **(c)** *N*-acetylglucosamine. **(d)** the Fischer structure and the Haworth projection of the α-anomer of galacturonate.

27. Some pathogenic bacteria release monosaccharides that can be detected by white blood cells. These monosaccharides are not found in humans, so the enzymes that catalyze their synthesis in bacteria are attractive drug targets. Using the name of the monosaccharide and examples provided in the text, draw the Haworth projection of a heptose produced in bacteria named 2-*O*-acetyl-6-deoxy-α-D-mannoheptopyranose. (*Hint*: See Solution 8 for the structure of mannose.)

11.2 Polysaccharides

28. Explain why lactose is a reducing sugar, whereas sucrose is not.

29. As described in the text, lactase is the enzyme that catalyzes the hydrolysis of lactose. The enzyme is expressed in newborns, but its activity typically decreases as the organism matures, and adults become *lactose-intolerant*. However, adult members of populations whose ancestors domesticated cows (Northern Europe) or goats (Africa) are lactose-tolerant. Explain why.

30. Cellobiose is a disaccharide composed of two glucose monomers linked by a β(1 → 4) glycosidic bond. Draw the structure of cellobiose. Is cellobiose a reducing sugar?

31. The structure of trehalose is shown. Is trehalose a reducing sugar?

32. Trehalase is an enzyme that catalyzes hydrolysis of the bond that links the two monosaccharide residues of trehalose (Problem 31). Draw the structures of the trehalase reaction products.

33. Trehalose (see Problem 31) accumulates in some plants under dehydrating conditions. The trehalose stabilizes dehydrated enzymes, proteins, and lipid membranes when the plant is desiccated. These plants are often called "resurrection plants" because the plants resume their normal metabolism when water becomes available. How does trehalose stabilize cellular molecules?

34. The disaccharide gentobiose is a component of amygdalin, a poisonous cyanogenic glycoside found in the kernels of apricot seeds. Draw the structure of gentobiose, which consists of two glucose residues linked by a β(1 → 6) glycosidic bond.

35. Propose a structure for the disaccharide maltose that is consistent with the following information: Complete hydrolysis yields only D-glucose; it reduces copper(II) to Cu_2O (see Problem 22), and it is hydrolyzed by α-glucosidase but not β-glucosidase.

36. An unknown disaccharide is subjected to methylation with CH_3I followed by acid hydrolysis (see Problem 23). The reaction yields 2,3,4,6-tetra-*O*-methyl-D-glucose and 2,3,4-tri-*O*-methyl-D-glucose. **(a)** Draw the structure of the unknown disaccharide. **(b)** How does this structure compare to the structure of gentobiose that you drew in Problem 34?

37. The sugar alcohol sorbitol (see Solution 21c) is sometimes used instead of sucrose to sweeten processed foods. Unfortunately, the consumption of large amounts of sorbitol may lead to gastrointestinal distress in susceptible individuals, especially children. Sorbitol-sweetened foods have a lower calorie count than sucrose-sweetened foods. Explain why.

38. A friend tells you that she plans to eat a lot of celery because she heard that the digestion of celery consumes more calories than the food contains, and this will help her to lose weight. How do you respond?

39. The presence of the oligosaccharide raffinose (below) in beans is blamed for causing flatulence when beans are consumed. Undigested

Raffinose

raffinose is acted upon by bacteria in the large intestine, which produce gas as a metabolic by-product. Why are humans unable to digest raffinose?

40. Nigerotriose consists of three glucose residues linked by $\alpha(1 \to 3)$ glycosidic bonds. Draw the structure of nigerotriose.

41. Scientists in Singapore injected soybeans with an enzyme extract from the fungus *R. oligosporus* before allowing the soybeans to germinate. After three days, the concentration of raffinose (see Problem 39) decreased dramatically. (As an added bonus, the concentration of cancer-fighting isoflavones in the soybeans increased.) What enzymes must have been present in the fungus?

42. In a homopolymer, all the monomers are the same, and in a heteropolymer, the monomers are different. Which of the polysaccharides discussed in this chapter are homopolymers and which are heteropolymers?

43. How many reducing ends are in a molecule of potato amylopectin that contains 100,000 residues with a branch every 50 residues?

44. Instead of starch, some plants produce inulin, which is a polymer of $\beta(2 \to 1)$-linked fructose residues. Chicory root is a particularly rich source of inulin.

(a) Draw the structure of an inulin disaccharide.

(b) Why do some food manufacturers add inulin to their products (yogurt, ice cream, and even drinks) to boost their fiber content?

45. When amylose is suspended in water in the presence of I_2, a blue color results due to the ability of iodine to occupy the interior of the helix. A drop of yellow iodine solution on a potato slice turns blue, but when a drop of iodine is placed on an apple slice, the color of the solution remains yellow. Explain why.

46. The repeating disaccharide unit of the polysaccharide heparin is shown below. An affinity chromatography method (see Section 4.6) of purifying DNA-binding proteins was developed by immobilizing heparin onto a solid support. DNA-binding proteins bind to heparin while proteins with no affinity for DNA elute from the column in the flow-through. Why do DNA-binding proteins have an affinity for heparin?

47. Explain why pectin (Box 11.A) is sometimes added to fruit extracts to make jams and jellies.

48. Hemicellulose, another polysaccharide produced in plants, is not a cellulose derivative but is a random heteropolymer of various monosaccharides. D-Xylose residues (see Box 11.A) are present in the largest amount and are linked with $\beta(1 \to 4)$ glycosidic bonds. Draw the structure of a hemicellulose disaccharide containing D-xylose.

49. Explain why a growing plant cell, which produces cellulose, must also produce cellulase.

50. Stonewashed cotton clothing, which appears faded and weathered, can be prepared by tumbling a garment with water and stones. Alternatively, the garment can be briefly soaked in a solution containing cellulase.

(a) Explain what cellulase does.

(b) What would happen if the cellulase treatment was prolonged?

51. The human genome codes for glycoside hydrolases that act on starch, glycogen, sucrose, and lactose. How many different types of glycosidic bonds must these enzymes break in order to generate monosaccharide products?

52. Brown algae are an attractive source of biofuels, since these organisms do not require land, fertilizer, or fresh water and lack lignin. One major polysaccharide in brown algae is laminarin, which consists of glucose residues linked by $\beta(1 \to 3)$ glycosidic bonds. Draw the structure of the disaccharide unit of laminarin.

53. Mycobacteria contain several unusual polymethylated polysaccharides. One consists of repeating units of 3-*O*-methylmannose linked with $\alpha(1 \to 4)$ glycosidic bonds. Draw the structure of the disaccharide unit of this polysaccharide.

11.3 Glycoproteins

54. Identify the *N*- and *O*-linked saccharides shown at the start of Section 11.3. Are the linkages α- or β-glycosidic bonds?

55. A common *O*-glycosidic attachment of an oligosaccharide to a glycoprotein is β-galactosyl-(1$\to$3)-α-*N*-acetylgalactosyl-Ser. Draw the structure of the oligosaccharide and its linkage to the glycoprotein.

56. A transferase enzyme highly expressed in cancer cells catalyzes the synthesis of a glycan that is found in cancer cells but not in normal cells, which makes the enzyme a potential drug target for cancer therapy. Draw the structure of the trisaccharide unit of the glycan: GlcNAcβ(1$\to$6)Manα(1$\to$6)Manβ1 (*Hint*: See Solution 8 for the structure of mannose.)

57. Identify the parent monosaccharides of the chondroitin sulfate disaccharide (Fig. 11.9) and identify the linkages between them.

58. Calculate the net charge of a chondroitin sulfate molecule containing 150 disaccharide units.

59. Collagen isolated from a deep-sea hydrothermal vent worm contains a glycosylated threonine residue in the "Y" position of the (Gly-X-Y)$_n$ repeating triplet. A galactose residue is covalently attached to the threonine via a β-glycosidic bond. Draw the structure of the galactosylated threonine residue.

60. In the past, it was believed that glycosylated proteins were found only in eukaryotic cells, but recently glycoproteins have been found in prokaryotic cells. The flagellin protein from the gram-positive bacterium *L. monocytogenes* is glycosylated with a single *N*-acetylglucosamine at up to six different serine or threonine residues. Draw the structure of this linkage.

61. (a) Identify the monosaccharide precursor of the repeating disaccharide in peptidoglycan.

(b) In one type of peptide cross-link in peptidoglycan, an alanine residue forms an amide bond with the repeating disaccharide. Identify the location of this linkage.

62. Lysozyme, an enzyme present in tears and mucus secretions, is a $\beta(1 \to 4)$ glycosidase. Explain how lysozyme helps prevent bacterial infections.

63. Transmembrane proteins destined for the plasma membrane are posttranslationally processed: Oligosaccharide chains are added in the ER to selected asparagine residues, then the oligosaccharide chains are processed in the Golgi apparatus (Fig. 11.7). One of the sugars added to the core oligosaccharide is *N*-acetylneuraminic acid, also called sialic acid (Problem 16). Tumor cells frequently overexpress sialic acid on their surface, which might contribute to a tumor cell's ability to detach from the tumor and travel through the bloodstream to form additional tumors. The sialic acid residues on tumor cells are typically not recognized by the immune system, which compounds the problem.

(a) How does the presence of sialic acid on the cell surface facilitate detachment?

(b) What strategies would you use to design therapeutic agents to kill these types of tumor cells?

64. Keratan sulfate is composed of repeating disaccharides of *N*-acetylglucosamine and galactose linked via a β(1→3) glycosidic bond. An enzyme catalyzes the addition of sulfate groups to both monosaccharides at the C6 positions in infants during the early postnatal period. Draw the structure of the repeating unit of keratan sulfate.

SELECTED READINGS

Frank, M. and Schloissnig, S., Bioinformatics and molecular modeling in glycobiology, *Cell. Mol. Life Sci.* **67**, 2749–2772 (2010). [Describes the databases and software used to study carbohydrates.]

Hart, G. W. and Copeland, R. J., Glycomics hits the big time, *Cell* **143**, 672–676 (2010). [Describes the biological roles of carbohydrates and some approaches used to study them.]

Kolter, R. and Greenberg, P., The superficial life of microbes, *Nature* **441**, 300–302 (2006). [A brief review of bacterial biofilms.]

Schwarz, F. and Aebi, M., Mechanism and principles of N-linked protein glycosylation, *Curr. Opin. Struct. Biol.* **21**, 576–582 (2011). [Reviews the synthesis and biological importance of oligosaccharides.]

Spiro, R. G., Protein glycosylation: nature, distribution, enzymatic formation, and disease implications of glycopeptide bonds, *Glycobiology* **12**, 43R–56R (2002). [Describes the different ways carbohydrates are linked to proteins.]

Vollmer, W., Blanot D., and de Pedro, M. A., Peptidoglycan structure and architecture, *FEMS Microbiol. Rev.* **32**, 149–167 (2008). [A review of data and models related to peptidoglycan structure.]

C H A P T E R **12**

METABOLISM AND BIOENERGETICS

Courtesy of Nicole Dubilier

Many microorganisms obtain raw materials and free energy in unfamiliar ways, but it is rare to find a larger organism as unusual as Olavius algarvensis. *This marine worm does not eat. It has no mouth or digestive system and lives entirely off the nutrients provided by the bacteria that live just underneath its skin.*

DO YOU REMEMBER?

- Living organisms obey the laws of thermodynamics (Section 1.3).
- Amino acids are linked by peptide bonds to form a polypeptide (Section 4.1).
- Allosteric regulators can inhibit or activate enzymes (Section 7.3).
- Lipids are predominantly hydrophobic molecules that can be esterified but cannot form polymers (Section 8.1).
- Monosaccharides can be linked by glycosidic bonds in various arrangements (Section 11.2).

The diversity of life-forms on earth makes it impossible to examine all the molecular components and chemical reactions that occur in each organism. But it turns out that organisms share some basic cellular structures, make the same types of molecules, and use similar enzymes to build and break down those molecules. In the coming chapters, we will focus on these common pathways that transform raw material and free energy in cells. We begin by introducing the notions of metabolic fuels and metabolic pathways and then turn to explore the meaning of free energy in biological systems.

12.1 | FOOD AND FUEL

The biochemical activity of cells can be summarized as shown in **Figure 12.1**. Cells break down or **catabolize** large molecules to release free energy and small molecules. The cells then use the free energy and small molecules to rebuild larger molecules, a process called **anabolism.** The set of all catabolic and anabolic activities constitutes an organism's **metabolism.**

Some microorganisms, known as **chemoautotrophs** (from the Greek *trophe*, "nourishment"), obtain virtually all their metabolic building materials and free energy from the simple inorganic compounds CO_2, N_2, H_2, and S_2. **Photoautotrophs,** such as the familiar green plants, need little more than CO_2, H_2O, a source of nitrogen, and sunlight. In contrast, **heterotrophs,** a group that includes animals, directly or indirectly obtain all their building materials and free energy from organic compounds produced by chemo- or photoautotrophs.

KEY CONCEPTS

- The macromolecules in food are hydrolyzed, and the monomeric products are absorbed by the intestine.
- Cells store fatty acids, glucose, and amino acids in the form of polymers.
- Metabolic fuels can be mobilized by breaking down glycogen, triacylglycerols, and proteins.

301

FIGURE 12.1

Catabolism and anabolism. Catabolic (degradative) reactions yield free energy and small molecules that can be used for anabolic (synthetic) reactions. Metabolism is the sum of all catabolic and anabolic processes.

As heterotrophs, mammals rely on food produced by other organisms. After food is digested and absorbed, it becomes a source of metabolic energy and materials to support the animal's growth and other activities. The human diet includes the four types of biological molecules introduced in Section 1.2 and described in more detail in subsequent chapters. These molecules are often present as macromolecular polymers, namely proteins, nucleic acids, polysaccharides, and triacylglycerols (technically, fats are not polymers since the monomeric units are not linked to each other but to glycerol). *Digestion reduces the polymers to their monomeric components: amino acids, nucleotides, monosaccharides, and fatty acids.* The breakdown of nucleotides does not yield significant amounts of metabolic free energy, so we will devote more attention to the catabolism of other types of biomolecules.

CELLS TAKE UP THE PRODUCTS OF DIGESTION

Digestion takes place extracellularly in the mouth, stomach, and small intestine and is catalyzed by hydrolytic enzymes (**Fig. 12.2**). For example, salivary amylase begins to break down starch, which consists of linear polymers of glucose residues (amylose) and branched polymers (amylopectin; Section 11.2). Gastric and pancreatic proteases (including trypsin, chymotrypsin, and elastase) degrade proteins to small peptides and amino acids. Lipases synthesized by the pancreas and secreted into the small intestine catalyze the release of fatty acids from triacylglycerols. Water-insoluble lipids do not freely mix with the other digested molecules but instead form micelles (Fig. 2.9).

The products of digestion are absorbed by the cells lining the intestine. Monosaccharides enter the cells via active transporters such as the Na⁺-glucose system diagrammed in Figure 9.18. Similar symport systems bring amino acids and di- and tripeptides into the cells. Some highly hydrophobic lipids diffuse through the cell membrane; others require transporters. Inside the

FIGURE 12.2

Digestion of biopolymers. These hydrolytic reactions are just a few of those that occur during food digestion. In each example, the bond to be cleaved is colored red. (a) The chains of glucose residues in starch are hydrolyzed by amylases. (b) Proteases catalyze the hydrolysis of peptide bonds in proteins. (c) Lipases hydrolyze the ester bonds linking fatty acids to the glycerol backbone of triacylglycerols.

? **Explain why the reactions shown here are thermodynamically favorable.**

cell, the triacylglycerol digestion products re-form triacylglycerols, and some fatty acids are linked to cholesterol to form cholesteryl esters, for example,

Cholesteryl stearate

Triacylglycerols and cholesteryl esters are packaged, together with certain proteins, to form **lipoproteins.** These particles, known specifically as chylomicrons, are released into the lymphatic circulation before entering the bloodstream for delivery to tissues.

Water-soluble substances such as amino acids and monosaccharides leave the intestinal cells and enter the portal vein, which drains the intestine and other visceral organs and leads directly to the liver. *The liver therefore receives the bulk of a meal's nutrients and catabolizes them, stores them, or releases them back into the bloodstream.* The liver also takes up chylomicrons and repackages the lipids with different proteins to form other lipoproteins, which circulate throughout the body, carrying cholesterol, triacylglycerols, and other lipids (lipoproteins are discussed in greater detail in Chapter 17). The allocation of resources following a meal varies with the individual's needs at that time and with the type of nutrients consumed. Fortunately, the body does this efficiently, regardless of what food was eaten (**Box 12.A**).

A CLOSER LOOK

Box 12.A Dietary Guidelines

Nutritionists have yet to come up with the ideal diet; the best they can do is identify the body's overall needs and roughly outline dietary requirements. For example, scientists have compiled lists of recommended daily intakes for various substances in terms of grams of the substance or the proportion of total energy intake contributed by that substance:

Distribution of Macronutrients for Adults

Carbohydrate	45–65%
Fat	20–35%
Protein	10–35%

However, few foods are composed of pure substances, so more practical guidelines focus on types of foods, with units that are more familiar to most consumers, such as ounces or cups. One source of information is the U.S. Department of Agriculture, which has published the following guidelines:

Food Group Choices

	Moderately active female, age 21–25	Moderately active male, age 21–25
Total calories	2200	2800
Fruits	2 cups	2.5 cups
Grains	7 oz	10 oz
Milk products	3 cups	3 cups
Meat, beans, nuts	6 oz	7 oz
Oils	6 tsp	8 tsp
Vegetables	3 cups	3.5 cups

[From www.cnpp.usda.gov/]

Even these recommendations are somewhat clumsy, since most individuals do not determine the volume or mass of what they

place on their plates. Nutrition educators strive to translate some formal quantities into yet more familiar units: A cup of rice or a medium apple is about the size of a baseball, and three ounces of meat is about the size of a deck of cards.

An additional drawback of dietary guidelines formulated like those above is that in the United States, recommendations are based on a traditional Western diet that includes meat and dairy products. Vegetarians (who do not consume meat), vegans (who avoid consuming any animal products), and those who do not drink milk must be more diligent in assessing whether the foods they consume meet the basic requirements for carbohydrates, proteins, and so on.

Finally, a serious challenge for anyone interested in tracking their nutrient consumption is that many foods are processed; that is, raw ingredients are combined, sometimes in unknown proportions, to generate a product that can be sold as a convenience item (think: instant soup). Such foods are typically accompanied by a nutrition facts label that lists, among other things, the serving size; calories per serving; and the quantities of carbohydrates, fats, and proteins (in grams) and their percentage of the recommended daily value.

The availability of different types of dietary guidelines, along with a plethora of advice (which may or may not be grounded in the scientific method), suggests that there is significant leeway regarding what humans can or should consume. Indeed, consideration of how eating patterns have varied across centuries and across continents indicates that the human body must be remarkably versatile in converting a variety of raw materials into the molecular building blocks and metabolic energy required to sustain life.

? **How would the recommended intake of protein vary from infancy to old age? Should intakes be adjusted according to body mass?**

FIGURE 12.3

Adipocytes. These cells, which make up adipose tissue, contain a small amount of cytoplasm surrounding a large globule of trlacylglycerols (fat). [© CNRI/ Phototake.]

MONOMERS ARE STORED AS POLYMERS

Immediately following a meal, the circulating concentrations of monomeric compounds are relatively high. All cells can take up these materials to some extent to fulfill their immediate needs, but *some tissues are specialized for the long-term storage of nutrients.* For example, fatty acids are used to build triacylglycerols, many of which travel in the form of lipoproteins to **adipose tissue.** Here, adipocytes take up the triacylglycerols and store them as intracellular fat globules. Because the mass of lipid is hydrophobic and does not interfere with activities in the aqueous cytoplasm, the fat globule can be enormous, occupying most of the volume of the adipocyte (**Fig. 12.3**).

Virtually all cells can take up monosaccharides and immediately catabolize them to produce free energy. Some tissues, primarily liver and muscle (which makes up a significant portion of the human body), use monosaccharides to synthesize glycogen, the storage polymer of glucose. Glycogen is a highly branched polymer with a compact shape. Several glycogen molecules may clump together to form granules that are visible by electron microscopy (**Fig. 12.4**). Glycogen's branched structure means that a single molecule can be expanded quickly, by adding glucose residues to its many branches, and degraded quickly, by simultaneously removing glucose from the ends of many branches. Glucose that does not become part of glycogen can be catabolized to two-carbon acetyl units and converted into fatty acids for storage as triacylglycerols.

Amino acids can be used to build polypeptides. A protein is not a dedicated storage molecule for amino acids, as glycogen is for glucose and triacylglycerols are for fatty acids, so excess amino acids cannot be saved for later. However, in certain cases, such as during starvation, proteins are catabolized to supply the body's energy needs. If the intake of amino acids exceeds the body's immediate protein-building needs, the excess amino acids can be broken down and converted to carbohydrate (which can be stored as glycogen) or converted to acetyl units (which can then be converted to fat).

Amino acids and glucose are both required to synthesize nucleotides. Aspartate, glutamine, and glycine supply some of the carbon and nitrogen atoms used to build the purine and pyrimidine bases (Section 18.3). The ribose-5-phosphate component of nucleotides is derived from glucose by a pathway that converts the six-carbon sugar to a five-carbon sugar (Section 13.4). In sum, the allocation of resources within a cell depends on the type of tissue and its need to build cellular structures, provide free energy, or stockpile resources in anticipation of future needs.

FUELS ARE MOBILIZED AS NEEDED

Amino acids, monosaccharides, and fatty acids are known as **metabolic fuels** *because they can be broken down by processes that make free energy available for the cell's activities.* After a meal, glucose and amino acids are catabolized to release their free energy. When these fuel supplies are exhausted, the body **mobilizes** its stored resources; that is, it converts its polysaccharide and triacylglycerol storage molecules (and sometimes proteins) to their respective monomeric units. Most of the body's tissues prefer to use glucose as their primary metabolic fuel, and the central nervous system can run on almost nothing else. In response to this demand, the liver mobilizes glucose by breaking down glycogen.

In general, depolymerization reactions are hydrolytic, but in the case of glycogen, the molecule that breaks the bonds between glucose residues is not water but phosphate. Thus, the degradation of glycogen is called **phosphorolysis** (**Fig. 12.5**). This reaction is catalyzed by glycogen phosphorylase, which releases residues from the ends of branches in the glycogen polymer.

(a)

(b)

FIGURE 12.4

Glycogen structure. (a) Schematic diagram of a glycogen molecule. Each hexagon represents a glucose monomer, and branches occur every 8 to 14 residues. (b) Electron micrograph of a liver cell showing glycogen granules (colored pink). Mitochondria are green, and a fat globule is yellow. [© CNRI/Science Photo Library/Photo Researchers.]

Glycogen

Glucose-1-phosphate

FIGURE 12.5

Glycogen phosphorolysis.
This mechanism of glycogen degradation generates glucose-1-phosphate.

The phosphate group of glucose-1-phosphate is removed before glucose is released from the liver into the circulation. Other tissues absorb glucose from the blood. In the disease **diabetes mellitus,** this does not occur, and the concentration of circulating glucose may become elevated.

Only when the supply of glucose runs low does adipose tissue mobilize its fat stores. A lipase hydrolyzes triacylglycerols so that fatty acids can be released into the bloodstream. These free fatty acids are not water-soluble and therefore bind to circulating proteins. Except for the heart, which uses fatty acids as its primary fuel, the body does not have a budget for burning fatty acids. In general, as long as dietary carbohydrates and amino acids can meet the body's energy needs, stored fat will not be mobilized, even if the diet includes almost no fat. This feature of mammalian fuel metabolism is a source of misery for many dieters!

Amino acids are not mobilized to generate energy except during a fast, when glycogen stores are depleted (in this situation, the liver can also convert some amino acids into glucose). However, cellular proteins are continuously degraded and rebuilt with the changing demand for particular enzymes, transporters, cytoskeletal elements, and so on. There are two major mechanisms for degrading unneeded proteins. In the first, the **lysosome,** an organelle containing proteases and other hydrolytic enzymes, breaks down proteins that are enclosed in a membranous vesicle. Membrane proteins and extracellular proteins taken up by endocytosis are degraded by this pathway, but intracellular proteins that become enclosed in vesicles can also be broken down by lysosomal enzymes.

A second pathway for degrading intracellular proteins requires a barrel-shaped structure known as a **proteasome.** The 700-kD core of this multiprotein complex encloses an inner chamber with multiple active sites that carry out peptide bond hydrolysis (**Fig. 12.6**). A protein can enter the proteasome only after it has been covalently tagged with a small protein called ubiquitin. This 76-residue protein is ubiquitous (hence its name) and highly conserved in eukaryotes (**Fig. 12.7**). Ubiquitin is attached to a protein by the action of a set of enzymes that link the C-terminus of ubiquitin to a lysine side chain. Additional ubiquitin molecules are then added to the first, each one linked via its C-terminus to a lysine side chain of the preceding ubiquitin. A chain of at least four ubiquitins is required to mark a protein for destruction by a proteasome.

The structural features that allow a protein to be ubiquitinated are not completely understood, but the system is sophisticated enough to allow unneeded or defective proteins to be destroyed while sparing essential proteins. A cap at the end of the proteasome barrel (not shown in Fig. 12.6) regulates the entry of ubiquitinated proteins into the inner chamber. The free energy of the ATP hydrolysis reaction drives conformational changes that apparently help the condemned protein to unfold so that it

FIGURE 12.6

Structure of the yeast proteasome core. This cutaway view shows the inner chamber, where proteolysis occurs. Additional protein complexes (not shown) assist the entry of proteins into the proteasome. The red structures mark the locations of three protease active sites. [*Courtesy Robert Huber, Max-Planck-Institut fur Biochemie, Germany.*]

FIGURE 12.7

Ubiquitin. Several copies of this 76-residue protein are linked to Lys residues in proteins that are to be degraded by a proteasome. Atoms are color-coded: C green, O red, N blue, and H white. [*Structure (pdb 1UBQ) determined by S. Vijay-Kumar, C. E. Bugg, and W. J. Cook.*]

 Draw the linkage between a protein's C-terminus and a ubiquitin Lys residue.

can be more easily hydrolyzed. The ubiquitin molecules are not degraded; instead they are detached and reused. The three protease active sites inside the proteasome cleave the unfolded polypeptide substrate, releasing peptides of about eight residues that can diffuse out of the proteasome (**Fig. 12.8**). These peptides are further broken down by cytosolic peptidases so that the amino acids can be catabolized or recycled.

CONCEPT CHECK

1. Explain the relationship between autotrophs and heterotrophs.
2. Review the steps by which nutrients from food molecules reach the body's tissues.
3. Name the major metabolic fuels and discuss how they are stored.
4. How are metabolic fuels mobilized?
5. Summarize the mechanisms for intracellular protein degradation.

12.2 METABOLIC PATHWAYS

KEY CONCEPTS
- A few metabolites appear in several metabolic pathways.
- Coenzymes such as NAD$^+$ and ubiquinone collect electrons from compounds that become oxidized.
- Metabolic pathways in cells are connected and are regulated.
- Many vitamins, substances that humans cannot synthesize, are components of coenzymes.

The interconversion of a biopolymer and its monomeric units is usually accomplished in just one or a few enzyme-catalyzed steps. In contrast, many steps are required to break down the monomeric compounds or build them up from smaller precursors. These series of reactions are known as **metabolic pathways.** A metabolic pathway can be considered from many viewpoints: as a series of **intermediates** or **metabolites,** as a set of enzymes that catalyze the reactions by which metabolites are interconverted, as an energy-producing or energy-requiring phenomenon, or as a dynamic process whose activity can be turned up or down. As we explore metabolic pathways in the coming chapters, we will use each of these perspectives.

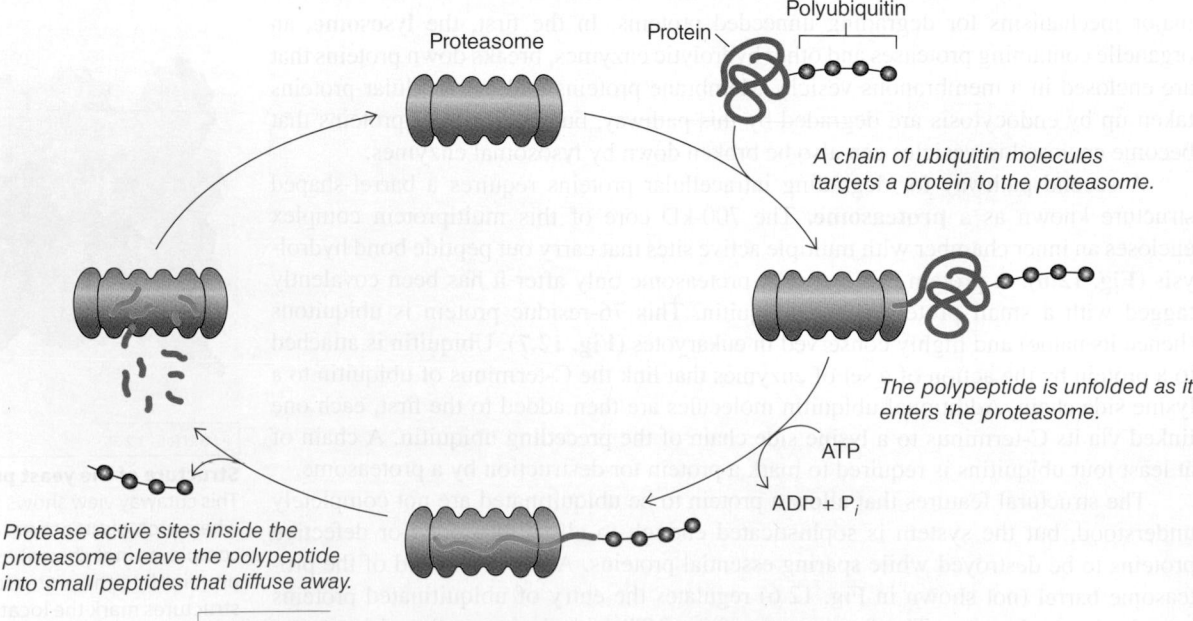

A chain of ubiquitin molecules targets a protein to the proteasome.

The polypeptide is unfolded as it enters the proteasome.

Protease active sites inside the proteasome cleave the polypeptide into small peptides that diffuse away.

FIGURE 12.8

Protein degradation by the proteasome.

Glucose $\Rightarrow$ (Glyceraldehyde-3-phosphate structure) $\Rightarrow$ (Pyruvate structure) $\Rightarrow$ (Acetyl-CoA structure)

Glyceraldehyde-3-phosphate   Pyruvate   Acetyl-CoA

FIGURE 12.9

Some intermediates resulting from glucose catabolism.

? Compare the oxidation states of the carbons in glyceraldehyde-3-phosphate and in pyruvate.

SOME MAJOR METABOLIC PATHWAYS SHARE A FEW COMMON INTERMEDIATES

One of the challenges of studying metabolism is dealing with the large number of reactions that occur in a cell—involving thousands of different intermediates. However, *a handful of metabolites appear as precursors or products in the pathways that lead to or from virtually all other types of biomolecules.* These intermediates are worth examining at this point, since they will reappear several times in the coming chapters.

In **glycolysis,** the pathway that degrades the monosaccharide glucose, the six-carbon sugar is phosphorylated and split in half, yielding two molecules of glyceraldehyde-3-phosphate (**Fig. 12.9**). This compound is then converted in several more steps to another three-carbon molecule, pyruvate. The decarboxylation of pyruvate (removal of a carbon atom as CO_2) yields acetyl-CoA, in which a two-carbon acetyl group is linked to the carrier molecule coenzyme A (CoA).

Glyceraldehyde-3-phosphate, pyruvate, and acetyl-CoA are key players in other metabolic pathways. For example, glyceraldehyde-3-phosphate is the metabolic precursor of the three-carbon glycerol backbone of triacylglycerols. In plants, it is also the entry point for the carbon "fixed" by photosynthesis; in this case, two molecules of glyceraldehyde-3-phosphate combine to form a six-carbon monosaccharide. Pyruvate can undergo a reversible amino-group transfer reaction to yield alanine. This makes pyruvate both a precursor of an amino acid and the degradation product of one. Pyruvate can also be carboxylated to yield oxaloacetate, a four-carbon precursor of several other amino acids:

(Pyruvate structure) + CO_2 $\longrightarrow$ (Oxaloacetate structure)

Pyruvate   Oxaloacetate

(Pyruvate ⇌ Alanine structures)

Pyruvate   Alanine

Fatty acids are built by the sequential addition of two-carbon units derived from acetyl-CoA; fatty acid breakdown yields acetyl-CoA. These relationships are summarized in **Figure 12.10**. If

FIGURE 12.10

Some of the metabolic roles of the common intermediates.

? Without looking at the text, draw the structures of glyceraldehyde-3-phosphate, pyruvate, and oxaloacetate.

not used to synthesize other compounds, two-carbon intermediates can be broken down to CO_2 by the **citric acid cycle,** a metabolic pathway essential for the catabolism of all metabolic fuels.

MANY METABOLIC PATHWAYS INCLUDE OXIDATION–REDUCTION REACTIONS

In general, *the catabolism of amino acids, monosaccharides, and fatty acids is a process of oxidizing carbon atoms, and the synthesis of these compounds involves carbon reduction.* Recall from Section 1.3 that **oxidation** is defined as the loss of electrons and **reduction** is the gain of electrons. Oxidation–reduction, or **redox,** reactions are like acid–base reactions: They involve pairs of reactants. As one compound becomes more oxidized (gives up electrons or loosens its hold on them), another compound becomes reduced (receives the electrons or tightens its grip on them).

In biological systems, unaccompanied electrons are rare. Keep in mind that electrons typically remain associated with molecules, usually as a pair of electrons that constitute a covalent bond. As in any chemical reaction—which is a matter of old bonds breaking and new bonds forming—we can see where electrons travel during a redox reaction by comparing the bonding arrangements in the reactant and product molecules.

For the metabolic reactions that we are concerned with, the oxidation of carbon atoms frequently appears as the replacement of C—H bonds (in which the C and H atoms share the bonding electrons equally) with C—O bonds (in which the more electronegative O atom "pulls" the electrons away from the carbon atom). Carbon has given up some of its electrons, even though the electrons are still participating in a covalent bond.

The transformation of methane to carbon dioxide represents the conversion of carbon from its most reduced state to its most oxidized state:

$$CH_4 \longrightarrow O{=}C{=}O$$

Similarly, oxidation occurs during the catabolism of a fatty acid, when saturated methylene (—CH_2—) groups are converted to CO_2 and when the carbons of a carbohydrate (represented as CH_2O) are converted to CO_2:

$$H{-}CH_2{-}H \longrightarrow O{=}C{=}O$$

$$H{-}CH{-}OH \longrightarrow O{=}C{=}O$$

The reverse of either of these processes—converting the carbons of CO_2 to the carbons of fatty acids or carbohydrates—is a reduction process (this is what occurs during photosynthesis, for example). In reduction processes, the carbon atoms regain electrons as C—O bonds are replaced by C—H bonds.

Turning CO_2 into carbohydrate (CH_2O) requires the input of free energy (think: sunlight). Therefore, *the reduced carbons of the carbohydrate represent a form of stored free energy.* This energy is recovered when cells break the carbohydrate back down to CO_2. Of course, such a metabolic conversion does not happen all at once but takes place through many enzyme-catalyzed steps.

In following metabolic pathways that include oxidation–reduction reactions, we can examine the redox state of the carbon atoms, and we can also trace the path of the electrons that are transferred. In some cases, this is straightforward, as when an oxidized metal ion such as iron gains an electron (represented as e^-) to become reduced.

$$Fe^{3+} + e^- \rightarrow Fe^{2+}$$

But in many cases, electron transfer means that one substance ends up with one less covalent bond and another substance ends up with one more bond (since each bond represents a pair of electrons). Sometimes, the electrons travel along with protons as H atoms, or a pair of electrons travels with a proton as a hydride ion (H^-).

When a metabolic fuel molecule is oxidized, its electrons may be transferred to a compound such as nicotinamide adenine dinucleotide (NAD^+) or nicotinamide adenine dinucleotide phosphate ($NADP^+$). The structure of these nucleotides is shown in Figure 3.2b. NAD^+ and $NADP^+$ are called **cofactors** or **coenzymes,** organic compounds that allow an enzyme to carry out a particular chemical reaction (Section 6.2). The redox-active portion of NAD^+ and $NADP^+$ is the nicotinamide group, which accepts a hydride ion to form NADH or NADPH.

NAD(P)$^+$ *(oxidized)* $\quad + \quad H^- \quad \rightleftharpoons \quad$ NAD(P)H *(reduced)*

This reaction is reversible, so the reduced cofactors can become oxidized by giving up a hydride ion. In general, NAD^+ participates in catabolic reactions and $NADP^+$ in anabolic reactions. Because these electron carriers are soluble in aqueous solution, they can travel throughout the cell, shuttling electrons from reduced compounds to oxidized compounds.

Many cellular oxidation–reduction reactions take place at membrane surfaces, for example, in the inner membranes of mitochondria and chloroplasts in eukaryotes and in the plasma membrane of prokaryotes. In these cases, a membrane-associated enzyme may transfer electrons from a substrate to a lipid-soluble electron carrier such as ubiquinone (coenzyme Q, abbreviated Q; see Section 8.1). Ubiquinone's hydrophobic tail, containing 10 five-carbon isoprenoid units in mammals, allows it to diffuse within the membrane. *Ubiquinone can take up one or two electrons* (in contrast to NAD^+, which is strictly a two-electron carrier). A one-electron reduction of ubiquinone (addition of an H atom) produces a semiquinone, a stable free radical (shown as QH·). A two-electron reduction (two H atoms) yields ubiquinol (QH_2):

Ubiquinone (Q) $\quad \xrightarrow{[H\cdot]} \quad$ Ubisemiquinone (QH·) $\quad \xrightarrow{[H\cdot]} \quad$ Ubiquinol (QH_2)

The reduced ubiquinol can then diffuse through the membrane to donate its electrons in another oxidation–reduction reaction.

Catabolic pathways, such as the citric acid cycle, generate considerable amounts of reduced cofactors. Some of them are reoxidized in anabolic reactions. The rest are reoxidized by a process that is accompanied by the synthesis of ATP from ADP and P_i. In mammals, the reoxidation of NADH and QH_2 and the concomitant production of ATP require the reduction of O_2 to H_2O. This pathway is known as **oxidative phosphorylation.**

reduced fuel molecule $\quad\longrightarrow\quad$ oxidized cofactor $\quad\longrightarrow\quad H_2O$

oxidized fuel molecule $\quad\longleftarrow\quad$ reduced cofactor $\quad\longleftarrow\quad O_2$

In effect, NAD^+ and ubiquinone collect electrons (and hence free energy) from reduced fuel molecules. When the electrons are ultimately transferred to O_2, this free energy is harvested in the form of ATP.

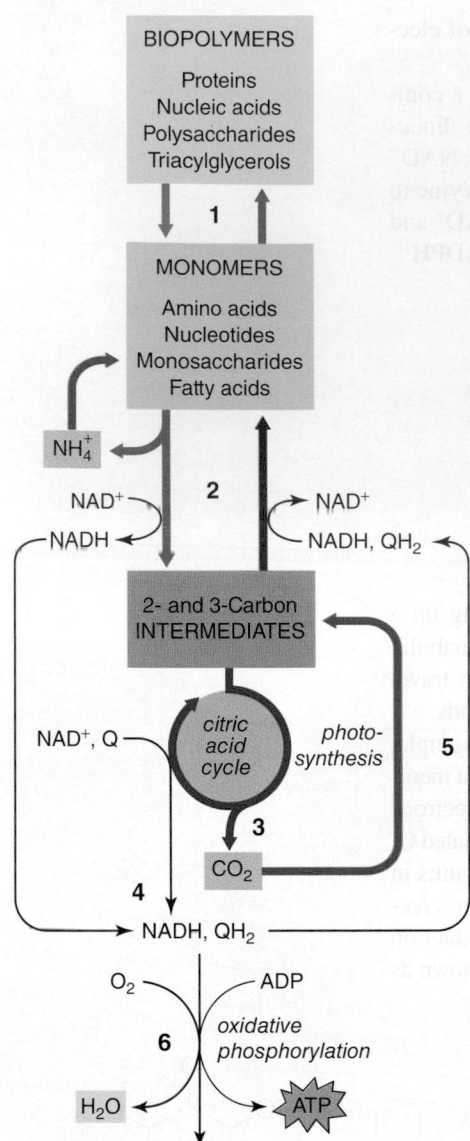

FIGURE 12.11

Outline of metabolism. In this composite diagram, downward arrows represent catabolic processes, and upward arrows represent anabolic processes. Red arrows indicate some major oxidation–reduction reactions. The major metabolic processes are highlighted: (1) Biological polymers (proteins, nucleic acids, polysaccharides, and triacylglycerols) are built from and are degraded to monomers (amino acids, nucleotides, monosaccharides, and fatty acids). (2) The monomers are broken down into two- and three-carbon intermediates such as glyceraldehyde-3-phosphate, pyruvate, and acetyl-CoA, which are also the precursors of many other biological compounds. (3) The complete degradation of biological molecules yields inorganic compounds such as NH_4^+, CO_2, and H_2O. These substances are returned to the pool of intermediates by processes such as photosynthesis. (4) Electron carriers (NAD^+ and ubiquinone) accept the electrons released by metabolic fuels (amino acids, monosaccharides, and fatty acids) as they are degraded and then completely oxidized by the citric acid cycle. (5) The reduced cofactors (NADH and QH_2) are required for many biosynthetic reactions. (6) The reoxidation of reduced cofactors drives the production of ATP from ADP + P_i (oxidative phosphorylation).

> **?** Is the citric acid cycle a process of carbon oxidation or reduction? Is photosynthesis a process of carbon oxidation or reduction?

METABOLIC PATHWAYS ARE COMPLEX

So far we have sketched the outlines of mammalian fuel metabolism, in which macromolecules are stored and mobilized so that their monomeric units can be broken down into smaller intermediates. These intermediates can be further degraded (oxidized) and their electrons collected by cofactors. We have also briefly mentioned anabolic (synthetic) reactions in which the common two- and three-carbon intermediates give rise to larger compounds. At this point, we can present this information in schematic form in order to highlight some important features of metabolism (**Fig. 12.11**).

1. *Metabolic pathways are all connected.* In a cell, a metabolic pathway does not operate in isolation; its substrates are the products of other pathways, and vice versa. For example, the NADH and QH_2 generated by the citric acid cycle are the starting materials for oxidative phosphorylation.

2. *Pathway activity is regulated.* Cells do not synthesize polymers when monomers are in short supply. Conversely, they do not catabolize fuels when the need for ATP is low. The **flux,** or flow of intermediates, through a metabolic pathway is regulated in various ways according to substrate availability and the cell's need for the pathway's products. The activity of one or more enzymes in a pathway may be controlled by allosteric effectors (Sections 5.1 and 7.3). These changes in turn may reflect extracellular signals that activate intracellular kinases, phosphatases, and second messengers (Section 10.1). Regulation of pathways is especially critical when the simultaneous operation of two opposing processes, such as fatty acid synthesis and degradation, would be wasteful.

3. *Not every cell carries out every pathway.* Figure 12.11 is a composite of a number of metabolic processes, and a given cell or organism may undertake only a subset of these. Mammals do not perform photosynthesis, and only the liver and kidney can synthesize glucose from noncarbohydrate precursors.

4. *Each cell has a unique metabolic repertoire.* In addition to the pathways outlined in Figure 12.11, which are centered on fuel metabolism, cells carry out a plethora of biosynthetic reactions that are not explicitly shown. Such pathways contribute to the unique metabolic capabilities of different cells and organisms (**Box 12.B**).

5. *Organisms may be metabolically interdependent.* Photosynthetic plants and the heterotrophs that consume them are an obvious example of metabolic complementarity, but there are numerous other examples, especially in the microbial world. Certain organisms that release methane as a waste product live in close proximity to methanotrophic species (which consume CH_4 as a fuel); neither organism can survive without the other.

Humans also exhibit interspecific cooperativity: Thousands of different microbial species, amounting to some 100 trillion cells, can live in or on the human body. Collectively, these organisms express millions of different genes and carry out a correspondingly rich set of metabolic activities.

A CLOSER LOOK

Box 12.B The Transcriptome, the Proteome, and the Metabolome

Modern biologists have developed research tools that use the power of computers to collect enormous data sets and analyze them. Such endeavors provide great insights but also have limitations. As we saw in Section 3.4, genomics, the study of an organism's complete set of genes, yields a glimpse of that organism's overall metabolic repertoire. But what the organism, or a single cell, is actually doing at a particular moment depends in part on which genes are active.

A cell's population of mRNA molecules represents genes that are turned on, or transcribed. The study of these mRNAs is known as **transcriptomics.** Identifying and quantifying all the mRNA transcripts (the **transcriptome**) from a single cell type can be done by assembling short strands of DNA with known sequences on a solid support, then allowing them to hybridize, or form double-stranded structures, with fluorescent-labeled mRNAs from a cell preparation. The strength of fluorescence indicates how much mRNA binds to a particular complementary DNA sequence. The collection of DNA sequences is called a **microarray** or **DNA chip** because thousands of sequences fit in a few square centimeters. The microarray may represent an entire genome or just a few selected genes. Each bright spot in the DNA chip shown here represents a DNA sequence to which a fluorescent mRNA molecule has bound.

[*Voker Steger/Science Photo Library/Photo Researchers.*]

Biologists use DNA chips to identify genes whose expression changes under certain conditions or at different developmental stages.

Unfortunately, the correlation between the amount of a particular mRNA and the amount of its protein product is not perfect; some mRNAs are rapidly degraded, whereas others are translated many times, yielding large quantities of the corresponding protein. Hence, a more reliable way to assess gene expression is through **proteomics**—by examining a cell's **proteome,** the complete set of proteins that are synthesized by the cell at a particular point in its life cycle. However, this approach is limited by the technical problems of detecting minute quantities of thousands of different proteins. Nucleic acids can be amplified by the polymerase chain reaction (see Section 3.5), but there is no comparable procedure for amplifying proteins.

Where genomics, transcriptomics, and proteomics fall short, **metabolomics** steps in, attempting to pin down the actual metabolic activity in a cell or tissue by identifying and quantifying all its metabolites, that is, its **metabolome.** This is no trivial task, as a cell may contain tens of thousands of different types of compounds, whose concentrations may range over many orders of magnitude. These substances include nonfood molecules such as toxins, preservatives, drugs, and their degradation products. Metabolites are typically detected through column chromatography, nuclear magnetic resonance (NMR) spectroscopy, or mass spectrometry (Section 4.6). In the example shown, approximately 20 metabolites are visible in an ^{1}H NMR spectrum of a 10-μL sample of rat brain.

Metabolite profile of rat brain. [*Courtesy Raghavendra Rao, University of Minnesota, Minneapolis.*]

As has been done for genomics and proteomics and other areas of bioinformatics, metabolomic data are deposited in publicly accessible databases for retrieval and analysis. One hope for metabolomics is that disease diagnosis could be streamlined by obtaining a complete metabolic profile of a patient's urine or blood. The unique "fingerprint" of a disease such as cancer could then be monitored to observe the patient's response to treatment. Industrial applications include monitoring biological processes such as winemaking and bioremediation (using microorganisms to detoxify contaminated environments).

The field of metabolomics also encompasses databases devoted to the relationships among metabolites, that is, the metabolic pathways in which the metabolites participate as intermediates. With hundreds of pathways, representing thousands of different organisms, such databases can be searched for specific compounds (including drugs), and links to enzyme and sequence databases can begin to paint a holistic view of the biochemical processes that occur in cells.

? **Compare the metabolomic complexity of a single-celled prokaryote and that of a multicellular eukaryote.**

TABLE 12.1 Some Essential Substances for Humans

AMINO ACIDS	FATTY ACIDS		OTHER	
Histidine	Linoleate	$CH_3(CH_2)_4(CH{=}CHCH_2)_2(CH_2)_6COO^-$	Choline	$(CH_3)_3N^+CH_2CH_2OH$
Isoleucine	Linolenate	$CH_3CH_2(CH{=}CHCH_2)_3(CH_2)_6COO^-$		
Leucine				
Lysine				
Methionine				
Phenylalanine				
Threonine				
Tryptophan				
Valine				

An overview such as Figure 12.11 does not convey the true complexity of cellular metabolism, which takes place in a milieu crowded with multiple substrates, competing enzymes, and layers of regulatory mechanisms. Moreover, Figure 12.11 does not include any of the reactions involved in transmitting and decoding genetic information (these topics are covered in the final section of this book). However, a diagram such as Figure 12.11 is a useful tool for mapping the relationships among metabolic processes, and we will refer back to it in the coming chapters. Online databases provide additional information about metabolic pathways, enzymes, intermediates, and metabolic diseases.

HUMAN METABOLISM DEPENDS ON VITAMINS

Humans lack many of the biosynthetic pathways that occur in plants and microorganisms and so rely on other species to provide certain raw materials. Some amino acids and unsaturated fatty acids are considered **essential** because the human body cannot synthesize them and must obtain them from food (**Table 12.1**). **Vitamins** likewise are compounds that humans need but cannot make. Presumably, the pathways for synthesizing these substances, which require many specialized enzymes, are not necessary for heterotrophic organisms and have been lost through evolution.

The word *vitamin* comes from *vital amine,* a term coined by Casimir Funk in 1912 to describe organic compounds that are required in small amounts for normal health. It turns out that most vitamins are not amines, but the name has stuck. **Table 12.2** lists the vitamins and their metabolic roles. Vitamins A, D, E, and K are lipids; their functions were described in Box 8.B. Many of the water-soluble vitamins are the precursors of coenzymes, which we will describe as we encounter them in the context of their particular metabolic reactions. Vitamins are a diverse group of compounds, whose discoveries and functional characterization have provided some of the more colorful stories in the history of biochemistry.

Many vitamins were discovered through studies of nutritional deficiencies. One of the earliest links between nutrition and disease was observed centuries ago in sailors suffering from scurvy, an illness caused by vitamin C deficiency (Box 5.B). A study of the disease beriberi led to the discovery of the first B vitamin. Beriberi, characterized by leg weakness and swelling, is caused by a deficiency of thiamine (vitamin B₁).

Thiamine (vitamin B$_1$)

TABLE 12.2 Vitamins and Their Roles

VITAMIN	COENZYME PRODUCT	BIOCHEMICAL FUNCTION	HUMAN DEFICIENCY DISEASE	TEXT REFERENCE
Water-Soluble				
Ascorbic acid (C)	Ascorbate	Cofactor for hydroxylation of collagen	Scurvy	Box 5.D
Biotin (B_7)	Biocytin	Cofactor for carboxylation reactions	*	Section 13.1
Cobalamin (B_{12})	Cobalamin coenzymes	Cofactor for alkylation reactions	Anemia	Section 17.2
Folic acid	Tetrahydrofolate	Cofactor for one-carbon transfer reactions	Anemia	Section 18.2
Lipoic acid	Lipoamide	Cofactor for acyl transfer reactions	*	Section 14.1
Nicotinamide (niacin, B_3)	Nicotinamide coenzymes (NAD^+, $NADP^+$)	Cofactor for oxidation–reduction reactions	Pellagra	Fig. 3.2, Section 12.2
Pantothenic acid (B_5)	Coenzyme A	Cofactor for acyl transfer reactions	*	Fig. 3.2, Section 12.3
Pyridoxine (B_6)	Pyridoxal phosphate	Cofactor for amino-group transfer reactions	*	Section 18.1
Riboflavin (B_2)	Flavin coenzymes (FAD, FMN)	Cofactor for oxidation–reduction reactions	*	Fig. 3.2
Thiamine (B_1)	Thiamine pyrophosphate	Cofactor for aldehyde transfer reactions	Beriberi	Sections 12.2, 14.1
Fat-Soluble				
Vitamin A (retinol)		Light-absorbing pigment	Blindness	Box 8.B
Vitamin D		Hormone that promotes Ca^{2+} absorption	Rickets	Box 8.B
Vitamin E (tocopherol)		Antioxidant	*	Box 8.B
Vitamin K (phylloquinone)		Cofactor for carboxylation of blood coagulation proteins	Bleeding	Box 8.B

*Deficiency in humans is rare or unobserved.

Thiamine acts as a prosthetic group in some essential enzymes, including the one that converts pyruvate to acetyl-CoA. Rice husks are rich in thiamine, and individuals whose diet consists largely of polished (huskless) rice can develop beriberi. The disease was originally thought to be infectious, until the same symptoms were observed in chickens and prisoners fed a diet of polished rice. Thiamine deficiency can occur in chronic alcoholics and others with a limited diet and impaired nutrient absorption.

Niacin, a component of NAD^+ and $NADP^+$, was first identified as the factor missing in the vitamin-deficiency disease pellagra.

Niacin

The symptoms of pellagra, including diarrhea and dermatitis, can be alleviated by boosting the intake of the essential amino acid tryptophan, which humans can convert to niacin. Niacin deficiency was once common in certain populations whose diet consisted largely of maize (corn). This grain is low in tryptophan and its niacin is covalently bound to other molecules; hence, it is not easily absorbed during digestion. In South America, where maize originated, the kernels are traditionally prepared by soaking or boiling them in an alkaline solution, a treatment that releases niacin and prevents pellagra. Unfortunately, this food-preparation custom did not spread to other parts of the world that adopted maize farming.

Most vitamins are readily obtained from a balanced diet, although poor nutrition, particularly in impoverished parts of the world, still causes vitamin-deficiency diseases. Intestinal bacteria, as well as plant- and animal-derived foods, are the natural sources of vitamins. However, plants do not contain cobalamin, so individuals who follow a strict vegetarian diet are at higher risk for developing a cobalamin deficiency.

 CONCEPT CHECK

1. Why are compounds such as glyceraldehyde-3-phosphate, pyruvate, and acetyl-CoA so important in metabolism?

2. Explain how cofactors such as NAD^+ and ubiquinone participate in metabolic reactions.

3. Explain the importance of reoxidizing NADH and QH_2 by molecular oxygen.

4. Summarize the main features of metabolic pathways.

5. Explain the relationship between vitamins and coenzymes.

12.3 FREE ENERGY CHANGES IN METABOLIC REACTIONS

KEY CONCEPTS

- The free energy change for a reaction depends on the equilibrium constant for the reaction and on the actual concentrations of the reacting species.
- A reaction with a large negative change in free energy can be coupled to another, unfavorable reaction.
- A reaction that breaks a phosphoanhydride bond in ATP occurs with a large change in free energy.
- Cells also use the free energy of other phosphorylated compounds, thioesters, reduced cofactors, and electrochemical gradients.
- Nonequilibrium reactions often serve as metabolic control points.

We have introduced the idea that catabolic reactions tend to release free energy and anabolic reactions tend to consume it (see Fig. 12.1), but, in fact, *all reactions in vivo occur with a net decrease in free energy; that is, ΔG is always less than zero* (free energy is discussed in Section 1.3). In a cell, metabolic reactions are not isolated but are linked, so the free energy of a thermodynamically favorable reaction can be used to pull a second, unfavorable reaction forward. How can free energy be transferred from one reaction to another? Free energy is not a substance or the property of a single molecule, so it is misleading to refer to a molecule or a bond within that molecule as having a large amount of free energy. Rather, *free energy is a property of a system, and it changes when the system undergoes a chemical reaction.*

THE FREE ENERGY CHANGE DEPENDS ON REACTANT CONCENTRATIONS

The change in free energy of a system is related to the concentrations of the reacting substances. When a reaction such as $A + B \rightleftharpoons C + D$ is at equilibrium, the concentrations of the four reactants define the **equilibrium constant, K_{eq},** for the reaction:

$$K_{eq} = \frac{[C]_{eq}[D]_{eq}}{[A]_{eq}[B]_{eq}} \qquad [12.1]$$

(the brackets indicate the molar concentration of each substance). Recall that at equilibrium, the rates of the forward and reverse reactions are balanced, so there is no net change in the concentration of any reactant. Equilibrium does *not* mean that the concentrations of the reactants and products are equal.

*When the system is not at equilibrium, the reactants experience a driving force to reach their equilibrium values. This force is the **standard free energy change** for the reaction, ΔG°′,* which is defined as

$$\Delta G^{\circ\prime} = -RT \ln K_{eq} \qquad [12.2]$$

R is the gas constant $(8.3145 \text{ J} \cdot \text{K}^{-1} \cdot \text{mol}^{-1})$ and T is the temperature in Kelvin. The larger the value of the free energy change, the farther the reaction is from equilibrium and the stronger the tendency for the reaction to proceed. Recall from Section 1.3 that free energy has units of joules per mole. One joule is equivalent to 0.239 calories, and 1 cal = 4.1484 J (1000 cal = 1 kcal = 1 Calorie). Equation 12.2 can be used to calculate $\Delta G^{\circ\prime}$ from K_{eq} and vice versa (see Sample Calculation 12.1).

SAMPLE CALCULATION 12.1

PROBLEM

Calculate $\Delta G^{\circ\prime}$ for a reaction at 25°C when $K_{eq} = 5.0$.

SOLUTION

Use Equation 12.2:

$$\Delta G^{\circ\prime} = -RT \ln K_{eq}$$
$$= -(8.3145 \text{ J} \cdot \text{K}^{-1} \cdot \text{mol}^{-1})(298 \text{ K}) \ln 5.0$$
$$= -4000 \text{ J} \cdot \text{mol}^{-1} = -4.0 \text{ kJ} \cdot \text{mol}^{-1}$$

PRACTICE PROBLEMS

1. Calculate $\Delta G^{\circ\prime}$ for a reaction at 25°C when $K_{eq} = 0.55$.
2. If the temperature for the reaction in Practice Problem 1 was raised to 37°C, how would $\Delta G^{\circ\prime}$ change?
3. If $\Delta G^{\circ\prime}$ for a reaction at 37°C is –5 kJ · mol^{-1}, what is K_{eq}?

By convention, measurements of standard free energy are valid under **standard conditions,** where the temperature is 25°C (298 K) and the pressure is 1 atm (these conditions are indicated by the degree symbol after ΔG). For a chemist, standard conditions specify an initial activity of 1 for each reactant (activity is the reactant's concentration corrected for its nonideal behavior). However, these conditions are impractical for biochemists since most biochemical reactions occur near neutral pH (where $[H^+] = 10^{-7}$ M rather than 1 M) and in aqueous solution (where $[H_2O] = 55.5$ M). The biochemical standard conditions are summarized in **Table 12.3**. Biochemists use a prime symbol to indicate the standard free energy change for a reaction under biochemical standard conditions. In most equilibrium expressions, $[H^+]$ and $[H_2O]$ are set to 1 so that these terms can be ignored. And because biochemical reactions typically involve dilute solutions of reactants, molar concentrations can be used instead of activities.

Like K_{eq}, $\Delta G^{\circ\prime}$ is a constant for a particular reaction. It may be a positive or negative value, and it indicates whether the reaction can proceed spontaneously ($\Delta G^{\circ\prime} < 0$) or not ($\Delta G^{\circ\prime} > 0$) under standard conditions. In a living cell, reactants and products are almost never present at standard-state concentrations and the temperature may not be 25°C, yet reactions do occur with some change in free energy. Thus, it is important to distinguish the standard free energy change of a reaction from its actual free energy change, ΔG. ΔG is a function of the actual concentrations of the reactants and the temperature (37°C or 310 K in humans). ΔG is related to the standard free energy change for the reaction:

$$\Delta G = \Delta G^{\circ\prime} + RT \ln \frac{[C][D]}{[A][B]}$$ [12.3]

Here, the bracketed quantities represent the actual, nonequilibrium concentrations of the reactants. The concentration term in Equation 12.3 is sometimes called the **mass action ratio.**

When the reaction is at equilibrium, $\Delta G = 0$ and

$$\Delta G^{\circ\prime} = -RT \ln \frac{[C]_{eq}[D]_{eq}}{[A]_{eq}[B]_{eq}}$$ [12.4]

which is equivalent to Equation 12.2. Note that Equation 12.3 shows that *the criterion for spontaneity for a reaction is ΔG, a property of the actual concentrations of the reactants, not the constant $\Delta G^{\circ\prime}$.* Thus, a reaction with a positive standard free energy change (a reaction that cannot occur when the reactants are present at standard concentrations) may proceed *in vivo*, depending on the concentrations of reactants in the cell (see Sample Calculation 12.2). Keep in mind that thermodynamic spontaneity does not imply a rapid reaction. Even a substance with a strong tendency to undergo reaction ($\Delta G \ll 0$) will usually not react until acted upon by an enzyme that catalyzes the reaction.

TABLE 12.3 Biochemical Standard State

Temperature	25°C (298 K)
Pressure	1 atm
Reactant concentration	1 M
pH	7.0 ($[H^+] = 10^{-7}$ M)
Water concentration	55.5 M

SAMPLE CALCULATION 12.2

PROBLEM

The standard free energy change for the reaction catalyzed by phosphoglucomutase is $-7.1 \text{ kJ} \cdot \text{mol}^{-1}$. Calculate the equilibrium constant for the reaction. Calculate ΔG at 37°C when the concentration of glucose-1-phosphate is 1 mM and the concentration of glucose-6-phosphate is 25 mM. Is the reaction spontaneous under these conditions?

Glucose-1-phosphate   Glucose-6-phosphate

SOLUTION

The equilibrium constant K_{eq} can be derived by rearranging Equation 12.2.

$$K_{eq} = e^{-\Delta G°'/RT}$$
$$= e^{-(-7100 \text{ J} \cdot \text{mol}^{-1})/(8.3145 \text{ J} \cdot \text{K}^{-1} \cdot \text{mol}^{-1})(298 \text{ K})}$$
$$= e^{2.87} = 17.6$$

At 37°C, $T = 310$ K.

$$\Delta G = \Delta G°' + RT \ln \frac{[\text{glucose-6-phosphate}]}{[\text{glucose-1-phosphate}]}$$
$$= -7100 \text{ J} \cdot \text{mol}^{-1} + (8.3145 \text{ J} \cdot \text{K}^{-1} \cdot \text{mol}^{-1})(310 \text{ K})\ln(0.025/0.001)$$
$$= -7100 \text{ J} \cdot \text{mol}^{-1} + 8300 \text{ J} \cdot \text{mol}^{-1}$$
$$= 1200 \text{ J} \cdot \text{mol}^{-1} = 1.2 \text{ kJ} \cdot \text{mol}^{-1}$$

The reaction is not spontaneous because ΔG is greater than zero.

PRACTICE PROBLEMS

4. Calculate ΔG for the reaction shown here when the initial concentration of glucose-1-phosphate is 4 mM and the initial concentration of glucose-6-phosphate is 20 mM. Is the reaction spontaneous under these conditions?

5. At equilibrium, the concentration of glucose-6-phosphate is 45 mM. What is the concentration of glucose-1-phosphate?

6. Calculate the ratio of the concentration of glucose-6-phosphate to the concentration of glucose-1-phosphate that gives a free energy change of $-4 \text{ kJ} \cdot \text{mol}^{-1}$.

UNFAVORABLE REACTIONS ARE COUPLED TO FAVORABLE REACTIONS

A biochemical reaction may at first seem to be thermodynamically forbidden because its free energy change is greater than zero. Yet the reaction can proceed *in vivo* when it is coupled to a second reaction whose value of ΔG is very large and negative so that the *net* change in free energy for the combined reactions is less than zero. *ATP is often involved in such coupled processes because its reactions occur with a relatively large negative change in free energy.*

Adenosine triphosphate (ATP) contains two phosphoanhydride bonds (**Fig. 12.12**). Cleavage of either of these bonds—that is, transfer of one or two of its phosphoryl groups to another molecule—is a reaction with a large negative standard free energy change (under physiological conditions, ΔG is even more negative). As a reference point, biochemists use the reaction in which a phosphoryl group is transferred to water—in other words, hydrolysis of the phosphoanhydride bond, such as

$$ATP + H_2O \rightarrow ADP + P_i$$

This is a spontaneous reaction with a $\Delta G°'$ value of $-30 \text{ kJ} \cdot \text{mol}^{-1}$.

FIGURE 12.12

Adenosine triphosphate. The three phosphate groups are sometimes described by the Greek letters α, β, and γ. The linkage between the first (α) and second (β) phosphoryl groups, and between the second (β) and third (γ), is a phosphoanhydride bond. A reaction in which one or two phosphoryl groups are transferred to another compound (a reaction in which a phosphoanhydride bond is cleaved) has a large negative value of $\Delta G^{\circ\prime}$.

🅿 **How does hydrolysis of a phosphoanhydride bond affect the net charge of a nucleotide?**

The following example illustrates the role of ATP in a coupled reaction. Consider the phosphorylation of glucose by inorganic phosphate (HPO_4^{2-} or P_i), a thermodynamically unfavorable reaction ($\Delta G^{\circ\prime} = +13.8$ kJ·mol^{-1}):

When this reaction is combined with the ATP hydrolysis reaction, the values of $\Delta G^{\circ\prime}$ for each reaction are added:

	$\Delta G^{\circ\prime}$
glucose + P_i → glucose-6-phosphate + H_2O	$+13.8$ kJ·mol^{-1}
ATP + H_2O → ADP + P_i	-30.5 kJ·mol^{-1}
glucose + ATP → glucose-6-phosphate + ADP	-16.7 kJ·mol^{-1}

The net chemical reaction, the phosphorylation of glucose, is thermodynamically favorable ($\Delta G < 0$). *In vivo*, this reaction is catalyzed by hexokinase (introduced in Section 6.3), and a phosphoryl group is transferred from ATP directly to glucose. The ATP is not actually hydrolyzed, and there is no free phosphoryl group floating around the enzyme. However, writing out the two coupled reactions, as shown above, makes it easier to see what's going on thermodynamically.

Some biochemical processes appear to occur with the concomitant hydrolysis of ATP to ADP + P_i, for example, the operation of myosin and kinesin (Section 5.4) or the Na,K-ATPase ion pump (Section 9.3). But a closer look reveals that in all these processes, ATP actually transfers a phosphoryl group to a protein. Later, the phosphoryl group is transferred to water, so the net reaction takes the form of ATP hydrolysis. The same ATP "hydrolysis" effect applies to some reactions in which the AMP moiety of ATP (rather than a phosphoryl group) is transferred to a substance, leaving inorganic pyrophosphate (PP_i). Cleavage of the phosphoanhydride bond of PP_i also has a large negative value of $\Delta G^{\circ\prime}$.

Because the ATP hydrolysis reaction appears to drive many thermodynamically unfavorable reactions, it is tempting to think of ATP as an agent that transfers packets of free energy around the cell. This is one reason why ATP is commonly called the energy currency of the cell. The general role of ATP in linking exergonic ATP-producing processes to endergonic ATP-consuming processes can be diagrammed as

In this scheme, it appears that the "energy" of the catabolized nutrient is transferred to ATP, then the "energy" of ATP is transferred to another product in a biosynthetic reaction. However, free energy is not a tangible item, and there is nothing magic about ATP. The two phospho-anhydride bonds of ATP are sometimes called "high-energy" bonds, but they are no different from other covalent bonds. What matters is that *a reaction in which phosphoryl groups are transferred to another molecule—breaking these bonds—is a process with a large negative free energy change.* Using the simple example of ATP hydrolysis, we can state that a large amount of free energy is released when ATP is hydrolyzed because the products of the reaction have less free energy than the reactants. It is worth examining two reasons why this is so.

1. *The ATP hydrolysis products are more stable than the reactants.* At physiological pH, ATP has three to four negative charges (its pK is close to 7), and the anionic groups repel each other. In the products ADP and P_i, separation of the charges relieves some of this unfavorable electrostatic repulsion.

2. *A compound with a phosphoanhydride bond experiences less resonance stabilization than its hydrolysis products.* **Resonance stabilization** reflects the degree of electron delocalization in a molecule and can be roughly assessed by the number of different ways of depicting the molecule's structure. There are fewer equivalent ways of arranging the bonds of the terminal phosphoryl group of ATP than there are in free P_i.

To summarize, *ATP functions as an energy currency because its hydrolysis reaction is highly exergonic ($\Delta G \ll 0$).* The favorable ATP reaction (ATP → ADP) can therefore pull another, unfavorable reaction with it, provided that the sum of the free energy changes for both reactions is less than zero. In effect, the cell "spends" ATP to make another process happen.

FREE ENERGY CAN TAKE DIFFERENT FORMS

ATP is not the only substance that functions as energy currency in the cell. Other compounds that participate in reactions with large negative changes in free energy can serve the same purpose. For example, a number of phosphorylated compounds other than ATP can give up their phosphoryl group to another molecule. **Table 12.4** lists the standard free energy changes for some of these reactions in which the phosphoryl group is transferred to water.

TABLE 12.4 Standard Free Energy Change for Phosphate Hydrolysis

COMPOUND	$\Delta G^{\circ\prime}$ (kJ · mol^{-1})
Phosphoenolpyruvate	−61.9
1,3-Bisphosphoglycerate	−49.4
ATP → AMP + PP$_i$	−45.6
Phosphocreatine	−43.1
ATP → ADP + P$_i$	−30.5
Glucose-1-phosphate	−20.9
PP$_i$ → 2 P$_i$	−19.2
Glucose-6-phosphate	−13.8
Glycerol-3-phosphate	−9.2

Although hydrolysis of the bond linking the phosphate group to the rest of the molecule could be a wasteful process (the product would be free phosphate, P_i), the values listed in the table are a guide to how such compounds would behave in a coupled reaction, such as the hexokinase reaction described above. For example, phosphocreatine has a standard free energy of hydrolysis of -43.1 kJ·mol^{-1}:

$$
\underset{\text{Phosphocreatine}}{
\begin{array}{c}
\overset{+}{H_2N} \quad\quad O \\
\| \quad\quad\quad \| \\
C-NH-P-O^- \\
| \quad\quad\quad\quad | \\
N-CH_3 \quad O^- \\
| \\
CH_2 \\
| \\
COO^-
\end{array}
}
\quad\xrightarrow{\;H_2O\quad P_i\;}\quad
\underset{\text{Creatine}}{
\begin{array}{c}
\overset{+}{H_2N} \\
\| \\
C-NH_2 \\
| \\
N-CH_3 \\
| \\
CH_2 \\
| \\
COO^-
\end{array}
}
$$

Creatine has lower free energy than phosphocreatine since it has two, rather than one, resonance forms; this resonance stabilization contributes to the large negative free energy change when phosphocreatine transfers its phosphoryl group to another compound. In muscles, phosphocreatine transfers a phosphoryl group to ADP to produce ATP (**Box 12.C**).

Like ATP, other nucleoside triphosphates have large negative standard free energies of hydrolysis. GTP rather than ATP serves as the energy currency for reactions that occur during cellular signaling (Section 10.2) and protein synthesis (Section 22.3). In the cell, nucleoside triphosphates are freely interconverted by reactions such as the one catalyzed by nucleoside diphosphate kinase, which transfers a phosphoryl group from ATP to a nucleoside diphosphate (NDP):

$$\text{ATP} + \text{NDP} \rightleftharpoons \text{ADP} + \text{NTP}$$

Because the reactants and products are energetically equivalent, $\Delta G^{\circ\prime}$ values for these reactions are near zero.

Another class of compounds that can release a large amount of free energy upon hydrolysis are **thioesters,** such as acetyl-CoA. Coenzyme A is a nucleotide derivative with a side chain ending in a sulfhydryl (SH) group (see Fig. 3.2a). An acyl or acetyl group (the "A" for which coenzyme A was named) is linked to the sulfhydryl group by a thioester bond. Hydrolysis of this bond has a $\Delta G^{\circ\prime}$ value of -31.5 kJ·mol^{-1}, comparable to that of ATP hydrolysis:

$$
\underset{\text{Acetyl-CoA}}{
\begin{array}{c}
\text{thioester} \\
\text{bond} \\
O \quad\swarrow \\
\| \\
CH_3-C-S-CoA
\end{array}
}
\quad\xrightarrow{\;H_2O\;}\quad
\begin{array}{c}
O \\
\| \\
CH_3-C-O^-
\end{array}
\;+\; CoA-SH
$$

Hydrolysis of a thioester is more exergonic than the hydrolysis of an ordinary (oxygen) ester because thioesters have less resonance stability than oxygen esters, owing to the larger size of an S atom relative to an O atom. An acetyl group linked to coenzyme A can be readily transferred to another molecule because formation of the new linkage is powered by the favorable free energy change of breaking the thioester bond.

We have already seen that in oxidation–reduction reactions, cofactors such as NAD$^+$ and ubiquinone can collect electrons. The reduced cofactors are a form of energy currency because their subsequent reoxidation by another compound occurs with a negative change in free energy. Ultimately, the transfer of electrons from one reduced cofactor to another and finally to oxygen, the final electron acceptor in many cells, releases enough free energy to drive the synthesis of ATP.

Keep in mind that free energy changes occur not just as the result of chemical changes such as phosphoryl-group transfer or electron transfer. As decreed by the first law of thermodynamics (Section 1.3), *energy can take many forms*. We will see that ATP production

Box 12.C Powering Human Muscles

In resting muscles, when the demand for ATP is low, creatine kinase catalyzes the transfer of a phosphoryl group from ATP to creatine to produce phosphocreatine:

$$\text{ATP} + \text{creatine} \rightleftharpoons \text{phosphocreatine} + \text{ADP}$$

This reaction runs in reverse when ADP concentrations rise, as they do when muscle contraction converts ATP to ADP + P_i. Phosphocreatine therefore acts as a sort of phosphoryl-group reservoir to maintain the supply of ATP. Cells cannot stockpile ATP; its concentration remains remarkably stable (between 2 and 5 mM in most cells) under widely varying levels of demand. Without phosphocreatine, a muscle would exhaust its ATP supply before it could be replenished by other, slower processes.

Different types of physical activity make different demands on a muscle's ATP-generating mechanisms. A single burst of activity is powered by the available ATP. Activities lasting up to a few seconds require phosphocreatine to maintain the ATP supply. Phosphocreatine itself is limited, so continued muscle contraction must rely on ATP produced by catabolizing glucose (obtained from the muscle's store of glycogen) via glycolysis. The end product of this pathway is lactate, the conjugate base of a weak acid, and muscle pain sets in as the acid accumulates and the pH begins to drop. Up to this point, the muscle functions anaerobically (without the participation of O_2). To continue its activity, it must switch to aerobic (O_2-dependent) metabolism and further oxidize glucose via the citric acid cycle. The muscle also catabolizes

fatty acids, whose products also enter the citric acid cycle. Recall that the citric acid cycle generates reduced cofactors that must be reoxidized by molecular oxygen. Aerobic metabolism of glucose and fatty acids is slower than anaerobic glycolysis, but it generates considerably more ATP. Some forms of physical activity and the systems that power them are diagrammed below.

A casual athlete can detect the shift from anaerobic to aerobic metabolism after about a minute and a half. In world-class athletes, the breakpoint occurs at about 150 to 170 seconds, which corresponds roughly to the finish line in a 1000-meter race.

The muscles of sprinters have a high capacity for anaerobic ATP generation, whereas the muscles of distance runners are better adapted to produce ATP aerobically. Such differences in energy metabolism are visibly manifest in the flight muscles of birds. Migratory birds such as geese, which power their long flights primarily with fatty acids, have large numbers of mitochondria to carry out oxidative phosphorylation. The reddish-brown color of the mitochondria gives the flight muscles a dark color. Birds that rarely fly, such as chickens, have fewer mitochondria and lighter-colored muscles. When these birds do fly, it is usually only a short burst of activity that is powered by anaerobic mechanisms.

? **Why do some athletes believe that creatine supplements boost their performance?**

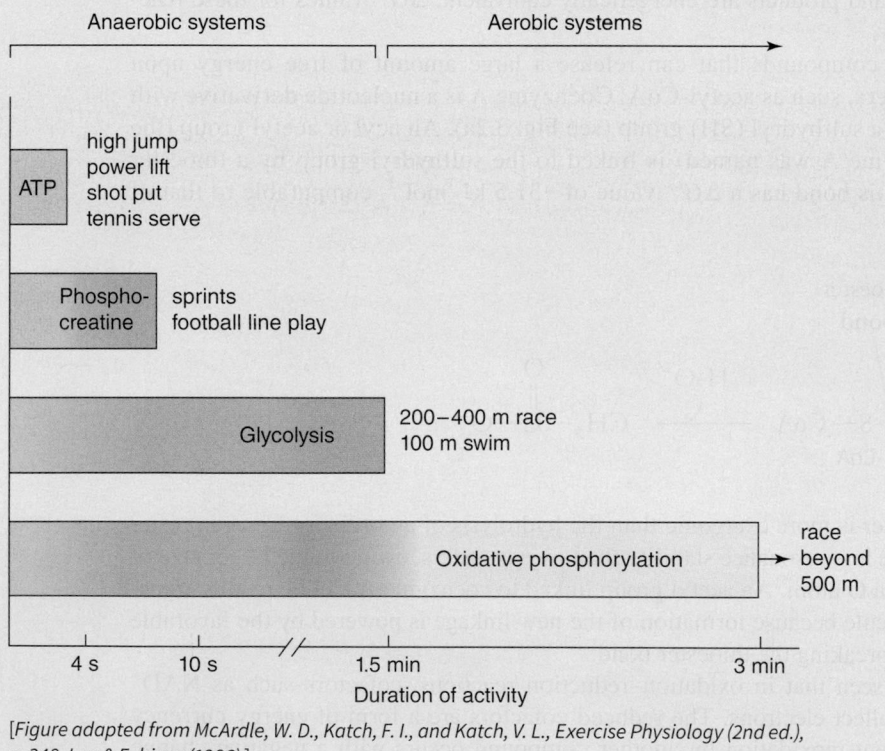

[Figure adapted from McArdle, W. D., Katch, F. I., and Katch, V. L., Exercise Physiology (2nd ed.), p. 348, Lea & Febiger (1986).]

in cells depends on the energy of an electrochemical gradient, that is, an imbalance in the concentration of a substance (in this case, protons) on the two sides of a membrane. The free energy change of dissipating this gradient (allowing the system to move toward equilibrium) is converted to the mechanical energy of an enzyme that synthesizes ATP. In photosynthetic cells, the chemical reactions required to generate carbohydrates are ultimately driven by the free energy changes of reactions in which light-excited molecules return to a lower-energy state.

REGULATION OCCURS AT THE STEPS WITH THE LARGEST FREE ENERGY CHANGES

In a series of reactions that make up a metabolic pathway, some reactions have ΔG values near zero. These near-equilibrium reactions are not subject to a strong driving force to proceed in either direction. Rather, flux can go forward or backward, according to slight fluctuations in the concentrations of reactants and products. When the concentrations of metabolites change, the enzymes that catalyze these near-equilibrium reactions tend to act quickly to restore the near-equilibrium state.

Reactions with large negative changes in free energy have a longer way to go to reach equilibrium; these are the reactions that experience the greatest "urge" to proceed forward. However, the enzymes that catalyze these reactions do not allow the reaction to reach equilibrium because they work too slowly. Often the enzymes are already saturated with substrate, so the reactions cannot go any faster (when $[S] \gg K_M$, $v \approx V_{max}$; Section 7.2). The rates of these far-from-equilibrium reactions limit flux through the entire pathway because the reactions function like dams.

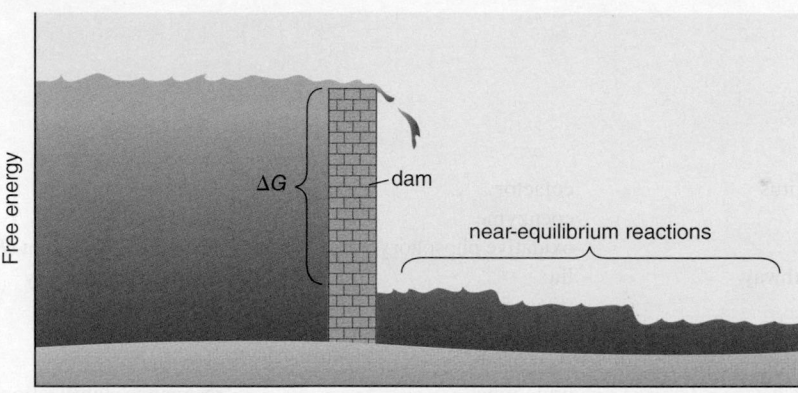

Cells can regulate flux through a pathway by adjusting the rate of a reaction with a large free energy change. This can be done by increasing the amount of enzyme that catalyzes that step or by altering the intrinsic activity of the enzyme through allosteric mechanisms (see Fig. 7.17). As soon as more metabolite has gotten past the dam, the near-equilibrium reactions go with the flow, allowing the pathway intermediates to move toward the final product. Most metabolic pathways do not have a single flow-control point, as the dam analogy might suggest. Instead, flux is typically controlled at several points to ensure that the pathway can work efficiently as part of the cell's entire metabolic network.

 CONCEPT CHECK

1. Explain why free energy changes must be negative for reactions *in vivo*.
2. What is the relationship between kinases and phosphatases?
3. Describe the difference between ΔG and $\Delta G°'$. How are they related?
4. Why is it misleading to refer to ATP as a high-energy molecule?
5. Explain why cleavage of one of ATP's phosphoanhydride bonds releases large amounts of free energy.
6. How do phosphorylated compounds, thioesters, and reduced cofactors appear to transfer free energy? What other forms of energy do cells use?
7. Describe the reasons why cells control metabolic reactions that have large negative free energy changes.

SUMMARY

12.1 Food and Fuel

• Polymeric food molecules such as starch, proteins, and triacylglycerols are broken down to their monomeric components (glucose, amino acids, and fatty acids), which are absorbed. These materials are stored as polymers in a tissue-specific manner.

• Metabolic fuels are mobilized from glycogen, fat, and proteins as needed.

12.2 Metabolic Pathways

• Series of reactions known as metabolic pathways break down and synthesize biological molecules. Several pathways make use of the same small molecule intermediates.

• During the oxidation of amino acids, monosaccharides, and fatty acids, electrons are transferred to carriers such as NAD^+ and ubiquinone. Reoxidation of the reduced cofactors drives the synthesis of ATP by oxidative phosphorylation.

• Metabolic pathways form a complex network, but not all cells or organisms carry out all possible metabolic processes. Humans rely on other organisms to supply vitamins and other essential materials.

12.3 Free Energy Changes in Metabolic Reactions

• The standard free energy change for a reaction is related to the equilibrium constant, but the actual free energy change is related to the actual cellular concentrations of reactants and products.

• A thermodynamically unfavorable reaction may proceed when it is coupled to a favorable process involving ATP, whose phosphoanhydride bonds release a large amount of free energy when cleaved.

• Other forms of cellular energy currency include phosphorylated compounds, thioesters, and reduced cofactors.

• Cells regulate metabolic pathways at the steps that are farthest from equilibrium.

KEY TERMS

catabolism	diabetes mellitus	cofactor	essential compound
anabolism	lysosome	coenzyme	vitamin
metabolism	proteasome	oxidative phosphorylation	equilibrium constant (K_{eq})
chemoautotroph	metabolic pathway	flux	standard free energy
photoautotroph	intermediate	transcriptomics	change ($\Delta G^{o\prime}$)
heterotroph	metabolite	transcriptome	standard conditions
lipoprotein	glycolysis	microarray (DNA chip)	mass action ratio
adipose tissue	citric acid cycle	proteomics	resonance stabilization
metabolic fuel	oxidation	proteome	thioester
mobilization	reduction	metabolomics	
phosphorolysis	redox reaction	metabolome	

BIOINFORMATICS

Brief Bioinformatics Exercises

12.1 Metabolism and the BRENDA and KEGG Databases

12.2 Cofactor Chemistry

Bioinformatics Projects

Metabolic Enzymes, Microarrays, and Proteomics

Metabolomics Databases and Tools

PROBLEMS

12.1 Food and Fuel

1. Classify the following organisms as chemoautotrophs, photoautotrophs, or heterotrophs:

(a) *Hydrogenobacter*, which converts molecular hydrogen and oxygen to water;

(b) *Arabidopsis thaliana*, a green plant;

(c) The nitrosifying bacteria, which oxidize NH_3 to nitrite;

(d) *Saccharomyces cerevisiae,* yeast;

(e) *Caenorhabditis elegans,* a nematode worm;

(f) The *Thiothrix* bacteria, which oxidize hydrogen sulfide;

(g) Cyanobacteria (erroneously termed "blue-green algae" in the past).

2. The purple nonsulfur bacteria obtain their cellular energy from a photosynthetic process that does not produce oxygen. These bacteria also require an organic carbon source. Using the terms in this chapter, coin a new term that describes the trophic strategy of this organism.

3. Digestion of carbohydrates begins in the mouth, where salivary amylases act on dietary starch. When the food is swallowed and enters the stomach, carbohydrate digestion ceases (it resumes in the small intestine). Why does carbohydrate digestion not occur in the stomach?

4. Pancreatic amylase, which is similar to salivary amylase, is secreted by the pancreas into the small intestine. The active site of pancreatic amylase accommodates five glucosyl residues and cleaves the glycosidic bond between the second and third residues. The enzyme cannot accommodate branched chains.
(a) What are the main products of amylase digestion?
(b) What are the products of amylopectin digestion?

5. Starch digestion is completed by the enzymes isomaltase (or α-dextrinase), which catalyzes the hydrolysis of $\alpha(1 \rightarrow 6)$ glycosidic bonds, and maltase, which hydrolyzes $\alpha(1 \rightarrow 4)$ bonds. Why are these enzymes needed in addition to α-amylase?

6. Monosaccharides, the products of polysaccharide and disaccharide digestion, enter the cells lining the intestine via a specialized transport system. What is the source of free energy for this transport process?

7. Unlike the monosaccharides described in Problem 6, sugar alcohols such as sorbitol (see Solution 11.17c) are absorbed via passive diffusion. Why? What process occurs more rapidly, passive diffusion or passive transport?

8. Use what you know about the properties of alcohol (ethanol) to describe how it is absorbed in both the stomach and the small intestine. What effect does the presence of food have on the absorption of ethanol?

9. Nucleic acids that are present in food are hydrolyzed by digestive enzymes. What type of mechanism most likely mediates the entry of the reaction products into intestinal cells?

10. Hydrolysis of proteins begins in the stomach, catalyzed by the hydrochloric acid secreted into the stomach by parietal cells. Draw the reaction that shows the hydrolysis of a peptide bond.

11. How does the low pH of the stomach affect protein structure in such a way that the proteins are prepared for hydrolytic digestion?

12. Like the serine proteases (see Section 6.4), pepsin is made as a zymogen and is inactive at its site of synthesis, where the pH is 7. Pepsin becomes activated when secreted into the stomach, where it encounters a pH of ~2. Pepsinogen contains a "basic peptide" that blocks its active site at pH 7. The basic peptide dissociates from the active site at pH 2 and is cleaved, resulting in the formation of the active form of the enzyme. What amino acids residues are found in the active site of pepsin? Why does the basic peptide bind tightly to the active site at pH 7 and why does it dissociate at the lower pH?

13. The cleavage of peptide bonds in the stomach is catalyzed both by hydrochloric acid (see Problem 10) and by the stomach enzyme pepsin (see Problem 12). Peptide bond cleavage continues in the small intestine, catalyzed by the pancreatic enzymes trypsin and chymotrypsin. At what pH does pepsin function optimally; that is, at what pH is the V_{max} for pepsin greatest? Is the pH optimum for pepsin different from that for trypsin and chymotrypsin? Explain.

14. Scientists have recently discovered why the botulinum toxin survives the acidic environment of the stomach. The toxin forms a complex with a second nontoxic protein that acts as a shield to protect the botulinum toxin from being digested by stomach enzymes. Upon entry into the small intestine, the two proteins dissociate and

the botulinum toxin is released. What is the likely interaction between the botulinum toxin and the nontoxic protein, and why does the complex form readily in the stomach but not in the small intestine?

15. Free amino acid transport from the intestinal lumen into intestinal cells requires Na^+ ions. Draw a diagram that illustrates amino acid transport into these cells.

16. In oral rehydration therapy (ORT), patients suffering from diarrhea are given a solution consisting of a mixture of glucose and electrolytes. Some formulations also contain amino acids. Why are electrolytes added to the mixture?

17. Triacylglycerol digestion begins in the stomach. Gastric lipase catalyzes hydrolysis of the fatty acid from the third glycerol carbon. (a) Draw the reactants and products of this reaction. (b) Conversion of the triacylglycerol to a diacylglycerol and a fatty acid promotes emulsification of fats in the stomach; that is, the products are more easily incorporated into micelles. Explain why.

18. Most of the fatty acids produced in the reaction described in Problem 17 form micelles and are absorbed as such, but a small percentage of fatty acids are free and are transported into the intestinal epithelial cells without the need for a transport protein. Explain why a transport protein is not required.

19. The cells lining the small intestine absorb cholesterol but not cholesteryl esters. Draw the reaction catalyzed by cholesteryl esterase that produces cholesterol from cholesteryl stearate.

20. Some cholesterol is converted back to cholesteryl esters in the epithelial cells lining the small intestine (the reverse of the reaction described in Problem 19). Both cholesterol and cholesteryl esters are packaged into particles called chylomicrons, which consist of lipid and protein. Use what you know about the physical properties of cholesterol and cholesteryl esters to describe their locations in the chylomicron particle.

21. (a) Consider the physical properties of a polar glycogen molecule and an aggregation of hydrophobic triacylglycerols. On a per-weight basis, why is fat a more efficient form of energy storage than glycogen? (b) Explain why there is an upper limit to the size of a glycogen molecule but there is no upper limit to the amount of triacylglycerols that an adipocyte can store.

22. Glycogen can be expanded quickly, by adding glucose residues to its many branches, and degraded quickly, by simultaneously removing glucose from the ends of these branches. Are the enzymes that catalyze these processes specific for the reducing or nonreducing ends of the glycogen polymer? Explain.

23. The phosphorolysis reaction that removes glucose residues from glycogen yields as its product glucose-1-phosphate. Glucose-1-phosphate is isomerized to glucose-6-phosphate; then the phosphate group is removed in a hydrolysis reaction. Why is it necessary to remove the phosphate group before the glucose exits the cell to enter the circulation?

24. Hydrolytic enzymes encased within the membrane-bound lysosomes all work optimally at pH ~5. This feature serves as a cellular "insurance policy" in the event of lysosomal enzyme leakage into the cytosol. Explain.

12.2 Metabolic Pathways

25. The common intermediates listed in the table—acetyl CoA, glyceraldehyde-3-phosphate (GAP), and pyruvate—appear as reactants or products in several pathways. Place a check mark in the box that indicates the appropriate pathway—glycolysis, citric acid cycle, fatty acid metabolism, triacylglycerol (TAG) synthesis, photosynthesis, and transamination—for each reactant.

	Acetyl-CoA	GAP	Pyruvate
Glycolysis			
Citric acid cycle			
Fatty acid metabolism			
TAG synthesis			
Photosynthesis			
Transamination			

26. For each of the (unbalanced) reactions shown below, tell whether the reactant is being oxidized or reduced.

(a) A reaction from the catabolic glycolytic pathway

(b) A reaction from the fatty acid synthesis pathway

(c) A reaction associated with the catabolic glycolytic pathway

(d) A reaction associated with the anabolic pentose phosphate pathway

27. For each of the reactions shown in Problem 26, identify the cofactor as NAD^+, $NADP^+$, NADH, or NADPH.

28. A potential way to reduce the concentration of methane, a greenhouse gas, is to take advantage of sulfate-reducing bacteria. **(a)** Complete the chemical equation for methane consumption by these organisms:

$$CH_4 + SO_4^{2-} \rightarrow \underline{\quad} + HS^- + H_2O$$

(b) Identify the reaction component that undergoes oxidation.
(c) Identify the component that undergoes reduction.

29. Vitamin B_{12} is synthesized by certain gastrointestinal bacteria and is also found in foods of animal origin such as meat, milk, eggs, and fish. When vitamin B_{12}–containing foods are consumed, the vitamin is released from the food and binds to a salivary vitamin B_{12}–binding protein called haptocorrin. The haptocorrin–vitamin B_{12} complex passes from the stomach to the small intestine, where the vitamin is released from the haptocorrin and then binds to intrinsic factor (IF). The IF–vitamin B_{12} complex then enters the cells lining the intestine by receptor-mediated endocytosis. Using this information, make a list of individuals most at risk for vitamin B_{12} deficiency.

30. Hartnup disease is a hereditary disorder caused by a defective transporter for nonpolar amino acids.

(a) The symptoms of the disease (photosensitivity and neurological abnormalities) can be prevented through dietary adjustments. What sort of diet would be effective?
(b) Patients with Hartnup disease often exhibit pellagra-like symptoms. Explain.

31. A vitamin K–dependent carboxylase enzyme catalyzes the γ-carboxylation of specific glutamate residues in blood coagulation proteins.

(a) Draw the structure of a γ-carboxyglutamate residue.
(b) Why does this post-translational modification assist coagulation proteins in binding the Ca^{2+} ions required for blood clotting?

32. Food scientists at Ohio State University gave volunteers salsa or salad with or without lipid-rich avocados, then drew blood samples periodically and measured levels of β-carotene (see Box 8-B). They found that serum β-carotene levels were 2 to 15 times greater when the volunteers consumed the food with avocado. Explain these results.

33. Refer to Table 12.2 and identify the vitamin required to accomplish each of the following reactions:

(a)

(b)

(c)

(d)

34. Why is niacin technically not a vitamin?

35. The microbial enzyme lactate racemase contains a nickel-based prosthetic group shown as follows. **(a)** To which common coenzyme is this prosthetic group related? **(b)** Which two amino acid side chains hold the group in place?

36. Prokaryotic species A and B live close together because they are metabolically interdependent.

(a) If Species A converts CH_4 to CO_2, what process would Species B need to undertake: the conversion of S^{2-} to SO_4^{2-} or the conversion of SO_4^{2-} to S^{2-}?

(b) If Species A converts CH_4 to CO_2, what process would Species B need to undertake: the conversion of Fe^{3+} to Fe^{2+} or the conversion of Fe^{2+} to Fe^{3+}?

12.3 Free Energy Changes in Metabolic Reactions

37. Consider two reactions: $A \rightleftharpoons B$ and $C \rightleftharpoons D$. K_{eq} for the $A \rightleftharpoons B$ reaction is 10, and K_{eq} for the $C \rightleftharpoons D$ reaction is 0.1. You place 1 mM A in Tube 1 and 1 mM C in Tube 2 and allow the reactions to reach equilibrium.

(a) Without doing any calculations, determine whether the concentration of B in Tube 1 will be greater than or less than the concentration of D in Tube 2.

(b) Calculate the $\Delta G^{\circ\prime}$ values for the reactions. Assume a temperature of 37°C.

38. Calculate $\Delta G^{\circ\prime}$ for a reaction at 25°C when $K_{eq} = 4$. **(a)** If the reaction were carried out at 37°C, how would $\Delta G^{\circ\prime}$ change? **(b)** Is the reaction spontaneous?

39. Consider the reaction $E \rightleftharpoons F$, $K_{eq} = 1$.
(a) Without doing any calculations, what can you conclude about the $\Delta G^{\circ\prime}$ value for the reaction?
(b) You place 1 mM F in a tube and allow the reaction to reach equilibrium. Determine the final concentrations of E and F.
(c) Determine the direction the reaction will proceed if you place 5 mM E and 2 mM F in a test tube. What are the final concentrations of E and F?

40. The standard free energy change for the reaction catalyzed by triose phosphate isomerase is 7.56 kJ·mol^{-1}.

Calculate ΔG at 37°C when the concentration of dihydroxyacetone phosphate is 10 mM and the concentration of glyceraldehyde-3-phosphate is 40 mM. Is the reaction spontaneous under these conditions?

41. Calculate the ratio of the concentration of glyceraldehyde-3-phosphate to the concentration of dihydroxyacetone phosphate that gives a free energy change of 2 kJ·mol^{-1}.

42. Calculate ΔG for the $A \rightleftharpoons B$ reaction described in Problem 37 when the concentrations of A and B are 0.8 mM and 0.2 mM, respectively. In which direction will the reaction proceed?

43. Calculate ΔG for the $C \rightleftharpoons D$ reaction described in Problem 37 when the concentrations of C and D are 0.8 mM and 0.2 mM, respectively. In which direction will the reaction proceed?

44. **(a)** The $\Delta G^{\circ\prime}$ value for a hypothetical reaction is 15 kJ·mol^{-1} at 25°C. Compare the K_{eq} for this reaction with the K_{eq} for a reaction whose $\Delta G^{\circ\prime}$ value is twice as large. **(b)** Carry out the same exercise for a hypothetical reaction whose $\Delta G^{\circ\prime}$ value is -15 kJ·mol^{-1}.

45. **(a)** Use the standard free energies provided in Table 12.4 to calculate $\Delta G^{\circ\prime}$ for the isomerization of glucose-1-phosphate to glucose-6-phosphate. Is this value consistent with the value of $\Delta G^{\circ\prime}$ shown in Sample Calculation 12.2? Is this reaction spontaneous under standard conditions?
(b) Is the reaction spontaneous at 37°C when the concentration of glucose-6-phosphate is 10 mM and the concentration of glucose-1-phosphate is 0.5 mM?

46. Calculate ΔG for the hydrolysis of ATP under cellular conditions, where [ATP] = 5 mM, [ADP] = 2 mM, and [P$_i$] = 10 mM.

47. The standard free energy change for the reaction catalyzed by triose phosphate isomerase is 7.9 kJ·mol^{-1}.

Glyceraldehyde-3-phosphate   Dihydroxyacetone phosphate

(a) Calculate the equilibrium constant for the reaction.
(b) Calculate ΔG at 37°C when the concentration of glyceraldehyde-3-phosphate (GAP) is 0.2 mM and the concentration of dihydroxyacetone phosphate (DHAP) is 0.6 mM.
(c) Is the reaction spontaneous under these conditions? Would the reverse reaction be spontaneous?

48. An apple contains about 72 Calories. Express this quantity in terms of ATP equivalents (that is, how many ATP → ADP + P$_i$ reactions?).

49. A large hot chocolate with whipped cream purchased at a national coffee chain contains 760 Calories. Express this quantity in terms of ATP equivalents (see Problem 48).

50. Construct a graph plotting free energy versus the reaction coordinate for the following reactions: **(a)** glucose + P$_i$ → glucose-6-phosphate; **(b)** ATP + H_2O → ADP + P$_i$; **(c)** the coupled reaction.

51. Some studies (but not all) show that creatine supplementation increases performance in high-intensity exercises lasting less than 30 seconds. Would you expect creatine supplements to affect endurance exercise?

52. The $\Delta G^{\circ\prime}$ for the hydrolysis of ATP under standard conditions at pH 7 and in the presence of magnesium ions is -30.5 kJ·mol^{-1}.
(a) How would this value change if ATP hydrolysis were carried out at a pH of less than 7? Explain.
(b) How would this value change if magnesium ions were not present?

53. The $\Delta G^{\circ\prime}$ for the formation of UDP–glucose from glucose-1-phosphate and UTP is about zero. Yet the production of UDP–glucose is highly favorable. What is the driving force for this reaction?

$$\text{glucose-1-phosphate} + UTP \rightleftharpoons UDP\text{–glucose} + PP_i$$

54. **(a)** The complete oxidation of glucose releases a considerable amount of energy. The $\Delta G^{\circ\prime}$ for the reaction shown is -2850 kJ·mol^{-1}.

$$C_6H_{12}O_6 + 6\,O_2 \longrightarrow 6\,CO_2 + 6\,H_2O$$

How many moles of ATP could be produced under standard conditions from the oxidation of one mole of glucose, assuming about 33% efficiency?

(b) The oxidation of palmitate, a 16-carbon saturated fatty acid, releases $9781 \text{ kJ} \cdot \text{mol}^{-1}$.

$$C_{16}H_{32}O_2 + 23 O_2 \rightarrow 16 CO_2 + 16 H_2O$$

How many moles of ATP could be produced under standard conditions from the oxidation of one mole of palmitate, assuming 33% efficiency?

(c) Calculate the number of ATP molecules produced per carbon for glucose and palmitate. Explain the reason for the difference.

55. A moderately active adult female weighing 125 pounds must consume 2200 Calories of food daily.

(a) If this energy is used to synthesize ATP, calculate the number of moles of ATP that would be synthesized each day under standard conditions (assuming 33% efficiency).

(b) Calculate the number of grams of ATP that would be synthesized each day. The molar mass of ATP is $505 \text{ g} \cdot \text{mol}^{-1}$. What is the mass of ATP in pounds? (2.2 kg = 1 lb)

(c) There is approximately 40 g of ATP in the adult 125-lb female. Considering this fact and your answer to part b, suggest an explanation that is consistent with these findings.

56. (a) How many apples (see Solution 48) would be required to provide the amount of ATP calculated in Problem 55? **(b)** How many large hot chocolate drinks (see Solution 49) would be required?

57. (a) Which of the compounds listed in Table 12.4 could be involved in a reaction coupled to the synthesis of ATP from ADP + P_i?
(b) Which of the compounds listed in Table 12.4 could be involved in a reaction coupled to the hydrolysis of ATP to ADP + P_i?

58. Citrate is isomerized to isocitrate in the citric acid cycle (Chapter 14). The reaction is catalyzed by the enzyme aconitase. The $\Delta G^{\circ\prime}$ of the reaction is $5 \text{ kJ} \cdot \text{mol}^{-1}$. The properties of the reaction are studied *in vitro*, where 1 M citrate and 1 M isocitrate are added to an aqueous solution of the enzyme at 25°C.

(a) What is the K_{eq} for the reaction?
(b) What are the equilibrium concentrations of the reactant and product?
(c) What is the preferred direction of the reaction under standard conditions?
(d) The aconitase reaction is the second step of an eight-step pathway and occurs in the direction shown in the figure. How can you reconcile these facts with your answer to part (c)?

Citrate — aconitase → Isocitrate

59. The equilibrium constant for the conversion of glucose-6-phosphate (G6P) to fructose-6-phosphate (F6P) is 0.41. The reaction is reversible and is catalyzed by the enzyme phosphoglucose isomerase.

Glucose-6-phosphate — phosphoglucose isomerase → Fructose-6-phosphate

(a) What is the $\Delta G^{\circ\prime}$ for this reaction? Would this reaction proceed in the direction written under standard conditions?

(b) What is the ΔG for this reaction at 37°C when the concentration of glucose-6-phosphate is 1.0 mM and the concentration of fructose-6-phosphate is 0.3 mM? Would the reaction proceed in the direction written under these cellular conditions?

60. The conversion of glutamate to glutamine is unfavorable. In order for this transformation to occur in the cell, it must be coupled to the hydrolysis of ATP. Consider two possible mechanisms:

Mechanism I: glutamate + $NH_3 \rightleftharpoons$ glutamine

$$ATP + H_2O \rightleftharpoons ADP + P_i$$

Mechanism II: glutamate + $ATP \rightleftharpoons \gamma$-glutamylphosphate + ADP

$$\gamma\text{-glutamylphosphate} + H_2O + NH_3 \rightleftharpoons \text{glutamine} + P_i$$

Write the overall equation for the reaction for each mechanism. Is one mechanism more likely than the other? Or are both mechanisms equally feasible for the conversion of glutamate to glutamine? Explain.

61. The phosphorylation of glucose to glucose-6-phosphate, the first step of glycolysis (Chapter 13), can be described by the equation glucose + $P_i \rightleftharpoons$ glucose-6-phosphate + H_2O.

(a) Calculate the equilibrium constant for this reaction.
(b) What would the equilibrium concentration of glucose-6-phosphate (G6P) be under cellular conditions (both glucose and phosphate concentrations are 5 mM) if glucose was phosphorylated according to this reaction? Does this reaction provide a feasible route for producing glucose-6-phosphate for glycolysis?
(c) One way to increase the amount of product is to increase the concentrations of the reactants. This would decrease the mass action ratio (see Equation 12.3) and would theoretically make the reaction as written more favorable. What concentration of glucose would be required to achieve a glucose-6-phosphate concentration of 250 μM? Is this strategy physiologically feasible, given that the solubility of glucose in aqueous medium is less than 1 M?
(d) Another way to promote the formation of glucose-6-phosphate from glucose is to couple the phosphorylation of glucose to the hydrolysis of ATP as shown in Section 12.3. Calculate K_{eq} for the reaction in which glucose is converted to glucose-6-phosphate with concomitant ATP hydrolysis.
(e) When the ATP-dependent phosphorylation of glucose is carried out, what concentration of glucose is needed to achieve a 250 μM intracellular concentration of glucose-6-phosphate when the concentrations of ATP and ADP are 5.0 mM and 1.25 mM, respectively?
(f) Which route is more feasible to accomplish the phosphorylation of glucose to glucose-6-phosphate: the direct phosphorylation by P_i or the coupling of this phosphorylation to ATP hydrolysis? Explain.

62. Fructose-6-phosphate is phosphorylated to fructose-1,6-bisphosphate in the third step of the glycolytic pathway (Chapter 14). The phosphorylation of fructose-6-phosphate is described by the following equation:

$$\text{fructose-6-phosphate} + P_i \rightleftharpoons \text{fructose-1,6-bisphosphate}$$

$$\Delta G^{\circ\prime} = 47.7 \text{ kJ} \cdot \text{mol}^{-1}$$

(a) What is the ratio of fructose-1,6-bisphosphate (F16BP) to fructose-6-phosphate (F6P) at equilibrium if the concentration of phosphate in the cell is 5 mM?
(b) Suppose that the phosphorylation of fructose-6-phosphate is coupled to the hydrolysis of ATP. Write the new equation that describes this process and calculate $\Delta G^{\circ\prime}$ for the reaction.
(c) What is the ratio of fructose-1,6-bisphosphate to fructose-6-phosphate at equilibrium for the reaction you wrote in part b if the equilibrium concentration of ATP = 3 mM and [ADP] = 1 mM?
(d) Compare your answers to part (a) and (c). What is the likely path for the cellular synthesis of fructose-1,6-bisphosphate?

(e) One can envision two mechanisms for coupling ATP hydrolysis to the phosphorylation of fructose-6-phosphate (F6P) to fructose-1,6-bisphosphate (F16BP), yielding the same overall reaction:

Mechanism I: ATP is hydrolyzed as F6P is transformed to F16BP:

$$F6P + P_i \rightleftharpoons F16BP$$
$$ATP + H_2O \rightleftharpoons ADP + P_i$$

Mechanism II: ATP transfers its γ phosphate directly to F6P in one step, producing F16BP.

$$F6P + ATP + H_2O \rightleftharpoons F16BP + ADP$$

Choose one of the above mechanisms as the more biochemically feasible and provide a rationale for your choice.

63. Glyceraldehyde-3-phosphate (GAP) is eventually converted to 3-phosphoglycerate (3PG) in the glycolytic pathway.

Glyceraldehyde-3-phosphate (GAP)
3-Phosphoglycerate (3PG)
1,3-Bisphosphoglycerate (1,3-BPG)

Consider these two scenarios:

I. GAP is oxidized to 1,3-BPG ($\Delta G^{\circ\prime} = 6.7$ kJ·mol^{-1}), which is subsequently hydrolyzed to yield 3PG ($\Delta G^{\circ\prime} = -49.3$ kJ·mol^{-1})

II. GAP is oxidized to 1,3BPG, which then transfers its phosphate to ADP, yielding ATP ($\Delta G^{\circ\prime} = -18.8$ kJ·mol^{-1}).

Write the overall equations for the two scenarios. Which is more likely to occur in the cell, and why?

64. Palmitate is activated in the cell by forming a thioester bond to coenzyme A. The $\Delta G^{\circ\prime}$ for the synthesis of palmitoyl-CoA from palmitate and coenzyme A is 31.5 kJ·mol^{-1}.

(a) What is the ratio of products to reactants at equilibrium for the reaction? Is the reaction favorable? Explain.
(b) Suppose the synthesis of palmitoyl-CoA were coupled with ATP hydrolysis. Write the new equation for the activation of palmitate when coupled with ATP hydrolysis to ADP. Calculate $\Delta G^{\circ\prime}$ for the reaction. What is the ratio of products to reactants at equilibrium for the reaction? Is the reaction favorable? Compare your answer to the answer you obtained in part (a).
(c) Suppose the reaction described in part a were coupled with ATP hydrolysis to AMP. Write the new equation for the activation of palmitate when coupled with ATP hydrolysis to AMP. Calculate $\Delta G^{\circ\prime}$ for the reaction. What is the ratio of products to reactants at equilibrium for the reaction? Is the reaction favorable? Compare your answer to the answer you obtained in part (b).
(d) Pyrophosphate, PP$_i$, is hydrolyzed to 2 P$_i$. The activation of palmitate, as described in part c, is coupled to the hydrolysis of pyrophosphate. Write the equation for this coupled reaction and calculate $\Delta G^{\circ\prime}$. What is the ratio of products to reactants at equilibrium for the reaction? Is the reaction favorable? Compare your answer to the answers you obtained in parts (b) and (c).

65. DNA containing broken phosphodiester bonds ("nicks") can be repaired by the action of a ligase enzyme. Formation of a new phosphodiester bond in DNA requires the free energy of ATP, which is hydrolyzed to AMP:

$$ATP + \text{nicked bond} \xrightarrow{\text{ligase}} AMP + PP_i + \text{phosphodiester bond}$$

The equilibrium constant expression for this reaction can be rearranged to define a constant, C, as follows:

$$K_{eq} = \frac{[\text{phosphodiester bond}][AMP][PP_i]}{[\text{nick}][ATP]}$$

$$\frac{[\text{nick}]}{[\text{phosphodiester bond}]} = \frac{[AMP][PP_i]}{K_{eq}[ATP]}$$

$$C = \frac{[PP_i]}{K_{eq}[ATP]}$$

$$\frac{[\text{nick}]}{[\text{phosphodiester bond}]} = C[AMP]$$

(a) The ratio of nicked bonds to phosphodiester bonds at various concentrations of AMP was determined. Using the data provided, construct a plot of [nick]/[phosphodiester bond] versus [AMP] and determine the value of C from the plot.

[AMP] (mM)	[Nick]/[Phosphodiester bond]
10	4.01×10^{-5}
20	5.47×10^{-5}
30	8.67×10^{-5}
40	9.30×10^{-5}
50	1.13×10^{-4}

(b) Determine the value of K_{eq} for the reaction in which the concentrations of PP$_i$ and ATP are 1.0 mM and 14 μM, respectively.
(c) What is the value of $\Delta G^{\circ\prime}$ for the reaction?
(d) What is the value of $\Delta G^{\circ\prime}$ for nicked bond → phosphodiester bond reaction? (*Note*: The $\Delta G^{\circ\prime}$ for the hydrolysis of ATP to AMP and PP$_i$ is -48.5 kJ·mol^{-1} under the conditions used in the experiment.)
(e) The $\Delta G^{\circ\prime}$ for the hydrolysis of a typical phosphomonoester bond is -13.8 kJ·mol^{-1}. Compare the stability of the phosphodiester bond in DNA to the stability of a typical phosphomonoester bond.

SELECTED READINGS

Falkowski, P. G., Fenchel, T., and Delong, E. F., The microbial engines that drive Earth's biogeochemical cycles, *Science* **320**, 1034–1038 (2008). [Discusses the diversity and interconnectedness of metabolic processes.]

Hanson, R. W., The role of ATP in metabolism, *Biochem. Ed.* **17**, 86–92 (1989). [Provides an excellent explanation of why ATP is an energy transducer rather than an energy store.]

Kim, M.-S. et al., A draft map of the human proteome, *Nature* **509**, 575–581 (2014). [Describes a dataset of 17,000 proteins representing 30 different human tissues.]

Smolin, L. A. and Grosvenor, M. B, *Nutrition: Science and Applications,* Wiley (2013). [Includes health-related information on macromolecular nutrients and vitamins.]

Wishart, D. S., Jewison, T., Guo, A. C., Wilson, M., Knox, C., et al., HMDB 3.0—the human metabolome database in 2013. *Nuc. Acids Res.* **41(D1),** D801–807. doi: 10.1093/nar/gks1065 (2013). [Describes the human metabolome database, with over 40,000 entries. Available at www.hmdb.ca.]

CHAPTER 13

GLUCOSE METABOLISM

Dmytro Pylypenko/shutterstock

Weddell seal pups are born with huge brains—75% of their adult size, compared to 25% for a human newborn—presumably to help them quickly learn to navigate under ice. Along with this extraordinarily large brain comes a high demand for glucose, the brain's primary fuel, which must be supplied by the mother seal's milk.

DO YOU REMEMBER?

- Enzymes accelerate chemical reactions using acid–base catalysis, covalent catalysis, and metal ion catalysis (Section 6.2).

- Glucose polymers include the fuel-storage polysaccharides starch and glycogen and the structural polysaccharide cellulose (Section 11.2).

- Coenzymes such as NAD$^+$ and ubiquinone collect electrons from compounds that become oxidized (Section 12.2).

- A reaction with a large negative change in free energy can be coupled to another unfavorable reaction (Section 12.3).

- A reaction that breaks a phosphoanhydride bond in ATP occurs with a large change in free energy (Section 12.3).

- Nonequilibrium reactions often serve as metabolic control points (Section 12.3).

Glucose occupies a central position in the metabolism of most cells. It is a major source of metabolic energy (in some cells, it is the only source), and it provides the precursors for the synthesis of other biomolecules. Recall that glucose is stored in polymeric form as starch in plants and as glycogen in animals (Section 11.2). The breakdown of these polymers provides glucose monomers that can be catabolized to release energy. In this chapter, we will examine the major metabolic pathways involving glucose, including the interconversion of the monosaccharide glucose with glycogen, the degradation of glucose to the three-carbon intermediate pyruvate, the synthesis of glucose from smaller compounds, and the conversion of glucose to the five-carbon monosaccharide ribose. For all the pathways, we will present the intermediates and some of the relevant enzymes. We will also examine the thermodynamics of reactions that release or consume large amounts of free energy and discuss how some of these reactions are regulated.

GLYCOLYSIS

Pathways dealing with carbohydrates, highlighted in **Figure 13.1**, are part of the larger metabolic scheme introduced in Figure 12.11. **Glycolysis,** the conversion of glucose to pyruvate, is a good place to begin a study of metabolic pathways. As a result of many years of research, we know a great deal about the pathway's intermediates and the enzymes that mediate their chemical transformations. We have also learned that glycolysis, along with other metabolic pathways, exhibits the following properties:

1. Each step of the pathway is catalyzed by a distinct enzyme.
2. The free energy consumed or released in certain reactions is transferred by molecules such as ATP and NADH.
3. The rate of the pathway can be controlled by altering the activity of individual enzymes.

If metabolic processes did not occur via multiple enzyme-catalyzed steps, cells would have little control over the amount and type of reaction products and no way to manage free energy. For example, the combustion of glucose and O_2 to CO_2 and H_2O— if allowed to occur in one grand explosion—would release about $2850 \text{ kJ} \cdot \text{mol}^{-1}$ of free energy all at once. In the cell, *glucose combustion requires many steps so that the cell can recover its free energy in smaller, more useful quantities.*

Glycolysis, representing the first ten steps of this process, appears to be an ancient metabolic pathway. The fact that it does not require molecular oxygen suggests that it evolved before photosynthesis increased the level of atmospheric O_2. Overall, glycolysis is a series of enzyme-catalyzed steps in which a six-carbon glucose molecule is broken down into two three-carbon pyruvate molecules. This catabolic pathway is accompanied by the phosphorylation of two molecules of ADP (to produce 2 ATP) and the reduction of two molecules of NAD^+. The net equation for the pathway (ignoring water and protons) is

$$\text{glucose} + 2\text{ NAD}^+ + 2\text{ ADP} + 2\text{ P}_i \rightarrow 2\text{ pyruvate} + 2\text{ NADH} + 2\text{ ATP}$$

It is convenient to divide the 10 reactions of glycolysis into two phases. In the first (Reactions 1–5), the hexose is phosphorylated and cleaved in half. In the second (Reactions 6–10), the three-carbon molecules are converted to pyruvate (**Fig. 13.2**).

As you examine each of the reactions of glycolysis described in the following pages, note how the reaction substrates are converted to products by the action of an enzyme (and note how the enzyme's name often reveals its purpose). Pay attention also to the free energy change of each reaction.

REACTIONS 1–5 ARE THE ENERGY-INVESTMENT PHASE OF GLYCOLYSIS

The first five reactions of glycolysis can be considered a preparatory phase for the second, energy-producing phase. In fact, the first phase requires the *investment* of free energy in the form of two ATP molecules.

FIGURE 13.1

Glucose metabolism in context. (1) The polysaccharide glycogen is degraded to glucose, which is then catabolized by the glycolytic pathway (2) to the three-carbon intermediate pyruvate. Gluconeogenesis (3) is the pathway for the synthesis of glucose from smaller precursors. Glucose can then be reincorporated into glycogen (4). The conversion of glucose to ribose, a component of nucleotides, is not shown in this diagram.

1. Hexokinase In the first step of glycolysis, the enzyme hexokinase transfers a phosphoryl group from ATP to the C6 OH group of glucose to form glucose-6-phosphate:

$$\text{Glucose} + \text{ATP} \xrightarrow{\text{hexokinase}} \text{Glucose-6-phosphate} + \text{ADP}$$

Glucose

ATP
1 hexokinase
ADP

Glucose-6-phosphate

2 phosphoglucose isomerase

Fructose-6-phosphate

ATP
3 phosphofructokinase
ADP

Fructose-1,6-bisphosphate

4 aldolase

Glyceraldehyde-3-phosphate + Dihydroxyacetone phosphate

5 triose phosphate isomerase

$2 P_i + 2 NAD^+$
6 glyceraldehyde-3-phosphate dehydrogenase
$2 NADH + 2 H^+$

2 1,3-Bisphosphoglycerate

2 ADP
7 phosphoglycerate kinase
2 ATP

2 3-Phosphoglycerate

8 phosphoglycerate mutase

2 2-Phosphoglycerate

9 enolase
$2 H_2O$

2 Phosphoenolpyruvate

2 ADP
10 pyruvate kinase
2 ATP

2 Pyruvate

Energy Investment Phase

Energy Payoff Phase

FIGURE 13.2

The reactions of glycolysis.
The substrates, products, and enzymes corresponding to the 10 steps of the pathway are shown. Shading indicates the substrates (purple) and products (green) of the pathway as a whole.

? Next to each reaction, write the term that describes the type of chemical change that occurs.

331

A **kinase** is an enzyme that transfers a phosphoryl group from ATP (or another nucleoside triphosphate) to another substance.

Recall from Section 6.3 that the hexokinase active site closes around its substrates so that a phosphoryl group is efficiently transferred from ATP to glucose. The standard free energy change for this reaction, which cleaves one of ATP's phosphoanhydride bonds, is -16.7 kJ·mol^{-1} (ΔG, the actual free energy change for the reaction inside a cell, has a similar value). The magnitude of this free energy change means that the reaction proceeds in only one direction; the reverse reaction is extremely unlikely since its standard free energy change would be $+16.7$ kJ·mol^{-1}. Consequently, hexokinase is said to catalyze a **metabolically irreversible reaction** that prevents glucose from backing out of glycolysis. *Many metabolic pathways have a similar irreversible step near the start that commits a metabolite to proceed through the pathway.*

2. Phosphoglucose Isomerase
The second reaction of glycolysis is an isomerization reaction in which glucose-6-phosphate is converted to fructose-6-phosphate:

Glucose-6-phosphate   phosphoglucose isomerase   Fructose-6-phosphate

Because fructose is a six-carbon ketose (Section 11.1), it forms a five-membered ring.

The standard free energy change for the phosphoglucose isomerase reaction is $+2.2$ kJ·mol^{-1}, but the reactant concentrations *in vivo* yield a ΔG value of about -1.4 kJ·mol^{-1}. A value of ΔG near zero indicates that the reaction operates close to equilibrium (at equilibrium, $\Delta G = 0$). *Such **near-equilibrium reactions** are considered to be freely reversible, since a slight excess of products can easily drive the reaction in reverse by mass action effects.* In a metabolically irreversible reaction, such as the hexokinase reaction, the concentration of product could never increase enough to compensate for the reaction's large value of ΔG.

3. Phosphofructokinase
The third reaction of glycolysis consumes a second ATP molecule in the phosphorylation of fructose-6-phosphate to yield fructose-1,6-bisphosphate.

Fructose-6-phosphate   phosphofructokinase   Fructose-1,6-bisphosphate

Phosphofructokinase operates in much the same way as hexokinase, and the reaction it catalyzes is irreversible, with a $\Delta G°'$ value of -17.2 kJ·mol^{-1}.

In cells, the activity of phosphofructokinase is regulated. We have already seen how the activity of a bacterial phosphofructokinase responds to allosteric effectors (Section 7.3). ADP binds to the enzyme and causes a conformational change that promotes fructose-6-phosphate binding, which in turn promotes catalysis. This mechanism is useful because the concentration of ADP in the cell is a good indicator of the need for ATP, which is a product of glycolysis. Phosphoenolpyruvate, the product of step 9 of glycolysis, binds to bacterial phosphofructokinase and causes it to assume a conformation that destabilizes fructose-6-phosphate binding, thereby diminishing catalytic activity. Thus, when the glycolytic pathway is producing plenty of phosphoenolpyruvate and ATP, the phosphoenolpyruvate can act as a feedback inhibitor to slow the pathway by decreasing the rate of the reaction catalyzed by phosphofructokinase (**Fig. 13.3a**). Citrate, an intermediate of the citric acid cycle (which completes glucose catabolism), is also a feedback inhibitor of phosphofructokinase.

(a)

Glucose

↓

Fructose-6-phosphate

Fructose-1,6-bisphosphate

↓

Phosphoenolpyruvate

↓

Pyruvate

ATP → P$_i$ → ADP

(b)

Glucose

↓

Fructose-6-phosphate → Fructose-2,6-bisphosphate

↓

Fructose-1,6-bisphosphate

↓

Pyruvate

FIGURE 13.3

Regulation of phosphofructokinase.
(a) Regulation in bacteria. ADP, produced when ATP is consumed elsewhere in the cell, stimulates the activity of phosphofructokinase (green arrow). Phosphoenolpyruvate, a late intermediate of glycolysis, inhibits phosphofructokinase (red symbol), thereby decreasing the rate of the entire pathway. (b) Regulation in mammals.

The most potent activator of phosphofructokinase in mammals is the compound fructose-2,6-bisphosphate, which is synthesized from fructose-6-phosphate by an enzyme known as phosphofructokinase-2. (The glycolytic enzyme is therefore sometimes called phosphofructokinase-1.)

Fructose-6-phosphate → (ATP → ADP, phosphofructokinase-2) → Fructose-2,6-bisphosphate

The activity of phosphofructokinase-2 is hormonally stimulated when the concentration of glucose in the blood is high. The resulting increase in fructose-2,6-bisphosphate concentration activates phosphofructokinase to increase the flux (rate of flow) of glucose through the glycolytic pathway (Fig. 13.3b).

The phosphofructokinase reaction is the primary control point for glycolysis. It is the slowest reaction of glycolysis, so the rate of this reaction largely determines the flux of glucose through the entire pathway. In general, a **rate-determining reaction**—such as the phosphofructokinase reaction—operates far from equilibrium; that is, it has a large negative free energy change and is irreversible under metabolic conditions. The rate of the reaction can be altered by allosteric effectors but not by fluctuations in the concentrations of its substrates or products. Thus, it acts as a one-way valve. In contrast, a near-equilibrium reaction—such as the phosphoglucose isomerase reaction—cannot serve as a rate-determining step for a pathway because it can respond to small changes in reactant concentrations by operating in reverse.

4. Aldolase Reaction 4 converts the hexose fructose-1,6-bisphosphate to two three-carbon molecules, each of which bears a phosphate group.

$$
\begin{array}{l}
\overset{1}{C}H_2OPO_3^{2-} \\
\overset{2}{C}=O \\
HO-\overset{3}{C}-H \\
H-\overset{4}{C}-OH \\
H-\overset{5}{C}-OH \\
\overset{6}{C}H_2OPO_3^{2-}
\end{array}
\quad \underset{\text{aldolase}}{\rightleftarrows} \quad
\begin{array}{l}
\overset{3}{C}H_2OPO_3^{2-} \\
\overset{2}{C}=O \\
HO-\overset{1}{C}H_2
\end{array}
\quad + \quad
\begin{array}{l}
O=\overset{1}{C}-H \\
H-\overset{2}{C}-OH \\
\overset{3}{C}H_2OPO_3^{2-}
\end{array}
$$

Fructose-1,6-bisphosphate

Dihydroxyacetone phosphate

Glyceraldehyde-3-phosphate

This reaction is the reverse of an aldol (aldehyde–alcohol) condensation, so the enzyme that catalyzes the reaction is called aldolase. It is worth examining its mechanism. The active site of mammalian aldolase contains two catalytically important residues: a lysine residue that forms a Schiff base (imine) with the substrate and an ionized aspartate residue that acts as a base catalyst (**Fig. 13.4**). (Bacterial aldolase uses lysine and an ionized tyrosine residue.)

1. Fructose-1,6-bisphosphate, shown in its linear form, binds to the enzyme.

2. The active-site Lys residue reacts with the substrate's carbonyl group to form a Schiff base.

3. The Asp side chain, acting as a base, accepts a proton from the substrate, and the bond between C3 and C4 breaks, releasing the first product, an aldehyde. The reaction is facilitated by the protonated Schiff base, which is a better electron-withdrawing group than the carbonyl group that was originally at C2.

4. The Asp residue gives up its proton to the remainder of the substrate, re-forming a Schiff base.

5. The Schiff base is hydrolyzed, releasing the second product and regenerating the enzyme.

Glyceraldehyde-3-phosphate

FIGURE 13.4

The aldolase reaction.

❓ Does this reaction follow an ordered or ping pong mechanism (see Section 7.2)?

Early studies of aldolase implicated a cysteine residue in catalysis because iodoacetate, a reagent that reacts with the cysteine side chain, also inactivates the enzyme:

$$
\begin{array}{c}
\text{C}{=}\text{O} \\
| \\
\text{HC}-\text{CH}_2-\text{SH} \; + \; \text{ICH}_2\text{COO}^- \; \xrightarrow{\;\;HI\;\;} \; \text{HC}-\text{CH}_2-\text{SCH}_2\text{COO}^- \\
| \\
\text{NH} \qquad\qquad \text{Iodoacetate} \\
| \\
\text{Cys}
\end{array}
$$

Researchers used iodoacetate to help identify the intermediates of glycolysis: In the presence of iodoacetate, fructose-1,6-bisphosphate accumulates because the next step is blocked. Acetylation of the cysteine residue, which turned out not to be part of the active site, probably interferes with a conformational change that is necessary for aldolase activity.

The $\Delta G^{\circ\prime}$ value for the aldolase reaction is $+22.8 \text{ kJ} \cdot \text{mol}^{-1}$, indicating that the reaction is unfavorable under standard conditions. However, the reaction proceeds *in vivo* (ΔG is actually less than zero) because the products of the reaction are quickly whisked away by subsequent reactions. In essence, the rapid consumption of glyceraldehyde-3-phosphate and dihydroxyacetone phosphate "pulls" the aldolase reaction forward.

5. Triose Phosphate Isomerase The products of the aldolase reaction are both phosphorylated three-carbon compounds, but only one of them—glyceraldehyde-3-phosphate—proceeds through the remainder of the pathway. Dihydroxyacetone phosphate is converted to glyceraldehyde-3-phosphate by triose phosphate isomerase:

$$
\begin{array}{ccc}
\begin{array}{c}
\text{H} \\
| \\
\text{H}-\text{C}-\text{OH} \\
| \\
\text{C}{=}\text{O} \\
| \\
\text{CH}_2\text{OPO}_3^{2-}
\end{array}
& \underset{\text{triose phosphate isomerase}}{\rightleftharpoons} &
\begin{array}{c}
\text{O}{\diagdown}_{\text{C}}{\diagup}^{\text{H}} \\
| \\
\text{H}-\text{C}-\text{OH} \\
| \\
\text{CH}_2\text{OPO}_3^{2-}
\end{array} \\
\text{Dihydroxyacetone} & & \text{Glyceraldehyde-} \\
\text{phosphate} & & \text{3-phosphate}
\end{array}
$$

Triose phosphate isomerase was introduced in Section 7.2 as an example of a catalytically perfect enzyme, one whose rate is limited only by the rate at which its substrates can diffuse to its active site. The catalytic mechanism of triose phosphate isomerase may involve low-barrier hydrogen bonds (which also help stabilize the transition state in serine proteases; see Section 6.3). In addition, the catalytic power of triose phosphate isomerase depends on a protein loop that closes over the active site (**Fig. 13.5**).

(a) (b)

FIGURE 13.5

Conformational changes in yeast triose phosphate isomerase.
(a) One loop of the protein, comprising residues 166–176, is highlighted in green. (b) When a substrate binds to the enzyme, the loop closes over the active site to stabilize the reaction's transition state. In this model, the transition state analog 2-phosphoglycolate (orange) occupies the active site. Triose phosphate isomerase is actually a homodimer; only one subunit is pictured here. [*Structure of the enzyme alone (pdb 1YPI) determined by T. Alber, E. Lolis, and G. A. Petsko; structure of the enzyme with the analog (pdb 2YPI) determined by E. Lolis and G. A. Petsko.*]

The standard free energy change for the triose phosphate isomerase reaction is slightly positive, even under physiological conditions ($\Delta G°' = +7.9 \cdot kJ \cdot mol^{-1}$ and $\Delta G = +4.4 \, kJ \cdot mol^{-1}$), but the reaction proceeds because glyceraldehyde-3-phosphate is quickly consumed in the next reaction, so more dihydroxyacetone phosphate is constantly being converted to glyceraldehyde-3-phosphate.

REACTIONS 6–10 ARE THE ENERGY-PAYOFF PHASE OF GLYCOLYSIS

So far, the reactions of glycolysis have consumed 2 ATP, but this investment pays off in the second phase of glycolysis when 4 ATP are produced, for a net gain of 2 ATP. All of the reactions of the second phase involve three-carbon intermediates, but keep in mind that *each glucose molecule that enters the pathway yields two of these three-carbon units.*

Some species convert glucose to glyceraldehyde-3-phosphate by different pathways than the one presented above. However, the second phase of glycolysis, which converts glyceraldehyde-3-phosphate to pyruvate, is the same in all organisms. This suggests that glycolysis may have evolved from the "bottom up"; that is, it first evolved as a pathway for extracting free energy from abiotically produced small molecules, before cells developed the ability to synthesize larger molecules such as hexoses.

6. Glyceraldehyde-3-Phosphate Dehydrogenase In the sixth reaction of glycolysis, glyceraldehyde-3-phosphate is both oxidized and phosphorylated:

Glyceraldehyde-3-phosphate   1,3-Bisphosphoglycerate

Unlike the kinases that catalyze Reactions 1 and 3, glyceraldehyde-3-phosphate dehydrogenase does not use ATP as a phosphoryl-group donor; it adds inorganic phosphate to the substrate. This reaction is also an oxidation–reduction reaction in which the aldehyde group of glyceraldehyde-3-phosphate is oxidized and the cofactor NAD^+ is reduced to NADH. In effect, glyceraldehyde-3-phosphate dehydrogenase catalyzes the removal of an H atom (actually, a hydride ion); hence the name "dehydrogenase." Note that the reaction product NADH must eventually be reoxidized to NAD^+, or else glycolysis will come to a halt. In fact, the reoxidation of NADH, which is a form of "energy currency," generates ATP (Chapter 15).

An active-site cysteine residue participates in the glyceraldehyde-3-phosphate dehydrogenase reaction (**Fig. 13.6**). The enzyme is inhibited by arsenate (AsO_4^{3-}), which competes with P_i (PO_4^{3-}) for binding in the enzyme active site.

7. Phosphoglycerate Kinase The product of Reaction 6, 1,3-bisphosphoglycerate, is an acyl phosphate.

An acyl phosphate

FIGURE 13.6

The glyceraldehyde-3-phosphate dehydrogenase reaction.

❓ Identify the reactant that undergoes oxidation and the reactant that undergoes reduction.

The subsequent removal of its phosphoryl group releases a large amount of free energy in part because the reaction products are more stable (the same principle contributes to the large negative value of ΔG for reactions involving cleavage of ATP's phosphoanhydride bonds; see Section 12.3). *The free energy released in this reaction is used to drive the formation of ATP, as 1,3-bisphosphoglycerate donates its phosphoryl group to ADP:*

Note that the enzyme that catalyzes this reaction is called a kinase since it transfers a phosphoryl group between ATP and another molecule.

The standard free energy change for the phosphoglycerate kinase reaction is $-18.8 \text{ kJ} \cdot \text{mol}^{-1}$. This strongly exergonic reaction helps pull the glyceraldehyde-3-phosphate dehydrogenase reaction forward, since its standard free energy change is greater than zero ($\Delta G^{\circ\prime} = +6.7 \text{ kJ} \cdot \text{mol}^{-1}$). This pair of reactions provides a good example of the coupling of a thermodynamically favorable and unfavorable reaction so that both proceed with a net decrease in free energy: $-18.8 \text{ kJ} \cdot \text{mol}^{-1} + 6.7 \text{ kJ} \cdot \text{mol}^{-1} = -12.1 \text{ kJ} \cdot \text{mol}^{-1}$. Under physiological conditions, ΔG for the paired reactions is close to zero.

8. Phosphoglycerate Mutase

In the next reaction, 3-phosphoglycerate is converted to 2-phosphoglycerate:

3-Phosphoglycerate   phosphoglycerate mutase   2-Phosphoglycerate

Although the reaction appears to involve the simple intramolecular transfer of a phosphoryl group, the reaction mechanism is a bit more complicated and requires an enzyme active site that contains a phosphorylated histidine residue. The phospho-His transfers its phosphoryl group to 3-phosphoglycerate to generate 2,3-bisphosphoglycerate, which then gives a phosphoryl group back to the enzyme, leaving 2-phosphoglycerate and phospho-His:

3-Phosphoglycerate   2-Phosphoglycerate

As can be guessed from its mechanism, the phosphoglycerate mutase reaction is freely reversible *in vivo*.

9. Enolase

Enolase catalyzes a dehydration reaction, in which water is eliminated:

2-Phosphoglycerate   enolase   Phosphoenolpyruvate

The enzyme active site includes an Mg^{2+} ion that apparently coordinates with the OH group at C3 and makes it a better leaving group. Fluoride ion and P_i can form a complex with the Mg^{2+} and thereby inhibit the enzyme. In early studies demonstrating the inhibition of glycolysis by F^-, 2-phosphoglycerate, the substrate of enolase, accumulated. The concentration of 3-phosphoglycerate also increased in the presence of F^- since phosphoglycerate mutase readily converted the excess 2-phosphoglycerate back to 3-phosphoglycerate.

10. Pyruvate Kinase

The tenth reaction of glycolysis is catalyzed by pyruvate kinase, which converts phosphoenolpyruvate to pyruvate and transfers a phosphoryl group to ADP to produce ATP:

Phosphoenolpyruvate   + ADP   pyruvate kinase   Pyruvate   + ATP

The reaction actually occurs in two parts. First, ADP attacks the phosphoryl group of phosphoenolpyruvate to form ATP and enolpyruvate:

Phosphoenolpyruvate   ADP   Enolpyruvate

Enolpyruvate   Pyruvate

Removal of phosphoenolpyruvate's phosphoryl group is not a particularly exergonic reaction: When written as a hydrolytic reaction (transfer of the phosphoryl group to water), the $\Delta G^{\circ\prime}$ value is -16 kJ $\cdot$ mol^{-1}. This is not enough free energy to drive the synthesis of ATP from ADP + P$_i$ (this reaction requires $+30.5$ kJ $\cdot$ mol^{-1}). However, the second half of the pyruvate kinase reaction is highly exergonic. This is the **tautomerization** (isomerization through the shift of an H atom) of enolpyruvate to pyruvate. $\Delta G^{\circ\prime}$ for this step is -46 kJ $\cdot$ mol^{-1}, so $\Delta G^{\circ\prime}$ for the net reaction (hydrolysis of phosphoenolpyruvate followed by tautomerization of enolpyruvate to pyruvate) is -61.9 kJ $\cdot$ mol^{-1}, more than enough free energy to drive the synthesis of ATP.

Three of the ten reactions of glycolysis (the reactions catalyzed by hexokinase, phosphofructokinase, and pyruvate kinase) have large negative values of ΔG. In theory, any of these far-from-equilibrium reactions could serve as a flux-control point for the pathway. The other seven reactions function near equilibrium ($\Delta G \approx 0$) and can therefore accommodate flux in either direction. The free energy changes for the ten reactions of glycolysis are shown graphically in **Figure 13.7**.

We have already discussed the mechanisms for regulating phosphofructokinase activity, the major control point for glycolysis. Hexokinase also catalyzes an irreversible reaction and is subject to inhibition by its product, glucose-6-phosphate. However, hexokinase cannot be the sole control point for glycolysis because glucose can also enter the pathway as glucose-6-phosphate, bypassing the hexokinase reaction. Finally, it would not be efficient for the pyruvate kinase reaction to be the primary regulatory step for glycolysis because it occurs at the very end of the 10-step pathway. Even so, pyruvate kinase activity can be adjusted. In some organisms, fructose-1,6-bisphosphate activates pyruvate kinase at an allosteric site. This is an example of **feed-forward activation**: Once a monosaccharide has entered glycolysis, fructose-1,6-bisphosphate helps ensure rapid flux through the pathway.

FIGURE 13.7

Graphical representation of the free energy changes of glycolysis. Three steps have large negative values of ΔG; the remaining steps are near equilibrium ($\Delta G \approx 0$). The height of each step corresponds to its ΔG value in heart muscle, and the numbers correspond to glycolytic enzymes. Keep in mind that ΔG values vary slightly among tissues. [Data from Newsholme, E. A. and Start, C., Regulation in Metabolism, p. 97, Wiley (1973).]

The glycolytic pathway can be considered to be a sort of pipe, with an entrance for glucose (or glucose-6-phosphate) and an exit for pyruvate. The control points for the pathway are like one-way valves that prevent backflow. Between those control points, intermediates can move in either direction. The pipe never runs dry because the pyruvate gets used up and more glucose is always available. In addition, intermediates can enter or leave the pathway at any point. Even the simplest cells contain many copies of each glycolytic enzyme, acting on a pool of millions of substrate molecules, so their collective behavior is what we refer to when we discuss flux through the pathway.

To sum up the second phase of glycolysis: Glyceraldehyde-3-phosphate is converted to pyruvate with the synthesis of 2 ATP (in Reactions 7 and 10). Since each molecule of glucose yields two molecules of glyceraldehyde-3-phosphate, the reactions of the second phase of glycolysis must be doubled, so the yield is 4 ATP. Two molecules of ATP are invested in phase 1, bringing the net yield to 2 ATP produced per glucose molecule. Two NADH are also generated for each glucose molecule. Monosaccharides other than glucose are metabolized in a similar fashion to yield ATP (**Box 13.A**).

A CLOSER LOOK

Box 13.A Catabolism of Other Sugars

A typical human diet contains many carbohydrates other than glucose and its polymers. For example, lactose, a disaccharide of glucose and galactose, is present in milk and food derived from it (Section 11.2). Lactose is cleaved in the intestine by the enzyme lactase, and the two monosaccharides are absorbed, transported to the liver, and metabolized. Galactose undergoes phosphorylation and isomerization and enters the glycolytic pathway as glucose-6-phosphate, so its energy yield is the same as that of glucose.

Sucrose, the other major dietary disaccharide, is composed of glucose and fructose (Section 11.2); it is present in a variety of foods of plant origin. Like lactose, sucrose is hydrolyzed in the small intestine, and its components glucose and fructose are absorbed. The monosaccharide fructose is also present in many foods, particularly fruits and honey. It tastes sweeter than sucrose, is more soluble, and is inexpensive to produce in the form of high-fructose corn syrup—all of which make fructose attractive to the manufacturers of soft drinks and other processed foods. This is the primary reason why the consumption of fructose in the United States rose dramatically from 1970 until about 2000 (it has been declining ever since).

Some researchers have proposed that the overconsumption of fructose is contributing to the obesity epidemic. One possible explanation relates to the catabolism of fructose, which differs somewhat from the catabolism of glucose. Fructose is metabolized primarily by the liver, but the form of hexokinase present in the liver (called glucokinase) has very low affinity for fructose. Fructose therefore enters glycolysis by a different route.

First, fructose is phosphorylated to yield fructose-1-phosphate. The enzyme fructose-1-phosphate aldolase then splits the six-carbon molecule into two three-carbon-molecules: glyceraldehyde and dihydroxyacetone phosphate (see diagram).

Dihydroxyacetone phosphate is converted to glyceraldehyde-3-phosphate by triose phosphate isomerase and can proceed through the second phase of glycolysis. The glyceraldehyde can be phosphorylated to glyceraldehyde-3-phosphate, but it can also be converted to glycerol-3-phosphate, a precursor of the backbone of triacylglycerols. This may contribute to an increase in fat deposition. A second potential hazard of the fructose catabolic pathway is that fructose catabolism bypasses the phosphofructokinase-catalyzed step of glycolysis, a major regulatory point, and thus may disrupt the entire body's fuel metabolism. However, fructose consumption has been decreasing for over 15 years, even while obesity rates rise. Furthermore, except in experimental settings, individuals typically consume about five times more glucose than fructose, so the actual fat-triggering culprit might simply be total sugar intake, not fructose in particular.

? **When its concentration is extremely high, fructose is converted to fructose-1-phosphate much faster than it can be cleaved by the aldolase. How would this affect the cell's ATP supply?**

Fructose — Fructose-1-phosphate — Glyceraldehyde — Dihydroxyacetone phosphate

FIGURE 13.8

Fates of pyruvate. Pyruvate may be converted to a two-carbon acetyl group linked to the carrier coenzyme A. Acetyl-CoA may be further broken down by the citric acid cycle or used to synthesize fatty acids. In muscle, pyruvate is reduced to lactate to regenerate NAD^+ for glycolysis. Yeast degrade pyruvate to ethanol and CO_2. Pyruvate can also be carboxylated to produce the four-carbon oxaloacetate.

? Beside each arrow, write the names of the enzymes that catalyze the process.

PYRUVATE IS CONVERTED TO OTHER SUBSTANCES

What happens to the pyruvate generated by the catabolism of glucose? It can be further broken down to acetyl-CoA or used to synthesize other compounds such as oxaloacetate. The fate of pyruvate depends on the cell type and the need for metabolic free energy and molecular building blocks. Some of the options are diagrammed in **Figure 13.8**.

During exercise, pyruvate may be temporarily converted to lactate. In a highly active muscle cell, glycolysis rapidly provides ATP to power muscle contraction, but the pathway also consumes NAD^+ at the glyceraldehyde-3-phosphate dehydrogenase step. The two NADH molecules generated for each glucose molecule catabolized can be reoxidized in the presence of oxygen. However, this process is too slow to replenish the NAD^+ needed for the rapid production of ATP by glycolysis. *To regenerate NAD^+, the enzyme lactate dehydrogenase reduces pyruvate to lactate:*

This reaction, sometimes called the eleventh step of glycolysis, allows the muscle to function anaerobically for a minute or so (see Box 12-C). The net reaction for anaerobic glucose catabolism is

$$\text{glucose} + 2\ \text{ADP} + 2\ \text{P}_i \rightarrow 2\ \text{lactate} + 2\ \text{ATP}$$

Lactate represents a sort of metabolic dead end: Its only options are to be eventually converted back to pyruvate (the lactate dehydrogenase reaction is reversible) or to be exported from the cell. The liver takes up lactate, oxidizes it back to pyruvate, and then converts the pyruvate to glucose. The fuel produced in this manner may eventually make its way back to the muscle to help power continued contraction. When the muscle is functioning aerobically, NADH produced by the glyceraldehyde-3-phosphate dehydrogenase reaction is reoxidized by oxygen and the lactate dehydrogenase reaction is not needed.

Organisms such as yeast growing under anaerobic conditions can regenerate NAD^+ by producing alcohol. In the mid-1800s, Louis Pasteur called this process **fermentation,** meaning life without air, although yeast also ferment sugars in the presence of O_2. Alcoholic fermentation is a two-step process. First, pyruvate decarboxylase (an enzyme not present in animals) catabolizes the removal of pyruvate's carboxylate group to produce acetaldehyde. Next, alcohol dehydrogenase reduces acetaldehyde to ethanol.

Pyruvate $\xrightarrow[\text{pyruvate decarboxylase}]{-CO_2}$ Acetaldehyde $\xrightarrow[\text{alcohol dehydrogenase}]{\text{NADH} \to \text{NAD}^+}$ Ethanol

Ethanol is considered to be a waste product of sugar metabolism; its accumulation is toxic to other organisms (**Box 13.B**), including the yeast that produce it. This is one reason why the alcohol content of yeast-fermented beverages such as wine is limited to about 13%. "Hard" liquor must be distilled to increase its ethanol content.

Although glycolysis is an oxidative pathway, its end product pyruvate is still a relatively reduced molecule. The further catabolism of pyruvate begins with its decarboxylation to form a two-carbon acetyl group linked to coenzyme A.

$$\text{CH}_3\text{—C(=O)—COO}^- \xrightarrow[\text{pyruvate dehydrogenase}]{\text{CoASH} \quad \text{CO}_2} \text{CH}_3\text{—C(=O)—SCoA}$$

Pyruvate → Acetyl-CoA

A CLOSER LOOK

Box 13.B Alcohol Metabolism

Unlike yeast, mammals do not produce ethanol, although it is naturally present in many foods and is produced in small amounts by intestinal microorganisms. The liver is equipped to metabolize ethanol, a small, weakly polar substance that is readily absorbed from the gastrointestinal tract and transported by the bloodstream. First, alcohol dehydrogenase converts ethanol to acetaldehyde. This is the reverse of the reaction yeast use to produce ethanol. A second reaction converts acetaldehyde to acetate:

Ethanol $\xrightarrow[\text{alcohol dehydrogenase}]{\text{NAD}^+ \quad \text{NADH, H}^+}$ Acetaldehyde

Acetaldehyde $\xrightarrow[\text{acetaldehyde dehydrogenase}]{\text{OH}^- \quad \text{NAD}^+ \quad \text{NADH, H}^+}$ Acetate

Note that both of these reactions require NAD^+, a cofactor used in many other oxidative cellular processes, including glycolysis. The liver uses the same two-enzyme pathway to metabolize the excess ethanol obtained from alcoholic beverages. Ethanol itself is mildly toxic, and the physiological effects of alcohol also reflect the toxicity of acetaldehyde and acetate in tissues such as the liver and brain.

Ethanol induces vasodilation, apparent as flushing (warming and reddening of the skin due to increased blood flow). At the same time, the heart rate and respiration rate become slightly lower. The kidneys increase the excretion of water as ethanol interferes with the ability of the hypothalamus (a region of the brain) to properly sense osmotic pressure.

Ethanol also stimulates signaling from certain neurotransmitter receptors that function as ligand-gated ion channels (Section 9.2) to inhibit neuronal signaling, producing a sedative effect. Sensory, motor, and cognitive functions are impaired, leading to delayed reaction time, loss of balance, and blurred vision. Some of the symptoms of ethanol intoxication can be experienced even at low doses, when the blood alcohol concentration is less than 0.05%. At high doses, usually at blood concentrations above 0.25%, ethanol can cause loss of consciousness, coma, and death.

The mostly pleasant responses to moderate ethanol consumption are followed by a period of recovery, when the concentrations of ethanol metabolites are relatively high. The unpleasant symptoms of a hangover in part reflect the chemistry of producing acetaldehyde and acetate. Their production in the liver consumes NAD^+, thereby lowering the cell's NAD^+:NADH ratio. Without sufficient NAD^+, the liver's ability to produce ATP by glycolysis is diminished (since NAD^+ is required for the glyceraldehyde-3-phosphate dehydrogenase reaction). Acetaldehyde itself can react with liver proteins, inactivating them. Acetate (acetic acid) production lowers blood pH.

Long-term, excessive alcohol consumption exacerbates the toxic effects of ethanol and its metabolites. For example, a shortage of liver NAD^+ slows fatty acid breakdown (like glycolysis, a process that requires NAD^+) and promotes fatty acid synthesis, leading to fat accumulation in the liver. Over time, cell death causes permanent loss of function in the central nervous system. The death of liver cells and their replacement by fibrous scar tissue causes liver cirrhosis.

? Normally, the liver converts lactate, produced mainly by muscles, back to pyruvate so that the pyruvate can be converted to glucose by gluconeogenesis (Section 13.2). How do the activities of alcohol dehydrogenase and acetaldehyde dehydrogenase contribute to hypoglycemia?

The resulting acetyl-CoA is a substrate for the citric acid cycle (Chapter 14), which converts the acetyl carbon atoms to CO_2. The complete oxidation of the six glucose carbons to CO_2 releases much more free energy than the conversion of glucose to lactate (**Table 13.1**). Much of this energy is recovered in the synthesis of ATP by the reactions of the citric acid cycle and oxidative phosphorylation (Chapter 15), pathways that require the presence of molecular oxygen.

Pyruvate is not always destined for catabolism. *Its carbon atoms provide the raw material for synthesizing a variety of molecules*, including, in the liver, more glucose (discussed in the following section). Fatty acids, the precursors of triacylglycerols and many membrane lipids, can be synthesized from the two-carbon units of acetyl-CoA derived from pyruvate. This is how fat is produced from excess carbohydrate.

Pyruvate is also the precursor of oxaloacetate, a four-carbon molecule that is an intermediate in the synthesis of several amino acids. It is also one of the intermediates of the citric acid cycle. Oxaloacetate is synthesized by the action of pyruvate carboxylase:

TABLE 13.1	Standard Free Energy Changes for Glucose Catabolism
CATABOLIC PROCESS	$\Delta G^{\circ\prime}$ $(kJ \cdot mol^{-1})$
$C_6H_{12}O_6 \rightarrow 2\ C_3H_5O_3^- + 2\ H^+$ (glucose)　(lactate)	-196
$C_6H_{12}O_6 + 6\ O_2 \rightarrow 6\ CO_2 + 6\ H_2O$ (glucose)	-2850

The pyruvate carboxylase reaction is interesting because of its unusual chemistry. The enzyme has a biotin prosthetic group that acts as a carrier of CO_2. Biotin is considered a vitamin, but a deficiency is rare because it is present in many foods and is synthesized by intestinal bacteria. The biotin group is covalently linked to an enzyme lysine residue:

The lysine side chain and its attached biotin group form a 14-Å-long flexible arm that swings between two active sites in the enzyme. In one active site, a CO_2 molecule is first "activated" by its reaction with ATP, then transferred to the biotin. The second active site transfers the carboxyl group to pyruvate to produce oxaloacetate (**Fig. 13.9**).

CONCEPT CHECK

1. Write the net equation for glycolysis.

2. Draw the structures of the substrates and products of the ten glycolytic reactions and name the enzyme that catalyzes each step.

3. List the glycolytic reactions that consume ATP or generate ATP.

4. Calculate the net yield of ATP and NADH per glucose molecule.

5. Identify the reactions that serve as flux-control points for glycolysis.

6. List the possible metabolic fates of pyruvate.

7. What is the metabolic function of lactate dehydrogenase?

1. *CO$_2$ (as bicarbonate, HCO$_3^-$) reacts with ATP such that some of the free energy released in the removal of ATP's phosphoryl group is conserved in the formation of the "activated" compound carboxyphosphate.*

Carboxyphosphate

2. *Like ATP, carboxyphosphate releases a large amount of free energy when its phosphoryl group is liberated. This free energy drives the carboxylation of biotin.*

P$_i$
Biotinyl-Lys

3. *The enzyme abstracts a proton from pyruvate, forming a carbanion.*

Pyruvate

4. *The carbanion attacks the carboxyl group attached to biotin, generating oxaloacetate.*

Biotinyl-Lys

Oxaloacetate

FIGURE 13.9

The pyruvate carboxylase reaction.

13.2 GLUCONEOGENESIS

KEY CONCEPTS

- Pyruvate is converted to glucose by glycolytic enzymes operating in reverse and by enzymes that bypass the irreversible steps of glycolysis.
- Gluconeogenic flux is regulated primarily by fructose-2,6-bisphosphate.

We have already alluded to the ability of the liver to synthesize glucose from non-carbohydrate precursors via the pathway of **gluconeogenesis**. This pathway, which also occurs to a limited extent in the kidneys, operates when the liver's supply of glycogen is exhausted. Certain tissues, such as the central nervous system and red blood cells, which burn glucose as their primary metabolic fuel, rely on the liver to supply them with newly synthesized glucose.

Gluconeogenesis is considered to be the reversal of glycolysis, that is, the conversion of two molecules of pyruvate to one molecule of glucose. Although some of the steps of gluconeogenesis are catalyzed by glycolytic enzymes operating in reverse, the gluconeogenic pathway contains several unique enzymes that bypass the three irreversible steps of glycolysis—the steps catalyzed by pyruvate kinase, phosphofructokinase, and hexokinase (**Fig. 13.10**). This principle applies to all pairs of opposing metabolic pathways: *The pathways may share some near-equilibrium reactions but cannot use the same enzymes to catalyze thermodynamically favorable irreversible reactions.* The three irreversible reactions of glycolysis are clearly visible in the "waterfall" diagram (see Fig. 13.7).

FOUR GLUCONEOGENIC ENZYMES PLUS SOME GLYCOLYTIC ENZYMES CONVERT PYRUVATE TO GLUCOSE

Pyruvate cannot be converted directly back to phosphoenolpyruvate because pyruvate kinase catalyzes an irreversible reaction (Reaction 10 of glycolysis). To get around this thermodynamic barrier, pyruvate is carboxylated by pyruvate carboxylase to yield the four-carbon compound oxaloacetate (the same reaction shown in Fig. 13.9). Next,

FIGURE 13.10

The reactions of gluconeogenesis. The pathway uses the seven glycolytic enzymes that catalyze reversible reactions. The three irreversible reactions of glycolysis are bypassed in gluconeogenesis by the four enzymes that are highlighted in blue.

? Compare the ATP yield of glycolysis with the ATP consumption of gluconeogenesis.

phosphoenolpyruvate carboxykinase catalyzes the decarboxylation of oxaloacetate to form phosphoenolpyruvate:

Note that the carboxylate group added in the first reaction is released in the second. The two reactions are energetically costly: Pyruvate carboxylase consumes ATP, and phosphoenolpyruvate carboxykinase consumes GTP (which is energetically equivalent to ATP). Cleavage of two phosphoanhydride bonds is required to supply enough free energy to "undo" the highly exergonic pyruvate kinase reaction.

Amino acids (except for leucine and lysine) are the main sources of gluconeogenic precursors because they can all be converted to oxaloacetate and then to phosphoenolpyruvate. Thus, during starvation, proteins can be broken down and used to produce glucose to fuel the central nervous system. In mammals, fatty acids cannot serve as gluconeogenic precursors because they cannot be converted to oxaloacetate. (However, the three-carbon glycerol backbone of triacylglycerols is a gluconeogenic precursor.)

Two molecules of phosphoenolpyruvate are converted to one molecule of fructose-1,6-bisphosphate in a series of six reactions that are all catalyzed by glycolytic enzymes (steps 4–9 in reverse order). These reactions are reversible because they are near equilibrium ($\Delta G \approx 0$), and the direction of flux is determined by the concentrations of substrates and products. Note that the phosphoglycerate kinase reaction consumes ATP when it operates in the direction of gluconeogenesis. NADH is also required to reverse the glyceraldehyde-3-phosphate dehydrogenase reaction.

The final three reactions of gluconeogenesis require two enzymes unique to this pathway. The first step undoes the phosphofructokinase reaction, the irreversible reaction that is the major control point of glycolysis. In gluconeogenesis, the enzyme fructose bisphosphatase hydrolyzes the C1 phosphate of fructose-1,6-bisphosphate to yield fructose-6-phosphate. This reaction is thermodynamically favorable, with a ΔG value of -8.6 kJ·mol^{-1}. Next, the glycolytic enzyme phosphoglucose isomerase catalyzes the reverse of step 2 of glycolysis to produce glucose-6-phosphate. Finally, the gluconeogenic enzyme glucose-6-phosphatase catalyzes a hydrolytic reaction that yields glucose and P$_i$. Note that the hydrolytic reactions catalyzed by fructose bisphosphatase and glucose-6-phosphatase undo the work of two kinases in glycolysis (phosphofructokinase and hexokinase).

GLUCONEOGENESIS IS REGULATED AT THE FRUCTOSE BISPHOSPHATASE STEP

Gluconeogenesis is energetically expensive. Producing 1 glucose from 2 pyruvate consumes 6 ATP, 2 each at the steps catalyzed by pyruvate carboxylase, phosphoenolpyruvate carboxykinase, and phosphoglycerate kinase. If glycolysis occurred simultaneously with gluconeogenesis, there would be a net consumption of ATP:

glycolysis	glucose + 2 ADP + 2 P$_i$ → 2 pyruvate + 2 ATP
gluconeogenesis	2 pyruvate + 6 ATP → glucose + 6 ADP + 6 P$_i$
net	4 ATP → 4 ADP + 4 P$_i$

To avoid this waste of metabolic free energy, gluconeogenic cells (mainly liver cells) carefully regulate the opposing pathways of glycolysis and gluconeogenesis according to the cell's energy needs. *The major regulatory point is centered on the interconversion of fructose-6-phosphate and fructose-1,6-bisphosphate.* We have already seen that fructose-2,6-bisphosphate is a potent allosteric activator of phosphofructokinase, which catalyzes step 3 of glycolysis. Not surprisingly, fructose-2,6-bisphosphate is a potent *inhibitor* of fructose bisphosphatase, which catalyzes the opposing gluconeogenic reaction.

This mode of allosteric regulation is efficient because *a single compound can control flux through two opposing pathways in a reciprocal fashion.* Thus, when the concentration of fructose-2,6-bisphosphate is high, glycolysis is stimulated and gluconeogenesis is inhibited, and vice versa.

Many cells that do not carry out gluconeogenesis do contain the gluconeogenic enzyme fructose bisphosphatase. What is the reason for this? When both fructose bisphosphatase (FBPase) and phosphofructokinase (PFK) are active, the net result is the hydrolysis of ATP:

PFK fructose-6-phosphate + ATP → fructose-1,6-bisphosphate + ADP

FBPase fructose-1,6-bisphosphate + H_2O → fructose-6-phosphate + P_i

net ATP + H_2O → ADP + P_i

This combination of metabolic reactions is called a **futile cycle** since it seems to have no useful result. However, Eric Newsholme realized that such futile cycles could actually provide a means for fine-tuning the output of a metabolic pathway. For example, flux through the phosphofructokinase step of glycolysis is diminished by the activity of fructose bisphosphatase. An allosteric compound such as fructose-2,6-bisphosphate modulates the activity of both enzymes so that as the activity of one enzyme increases, the activity of the other one decreases. This dual regulatory effect results in a greater possible range of net flux than if the regulator merely activated or inhibited a single enzyme. Similarly, a car's speed is easier to control if it has both an accelerator and a brake.

CONCEPT CHECK

1. List the reactions of gluconeogenesis that are catalyzed by glycolytic enzymes.
2. Why are some enzymes unique to gluconeogenesis?
3. Describe the fructose-6-phosphate futile cycle and explain its purpose.

GLYCOGEN SYNTHESIS AND DEGRADATION

In animals, dietary glucose and the glucose produced by gluconeogenesis are stored in the liver and other tissues as glycogen. Later, glucose units can be removed from the glycogen polymer by phosphorolysis (see Section 12.1). Because glycogen degradation is thermodynamically spontaneous, glycogen synthesis requires the input of free energy. The two opposing pathways use different sets of enzymes so that each process can be thermodynamically favorable under cellular conditions.

KEY CONCEPTS
- The substrate for glycogen synthase is UDP–glucose, whose production costs the free energy of one phosphoanhydride bond.
- Glycogen is phosphorolyzed to produce glucose that can exit the cell or be catabolized by glycolysis.

GLYCOGEN SYNTHESIS CONSUMES THE FREE ENERGY OF UTP

The monosaccharide unit that is incorporated into glycogen is glucose-1-phosphate, which is produced from glucose-6-phosphate (the penultimate product of gluconeogenesis) by the action of the enzyme phosphoglucomutase:

Glucose-6-phosphate ⇌ (phosphoglucomutase) Glucose-1-phosphate

In mammalian cells, glucose-1-phosphate is then "activated" by reacting with UTP to form UDP–glucose (like GTP, UTP is energetically equivalent to ATP).

Glucose-1-phosphate

UDP–glucose pyrophosphorylase → PP$_i$ → (inorganic pyrophosphatase) 2 P$_i$

UDP–glucose

This process is a reversible phosphoanhydride exchange reaction ($\Delta G \approx 0$). Note that the two phosphoanhydride bonds of UTP are conserved, one in the product PP$_i$ and one in UDP–glucose. However, the PP$_i$ is rapidly hydrolyzed by inorganic pyrophosphatase to 2 P$_i$ in a highly exergonic reaction ($\Delta G^{\circ\prime} = -19.2$ kJ · mol^{-1}). Thus, cleavage of a phosphoanhydride bond makes the formation of UDP–glucose an exergonic, irreversible process—that is, *PP$_i$ hydrolysis "drives" a reaction that would otherwise be near equilibrium.* The hydrolysis of PP$_i$ by inorganic pyrophosphatase is a common strategy in biosynthetic reactions;

we will see this reaction again in the synthesis of other polymers, namely DNA, RNA, and polypeptides.

Finally, glycogen synthase transfers the glucose unit to the C4 OH group at the end of one of glycogen's branches to extend the linear polymer of $\alpha(1 \rightarrow 4)$-linked residues.

A separate enzyme, called a transglycosylase or branching enzyme, cleaves off a seven-residue segment and reattaches it to a glucose C6 OH group to create an $\alpha(1 \rightarrow 6)$ branch point.

The steps of glycogen synthesis can be summarized as follows:

UDP–glucose pyrophosphorylase	glucose-1-phosphate + UTP $\rightleftharpoons$ UDP–glucose + PP$_i$
pyrophosphatase	PP$_i$ + H$_2$O $\rightarrow$ 2 P$_i$
glycogen synthase	UDP–glucose + glycogen (n residues) $\rightarrow$ glycogen (n +1 residues) + UDP
net	glucose-1-phosphate + glycogen + UTP + H$_2$O $\rightarrow$ glycogen + UDP + 2 P$_i$

The energetic price for adding one glucose unit to glycogen is the cleavage of one phospho-anhydride bond of UTP. Nucleotides are also required for the synthesis of other saccharides. For example, lactose is synthesized from glucose and UDP–galactose. In plants, starch is synthesized using ADP–glucose, and cellulose is synthesized using CDP–glucose as starting materials.

GLYCOGEN PHOSPHORYLASE CATALYZES GLYCOGENOLYSIS

Glycogen breakdown follows a different set of steps than glycogen synthesis. *In glycogenolysis, glycogen is phosphorolyzed, not hydrolyzed, to yield glucose-1-phosphate.* However, a debranching enzyme can remove $\alpha(1 \rightarrow 6)$-linked residues by hydrolysis. In the liver, phosphoglucomutase converts glucose-1-phosphate to glucose-6-phosphate, which is then hydrolyzed by glucose-6-phosphatase to release free glucose.

glycogen $\xrightarrow[\substack{\text{glycogen} \\ \text{phosphorylase}}]{P_i}$ glucose-1-phosphate $\xrightleftharpoons[\substack{\text{phosphogluco-} \\ \text{mutase}}]{}$ glucose-6-phosphate $\xrightarrow[\substack{\text{glucose-6-} \\ \text{phosphatase}}]{H_2O \quad P_i}$ glucose

This glucose leaves the cell and enters the bloodstream. Only gluconeogenic tissues such as the liver can make glucose available to the body at large. Other tissues that store glycogen, such as muscle, lack glucose-6-phosphatase and so break down glycogen only for their own needs. In these tissues, the glucose-1-phosphate liberated by phosphorolysis of glycogen is converted to glucose-6-phosphate, which then enters glycolysis at the phosphoglucose isomerase reaction (step 2). The hexokinase reaction (step 1) is skipped, thereby sparing the consumption of ATP. Consequently, *glycolysis using glycogen-derived glucose has a higher net yield of ATP than glycolysis using glucose supplied by the bloodstream.*

Because the mobilization of glucose must be tailored to meet the energy demands of a particular tissue or the entire body, the activity of glycogen phosphorylase is carefully regulated by a variety of mechanisms linked to hormonal signaling. Likewise, the activity of glycogen synthase is subject to hormonal control. In Chapter 19 we will examine some of the mechanisms for regulating different aspects of fuel metabolism, including glycogen synthesis and degradation. Some disorders of glycogen metabolism are discussed in Section 13.5.

CONCEPT CHECK

1. Describe the role of UTP in glycogen synthesis.

2. What is the advantage of breaking down glycogen by phosphorolysis rather than hydrolysis?

3. Explain why only some tissues contain glucose-6-phosphatase.

THE PENTOSE PHOSPHATE PATHWAY

We have already seen that glucose catabolism can lead to pyruvate, which can be further oxidized to generate more ATP or used to synthesize amino acids and fatty acids. Glucose is also a precursor of the ribose groups used for nucleotide synthesis. The **pentose phosphate pathway,** which converts glucose-6-phosphate to ribose-5-phosphate, is an oxidative pathway that occurs in all cells. But unlike glycolysis, the pentose phosphate pathway generates NADPH rather than NADH. The two cofactors are not interchangeable and are easily distinguished by degradative enzymes (which generally use NAD^+) and biosynthetic enzymes (which generally use $NADP^+$). The pentose phosphate pathway is by no means a minor feature of glucose metabolism. As much as 30% of glucose in the liver may be catabolized by the pentose phosphate pathway. This pathway can be divided into two phases: a series of oxidative reactions followed by a series of reversible interconversion reactions.

THE OXIDATIVE REACTIONS OF THE PENTOSE PHOSPHATE PATHWAY PRODUCE NADPH

The starting point of the pentose phosphate pathway is glucose-6-phosphate, which can be derived from free glucose, from the glucose-1-phosphate produced by glycogen phosphorolysis, or from gluconeogenesis. In the first step of the pathway, glucose-6-phosphate dehydrogenase catalyzes the metabolically irreversible transfer of a hydride ion from glucose-6-phosphate to $NADP^+$, forming a lactone and NADPH:

Glucose-6-phosphate   6-Phosphoglucono-δ-lactone

The lactone intermediate is hydrolyzed to 6-phosphogluconate by the action of 6-phosphogluconolactonase, although this reaction can also occur in the absence of the enzyme:

6-Phosphoglucono-δ-lactone   6-Phosphogluconate

In the third step of the pentose phosphate pathway, 6-phosphogluconate is oxidatively decarboxylated in a reaction that converts the six-carbon sugar to a five-carbon sugar and reduces a second $NADP^+$ to NADPH:

6-Phosphogluconate   Ribulose-5-phosphate

The two molecules of NADPH produced for each glucose molecule that enters the pathway are used primarily for biosynthetic reactions, such as fatty acid synthesis and the synthesis of deoxynucleotides.

ISOMERIZATION AND INTERCONVERSION REACTIONS GENERATE A VARIETY OF MONOSACCHARIDES

The ribulose-5-phosphate product of the oxidative phase of the pentose phosphate pathway can isomerize to ribose-5-phosphate:

Ribulose-5-phosphate   Ribose-5-phosphate

Ribose-5-phosphate is the precursor of the ribose unit of nucleotides. In many cells, this marks the end of the pentose phosphate pathway, which has the net equation

$$\text{glucose-6-phosphate} + 2\,NADP^+ + H_2O \rightarrow \text{ribose-5-phosphate} + 2\,NADPH + CO_2 + 2\,H^+$$

Not surprisingly, the activity of the pentose phosphate pathway is high in rapidly dividing cells that must synthesize large amounts of DNA. In fact, the pentose phosphate pathway not only produces ribose, it also provides a reducing agent (NADPH) required for the reduction of ribose to deoxyribose. Ribonucleotide reductase carries out the reduction of nucleoside diphosphates (NDPs):

NDP   dNDP

The enzyme, which is oxidized in the process, is restored to its original state by a series of reactions in which NADPH is reduced (Section 18.5).

In some cells, however, the need for NADPH for other biosynthetic reactions is greater than the need for ribose-5-phosphate. In this case, *the excess carbons of the pentose are*

$$C_5 \quad C_5 \quad C_5 \qquad C_3 \quad C_7 \quad C_5 \qquad C_6 \quad C_4 \quad C_5 \qquad C_6 \quad C_6 \quad C_3$$

FIGURE 13.11

Rearrangements of the products of the pentose phosphate pathway. Three of the five-carbon products of the oxidative phase of the pentose phosphate pathway are converted to two fructose-6-phosphate and one glyceraldehyde-3-phosphate by reversible reactions involving the transfer of two- and three-carbon units. Each square represents a carbon atom in a monosaccharide. This pathway also allows ribose carbons to be used in glycolysis and gluconeogenesis.

? **Draw the structures of the carbohydrates corresponding to each shape.**

recycled into intermediates of the glycolytic pathway so that they can be degraded to pyruvate or used in gluconeogenesis, depending on the cell type and its metabolic needs.

A set of reversible reactions transform five-carbon ribulose units into six-carbon units (fructose-6-phosphate) and three-carbon units (glyceraldehyde-3-phosphate). These transformations are accomplished mainly by the enzymes transketolase and transaldolase, which transfer two- and three-carbon units among various intermediates to produce a set of sugars containing three, four, five, six, or seven carbons (the reaction catalyzed by transketolase was introduced in Section 7.2). **Figure 13.11** is a schematic view of this process. Because all these interconversions are reversible, *glycolytic intermediates can also be siphoned from glycolysis or gluconeogenesis to synthesize ribose-5-phosphate.* Thus, the cell can use some or all of the steps of the pentose phosphate pathway to generate NADPH, to produce ribose, and to interconvert other monosaccharides.

A SUMMARY OF GLUCOSE METABOLISM

Although our coverage of glucose metabolism is far from exhaustive, this chapter describes quite a few enzymes and reactions, which are compiled in **Figure 13.12**. As you examine this diagram, keep in mind the following points, which also apply to the metabolic pathways we will encounter in subsequent chapters:

1. A metabolic pathway is a series of enzyme-catalyzed reactions, so the pathway's substrate is converted to its product in discrete steps.
2. A monomeric compound such as glucose is interconverted with its polymeric form (glycogen), with other monosaccharides (fructose-6-phosphate and ribose-5-phosphate, for example), and with smaller metabolites such as the three-carbon pyruvate.
3. Although anabolic and catabolic pathways may share some steps, their irreversible steps are catalyzed by enzymes unique to each pathway.
4. Certain reactions consume or produce free energy in the form of ATP. In most cases, these are phosphoryl-group transfer reactions.
5. Some steps are oxidation–reduction reactions that require or generate a reduced cofactor such as NADH or NADPH.

 CONCEPT CHECK

1. List the products of the pentose phosphate pathway and describe how the cell uses them.
2. How does the cell catabolize excess ribose groups?

FIGURE 13.12

Summary of glucose metabolism. This diagram includes the pathways of glycogen synthesis and degradation, glycolysis, gluconeogenesis, and the pentose phosphate pathway. Dotted lines are used where the individual reactions are not shown. Filled gold symbols indicate ATP production; shadowed gold symbols indicate ATP consumption. Filled and shadowed red symbols represent the production and consumption of the reduced cofactors NADH and NADPH.

CLINICAL CONNECTION: DISORDERS OF CARBOHYDRATE METABOLISM

Although there is not enough space to describe all the known enzyme deficiencies that affect carbohydrate metabolic pathways in humans, a few disorders are worth highlighting. As with many metabolic diseases, the discovery and study of carbohydrate metabolic disorders has helped shed light on how the normal metabolic pathways function. Keep in mind that an enzyme deficiency may result from a genetic variation that limits production of the protein, directly impacts its catalytic power, or affects its regulation.

Deficiencies of glycolytic enzymes usually have severe consequences, particularly in tissues that rely heavily on glycolysis for ATP production. For example, red blood cells, which lack mitochondria to produce ATP by oxidative phosphorylation, use the ATP generated by glycolysis to power the Na,K-ATPase that maintains the cells' ion concentration gradients (Section 9.3). Low glycolytic activity reduces the supply of ATP, and the resulting ion imbalance leads to osmotic swelling and bursting of the red blood cells.

In addition to anemia (the loss of red blood cells), other abnormalities may develop when glycolytic enzymes are defective. In a pyruvate kinase deficiency, several of the intermediates

KEY CONCEPT
- Glycolytic enzyme deficiencies cause defects in carbohydrate metabolism.

upstream of phosphoenolpyruvate (the pyruvate kinase substrate) accumulate because they are in equilibrium. Red blood cells normally convert one of these—1,3-bisphosphoglycerate—to 2,3-bisphosphoglycerate (BPG), which binds to hemoglobin to decrease its oxygen affinity (Section 7.1). The elevated concentrations of glycolytic intermediates resulting from a pyruvate kinase deficiency actually boost BPG production, allowing red blood cells to deliver O_2 more efficiently, which helps offset the anemia caused by the enzyme deficiency. A shortage of hexokinase, however, slows the entire glycolytic pathway, which limits the production of BPG in red blood cells and thereby reduces the amount of O_2 delivered to tissues.

Defects in pathways that metabolize sugars other than glucose have variable effects. Individuals with fructose intolerance lack fructose-1-phosphate aldolase (see Box 13.A). The resulting accumulation of fructose-1-phosphate ties up the liver's phosphate supply, which hinders ATP production from ADP. As in the disorders already described, one of the first casualties is the Na,K-ATPase, whose inadequate activity leads to cell death.

An inability to convert galactose to glucose can be deadly, especially in infancy, when the major carbohydrate source is lactose, a disaccharide of glucose and galactose. High concentrations of galactose that cannot be metabolized contribute to side reactions such as the addition of sugar derivatives to proteins. Damage to proteins in nerve cells causes growth retardation and abnormal brain development. Fortunately, an early diagnosis and a galactose-free diet can avoid such damage.

A deficiency of glucose-6-phosphate dehydrogenase, the first enzyme of the pentose phosphate pathway, is the most common human enzyme deficiency. This defect decreases the cellular production of the reducing agent NADPH, which participates in certain oxidation–reduction processes and helps protect cells from oxidative damage. Red blood cells, where the O_2 concentration is high, are most at risk. Glucose-6-phosphate dehydrogenase deficiency affects about 500 million people, mostly in Africa, tropical South America, and southeast Asia—areas with historically high rates of malaria. The enzyme defect causes anemia, but the release of heme from damaged red blood cells triggers anti-inflammatory responses that increase an individual's chances of surviving malaria. The same effect is observed in individuals carrying the hemoglobin S variant (Section 5.2).

GLYCOGEN STORAGE DISEASES AFFECT LIVER AND MUSCLE

The **glycogen storage diseases** are a set of inherited disorders of glycogen metabolism, not all of which result in glycogen accumulation, as the name might suggest. The symptoms of the glycogen storage diseases vary, depending on whether the affected tissue is liver or muscle or both. In general, the disorders that affect the liver cause hypoglycemia (too little glucose in the blood) and an enlarged liver. Glycogen storage diseases that affect primarily muscle are characterized by muscle weakness and cramps. The incidence of glycogen storage diseases is estimated to be as high as 1 in 20,000 births, although some disorders are not apparent until adulthood. Twelve types of glycogen storage diseases have been described, and the defect in each is listed in **Table 13.2**.

A defect of glucose-6-phosphatase (type I glycogen storage disease) affects both gluconeogenesis and glycogenolysis, since the phosphatase catalyzes the final step of gluconeogenesis and makes free glucose available from glycogenolysis. The enlarged liver and hypoglycemia can lead to a host of other symptoms, including irritability, lethargy, and, in severe cases, death. A related defect is the deficiency of the transport protein that imports glucose-6-phosphate into the endoplasmic reticulum, where the phosphatase is located.

Type III glycogen storage disease results from a deficiency of the glycogen debranching enzyme. This condition accounts for about one-quarter of all cases of glycogen storage disease and usually affects both liver and muscle. The symptoms include muscle weakness and liver enlargement due to the accumulation of glycogen that cannot be efficiently broken down. The symptoms of type III glycogen storage disease often improve with age and disappear by early adulthood.

TABLE 13.2

TYPE	ENZYME DEFICIENCY
I	Glucose-6-phosphatase
II	α-1,4-Glucosidase
III	Amylo-1,6-glucosidase (debranching enzyme)
IV	Amylo-(1,4 → 1,6)-transglycosylase (branching enzyme)
V	Muscle glycogen phosphorylase
VI	Liver glycogen phosphorylase
VII	Phosphofructokinase
VIII, IX, X	Phosphorylase kinase
XI	GLUT2 transporter
0	Glycogen synthase

The most common type of glycogen storage disease is type IX. In this disorder, the kinase that activates glycogen phosphorylase is defective. Symptoms range from severe to mild and may fade with time. The complexity of this disease reflects the fact that the phosphorylase kinase consists of four subunits, with isoforms that are differentially expressed in the liver and other tissues.

In the past, glycogen storage diseases were diagnosed on the basis of symptoms, blood tests, and painful biopsies of liver or muscle to assess its glycogen content. Current diagnostic methods are centered on analyzing the relevant genes for mutations, a noninvasive approach. Treatment of glycogen storage diseases typically includes a regimen of frequent, small, carbohydrate-rich meals to alleviate hypoglycemia. However, because dietary therapy does not completely eliminate the symptoms of some glycogen storage diseases, and because the metabolic abnormalities, such as chronic hypoglycemia and liver damage, can severely impair physical growth as well as cognitive development, liver transplant has proved to be an effective treatment. The glycogen storage diseases are single-gene defects, which makes them attractive targets for gene therapy (see Section 3.5).

 CONCEPT CHECK

1. Make a list of enzyme deficiencies and their physiological effects.

2. Compare the impact of various disorders on red blood cells, liver cells, and muscle cells.

SUMMARY

13.1 Glycolysis

• The pathway of glucose catabolism, or glycolysis, is a series of enzyme-catalyzed steps in which free energy is conserved as ATP or NADH.

• The 10 reactions of glycolysis convert the six-carbon glucose to two molecules of pyruvate and produce two molecules of NADH and two molecules of ATP. The first phase (reactions catalyzed by hexokinase, phosphoglucose isomerase, phosphofructokinase, aldolase, and triose phosphate isomerase) requires the investment of two ATP. The irreversible reaction catalyzed by phosphofructokinase is the rate-determining step and the major control point for glycolysis. The second phase of the pathway (reactions catalyzed by glyceraldehyde-3-phosphate dehydrogenase, phosphoglycerate kinase, phosphoglycerate mutase, enolase, and pyruvate kinase) generates four ATP per glucose.

• Pyruvate may be reduced to lactate or ethanol, further oxidized by the citric acid cycle, or converted to other compounds.

13.2 Gluconeogenesis

• The pathway of gluconeogenesis converts two molecules of pyruvate to one molecule of glucose at a cost of six ATP. The pathway uses seven glycolytic enzymes, and the activities of pyruvate carboxylase, phosphoenolpyruvate carboxykinase, fructose bisphosphatase, and glucose-6-phosphatase bypass the three irreversible steps of glycolysis.

• A futile cycle involving phosphofructokinase and fructose bisphosphatase helps regulate the flux through glycolysis and gluconeogenesis.

13.3 Glycogen Synthesis and Degradation

• Glucose residues are incorporated into glycogen after first being activated by attachment to UDP.

• Phosphorolysis of glycogen produces phosphorylated glucose that can enter glycolysis. In the liver, this glucose is dephosphorylated and exported.

13.4 The Pentose Phosphate Pathway

• The pentose phosphate catabolic pathway for glucose yields NADPH and ribose groups. The five-carbon sugar intermediates can be converted to glycolytic intermediates.

13.5 Clinical Connection: Disorders of Carbohydrate Metabolism

• Disorders are associated with deficiencies of glycolytic enzymes and enzymes that metabolize fructose and galactose.

• Glycogen storage diseases cause hypoglycemia, muscle weakness, and liver damage.

KEY TERMS

glycolysis	near-equilibrium reaction	fermentation	pentose phosphate pathway
kinase	rate-determining reaction	gluconeogenesis	glycogen storage disease
metabolically irreversible reaction	tautomerization	futile cycle	
	feed-forward activation	glycogenolysis	

Brief Bioinformatics Exercises

13.1 Hexokinase Structure and Ligand Binding

13.2 Glycolysis and the KEGG Database

13.1 Glycolysis

1. Identify which of the ten reactions of glycolysis are (a) phosphorylations; (b) isomerizations; (c) oxidation–reductions; (d) dehydrations; (e) carbon–carbon bond cleavages.

2. Which reactions of glycolysis can be reversed? Which reactions are irreversible? What is the significance of the metabolically irreversible reactions?

3. The $\Delta G^{\circ\prime}$ value for the hexokinase reaction is -16.7 kJ·mol^{-1}, while the ΔG value under cellular conditions is similar.

(a) What is the ratio of glucose-6-phosphate to glucose under standard conditions if the ratio of [ATP] to [ADP] is 15:1?

(b) How high would the ratio of glucose-6-phosphate to glucose have to be in order to reverse the hexokinase reaction by mass action?

4. (a) What is the ratio of fructose-6-phosphate (F6P) to glucose-6-phosphate (G6P) under standard conditions? (b) Under cellular conditions? In which direction does the reaction proceed under cellular conditions?

5. Except during starvation, the brain burns glucose as its sole metabolic fuel and consumes up to 40% of the body's circulating glucose. Brain hexokinase has a K_M for glucose that is 100 times lower than the concentration of circulating glucose (5 mM). What is the advantage of this low K_M?

6. The V_{max} and K_M values for an unusual hexokinase found in *Trypansoma cruzi* (the causative agent of Chagas disease) are shown in the presence and absence of a bisphonate inhibitor (structure shown).

	Without inhibitor	With inhibitor
K_M (mM)	90	125
V_{max} (μmol·min^{-1}·mL^{-1})	0.30	0.12

A bisphonate compound

(a) What type of inhibitor is bisphonate?

(b) The parasite hexokinase, unlike the mammalian enzyme, is not inhibited by glucose-6-phosphate but is inhibited by pyrophosphate (PP$_i$). Is this observation consistent with your answer to part a?

(c) Might bisphonate be a good candidate as a drug to treat the disease?

7. The radioactively labeled compound [^{18}F]fluorodeoxyglucose (FDG) is used to measure glucose uptake in cells. FDG, a derivative of glucose in which the C2 hydroxyl is replaced with fluorine, is phosphorylated to FDG-6-phosphate upon entering the cell but cannot proceed any further through glycolysis. The phosphorylated FDG remains in the cell, and its presence can be detected and quantitated, providing information on the rate of glucose uptake. (Cancer cells take up glucose particularly rapidly.)

(a) Draw the structure of FDG.

(b) Why does phosphorylation trap the FDG in the cell?

(c) Why is FDG unable to proceed though glycolysis past the hexokinase reaction? (*Hint:* When glucose-6-phosphate is converted to fructose-6-phosphate, the ring opens, the isomerization reaction takes place, and then the ring closes again.)

(d) Is FDG taken up only by cancer cells?

8. ADP stimulates the activity of phosphofructokinase (PFK), yet it is a product of the reaction and not a reactant. Explain this apparently contradictory regulatory strategy.

9. We can apply the 'T' and 'R' nomenclature used to describe the low- and high-affinity forms of hemoglobin (see Section 5.1) can also be used to describe the conformational changes that occur in allosteric enzymes like phosphofructokinase (PFK). Allosteric inhibitors stabilize the T form, which has low affinity for its substrate, while activators stabilize the high-affinity R form. Do the following allosteric effectors stabilize the T form of PFK or the R form? (a) ADP (bacteria); (b) phosphoenolpyruvate (PEP, bacteria); (c) fructose-2,6-bisphosphate (mammals).

10. PFK isolated from the bacterium *Bacillus stearothermophilus* is a tetramer that binds fructose-6-phosphate with hyperbolic kinetics and a K_M of 23 μM. What happens to the K_M in the presence of phosphoenolpyruvate (PEP; see Fig. 7.15)? Use the T $\rightarrow$ R terminology (see Problem 9) to explain what happens.

11. Refer to Figure 7.16. Why does the conformational change that results when Arg 162 changes places with Glu 161 result in a form of phosphofructokinase that has a low affinity for its substrate?

12. Researchers isolated a yeast mutant that was deficient in phosphofructokinase. The mutant yeast was able to grow on glycerol as an energy source, but not glucose. Explain why.

13. Researchers isolated a yeast phosphofructokinase mutant in which a serine at the fructose-2,6-bisphosphate (F26BP) binding site was replaced with an aspartate residue. The amino acid substitution completely abolished the binding of F26BP to PFK. There was a dramatic decline in glucose consumption and ethanol production in the mutant compared to control yeast.

(a) Propose a hypothesis that explains why the mutant PFK cannot bind F26BP.

(b) What does the decline of glucose consumption and ethanol production in the yeast reveal about the role of F26BP in glycolysis?

14. Refer to the mechanism of aldolase shown in Figure 13.4.

(a) Is the pK of the Lys side chain higher, lower, or the same as it would be in the free amino acid? What is the role of the Lys side chain in catalysis?

(b) Does the pK of the Asp side chain change after formation of the Schiff base? What is the role of the Asp side chain in catalysis?

(c) Does the aldolase enzyme mechanism use acid catalysis, base catalysis, covalent catalysis, or some combination of these strategies (see Section 6.2)? Explain.

15. Explain why iodoacetate was useful for determining the order of intermediates in glycolysis but provided misleading information about an enzyme's active site.

16. Biochemists use transition state analogs to determine the structure of a short-lived intermediate in an enzyme-catalyzed reaction. Because an enzyme binds tightly to the transition state, a compound that resembles the transition state should be a potent competitive inhibitor. Phosphoglycohydroxamate binds 150 times more tightly than dihydroxyacetone phosphate to triose phosphate isomerase. Based on this information, propose a structure for the intermediate of the triose phosphate isomerase reaction.

$$
\begin{array}{c}
\text{OH} \\
| \\
\text{N} \\
\parallel \\
\text{C} \longrightarrow \text{O}^- \\
| \\
\text{CH}_2\text{OPO}_3^{2-}
\end{array}
$$

Phosphoglycohydroxamate

17. What is the ratio of glyceraldehyde-3-phosphate (GAP) to dihydroxyacetone phosphate (DHAP) in cells at 37°C under nonequilibrium conditions? Considering your answer to this question, how do you account for the fact that the conversion of DHAP to GAP occurs readily in cells?

18. Cancer cells have elevated levels of glyceraldehyde-3-phosphate dehydrogenase (GAPDH), which may account for the high rate of glycolysis in these cells. The compound methylglyoxal has been shown to inhibit GAPDH in cancer cells but not in normal cells. This observation may lead to the development of drugs for treating cancer.
(a) Propose a hypothesis to explain why GAPDH levels in cancer cells are elevated.
(b) Why might methylglyoxal inhibit GAPDH in cancer cells but not in normal cells?

19. Arsenate, AsO_4^{3-}, acts as a phosphate analog and can replace phosphate in the GAPDH reaction. The product of this reaction is 1-arseno-3-phosphoglycerate, which is unstable and spontaneously hydrolyzes to form 3-phosphoglycerate, as shown. What is the effect of arsenate on cells undergoing glycolysis?

$$
\begin{array}{c}
\text{O} \\
\parallel \\
\text{CH} \\
| \\
\text{HCOH} \quad + \quad AsO_4^{3-} \\
| \\
\text{H}_2\text{COPO}_3^{2-}
\end{array}
\xrightarrow[\text{GAPDH}]{\text{NAD}^+ \quad \text{NADH} + \text{H}^+}
\begin{array}{c}
\text{O} \\
\parallel \\
\text{C} \longrightarrow \text{OAsO}_3^{2-} \\
| \\
\text{HCOH} \\
| \\
\text{H}_2\text{COPO}_3^{2-}
\end{array}
$$

$$
\begin{array}{c}
\text{O} \\
\parallel \\
\text{C} \longrightarrow \text{OAsO}_3^{2-} \\
| \\
\text{HCOH} \\
| \\
\text{H}_2\text{COPO}_3^{2-}
\end{array}
\xrightarrow[\quad]{\text{H}_2\text{O} \qquad AsO_4^{3-}}
\begin{array}{c}
\text{O} \\
\parallel \\
\text{C} \longrightarrow \text{O}^- \\
| \\
\text{HCOH} \\
| \\
\text{H}_2\text{COPO}_3^{2-}
\end{array}
$$

20. In several species of bacteria, GAPDH activity is controlled by the $NADH/NAD^+$ ratio. Does the activity of GAPDH increase or decrease when the $NADH/NAD^+$ ratio increases? Explain. Assume that only the forward direction of the reaction is relevant.

21. The thermophilic archaebacterium *Thermoproteus tenax* expresses two GAPDH enzymes; one that carries out the reaction shown in the text (although $NADP^+$ is the reactant rather than NAD^+) and a second GAPDH enzyme that catalyzes the reaction shown here.

$$
\text{Glyceraldehyde-3-phosphate} + \text{NAD}^+ \xrightarrow[\text{GAPDH}]{\textit{T. tenax}}
$$
$$
\text{3-phosphoglycerate} + \text{NADH} + \text{H}^+
$$

(a) How do the properties of the two enzymes differ?
(b) In an experiment, the activity of the second *T. tenax* GAPDH enzyme was measured in the presence of various metabolites. The

results are shown in the figure. Estimate the K_M values for the enzyme in the presence and absence of effectors and classify the effectors as activators or inhibitors of the enzyme.
(c) Propose a hypothesis that explains why these effectors act as either activators or inhibitors.

22. Phosphoglycerate kinase in red blood cells is bound to the plasma membrane. This allows the kinase reaction to be coupled to the Na,K-ATPase pump. How does the proximity of the enzyme to the membrane facilitate the action of the pump?

23. Vanadate, VO_4^{3-}, inhibits GAPDH, not by acting as a phosphate analog, but by interacting with essential—SH groups on the enzyme. What happens to cellular levels of phosphate, ATP, and 2,3-bisphosphoglycerate (see pathway below) when red blood cells are incubated with vanadate?

$$
\begin{array}{c}
\text{Glyceraldehyde-3-phosphate} \\
\updownarrow \text{GAPDH} \\
\text{1,3-Bisphosphoglycerate} \\
\updownarrow \text{PGK} \searrow \text{2,3-Bisphosphoglycerate} \\
\text{3-Phosphoglycerate} \swarrow
\end{array}
$$

24. The mechanism of plant phosphoglycerate mutase is different from the mechanism of mammalian phosphoglycerate mutase presented in the text. 3-Phosphoglycerate (3PG) binds to the plant enzyme, transfers its phosphate to the enzyme, and then the enzyme transfers the phosphate group back to the substrate to form 2-phosphoglycerate (2PG). When [^{32}P]-labeled 3PG is added to cultured **(a)** hepatocytes or **(b)** plant cells, what is the fate of the [^{32}P] label?

25. Assume a standard free energy change of 30.5 kJ·mol^{-1} for the synthesis of ATP from ADP and P_i.
(a) How many molecules of ATP could be theoretically produced by the catabolism of glucose to 0(i) lactate and (ii) CO_2 (see Table 13.1), assuming 33% efficiency?
(b) Does this explain the Pasteur effect (the observation, first made by Louis Pasteur, that glucose consumption in yeast dramatically decreases in the presence of oxygen)?

26. Organisms such as yeast growing under anaerobic conditions can convert pyruvate to alcohol in a process called fermentation, as described in the text. Instead of being converted to lactate, pyruvate is converted to ethanol in a two-step reaction. Why is the second step of this process essential to the yeast cell?

27. Several studies have shown that aluminum inhibits phosphofructokinase in liver cells.
(a) Compare the production of pyruvate by perfused livers in control and aluminum-treated rats using fructose as an energy source.
(b) What would the experimental results be if glucose were used instead of fructose?

28. (a) Explain why alcohol consumption is associated with increased risk of developing hypothermia.

(b) Drinking a glass of water for each alcoholic drink is a popular hangover-prevention strategy. Explain how increased water consumption might relieve some of the negative effects of alcohol consumption.

29. Drinking methanol can cause blindness and death, depending on the dosage. The causative agent is formaldehyde derived from methanol.
(a) Draw the balanced chemical reaction for the conversion of methanol to formaldehyde.
(b) Why would administering whiskey (ethanol) to a person poisoned with methanol be a good antidote?

30. About 15% of ingested ethanol is metabolized by a cytochrome P450 (see Section 7.4), and chronic alcohol consumption induces the expression of this enzyme. Explain how this would change the effectiveness of therapeutic drugs.

31. The term "turbo design" has been used to describe pathways such as glycolysis that have one or more ATP-consuming steps followed by one or more ATP-producing steps with a net yield of ATP production for the pathway overall. Mathematical models have shown that "turbo" pathways have the risk of substrate-accelerated death unless there is a "guard at the gate," that is, a mechanism for inhibiting an early step of the pathway. In yeast, hexokinase is inhibited by a complex mechanism mediated by trehalose-6-phosphate synthase (TPSI). Mutant yeast in which TPSI is defective (there is no "guard at the gate") die if grown under conditions of high glucose concentration. Explain why.

32. Studies have shown that the halophilic organism *Halococcus saccharolyticus* degrades glucose via the Entner–Doudoroff pathway rather than by the glycolytic pathway presented in the text. A modified scheme for the Entner–Doudoroff pathway is shown here.
(a) What is the ATP yield per mole of glucose for this pathway?
(b) Describe (in general) what kinds of reactions would need to follow the Entner–Doudoroff pathway in this organism.

Glucose

↓ NADP⁺ → NADPH

Gluconate

↓ H₂O

2-Keto-3-deoxygluconate (KDG)

↓ ATP → ADP

2-Keto-3-deoxy-6-phosphogluconate (KDPG)

↓ ↓

Pyruvate   Glyceraldehyde-3-phosphate

↓

1,3-BPG

↓
↓
↓

Pyruvate

33. Trypanosomes living in the bloodstream obtain all their free energy from glycolysis. They take up glucose from the host's blood and excrete pyruvate as a waste product. In this part of their life cycle, trypanosomes do not carry out any oxidative phosphorylation, but they do use another oxygen-dependent pathway, which is absent in mammals, to oxidize NADH.
(a) Why is this other pathway necessary?
(b) Would the pathway be necessary if the trypanosome excreted lactate rather than pyruvate?
(c) Why would this pathway be a good target for antiparasitic drugs?

13.2 Gluconeogenesis

34. Flux through the opposing pathways of glycolysis and gluconeogenesis is controlled in several ways.

(a) Explain how the activation of pyruvate carboxylase by acetyl-CoA affects glucose metabolism.

(b) Pyruvate can undergo a reversible amino-group transfer reaction to yield alanine (see Section 12.2). Alanine is an allosteric effector of pyruvate kinase. Would you expect alanine to stimulate or inhibit pyruvate kinase? Explain.

35. A physician diagnoses an infant patient with a pyruvate carboxylase deficiency in part by measuring the patient's blood levels of lactate and pyruvate. How would the patient's [lactate]/[pyruvate] ratio compare to the ratio in a normal infant?

36. A liver biopsy of a four-year-old boy indicated that fructose-1,6-bisphosphatase enzyme activity was 20% of normal. The patient's blood glucose levels were normal at the beginning of a fast but then decreased suddenly. Pyruvate and alanine concentrations were also elevated, as was the glyceraldehyde-3-phosphate/dihydroxyacetone phosphate ([GAP]/[DHAP]) ratio. Explain the reason for these symptoms.

37. Insulin is one of the major hormones that regulates gluconeogenesis. Insulin acts in part by decreasing the transcription of genes coding for certain gluconeogenic enzymes. For which genes would you expect insulin to suppress transcription?

38. Type 2 diabetes is characterized by insulin resistance, in which insulin is unable to perform its many functions. What symptoms would you expect in a Type 2 diabetic patient if insulin is unable to perform the function described in Problem 37?

39. The concentration of fructose-2,6-bisphosphate (F26BP) is regulated in the cell by a homodimeric enzyme with two catalytic activities: a kinase that phosphorylates fructose-6-phosphate to form F26BP, and a phosphatase that catalyzes the hydrolysis of the phosphate group.

(a) Which enzyme activity, the kinase or the phosphatase, would you expect to be active under fasting conditions? Explain.

(b) Which hormone is likely to be responsible for inducing this activity?

(c) Consult Section 10.2 and propose a mechanism for this induction.

40. The "carbon skeletons" of most amino acids can be converted to glucose, a process that may require many enzymatic steps. Which amino acids can enter the gluconeogenic pathway directly after undergoing deamination (a reaction in which the carbon with the amino group becomes a ketone)?

41. Brazilin, a compound found in aqueous extracts of sappan wood, has been used to treat diabetics in Korea. Brazilin increases the activity of the kinase enzyme that produces F26BP (see Problem 39), and the compound also stimulates the activity of pyruvate kinase.
(a) What is the effect of adding brazilin to hepatocytes in culture?
(b) Why would brazilin be an effective treatment for diabetes?

42. Metformin is a drug that decreases the expression of phosphoenolpyruvate carboxykinase. Explain why metformin would be helpful in treating diabetes.

---

43. Draw a diagram that illustrates how.
(a) Lactate released from the muscle is converted back to glucose in the liver. What is the cost (in ATP) of running this cycle?
(b) Alanine released from the muscle is converted back to glucose in the liver. What is the physiological cost if this cycle runs for a prolonged period of time?

13.3   Glycogen Synthesis and Degradation

44. Beer is produced from raw materials such as wheat and barley. Explain why the grains are allowed to sprout, a process in which their starch is broken down to glucose, before fermentation begins.

45. Some bread manufacturers add amylase to bread dough prior to the fermentation process. What role does this enzyme (see Section 12.1) play in the bread-making process?

46. Glycogen is degraded via a phosphorolysis process, which produces glucose-1-phosphate. What advantage does this process have over a simple hydrolysis, which would produce glucose instead of phosphorylated glucose?

47. The equation for the degradation of glycogen is shown below.
(a) What is the ratio of $[P_i]/[G1P]$ under standard conditions?
(b) What is the value of ΔG under cellular conditions when the $[P_i]/[G1P]$ ratio is 50/1?
(c) What advantage does degradation by phosphorolysis have over a simple hydrolysis, which would produce glucose instead of glucose-1-phosphate?

$$\text{glycogen } (n \text{ residues}) + P_i \underset{\phantom{\text{phosphorylase}}}{\overset{\text{phosphorylase}}{\rightleftharpoons}}$$
$$\text{glycogen } (n - 1 \text{ residues}) + \text{G1P} \quad \Delta G^{\circ\prime} = +3.1 \text{ kJ} \cdot \text{mol}^{-1}$$

48. The mechanism of the phosphoglucomutase enzyme is similar to that of the plant mutase described in Problem 24 and is shown below. On occasion, the glucose-1,6-bisphosphate dissociates from the enzyme. Why does the dissociation of glucose-1,6-bisphosphate inhibit the enzyme?

Glucose-6-phosphate   Glucose-1,6-bisphosphate   Glucose-1-phosphate

49. Trehalose, a disaccharide consisting of two glucose residues (see Problems 11.32 through 11.34), is one of the major sugars in insect hemolymph (the fluid that circulates through the insect's body). Trehalose serves as a storage form of glucose and also helps protect the insect from desiccation and freezing. Its concentration in the hemolymph must be closely regulated. Trehalose is synthesized in the insect fat body, which plays a role in metabolism analogous to the vertebrate liver. Studies of the insect *Manduca sexta* have shown that during starvation, hemolymph glucose concentration decreases, which results in an increase in fat body glycogen phosphorylase activity and a decrease in the concentration of fructose-2,6-bisphosphate. What effect do these changes have on hemolymph trehalose concentration in the fasted insect?

50. The glycolytic pathway in the thermophilic archaebacterium *Thermoproteus tenax* differs from the pathway presented in this chapter. The phosphofructokinase (PFK) reaction in *T. tenax* is reversible and depends on pyrophosphate rather than ATP. In addition, *T. tenax* has two glyceraldehyde-3-phosphate dehydrogenase (GADPH) isozymes. The "phosphorylating GAPDH" is similar to the enzyme described in this chapter. The second isozyme is the irreversible "nonphosphorylating GAPDH," which catalyzes the reaction described in Problem 21. *T. tenax* relies on glycogen stores as a source of energy. What is the ATP yield for one mole of glucose oxidized by the pathway that uses the nonphosphorylating GAPDH enzyme?

13.4   The Pentose Phosphate Pathway

51. Most metabolic pathways include an enzyme-catalyzed reaction that commits a metabolite to continue through the pathway.
(a) Identify the first committed step of the pentose phosphate pathway. Explain your reasoning.
(b) Hexokinase catalyzes an irreversible reaction at the start of glycolysis. Does this step commit glucose to continue through glycolysis?

52. A given metabolite may follow more than one metabolic pathway. List all the possible fates of glucose-6-phosphate in **(a)** a liver cell and **(b)** a muscle cell.

53. Write a mechanism for the nonenzymatic hydrolysis of 6-phosphogluconolactone to 6-phosphogluconate.

54. Enzymes in the soil fungus *Aspergillus nidulans* use NADPH as a coenzyme when converting nitrate to ammonium ions. When the fungus was cultured in a growth medium containing nitrate, the activities of several enzymes involved in glucose metabolism increased. What enzymes are good candidates for regulation under these conditions? Explain.

55. Several studies have shown that the metabolite glucose-1,6-bisphosphate (G16BP) regulates several pathways of carbohydrate metabolism by inhibiting or activating key enzymes. The effect of G16BP on several enzymes is summarized in the table below. What pathways are active when G16BP is present? What pathways are inactive? What is the overall effect? Explain.

Enzyme	Effect of G16BP
Hexokinase	Inhibits
Phosphofructokinase (PFK)	Activates
Pyruvate kinase (PK)	Activates
Phosphoglucomutase	Activates
6-Phosphogluconate dehydrogenase	Inhibits

56. Xylulose-5-phosphate (an intermediate of the pentose phosphate pathway) acts as an intracellular signaling molecule that activates kinases and phosphatases in liver cells. As a result of this signaling, there is an increase in the activity of the enzyme that produces fructose-2,6-bisphosphate, and the expression of genes for lipid synthesis is increased. What is the net effect of these responses?

13.5   Clinical Connection: Disorders of Carbohydrate Metabolism

57. Type VII glycogen storage disease is actually a deficiency of muscle phosphofructokinase. Patients with this genetic disease may have muscle PFK levels that are 1–5% of normal. Why do these patients suffer from myoglobinuria (myoglobin in the urine) and muscle cramping during exercise?

58. Red blood cells synthesize and degrade 2,3-bisphosphoglycerate (BPG) as a detour from the glycolytic pathway, as shown in Problem 23. BPG decreases the oxygen affinity of hemoglobin by binding in the central cavity of the deoxygenated form of hemoglobin (see Fig. 5.11). This encourages delivery of oxygen to tissues. A defect in one of the glycolytic enzymes may affect levels of BPG. The plot shows oxygen-binding curves for normal erythrocytes and for hexokinase-deficient and pyruvate kinase–deficient erythrocytes. Identify which curve corresponds to which enzyme deficiency.

59. (a) What happens to the [ADP]/[ATP] and [NAD$^+$]/[NADH] ratios in red blood cells with a pyruvate kinase deficiency (see Problem 58)? **(b)** One of the symptoms of a pyruvate kinase deficiency is hemolytic anemia, in which red blood cells swell and eventually lyse. Explain why the enzyme deficiency brings about this symptom.

60. During even mild exertion, individuals with McArdle's disease experience painful muscle cramps due to a genetic defect in glycogen phosphorylase, the enzyme that breaks down glycogen. Yet the muscles in these individuals contain normal amounts of glycogen. What did this observation tell researchers about the pathways for glycogen degradation and glycogen synthesis?

61. Patients with McArdle's disease have normal liver glycogen content and structure. Identify the type of glycogen storage disease as listed in Table 13.2.

62. A patient with McArdle's disease (see Problem 61) performs ischemic (anaerobic) exercise for as long as he is able to do so. The patient's blood is withdrawn every few minutes during the exercise period and tested for lactate. The patient's samples are compared with control samples from a patient who does not suffer from a glycogen storage disease. The results are shown in the figure. Why does the lactate concentration increase in the normal patient? Why is there no corresponding increase in the patient's lactate concentration?

63. Patients with von Gierke's disease (type I glycogen storage disease) have a deficiency of glucose-6-phosphatase. One of the most prominent symptoms of the disease is a protruding abdomen due to an enlarged liver.
(a) Explain why the liver is enlarged in patients with von Gierke's disease.
(b) Some patients with von Gierke's disease also have enlarged kidneys. Explain why.

64. (a) Does a patient with McArdle's disease (see Problems 60–62) suffer from hypoglycemia, hyperglycemia, or neither? **(b)** Does a patient with von Gierke's disease (see Problem 63) suffer from hypoglycemia, hyperglycemia, or neither?

65. Individuals with fructose intolerance lack fructose-1-phosphate aldolase, a liver enzyme essential for catabolizing fructose. In the absence of fructose-1-phosphate aldolase, fructose-1-phosphate accumulates in the liver and inhibits glycogen phosphorylase and fructose-1,6-bisphosphatase.
(a) Explain why individuals with fructose intolerance exhibit hypoglycemia (low blood sugar).
(b) Administering glycerol and dihydroxyacetone phosphate does not alleviate the hypoglycemia, but administering galactose does relieve the hypoglycemia. Explain.

66. Reduced glutathione, a tripeptide containing a Cys residue, is found in red blood cells, where it reduces organic peroxides formed in cellular structures exposed to high concentrations of reactive oxygen.

$$2\ \gamma\text{-Glu}-\text{Cys}-\text{Gly}\ +\ \ \ \ R-O-OH \longrightarrow$$
$$\ |$$
$$\ SH$$
Reduced glutathione   Organic peroxide

γ-Glu — Cys — Gly
        |
        S
        | + R — OH + H_2O
        S
        |
γ-Glu — Cys — Gly
Oxidized glutathione

Reduced glutathione also plays a role in maintaining normal red blood cell structure and keeping the iron ion of hemoglobin in the +2 oxidation state. Glutathione is regenerated as shown in the following reaction:

γ-Glu — Cys — Gly
        |
        S
        | + NADPH + H$^+$ ⟶
        S
        |
γ-Glu — Cys — Gly

$$2\ \gamma\text{-Glu}-\text{Cys}-\text{Gly}\ +\ NADP^+$$
$$\ \ \ \ \ \ \ \ \ \ \ \ \ \ \ \ \ \ |$$
$$\ \ \ \ \ \ \ \ \ \ \ \ \ \ \ \ \ \ SH$$

Use this information to predict the physiological effects of a glucose-6-phosphate dehydrogenase deficiency.

67. Experiments were carried out in cultured cells to determine the relationship between glucose-6-phosphate dehydrogenase (G6PD) activity and rates of cell growth. Cells were cultured in a medium supplemented with serum, which contains growth factors that stimulate

G6PD activity. Predict how the cellular NADPH/NADP$^+$ ratio would change under the following circumstances:

(a) Serum is withdrawn from the medium.

(b) DHEA, an inhibitor of glucose-6-phosphate dehydrogenase, is added.

(c) The oxidant H_2O_2 is added.

(d) Serum is withdrawn and H_2O_2 is added.

68. Would a feeding of cornstarch administered at bedtime help relieve the symptoms of type 0 glycogen storage disease? Explain why or why not.

SELECTED READINGS

Brosnan, J. T., Comments on metabolic needs for glucose and the role of gluconeogenesis, *Eur. J. Clin. Nutr.* **53,** S107–S111 (1999). [A very readable review that discusses possible reasons why carbohydrates are used universally as metabolic fuels, why glucose is stored as glycogen, and why the pentose phosphate pathway is important.]

Lenzen, S., A fresh view of glycolysis and glucokinase regulation: History and current status, *J. Biol. Chem.* **289,** 12189–12194 (2014). [Includes some history of glycolysis research, with an emphasis on regulation of Steps 1 and 3 of glycolysis.]

Özen, H., Glycogen storage diseases: New perspectives, *World J. Gastroenterology* **13,** 2541–2553 (2007). [Describes the symptoms, biochemistry, and treatment of the major forms of these diseases.]

Patra, K. C. and Hay, N., The pentose phosphate pathway and cancer, *Trends Biochem. Sci.* **39,** 347–354 (2014). [Reviews the reactions of the pentose phosphate pathway.]

Roach, P. J., Depaoli-Roach, A. A., Hurley, T. D., and Tagliabracci, V. S., Glycogen and its metabolism: Some new developments and old themes, *Biochem. J.* **441,** 763–787 (2012). [Includes discussions of the hormone-mediated regulation of key enzymes of glycogen synthesis and breakdown.]

CHAPTER 14

THE CITRIC ACID CYCLE

Metabolism converts the stored energy of fuel molecules into the chemical energy of ATP, but some energy is always lost as heat, which is one way animals can keep warm. Until recently, birds and mammals were the only organisms known to adjust their metabolism to maintain high body temperatures. However, the Tegu lizard can also generate heat during its reproductive phase, boosting its nest temperature as much as 10°C higher than the ambient temperature.

Paul Starosta/Getty Images, Inc.

DO YOU REMEMBER?

- Enzymes accelerate chemical reactions using acid–base catalysis, covalent catalysis, and metal ion catalysis (Section 6.2).

- Coenzymes such as NAD^+ and ubiquinone collect electrons from compounds that become oxidized (Section 12.2).

- Metabolic pathways in cells are connected and are regulated (Section 12.2).

- Many vitamins, substances that humans cannot synthesize, are components of coenzymes (Section 12.2).

- Pyruvate can be converted to lactate, acetyl-CoA, or oxaloacetate (Section 13.1).

The citric acid cycle logically follows glycolysis in an overview of cellular energy metabolism, but the cycle does much more than just continue the breakdown of glucose. Occupying a central place in the metabolism of most cells, the citric acid cycle processes the remnants of all types of metabolic fuels, including fatty acids and amino acids, so that their energy can be used to synthesize ATP. The citric acid cycle also operates anabolically, supplying the precursors for biosynthetic pathways. We will use pyruvate, the end product of glycolysis, as the starting point for our study of the citric acid cycle. We will then examine the eight steps of the citric acid cycle and discuss how this sequence of reactions might have evolved. Finally, we will consider the citric acid cycle as a multifunctional pathway with links to other metabolic processes.

14.1 THE PYRUVATE DEHYDROGENASE REACTION

KEY CONCEPT
- The pyruvate dehydrogenase complex includes three types of enzymes that collectively remove a carboxylate group from pyruvate and produce acetyl-CoA and NADH.

The end product of glycolysis is the three-carbon compound pyruvate. In aerobic organisms, these carbons are ultimately oxidized to 3 CO_2 (although the oxygen atoms come not from molecular oxygen but from water and phosphate). The first molecule of CO_2 is released when pyruvate is decarboxylated to an acetyl unit. The second and third CO_2 molecules are products of the citric acid cycle.

THE PYRUVATE DEHYDROGENASE COMPLEX CONTAINS MULTIPLE COPIES OF THREE DIFFERENT ENZYMES

The decarboxylation of pyruvate is catalyzed by the pyruvate dehydrogenase complex. In eukaryotes, this enzyme complex, and the enzymes of the citric acid cycle itself, are located inside the mitochondrion (an organelle surrounded by a double membrane and whose interior is called the **mitochondrial matrix**). Accordingly, pyruvate produced by glycolysis in the cytosol must first be transported into the mitochondria.

For convenience, the three kinds of enzymes that make up the pyruvate dehydrogenase complex are called E1, E2, and E3. Together *they catalyze the oxidative decarboxylation of pyruvate and the transfer of the acetyl unit to coenzyme A*:

$$\text{pyruvate} + \text{CoA} + \text{NAD}^+ \rightarrow \text{acetyl-CoA} + \text{CO}_2 + \text{NADH}$$

The structure of coenzyme A, a nucleotide derivative containing the vitamin pantothenate, is shown in Figure 3.2a.

In some bacteria, the 4600-kD pyruvate dehydrogenase complex consists of a cubic core of 24 E2 subunits (**Fig. 14.1a**), which are surrounded by an outer shell of 24 E1 and 12 E3 subunits. In mammals and some other bacteria, the enzyme complex is even larger, with 42–48 E1, 60 E2, and 6–12 E3 plus additional proteins that hold the complex together and regulate its enzymatic activity. The 60-subunit E2 core of the pyruvate dehydrogenase complex from *Bacillus stearothermophilus* is shown in Figure 14.1b.

FIGURE 14.1

Models of the E2 core of the pyruvate dehydrogenase complex. (a) In the *Azotobacter vinelandii* complex, 24 E2 polypeptides are arranged in a cube. (b) The 60 subunits of the E2 core from *B. stearothermophilus* form a dodecahedron, a shape with 12 pentagonal faces. [*Structure of the A.* vinelandii *core (pdb 1EAA) determined by A. Mattevei and W. G. J. Hol. Structure of the* B. stearothermophilus *core (pdb 1B5S) determined by T. Izard, A. Aevarsson, M. D. Allen, A. H. Westphal, R. N. Perham, A. De Kok, and W. G. Hol.*]

PYRUVATE DEHYDROGENASE CONVERTS PYRUVATE TO ACETYL-CoA

The operation of the pyruvate dehydrogenase complex requires several coenzymes, whose functional roles in the five-step reaction are described below.

1. In the first step, which is catalyzed by E1 (also called pyruvate dehydrogenase), pyruvate is decarboxylated. This reaction requires the cofactor thiamine pyrophosphate (TPP; **Fig. 14.2**). TPP attacks the carbonyl carbon of pyruvate, and the departure of CO_2 leaves a hydroxyethyl group attached to TPP. This carbanion is stabilized by the positively charged thiazolium ring group of TPP:

Pyruvate → Hydroxyethyl–TPP

FIGURE 14.2

Thiamine pyrophosphate (TPP). This cofactor is the phosphorylated form of thiamine, also known as vitamin B_1 (see Section 12.2). The central thiazolium ring (blue) is the active portion. An acidic proton (red) dissociates, and the resulting carbanion is stabilized by the nearby positively charged nitrogen. TPP is a cofactor for several different decarboxylases.

Thiamine pyrophosphate

2. The hydroxyethyl group is then transferred to E2 of the pyruvate dehydrogenase complex. The hydroxyethyl acceptor is a lipoamide prosthetic group (**Fig. 14.3**).

FIGURE 14.3

Lipoamide. This prosthetic group consists of lipoic acid (a vitamin) linked via an amide bond to the ε-amino group of a protein lysine residue. The active portion of the 14-Å-long lipoamide is the disulfide bond (red), which can be reversibly reduced.

The transfer reaction regenerates the TPP cofactor of E1 and oxidizes the hydroxyethyl group to an acetyl group:

3. Next, E2 transfers the acetyl group to coenzyme A, producing acetyl-CoA and leaving a reduced lipoamide group.

Recall that acetyl-CoA is a thioester, a form of energy currency (see Section 12.3). Some of the free energy released in the oxidation of the hydroxyethyl group to an acetyl group is conserved in the formation of acetyl-CoA.

4. The final two steps of the reaction restore the pyruvate dehydrogenase complex to its original state. E3 reoxidizes the lipoamide group of E2 by transferring electrons to a Cys–Cys disulfide group in the enzyme.

5. Finally, NAD^+ reoxidizes the reduced cysteine sulfhydryl groups. This electron-transfer reaction is facilitated by an FAD prosthetic group (the structure of FAD, a nucleotide derivative, is shown in Fig. 3.2c).

During the five-step reaction (summarized in **Fig. 14.4**), the long lipoamide group of E2 acts as a swinging arm that visits the active sites of E1, E2, and E3 within the multienzyme complex. The arm picks up an acetyl group from an E1 subunit and transfers it to coenzyme A in an E2 active site. The arm then swings to an E3 active site, where it is reoxidized. Some other multienzyme complexes also include swinging arms, often attached to hinged protein domains to maximize their mobility.

FIGURE 14.4

Reactions of the pyruvate dehydrogenase complex. In these five reactions, an acetyl group from pyruvate is transferred to CoA, CO_2 is released, and NAD^+ is reduced to NADH.

? **Without looking at the text, write the net equation for the reactions shown here. How many vitamin-derived cofactors are involved?**

A **multienzyme complex** such as the pyruvate dehydrogenase complex can carry out a multistep reaction sequence efficiently because *the product of one reaction can quickly become the substrate for the next reaction without diffusing away or reacting with another substance.* There is also evidence that the individual enzymes of glycolysis and the citric acid cycle associate loosely with each other so that the close proximity of their active sites can increase flux through their respective pathways.

Flux through the pyruvate dehydrogenase complex is regulated by product inhibition: Both NADH and acetyl-CoA act as inhibitors. The activity of the complex is also regulated by hormone-controlled phosphorylation and dephosphorylation, which suits its function as the gatekeeper for the entry of a metabolic fuel into the citric acid cycle.

◆ **CONCEPT CHECK**

1. Describe the functional importance of the coenzymes that participate in the reactions carried out by the pyruvate dehydrogenase complex.

2. What are the advantages of a multienzyme complex?

14.2 THE EIGHT REACTIONS OF THE CITRIC ACID CYCLE

The starting material for the **citric acid cycle** is an acetyl-CoA molecule that may be derived from a carbohydrate via pyruvate, as just described, or from another metabolic fuel. The carbon skeletons of amino acids are broken down to either pyruvate or acetyl-CoA, and fatty acids are broken down to acetyl-CoA. In some tissues, the bulk of acetyl-CoA entering the citric acid cycle comes from fatty acids rather than carbohydrates or amino acids. Whatever their source, the citric acid cycle converts all these two-carbon acetyl groups into CO_2 and therefore represents the final stage in fuel oxidation (**Fig. 14.5**). As the carbons become fully oxidized to CO_2, their energy is conserved and subsequently used to produce ATP.

KEY CONCEPTS

- The citric acid cycle is a set of eight reactions in which an acetyl group is condensed with oxaloacetate, two CO_2 are lost, and oxaloacetate is regenerated.

- Each round of the citric acid cycle generates three NADH, one QH_2, and one GTP or ATP.

FIGURE 14.5

The citric acid cycle in context. The citric acid cycle is a central metabolic pathway whose starting material is two-carbon acetyl units derived from amino acids, monosaccharides, and fatty acids. These are oxidized to the waste product CO_2, with the reduction of the cofactors NAD^+ and ubiquinone (Q).

The eight reactions of the citric acid cycle take place in the cytosol of prokaryotes and in the mitochondria of eukaryotes. Unlike a linear pathway such as glycolysis (see Fig. 13.2) or gluconeogenesis (see Fig. 13.10), the citric acid cycle always returns to its starting position, essentially behaving as a multistep catalyst.

The cycle as a whole is highly exergonic, and free energy is conserved at several steps in the form of a nucleotide triphosphate (GTP) and reduced cofactors. *For each acetyl group that enters the citric acid cycle, two molecules of fully oxidized CO_2 are produced, representing a loss of four pairs of electrons.* These electrons are transferred to 3 NAD^+ and 1 ubiquinone (Q) to produce 3 NADH and 1 QH_2. The net equation for the citric acid cycle is therefore

$$acetyl\text{-}CoA + GDP + P_i + 3\,NAD^+ + Q \rightarrow$$
$$2\,CO_2 + CoA + GTP + 3\,NADH + QH_2$$

In this section we examine the sequence of eight enzyme-catalyzed reactions of the citric acid cycle, focusing on a few interesting reactions. The entire pathway is summarized in **Figure 14.6**.

1. CITRATE SYNTHASE ADDS AN ACETYL GROUP TO OXALOACETATE

In the first reaction of the citric acid cycle, the acetyl group of acetyl-CoA condenses with the four-carbon compound oxaloacetate to produce the six-carbon compound citrate:

$$
\begin{array}{ccc}
\text{COO}^- & & \text{COO}^- \\
| & & | \\
\text{C=O} & \quad\quad \text{O} \quad\quad \xrightarrow[\text{synthase}]{\text{citrate}} & \text{CH}_2 \\
| & + \;\; \text{CH}_3-\overset{\|}{\text{C}}-\text{SCoA} \quad \overset{\text{H}_2\text{O} \;\; \text{HSCoA}}{\longrightarrow} & \text{HO}-\text{C}-\text{COO}^- \\
\text{CH}_2 & & | \\
| & & \text{CH}_2 \\
\text{COO}^- & & | \\
& & \text{COO}^- \\
\text{Oxaloacetate} & \text{Acetyl-CoA} & \text{Citrate}
\end{array}
$$

Citrate synthase, a dimer, undergoes a large conformational change upon substrate binding (**Fig. 14.7**).

Citrate synthase is one of the few enzymes that can synthesize a carbon–carbon bond without using a metal ion cofactor. Its mechanism is shown in **Figure 14.8**. The first reaction intermediate may be stabilized by the formation of low-barrier hydrogen bonds, which are stronger than ordinary hydrogen bonds (see Section 6.3). The coenzyme A released during the final step can be reused by the pyruvate dehydrogenase complex or used later in the citric acid cycle to synthesize the intermediate succinyl-CoA.

The reaction catalyzed by citrate synthase is highly exergonic ($\Delta G^{\circ\prime} = -31.5$ kJ·mol^{-1}, equivalent to the free energy of hydrolyzing the thioester bond of acetyl-CoA). We will see later why the efficient operation of the citric acid cycle requires that this step have a large negative free energy change.

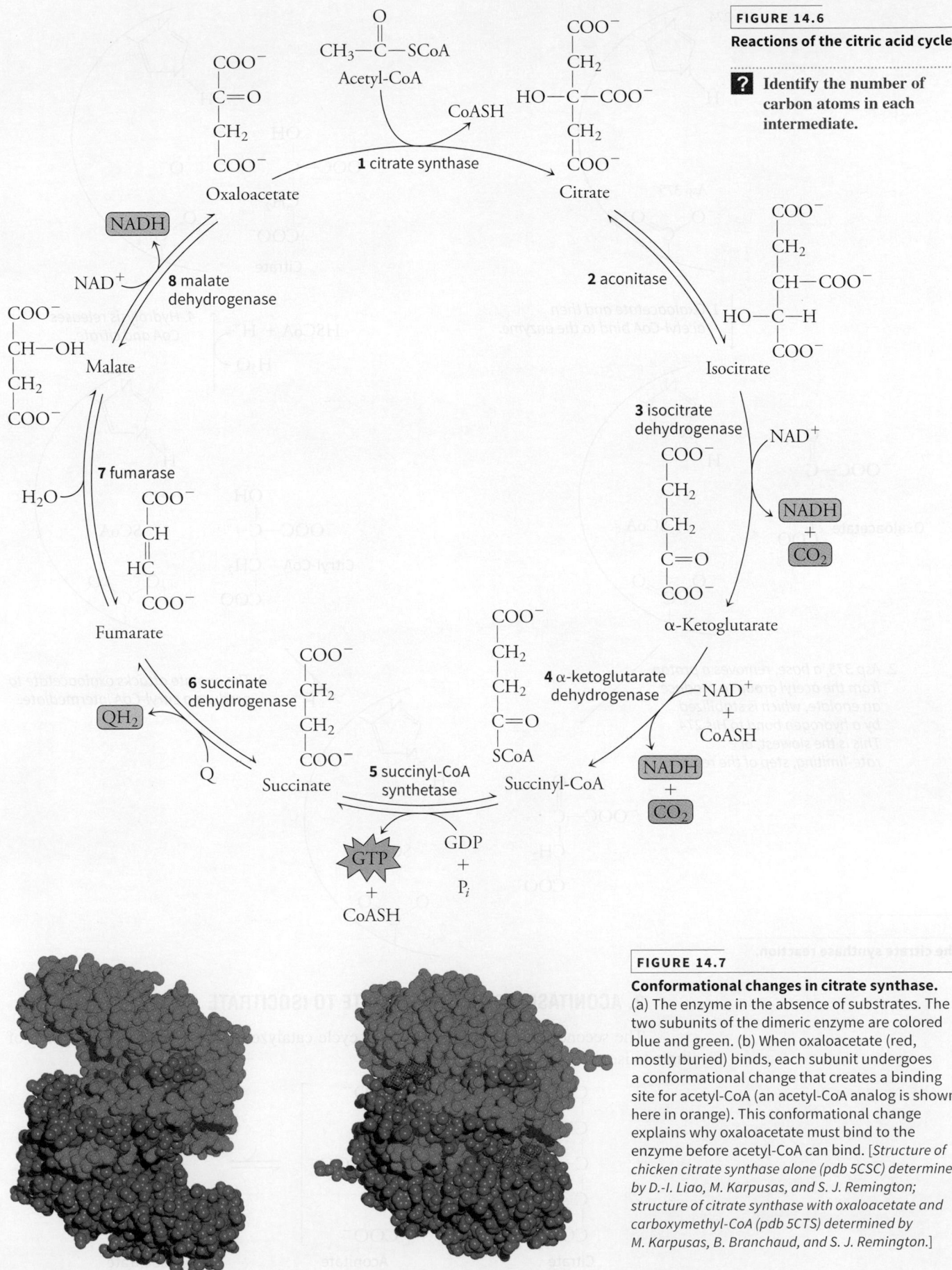

FIGURE 14.6

Reactions of the citric acid cycle.

? Identify the number of carbon atoms in each intermediate.

Oxaloacetate

Acetyl-CoA $CH_3-\overset{\overset{\displaystyle O}{\|}}{C}-SCoA$

CoASH

1 citrate synthase

Citrate

2 aconitase

Isocitrate

3 isocitrate dehydrogenase

NAD^+

NADH + CO_2

α-Ketoglutarate

4 α-ketoglutarate dehydrogenase

NAD^+ + CoASH

NADH + CO_2

Succinyl-CoA

5 succinyl-CoA synthetase

GDP + P_i

GTP + CoASH

Succinate

6 succinate dehydrogenase

Q

QH_2

Fumarate

7 fumarase

H_2O

Malate

8 malate dehydrogenase

NAD^+

NADH

FIGURE 14.7

Conformational changes in citrate synthase.
(a) The enzyme in the absence of substrates. The two subunits of the dimeric enzyme are colored blue and green. (b) When oxaloacetate (red, mostly buried) binds, each subunit undergoes a conformational change that creates a binding site for acetyl-CoA (an acetyl-CoA analog is shown here in orange). This conformational change explains why oxaloacetate must bind to the enzyme before acetyl-CoA can bind. [*Structure of chicken citrate synthase alone (pdb 5CSC) determined by D.-I. Liao, M. Karpusas, and S. J. Remington; structure of citrate synthase with oxaloacetate and carboxymethyl-CoA (pdb 5CTS) determined by M. Karpusas, B. Branchaud, and S. J. Remington.*]

(a)   (b)

His 274

Asp 375

1. Oxaloacetate and then acetyl-CoA bind to the enzyme.

OH

$^-$OOC$-$C$-$CH$_2$$-COO^-$

CH$_2$

COO$^-$

Citrate

4. Hydrolysis releases CoA and citrate.

HSCoA + H$^+$

H$_2$O

$^-$OOC$-$C

CH$_2$

COO$^-$

Oxaloacetate

O Acetyl-CoA

H$_2$C$-$C$-$SCoA

H

O

C

O

2. Asp 375, a base, removes a proton from the acetyl group to produce an enolate, which is stabilized by a hydrogen bond to His 274. This is the slowest, or rate-limiting, step of the reaction.

OH

$^-$OOC$-$C$-$CH$_2$$-C-$SCoA

CH$_2$

COO$^-$

Citryl-CoA

HO C O

3. The enolate attacks oxaloacetate to produce a citryl-CoA intermediate.

H$^+$

$^-$OOC$-$C

CH$_2$

COO$^-$

O

H$_2$C$=$C$-$SCoA

HO C O

FIGURE 14.8

The citrate synthase reaction.

2. ACONITASE ISOMERIZES CITRATE TO ISOCITRATE

The second enzyme of the citric acid cycle catalyzes the reversible isomerization of citrate to isocitrate:

COO$^-$
|
CH$_2$
|
HO$-$C$-$COO$^-$
|
CH$_2$
|
COO$^-$

Citrate

$\rightleftharpoons$ H$_2$O

$\left[\begin{array}{c} \text{COO}^- \\ | \\ \text{CH}_2 \\ | \\ \text{C}-\text{COO}^- \\ || \\ \text{CH} \\ | \\ \text{COO}^- \end{array}\right]$

Aconitate

$\rightleftharpoons$ H$_2$O

COO$^-$
|
CH$_2$
|
H$-$C$-$COO$^-$
|
HO$-$C$-$H
|
COO$^-$

Isocitrate

The enzyme is named after the reaction intermediate.

Citrate is a symmetrical molecule, yet only one of its two carboxymethyl arms (—CH_2—COO^-) undergoes dehydration and rehydration during the aconitase reaction. This stereochemical specificity long puzzled biochemists, including Hans Krebs, who first described the citric acid cycle (also known as the Krebs cycle). Eventually, Alexander Ogston pointed out that although citrate is symmetrical, its two carboxymethyl groups are no longer identical when it is bound to an asymmetrical enzyme (**Fig. 14.9**). In fact, a three-point attachment is not even necessary for an enzyme to distinguish two groups in a molecule such as citrate, which are related by mirror symmetry. You can prove this yourself with a simple organic chemistry model kit. By now you should appreciate that biological systems, including enzyme, are inherently chiral (also see Section 4.1).

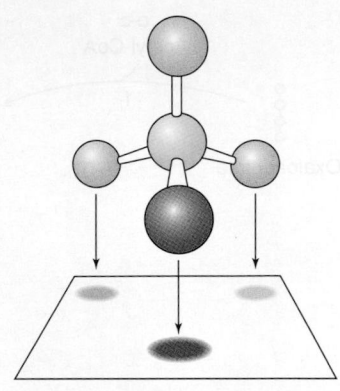

FIGURE 14.9

Stereochemistry of aconitase. The three-point attachment of citrate to the enzyme allows only one carboxymethyl group (shown in green) to react.

3. ISOCITRATE DEHYDROGENASE RELEASES THE FIRST CO_2

The third reaction of the citric acid cycle is the oxidative decarboxylation of isocitrate to α-ketoglutarate. The substrate is first oxidized in a reaction accompanied by the reduction of NAD^+ to NADH. Then the carboxylate group β to the ketone function (that is, two carbon atoms away from the ketone) is eliminated as CO_2. An Mn^{2+} ion in the active site helps stabilize the negative charges of the reaction intermediate.

The CO_2 molecules generated by isocitrate dehydrogenase—along with the CO_2 generated in the following reaction and the CO_2 produced by the decarboxylation of pyruvate—diffuse out of the cell and are carried in the bloodstream to the lungs, where they are breathed out. Note that these CO_2 molecules are produced through oxidation–reduction reactions: The carbons are oxidized, while NAD^+ is reduced. O_2 is not directly involved in this process.

4. α-KETOGLUTARATE DEHYDROGENASE RELEASES THE SECOND CO_2

α-Ketoglutarate dehydrogenase, like isocitrate dehydrogenase, catalyzes an oxidative decarboxylation reaction. It also transfers the remaining four-carbon fragment to CoA:

The free energy of oxidizing α-ketoglutarate is conserved in the formation of the thioester succinyl-CoA. α-Ketoglutarate dehydrogenase is a multienzyme complex that resembles the pyruvate dehydrogenase complex in both structure and enzymatic mechanism. In fact, the same E3 enzyme is a member of both complexes.

Acetyl-CoA

Oxaloacetate

1

Citrate

2

Isocitrate

3

α-Ketoglutarate

4

Succinyl-CoA

Fates of carbon atoms in the citric acid cycle. The two carbon atoms that are lost as CO_2 in the reactions catalyzed by isocitrate dehydrogenase (step 3) and α-ketoglutarate dehydrogenase (step 4) are not the same carbons that entered the cycle as acetyl-CoA (red). The acetyl carbons become part of oxaloacetate and are lost in subsequent rounds of the cycle.

The succinyl-CoA synthetase reaction.

⁇ **What is the fate of the free CoA molecule?**

The isocitrate dehydrogenase and α-ketoglutarate dehydrogenase reactions both release CO_2. These two carbons are not the ones that entered the citric acid cycle as acetyl-CoA; those acetyl carbons are released in subsequent rounds of the cycle (**Fig. 14.10**). *However, the net result of each round of the citric acid cycle is the loss of two carbons as CO_2 for each acetyl-CoA that enters the cycle.*

5. SUCCINYL-CoA SYNTHETASE CATALYZES SUBSTRATE-LEVEL PHOSPHORYLATION

The thioester succinyl-CoA releases a large amount of free energy when it is hydrolyzed ($\Delta G^{\circ\prime} = -32.6$ kJ·mol^{-1}). This is enough free energy to drive the synthesis of a nucleoside triphosphate from a nucleoside diphosphate and P_i ($\Delta G^{\circ\prime} = 30.5$ kJ·mol^{-1}). The change in free energy for the net reaction is near zero, so the reaction is reversible. In fact, the enzyme is named for the reverse reaction. Succinyl-CoA synthetase in the mammalian citric acid cycle generates GTP, whereas the plant and bacterial enzymes generate ATP (recall that GTP is energetically equivalent to ATP). An exergonic reaction coupled to the transfer of a phosphoryl group to a nucleoside diphosphate is termed **substrate-level phosphorylation** to distinguish it from oxidative phosphorylation (Section 15.4) and photophosphorylation (Section 16.2), which are more indirect ways of synthesizing ATP.

How does succinyl-CoA synthetase couple thioester cleavage to the synthesis of a nucleoside triphosphate? The reaction is a series of phosphoryl-group transfers that involve an active-site histidine residue (**Fig. 14.11**). The phospho-His reaction intermediate must move a large distance to shuttle the phosphoryl group between the succinyl group and the nucleoside diphosphate (**Fig. 14.12**).

6. SUCCINATE DEHYDROGENASE GENERATES UBIQUINOL

The final three reactions of the citric acid cycle convert succinate back to the cycle's starting substrate, oxaloacetate. Succinate dehydrogenase catalyzes the reversible dehydrogenation

1. *A phosphate group displaces CoA in succinyl-CoA. The product, succinyl phosphate, is an acyl phosphate, which releases a large amount of free energy when hydrolyzed.*

2. *Succinyl phosphate donates its phosphoryl group to a His residue on the enzyme, producing a phospho-His intermediate and releasing succinate.*

3. *The phosphoryl group is then transferred to GDP to form GTP.*

of succinate to fumarate. This oxidation–reduction reaction requires an FAD prosthetic group, which is reduced to $FADH_2$ during the reaction:

$$\underset{\text{Succinate}}{\begin{array}{c} COO^- \\ | \\ H-C-H \\ | \\ H-C-H \\ | \\ COO^- \end{array}} \xrightleftharpoons[\text{succinate dehydrogenase}]{\overset{\text{Enz-FAD} \quad \text{Enz-FADH}_2}{\curvearrowright}} \underset{\text{Fumarate}}{\begin{array}{c} H \quad\quad COO^- \\ \backslash \quad / \\ C \\ \| \\ C \\ / \quad\quad \backslash \\ {}^-OOC \quad\quad H \end{array}}$$

To regenerate the enzyme, the $FADH_2$ group must be reoxidized. Since succinate dehydrogenase is embedded in the inner mitochondrial membrane (it is the only one of the eight citric acid cycle enzymes that is not soluble in the mitochondrial matrix), it can be reoxidized by the lipid-soluble electron carrier ubiquinone (see Section 12.2) rather than by the soluble cofactor NAD^+. Ubiquinone (abbreviated Q) acquires two electrons to become ubiquinol (QH_2).

$$\text{Enzyme-FADH}_2 \xrightleftharpoons[]{\overset{\text{Q} \quad\quad \text{QH}_2}{\curvearrowright}} \text{Enzyme-FAD}$$

7. FUMARASE CATALYZES A HYDRATION REACTION

In the seventh reaction, fumarase (also known as fumarate hydratase) catalyzes the reversible hydration of a double bond to convert fumarate to malate:

$$\underset{\text{Fumarate}}{\begin{array}{c} COO^- \\ | \\ CH \\ \| \\ HC \\ | \\ COO^- \end{array}} \xrightleftharpoons[\text{fumarase}]{H_2O} \underset{\text{Malate}}{\begin{array}{c} COO^- \\ | \\ H-C-OH \\ | \\ H-C-H \\ | \\ COO^- \end{array}}$$

8. MALATE DEHYDROGENASE REGENERATES OXALOACETATE

The citric acid cycle concludes with the regeneration of oxaloacetate from malate in an NAD^+-dependent oxidation reaction:

$$\underset{\text{Malate}}{\begin{array}{c} COO^- \\ | \\ H-C-OH \\ | \\ CH_2 \\ | \\ COO^- \end{array}} \xrightleftharpoons[\text{malate dehydrogenase}]{\overset{\text{NAD}^+ \quad \text{NADH} + \text{H}^+}{\curvearrowright}} \underset{\text{Oxaloacetate}}{\begin{array}{c} COO^- \\ | \\ C=O \\ | \\ CH_2 \\ | \\ COO^- \end{array}}$$

The standard free energy change for this reaction is +29.7 kJ·mol⁻¹, indicating that the reaction has a low probability of occurring as written. However, the product oxaloacetate is a substrate for the next reaction (Reaction 1 of the citric acid cycle). The highly exergonic—and therefore highly favorable—citrate synthase reaction helps pull the malate dehydrogenase reaction

FIGURE 14.12

Substrate binding in succinyl-CoA synthetase. Succinyl-CoA (represented by coenzyme A, red) binds to the enzyme, and its succinyl group is phosphorylated. The succinyl phosphate then transfers its phosphoryl group to the His 246 side chain (green). A protein loop containing the phospho-His side chain must undergo a large movement because the nucleoside diphosphate awaiting phosphorylation (ADP, orange) binds to a site about 35 Å away. [*Structure of* E. coli *succinyl-CoA synthetase (pdb 1CQI) determined by M. A. Joyce, M. E. Fraser, M. N. G. James, W. A. Bridger, and W. T. Wolodko.*]

forward. This is the reason for the apparent waste of free energy released by cleaving the thioester bond of acetyl-CoA in the first reaction of the citric acid cycle.

 CONCEPT CHECK

1. List the sources of the acetyl groups that enter the citric acid cycle.

2. Write the net equation for the citric acid cycle.

3. Draw the structures of the substrates and products of the eight reactions and name the enzyme that catalyzes each step.

4. Identify the steps that generate ATP, CO_2, and reduced cofactors.

14.3 THERMODYNAMICS OF THE CITRIC ACID CYCLE

KEY CONCEPTS
- Flux through the citric acid cycle is regulated primarily by feedback inhibition at three steps.
- The citric acid cycle likely evolved by the combination of oxidative and reductive pathways.

Because the eighth reaction of the citric acid cycle returns the system to its original state, *the entire pathway acts in a catalytic fashion to dispose of carbon atoms derived from amino acids, carbohydrates, and fatty acids.* Albert Szent-Györgyi discovered the catalytic nature of the pathway by observing that small additions of organic compounds such as succinate, fumarate, and malate stimulated O_2 uptake in a tissue preparation. Because the O_2 consumption was much greater than would be required for the direct oxidation of the added substances, he inferred that the compounds acted catalytically.

THE CITRIC ACID CYCLE IS AN ENERGY-GENERATING CATALYTIC CYCLE

We now know that oxygen is consumed during oxidative phosphorylation, the process that reoxidizes the reduced cofactors (NADH and QH_2) that are produced by the citric acid cycle. Although the citric acid cycle generates one molecule of GTP (or ATP), considerably more ATP is generated when the reduced cofactors are reoxidized by O_2. Each NADH yields approximately 2.5 ATP, and each QH_2 yields approximately 1.5 ATP (we will see in Section 15.4 why these values are not whole numbers). *Every acetyl unit that enters the citric acid cycle can therefore generate a total of 10 ATP equivalents.* The energy yield of a molecule of glucose, which generates two acetyl units, can be calculated (**Fig. 14.13**).

A muscle operating anaerobically produces only 2 ATP per glucose, but under aerobic conditions when the citric acid cycle is fully functional, each glucose molecule generates about 32 ATP equivalents. This general phenomenon is called the **Pasteur effect,** after Louis Pasteur, who first observed that the rate of glucose consumption by yeast cells decreased dramatically when the cells were shifted from anaerobic to aerobic growth conditions.

THE CITRIC ACID CYCLE IS REGULATED AT THREE STEPS

Flux through the citric acid cycle is regulated primarily at the cycle's three metabolically irreversible steps: those catalyzed by citrate synthase (Reaction 1), isocitrate dehydrogenase (Reaction 3), and α-ketoglutarate dehydrogenase (Reaction 4). The major regulators are shown in **Figure 14.14**.

Neither acetyl-CoA nor oxaloacetate is present at concentrations high enough to saturate citrate synthase, so flux through the first step of the citric acid cycle depends largely on the substrate concentrations. The product of the reaction, citrate, inhibits citrate synthase (citrate also inhibits phosphofructokinase, thereby decreasing the supply of acetyl-CoA produced by glycolysis). Succinyl-CoA, the product of Reaction 4, inhibits the enzyme that produces it. It also acts as a feedback inhibitor by competing with acetyl-CoA in Reaction 1.

FIGURE 14.13

ATP yield from glucose. Two rounds of the citric acid cycle are required to fully oxidize one molecule of glucose.

? Trace the fate of the six glucose carbon atoms.

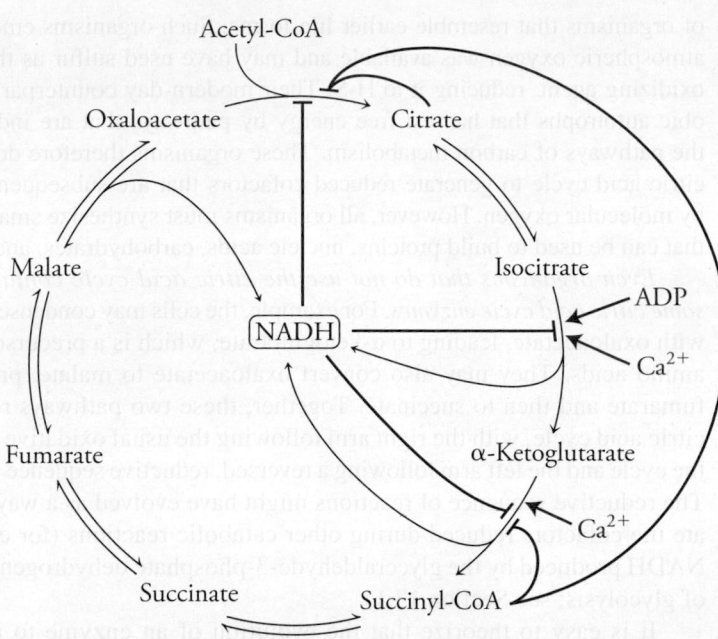

FIGURE 14.14

Regulation of the citric acid cycle. Inhibition is represented by red symbols, activation by green symbols.

The activity of isocitrate dehydrogenase is inhibited by its reaction product, NADH. NADH also inhibits α-ketoglutarate dehydrogenase and citrate synthase. Both dehydrogenases are activated by Ca^{2+} ions, which generally signify the need to generate cellular free energy. ADP, also representing the need for more ATP, activates isocitrate dehydrogenase.

Changes in enzyme reaction rates also regulate the flow of acetyl carbons through the cycle by altering the concentrations of cycle intermediates. Because the entire cycle acts as a catalyst, more intermediates means that more acetyl groups can be processed, just as a city can move more commuters by adding buses during rush hour. Not surprisingly, citric acid cycle defects have serious consequences (**Box 14.A**).

THE CITRIC ACID CYCLE PROBABLY EVOLVED AS A SYNTHETIC PATHWAY

A circular pathway such as the citric acid cycle must have evolved from a linear set of preexisting biochemical reactions. Clues to its origins can be found by examining the metabolism

A CLOSER LOOK

Box 14.A Mutations in Citric Acid Cycle Enzymes

Possibly because the citric acid cycle is a central metabolic pathway, severe defects in any of its components are expected to be incompatible with life. However, researchers have documented mutations in the genes for several of the cycle's enzymes, including α-ketoglutarate dehydrogenase, succinyl-CoA synthetase, and succinate dehydrogenase. These defects, which are all rare, typically affect the central nervous system, causing symptoms such as movement disorders and neurodegeneration. A rare form of fumarase deficiency results in brain malformation and developmental disabilities.

Some citric acid cycle enzyme mutations are linked to cancer. One possible explanation is that a defective enzyme contributes to **carcinogenesis** (the development of cancer, or uncontrolled cell growth) by causing the accumulation of particular metabolites, which are responsible for altering the cell's activities. For example, normal cells respond to a drop in oxygen availability (hypoxia) by activating transcription factors known as hypoxia-inducible factors (HIFs). These proteins interact with DNA to turn on the expression of genes for glycolytic enzymes and a growth factor that promotes the development of new blood vessels. When the fumarase gene is defective, fumarate accumulates and inhibits a protein that destabilizes HIFs. As a result, the fumarase deficiency promotes glycolysis (an anaerobic pathway) and the growth of blood vessels. These two adaptations favor tumors, whose growth, although characteristically rapid, may be limited by the availability of oxygen and other nutrients delivered by the bloodstream.

Defects in isocitrate dehydrogenase also promote cancer in an indirect fashion. Many cancerous cells exhibit a mutation in one of the two genes for the enzyme, suggesting that the unaltered copy is necessary for maintaining the normal activity of the citric acid cycle, while the mutated copy plays a role in carcinogenesis. The mutated isocitrate dehydrogenase no longer carries out the usual reaction (converting isocitrate to α-ketoglutarate) but instead converts α-ketoglutarate to 2-hydroxyglutarate in an NADPH-dependent manner. The mechanism whereby 2-hydroxyglutarate contributes to carcinogenesis is not clear, but its involvement is bolstered by the observation that individuals who harbor other mutations that lead to 2-hydroxyglutarate accumulation have an increased risk of developing brain tumors.

? **How would a fumarase deficiency affect the levels of pyruvate, fumarate, and malate?**

FIGURE 14.15

Pathways that might have given rise to the citric acid cycle. The pathway starting from oxaloacetate and proceeding to the right is an oxidative biosynthetic pathway, whereas the pathway that proceeds to the left is a reductive pathway. The modern citric acid cycle may have evolved by connecting these pathways.

FIGURE 14.16

A proposed reductive biosynthetic pathway based on the citric acid cycle. This pathway might have operated to incorporate CO_2 into biological molecules.

of organisms that resemble earlier life-forms. Such organisms emerged before atmospheric oxygen was available and may have used sulfur as their ultimate oxidizing agent, reducing it to H_2S. Their modern-day counterparts are anaerobic autotrophs that harvest free energy by pathways that are independent of the pathways of carbon metabolism. These organisms therefore do not use the citric acid cycle to generate reduced cofactors that are subsequently oxidized by molecular oxygen. However, all organisms must synthesize small molecules that can be used to build proteins, nucleic acids, carbohydrates, and so on.

Even organisms that do not use the citric acid cycle contain genes for some citric acid cycle enzymes. For example, the cells may condense acetyl-CoA with oxaloacetate, leading to α-ketoglutarate, which is a precursor of several amino acids. They may also convert oxaloacetate to malate, proceeding to fumarate and then to succinate. Together, these two pathways resemble the citric acid cycle, with the right arm following the usual oxidative sequence of the cycle and the left arm following a reversed, reductive sequence (**Fig. 14.15**). The reductive sequence of reactions might have evolved as a way to regenerate the cofactors reduced during other catabolic reactions (for example, the NADH produced by the glyceraldehyde-3-phosphate dehydrogenase reaction of glycolysis; see Section 13.1).

It is easy to theorize that the evolution of an enzyme to interconvert α-ketoglutarate and succinate could have created a cyclic pathway similar to the modern citric acid cycle. Interestingly, *E. coli*, which uses the citric acid cycle under aerobic growth conditions, uses an interrupted citric acid cycle like the one diagrammed in Figure 14.15 when it is growing anaerobically.

Since the final four reactions of the modern citric acid cycle are metabolically reversible, the primitive citric acid cycle might easily have accommodated one-way flux in the clockwise direction, forming an oxidative cycle. If the complete cycle proceeded in the counterclockwise direction, the result would have been a reductive biosynthetic pathway (**Fig. 14.16**). This pathway, which would incorporate, or "fix," atmospheric CO_2 into biological molecules, may have preceded the modern CO_2-fixing pathway found in green plants and some photosynthetic bacteria (described in Section 16.3).

CONCEPT CHECK

1. Identify the products of the citric acid cycle that represent forms of energy currency for the cell.

2. Compare the ATP yield for glucose degradation via glycolysis and via glycolysis plus the citric acid cycle.

3. Which substrates and products of the citric acid cycle regulate flux through the pathway?

4. Explain how primitive oxidative and reductive biosynthetic pathways might have combined to generate a circular metabolic pathway.

14.4 ANABOLIC AND CATABOLIC FUNCTIONS OF THE CITRIC ACID CYCLE

KEY CONCEPTS
- The citric acid cycle supplies precursors for the synthesis of other compounds.
- Citric acid cycle intermediates can be replenished.

The citric acid cycle does not operate like a simple pipeline, where one substance enters at one end and another emerges from the opposite end. In mammals, six of the eight citric acid cycle intermediates (all except isocitrate and succinate) are the precursors or products of other pathways. For this reason, it is impossible to designate the citric acid cycle as a purely catabolic or anabolic pathway.

FIGURE 14.17

Citric acid cycle intermediates as biosynthetic precursors.

CITRIC ACID CYCLE INTERMEDIATES ARE PRECURSORS OF OTHER MOLECULES

Intermediates of the citric acid cycle can be siphoned off to form other compounds (**Fig. 14.17**). For example, succinyl-CoA is used for the synthesis of heme. The five-carbon α-ketoglutarate (sometimes called 2-oxoglutarate) can undergo reductive amination by glutamate dehydrogenase to produce the amino acid glutamate:

Glutamate is a precursor of the amino acids glutamine, arginine, and proline. Glutamine in turn is a precursor for the synthesis of purine and pyrimidine nucleotides. We have already seen that oxaloacetate is a precursor of monosaccharides (Section 13.2). Consequently, *any of the citric acid cycle intermediates, which can be converted by the cycle to oxaloacetate, can ultimately serve as gluconeogenic precursors.*

Citrate produced by the condensation of acetyl-CoA with oxaloacetate can be transported out of the mitochondria to the cytosol. ATP-citrate lyase then catalyzes the reaction

$$\text{ATP} + \text{citrate} + \text{CoA} \rightarrow \text{ADP} + P_i + \text{oxaloacetate} + \text{acetyl-CoA}$$

The resulting acetyl-CoA is used for fatty acid and cholesterol synthesis, which take place in the cytosol. Note that the ATP-citrate lyase reaction undoes the work of the exergonic citrate synthase reaction. This seems wasteful, but cytosolic ATP-citrate lyase is essential because acetyl-CoA, which is produced in the mitochondria, cannot cross the mitochondrial membrane to reach the cytosol, whereas citrate can. The oxaloacetate product of the ATP-citrate

FIGURE 14.18

The citrate transport system.
Both citrate and pyruvate cross the inner mitochondrial membrane via specific transport proteins. This system allows carbon atoms from mitochondrial acetyl-CoA to be transferred to the cytosol for the synthesis of fatty acids and cholesterol.

? Can the same CO_2 molecule released in the malic enzyme reaction be used in the pyruvate carboxylase reaction?

lyase reaction can be converted to malate by a cytosolic malate dehydrogenase operating in reverse. Malate is then decarboxylated by the action of malic enzyme to produce pyruvate:

$$\begin{array}{c} COO^- \\ | \\ CH-OH \\ | \\ CH_2 \\ | \\ COO^- \end{array} \quad \underset{\text{malic enzyme}}{\overset{NADP^+ \quad NADPH}{\rightleftharpoons}} \quad \begin{array}{c} COO^- \\ | \\ C=O \\ | \\ CH_3 \end{array} \;+\; CO_2$$

Malate   Pyruvate

Pyruvate can reenter the mitochondria and be converted back to oxaloacetate to complete the cycle shown in **Figure 14.18**. In plants, isocitrate is diverted from the citric acid cycle in a biosynthetic pathway known as the **glyoxylate pathway** (**Box 14.B**).

ANAPLEROTIC REACTIONS REPLENISH CITRIC ACID CYCLE INTERMEDIATES

Intermediates that are diverted from the citric acid cycle for other purposes can be replenished through **anaplerotic reactions** (from the Greek *ana*, "up," and *plerotikos*, "to fill"; **Fig. 14.19**). One of the most important of these reactions is catalyzed by pyruvate carboxylase (this is also the first step of gluconeogenesis; Section 13.2):

$$\text{pyruvate} + CO_2 + ATP + H_2O \rightarrow \text{oxaloacetate} + ADP + P_i$$

Acetyl-CoA activates pyruvate carboxylase, so when the activity of the citric acid cycle is low and acetyl-CoA accumulates, more oxaloacetate is produced. The concentration of oxaloacetate is normally low since the malate dehydrogenase reaction is thermodynamically unfavorable and the citrate synthase reaction is highly favorable. The replenished oxaloacetate is converted to citrate, isocitrate, α-ketoglutarate, and so on, so the concentrations of all the citric

A CLOSER LOOK

Box 14.B The Glyoxylate Pathway

Plants and some bacterial cells contain certain enzymes that act together with some citric acid cycle enzymes to convert acetyl-CoA to oxaloacetate, a gluconeogenic precursor. Animals lack the enzymes to do this and therefore cannot undertake the net synthesis of carbohydrates from two-carbon precursors. In plants, the glyoxylate pathway includes reactions that take place in the mitochondria and the **glyoxysome,** an organelle that, like the peroxisome, contains enzymes that carry out some essential metabolic processes.

In the glyoxysome, acetyl-CoA condenses with oxaloacetate to form citrate, which is then isomerized to isocitrate, as in the citric acid cycle. However, the next step is not the isocitrate dehydrogenase reaction but a reaction catalyzed by the glyoxysome enzyme isocitrate lyase, which converts isocitrate to succinate and the two-carbon compound glyoxylate. Succinate continues as usual through the mitochondrial citric acid cycle to regenerate oxaloacetate.

In the glyoxysome, the glyoxylate condenses with a second molecule of acetyl-CoA in a reaction catalyzed by the glyoxysome enzyme malate synthase to form the four-carbon compound malate. Malate can then be converted to oxaloacetate for gluconeogenesis. The two reactions that are unique to the glyoxylate pathway are shown in green in the figure; reactions that are identical to those of the citric acid cycle are shown in blue.

In essence, the glyoxylate pathway bypasses the two CO_2-generating steps of the citric acid cycle (catalyzed by isocitrate dehydrogenase and α-ketoglutarate dehydrogenase) and incorporates a second acetyl unit (at the malate synthase step). The net result of the glyoxylate pathway is the production of a four-carbon compound that can be used to synthesize glucose. This pathway is highly active in germinating seeds, where stored oils (triacylglycerols) are broken down to acetyl-CoA. The glyoxylate pathway thus provides a route for synthesizing glucose from fatty acids. Because animals lack isocitrate lyase and malate synthase, they cannot convert excess fat to carbohydrate.

? **Write the net equation for the glyoxylate cycle as shown here.**

acid cycle intermediates increase and the cycle can proceed more quickly. *Since the citric acid cycle acts as a catalyst, increasing the concentrations of its components increases flux through the pathway.*

The degradation of fatty acids with an odd number of carbon atoms yields the citric acid cycle intermediate succinyl-CoA. Other anaplerotic reactions include the pathways for the degradation of some amino acids, which produce α-ketoglutarate, succinyl-CoA, fumarate, and oxaloacetate. Some of these reactions are transaminations, such as

FIGURE 14.19

Anaplerotic reactions of the citric acid cycle.

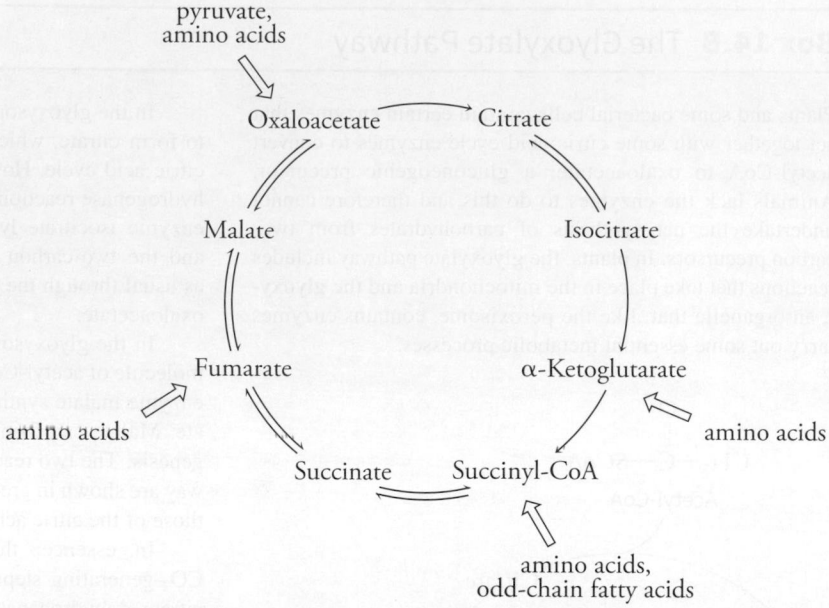

Because transamination reactions have ΔG values near zero, the direction of flux into or out of the pool of citric acid cycle intermediates depends on the relative concentrations of the reactants.

In vigorously exercising muscle, the concentrations of citric acid cycle intermediates increase about three- to fourfold within a few minutes. This may help boost the energy-generating activity of the citric acid cycle, but it cannot be the sole mechanism, since flux through the citric acid cycle actually increases as much as 100-fold due to the increased activity of the three enzymes at the control points: citrate synthase, isocitrate dehydrogenase, and α-ketoglutarate dehydrogenase. The increase in citric acid cycle intermediates may actually be a mechanism for accommodating the large increase in pyruvate that results from rapid glycolysis at the start of exercise. Not all of this pyruvate is converted to lactate (Section 13.1); some is shunted into the pool of citric acid cycle intermediates via the pyruvate carboxylase reaction. Some pyruvate also undergoes a reversible reaction catalyzed by alanine aminotransferase:

The resulting α-ketoglutarate then augments the pool of citric acid cycle intermediates, thereby increasing the ability of the cycle to oxidize the extra pyruvate.

Note that any compound that enters the citric acid cycle as an intermediate is not itself oxidized; it merely boosts the catalytic activity of the cycle, whose net reaction is still the oxidation of the two carbons of acetyl-CoA.

CONCEPT CHECK

1. List the citric acid cycle intermediates that are used for the synthesis of amino acids, glucose, and fatty acids.
2. What does the ATP-citrate lyase reaction accomplish?
3. Explain why the concentration of oxaloacetate remains low.
4. Why does synthesizing more oxaloacetate increase flux through the cycle?
5. Write equations for all the anaplerotic reactions.

SUMMARY

14.1 The Pyruvate Dehydrogenase Reaction

• In order for pyruvate, the product of glycolysis, to enter the citric acid cycle, it must undergo oxidative decarboxylation catalyzed by the multienzyme pyruvate dehydrogenase complex, which yields acetyl-CoA, CO_2, and NADH.

14.2 The Eight Reactions of the Citric Acid Cycle

• The eight reactions of the citric acid cycle function as a multistep catalyst to convert the two carbons of acetyl-CoA to 2 CO_2.

14.3 Thermodynamics of the Citric Acid Cycle

• The electrons released in the oxidative reactions of the citric acid cycle are transferred to 3 NAD$^+$ and to ubiquinone. The reoxidation of the reduced cofactors generates ATP by oxidative phosphorylation.

In addition, succinyl-CoA synthetase yields one molecule of GTP or ATP.

• The regulated reactions of the citric acid cycle are its irreversible steps, catalyzed by citrate synthase, isocitrate dehydrogenase, and α-ketoglutarate dehydrogenase.

• The citric acid cycle most likely evolved from biosynthetic pathways leading to α-ketoglutarate or succinate.

14.4 Anabolic and Catabolic Functions of the Citric Acid Cycle

• Six of the eight citric acid cycle intermediates serve as precursors of other compounds, including amino acids, monosaccharides, and lipids. Anaplerotic reactions convert other compounds into citric acid cycle intermediates, thereby allowing increased flux of acetyl carbons through the pathway.

KEY TERMS

mitochondrial matrix
multienzyme complex
citric acid cycle
substrate-level phosphorylation
Pasteur effect
carcinogenesis
glyoxylate pathway
glyoxysome
anaplerotic reaction

BIOINFORMATICS

Brief Bioinformatics Exercises

14.1 Viewing and Analyzing the Pyruvate Dehydrogenase Complex
14.2 The Citric Acid Cycle and the KEGG Database

PROBLEMS

14.1 The Pyruvate Dehydrogenase Reaction

1. What are four possible transformations of pyruvate in mammalian cells?

2. Determine which one of the five steps of the pyruvate dehydrogenase complex reaction is metabolically irreversible and explain why.

3. The product of the pyruvate dehydrogenase complex, acetyl-CoA, is released in step 3 of the overall reaction. What is the purpose of steps 4 and 5?

4. Beriberi is a disease that results from a dietary lack of thiamine, the vitamin that serves as the precursor for thiamine pyrophosphate (TPP). There are two metabolites that accumulate in individuals with beriberi, especially after ingestion of glucose. Which metabolites accumulate and why?

5. Arsenite is toxic in part because it binds to sulfhydryl compounds such as dihydrolipoamide, as shown in the figure. What effect would the presence of arsenite have on the citric acid cycle?

6. Using the pyruvate dehydrogenase complex reaction as a model, reconstruct the TPP-dependent yeast pyruvate decarboxylase reaction in alcoholic fermentation (see Box 13.B).

7. How is the activity of the pyruvate dehydrogenase complex affected by (a) a high [NADH]/[NAD$^+$] ratio or (b) a high [acetyl-CoA]/[CoASH] ratio?

8. The activity of the pyruvate dehydrogenase complex is also controlled by phosphorylation. Pyruvate dehydrogenase kinase (PDH kinase) catalyzes the phosphorylation of a specific Ser residue on the E1 subunit of the enzyme, rendering it inactive. Pyruvate dehydrogenase phosphatase (PDH phosphatase) reverses the inhibition by catalyzing the removal of this phosphate group. The PDH kinase is highly regulated and its activity is influenced by various cellular metabolites. Indicate whether the following would activate or inhibit PDH kinase: (a) NAD$^+$; (b) NADH; (c) coenzyme A; (d) acetyl-CoA; (e) ADP.

9. The PDH kinase and PDH phosphatase enzymes (see Problem 8) are controlled by cytosolic Ca^{2+} levels. In the muscle, Ca^{2+} levels rise when the muscle contracts. Which of these two enzymes is inhibited by Ca^{2+} and which is activated by Ca^{2+}?

10. Most cases of pyruvate dehydrogenase deficiency disease that have been studied to date involve a mutation in the E1 subunit of the enzyme. The disease is extremely difficult to treat successfully.
(a) Physicians who identify patients with a pyruvate dehydrogenase deficiency administer thiamine as a first course of treatment. Explain why.
(b) A second strategy to treat a pyruvate dehydrogenase deficiency disease involves administering dichloroacetate, a compound that inhibits pyruvate dehydrogenase kinase (see Problem 8). How might this strategy be effective?

14.2 The Eight Reactions of the Citric Acid Cycle

11. Does the citrate synthase enzyme mechanism (Fig. 14.8) use an acid catalysis strategy, a base catalysis strategy, a covalent catalysis strategy, or some combination of these strategies (see Section 6.2)? Explain.

12. Animals that have ingested the leaves of the poisonous South African plant *Dichapetalum cymosum* exhibit a 10-fold increase in levels of cellular citrate. The plant contains fluoroacetate, which is converted to fluoroacetyl-CoA. Describe the mechanism that leads to increased levels of citrate in animals that have ingested this poisonous plant. (*Note:* Fluoroacetyl-CoA is not an inhibitor of citrate synthase.)

13. Site-directed mutagenesis techniques were used to synthesize a mutant citrate synthase enzyme in which the active site histidine was converted to an alanine. Why did the mutant citrate synthase enzyme exhibit decreased catalytic activity?

14. Investigators interested in studying the effect of acetyl-CoA analogs on citrate synthase activity synthesized the compound S-acetonyl-CoA from 1-bromoacetone and coenzyme A.
(a) Write the reaction for the formation of S-acetonyl-CoA.
(b) The Lineweaver–Burk plot for the citrate synthase reaction with and without S-acetonyl-CoA is shown. What type of inhibitor is S-acetonyl-CoA? Explain.

$$H_3C-\overset{\overset{\displaystyle O}{\|}}{C}-CH_2-S-CoA$$

S-Acetonyl-CoA

15. The compound carboxymethyl-CoA (shown below) is a competitive inhibitor of citrate synthase and is a proposed transition state analog. Propose a structure for the reaction intermediate derived from acetyl-CoA in the rate-limiting step of the reaction, just prior to its reaction with oxaloacetate.

$$CoA-S-CH_2-\overset{\overset{\displaystyle OH}{\diagup}}{\underset{\diagdown O}{C}}$$

Carboxymethyl-CoA
(transition state analog)

16. Kinetic studies with aconitase revealed that *trans*-aconitate is a competitive inhibitor of the enzyme if *cis*-aconitate is used as the substrate. But if citrate is used as the substrate, *trans*-aconitate is a noncompetitive inhibitor. Propose a hypothesis that explains this observation.

17. In a yeast mutant, the gene for aconitase is nonfunctional. What are the consequences for the cell, particularly with regard to energy production?

18. The $\Delta G^{\circ\prime}$ value for the isocitrate dehydrogenase reaction is -21 kJ·mol^{-1}. What is K_{eq} for this reaction? Assume $T = 25$°C.

19. The crystal structure of isocitrate dehydrogenase shows that there is a cluster of highly conserved amino acids in the substrate binding pocket—three arginines, a tyrosine, and a lysine. Why are these residues conserved and what is a possible role for their side chains in substrate binding?

20. Using the pyruvate dehydrogenase complex reaction as a model, draw the intermediates of the α-ketoglutarate dehydrogenase reaction. Describe what happens in each of the five reaction steps.

21. Using the mechanism you drew for Problem 20, explain how succinyl phosphonate (below) inhibits α-ketoglutarate dehydrogenase.

$$
\begin{array}{c}
O^- \\
| \\
^-O-P=O \\
| \\
C=O \\
| \\
CH_2 \\
| \\
CH_2 \\
| \\
COO^-
\end{array}
$$

Succinyl phosphonate

22. Succinyl-CoA synthetase is also called succinate thiokinase. Why is the enzyme considered to be a kinase?

23. Malonate is a competitive inhibitor of succinate dehydrogenase. What citric acid cycle intermediates accumulate if malonate is present in a preparation of isolated mitochondria?

24. Succinate dehydrogenase is not considered to be part of the glyoxylate pathway (see Box 14.B), yet it is vital to the proper functioning of the pathway. Why?

25. The $\Delta G^{\circ\prime}$ for the fumarase reaction is $-3.4\,kJ\cdot mol^{-1}$, but the ΔG value is close to zero. What is the ratio of fumarate to malate under cellular conditions at 37°C? Is this reaction likely to be a control point for the citric acid cycle?

26. A mutant bacterial fumarase was constructed by replacing the Glu (E) at position 315 with Gln (Q). The kinetic parameters of the mutant and wild-type enzymes are shown in the table. Explain the significance of the changes.

	Wild type enzyme	E315Q mutant enzyme
V_{max} ($\mu mol\cdot min^{-1}\cdot mg^{-1}$)	345	32
K_M (mM)	0.21	0.25
k_{cat}/K_M ($M^{-1}\cdot s^{-1}$)	5.6×10^6	4.3×10^5

27. **(a)** Oxaloacetate labeled at C4 with ^{14}C is added to a suspension of respiring mitochondria. What is the fate of the labeled carbon?
(b) Acetyl-CoA labeled at C1 with ^{14}C is added to a suspension of respiring mitochondria. What is the fate of the labeled carbon?

28. The complex metabolic pathways in the parasite *Trypanosoma brucei* (the causative agent of sleeping sickness) were elucidated in part by adding radiolabeled metabolites to cultured parasites. In the parasite, glucose is converted to phosphoenolpyruvate (PEP) in the cytosol. PEP then enters an organelle called the glycosome and is converted to oxaloacetate (OAA); OAA is then converted to malate, and malate to fumarate. Fumarate reductase catalyzes the conversion of fumarate to succinate; the succinate is then secreted from the glycosome.
(a) If C1 of glucose is labeled, what carbons in succinate are labeled?
(b) If citrate becomes radioactively labeled, what can you conclude about the connection between glycosomal and mitochondrial pathways in the parasite?

14.3 Thermodynamics of the Citric Acid Cycle

29. Acetyl-CoA acts as an allosteric activator of pyruvate carboxylase. S-acetonyl-CoA (see Problem 14) does not activate pyruvate carboxylase, and it cannot compete with acetyl-CoA for binding to the enzyme. What does this tell you about the binding requirements for an allosteric activator of pyruvate carboxylase?

30. Why is it advantageous for citrate, the product of Reaction 1 of the citric acid cycle, to inhibit phosphofructokinase, which catalyzes the third reaction of glycolysis?

31. Citrate competes with oxaloacetate for binding to citrate synthase. Isocitrate dehydrogenase is activated by Ca^{2+} ions, which are released when muscle contracts. How do these two regulatory strategies assist the cell in making the transition from the rested state (low citric acid cycle activity) to the exercise state (high citric acid cycle activity)?

32. Administering high concentrations of oxygen (hyperoxia) is effective in treating lung injuries but at the same time can also be quite damaging.
(a) Lung aconitase activity is dramatically decreased during hyperoxia. How would this affect the concentration of citric acid cycle intermediates?
(b) The decreased aconitase activity and decreased mitochondrial respiration in hyperoxia are accompanied by elevated rates of glycolysis and the pentose phosphate pathway. Explain why.

33. The scientists who carried out the hyperoxia experiments described in Problem 32 noted that they could mimic this effect by administering either fluoroacetate or fluorocitrate to cells in culture. Explain. (*Hint*: Fluoroacetate can react with coenzyme A to form fluoroacetyl-CoA.)

34. In bacteria, isocitrate dehydrogenase is regulated by phosphorylation of a specific Ser residue in the enzyme active site. X-ray

structures of the phosphorylated and the non-phosphorylated enzyme show no significant conformational differences.
(a) How does phosphorylation regulate isocitrate dehydrogenase activity?
(b) To confirm their hypothesis, investigators constructed a mutant enzyme in which the Ser residue was replaced with Asp. The mutant was unable to bind isocitrate. Are these results consistent with the hypothesis you proposed in part (a)?

35. The expression of several enzymes changes when yeast grown on glucose are abruptly shifted to a 2-carbon food source such as acetate
(a) Why does the level of expression of isocitrate dehydrogenase increase when the yeast are shifted from glucose to acetate?
(b) The metabolism of a yeast mutant with a nonfunctional isocitrate dehydrogenase enzyme was compared to that of a wild-type yeast. The yeast were grown on glucose and then abruptly shifted to acetate as the sole carbon source. The [NAD⁺]/[NADH] ratio was measured over a period of 48 hours. The results are shown below. Why does the ratio increase slightly at 36 hours for the wild-type yeast? Why is there a more dramatic increase in the ratio for the mutant?

36. Succinyl-CoA inhibits both citrate synthase and α-ketoglutarate dehydrogenase. How is succinyl-CoA able to inhibit both enzymes?

37. A patient with an α-ketoglutarate deficiency exhibits a small increase in blood pyruvate level and a large increase in blood lactate level, resulting in a [lactate]/[pyruvate] ratio that is many times greater than normal. Explain the reason for these symptoms.

38. Succinyl-CoA synthetase is a dimer composed of an α subunit and a β subunit. A single gene codes for the α subunit protein. Two genes code for two different β subunit proteins—one subunit is specific for GDP, and one is specific for ADP.
(a) The β subunit specific for ADP is expressed in "catabolic tissues" such as brain and muscles, whereas the β subunit specific for GDP is expressed in "anabolic tissues" such as liver and kidneys. Propose a hypothesis to explain this observation.
(b) Individuals who are born with a mutation in the gene coding for the α subunit of the enzyme experience severe lactic acidosis and usually die within a few days of birth. Why is this mutation so deleterious?
(c) Individuals who are born with a mutation in the gene coding for the ADP-specific β subunit of the enzyme experience normal to moderately elevated concentrations of lactate and usually survive to their early 20s. Why is the prognosis for these patients better than for patients with a mutation in the gene for the α subunit?

39. Reactions 8 and 1 of the citric acid cycle can be considered to be coupled because the exergonic hydrolysis of the thioester bond of acetyl-CoA in Reaction 1 drives the regeneration of oxaloacetate in Reaction 8.

(a) Write the equation for the overall coupled reaction and calculate its $\Delta G^{\circ\prime}$.

(b) What is the equilibrium constant for the coupled reaction? Compare this equilibrium constant with the equilibrium constant of Reaction 8 alone.

40. Malate dehydrogenase is more active in cells oxidizing glucose aerobically than in cells oxidizing glucose anaerobically. Explain why.

41. Flux through the citric acid cycle is regulated by the simple mechanisms of **(a)** substrate availability, **(b)** product inhibition, and **(c)** feedback inhibition. Give examples of each.

42. Why would a deficiency of succinate dehydrogenase lead to a shortage of coenzyme A?

43. Certain microorganisms with an incomplete citric acid cycle decarboxylate α-ketoglutarate to produce succinate semialdehyde. A dehydrogenase then converts succinate semialdehyde to succinate. These reactions can be combined with other standard citric acid cycle reactions to create a pathway from citrate to oxaloacetate. How does this alternative pathway compare to the standard citric acid cycle in its ability to make free energy available to the cell?

14.4 Anabolic and Catabolic Functions of the Citric Acid Cycle

44. (a) Why is the reaction catalyzed by pyruvate carboxylase the most important anaplerotic reaction of the citric acid cycle?

(b) Why is the activation of pyruvate carboxylase by acetyl-CoA a good regulatory strategy?

45. Many amino acids are broken down to intermediates of the citric acid cycle.

(a) Why can't these amino acid "remnants" be completely oxidized to CO_2 by the citric acid cycle?

(b) Explain why amino acids that are broken down to pyruvate can be completely oxidized by the citric acid cycle.

46. Describe how the transamination reaction below could function as an anaplerotic reaction for the citric acid cycle.

47. Is net synthesis of glucose in mammals possible from the following compounds?

(a) The fatty acid palmitate (16:0), which is degraded to eight acetyl-CoA.

(b) The fatty acid pentadecanoate (15:0), which is degraded to six acetyl-CoA and one propionyl-CoA.

(c) Glyceraldehyde-3-phosphate.

(d) Leucine, which is degraded to acetyl-CoA and acetoacetate (a compound that is metabolically equivalent to two acetyl-CoA groups).

(e) Tryptophan, which is degraded to alanine and acetoacetate.

(f) Phenylalanine, which is degraded to acetoacetate and fumarate.

48. Pancreatic islet cells cultured in the presence of 1–20 mM glucose showed increased activities of pyruvate carboxylase and the E1 subunit of the pyruvate dehydrogenase complex proportional to the increase in glucose concentration. Explain why.

49. A physician is attempting to diagnose a neonate with a pyruvate carboxylase deficiency. An injection of alanine normally leads to a gluconeogenic response, but in the patient no such response occurs. Explain.

50. The physician treats the patient described in Problem 49 by administering glutamine. Explain why glutamine supplements are effective in treating the disease.

51. Physicians often attempt to treat a pyruvate carboxylase deficiency by administering biotin. Explain why this strategy might be effective.

52. Patients with a pyruvate dehydrogenase deficiency and patients with a pyruvate carboxylase deficiency (see Problems 49–51) both have high blood levels of pyruvate and lactate. Explain why.

53. The udder of a ewe uses the majority of the glucose synthesized by the ewe, both for production of the milk sugar lactose and for production of triacylglycerols. During the winter, milk production decreases due to a poor food supply. Administration of propionate (which is converted to succinyl-CoA in ruminants) is an appropriate remedy. How does this treatment work?

54. Metabolites in rat muscle were measured before and after exercising. After exercise, the rat muscle showed an increase in oxaloacetate concentration, a decrease in phosphoenolpyruvate concentration, and no change in pyruvate concentration. Explain.

55. Oxygen does not appear as a reactant in any of the citric acid cycle reactions; yet it is essential for the proper functioning of the cycle. Explain why.

56. The activity of isocitrate dehydrogenase in *E. coli* is regulated by the covalent attachment of a phosphate group, which inactivates the enzyme. When acetate is the food source for a culture of *E. coli*, isocitrate dehydrogenase is phosphorylated.

(a) Draw a diagram showing how acetate is metabolized in *E. coli*.

(b) When glucose is added to the culture, the phosphate group is removed from isocitrate dehydrogenase. How does flux through the metabolic pathways change in *E. coli* when glucose is the food source instead of acetate?

57. Yeast are unusual in their ability to use ethanol as a gluconeogenic substrate. Ethanol is converted to glucose with the assistance of the glyoxylate pathway. Describe how the ethanol → glucose conversion takes place.

58. Animals lack a glyoxylate pathway and cannot convert fats to carbohydrates. If an animal is fed a fatty acid with all of its carbons replaced by the isotope ^{14}C, some of the labeled carbons later appear in glucose. How is this possible?

59. A bacterial mutant with low levels of isocitrate dehydrogenase is able to grow normally when the culture medium is supplemented with glutamate. Explain why.

60. The activity of the purine nucleotide cycle (shown below) in muscles increases during periods of high activity. Explain how the cycle contributes to the ability of the muscle cell to generate energy during intense exercise. IMP is inosine monophosphate.

61. The plant metabolite hydroxycitrate is advertised as an agent that prevents fat buildup.

(a) How does this compound differ from citrate?
(b) Hydroxycitrate inhibits the activity of ATP-citrate lyase. What kind of inhibition is likely to occur?
(c) Why might inhibition of ATP-citrate lyase block the conversion of carbohydrates to fats?
(d) The synthesis of what other compounds would be inhibited by hydroxycitrate?

$$CH_2-COO^-$$
$$|$$
$$HO-C-COO^-$$
$$|$$
$$HO-CH-COO^-$$
Hydroxycitrate

62. *Helicobacter pylori* is a bacterium that colonizes the upper gastrointestinal tract in humans and is the causative agent of chronic gastritis, ulcers, and possibly gastric cancer. Knowledge of the intermediary metabolism of this organism will be helpful in the development of effective drug therapies to treat these diseases. The citric acid "cycle" in *H. pylori* is a noncyclic, branched pathway that is used to produce biosynthetic intermediates instead of metabolic energy. Succinate is produced in the "reductive branch," whereas α-ketoglutarate is produced in the "oxidative branch." The two branches are linked by the α-ketoglutarate oxidase reaction. The pathway is shown in the diagram below.

Glyoxylate
+
Acetyl-CoA
| malate synthase
↓ NAD^+ NADH
Malate ⇌ Oxaloacetate
         malate dehydrogenase   citrate synthase ⟵ Acetyl-CoA
H_2O ⟵ fumarase
Fumarate   Citrate
QH_2 ⟍ fumarate reductase   aconitase
Q ⟋
Succinate   Isocitrate
CO_2, ⟵ α-keto-glutarate oxidase   isocitrate dehydro-genase ⟵ $NADP^+$
QH_2
Q NADPH
α-Ketoglutarate

(a) Compare and contrast the citric acid cycle in *H. pylori* with the citric acid cycle in mammals.
(b) The K_M values for the enzymes listed in the table below are higher than the K_M values for the corresponding enzymes in other species of bacteria. What does this tell you about the conditions under which the citric acid cycle operates in *H. pylori*?
(c) Compare the properties of *H. pylori* citrate synthase with mammalian citrate synthase.
(d) What enzymes might serve to regulate the citric acid cycle in *H. pylori*?
(e) What enzymes might be used as drug targets for persons suffering from gastritis, ulcers, or gastric cancer?

Enzyme	Substrate	Inhibitors	Activators
Citrate synthase	Acetyl-CoA, oxaloacetate	ATP	
Aconitase	Citrate		
Isocitrate dehydrogenase (NADP⁺-dependent)	Isocitrate, NADP⁺	Higher concentrations of NADP⁺, isocitrate	AMP (slight)
α-ketoglutarate oxidase	α-ketoglutarate, FAD		CoASH
Malate dehydrogenase	Oxaloacetate, NADH		
Fumarase	Malate		
Fumarate reductase	Fumarate, QH_2		
Malate synthase	Glyoxylate, acetyl-CoA		

63. *H. pylori* (see Problem 62) uses amino acids and fatty acids present in the gastrointestinal tract as a source of biosynthetic intermediates.
(a) Describe how *H. pylori* uses the acetyl-CoA derived from fatty acid breakdown to synthesize glucose and glutamate.
(b) Describe how *H. pylori* converts aspartate to glutamate.

64. Yeast cells that are grown on nonfermentable substrates and then abruptly switched to glucose exhibit substrate-induced inactivation of several enzymes. Which enzymes would glucose cause to be inactivated, and why?

65. Phagocytes protect the host against damage caused by invading microorganisms by engulfing a foreign microbe and forming a membrane-bound structure around it called a phagosome. The phagosome then fuses with a lysosome, a cellular organelle that contains proteolytic enzymes that destroy the pathogen. But some microbes, such as *Mycobacterium tuberculosis,* survive dormant inside the phagosome for a prolonged period of time. In this scenario, levels of bacterial isocitrate lyase, malate synthase, citrate synthase, and malate dehydrogenase increase inside the phagosome to levels as much as 20 times above normal.
(a) What pathway(s) does *M. tuberculosis* employ while in the phagosome and why are these pathways essential to its survival?
(b) What might be good drug targets for treating a patient infected with *M. tuberculosis*?

66. Bacteria and plants (but not animals) possess the enzyme phosphoenolpyruvate carboxylase (PPC), which catalyzes the reaction shown.
(a) What is the importance of this reaction to the organism?
(b) PPC is allosterically activated by both acetyl-CoA and fructose-1,6-bisphosphate. Explain these regulatory strategies.

$$
\underset{\text{Phosphoenolpyruvate}}{
\begin{array}{c}
CH_2 \\
\| \\
C-O-PO_3^{2-} \\
| \\
COO^-
\end{array}}
+ HCO_3^-
\xrightarrow{\text{PPC}}
\underset{\text{Oxaloacetate}}{
\begin{array}{c}
COO^- \\
| \\
CH_2 \\
| \\
C=O \\
| \\
COO^-
\end{array}}
+ P_i
$$

67. The pharmaceutical, cosmetics, and food industries synthesize succinic acid by "green" or environmentally responsible methods that involve bacteria instead of petrochemicals. Industrial succinate production by bacteria occurs under anaerobic conditions, in which malate dehydrogenase activity increases.

(a) Draw a reaction scheme outlining how phosphoenolpyruvate is converted to succinate. Include the names of all reactants, products, and enzymes.

(b) Why is it essential that the production of succinate take place under anaerobic conditions?

68. Experiments with cancer cells grown in culture show that glutamine is consumed at a high rate and used for biosynthetic reactions, aside from protein synthesis. One possible pathway involves the conversion of glutamine to glutamate and then to α-ketoglutarate. The α-ketoglutarate can then be used to produce pyruvate for gluconeogenesis.

(a) Describe the types of reactions that convert glutamine to α-ketoglutarate.

(b) Give the sequence of enzymes that can convert α-ketoglutarate to pyruvate.

69. Many cancer cells carry out glycolysis at a high rate but convert most of the resulting pyruvate to lactate rather than to acetyl-CoA. Acetyl-CoA, however, is required for the synthesis of fatty acids, which are needed in large amounts by rapidly growing cancer cells. In these cells, the isocitrate dehydrogenase reaction apparently operates in reverse. Explain why this reaction could facilitate the conversion of amino acids such as glutamate into fatty acids.

SELECTED READINGS

Barry, M. J., Enzymes and symmetrical molecules, *Trends Biochem. Sci.* **22,** 228–230 (1997). [Recounts how experiments and insight revealed that the symmetrical citrate molecule can react asymmetrically.]

Brière, J.-J., Favier, J., Giminez-Roqueplo, A.-P., and Rustin, P., Tricarboxylic acid cycle dysfunction as a cause of human disease and tumor formation, *Am. J. Physiol. Cell Physiol.* **291,** C1114–C1120 (2006). [Includes discussions of the metabolic role of the citric acid cycle and its location.]

Gray, L. R., Tompkins, S. C., and Taylor, E. B., Regulation of pyruvate metabolism and human disease. *Cell Mol. Life Sci.* **71,** 2577–2604 (2014). [Includes useful diagrams of pathways involving pyruvate and discusses diseases related to them.]

Owen, O. E., Kalhan, S. C., and Hanson, R. W., The key role of anaplerosis and cataplerosis for citric acid cycle function, *J. Biol. Chem.* **277,** 30409–30412 (2002). [Describes the addition (anaplerosis) and removal (cataplerosis) of citric acid cycle intermediates in different organ systems.]

Patel, M. S., Nemeria, N. S., Furey, W., and Jordan, F., The pyruvate dehydrogenase complexes: structure-based function and regulation, *J. Biol. Chem.* **289,** 16615–16623 (2014). [Compares the structure and chemical reactions of the human and *E. coli* enzyme complexes.]

OXIDATIVE PHOSPHORYLATION

Courtesy Kathryn Tosney

Loricifera, a group of tiny (< 1 mm) organisms that resemble armored jellyfish, live in marine sediments. Remarkably—for a multicellular organism—they are entirely anaerobic and lack mitochondria, the site of aerobic metabolism in all other animals.

DO YOU REMEMBER?

- Living organisms obey the laws of thermodynamics (Section 1.3).
- Transporters obey the laws of thermodynamics, providing a way for solutes to move down their concentration gradients or using ATP to move substances against their gradients (Section 9.1).
- Coenzymes such as NAD^+ and ubiquinone collect electrons from compounds that become oxidized (Section 12.2).
- A reaction that hydrolyzes a phosphoanhydride bond in ATP occurs with a large negative change in free energy (Section 12.3).

The early stages of oxidation of metabolic fuels such as glucose, fatty acids, and amino acids, as well as the oxidation of acetyl carbons to CO_2 via the citric acid cycle, yield the reduced cofactors NADH and ubiquinol (QH_2). These compounds are forms of energy currency (see Section 12.3) not because they are chemically special but because their reoxidation—ultimately by molecular oxygen in aerobic organisms—is an exergonic process. That free energy is harvested to synthesize ATP, a phenomenon called oxidative phosphorylation. To understand oxidative phosphorylation, we need to first examine why and how electrons flow from reduced cofactors to O_2. Then we can explore how the free energy of the redox reactions is conserved in the formation of a transmembrane gradient of protons, another type of free energy that drives the rotation of ATP synthase so that it can build ATP from ADP and P_i.

15.1 THE THERMODYNAMICS OF OXIDATION–REDUCTION REACTIONS

In the scheme introduced in Figure 12.11, **oxidative phosphorylation** represents the final phase of the catabolism of metabolic fuels and the major source of the cell's ATP (**Fig. 15.1**). Oxidative phosphorylation differs from the conventional biochemical reactions

KEY CONCEPTS

- The standard reduction potential indicates a substance's tendency to become reduced; the actual reduction potential depends on the concentrations of reactants.
- Electrons are transferred from a substance with a lower reduction potential to a substance with a higher reduction potential.
- The free energy change for an oxidation–reduction reaction depends on the change in reduction potential.

FIGURE 15.1

Oxidative phosphorylation in context. The reduced cofactors NADH and QH$_2$, which are generated in the oxidative catabolism of amino acids, monosaccharides, and fatty acids, are reoxidized by molecular oxygen. The free energy of this process is conserved in a manner that powers the synthesis of ATP from ADP + P$_i$.

we have focused on in the last two chapters. In particular, ATP synthesis is not directly coupled to a single discrete chemical reaction, such as a kinase-catalyzed reaction. Rather, oxidative phosphorylation is a more indirect process of free energy transformation.

The flow of electrons from reduced compounds such as NADH and QH$_2$ to an oxidized compound such as O$_2$ is a thermodynamically favorable process. The free energy changes for the movements of electrons through a series of electron carriers can be quantified by considering the reduction potentials of the chemical species involved in each transfer.

Oxidation–reduction reactions (or redox reactions, introduced in Section 12.2) are similar to other chemical reactions in which a portion of a molecule—electrons in this case—is transferred. In any oxidation–reduction reaction, one reactant (called the **oxidizing agent** or **oxidant**) is reduced as it gains electrons. The other reactant (called the **reducing agent** or **reductant**) is oxidized as it gives up electrons:

$$A_{oxidized} + B_{reduced} \rightleftharpoons A_{reduced} + B_{oxidized}$$

For example, in the succinate dehydrogenase reaction (step 6 of the citric acid cycle; see Section 14.2), the two electrons of the reduced FADH$_2$ prosthetic group of the enzyme are transferred to ubiquinone (Q) so that FADH$_2$ is oxidized and ubiquinone is reduced:

$$\underset{(reduced)}{FADH_2} + \underset{(oxidized)}{Q} \rightleftharpoons \underset{(oxidized)}{FAD} + \underset{(reduced)}{QH_2}$$

In this reaction, the two electrons are transferred as H atoms (an H atom consists of a proton and an electron, or H$^+$ and e^-). In oxidation–reduction reactions involving the cofactor NAD$^+$, the electron pair takes the form of a hydride ion (H$^-$, a proton with two electrons). In biological systems, electrons usually travel in pairs, although, as we will see, they may also be transferred one at a time. Note that the change in oxidation state of a reactant may be obvious, such as when Fe^{3+} is reduced to Fe^{2+}, or it may require closer inspection of the molecule's structure, such as when succinate is oxidized to fumarate (Section 14.2).

REDUCTION POTENTIAL INDICATES A SUBSTANCE'S TENDENCY TO ACCEPT ELECTRONS

The tendency of a substance to accept electrons (to become reduced) or to donate electrons (become oxidized) can be quantified. Although an oxidation–reduction reaction necessarily requires both an oxidant and a reductant, it is helpful to consider just one substance at a time, that is, a **half-reaction.** Using the example above, the half-reaction for ubiquinone (by convention, written as a reduction reaction) is

$$Q + 2H^+ + 2e^- \rightleftharpoons QH_2$$

(the reverse reaction would describe an oxidation half-reaction).

The affinity of a substance such as ubiquinone for electrons is its **standard reduction potential ($\mathcal{E}°'$),** which has units of volts (note that the degree and prime symbols indicate a value under standard biochemical conditions where the pressure is 1 atm, the temperature is

25°C, the pH is 7.0, and all species are present at concentrations of 1 M). *The greater the value of $\mathcal{E}^{\circ\prime}$, the greater the tendency of the oxidized form of the substance to accept electrons and become reduced.* The standard reduction potentials of some biological substances are given in **Table 15.1**.

Like a ΔG value, *the actual reduction potential depends on the actual concentrations of the oxidized and reduced species.* The actual reduction potential ($\mathcal{E}$) is related to the standard reduction potential ($\mathcal{E}^{\circ\prime}$) by the **Nernst equation:**

$$\mathcal{E} = \mathcal{E}^{\circ\prime} - \frac{RT}{n\mathcal{F}} \ln \frac{[A_{reduced}]}{[A_{oxidized}]} \qquad [15.1]$$

R (the gas constant) has a value of $8.3145 \text{ J} \cdot \text{K}^{-1} \cdot \text{mol}^{-1}$, T is the temperature in Kelvin, n is the number of electrons transferred (one or two in most of the reactions we will encounter), and $\mathcal{F}$ is the **Faraday constant** ($96,485 \text{ J} \cdot \text{V}^{-1} \cdot \text{mol}^{-1}$; it is equivalent to the electrical charge of one mole of electrons). At 25°C (298 K), the Nernst equation reduces to

$$\mathcal{E} = \mathcal{E}^{\circ\prime} - \frac{0.026 \text{ V}}{n} \ln \frac{[A_{reduced}]}{[A_{oxidized}]} \qquad [15.2]$$

In fact, for many substances in biological systems, the concentrations of the oxidized and reduced species are similar, so the logarithmic term is small (recall that $\ln 1 = 0$) and $\mathcal{E}$ is close to $\mathcal{E}^{\circ\prime}$ (Sample Calculation 15.1).

TABLE 15.1 Standard Reduction Potentials of Some Biological Substances

HALF-REACTION	$\mathcal{E}^{\circ\prime}$ (V)
$\frac{1}{2} O_2 + 2 H^+ + 2 e^- \rightleftharpoons H_2O$	0.815
$SO_4^{2-} + 2 H^+ + 2 e^- \rightleftharpoons SO_3^{2-} + H_2O$	0.48
$NO_3^- + 2 H^+ + 2e^- \rightleftharpoons NO_2^- + H_2O$	0.42
Cytochrome a_3 (Fe^{3+}) + $e^- \rightleftharpoons$ cytochrome a_3 (Fe^{2+})	0.385
Cytochrome a (Fe^{3+}) + $e^- \rightleftharpoons$ cytochrome a (Fe^{2+})	0.29
Cytochrome c (Fe^{3+}) + $e^- \rightleftharpoons$ cytochrome c (Fe^{2+})	0.235
Cytochrome c_1 (Fe^{3+}) + $e^- \rightleftharpoons$ cytochrome c_1 (Fe^{2+})	0.22
Cytochrome b (Fe^{3+}) + $e^- \rightleftharpoons$ cytochrome b (Fe^{2+})(*mitochondrial*)	0.077
Ubiquinone + 2 H^+ + 2 $e^- \rightleftharpoons$ ubiquinol	0.045
Fumarate$^-$ + 2 H^+ + 2 $e^- \rightleftharpoons$ succinate$^-$	0.031
FAD + 2 H^+ + 2 $e^- \rightleftharpoons$ FADH$_2$ (*in flavoproteins*)	~ 0.
Oxaloacetate$^-$ + 2 H^+ + 2 $e^- \rightleftharpoons$ malate$^-$	− 0.166
Pyruvate$^-$ + 2 H^+ + 2 $e^- \rightleftharpoons$ lactate$^-$	− 0.185
Acetaldehyde + 2 H^+ + 2 $e^- \rightleftharpoons$ ethanol	− 0.197
S + 2 H^+ + 2 $e^- \rightleftharpoons$ H$_2$S	− 0.23
Lipoic acid + 2 H^+ + 2 $e^- \rightleftharpoons$ dihydrolipoic acid	− 0.29
NAD$^+$ + H^+ + 2 $e^- \rightleftharpoons$ NADH	− 0.315
NADP$^+$ + H^+ + 2 $e^- \rightleftharpoons$ NADPH	− 0.320
Acetoacetate$^-$ + 2 H^+ + 2 $e^- \rightleftharpoons$ 3-hydroxybutyrate$^-$	− 0.346
Acetate$^-$ + 3 H^+ + 2 $e^- \rightleftharpoons$ acetaldehyde + H$_2$O	− 0.581

Source: Mostly from Loach, P. A., in Fasman, G. D. (ed.), *Handbook of Biochemistry and Molecular Biology* (3rd ed.), Physical and Chemical Data, Vol. I, pp. 123–130, CRC Press (1976).

SAMPLE CALCULATION 15.1

PROBLEM Calculate the reduction potential of fumarate ($\mathcal{E}^{\circ\prime} = 0.031$ V) at 25°C when [fumarate] = 40 μM and [succinate] = 200 μM.

SOLUTION Use Equation 15.2. Fumarate is the oxidized compound and succinate is the reduced compound.

$$\mathcal{E} = \mathcal{E}^{\circ\prime} - \frac{0.026 \text{ V}}{n} \ln \frac{[A_{reduced}]}{[A_{oxidized}]}$$

$$= 0.031 \text{ V} - \frac{0.026 \text{ V}}{2} \ln \frac{(2 \times 10^{-4})}{(4 \times 10^{-5})}$$

$$= 0.031 \text{ V} - 0.021 \text{ V} = 0.010 \text{ V}$$

PRACTICE PROBLEMS

1. Calculate the reduction potential of fumarate at 37°C when [fumarate] = 400 μM and [succinate] = 500 μM.

2. Calculate the standard reduction potential of substance A when $\mathcal{E} = 0.2$ V at 25°C, $[A_{reduced}] = 2.5$ μM, and $[A_{oxidized}] = 100$ μM. Assume that $n = 2$.

THE FREE ENERGY CHANGE CAN BE CALCULATED FROM THE CHANGE IN REDUCTION POTENTIAL

Knowing the reduction potentials of different substances is useful for predicting the movement of electrons between the two substances. When the substances are together in solution or connected by wire in an electrical circuit, *electrons flow spontaneously from the substance with the lower reduction potential to the substance with the higher reduction potential.* For example, in a system containing Q/QH_2 and $NAD^+/NADH$, we can predict whether electrons will flow from QH_2 to NAD^+ or from NADH to Q. Using the standard reduction potentials given in Table 15.1, we note that $\mathcal{E}^{\circ\prime}$ for NAD^+ (–0.315 V) is lower than $\mathcal{E}^{\circ\prime}$ for ubiquinone (0.045 V). Therefore, NADH will tend to transfer its electrons to ubiquinone; that is, NADH will be oxidized and Q will be reduced.

A complete oxidation–reduction reaction is just a combination of two half-reactions. For the NADH–ubiquinone reaction, the net reaction is the ubiquinone reduction half-reaction (the half-reaction as listed in Table 15.1) combined with the NADH oxidation half-reaction (the reverse of the half-reaction listed in Table 15.1). Note that because the NAD^+ half-reaction has been reversed to indicate oxidation, we have also reversed the sign of its $\mathcal{E}^{\circ\prime}$ value:

$$\text{NADH} \rightleftharpoons \text{NAD}^+ + \text{H}^+ + 2\,e^- \qquad\qquad \mathcal{E}^{\circ\prime} = +0.315 \text{ V}$$
$$\underline{\text{Q} + 2\,\text{H}^+ + 2\,e^- \rightleftharpoons \text{QH}_2 \qquad\qquad\qquad\quad \mathcal{E}^{\circ\prime} = 0.045 \text{ V}}$$
$$\textit{net}: \text{NADH} + \text{Q} + \text{H}^+ \rightleftharpoons \text{NAD}^+ + \text{QH}_2 \qquad \Delta\mathcal{E}^{\circ\prime} = +0.360 \text{ V}$$

When the two half-reactions are added, their reduction potentials are also added, yielding a $\Delta\mathcal{E}^{\circ\prime}$ value. Keep in mind that the reduction potential is a property of the half-reaction and is independent of the direction in which the reaction occurs. Reversing the sign of $\mathcal{E}^{\circ\prime}$, as shown above, is just a shortcut to simplify the task of calculating $\Delta\mathcal{E}^{\circ\prime}$. Another method for calculating $\Delta\mathcal{E}^{\circ\prime}$ uses the following equation:

$$\Delta\mathcal{E}^{\circ\prime} = \mathcal{E}^{\circ\prime}_{(e^- \text{ acceptor})} - \mathcal{E}^{\circ\prime}_{(e^- \text{ donor})} \qquad\qquad\qquad \textbf{[15.3]}$$

Not surprisingly, *the larger the difference in $\mathcal{E}$ values (the greater the $\Delta\mathcal{E}$ value), the greater the tendency of electrons to flow from one substance to the other, and the greater the change in free energy of the system.* ΔG is related to $\Delta\mathcal{E}$ as follows:

$$\Delta G^{\circ\prime} = -n\mathcal{F}\Delta\mathcal{E}^{\circ\prime} \quad \text{or} \quad \Delta G = -n\mathcal{F}\Delta\mathcal{E} \qquad\qquad\qquad \textbf{[15.4]}$$

Overview of mitochondrial electron transport. The reduction potentials of the key electron carriers are indicated. The oxidation–reduction reactions mediated by Complexes I, III, and IV release free energy.

 Determine the total free energy change for the oxidation of NADH by O_2.

Accordingly, an oxidation–reduction reaction with a large positive $\Delta\mathcal{E}$ value has a large negative value of ΔG (see Sample Calculation 15.2). Depending on the relevant reduction potentials, an oxidation–reduction reaction can release considerable amounts of free energy. This is what happens in the mitochondria, where the reduced cofactors generated by the oxidation of metabolic fuels are reoxidized. The free energy released in this process powers ATP synthesis by oxidative phosphorylation. **Figure 15.2** shows the major mitochondrial electron transport components arranged by their reduction potentials. Each stage of electron transfer, from NADH to O_2, the final electron acceptor, occurs with a negative change in free energy.

CONCEPT CHECK

1. Explain why an oxidation–reduction reaction must include both an oxidant and a reductant.

2. When two reactants are mixed together, explain how you can predict which one will become reduced and which one will become oxidized.

3. Explain how adding the $\mathcal{E}^{\circ\prime}$ values for two half-reactions yields a value of $\Delta\mathcal{E}^{\circ\prime}$ and $\Delta G^{\circ\prime}$ for an oxidation–reduction reaction.

4. Select the two half-reactions from Table 15.1 that would be most likely to form a freely reversible (near-equilibrium) redox reaction.

SAMPLE CALCULATION 15.2

PROBLEM Calculate the standard free energy change for the oxidation of malate by NAD^+. Is the reaction spontaneous under standard conditions?

SOLUTION

Method 1
Write the relevant half-reactions, reversing the malate half-reaction (so that it becomes an oxidation reaction) and reversing the sign of its $\mathcal{E}^{\circ\prime}$:

malate $\rightarrow$ oxaloacetate $+ 2\,H^+ + e^-$	$\mathcal{E}^{\circ\prime} = +0.166\ V$
$NAD^+ + H^+ + 2e^- \rightarrow NADH$	$\mathcal{E}^{\circ\prime} = -0.315\ V$

net: malate $+ NAD^+ \rightarrow$ oxaloacetate $+ NADH + H^+$ $\Delta\mathcal{E}^{\circ\prime} = -0.149\ V$

Method 2
Identify the electron acceptor (NAD^+) and electron donor (malate). Substitute their standard reduction potentials into Equation 15.3:

$$\Delta\mathcal{E}^{\circ\prime} = \mathcal{E}^{\circ\prime}_{(e^-\ acceptor)} - \mathcal{E}^{\circ\prime}_{(e^-\ donor)}$$
$$= -0.315\ V - (-0.166\ V)$$
$$= -0.149\ V$$

Both Methods

The $\Delta \mathcal{E}^{\circ\prime}$ for the net reaction is -0.149 V. Use Equation 15.4 to calculate $\Delta G^{\circ\prime}$:

$$\Delta G^{\circ\prime} = -n\mathcal{F}\Delta \mathcal{E}^{\circ\prime}$$
$$= -(2)(96{,}485 \text{ J}\cdot\text{V}^{-1}\cdot\text{mol}^{-1})(-0.149 \text{ V})$$
$$= +28{,}750 \text{ J}\cdot\text{mol}^{-1} = +28.8 \text{ kJ}\cdot\text{mol}^{-1}$$

The reaction has a positive value of $\Delta G^{\circ\prime}$ and so is not spontaneous. (*In vivo*, this endergonic reaction occurs as step 8 of the citric acid cycle and is coupled to step 1, which is exergonic.)

PRACTICE PROBLEMS

3. Calculate the standard free energy change for the oxidation of malate by ubiquinone. Is the reaction spontaneous under standard conditions?

4. In yeast, alcohol dehydrogenase reduces acetaldehyde to ethanol (Section 13.1). Calculate the free energy change for this reaction under standard conditions.

5. In cells, cytochrome c oxidizes cytochrome c_1. Calculate the change in free energy for the reverse reaction.

15.2 MITOCHONDRIAL ELECTRON TRANSPORT

KEY CONCEPTS

- The inner mitochondrial membrane encloses the matrix and includes specific transport proteins.
- Complex I transfers electrons from NADH to ubiquinone.
- The citric acid cycle, fatty acid oxidation, and other processes also generate mitochondrial ubiquinol.
- The Q cycle mediated by Complex III reduces cytochrome c.
- Complex IV uses electrons from cytochrome c to reduce O_2 to H_2O.

In aerobic organisms, the NADH and ubiquinol produced by glycolysis, the citric acid cycle, fatty acid oxidation, and other metabolic pathways are ultimately reoxidized by molecular oxygen, a process called **cellular respiration.** The standard reduction potential of $+0.815$ V for the reduction of O_2 to H_2O indicates that O_2 is a more effective oxidizing agent than any other biological compound (see Table 15.1). The oxidation of NADH by O_2, that is, the transfer of electrons from NADH directly to O_2, would release a large amount of free energy, but this reaction does not occur in a single step. Instead, *electrons are shuttled from NADH to O_2 in a multistep process that offers several opportunities to conserve the free energy of oxidation.* In eukaryotes, all the steps of oxidative phosphorylation are carried out by a series of electron carriers that include small molecules as well as the prosthetic groups of large integral membrane proteins in mitochondria (in prokaryotes, similar electron carriers are located in the plasma membrane). The following sections describe how electrons flow through this respiratory **electron transport chain** from reduced cofactors to oxygen.

MITOCHONDRIAL MEMBRANES DEFINE TWO COMPARTMENTS

In accordance with its origin as a bacterial symbiont, the **mitochondrion** (plural, *mitochondria*) has two membranes. The outer membrane, analogous to the outer membrane of some bacteria, is relatively porous due to the presence of porin-like proteins that permit the transmembrane diffusion of substances with masses up to about 10 kD (see Section 9.2 for an example of porin structure and function). The inner membrane has a convoluted architecture that encloses a space called the **mitochondrial matrix.** Because the inner mitochondrial membrane prevents the transmembrane movements of ions and small molecules (except via specific transport proteins), the composition of the matrix differs from that of the space between the inner and outer membranes. In fact, the ionic composition of the **intermembrane space** is considered to be equivalent to that of the cytosol due to the presence of the porins in the outer mitochondrial membrane (**Fig. 15.3**).

Mitochondria are customarily shown as kidney-shaped organelles with the inner mitochondrial membrane forming a system of baffles called **cristae** (**Fig. 15.4a**). However, **electron tomography,** a technique for visualizing cellular structures in three dimensions by analyzing micrographs of sequential cell slices, reveals that mitochondria are highly variable structures. For example, the cristae may be irregular and bulbous rather than planar and may make several tubular connections with the rest of the inner mitochondrial membrane (Fig. 15.4b). Moreover, a cell may contain hundreds to thousands of discrete bacteria-shaped mitochondria, or a single tubular organelle may take the form of an extended network with many branches and interconnections (Fig. 15.4c). Individual mitochondria can move around the cell and undergo fusion (joining) and fission (separating).

Outer membrane
Inner membrane
Intermembrane space
Matrix

FIGURE 15.3

Model of mitochondrial structure. The relatively impermeable inner mitochondrial membrane encloses the protein-rich matrix. The intermembrane space has an ionic composition similar to that of the cytosol because the outer mitochondrial membrane is permeable to substances with masses of less than about 10 kD.

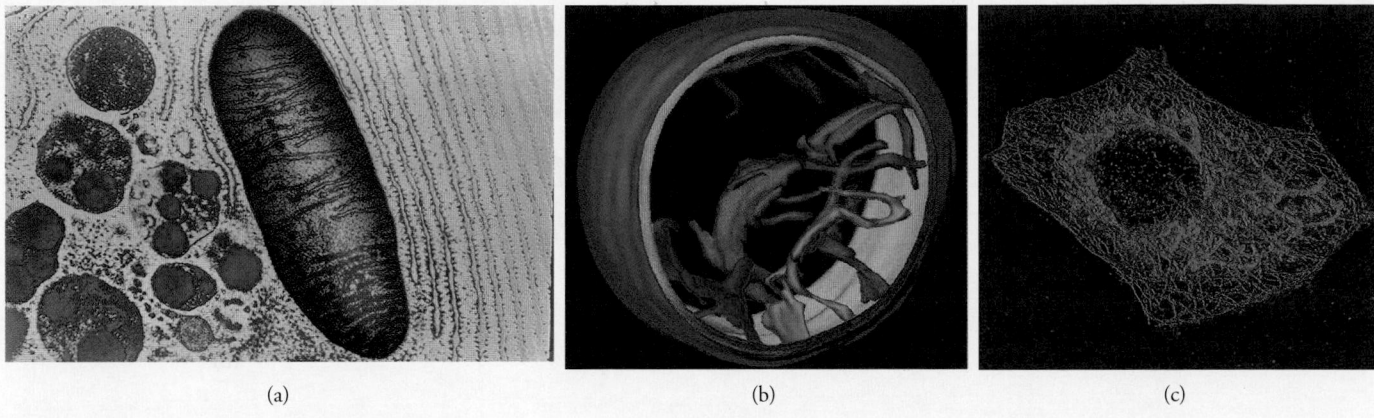

(a) (b) (c)

FIGURE 15.4

Images of mitochondria. (a) Conventional electron micrograph showing cristae as a system of planar baffles. [*K. Porter/Photo Researchers.*] (b) Three-dimensional reconstruction of a mitochondrion by electron tomography, showing irregular tubular cristae. [*Courtesy Carmen Mannella, Wadsworth Center, Albany, New York.*] (c) Electron micrograph of a mammalian fibroblast, showing a network of tubular mitochondria (labeled with a green fluorescent dye). The remainder of the cytosol is delineated by microtubules (labeled with a red fluorescent dye). [*Courtesy Michael P. Yaffe. From* Science *283, 1493–1497 (1991).*]

Reflecting its ancient origin as a free-living organism, the mitochondrion has its own genome and protein-synthesizing machinery consisting of mitochondrially encoded rRNA and tRNA. The mitochondrial genome encodes 13 proteins, all of which are components of the respiratory chain complexes. This is only a small subset of the approximately 1500 proteins required for mitochondrial function; the other respiratory chain proteins, matrix enzymes, transporters, and so on are encoded by the cell's nuclear genome, synthesized in the cytosol, and imported into the mitochondria (across one or both membranes) by special mechanisms.

Much of the cell's NADH and QH$_2$ is generated by the citric acid cycle inside mitochondria. Fatty acid oxidation also takes place largely inside mitochondria and yields NADH and QH$_2$. These reduced cofactors transfer their electrons to the protein complexes of the respiratory electron transport chain, which are tightly associated with the inner mitochondrial membrane. However, NADH produced by glycolysis and other oxidative processes in the cytosol cannot directly reach the respiratory chain. There is no transport protein that can ferry NADH across the inner mitochondrial membrane. Instead, "reducing equivalents" are imported into the matrix by the chemical reactions of systems such as the malate–aspartate shuttle system (**Fig. 15.5**).

FIGURE 15.5

The malate–aspartate shuttle system. Cytosolic oxaloacetate is reduced to malate for transport into mitochondria. Malate is then reoxidized in the matrix. The net result is the transfer of "reducing equivalents" from the cytosol to the matrix. Mitochondrial oxaloacetate can be exported back to the cytosol after being converted to aspartate by an aminotransferase.

FIGURE 15.6

Mitochondrial transport systems. (a) The adenine nucleotide translocase binds either ATP or ADP and changes its conformation to release the nucleotide on the opposite side of the inner mitochondrial membrane. This transporter can therefore export ATP and import ADP. (b) A P_i–H^+ symport protein permits the simultaneous movement of inorganic phosphate and a proton into the mitochondrial matrix.

? **Does the activity of either of these transporters contribute to the mitochondrial membrane potential?**

Mitochondria also need a mechanism to export ATP and to import ADP and P_i, since most of the cell's ATP is generated in the matrix by oxidative phosphorylation and is consumed in the cytosol. A transport protein called the adenine nucleotide translocase exports ATP and imports ADP, binding one or the other and changing its conformation to release the bound nucleotide on the other side of the membrane (**Fig. 15.6a**). Inorganic phosphate, a substrate for oxidative phosphorylation, is imported from the cytosol in symport with H^+ (Fig. 15.6b).

The protein complexes that carry out electron transport and ATP synthesis are oriented in the inner mitochondrial membrane so that they can bind the NADH, ADP, and P_i present in the matrix. Electron microscopy studies show that Complexes I, III, and IV form a "supercomplex," which likely increases the efficiency of electron transfer between them.

COMPLEX I TRANSFERS ELECTRONS FROM NADH TO UBIQUINONE

The path electrons travel through the respiratory chain begins with Complex I, also called NADH:ubiquinone oxidoreductase or NADH dehydrogenase. This enzyme catalyzes the transfer of a pair of electrons from NADH to ubiquinone:

$$NADH + H^+ + Q \rightleftharpoons NAD^+ + QH_2$$

Complex I is the largest of the electron transport proteins in the mitochondrial respiratory chain, with 44 different subunits and a total mass of about 980 kD in mammals. The crystal structure of a smaller (536-kD) bacterial Complex I reveals an L-shaped protein with numerous transmembrane helices and a peripheral arm (**Fig. 15.7**). Electron transport takes place in the peripheral arm, which includes several prosthetic groups that undergo reduction as they receive electrons and become oxidized as they give up their electrons to the next group. All these groups, or **redox centers,** appear to have reduction potentials approximately between the reduction potentials of NAD^+ ($\mathcal{E}^{\circ\prime} = -0.315$ V) and ubiquinone ($\mathcal{E}^{\circ\prime} = +0.045$ V). This allows them to form a chain where the electrons travel a path of increasing reduction potential. The redox centers do not need to be in intimate contact with each other, as they would be if the transferred group were a larger chemical entity. An electron can move between redox centers up to 14 Å apart by "tunneling" through the covalent bonds of the protein.

The two electrons donated by NADH are first picked up by flavin mononucleotide (FMN; **Fig. 15.8**) near the far end of the Complex I arm. This noncovalently bound prosthetic group, which is similar to FAD, then transfers the electrons, one at a time, to a second type of redox center, an iron–sulfur (Fe–S) cluster. Depending on the species, Complex I bears 8 to 10 of

(a)

(b)

FIGURE 15.7

Structure of bacterial Complex I. (a) The 16 subunits are shown in different colors, with the redox centers in red (FMN) and orange (Fe–S clusters). The horizontal portion is embedded in the membrane (represented by black lines) and the "arm" projects into the cytosol in bacteria (or mitochondrial matrix in eukaryotes). (b) Arrangement of the redox centers in Complex I. Atoms are color coded: C gray, N blue, O red, P orange, Fe gold, and S yellow. Electrons flow through the groups from upper left to lower right. [*Structure of* Thermus thermophilus *Complex I (pdb 4HEA) determined by R. Baradaran, J. M. Berrisford, G. S. Minhas, and L. A. Sazanov.*]

? **Identify the FMN group, the 4Fe–4S clusters, and the 2Fe–2S clusters in part (b).**

these prosthetic groups, which contain equal numbers of iron and sulfide ions (**Fig. 15.9**). Unlike the electron carriers we have introduced so far, Fe–S clusters are one-electron carriers. They have an oxidation state of either +3 (oxidized) or +2 (reduced), regardless of the number of Fe atoms in the cluster (each cluster is a conjugated structure that functions as a single unit). Electrons travel between several Fe–S clusters before reaching ubiquinone. Like FMN, ubiquinone is a two-electron carrier (see Section 12.2), but it accepts one electron at a time from an Fe–S donor. Iron–sulfur clusters may be among the most ancient of electron carriers, dating from a time when the earth's abundant iron and sulfur were major players in prebiotic chemical reactions. The ubiquinone binding site is located in the Complex I arm not far from the membrane surface.

As electrons are transferred from NADH to ubiquinone, Complex I transfers four protons from the matrix to the intermembrane space. Comparisons with other transport proteins

$$2 H^+, 2 e^-$$

Flavin mononucleotide (FMN)

FMNH$_2$

FIGURE 15.8

Flavin mononucleotide (FMN). This prosthetic group resembles flavin adenine dinucleotide (FAD; see Fig. 3.2c) but lacks the AMP group of FAD. The transfer of two electrons and two protons to FMN yields FMNH$_2$.

2Fe–2S

4Fe–4S

FIGURE 15.9

Iron–sulfur clusters. Although some Fe–S clusters contain up to eight Fe atoms, the most common are the 2Fe–2S and 4Fe–4S clusters. In all cases, the iron–sulfur clusters are coordinated by the S atoms of cysteine side chains. These prosthetic groups undergo one-electron redox reactions.

and detailed analysis of the crystal structure indicate the presence of four proton-translocating "channels" in the membrane-embedded arm of Complex I. When redox groups in the peripheral arm are transiently reduced and reoxidized, the protein undergoes conformational changes that are transmitted from the peripheral arm to the membrane arm in part by a horizontally oriented helix that lies within the membrane portion of the complex. These conformational changes do not open passageways, as occurs in Na^+ and K^+ transporters (Section 9.2). Instead, each proton passes from one side of the membrane to the other via a **proton wire,** a series of hydrogen-bonded protein groups plus water molecules that form a chain through which a proton can be rapidly relayed. (Recall from Fig. 2.14 that protons readily jump between water molecules.) Note that in this relay mechanism, the protons taken up from the matrix are not the same ones that are released into the intermembrane space. The reactions of Complex I are summarized in **Figure 15.10**.

OTHER OXIDATION REACTIONS CONTRIBUTE TO THE UBIQUINOL POOL

The reduced quinone product of the Complex I reaction joins a pool of quinones that are soluble in the inner mitochondrial membrane by virtue of their long hydrophobic isoprenoid tails (see Section 12.2). *The pool of reduced quinones is augmented by the activity of other oxidation–reduction reactions.* One of these is catalyzed by succinate dehydrogenase, which carries out step 6 of the citric acid cycle (see Section 14.2).

$$\text{succinate} + Q \rightleftharpoons \text{fumarate} + QH_2$$

Succinate dehydrogenase is the only one of the citric acid cycle enzymes that is not soluble in the mitochondrial matrix; it is embedded in the inner membrane. Like the other respiratory complexes, it contains several redox centers, including an FAD group. Succinate dehydrogenase is also called Complex II of the mitochondrial respiratory chain. However, it is more like a tributary because it does not undertake proton translocation and therefore does not directly contribute the free energy of its oxidation–reduction reaction toward ATP synthesis. Nevertheless, it does feed reducing equivalents as ubiquinol into the electron transport chain **(Fig. 15.11a)**.

A major source of ubiquinol is fatty acid oxidation, another energy-generating catabolic pathway that takes place in the mitochondrial matrix. A membrane-bound fatty acyl-CoA dehydrogenase catalyzes the oxidation of a C—C bond in a fatty acid attached to coenzyme A. The electrons removed in this dehydrogenation reaction are transferred to ubiquinone (Fig. 15.11b). As we will see in Section 17.2, the complete oxidation of a fatty acid also produces NADH that is reoxidized by the mitochondrial electron transport chain, starting with Complex I.

Electrons from cytosolic NADH can also enter the mitochondrial ubiquinol pool through the actions of a cytosolic and a mitochondrial glycerol-3-phosphate dehydrogenase (Fig. 15.11c). This system, which shuttles electrons from NADH to ubiquinol, bypasses Complex I.

COMPLEX III TRANSFERS ELECTRONS FROM UBIQUINOL TO CYTOCHROME *c*

Ubiquinol is reoxidized by Complex III, an integral membrane protein with 11 subunits in each of its two monomeric units. Complex III, also called ubiquinol:cytochrome *c* oxidoreductase or cytochrome bc_1, transfers electrons to the peripheral membrane protein cytochrome *c*. **Cytochromes** are proteins with heme prosthetic groups. The name *cytochrome* literally means "cell color"; cytochromes are largely responsible for the purplish-brown color of mitochondria. Cytochromes are commonly named with a letter (*a*, *b*, or *c*) indicating the exact structure of the porphyrin ring of their heme group (**Fig. 15.12**). The structure of the heme group and the surrounding protein microenvironment influence the protein's absorption spectrum.

FIGURE 15.10

Complex I function. As two electrons from the water-soluble NADH are transferred to the lipid-soluble ubiquinone, four protons are translocated from the matrix into the intermembrane space.

MATRIX

INTERMEMBRANE
SPACE

Succinate → Fumarate

succinate
dehydrogenase
(Complex II)

$2e^-$ Q

QH$_2$

(a)

acyl-CoA
dehydrogenase

$2e^-$ Q

QH$_2$

(b)

mitochondrial
dehydrogenase

$2e^-$ Q

QH$_2$

Dihydroxyacetone
phosphate

Glycerol-3-
phosphate

cytosolic
dehydrogenase

NADH + H$^+$ NAD$^+$

(c)

FIGURE 15.11

Reactions that contribute to the ubiquinol pool. (a) The succinate dehydrogenase (Complex II) reaction transfers electrons to the pool of reduced ubiquinone in the inner mitochondrial membrane. (b) The acyl-CoA dehydrogenase reaction, which is one step of the fatty acid oxidation pathway, generates ubiquinol. R represents the hydrocarbon tail of the fatty acid. (c) In the glycerol-3-phosphate shuttle system, electrons from cytosolic NADH are used by a cytosolic glycerol-3-phosphate dehydrogenase to reduce dihydroxyacetone phosphate to glycerol-3-phosphate. The mitochondrial enzyme, embedded in the inner membrane, then reoxidizes the glycerol-3-phosphate, ultimately transferring the two electrons to the membrane ubiquinone pool.

They also determine the reduction potentials of cytochromes, which range from about -0.080 V to about $+0.385$ V.

Unlike the heme prosthetic groups of hemoglobin and myoglobin, the heme groups of cytochromes undergo reversible one-electron reduction, with the central Fe atom cycling between the Fe^{3+} (oxidized) and Fe^{2+} (reduced) states. Consequently, the net reaction for Complex III, in which two electrons are transferred, includes two cytochrome c proteins:

$$QH_2 + 2 \text{ cytochrome } c \text{ (Fe}^{3+}) \rightleftharpoons Q + 2 \text{ cytochrome } c \text{ (Fe}^{2+}) + 2 \text{ H}^+$$

Complex III itself contains two cytochromes (cytochrome b and cytochrome c_1) that are integral membrane proteins. These two proteins, along with an iron–sulfur protein (also called the Rieske protein), form the functional core of Complex III (these same three subunits are the only ones that have homologs in the corresponding bacterial respiratory complex). Altogether, each monomer of Complex III is anchored in the membrane by 14 transmembrane α helices (**Fig. 15.13**).

The flow of electrons through Complex III is complicated, in part because the two electrons donated by ubiquinol must split up in order to travel through a series of one-electron carriers that includes the 2Fe–2S cluster of the iron–sulfur protein, cytochrome c_1, and cytochrome b (which actually contains two heme groups with slightly different reduction potentials). Except for the 2Fe–2S cluster, all the redox centers are arranged in such a way that electrons can tunnel from one to another. The iron–sulfur protein must change its conformation by rotating and moving about 22 Å in order to pick up and deliver an electron. Further complicating the picture is the fact that each monomeric unit of Complex III has two active sites where quinone cofactors undergo reduction and oxidation.

The circuitous route of electrons from ubiquinol to cytochrome c is described by the two-round **Q cycle,** diagrammed in **Figure 15.14**. *The net result of the Q cycle is that two electrons from QH$_2$ reduce two molecules of cytochrome c. In addition, four*

Heme b

FIGURE 15.12

The heme group of a b cytochrome. The planar porphyrin ring surrounds a central Fe atom, shown here in its oxidized (Fe^{3+}) state. The heme substituent groups that are colored blue differ in the a and c cytochromes (the heme group of hemoglobin and myoglobin has the b structure; see Section 5.1).

MATRIX

INTERMEMBRANE
SPACE

(a)   (b)

FIGURE 15.13

Structure of mammalian Complex III. (a) Backbone model. Eight transmembrane helices in each monomer of the dimeric complex are contributed by cytochrome *b* (light blue with heme groups dark blue). The iron–sulfur protein (green with Fe–S clusters orange) and cytochrome *c*₁ (pink with heme groups purple) project into the intermembrane space. (b) Arrangement of prosthetic groups. The two heme groups of each cytochrome *b* (blue) and the heme group of cytochrome *c*₁ (purple), along with the iron–sulfur clusters (Fe atoms orange), provide a pathway for electrons between ubiquinol (in the membrane) and cytochrome *c* (in the intermembrane space). [*Structure (pdb 1BE3) determined by S. Iwata, J. W. Lee, K. Okada, J. K. Lee, M. Iwata, S. Ramaswamy, and B. K. Jap.*]

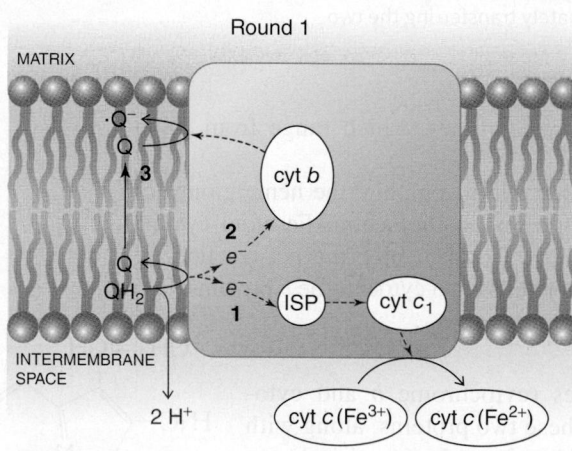

1. *In the first round, QH₂ donates one electron to the iron–sulfur protein (ISP). The electron then travels to cytochrome c₁ and then to cytochrome c.*

2. *QH₂ donates its other electron to cytochrome b. The two protons from QH₂ are released into the intermembrane space.*

3. *The oxidized ubiquinone diffuses to another quinone-binding site, where it accepts the electron from cytochrome b, becoming a half-reduced semiquinone (·Q⁻).*

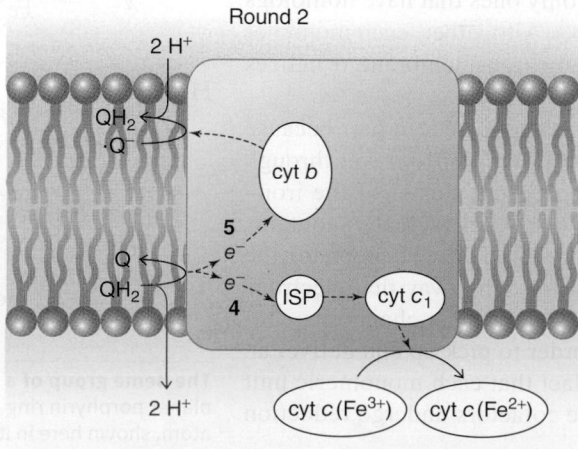

4. *In the second round, a second QH₂ surrenders its two electrons to Complex III and its two protons to the intermembrane space. One electron goes to reduce cytochrome c.*

5. *The other electron goes to cytochrome b and then to the waiting semiquinone produced in the first part of the cycle. This step regenerates QH₂, using protons from the matrix.*

FIGURE 15.14

The Q cycle.

 Write an equation for round 1 and for round 2.

FIGURE 15.15

Complex III function. For every two electrons that pass from ubiquinol to cytochrome *c*, four protons are translocated to the intermembrane space.

? How does the proton-translocating mechanism of Complex III differ from the one in Complex I?

protons are translocated to the intermembrane space, two from QH$_2$ in the first round of the Q cycle and two from QH$_2$ in the second round. This proton movement contributes to the transmembrane proton gradient. The reactions of Complex III are summarized in **Figure 15.15**.

COMPLEX IV OXIDIZES CYTOCHROME *c* AND REDUCES O$_2$

Just as ubiquinone ferries electrons from Complex I and other enzymes to Complex III, cytochrome *c* ferries electrons between Complexes III and IV. Unlike ubiquinone and the other proteins of the respiratory chain, cytochrome *c* is soluble in the intermembrane space (**Fig. 15.16**). Because this small peripheral membrane protein is central to the metabolism of many organisms, analysis of its sequence played a large role in elucidating evolutionary relationships.

Complex IV, also called cytochrome *c* oxidase, is the last enzyme to deal with the electrons derived from the oxidation of metabolic fuels. Four electrons delivered by cytochrome *c* are consumed in the reduction of molecular oxygen to water:

$$4 \text{ cytochrome } c \text{ (Fe}^{2+}) + O_2 + 4 \text{ H}^+ \longrightarrow 4 \text{ cytochrome } c \text{ (Fe}^{3+}) + 2 \text{ H}_2\text{O}$$

The redox centers of mammalian Complex IV include heme groups and copper ions situated among the 13 subunits in each half of the dimeric complex (**Fig. 15.17**).

FIGURE 15.16

Cytochrome *c*. The protein is shown as a gray transparent surface over its ribbon backbone. The heme group (pink) lies in a deep pocket. Cytochrome *c* transfers one electron at a time from Complex III to Complex IV. [*Structure of tuna cytochrome c (pdb 5CYT) determined by T. Takano.*]

FIGURE 15.17

Structure of cytochrome *c* oxidase. (a) The 13 subunits in each monomeric half of the mammalian complex comprise 28 transmembrane α helices. (b) The heme groups (C atoms gray, N blue, O red, and Fe gold) and copper ions (brown) from one half of the complex are shown in space-filling form. [*Structure (pdb 2OCC) determined by T. Tsukihara and M. Yao.*]

? Locate the copper ion and heme iron of the binuclear center where O$_2$ is reduced.

$Fe^{2+} \text{---} Cu^+$

O_2

$Fe^{2+} \text{---} Cu^+$

O_2

H^+, e^-

$Fe^{4+} \text{---} Cu^{2+}$

$O^{2-} \quad OH^-$

H^+

$\rightarrow H_2O$

$Fe^{4+} \text{---} Cu^{2+}$

O^{2-}

H^+, e^-

$Fe^{3+} \text{---} Cu^{2+}$

OH^-

$H^+, 2\,e^-$

$\rightarrow H_2O$

$Fe^{2+} \text{---} Cu^+$

FIGURE 15.18

A proposed model for the cytochrome c oxidase reaction. Although the exact sequence of proton and electron transfers is not known, the reaction intermediates shown here are inferred from spectroscopic and other evidence. An enzyme tyrosine radical (not shown) plays a role in electron transfer.

Each electron travels from cytochrome c to the Cu_A redox center, which has two copper ions, and then to a heme a group. From there it travels to a binuclear center consisting of the iron atom of heme a_3 and a copper ion (Cu_B). The four-electron reduction of O_2 occurs at the Fe–Cu binuclear center. Note that the chemical reduction of O_2 to H_2O consumes four protons from the mitochondrial matrix. One possible sequence of reaction intermediates is shown in **Figure 15.18**. The incomplete reduction of O_2 to H_2O is believed to generate free radicals that can damage mitochondria (**Box 15.A**).

Cytochrome c oxidase also relays four additional protons from the matrix to the intermembrane space (two protons for every pair of electrons). The protein complex appears to harbor two proton wires. One delivers H^+ ions from the matrix to the oxygen-reducing active site. The other one spans the 50-Å distance between the matrix and intermembrane faces of the protein. Protons are relayed through the proton wires when the protein changes its conformation in response to changes in its oxidation state. *The production of water and the proton relays both deplete the matrix proton concentration and thereby contribute to the formation of a proton gradient across the inner mitochondrial membrane* (**Fig. 15.19**).

As the reduced cofactors NADH and ubiquinol deliver their electrons to the electron transport chain, they become oxidized. The cofactors rejoin the pool of oxidized cofactors in the cell, ready to accept additional electrons by participating in redox reactions such as those of the citric acid cycle. In this way, NAD^+ and ubiquinone function as shuttles, accepting electrons (becoming reduced) then giving them up (becoming oxidized) over and over. Because the electron carriers are regenerated with each reaction cycle, electron flow continues, provided there is plenty of metabolic fuel (the source of electrons) and O_2 (the final electron acceptor).

A CLOSER LOOK

Box 15.A Free Radicals and Aging

The oxidizing power of molecular oxygen allows aerobic metabolism—a far more efficient strategy than anaerobic metabolism—but comes at a cost. Partial reduction of O_2 by Complex IV, or possibly via side reactions carried out in Complexes I and III, can produce the superoxide free radical, $\cdot O_2^-$.

$$O_2 + e^- \rightarrow \cdot O_2^-$$

A **free radical** is an atom or molecule with a single unpaired electron and is highly reactive as it seeks another electron to form a pair. Such reactivity means that a free radical, although extremely short-lived (the half-life of $\cdot O_2^-$ is 1×10^{-6} seconds), can chemically alter nearby molecules. Presumably, the most damage is felt by mitochondria, whose proteins, lipids, and DNA are all susceptible to oxidation as superoxide steals an electron.

As the damage accumulates, the mitochondria become less efficient and eventually nonfunctional, at which point the cell self-destructs. According to the free radical theory of aging, oxidative damage mediated by $\cdot O_2^-$ and other free radicals is responsible for the degeneration of tissues that occurs with aging.

Several lines of evidence support the link between free radicals and aging. First, the tissues of some individuals with progeria, a

form of accelerated aging, appear to produce higher than normal levels of oxygen free radicals. Second, cells of all kinds are equipped with antioxidant mechanisms, suggesting that these components perform an essential function. For example, the enzyme superoxide dismutase converts superoxide to a less toxic product, peroxide:

$$2 \cdot O_2^- + 2\,H^+ \rightarrow H_2O_2 + O_2$$

Other cellular components, such as ascorbate (see Box 5.C) and α-tocopherol (see Box 8.B) may protect cells from oxidative damage by scavenging free radicals. Finally, animal experiments suggest that caloric restriction, which extends life spans, generates fewer free radicals by decreasing the availability of fuel molecules that undergo oxidative metabolism. Unfortunately, studies in humans have not yielded conclusive evidence that consuming particular antioxidants or decreasing fuel consumption diminishes the degeneration that normally accompanies aging.

? **Free radicals have been identified as hormone-like signaling molecules in both animals and plants. How might this information alter the free radical theory of aging?**

FIGURE 15.19

Complex IV function. For every two electrons donated by cytochrome c, two protons are translocated to the intermembrane space. Two protons from the matrix are also consumed in the reaction $\frac{1}{2} O_2 \rightarrow H_2O$ (the full reduction of O_2 requires four electrons).

CONCEPT CHECK

1. Describe the compartments of a mitochondrion.

2. What transport proteins occur in the inner mitochondrial membrane?

3. List the different types of redox centers in the respiratory electron transport chain. Which are one-electron carriers and which are two-electron carriers?

4. How are electrons delivered to complexes I, III and IV?

5. Explain why O_2 is the final electron acceptor in the chain.

6. Describe the operation of a proton wire.

7. Write an equation to describe the overall redox reaction carried out by each mitochondrial complex.

CHEMIOSMOSIS

The electrons collected from metabolic fuels during their oxidation are now fully disposed of in the reduction of O_2 to H_2O. However, their free energy has been conserved. How much free energy is potentially available? Using the ΔG values calculated from the standard reduction potentials of the substrates and products of Complexes I, III, and IV (presented graphically in Fig. 15.2), we can see that each of the three respiratory complexes theoretically releases enough free energy to drive the endergonic phosphorylation of ADP to form ATP ($\Delta G^{\circ\prime} = +30.5 \text{ kJ} \cdot \text{mol}^{-1}$).

Complex I: NADH $\rightarrow$ QH$_2$	$\Delta G^{\circ\prime} = -69.5 \text{ kJ} \cdot \text{mol}^{-1}$
Complex III: QH$_2$ $\rightarrow$ cytochrome c	$\Delta G^{\circ\prime} = -36.7 \text{ kJ} \cdot \text{mol}^{-1}$
Complex IV: cytochrome c $\rightarrow$ O$_2$	$\Delta G^{\circ\prime} = -112.0 \text{ kJ} \cdot \text{mol}^{-1}$
NADH $\rightarrow$ O$_2$	$\Delta G^{\circ\prime} = -218.2 \text{ kJ} \cdot \text{mol}^{-1}$

Recall that energy cannot be created or destroyed, but it can be transformed. Understanding oxidative phosphorylation requires recognizing energy in several different forms along the way from metabolic fuels to ATP.

KEY CONCEPTS
- The formation of a transmembrane proton gradient during electron transport provides the free energy to synthesize ATP.
- Both concentration and charge contribute to the free energy of the proton gradient.

FIGURE 15.20

Generation of a proton gradient. During the oxidation–reduction reactions catalyzed by mitochondrial Complexes I, III, and IV, protons (represented by positive charges) are translocated out of the matrix into the intermembrane space. This creates an imbalance in both proton concentration and electrical charge.

CHEMIOSMOSIS LINKS ELECTRON TRANSPORT AND OXIDATIVE PHOSPHORYLATION

Until the 1960s, the connection between respiratory electron transport (measured as O_2 consumption) and ATP synthesis was a mystery. Credit for discovering the connection belongs primarily to Peter Mitchell, who was inspired by his work on mitochondrial phosphate transport and recognized the importance of compartmentation in biological systems. Mitchell's **chemiosmotic theory** proposed that the proton-translocating activity of the electron transport complexes in the inner mitochondrial membrane generates a proton gradient across the membrane. The protons cannot diffuse back into the matrix because the membrane is impermeable to ions. *The imbalance of protons represents a source of free energy, also called a **protonmotive force,** that can drive the activity of an ATP synthase.*

We now know that for each pair of electrons that flow through Complexes I, III, and IV, 10 protons are translocated from the matrix to the intermembrane space (which is ionically equivalent to the cytosol). In bacteria, electron transport complexes in the plasma membrane translocate protons from the cytosol to the cell exterior. Mitchell's theory of chemiosmosis actually explains more than just aerobic respiration. It also applies to systems where the energy from sunlight is used to generate a transmembrane proton gradient (this aspect of photosynthesis is described in Section 16.2).

THE PROTON GRADIENT IS AN ELECTROCHEMICAL GRADIENT

When the mitochondrial complexes translocate protons across the inner mitochondrial membrane, the concentration of H^+ outside increases and the concentration of H^+ inside decreases (**Fig. 15.20**). *This imbalance of protons, a nonequilibrium state, has an associated free energy (the force that would restore the system to equilibrium).* The free energy of the proton gradient has two components, reflecting the difference in the concentration of the chemical species and the difference in electrical charge of the positively charged protons (for this reason, the mitochondrial proton gradient is referred to as an electrochemical gradient rather than a simple concentration gradient). The free energy change for generating the chemical imbalance of protons is

$$\Delta G = RT \ln \frac{[H^+]_{out}}{[H^+]_{in}} \qquad [15.5]$$

The pH ($-\log [H^+]$) of the intermembrane space (*out*) is typically about 0.75 units less than the pH of the matrix (*in*).

The free energy change for generating the electrical imbalance of protons is

$$\Delta G = Z\mathcal{F}\Delta\psi \qquad [15.6]$$

where Z is the ion's charge (+1 in this case) and $\Delta\psi$ is the membrane potential caused by the imbalance in positive charges (see Section 9.1). For mitochondria, $\Delta\psi$ is positive, usually 150 to 200 mV. This value indicates that the intermembrane space or cytosol is more positive than the matrix (recall from Section 9.1 that for a whole cell, the cytosol is more negative than the extracellular space and $\Delta\psi$ is negative).

Combining the chemical and electrical effects gives an overall free energy change for transporting protons from the matrix (*in*) to the intermembrane space (*out*):

$$\Delta G = RT \ln \frac{[H^+]_{out}}{[H^+]_{in}} + Z\mathcal{F}\Delta\psi \qquad [15.7]$$

Typically, the free energy change for translocating one proton out of the matrix is about $+20 \text{ kJ} \cdot \text{mol}^{-1}$ (see Sample Calculation 15.3 for a detailed application of Equation 15.7). This is a thermodynamically costly event. Passage of the proton back *into* the matrix, following its electrochemical gradient, would have a free energy change of about $-20 \text{ kJ} \cdot \text{mol}^{-1}$. This event is thermodynamically favorable, but it does not provide enough free energy to drive the synthesis of ATP. However, the 10 protons translocated for each pair of electrons transferred from NADH to O_2 have an associated protonmotive force of over $200 \text{ kJ} \cdot \text{mol}^{-1}$, enough to drive the phosphorylation of several molecules of ADP.

SAMPLE CALCULATION 15.3

PROBLEM

Calculate the free energy change for translocating a proton out of the mitochondrial matrix, where $pH_{matrix} = 7.8$, $pH_{cytosol} = 7.15$, $\Delta\psi = 170$ mV, and $T = 25°C$.

SOLUTION

Since $pH = -\log[H^+]$ (Equation 2.4), the logarithmic term of Equation 15.7 can be rewritten. Equation 15.7 then becomes

$$\Delta G = 2.303\, RT\,(pH_{in} - pH_{out}) + Z\mathcal{F}\Delta\psi$$

Substituting known values gives

$$\Delta G = 2.303(8.3145\ \text{J}\cdot\text{K}^{-1}\cdot\text{mol}^{-1})(298\ \text{K})(7.8 - 7.15)$$

$$+ (1)(96,485\ \text{J}\cdot\text{V}^{-1}\cdot\text{mol}^{-1})(0.170\ \text{V})$$

$$= 3700\ \text{J}\cdot\text{mol}^{-1} + 16{,}400\ \text{J}\cdot\text{mol}^{-1}$$

$$= +20.1\ \text{kJ}\cdot\text{mol}^{-1}$$

PRACTICE PROBLEMS

6. Calculate the free energy change for transporting a proton out of the mitochondrial matrix, where $pH_{matrix} = 7.5$, $pH_{cytosol} = 7.25$, $\Delta\psi = 170$ mV, and $T = 37°C$.

7. What size pH gradient (the difference between pH_{matrix} and $pH_{cytosol}$) would correspond to a free energy change of 30.5 kJ · mol^{-1}? Assume that $\Delta\psi = 170$ mV and $T = 25°C$.

CONCEPT CHECK

1. Describe the importance of mitochondrial structure for generating the protonmotive force.

2. Identify the source of the protons that form the gradient across the inner mitochondrial membrane.

3. Why does the free energy of the proton gradient have a chemical and an electrical component?

KEY CONCEPTS
- Proton translocation drives the rotation of a portion of ATP synthase.
- Rotation-induced conformational changes allow ATP synthase to bind ADP and P_i, to phosphorylate ADP, and to release ATP.
- Because ATP synthesis is indirectly linked to electron transport, the P:O ratio is not a whole number.
- The supply of reduced cofactors determines the rate of oxidative phosphorylation.

15.4 ATP SYNTHASE

The protein that taps the electrochemical proton gradient to phosphorylate ADP is known as the F-ATP synthase (or Complex V). One part of the protein, called F_0, functions as a transmembrane channel that permits H^+ to flow back into the matrix, following its gradient. The F_1 component catalyzes the reaction ADP + $P_i \rightarrow$ ATP + H_2O (**Fig. 15.21**). This section describes the structures of the two components of ATP synthase and shows how their activities are linked so that exergonic H^+ transport can be coupled to endergonic ATP synthesis.

ATP SYNTHASE ROTATES AS IT TRANSLOCATES PROTONS

Not surprisingly, the overall structure of ATP synthase is conserved among different species. The F_1 component consists of three α and three β subunits surrounding a central shaft. The membrane-embedded portion of ATP synthase includes an a subunit, two b subunits that extend upward to interact with the F_1 component, and a ring of c subunits (**Fig. 15.22**). The exact number of c subunits varies with the source; bovine mitochondrial ATP synthase, for example, has 8 c subunits, while some bacterial enzymes have 15 c subunits.

FIGURE 15.21

ATP synthase function. As protons flow through the F_0 component from the intermembrane space to the matrix, the F_1 component catalyzes the synthesis of ATP from ADP + P_i.

FIGURE 15.22

Structure of ATP synthase. (a) Model of the mammalian enzyme with individual subunits labeled. A spherical structure consisting of three α and three β subunits is connected via a central stalk (subunits γ, δ, and ε) to the membrane-embedded *c* ring. The *a* subunit is closely associated with the *c* ring, and a peripheral stalk containing several subunits (including *b*) links subunit *a* to the catalytic domain. (b) X-Ray structure of bovine ATP synthase at 3.5-Å resolution. Some subunits are not visible in the crystal structure. [*Structure (pdb 2XND) determined by I. N. Watt, M. G. Montgomery, M. J. Runswick, A. G. W. Leslie, and J. E. Walker.*]

FIGURE 15.23

Mechanism of proton transport by ATP synthase. When a *c* subunit (pink) binds a proton from one side of the membrane, it moves away from the *a* subunit (blue). Because the *c* subunits form a ring, rotation brings another *c* subunit toward the *a* subunit, where it releases its bound proton to the opposite side of the membrane. In mammalian ATP synthase (shown here), one complete rotation of the *c* ring corresponds to the translocation of eight protons. The γ subunit in the center of the *c* ring rotates along with it.

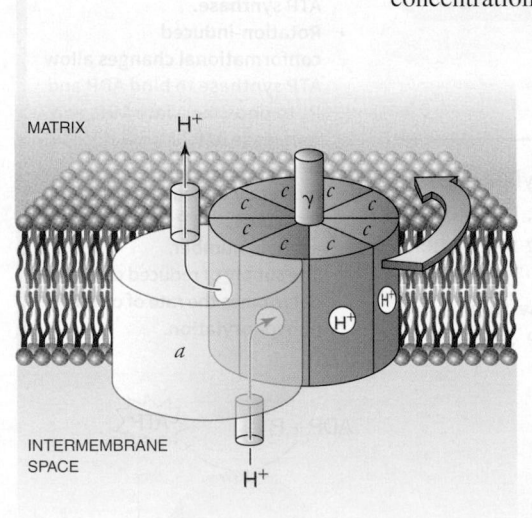

In all species, proton transport through ATP synthase involves the rotation of the *c* ring past the stationary *a* subunit. The carboxylate side chain of a highly conserved aspartate or glutamate residue on each *c* subunit serves as a proton binding site (**Fig. 15.23**). When properly positioned at the *a* subunit, a *c* subunit can take up a proton from the intermembrane space. A slight rotation of the *c* ring brings another *c* subunit into position so that it can release its bound proton into the matrix. The favorable thermodynamics of proton translocation force the *c* ring to keep moving in one direction. Experiments show that depending on the relative concentrations of protons on the two sides of the membrane, the *c* ring can actually spin in either direction. Related proteins, known as P- and V-ATPases, in fact function as active transporters that use the free energy of the ATP hydrolysis reaction to drive ion movement across the membrane.

Attached to the *c* ring and rotating along with it are the γ, δ, and ε subunits (Fig. 15.22). The δ and ε subunits are relatively small, but the γ subunit consists of two long α helices arranged as a bent coiled coil that protrudes into the center of the globular F_1 structure. The three α and three β subunits of F_1 have similar tertiary structures and are arranged like the sections of an orange around the γ subunit (**Fig. 15.24**). Although all six subunits can bind adenine nucleotides, only the β subunits have catalytic activity (nucleotide binding to the α subunits may play a regulatory role).

A close examination of the F_1 assembly reveals that the γ subunit interacts asymmetrically with the three pairs of αβ units. In fact, each αβ unit has a slightly different conformation, and model-building indicates that for steric reasons, the three units cannot simultaneously adopt the same conformation. *The three αβ pairs change their conformations as the γ subunit rotates* (it is like a shaft driven by the *c* ring "rotor"). The αβ hexamer itself does not rotate, since it is held in place by the peripheral arm that is anchored to the *a* subunit (see Fig. 15.22a).

For an ATP synthase containing 8 *c* subunits, the transmembrane movement of each proton could potentially turn the γ shaft by 45° (360° ÷ 8). However, videomicroscopy experiments indicate that the γ subunit rotates in steps of 120°, interacting successively with each of the three αβ pairs in one full rotation of 360°. Electrostatic interactions between the γ and β subunits apparently act as a catch that holds the γ subunit in place while translocation of two to three protons builds up strain. Translocation of the next proton causes the γ subunit to suddenly snap into position at the next β subunit, a movement of

120°. This mechanism accounts for the variation in the number of c subunits in ATP synthases from different sources. The c ring spins in small increments (24° to 45°, depending on the number of c subunits), but the γ subunit makes just three large shifts of 120°.

THE BINDING CHANGE MECHANISM EXPLAINS HOW ATP IS MADE

At the start of the chapter, we pointed out that ATP synthase catalyzes a highly endergonic reaction ($\Delta G^{\circ\prime} = +30.5$ kJ·mol^{-1}) in order to produce the bulk of a cell's ATP supply. This enzyme operates in an unusual fashion, using mechanical energy (rotation) to form a chemical bond (the attachment of a phosphoryl group to ADP). In other words, the enzyme converts mechanical energy to the chemical energy of ATP. The interaction between the γ subunit and the $\alpha\beta$ hexamer explains this energy transduction.

According to the **binding change mechanism** described by Paul Boyer, *rotation-driven conformational changes alter the affinity of each catalytic β subunit for an adenine nucleotide*. At any moment, each catalytic site has a different conformation (and binding affinity), referred to as the open, loose, or tight state. ATP synthesis occurs as follows (**Fig. 15.25**):

1. The substrates ADP and P$_i$ bind to a β subunit in the loose state.

2. The substrates are converted to ATP as rotation of the γ subunit causes the β subunit to shift to the tight conformation.

3. The product ATP is released after the next rotation, when the β subunit shifts to the open conformation.

Because the three β subunits of ATP synthase act cooperatively, they all change their conformations simultaneously as the γ subunit turns. A full rotation of 360° is required to restore the enzyme to its initial state, but each rotation of 120° results in the release of ATP from one of the three active sites.

Experiments with the isolated F$_1$ component of ATP synthase show that in the absence of F$_0$, F$_1$ functions as an ATPase, hydrolyzing ATP to ADP + P$_i$ (a thermodynamically favorable reaction). In the intact ATP synthase, dissipation of the proton gradient is tightly coupled to ATP synthesis with near 100% efficiency. Consequently, *in the absence of a proton gradient, no ATP is synthesized because there is no free energy to drive the rotation of the γ subunit*. Agents that dissipate the proton gradient can therefore "uncouple" ATP synthesis from electron transport, the source of the proton gradient (**Box 15.B**).

THE P:O RATIO DESCRIBES THE STOICHIOMETRY OF OXIDATIVE PHOSPHORYLATION

Since the γ shaft of ATP synthase is attached to the c-subunit rotor, 3 ATP molecules are synthesized for every complete c-ring rotation. However, the number of protons translocated per ATP depends on the number of c subunits. For mammalian ATP synthase, which has 8 c subunits, the stoichiometry is 8 H$^+$ per 3 ATP, or 2.7 H$^+$ per ATP. Such nonintegral values would be difficult to reconcile with most biochemical reactions, but they are consistent with the chemiosmotic theory: *Chemical energy (from*

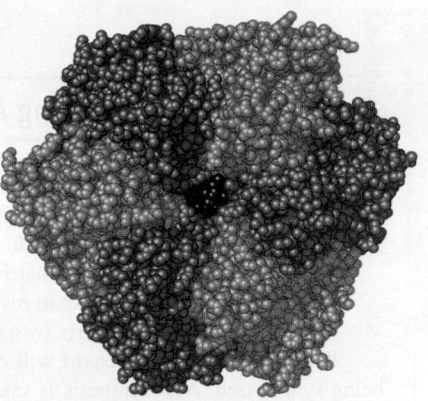

FIGURE 15.24

Structure of the F$_1$ component of ATP synthase. The alternating α (blue) and β (green) subunits form a hexamer around the end of the γ shaft (purple). This view is looking from the matrix down onto the top of the ATP synthase structure shown in Figure 15.22b. [*Structure (pdb 1E79) determined by C. Gibbons, M. G. Montgomery, A. G. W. Leslie, and J. E. Walker.*]

FIGURE 15.25

The binding change mechanism. The diagram shows the catalytic (β) subunits of the F$_1$ component of ATP synthase from the same perspective as in Fig. 15.24. Each of the three β subunits adopts a different conformation: open (O), loose (L), or tight (T). The substrates ADP and P$_i$ bind to a loose site, ATP is synthesized when the site becomes tight, and ATP is released when the subunit becomes open. The conformational shifts are triggered by the 120° rotation of the γ subunit, arbitrarily represented by the purple shape. Because each of the three catalytic sites cycles through the three conformational states, ATP is released from one of the three β subunits with each 120° rotation of the γ subunit.

Box 15.B Uncoupling Agents Prevent ATP Synthesis

When the metabolic need for ATP is low, the oxidation of reduced cofactors proceeds until the transmembrane proton gradient builds up enough to halt further electron transport. When the protons reenter the matrix via the F_0 component of ATP synthase, electron transport resumes. However, if the protons leak back into the matrix by a route other than ATP synthase, then electron transport will continue without any ATP being synthesized. ATP synthesis is said to be "uncoupled" from electron transport, and the agent that allows the proton gradient to dissipate in this way is called an **uncoupler.** Some small molecules that act as uncouplers are poisons, but physiological uncoupling does occur. Dissipating a proton gradient prevents ATP synthesis, but it allows oxidative metabolism to continue at a high rate. The by-product of this metabolic activity is heat.

Uncoupling for **thermogenesis** (heat production) occurs in specialized adipose tissue known as brown fat (its dark color is

due to the relatively high concentration of cytochrome-containing mitochondria; ordinary adipose tissue is lighter). The inner membrane of the mitochondria in brown fat contains a transmembrane proton channel called a UCP (uncoupling protein). Protons translocated to the intermembrane space during respiration can reenter the mitochondrial matrix via the uncoupling protein, bypassing ATP synthase. The free energy of respiration is therefore given off as heat rather than used to synthesize ATP. Brown fat is abundant in hibernating mammals and newborn humans, and the activity of the UCP is under the control of hormones that also mobilize the stored fatty acids to be oxidized in the brown fat mitochondria.

? **Why would increasing the activity of UCP promote weight loss?**

the respiratory oxidation–reduction reactions) is transduced to a protonmotive force, then to the mechanical movement of a rotary engine (the c ring and its attached γ shaft), and finally back to chemical energy in the form of ATP.

The relationship between respiration (the activity of the electron transport complexes) and ATP synthesis is traditionally expressed as a **P:O ratio,** that is, the number of phosphorylations of ADP relative to the number of oxygen atoms reduced. For example, the oxidation of NADH by O_2 (carried out by the sequential activities of Complexes I, III, and IV) translocates 10 protons into the intermembrane space. The movement of these 10 protons back into the matrix via the F_0 component would theoretically drive the synthesis of about 3.7 ATP since 1 ATP can be made for every 2.7 protons translocated, at least in mammalian mitochondria:

$$\frac{1\ \text{ATP}}{2.7\ \text{H}^+} \times 10\ \text{H}^+ = 3.7$$

Thus, the P:O ratio would be about 3.7 (3.7 ATP per $\frac{1}{2} O_2$ reduced). For an electron pair originating as QH_2, only 6 protons would be translocated (by the activities of Complexes III and IV), and the P:O ratio would be approximately 2.2:

$$\frac{1\ \text{ATP}}{2.7\ \text{H}^+} \times 6\ \text{H}^+ = 2.2$$

In vivo, the P:O ratios are actually a bit lower than the theoretical values, because some of the protons translocated during electron transport do leak across the membrane or are consumed in other processes, such as the transport of P_i into the mitochondrial matrix (see Fig. 15.6). Consequently, experimentally determined P:O ratios are closer to 2.5 when NADH is the source of electrons and 1.5 for ubiquinol. These values are the basis for our tally of the ATP yield for the complete oxidation of glucose by glycolysis and the citric acid cycle (see Fig. 14.13).

THE RATE OF OXIDATIVE PHOSPHORYLATION DEPENDS ON THE RATE OF FUEL CATABOLISM

In most metabolic pathways, control is exerted at highly exergonic (irreversible) steps. In oxidative phosphorylation, this step would be the reaction catalyzed by cytochrome *c* oxidase (Complex IV; see Fig. 15.2). However, there are no known effectors of cytochrome *c* oxidase activity. Apparently, *the close coupling between generation of the proton gradient and ATP synthesis allows oxidative phosphorylation to be regulated primarily by the availability of reduced cofactors (NADH and QH_2) produced by other metabolic processes.*

Less important regulatory mechanisms may involve the availability of the substrates ADP and P_i (which depend on the activity of their respective transport proteins). Experiments with ATP synthase show that when ADP and P_i are absent, the β subunits cannot undergo the conformational changes required by the binding change mechanism. The γ subunit therefore remains immobile, and no protons are translocated through the c ring. This tight coupling between ATP synthesis and proton translocation prevents the waste of the free energy of the proton gradient.

There is also evidence that mitochondria contain a regulatory protein that binds to ATP synthase to inhibit its rate of ATP hydrolysis. The inhibitor is sensitive to pH, so it does not bind to ATP synthase when the matrix pH is high (as it is when electron transport is occurring). However, if the matrix pH drops as a result of a momentary disruption of the proton gradient, the inhibitor binds to ATP synthase. This regulatory mechanism prevents ATP synthase from operating in reverse as an ATPase.

The processes of electron transport and oxidative phosphorylation, like other metabolic pathways, maintain a steady state. The proton gradient does not build up beyond the typical 0.75-pH-unit difference across the membrane, because as soon as protons are pumped into the intermembrane space, they return to the matrix via the c ring of ATP synthase. Similarly, the cell's concentration of ATP remains more or less constant. Note that ATP synthase does not build an ATP molecule from scratch; it simply forms a bond between the second and third phosphate groups. When free energy–requiring cellular reactions consume ATP, this same bond is most often the one that is broken. The rates of ATP consumptions and synthesis may fluctuate dramatically, depending on the cell's activity level, but they must be balanced. The constant cleavage and regeneration of ATP continues as long as the cell has the fuel and oxygen to support oxidative phosphorylation.

 CONCEPT CHECK

1. Draw a simple diagram of ATP synthase and indicate which parts are stationary and which rotate.

2. Explain how ATP synthase dissipates the proton gradient.

3. Describe how the three conformational states of the β subunits of ATP synthase are involved in ATP synthesis.

4. How does the binding change mechanism account for ATP hydrolysis by ATP synthase?

5. Why does the number of protons translocated per ATP synthesized vary among species?

6. Explain why the availability of reduced substrates is the primary mechanism for regulating oxidative phosphorylation.

7. What is a P:O ratio and why is it nonintegral?

SUMMARY

15.1 The Thermodynamics of Oxidation–Reduction Reactions

• The electron affinity of a substance participating in an oxidation–reduction reaction, which involves the transfer of electrons, is indicated by its reduction potential, $\mathcal{E}°'$.

• The difference in reduction potential between substances undergoing oxidation and reduction is related to the free energy change for the reaction.

15.2 Mitochondrial Electron Transport

• Oxidation of reduced cofactors generated by metabolic reactions takes place in the mitochondrion. Shuttle systems and transport proteins allow the transmembrane movement of reducing equivalents, ATP, ADP, and P_i.

• The electron transport chain consists of a series of integral membrane protein complexes that contain multiple redox groups, including iron–sulfur clusters, flavins, cytochromes, and copper ions, and that are linked by mobile electron carriers. Starting from NADH, electrons travel a path of increasing reduction potential through Complex I, ubiquinone, Complex III, cytochrome c, and then to Complex IV, where O_2 is reduced to H_2O.

• As electrons are transferred, protons are translocated to the intermembrane space via proton wires in Complexes I and IV and by the action of the Q cycle associated with Complex III.

15.3 Chemiosmosis

• The chemiosmotic theory describes how proton translocation during mitochondrial electron transport generates an electrochemical gradient whose free energy drives ATP synthesis.

15.4 ATP Synthase

• The energy of the proton gradient is tapped as protons spontaneously flow through ATP synthase. Proton transport allows rotation of a ring of integral membrane c subunits. The linked γ subunit thereby rotates, triggering conformational changes in the F_1 portion of ATP synthase.

• According to the binding change mechanism, the three functional units of the F_1 portion cycle through three conformational states to

sequentially bind ADP and P_i, convert the substrates to ATP, and release ATP.

• The P:O ratio quantifies the link between electron transport and oxidative phosphorylation in terms of the ATP synthesized and the O_2 reduced. Because these processes are coupled, the rate of oxidative phosphorylation is controlled primarily by the availability of reduced cofactors.

KEY TERMS

oxidative phosphorylation	$\mathcal{F}$	electron tomography	protonmotive force
oxidizing agent (oxidant)	cellular respiration	redox center	Z
reducing agent (reductant)	electron transport chain	proton wire	binding change mechanism
half-reaction	mitochondrion	cytochrome	uncoupler
$\mathcal{E}^{\circ\prime}$	mitochondrial matrix	Q cycle	thermogenesis
$\mathcal{E}$	intermembrane space	free radical	P:O ratio
Nernst equation	cristae	chemiosmotic theory	

BIOINFORMATICS

Brief Bioinformatics Exercises

15.1 Viewing and Analyzing Complexes I–IV

15.2 Diversity of the Electron-Transport Chain and the KEGG Database

---

PROBLEMS

15.1 The Thermodynamics of Oxidation–Reduction Reactions

1. Calculate the standard free energy change for the reduction of
(a) Pyruvate by NADH. (b) Oxygen by cytochrome a_3.

Consult Table 15.1 for the relevant half-reactions. Are these reactions spontaneous under standard conditions?

2. Identify the oxidized and reduced forms from the following pairs:
(a) malate/oxaloacetate; (b) pyruvate/lactate; (c) NADH/NAD$^+$;
(d) fumarate/succinate; (e) Fe(CN)$_6^{3-}$/Fe(CN)$_6^{2-}$; (f) H$_2$O$_2$/O$_2$;
(g) FMN/FMNH$_2$ (see Fig. 15.8).

3. Calculate the reduction potential of
(a) Fumarate at 37°C when [fumarate] = 60 μM and [succinate] = 90 μM.
(b) Ubiquinone (Q) at 37°C when the concentration of Q is 30 μM and the concentration of ubiquinol (QH$_2$) is 6 μM.

4. (a) Calculate the standard reduction potential of substance A when $\mathcal{E}$ = 0.49 V at 25°C; [A$_{reduced}$] = 10 μM, and [A$_{oxidized}$] = 250 μM. Assume that n = 2.
(b) Calculate the standard reduction potential of substance B when $\mathcal{E}$ = –0.22 V at 25°C; [B$_{reduced}$] = 60 μM, and [B$_{oxidized}$] = 3 μM. Assume that n = 2.

(c) Consult Table 15.1. What are the possible identities of substance A and B?

5. Calculate the standard free energy change for the oxidation of malate by ubiquinone. Is the reaction spontaneous under standard conditions?

6. In one of the final steps of the pyruvate dehydrogenase reaction (see Section 14.1), E3 reoxidizes the lipoamide group of E2, then NAD$^+$ reoxidizes E3. Calculate $\Delta G^{\circ\prime}$ for the electron transfer from dihydrolipoic acid to NAD$^+$.

7. Each electron from cytochrome c is donated to a Cu$_A$ redox center in Complex IV. The $\mathcal{E}^{\circ\prime}$ value for the Cu$_A$ redox center is 0.245 V. Calculate $\Delta G^{\circ\prime}$ for this electron transfer.

8. Acetaldehyde may be oxidized to acetate. Would NAD$^+$ be an effective oxidizing agent? Explain.

9. Acetoacetate may be reduced to 3-hydroxybutyrate. What serves as a better reducing agent, NADH or FADH$_2$? Explain.

10. For every two QH$_2$ that enter the Q cycle, one is regenerated and the other passes its two electrons to two cytochrome c_1 centers. The overall equation is

$$\text{QH}_2 + 2 \text{ cyt } c_1 \text{ (Fe}^{3+}) + 2 \text{ H}^+ \rightarrow \text{Q} + 2 \text{ cyt } c_1 \text{ (Fe}^{2+}) + 4 \text{ H}^+$$

Calculate the free energy change associated with the Q cycle.

11. Why is succinate oxidized by FAD instead of by NAD$^+$?

12. (a) What is the $\Delta\mathcal{E}$ value for the oxidation of ubiquinol by cytochrome c when the ratio of QH_2/Q is 10 and the ratio of cyt c (Fe^{3+})/ cyt c (Fe^{2+}) is 5?

(b) Calculate ΔG for the reaction in part (a). Assume $T = 25°C$.

13. An iron–sulfur protein in Complex III donates an electron to cytochrome c_1. The reduction half-reactions and $\mathcal{E}°'$ values are shown below.

(a) Write the balanced equation for the reaction and calculate the standard free energy change.

$$FeS\ (ox) + e^- \rightarrow FeS\ (red) \qquad \mathcal{E}°' = 0.280\ V$$
$$\text{cyt } c_1\ (Fe^{3+}) + e^- \rightarrow \text{cyt } c_1\ (Fe^{2+}) \qquad \mathcal{E}°' = 0.215\ V$$

(b) How can you account for the fact that this reaction occurs spontaneously in the cell?

(c) If the ΔG value for the reaction is $-6.0\ kJ \cdot mol^{-1}$, what is the value of $\Delta\mathcal{E}$?

14. Calculate the overall efficiency of oxidative phosphorylation, assuming standard conditions, by comparing the free energy potentially available from the oxidation of NADH by O_2 and the free energy required to synthesize 2.5 ATP from 2.5 ADP.

15. Using the percent efficiency calculated in Problem 14, calculate the number of ATP generated by:

(a) Complex I (where NADH is oxidized by ubiquinone).

(b) Complex III (where ubiquinol is oxidized by cytochrome c).

(c) Complex IV (where cytochrome c is oxidized by molecular oxygen).

16. Calculate $\Delta\mathcal{E}°'$ and $\Delta G°'$ for the succinate dehydrogenase (Complex II) reaction. Does this reaction provide sufficient free energy to drive ATP synthesis under standard conditions? Explain.

15.2 Mitochondrial Electron Transport

17. The sequence of events in electron transport was elucidated in part by the use of inhibitors that block electron transfer at specific points along the chain. For example, adding rotenone (a plant toxin) or amytal (a barbiturate) blocks electron transport in Complex I; antimycin A (an antibiotic) blocks electron transport in Complex III; and cyanide (CN^-) blocks electron transport in Complex IV by binding to the Fe^{2+} in the Fe–Cu binuclear center.

(a) What happens to oxygen consumption when these inhibitors are added to a suspension of respiring mitochondria?

(b) What is the redox state of the electron carriers in the electron transport chain when each of the inhibitors is added separately to the mitochondrial suspension?

18. What is the effect of added succinate on rotenone-blocked, antimycin A–blocked, or cyanide-blocked mitochondria (see Problem 17)? In other words, can succinate help "bypass" the block? Explain.

19. (a) The compound tetramethyl-p-phenylenediamine (TMPD) donates a pair of electrons directly to Complex IV. Can TMPD act as a bypass for the rotenone-blocked, antimycin A–blocked, or cyanide-blocked mitochondria described in Problem 17?

(b) Ascorbate (vitamin C) can donate a pair of electrons to cytochrome c. Can ascorbate act as a bypass for the rotenone-blocked, antimycin A–blocked, or cyanide-blocked mitochondria described in Problem 17?

20. When the antifungal agent myxothiazol is added to a suspension of respiring mitochondria, the QH_2/Q ratio increases. Where in the electron transport chain does myxothiazol inhibit electron transfer?

21. If cyanide poisoning (see Problem 17) is diagnosed immediately, it can be treated by administering nitrites that can oxidize the Fe^{2+} in hemoglobin to Fe^{3+}. Why is this treatment effective?

22. The effect of the drug fluoxetine (Prozac) on isolated rat brain mitochondria was examined by measuring the rate of electron transport (units not given) in the presence of various combinations of substrates and inhibitors (see Problems 17–19).

(a) How do pyruvate and malate serve as substrates for electron transport?

(b) What is the effect of fluoxetine on electron transport? Explain.

(c) Fluoxetine can also inhibit ATP synthase. Why might long-term use of fluoxetine be a concern?

Rate of Electron Transport

[Fluoxetine], (mM)	Pyruvate + malate	Succinate + rotenone	Ascorbate + TMPD
0	163 ± 15.1	145 ± 14.2	184 ± 22.2
0.15	77 ± 7.3	131 ± 13.5	116 ± 13.9

23. Complex I, succinate dehydrogenase, acyl-CoA dehydrogenase, and glycerol-3-phosphate dehydrogenase (see Fig. 15.11) are all flavoproteins; that is, they contain an FMN or FAD prosthetic group. Explain the function of the flavin group in these enzymes. Why are the flavoproteins ideally suited to transfer electrons to ubiquinone?

24. What side chains would you expect to find as part of a proton wire in a proton-translocating membrane protein?

25. Ubiquinone is not anchored in the mitochondrial membrane but is free to diffuse laterally throughout the membrane among the electron transport chain components. What aspects of its structure account for this behavior?

26. Explain why the ubiquinone-binding site of Complex I (Fig. 15.7) is located at the end of the peripheral arm closest to the membrane.

27. Cytochrome c is easily dissociated from isolated mitochondrial membrane preparations, but the isolation of cytochrome c_1 requires the use of strong detergents. Explain why.

28. Release of cytochrome c from the mitochondrion to the cytosol is one of the signals that induces apoptosis, a form of programmed cell death. What structural features of cytochrome c allow it to play this role?

29. In coastal marine environments, high concentrations of nutrients from terrestrial runoff may lead to algal blooms. When the nutrients are depleted, the algae die and sink and are degraded by other microorganisms. The algal die-off may be followed by a sharp drop in oxygen in the depths, which can kill fish and bottom-dwelling invertebrates. Why do these "dead zones" form?

30. Chromium is most toxic and highly soluble in its oxidized Cr(VI) state but is less toxic and less soluble in its more reduced Cr(III) state. Efforts to detoxify Cr-contaminated groundwater have involved injecting chemical reducing agents underground. Another approach is bioremediation, which involves injecting molasses or cooking oil into the contaminated groundwater. Explain how these substances would promote the reduction of Cr(VI) to Cr(III).

31. At one time, it was believed that myoglobin functioned simply as an oxygen-storage protein. New evidence suggests that myoglobin plays a much more active role in the muscle cell. The phrase *myoglobin-facilitated oxygen diffusion* describes myoglobin's role in transporting oxygen from the muscle cell sarcolemma to the mitochondrial membrane surface. Mice in which the myoglobin gene was knocked out had higher tissue capillary density, elevated red blood cell counts, and increased coronary blood flow. Explain the reasons for these compensatory mechanisms in the knockout mice.

32. The myoglobin and cytochrome c oxidase content were determined in several animals, as shown in the table. What is the relationship between the two proteins? Explain.

	Myoglobin content, mmol · kg⁻¹	Cytochrome c oxidase activity
Hare	0.1	900
Sheep	0.19	950
Ox	0.31	1200
Horse	0.38	1800

33. It has recently been found that myoglobin distribution is not confined to muscle cells. Tumor cells, which generally exist in hypoxic (low-oxygen) conditions because of limited blood flow, have been found to express myoglobin. How does this adaptation increase the chances of tumor cell survival?

34. Cancer cells, even when sufficient oxygen is available, produce large amounts of lactate. It has been observed that the concentration of fructose-2,6-bisphosphate is much higher in cancer cells than in normal cells. Why would this result in anaerobic metabolism being favored, even when oxygen is available?

35. A group of elderly patients who did not exercise regularly were asked to participate in a 12-week exercise program. Data collected from the patients are shown in the table. What was the result of the exercise intervention and why did this occur?

	Prior to exercise intervention	Post 12-week exercise intervention
Total mitochondrial DNA (copies per diploid genome)	1300	1900
Complex II activity	0.13	0.20
Complex I–IV activity	0.51	1.00

36. Amyotrophic lateral sclerosis (ALS) is a neurodegenerative disease that causes muscle paralysis and eventually death. Researchers measured the activity of the electron transport chain complexes in various regions of the nervous system in patients with ALS. In a certain region of the spinal cord, Complex I showed decreased activity but not decreased concentration. How does this contribute to progression of the disease?

15.3 Chemiosmosis

37. What is the free energy change for generating the electrical imbalance of protons: **(a)** In neuroblastoma cells, where $\Delta\psi$ is 81 mV? **(b)** In respiring mitochondria in culture, where $\Delta\psi$ is 150 mV?

38. Calculate the free energy change for translocating a proton out of the mitochondrial matrix, where $pH_{matrix} = 7.6$, $pH_{cytosol} = 7.2$, $\Delta\psi = 200$ mV, and $T = 37°C$.

39. What size pH gradient (the difference between pH_{matrix} and $pH_{cytosol}$) would correspond to a free energy change of 29.2 kJ·mol^{-1} (Assume that $\Delta\psi = 170$ mV and $T = 37°C$).

40. Several key experimental observations were important in the development of the chemiosmotic theory. Explain how each of these observations is consistent with the chemiosmotic theory as described by Peter Mitchell.

(a) The pH of the intermembrane space is lower than the pH of the mitochondrial matrix.

(b) Oxidative phosphorylation does not occur in mitochondrial preparations to which detergents have been added.

41. Mitchell's original chemiosmotic hypothesis relies on the impermeability of the inner mitochondrial membrane to ions other than H$^+$, such as Na$^+$ and Cl$^-$.

(a) Why was this thought to be important?

(b) Could ATP still be synthesized if the membrane were permeable to other ions?

42. Nigericin is an antibiotic that integrates into membranes and functions as a K$^+$/H$^+$ antiporter. Another antibiotic, valinomycin, is similar, but it allows the passage of K$^+$ ions. When both antibiotics are added simultaneously to suspensions of respiring mitochondria, the electrochemical gradient completely collapses.

(a) Draw a diagram of a mitochondrion in which nigericin and valinomycin have integrated into the inner mitochondrial membrane, in a manner that is consistent with the experimental results.

(b) Explain why the electrochemical gradient dissipates. What happens to ATP synthesis?

43. How does transport of inorganic phosphate from the intermembrane space to the mitochondrial matrix affect the pH difference across the inner mitochondrial membrane?

44. Pioglitazone, a drug used to treat diabetes, causes some membrane- embedded portions of mitochondrial Complex I to separate from the rest of the protein that includes the matrix "arm." Predict the effect of pioglitazone on electron transport and ATP production.

15.4 ATP Synthase

45. How much ATP can be obtained by the cell from the complete oxidation of one mole of glucose? Compare this value with the amount of ATP obtained when glucose is anaerobically converted to lactate or ethanol. Do organisms that can completely oxidize glucose have an advantage over organisms that cannot?

46. The glycerol-3-phosphate shuttle can transport cytosolic NADH equivalents into the mitochondrial matrix (see Fig. 15.11c). In this shuttle, the protons and electrons are donated to FAD, which is reduced to FADH$_2$. These protons and electrons are subsequently donated to coenzyme Q in the electron transport chain. How much ATP is generated per mole of glucose when the glycerol-3-phosphate shuttle is used?

47. The complete oxidation of glucose yields -2850 kJ·mol^{-1} of free energy. Incomplete oxidation by conversion to lactate yields -196 kJ·mol^{-1} and by alcoholic fermentation yields -235 kJ·mol^{-1}.

(a) Calculate the overall efficiencies of glucose oxidation by these three processes.

(b) Do organisms that can completely oxidize glucose have an advantage over organisms that cannot?

48. A culture of yeast grown under anaerobic conditions is exposed to oxygen, resulting in a dramatic decrease in glucose consumption by the cells. This phenomenon is referred to as the Pasteur effect.

(a) Explain the Pasteur effect.

(b) The [NADH]/[NAD$^+$] and [ATP]/[ADP] ratios also change when an anaerobic culture is exposed to oxygen. Explain how these ratios change, and what effect this has on glycolysis and the citric acid cycle in the yeast.

49. Experiments in the late 1970s attributed the Pasteur effect (see Problem 48) to the stimulation of hexokinase and phosphofructokinase under anaerobic conditions. Upon exposure to oxygen, the stimulation of these enzymes ceases. Why are these enzymes more active in the absence of oxygen?

50. Consider the adenine nucleotide translocase and the $P_i - H^+$ symport protein that import ADP and P_i, the substrates for oxidative phosphorylation, into the mitochondrion (see Fig. 15.6).

(a) How does the activity of the adenine nucleotide translocase affect the electrochemical gradient across the mitochondrial membrane?

(b) How does the activity of the $P_i - H^+$ symport protein affect the gradient?

(c) What can you conclude about the thermodynamic force that drives the two transport systems?

51. The compounds atractyloside and bongkrekic acid both bind tightly to and inhibit the adenine nucleotide translocase. How do these compounds affect ATP synthesis? Electron transport?

52. The compound tetramethyl-*p*-phenylenediamine (TMPD) donates a pair of electrons directly to complex IV (see Problem 19a). What is the P:O ratio of TMPD?

53. Ascorbate (vitamin C) can donate a pair of electrons to cytochrome *c* (see Problem 19b). What is the P:O ratio for ascorbate?

54. Dicyclohexylcarbodiimide (DCCD) is a reagent that reacts with Asp or Glu residues. Explain why the reaction of DCCD with just one *c* subunit completely blocks both the ATP-synthesizing and ATP-hydrolyzing activity of ATP synthase.

55. Dinitrophenol was introduced as a "diet pill" in the 1920s. Its use was discontinued because the side effects were fatal in some cases. What was the rationale for believing that DNP would be an effective diet aid?

56. Oligomycin is an antibiotic that blocks proton transfer through the F_0 proton channel of ATP synthase. What is the effect on **(a)** ATP synthesis, **(b)** electron transport, and **(c)** oxygen consumption when oligomycin is added to a suspension of respiring mitochondria? **(d)** What changes occur when dinitrophenol (see Problem 55) is then added to the suspension?

57. The compound carbonylcyanide-*p*-trifluoromethoxy phenylhydrazone (FCCP) is an uncoupler similar to DNP (see Problem 55). Describe how FCCP acts as an uncoupler.

FCCP

58. In the 1950s, experiments with isolated mitochondria showed that organic compounds are oxidized and O_2 is consumed only when ADP is included in the preparation. When the ADP supply runs out, oxygen consumption halts. Explain these results.

59. A patient seeks treatment because her metabolic rate is twice normal and her temperature is elevated. A biopsy reveals that her muscle mitochondria are structurally unusual and not subject to normal respiratory controls. Electron transport takes place regardless of the concentration of ADP.

(a) What is the P:O ratio (compared to normal) of NADH that enters the electron transport chain in the mitochondria of this patient?
(b) Why are the patient's metabolic rate and temperature elevated?
(c) Will this patient be able to carry out strenuous exercise?

60. In experimental systems, the F_0 component of ATP synthase can be reconstituted into a membrane. F_0 can then act as a proton channel that is blocked when the F_1 component is added to the system. What molecule must be added to the system in order to restore the proton-translocating activity of F_0? Explain.

61. Calculate the ratio of protons translocated to ATP synthesized for yeast ATP synthase, which has 10 *c* subunits, and for spinach chloroplast ATP synthase, which has 14 *c* subunits.

62. A bacterial ATP synthase has 10 *c* subunits, and a chloroplast ATP synthase has 14 *c* subunits. Would you expect the bacterium or the chloroplast to have a higher P:O ratio?

63. Experiments indicate that the *c* ring of ATP synthase spins at a rate of 6000 rpm. How many ATP molecules are generated each second?

64. Mutations that impair ATP synthase function are rare. Laboratory studies indicate that adding α-ketoglutarate boosts ATP production in ATP synthase–deficient cells, but only when aspartate is also added to the cells. Explain.

65. In yeast, pyruvate may be converted to ethanol in a two-step pathway catalyzed by pyruvate decarboxylase and alcohol dehydrogenase (see Section 13.1). Pyruvate may also be converted to acetyl-CoA by pyruvate dehydrogenase. Yeast mutants in which the pyruvate

decarboxylase gene is missing (*pdc*–) are useful for studying the regulation of the pyruvate dehydrogenase enzyme. When wildtype yeast were pulsed with glucose, glycolytic flux increased dramatically and the rate of respiration increased. But when the same experiments were performed with *pdc*– mutants, only a small increase in glycolysis was observed, and pyruvate was the main product excreted by the yeast cells. Explain these results.

66. During anaerobic fermentation in yeast, the majority of the available glucose is oxidized via the glycolytic pathway and the rest enters the pentose phosphate pathway to generate NADPH and ribose. This occurs during aerobic respiration as well, except that the percentage of glucose entering the pentose phosphate pathway is much greater in aerobic respiration than during anaerobic fermentation. Explain why.

67. UCP1 is an uncoupling protein in brown fat (Box 15.B). Experiments using UCP1-knockout mice (animals missing the gene for UCP1) resulted in the discovery of a second uncoupling protein named UCP2.

(a) Oxygen consumption increased over twofold when a β_3 adrenergic agonist that stimulates UCP1 was injected into normal mice. This was not observed when the agonist was injected into the knockout mice. Explain these results.
(b) In one experiment, normal mice and UCP1-knockout mice were placed in a cold (5°C) room overnight. The normal mice were able to maintain their body temperature at 37°C even after 24 hours in the cold. But the body temperatures of the cold-exposed knockout mice decreased 10°C or more. Explain.

68. The Eastern skunk cabbage can maintain its temperature 15–35°C higher than ambient temperature during the months of February and March, when ambient temperatures range from –15 to +15°C. Thermogenesis in the skunk cabbage is critical to the survival of the plant since the spadix (a flower component) is not frost-resistant. An uncoupling protein is responsible for the observed thermogenesis.

(a) The spadix relies on the skunk cabbage's massive root system, which stores appreciable quantities of starch. Why is a large quantity of starch required for the skunk cabbage to carry out sustained thermogenesis for weeks rather than hours?
(b) Oxygen consumption by the skunk cabbage increases as the temperature decreases, nearly doubling with every 10°C drop in ambient temperature. Oxygen consumption was observed to decrease during the day, when temperatures were close to 30°C, and increase at night. What is the biochemical explanation for these observations?

69. The gene that codes for an uncoupling protein (see Problem 68) in potatoes was isolated. The results of a Northern blot analysis (which detects mRNA) are shown below. What is your interpretation of these results? How does the mRNA level affect thermogenesis in the potato?

	20°C				4°C		
Days:	0	1	2	3	0	1	2	3

70. Glutamate can be used as an artificial substrate for mitochondrial respiration, as shown in the diagram. When ceramide is added to a mitochondrial suspension respiring in the presence of glutamate, respiration decreases, leading scientists to hypothesize that ceramide might regulate mitochondrial function *in vivo*.

(a) How does glutamate act as a substrate for mitochondrial respiration?
(b) Ceramide-induced inhibition of respiration could be due to several different factors. List several possibilities.
(c) Mitochondria treated with ceramide were exposed to an uncoupler, but the respiration rate did not increase. What site(s) of inhibition can be ruled out?

(d) In another experiment, mitochondria were subjected to a freeze–thaw cycle that rendered the inner mitochondrial membrane permeable to NADH. NADH could then be added to a mitochondrial suspension as a substrate for electron transport. When NADH was added, ceramide decreased the respiration rate to the same extent as when glutamate was the substrate. What site(s) of inhibition can be ruled out?

71. When cells cannot carry out oxidative phosphorylation, they synthesize ATP through substrate-level phosphorylation.

(a) Which enzymes of glycolysis and the citric acid cycle catalyze substrate-level phosphorylation?

(b) The O_2 that we breathe in is not directly converted to the CO_2 that we breathe out. Write a balanced equation for the complete combustion of glucose and oxygen.

72. Rapidly growing bacterial cells tend to rely on glycolysis and fermentation rather than oxidative phosphorylation to generate ATP, even when O_2 is abundant. One group of researchers noted that glycolysis/fermentation requires fewer proteins than oxidative phosphorylation. Could this observation explain why rapidly growing cells prefer glycolysis over more-efficient oxidative phosphorylation?

SELECTED READINGS

Boekema, E. J. and Braunm H.-P., Supramolecular structure of the mitochondrial oxidative phosphorylation system, *J. Biol. Chem.* **282**, 1–4 (2007). [Presents evidence that Complexes I through IV form "supercomplexes."]

Boyer, P. D., Catalytic site forms and controls in ATP synthase catalysis, *Biochim. Biophys. Acta* **1458**, 252–262 (2000). [Reviews the steps of ATP synthesis and hydrolysis, along with experimental evidence and alternative explanations; by the originator of the binding change mechanism.]

Friedman, J. R. and Nunnari, J., Mitochondrial form and function, *Nature* **505**, 335–343 (2014). [Descibes some attributes of mitochondria as semi-independent cellular organelles.]

Watt, I. N., Montgomery, M. G., Runswick, M. J., Leslie, A. G. W., and Walker, J. E., Bioenergetic cost of making an adenosine triphosphate molecule in animal mitochondria, *Proc. Nat. Acad. Sci.* **107**, 16823–16827 (2010). [Includes details of ATP synthase structure.]

Zickerman, V., Wirth, C., Nasiri, H., Siegmund, K., Schwalbe, H., Hunte, C., and Brandt, U., Mechanistic insight from the crystal structure of mitochondrial complex I, *Science* **347**, 44–49 (2015). [Describes the structure of a eukaryotic Complex I and a possible mechanism for linking electron transport to proton pumping.]

C H A P T E R 16

PHOTOSYNTHESIS

The bright green color of the anemone Anthopleura xanthogrammica *is due in part to the presence of symbiotic algae in its tentacles. The algae obtain nutrients from the anemone and in turn supply it with O_2, a by-product of photosynthesis.*

DO YOU REMEMBER?

- Glucose polymers include the fuel-storage polysaccharides starch and glycogen and the structural polysaccharide cellulose (Section 11.2).
- Coenzymes such as NAD^+ and ubiquinone collect electrons from compounds that become oxidized (Section 12.2).
- Electrons are transferred from a substance with a lower reduction potential to a substance with a higher reduction potential (Section 15.1).
- The formation of a transmembrane proton gradient during electron transport provides the free energy to synthesize ATP (Section 15.3).

Every year, plants and bacteria convert an estimated 6×10^{16} grams of carbon to organic compounds by photosynthesis. About half of this activity occurs in forests and savannas, and the rest occurs in the ocean and under ice—wherever water, carbon dioxide, and light are available. The organic materials produced by photosynthetic organisms sustain them as well as the organisms that feed on them. We will begin by examining the absorption of light energy and then look at the electron transport complexes that convert solar energy to biologically useful forms of free energy such as ATP and the reduced cofactor NADPH. Finally, we will see how plants use these energy currencies to synthesize carbohydrates.

16.1 CHLOROPLASTS AND SOLAR ENERGY

The ability to use sunlight as an energy source evolved about 3.5 billion years ago. Before that, cellular metabolism probably centered around the inorganic reductive reactions associated with hydrothermal vents. The first **photosynthetic** organisms produced various pigments (light-absorbing molecules) to capture solar energy and thereby drive the reduction of metabolites. The descendants of some of these organisms are known today as purple bacteria and green sulfur bacteria. By about 2.5 billion years ago, the cyanobacteria had evolved.

KEY CONCEPTS
- Photosynthetic pigments absorb different wavelengths of light to become excited.
- Light-harvesting complexes direct light energy to the reaction center.

Mark Newman/Getty Images

BIOPOLYMERS

Proteins
Nucleic acids
Polysaccharides
Triacylglycerols

MONOMERS

Amino acids
Nucleotides
Monosaccharides
Fatty acids

NH_4^+

NAD^+

NADH

NAD^+

NADH, QH_2

2- and 3-Carbon
INTERMEDIATES

NAD^+, Q

citric
acid
cycle

photo-
synthesis

CO_2

NADH, QH_2

O_2

ADP

oxidative
phosphorylation

H_2O

ATP

NAD^+, Q

FIGURE 16.1

Photosynthesis in context. Photosynthetic organisms incorporate atmospheric CO_2 into three-carbon compounds that are the precursors of biological molecules such as carbohydrates and amino acids. Photosynthesis requires light energy to drive the production of the ATP and NADPH consumed in biosynthetic reactions.

These organisms absorb enough solar energy to undertake the energetically costly oxidation of water to molecular oxygen. In fact, the dramatic increase in the level of atmospheric oxygen (from an estimated 1% to the current level of about 20%) around 2.1 to 2.4 billion years ago is attributed to the rise of cyanobacteria. Modern plants are the result of the symbiosis of early eukaryotic cells with cyanobacteria.

Although the apparatus and reactions of photosynthesis are not found in all organisms, they can be placed in the context of the metabolic scheme outlined in Chapter 12 (**Fig. 16.1**). As you examine the harvest of solar energy and its use in incorporating CO_2 into three-carbon compounds, you will see that significant portions of these processes resemble metabolic pathways that you have already encountered.

Photosynthesis in green plants takes place in **chloroplasts,** discrete organelles that are descended from cyanobacteria. Like mitochondria, chloroplasts contain their own DNA, in this case coding for 100 to 200 chloroplast proteins. DNA in the cell's nucleus contains close to a thousand more genes whose products are essential for photosynthesis.

The chloroplast is enclosed by a porous outer membrane and an ion-impermeable inner membrane (**Fig. 16.2**). The inner compartment, called the **stroma,** is analogous to the mitochondrial matrix and is rich in enzymes, including those required for carbohydrate synthesis. Within the stroma is a membranous structure called the **thylakoid.** Unlike the planar or tubular mitochondrial cristae (see Fig. 15.4), the thylakoid membrane folds into stacks of flattened vesicles and encloses a compartment called the thylakoid lumen. The energy-transducing reactions of photosynthesis take place in the thylakoid membrane. The analogous reactions in photosynthetic bacteria typically take place in folded regions of the plasma membrane.

PIGMENTS ABSORB LIGHT OF DIFFERENT WAVELENGTHS

Light can be considered as both a wave and a particle, the **photon.** The energy (E) of a photon depends on its wavelength, as expressed by **Planck's law:**

$$E = \frac{hc}{\lambda} \qquad [16.1]$$

where h is Planck's constant (6.626×10^{-34} J·s), c is the speed of light (2.998×10^8 m·s^{-1}), and λ is the wavelength (about 400 to 700 nm for visible light; see Sample Calculation 16.1). *This energy is absorbed by the photosynthetic apparatus of the chloroplast and transduced to chemical energy.*

FIGURE 16.2

The chloroplast. (a) Electron micrograph of a chloroplast from tobacco. [*Dr. Jeremy Burgess/Photo Researchers, Inc.*] (b) Model. The stacked thylakoid membranes are known as grana (singular, *granum*).

? **Compare these images to the images of a mitochondrion in Figure 15.4.**

(a)

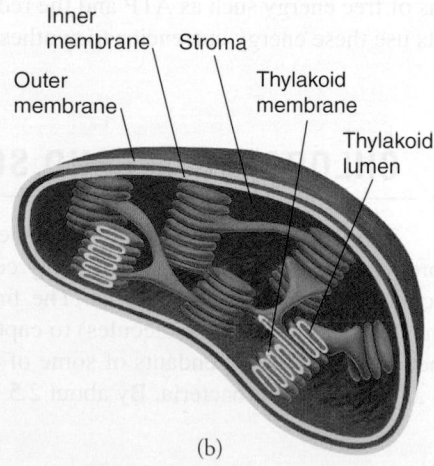

Inner membrane

Stroma

Outer membrane

Thylakoid membrane

Thylakoid lumen

(b)

PROBLEM

Calculate the energy of a photon with a wavelength of 550 nm.

SOLUTION

$$E = \frac{hc}{\lambda}$$

$$E = \frac{(6.626 \times 10^{-34}\,\text{J}\cdot\text{s})(2.998 \times 10^{8}\,\text{m}\cdot\text{s}^{-1})}{550 \times 10^{-9}\,\text{m}}$$

$$E = 3.6 \times 10^{-19}\,\text{J}$$

PRACTICE PROBLEMS

1. Calculate the energy of one mole of photons (6.022×10^{23} mol^{-1}) with a wavelength of 550 nm.
2. At what wavelength would a mole of photons have an energy of 300 kJ?

Chloroplasts contain a variety of light-absorbing groups called pigments or **photoreceptors** (**Fig. 16.3**). Chlorophyll is the principal photoreceptor. It appears green because it absorbs both blue and red light. The second most common pigments are the red carotenoids, which absorb blue light. Pigments such as phycocyanin, which absorb longer-wavelength red light, are common in aquatic systems because water absorbs blue light. Together, these types of pigments absorb all the wavelengths of visible light (**Fig. 16.4**).

(a)

Chlorophyll *a*

(b)

β-Carotene

(c)

Phycocyanin

FIGURE 16.3

Some common chloroplast photoreceptors. (a) Chlorophyll *a*. In chlorophyll *b*, a methyl group (blue) is replaced by an aldehyde group. Chlorophyll resembles the heme groups of hemoglobin and cytochromes (see Fig. 15.12), but it has a central Mg^{2+} rather than an Fe^{2+} ion; it includes a fused cyclopentane ring, and it has a long lipid side chain. (b) The carotenoid β-carotene, a precursor of vitamin A (see Box 8.B). (c) Phycocyanin, a linear tetrapyrrole. It resembles an unfolded chlorophyll molecule.

FIGURE 16.4

Visible light absorption by some photosynthetic pigments. The wavelengths of absorbed light correspond to the peak of the solar energy that reaches the earth.

? Use this diagram to explain the color of each type of pigment molecule.

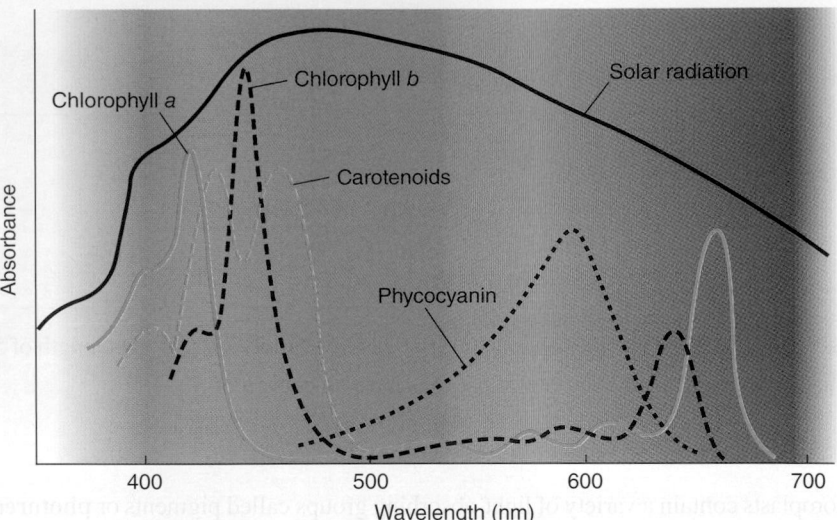

Each photosynthetic pigment is a highly conjugated molecule. When it absorbs a photon of the appropriate wavelength, one of its delocalized electrons is promoted to a higher-energy orbital, and the molecule is said to be excited. The excited molecule can return to its low-energy, or ground, state by several mechanisms (**Fig. 16.5**):

1. The absorbed energy can be lost as heat.

2. The energy can be given off as light, or **fluorescence.** For thermodynamic reasons, the emitted photon has a lower energy (longer wavelength) than the absorbed photon.

3. The energy can be transferred to another molecule. This process is called **exciton transfer** (an exciton is the packet of transferred energy) or resonance energy transfer, since the molecular orbitals of the donor and recipient groups must be oscillating in a coordinated manner in order to transfer energy.

4. An electron from the excited molecule can be transferred to a recipient molecule. In this process, called **photooxidation,** the excited molecule becomes oxidized and the acceptor molecule becomes reduced. Another electron-transfer reaction is required to restore the photooxidized molecule to its original reduced state.

All of these energy-transferring processes occur in chloroplasts to some extent, but *exciton transfer and photooxidation are the most important for photosynthesis.*

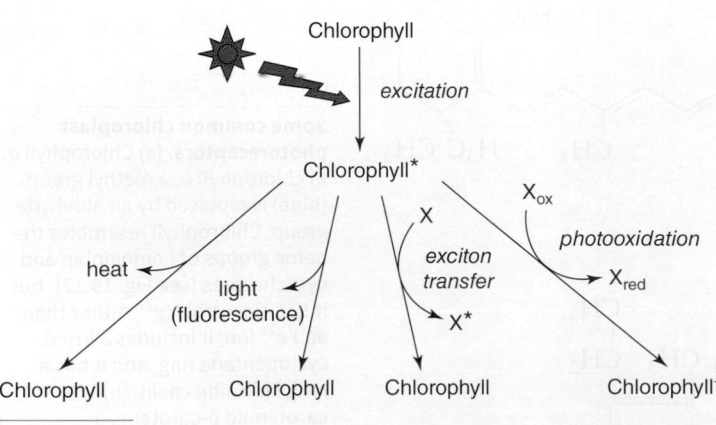

FIGURE 16.5

Dissipation of energy in a photoexcited molecule. A pigment molecule such as chlorophyll is excited by absorbing a photon. The excited molecule (chlorophyll*) can return to its ground state by one of several mechanisms.

LIGHT-HARVESTING COMPLEXES TRANSFER ENERGY TO THE REACTION CENTER

The primary reactions of photosynthesis occur at specific chlorophyll molecules called **reaction centers.** However, chloroplasts contain many more chlorophyll molecules and other pigments than reaction centers. *Many of these extra, or **antenna,** pigments are located in membrane proteins called **light-harvesting complexes.*** Over 30 different kinds of light-harvesting complexes have been characterized, and they are remarkable for their regular geometry. For example, one light-harvesting complex in purple photosynthetic bacteria consists of 18 polypeptide chains holding two concentric rings of chlorophyll molecules, plus carotenoids (**Fig. 16.6**). This artful arrangement of light-absorbing groups is essential for the function of the light-harvesting complex.

The protein microenvironment of each photoreceptor influences the wavelength (and therefore the energy) of the photon it can absorb (just as cytochrome protein structure

(a) (b) (c)

FIGURE 16.6

A light-harvesting complex from *Rhodopseudomonas acidophila*. The nine pairs of subunits (light and dark gray) are mostly buried in the membrane and form a scaffold for two rings of chlorophyll molecules (yellow and green) and carotenoids (red). The pigments are all within a few angstroms of each other. (a) Side view. The extracellular side is at the top. (b) Top view. (c) Top view showing only the chlorophyll molecules. The 18 chlorophyll molecules in the inner ring (green) overlap so that excitation energy may be delocalized over the entire ring. [*Structure (pdb 1KZU) determined by R. J. Cogdell, A. A. Freer, N. W. Isaacs, A. M. Hawthornthwaite-Lawless, G. McDermott, M. Z. Papiz, and S. M. Prince.*]

influences the reduction potential of its heme group; see Section 15.2). Consequently, *the various light-harvesting complexes with their multiple pigments can absorb light of many different wavelengths.* Within a light-harvesting complex, the precisely aligned pigment molecules can quickly transfer their energy to other pigments. *Exciton transfer eventually brings the energy to the chlorophyll at the reaction center* (**Fig. 16.7**). Without light-harvesting complexes to collect and concentrate light, the reaction center chlorophyll could collect only a small fraction of the incoming solar radiation. Even so, a leaf captures only about 1% of the available solar energy.

During periods of high light intensity, some accessory pigments may function to dissipate excess solar energy as heat so that it does not damage the photosynthetic apparatus by inappropriate photooxidation. Various pigment molecules may also act as photosensors to regulate the plant's growth rate and shape and to coordinate the plant's activities—such as germination, flowering, and dormancy—according to daily or seasonal light levels.

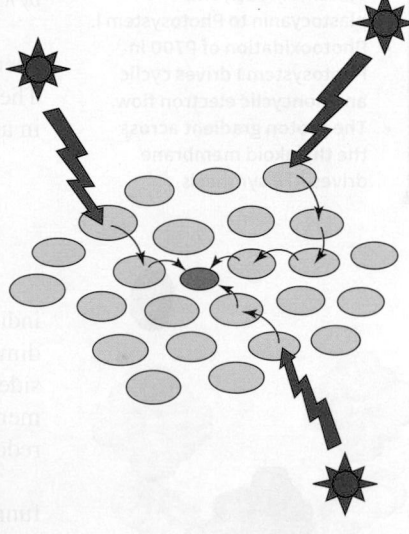

FIGURE 16.7

Function of light-harvesting complexes. A typical photosynthetic system consists of a reaction center (dark green) surrounded by light-harvesting complexes (light green), whose multiple pigments absorb light of different wavelengths. Exciton transfer funnels this captured solar energy to the chlorophyll at the reaction center.

CONCEPT CHECK

1. Correlate a photon's energy to its wavelength.
2. Why is it advantageous for photosynthetic pigments to absorb different colors of light?
3. Use nonscientific terms to explain why leaves are green.
4. Describe the four mechanisms by which a photoexcited molecule can return to its ground state.
5. Explain the function of a light-harvesting complex.

THE LIGHT REACTIONS

In plants and cyanobacteria, the energy captured by the antenna pigments of the light-harvesting complexes is funneled to two photosynthetic reaction centers. *Excitation of the reaction centers drives a series of oxidation–reduction reactions whose net results are the oxidation of water, the reduction of NADP$^+$, and the generation of a transmembrane proton gradient that powers ATP synthesis.* These events are known as the **light reactions** of photosynthesis. (Most photosynthetic bacteria undertake similar reactions but have a single reaction center and do not produce oxygen.) The two photosynthetic reaction centers that mediate light

KEY CONCEPTS

- The P680 reaction center of Photosystem II undergoes photooxidation.
- Photosystem II splits water to replace the lost P680 electron and generate O$_2$.

(continued)

FIGURE 16.8

Structure of cyanobacterial Photosystem II. The proteins are shown as gray ribbons, and the various prosthetic groups and cofactors are shown as stick models and color-coded: chlorophyll, green; pheophytin, orange; β-carotene, red; heme, purple; and quinone, blue. The stroma is at the top and the thylakoid lumen at the bottom. [*Structure of Photosystem II from* Synechococcus elongatus *(1S5L) determined by K. N. Ferreira, T. M. Iverson, K. Maghlaoui, J. Barber, and S. Iwata.*]

KEY CONCEPTS (*continued*)

- Electrons from Photosystem II travel via plastoquinone, cytochrome b_6f, and plastocyanin to Photosystem I.
- Photooxidation of P700 in Photosystem I drives cyclic and noncyclic electron flow.
- The proton gradient across the thylakoid membrane drives ATP synthesis.

FIGURE 16.9

Arrangement of prosthetic groups in Photosystem II. The green chlorophyll groups constitute the photooxidizable P680. The two "accessory" chlorophyll groups (yellow) do not undergo oxidation or reduction. An electron from P680 travels to one of the pheophytin groups (orange), which are essentially chlorophyll molecules without the central Mg^{2+} ion. Next, the electron is transferred to a tightly bound plastoquinone molecule (blue) and then to a loosely bound plastoquinone (not shown). An iron atom (red) may assist the final electron transfer. The lipid tails of the prosthetic groups are not shown.

energy transduction are part of protein complexes called Photosystem I and Photosystem II. These, along with other integral and peripheral proteins of the thylakoid membrane, operate in a series, much like the mitochondrial electron transport chain.

PHOTOSYSTEM II IS A LIGHT-ACTIVATED OXIDATION–REDUCTION ENZYME

In plants and cyanobacteria, the light reactions begin with Photosystem II (the number indicates that it was the second to be discovered). This integral membrane protein complex is dimeric, with more bulk on the lumenal side of the thylakoid membrane than on the stromal side. The cyanobacterial Photosystem II contains at least 19 subunits (14 of them integral membrane proteins). Its numerous prosthetic groups include light-absorbing pigments and redox-active cofactors (**Fig. 16.8**).

Several dozen chlorophyll molecules in Photosystem II function as internal antennas, funneling energy to the two reaction centers, each of which includes a pair of chlorophyll molecules known as P680 (680 nm is the wavelength of one of their absorption peaks). The reaction center chlorophylls overlap so that they are electronically coupled and function as a single unit. When P680 is excited, as indicated by the notation P680*, it quickly gives up an electron, dropping to a lower-energy state, P680⁺. In other words, *light has oxidized P680*. The photooxidized chlorophyll molecule must be reduced in order to return to its original state.

The two P680 groups are located near the lumenal side of Photosystem II. The electron given up by each photooxidized P680 travels through several redox groups (**Fig. 16.9**).

Although the prosthetic groups in Photosystem II are arranged more or less symmetrically, they do not all directly participate in electron transfer. The electron eventually reaches a plastoquinone molecule on the stromal side of Photosystem II. Plastoquinone (PQ) is similar to mammalian mitochondrial ubiquinone (see Section 12.2).

Plastoquinone

FIGURE 16.10

Structure of the Mn₄CaO₅ cluster. Atoms are color-coded: Mn purple, Ca green, and O red. One or more of the four H₂O molecules associated with the cluster (oxygen atoms in blue) may be substrates for the water-splitting reaction. The Asp, Glu, and His side chains that hold the cluster in place are not shown. [*Structure of PSII (pdb 3WU2) determined by Y. Umena, K. Kawakami, J.-R. Shen, and N. Kamiya.*]

It functions in the same way as a two-electron carrier. The fully reduced plastoquinol (PQH_2) joins a pool of plastoquinones that are soluble in the thylakoid membrane. Two electrons (two photooxidations of P680) are required to fully reduce plastoquinone to PQH_2. This reaction also consumes two protons, which are taken from the stroma.

THE OXYGEN-EVOLVING COMPLEX OF PHOTOSYSTEM II OXIDIZES WATER

O_2, a waste product of photosynthesis, is generated from H_2O by a lumenal portion of Photosystem II called the oxygen-evolving center. This reaction can be written as

$$2 H_2O \rightarrow O_2 + 4 H^+ + 4 e^-$$

The electrons derived from H_2O are used to restore photooxidized P680 to its reduced state.

The catalyst for the water-splitting reaction is a cofactor with the composition Mn_4CaO_5 (**Fig. 16.10**). This unusual inorganic cofactor occurs in all Photosystem II complexes, which suggests a unique chemistry that has remained unaltered for about 2.5 billion years. No synthetic catalyst can match the manganese cluster in its ability to extract electrons from water to form O_2. The water-splitting reaction is rapid, with about 50 O_2 produced per second per Photosystem II, and generates most of the earth's atmospheric O_2.

During water oxidation, the manganese cluster undergoes multiple changes in its oxidation state, somewhat reminiscent of the changes in the Fe–Cu binuclear center of cytochrome *c* oxidase (mitochondrial Complex IV; see Fig. 15.18), which carries out the reverse reaction. The four water-derived protons are released into the thylakoid lumen, contributing to a drop in pH relative to the stroma. A tyrosine radical (Y·) in Photosystem II transfers each of the four water-derived electrons to $P680^+$ (a tyrosine radical also plays a role in electron transfer in cytochrome *c* oxidase; see Section 15.2).

Tyrosine radical

The oxidation of water is a thermodynamically demanding reaction because O_2 has an extremely high reduction potential (+0.815 V) and electrons spontaneously flow from a group with a lower reduction potential to a group with a higher reduction potential (see Section 15.1). In fact, photooxidized P680 is the most powerful biological oxidant, with a reduction potential of about +1.15 V.

Upon photoexcitation, the reduction potential of P680 (now P680*) is dramatically diminished, to about –0.8 V. *This low reduction potential allows P680* to surrender an electron to a series of groups with increasingly positive reduction potentials* (**Fig. 16.11**). Recall that the lower the reduction potential, the higher the free energy. The overall result is that the input of solar energy allows an electron to travel a thermodynamically favorable path from water to plastoquinone. *Four photooxidation events in Photosystem II are required to oxidize two H₂O molecules and produce one O₂ molecule.* **Figure 16.12** summarizes the functions of Photosystem II.

FIGURE 16.11

Reduction potential and electron flow in Photosystem II. Electrons flow spontaneously from a group with a lower reduction potential to a group with a higher reduction potential. The transfer of electrons from H_2O to plastoquinone is made possible by the excitation of P680 (wavy arrow), which dramatically lowers its reduction potential.

> **?** **Will pheophytin be in the oxidized or reduced state when it is dark?**

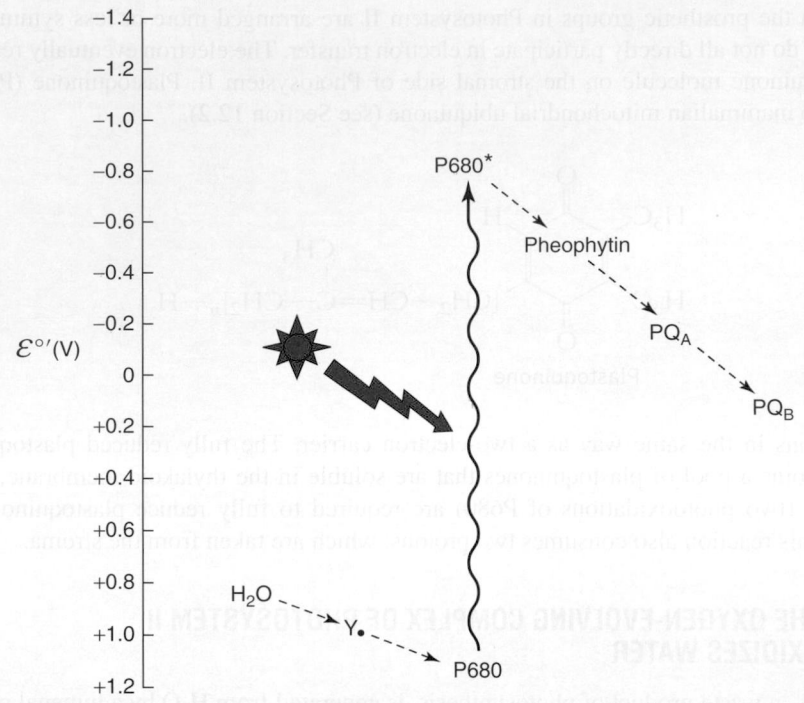

CYTOCHROME $b_6 f$ LINKS PHOTOSYSTEMS I AND II

After they leave Photosystem II as plastoquinol, electrons reach a second membrane-bound protein complex known as cytochrome $b_6 f$. This complex resembles mitochondrial Complex III (also called cytochrome bc_1)—from the entry of electrons in the form of a reduced quinone, through the circular flow of electrons among its redox groups, to the final transfer of electrons to a mobile electron carrier.

The cytochrome $b_6 f$ complex contains eight subunits in each of its monomeric halves (**Fig. 16.13**). Three subunits bear electron-transporting prosthetic groups. One of these subunits is cytochrome b_6, which is homologous to mitochondrial cytochrome b. The second is cytochrome f, whose heme group is actually of the c type. Although it shares no sequence homology with mitochondrial cytochrome c_1, it functions similarly. The chloroplast complex also contains a Rieske iron–sulfur protein with a 2Fe–2S group that behaves like its mitochondrial counterpart. However, the cytochrome $b_6 f$ complex also contains subunits with prosthetic groups that are absent in the mitochondrial complex: a chlorophyll molecule and a β-carotene. These

FIGURE 16.12

Photosystem II function. For every oxygen molecule evolved, two plastoquinone molecules are reduced.

FIGURE 16.13

Structure of cyanobacterial cytochrome $b_6 f$. Each subunit of the dimeric complex is a different color. The prosthetic groups are not shown. [*Structure (pdb 1VF5) determined by G. Kurisu, H. Zhang, J. L. Smith, and W. A. Cramer.*]

> **?** **Compare this structure to the functionally similar mitochondrial cytochrome bc_1 (Complex III) in Figure 15.13.**

light-absorbing molecules do not appear to participate in electron transfer and may instead help regulate the activity of cytochrome b_6f by registering the amount of available light.

Electron flow in the cytochrome b_6f complex follows a cyclic pattern that is probably identical to the Q cycle in mitochondrial Complex III (see Fig. 15.14). However, in chloroplasts, the final electron acceptor is not cytochrome c but plastocyanin, a small protein with an active-site copper ion (**Fig. 16.14**). Plastocyanin functions as a one-electron carrier by cycling between the Cu^+ and Cu^{2+} oxidation states. Like cytochrome c, plastocyanin is a peripheral membrane protein; it picks up electrons at the lumenal surface of cytochrome b_6f and delivers them to another integral membrane protein complex, in this case Photosystem I.

The net result of the cytochrome b_6f Q cycle is that for every two electrons emanating from Photosystem II, four protons are released into the thylakoid lumen. Since the oxidation of $2 H_2O$ is a four-electron reaction, *the production of one molecule of O_2 causes the cytochrome b_6f complex to produce eight lumenal H^+* (**Fig. 16.15**). The resulting pH gradient between the stroma and the lumen is a source of free energy that drives ATP synthesis, as described below.

A SECOND PHOTOOXIDATION OCCURS AT PHOTOSYSTEM I

Photosystem I, like Photosystem II, is a large protein complex containing multiple pigment molecules. The Photosystem I in the cyanobacterium *Synechococcus* is a symmetric trimer with 12 proteins in each monomer (**Fig. 16.16**). Ninety-six chlorophyll molecules and 22 carotenoids operate as a built-in light-harvesting complex for Photosystem I.

In the core of each monomer, a pair of chlorophyll molecules constitute the photoactive group known as P700 (it has a slightly longer-wavelength absorbance maximum than P680). Like P680, P700 undergoes exciton transfer from an antenna pigment. P700* gives up an electron to achieve a low-energy oxidized state, P700$^+$. The group is then reduced by accepting an electron donated by plastocyanin.

FIGURE 16.14

Plastocyanin. The redox-active copper ion (green) is coordinated by a Cys, a Met, and two His residues (yellow). [*Structure of plastocyanin from poplar leaves (pdb 1PLC) determined by J. M. Guss and H. C. Freeman.*]

FIGURE 16.15

Cytochrome b_6f function. The stoichiometry shown for the cytochrome b_6f Q cycle reflects the four electrons released by the oxygen-evolving complex of Photosystem II.

P700 is not a particularly good reducing agent (its reduction potential is relatively high, about +0.45 V). However, excited P700 (P700*) has an

FIGURE 16.16

Structure of cyanobacterial Photosystem I. The protein is shown as a gray ribbon, and the various prosthetic groups are color-coded: chlorophyll, green; β-carotene, red; phylloquinone, blue; and Fe–S clusters, orange. Only one monomer of the trimeric complex is shown. The stroma is at the top. [*Structure of the Synechococcus Photosystem I (pdb 1JB0) determined by P. Jordan, P. Fromme, H. T. Witt, O. Klukas, W. Saenger, and N. Krauss.*]

FIGURE 16.17

Prosthetic groups in Photosystem I. The groups include P700 (the green chlorophyll molecules), "accessory" chlorophylls (yellow), quinones (marked by blue spheres), and 4Fe–4S clusters (orange). The lipid tails of the prosthetic groups are not shown.

extremely low $\mathcal{E}^{\circ\prime}$ value (about –1.3 V), so electrons can spontaneously flow from P700* to the other redox groups of Photosystem I. These groups include four additional chlorophyll molecules, quinones, and iron–sulfur clusters of the 4Fe–4S type (**Fig. 16.17**). As in Photosystem II, these prosthetic groups are arranged with approximate symmetry. However, in Photosystem I, all the redox groups appear to undergo oxidation and reduction.

Each electron given up by photooxidized P700 eventually reaches ferredoxin, a small peripheral protein on the stromal side of the thylakoid membrane. Ferredoxin undergoes a one-electron reduction at a 2Fe–2S cluster (**Fig. 16.18**). Reduced ferredoxin participates in two different electron transport pathways in the chloroplast, which are known as noncyclic and cyclic electron flow.

In **noncyclic electron flow,** ferredoxin serves as a substrate for ferredoxin–NADP$^+$ reductase. This stromal enzyme uses two electrons (from two separate ferredoxin molecules) to reduce NADP$^+$ to NADPH (**Fig. 16.19**). *The net result of noncyclic electron flow is therefore the transfer of electrons from water, through Photosystem II, cytochrome b_6f, Photosystem I, and then on to NADP$^+$.* Photosystem I does not contribute to the transmembrane proton gradient except by consuming stromal protons in the reduction of NADP$^+$ to NADPH.

When plotted according to reduction potential, the electron-carrying groups of the pathway from water to NADP$^+$ form a diagram called the **Z-scheme** of photosynthesis (**Fig. 16.20**). The zigzag pattern is due to the two photooxidation events, which markedly decrease the reduction potentials of P680 and P700. Note that the four-electron process of producing one O_2 and two NADPH is accompanied by the absorption of eight photons (four each at Photosystem II and Photosystem I).

In **cyclic electron flow,** electrons from Photosystem I do not reduce NADP$^+$ but instead return to the cytochrome b_6f complex. There, the electrons are transferred to plastocyanin and flow back to Photosystem I to reduce photooxidized P700$^+$. Meanwhile, plastoquinol molecules circulate between the two quinone-binding sites of cytochrome b_6f so that protons are translocated from the stroma to the lumen, in the Q cycle (**Fig. 16.21**). Cyclic electron flow requires the input of light energy at Photosystem I but not Photosystem II. During cyclic flow, no free energy is recovered in the form of the reduced cofactor NADPH, but free energy is conserved in the formation of a transmembrane proton gradient by the activity of the cytochrome b_6f complex. Consequently, *cyclic electron flow augments ATP generation by chemiosmosis* (in some bacteria with just a single reaction center, electrons flow through a similar pathway that does not produce O_2 or NADPH). By varying the proportion of electrons that follow the noncyclic and cyclic pathways through Photosystem I, a photosynthetic cell can vary the proportions of ATP and NADPH produced by the light reactions.

FIGURE 16.18

Ferredoxin. The 2Fe–2S cluster is shown in orange. [*Structure of ferredoxin from the cyanobacterium* Anabaena *(pdb 1CZP) determined by R. Morales, M. H. Charon, and M. Frey.*]

FIGURE 16.19

Noncyclic electron flow through Photosystem I. Electrons donated by plastocyanin are transferred to ferredoxin and used to reduce NADP$^+$. The stoichiometry reflects the four electrons released by the oxidation of 2 H_2O in Photosystem II. Therefore, 2 NADPH are produced for every molecule of O_2.

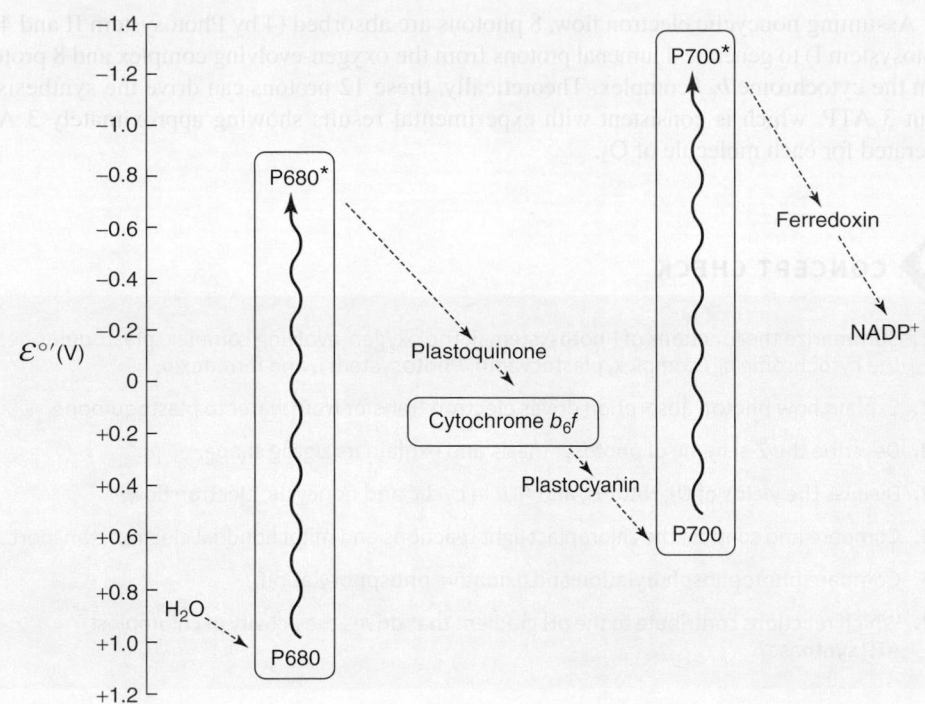

FIGURE 16.20

The Z-scheme of photosynthesis. The major components are positioned according to their reduction potentials (the individual redox groups within Photosystem II, cytochrome $b_6 f$, and Photosystem I are not shown). Excitation of P680 and P700 ensures that electrons follow a thermodynamically favorable pathway to groups with increasing reduction potential.

? Compare the redox changes depicted here with those of the mitochondrial electron transport chain in Figure 15.2.

CHEMIOSMOSIS PROVIDES THE FREE ENERGY FOR ATP SYNTHESIS

Chloroplasts and mitochondria use the same mechanism to synthesize ATP: *They couple the dissipation of a transmembrane proton gradient to the phosphorylation of ADP.* In photosynthetic organisms, this process is called **photophosphorylation.** Chloroplast ATP synthase is highly homologous to mitochondrial and bacterial ATP synthases. The CF_1CF_0 complex ("C" indicates chloroplast) consists of a proton-translocating integral membrane component (CF_0) mechanically linked to a soluble CF_1 component where ATP synthesis occurs by a binding change mechanism (as described in Fig. 15.25). The movement of protons from the thylakoid lumen to the stroma provides the free energy to drive ATP synthesis (**Fig. 16.22**).

As in mitochondria, the proton gradient has both chemical and electrical components. In chloroplasts, the pH gradient (about 3.5 pH units) is much larger than in mitochondria (about 0.75 units). However, in chloroplasts, the electrical component is less than in mitochondria because of the permeability of the thylakoid membrane to ions such as Mg^{2+} and Cl^-. Diffusion of these ions tends to minimize the difference in charge due to protons.

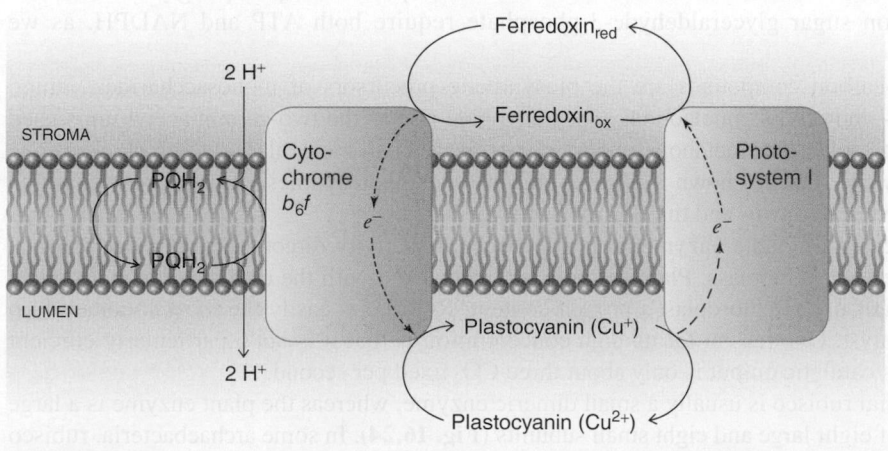

FIGURE 16.21

Cyclic electron flow. Electrons circulate between Photosystem I and the cytochrome $b_6 f$ complex. No NADPH or O_2 is produced, but the activity of cytochrome $b_6 f$ builds up a proton gradient that drives ATP synthesis.

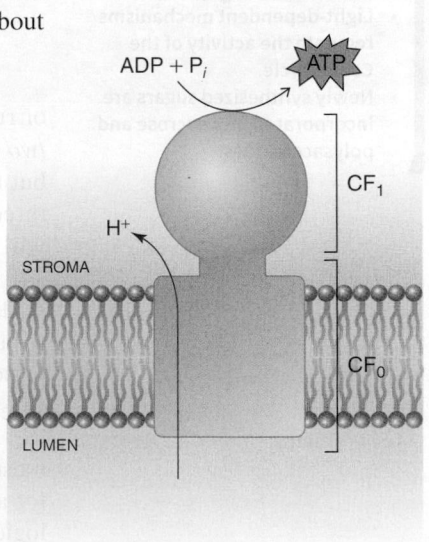

FIGURE 16.22

Photophosphorylation. As protons traverse the CF_0 component of chloroplast ATP synthase (following their concentration gradient from the lumen to the stroma), the CF_1 component carries out ATP synthesis.

Assuming noncyclic electron flow, 8 photons are absorbed (4 by Photosystem II and 4 by Photosystem I) to generate 4 lumenal protons from the oxygen-evolving complex and 8 protons from the cytochrome b_6f complex. Theoretically, these 12 protons can drive the synthesis of about 3 ATP, which is consistent with experimental results showing approximately 3 ATP generated for each molecule of O_2.

 CONCEPT CHECK

1. Summarize the functions of Photosystem II, the oxygen-evolving complex, plastoquinone, the cytochrome $b_6 f$ complex, plastocyanin, Photosystem I, and ferredoxin.

2. Explain how photon absorption drives electron transfer from water to plastoquinone.

3. Describe the Z-scheme of photosynthesis and explain its zigzag shape.

4. Discuss the yields of O_2, NADPH, and ATP in cyclic and noncyclic electron flow.

5. Compare and contrast the chloroplast light reactions and mitochondrial electron transport.

6. Compare photophosphorylation and oxidative phosphorylation.

7. Which reactions contribute to the pH gradient that drives the activity of chloroplast ATP synthase?

16.3 CARBON FIXATION

KEY CONCEPTS
- Rubisco catalyzes carbon fixation by adding CO_2 to a five-carbon acceptor molecule.
- The Calvin cycle shuffles sugars for the net conversion of three CO_2 to one glyceraldehyde-3-phosphate.
- Light-dependent mechanisms regulate the activity of the Calvin cycle.
- Newly synthesized sugars are incorporated into sucrose and polysaccharides.

The production of ATP and NADPH by the photoactive complexes of the thylakoid membrane (or bacterial plasma membrane) is only part of the story of photosynthesis. The rest of this chapter focuses on the use of the products of the light reactions in the so-called **dark reactions.** These reactions, which occur in the chloroplast stroma, incorporate atmospheric carbon dioxide in biologically useful organic molecules, a process called **carbon fixation.**

RUBISCO CATALYZES CO$_2$ FIXATION

Carbon dioxide is fixed by the action of ribulose bisphosphate carboxylase/oxygenase, or rubisco. *This enzyme adds CO$_2$ to a five-carbon sugar and then cleaves the product to two three-carbon units* (**Fig. 16.23**). This reaction itself does not require ATP or NADPH, but the reactions that transform the rubisco reaction product, 3-phosphoglycerate, to the three-carbon sugar glyceraldehyde-3-phosphate require both ATP and NADPH, as we will see.

Three-carbon compounds are the biosynthetic precursors of monosaccharides, amino acids, and—indirectly—nucleotides. They also give rise to the two-carbon acetyl units used to build fatty acids. The metabolic importance of these small molecular building blocks is one reason why the scheme shown in Figure 16.1 presents photosynthesis as a process in which CO_2 is converted to two- and three-carbon intermediates.

Rubisco is a notable enzyme, in part because its activity directly or indirectly sustains most of the earth's biomass. Plant chloroplasts are packed with the enzyme, which accounts for about half of the chloroplast's protein content. Rubisco is easily the most abundant biological catalyst. One reason for its high concentration is that it is not a particularly efficient enzyme. Its catalytic output is only about three CO_2 fixed per second.

Bacterial rubisco is usually a small dimeric enzyme, whereas the plant enzyme is a large multimer of eight large and eight small subunits (**Fig. 16.24**). In some archaebacteria, rubisco has ten identical subunits. Enzymes with multiple catalytic sites typically exhibit cooperative behavior and are regulated allosterically, but this does not seem to be true for plant rubisco, whose eight active sites operate independently. Multimerization may simply be an efficient way to pack more active sites into the limited space of the chloroplast.

FIGURE 16.23

1. The enzyme abstracts a proton from C3 of ribulose-1,5-bisphosphate. An active-site Mg^{2+} ion may help stabilize the developing negative charge.

Ribulose-1,5-bisphosphate

2. The enediolate intermediate nucleophilically attacks CO_2.

Enediolate

CO_2

3. H_2O attacks C3.

4. The 6-carbon product splits to yield two molecules of 3-phosphoglycerate. This step provides much of the free energy for the reaction, since it yields two resonance-stabilized carboxylate products.

2 3-Phosphoglycerate

The rubisco carboxylation reaction.

? Before this reaction was understood, scientists believed that carbon fixation involved the reaction of CO_2 with a two-carbon molecule. Explain.

Despite its metabolic importance, rubisco is not a highly specific enzyme. It also acts as an oxygenase (as reflected in its name) by reacting with O_2, which chemically resembles CO_2. The products of the oxygenase reaction are a three-carbon and a two-carbon compound:

Ribulose bisphosphate $+ O_2 \xrightarrow{\text{rubisco}}$ H_2O 2-Phosphoglycolate $+$ 3-Phosphoglycerate

The 2-phosphoglycolate product of the rubisco oxygenation reaction is subsequently metabolized by a pathway that consumes ATP and NADPH and produces CO_2. This process, called **photorespiration,** uses the products of the light reactions and therefore wastes some of the free energy of captured photons.

Oxygenase activity is a feature of all known rubisco enzymes and must play an essential role that has been conserved throughout plant evolution. *Photorespiration apparently provides a mechanism for plants to dissipate excess free energy under conditions where the CO_2 supply is insufficient for carbon fixation.* Photorespiration may consume significant amounts of ATP and NADPH at high temperatures, which favor oxygenase activity. Some plants have evolved a mechanism, called the **C₄ pathway,** to minimize photorespiration (**Box 16.A**).

FIGURE 16.24

Spinach rubisco. The complex has a mass of approximately 550 kD. The eight catalytic sites are located in the large subunits (dark colors). Only four of eight small subunits (light colors) are visible in this image. [*Structure (pdb 1RCX) determined by T. C. Taylor and I. Anderson.*]

Box 16.A The C$_4$ Pathway

On hot, bright days, the light reactions produce O$_2$, the substrate for photorespiration, and CO$_2$ supplies are low as plants close their stomata (pores in the leaf surface) to avoid evaporative water loss. This combination of events can bring photosynthesis to a halt. Some plants avoid this possibility by stockpiling CO$_2$ in four-carbon molecules so that photosynthesis can proceed even while stomata are closed.

The mechanism for storing carbon begins with the condensation of bicarbonate (HCO$_3^-$) with phosphoenolpyruvate to yield oxaloacetate, which is then reduced to malate. These four-carbon acids give the C$_4$ pathway its name. The subsequent oxidative decarboxylation of malate regenerates CO$_2$ and NADPH to be used in the Calvin cycle. The three-carbon remnant, pyruvate, is recycled back to phosphoenolpyruvate.

Because the C$_4$ pathway and the rubisco reaction compete for CO$_2$, they take place in different types of cells or at different times of day. For example, in some plants, carbon accumulates in mesophyll cells, which are near the leaf surface and lack rubisco. The C$_4$ compounds then enter bundle sheath cells in the leaf interior, which contain abundant rubisco. In other plants, the C$_4$ pathway occurs at night, when the stomata are open and water loss is minimal, and carbon is fixed by rubisco during the day.

The C$_4$ pathway is energetically expensive, so it requires lots of sunlight. Consequently, C$_4$ plants grow more slowly than conventional, or C$_3$, plants when light is limited, but they have the advantage in hot, dry climates. About 5% of the earth's plants, including the economically important maize (corn), sugarcane, and sorghum, use the C$_4$ pathway.

The recognition of global warming has led to predictions that C$_4$ "weeds" may overtake economically important C$_3$ plants as temperatures increase. In fact, the increase in atmospheric CO$_2$ that is driving the warming trend appears to promote the growth of C$_3$ plants, which can obtain CO$_2$ more easily without losing too much water via their stomata. However, if water is limited, C$_4$ plants may still have a competitive edge, as they are adapted not just for hot environments but for arid ones.

? **Operation of the C$_4$ pathway consumes ATP. Using the pathway shown here, indicate where this occurs.**

THE CALVIN CYCLE REARRANGES SUGAR MOLECULES

If rubisco is responsible for fixing CO$_2$, what is the origin of its other substrate, ribulose bisphosphate? The answer—elucidated over many years by Melvin Calvin, James Bassham, and Andrew Benson—is a metabolic pathway known as the **Calvin cycle.** Early experiments to study the fate of ^{14}C-labeled CO$_2$ in algae showed that within a few minutes, the cells had synthesized a complex mixture of sugars, all containing the radioactive label. Rearrangements among these sugar molecules generate the five-carbon substrate for rubisco.

The Calvin cycle actually begins with a sugar monophosphate, ribulose-5-phosphate, which is phosphorylated in an ATP-dependent reaction (**Fig. 16.25**). The resulting bisphosphate is a substrate for rubisco, as we have already seen. Each 3-phosphoglycerate product of the rubisco reaction is then phosphorylated, again at the expense of ATP. This phosphorylation reaction (step 3 of Fig. 16.25) is identical to the phosphoglycerate kinase reaction of glycolysis (see Section 13.1). Next, bisphosphoglycerate is reduced by the chloroplast enzyme glyceraldehyde-3-phosphate dehydrogenase, which resembles the glycolytic enzyme but uses NADPH rather than NADH. The NADPH is the product of the light reactions of photosynthesis.

The initial reactions of the Calvin cycle are shown in the following diagram:

Ribulose-5-phosphate → (ATP ADP, phosphoribulokinase, **1**) → Ribulose-1,5-bisphosphate → (CO_2, rubisco, **2**) → 2 3-Phosphoglycerate → (ATP/ADP, phosphoglycerate kinase, **3**) → 2 1,3-Bisphosphoglycerate → (glyceraldehyde-3-phosphate dehydrogenase, P_i + $NADP^+$ → NADPH, **4**) → 2 Glyceraldehyde-3-phosphate

FIGURE 16.25

Initial reactions of the Calvin cycle. Note that ATP and NADPH, products of the light-dependent reactions, are consumed in the process of converting CO_2 to glyceraldehyde-3-phosphate.

Some glyceraldehyde-3-phosphate is siphoned from the Calvin cycle for metabolic fates such as glucose or amino acid synthesis. Recall from Section 13.2 that the pathway from glyceraldehyde-3-phosphate to glucose consists of reactions that require no further input of ATP. Glyceraldehyde-3-phosphate can also be converted to pyruvate and then to oxaloacetate, both of which can undergo transamination to generate amino acids. Additional reactions lead to other metabolites.

The glyceraldehyde-3-phosphate that is not used for biosynthetic pathways remains part of the Calvin cycle and enters a series of isomerization and group-transfer reactions that regenerate ribulose-5-phosphate. These interconversion reactions are similar to those of the pentose phosphate pathway (Section 13.4). We can represent them simply by showing how carbon atoms are shuffled among three- to seven-carbon sugars:

$$C_3 + C_3 \rightarrow C_6$$
$$C_3 + C_6 \rightarrow C_4 + C_5$$
$$C_3 + C_4 \rightarrow C_7$$
$$C_3 + C_7 \rightarrow C_5 + C_5$$
$$net\ reaction: 5\ C_3 \rightarrow 3\ C_5$$

Consequently, if the Calvin cycle starts with three five-carbon ribulose molecules, so that three CO_2 molecules are fixed, the products are six three-carbon glyceraldehyde-3-phosphate molecules, five of which are recycled to form three ribulose molecules, leaving the sixth (representing the three fixed CO_2) as its net product.

The net equation for the Calvin cycle, including the ATP and NADPH cofactors, is

$$3 \text{ CO}_2 + 9 \text{ ATP} + 6 \text{ NADPH} \rightarrow \text{glyceraldehyde-3-phosphate} + 9 \text{ ADP} + 8 \text{ P}_i + 6 \text{ NADP}^+$$

Fixing a single CO$_2$ therefore requires 3 ATP and 2 NADPH—approximately the same quantity of ATP and NADPH produced by the absorption of eight photons. The relationship between the number of photons absorbed and the amount of carbon fixed or oxygen released is known as the **quantum yield** of photosynthesis. Keep in mind that the exact number of carbons fixed per photon absorbed depends on factors such as the number of protons translocated per ATP synthesized by the chloroplast ATP synthase and the ratio of cyclic to noncyclic electron flow in Photosystem I.

THE AVAILABILITY OF LIGHT REGULATES CARBON FIXATION

Plants must coordinate the light reactions with carbon fixation. During the day, both processes occur. At night, when the photosystems are inactive, the plant turns off the "dark" reactions to conserve ATP and NADPH while it turns on pathways to regenerate these cofactors by metabolic pathways such as glycolysis and the pentose phosphate pathway. It would be wasteful for these catabolic processes to proceed simultaneously with the Calvin cycle. Thus, the "dark" reactions do not actually occur in the dark!

All the mechanisms for regulating the Calvin cycle are directly or indirectly linked to the availability of light energy. Some of the regulatory mechanisms are highlighted here. For example, a catalytically essential Mg^{2+} ion in the rubisco active site is coordinated in part by a carboxylated lysine side chain that is produced by the reaction of CO$_2$ with the ε-amino group:

$$-(\text{CH}_2)_4-\text{NH}_2 + \text{CO}_2 \rightleftharpoons -(\text{CH}_2)_4-\text{NH}-\text{COO}^- + \text{H}^+$$
$$\text{Lys}$$

By forming the Mg^{2+}-binding site, this "activating" CO$_2$ molecule promotes the ability of rubisco to fix additional substrate CO$_2$ molecules. The carboxylation reaction is favored at high pH, a signal that the light reactions are working (depleting the stroma of protons) and that ATP and NADPH are available for the Calvin cycle.

Magnesium ions also directly activate rubisco and several of the Calvin cycle enzymes. During the light reactions, the rise in stromal pH triggers the flux of Mg^{2+} ions from the lumen to the stroma (this ion movement helps balance the charge of the protons that are translocated in the opposite direction). Some of the Calvin cycle enzymes are also activated when the ratio of reduced ferredoxin to oxidized ferredoxin is high, another signal that the photosystems are active.

CALVIN CYCLE PRODUCTS ARE USED TO SYNTHESIZE SUCROSE AND STARCH

Many of the three-carbon sugars produced by the Calvin cycle are converted to sucrose or starch. The polysaccharide starch is synthesized in the chloroplast stroma as a temporary storage depot for glucose. It is also synthesized as a long-term storage molecule elsewhere in the plant, including leaves, seeds, and roots. In the first stage of starch synthesis, two molecules of glyceraldehyde-3-phosphate are converted to glucose-6-phosphate by reactions analogous to those of mammalian gluconeogenesis (see Fig. 13.10). Phosphoglucomutase then carries out an isomerization reaction to produce glucose-1-phosphate. Next, this sugar is "activated" by its reaction with ATP to form ADP–glucose:

(Recall from Section 13.3 that glycogen synthesis uses the chemically related nucleotide sugar UDP–glucose.) Starch synthase then transfers the glucose residue to the end of a starch polymer, forming a new glycosidic linkage.

The overall reaction is driven by the exergonic hydrolysis of the PP$_i$ released in the formation of ADP–glucose. Thus, one phosphoanhydride bond is consumed in lengthening a starch molecule by one glucose residue.

Sucrose, a disaccharide of glucose and fructose, is synthesized in the cytosol. Glyceraldehyde-3-phosphate or its isomer dihydroxyacetone phosphate is transported out of the chloroplast by an antiport protein that exchanges phosphate for a phosphorylated three-carbon sugar. Two of these sugars combine to form fructose 6-phosphate, and two others combine to form glucose-1-phosphate, which is subsequently activated by UTP. Next, fructose-6-phosphate reacts with UDP–glucose to produce sucrose-6-phosphate. Finally, a phosphatase converts the phosphorylated sugar to sucrose:

Sucrose can then be exported to other plant tissues. This disaccharide probably became the preferred transport form of carbon in plants because its glycosidic linkage is insensitive to amylases (starch-digesting enzymes) and other common hydrolases. Also, its two anomeric carbons are tied up in the glycosidic bond and therefore cannot react nonenzymatically with other substances.

Cellulose, the other major polysaccharide of plants, is also synthesized from UDP–glucose (cellulose is described in Section 11.2). Plant cell walls consist of almost-crystalline cables, each containing approximately 36 cellulose polymers, and all embedded in an amorphous matrix of other polysaccharides (see Box 11.A; synthetic materials such as fiberglass are built on the same principle). Unlike starch in plants or glycogen in mammals, cellulose is synthesized by enzyme complexes in the plant plasma membrane and is extruded into the extracellular space.

 CONCEPT CHECK

1. List the reactants and products of the two reactions catalyzed by rubisco.

2. Compare the physiological implications of carbon fixation and photorespiration.

3. Explain how the carbon from a molecule of fixed CO_2 become incorporated into other compounds such as monosaccharides.

4. Identify the source of the ribulose-1,5-bisphosphate used for carbon fixation by rubisco.

5. Describe some of the mechanisms for regulating the activity of the "dark" reactions.

6. What is the role of nucleotides in the synthesis of starch and sucrose?

SUMMARY

16.1   Chloroplasts and Solar Energy

• Plant chloroplasts contain pigments that absorb photons and release the energy, primarily by transferring it to another molecule (exciton transfer) or giving up an electron (photooxidation). Light-harvesting complexes act to capture and funnel light energy to the photosynthetic reaction centers.

16.2   The Light Reactions

• In the so-called light reactions of photosynthesis, electrons from the photooxidized P680 reaction center of Photosystem II pass through several prosthetic groups and then to plastoquinone. The P680 electrons are replaced when the oxygen-evolving complex of Photosystem II converts water to O_2, a four-electron oxidation reaction.

• Electrons flow next to a cytochrome $b_6 f$ complex that carries out a proton-translocating Q cycle, and then to the protein plastocyanin.

• A second photooxidation at P700 of Photosystem I allows electrons to flow to the protein ferredoxin and finally to $NADP^+$ to produce NADPH.

• The free energy of light-driven electron flow, particularly cyclic flow, is also conserved in the formation of a transmembrane proton gradient that drives ATP synthesis in the process called photophosphorylation.

16.3   Carbon Fixation

• The enzyme rubisco "fixes" CO_2 by catalyzing the carboxylation of a five-carbon sugar. Rubisco also acts as an oxygenase in the process of photorespiration.

• The reactions of the Calvin cycle use the products of the light reactions (ATP and NADPH) to convert the product of the rubisco reaction to glyceraldehyde-3-phosphate and to regenerate the five-carbon carboxylate receptor. These "dark" reactions are regulated according to the availability of light energy.

• Chloroplasts convert the glyceraldehyde-3-phosphate product of photosynthesis into glucose residues for incorporation into starch, sucrose, and cellulose.

KEY TERMS

photosynthesis	photoreceptor	light-harvesting complex	dark reactions
chloroplast	fluorescence	light reactions	carbon fixation
stroma	exciton transfer	noncyclic electron flow	photorespiration
thylakoid	photooxidation	Z-scheme	C_4 pathway
photon	reaction center	cyclic electron flow	Calvin cycle
Planck's law	antenna pigment	photophosphorylation	quantum yield

BIOINFORMATICS

Brief Bioinformatics Exercises

16.1 Viewing and Analyzing Photosystems I and II

16.2 Photosynthesis and the KEGG Database

PROBLEMS

16.1 Chloroplasts and Solar Energy

1. Indicate with a C or an M whether the following occur in chloroplasts, mitochondria, or both: **(a)** proton translocation; **(b)** photophosphorylation; **(c)** photooxidation; **(d)** quinones; **(e)** oxygen reduction; **(f)** water oxidation; **(g)** electron transport; **(h)** oxidative phosphorylation; **(i)** carbon fixation; **(j)** NADH oxidation; **(k)** Mn cofactor; **(l)** heme groups; **(m)** binding change mechanism; **(n)** iron–sulfur clusters; **(o)** NADP$^+$ reduction.

2. Compare and contrast the structures of chloroplasts and mitochondria.

3. The thylakoid membrane contains some unusual lipids. One of these is galactosyl diacylglycerol; a β-galactose residue is attached to the first glycerol carbon. Draw the structure of galactosyl diacylglycerol.

4. Thylakoid membranes contain lipids with a high degree of unsaturation. What does this tell you about the character of the thylakoid membrane?

5. Calculate the energy per photon and per mole of photons with a wavelength of **(a)** 680 nm **(b)** 450 nm and **(c)** 600 nm. What is the relationship between wavelength and energy?

6. At what wavelength would a mole of photons have an energy of **(a)** 350 kJ and **(b)** 500 kJ?

7. Assuming 100% efficiency, calculate the number of moles of ATP that could be generated per mole of photons with the energy calculated in Problem 5.

8. Red tides are caused by algal blooms that cause seawater to become visibly red. In the photosynthetic process, red algae take advantage of wavelengths not absorbed by other organisms. Describe the photosynthetic pigments of the red algae.

9. Some photosynthetic bacteria live in murky ponds where visible light does not penetrate easily. What wavelengths might the photosynthetic pigments in these organisms absorb?

10. Compare the structures of chlorophyll *a* (Fig. 16.3) and the reduced form of heme *b* (Fig. 15.12).

11. You are investigating the functional similarities of chloroplast cytochrome *f* and mitochondrial cytochrome c_1.

(a) Which would provide more useful information: the amino acid sequences of the proteins or models of their three-dimensional shapes? Explain.

(b) Would it be better to examine high-resolution models of the two apoproteins (the polypeptides without their heme groups) or low-resolution models of the holoproteins (polypeptides plus heme groups)?

12. Under conditions of very high light intensity, excess absorbed solar energy is dissipated by the action of "photoprotective" proteins in the thylakoid membrane. Explain why it is advantageous for these proteins to be activated by a buildup of a proton gradient across the membrane.

13. Of the four mechanisms for dissipating light energy shown in Figure 16.5, which would be best for "protecting" the photosystems from excess light energy?

16.2 The Light Reactions

14. Use Equation 15.4 to calculate the free energy change for transforming one mole of P680 to P680*.

15. Determine the wavelength of the photons whose absorption would supply the free energy to transform one mole of P680 to P680* (see Problem 14).

16. The three electron-transporting complexes of the thylakoid membrane can be called plastocyanin–ferredoxin oxidoreductase, plastoquinone–plastocyanin oxidoreductase, and water–plastoquinone oxidoreductase. What are the common names of these enzymes and in what order do they act?

17. Photosystem II is located mostly in the tightly stacked regions of the thylakoid membrane, whereas Photosystem I is located mostly in the unstacked regions (see Fig. 16.2). Why might it be important for the two photosystems to be separated?

18. The herbicide 3-(3,4-dichlorophenyl)-1,1-dimethylurea (DCMU) blocks electron flow from Photosystem II to Photosystem I. What is the effect on oxygen production and photophosphorylation when DCMU is added to plants?

19. When the antifungal agent myxothiazol is added to a suspension of chloroplasts, the QH$_2$/Q ratio increases. Where does myxothiazol inhibit electron transfer?

20. Plastoquinone is not firmly anchored to any thylakoid membrane component but is free to diffuse laterally throughout the membrane among the photosynthetic components. What aspects of its structure account for this behavior?

21. The Photosystem II reaction center contains a protein called D1 that contains the PQ$_B$ binding site. The D1 protein in the single-celled alga *Chlamydomonas reinhardtii* is predicted to have five hydrophobic membrane-spanning helical regions. A loop between the fourth and fifth segments is located in the stroma. This loop is believed to lie along the membrane surface, and it contains several highly conserved amino acid residues. D1 proteins with mutations at the Ala 251 position were evaluated for photosynthetic activity and herbicide susceptibility. The results are shown in the table. What are the essential properties of the amino acid at position 251 in the D1 protein?

Amino acid at position 251	Characteristics
Ala	Wild-type
Cys	Similar to wild-type
Gly, Pro, Ser	Impaired in photosynthesis
Ile, Lau, Val	Impaired in photoautotrophic growth
	Impaired in photosynthesis
	Resistant to herbicides
Arg, Glan, Glu, His, and Asp	Not photosynthetically competent

22. Calculate the free energy of translocating a proton out of the stroma when the lumenal pH is 3.5 units lower than the stromal pH and $\Delta\psi$ is –50 mV.

23. Compare the free energy of proton translocation calculated in Problem 22 to the free energy of translocating a proton out of a mitochondrion where the pH difference is 0.75 units and $\Delta\psi$ is 200 mV. Compare both types of translocation. Are the processes exergonic or endergonic? Which contributes a larger component of the free energy for each process, the pH difference or the membrane potential?

24. Calculate the standard free energy change for the oxidation of one molecule of water by NADP$^+$.

25. Calculate the energy available in two photons of wavelength 700 nm. Compare this value with the standard free energy changes you calculated in Problem 24. Do two photons supply enough energy to drive the oxidation of one molecule of water by NADP$^+$?

26. Photophosphorylation in chloroplasts is similar to oxidative phosphorylation in mitochondria. What is the final electron acceptor in photosynthesis? What is the final electron acceptor in mitochondrial electron transport?

27. If radioactively labeled water ($H_2^{18}O$) is provided to a plant, where does the label appear?

28. Predict the effect of an uncoupler such as dinitrophenol (see Box 15.B) on production of **(a)** ATP and **(b)** NADPH by a chloroplast.

29. Antimycin A (an antibiotic) blocks electron transport in Complex III of the electron transport chain in mitochondria. How would the addition of antimycin A to chloroplasts affect chloroplast ATP synthesis and NADPH production?

30. Does the quantum yield of photosynthesis increase or decrease for systems where **(a)** the CF_0 component of ATP synthase contains more c subunits or **(b)** when the proportion of cyclic electron flow through Photosystem I increases?

31. Oligomycin inhibits the proton channel (F_0) of the ATP synthase enzyme in mitochondria but does not inhibit CF_0. When oligomycin is added to plant cells undergoing photosynthesis, the cytosolic ATP/ADP ratio decreases whereas the chloroplastic ATP/ADP ratio is unchanged or even increases. Explain these results.

16.3 Carbon Fixation

32. Defend or refute this statement: The "dark" reactions are so named because these reactions occur only at night.

33. Examine the net equation for the light and "dark" reactions of photosynthesis—that is, the incorporation of one molecule of CO_2 into carbohydrate, which has the chemical formula $(CH_2O)_n$.

$$CO_2 + 2\ H_2O \rightarrow (CH_2O) + O_2 + H_2O$$

How would this equation differ for a bacterial photosynthetic system in which H_2S rather than H_2O serves as a source of electrons?

34. Melvin Calvin and his colleagues noted that when $^{14}CO_2$ was added to algal cells, a single compound was radiolabeled within 5 seconds of exposure. What is the compound, and where does the radioactive label appear?

35. As described in the text, rubisco is not a particularly specific enzyme. Scientists have wondered why millions of years of evolution failed to produce a more specific enzyme. What advantage would be conferred upon a plant that evolved a rubisco enzyme that was able to react with CO_2 but not oxygen?

36. A tiny acorn grows into a massive oak tree. Using what you know about photosynthesis, what accounts for the increase in mass?

37. The $\Delta G^{\circ\prime}$ value for the rubisco reaction is -35.1 kJ·mol^{-1} and the ΔG is -41.0 kJ·mol^{-1}. What is the ratio of products to reactants under normal cellular conditions? Assume a temperature of 25°C.

38. Efforts to engineer a more efficient rubisco, one that could fix CO_2 more quickly than three per second, could improve farming by allowing plants to grow larger and/or faster. Explain why the engineered rubisco might also decrease the need for nitrogen-containing fertilizers.

39. Some plants synthesize the sugar 2-carboxyarabinitol-1-phosphate. This compound inhibits the activity of rubisco.

(a) What is the probable mechanism of action of the inhibitor?

(b) Why do plants synthesize the inhibitor at night and break it down during the day?

$$
\begin{array}{c}
CH_2OPO_3^{2-} \\
| \\
HO-C-COO^- \\
| \\
H-C-OH \\
| \\
H-C-OH \\
| \\
CH_2OH
\end{array}
$$

2-Carboxyarabinitol-
1-phosphate

40. An "activating" CO_2 reacts with a lysine side chain on rubisco to carboxylate it.

(a) Explain why the carboxylation reaction is favored at high pH.

(b) How does this activation step help link carbon fixation to the light reactions?

41. Chloroplast phosphofructokinase (PFK) is inhibited by ATP and NADPH. How does this observation link the light reactions to the regulation of glycolysis?

42. Crabgrass, a C_4 plant, remains green during a long spell of hot dry weather when C_3 grasses turn brown. Explain this observation.

43. Chloroplasts contain thioredoxin, a small protein with two cysteine residues that can form an intramolecular disulfide bond. The sulfhydryl/disulfide interconversion in thioredoxin is catalyzed by an enzyme known as ferredoxin–thioredoxin reductase. This enzyme, along with some of the Calvin cycle enzymes, also includes two Cys residues that undergo sulfhydryl/disulfide transitions. Show how disulfide interchange reactions involving thioredoxin could coordinate the activity of Photosystem I with the activity of the Calvin cycle.

44. The inner chloroplast membrane is impermeable to large polar and ionic compounds such as NADH and ATP. However, the membrane has an antiport protein that facilitates the passage of dihydroxyacetone phosphate or 3-phosphoglycerate in exchange for P_i. This system permits the entry of P_i for photophosphorylation and the exit of the products of carbon fixation. Show how the same antiport could "transport" ATP and reduced cofactors from the chloroplast to the cytosol.

45. The sedoheptulose bisphosphatase (SBPase) enzyme in the Calvin cycle catalyzes the removal of a phosphate group from C1 of sedoheptulose-1,7-bisphosphate (SBP) to produce sedoheptulose-7-phosphate (S7P). The $\Delta G^{\circ\prime}$ for this reaction is -14.2 kJ·mol^{-1} and the ΔG is -29.7 kJ·mol^{-1}. What is the ratio of products to reactants under normal cellular conditions? Is this enzyme likely to be regulated in the Calvin cycle? Assume a temperature of 25°C.

46. Phosphoenolpyruvate carboxylase (PEPC) catalyzes the carboxylation of phosphoenolpyruvate (PEP) to oxaloacetate (OAA). The enzyme is commonly found in plants but is absent in animals. The reaction is shown.

(a) Why is PEPC referred to as an anaplerotic enzyme?

(b) Acetyl-CoA is an allosteric regulator of PEPC. Does acetyl-CoA activate or inhibit PEPC? Explain.

$$
\begin{array}{c}
COO^- \\
| \\
C-OPO_3^{2-} \\
\| \\
CH_2 \\
PEP
\end{array}
+ HCO_3^- \xrightarrow{\text{PEPC}}
\begin{array}{c}
COO^- \\
| \\
C=O \\
| \\
CH_2 \\
| \\
COO^- \\
OAA
\end{array}
+ HPO_4^{2-}
$$

47. In germinating oil seeds, triacylglycerols are rapidly converted to sucrose and protein. What is the role of PEPC (see Problem 46) in this process?

48. The use of transgenic plants is increasing. These plants are constructed by inserting a new gene into the plant genome to give the plant a desirable characteristic, such as resistance to pesticides or frost, or to confer a nutritional benefit. Transformed plants contain the gene of interest, a promoter (to induce expression of the gene), and a terminator. The cauliflower mosaic virus (CaMV) is often used as a promoter in transgenic plants. A partial DNA sequence is shown below. Design a set of 18-bp primers (see Figure 3.18) that you could use to detect the presence of this promoter in a transgenic plant.

GTAGTGGGATTGTGCGTCATCCCTTACGTCAGT···
(112 bases)··· TCAACGATGGCCTTTCCTTTATCGCAATGAT-
GGCATTTGTAGGAGC

49. Novartis has constructed a corn cultivar that contains a gene from the bacterium *Bacillus thuringiensis* (Bt) that codes for an endotoxin protein. When insects of the order Lepidoptera (which includes the European corn borer but also unfortunately includes the Monarch butterfly) eat the corn, the ingested endotoxin enters the high pH environment of the insect's midgut. Under these conditions, the endotoxin forms a pore in the membrane of the cells lining the midgut, causing ions to flow into the cell and ultimately resulting in the death of the organism. Is the Bt endotoxin toxic to humans? Explain why or why not.

50. The Monsanto Company has constructed a transgenic "Roundup Ready" soybean cultivar in which the bacterial gene for the enzyme EPSPS is inserted into the plant genome. EPSPS catalyzes an important step in the synthesis of aromatic amino acids, as shown in the diagram. The herbicide Roundup® contains glyphosate, a compound that competitively inhibits plant EPSPS but not the bacterial form of the enzyme. Explain the strategy for using glyphosate-containing herbicide on a soybean crop to kill weeds.

51. Is Roundup® (see Problem 50) toxic to humans? Explain why or why not.

SELECTED READINGS

Allen, J. F. and Martin, W., Out of thin air, *Nature* **445,** 610–612 (2007). [A short discussion of the cyanobacterial shift to oxygenic photosynthesis billions of years ago.]

Andersson, I., Catalysis and regulation in rubisco, *J. Exp. Botany* **59,** 1555–1568 (2008). [Discusses the structure, chemistry, and regulation of rubisco.]

Busch A. and Hippler M., The structure and function of eukaryotic photosystem I, *Biochim. Biophys. Acta.* **1807,** 864–77 (2011).

[Discusses electron transfer between the light-harvesting groups and the core of Photosystem I.]

Evans, J. R., Improving photosynthesis, *Plant Physiol.* **162,** 1780–93 (2013). [Focuses on carbon fixation and how the process could be engineered for greater efficiency.]

Suga, M., et al., Native structure of photosystem II at 1.95 Å resolution viewed by femtosecond X-ray pulses, *Nature* **517,** 99–103 (2015). [Provide details on the fine structure of Photosystem II.]

LIPID METABOLISM

© Nature Photographers Ltd/Alamy Stock Photo

Although many bird species migrate long distances, some even from the Arctic to the Antarctic, the record for the longest non-stop flight belongs to the bar-tailed godwit, which migrates from Alaska to New Zealand (and back again) without refueling along the way. In addition to having a size and shape adapted for efficient flight, the godwit powers its long migration by catabolizing significant amounts of stored fat.

DO YOU REMEMBER?

- Lipids are predominantly hydrophobic molecules that can be esterified but cannot form polymers (Section 8.1).
- Metabolic fuels can be mobilized by breaking down glycogen, triacylglycerols, and proteins (Section 12.1).
- A few metabolites appear in several metabolic pathways (Section 12.2).
- Cells also use the free energy of other phosphorylated compounds, thioesters, reduced cofactors, and electrochemical gradients (Section 12.3).

Like carbohydrates, lipids are metabolic fuels and therefore can be examined in terms of their synthesis, storage, mobilization, and catabolism—pathways that intersect with the processes we have already studied. In this chapter, we will investigate the breakdown and synthesis of fatty acids and related molecules. But unlike other classes of biological molecules, lipids are insoluble in water, so we will begin by looking at how they travel between organs.

17.1 | LIPID TRANSPORT

Approximately half of all deaths in the United States are linked to the vascular disease **atherosclerosis** (a term derived from the Greek *athero,* "paste," and *sclerosis,* "hardness"). Atherosclerosis is a slow progressive disease that begins with the accumulation of lipids in the walls of large blood vessels. The trapped lipids initiate inflammation by triggering the production of chemical signals that attract white blood cells, particularly macrophages. These cells engorge themselves by taking up the accumulated lipids and continue to recruit more macrophages, thereby perpetuating the inflammation.

The damaged vessel wall forms a plaque with a core of cholesterol, cholesteryl esters, and remnants of dead macrophages, surrounded by proliferating smooth muscle cells that may undergo calcification, as occurs in bone formation. This accounts for the "hardening" of the arteries. Although a very large plaque can occlude the lumen of the artery (**Fig. 17.1**), blood flow is usually not completely blocked unless the plaque ruptures, triggering formation of a blood clot that can prevent circulation to the heart (a heart attack) or brain (a stroke).

What is the source of the lipids that accumulate in vessel walls? They are deposited by lipoproteins known as LDL (for low-density lipoproteins). **Lipoproteins** (particles consisting of lipids and specialized proteins) are the primary form of circulating lipid in animals (**Fig. 17.2**). Recall from Section 12.1 that dietary lipids travel from the intestine to other tissues as chylomicrons. These lipoproteins are relatively large (1000 to 5000 Å in diameter) with a protein content of only 1% to 2%. Their primary function is to transport dietary triacylglycerols to adipose tissue and cholesterol to the liver. The liver repackages the cholesterol and other lipids—including triacylglycerols, phospholipids, and cholesteryl esters—into other lipoproteins known as VLDL (very-low-density lipoproteins). VLDL have a triacylglycerol content of about 50% and a diameter of about 500 Å. As they circulate in the bloodstream, VLDL give up triacylglycerols to the tissues, becoming smaller, denser, and richer in cholesterol and cholesteryl esters. After passing through an intermediate state (IDL, or intermediate-density lipoproteins), they become LDL, about 200 Å in diameter and about 45% cholesteryl ester (**Table 17.1**).

High concentrations of circulating LDL, measured as serum cholesterol (popularly called "bad cholesterol"), are a major factor in atherosclerosis. Some high-fat diets (especially those rich in saturated fats) may contribute to atherosclerosis by boosting LDL levels, but genetic factors, smoking, and infection also increase the risk of atherosclerosis. The disease is less likely to occur in individuals who consume low-cholesterol diets and who have high levels of HDL (high-density lipoproteins, sometimes called "good" cholesterol). HDL particles are even smaller and denser than LDL (see Table 17.1), and their primary function is to transport the body's excess cholesterol back to the liver. HDL therefore counter the atherogenic tendencies of LDL. The roles of the various lipoproteins are summarized in **Figure 17.3**.

All cells in the body can synthesize cholesterol (Section 17.4), which is an essential membrane component, but LDL are also a major source of this lipid. When LDL proteins dock with the LDL receptor on the cell surface, the lipoprotein–receptor complex undergoes endocytosis (Section 9.4). Inside the cell, the lipoprotein is degraded and cholesterol enters the cytosol. The LDL receptor itself is not degraded but cycles back to the cell surface, ready for another round of receptor-mediated endocytosis of an LDL particle.

FIGURE 17.1

An atherosclerotic plaque in an artery. Note the thickening of the vessel wall. [*James Cavallini/BSIP/Phototake.*]

FIGURE 17.2

Structure of a lipoprotein. This image is based on small angle neutron scattering of HDL particles. Three copies of apolipoprotein A1 (orange) wrap around a core containing phospholipids, cholesterol, and cholesteryl esters. The particle has dimensions of about 110 Å × 96 Å. The proteins help target the particle to cell surfaces and modulate the activities of enzymes that act on the component lipids. The various types of lipoproteins differ in size, lipid composition, protein composition, and density (a function of the relative proportions of lipid and protein). [*From Wu et al., J. Biol. Chem. **286,** 12495–12508 (2011). Reproduced with permission.*]

LDL   cholesterol

LDL receptor

CELL

> ❓ **Explain why the protein is amphipathic.**

TABLE 17.1 Characteristics of Lipoproteins

LIPOPROTEIN	DIAMETER (Å)	DENSITY (g · cm⁻³)	% PROTEIN	% TRIACYLGLYCEROL	% CHOLESTEROL AND CHOLESTERYL ESTER
Chylomicrons	1000–5000	<0.95	1–2	85–90	4–8
VLDL	300–800	0.95–1.006	5–10	50–65	15–25
IDL	250–350	1.006–1.019	10–20	20–30	40–45
LDL	180–250	1.019–1.063	20–25	7–15	45–50
HDL	50–120	1.063–1.210	40–55	3–10	15–20

FIGURE 17.3

Lipoprotein function. Large chylomicrons, which are mostly lipid, transport dietary lipids to the liver and other tissues. The liver produces triacylglycerol-rich very-low-density lipoproteins (VLDL). As they circulate in the tissues, VLDL give up their triacylglycerols, becoming cholesterol-rich low-density lipoproteins (LDL), which are taken up by tissues. High-density lipoproteins (HDL), the smallest and densest of the lipoproteins, transport cholesterol from the tissues back to the liver.

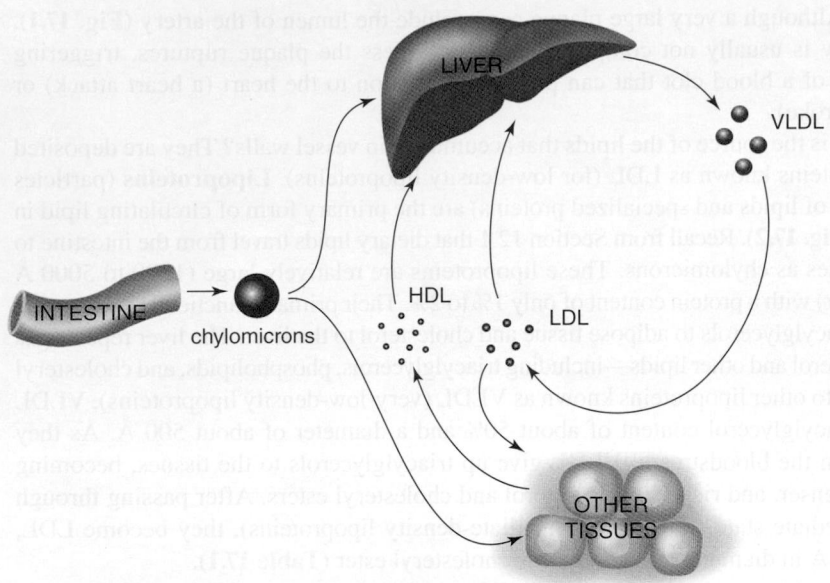

The role of LDL in delivering cholesterol to cells is dramatically illustrated by the disease familial hypercholesterolemia, which is due to a genetic defect in the LDL receptor. The cells of homozygotes are unable to take up LDL, so the concentration of serum cholesterol is about three times higher than normal. This contributes to atherosclerosis, and many individuals die of the disease before age 30. Some other cases of **hypercholesterolemia** (high concentrations of cholesterol in the blood) can be treated with drugs that block the activity of a protein called PCSK9. This protein normally binds to the LDL receptor, making it more likely to be degraded than recycled to the cell surface. Inhibiting PCSK9 leads to more LDL receptors on the cell surface and more uptake of circulating cholesterol-rich LDL.

High-density lipoproteins (HDL) are essential for removing excess cholesterol from cells. The efflux of cholesterol requires the close juxtaposition of the cell membrane and an HDL particle as well as specific cell-surface proteins. One of these is an ABC transporter (Section 9.3) that acts as a flippase to move cholesterol from the cytosolic leaflet to the extracellular leaflet, from which it can diffuse into the HDL particle.

Defects in the gene for the transporter cause Tangier disease, which is characterized by accumulations of cholesterol in tissues and a high risk of heart attack.

 CONCEPT CHECK

1. Draw a diagram showing how cholesterol and other lipids are transported in the body.
2. Describe the relationship between a lipoprotein's protein and lipid content and its density.
3. Explain why both LDL and HDL are needed for cholesterol homeostasis.

17.2 FATTY ACID OXIDATION

The opposing actions of LDL and HDL are just one part of the body's efforts to regulate lipid metabolism, which includes multiple pathways. For example, lipids are synthesized from smaller precursors as well as obtained from food. Cells use lipids as a source of free energy, as building materials, and as signaling molecules, Lipids may be catabolized quickly or stored long-term in adipose tissue. For the most part, the pathways responsible for synthesizing and degrading lipids fall within the shaded portions of the metabolic map shown in **Figure 17.4**.

The degradation (oxidation) of fatty acids is a source of metabolic free energy. In this section we describe how cells obtain, activate, and oxidize fatty acids. In humans, *dietary triacylglycerols are the primary source of fatty acids used as metabolic fuel*. The triacylglycerols are carried by lipoproteins to tissues, where hydrolysis releases their fatty acids from the glycerol backbone.

KEY CONCEPTS

- Fatty acids to be degraded are linked to coenzyme A and then transported into mitochondria.
- The four reactions of each cycle of β oxidation produce acetyl-CoA, QH_2, and NADH.
- Additional enzymes are required to break down unsaturated fatty acids.
- Fatty acids with an odd number of carbons yield propionyl-CoA that is ultimately converted to acetyl-CoA.
- Peroxisomes oxidize long-chain and branched fatty acids, producing H_2O_2.

CH$_2$—CH—CH$_2$ ... (Triacylglycerol) $\xrightarrow[\text{lipase}]{\text{3 H}_2\text{O}}$ CH$_2$—CH—CH$_2$ (Glycerol) + 3 R—C—O$^-$ (Fatty acyl groups)

Hydrolysis occurs extracellularly, catalyzed by lipoprotein lipase, an enzyme associated with the outer surface of cells.

Triacylglycerols that are stored in adipose tissue are mobilized (their fatty acids are released to be used as fuel) by an intracellular hormone-sensitive lipase. The mobilized fatty acids travel through the bloodstream, not as part of lipoproteins, but bound to albumin, a 66-kD protein that accounts for about half of the serum protein (it also binds metal ions and hormones, serving as an all-purpose transport protein).

The concentration of free fatty acids in the body is very low because these molecules are detergents (which form micelles; see Section 2.2) and can disrupt cell membranes. After they enter cells, probably with the assistance of proteins, the fatty acids are either broken down for energy or re-esterified to form triacylglycerols or other complex lipids (as described in Section 17.4). Many free fatty acids are deployed to the liver and muscle cells, especially heart muscle, which prefers to burn fatty acids even when carbohydrate fuels are available.

FATTY ACIDS ARE ACTIVATED BEFORE THEY ARE DEGRADED

To be oxidatively degraded, a fatty acid must first be activated. Activation is a two-step reaction catalyzed by acyl-CoA synthetase. First, the fatty acid displaces the diphosphate group of ATP, then coenzyme A (HSCoA) displaces the AMP

FIGURE 17.4

Lipid metabolism in context. Triacylglycerols, the "polymeric" form of fatty acids, are hydrolyzed to release fatty acids (1) that are oxidatively degraded to the two-carbon intermediate acetyl-CoA (2). Acetyl-CoA is also the starting material for the reductive biosynthesis of fatty acids (3), which can then be stored as triacylglycerols (4) or used in the synthesis of other lipids. Acetyl-CoA is also the precursor of lipids that are not built from fatty acids (these pathways are not shown here).

group to form an acyl-CoA:

$$\text{R}-\overset{\displaystyle \overset{O}{\|}}{\text{C}}\underset{\text{O}^-}{\underset{|}{}}\ \ \text{Fatty acid} \quad + \quad \text{$^-$O}-\overset{\displaystyle \overset{O}{\|}}{\underset{\text{O}^-}{\underset{|}{\text{P}}}}-\text{O}-\overset{\displaystyle \overset{O}{\|}}{\underset{\text{O}^-}{\underset{|}{\text{P}}}}-\text{O}-\overset{\displaystyle \overset{O}{\|}}{\underset{\text{O}^-}{\underset{|}{\text{P}}}}-\text{O}-\text{Adenosine} \quad \text{ATP}$$

$$\downarrow \quad \text{PP}_i \xrightarrow{\ \ \text{H}_2\text{O}\ \ }_{\substack{\text{inorganic}\\\text{pyrophosphatase}}} 2\,\text{P}_i$$

$$\text{R}-\overset{\displaystyle \overset{O}{\|}}{\text{C}}-\text{O}-\overset{\displaystyle \overset{O}{\|}}{\underset{\text{O}^-}{\underset{|}{\text{P}}}}-\text{O}-\text{Adenosine}$$

Acyladenylate

$$\downarrow \quad \text{H}-\text{SCoA}$$

$$\text{R}-\overset{\displaystyle \overset{O}{\|}}{\text{C}}-\text{SCoA} \quad + \quad \text{$^-$O}-\overset{\displaystyle \overset{O}{\|}}{\underset{\text{O}^-}{\underset{|}{\text{P}}}}-\text{O}-\text{Adenosine}$$

Acyl-CoA   AMP

The acyladenylate product of the first step has a large free energy of hydrolysis so its formation conserves the free energy of the cleaved phosphoanhydride bond in ATP. The second step, transfer of the acyl group to CoA (the same molecule that carries acetyl groups as acetyl-CoA), likewise conserves free energy in the formation of a thioester bond (see Section 12.3). Consequently, the overall reaction

$$\text{fatty acid} + \text{CoA} + \text{ATP} \rightleftharpoons \text{acyl-CoA} + \text{AMP} + \text{PP}_i$$

has a free energy change near zero. However, subsequent hydrolysis of the product PP_i (by the ubiquitous enzyme inorganic pyrophosphatase) is highly exergonic, and this reaction makes the formation of acyl-CoA spontaneous and irreversible. This sequence of steps is another example of cells using the free energy of ATP hydrolysis to drive a reaction (the attachment of the fatty acid to coenzyme A) that would not occur on its own. Note, again, that ATP is not simply hydrolyzed in isolation; it participates in the reaction via formation of the acyl-adenylate intermediate.

Most cells contain a set of acyl-CoA synthetases specific for fatty acids that are short (C_2–C_3), medium (C_4–C_{12}), long ($\geq\text{C}_{12}$), or very long ($\geq\text{C}_{22}$). The enzymes that are specific for the longest acyl chains may function in cooperation with a membrane transport protein so that the fatty acid is activated as it enters the cell. Once the large and polar coenzyme A is attached, the fatty acid is unable to diffuse back across the membrane and remains inside the cell to be metabolized.

In animals, fatty acids are activated in the cytosol, but the rest of the oxidation pathway occurs in the mitochondria. Because there is no transport protein for CoA adducts, acyl groups must enter the mitochondria via a shuttle system involving the small molecule carnitine (**Fig. 17.5**). The acyl group is now ready to be oxidized.

EACH ROUND OF β OXIDATION HAS FOUR REACTIONS

The pathway known as **β oxidation** degrades an acyl-CoA in a way that produces acetyl-CoA molecules for further oxidation and energy production by the citric acid cycle. In fact, in some tissues or under certain conditions, β oxidation supplies far more acetyl groups to the citric acid cycle than does glycolysis. β Oxidation also feeds electrons directly into the mitochondrial electron transport chain, which generates ATP by oxidative phosphorylation.

FIGURE 17.5

The carnitine shuttle system.
(1) A cytosolic carnitine acyltransferase transfers an acyl group from CoA to carnitine.
(2) The carnitine transporter allows the acyl-carnitine to enter the mitochondrial matrix.
(3) A mitochondrial carnitine acyltransferase transfers the acyl group to a mitochondrial CoA molecule. (4) Free carnitine returns to the cytosol via the transport protein.

? Why do acyl groups move into the mitochondrion, not out of it?

β Oxidation is a spiral pathway. *Each round consists of four enzyme-catalyzed steps that yield one molecule of acetyl-CoA and an acyl-CoA shortened by two carbons, which becomes the starting substrate for the next round.* Seven rounds of β oxidation degrade a C_{16} fatty acid to eight molecules of acetyl-CoA:

CH$_3$—(CH$_2$)$_n$—C—C—C—SCoA

Fatty acyl-CoA

1 FAD / acyl-CoA dehydrogenase / FADH$_2$ ⇌ QH$_2$ / Q

CH$_3$—(CH$_2$)$_n$—C=C—C—SCoA

Enoyl-CoA

2 H$_2$O / enoyl-CoA hydratase

CH$_3$—(CH$_2$)$_n$—C—CH$_2$—C—SCoA

3-Hydroxyacyl-CoA

3 NAD$^+$ / 3-hydroxyacyl-CoA dehydrogenase / NADH + H$^+$

CH$_3$—(CH$_2$)$_n$—C—CH$_2$—C—SCoA

Ketoacyl-CoA

4 CoASH / thiolase

CH$_3$—(CH$_2$)$_n$—C—SCoA + CH$_3$—C—SCoA

Fatty acyl-CoA (2 C atoms shorter)   Acetyl-CoA

1. *Oxidation of acyl-CoA at the 2,3 position is catalyzed by an acyl-CoA dehydrogenase to yield a 2,3-enoyl-CoA. The two electrons removed from the acyl group are transferred to an FAD prosthetic group. A series of electron-transfer reactions eventually transfers the electrons to ubiquinone (Q).*

2. *The second step is catalyzed by a hydratase, which adds the elements of water across the double bond produced in the first step.*

3. *The hydroxyacyl-CoA is oxidized by another dehydrogenase. In this case, NAD$^+$ is the cofactor.*

4. *The final step, thiolysis, is catalyzed by a thiolase and releases acetyl-CoA. The remaining acyl-CoA, two carbons shorter than the starting substrate, undergoes another round of the four reactions (dotted line).*

FIGURE 17.6

The reactions of β oxidation.

? **Which steps of the pathway are redox reactions?**

Figure 17.6 shows the reactions of β oxidation. The β indicates that oxidation occurs at the β position, which is the carbon atom that is two away from the carbonyl carbon (C3 is the β carbon). Note that acetyl units are not lost from the methyl end of the fatty acid but from the activated, CoA end.

The oxidation of fatty acids by the successive removal of two-carbon units was discovered over 100 years ago, and the enzymatic steps were elucidated about 50 years ago. But β oxidation still offers surprises in the details. For example, many enzymes are required to fully degrade an acyl-CoA to acetyl-CoA. Each of the four steps shown in Figure 17.6 appears to be catalyzed by two to five different enzymes with different chain-length specificities, as in the acyl-CoA synthetase reaction. The existence of some of these isozymes was inferred from studies of patients with disorders of fatty acid oxidation. One of these often-fatal diseases is due to a deficiency of medium-chain acyl-CoA dehydrogenase; affected individuals cannot degrade acyl-CoAs having 4 to 12 carbons, and derivatives of these molecules accumulate in the liver and are excreted in the urine.

β Oxidation is a major source of cellular free energy, especially during a fast, when carbohydrates are not available. Each round of β oxidation produces one QH$_2$, one NADH, and one acetyl-CoA. The citric acid cycle oxidizes the acetyl-CoA to produce an additional three NADH, one QH_2, and one GTP. Oxidation of all the reduced cofactors yields approximately 13 ATP: 3 from the two QH_2 and 10 from the four NADH (recall from the discussion of P:O ratios that oxidative phosphorylation is an indirect process, so the amount of ATP produced per pair of electrons entering the electron transport chain is not a whole number; see Section 15.4). A total of 14 ATP are generated from each round of β oxidation:

One round of β oxidation	Citric acid cycle	Oxidative phosphorylation
1 QH_2	⟶	→ 1.5 ATP
1 NADH	⟶	→ 2.5 ATP
1 Acetyl-CoA ⟶	3 NADH	→ 7.5 ATP
	1 QH_2	→ 1.5 ATP
	1 GTP	→ 1 ATP

Total 14 ATP

β Oxidation is regulated primarily by the availability of free CoA (to make acyl-CoA) and by the ratios of NAD^+/NADH and Q/QH_2 (these reflect the state of the oxidative phosphorylation system). Some individual enzymes are also regulated by product inhibition.

DEGRADATION OF UNSATURATED FATTY ACIDS REQUIRES ISOMERIZATION AND REDUCTION

Common fatty acids such as oleate and linoleate contain *cis* double bonds that present obstacles to the enzymes that catalyze β oxidation.

For linoleate, the first three rounds of β oxidation proceed as usual. But the acyl-CoA that begins the fourth round has a 3,4 double bond (originally the 9,10 double bond). Furthermore, this molecule is a *cis* enoyl-CoA, but enoyl-CoA hydratase (the enzyme that catalyzes step 2 of β oxidation) recognizes only the *trans* configuration. This metabolic obstacle is removed by the enzyme enoyl-CoA isomerase, which converts the *cis* 3,4 double bond to a *trans* 2,3 double bond so that β oxidation can continue.

A second obstacle for linoleate oxidation arises after the first reaction of the fifth round of β oxidation. Acyl-CoA dehydrogenase introduces a 2,3 double bond as usual, but the original 12,13 double bond of linoleate is now at the 4,5 position. The resulting dienoyl-CoA is not a good substrate for the next enzyme, enoyl-CoA hydratase. The dienoyl-CoA must undergo an NADPH-dependent reduction to convert its two double bonds to a single *trans* 3,4 double bond. This product must then be isomerized to produce the *trans* 2,3 double bond that is recognized by enoyl-CoA hydratase.

Carbon compounds with double bonds are slightly more oxidized than saturated compounds (see Table 1.3), so less energy is released in converting them to CO_2. Accordingly, a diet rich in unsaturated fatty acids contains fewer calories than a diet rich in saturated fatty acids. The bypass reactions described above provide the molecular explanation why *unsaturated fatty acids yield less free energy than saturated fatty acids*. First, the enoyl-CoA isomerase reaction bypasses the QH_2-producing acyl-CoA dehydrogenase step, so 1.5 fewer ATP are produced. Second, the NADPH-dependent reductase consumes 2.5 ATP equivalents because NADPH is energetically equivalent to NADH.

OXIDATION OF ODD-CHAIN FATTY ACIDS YIELDS PROPIONYL-CoA

Most fatty acids have an even number of carbon atoms (this is because they are synthesized by the addition of two-carbon acetyl units, as we will see later in this chapter). However, some plant and bacterial fatty acids that make their way into the human system have an odd number of carbon atoms. *The final round of β oxidation of these molecules leaves a three-carbon fragment, propionyl-CoA, rather than the usual acetyl-CoA.*

Propionyl-CoA

This intermediate can be further metabolized by the sequence of steps outlined in **Figure 17.7**. At first, this pathway seems longer than necessary. For example, adding a carbon to C3 of the propionyl group would immediately generate succinyl-CoA. However, such a reaction is not chemically favored, because C3 is too far from the electron-delocalizing effects of the CoA thioester. Consequently, propionyl-CoA carboxylase must add a carbon to C2, and then methylmalonyl-CoA mutase must rearrange the carbon skeleton to produce succinyl-CoA. Note that succinyl-CoA is not the end point of the pathway. Because it is a citric acid cycle intermediate, it acts catalytically and is not consumed by the cycle (see Section 14.2). *The complete catabolism of the carbons derived from propionyl-CoA requires that the succinyl-CoA be converted to pyruvate and then to acetyl-CoA, which enters the citric acid cycle as a substrate.*

FIGURE 17.7
Catabolism of propionyl-CoA.

? How many ATP equivalents can be produced from propionyl-CoA?

1. *Propionyl-CoA carboxylase adds a carboxyl group at C2 of the propionyl group to form a four-carbon methylmalonyl group.*

2. *A racemase interconverts the two different methylmalonyl-CoA stereoisomers (the two configurations are indicated by the R and S symbols).*

3. *Methylmalonyl-CoA mutase rearranges the carbon skeleton to generate succinyl-CoA.*

4–6. *Succinyl-CoA, a citric acid cycle intermediate, is converted to malate by reactions 5–7 of the citric acid cycle (see Fig. 14.6).*

7. *After being exported from the mitochondria, malate is decarboxylated by malic enzyme to produce pyruvate in the cytosol.*

8. *Pyruvate, imported back into the mitochondria, can then be converted to acetyl-CoA by the pyruvate dehydrogenase complex.*

The cobalamin-derived cofactor. The prosthetic group of methylmalonyl-CoA mutase is a derivative of the vitamin cobalamin. The structure includes a hemelike ring structure with a central cobalt ion. Note that one of the Co ligands is a carbon atom, an extremely rare instance of a carbon–metal bond in a biological system.

Methylmalonyl-CoA mutase, which catalyzes step 3 of Figure 17.7, is an unusual enzyme because it mediates a rearrangement of carbon atoms and requires a prosthetic group derived from the vitamin cobalamin (vitamin B_{12}; **Fig. 17.8**). Only about a dozen enzymes are known to use cobalamin cofactors. The small amounts of cobalamin required for human health are usually easily obtained from a diet that contains animal products. Vegans, however, are advised to consume B_{12} supplements. A disorder of vitamin B_{12} absorption causes the disease pernicious anemia.

SOME FATTY ACID OXIDATION OCCURS IN PEROXISOMES

The majority of a mammalian cell's fatty acid oxidation occurs in mitochondria, but a small percentage is carried out in organelles known as **peroxisomes** (**Fig. 17.9**). In plants, all fatty acid oxidation occurs in peroxisomes and glyoxysomes. Peroxisomes are enclosed by a single membrane and contain a variety of degradative and biosynthetic enzymes. The peroxisomal β oxidation pathway differs from the mitochondrial pathway in the first step. An acyl-CoA oxidase catalyzes the reaction

FIGURE 17.9

Peroxisomes. Nearly all eukaryotic cells contain these single-membrane-bound organelles (dark structures), which are similar to plant glyoxysomes (see Box 14.B). [*Don W. Fawcett/Photo Researchers, Inc.*]

$$CH_3-(CH_2)_n-\overset{H}{\underset{H}{C}}-\overset{H}{\underset{H}{C}}-\overset{O}{C}-SCoA$$
Acyl-CoA

acyl-CoA oxidase, FAD → FADH$_2$, O$_2$ → H$_2$O$_2$

$$CH_3-(CH_2)_n-\overset{H}{C}=\overset{}{\underset{H}{C}}-\overset{O}{C}-SCoA$$
Enoyl-CoA

The enoyl-CoA product of the reaction is identical to the product of the mitochondrial acyl-CoA dehydrogenase reaction (see Fig. 17.6), but the electrons removed from the acyl-CoA are transferred not to ubiquinone but directly to molecular oxygen to produce hydrogen peroxide, H_2O_2. This reaction product, which gives the peroxisome its name, is subsequently broken down by the peroxisomal enzyme catalase:

$$2\ H_2O_2 \rightarrow 2\ H_2O + O_2$$

The second, third, and fourth reactions of fatty acid oxidation are the same as in mitochondria.

Because the peroxisomal oxidation enzymes are specific for very-long-chain fatty acids (such as those containing over 20 carbons) and bind short-chain fatty acids with low affinity, *the peroxisome serves as a chain-shortening system*. The partially degraded fatty acyl-CoAs then make their way to the mitochondria for complete oxidation.

The peroxisome is also responsible for degrading some branched-chain fatty acids, which are not recognized by the mitochondrial enzymes. One such nonstandard fatty acid is phytanate,

Phytanate

which is derived from the side chain of chlorophyll molecules (see Fig. 16.3) and is present in all plant-containing diets. Phytanate must be degraded by peroxisomal enzymes because the methyl group at C3 prevents dehydrogenation by 3-hydroxyacyl-CoA dehydrogenase (step 3 of the standard β oxidation pathway). A deficiency of any of the phytanate-degrading enzymes results in Refsum's disease, a degenerative neuronal disorder characterized by an accumulation of phytanate in the tissues. The importance of peroxisomal enzymes in lipid metabolism (both catabolic and anabolic) is confirmed by the fatal outcome of most diseases stemming from deficient peroxisomal enzymes or improper synthesis of the peroxisomes themselves.

CONCEPT CHECK

1. Summarize the role of lipoproteins in making fatty acids available for oxidation.

2. How is a fatty acid activated? What is the cost in free energy?

3. Identify the energy-generating steps of fatty acid oxidation and explain how they contribute to ATP synthesis.

4. Compare the contributions of the cytosol, mitochondria, and peroxisomes to fatty acid catabolism.

5. What is the role of peroxisomes in fatty acid catabolism?

FATTY ACID SYNTHESIS

At first glance, fatty acid synthesis appears to be the exact reverse of fatty acid oxidation. For example, fatty acyl groups are built and degraded two carbons at a time, and several of the reaction intermediates in the two pathways are similar or identical. However, *the pathways for fatty acid synthesis and degradation must differ for thermodynamic reasons*, as we saw for glycolysis and gluconeogenesis. Since fatty acid oxidation is a thermodynamically favorable process, simply reversing the steps of this pathway would be energetically unfavorable.

In mammalian cells, the opposing metabolic pathways of fatty acid synthesis and degradation are entirely separate. β Oxidation takes place in the mitochondrial matrix, and synthesis occurs in the cytosol. Furthermore, the two pathways use different cofactors. In β oxidation, the acyl group is attached to coenzyme A, but a growing fatty acyl chain is bound by an acyl-carrier protein (ACP; **Fig. 17.10**). β Oxidation funnels electrons to ubiquinone and NAD^+, but in fatty acid synthesis, NADPH is the source of reducing power. Finally, β oxidation requires two ATP equivalents (two phosphoanhydride bonds) to "activate" the acyl group, but the biosynthetic pathway consumes one ATP for every two carbons incorporated

KEY CONCEPTS

- Fatty acid synthesis begins with the carboxylation of acetyl-CoA in the cytosol.
- Fatty acid synthase catalyzes seven separate reactions to extend a fatty acid by two carbons.
- Elongases and desaturases modify newly synthesized fatty acids.
- Various metabolites contribute to the regulation of fatty acid synthesis.
- Ketogenesis converts acetyl-CoA to small soluble ketone bodies.

FIGURE 17.10

Acyl-carrier protein and coenzyme A. Both acyl-carrier protein (ACP) and coenzyme A (CoA) include a pantothenate (vitamin B_5) derivative ending with a sulfhydryl group that forms a thioester with an acyl or acetyl group. In CoA, the pantothenate derivative is esterified to an adenine nucleotide; in ACP, the group is esterified to a serine OH group of a polypeptide (in mammals, ACP is part of a larger multifunctional protein, fatty acid synthase).

into a fatty acid. In this section, we focus on the reactions of fatty acid synthesis, comparing and contrasting them to β oxidation.

ACETYL-CoA CARBOXYLASE CATALYZES THE FIRST STEP OF FATTY ACID SYNTHESIS

The starting material for fatty acid synthesis is acetyl-CoA, which may be generated in the mitochondria by the action of the pyruvate dehydrogenase complex (Section 14.1). But just as cytosolic acyl-CoA cannot directly enter the mitochondria to be oxidized, mitochondrial acetyl-CoA cannot exit to the cytosol for biosynthetic reactions. *The transport of acetyl groups to the cytosol involves citrate, which has a transport protein.* Citrate synthase (the enzyme that catalyzes the first step of the citric acid cycle; see Fig. 14.6) combines acetyl-CoA with oxaloacetate to produce citrate, which then leaves the mitochondria. ATP-citrate lyase "undoes" the citrate synthase reaction to produce acetyl-CoA and oxaloacetate in the cytosol (**Fig. 17.11**). Note that ATP is consumed in the ATP-citrate lyase reaction to drive the formation of a thioester bond.

The first step of fatty acid synthesis is the carboxylation of acetyl-CoA, an ATP-dependent reaction carried out by acetyl-CoA carboxylase. This enzyme catalyzes the rate-controlling step for the fatty acid synthesis pathway. The acetyl-CoA carboxylase mechanism is similar to that of propionyl-CoA carboxylase (step 1 in Fig. 17.7) and pyruvate carboxylase (see Fig. 13.9). First,

FIGURE 17.11

The citrate transport system. The citrate transport protein, along with mitochondrial citrate synthase and cytoplasmic ATP-citrate lyase, provides a route for transporting acetyl units from the mitochondrial matrix to the cytosol.

? Add to the diagram the steps by which oxaloacetate carbons are returned to the matrix via the pyruvate transporter (see Fig. 14.18).

CO_2 (as bicarbonate, HCO_3^-) is "activated" by its attachment to a biotin prosthetic group in a reaction that converts ATP to ADP + P_i:

$$biotin + HCO_3^- + ATP \rightarrow biotin-COO^- + ADP + P_i$$

Next, the carboxybiotin prosthetic group transfers the carboxylate group to acetyl-CoA to form the three-carbon malonyl-CoA and regenerate the enzyme:

$$\underset{\text{Acetyl-CoA}}{biotin-COO^- + CH_3-\overset{\overset{\displaystyle O}{\|}}{C}-SCoA} \longrightarrow \underset{\text{Malonyl-CoA}}{^-OOC-CH_2-\overset{\overset{\displaystyle O}{\|}}{C}-SCoA + biotin}$$

Malonyl-CoA is the donor of the two-carbon acetyl units that are used to build a fatty acid. The carboxylate group added by the carboxylation reaction is lost in a subsequent decarboxylation reaction. This sequence of carboxylation followed by decarboxylation also occurs in the conversion of pyruvate to phosphoenolpyruvate in gluconeogenesis (see Section 13.2). Note that fatty acid synthesis requires a C_3 intermediate, whereas β oxidation involves only two-carbon acetyl units.

FATTY ACID SYNTHASE CATALYZES SEVEN REACTIONS

The protein that carries out the main reactions of fatty acid synthesis in animals is a 540-kD **multifunctional enzyme** made of two identical polypeptides (**Fig. 17.12**). Each polypeptide of this fatty acid synthase has six active sites to carry out seven discrete reactions, which are summarized in **Figure 17.13**. In plants and most bacteria, the reactions are catalyzed by separate polypeptides, but the chemistry is the same.

FIGURE 17.12

Mammalian fatty acid synthase. (a) Domain organization of the fatty acid synthase polypeptide. The domains labeled KS, MAT, DH1/DH2, ER, KR, and TE are the six enzymes. ACP is the acyl-carrier protein, whose pantothenate arm swings between the active sites. The domains labeled ψME and ψKR have no enzymatic activity. (b) Three-dimensional structure of the fatty acid synthase dimer, with the domains of each monomer colored as in part (a). The ACP and TE (thioesterase) modules are missing in this model. The NADP$^+$ molecules are shown in blue, and black spheres represent the anchor points for the acyl-carrier protein domains. [*From T. Maier, M. Leibundgut, N. Ban, Science 321: 5894, 2008. Reprinted with permission from AAAS.*]

? Provide an example of another multifunctional enzyme.

1. The two-carbon acetyl group that will be lengthened is transferred from CoA to a Cys side chain of fatty acid synthase.

2. The malonyl group that will donate an acetyl group to the growing fatty acyl chain is transferred from CoA to the ACP domain of the enzyme.

3. In this condensation reaction, the malonyl group is decarboxylated and the resulting two-carbon fragment attacks the acetyl group to form a four-carbon product.

4. The 3-ketoacyl product of step 3 is reduced.

5. A dehydration introduces a 2,3 double bond.

6. A second NADPH-dependent reduction completes the conversion of the condensation product to an acyl group.

7. The acyl group is transferred from ACP to the enzyme Cys group, and another malonyl group is loaded onto the free ACP, ready for another condensation reaction.

8. Steps 3 – 6 are repeated six times to build a C_{16} fatty acid.

9. A thioesterase hydrolyzes the thioester bond to release palmitate.

FIGURE 17.13

Fatty acid synthesis. The steps show how fatty acid synthase carries out the synthesis of the C_{16} fatty acid palmitate, starting from acetyl-CoA. The abbreviation next to each enzyme corresponds to a structural domain as shown in Figure 17.12.

Reactions 1 and 2 are transacylation reactions that serve to prime or load the enzyme with the reactants for the condensation reaction (step 3). In the condensation reaction, decarboxylation of the malonyl group allows C2 to attack the acetyl thioester group to form acetoacetyl-ACP:

Malonyl-ACP   Acetyl-Cys   Acetoacetyl-ACP

This chemistry explains the necessity for carboxylating an acetyl group to a malonyl group: C2 of an acetyl group would not be sufficiently reactive.

The hydroxyacyl product of Reaction 4 is chemically similar to the hydroxyacyl product of step 2 of β oxidation, but the intermediates of the two pathways have opposite configurations (see Section 4.1):

3-Hydroxyacyl-ACP intermediate of
fatty acid synthesis
(D configuration)

3-Hydroxyacyl-ACP intermediate of
β oxidation
(L configuration)

Note also that growth of the acyl chain—like chain shortening in β oxidation—occurs at the thioester end of the molecule, not at the methyl end.

The NADPH required for the two reduction steps of fatty acid synthesis (steps 4 and 6) is supplied mostly by the pentose phosphate pathway (see Section 13.4). The synthesis of one molecule of palmitate (the usual product of fatty acid synthase) requires the production of 7 malonyl-CoA, at a cost of 7 ATP. The seven rounds of fatty acid synthesis consume 14 NADPH, which is equivalent to 14×2.5, or 35, ATP, bringing the total cost to 42 ATP. Still, this energy investment is much less than the free energy yield of oxidizing palmitate.

During fatty acid synthesis, the long flexible arm of the pantothenate derivative in ACP (see Fig. 17.10) shuttles intermediates between the active sites of fatty acid synthase (the lipo-amide group in the pyruvate dehydrogenase complex functions similarly; see Section 14.1). In the fatty acid synthase dimer, two fatty acids can be built simultaneously.

Packaging several enzyme activities into one multifunctional protein like mammalian fatty acid synthase allows the enzymes to be synthesized and controlled in a coordinated fashion. Also, the product of one reaction can quickly diffuse to the next active site. Bacterial and plant fatty acid synthase systems may lack the efficiency of a multifunctional protein, but because the enzymes are not locked together, a wider variety of fatty acid products can be more easily made. In mammals, fatty acid synthase produces mostly the 16-carbon saturated fatty acid palmitate.

OTHER ENZYMES ELONGATE AND DESATURATE NEWLY SYNTHESIZED FATTY ACIDS

Some sphingolipids contain C_{22} and C_{24} fatty acyl groups. *These and other long-chain fatty acids are generated by enzymes known as elongases,* which extend the C_{16} fatty acid produced by fatty acid synthase. Elongation can occur in either the endoplasmic reticulum or mitochondria. The endoplasmic reticulum reactions use malonyl-CoA as the acetyl-group

A CLOSER LOOK

Box 17.A Fats, Diet, and Heart Disease

Years of study have established a link between elevated LDL levels and atherosclerosis and indicate that certain diets contribute to the formation of fatty deposits that clog arteries and cause cardiovascular disease. Considerable research has been devoted to showing how dietary lipids influence serum lipid levels. For example, early studies showed that diets rich in saturated fats increased blood cholesterol (that is, LDL), whereas diets in which unsaturated vegetable oils replaced the saturated fats had the opposite effect. These and other findings led to recommendations that individuals at risk for atherosclerosis avoid butter, which is rich in saturated fat as well as cholesterol, and instead use margarine, which is prepared from cholesterol-free vegetable oils.

The production of semisolid margarine from liquid plant oils (triacylglycerols containing unsaturated fatty acids) often includes a hydrogenation step to chemically saturate the carbons of the fatty acyl chains. In this process, some of the original *cis* double bonds are converted to *trans* double bonds. In clinical studies, *trans* fatty acids are comparable to saturated fatty acids in their tendency to increase LDL levels and decrease HDL levels. Dietary guidelines now warn against the excessive intake of *trans* fatty

acids in the form of hydrogenated vegetable oils (small amounts of *trans* fatty acids also occur naturally in some animal fats). This would mean avoiding processed foods whose list of ingredients includes "partially hydrogenated vegetable oil."

So should you consume butter or margarine? Linking specific types of dietary fats to human health and disease has always been a risky venture because quantitative information comes mainly from epidemiological and clinical studies, which are typically time-consuming and often inconclusive or downright contradictory. Scientists still do not fully understand *how* the consumption of specific fatty acids—saturated or unsaturated, *cis* or *trans*—influences lipoprotein metabolism. Other dietary factors also play a role. For example, one consequence of low-fat diets is that individuals consume relatively more carbohydrates. And when they reduce their meat intake (an obvious source of fat), people eat more fruits and vegetables, which may have health-enhancing effects of their own.

? **Rank the "healthiness" of the following sources of fatty acids: animal fat, olive oil, and hydrogenated soybean oil.**

donor and are chemically similar to those of fatty acid synthase. In the mitochondria, fatty acids are elongated by reactions that more closely resemble the reversal of β oxidation but use NADPH.

Desaturases introduce double bonds into saturated fatty acids. These reactions take place in the endoplasmic reticulum, catalyzed by membrane-bound enzymes. The electrons removed in the dehydrogenation of the fatty acid are eventually transferred to molecular oxygen to produce H_2O. The most common unsaturated fatty acids in animals are palmitoleate (a C_{16} molecule) and oleate (a C_{18} fatty acid; see Section 8.1), both with one *cis* double bond at the 9,10 position. *Trans* fatty acids are relatively rare in plants and animals, but they are abundant in some prepared foods, which has produced confusion among individuals concerned with eating the "right" kinds of fats (**Box 17.A**).

Elongation can follow desaturation (and vice versa), so animals can synthesize a variety of fatty acids with different chain lengths and degrees of unsaturation. However, mammals cannot introduce double bonds at positions beyond C9 and therefore cannot synthesize fatty acids such as linoleate and linolenate. These molecules are precursors of the C_{20} fatty acid arachidonate and other lipids with specialized biological activities (**Fig. 17.14**). *Mammals*

FIGURE 17.14

Synthesis of arachidonate. Linoleate (or linolenate) is elongated and desaturated to produce arachidonate, a C_{20} fatty acid with four double bonds.

Linoleate

desaturation

Linolenate

elongation

desaturation

Arachidonate

must therefore obtain linoleate and linolenate from their diet. These **essential fatty acids** are abundant in fish and plant oils. Unsaturated fatty acids with a double bond three carbons from the end, omega-3 fatty acids, may have health benefits (Box 8.A). A deficiency of essential fatty acids resulting from a very-low-fat diet may elicit symptoms such as slow growth and poor wound healing.

FATTY ACID SYNTHESIS CAN BE ACTIVATED AND INHIBITED

Under conditions of abundant metabolic fuel, the products of carbohydrate and amino acid catabolism are directed toward fatty acid synthesis, and the resulting fatty acids are stored inside cells as triacylglycerols. These form a lipid droplet that is coated with a phospholipid monolayer.

The rate of fatty acid synthesis is controlled by acetyl-CoA carboxylase, which catalyzes the first step of the pathway. This enzyme is inhibited by a pathway product (palmitoyl-CoA) and is allosterically activated by citrate (which signals abundant acetyl-CoA). The enzyme is also subject to allosteric regulation by hormone-stimulated phosphorylation and dephosphorylation.

The concentration of malonyl-CoA is also critical for preventing the wasteful simultaneous activity of fatty acid synthesis and fatty acid oxidation. Malonyl-CoA is the source of acetyl groups that are incorporated into fatty acids, and it also blocks β oxidation by inhibiting carnitine acyltransferase, the enzyme involved in shuttling acyl groups into the mitochondria (see Fig. 17.5). Consequently, *when fatty acid synthesis is under way, no acyl groups are transported into the mitochondria for oxidation.* Some of the mechanisms that regulate fatty acid metabolism are summarized in **Figure 17.15**.

There are both natural and synthetic inhibitors of fatty acid synthase, such as the widely used antibacterial agent triclosan and drugs that target pathogens more specifically (**Box 17.B**). Fatty acid synthase inhibitors are of great scientific and popular interest, given that excess body weight (due to fat) is a major health problem, affecting about two-thirds of the population of the United States. And because many tumors sustain high levels of fatty acid synthesis, fatty acid synthase inhibitors may also be useful for treating cancer.

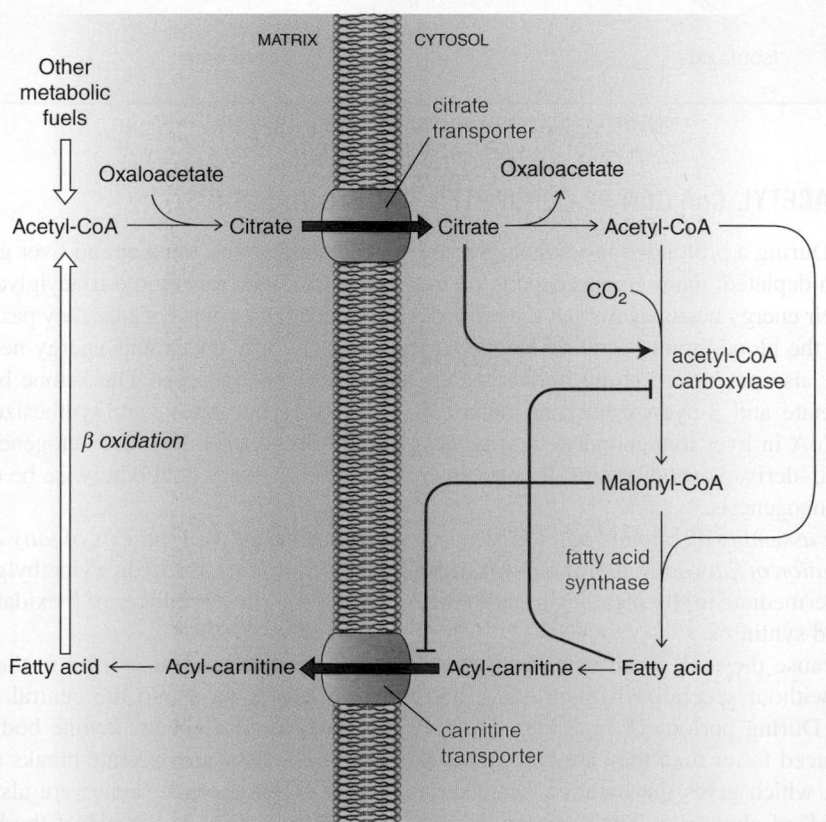

FIGURE 17.15

Some control mechanisms in fatty acid metabolism. Red symbols indicate inhibition, and the green symbol indicates activation.

? Name all the enzymes required to convert excess glucose into fatty acids.

A CLOSER LOOK

Box 17.B Inhibitors of Fatty Acid Synthesis

Because fatty acid synthesis is an essential metabolic activity, inhibiting the process in pathogenic organisms—but not in their mammalian hosts—is a useful strategy for preventing or curing certain infectious diseases. For example, many cosmetics, toothpastes, antiseptic soaps, and even plastic toys and kitchenware contain the compound 5-chloro-2-(2,4-dichlorophenoxy)-phenol, better known as triclosan.

Triclosan was long believed to act as a general microbicide, an agent that kills nonspecifically, much like household bleach or ultraviolet light. Such microbicides are effective because it is difficult for bacteria to evolve specific resistance mechanisms. However, triclosan actually operates more like an antibiotic with a specific biochemical target, in this case, enoyl-ACP reductase, which catalyzes step 6 of fatty acid synthesis (see Fig. 17.13).

The enzyme's natural substrate has a K_M of about 22 μM, but the dissociation constant for the inhibitor is 20 to 40 pM, indicating extremely tight binding. In the active site, one of the phenyl rings of triclosan, whose structure mimics the structure of the reaction intermediate, stacks on top of the nicotinamide ring of the NADH cofactor. Triclosan also binds through van der Waals interactions and hydrogen bonds with amino acid residues in the active site.

The antibiotic isoniazid has been used to treat *Mycobacterium tuberculosis* infections. Inside the bacterial cell, isoniazid is oxidized,

and the reaction product combines with NAD^+ to generate a compound that inhibits one of the cell's enoyl-ACP reductases. The target enzyme is specific for extremely long-chain fatty acids, which are incorporated into mycolic acids, the waxlike components of the mycobacterial cell wall.

Drugs such as isoniazid must be taken for many months, since mycobacteria replicate very slowly and can remain dormant inside host cells, protected from drugs as well as the host's immune system.

Some fungal species are susceptible to cerulenin, which inhibits 3-ketoacyl-ACP synthase (step 3 of fatty acid synthesis; see Fig. 17.13) by blocking the reaction of malonyl-ACP, that is, the condensation step. Cerulenin is also effective against *M. tuberculosis*, inhibiting the production of long-chain fatty acids required for cell-wall synthesis. The drug, which contains a reactive epoxide group, reacts irreversibly with the enzyme's active-site cysteine residue, forming a C2—S covalent bond. Cerulenin's hydrocarbon tail occupies the site that would normally accommodate the growing fatty acyl chain.

? **What types of mutations might allow bacteria to become resistant to triclosan?**

Triclosan   Isoniazid   Cerulenin

ACETYL-CoA CAN BE CONVERTED TO KETONE BODIES

During a prolonged fast, when glucose is unavailable from the diet and liver glycogen has been depleted, many tissues depend on fatty acids released from stored triacylglycerols to meet their energy needs. However, the brain does not burn fatty acids because they pass poorly through the blood–brain barrier. Gluconeogenesis helps supply the brain's energy needs, but the liver also produces **ketone bodies** to supplement gluconeogenesis. The ketone bodies—acetoacetate and 3-hydroxybutyrate (also called β-hydroxybutyrate)—are synthesized from acetyl-CoA in liver mitochondria by a process called **ketogenesis.** Because ketogenesis uses fatty acid–derived acetyl groups, it helps spare amino acids that would otherwise be diverted to gluconeogenesis.

The assembly of ketone bodies is somewhat reminiscent of the synthesis of fatty acids or the oxidation of fatty acids in two-carbon steps (**Fig. 17.16**). In fact, the hydroxymethylglutaryl-CoA intermediate is chemically similar to the 3-hydroxyacyl intermediates of β oxidation and fatty acid synthesis.

Because they are small and water-soluble, ketone bodies are transported in the bloodstream without specialized lipoproteins, and they can easily pass into the central nervous system. During periods of high ketogenic activity, such as in diabetes, ketone bodies may be produced faster than they are consumed. Some of the excess acetoacetate breaks down to acetone, which gives the breath a characteristic sweet smell. Ketone bodies are also acids, with a pK of about 3.5. Their overproduction can lead to a drop in the pH of the blood, a

$$CH_3-\overset{\overset{\displaystyle O}{\|}}{C}-SCoA + CH_3-\overset{\overset{\displaystyle O}{\|}}{C}-SCoA$$

Acetyl-CoA Acetyl-CoA

H—SCoA ← **1** thiolase

$$CH_3-\overset{\overset{\displaystyle O}{\|}}{C}-CH_2-\overset{\overset{\displaystyle O}{\|}}{C}-SCoA$$

Acetoacetyl-CoA

$$CH_3-\overset{\overset{\displaystyle O}{\|}}{C}-SCoA$$

2 HMG-CoA synthase

H—SCoA ←

$$^-OOC-CH_2-\overset{\overset{\displaystyle OH}{|}}{\underset{\underset{\displaystyle CH_3}{|}}{C}}-CH_2-\overset{\overset{\displaystyle O}{\|}}{C}-SCoA$$

3-Hydroxy-3-methylglutaryl-CoA (HMG-CoA)

$$CH_3-\overset{\overset{\displaystyle O}{\|}}{C}-SCoA$$ ← **3** HMG-CoA lyase

$$^-OOC-CH_2-\overset{\overset{\displaystyle O}{\|}}{C}-CH_3$$

Acetoacetate

NADH + H⁺

NAD⁺ ← **4** 3-hydroxybutyrate dehydrogenase

5 CO_2

$$^-OOC-CH_2-\overset{\overset{\displaystyle OH}{|}}{C}H-CH_3$$

3-Hydroxybutyrate

$$CH_3-\overset{\overset{\displaystyle O}{\|}}{C}-CH_3$$

Acetone

1. Two molecules of acetyl-CoA condense to form acetoacetyl-CoA. The reaction is catalyzed by a thiolase, which breaks a thioester bond.

2. The four-carbon acetoacetyl group condenses with a third molecule of acetyl-CoA to form the six-carbon 3-hydroxymethylglutaryl-CoA (HMG-CoA).

3. HMG-CoA is then degraded to the ketone body acetoacetate and acetyl-CoA.

4. Acetoacetate undergoes reduction to produce another ketone body, 3-hydroxybutyrate.

5. Some acetoacetate may also undergo nonenzymatic decarboxylation to acetone and CO_2.

FIGURE 17.16

Ketogenesis. The ketone bodies are boxed.

condition called ketoacidosis. Mild symptoms may also develop in some individuals following a high-protein, low-carbohydrate diet, when ketogenesis increases to offset the shortage of dietary carbohydrates.

Ketone bodies produced by the liver are used by other tissues as metabolic fuels after being converted back to acetyl-CoA (**Fig. 17.17**). The liver itself cannot catabolize ketone bodies because it lacks one of the required enzymes, 3-ketoacyl-CoA transferase.

 CONCEPT CHECK

1. Explain why the pathways of fatty acid synthesis and degradation must differ.

2. Compare fatty acid synthesis and oxidation with respect to cellular location, cofactors, use of ATP, acyl group carrier, and role of acetyl-CoA.

3. What is the role of malonyl-CoA in fatty acid synthesis?

4. What is the function of elongases and desaturases?

5. List the metabolites that regulate fatty acid synthesis.

6. Describe the conditions under which ketone bodies are made and used.

FIGURE 17.17

Catabolism of ketone bodies.

? Compare this pathway to ketogenesis. Which steps are similar?

1. 3-Hydroxybutyrate is oxidized back to acetoacetate (this is the reverse of Reaction 4 in Fig. 17.16).

2. Succinyl-CoA donates its CoA group to produce acetoacetyl-CoA.

3. Thiolase then uses a free CoA group to cleave the four-carbon unit to two molecules of acetyl-CoA.

3-Hydroxybutyrate

NAD$^+$
3-hydroxybutyrate dehydrogenase
NADH + H$^+$

Acetoacetate

Succinyl-CoA
3-ketoacyl-CoA transferase

Succinate

Acetoacetyl-CoA

H—SCoA
thiolase

2 Acetyl-CoA

17.4

SYNTHESIS OF OTHER LIPIDS

KEY CONCEPTS
- Acyl groups are transferred from CoA to a glycerol backbone to generate triacylglycerols and phospholipids.
- Cholesterol is synthesized from acetyl-CoA.
- Cholesterol can be used both inside and outside the cell.

Lipid metabolism encompasses many chemical reactions involving fatty acids, which are structural components of other lipids such as triacylglycerols, glycerophospholipids, and sphingolipids. Fatty acids such as arachidonate are also the precursors of eicosanoids that function as signaling molecules (Section 10.4). This section covers the biosynthesis of some of the major types of lipids, including the synthesis of cholesterol from acetyl-CoA.

TRIACYLGLYCEROLS AND PHOSPHOLIPIDS ARE BUILT FROM ACYL-CoA GROUPS

Cells have a virtually unlimited capacity for storing fatty acids in the form of triacylglycerols, which aggregate in the cytoplasm to form droplets surrounded by a layer of amphipathic phospholipids. *Triacylglycerols are synthesized by attaching fatty acyl groups to a glycerol backbone derived from phosphorylated glycerol or from glycolytic intermediates,* for example, dihydroxyacetone phosphate:

Dihydroxyacetone phosphate

NADH + H$^+$ NAD$^+$

glycerol-3-phosphate dehydrogenase

Glycerol-3-phosphate

FIGURE 17.18

Triacylglycerol synthesis. An acyltransferase appends a fatty acyl group to C1 of glycerol-3-phosphate. A second acyltransferase reaction adds an acyl group to C2, yielding phosphatidate. A phosphatase removes P_i to produce diacylglycerol. The addition of a third acyl group yields a triacylglycerol.

The fatty acyl groups are first activated to CoA thioesters in an ATP-dependent manner:

$$\text{fatty acid} + \text{CoA} + \text{ATP} \rightleftharpoons \text{acyl-CoA} + \text{AMP} + \text{PP}_i$$

This reaction is catalyzed by acyl-CoA synthetase, the same enzyme that activates fatty acids for oxidation. Triacylglycerols are assembled as shown in **Figure 17.18**. The acyltransferases that add fatty acids to the glycerol backbone are not highly specific with respect to chain length or degree of unsaturation of the fatty acyl group, but human triacylglycerols usually contain palmitate at C1 and unsaturated oleate at C2.

The triacylglycerol biosynthetic pathway also provides the precursors for glycerophospholipids. *These amphipathic phospholipids are synthesized from phosphatidate or diacylglycerol by pathways that include an activating step in which the nucleotide cytidine triphosphate (CTP) is cleaved.* In some cases, the phospholipid head group is activated; in other cases, the lipid tail portion is activated.

Figure 17.19 shows how the head groups ethanolamine and choline are activated before being added to diacylglycerol to produce phosphatidylethanolamine and phosphatidylcholine. Similar chemistry involving nucleotide sugars is used in the synthesis of glycogen from UDP–glucose (see Section 13.3) and starch from ADP–glucose (see Section 16.3).

Phosphatidylserine is synthesized from phosphatidylethanolamine by a head-group exchange reaction in which serine displaces the ethanolamine head group:

$$HO—CH_2—CH_2—NX_3^+$$

Ethanolamine/Choline

1. *ATP phosphorylates the OH group of ethanolamine or choline (X is H in ethanolamine, and CH₃ in choline).*

ATP
ADP

$$^-O—\overset{\displaystyle O}{\underset{\displaystyle O^-}{\overset{\|}{P}}}—O—CH_2—CH_2—NX_3^+$$

Phosphoethanolamine/
Phosphocholine

2. *The phosphoryl group attacks CTP to form CDP–ethanolamine or CDP–choline. The PP$_i$ product is hydrolyzed.*

CTP
$PP_i \longrightarrow 2\ P_i$

Diacylglycerol

CMP

$$Cytidine—\overset{\displaystyle O}{\underset{\displaystyle O^-}{\overset{\|}{P}}}—O—\overset{\displaystyle O}{\underset{\displaystyle O^-}{\overset{\|}{P}}}—O—CH_2—CH_2—NX_3^+$$

CDP–ethanolamine/
CDP–choline

3. *The C3 OH group of diacylglycerol displaces CMP to generate the glycerophospholipid.*

Phosphatidylethanolamine/Phosphatidylcholine

FIGURE 17.19

Synthesis of phosphatidylethanolamine and phosphatidylcholine.

In the synthesis of phosphatidylinositol, the diacylglycerol component is activated, rather than the head group, so that the inositol head group adds to CDP–diacylglycerol (**Fig. 17.20**).

Glycerophospholipids (and sphingolipids) are components of cellular membranes. New membranes are formed by inserting proteins and lipids into preexisting membranes, mainly in the endoplasmic reticulum. The newly synthesized membrane components reach their final cellular destinations primarily via vesicles that bud off the endoplasmic reticulum and, in some cases, by diffusing at points where two membranes make physical contact. Glycerophospholipids may undergo remodeling through the action of phospholipases and acyltransferases that remove and reattach different fatty acyl groups.

CHOLESTEROL SYNTHESIS BEGINS WITH ACETYL-CoA

Cholesterol molecules, like fatty acids, are built from two-carbon acetyl units. In fact, *the first steps of cholesterol synthesis resemble those of ketogenesis.* However, ketone bodies are synthesized in the mitochondria (and only in the liver), and cholesterol is synthesized in the cytosol. The reactions of cholesterol biosynthesis and ketogenesis diverge after the production of HMG-CoA. In ketogenesis, this compound is cleaved to produce acetoacetate (see Fig. 17.16). In cholesterol synthesis, the thioester group of HMG-CoA is reduced to an alcohol, releasing the six-carbon compound mevalonate (**Fig. 17.21**).

In the next four steps of cholesterol synthesis, mevalonate acquires two phosphoryl groups and is decarboxylated to produce the five-carbon compound isopentenyl pyrophosphate:

$$CH_2{=}\overset{\displaystyle }{\underset{\displaystyle CH_3}{C}}—CH_2—CH_2—O—\overset{\displaystyle O}{\underset{\displaystyle O^-}{\overset{\|}{P}}}—O—\overset{\displaystyle O}{\underset{\displaystyle O^-}{\overset{\|}{P}}}—O^-$$

Isopentenyl pyrophosphate

FIGURE 17.20

Phosphatidylinositol synthesis.

Phosphatidate

1. Phosphatidate attacks CTP to form CDP–diacylglycerol.

$2 P_i \longleftarrow PP_i$

CDP–Diacylglycerol

Inositol

CMP

2. An inositol group displaces CMP to produce phosphatidylinositol.

Phosphatidylinositol

This isoprene derivative is the precursor of cholesterol as well as other isoprenoids, such as ubiquinone, the C_{15} farnesyl group that is attached to some lipid-linked membrane proteins, and pigments such as β-carotene. Isoprenoids are an extremely diverse group of compounds, particularly in plants, with about 25,000 characterized to date.

In cholesterol synthesis, six isoprene units condense to form the C_{30} compound squalene. Cyclization of this linear molecule leads to a structure with four rings, resembling cholesterol (**Fig. 17.22**). A total of 21 reactions are required to convert squalene to cholesterol. NADH or NADPH is required for several steps.

The rate-determining step of cholesterol synthesis (a pathway with over 30 steps) and the major control point is the conversion of HMG-CoA to mevalonate by HMG-CoA reductase. This enzyme is one of the most highly regulated enzymes known. For example, the rates of its synthesis and degradation are tightly controlled, and the enzyme is subject to inhibition by phosphorylation of a serine residue.

Synthetic inhibitors known as statins bind extremely tightly to HMG-CoA reductase, with K_I values in the nanomolar range (Section 7.3). The substrate HMG-CoA has a K_M of about 4 μM. All the statins have an HMG-like group that acts as a competitive inhibitor of HMG-CoA binding to the enzyme (**Fig. 17.23**). Their rigid hydrophobic groups also prevent the enzyme

Acetyl-CoA + Acetyl-CoA

$H-SCoA$ thiolase

Acetoacetyl-CoA

$CH_3-C-SCoA$ HMG-CoA synthase

$H-SCoA$

3-Hydroxy-3-methylglutaryl-CoA (HMG-CoA)

2 NADPH HMG-CoA reductase

$2 NADP^+ + HSCoA$

Mevalonate

FIGURE 17.21

The first steps of cholesterol biosynthesis. Note the resemblance of this pathway to ketogenesis (Fig. 17.16) through the production of HMG-CoA. HMG-CoA reductase then catalyzes a four-electron reductive deacylation to yield mevalonate.

Squalene ⟶

Folded squalene   Cholesterol

FIGURE 17.22

Conversion of squalene to cholesterol. The six isoprene units of squalene are shown in different colors. The molecule folds and undergoes cyclization. Additional reactions convert the C_{30} squalene to cholesterol, a C_{27} molecule.

from forming a structure that would accommodate the pantothenate moiety of CoA. The physiological effect of the statins is to lower serum cholesterol levels by blocking mevalonate synthesis. Cells must then obtain cholesterol from circulating lipoproteins. But since mevalonate is also the precursor of other isoprenoids such as ubiquinone, the long-term use of statins can have negative side effects.

Newly synthesized cholesterol has several fates:

1. It can be incorporated into a cell membrane.

2. It may be acylated to form a cholesteryl ester for storage or, in the liver, for packaging in VLDL.

Cholesterol   Cholesteryl ester

3. It is a precursor of steroid hormones such as testosterone and estrogen in the appropriate tissues.

FIGURE 17.23

Some statins. These inhibitors of HMG-CoA reductase have a bulky hydrophobic group plus an HMG-like group (colored red).

Lovastatin (Mevacor®)   Atorvastatin (Lipitor®)   HMG-CoA

4. It is a precursor of bile acids such as cholate:

Cholate

Bile acids are synthesized in the liver, stored in the gallbladder, and secreted into the small intestine. There, they aid digestion by acting as detergents to solubilize dietary fats and make them more susceptible to lipases. Although bile acids are mostly reabsorbed and recycled through the liver for reuse, some are excreted from the body. *This is virtually the only route for cholesterol disposal.*

Because cells do not break down cholesterol and because the accumulation of cholesterol is potentially toxic (it could disrupt membrane structure), the body must coordinate cholesterol synthesis and transport among tissues. For example, cholesterol shuts down its own synthesis by inhibiting the synthesis of enzymes such as HMG-CoA reductase. Cellular cholesterol also represses transcription of the gene for the LDL receptor.

A SUMMARY OF LIPID METABOLISM

The processes of breaking down and synthesizing lipids illustrate some general principles related to how the cell carries out opposing metabolic pathways. The diagram in **Figure 17.24** includes the major lipid metabolic pathways covered in this chapter. Several features are worth noting:

1. The pathways for fatty acid catabolism and synthesis as well as the synthesis of some other compounds all converge at the common intermediate acetyl-CoA, which is also a product of carbohydrate metabolism (see Section 14.1) and a key player in amino acid metabolism (which will be covered in Chapter 18).

2. The pathways for fatty acid degradation and fatty acid synthesis have a certain degree of symmetry, with similar intermediates and a role for thioesters, but the pathways have very different free energy considerations. β Oxidation produces reduced cofactors and requires only two ATP equivalents; fatty acid synthesis consumes NADPH and requires the input of ATP in each round. Other metabolic pathways, including cholesterol synthesis, consume reduced cofactors generated by catabolic reactions.

3. The catabolic pathway of β oxidation, the conversion of acetyl-CoA to ketone bodies, the oxidation of acetyl-CoA via the citric acid cycle, and the reoxidation of reduced cofactors occur in mitochondria (although some lipid metabolic reactions also occur in peroxisomes). In contrast, many lipid biosynthetic reactions take place in the cytosol or in association with the endoplasmic reticulum. Various pathways therefore require transmembrane transport systems and/or separate pools of substrates and cofactors.

4. Although the central pathways of lipid metabolism, as outlined in Figure 17.24, comprise just a few different reactions, complexity is introduced in the form of isozymes with different acyl-chain-length specificities; in additional enzymes to deal with odd-chain, branched, and unsaturated fatty acids; and in tissue-specific reactions leading to particular products such as eicosanoids or isoprenoids.

FIGURE 17.24

Summary of lipid metabolism.
Only the major pathways covered in this chapter are included. Open gold symbols indicate ATP consumption; filled gold symbols indicate ATP production. Open and filled red symbols represent the consumption and production of reduced cofactors (NADH, NADPH, and QH$_2$). The shaded portions of the diagram indicate reactions that occur in mitochondria.

 CONCEPT CHECK

1. What is the role of coenzyme A in triacylglycerol biosynthesis?
2. Describe how CTP can activate either diacylglycerol or a lipid head group.
3. Compare cholesterol synthesis and ketogenesis.
4. List the metabolic fates of newly synthesized cholesterol.
5. Summarize the roles of LDL and HDL in cholesterol metabolism.

SUMMARY

17.1 Lipid Transport

• Lipoproteins transport lipids, including cholesterol, in the blood-stream. High levels of LDL are associated with the development of atherosclerosis.

17.2 Fatty Acid Oxidation

• Fatty acids released from triacylglycerols by the action of lipases are activated by their attachment to CoA in an ATP-dependent reaction.

• In the process of β oxidation, a series of four enzymatic reactions degrades a fatty acyl-CoA two carbons at a time, producing one QH_2, one NADH, and one acetyl-CoA, which can be further oxidized by the citric acid cycle. Reoxidation of the reduced cofactors generates considerable ATP.

• Oxidation of unsaturated and odd-chain fatty acids requires additional enzymes. Very-long-chain and branched fatty acids are oxidized in peroxisomes.

17.3 Fatty Acid Synthesis

• Fatty acids are synthesized by a pathway that resembles the reverse of β oxidation. In the first step of fatty acid synthesis, acetyl-CoA carboxylase catalyzes an ATP-dependent reaction that converts acetyl-CoA to malonyl-CoA, which becomes the donor of two-carbon groups.

• Mammalian fatty acid synthase is a multifunctional enzyme in which the growing fatty acyl chain is attached to acyl-carrier protein rather than CoA. Elongases and desaturases may modify newly synthesized fatty acids.

• The liver can convert acetyl-CoA to ketone bodies to be used as metabolic fuels in other tissues.

17.4 Synthesis of Other Lipids

• Triacylglycerols are synthesized by attaching three fatty acyl groups to a glycerol backbone. Intermediates of the triacylglycerol pathway are the starting materials for the synthesis of phospholipids.

• Cholesterol is synthesized from acetyl-CoA. The rate-determining step of this pathway is the target of drugs known as statins.

KEY TERMS

atherosclerosis	β oxidation	essential fatty acid	bile acid
lipoprotein	peroxisome	ketone bodies	
hypercholesterolemia	multifunctional enzyme	ketogenesis	

BIOINFORMATICS

Brief Bioinformatics Exercises

17.1 Viewing and Analyzing Fatty Acid Synthase

17.2 Lipid Metabolism and the KEGG Database

Bioinformatics Project

Drug Design and Cholesterol Medications

PROBLEMS

17.1 Lipid Transport

1. Use the information in Table 17.1 to explain the density rankings of the lipoproteins.

2. What amino acid side chains of apolipoprotein A1 would be expected to be in contact with the lipid core of the lipoprotein shown in Figure 17.2?

3. Which of the lipid components of lipoproteins are amphipathic and which are completely nonpolar? **(a)** cholesterol; **(b)** cholesteryl esters; **(c)** triacylglycerols; and **(d)** phospholipids. Based on the structures of the lipid components of lipoproteins, describe how the lipids are arranged in the lipoprotein particle.

4. Physicians inform their patients of the total cholesterol concentration in the blood, but the HDL:LDL cholesterol ratio is also reported to the patient. Why is the HDL:LDL cholesterol ratio more informative? Which is more desirable, a high HDL:LDL cholesterol ratio or a low ratio?

5. Premenopausal women typically have higher HDL levels than men.

(a) Why would this tend to decrease the risk of heart disease in these women?

(b) Why is HDL level alone not a good indicator of the risk of developing heart disease?

6. Individuals who are unable to produce chylomicrons display symptoms consistent with vitamin A deficiency. Explain.

17.2 Fatty Acid Oxidation

7. (a) Biodiesel, a fuel typically derived from plant oils, can be manufactured by treating the oil with a methanol/KOH mixture to produce fatty acid methyl esters. Indicate the structures of the products generated by treating the triacylglycerol 1-palmitoyl-2,3-dioleoylglycerol with methanol/KOH.

(b) The methanol/KOH reaction described must be carried out in "dry" methanol. If any H_2O were present, what would happen?

8. The overall reaction for the activation of a fatty acid to fatty acyl-CoA, with concomitant hydrolysis of ATP to AMP, has a free energy change of about zero. The reaction is favorable because of subsequent hydrolysis of pyrophosphate to orthophosphate (the reaction has a $\Delta G^{\circ\prime}$ value of $-19.2 \text{ kJ} \cdot \text{mol}^{-1}$). **(a)** Write the equation for the coupled reaction and calculate $\Delta G^{\circ\prime}$ (assume a temperature of 25°C). **(b)** Calculate the equilibrium constant for the reaction.

9. Fatty acid activation catalyzed by acyl-CoA synthetase begins with nucleophilic attack by the negatively charged carboxylate oxygen of the fatty acid on the α-phosphate (the innermost phosphate) of ATP. An acyladenylate mixed anhydride is formed. Write the mechanism of the reaction.

10. The reactions in Figure 17.5 are all reversible. Why do the reactions tend to run in one direction (i. e., favoring the delivery of acyl-CoA to the mitochondrial matrix)?

11. Explain why one symptom of a deficiency of liver cytosolic carnitine acyltransferase (see Fig. 17.5) is hypoglycemia. What would you suggest as an effective treatment for this disorder?

12. A deficiency of carnitine results in muscle cramps, which are exacerbated by fasting or exercise. Give a biochemical explanation for the muscle cramping, and explain why cramping increases during fasting and exercise.

13. Muscle biopsy and enzyme assays of a carnitine-deficient individual show that medium-chain (C_8–C_{10}) fatty acids can be metabolized normally, despite the carnitine deficiency. What does this tell you about the role of carnitine in fatty acid transport across the inner mitochondrial membrane?

14. Some individuals suffer from medium-chain acyl-CoA dehydrogenase (MCAD) deficiency. Which intermediates accumulate in individuals with MCAD deficiency?

15. How should a patient with a medium-chain acyl-CoA dehydrogenase deficiency be treated?

16. The first three reactions of β oxidation are similar to three reactions of the citric acid cycle. Which reactions are these, and why are they similar?

17. During β oxidation, methylene (—CH_2—) groups in a fatty acid are oxidized to carbonyl (C=O) groups, yet no oxygen is consumed by the reactions of β oxidation. How is this possible?

18. The β oxidation pathway was elucidated in part by Franz Knoop in 1904. He fed dogs fatty acid phenyl derivatives and then analyzed their urine for the resulting metabolites. What metabolite was produced when the dogs were fed (a) phenylpropionate or (b) phenylbutyrate?

Phenylpropionate

Phenylbutyrate

19. A deficiency of phytanate-degrading enzymes in peroxisomes results in Refsum's disease, a neuronal disorder caused by phytanate accumulation. Patients with Refsum's disease cannot convert phytanate to pristanate because they lack the enzymes involved in the α oxidation reaction (shown below). Show how pristanate is oxidized via β oxidation, and list the products of pristanate oxidation.

Phytanate

α oxidation → CO_2

Pristanate

20. How many molecules of ATP are generated when (a) palmitate or (b) stearate is completely oxidized via β oxidation in mitochondria?

21. How many molecules of ATP are generated when (a) oleate or (b) linoleate is completely oxidized via β oxidation?

22. How many molecules of ATP are generated when a fully saturated 17-carbon fatty acid is oxidized via β oxidation?

23. How many molecules of ATP are generated when oxidation of a fully saturated C_{24} fatty acid begins in the peroxisome and is completed by the mitochondrion when 12 carbons remain?

24. A vitamin B_{12} deficiency leads to pernicious anemia. Often the disease is caused not by the lack of the vitamin itself but by the lack of a protein called intrinsic factor, which is secreted by gastric parietal cells. Intrinsic factor binds to vitamin B_{12} and facilitates its absorption into the small intestine. Use this information to devise a treatment for a patient diagnosed with pernicious anemia.

25. If you were a physician and wanted to test a patient for pernicious anemia (see Problem 24), what metabolite would you measure in the patient's blood or urine, and why?

26. Both fatty acid oxidation and glucose oxidation by glycolysis generate large amounts of ATP. Explain why a cell preparation containing all the enzymes required for either pathway cannot generate ATP when a fatty acid or glucose is added, unless a small amount of ATP is also added.

27. The complete oxidation to CO_2 of glucose and palmitate releases considerable free energy: $\Delta G^{\circ\prime} = -2850$ kJ·mol^{-1} for glucose oxidation, and $\Delta G^{\circ\prime} = -9781$ kJ·mol^{-1} for palmitate. For each fuel molecule, compare the ATP yield per carbon atom (a) in theory and (b) *in vivo*. (c) What do these results tell you about the relative efficiency of oxidizing carbohydrates and fatty acids?

17.3 Fatty Acid Synthesis

28. Compare fatty acid degradation and fatty acid synthesis by filling in the table below.

	Fatty acid degradation	Fatty acid synthesis
Cellular location		
Acyl-group carrier		
Electron carrier(s)		
ATP requirement		
Unit product/unit donor		
Configuration of hydroxyacyl intermediate		
Shortening/growth occurs at which end of the fatty acyl chain?		

29. Mice that are deficient in acetyl-CoA carboxylase are thinner than normal and exhibit continuous fatty acid oxidation. Explain these observations.

30. Write the mechanism for the carboxylation of acetyl-CoA to malonyl-CoA catalyzed by acetyl-CoA carboxylase.

31. During fatty acid synthesis, why is the condensation of an acetyl group and a malonyl group energetically favorable, whereas the condensation of two acetyl groups would be unfavorable?

32. What do acetyl-CoA carboxylase, pyruvate carboxylase, and propionyl-CoA carboxylase have in common?

33. The activity of acetyl-CoA carboxylase is regulated by hormone-controlled phosphorylation and dephosphorylation. Based on what you know about signaling via epinephrine (Section 10.2), describe the effect of epinephrine on acetyl-CoA carboxylase and fatty acid metabolism. Is this consistent with epinephrine's effect on glycogen metabolism?

34. Is the K_I (see Equation 7.30) for inhibition of acetyl-CoA carboxylase by palmitoyl-CoA higher or lower when the enzyme is phosphorylated (see Problem 33)?

35. Why is a high cytosolic citrate concentration correlated with a high rate of fatty acid synthesis?

36. Would higher or lower concentrations of citrate be required to activate acetyl-CoA carboxylase when the enzyme is phosphorylated (see Problem 33)?

37. What is the cost of synthesizing palmitate from acetyl-CoA?

38. On what carbon atoms does the $^{14}CO_2$ used to synthesize malonyl-CoA from acetyl-CoA appear in palmitate?

39. Why does triclosan inhibit bacterial fatty acid synthase but not mammalian fatty acid synthase?

40. Cancer cells have a greater-than-normal level of fatty acid synthesis, which has been attributed to enhanced expression of fatty acid synthase. Rapid fatty acid synthesis is required for tumor growth. This observation has led cancer researchers to investigate the use of fatty acid synthesis inhibitors as antitumor agents.

(a) A series of potential inhibitors of fatty acid synthase were synthesized. All the inhibitors had the structure shown in the figure; only the alkyl chain (R) varied in length. Why were these compounds effective as inhibitors of fatty acid synthase?

(b) Each compound was tested for its ability to inhibit fatty acid synthase activity in both normal cells and breast cancer cells. ID_{50} values were determined (this is the concentration of inhibitor required to inhibit cell growth by 50%). The results are shown in the table. Which inhibitors are most effective? What are the characteristics of the effective inhibitors? (*Hint:* Calculate the ratio of the ID_{50} values for normal and breast cancer cells.) An effective inhibitor must also be soluble in aqueous solution, so consider solubility as an additional factor in your answer.

Compound	Alkyl side chain (R)	Breast cancer cells ID_{50} (mg/mL)	Normal cells ID_{50} (mg/mL)
A	$-C_{13}H_{27}$	3.9	10.6
B	$-C_{11}H_{23}$	4.8	29.0
C	$-C_9H_{19}$	5.2	12.8
D	$-C_8H_{17}$	5.0	21.3
E	$-C_7H_{15}$	4.8	21.7
F	$-C_6H_{13}$	8.4	12.4

41. Many bacteria can convert saturated fatty acids to unsaturated fatty acids by a dehydration reaction that does not require O_2. Explain why this reaction might be a potential target for antibiotics.

42. As described in Box 17.B, cerulenin inhibits the production of long-chain fatty acids required for bacterial cell wall synthesis. A mycolic acid produced by *M. tuberculosis* consists of a 50-carbon β-hydroxy fatty acid with a 26-carbon α-alkyl group. Draw the abbreviated structure of this molecule.

43. A tuberculosis patient typically harbors 10^{12} *M. tuberculosis* cells. **(a)** If mutations that confer resistance to isoniazid (see Box 17.B) occur with a frequency of 1 in 10^8, approximately how many bacterial cells will be resistant to isoniazid?

(b) Assuming a similar rate for mutations that confer resistance to other antibiotics, approximately how many bacterial cells will be resistant to a combination of three drugs?

44. Draw the structures of the following fatty acids, given the shorthand form described in Problem 8.1. Which are essential for humans? **(a)** Oleate (18:1n-9); **(b)** Linoleate (18:2n-6); **(c)** α-Linolenate (18:3n-3); **(d)** Palmitoleate (16:1n-6).

45. Why is docosahexaenoic acid (DHA, 22:6n-3), a fatty acid commonly found in fish, added to baby formula?

46. Compare the carboxylation/decarboxylation sequence of reactions in gluconeogenesis (Fig. 13.10) and fatty acid synthesis. Discuss the source of free energy for these steps in each pathway.

47. Does fatty acid synthase activity increase or decrease under the following conditions? Explain. **(a)** High-carbohydrate diet (liver fatty acid synthase); **(b)** High-fat diet (liver fatty acid synthase); **(c)** Mid to late pregnancy (mammary gland fatty acid synthase).

48. Isolated heart cells undergo contraction even in the absence of glucose and fatty acids if they are supplied with acetoacetate.

(a) How does this compound act as a metabolic fuel?
(b) Even with plentiful acetoacetate, the rate of flux through the citric acid cycle gradually drops off unless pyruvate is added to the cells. Explain.

49. When glucose is unavailable, the liver begins to break down fatty acids to supply the rest of the body with metabolic fuel. Explain why fatty acid–derived acetyl-CoA is not catabolized by the citric acid cycle but is instead diverted to ketogenesis when no glucose is available.

50. Discuss the energetic costs of converting two acetyl-CoA to the ketone body 3-hydroxybutyrate in the liver and then converting the 3-hydroxybutyrate back to two acetyl-CoA in the muscle.

51. It has been said that "fats burn in the flame of carbohydrates." Give a biochemical explanation for this statement.

52. Two siblings born two years apart were separately diagnosed with a pyruvate carboxylase deficiency. Both neonates died less than a month after they were born. Blood samples taken before the death of the infants showed a high concentration of ketone bodies in the blood. Explain why the ketone body concentration was elevated.

53. A symptom of a pyruvate carboxylase deficiency (see Problem 52) is a decreased β-hydroxybutyrate:acetoacetate ratio. Explain.

54. The activity of phosphofructokinase (PFK; Section 13.1) is inhibited by long chain fatty acids. Inhibition involves the acylation of several amino acid side chains in PFK and is reversed if a thioesterase is added.

(a) What amino acid is acylated by the long chain fatty acids? Draw the structure of the linkage.
(b) Propose a hypothesis to explain why acylation inhibits PFK.
(c) What happens to the concentration of citrate under these conditions?

17.4 Synthesis of Other Lipids

55. The glycerol-3-phosphate required for the first step of triacylglycerol synthesis can be obtained either from glucose or pyruvate. Explain how this occurs. Can all cells obtain glycerol-3-phosphate from either precursor?

56. A fatty acid inhibitor called Compound D (see Problem 40) was tested for its ability to inhibit triacylglycerol synthesis in leukemia cells. The cells were incubated with ^{14}C-labeled acetate, and the amount of radioactivity in various types of cellular lipids was measured. What is your interpretation of the results shown in the graph?

57. In a site-directed mutagenesis experiment, an essential serine of HMG-CoA reductase was replaced with alanine. In normal cells, HMG-CoA reductase levels decrease when the cells are incubated with media containing LDL particles, but in cells expressing the mutant enzyme, enzyme activity did not change. What do these results indicate regarding the regulation of HMG-CoA reductase?

58. Fumonisins are mycotoxins isolated from fungi commonly found on corn and other grains. They are both toxic and carcinogenic and can cause disease in animals grazing on fungus-contaminated grain. The structure of fumonisin B_1 is shown. Note the structural similarity to sphingosine.

OR

OH OH

CH₃

CH₃ OR CH₃ OH

NH₃⁺

Fumonisin B_1

R=

COO⁻

COO⁻

Fumonisins inhibit one of the enzymes in the ceramide synthetic pathway, which is outlined below. Ceramide is an important cell-signaling molecule, and its regulation is critical to cell survival.

Palmitoyl-CoA + Serine

↓ serine palmitoyl transferase

3-Ketosphinganine

↓ 3-ketosphinganine reductase

Sphinganine

↓ ceramide synthase

Dihydroceramide

↓ desaturase

Ceramide

(a) Deduce which enzyme in the pathway is inhibited by fumonisin, given the following clues: (1) The addition of fumonisin B_1 to rat

hepatocytes almost completely inhibited ceramide synthesis. The synthesis of other phospholipids was not affected. (2) Addition of fumonisin B_1 to the cultured cells did not significantly change the rate of formation of 3-ketosphinganine. (3) There was no accumulation of 3-ketosphinganine. (4) When radioactively labeled serine was added to culture medium containing fumonisin B_1, the amount of label in sphinganine increased compared to controls.

(b) How does fumonisin inhibit the target enzyme identified in part (a)?

59. How many phosphoanhydride bonds must be cleaved to synthesize phosphatidylcholine from choline and diacylglycerol?

60. Manufacturers of cooking oil can chemically convert triacylglycerols to diacylglycerols. What kind of chemical reaction occurs and what is its purpose?

61. Cancer cells appear to increase the expression of choline kinase. Why is this enzyme necessary for phospholipid synthesis?

62. A recently developed magnetic resonance spectroscopy technique to quantitate choline and choline derivatives allowed cancer researchers to observe that tumor cells accumulate phosphocholine derivatives. In addition to greater expression of choline kinase (see Solution 61), researchers hypothesized that increased expression of phospholipases C and D could contribute to the higher concentration of phosphocholine. Explain.

63. The malaria parasite can synthesize large amounts of phosphatidylcholine even in the absence of choline. To do this, it relies on an enzyme called phosphoethanolamine methyltransferase.

(a) Describe the reaction catalyzed by this enzyme.

(b) What information would you need in order to assess whether this methyltransferase would be a suitable antimalarial drug target?

64. HMG-CoA reductase is phosphorylated by the same kinase that phosphorylates acetyl-CoA carboxylase. How does this strategy assist the cell in regulating the two biosynthetic pathways?

65. Fungi produce ergosterol rather than cholesterol. List the ways that ergosterol differs from cholesterol.

CH₃

CH₃

CH₃

CH₃

CH₃

H₃C

HO

Ergosterol

66. Cholesterol is poorly soluble in aqueous solution, yet cells must be able to sense the cholesterol level in order to regulate uptake and biosynthesis, in part by altering the expression of genes for HMG-CoA reductase and the LDL receptor. The cellular cholesterol sensors are proteins called SREBPs (sterol regulatory element binding proteins). In the absence of cholesterol, an SREBP residing in the endoplasmic reticulum is proteolytically cleaved to release a large soluble N-terminal domain that includes a structural motif found in many DNA-binding proteins.

(a) Why is it important that the SREBP be an integral membrane protein?

(b) Why is proteolysis of the SREBP required?

(c) How might the SREBP regulate the transcription of enzymes related to cholesterol metabolism?

67. Individuals with a mutation in the gene for apolipoprotein B-100 produce very low levels of this protein, which is a component of LDL.

(a) Explain why these individuals exhibit accumulation of fat in the liver.

(b) Would such individuals exhibit hypercholesterolemia or hypocholesterolemia?

SELECTED READINGS

Calabresi, L., Gomaraschi, M., Simonelli, S., Bernini, F., and Franceschini, G., HDL and atherosclerosis: Insights from inherited HDL disorders, *Biochim. Biophys. Acta* **1851,** 13–18 (2015). [Includes a summary of the roles of the different types of lipoproteins.]

Houten, S. M. and Wanders, R. J. A., A general introduction to the biochemistry of mitochondrial fatty acid β-oxidation, *J. Inherit. Metab. Dis.* **33,** 469–477 (2010). [Includes a review of the relevant chemical reaction and diseases resulting from enzyme deficiencies.]

Maier, T., Leibundgut, M., and Ban, N., The crystal structure of a mammalian fatty acid synthase, *Science* **321,** 1315–1322 (2008).

[Presents the structure of all but two domains of the multifunctional protein.]

Mizuno, Y., Jacob, R. F., and Mason, R. P., Inflammation and the development of atherosclerosis, *J. Atheroscler. Thromb.* **18,** 351–358 (2011). [Describes the pathology of atherosclerosis, the role of lipoproteins, and drug treatments.]

Zhang, Y.-M., White, S. W., and Rock, C. O., Inhibiting bacterial fatty acid synthesis, *J. Biol. Chem.* **281,** 17541–17544 (2006). [Describes each enzyme of bacterial pathway along with inhibitors.]

NITROGEN METABOLISM

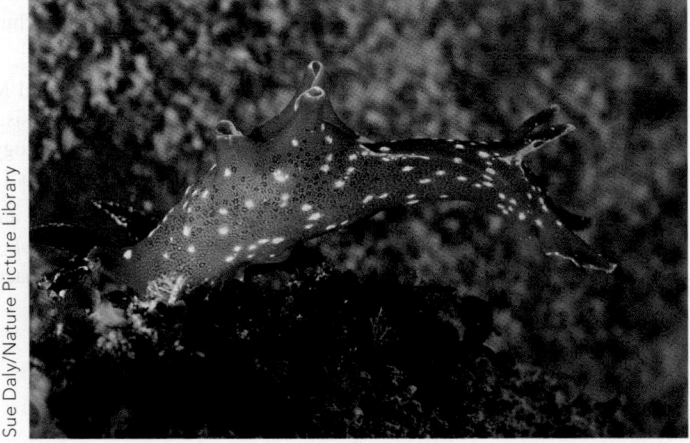

Sue Daly/Nature Picture Library

The sea hare Aplysia, *like many other shell-less mollusks, secretes a solution of "ink" when attacked. In addition to visually distracting the predator, the ink cloud contains high concentrations of amino acids that trigger feeding behavior in the misguided predator while the slow-moving* Aplysia *makes its escape unnoticed.*

DO YOU REMEMBER?

- Nucleotides consist of a purine or pyrimidine base, deoxyribose or ribose, and phosphate (Section 3.1).
- The 20 amino acids differ in the chemical characteristics of their R groups (Section 4.1).
- A few metabolites appear in several metabolic pathways (Section 12.2).
- Many vitamins, substances that humans cannot synthesize, are components of coenzymes (Section 12.2).
- The citric acid cycle supplies precursors for the synthesis of other compounds (Section 14.4).

The metabolic pathways we have examined so far are centered on carbon. In this chapter, we shift focus to another essential element, the nitrogen that is a component of amino acids and nucleotides. In addition to examining how amino groups, the most common biological form of nitrogen, are acquired and disposed of, we will survey the pathways for synthesizing and breaking down amino acids and nucleotides.

NITROGEN FIXATION AND ASSIMILATION

KEY CONCEPTS
- Nitrogen fixation by the activity of nitrogenase is part of the nitrogen cycle.
- Other enzymes incorporate amino groups into glutamine and glutamate.
- Transaminases transfer amino groups to interconvert amino acids and α-keto acids.

Approximately 80% of the air we breathe is nitrogen (N_2), but we cannot use this form of nitrogen for the synthesis of amino acids, nucleotides, and other nitrogen-containing biomolecules. Instead, we—along with most macroscopic and many microscopic life-forms—depend on the activity of a few types of microorganisms that can "fix" gaseous N_2 by transforming it into biologically useful forms. The low availability of fixed nitrogen—mainly as nitrate and ammonia—is believed to limit the biological productivity in much of the world's oceans. It also limits the growth of terrestrial organisms, which is why farmers use fertilizer (a source of fixed nitrogen, among other things) to promote crop growth.

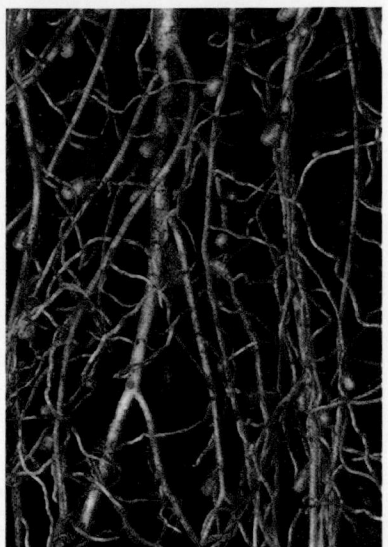

NITROGENASE CONVERTS N₂ TO NH₃

The known **nitrogen-fixing** organisms, or **diazotrophs,** include certain marine cyanobacteria and bacteria that colonize the root nodules of leguminous plants (**Fig. 18.1**). These bacteria make the enzyme nitrogenase, which carries out the energetically expensive reduction of N_2 to NH_3. Nitrogenase is a metalloprotein containing iron–sulfur centers and a cofactor with both iron and molybdenum, which resembles an elaborate Fe–S cluster (**Fig. 18.2**). The industrial fixation of nitrogen also involves metal catalysts, but this nonbiological process requires temperatures of 300 to 500°C and pressures of over 300 atm in order to break the triple bond between the two nitrogen atoms.

Biological N_2 reduction consumes large amounts of ATP and requires a strong reducing agent such as ferredoxin (see Section 16.2) to donate electrons. The net reaction is

$$N_2 + 8\,H^+ + 8\,e^- + 16\,ATP + 16\,H_2O \rightarrow 2\,NH_3 + H_2 + 16\,ADP + 16\,P_i$$

Note that eight electrons are required for the nitrogenase reaction, although N_2 reduction formally requires only six electrons; the two extra electrons are used to produce H_2. *In vivo*, the inefficiency of the reaction boosts the ATP toll to about 20 or 30 per N_2 reduced. Oxygen inactivates nitrogenase, so many nitrogen-fixing bacteria are confined to anaerobic habitats or carry out nitrogen fixation when O_2 is scarce. Some plant leaves contain nitrogen-fixing bacteria that limit their oxygen exposure by forming a biofilm (Section 11.2); these organisms may be an unrecognized source of fixed nitrogen for the plants.

Biologically useful nitrogen also originates from nitrate (NO_3^-), which is naturally present in water and soils. *Nitrate is reduced to NH_3 by plants, fungi, and many bacteria.* First, nitrate reductase catalyzes the two-electron reduction of nitrate to nitrite (NO_2^-):

$$NO_3^- + 2\,H^+ + 2\,e^- \rightarrow NO_2^- + H_2O$$

Next, nitrite reductase converts nitrite to ammonia:

$$NO_2^- + 8\,H^+ + 6\,e^- \rightarrow NH_4^+ + 2\,H_2O$$

Under physiological conditions, ammonia exists primarily in the protonated form, NH_4^+ (the ammonium ion), which has a pK of 9.25.

Nitrate is also produced by certain bacteria that oxidize NH_4^+ to NO_2^- and then NO_3^-, a process called **nitrification.** Still other organisms convert nitrate back to N_2, which is called **denitrification.** All the reactions we have discussed so far constitute the earth's **nitrogen cycle** (**Fig. 18.3**).

FIGURE 18.2

Model of the FeMo cofactor of nitrogenase. This prosthetic group in the enzyme nitrogenase consists of iron atoms (orange), sulfur atoms (yellow), and a molybdenum atom (cyan). A carbon atom (green) is liganded to six iron atoms. The manner in which N_2 interacts with the FeMo cofactor is not understood. [*Structure of the FeMo cofactor in nitrogenase (pdb 1M1N) determined by O. Einsle, F. A. Tezcan, S. L. Andrade, B. Schmid, M. Yoshida, J. B. Howard, and D. C. Rees.*]

FIGURE 18.3

The nitrogen cycle. Nitrogen fixation converts N_2 to the biologically useful NH_4^+. Nitrate can also be converted to NH_4^+. Ammonia is transformed back to N_2 by nitrification followed by denitrification.

? Indicate which processes are oxidations and which are reductions.

AMMONIA IS ASSIMILATED BY GLUTAMINE SYNTHETASE AND GLUTAMATE SYNTHASE

The enzyme glutamine synthetase is found in all organisms. In microorganisms, it is a metabolic entry point for fixed nitrogen. In animals, it helps mop up excess ammonia, which is toxic. In the first step of the reaction, ATP donates a phosphoryl group to glutamate. Then ammonia reacts with the reaction intermediate, displacing P_i to produce glutamine:

The name *synthetase* indicates that ATP is consumed in the reaction.

Glutamine, along with glutamate, is usually present in organisms at much higher concentrations than the other amino acids, which is consistent with its role as a carrier of amino groups. Not surprisingly, *the activity of glutamine synthetase is tightly regulated to maintain a supply of accessible amino groups*. For example, the dodecameric glutamine synthetase from *E. coli* is regulated allosterically and by covalent modification (**Fig. 18.4**).

The glutamine synthetase reaction that introduces fixed nitrogen (ammonia) into biological compounds requires a nitrogen-containing compound (glutamate) as a substrate. So what is the source of the nitrogen in glutamate? In bacteria and plants, the enzyme glutamate synthase catalyzes the reaction (a reaction catalyzed by a synthase does not require ATP). The net result of the glutamine synthetase and glutamate synthase reactions is

$$\alpha\text{-ketoglutarate} + NH_4^+ + NADPH + ATP \rightarrow glutamate + NADP^+ + ADP + P_i$$

In other words, *the combined action of these two enzymes assimilates fixed nitrogen (NH_4^+) into an organic compound (α-ketoglutarate, a citric acid cycle intermediate) to produce an amino acid (glutamate).* Mammals lack glutamate synthase, but glutamate concentrations are relatively high because glutamate is produced by other reactions.

TRANSAMINATION MOVES AMINO GROUPS BETWEEN COMPOUNDS

Because reduced nitrogen is so precious but free ammonia is toxic, amino groups are transferred from molecule to molecule, with glutamate often serving as an amino-group donor. We saw some of these **transamination** reactions in Section 14.4 when we examined how citric acid cycle intermediates participate in other metabolic pathways.

A transaminase (also called an aminotransferase) catalyzes the transfer of an amino group to an α-keto acid. For example,

Glutamate (amino acid) + Pyruvate (α-keto acid) ⇌ (transaminase) α-Ketoglutarate (α-keto acid) + Alanine (amino acid)

FIGURE 18.4

***E. coli* glutamine synthetase.** The 12 identical subunits of this enzyme are arranged in two stacked rings of 6 subunits (only the upper ring is visible here). The symmetrical arrangement of subunits is a general feature of enzymes that are regulated by allosteric effectors: Changes in activity at one of the active sites can be efficiently communicated to the other active sites. [*Structure (pdb 2GLS) determined by D. Eisenberg, R. J. Almassy, and M. M. Yamashita.*]

During such an amino-group transfer reaction, the amino group is transiently attached to a prosthetic group of the enzyme. This group is pyridoxal-5'-phosphate (PLP), a derivative of pyridoxine (an essential nutrient also known as vitamin B_6):

Pyridoxal-5'-phosphate (PLP)

Pyridoxine (vitamin B_6)

PLP is covalently attached to the enzyme via a Schiff base (imine) linkage to the ε-amino group of a lysine residue. The amino acid substrate of the transaminase displaces this Lys amino group, which then acts as an acid–base catalyst. The steps of the reaction are diagrammed in **Figure 18.5**.

The transamination reaction is freely reversible, so transaminases participate in pathways for amino acid synthesis as well as degradation. Note that if the α-keto acid produced in step 4 reenters the active site, then the amino group that was removed from the starting amino acid is restored. However, most transaminases accept only α-ketoglutarate or oxaloacetate as the α-keto acid substrate for the second part of the reaction (steps 5 to 7).

Enzyme–PLP Schiff base

1. The α-amino group of an amino acid attacks the enzyme–PLP Schiff base. This transamination reaction forms an amino acid–PLP Schiff base and releases the enzyme's Lys ε-amino group.

2. The Lys amino group, acting as a base, removes the hydrogen from the substrate amino acid's α carbon. The negative charge of the resulting carbanion is stabilized by the PLP group, which acts as an electron sink.

3. The protonated Lys residue, now acting as an acid, donates the proton to the PLP group, generating a ketimine. The molecular rearrangement resulting from the movement of an H atom is known as tautomerization.

4. Hydrolysis frees the α-keto acid and leaves the amino group bound to the PLP group.

5. Another α-keto acid enters the active site to reform a ketimine (this is the reverse of step 4).

6. Lysine-catalyzed tautomerization yields an amino acid–Schiff base (the reverse of steps 2 and 3).

7. In a transamination reaction, the ε-amino group of the Lys residue displaces the amino acid and regenerates the enzyme–PLP Schiff base (the reverse of step 1).

FIGURE 18.5

PLP-catalyzed transamination.

Box 18.A Transaminases in the Clinic

Assays of transaminase activity in the blood are the basis of the widely used clinical measurements known as AST (aspartate aminotransferase; also known as serum glutamate–oxaloacetate transaminase, or SGOT) and ALT (alanine transaminase; also known as serum glutamate–pyruvate transaminase, or SGPT). In clinical lab tests, blood samples are added to a mixture of the enzymes' substrates. The reaction products, whose concentrations are proportional to the amount of enzyme present, are then detected by secondary reactions that generate colored products easily quantified by spectrophotometry. Prepackaged kits give reliable results in a matter of minutes.

The concentration of AST in the blood increases after a heart attack, when damaged heart muscle leaks its intracellular contents. Typically, AST concentrations rise in the first hours after a heart attack, peak in 24 to 36 hours, and return to normal within a few days. However, since many tissues contain AST, monitoring cardiac muscle damage more commonly relies on measurements of cardiac troponins (proteins specific to heart muscle). ALT is primarily a liver enzyme, so it is useful as a marker of liver damage resulting from infection, trauma, or chronic alcohol abuse. Certain drugs, including the cholesterol-lowering statins (Section 17.4), sometimes increase AST and ALT levels to such an extent that the drugs must be discontinued.

? **Identify the substrates and products for the AST and ALT reactions.**

This means that most transaminases generate glutamate or aspartate. Lysine is the only amino acid that cannot be transaminated. The presence of transaminases in muscle and liver cells makes them useful markers of tissue damage (**Box 18.A**).

◆ CONCEPT CHECK

1. What does the nitrogenase reaction accomplish?

2. List the processes that can generate ammonia.

3. Write equations for the reactions catalyzed by glutamine synthetase and glutamate synthase.

4. Describe the function of the PLP cofactor.

5. Explain why transaminases catalyze reversible reactions.

6. Draw a diagram to trace a nitrogen atom from N_2 to the amino group of threonine.

18.2 AMINO ACID BIOSYNTHESIS

Amino acids are synthesized from intermediates of glycolysis, the citric acid cycle, and the pentose phosphate pathway. Their amino groups are derived from the nitrogen carrier molecules glutamate and glutamine. Using the metabolic scheme introduced in Chapter 12, we can show how amino acid biosynthesis and other reactions of nitrogen metabolism are related to the other pathways we have examined (**Fig. 18.6**).

Humans can synthesize only some of the 20 amino acids that are commonly found in proteins. These are known as **nonessential** amino acids. The other amino acids are said to be **essential** because humans cannot synthesize them and must obtain them from their food. The ultimate sources of the essential amino acids are plants and microorganisms, which produce all the enzymes necessary to undertake the synthesis of these compounds. The essential and nonessential amino acids for humans are listed in **Table 18.1**. This classification scheme can be somewhat confusing. For example, some nonessential amino acids, such as arginine, may be essential for young children; that is, dietary sources must supplement what the body can produce on its own. Human cells cannot synthesize histidine, so it is classified as an essential amino acid, even though a dietary requirement has never been defined (probably because sufficient quantities are naturally supplied by intestinal microorganisms). Tyrosine can be considered essential in that it is synthesized directly from the essential amino acid phenylalanine. Likewise, cysteine synthesis depends on the availability of sulfur provided by the essential amino acid methionine.

KEY CONCEPTS

- Alanine, arginine, asparagine, aspartate, glutamate, glutamine, glycine, proline, and serine are synthesized from intermediates of glycolysis and the citric acid cycle.

- Bacteria and plants synthesize amino acids with sulfur (cysteine and methionine), branched chains (isoleucine, leucine, and valine), and aromatic groups (phenylalanine, tryptophan, and tyrosine) as well as histidine, lysine, and threonine.

- Glutamate and tyrosine are modified to generate neurotransmitters and hormones.

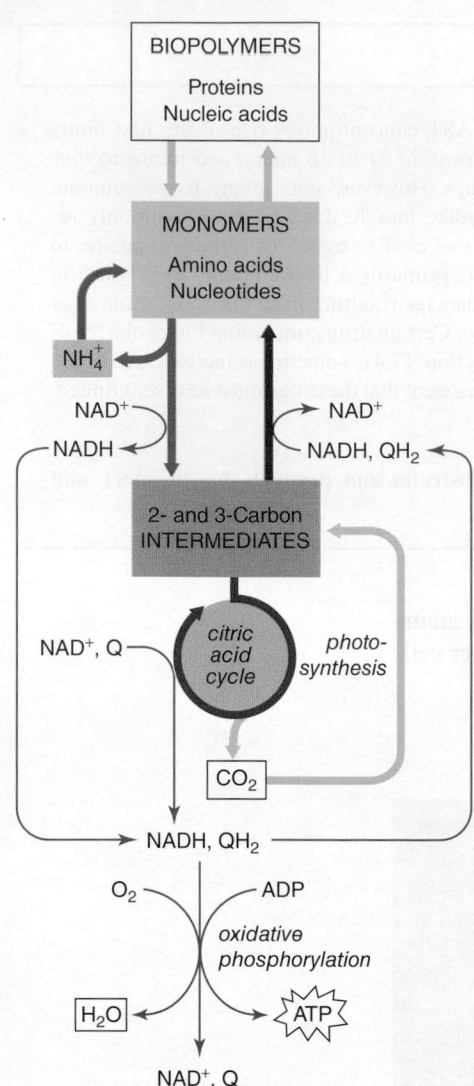

FIGURE 18.6

Nitrogen metabolism in context.
Amino acids are synthesized
mostly from three-carbon
intermediates of glycolysis
and from intermediates of the
citric acid cycle. Amino acid
catabolism yields some of the
same intermediates, as well as
the two-carbon acetyl-CoA. Amino
acids are also the precursors
of nucleotides. Both types of
molecules contain nitrogen,
so a discussion of amino acid
metabolism includes pathways for
obtaining, using, and disposing of
amino groups.

TABLE 18.1 Essential and Nonessential Amino Acids

ESSENTIAL	NONESSENTIAL
Histidine	Alanine
Isoleucine	Arginine
Leucine	Asparagine
Lysine	Aspartate
Methionine	Cysteine
Phenylalanine	Glutamate
Threonine	Glutamine
Tryptophan	Glycine
Valine	Proline
	Serine
	Tyrosine

SEVERAL AMINO ACIDS ARE EASILY SYNTHESIZED FROM COMMON METABOLITES

We have already seen that *some amino acids can be produced by transamination reactions*. In this way, alanine is produced from pyruvate, aspartate from oxaloacetate, and glutamate from α-ketoglutarate. Glutamine synthetase catalyzes the amidation of glutamate to produce glutamine. Asparagine synthetase, which uses glutamine as an amino-group donor rather than ammonia, converts aspartate to asparagine:

To summarize, three common metabolic intermediates (pyruvate, oxaloacetate, and α-ketoglutarate) give rise to five nonessential amino acids by simple transamination and amidation reactions.

Slightly longer pathways convert glutamate to proline and arginine, which each have the same five-carbon core:

Serine is derived from the glycolytic intermediate 3-phosphoglycerate in three steps:

3-Phosphoglycerate   3-Phospho-hydroxypyruvate   3-Phosphoserine   Serine

Serine, a three-carbon amino acid, gives rise to the two-carbon glycine in a reaction catalyzed by serine hydroxymethyltransferase (the reverse reaction converts glycine to serine). This enzyme uses a PLP-dependent mechanism to remove the hydroxymethyl (—CH_2OH) group attached to the α carbon of serine; this one-carbon fragment is then transferred to the cofactor tetrahydrofolate:

Serine   Glycine

Tetrahydrofolate functions as a carrier of one-carbon units in several reactions of amino acid and nucleotide metabolism (**Fig. 18.7**). Mammals cannot synthesize folate (the oxidized form of tetrahydrofolate) and must therefore obtain it as a vitamin from their diet. Folate is abundant in foods such as fortified cereal, fruits, and vegetables. The requirement for folate increases during the first few weeks of pregnancy, when the fetal nervous system begins to develop. Supplemental folate appears to prevent certain neural tube defects such as spina bifida, in which the spinal cord remains exposed.

AMINO ACIDS WITH SULFUR, BRANCHED CHAINS, OR AROMATIC GROUPS ARE MORE DIFFICULT TO SYNTHESIZE

We have just described how a few metabolites—pyruvate, 3-phosphoglycerate, oxaloacetate, and α-ketoglutarate—are converted in a few enzyme-catalyzed steps to nine different amino acids. Synthesis of the other amino acids (the essential amino acids and those derived directly from them) also begins with common metabolites. However, these biosynthetic pathways tend to be more complicated. At some point in their evolution, animals lost the ability to synthesize these amino acids, probably because the pathways were energetically expensive and the compounds were already available in food. In general, humans cannot synthesize branched-chain amino acids or aromatic amino acids and cannot incorporate sulfur into compounds such as methionine. In this section, we will focus on a few interesting points related to the synthesis of essential amino acids.

(a)

2-Amino-4-oxo-6-methylpterin   *p*-Aminobenzoate   Glutamates ($n = 1$–6)

Tetrahydrofolate

(b)

N^5,N^{10}-Methylenetetrahydrofolate

FIGURE 18.7

Tetrahydrofolate. (a) This cofactor consists of a pterin derivative, a *p*-aminobenzoate residue, and up to six glutamate residues. It is a reduced form of the vitamin folate. The four H atoms of the tetrahydro form are colored red. (b) In the conversion of serine to glycine, a methylene group (blue) becomes attached to both N5 and N10 of tetrahydrofolate. Tetrahydrofolate can carry carbon units of different oxidation states. For example, a methyl group can attach to N5, and a formyl group (—HCO) can attach at N5 or N10.

The bacterial pathway for producing sulfur-containing amino acids begins with serine and uses sulfur that comes from inorganic sulfide:

Serine → O-Acetylserine → Cysteine

Cysteine can then donate its sulfur atom to a four-carbon compound derived from aspartate, forming the nonstandard amino acid homocysteine. The final step of methionine synthesis is catalyzed by methionine synthase, which adds to homocysteine a methyl group carried by tetrahydrofolate:

Homocysteine → Methionine

In humans, serine reacts with homocysteine to yield cysteine:

Serine + Homocysteine → Cysteine + α-Ketobutyrate

This pathway is the reason why cysteine is considered a nonessential amino acid, although its sulfur atom must ultimately come from methionine, an essential amino acid.

High levels of homocysteine in the blood are associated with cardiovascular disease. The link was first discovered in individuals with homocystinuria, a disorder in which excess homocysteine is excreted in the urine. These individuals develop atherosclerosis as children, probably because the homocysteine directly damages the walls of blood vessels even in the absence of elevated LDL levels (see Section 17.1). Increasing the intake of folate, the vitamin precursor of tetrahydrofolate, helps decrease the level of homocysteine by promoting its conversion to methionine.

Aspartate, the precursor of methionine, is also the precursor of the essential amino acids threonine and lysine. Since these amino acids are derived from another amino acid, they already have an amino group. The branched-chain amino acids (valine, leucine, and isoleucine) are synthesized by pathways that use pyruvate as the starting substrate. These amino acids require a step catalyzed by a transaminase (with glutamate as a substrate) to introduce an amino group.

In plants and bacteria, the pathway for synthesizing the aromatic amino acids (phenylalanine, tyrosine, and tryptophan) begins with the condensation of the C_3 compound phosphoenolpyruvate (a glycolytic intermediate) and erythrose-4-phosphate (a four-carbon intermediate of the pentose phosphate pathway). The seven-carbon reaction product then cyclizes and undergoes additional modifications, including the addition of three more carbons from phosphoenolpyruvate, before becoming chorismate, the last common intermediate in the synthesis of the three aromatic amino acids. *Because animals do not synthesize chorismate, this pathway is an obvious target for agents that can inhibit plant metabolism without affecting animals* (**Box 18.B**).

A CLOSER LOOK

Box 18.B Glyphosate, the Most Popular Herbicide

Glycine phosphonate, also known as glyphosate or Roundup (its trade name), competes with the second phosphoenolpyruvate in the pathway leading to chorismate:

Phosphoenolpyruvate   Glyphosate

Because plants cannot manufacture aromatic amino acids without chorismate, glyphosate acts as an herbicide. Used widely in agriculture as well as home gardens, it has become the most popular herbicide in the United States, replacing other, more toxic compounds. Glyphosate that is not directly absorbed by the plant appears to bind tightly to soil particles and then is rapidly broken down by bacteria. Consequently, glyphosate has less potential to contaminate water supplies than do more stable compounds.

Farmers can take advantage of glyphosate's weed-killing properties by planting glyphosate-resistant crops and then spraying the field with glyphosate when weeds emerge and begin to compete with the crop plants. Such "Roundup-Ready" species include soybeans, corn (maize), and cotton. These plants have been genetically engineered to express a bacterial version of the enzyme that uses phosphoenolpyruvate but is not inhibited by glyphosate. Predictably, the use of glyphosate selects for herbicide resistance, so many types of weeds have already evolved resistance to glyphosate.

? **In addition to plants, what other types of organisms synthesize chorismate as a precursor of aromatic amino acids? How would glyphosate affect them?**

Phosphoenolpyruvate

Erythrose-4-phosphate

Chorismate

Phenylalanine and tyrosine are derived from chorismate by diverging pathways. In humans, tyrosine is generated by hydroxylating phenylalanine, which is why tyrosine is not considered an essential amino acid.

The final two reactions of the tryptophan biosynthetic pathway (which has 13 steps altogether) are catalyzed by tryptophan synthase, a bifunctional enzyme with an $\alpha_2\beta_2$ quaternary structure. The α subunit cleaves indole-3-glycerol phosphate to indole and glyceraldehyde-3-phosphate, then the β subunit adds serine to indole to produce tryptophan:

Phenylalanine   Tyrosine

Indole-3-glycerol phosphate   Indole   Tryptophan

FIGURE 18.8

Tryptophan synthase. Only one α subunit (blue and tan) and one β subunit (yellow, orange, and tan) are shown. Indolepropanol phosphate (IPP; red) marks the active site of the α subunit. The β active site is marked by its PLP cofactor (yellow). The surface of the tunnel between the two active sites is outlined with yellow dots. Several indole molecules (green) are included in the model to show how this intermediate can pass between the active sites. [*Courtesy of Craig Hyde, National Institutes of Health.*]

Indole, the product of the α-subunit reaction and the substrate for the β-subunit reaction, never leaves the enzyme. Instead, it diffuses directly from one active site to the other without entering the surrounding solvent. The X-ray structure of the enzyme reveals that the active sites in adjacent α and β subunits are 25 Å apart but are connected by a tunnel through the protein that is large enough to accommodate indole (**Fig. 18.8**). *The movement of a reactant between two active sites is called* **channeling,** *and it increases the rate of a metabolic process by preventing the loss of intermediates.* Channeling is known to occur in a few other multifunctional enzymes.

All but one of the 20 standard amino acids are synthesized entirely from precursors produced by the main carbohydrate-metabolizing pathways. The exception is histidine, to which ATP provides one nitrogen and one carbon atom. Glutamate and glutamine donate the other two nitrogen atoms, and the remaining five carbons are derived from a phosphorylated monosaccharide, 5-phosphoribosyl pyrophosphate (PRPP):

5-Phosphoribosyl pyrophosphate is also the source of the ribose group of nucleotides. This suggests that histidine might have been one of the first amino acids synthesized by an early life-form making the transition from an all-RNA metabolism to an RNA-and-protein-based metabolism.

AMINO ACIDS ARE THE PRECURSORS OF SOME SIGNALING MOLECULES

Many amino acids that are ingested or built from scratch are used to manufacture a cell's proteins, but some also have essential functions as precursors of other compounds, including **neurotransmitters.** Communication in the complex neuronal circuitry of the nervous system relies on small chemical signals that are released by one neuron and taken up by another (see Section 9.4). Common neurotransmitters include the amino acids glycine and glutamate and a glutamate derivative (its carboxylate group has been removed) known as γ-aminobutyric acid (GABA) or γ-aminobutyrate.

$$
\overset{+}{H_3N}-CH_2 \\
| \\
CH_2 \\
| \\
CH_2 \\
| \\
COO^-
$$

γ-Aminobutyrate

Several other amino acid derivatives also function as neurotransmitters. For example, tyrosine gives rise to dopamine, norepinephrine, and epinephrine. These compounds are called catecholamines, reflecting their resemblance to catechol.

Tyrosine · Dopamine · Norepinephrine · Epinephrine · Catechol

A deficiency of dopamine produces the symptoms of Parkinson's disease: tremor, rigidity, and slow movements. As we saw in Section 10.2, catecholamines are also produced by other tissues and function as hormones.

Tryptophan is the precursor of the neurotransmitter serotonin:

Tryptophan · Serotonin

Low levels of serotonin in the brain have been linked to conditions such as depression, aggression, and hyperactivity. The antidepressive effect of drugs such as Prozac® results from their ability to increase serotonin levels by blocking the reabsorption of the released neurotransmitter (see Box 9.B). Serotonin is the precursor of melatonin. This tryptophan derivative is synthesized in the pineal gland and retina. Its concentration is low during the day, rising during darkness. Because melatonin appears to govern the synthesis of some other neurotransmitters that control circadian (daily) rhythms, it has been touted as a cure for sleep disorders and jet lag.

Arginine is also the precursor of a signaling molecule that was discovered to be the free radical gas nitric oxide (NO; **Box 18.C**).

Melatonin

Box 18.C Nitric Oxide

In the 1980s, vascular biologists were investigating the nature of an endothelial cell–derived "relaxation factor" that caused blood vessels to dilate. This substance diffused quickly, acted locally, and disappeared within seconds. To the surprise of many, the mysterious factor turned out to be the free radical nitric oxide (·NO). Although NO was known to elicit vasodilation, it had not been considered a good candidate for a biological signaling molecule because its unpaired electron makes it extremely reactive and it breaks down to yield the corrosive nitric acid.

NO is a signaling molecule in a wide array of tissues. At low concentrations it induces blood vessel dilation; at high concentrations (along with oxygen radicals) it kills pathogens. NO is synthesized from arginine by nitric oxide synthase, an enzyme whose cofactors include FMN, FAD, tetrahydrobiopterin (discussed in Section 18.3), and a heme group. The first step of NO production is a hydroxylation reaction. In the second step, one electron oxidizes N-hydroxyarginine.

NO is unusual among signaling molecules for several reasons: It cannot be stockpiled for later release; it diffuses into cells, so it does not need a cell-surface receptor; and it needs no

degradative enzyme because it breaks down on its own. NO is produced only when and where it is needed. A free radical gas such as NO cannot be directly introduced into the body, but an indirect source of NO has been clinically used for over a century. Individuals who suffer from angina pectoris, a painful condition caused by obstruction of the coronary blood vessels, can relieve their symptoms by taking nitroglycerin:

$$CH_2 - CH - CH_2$$
$$\ \ |\qquad\ |\qquad\ |$$
$$\ \ O\qquad O\qquad O$$
$$\ \ |\qquad\ |\qquad\ |$$
$$NO_2\quad NO_2\quad NO_2$$

Nitroglycerin

In vivo, nitroglycerin yields NO, which rapidly stimulates vasodilation, temporarily relieving the symptoms of angina.

? **Explain why blood vessels express constant amounts of nitric oxide synthase whereas white blood cells must be induced to produce the enzyme.**

COO⁻ structures — Arginine, N-Hydroxyarginine, Citrulline:

$$\begin{array}{c}
COO^- \\
| \\
H_3\overset{+}{N}-C-H \\
| \\
CH_2 \\
| \\
CH_2 \\
| \\
CH_2 \\
| \\
NH \\
| \\
C \\
H_2N \diagup \ \diagdown \overset{+}{N}H_2
\end{array}
\quad
\xrightarrow[\begin{array}{c}O_2\ \ \ \ \ H_2O\end{array}]{\begin{array}{c}NADPH\ \ NADP^+\end{array}}
\quad
\begin{array}{c}
COO^- \\
| \\
H_3\overset{+}{N}-C-H \\
| \\
CH_2 \\
| \\
CH_2 \\
| \\
CH_2 \\
| \\
NH \\
| \\
C \\
H_2N \diagup \ \diagdown N-OH
\end{array}
\quad
\xrightarrow[\begin{array}{c}O_2\ \ \ \ \ H_2O\end{array}]{\begin{array}{c}\frac{1}{2}NADPH\ \ \frac{1}{2}NADP^+\end{array}}
\quad
\begin{array}{c}
COO^- \\
| \\
H_3\overset{+}{N}-C-H \\
| \\
CH_2 \\
| \\
CH_2 \\
| \\
CH_2 \\
| \\
NH \\
| \\
C \\
H_2N \diagup \ \diagdown O
\end{array}
\ +\ \cdot NO$$

Arginine N-Hydroxyarginine   Citrulline

CONCEPT CHECK

1. List the metabolites that are used as precursors for the nonessential amino acids.
2. Make a list of amino acids that cannot be synthesized by humans.
3. Identify the amino acids that are synthesized by simple transamination reactions.
4. Describe the role of tetrahydrofolate in amino acid biosynthesis.
5. Explain how histidine synthesis differs from the synthesis of other amino acids.
6. List some amino acids that give rise to neurotransmitters and signaling molecules.
7. Why do herbicides target the pathway for synthesizing aromatic amino acids?

KEY CONCEPT
- Degradation of the carbon skeletons of amino acids produces acetyl-CoA and precursors for gluconeogenesis.

18.3 AMINO ACID CATABOLISM

Like monosaccharides and fatty acids, *amino acids are metabolic fuels that can be broken down to release free energy*. In fact, amino acids, not glucose, are the major fuel for the cells lining the small intestine. These cells absorb dietary amino acids and break down almost all of the available glutamate and aspartate and a good portion of the glutamine supply (note that these are all nonessential amino acids).

Other tissues, mainly the liver, also catabolize amino acids originating from the diet and from the normal turnover of intracellular proteins. During periods when dietary amino acids are not available, such as during a prolonged fast, amino acids are mobilized through the breakdown of muscle tissue, which accounts for about 40% of the total protein in the body. The amino acids undergo transamination reactions to remove their α-amino groups, and their carbon skeletons then enter the central pathways of energy metabolism (principally the citric acid cycle). However, *the catabolism of amino acids in the liver is not complete*. There is simply not enough oxygen available for the liver to completely oxidize all the carbon to CO_2. And even if there were, the liver would not need all the ATP that would be produced as a result. Instead, the amino acids are partially oxidized to substrates for gluconeogenesis (or ketogenesis). Glucose can then be exported to other tissues or stored as glycogen.

The reactions of amino acid catabolism, like those of amino acid synthesis, are too numerous to describe in full here, and the catabolic pathways do not necessarily mirror the anabolic pathways, as they do in carbohydrate and fatty acid metabolism. In this section, we will focus on some general principles and a few interesting chemical aspects of amino acid catabolism. In the following section we will see how organisms dispose of the nitrogen component of catabolized amino acids.

AMINO ACIDS ARE GLUCOGENIC, KETOGENIC, OR BOTH

It is useful to classify amino acids in humans as **glucogenic** (giving rise to gluconeogenic precursors such as citric acid cycle intermediates) or **ketogenic** (giving rise to acetyl-CoA, which can be used for ketogenesis or fatty acid synthesis, but not gluconeogenesis). As shown in **Table 18.2**, *all but leucine and lysine are at least partly glucogenic, most of the nonessential amino acids are glucogenic, and the large skeletons of the aromatic amino acids are both glucogenic and ketogenic*.

Three amino acids are converted to gluconeogenic substrates by simple transamination (the reverse of their biosynthetic reactions): alanine to pyruvate, aspartate to oxaloacetate, and glutamate to α-ketoglutarate. Glutamate can also be **deaminated** in an oxidation reaction

TABLE 18.2 Catabolic Fates of Amino Acids

GLUCOGENIC	BOTH GLUCOGENIC AND KETOGENIC	KETOGENIC
Alanine	Isoleucine	Leucine
Arginine	Phenylalanine	Lysine
Asparagine	Threonine	
Aspartate	Tryptophan	
Cysteine	Tyrosine	
Glutamate		
Glutamine		
Glycine		
Histidine		
Methionine		
Proline		
Serine		
Valine		

that we will examine in the following section. Asparagine undergoes a simple hydrolytic deamidation to aspartate, which is then transaminated to oxaloacetate:

Asparagine → (H₂O, NH₄⁺, asparaginase) → Aspartate → (α-Ketoglutarate, Glutamate, transaminase) → Oxaloacetate

Similarly, glutamine is deamidated by a glutaminase to glutamate, and the glutamate dehydrogenase reaction yields α-ketoglutarate. Serine is converted to pyruvate:

Serine → (NH₄⁺) → Pyruvate

Note that in this reaction and in the conversion of asparagine and glutamine to their acid counterparts, the amino group is released as NH_4^+ rather than being transferred to another compound.

Arginine and proline (which are synthesized from glutamate) as well as histidine are catabolized to glutamate, which is then converted to α-ketoglutarate. Amino acids of the glutamate "family," namely arginine, glutamine, histidine, and proline, constitute about 25% of dietary amino acids, so their potential contribution to energy metabolism is significant.

Cysteine is converted to pyruvate by a process that releases ammonia as well as sulfur:

Cysteine → (H₂O, H₂S, NH₄⁺) → Pyruvate

The products of the reactions listed so far—pyruvate, oxaloacetate, and α-ketoglutarate—are all gluconeogenic precursors. Threonine is both glucogenic and ketogenic because it is broken down to acetyl-CoA and glycine:

Threonine → (NAD⁺, NADH + H⁺) → α-Amino-β-ketobutyrate → (CoASH) → Glycine + Acetyl-CoA

The acetyl-CoA is a precursor of ketone bodies (see Section 17.3), and the glycine is potentially glucogenic—if it is first converted to serine by the action of serine hydroxymethyltransferase. The major route for glycine disposal, however, is catalyzed by a multiprotein complex known as the glycine cleavage system:

Glycine + tetrahydrofolate → (NAD⁺, NADH, glycine cleavage system) → methylene-tetrahydrofolate + NH_4^+ + CO_2

FIGURE 18.9

The initial steps of valine degradation.

❓ **List all the cofactors involved in this process. (*Hint:* Step 2 is carried out by a multienzyme complex.)**

The degradation pathways for the remaining amino acids are more complicated. For example, the branched-chain amino acids—valine, leucine, and isoleucine—undergo transamination to their α-keto-acid forms and are then linked to coenzyme A in an oxidative decarboxylation reaction. This step is catalyzed by the branched-chain α-keto-acid dehydrogenase complex, a multienzyme complex that resembles the pyruvate dehydrogenase complex (see Section 14.1) and even shares some of the same subunits.

The initial reactions of valine catabolism are shown in **Figure 18.9**. Subsequent steps yield the citric acid cycle intermediate succinyl-CoA. Isoleucine is degraded by a similar pathway that yields succinyl-CoA and acetyl-CoA. Leucine degradation yields acetyl-CoA and the ketone body acetoacetate. Lysine degradation, which follows a different pathway from the branched-chain amino acids, also yields acetyl-CoA and acetoacetate. The degradation of methionine produces succinyl-CoA.

Finally, the cleavage of the aromatic amino acids—phenylalanine, tyrosine, and tryptophan—yields the ketone body acetoacetate as well as a glucogenic compound (alanine or fumarate). The first step of phenylalanine degradation is a hydroxylation reaction that produces tyrosine, as we have already seen. This reaction is worth noting because it uses the cofactor tetrahydrobiopterin (which, like folate, contains a pterin group).

Box 18.D Diseases of Amino Acid Metabolism

Inherited diseases result from defective genes, and malfunctioning genes underlie many noninherited diseases as well. The link between genes and disease was first recognized by the physician Archibald Garrod, who coined the term *inborn error of metabolism* in 1902. Garrod's insights came from his studies of individuals with alcaptonuria. Their urine turned black upon exposure to air because it contained homogentisate, a product of tyrosine catabolism. Garrod concluded that this inherited condition resulted from the lack of a specific enzyme. We now know that homogentisate is excreted because the enzyme that breaks it down, homogentisate dioxygenase, is missing or defective.

Garrod also described a number of other inborn errors of metabolism, including albinism, cystinuria (excretion of cysteine in the urine), and several other disorders that were not life-threatening and left easily detected clues in the patient's urine. Of course, many "inborn errors" are catastrophic. For example, phenylketonuria (PKU) results from a deficiency of phenylalanine hydroxylase (see the diagram). Without this enzyme, phenylalanine cannot be broken down, although it can

undergo transamination. The resulting α-keto-acid derivative phenylpyruvate accumulates and is excreted in the urine, giving it a mousy odor. If not treated, PKU causes mental retardation. Fortunately, the disease can be detected in newborns. Afflicted individuals develop normally if they consume a diet that is low in phenylalanine.

Maple syrup urine disease results from a deficiency of the branched-chain α-keto-acid dehydrogenase complex. The deaminated forms of the branched-chain amino acids accumulate and are excreted in the urine, giving it a smell like maple syrup. Like PKU, this disorder can be treated by a controlled diet.

Defects in the glycine cleavage system cause hyperglycinemia and lead to severe neurological abnormalities, perhaps due to glycine's role as a neurotransmitter. It cannot be treated by dietary adjustments.

? **Individuals with PKU must consume a certain amount of phenylalanine. Describe the three metabolic fates of this amino acid.**

Phenylalanine → (phenylalanine hydroxylase, Missing in phenylketonuria) → Tyrosine → (transaminase) → p-Hydroxyphenylpyruvate → (p-hydroxyphenylpyruvate dioxygenase) → Homogentisate → (homogentisate dioxygenase, Missing in alcaptonuria) → 4-Maleylacetoacetate

The tetrahydrobiopterin is oxidized to dihydrobiopterin in the phenylalanine hydroxylase reaction. This cofactor must be subsequently reduced to the tetrahydro form by a separate NADH-dependent enzyme. Another step of the phenylalanine (and tyrosine) degradation pathway is also notable because a deficiency of the enzyme was one of the first-characterized metabolic diseases (**Box 18.D**).

CONCEPT CHECK

1. List the end-products of catabolism of the 20 amino acids.
2. What is the role of the liver in catabolizing amino acids?
3. List the cofactors involved in amino acid breakdown.
4. Distinguish the glucogenic and ketogenic amino acids.

18.4 NITROGEN DISPOSAL: THE UREA CYCLE

When the supply of amino acids exceeds the cell's immediate needs for protein synthesis or other amino acid–consuming pathways, the carbon skeletons are broken down and the nitrogen disposed of. All amino acids except lysine can be deaminated by the action of transaminases, but this merely transfers the amino group to another molecule; it does not eliminate it from the body.

Some catabolic reactions do release free ammonia, which can be excreted as a waste product in the urine. In fact, the kidney is a major site of glutamine catabolism, and the resulting NH_4^+ facilitates the excretion of metabolic acids such as H_2SO_4 that arise from the catabolism of methionine and cysteine. However, ammonia production is not feasible for disposing of large amounts of excess nitrogen. First, high concentrations of NH_4^+ in the blood cause alkalosis. Second, ammonia is highly toxic. It easily enters the brain, where it activates the NMDA receptor, whose normal agonist is the neurotransmitter glutamate. The activated receptor is an ion channel that normally opens to allow Ca^{2+} and Na^+ ions to enter the cell and K^+ ions to exit the cell. However, the large Ca^{2+} influx triggered by ammonia binding to the receptor results in neuronal cell death, a phenomenon called excitotoxicity. Humans and many other organisms have therefore evolved safer ways to deal with excess amino groups.

Approximately 80% of the body's excess nitrogen is excreted in the form of urea,

$$H_2N-\overset{\overset{\displaystyle O}{\|}}{C}-NH_2$$

Urea

*which is produced in the liver by the reactions of the **urea cycle**.* This catabolic cycle was elucidated in 1932 by Hans Krebs and Kurt Henseleit; Krebs went on to outline another circular pathway—the citric acid cycle—in 1937.

GLUTAMATE SUPPLIES NITROGEN TO THE UREA CYCLE

Because many transaminases use α-ketoglutarate as the amino-group acceptor, glutamate is one of the most abundant amino acids inside cells. Glutamate can be deaminated to regenerate α-ketoglutarate and release NH_4^+ in an oxidation–reduction reaction catalyzed by glutamate dehydrogenase:

$$^-OOC-CH_2-CH_2-\overset{\overset{\displaystyle NH_3^+}{|}}{\underset{\underset{\displaystyle H}{|}}{C}}-COO^-$$

Glutamate

$$\xrightarrow[\quad NAD(P)^+ \quad NAD(P)H + H^+ \quad]{}$$

$$\left[^-OOC-CH_2-CH_2-\overset{\overset{\displaystyle NH_2^+}{\|}}{C}-COO^- \right]$$

$$\xrightarrow[\quad H_2O \quad NH_4^+ \quad]{}$$

$$^-OOC-CH_2-CH_2-\overset{\overset{\displaystyle O}{\|}}{C}-COO^-$$

α-Ketoglutarate

This mitochondrial enzyme is unusual: It is one of only a few enzymes that can use either NAD^+ or $NADP^+$ as a cofactor. The glutamate dehydrogenase reaction is a major route for feeding amino acid–derived amino groups into the urea cycle and, not surprisingly, is subject to allosteric activation and inhibition.

The starting substrate for the urea cycle is an "activated" molecule produced by the condensation of bicarbonate and ammonia, as catalyzed by carbamoyl phosphate synthetase (**Fig. 18.10**). The NH_4^+ may be contributed by the glutamate dehydrogenase reaction or another process that releases ammonia. The bicarbonate is the source of the

1. ATP activates bicarbonate by transferring a phosphoryl group.

2. Ammonia attacks the resulting carboxyphosphate intermediate, displacing the phosphate.

3. A second ATP phosphorylates carbamate to generate carbamoyl phosphate.

Carboxyphosphate

Carbamate

Carbamoyl phosphate

FIGURE 18.10

The carbamoyl phosphate synthetase reaction.

? Bicarbonate (as CO_2) and ammonia can easily diffuse into mitochondria. Why doesn't the reaction product diffuse out?

urea carbon. Note that the phosphoanhydride bonds of two ATP molecules are consumed in the energetically costly production of carbamoyl phosphate.

THE UREA CYCLE CONSISTS OF FOUR REACTIONS

The four enzyme-catalyzed reactions of the urea cycle proper are shown in **Figure 18.11**. The cycle also provides a means for synthesizing arginine: The five-carbon ornithine is derived from glutamate, and the urea cycle converts it to arginine. However, the arginine needs of children exceed the biosynthetic capacity of the urea cycle, so arginine is classified as an essential amino acid.

The fumarate generated in step 3 of the urea cycle is converted to malate and then oxaloacetate, which is used for gluconeogenesis. The aspartate substrate for Reaction 2 may represent oxaloacetate that has undergone transamination. Combining these ancillary reactions with those of the urea cycle, the carbamoyl phosphate synthetase reaction, and the glutamate dehydrogenase reaction yields the pathway outlined in **Figure 18.12**. *The overall effect is that transaminated amino acids donate amino groups, via glutamate and aspartate, to urea synthesis.* Because the liver is the only tissue that can carry out urea synthesis, amino groups to be eliminated travel through the blood to the liver mainly as glutamine, which accounts for up to one-quarter of circulating amino acids.

Like many other metabolic loops, the urea cycle involves enzymes located in both the mitochondria and cytosol. Glutamate dehydrogenase, carbamoyl phosphate synthetase, and ornithine transcarbamoylase are mitochondrial, whereas argininosuccinate synthetase, argininosuccinase, and arginase are cytosolic. Consequently, citrulline is produced in the mitochondria but must be transported to the cytosol for the next step, and ornithine produced in the cytosol must be imported into the mitochondria to begin a new round of the cycle.

The carbamoyl phosphate synthetase reaction and the argininosuccinate synthetase reactions each consume 2 ATP equivalents, so the cost of the urea cycle is 4 ATP per urea. However, when considered in context, operation of the urea cycle is often accompanied by ATP synthesis. The glutamate dehydrogenase reaction produces NADH (or NADPH), whose free energy is conserved in the synthesis of 2.5 ATP by oxidative phosphorylation. Catabolism of the carbon skeletons of the amino acids that donated their amino groups via transamination also yields ATP.

The rate of urea production is controlled largely by the activity of carbamoyl phosphate synthetase. This enzyme is allosterically activated by *N*-acetylglutamate, which is synthesized from glutamate and acetyl-CoA:

Glutamate   Acetyl-CoA   *N*-Acetylglutamate

When amino acids are undergoing transamination and being catabolized, the resulting increases in the cellular glutamate and acetyl-CoA concentrations boost production of *N*-acetylglutamate. This stimulates carbamoyl phosphate synthetase activity, and flux through the urea cycle increases. Such a regulatory system allows the cell to efficiently dispose of the nitrogen released from amino acid degradation.

1. *Ornithine transcarbamoylase produces citrulline by transferring a carbamoyl group to ornithine.*

2. *Argininosuccinate synthetase adds aspartate to citrulline in an ATP-dependent condensation reaction. The products of the reaction include AMP and PP_i; the PP_i is subsequently hydrolyzed to $2 P_i$. Consequently, this step consumes 2 ATP equivalents.*

3. *Argininosuccinase releases fumarate (which represents the carbon skeleton of aspartate) from argininosuccinate.*

4. *Arginase hydrolyzes arginine to generate urea and regenerate ornithine for another round of the urea cycle.*

FIGURE 18.11

The four reactions of the urea cycle.

? **Which steps are likely to be irreversible *in vivo*?**

FIGURE 18.12

The urea cycle and related reactions. Two routes for the disposal of amino groups are highlighted. The blue pathway shows how an amino group from an amino acid enters the urea cycle via glutamate and carbamoyl phosphate. The red pathway shows how an amino group from an amino acid enters via aspartate.

? Identify all the products of this metabolic scheme. Which ones are recycled?

Urea is relatively nontoxic and easily transported through the bloodstream to the kidneys for excretion in the urine. However, the polar urea molecule requires large amounts of water for its efficient excretion. This presents a problem for flying vertebrates such as birds and for reptiles that are adapted to arid habitats. These organisms deal with waste nitrogen by converting it to uric acid via purine synthesis (Section 18.5). The relatively insoluble uric acid is excreted as a semisolid paste, which conserves water.

Bacteria, fungi, and some other organisms use an enzyme called urease to break down urea:

$$H_2N-\overset{\overset{\displaystyle O}{\|}}{C}-NH_2 \ + \ H_2O \ \xrightarrow{urease} \ 2\,NH_3 \ + \ CO_2$$

Urease has the distinction of being the first enzyme to be crystallized (in 1926). It helped promote the theory that catalytic activity was a property of proteins. This premise is only partly true, as we have seen, since many enzymes contain metal ions or inorganic cofactors (urease itself contains two catalytic nickel atoms).

◆ CONCEPT CHECK

1. Explain the importance of glutamate dehydrogenase and carbamoyl phosphate synthetase for nitrogen disposal.

2. Summarize the reactions that occur in the four steps of the urea cycle.

3. Compare the solubility and toxicity of ammonia, urate, and urea.

4. Describe some situations when flux through the urea cycle would be very high or very low.

5. Why do mammals eliminate waste nitrogen as urea rather than as ammonia? Why do some organisms excrete uric acid?

6. How are amino groups from amino acids incorporated into urea?

18.5

NUCLEOTIDE METABOLISM

A discussion of nitrogen metabolism would not be complete without considering nucleotides, whose *nitrogenous bases are synthesized mainly from amino acids.* The human body can also recycle nucleotides from nucleic acids and nucleotide cofactors that are broken down. Although food supplies nucleotides, the biosynthetic and recycling pathways are so efficient that there is no true dietary requirement for purines and pyrimidines. In this section we will take a brief look at the biosynthesis and degradation of purine and pyrimidine nucleotides in mammals.

PURINE NUCLEOTIDE SYNTHESIS YIELDS IMP AND THEN AMP AND GMP

Purine nucleotides (AMP and GMP) are synthesized by building the purine base onto a ribose-5-phosphate molecule. In fact, the first step of the pathway is the production of 5-phosphoribosyl pyrophosphate (which is also a precursor of histidine):

Ribose-5-phosphate   5-Phosphoribosyl pyrophosphate

The subsequent ten steps of the pathway require as substrates glutamine, glycine, aspartate, and bicarbonate, plus one-carbon formyl (—HC=O) groups donated by tetrahydrofolate. The product is inosine monophosphate (IMP), a nucleotide whose base is the purine hypoxanthine:

Inosine monophosphate (IMP)

IMP is the substrate for two short pathways that yield AMP and GMP. In AMP synthesis, an amino group from aspartate is transferred to the purine; in GMP synthesis, glutamate is the source of the amino group (**Fig. 18.13**). Kinases then catalyze phosphoryl-group transfer reactions to convert the nucleoside monophosphates to diphosphates and then triphosphates (ATP and GTP).

Figure 18.13 indicates that GTP participates in AMP synthesis and ATP participates in GMP synthesis. High concentrations of ATP therefore promote GMP production, and high concentrations of GTP promote AMP production. *This reciprocal relationship is one mechanism for balancing the production of adenine and guanine nucleotides.* (Because most nucleotides are destined for DNA or RNA synthesis, they are required in roughly equal amounts.)

FIGURE 18.13

AMP and GMP synthesis from IMP.

The pathway leading to AMP and GMP is also regulated by feedback inhibition at several points, including the first step, the production of 5-phosphoribosyl pyrophosphate from ribose-5-phosphate, which is inhibited by both ADP and GDP.

PYRIMIDINE NUCLEOTIDE SYNTHESIS YIELDS UTP AND CTP

In contrast to purine nucleotides, pyrimidine nucleotides are synthesized as a base that is subsequently attached to 5-phosphoribosyl pyrophosphate to form a nucleotide. The six-step pathway that yields uridine monophosphate (UMP) requires glutamine, aspartate, and bicarbonate.

Uridine monophosphate (UMP)

UMP is phosphorylated to yield UDP and then UTP. CTP synthetase catalyzes the amination of UTP to CTP, using glutamine as the donor:

The UMP synthetic pathway in mammals is regulated primarily through feedback inhibition by UMP, UDP, and UTP. ATP activates the enzyme that catalyzes the first step; this helps balance the production of purine and pyrimidine nucleotides.

RIBONUCLEOTIDE REDUCTASE CONVERTS RIBONUCLEOTIDES TO DEOXYRIBONUCLEOTIDES

So far, we have accounted for the synthesis of ATP, GTP, CTP, and UTP, which are substrates for the synthesis of RNA. DNA, of course, is built from deoxynucleotides. In deoxynucleotide synthesis, each of the four nucleoside triphosphates (NTPs) is converted to its diphosphate (NDP) form, ribonucleotide reductase replaces the 2′ OH group with H, and then the resulting deoxynucleoside diphosphate (dNDP) is phosphorylated to produce the corresponding triphosphate (dNTP).

Ribonucleotide reductase is an essential enzyme that carries out a chemically difficult reaction using a mechanism that involves free radicals. Three types of ribonucleotide reductases, which differ in their catalytic groups, have been described. Class I enzymes (the type that occurs in mammals and most bacteria) have two Fe^{3+} ions and an unusually stable tyrosine radical (most free radicals, which have one unpaired electron, are highly reactive and short-lived).

Tyrosine radicals are also features of the active sites of cytochrome c oxidase (mitochondrial Complex IV) and Photosystem II in plants. A possible reaction mechanism for ribonucleotide reductase is shown in **Figure 18.14**.

The final step of the reaction, which regenerates the enzyme, requires the small protein thioredoxin. The oxidized thioredoxin must then undergo reduction to return to its original state. *This reaction uses NADPH, which is therefore the ultimate source of reducing power for the synthesis of deoxyribonucleotides.* Recall that the pentose phosphate pathway, which provides the ribose-5-phosphate for nucleotide synthesis, also generates NADPH (Section 13.4).

Not surprisingly, the activity of ribonucleotide reductase is tightly regulated so that the cell can balance the levels of ribo- and deoxyribonucleotides as well as the proportions of each of the four deoxyribonucleotides. Control of the enzyme involves two regulatory sites that are distinct from the substrate-binding site. For example, ATP binding to the so-called activity site activates the enzyme. Binding of the deoxyribonucleotide dATP decreases enzyme activity. Several nucleotides bind to the so-called substrate specificity site. Here, ATP binding induces the enzyme to act on pyrimidine nucleotides, and dTTP binding causes the enzyme to prefer GDP as a substrate. These mechanisms, in concert

1. The Cys free radical (—S·) reacts with the ribose of the NDP substrate to create a radical at C3'.

2. An enzyme SH group donates a proton to the oxygen at C2', which leaves as H_2O, and the radical migrates to C2'.

3. Proton transfer from the enzyme leads to formation of a disulfide bond between two enzyme Cys residues.

4. A second proton and electron transfer (the reverse of step 1) generates the deoxyribose group and regenerates the thiyl radical.

5. The oxidized Cys groups of the enzyme are reduced in a disulfide exchange reaction with thioredoxin.

Enzyme

NDP

+ dNDP

Thioredoxin

FIGURE 18.14

Proposed mechanism for ribonucleotide reductase. Only part of the nucleotide's ribose ring is shown.

with other mechanisms for balancing the amounts of the various nucleotides, help make all four deoxynucleotides available for DNA synthesis.

THYMIDINE NUCLEOTIDES ARE PRODUCED BY METHYLATION

The ribonucleotide reductase reaction, followed by kinase-catalyzed phosphorylation, generates dATP, dCTP, dGTP, and dUTP. However, dUTP is not used for DNA synthesis. Instead, *it is rapidly converted to thymine nucleotides* (which helps prevent the accidental incorporation of uracil into DNA). First, dUTP is hydrolyzed to dUMP. Next, thymidylate synthase adds a methyl group to dUMP to produce dTMP, using methylene-tetrahydrofolate as a one-carbon donor.

dUMP

N^5,N^{10}-Methylene-tetrahydrofolate

thymidylate synthase

dTMP

Dihydrofolate

The serine hydroxymethyltransferase reaction, which converts serine to glycine (Section 18.2), is the main source of methylene-tetrahydrofolate.

In converting the methylene group ($-CH_2-$) of the cofactor to the methyl group ($-CH_3$) attached to thymine, thymidylate synthase oxidizes the tetrahydrofolate cofactor to dihydrofolate. An NADPH-dependent enzyme called dihydrofolate reductase must then regenerate the reduced tetrahydrofolate cofactor. Finally, dTMP is phosphorylated to produce dTTP, the substrate for DNA polymerase.

Because cancer cells undergo rapid cell division, the enzymes of nucleotide synthesis, including thymidylate synthase and dihydrofolate reductase, are highly active. Compounds that inhibit either of these reactions can therefore act as anticancer agents. For example, the dUMP analog 5-fluorodeoxyuridylate, introduced in Section 7.3, inactivates thymidylate synthase. "Antifolates" such as methotrexate are competitive inhibitors of dihydrofolate reductase because they compete with dihydrofolate for binding to the enzyme. In the presence of methotrexate, a cancer cell cannot regenerate the tetrahydrofolate required for dTMP production, and the cell dies. Most noncancerous cells, which grow much more slowly, are not as sensitive to the effect of the drug.

Methotrexate

NUCLEOTIDE DEGRADATION PRODUCES URIC ACID OR AMINO ACIDS

Nucleotides that are obtained from food or synthesized by cells can be broken down, releasing ribose groups and a purine or pyrimidine that can be further catabolized and excreted (purines) or used as a metabolic fuel (pyrimidines). At several points in the degradation pathways, intermediates may be redirected toward the synthesis of new nucleotides by so-called **salvage pathways.** For example, a free adenine base can be reattached to ribose by the reaction

$$\text{adenine} + \text{5-phosphoribosyl pyrophosphate} \rightleftharpoons \text{AMP} + \text{PP}_i$$

Degradation of a nucleoside monophosphate begins with dephosphorylation to produce a nucleoside. In a subsequent step, a phosphorylase breaks the glycosidic bond between the base and the ribose by adding phosphate (a similar phosphorolysis reaction occurs during glycogenolysis; Section 13.3).

Uridine + HPO_4^{2-} ⟶ Uracil + Ribose-1-phosphate

The phosphorylated ribose can be catabolized or salvaged and converted to 5-phosphoribosyl pyrophosphate for synthesis of another nucleotide. The fate of the base depends on whether it is a purine or a pyrimidine.

The purine bases are eventually converted to uric acid in a process that may require deamination and oxidation, depending on whether the original base was adenine, guanine, or hypoxanthine. Uric acid has a pK of 5.4, so it exists mainly as urate.

Uric acid ⇌ Urate + H^+

In humans, urate, a poorly soluble compound, is excreted in the urine. Excess urate may precipitate as crystals of sodium urate in the kidneys (kidney "stones"). Deposits of urate in the joints, primarily the knees and toes, cause a painful condition called gout. Other organisms can further catabolize urate to generate more soluble waste products such as urea and ammonia.

The pyrimidines cytosine, thymine, and uracil undergo deamination and reduction, after which the pyrimidine ring is opened. Further catabolism produces the nonstandard amino acid β-alanine (from cytosine and uracil) or β-aminoisobutyrate (from thymine), both of which feed into other metabolic pathways.

Cytosine, Uracil ⟶ β-Ureidopropionate $\xrightarrow{H_2O \quad CO_2 + NH_4^+}$ β-Alanine

Thymine ⟶ β-Ureidoisobutyrate $\xrightarrow{H_2O \quad CO_2 + NH_4^+}$ β-Aminoisobutyrate

Consequently, pyrimidine catabolism contributes to the pool of cellular metabolites for both anabolic and catabolic processes. In contrast, purine catabolism generates a waste product that is excreted from the body.

CONCEPT CHECK

1. List the molecules that are used to build purine and pyrimidine nucleotides.

2. Why don't humans require purines and pyrimidines in their diet?

3. Compare the metabolic fates of IMP and UTP.

4. What does ribonucleotide reductase do?

5. Explain why dihydrofolate reductase is as important as thymidylate synthase.

6. Describe the metabolic fates of ribose, purines, and pyrimidines.

SUMMARY

18.1 Nitrogen Fixation and Assimilation

• Nitrogen-fixing organisms convert N_2 to NH_3 in the ATP-consuming nitrogenase reaction. Nitrate and nitrite can also be reduced to NH_3.

• Ammonia is incorporated into glutamine by the action of glutamine synthetase.

• Transaminases use a PLP prosthetic group to catalyze the reversible interconversion of α-amino acids and α-keto acids.

18.2 Amino Acid Biosynthesis

• In general, the nonessential amino acids are synthesized from common metabolic intermediates such as pyruvate, oxaloacetate, and α-ketoglutarate.

• The essential amino acids, which include the sulfur-containing, branched-chain, and aromatic amino acids, are synthesized by more elaborate pathways in bacteria and plants.

• Amino acids are the precursors of some neurotransmitters and hormones.

18.3 Amino Acid Catabolism

• Following removal of their amino groups by transamination, amino acids are broken down to intermediates that can be converted

to glucose or acetyl-CoA for use in the citric acid cycle, fatty acid synthesis, or ketogenesis.

18.4 Nitrogen Disposal: The Urea Cycle

• In mammals, excess amino groups are converted to urea for disposal. The urea cycle is regulated at the carbamoyl phosphate synthetase step, an entry point for ammonia. Other organisms convert excess nitrogen to compounds such as uric acid.

18.5 Nucleotide Metabolism

• The synthesis of nucleotides requires glutamate, glycine, and aspartate as well as ribose-5-phosphate. The pathways for purine and pyrimidine biosynthesis are regulated to balance the production of the various nucleotides.

• A ribonucleotide reductase uses a free radical mechanism to convert nucleotides to deoxynucleotides.

• Thymidine production requires a methyl group donated by the cofactor tetrahydrofolate.

• In humans, purines are degraded to uric acid for excretion, and pyrimidines are converted to β-amino acids.

KEY TERMS

nitrogen fixation	nitrogen cycle	channeling	deamination
diazotroph	transamination	neurotransmitter	urea cycle
nitrification	nonessential amino acid	glucogenic amino acid	salvage pathway
denitrification	essential amino acid	ketogenic amino acid	

BIOINFORMATICS

Brief Bioinformatics Exercises

18.1 Viewing and Analyzing Pyridoxal Phosphate Enzymes

18.2 Amino Acid Metabolism and the KEGG Database

PROBLEMS

18.1   Nitrogen Fixation and Assimilation

1. $\Delta\mathcal{E}^{\circ\prime}$ for the half-reaction

$$N_2 + 6\,H^+ + 6\,e^- \rightarrow 2\,NH_3$$

is -0.34 V. The reduction potential of the nitrogenase component that donates electrons for nitrogen reduction is about -0.29 V. ATP hydrolysis apparently induces a conformational change in the protein that alters its reduction potential by about 0.11 V. Does this change increase or decrease $\mathcal{E}^{\circ\prime}$ of the electron donor and why is this change necessary?

2. Why is it a good agricultural practice to plant a field with alfalfa every few years?

3. Plants whose root nodules contain nitrogen-fixing bacterial symbionts synthesize a heme-containing protein, called leghemoglobin, which structurally resembles myoglobin. What is the function of this protein in the root nodules?

4. Photosynthetic cyanobacteria carry out nitrogen fixation in specialized cells that have Photosystem I but lack Photosystem II. Explain the reason for this strategy.

5. Over the past 20 years, scientists have discovered prokaryotic species that can oxidize ammonia to N_2 and convert ammonia directly to nitrate.
(a) Draw a diagram that includes these processes along with those illustrated in Figure 18.3.
(b) If each process in your diagram corresponds to one organism, which organism(s) could be added to farm plots to avoid the need for chemical fertilizers?

6. The highly versatile prokaryotic cell can incorporate ammonia into amino acids using two different mechanisms, depending on the concentration of available ammonia.

(a) One method involves coupling glutamine synthetase, glutamate synthase, and transamination reactions. Write the overall balanced equation for this process.
(b) A second method involves coupling the freely reversible glutamate dehydrogenase reaction and a transamination reaction. Write the overall balanced equation for this process.

7. Refer to your answer to Problem 6.

(a) Which process is used when the concentration of available ammonia is low?
(b) When the concentration of ammonia is high? (*Hint:* The K_M of glutamine synthetase for the ammonium ion is lower than the K_M of glutamate dehydrogenase for the ammonium ion.)
(c) Why is the prokaryotic cell at a disadvantage when the concentration of ammonia is low? Explain.

8. Cancer cells show a much higher rate of glutamine utilization than normal cells. What are two strategies that cancer cells could employ to increase the cellular glutamine content? How could you use this information to devise a therapeutic agent for treating cancer?

9. *E. coli* glutamine synthetase is regulated by adenylylation; that is, an AMP group is covalently attached to a tyrosine side chain. The enzyme is less active in the adenylylated form. The reaction is catalyzed by an adenylyltransferase (ATase) enzyme.

(a) Write a balanced equation for this reaction and show the structure of the adenylylated tyrosine side chain.
(b) Would you expect α-ketoglutarate to stimulate or inhibit the adenylylation of the enzyme?

10. Draw the products of the following transamination reactions:

(a) glycine + α-ketoglutarate $\rightarrow$ glutamate + _____
(b) arginine + α-ketoglutarate $\rightarrow$ glutamate + _____

(c) serine + α-ketoglutarate $\rightarrow$ glutamate + _____
(d) phenylalanine + α-ketoglutarate $\rightarrow$ glutamate + _____

11. Draw the products of the following transamination reactions. What do all of the products have in common?

(a) aspartate + α-ketoglutarate $\rightarrow$ glutamate + _____
(b) alanine + α-ketoglutarate $\rightarrow$ glutamate + _____
(c) glutamate + oxaloacetate $\rightarrow$ aspartate + _____

12. The transamination reaction is an example of an enzyme that uses a ping pong mechanism (see Section 7.2). Explain why.

13. Which amino acid generates each of the following products in a transamination reaction with α-ketoglutarate?

14. Serine hydroxymethyltransferase catalyzes the conversion of threonine to glycine in a PLP-dependent reaction. The mechanism is slightly different from that shown in the transamination reaction in Figure 18.5. The degradation of threonine to glycine begins with a threonine C_α—C_β bond cleavage. Draw the structure of the threonine–Schiff base intermediate that forms in this reaction and show how C_α—C_β bond cleavage occurs.

15. Alcoholics may suffer from malnutrition and vitamin deficiencies. How might this affect the results of the ALT test (see Box 18.A)? If you were the clinician running the test, how could you alter the assay procedure to account for this?

18.2   Amino Acid Biosynthesis

16. Glutamine synthetase and asparagine synthetase catalyze reactions to produce glutamine and asparagine, respectively. It might be reasonable to expect that these enzymes are similar, but they catalyze the formation of their amino acid products very differently. Compare and contrast the two enzymes.

17. Asparagine synthetase can use a similar mechanism to catalyze the synthesis of asparagine from aspartate using ammonium ions rather than glutamate as the nitrogen donor (see Problem 16).
(a) Write a balanced equation for this reaction. (b) Compare this reaction to the glutamine synthetase reaction.

18. The expression of phosphoglycerate dehydrogenase, which catalyzes the first step of serine biosynthesis, is dramatically elevated in some cancers, which are highly metabolically active. Experiments suggest that in addition to supplying extra serine, the elevated dehydrogenase activity increases flux through the citric acid cycle. Explain. (*Hint:* 3-phosphohydroxypyruvate is an α-keto acid.)

19. Serine hydroxymethyltransferase catalyzes the reaction shown in the text, in which serine and tetrahydrofolate are converted to glycine and methylene-tetrahydrofolate. The methylene-tetrahydrofolate is then used as a reactant in nucleotide synthesis. What types of cells have a high expression of serine hydroxymethyltransferase?

20. Researchers have identified several metabolites whose concentrations in urine may be elevated in prostate cancer. One of these metabolites is sarcosine (*N*-methylglycine).

(a) Draw its structure.

(b) Folate deficiency causes sarcosine levels in the blood to increase. Propose an explanation for this observation.

21. All of the nonessential amino acids, with the exception of tyrosine, can be synthesized from four metabolites: pyruvate, oxaloacetate, α-ketoglutarate, and 3-phosphoglycerate. Draw a diagram that shows how the 10 amino acids are obtained from these metabolites.

22. In many bacteria, the pathway for synthesizing cysteine and methionine consists of enzymes that contain relatively few Cys and Met residues. Explain why this is an advantage.

23. Taurine, a naturally occurring compound that is added to some energy drinks, is used in the synthesis of bile salts (Section 17.4); it may also help regulate cardiovascular function and lipoprotein metabolism. From what amino acid is taurine derived, and what types of reactions are required to convert this amino acid to taurine?

$$^{+}H_3N-CH_2-CH_2-\overset{\overset{\textstyle O}{\|}}{\underset{\underset{\textstyle O}{\|}}{S}}-O^-$$

Taurine

24. Sulfonamides (sulfa drugs) act as antibiotics by inhibiting the synthesis of folate in bacteria.

(a) Which portion of the folate molecule does the sulfonamide resemble?

(b) Why do sulfonamides kill bacteria without harming their mammalian host?

$$H_2N-\!\!\!\!\bigcirc\!\!\!\!-\overset{\overset{\textstyle O}{\|}}{\underset{\underset{\textstyle O}{\|}}{S}}-NH-R$$

A sulfonamide

25. A person whose diet is poor in just one of the essential amino acids may enter a state of negative nitrogen balance, in which nitrogen excretion is greater than nitrogen intake. Explain why this occurs, even when the supply of other amino acids is high.

26. A dilute solution of gelatin, which is derived from the protein collagen, is sometimes given to ill children who have not been able to consume solid food for several days.

(a) Explain why gelatin is not a good source of essential amino acids.

(b) What is the advantage of giving gelatin, rather than a sugar solution, to someone who has not eaten for several days?

27. (a) Is tyrosine an essential amino acid?

(b) Would tyrosine be an essential amino acid in a patient with a phenylalanine hydroxylase deficiency?

28. Threonine deaminase catalyzes the committed step of the biosynthetic pathway leading to the branched-chain amino acid isoleucine. The enzyme catalyzes the dehydration and deamination of threonine to α-ketobutyrate. This pathway is linked to pathways that produce valine and leucine as shown in the figure.

Threonine deaminase is a tetramer and has a low affinity, or "T," form and a high affinity, or "R," form. The enzyme is allosterically regulated. The activity of threonine deaminase was measured at increasing concentrations of its substrate, threonine (curve A). Measurements were also made in the presence of isoleucine (curve B) and valine (curve C). The kinetic constants obtained from the graphs are shown in the table.

	No modulators present (Thr only)	Isoleucine added	Valine added
V_{max} ($\mu mol \cdot mg^{-1} \cdot min^{-1}$)	214	180	225
K_M (mM)	8.0	74	5.7

(a) What can you determine about threonine deaminase from these data?

(b) What is the effect of isoleucine on threonine deaminase activity? To which form of the enzyme does isoleucine bind?

(c) What is the effect of valine (a product of a parallel pathway) on threonine deaminase activity? To which form of the enzyme does valine bind?

29. Bacteria synthesize the essential amino acids lysine, methionine, and threonine using aspartate as a substrate, as summarized below. Certain enzymes in this pathway serve as regulatory points so that the cell can maintain appropriate concentrations of each amino acid. The amino acids themselves serve as allosteric enzyme regulators.

Which enzyme(s) is(are) good candidates for regulating the synthesis of **(a)** lysine, **(b)** methionine, and **(c)** threonine?

18.3 Amino Acid Catabolism

30. (a) From what pathway is 3-phosphoglycerate derived in order to synthesize serine?
(b) What type of enzyme catalyzes the second reaction of the serine biosynthetic pathway?

31. Peptides containing citrulline residues are often detected in inflammation, such as occurs in rheumatoid arthritis.

(a) From which amino acid are citrulline residues derived, and what kind of reaction generates them?
(b) Why are citrullinated peptides likely to trigger an autoimmune response?

32. Urinary 3-methylhistidine (a modified amino acid found primarily in actin) is used as an indicator of the rate of muscle degradation. When actin is degraded, 3-methylhistidine is excreted because it cannot be re-used for protein synthesis. Why? Explain why monitoring 3-methylhistidine levels offers only an approximation of the rate of muscle degradation.

33. The catabolic pathways for the 20 amino acids vary considerably, but all amino acids are degraded to one of seven metabolites: pyruvate, α-ketoglutarate, succinyl-CoA, fumarate, oxaloacetate, acetyl-CoA, or acetoacetate. What is the fate of each of these metabolites?

34. Although amino acids are classified as glucogenic, ketogenic, or both, it is possible for all their carbon skeletons to be broken down to acetyl-CoA. Explain.

35. Mouse embryonic stem cells are small but divide extremely rapidly. To maintain high metabolic flux, these cells require a high concentration of threonine and express high levels of threonine dehydrogenase, which catalyzes the first step of threonine breakdown. Explain how threonine catabolism contributes to citric acid cycle activity and nucleotide biosynthesis.

36. Isoleucine is degraded to acetyl-CoA and propionyl-CoA by a pathway in which the first steps are identical to those of valine degradation (Fig. 18.9) and the last steps are identical to those of fatty acid oxidation.

(a) Draw the intermediates of isoleucine degradation and indicate the enzyme that catalyzes each step.
(b) Which reaction in the degradation scheme is analogous to the reaction catalyzed by pyruvate dehydrogenase?

(c) What reaction is analogous to the reaction catalyzed by acyl-CoA dehydrogenase in fatty acid oxidation?

37. What is the fate of the propionyl-CoA that is produced by degradation of isoleucine?

38. Leucine is degraded to acetyl-CoA and acetoacetate by a pathway whose first two steps are identical to those of valine degradation (Fig. 18.9). The third step is the same as the first step in fatty acid oxidation. The fourth step involves an ATP-dependent carboxylation, the fifth step is a hydration reaction, and the last step is a cleavage reaction catalyzed by a lyase enzyme that releases the products. Draw the intermediates of leucine degradation and indicate the enzyme that catalyzes each step.

39. (a) What is the metabolic fate of the products of leucine degradation (see Solution 38) and how does this differ from the metabolic fate of the products of isoleucine degradation (see Solution 37)?
(b) The last reaction in the leucine degradation pathway is catalyzed by HMG-CoA lyase (see Solution 38). Why do patients who are missing this enzyme need to restrict not just leucine but fat from their diets? [*Hint*: see Fig. 17.16.]

40. Maple syrup urine disease (MSUD) is an inborn error of metabolism that results in the excretion of α-keto acids derived from leucine, isoleucine, and valine. What enzyme is nonfunctional in these patients? Severe neurological symptoms develop if the disease is not treated, but if treated, patients can live a fairly normal life. How would you treat the disease?

41. The effect of the hormones glucagon and insulin on the activity of phenylalanine hydroxylase was investigated, with and without preincubation with phenylalanine (phenylalanine converts the enzyme from the dimeric to the tetrameric form). The results are shown below.

(a) What hormone stimulates the enzyme?
(b) What is the role of phenylalanine in regulating enzyme activity?
(c) A separate experiment showed that the amount of radioactively labeled phosphate incorporated into the enzyme with glucagon treatment was nearly seven-fold greater than in controls. Explain this observation.
(d) Is phenylalanine hydroxylase more active in the fed or fasting state?

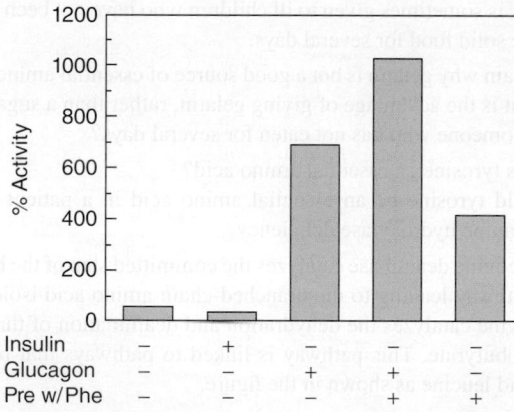

42. Phenylketonuria (PKU) is an inherited disease that results from the lack of phenylalanine hydroxylase (see Problem 41), which catalyzes the first step in phenylalanine degradation. In PKU patients, phenylalanine accumulates in the blood and is transaminated to phenylpyruvate, a phenylketone. The accumulation of phenylpyruvate causes irreversible brain damage if the disease is not treated.

(a) Draw the structure of phenylpyruvate.
(b) Why do children with a deficiency of tetrahydrobiopterin excrete large quantities of the phenylketone?

(c) Why are patients with PKU given a low phenylalanine diet and not a phenylalanine-free diet?

(d) Why should patients with PKU avoid the artificial sweetener aspartame (see Problem 4.9)?

(e) Why should PKU patients increase their dietary tyrosine?

43. Nonketotic hyperglycinemia (NKH) is an inborn error of metabolism characterized by high levels of glycine in the blood, urine, and cerebrospinal fluid. Babies with this disease suffer from hypotonia, seizures, and intellectual disability. What enzyme is most likely to be nonfunctional in patients with NKH?

44. In mammals, metabolic fuels can be stored: glucose as glycogen and fatty acids as triacylglycerols. What type of molecule could be considered as a sort of storage depot for amino acids? How does it differ from other fuel-storage molecules?

18.4 Nitrogen Disposal: The Urea Cycle

45. At one time, ammonia's toxicity was believed to result from its participation in the glutamate dehydrogenase reaction, which is reversible. Explain how this reaction could affect the brain's energy metabolism.

46. Glutamate dehydrogenase is allosterically regulated by a variety of metabolites. Predict the effect of each of the following on glutamate dehydrogenase activity: **(a)** GTP; **(b)** ADP; **(c)** NADH.

47. *N*-acetylglutamate synthase, which catalyzes the synthesis of the allosteric regulator *N*-acetylglutamate, is itself allosterically regulated by arginine. Is arginine an allosteric activator or inhibitor of *N*-acetylglutamate synthase? Explain.

48. Recall that ketone bodies are produced during starvation (Section 17.3). Under these conditions, the kidneys increase their uptake of glutamine. Explain how this helps counteract the effects of ketosis.

49. Identify the source of the two nitrogen atoms in urea.

50. List all the reactions shown in this chapter that generate free ammonia.

51. Which three mammalian enzymes can potentially "mop up" excess NH_4^+?

52. Which of the metabolites listed are consumed in the urea cycle? Which are not consumed? **(a)** NH_4^+; **(b)** CO_2; **(c)** ATP; **(d)** ornithine; **(e)** citrulline; **(f)** argininosuccinate; **(g)** arginine; **(h)** aspartate.

53. A complete deficiency of a urea cycle enzyme usually causes death soon after birth, but a partial deficiency may be tolerated.

(a) Explain why hyperammonemia (high levels of ammonia in the blood) accompanies a urea cycle enzyme deficiency.

(b) What dietary adjustments might minimize ammonia toxicity?

54. The drug Ucephan, a mixture of sodium salts of phenylacetate and benzoate (shown below), is used to treat the hyperammonemia associated with a urea cycle enzyme deficiency (see Problem 53). Phenylacetate reacts with glutamine, and benzoate reacts with glycine; the reaction products are excreted in the urine. Draw the structures of the products. What is the biochemical rationale for this treatment?

Benzoate   Phenylacetate

55. An inborn error of metabolism results in the deficiency of argininosuccinase. What could be added to the diet to boost urea production?

56. Production of the enzymes that catalyze the reactions of the urea cycle can increase or decrease according to the metabolic needs of the organism. High levels of these enzymes are associated with high-protein diets as well as starvation. Explain this paradox.

57. (a) Describe how the reversible glutamate dehydrogenase reaction contributes to amino acid biosynthesis.

(b) How does it function as an anaplerotic reaction for the citric acid cycle?

58. The kidneys play a role in regulating acid–base balance in humans by releasing amino groups from glutamine so that the resulting ammonium ions can neutralize metabolic acids (Section 2.5). Which two kidney enzymes are responsible for removing glutamine's amino groups? Write the reaction catalyzed by each enzyme, and write the net reaction.

59. Vigorous exercise is known to break down muscle proteins. What is the probable metabolic fate of the resulting free amino acids?

60. (a) As described in the text, ammonia activation of the NMDA receptor in the brain leads to neuronal cell death. One cause of death is depletion of ATP in the post-synaptic cell. Explain why over-stimulation of the NMDA receptor leads to loss of ATP. (*Hint*: The Ca^{2+} ions that entered the cytosol upon stimulation must be returned to the mitochondrial matrix and Na^+ ions must be ejected from the cell.)

(b) Ammonia-induced activation of NMDA receptors in neurons results in the formation of NO (see Box 18.C), which inhibits glutamine synthetase. What effect does this have on the neurons?

61. Gastric ulcers result from infection by *Helicobacter pylori*. To survive in the extreme acidity of the stomach, the bacteria express high levels of the enzyme urease.

(a) Why is urease activity essential for *H. pylori* survival?

(b) Why is it important for at least some urease to be associated with the bacterial cell surface?

62. The citric acid "cycle" in *H. pylori* is a noncyclic, branched pathway, as shown in Problem 14.62. Succinate is produced in the "reductive branch," whereas α-ketoglutarate is produced in the "oxidative branch." How might this pathway be linked to amino acid metabolism in the bacterium?

18.5 Nucleotide Metabolism

63. Although methotrexate is still used for cancer chemotherapy, it is also prescribed to treat some autoimmune diseases such as rheumatoid arthritis. Explain.

64. Multifunctional enzymes are common in eukaryotic metabolism. Explain why it would be an advantage for dihydrofolate reductase (DHFR) and thymidylate synthase (TS) activities to be part of the same protein.

65. Class I ribonucleotide reductases use a tyrosine radical, class II enzymes use adenosylcobalamin (Section 17.2), and class III enzymes use a glycyl radical to generate the cysteine free radical required in the first step of the reaction mechanism (see Fig. 18.14). Is this an example of convergent or divergent evolution?

66. Purine nucleotide synthesis, outlined below, is a highly regulated process. The main objective is to provide the cell with approximately equal concentrations of ATP and GTP for DNA synthesis.

(a) How do ADP and GDP regulate ribose phosphate pyrophosphokinase?

(b) Amidophosphoribosyl transferase catalyzes the committed step of the IMP synthetic pathway. How might 5-phosphoribosyl pyrophosphate (PRPP), AMP, ADP, ATP, GMP, GDP, and GTP affect the activity of this enzyme? Explain your reasoning.

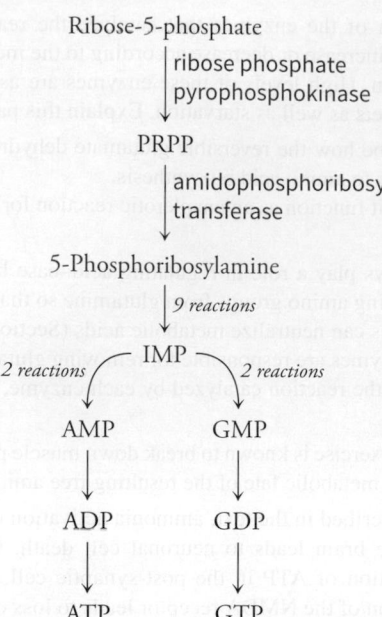

Ribose-5-phosphate
↓ ribose phosphate pyrophosphokinase
PRPP
↓ amidophosphoribosyl transferase
5-Phosphoribosylamine
↓ *9 reactions*
IMP
2 reactions ↙   ↘ *2 reactions*
AMP   GMP
↓   ↓
ADP   GDP
↓   ↓
ATP   GTP

67. The purine nucleotide synthesis pathway shown in Problem 64 is the *de novo* synthetic pathway and is virtually identical in all organisms. Most organisms have additional salvage pathways in which purines released from degradative processes are recycled to form their respective nucleotides. (Some organisms that do not have the *de novo* pathway rely exclusively on a salvage pathway to synthesize purine nucleotides.) One salvage pathway involves the conversion of adenine to AMP, catalyzed by the enzyme adenine phosphoribosyltransferase (APRT). If adenine is present in large amounts, it is converted to dihydroxyadenine, which can form kidney stones.

(a) A mutation in the APRT gene results in a 10-fold increase in the K_M value for one of the APRT substrates. What is the consequence of this mutation?

(b) How would you treat the condition resulting from this mutation?

adenine + 5-phosphoribosyl pyrophosphate $\xrightleftharpoons{\text{APRT}}$ AMP + PP$_i$

↓ xanthine dehydrogenase

dihydroxyadenine

68. A second purine salvage pathway (see Problem 65) is catalyzed by hypoxanthine-guanine phosphoribosyl transferase (HGPRT), which catalyzes the following reactions:

hypoxanthine + 5-phosphoribosyl pyrophosphate $\xrightleftharpoons{\text{HGPRT}}$
IMP + PP$_i$

guanine + 5-phosphoribosyl pyrophosphate $\xrightleftharpoons{\text{HGPRT}}$
GMP + PP$_i$

Some intracellular protozoan parasites have high levels of HGPRT. Inhibitors of this enzyme are being studied for their effectiveness in blocking parasite growth. What is the metabolic effect of inhibiting the parasite's HGPRT, and what does this tell you about the parasite's metabolic capabilities?

69. Lesch–Nyhan syndrome is a disease caused by a severe deficiency in HGPRT activity (see Problem 68). The disease is characterized by the accumulation of excessive amounts of uric acid, a product of nucleotide degradation, which causes neurological abnormalities and destructive behavior, including self-mutilation. Explain why the absence of HGPRT causes uric acid to accumulate.

70. Antibody-producing B lymphocytes use both the *de novo* (see Problem 66) and salvage pathways (see Problem 67) to synthesize nucleotides, but their survival rate in culture is only 7–10 days. Myeloma cells lack the HGPRT enzyme (see Problem 68) and survive indefinitely in culture. The preparation of long-lived antibody-producing hybridoma cells involves fusing a lymphocyte and a myeloma cell to produce a hybridoma, then selecting for hybridomas that can grow in HAT medium. The medium contains hypoxanthine, aminopterin (an antibiotic that inhibits enzymes of the *de novo* nucleotide synthetic pathway), and thymidine. How does the HAT medium select for hybridoma cells?

71. The compound 5-fluorouracil (shown below) is often used topically to treat minor skin cancers. When 5-fluorouracil is added to cells in culture, the concentration of dUTP increases, while dTTP is depleted. How do you account for these observations, and how does 5-fluorouracil kill cancer cells?

5-Fluorouracil

SELECTED READINGS

Brosnan, J. T., Glutamate, at the interface between amino acid and carbohydrate metabolism, *J. Nutr.* **130**, 988S–990S (2000). [Summarizes the metabolic roles of glutamate as a fuel and as a participant in deamination and transamination reactions.]

Gropman, A. L., Summar, M., and Leonard, J. V., Neurological implications of urea cycle disorders, *J. Inherit. Metab. Dis.* **30**, 865–879 (2007). [Reviews the urea cycle and discusses the clinical consequences of urea cycle enzyme deficiencies.]

Lane, A. N. and Fan, T. W.-M., Regulation of mammalian nucleotide metabolism and biosynthesis, *Nuc. Acids Res.* **43**, 2466–2485 (2015). [Includes overviews of the major metabolic pathways for synthesizing nucleotides.]

Reichard, P., Ribonucleotide reductases: Substrate specificity by allostery, *Biochem. Biophys. Res. Commun.* **396**, 19–23 (2010). [Includes a discussion of how these enzymes are regulated.]

Toney, M. D., Controlling reaction specificity in pyridoxal phosphate enzymes, *Biochim. Biophys Acta,* **1814,** 1407–1418 (2011). [Discusses how these enzymes use the same cofactor for a variety of reactions.]

Williams, R. A., Mamotte, C. D. S., and Burnett, J. R., Phenylketonuria: An inborn error of phenylalanine metabolism, *Clin. Biochem. Rev.* **29**, 31–41 (2008). [Includes the history, biochemistry, diagnosis, and treatment of the disease.]

MAMMALIAN FUEL METABOLISM: REGULATION

John Carnemolla/ShutterstocK

Because a koala's diet consists almost entirely of fiber-rich, nutrient-poor eucalyptus leaves, this marsupial sleeps as much as 22 hours each day and has a relatively slow metabolic rate that minimizes energy expenditure and maximizes the time for gut bacteria to digest food.

DO YOU REMEMBER?

- Allosteric regulators can inhibit or activate enzymes (Section 7.3).

- G protein–coupled receptors and receptor tyrosine kinases are the two major types of receptors that transduce extracellular signals to the cell interior (Section 10.1).

- Metabolic fuels can be mobilized by breaking down glycogen, triacylglycerols, and proteins (Section 12.1).

- Metabolic pathways in cells are connected and are regulated (Section 12.2).

As a machine obeying the laws of thermodynamics, the human body is remarkably flexible in managing its resources. Humans and other mammals rely on different organs that are specialized for using, storing, and interconverting metabolic fuels. The exchange of materials between organs and communication between organs allows the body to operate as a unified whole. But for the same reasons, disorders in one aspect of fuel metabolism can have body-wide consequences. We begin this chapter by reviewing the roles of different organs. Then we can explore how metabolic processes are coordinated and how they are disrupted in disease.

INTEGRATION OF FUEL METABOLISM

In examining the various pathways for the catabolism and anabolism of the major metabolic fuels and building blocks in mammals—carbohydrates, fatty acids, amino acids, and nucleotides—we have seen that biosynthetic and degradative pathways differ for thermodynamic reasons. These pathways are also regulated so that their simultaneous operation does not waste resources. One form of regulation is the compartmentation of opposing processes. For example, fatty acid oxidation takes place in mitochondria, whereas fatty acid synthesis takes place in the cytosol. The locations of the major metabolic pathways are shown in **Figure 19.1**. The movement of materials between cellular compartments requires an extensive set of membrane transporters, some of which are included in Figure 19.1.

ORGANS ARE SPECIALIZED FOR DIFFERENT FUNCTIONS

Compartmentation also takes the form of organ specialization: *Different tissues have different roles in energy storage and use.* For example, the liver carries out most metabolic processes as well as liver-specific functions such as gluconeogenesis, ketogenesis, and urea production. Adipose tissue is specialized to store about 95% of the body's triacylglycerols. Some tissues, such as red blood cells, do not store glycogen or fat and rely primarily on glucose supplied by the liver.

The functions of some organs as fuel depositories or fuel sources depend on whether the body is experiencing abundance (for example, immediately following a meal) or deprivation (after many hours of fasting). The major metabolic functions of some organs, including their reciprocating roles in storing and mobilizing fuels, are diagrammed in **Figure 19.2**.

Following a meal, the liver takes up glucose and converts it to glycogen for storage. Excess glucose and amino acids are catabolized to acetyl-CoA, which is used to synthesize fatty acids. The fatty acids are esterified to glycerol, and the resulting triacylglycerols, along with dietary triacylglycerols, are exported to other tissues. During a fast, the liver mobilizes glucose from glycogen stores and releases it into the circulation for other tissues to use. Triacylglycerols are broken down to acetyl-CoA, which can be converted to ketone bodies to power the brain and heart when glucose is in short supply. Amino acids derived from proteins can be converted to glucose by gluconeogenesis (the nonglucogenic amino acids can be converted to ketone bodies). The liver also deals with lactate and alanine produced by muscle activity, converting these molecules to glucose and disposing of amino groups through urea synthesis.

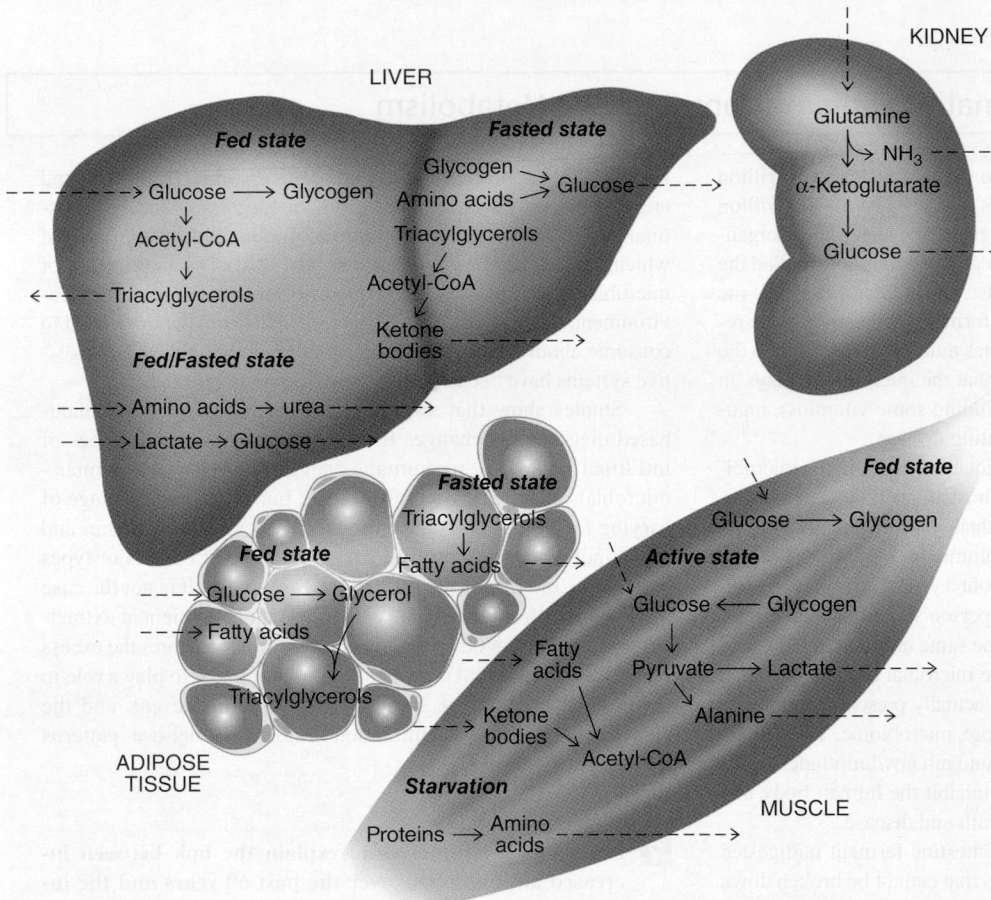

FIGURE 19.2

The major metabolic roles of the liver, kidney, muscle, and adipose tissue. The liver is the most metabolically active organ in the body, followed by adipose tissue, and then muscle.

? Identify the processes that are mainly catabolic and mainly anabolic.

Muscle cells take up glucose when it is available and store it as glycogen, although there is a limit to how much glycogen can be stockpiled. During exercise, the glycogen is quickly broken down for glycolysis, a pathway for the rapid—if inefficient—production of ATP. Muscle cells can also burn fatty acids and ketone bodies. Heart muscle, which maintains a near-constant level of activity, is specialized to burn fatty acids as its primary fuel and is rich in mitochondria to carry out this aerobic activity. During intense prolonged activity, muscle cells export lactate and alanine (see below). Muscle protein can be tapped as a source of metabolic fuel during starvation, when amino acids are needed to generate glucose.

Adipocytes take up glucose and convert it to glycerol; this plus fatty acids taken up from the circulation are the raw materials for triacylglycerols that are stored as a fat globule inside the cell. Fatty acids are mobilized in times of need and released from adipose tissue into the circulation.

In addition to eliminating wastes and maintaining acid–base balance (see Section 2.5), the kidneys play a minor role in fuel metabolism. The removal of amino groups from glutamine leaves α-ketoglutarate, which can be converted to glucose (the liver and kidney are the only organs that carry out gluconeogenesis).

METABOLITES TRAVEL BETWEEN ORGANS

The body's organs are connected to each other by the circulatory system so that metabolites synthesized by one organ, such as glucose produced by the liver, can easily reach other tissues. Amino acids released from various tissues travel to the liver or kidney for disposal of their amino groups. Materials are also exchanged between the body's organs and the microorganisms that inhabit the intestines (**Box 19.A**).

Some metabolic pathways are circuits that include interorgan transport. For example, the **Cori cycle** (named after Carl and Gerty Cori, who first described it) is a metabolic pathway involving the muscles and liver. During periods of high activity, muscle glycogen is broken down to glucose, which undergoes glycolysis to produce the ATP required for muscle contraction.

Box 19.A The Intestinal Microbiome Contributes to Metabolism

The human body is estimated to contain at least 10 trillion (10^{13}) cells, and there may be as many as 40 to 100 trillion microorganisms, primarily living in the intestine. These organisms, mostly bacteria, form an integrated community called the **microbiome.** At one time, the existence of microbes inside the human host was believed to be a form of commensalism, a relationship in which neither party has much to gain or lose in the partnership. It is now clear, however, that the microbiome plays an active role in providing nutrients (including some vitamins), regulating fuel use and storage, and preventing disease.

DNA sequencing, the main technique for studying the microbiome, suggests that the human gut can host about 1000 different microbial species, although an individual typically harbors only 100–200 of these. Establishing this community of microbes begins at birth and is mostly complete after about a year. The mix of species remains fairly constant throughout the person's lifetime but can vary markedly among individuals, even in the same household. However, the overall metabolic capabilities of the microbial community seem to matter more than which species are actually present. Contrary to expectations, there is no common, or core, microbiome. The Human Microbiome Project (https://commonfund.nih.gov/hmp/index) aims to better characterize the species that inhabit the human body and assess their contribution to human health and disease.

Bacteria and fungi in the small intestine ferment undigested carbohydrates, mainly polysaccharides that cannot be broken down by human digestive enzymes, and produce acetate, propionate, and butyrate. These short-chain fatty acids are absorbed by the host and are transformed into triacylglycerols for long-term storage. Intestinal bacteria also produce vitamin K, biotin, and folate, some of which can be taken up and used by the host. The importance of microbial digestion is illustrated by mice grown in a germ-free environment. Without the normal bacterial partners, the mice need to consume about 30% more food than normal animals whose digestive systems have been colonized by microorganisms.

Studies show that switching between meat-rich and plant-based diets rapidly changes the proportions of different types of intestinal microbes in humans, suggesting that the human–microbial ecosystem has evolved to let humans take advantage of varying food availability. But there is also evidence that thin and obese individuals harbor different proportions of two major types of bacteria, the Bacteroidetes and the Firmicutes. It is not the case that microbes from obese individuals are more efficient at extracting nutrients from food so that the host absorbs and stores the excess as fat. Rather, different classes of microbes appear to play a role in promoting or inhibiting immune responses in the gut, and the resulting degree of inflammation influences fuel-use patterns throughout the host.

? **Formulate a hypothesis to explain the link between increased antibiotic use over the past 60 years and the increasing incidence of obesity in humans.**

The rapid catabolism of muscle glucose exceeds the capacity of the mitochondria to reoxidize the resulting NADH and so generates lactate as an end product. This three-carbon molecule is excreted from the muscle cells and travels via the bloodstream to the liver, where it serves as a substrate for gluconeogenesis. The newly synthesized glucose can then return to the muscle cells to sustain ATP production even after the muscle glycogen has been depleted (**Fig. 19.3**). The free energy to drive gluconeogenesis in the liver is derived from ATP produced by the oxidation of fatty acids. In effect, *the Cori cycle transfers free energy from the liver to the muscles.*

FIGURE 19.3

The Cori cycle. The product of muscle glycolysis is lactate, which travels to the liver. Lactate dehydrogenase converts the lactate back to pyruvate, which can then be used to synthesize glucose by gluconeogenesis. The input of free energy in the liver (in the form of ATP) is recovered when glucose returns to the muscles to be catabolized.

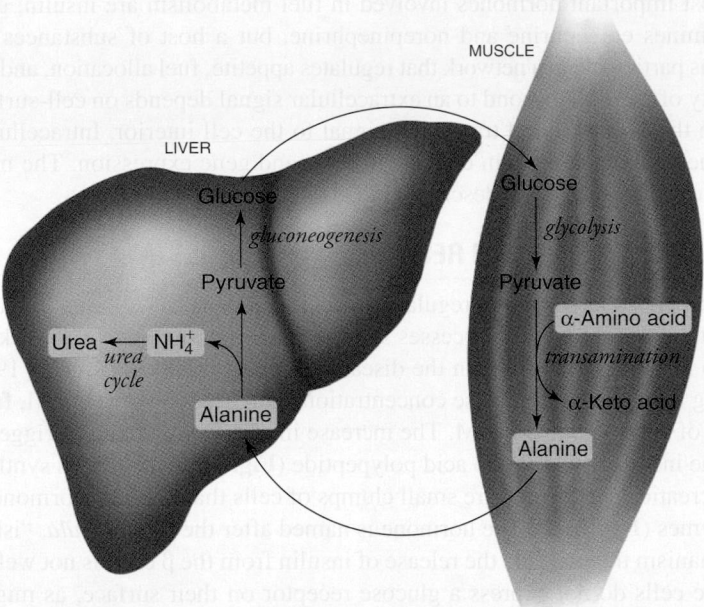

FIGURE 19.4

The glucose–alanine cycle.
The pyruvate produced by muscle glycolysis undergoes transamination to alanine, which delivers amino groups to the liver. The carbon skeleton of alanine is converted back to glucose to be used by the muscles, and the nitrogen is converted to urea for disposal.

 What is the fate of the α-keto acids in muscle?

A second interorgan pathway, the **glucose–alanine cycle,** also links the muscles and liver. During vigorous exercise, muscle proteins break down, and the resulting amino acids undergo transamination to generate intermediates to boost the activity of the citric acid cycle (Section 14.3). Transamination reactions convert pyruvate, a product of glycolysis, to alanine, which travels by the blood to the liver. There, the amino group is used for urea synthesis (Section 18.4) and the resulting pyruvate is converted back to glucose by the reactions of gluconeogenesis. As in the Cori cycle, the glucose returns to the muscle cells to complete the metabolic loop (**Fig. 19.4**). *The net effect of the glucose–alanine cycle is to transport nitrogen from muscles to the liver.*

 CONCEPT CHECK

1. Explain the importance of intracellular transport systems in coordinating metabolic activities.

2. Summarize the role of the liver, muscle, adipose tissue, and kidney in storing and mobilizing metabolic fuel.

3. What metabolic pathways are unique to the liver?

4. Draw a diagram showing fuel distribution to each organ in the fed state.

5. Draw a diagram showing fuel mobilization in each organ in the fasted state.

6. Describe the individual steps and the net effect of the Cori cycle and the glucose–alanine cycle.

HORMONAL CONTROL OF FUEL METABOLISM

Individual cells or organs must regulate the activities of their respective pathways according to their metabolic needs and the availability of fuel and building materials, which are supplied intermittently. Metabolic processes are not at equilibrium but can be continually adjusted to maintain a steady state. The human body buffers itself against fluctuations in the fuel supply by storing metabolic fuels, mobilizing them as needed, and replenishing them after the next meal. Maintaining a steady supply of glucose is especially critical for the brain, which exerts a large and relatively constant demand for glucose, regardless of how the dietary intake of carbohydrates varies or how much carbohydrate is oxidized to support other activities.

How does the body control the level of glucose and other fuels from hour to hour or day to day? The activities of organs that store and release fuels are coordinated by **hormones,** which are substances produced by one tissue that affect the functions of other tissues throughout the

KEY CONCEPTS
- Pancreatic cells release insulin when circulating glucose increases.
- Insulin stimulates the uptake of glucose and the storage of metabolic fuel.
- Glucagon and epinephrine promote glycogenolysis and lipolysis.
- Hormones produced by adipose tissue and digestive organs regulate appetite and fuel metabolism.
- AMP-dependent protein kinase activates ATP-producing pathways and inhibits ATP-consuming pathways.

FIGURE 19.5

Structure of human insulin. This two-chain hormone is colored by atom type: C gray, O red, N blue, H white, and S yellow. [*Structure (pdb 1AI0) determined by X. Chang, A. M. M. Jorgensen, P. Bardrum, and J. J. Led.*]

body. The most important hormones involved in fuel metabolism are insulin, glucagon, and the catecholamines epinephrine and norepinephrine, but a host of substances produced by multiple organs participate in a network that regulates appetite, fuel allocation, and body weight.

The ability of a cell to respond to an extracellular signal depends on cell-surface receptors that recognize the hormone and transmit a signal to the cell interior. Intracellular responses to the hormone include changes in enzyme activity and gene expression. The major types of signal transduction pathways are described in Chapter 10.

INSULIN IS RELEASED IN RESPONSE TO GLUCOSE

Insulin plays a large role in regulating fuel metabolism by stimulating activities such as glucose uptake and inhibiting processes such as glycogen breakdown. A lack of insulin or an inability to respond to it results in the disease **diabetes mellitus** (Section 19.3). Immediately following a meal, blood glucose concentrations may rise to about 8 mM, from a normal concentration of about 3.6 to 5.8 mM. The increase in circulating glucose triggers the release of the hormone insulin, a 51–amino acid polypeptide (**Fig. 19.5**). Insulin is synthesized in the β cells of pancreatic islets, which are small clumps of cells that produce hormones rather than digestive enzymes (**Fig. 19.6**). The hormone is named after the Latin *insula,* "island."

The mechanism that triggers the release of insulin from the β cells is not well understood. The pancreatic cells do not express a glucose receptor on their surface, as might be expected. Instead; *the cellular metabolism of glucose itself seems to generate the signal to release insulin.* In liver and pancreatic β cells, the glycolytic degradation of glucose begins with a reaction catalyzed by glucokinase (an isozyme of hexokinase; see Section 13.1):

$$\text{glucose} + \text{ATP} \rightarrow \text{glucose-6-phosphate} + \text{ADP}$$

The hexokinases in other cell types have a relatively low K_M for glucose (less than 0.1 mM), which means that the enzymes are saturated with substrate at physiological glucose concentrations. Glucokinase, in contrast, has a high K_M of 5–10 mM, *so it is never saturated and its activity is maximally sensitive to the concentration of available glucose* (**Fig. 19.7**).

Interestingly, the velocity versus substrate curve for glucokinase is not hyperbolic, as might be expected for a monomeric enzyme such as glucokinase. Instead, the curve is sigmoidal, which is typical of allosteric enzymes with multiple active sites operating cooperatively (see Section 7.2). The sigmoidal kinetics of glucokinase, which has only one active site, may be due to a substrate-induced conformational change such that at the end of the catalytic cycle, the enzyme briefly maintains a high affinity for the next glucose molecule. At high glucose concentrations, this would mean a high reaction velocity; at low glucose concentrations, the enzyme

FIGURE 19.6

Pancreatic islet cells. The pancreatic islets of Langerhans (named for their discoverer) consist of two types of cells. The β cells produce insulin, and the α cells produce glucagon. Most other pancreatic cells produce digestive enzymes. [*Carolina Biological Supply Co./Phototake.*]

FIGURE 19.7

Activities of glucokinase and hexokinase. Both enzymes catalyze the ATP-dependent phosphorylation of glucose as the first step of glycolysis. Glucokinase has a high K_M, so its reaction velocity changes in response to changes in glucose concentrations. In contrast, hexokinase is saturated with glucose at physiological concentrations (shaded region).

would operate more slowly because it reverts to a low-affinity conformation before binding another glucose substrate.

The role of glucokinase as a pancreatic glucose sensor is supported by the fact that mutations in the glucokinase gene cause a rare form of diabetes. However, other cellular factors may be involved, particularly in the mitochondria of the β cells. The glucose sensor responsible for triggering insulin release may also depend on the mitochondrial NAD$^+$/NADH or ADP/ATP ratios. For this reason, age-related declines in mitochondrial function may be a factor in the development of diabetes in the elderly.

Once released into the bloodstream, insulin can bind to receptors on cells in muscle and other tissues. Insulin binding to its receptor stimulates the tyrosine kinase activity of the receptor's intracellular domains (Section 10.3). These kinases phosphorylate each other as well as tyrosine residues in other proteins, including IRS-1 and IRS-2 (insulin receptor substrates 1 and 2). The IRS proteins then trigger additional events in the cell, not all of which have been fully characterized.

TABLE 19.1 Summary of Insulin Action

TARGET TISSUE	METABOLIC EFFECT
Muscle and other tissues	Promotes glucose transport into cells Stimulates glycogen synthesis Suppresses glycogen breakdown
Adipose tissue	Activates extracellular lipoprotein lipase Increases level of acetyl-CoA carboxylase Stimulates triacylglycerol synthesis Suppresses lipolysis
Liver	Promotes glycogen synthesis Promotes triacylglycerol synthesis Suppresses gluconeogenesis

INSULIN PROMOTES FUEL USE AND STORAGE

Only cells that bear insulin receptors can respond to the hormone, and the cells' response is tissue specific. In general, *insulin signals fuel abundance: It decreases the metabolism of stored fuel while promoting fuel storage.* The effects of insulin on various tissues are summarized in **Table 19.1**.

In tissues such as muscle and adipose tissue, insulin stimulates glucose transport into cells by several fold. The $V_{\max}$ for glucose transport increases, not because insulin alters the intrinsic catalytic activity of the transporter but because insulin increases the number of transporters at the cell surface. These transporters, named GLUT4, are situated in the membranes of intracellular vesicles. When insulin binds to the cell, the vesicles fuse with the plasma membrane. *This translocation of transporters to the cell surface increases the rate at which glucose enters the cell* (**Fig. 19.8**). GLUT4 is a passive transporter, operating as shown in Figure 9.12. When the insulin stimulus is removed, endocytosis returns the transporters to intracellular vesicles.

Insulin stimulates fatty acid uptake as well as glucose uptake. When the hormone binds to its receptors in adipose tissue, it activates the extracellular protein lipoprotein lipase, which hydrolyzes triacylglycerols in circulating lipoproteins so that the released fatty acids can be taken up for storage by adipocytes.

The insulin signaling pathway also alters the activity of glycogen-metabolizing enzymes. Glycogen metabolism is characterized by a balance between glycogen synthesis and glycogen degradation. Synthesis is carried out by the enzyme glycogen synthase, which adds glucose units donated by UDP–glucose to the ends of the branches of a glycogen polymer (see Section 13.3):

$$\text{UDP–glucose} + \text{glycogen}_{(n \text{ residues})} \rightarrow \text{UDP} + \text{glycogen}_{(n+1 \text{ residues})}$$

Glycogen phosphorylase mobilizes glucose residues from glycogen by phosphorolysis (cleavage through addition of a phosphoryl group rather than water):

$$\text{glycogen}_{(n \text{ residues})} + P_i \rightarrow \text{glycogen}_{(n-1 \text{ residues})} + \text{glucose-1-phosphate}$$

This reaction, followed by an isomerization reaction, yields glucose-6-phosphate, the first intermediate of glycolysis.

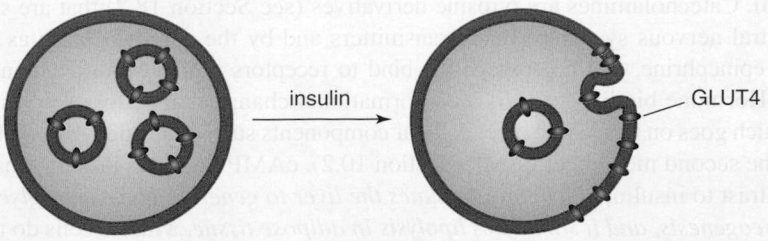

FIGURE 19.8

Effect of insulin on GLUT4. Insulin triggers vesicle fusion so that the glucose transport protein GLUT4 is translocated from intracellular vesicles to the plasma membrane. This increases the rate at which the cells take up glucose.

FIGURE 19.9

The reciprocal regulation of glycogen synthase and glycogen phosphorylase. Phosphorylation (transfer of a phosphoryl group from ATP to the enzyme) deactivates glycogen synthase and activates glycogen phosphorylase. Dephosphorylation has the opposite effect. The more active form of each enzyme is known as the *a* form (indicated in green), and the less active form is known as the *b* form (in red).

 Would insulin signaling lead to activation of the kinases or of the phosphatases in this diagram?

Glycogen synthase is a homodimer, and glycogen phosphorylase is a heterodimer. Both enzymes are regulated by allosteric effectors. For example, glycogen synthase is activated by glucose-6-phosphate. AMP activates glycogen phosphorylase and ATP inhibits it. These effects are consistent with the role of glycogen phosphorylase in making glucose available to boost cellular ATP production. However, *the primary mechanism for regulating glycogen synthase and glycogen phosphorylase is through covalent modification (phosphorylation and dephosphorylation) that is under hormonal control.* Both enzymes undergo reversible phosphorylation at specific serine residues. Phosphorylation deactivates glycogen synthase and activates glycogen phosphorylase. Removal of the phosphoryl groups has the opposite effect: Dephosphorylation activates glycogen synthase and deactivates glycogen phosphorylase (**Fig. 19.9**).

Covalent modification is a type of allosteric regulation (Section 7.3). The attachment or removal of the highly anionic phosphoryl group triggers a conformational shift between a more active (*a* or R) state and a less active (*b* or T) state. *The reciprocal regulation of glycogen synthase and glycogen phosphorylase promotes metabolic efficiency, since the two enzymes catalyze key reactions in opposing metabolic pathways.* The advantage of this regulatory system is that a single kinase can tip the balance between glycogen synthesis and degradation. Similarly, a single phosphatase can tip the balance in the opposite direction. Covalent modifications, such as phosphorylation and dephosphorylation, permit a much wider range of enzyme activities than could be accomplished solely through the allosteric effects of metabolites whose cellular concentrations do not vary much. Insulin signaling activates phosphatases that dephosphorylate (activate) glycogen synthase and dephosphorylate (deactivate) glycogen phosphorylase. As a result, glycogen synthesis accelerates and the rate of glycogenolysis decreases when glucose is abundant.

GLUCAGON AND EPINEPHRINE TRIGGER FUEL MOBILIZATION

Within hours of a meal, dietary glucose has mostly been taken up by cells and consumed as fuel, stored as glycogen, or converted to fatty acids for long-term storage. At this point, the liver must begin mobilizing glucose in order to keep the blood glucose concentration constant. This phase of fuel metabolism is governed not by insulin but by other hormones, mainly glucagon and the catecholamines epinephrine and norepinephrine.

Glucagon, a 29-residue peptide hormone, is synthesized and released by the α cells of pancreatic islets when the blood glucose concentration begins to drop below about 5 mM (**Fig. 19.10**). Catecholamines are tyrosine derivatives (see Section 18.2) that are synthesized by the central nervous system as neurotransmitters and by the adrenal glands as hormones. Glucagon, epinephrine, and norepinephrine bind to receptors with seven membrane-spanning segments. Hormone binding triggers a conformational change that activates an associated G protein, which goes on to activate other cellular components such as an adenylate cyclase, which produces the second messenger cAMP (Section 10.2). cAMP activates protein kinase A.

In contrast to insulin, *glucagon stimulates the liver to generate glucose by glycogenolysis and gluconeogenesis, and it stimulates* **lipolysis** *in adipose tissue.* Muscle cells do not express

FIGURE 19.10

Structure of glucagon. The atoms of the 29-residue peptide are colored by type: C gray, O red, N blue, H white, and S yellow. [*Structure (pdb 1GCN) determined by T. L. Blundell, K. Sasaki, S. Dockerill, and I. J. Tickle.*]

a glucagon receptor but do respond to catecholamines, which elicit the same overall effects as glucagon. Thus, epinephrine stimulation of muscle cells activates glycogenolysis, which makes more glucose available to power muscle contraction.

One of the intracellular targets of protein kinase A is phosphorylase kinase, the enzyme that phosphorylates (deactivates) glycogen synthase and phosphorylates (activates) glycogen phosphorylase. Consequently, hormones such as glucagon and epinephrine, which lead to cAMP production, promote glycogenolysis and inhibit glycogen synthesis. Although phosphorylase kinase is activated by protein kinase A, it is maximally active when Ca^{2+} ions are also present. Ca^{2+} concentrations increase during signaling via the phosphoinositide pathway, which responds to the catecholamine hormones (Section 10.2).

In adipocytes, protein kinase A phosphorylates an enzyme known as hormone-sensitive lipase, thereby activating it. This lipase catalyzes the rate-limiting step of lipolysis, the conversion of stored triacylglycerols to diacylglycerols and then to monoacylglycerols, which releases fatty acids. *Hormone stimulation not only increases the lipase catalytic activity, it also relocates the lipase from the cytosol to the fat droplet of the adipocyte.* Co-localization with its substrate, possibly by interacting with a lipid-binding protein, boosts the rate at which fatty acids are mobilized. Thus, glucagon and epinephrine promote the breakdown of both glycogen and fat. These responses are summarized in **Figure 19.11**.

ADDITIONAL HORMONES INFLUENCE FUEL METABOLISM

In addition to well-known endocrine organs such as the pancreas (the source of insulin and glucagon) and adrenal glands (the source of epinephrine and norepinephrine), many other tissues produce hormones that help regulate all aspects of food acquisition and use (**Table 19.2**). In fact, adipose tissue, once thought to be a relatively inert fat-storage site, actively communicates with the rest of the body.

Adipose tissue produces the hormone leptin, a 146-residue polypeptide that functions as a signal of satiety (satisfaction, or fullness). It acts on the hypothalamus, a part of the brain, to suppress appetite. The level of leptin is proportional to the amount of adipose tissue: The more fat that accumulates in the body, the stronger the appetite-suppressing signal.

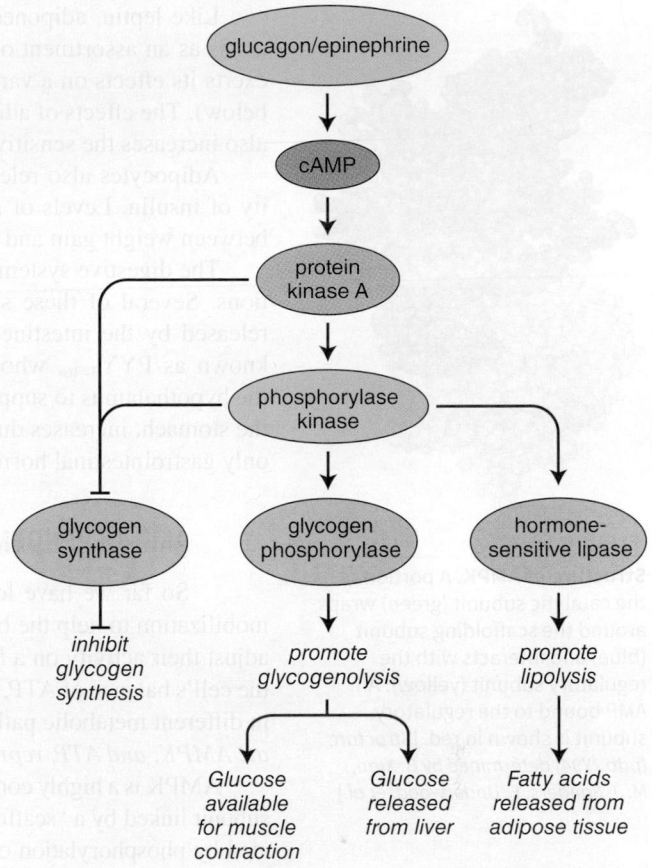

FIGURE 19.11

Effect of glucagon and epinephrine on fuel metabolism. Green arrows represent activation events, and red symbols represent inhibition. Both glucagon and epinephrine inhibit glycogen synthesis and promote the mobilization of glucose and fatty acids.

? Explain how these events account for the fight-or-flight response triggered by epinephrine (adrenaline).

TABLE 19.2 Some Hormones that Regulate Fuel Metabolism

HORMONE	SOURCE	ACTION
Adiponectin	Adipose tissue	Activates AMPK (promotes fuel catabolism)
Leptin	Adipose tissue	Signals satiety
Resistin	Adipose tissue	Blocks insulin activity
Neuropeptide Y	Hypothalamus	Stimulates appetite
Cholecystokinin	Intestine	Suppresses appetite
Incretin	Intestine	Promotes insulin release; inhibits glucagon release
PYY_{3-36}	Intestine	Suppresses appetite
Amylin	Pancreas	Signals satiety
Ghrelin	Stomach	Stimulates appetite

FIGURE 19.12

Structure of AMPK. A portion of the catalytic subunit (green) wraps around the scaffolding subunit (blue) and interacts with the regulatory subunit (yellow). An AMP bound to the regulatory subunit is shown in red. [*Structure (pdb 2Y94) determined by B. Xiao, M. J. Sanders, E. Underwood, et al.*]

Like leptin, adiponectin is released by adipose tissue, but this 247-residue polypeptide exists as an assortment of multimers with different receptor-binding properties. Adiponectin exerts its effects on a variety of tissues by activating an AMP-dependent protein kinase (see below). The effects of adiponectin include increased combustion of glucose and fatty acids. It also increases the sensitivity of tissues to insulin.

Adipocytes also release a 108-residue hormone called resistin, which blocks the activity of insulin. Levels of resistin increase during obesity, which would help explain the link between weight gain and decreased responsiveness to insulin (Section 19.3).

The digestive system produces at least 20 different peptide hormones with varying functions. Several of these signal that a meal has been consumed. For example, incretins are released by the intestine and enhance insulin secretion by the pancreas. The oligopeptide known as PYY_{3-36}, whose release is triggered especially by a high-protein meal, acts on the hypothalamus to suppress appetite. The level of ghrelin, a 28-residue peptide produced by the stomach, increases during fasting and decreases immediately following a meal; this is the only gastrointestinal hormone that stimulates appetite.

AMP-DEPENDENT PROTEIN KINASE ACTS AS A FUEL SENSOR

So far we have looked at a variety of signals that regulate fuel intake, storage, and mobilization to help the body maintain homeostasis. Individual cells also have a fuel gauge to adjust their activity on a finer scale. The AMP-dependent protein kinase (AMPK) responds to the cell's balance of ATP, ADP, and AMP to activate and inhibit a number of enzymes involved in different metabolic pathways. *AMP and ADP, representing the cell's need for energy, activate AMPK, and ATP, representing a state of energy sufficiency, inhibits the kinase.*

AMPK is a highly conserved Ser/Thr kinase consisting of a catalytic subunit and a regulatory subunit linked by a "scaffolding" subunit (**Fig. 19.12**). Like many other kinases, AMPK is activated by phosphorylation of a specific threonine residue. ADP binding to the regulatory subunit of AMPK prevents dephosphorylation of this residue, thereby maintaining the kinase in an active state (about 200 times more active than its dephospho form). AMP also acts as an allosteric activator of the kinase so that overall its activity increases about 2000-fold. ATP, which competes with AMP and ADP for binding to the regulatory subunit, inhibits AMPK. This multipart regulatory scheme allows AMPK to respond to a wide range of cellular energy states.

In addition to responding to intracellular energy deficits, AMPK responds to hormones such as leptin and adiponectin. *As a result of AMPK activity, the cell switches off ATP-consuming anabolic pathways and switches on ATP-generating catabolic pathways.* For example, in exercising muscle, AMPK phosphorylates and activates the enzyme that produces fructose-2,6-bisphosphate, an allosteric activator of phosphofructokinase, so that glycolytic flux increases (Section 13.1). In adipose tissue, AMPK phosphorylates and inactivates acetyl-CoA carboxylase, the enzyme that generates malonyl-CoA, to suppress fatty acid synthesis. Since malonyl-CoA inhibits fatty acid transport into mitochondria, AMPK increases the rate of mitochondrial β oxidation in tissues such as muscle. AMPK activation also promotes the production of new mitochondria. Some metabolic effects of AMPK are listed in **Table 19.3**.

TABLE 19.3 Effects of AMP-Dependent Protein Kinase	
TISSUE	**RESPONSE**
Hypothalamus	Increases food intake
Liver	Increases glycolysis Increases fatty acid oxidation Decreases glycogen synthesis Decreases gluconeogenesis
Muscle	Increases fatty acid oxidation Increases mitochondrial biogenesis
Adipose tissue	Decreases fatty acid synthesis Increases lipolysis

CONCEPT CHECK

1. Summarize the metabolic effects of insulin signaling on muscle cells and adipocytes.

2. How does glucokinase and hexokinase differ from each other?

3. How does insulin increase the rate of glucose entry into cells?

4. Explain how phosphorylation and dephosphorylation reciprocally regulate glycogen synthase and glycogen phosphorylase.

5. Summarize the metabolic effects of glucagon and epinephrine on liver cells and adipocytes.

6. Summarize the metabolic effects of hormones produced by adipose tissue and the digestive system.

7. How does AMPK function as a cell's energy sensor?

8. List the metabolic activities that are stimulated or suppressed by AMPK.

DISORDERS OF FUEL METABOLISM

The multifaceted regulation of mammalian fuel metabolism offers many opportunities for things to go wrong. Excessive intake and storage of fuel can cause obesity. Starvation results from insufficient food. The faulty regulation of carbohydrate and lipid metabolism can lead to diabetes. In this section we examine some of the biochemistry behind these conditions.

THE BODY GENERATES GLUCOSE AND KETONE BODIES DURING STARVATION

Most tissues in the body use glucose as their preferred fuel and turn to fatty acids only when the glucose supply diminishes. Except in the intestine, amino acids are not a primary fuel. But when no food is available for an extended period, the body must make adjustments to mobilize different types of fuels. An average adult can survive a famine lasting up to a few months, an adaptation likely shaped by seasonal food shortages during human evolution. Starvation in children, of course, may severely impact development (**Box 19.B**).

The liver and muscles store less than a day's supply of glucose in the form of glycogen. As glycogen stores are depleted, muscle switches from burning glucose to burning fatty acids. Insulin secretion ceases with the drop in circulating glucose, so insulin-responsive tissues are not stimulated to take up glucose. This means that more glucose will be available for tissues such as the brain, which stores very little glycogen and cannot use fatty acids as fuel.

The liver and kidneys respond to the continued demand for glucose by increasing the rate of gluconeogenesis, using noncarbohydrate precursors such as amino acids (derived from protein degradation) and glycerol (from fatty acid breakdown). After several days, the liver begins to convert mobilized fatty acids to acetyl-CoA and then to ketone bodies. These small water-soluble fuels are used by a variety of tissues, including the heart and brain. The gradual switch from glucose to ketone bodies prevents the body from using up its proteins to supply gluconeogenic precursors. During a 40-day fast, the concentration of circulating fatty acids varies about 15-fold, and the concentration of ketone bodies increases about 100-fold. In contrast, the concentration of glucose in the blood varies by no more than threefold. These patterns of fuel use are summed up in **Table 19.4**.

KEY CONCEPTS

- The body breaks down glycogen, fats, and proteins to generate glucose, fatty acids, and ketone bodies during starvation.
- Obesity appears to result from metabolic, environmental, and genetic factors.
- In diabetes, either a lack of insulin or the inability to respond to it leads to hyperglycemia.
- The metabolic syndrome is characterized by obesity and insulin resistance.

TABLE 19.4 Source of Metabolic Fuels Under Different Conditions

	CARBOHYDRATES (%)	FATTY ACIDS (%)	AMINO ACIDS (%)
Immediately after a meal	50	33	17
After an overnight fast	12	70	18
After a 40-day fast	0	95[a]	5

[a]This value reflects a high concentration of fatty acid–derived ketone bodies.

Box 19.B Marasmus and Kwashiorkor

Chronic malnutrition takes a toll on human life in many ways. For example, it exacerbates infectious diseases that would not necessarily be fatal in well-nourished individuals. Severely malnourished children also fail to reach their full potential in terms of body size and cognitive development, even if their food intake later increases to normal levels. Plentiful glucose for the brain is especially critical in infancy, when a child's brain is relatively large and its liver (where glycogen is stored) is relatively small.

There are two major forms of severe malnutrition, **marasmus** and **kwashiorkor,** which may also occur in combination. In marasmus, inadequate intake of metabolic fuels of all types causes wasting. Individuals with this condition are emaciated, with very little muscle mass and essentially no subcutaneous fat. Similar symptoms also develop in some chronic diseases such as cancer, tuberculosis, and AIDS.

Kwashiorkor results from inadequate protein intake, which may or may not be accompanied by inadequate energy intake. A child with kwashiorkor typically has thin limbs, reddish hair, and a swollen belly. Without an adequate supply of amino acids, the liver makes too little albumin, a protein that helps retain fluid inside blood vessels. When the concentration of albumin drops, fluid enters tissues by osmosis. This swelling (edema) also occurs in other diseases that impair liver function. The liver is enlarged in kwashiorkor due to the deposition of fat. Depigmentation of the hair and skin occurs because tyrosine, which is derived from the essential amino acid phenylalanine, is also the precursor of melanin, a brown pigment molecule.

 Explain why kwashiorkor sometimes develops in infants who are fed rice milk rather than breast milk.

OBESITY HAS MULTIPLE CAUSES

Obesity has become an enormous public health problem. In addition to its impact on the quality of life, it is physiologically costly: Masses of fat prevent the lungs from fully expanding; the heart must work harder to circulate blood through a larger body; and the additional weight stresses hip, knee, and ankle joints. Obesity also increases the risk of cardiovascular disease, diabetes, and cancer. And it is not an exaggeration to describe obesity as an epidemic, since it affects an estimated one-third of the adult population of the United States.

Like many conditions, obesity has no single cause. It is a complex disorder involving appetite and metabolism and reflecting environmental as well as genetic factors. Despite a high degree of heritability for obesity, only a few rare cases can be explained by a single defective gene. Although obesity is often considered to be a "lifestyle" disease, simple overeating or lack of exercise cannot account for more than about a 15-pound weight gain. Instead, obesity seems to result from the readjustment of multiple regulatory mechanisms and can be considered to be a chronic illness rather than a lifestyle choice.

The human body appears to have a **set-point** for body weight that remains constant and relatively independent of energy intake and expenditure even over many decades. The hormone leptin may help establish the set-point, since the absence of leptin causes severe obesity in rodents and humans (**Fig. 19.13**). However, the majority of obese humans do not appear to lack leptin, so they may instead be suffering from leptin resistance due to a defect somewhere in the leptin signaling pathway. When leptin is less effective at suppressing the appetite, the individual gains weight. Eventually, the increase in leptin concentration resulting from the increase in adipose tissue mass succeeds in signaling satiety, but the result is a high set-point (a higher body weight that must be maintained). This may be one reason why overweight people who manage to shed a few pounds often regain the lost weight and return to the original set-point.

It turns out that humans have several types of fat, including subcutaneous fat (beneath the skin), visceral fat (surrounding the abdominal organs), and brown fat. **Brown adipose tissue,** named for its high mitochondrial content, is specialized for generating heat to maintain body temperature. Brown fat is prominent in newborns and hibernating mammals (see Box 15.B), but it also occurs at least in small amounts in adult humans, mainly in the neck and in the body cavity. Developmentally and metabolically, brown adipose tissue more closely resembles muscle than ordinary white adipose tissue (**Fig. 19.14**). Instead of one

FIGURE 19.13

Normal and obese mice. The mouse on the left lacks a functional gene for leptin and is several times the size of a normal mouse (right). [*The Rockefeller University/AP/© Wide World Photos.*]

(a)

(b)

FIGURE 19.14

Brown and white adipose tissue. (a) In brown adipose tissue, cells contain relatively more mitochondria, and triacylglycerols are present as numerous small globules in the cytoplasm. (b) In white adipose tissue, each cell is occupied mostly by a single large fat globule, and there is little cytoplasm. [*Biophoto Associates/ Photo Researchers, Inc.*]

large fat globule, brown adipose tissue contains many small fat droplets, which are a source of fatty acids that are oxidized to generate heat.

The hormone norepinephrine binds to receptors on brown adipocytes, and signal transduction via protein kinase A activates a lipase that liberates fatty acids from triacylglycerols. The uncoupling protein (UCP) is expressed in the mitochondria of brown adipose tissue, so fuel oxidation occurs without ATP synthesis. A compelling hypothesis is that lean individuals have a higher capacity to burn off excess fuel in this manner instead of storing it in white adipose tissue. In fact, the amount of brown fat appears to be inversely proportional to the degree of obesity. Intriguingly, hormonal stimulation of white fat causes some cells (called beige fat) to develop the characteristics of brown fat, which suggests that there may be a way to manipulate the body's fat deposits to treat severe obesity.

DIABETES IS CHARACTERIZED BY HYPERGLYCEMIA

Another well-characterized disorder of fuel metabolism is diabetes mellitus, which affects about 10% of the population of the United States. Worldwide, the disease affects about 350 million people, killing about 3.5 million each year. The highest rates of diabetes are observed in middle-income countries, where around 20% of the population is affected.

The words *diabetes* (meaning "to run through") and *mellitus* ("honey") describe an obvious symptom of the disease. Diabetics excrete large amounts of urine containing high concentrations of glucose (the kidneys work to eliminate excess circulating glucose by excreting it in urine, a process that requires large amounts of water).

Type 1 diabetes (juvenile-onset or insulin-dependent diabetes) is an autoimmune disease in which the immune system destroys pancreatic β cells. Symptoms first appear in childhood as insulin production begins to drop off. At one time, the disease was invariably fatal. This changed dramatically in 1922, when Frederick Banting and Charles Best administered a pancreatic extract to save the life of a severely ill diabetic boy (**Fig. 19.15**). Since then, the treatment of type 1 diabetes with purified insulin has been refined, with delivery options including preloaded syringes and small pumps.

An ongoing challenge lies in tailoring the delivery of insulin to the body's needs over the course of a typical 24-hour cycle of eating and fasting. Diabetics typically measure the glucose concentration in a tiny sample of blood—often less than a microliter—several times a day. Devices for noninvasive, continual glucose monitoring are being tested.

Freeing patients from frequent needle sticks, whether for monitoring blood glucose or injecting insulin, is the goal of pancreatic islet cell transplants, which have seen some success. Gene therapy (Section 3.5) to treat diabetes is an elusive goal because the insulin gene must be introduced into the body in such a way that the gene's expression is glucose-sensitive.

By far the most common form of diabetes, accounting for up to 95% of all cases, is type 2 diabetes (also known as adult-onset or non-insulin-dependent diabetes). These cases are characterized by **insulin resistance,** which is the failure of the body to respond to normal or even elevated concentrations of the hormone. Only a small fraction of patients with type 2 diabetes bear genetic defects in the insulin receptor, as might be expected; in the majority of cases, the underlying cause is not known.

The primary feature of untreated diabetes is chronic **hyperglycemia** (high levels of glucose in the blood). *The loss of responsiveness of tissues to insulin means that cells fail to take up glucose.* The body's metabolism responds as if no glucose were available, so liver gluconeogenesis increases, further promoting hyperglycemia. Glucose circulating at high concentrations can participate in nonenzymatic glycosylation of proteins. This process is slow, but the modified

FIGURE 19.15

Banting and Best. Frederick Banting (right) and Charles Best (left) surgically removed the pancreas of dogs to induce diabetes. When preparations of the pancreatic tissue were administered to the animals, their symptoms improved. This work laid the foundation for treating human diabetes with pancreatic extracts, which contain insulin. [*Hulton Archive/Getty Images, Inc.*]

FIGURE 19.16

Photo of a diabetic cataract. The accumulation of sorbitol in the lens leads to swelling and precipitation of lens proteins. The resulting opacification can cause blurred vision or complete loss of sight. [*Courtesy Dr. Manuel Datiles III, Cataract and Cornea Section, OGCSB, National Eye Institute, National Institutes of Health.*]

proteins may gradually accumulate and damage tissues with low turnover rates, such as neurons.

Tissue damage also results from the metabolic effects of hyperglycemia. Since muscle and adipose tissue are unable to increase their uptake of glucose in response to insulin, glucose tends to enter other tissues. Inside these cells, aldose reductase catalyzes the conversion of glucose to sorbitol:

$$NADPH + H^+ \longrightarrow NADP^+$$

Glucose → aldose reductase → Sorbitol

Because aldose reductase has a relatively high K_M for glucose (about 100 mM), flux through this reaction is normally very low. But under hyperglycemic conditions, sorbitol accumulates and may alter the cell's osmotic balance. This may alter kidney function and may trigger protein precipitation in other tissues. Aggregation of lens proteins leads to cataracts (**Fig. 19.16**). Neurons and cells lining blood vessels may be similarly damaged, increasing the likelihood of neuropathies and circulatory problems that in severe cases result in kidney failure, heart attack, stroke, or the amputation of extremities.

Although commonly considered a disorder of glucose metabolism, *diabetes is also a disorder of fat metabolism,* since insulin normally stimulates triacylglycerol synthesis and suppresses lipolysis in adipocytes. Uncontrolled diabetics tend to metabolize fatty acids rather than carbohydrates, and the resulting production of ketone bodies may give the breath a sweet odor. Overproduction of ketone bodies leads to diabetic ketoacidosis (Section 2.5).

A variety of drugs have been developed to help compensate for the physiological effects of insulin resistance; within each class of drugs there are multiple options with slightly different pharmacokinetics (**Table 19.5**). For example, metformin improves diabetic symptoms by activating AMPK in liver and other tissues. Liver glucose production is suppressed by the decreased expression of the gluconeogenic enzymes phosphoenolpyruvate carboxykinase and glucose-6-phosphatase (Section 13.2). Metformin also increases glucose uptake and fatty acid oxidation in muscle.

Drugs of the thiazolidinedione class, such as rosiglitazone (Avandia®), act via intracellular receptors known as peroxisome proliferator–activated receptors. These receptors, which normally respond to lipid signals, are transcription factors that alter gene expression (Section 10.4). Thiazolidinediones increase adiponectin levels and decrease resistin levels (in fact, research on the pharmacology of these drugs led to the discovery of resistin). The net result is an increase in insulin sensitivity.

Many diabetic patients use a combination of drugs to help lower blood glucose levels. The most widely prescribed drugs can be taken orally, although they all have side effects. For example, the increased risk of heart attacks has severely restricted the use of rosiglitazone.

TABLE 19.5 Some Antidiabetic Drugs

CLASS	EXAMPLE	MECHANISM OF ACTION
Biguanides	Metformin (Glucophage®)	Stimulates AMPK; reduces glucose release from liver; increases glucose uptake by muscle
Sulfonylureas	Glipizide (Glucotrol®)	Blocks a K^+ channel in β cells, leading to increased production and secretion of insulin
Thiazolidinediones	Rosiglitazone (Avandia®)	Binds to peroxisome proliferator–activated receptors to activate gene transcription; increases insulin sensitivity

THE METABOLIC SYNDROME LINKS OBESITY AND DIABETES

In diabetes, the body behaves as if it were starving. Paradoxically, about 80% of patients with type 2 diabetes are obese, and obesity—particularly when abdominal fat deposits are large—is strongly correlated with the development of the disease. Some researchers use the term **metabolic syndrome** to refer to a set of symptoms, including obesity and insulin resistance, that appear to be related. As many as 40% of Americans over age 60 meet the criteria for a diagnosis of metabolic syndrome. Individuals with this disorder often develop type 2 diabetes, they may have atherosclerosis and hypertension (high blood pressure) that put them at risk for a heart attack, and they have a higher incidence of cancer. Several factors appear to underlie metabolic syndrome and to link obesity with diabetes.

Individuals with metabolic syndrome tend to have a relatively high proportion of visceral fat (assessed as a high waist-to-hip ratio). This type of fat exhibits a different hormone profile than subcutaneous fat. For example, visceral fat produces less leptin and adiponectin (hormones that increase insulin sensitivity) and more resistin (which promotes insulin resistance). Visceral fat also produces a hormone called tumor necrosis factor α (TNFα), which is a powerful mediator of inflammation, a normal part of the body's immune defenses. Chronic inflammation triggered by the visceral fat–derived TNFα may be responsible for some of the symptoms, such as atherosclerosis, that characterize metabolic syndrome. The TNFα signaling pathway in cells may lead to phosphorylation of IRS-1, a modification that prevents its activation by the insulin receptor kinase. This would explain the insulin resistance of metabolic syndrome and may also explain why disturbances in the intestinal microbiome that lead to inflammation (see Box 19.A) may trigger insulin resistance.

Another possible cause of metabolic syndrome, which may operate in concert with inflammation, is fat toxicity. High levels of dietary fatty acids promote fat accumulation in muscle tissue in addition to adipose tissue, impairing GLUT4 translocation and impeding glucose uptake. Circulating fatty acids also trigger gluconeogenesis in the liver, contributing to hyperglycemia. Pancreatic β cells respond to the hyperglycemia by increasing insulin secretion, which may stress the cells to the point of death, resulting in "β cell exhaustion."

Whatever its biochemical basis, the link between obesity and metabolic syndrome is underscored by the improvement of symptoms when the individual loses weight. If lifestyle changes related to diet and exercise are not effective, metabolic syndrome can be treated by the same drugs used to treat type 2 diabetes, since these increase insulin sensitivity.

 CONCEPT CHECK

1. Summarize the metabolic changes that occur during starvation.

2. Explain how a signaling molecule such as leptin could help determine the set-point for body weight.

3. What are the differences between type 1 and type 2 diabetes?

4. Explain why hyperglycemia is a symptom of diabetes.

5. Describe the action of some antidiabetic drugs.

6. How is obesity related to type 2 diabetes?

CLINICAL CONNECTION: CANCER METABOLISM

The typical patterns of fuel use described in Fig. 19.2 are altered in cancer cells, whose metabolism must support rapid growth and cell division. Normal differentiated cells grow slowly, if at all, and rely on oxidative phosphorylation to meet their energy needs. In contrast, most cancer cells are characterized by rapid, uncontrolled proliferation and carry out glycolysis at a high rate. At first, this activity was attributed to a lack of oxygen in tumors (glycolysis

KEY CONCEPT
- Cancer cells are characterized by rapid, uncontrolled proliferation and increased metabolic rate.

is an anaerobic pathway), but cancer cells actually consume large quantities of glucose even when oxygen is plentiful.

Heightened glucose uptake by cancer cells is the basis for the PET (positron emission tomography) scan used to locate tumors and monitor their growth. About an hour before entering the scanner, a patient is given an injection of 2-deoxy-2-[^{18}F]fluoroglucose (fluorodeoxyglucose, or FDG), which is taken up by all glucose-metabolizing cells. Decay of the ^{18}F isotope emits a positron (which is like an electron with a positive charge) that is ultimately detected as a flash of light. The PET scanner generates a two- or three-dimensional map showing the location of the tracer, that is, tissues with a high rate of glucose uptake (**Fig. 19.17**).

FIGURE 19.17

PET scan. The dark areas of the PET scan indicate tissues that rely heavily on glucose metabolism (brain, heart) or accumulate the radioactive tracer (bladder). The small dark areas are tumors. [*Living Art Enterprises/Photo Researchers.*]

AEROBIC GLYCOLYSIS SUPPORTS BIOSYNTHESIS

Aerobic glycolysis, known as the **Warburg effect,** has puzzled biochemists since Otto Warburg described it in the 1920s. One would expect that cancer cells would outcompete normal cells by burning fuel and generating ATP using more efficient pathways. In fact, cancer cells do perform oxidative phosphorylation, yet the cells still consume large amounts of glucose and eliminate waste carbons in the form of lactate rather than CO_2. Why do cancer cells use glucose in an apparently inefficient manner? Glycolysis is not just an ATP-generating pathway; *it also converts glucose carbons to the precursors for anabolic processes such as the synthesis of fatty acids, amino acids, and nucleotides, all of which are needed in large amounts as cells divide.*

The need for biosynthetic precursors also seems to explain why—despite the high rate of glucose catabolism—many cancer cells express a variant pyruvate kinase that has lower enzymatic activity. The resulting bottleneck in glycolytic flux may help divert some glucose carbons into other pathways. For example, the glycolytic intermediate 3-phosphoglycerate can be diverted to generate serine, which, when converted to glycine, supplies one-carbon groups for other metabolic processes, including thymidine synthesis, that support cell growth and division. Glycine itself is a precursor of adenine and guanine nucleotides. Serine allosterically activates the pyruvate kinase variant, so that glucose is converted to pyruvate when serine is plentiful.

Glycolytically generated pyruvate is a precursor of alanine, which is, after leucine, the most abundant amino acid in proteins. Pyruvate can also be processed to acetyl-CoA, which combines with oxaloacetate to form citrate. But rather than continuing through the citric acid cycle, the citrate is bled off and converted by the action of ATP-citrate lyase back to acetyl groups destined for fatty acid synthesis. Both citrate and fatty acyl-CoA molecules inhibit the activity of phosphofructokinase, thereby directing relatively more glucose through the pentose phosphate pathway, which yields the ribose and NADPH required for nucleotide synthesis.

CANCER CELLS CONSUME LARGE AMOUNTS OF GLUTAMINE

Glutamine, the most abundant amino acid in the body, supports the growth of cancer cells by providing a source of nitrogen for purine and pyrimidine synthesis. In addition, glutamate derived from glutamine can be deaminated by glutamate dehydrogenase to produce α-ketoglutarate. Although the increased α-ketoglutarate can increase flux through the citric acid cycle for the purpose of oxidative phosphorylation, some citric acid cycle intermediates have other fates. For example, malate can be converted by malic enzyme to pyruvate; this reaction also generates NADPH for biosynthetic pathways. Oxaloacetate can be transaminated to aspartate, which is a precursor for purine nucleotides. Oxaloacetate can also condense with acetyl-CoA to generate citrate to supply cytosolic acetyl groups for fatty acid synthesis.

Glutamate dehydrogenase is a major control point for cancer cell metabolism. The enzyme is activated by ADP and inhibited by GTP, which helps tie amino acid metabolism to the cell's energy budget. Leucine stimulates glutamate dehydrogenase activity, which could help balance the amino acid supply in cancer cells. Palmitoyl-CoA inhibits glutamate dehydrogenase activity, so that when fatty acid synthesis rates are high and acyl-CoA groups accumulate, less α-ketoglutarate is added to the citric acid cycle.

FIGURE 19.18

Cancer metabolism. Some of the biosynthetic pathways that require glucose and glutamine are highlighted.

When fatty acid oxidation rates are high, the glutamate dehydrogenase inhibition is relieved and glutamate can replenish α-ketoglutarate to increase the flux of fat-derived acetyl-CoA through the citric acid cycle.

Some of the metabolic patterns in cancer cells are summarized in **Figure 19.18**. Similar metabolic adjustments occur in some other types of rapidly dividing cells, including white blood cells responding to an infection.

Eighty percent of individuals with advanced cancer develop **cachexia,** the extreme loss of weight and muscle mass that is also common in the late stages of some other chronic illnesses such as tuberculosis and AIDS. Cachexia cannot be reversed by consuming more food, suggesting that the high rate of protein catabolism that results in muscle wasting is the result of reprogramming the entire body's metabolism, which includes adipose tissue and the brain's appetite-control center. During cachexia, white fat undergoes "browning" so that the body burns up its stored fuel. Halting this process would be one way to improve the quality of life in patients with end-stage illness, even if the underlying disease cannot be cured.

 CONCEPT CHECK

1. Explain how aerobic glycolysis differs from anaerobic glycolysis.

2. Summarize how the metabolism of glucose and glutamine supports the synthesis of nucleotides, lipids, and amino acids.

SUMMARY

19.1 Integration of Fuel Metabolism

• The liver is specialized to store glucose as glycogen, to synthesize triacylglycerols, to carry out gluconeogenesis, and to synthesize ketone bodies and urea. The muscles synthesize glycogen and can use glucose, fatty acids, and ketone bodies as fuel. Adipose tissue stores fatty acids as triacylglycerols.

• Pathways such as the Cori cycle and the glucose–alanine cycle link different organs.

19.2 Hormonal Control of Fuel Metabolism

• Insulin, which is synthesized by the pancreas in response to glucose, binds to a receptor tyrosine kinase. Cellular responses to insulin include increased uptake of glucose and fatty acids.

• The balance between glycogen synthesis and degradation depends on the relative activities of glycogen synthase and glycogen phosphorylase, which are controlled by hormone-triggered phosphorylation and dephosphorylation.

• Glucagon and catecholamines lead to the activation of cAMP-dependent protein kinase, which promotes glycogenolysis in liver and muscle, and lipolysis in adipose tissue.

• Adipose tissue is the source of the hormones leptin, adiponectin, and resistin, which help regulate appetite, fuel combustion, and insulin resistance. The stomach, intestines, and other organs also produce hormones that regulate appetite.

• AMP is an allosteric activator of AMPK, whose activity switches on pathways such as glycolysis and fatty acid oxidation.

19.3 Disorders of Fuel Metabolism

• In starvation, glycogen stores are depleted, but the liver makes glucose from amino acids and converts fatty acids into ketone bodies.

• The cause of obesity is not clear but may involve a failure in leptin signaling that raises the body weight set-point.

• The most common form of diabetes is characterized by insulin resistance, the inability to respond to insulin. The resulting hyperglycemia can lead to tissue damage.

• Metabolic disturbances resulting from obesity may lead to insulin resistance, a condition termed metabolic syndrome.

19.4 Clinical Connection: Cancer Metabolism

Cancer cell metabolism is characterized by increased consumption of glucose for aerobic glycolysis, and increased consumption of glutamine. In cancer cells, metabolic flux is directed toward the biosynthesis of amino acids, nucleotides, and fatty acids to support rapid cell growth and division.

KEY TERMS

Cori cycle	diabetes mellitus	set-point	metabolic syndrome
microbiome	lipolysis	brown adipose tissue	Warburg effect
glucose–alanine cycle	marasmus	insulin resistance	cachexia
hormone	kwashiorkor	hyperglycemia	

BIOINFORMATICS

Brief Bioinformatics Exercises

19.1 Insulin Secretion and the KEGG Database

19.2 Insulin Control of Glucose Metabolism and the KEGG Database

19.3 Glucagon Control of Glycogen and Glucose Metabolism

PROBLEMS

19.1 Integration of Fuel Metabolism

1. Name the two small metabolites at the "crossroads" of metabolism. How are these metabolites connected to the metabolic pathways we have studied?

2. Glucose-6-phosphate (G6P) is a metabolite that is linked to several metabolic pathways in carbohydrate metabolism. Describe how glucose-6-phosphate is linked to these pathways.

3. Incubating brain slices in a medium containing ouabain (a Na,K-ATPase inhibitor) decreases respiration by 50%. What does this tell you about ATP use in the brain? What pathways are involved in producing ATP in the brain?

4. Red blood cells lack mitochondria. Describe the metabolic pathways involved in ATP production in the red blood cell. What is the ATP yield per glucose molecule?

5. Adenylate kinase catalyzes the reaction

$$ATP + AMP \rightleftharpoons 2\ ADP$$

(a) How can you tell whether this reaction is likely to be a near-equilibrium reaction?

(b) Explain why muscle adenylate kinase would be very active during vigorous exercise.

6. During exercise, the concentration of AMP in muscle cells increases (see Solution 5). AMP is a substrate for the adenosine deaminase reaction:

$$AMP + H_2O \rightarrow IMP + NH_4^+$$

AMP is subsequently regenerated by a process in which the amino group of aspartate becomes attached to the purine ring of IMP and fumarate is released in a series of reactions known as the purine nucleotide cycle.

(a) What is the likely fate of the fumarate product?

(b) Why doesn't the muscle cell increase the concentration of citric acid cycle intermediates by converting aspartate to oxaloacetate by a simple transamination reaction?

(c) Ammonium ions stimulate the activity of phosphofructokinase and pyruvate kinase. Explain how adenosine deaminase activity could promote ATP production in active muscle.

7. How can glucose provide all of the substrates needed for fatty acid biosynthesis?

8. What is the "energy cost" in ATP of running the Cori cycle? How is the ATP obtained?

9. In the Cori cycle, the muscle converts pyruvate into lactate, which then diffuses out of the muscle and travels via the bloodstream to the

liver, where the reaction is reversed and lactate is converted back to pyruvate. Why is this extra step necessary? Why doesn't the muscle simply release pyruvate for uptake by the liver?

10. Explain how the reactions of the glucose–alanine cycle would operate during starvation.

11. What happens to plasma alanine levels in patients with inherited diseases of pyruvate or lactate metabolism that result in elevated plasma pyruvate levels? Explain.

12. An infant seemed normal at birth but was diagnosed at the age of three months with a pyruvate carboxylase deficiency. She suffered from lactic acidosis and ketosis. She had poor muscle tone and was experiencing seizures.

(a) Which metabolites would be elevated in this patient? Which metabolites would be deficient?

(b) Why does the patient suffer from lactic acidosis and ketosis?

(c) The poor muscle tone results from a lack of the neurotransmitter amino acids glutamate, aspartate, and γ-aminobutyric acid (GABA). Why would a pyruvate carboxylase deficiency decrease synthesis of these neurotransmitters?

(d) Acetyl-CoA was added to the patient's cultured fibroblasts to see whether pyruvate carboxylase activity could be detected. What was the rationale behind this experiment?

13. Physicians treating an infant with a pyruvate carboxylase deficiency (see Problem 12) noted that the patient suffered from hyperammonemia and that plasma levels of citrulline were elevated. Provide an explanation for this observation.

14. Treating farm animals such as cows and chickens with low doses of antibiotics promotes weight gain.

(a) Propose an explanation for this observation.

(b) The widespread use of antibiotics in animals has been implicated in the rise of antibiotic resistance in human pathogenic bacteria. How might this happen?

15. The caterpillar (larva) of the blue *Morpho* butterfly contains about 20 mg of fatty acids, compared to about 7 mg in the adult butterfly. What does this tell you about the energy source for metamorphosis (when the insect does not eat)?

16. During early lactation, cows cannot eat enough to supply the nutrients required for milk production. A recent study showed that enzymes involved in glycogen synthesis and the citric acid cycle were decreased during the early lactation period, whereas glycolytic enzyme activity, lactate production, and fatty acid degradation activity increased. What metabolic strategies are used by the cows in order to obtain the nutrients required for milk production in the absence of sufficient food intake? Be sure to specify which tissues are involved.

19.2 Hormonal Control of Fuel Metabolism

17. Glycogen phosphorylase cleaves glucose residues from glycogen via a phosphorolytic rather than a hydrolytic cleavage.

(a) Write an equation for each process.

(b) What is the metabolic advantage of phosphorolytic cleavage?

18. Estimate the K_M values for hexokinase and glucokinase from Figure 19.7. Compare these values to the normal blood glucose concentration, which ranges from 3.6–5.8 mM but can reach over 8 mM immediately after a meal.

19. Why does the sigmoidal behavior of glucokinase, the liver isozyme of hexokinase, help the liver to adjust its metabolic activities to the amount of available glucose?

20. Explain why a tyrosine phosphatase might be involved in limiting the signaling effect of insulin.

21. Why is insulin required for triacylglycerol synthesis in adipocytes?

22. Glycogen synthase kinase 3 (GSK3) can phosphorylate glycogen synthase in muscle cells. Activation of the insulin receptor leads to the activation of protein kinase B (Akt; see Section 10.2), which phosphorylates GSK3. How does insulin affect glycogen metabolism through GSK3?

23. In addition to its role in activating AMPK, adiponectin blocks the phosphorylation of glycogen synthase kinase 3 (GSK3; see Problem 22). What is the effect of adiponectin on glycogen metabolism?

24. Inexperienced athletes might consume a meal high in glucose just before a race, but veteran marathon runners know that doing so would impair their performance. Explain.

25. The results of one study showed that glucagon leads to an increase in the rate of glucose-6-phosphate hydrolysis.

(a) How could this explain the following results: Phosphoenolpyruvate concentration increased twofold, glucose-6-phosphate concentration decreased by 60%, and hepatic glucose concentration increased twofold in the presence of glucagon and exogenously administered dihydroxyacetone phosphate.

(b) Inhibition of glucose-6-phosphate hydrolysis resulted in both activation of gluconeogenesis and inhibition of glycolysis. Explain.

26. A 15-year-old male patient sees a physician because his parents are concerned about his inability to perform any kind of strenuous exercise without suffering painful muscle cramps. His liver is normal in size, but his muscles are flabby and poorly developed. Liver and muscle biopsies reveal that glycogen content in the liver is normal, but muscle glycogen content is elevated. The biochemical structure of glycogen in both tissues appears to be normal. A fasting glucose test shows that he is neither hypo- nor hyperglycemic. The patient's response to glucagon is tested by injecting a high dose of glucagon intravenously and then drawing samples of blood periodically and measuring the glucose content.

(a) After the glucagon injection, the patient's blood sugar rises dramatically. Is this the response you would expect in a normal person? Explain.

(b) Suggest an explanation for the results obtained from the liver and muscle biopsies. What type of glycogen storage disease does this patient have?

(c) The patient performs 30 minutes of ischemic (anaerobic) exercise and blood is withdrawn every few minutes and analyzed for alanine. In a normal person, the concentration of alanine in blood increases during ischemic exercise. But in the patient, alanine decreases during exercise, leading you to believe that his muscle cells are taking up alanine rather than releasing it. Why would blood alanine concentrations increase in a normal person? Why do blood alanine concentrations decrease in the patient?

(d) Explain why the patient does not suffer from either hypo- or hyperglycemia.

(e) The patient is advised to avoid strenuous exercise. If he does wish to perform light or moderate exercise, he is advised to consume sports drinks containing glucose or fructose frequently while exercising. Why would this help alleviate the muscle cramps suffered during exercise?

27. Phosphorylase kinase is one of the most complex enzymes known. It consists of four copies each of four different subunits, denoted as $\alpha_4\beta_4\gamma_4\delta_4$. The γ subunit contains the catalytic site. The α and β subunits can be phosphorylated. The δ subunit is calmodulin (see Section 10.2). What does this information tell you about the regulation of this enzyme's activity?

28. Hyperthyroidism is a condition that occurs when the thyroid secretes an excess of hormones, leading to an increase in the metabolic rate of all the cells in the body. Hyperthyroidism increases the demand for glucose and for substrates for oxidative phosphorylation. Which metabolic pathways are active in the fasted state in the liver, muscle, and adipose tissue in order to meet this demand?

29. AMPK affects gene transcription. Predict the effect of AMPK activation on the expression of (a) muscle GLUT4 and (b) liver glucose-6-phosphatase.

30. How would phosphorylation by AMPK affect the activity of the following enzymes. (a) glycogen synthase (b) phosphorylase kinase (c) HMG-CoA reductase and (d) hormone-sensitive lipase?

31. Insulin activates cAMP phosphodiesterase. Explain how this augments insulin's metabolic effect.

32. Why is the liver sometimes referred to as the body's "glucose buffer"?

19.3 Disorders of Fuel Metabolism

33. Explain why fasting increases the liver concentrations of phosphoenolpyruvate carboxykinase and glucose-6-phosphatase.

34. After several days of starvation, the ability of the liver to metabolize acetyl-CoA via the citric acid cycle is severely compromised. Explain why.

35. During a 24-hour fast, a person utilizes protein at a rate of 75 g/day. If a non-obese person has 6000 g of protein reserves and if death occurs when 50% of the protein reserves have been utilized, how prolonged can the fast be before death occurs?

36. In fact, during starvation, protein utilization does not progress at a rate of 75 g/day (Problem 35) but dramatically slows down to 20 g/day as the fast increases in duration. What body fuels are utilized during a prolonged fast in order to conserve body protein?

37. Why do dieters who follow the Atkins diet (a diet high in fat and protein and very low in carbohydrate) sometimes suffer from bad breath? (*Hint*: The odorous component of the breath is acetone.)

38. Individuals who are trying to lose weight are advised to consume fewer calories as well as to exercise. Why would exercising (keeping muscles active) help promote the loss of stored fat from adipose tissue?

39. Adipocytes secrete leptin, a hormone that suppresses appetite. Leptin exerts its effects through the central nervous system and also directly on target tissues by binding to specific receptors. Leptin can inhibit insulin secretion but can also act as an insulin mimic by activating some of the same intracellular signaling components as insulin. For example, leptin can induce tyrosine phosphorylation of insulin receptor substrate-1 (IRS-1). Using this information, predict leptin's effect on the following:

(a) Glucose uptake by skeletal muscle;

(b) Hepatic glycogenolysis and liver glycogen phosphorylase activity;

(c) cAMP phosphodiesterase activity.

40. What would happen if IRS-1 were overexpressed in rat muscle cells in culture?

41. Insulin resistance is characterized by the failure of insulin-sensitive tissues to respond to the hormone, which normally results in the uptake of glucose and its subsequent conversion to a storage form, such as glycogen or triacylglycerols. Why might inhibiting glycogen synthase kinase 3 (GSK3, see Problem 22) be useful for treating diabetes?

42. As described in Problem 23, adiponectin blocks the phosphorylation of glycogen synthase kinase 3 (GSK3; see Problem 22). How does the lack of adiponectin in an obese person predispose the individual to insulin resistance?

43. Adults have deposits of brown adipose tissue located mainly in the muscles of the lower neck and collarbone.

(a) Brown adipose tissue expresses more cytochrome *c* than white adipose tissue. What is the purpose of the elevated cytochrome *c*?

(b) Investigators measured the uptake of labeled glucose into brown fat in volunteers under room-temperature conditions and while the volunteers placed one foot in 7–9°C water. There was a 15-fold increase in uptake of labeled glucose by brown fat when the subjects were exposed to the colder temperature. Explain.

44. Decreased carnitine acyltransferase activity along with reduced activity of the mitochondrial electron transport chain have been observed in obese individuals. Explain the significance of these observations.

45. The activity of acetyl-CoA carboxylase is stimulated by a fat-free diet and inhibited in starvation and diabetes. Explain.

46. The properties of acetyl-CoA carboxylase (ACC) were studied to see whether the enzyme might be a possible drug target to treat obesity. Mammals have two forms of acetyl-CoA carboxylase, termed ACC1 and ACC2, whose properties are summarized in the table.

	ACC1	ACC2
Molecular mass (D)	265,000	280,000
Tissue expression	Liver and adipose tissue	Heart and muscle
Cellular location	Cytosol	Mitochondrial matrix
Sensitive to regulation by malonyl-CoA	Yes	Yes

(a) How does malonyl-CoA regulate the activity of acetyl-CoA carboxylase?

(b) In ACC2 knockout mice, the ACC2 gene is missing, but the ACC1 gene is still expressed normally. The ACC2-knockout mice showed a 20% reduction in liver glycogen compared to control mice. Explain why.

(c) In the knockout mice, the concentration of fatty acids in the blood was lower but the concentration of triacylglycerols was higher than in the wild-type mice. Explain.

(d) Fatty acid oxidation was measured in muscle tissue samples collected from both knockout and control mice. Administration of insulin caused a 45% decrease in palmitate oxidation in muscle tissue from normal mice, but there was no change in the rate of palmitate oxidation in the knockout mice. Explain.

(e) Both knockout and normal mice were allowed access to as much food as they cared to eat. At the end of a 27-week period, the knockout mice had consumed 20–30% more food than the wild-type mice. Interestingly, despite the increased food intake, the knockout mice weighed about 10% less and accumulated less fat in the adipose tissue than normal mice. Explain.

(f) How might you design the next new "diet pill," based on these results?

47. A fatty acid synthase (FAS) inhibitor ("Compound D" in Problem 17.40) was investigated as a possible candidate for a weight-loss drug.

(a) Mice were injected intraperitoneally with Compound D and radioactively labeled acetate. What is the fate of the label?

(b) Mice receiving intraperitoneal injections of Compound D reduced their food intake by more than 90% and lost nearly one-third of their body weight, although they gained back the weight when the drug was withdrawn. The investigators measured brain concentrations of neuropeptide Y (NPY), a compound known to act on the hypothalamus to increase appetite during starvation. Predict the effect of Compound D on brain levels of NPY.

(c) Because hepatic malonyl-CoA levels were high in the inhibitor-treated mice but not in control mice, the investigators hypothesized that malonyl-CoA inhibits feeding. If their hypothesis is correct, predict what would happen if mice were pretreated with an acetyl-CoA carboxylase inhibitor prior to injection with Compound D.

(d) What other cellular metabolites accumulate when concentrations of malonyl-CoA rise? (These molecules are candidates for signaling molecules, which might stimulate a biochemical pathway that decreases appetite.)

48. Explain why some drugs used to treat type 1 diabetes are compounds that diffuse into the cell and activate tyrosine kinases.

49. PTP-1B is a phosphatase that dephosphorylates the insulin receptor and might also dephosphorylate IRS-1.

(a) After feeding, mice deficient in PTP-1B reduce circulating blood glucose levels, using half as much insulin as normal mice. Explain this observation.

(b) What intracellular changes are observed in muscle cells when insulin is injected into the PTP-1B–deficient mice?

(c) How would you use this information to design a drug to treat diabetes? Are there any concerns involved in using the drug you have designed?

50. Several studies have shown that the insulin insufficiency of type 1 diabetes is accompanied by a hypersecretion of glucagon. The excess glucagon leads to release of glucose from the liver, which exacerbates the hyperglycemia characteristic of untreated diabetes. This observation has led to the suggestion that a diabetic's treatment regimen should include the administration of a glucagon antagonist along with insulin. A glucagon antagonist binds to receptors on the surface of liver cells but cannot initiate signal transduction. The antagonist prevents endogenous glucagon from binding and stimulating glycogenolysis.

Construction of a glucagon antagonist involves modifying the peptide hormone in such a way that amino acid residues important for binding are retained, while residues important for signal transduction are modified. There is an aspartate residue in the glucagon receptor that is essential for glucagon binding.

Glucagon analogs were synthesized in which amino acid residues were modified at several positions. The resulting analogs were tested for their ability to bind to liver membrane receptors and to initiate signal transduction (determined by measuring an increase in cAMP). A true antagonist binds to receptors without eliciting a response. Analogs capable of binding with diminished activity are referred to as partial agonists.

Sequence of human glucagon

1	2	3	4	5	6	7	8	9	10
His	Ser	Gln	Gly	Thr	Phe	Thr	Ser	Asp	Tyr

11	12	13	14	15	16	17	18	19	20
Ser	Lys	Tyr	Leu	Asp	Ser	Arg	Arg	Ala	Gln

21	22	23	24	25	26	27	28	29
Asp	Phe	Val	Gln	Trp	Leu	Met	Asn	Thr

(a) Glucagon carries out its biological function by binding to hepatic receptors and putting into motion a series of events that leads to glycogenolysis. Draw a diagram showing the steps of this process.

(b) Why did the investigators choose to modify the amino acids at positions 1, 12, and 18?

(c) Some of the synthetic glucagon analogs are listed in the table below. The ability of the glucagon analogs to both bind to receptors and elicit a biological response was measured and compared to native glucagon. What is the effect of substituting or eliminating the amino acid at position 9? Of replacement or modification at position 12? Of replacement at position 18? What is the role of the histidine at position 1?

(d) Of the glucagon analogs presented here, which is the best glucagon antagonist?

Glucagon analog (The *des* prefix indicates a deleted amino acid residue)	**Binding affinity (%)**	**% of maximum activity**
Glucagon	100	100
Des-Asp^9	45	8.3
Lys^9	54	0
Ala^{12}	17.3	59.7
Glu^{12}	1.0	80.4
Ala^{18}	13	94.4
Leu^{18}	56	95
Glu^{18}	6.2	100
Des-His^1	63	44
Des-His^1-Des-Asp^9	7	0
Des-His^1-Lys^9	70	0

51. Some obese patients with type 2 diabetes have undergone gastric bypass surgery, in which the upper part of the stomach is reconnected to the lower part of the small intestine. In some patients, the surgery appears to cure the symptoms of diabetes even before the patient has lost any weight. Propose an explanation for this observation.

52. One target of AMPK is phosphofructokinase-2, the enzyme that catalyzes the synthesis of fructose-2,6-bisphosphate (see Section 13.1). How does stimulating AMPK assist in the treatment of diabetes?

53. There is convincing evidence that AMPK can phosphorylate acetyl-CoA carboxylase. It's also possible that AMPK phosphorylates protein kinase B (see Section 10.2), which increases the translocation of GLUT4 vesicles to the plasma membrane. Given this information, how does metformin (Table 19.5) treat the symptoms of metabolic syndrome?

19.4 Clinical Connection: Cancer Metabolism

54. The compound 2-deoxy-D-glucose is structurally similar to glucose and can be taken up by cells via glucose transporters. **(a)** Once inside the cells, the compound is converted to 2-deoxy-D-glucose-6-phosphate by hexokinase. Draw the structures of the substrate and product of the reaction. **(b)** If 2-deoxy-D-glucose is added to cultured cancer cells, the intracellular concentration of ATP rapidly decreases. In a separate experiment, antimycin A (which prevents the transfer of electrons to cytochrome *c* in the electron transport chain) is added to the cells, but in this case there is no effect on ATP production. Explain these results.

55. Which glycolytic intermediates are used for amino acid biosynthesis in humans?

56. Positron emission tomography (PET) using glucose as a tracer is not useful for visualizing brain tumors because the background level of glucose uptake by the brain is already high.

(a) Based on what you know about cancer biology, propose an alternative substance that could be used to visualize brain tumors.

(b) Explain your choice of tracer.

57. In cancer cells with low pyruvate kinase activity, phosphoenolpyruvate accumulates and can donate its phosphoryl group to the active-site histidine in phosphoglycerate mutase. The phospho-His then spontaneously hydrolyzes from the enzyme. What is accomplished by this alternative pathway for phosphoenolpyruvate?

58. In addition to glucose, cancer cells consume relatively large amounts of glutamine. Explain

59. 2-deoxy-2-fluoroglucose (FDG) is taken up by cancer cells and phosphorylated by hexokinase.

(a) What prevents the reaction product from leaving the cells?

(b) The reaction product accumulates in the cancer cells. Explain why it cannot be further metabolized.

(c) Whole-body PET scans show high levels of FDG in the brain and bladder. Explain

60. Lactate stimulates growth in the cells that form new blood vessels. Does this observation help explain tumor growth?

61. The compound 5-aminoimidazole-4-carboxamide ribonucleoside (AICAR) activates AMPK. Adding AICAR to cancer cells in culture increases the concentration of reactive oxygen species (see Box 15.A). Explain why.

62. A type of renal cancer is characterized by a deficiency in fumarase.

(a) How would a fumarase deficiency affect the levels of pyruvate, fumarate, and malate?

(b) Why does a patient with a fumarase deficiency have the same symptoms as a patient with a succinate dehydrogenase deficiency?

63. Cancer researchers examined the relationship between fumarase and lactate dehydrogenase (LDH) activities and cell survival. They isolated cells from fumarase-deficient renal cancer patients (FH63 cells), then constructed a cell line in which they "knocked down" the expression of LDH in these cells (indicated as ΔLDH). The survival of the FH63 cells—with and without LDH—was compared to control cells and cells missing the LDH gene. The results are shown below.

(a) How does the percentage of apoptosis (programmed cell death) vary among the different types of cells?

(b) What strategies do the different types of cells use to survive?

64. A study of glioma tumor cells revealed an Arg → His mutation in an isozyme of isocitrate dehydrogenase. What single nucleotide change could result in this mutation?

65. The mutant isocitrate dehydrogenase described in Problem 64 produces 2-hydroxyglutarate, which accumulates in the cancer cell.

(a) Draw the structure of this metabolite.

(b) How does the reaction catalyzed by the mutant enzyme differ from the reaction catalyzed by the normal enzyme?

(c) How does operation of the mutant enzyme affect the cell's supply of reduced cofactors?

SELECTED READINGS

Betz, M. J. and Enerbäck, S., Human brown adipose tissue: what we have learned so far, *Diabetes* **64**, 2352–2360 (2015). [Summarizes research that links brown fat to obesity and diabetes and to potential treatments for these conditions.]

Cairns, R. A., Drivers of the Warburg phenotype, *Cancer J.* **21**, 56–61 (2015). [Describes the molecular mechanisms behind the metabolic changes that occur in tumors.]

Friedman, J. M., Causes and control of excess body fat, *Nature* **459**, 340–342 (2009). [A short question-and-answer summary of obesity and the role of leptin.]

Hardie, D. G., AMP-activated protein kinase—an energy sensor that regulates all aspects of cell function, *Genes Dev.* **25**, 1895–1908

(2011). [An extensive review of AMPK's effects on fuel metabolism and other aspects of cell growth.]

Odegaard, J. I. and Chawla, A., Pleiotropic actions of insulin resistance and inflammation in metabolic homeostasis, *Science* **339**, 172–177 (2013). [Discusses the links between insulin resistance and obesity.]

Rosen, E. D. and Spiegelman, B. M., Adipocytes as regulators of energy balance and glucose homeostasis, *Nature* **444**, 847–853 (2006). [Includes a summary of the actions of adipocyte hormones.]

Rui, J., Energy metabolism in the liver, *Compr. Physiol.* **4**, 177–197 (2014). [A detailed review of carbohydrate and lipid metabolism in the fed and fasted states.]

C H A P T E R **20**

DNA REPLICATION AND REPAIR

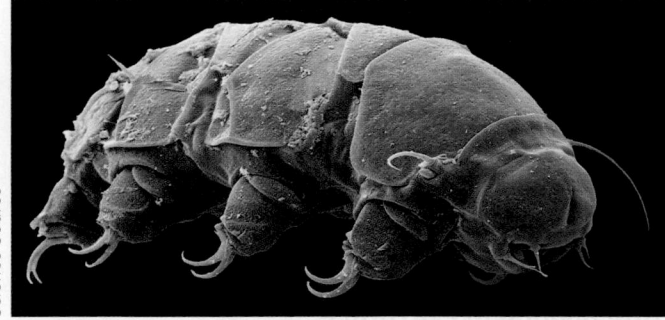

© Steve Gschmeissner/
Science Source

Tardigrades can survive extreme environmental conditions such as radiation, icy temperatures, and lack of oxygen or water. Their remarkable ability to survive many years in a desiccated state reflects a number of adaptations, including, in some species, resistance to ultraviolet light–induced DNA damage while desiccated and a high level of DNA repair activity following rehydration.

DO YOU REMEMBER?

- A DNA molecule contains two antiparallel strands that wind around each other to form a double helix in which A and T bases in opposite strands, and C and G bases in opposite strands, pair through hydrogen bonding (Section 3.1).

- Double-stranded nucleic acids are denatured at high temperatures; at lower temperatures, complementary polynucleotides anneal (Section 3.1).

- A DNA molecule can be sequenced or amplified by using DNA polymerase to make a copy of a template strand (Section 3.5).

- A reaction that hydrolyzes a phosphoanhydride bond in ATP occurs with a large change in free energy (Section 12.3).

A human cell contains 46 separate DNA molecules—chromosomes—comprising over 6 billion base pairs. Lined up end-to-end, these molecules would be slightly longer than 2 m, but the average diameter of a mammalian nucleus is only 6 μm (0.000006 m). Fitting all the DNA into the nucleus would be like stuffing a 100-km strand of hair into a backpack. Not surprisingly, cells have elaborate mechanisms for keeping DNA neatly packaged as well as unpacking it so that it can be copied. DNA replication can be considered to be part of the central dogma of molecular biology: the information in a parent DNA molecule must be copied to produce two identical DNA molecules that are passed on to daughter cells when the parent cell divides. In this chapter, we'll look at the process of DNA replication in detail, as well as some of the challenges cells face in accurately replicating DNA, repairing damaged DNA, and storing it safely.

20.1 THE DNA REPLICATION MACHINERY

When Watson and Crick described the complementary, double-stranded nature of DNA in 1953, they recognized that DNA could be duplicated by a process involving separation of the strands followed by the assembly of two new complementary strands. This mechanism of copying, or **replication,** was elegantly demonstrated by Matthew Meselson and Franklin Stahl in 1958. They grew bacteria in a medium containing the heavy isotope ^{15}N in order to label the cells' DNA. The bacteria were then transferred to fresh

KEY CONCEPTS

- DNA replication is carried out by stationary protein complexes.
- The proteins required for DNA replication include helicase, single-strand binding protein, primase, DNA polymerase, a sliding clamp, RNase, and DNA ligase.
- DNA polymerase synthesizes the leading strand continuously and the lagging strand as a series of Okazaki fragments.
- The accuracy of DNA polymerase is enhanced by its ability to sense correctly paired nucleotides and to hydrolytically excise mispaired nucleotides.

(a) (b)

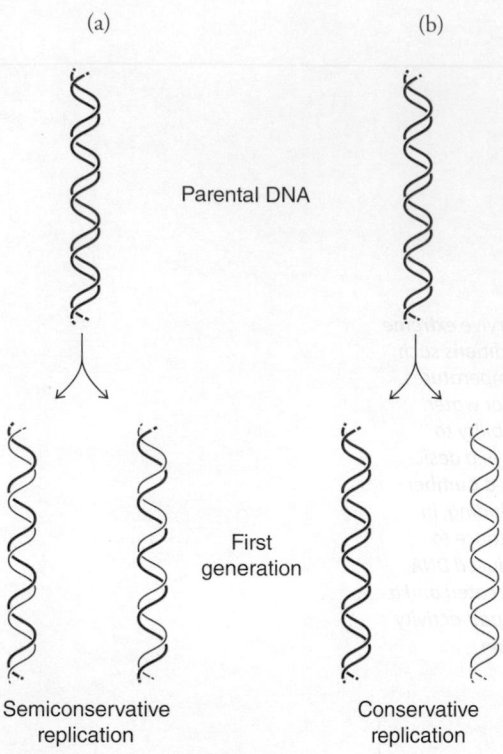

Parental DNA

First generation

Semiconservative replication

Conservative replication

FIGURE 20.1

Semiconservative versus conservative DNA replication. (a) Experiments performed by Meselson and Stahl demonstrated that new DNA molecules contain one parental (heavy) and one new (light) polynucleotide strand. Thus, DNA replication is semiconservative. (b) If DNA replication were conservative, the parental DNA (both strands heavy) would persist, while new DNA would consist of two light strands.

❓ **What would the second generation of DNA look like for each mode of replication?**

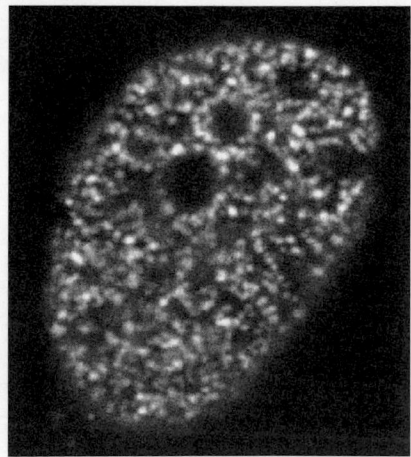

FIGURE 20.2

Replication foci. The fluorescent patches (foci) in the nucleus of a eukaryotic cell mark the presence of newly synthesized DNA. These sites of DNA replication remain stationary, probably due to attachment of the replication machinery to the nucleoskeleton. [*Courtesy A. Pombo. From Science 284, 1790–1795 (1999).*]

medium containing only ^{14}N, and the newly synthesized DNA was isolated and sedimented according to its density in an ultracentrifuge. Meselson and Stahl found that the first generation of replicated DNA had a lower density than the parental DNA but a higher density than DNA containing only ^{14}N. From this, they concluded that newly synthesized DNA is a hybrid containing one parental (heavy) strand and one new (light) strand. In other words, DNA is replicated **semiconservatively.** Because Meselson and Stahl did not observe any all-heavy DNA in the first generation, they were able to discount the possibility that DNA was copied in a way that left intact—or conserved—the original double-stranded molecule (**Fig. 20.1**).

DNA polymerase, the enzyme that catalyzes the polymerization of deoxynucleotides, is just one of the proteins involved in replicating double-stranded DNA. The entire process—separating the two template strands, initiating new complementary polynucleotide chains, and extending them—is carried out by a complex of enzymes and other proteins. In this section we examine the structures and functions of the major players in DNA replication in bacteria and in eukaryotes.

REPLICATION OCCURS IN FACTORIES

In the circular chromosomes of bacteria, DNA replication begins at a particular site called the origin. Here, proteins bind to the DNA and melt it open in an ATP-dependent manner. Polymerization then proceeds in both directions from this point until the entire chromosome (4.6×10^6 bp in *E. coli*) has been replicated. The point where the parental strands separate and the new strands are synthesized is known as the **replication fork.**

Parental strands

New strands

Replication fork

The much larger chromosomes of eukaryotes have multiple origins of replication. In yeast, these sites are about 40 kb apart and have a characteristic sequence. In mammals, the origins are farther apart and their locations appear to be determined by features of chromosome organization rather than sequence.

At one time, complexes of DNA polymerase and other replication proteins were thought to move along DNA like a train on tracks. This "locomotive" model of DNA replication requires that the large replication proteins move along the relatively thin template strand, rotating around it while generating a double-stranded helical product. In fact, cytological studies indicate that DNA replication (as well as transcription) occurs in "factories" at discrete sites. For example, in bacteria, DNA polymerase and associated factors appear to be immobilized in one or two complexes near the plasma membrane. In the eukaryotic nucleus, newly synthesized DNA appears at 100 to 150 spots, each one representing several hundred replication forks (**Fig. 20.2**). *According to the factory model of replication, the protein machinery is stationary and the DNA is reeled through it.* In eukaryotes, this organization presumably facilitates the synchronous elongation of many DNA segments, which allows for efficient replication of enormous eukaryotic genomes.

HELICASES CONVERT DOUBLE-STRANDED DNA TO SINGLE-STRANDED DNA

In order to replicate a DNA molecule, its two strands, which are coiled around each other, must be separated so that each one can serve as a template for DNA synthesis. One of the first proteins to bind to the origin of replication is an enzyme called a **helicase,** which catalyzes unwinding of the strands of the double helix. In eukaryotes, an inactive form of the helicase binds to the DNA at each origin well before replication begins. When the appropriate signals are received from internal as well as external signaling pathways, the cell commits to replicating its DNA in preparation for cell division. In response to kinase-catalyzed phosphorylation events and the addition of more proteins at each origin, each helicase becomes activated. After this point, no additional helicases can bind to the DNA. It is believed that by separating the helicase loading and activation steps, the cell can ensure that all its DNA is replicated more or less simultaneously (since all the helicases are activated at the same time) but only once per cell cycle (since only the helicases already poised at the origins are activated).

Most helicases are hexameric proteins shaped roughly like a donut that circles one DNA strand (**Fig. 20.3**). Helicase is a motor protein (see Section 5.4) that uses the free energy of ATP hydrolysis to move along the DNA strand, pushing away the complementary DNA strand to open up the helix. For each ATP hydrolyzed, up to five base pairs of DNA are separated. The helicase appears to operate in a rotary fashion, with conformational changes driven by ATP hydrolysis—somewhat reminiscent of the binding change mechanism of the F_1 component of ATP synthase (Section 15.4).

As replication proceeds, the parental DNA strands are continuously separated by the action of a helicase at each replication fork. A second helicase, which may be a dimer rather than a hexamer, may bind to the other strand of DNA to help open up the DNA helix. However, strand separation forces the helix to wind more tightly ahead of the replication fork (see diagram).

The strain generated by the extra twisting is relieved by a **topoisomerase,** an enzyme that cuts the DNA and allows it to "relax" before the cuts are sealed. These enzymes are described in more detail in Section 20.5.

As single-stranded DNA is exposed at the replication fork, it associates with a protein known as single-strand binding protein (SSB). SSB coats the DNA strands to protect them from nucleases and to prevent them from reannealing or forming secondary structures that might impede replication. *E. coli* SSB is a tetramer with a positively charged cleft that accommodates loops of single-stranded—but not double-stranded—DNA. The eukaryotic SSB, called replication protein A, is a larger protein that includes four DNA-binding domains separated by flexible regions (**Fig. 20.4**). Both prokaryotic and eukaryotic SSBs can adopt different conformations and bind DNA in several different ways. Studies of replication protein A suggest that DNA binding occurs in stages such that the first protein–DNA interaction is relatively weak (with a K_d in the µM range) but allows the other domains

FIGURE 20.3

Structure of a hexameric helicase. This helicase, from the bacteriophage T7, forms a hexameric ring around a single strand of DNA and pushes double-stranded DNA apart. [*Structure (pdb 1E0J) determined by M. R. Singleton, M. R. Sawaya, T. Ellenberger, and D. B. Wigley.*]

> ❓ **What would happen if an activated helicase could dissociate from DNA and re-bind to another origin?**

(a)   (b)

FIGURE 20.4

DNA-binding domains of replication protein A. Two of the four DNA-binding domains are shown in yellow and green in this model, with a bound octanucleotide (polydeoxycytidine, shown in purple). (a) Front view. (b) Side view. [*Structure (pdb 1JMC) determined by A. Bochkarev, R. Pfuetzner, A. Edwards, and L. Frappier.*]

> ❓ **Which types of amino acids would you expect to find in the DNA-binding site of this protein?**

to "unroll" along the DNA to form a stable complex that protects about 30 nucleotides and has an overall dissociation constant of about 10^{-9} M. As the template DNA is reeled through the polymerase, SSB is displaced and is presumably redeployed as helicase exposes more single-stranded DNA.

DNA POLYMERASE FACES TWO PROBLEMS

The mechanism of DNA polymerase and the double-stranded structure of DNA present two potential obstacles to the efficient replication of DNA. First, *DNA polymerase can only extend a preexisting chain; it cannot initiate polynucleotide synthesis.* However, an RNA polymerase can do this, so DNA chains *in vivo* begin with a short stretch of RNA that is later removed and replaced with DNA:

This stretch of RNA, usually about 12 nucleotides long, is known as a **primer,** and the enzyme that produces it during DNA replication is known as a **primase.** As we will see, primase is required throughout DNA replication, not just at the start. The active site of the primase is narrow (about 9 Å in diameter) at one end, where the single-stranded DNA template is threaded through. The other end of the active site is wide enough to accommodate a DNA–RNA hybrid helix (which has an A-DNA–like conformation; see Fig. 3.5).

The second problem facing DNA polymerase is that the antiparallel template DNA strands are replicated simultaneously by a pair of polymerase enzymes. But each DNA polymerase catalyzes a reaction in which the 3′ OH group at the end of the growing DNA chain attacks the phosphate group of a free nucleotide that base pairs with the template DNA strand (**Fig. 20.5**). For this reason, *a polynucleotide chain is said to be synthesized in the 5′ → 3′ direction.* Because the two template DNA strands are antiparallel, the synthesis of two new DNA strands would require that the template strands be pulled in opposite directions through the replication machinery so that the DNA polymerases could continually add nucleotides to the 3′ end of each new strand.

This awkward situation does not occur in cells. Instead, *the two polymerases work side-by-side, and one template DNA strand periodically loops out.* In this scenario, one strand of DNA, called the **leading strand,** can be synthesized in one continuous piece. It is initiated by the action of a primase, then extended in the 5′ → 3′ direction by the action of one DNA polymerase enzyme. The other strand, called the **lagging strand,** is synthesized in pieces, or **discontinuously.** Its template is repeatedly looped out so that its polymerase can also operate in the 5′ → 3′ direction. Thus, *the lagging strand consists of a series of polynucleotide segments, which are called **Okazaki fragments** after their discoverer* (**Fig. 20.6**).

Bacterial Okazaki fragments are about 500 to 2000 nucleotides long; in eukaryotes, they are about 100 to 200 nucleotides long. Each Okazaki fragment has a short stretch of RNA at its 5′ end, since each segment is initiated by a separate priming event. This explains why primase is required throughout replication: Although the leading strand theoretically requires only one priming event, the discontinuously synthesized lagging strand requires multiple primers.

The mechanism for continuous synthesis of the leading strand and discontinuous synthesis of the lagging strand means that the lagging-strand template must periodically be repositioned. Each time an Okazaki fragment is completed, a polymerase begins extending the RNA primer of the next Okazaki fragment. Other protein components of the replication complex assist in this repositioning in order to coordinate the activities of the two DNA polymerases at the replication fork. In *E. coli* and possibly other bacteria, three polymerases localize to the replication fork. One synthesizes the leading strand, and the other two apparently take turns synthesizing Okazaki fragments; sharing the work appears to increase replication efficiency.

DNA POLYMERASES SHARE A COMMON STRUCTURE AND MECHANISM

Most DNA polymerases are shaped somewhat like a hand, with domains corresponding to palm, fingers, and thumb. These structures are likely the result of convergent evolution, as only

FIGURE 20.6

A model for DNA replication. Two DNA polymerase enzymes (green) are positioned at the replication fork to make two complementary strands of DNA. The leading and lagging strands both start with RNA primers (red) and are extended by DNA polymerase in the 5′ → 3′ direction. The replication machinery is stationary, and the template DNA is reeled through it. Because the two template strands are antiparallel, the lagging-strand template loops out (the single-stranded DNA becomes coated with SSB). The leading strand is therefore synthesized continuously, while the lagging strand is synthesized as a series of Okazaki fragments.

FIGURE 20.7

***E. coli* DNA polymerase I.** This model shows the so-called Klenow fragment of DNA polymerase I (residues 324 to 928). The palm, fingers, and thumb domains are labeled. A loop at the end of the thumb is missing in this model. [*Structure (pdb 1KFD) determined by L. S. Beese, J. M. Friedman, and T. A. Steitz.*]

the palm domains exhibit strong homology. One of the best known polymerases is *E. coli* DNA polymerase I, the first such enzyme to be characterized (**Fig. 20.7**; this is the enzyme used for sequencing DNA; Section 3.5). The template strand and the newly synthesized DNA strand, which form a double helix, lie across the palm, in a cleft lined with basic residues.

The polymerase active site, at the bottom of the cleft, contains two Mg^{2+} ions about 3.6 Å apart. These metal ions are coordinated by aspartate side chains of the enzyme and by the phosphate groups of the substrate nucleoside triphosphate. One of the Mg^{2+} ions interacts with the 3′ O atom at the end of the primer or growing DNA strand to enhance its nucleophilicity as it attacks the incoming nucleotide:

The deoxy carbon at the 2′ position of the nucleotide lies in a hydrophobic pocket. This binding site allows the polymerase to discriminate against ribonucleotides, which bear a 2′ OH group.

After each polymerization event, the enzyme must advance the template strand by one nucleotide. Most DNA polymerases are **processive** enzymes, which means that they undergo several catalytic cycles (about 10 to 15 for *E. coli* DNA polymerase *in vitro*) before dissociating from their substrates. *E. coli* DNA polymerase is even more processive *in vivo*, polymerizing as many as 5000 nucleotides before releasing the DNA. *This enhanced processivity is due to an accessory protein that forms a sliding clamp around the DNA and helps hold DNA polymerase in place.* In *E. coli* the clamp is a dimeric protein, and in eukaryotes the clamp is a trimer. Both types of protein have a hexagonal ring structure and similar dimensions, consistent with a common function (**Fig. 20.8**). Additional proteins help assemble the clamp around the DNA. For the lagging-strand polymerase, the clamp must be reloaded at the start of each Okazaki fragment. This occurs about once every second in *E. coli*.

E. coli has five different DNA polymerase enzymes (I–V), and there are at least 14 eukaryotic DNA polymerases (designated by Greek letters), not including those found in mitochondria and chloroplasts. Why so many? First, DNA replication

(a)

(b)

FIGURE 20.8

DNA polymerase-associated clamps. In *E. coli*, the β subunit of DNA polymerase III forms a dimeric clamp. (b) In humans, the clamp is a trimer, called the proliferating cell nuclear antigen (PCNA). The inner surface of each clamp is positively charged and encloses a space with a diameter of about 35 Å, more than large enough to accommodate double-stranded DNA or a DNA–RNA hybrid helix with a diameter of 26 Å. Both structures enhance the processivity of their respective DNA polymerases, thereby increasing the efficiency of DNA replication. [*Structure of the β clamp (pdb 2POL) determined by X.-P. Kong and J. Kuriyan; structure of PCNA (pdb 1AXC) determined by J. M. Gulbis and J. Kuriyan.*]

? **What structural changes must occur each time the clamp is loaded onto DNA?**

requires polymerases with different activities to completely synthesize the leading and lagging strands. Other polymerases perform specialized roles in DNA repair pathways, most of which involve the excision and replacement of damaged DNA (Section 20.3).

For example, in *E. coli*, DNA polymerase III is the main replication polymerase for the leading and lagging strands, and DNA polymerase I participates in replacing RNA primers (described below). Polymerases II, IV, and V are used for DNA repair. DNA polymerase III is fast and highly processive, polymerizing about 1 kb of DNA per second at each replication fork. At that rate, the *E. coli* chromosome can be completely copied in 30 minutes.

In humans, DNA synthesis is about 20 times slower than in *E. coli*. And even with multiple origins of replication, it takes an estimated 8 hours to copy the largest chromosome of 250 Mb (Mb = millions of base pairs). Synthesis of both the leading and lagging strands begins with DNA polymerase α, which consists of two DNA polymerase subunits and two primase subunits. After building a primer of about 7 to 12 ribonucleotides, polymerase α extends the primer with another 10 to 25 deoxyribonucleotides. At this point, DNA polymerase δ takes over synthesis of the lagging strand, building each Okazaki fragment in turn. After the initial work of polymerase α, DNA polymerase ε completes the leading strand. Unlike other polymerases, DNA polymerase ε does not need an external sliding clamp to be processive, because its structure includes an extra loop of protein that circles the DNA as a built-in clamp.

DNA POLYMERASE PROOFREADS NEWLY SYNTHESIZED DNA

During polymerization, the incoming nucleotide base pairs with the template DNA so that the new strand will be complementary to the original. The polymerase accommodates base pairs snugly—recall that all possible pairings (A:T, T:A, C:G, and G:C) have the same overall geometry (see Section 3.2). The tight fit minimizes the chance of mispairings. In fact, structural studies indicate that DNA polymerase shifts between an "open" conformation and a "closed" conformation, analogous to the fingers moving closer to the thumb, when a nucleotide substrate is bound (**Fig. 20.9**). This conformational change may facilitate the polymerization of a tightly bound (correctly paired) nucleotide, or it could reflect a mechanism for quickly releasing a mismatched nucleotide before catalysis occurs.

If the wrong nucleotide does become covalently linked to the growing chain, the polymerase can detect the distortion it creates in the newly generated double helix. Many DNA polymerases contain a second active site that catalyzes hydrolysis of the nucleotide at the 3′ end of the growing DNA strand. *This 3′ → 5′ exonuclease excises misincorporated nucleotides, thereby acting as a* **proofreader** *for DNA polymerase* (**Fig. 20.10**). Recall that an exonuclease removes residues from the end of a polymer; an endonuclease cleaves within the polymer. In *E. coli* DNA polymerase I, the 3′ → 5′ exonuclease active site is located about 25 Å from the polymerase active site, indicating that the enzyme–DNA complex must undergo a large conformational change in order to shift from polymerization to nucleotide hydrolysis.

Proofreading during polymerization limits the error rate of DNA polymerase to about one in 10^6 bases. Misincorporated bases can also be removed following replication, through various DNA repair mechanisms, which further reduces the error rate of replication. This high degree of fidelity is absolutely essential for the accurate transmission of biological information from one generation to the next.

AN RNASE AND A LIGASE ARE REQUIRED TO COMPLETE THE LAGGING STRAND

In bacteria as well as eukaryotes, replication forks advance until they essentially run into each other. In a circular bacterial chromosome, a set of *Ter* sequences (*Ter* for termination) are located roughly opposite the replication origin. The first replication fork that arrives at a *Ter* sequence pauses until the other replication fork meets it. In linear eukaryotic chromosomes, two approaching replication forks merge so that

FIGURE 20.9

Open and closed conformations of a DNA polymerase. The structure of the polymerase from *Thermus aquaticus* was determined in the absence (magenta trace) and presence (green trace) of a nucleotide substrate analog (shown in space-filling form, purple). The models include a segment of DNA representing the template and primer strands (purple). [*Structures of the open conformation (pdb 2KTQ) and closed conformation (pdb 3KTQ) determined by Y. Li and G. Waksman.*]

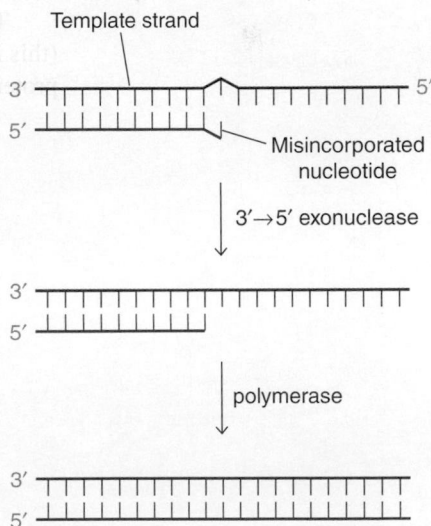

FIGURE 20.10

Proofreading during polymerization. DNA polymerase detects a distortion in double-stranded DNA that results from the incorporation of a mismatched nucleotide. The 3′ → 5′ exonuclease activity hydrolyzes the nucleotide at the 3′ end of the new strand. The polymerase then resumes its activity, generating an accurately base-paired DNA product.

each leading-strand polymerase continues adding nucleotides until it runs into the lagging strand of the other replication fork:

However, replication is not complete until the lagging strand—which is synthesized one Okazaki fragment at a time—is made whole. *As the polymerase finishes each Okazaki fragment, its RNA primer, along with some of the adjoining DNA, is hydrolytically removed and replaced with DNA; then the backbone is sealed to generate a continuous DNA strand.* This process also increases the accuracy of DNA replication: Primase has low fidelity, so the RNA primers tend to contain errors, as do the first few deoxynucleotides added to the primer by the action of the DNA polymerase. For example, in humans, DNA polymerase α lacks exonuclease activity and therefore cannot proofread its work.

In many cells, an exonuclease known as RNase H (H stands for hybrid) operates in the 5′ → 3′ direction to excise nucleotides at the primer end of an Okazaki fragment. Nucleotide hydrolysis may continue until DNA polymerase, now in the process of completing another Okazaki fragment, "catches up" with RNase H (the polymerase is faster than the exonuclease). In extending the newer Okazaki fragment, the polymerase replaces the excised ribonucleotides of the older Okazaki fragment with deoxyribonucleotides, leaving a single-strand **nick** between the two lagging-strand segments (**Fig. 20.11**).

The *E. coli* DNA polymerase I polypeptide actually includes a 5′ → 3′ exonuclease activity (this is in addition to the 3′ → 5′ proofreading endonuclease), so that, at least *in vitro*, a single protein can remove ribonucleotides from the previous Okazaki fragment as it extends the next

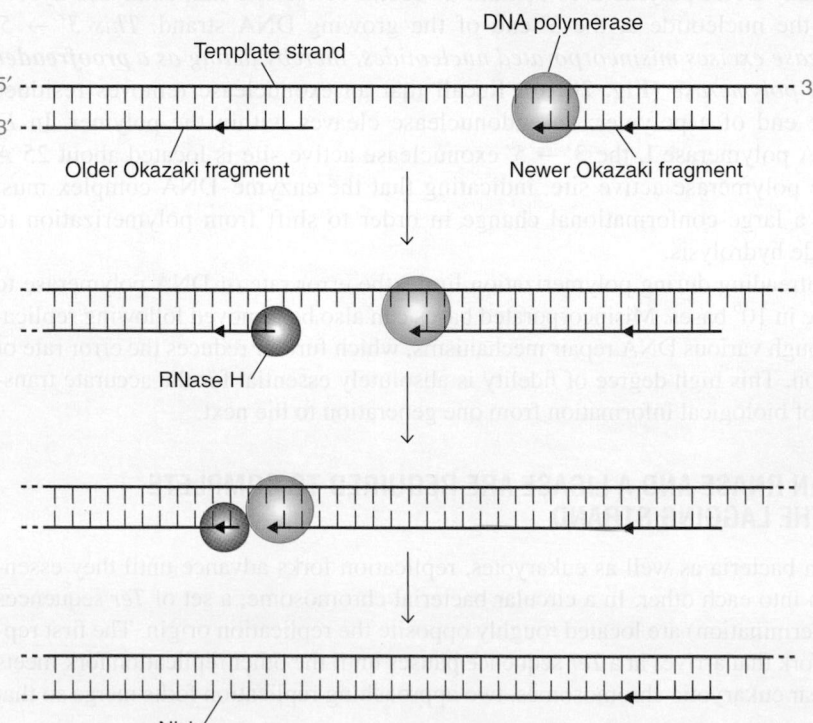

FIGURE 20.11

Primer excision. RNase H removes the RNA primer and some of the adjoining DNA in the older Okazaki fragment, allowing DNA polymerase to accurately replace these nucleotides. The nick can then be sealed.

Okazaki fragment. The combined activities of removing and replacing nucleotides have the net effect of moving the nick in the $5' \rightarrow 3'$ direction. This phenomenon is known as **nick translation.** DNA polymerase cannot seal the nick; this is the function of yet another enzyme.

The discontinuous segments of the lagging strand are joined by the action of **DNA ligase.** The reaction, which results in the formation of a phosphodiester bond, consumes the free energy of a similar bond in a nucleotide cofactor. Prokaryotes use NAD^+ for the reaction, yielding as products AMP and nicotinamide mononucleotide; eukaryotes use ATP and produce AMP and PP_i.

Ligation of Okazaki fragments yields a continuous lagging strand, completing the process of DNA replication.

Figure 20.12 is a composite, showing the major proteins that participate in DNA replication in *E. coli*. Other proteins are also present, including the machinery that reloads the sliding clamp onto the lagging-strand template each time a new Okazaki fragment begins. In a cell, the necessary proteins (dozens in eukaryotes) form a large complex called a **replisome** at each replication fork. All of the activities of building primers, extending

FIGURE 20.12

Overview of replication in *E. coli*.

1. Helicase unwinds the two strands of parental DNA.
2. Topoisomerase relieves the overwinding ahead of the replication fork.
3. SSB coats the exposed single strands.
4. One DNA polymerase III extends a primer continuously to build the leading strand. The sliding clamp increases the polymerase's processivity.
5. The lagging-strand template loops out to allow primase to synthesize a new RNA primer.
6. A second DNA polymerase III extends each RNA primer to complete an Okazaki fragment.
7. When DNA polymerase III reaches the previous Okazaki fragment, it is replaced by DNA polymerase I, which removes the RNA primer of the older Okazaki fragment and extends the newer Okazaki fragment with DNA.
8. DNA ligase seals the remaining nick between Okazaki fragments to generate a continuous lagging strand.

them, replacing RNA, and sealing nicks take place in a small area. The functions of each enzyme are coordinated so that the steps that are separated in Figure 20.12 actually occur simultaneously.

 CONCEPT CHECK

1. Describe how two DNA polymerases, operating in the $5' \rightarrow 3'$ direction, replicate the two antiparallel template strands.

2. Explain why the factory model for DNA replication is superior to the locomotive model.

3. Summarize the roles of helicase and SSB during replication.

4. Explain why DNA polymerization must be primed by RNA.

5. What protein enhances the processivity of DNA polymerase?

6. How does the polymerase proofread its mistakes?

7. Explain why an RNase, a polymerase, and a ligase are required to complete lagging-strand synthesis.

20.2 TELOMERES

KEY CONCEPT
• The ends of eukaryotic chromosomes are extended by telomerase to form telomeres.

Bacterial DNA replication produces two identical circular DNA molecules. Eukaryotic DNA replication yields two identical linear DNA molecules that remain attached at their **centromeres,** giving rise to the familiar X-shaped chromosome that becomes visible during cell division (**Fig. 20.13**).

It seems reasonable to assume that eukaryotic DNA polymerases proceed to the very ends of the linear chromosomes. In fact, the leading-strand DNA polymerase can copy the $5'$ end of the parental DNA template, since it extends a new complementary strand in the $5' \rightarrow 3'$ direction.

However, replication of the extreme $3'$ ends of the parental DNA strands presents a problem, for the same reason. Even if an RNA primer were paired with the $3'$ end of a template strand, DNA polymerase would not be able to replace the ribonucleotides with deoxynucleotides.

FIGURE 20.13

A replicated eukaryotic chromosome. The X shape is two identical DNA molecules linked at the centromere, where each molecule consists of a parental DNA strand and a newly synthesized strand. During most of the cell cycle, chromosomes are not visible. They become highly condensed during cell division, and the two DNA molecules separate so that each daughter cell receives one. [*Biophoto Associates/ Science Source/Getty Images*]

The $3'$ end of each parental (template) DNA strand would then extend past the end of each new strand and would be susceptible to nucleases. Consequently, *each round of DNA replication would lead to chromosome shortening:*

FIGURE 20.14

Telomerase. This model of an insect telomerase includes a segment of the RNA template (green) and telomeric DNA (gold). The built-in RNA directs the telomerase to extend the 3′ end of a DNA strand by repeatedly adding 6-nucleotide segments. [*Structure (pdb 3KYL) determined by E. Skordalakes.*]

Eukaryotic cells counteract the potential loss of genetic information by actively extending the ends of chromosomes. The resulting structure, which consists of short tandem (side-by-side) repeats, is called a **telomere.** The proteins associated with the telomeric DNA protect the end of the chromosome from nucleolytic degradation and prevent the joining of two chromosomes through end-to-end ligation (a reaction that normally occurs to repair broken DNA molecules).

TELOMERASE EXTENDS CHROMOSOMES

The telomere proteins include an enzyme complex known as **telomerase,** which was first described by Elizabeth Blackburn. *Telomerase repeatedly adds a sequence of six nucleotides to the 3′ end of a DNA strand, using an enzyme-associated RNA molecule as a template* (**Fig. 20.14**). The catalytic subunit of telomerase is a **reverse transcriptase,** a homolog of a viral enzyme that copies the viral RNA genome into DNA (**Box 20.A**).

Although the telomerase catalytic core is highly conserved among eukaryotes, its RNA subunit varies from ~150 to ~1300 nucleotides in different species. In humans, the RNA is a sequence of 451 bases, including the six that serve as a template for the addition of DNA repeats with the sequence TTAGGG. Similar G-rich sequences extend the 3′ ends of chromosomes in all eukaryotes. In order to synthesize telomeric DNA, the template RNA must be repeatedly realigned with the end of the preceding hexanucleotide extension. The precise alignment depends on base pairing between telomeric DNA and a nontemplate portion of the telomerase RNA. When the 3′ end of a DNA strand has been extended by telomerase, it can then serve as a template for the conventional synthesis of a C-rich extension of the complementary DNA strand (**Fig. 20.15**).

In humans, telomeric DNA is 2 to 10 kb long, depending on the tissue and the age of the individual, and includes a 3′ single-strand overhang of 100–300 nucleotides. This single strand

FIGURE 20.15

Synthesis of telomeric DNA. Telomerase extends the 3′ end of a DNA strand by adding TTAGGG repeats (yellow segments). DNA polymerase can then extend the complementary strand by the normal mechanism for lagging-strand synthesis (orange segment). Note that this process still leaves a 3′ overhang (up to 300 nucleotides in humans), but the chromosome has been lengthened.

? Explain why primase, RNase H, and DNA ligase are required to complete the lagging strand.

Box 20.A HIV Reverse Transcriptase

The human immunodeficiency virus (HIV, introduced in Box 7.A) is a retrovirus, a virus whose RNA genome must be copied to DNA inside the host cell. After entering a cell, the HIV particle disassembles. The 9-kb viral RNA is then transcribed into DNA by the action of the viral enzyme reverse transcriptase. Another viral enzyme, an integrase, incorporates the resulting DNA into the host genome. Expression of the viral genes produces 15 different proteins, some of which must be processed by HIV protease to achieve their mature forms. Eventually, new viral particles are assembled and bud off from the host cell, which dies. Because HIV preferentially infects cells of the immune system, cell death leads to an almost invariably fatal immunodeficiency.

HIV reverse transcriptase, which synthesizes DNA from an RNA template (a contradiction of the central dogma outlined in Section 3.3), resembles other polymerases in having fingers, thumb, and palm domains. These parts of the protein (colored red in the model shown here) comprise a polymerase active site that can use either DNA or RNA as a template (no other enzyme has this dual specificity). A separate domain (green) contains an RNase active site that degrades the RNA template.

[*Structure of HIV reverse transcriptase (pdb 1BQN) determined by Y. Hsiou, K. Das, and E. Arnold.*]

Reverse transcription occurs as follows: The enzyme binds to the RNA template and generates a complementary DNA strand. In a host cell, DNA synthesis is primed by a transfer RNA molecule. As polymerization proceeds, the RNase active site degrades the RNA strand of the RNA–DNA hybrid molecule, leaving a single DNA strand that then serves as a template for the reverse transcriptase polymerase active site to use in synthesizing a second strand of DNA. The result is a double-stranded DNA molecule (see diagram).

Viral reverse transcriptase has proved to be more than a biological curiosity. It has become a valuable laboratory tool, allowing researchers to purify messenger RNA transcripts from cells, transform them to DNA (called **cDNA** for complementary DNA), and then quantify them, sequence them, or use them to direct protein synthesis.

Reverse transcriptase activity can be blocked by two different types of drugs. Nucleoside analogs such as AZT and ddC readily enter cells and are phosphorylated. The resulting nucleotides bind in the reverse transcriptase active site and are linked, via their 5′ phosphate group, to the growing DNA chain. However, because they lack a 3′ OH group, further addition of nucleotides is impossible.

Reverse transcriptase can also be inhibited by non-nucleoside analogs such as nevirapine, a noncompetitive inhibitor that binds to a hydrophobic patch on the surface of reverse transcriptase near the base of the thumb domain. This does not interfere with RNA or nucleotide binding, but it does inhibit polymerase activity, probably by restricting thumb movement. HIV infections are typically treated with a "cocktail" of drugs that often includes a reverse transcriptase inhibitor along with a protease inhibitor (see Box 7.A).

? **Explain why the drugs described here interfere only minimally with nucleic acid metabolism in the human host.**

HOCH₂ O Thymine

AZT
(3′-Azido-2′,3′-dideoxythymidine, Zidovudine)

HOCH₂ O Cytosine

ddC
(2′,3′-Dideoxycytidine, Zalcitabine)

of DNA appears to fold back on itself to form a structure called a T-loop (**Fig. 20.16**). Multiple copies of six different proteins, collectively known as shelterin, bind to the telomere. These proteins play a critical role in regulating telomere length, which may be an indicator of the cells' longevity.

FIGURE 20.16

A T-loop. Telomeric DNA folds back on itself, and the G-rich single strand invades the double helix. Shelterin proteins (not shown here) associate with both single-stranded and double-stranded DNA to form a cap protecting the chromosome end.

IS TELOMERASE ACTIVITY LINKED TO CELL IMMORTALITY?

Cells that normally undergo a limited number of cell divisions appear to contain no active telomerase. Consequently, the size of the telomeres decreases with each replication cycle, until the cells reach a senescent stage and no longer divide. For most human cells, this point is reached after about 35–50 cell divisions. In contrast, cells that are "immortal"—such as unicellular organisms, stem cells, and the reproductive cells of multicellular organisms—appear to have active telomerase. These findings are consistent with the role of telomerase in maintaining the ends of chromosomes over many rounds of replication.

Telomerase appears to be activated in cancer cells derived from tissues that are normally senescent (no longer dividing). This suggests that shutting down telomerase activity might be an effective treatment for cancer. However, for many cells, telomerase activity by itself is not an indicator of the cells' potential for immortality. Rather, the ability of the cells to undergo repeated divisions without chromosome shortening appears to be a more complicated function of telomerase activity, telomere length, and the integrity of the telomere, which, like other portions of DNA, is susceptible to damage.

CONCEPT CHECK

1. Explain why DNA replication leads to chromosome shortening.
2. Describe the structure of telomeric DNA.
3. What is the function of the RNA component of telomerase?
4. Explain the relationship between cell immortality and telomerase activity.

20.3 DNA DAMAGE AND REPAIR

An alteration in a cell's DNA, whether from a polymerization error or another cause, becomes permanent—that is, a **mutation**—unless it is repaired. In a unicellular organism, altered DNA is replicated and passed on to the daughter cells when the cell divides. In a multicellular organism, a mutation is passed to offspring only if the altered DNA is present in the reproductive cells. Mutations that arise in other types of cells affect only the progeny of those cells within the parent organism.

A genetic change can affect the expression of genes positively or negatively (or it may have no measurable effect on the organism). The accumulation of genetic changes over an individual's lifetime may contribute to the gradual loss of functionality associated with aging. In this section, we examine different types of DNA damage and the mechanisms cells use to restore DNA integrity.

KEY CONCEPTS
- A mutation in DNA can arise from DNA replication errors, from nonenzymatic cellular processes, and from environmental factors.
- A variety of enzymes, including nucleases, polymerases, and ligases, participate in DNA repair pathways.

DNA DAMAGE IS UNAVOIDABLE

DNA damage is a fact of life. Even with its proofreading activity, DNA polymerase makes mistakes by introducing mismatched nucleotides or incorporating uridine rather than thymidine. DNA polymerase may occasionally add or delete nucleotides, producing bulges or other irregularities in the DNA that lead to insertion or deletion mutations.

Cellular metabolism itself exposes DNA to the damaging effects of reactive oxygen species (for example, the superoxide anion $\cdot O_2^-$, the hydroxyl radical $\cdot OH$, or H_2O_2) that are normal by-products of oxidative metabolism. Over 100 different oxidative modifications of DNA have been catalogued. For example, guanine can be oxidized to 8-oxoguanine (oxoG):

Guanine → 8-Oxoguanine (oxoG)

When the modified DNA strand is replicated, the oxoG can base pair with either an incoming C or A. Ultimately, the original G:C base pair can become a T:A base pair. A nucleotide substitution such as this is called a **point mutation.** Changing a purine (or pyrimidine) to another purine (or pyrimidine) is known as a **transition mutation;** a **transversion** occurs when a purine replaces a pyrimidine or vice versa.

Other nonenzymatic reactions disintegrate DNA under physiological conditions. For example, hydrolysis of the *N*-glycosidic bond linking a base to a deoxyribose group yields an **abasic site** (also called an apurinic or apyrimidinic or AP site).

Deamination reactions can alter the identities of bases. This is particularly dangerous in the case of cytosine, since deamination (actually an oxidative deamination) yields uracil:

Cytosine → Uracil

Recall that uracil has the same base-pairing propensies as thymine, so the original C:G base pair could give rise to a T:A base pair after DNA replication. Since DNA has evolved to contain thymine rather than uracil, a uracil base resulting from cytosine deamination can be recognized and corrected before the change becomes permanent.

In addition to the cellular factors described above, *environmental agents such as ultraviolet light, ionizing radiation, and certain chemicals can physically damage DNA.* For example, ultraviolet (UV) light induces covalent linkages between adjacent thymine bases:

Thymine residues → Thymine dimer

This brings the bases closer together, which distorts the helical structure of DNA. Thymine dimers can thereby interfere with normal replication and transcription.

Ionizing radiation also damages DNA either through its direct action on the DNA molecule or indirectly by inducing the formation of free radicals, particularly the hydroxyl radical, in the surrounding medium. This can lead to strand breakage. Many thousands of chemicals, both natural and man-made, can potentially react with the DNA molecule and cause mutations. Such compounds are known as **mutagens,** or as **carcinogens** if the mutations lead to cancer.

The unavoidable nature of many DNA lesions has driven the evolution of mechanisms to detect and remedy errors. A cell containing a point mutation or a small insertion or deletion may suffer no ill effects, particularly if the mutation is in a part of the genome that does not contain an essential gene. However, more serious lesions, such as single- or double-strand breaks, usually bring replication or translation to a halt, and the cell must deal with these show-stoppers immediately.

REPAIR ENZYMES RESTORE SOME TYPES OF DAMAGED DNA

In a few cases, repair of damaged DNA is a simple process involving one enzyme. For example, in bacteria and some other organisms (but not mammals), UV-induced thymine dimers can be restored to their monomeric form by the action of a light-activated enzyme called DNA photolyase.

Mammals can reverse other simple forms of DNA damage, such as the methylation of a guanine residue, which yields O^6-methylguanine (this modified base can pair with either cytosine or thymine):

O^6-Methylguanine

A methyltransferase removes the offending methyl group, transferring it to one of its cysteine residues. This permanently inactivates the protein. Apparently, the expense of sacrificing the methyltransferase is justified by the highly mutagenic nature of O^6-methylguanine.

In bacteria as well as eukaryotes, *nucleotide mispairings are corrected shortly after DNA replication by a **mismatch repair** system.* A protein, called MutS in bacteria, monitors newly synthesized DNA and binds to the mispair. Although it binds only 20 times more tightly to the mispair than to a normal base pair, MutS undergoes a conformational change and causes the DNA to bend (**Fig. 20.17**). These changes apparently induce an endonuclease to cleave the strand with the incorrect base at a site as far as 1000 bases away. A third protein then unwinds the helix so that the defective segment of DNA can be destroyed and replaced with accurately paired nucleotides by DNA polymerase. How does the endonuclease know which strand contains the incorrect base? Cellular DNA is normally methylated (Section 20.5), and the endonuclease can select the newly synthesized strand because it has not yet undergone methylation. The mismatch repair system lowers the error rate of DNA replication about 1000-fold.

BASE EXCISION REPAIR CORRECTS THE MOST FREQUENT DNA LESIONS

*Modified bases that cannot be directly repaired can be removed and replaced in a process known as **base excision repair.*** This pathway begins with a glycosylase, which removes the damaged base. An endonuclease then cleaves the backbone, and the gap is filled in by DNA polymerase (**Fig. 20.18**).

The structures and mechanisms of several DNA glycosylases have been described in detail. For example, there is a glycosylase that recognizes oxoG. Uracil-DNA glycosylase recognizes and removes the uracil bases that are mistakenly incorporated into DNA during replication or that result from cytosine deamination. When a glycosylase binds DNA, the damaged base typically flips out from the helix so that it can bind in

(a)

(b)

FIGURE 20.17

The mismatch repair protein MutS bound to DNA. Two subunits of the comma-shaped MutS protein encircle the DNA at the site of a mispaired nucleotide. Binding causes the DNA to bend sharply, compressing the major groove and widening the minor groove. (a) Front view. (b) Side view. [*Courtesy Wei Yang/NIH. From* Nature *407, 703–710 (2000).*]

damaged base

DNA glycosylase

abasic site

endonuclease

DNA polymerase
DNA ligase

FIGURE 20.18

Base excision repair.

FIGURE 20.19

Uracil-DNA glycosylase bound to DNA. The enzyme is shown in gray, and the DNA substrate is shown with blue and purple strands. The flipped-out uracil (whose glycosidic bond has already been hydrolyzed) is shown in red. An Arg side chain that takes the place of the uracil is shown in orange. [*Structure (pdb 4SKN) determined by G. Slupphaug, C. D. Mol, B. Kavil, A. S. Arvai, H. E. Krokan, and J. A. Tainer.*]

a cavity on the protein surface, and various protein side chains take the place of the damaged base and form hydrogen bonds with its complement on the other DNA strand (**Fig. 20.19**).

After excising the offending base, DNA glycosylases appear to remain bound to the DNA at the abasic site, possibly to help recruit the next enzyme of the repair pathway. This enzyme, usually called the AP endonuclease (APE1 in humans), nicks the DNA backbone at the 5′ side of the abasic ribose. The nuclease inserts two protein loops into the major and minor grooves of the DNA and bends the DNA by about 35° to expose the abasic site. A backbone structure with a base attached cannot enter the active-site pocket. During the hydrolysis reaction, an Mg^{2+} ion in the active site stabilizes the anionic leaving group (**Fig. 20.20**). Like the glycosylases, the AP endonuclease remains with its product. For most enzymes,

FIGURE 20.20

The AP endonuclease reaction.

rapid product dissociation is the rule, but in DNA repair, the continued association of the endonuclease with the broken DNA strand may be favored because it prevents unwanted side reactions.

In the final step of base excision repair, a DNA polymerase (such as DNA polymerase β in eukaryotes) fills in the one-nucleotide gap, and a DNA ligase seals the nick. In some cases, DNA polymerase may replace as many as 10 nucleotides. The displaced single strand can then be cleaved off by an endonuclease (Section 20.1).

NUCLEOTIDE EXCISION REPAIR TARGETS THE SECOND MOST COMMON FORM OF DNA DAMAGE

Nucleotide excision repair, as its name suggests, is similar to base excision repair but mainly targets DNA damage resulting from insults such as ultraviolet light or oxidation. In nucleotide excision repair, a segment containing the damaged nucleotide and about 30 of its neighbors is removed, and the resulting gap is filled in by a DNA polymerase that uses the intact complementary strand as a template (**Fig. 20.21**). Many of the 30 or so proteins that are involved in this pathway in humans have been identified through mutations that are manifest as two genetic diseases.

The rare hereditary disease Cockayne syndrome is characterized by neural underdevelopment, failure to grow, and sensitivity to sunlight. It results from a mutation in any of several genes that encode proteins participating in a pathway for recognizing an RNA polymerase that has stalled in the act of transcribing a gene into messenger RNA. Stalling occurs when the DNA template is damaged and distorted so that it blocks the progress of the RNA polymerase. The polymerase must be removed so that the damage can be addressed by the nucleotide excision repair system. Defects in the Cockayne syndrome genes prevent the cell from recognizing and removing the stalled RNA polymerase. Consequently, *the DNA never has a chance to be repaired, and the cell dies*. The death of transcriptionally active cells may account for the developmental symptoms of Cockayne syndrome.

Like Cockayne syndrome, the disease xeroderma pigmentosum is characterized by high sensitivity to sunlight, but individuals with xeroderma pigmentosum are about 1000 times more likely to develop skin cancer and do not suffer from developmental problems. Xeroderma pigmentosum is caused by a mutation in one of the genes that participate directly in nucleotide excision repair. Whereas Cockayne syndrome gene products appear to detect DNA damage that prevents transcription, the xeroderma pigmentosum proteins are responsible for repairing the damage. The failure to repair UV-induced lesions explains the high incidence of skin cancer.

When the cell attempts to replicate damaged DNA that has not been repaired, it may rely on one of its nonstandard DNA polymerases. For example, eukaryotic DNA polymerase η can bypass DNA lesions such as UV-induced thymine dimers by incorporating two adenine bases in the new strand. Although it is useful as a translesion polymerase, DNA polymerase η is relatively inaccurate and has no proofreading exonuclease activity. It inserts an incorrect base on average every 30 nucleotides. This may not be problematic, as the errors can be detected and corrected by the mismatch repair system described above.

The existence of error-prone polymerases provides a fail-safe mechanism for replicating stretches of DNA that cannot be navigated by the standard replication machinery. In fact, synthesis of these alternative polymerases increases when bacterial cells experience DNA damage. Evidently, the possibility of introducing small errors during replication is acceptable when the only other option is cell death.

DOUBLE-STRAND BREAKS CAN BE REPAIRED BY JOINING THE ENDS

Segments of a DNA double helix that have been completely severed by the effects of radiation or free radicals can be rejoined by **nonhomologous end-joining,** a process that does not require the presence of a homologous DNA molecule. In mammals, nearly all double-strand breaks are repaired by nonhomologous end-joining.

FIGURE 20.21

Nucleotide excision repair.

? Compare this process to base excision repair (Fig. 20.18).

FIGURE 20.22

FIGURE 20.22

Ku bound to DNA. The two subunits of the Ku heterodimer are shown in light and dark green, and the DNA strands are shown in blue and purple. [*Structure (pdb 1JEY) determined by J. R. Walker, R. A. Corpina, and J. Goldberg.*]

The first step in this repair pathway is the recognition of the broken DNA ends by a dimeric protein called Ku (**Fig. 20.22**). When Ku binds the cut DNA, it undergoes a conformational change so that it can recruit a nuclease, which trims up to 10 residues from the ends of the DNA molecule. The protein–DNA complex may then be joined by a DNA polymerase such as polymerase μ, which can extend the ends of the DNA either with or without a template. Template-independent polymerization, along with a tendency for polymerase μ to slip, means that the break site may end up with additional nucleotides that were not present in the DNA before it broke. A DNA ligase finishes the repair job by joining the two backbones of the DNA segment (**Fig. 20.23**).

Two Ku–DNA complexes associate with each other so that, ideally, the proper halves of a broken DNA molecule can be stitched back together. However, nonhomologous end-joining is also responsible for combining DNA segments that do not belong together, leading to chromosomal rearrangements. In addition, the Ku–DNA complex can interact with the nuclease, polymerase, and ligase in any order, so a DNA break could potentially be repaired in many different ways, with or without the addition and removal of nucleotides. Consequently, *nonhomologous end-joining is inherently mutagenic*, but, as in other forms of DNA repair, this may be a small price to pay for restoring a continuous double-stranded DNA. The imperfect nature of repair by end-joining is the basis for inactivating a gene using the CRISPR-Cas9 nuclease system described in Section 3.5.

RECOMBINATION ALSO RESTORES BROKEN DNA MOLECULES

In some organisms, double-strand breaks can be repaired through **recombination,** a process that also occurs in the absence of DNA damage as a mechanism for shuffling genes between homologous chromosomes during meiosis. Recombination repair of a broken chromosome can occur at any time in a diploid organism (which has two sets of homologous chromosomes) but can occur only after DNA replication in an organism with only one chromosome. Recombination requires another intact homologous double-stranded molecule as well as nucleases, polymerases, ligases, and other proteins (**Fig. 20.24**).

In recombination repair, a single strand from the damaged DNA molecule changes places with a homologous strand in another DNA molecule. In order for a single strand of DNA to "invade" a double-stranded DNA molecule (step 2 in Fig. 20.24), the single strand must first be coated with an ATP-binding protein, called RecA in *E. coli* and Rad51 in humans. RecA binds to a DNA strand in a cooperative fashion, beginning at the break point. This binding unwinds and stretches the DNA by about 50%, but not uniformly: Sets of three nucleotides retain a near-standard conformation (with about 3.4 Å between bases) but are separated from the next triplet by about 7.8 Å (**Fig. 20.25**). During recombination, the RecA–DNA filament aligns with double-stranded DNA containing a complementary strand. The elongated structure of the RecA filament may induce a similar shape change in the double-stranded DNA so that base pairs in the double-stranded DNA are disrupted and bases become unstacked. These changes would facilitate the strand swapping that occurs during recombination. At this point, hydrolysis of the ATP bound to RecA allows the protein to release the displaced single strand and a new double-stranded DNA. When the CRISPR-Cas9 system (Section 3.5) is used to edit rather than inactivate a gene, a DNA segment with the desired gene sequence is included so that it can serve as a template for recombination repair.

The mechanism of recombination is highly conserved among prokaryotes and eukaryotes, which argues for its essential role in maintaining the integrity of DNA as a vehicle of genetic information. The proteins that carry out recombination appear

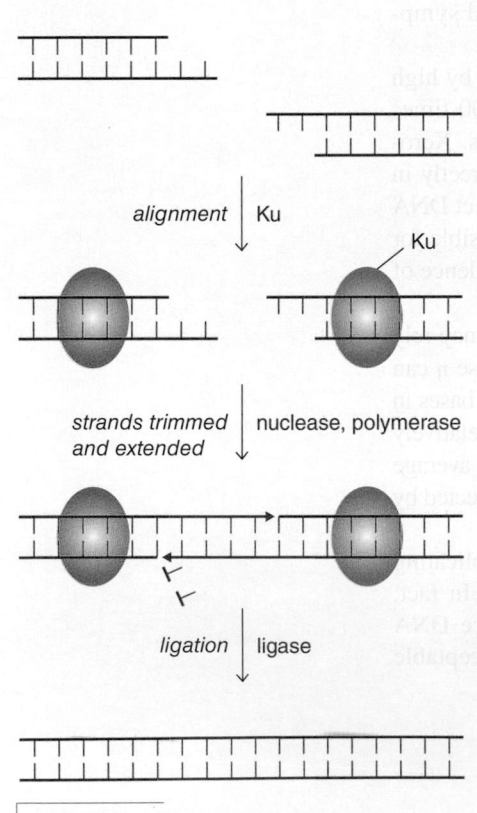

alignment | Ku

Ku

strands trimmed and extended | nuclease, polymerase

ligation | ligase

FIGURE 20.23

Nonhomologous end-joining. Ku recognizes the ends of broken DNA and aligns them. The activities of a nuclease, polymerase, and ligase generate an unbroken DNA molecule that may differ in sequence from the original.

FIGURE 20.24

Homologous recombination to repair a double-strand break.

1. An exonuclease trims the 5′ ends at the break.

Homologous DNA

2. One of the free 3′ ends invades a homologous segment of double-stranded DNA.

3. DNA polymerase extends the 3′ ends.

4. Resolution of the crossed-over structure yields two DNA segments with no breaks.

to function **constitutively** (that is, the genes are always expressed), which is consistent with proposals that DNA strand breaks are unavoidable. However, *many DNA repair mechanisms are induced only when the relevant form of DNA damage is detected*. This makes sense, since the repair enzymes might otherwise interfere with normal replication. In fact, activation

FIGURE 20.25

Conformation of DNA bound to RecA. The single strand of DNA is shown in space-filling form with atoms color coded: C gray, N blue, O red, and P orange. Brackets indicate sets of three nucleotides that retain a near–B-DNA conformation. The DNA backbone stretches in between these triplets so that the strand extends to about 1.5 times its original length. The RecA protein subunits are not shown. [*Structure (pdb 3CMW) determined by N. P. Pavletich.*]

of the repair pathways usually halts DNA synthesis, an advantage when error-prone DNA polymerases—which would be a liability in normal replication—are active.

 CONCEPT CHECK

1. Describe how replication errors, nonenzymatic cellular processes, and environmental factors can damage DNA.

2. Explain why DNA damage may not result in a mutation.

3. Describe the advantages and disadvantages of error-prone DNA polymerases.

4. Describe the general pathways and enzymes required for mismatch repair, base excision repair, nucleotide excision repair, end-joining and recombination.

5. Compare end-joining and recombination as mechanisms for repairing double-strand breaks.

CLINICAL CONNECTION: CANCER AS A GENETIC DISEASE

KEY CONCEPTS

- Several genetic events are necessary for carcinogenesis.
- Faulty DNA repair is closely linked to cancer.

Cells with badly damaged DNA tend to be so impaired that they undergo **apoptosis,** or programmed cell death, making room for their replacement by healthy cells. But in some cases, cells with damaged DNA do not die but instead escape the normal growth-control mechanisms and proliferate excessively, resulting in cancer. Some of the metabolic capabilities of cancer cells are outlined in Section 19.4.

Cancer is one of the most common diseases, affecting one in three people and killing one in four. The clinical picture of cancer varies tremendously, depending on the tissue affected, but all cancer cells share an ability to proliferate uncontrollably and ignore the usual signals to differentiate or undergo apoptosis. Eventually, cancer may kill by invading and eroding the surrounding normal tissue (**Fig. 20.26**).

There are believed to be three causes of cancer: inherited genetic variations, environmental factors, and random mutations. Only a few percent of cancers are classified as hereditary, but there are over 20 different kinds of these diseases, and they shed considerable light on specific molecular mechanisms of **carcinogenesis** (cancer development). The linkage between environmental factors and cancer is supported by epidemiological studies, which have shown, for example, that sunlight increases the risk of cutaneous melanoma (skin cancer) and smoking and asbestos exposure promote the development of lung cancer. Viral infections contribute to certain types of cancer, for example, liver cancer from hepatitis B virus and cervical cancer from human papillomaviruses. Chronic bacterial infections may also lead to cancer.

The analysis of tumor genomes has revealed the importance of genetic mutations. A typical cancer contains a few dozen genetic mutations that may or may not be present in other cells in the body. Melanomas and lung tumors contain an average of about 200 mutations, consistent with the known role of DNA-damaging events in these cancers (ultraviolet light and smoking). The link between cancer and DNA changes means that cancer can be considered as a disease of the genes. And replication errors occur every time a cell divides, so mutations tend to occur more frequently in tissues where stem cells (the cells that give rise to other cells) divide more frequently.

FIGURE 20.26

Tumors in human liver. The white bulges are masses of cancerous cells. [*CNRI/Science Photo Library/Photo Researchers.*]

TUMOR GROWTH DEPENDS ON MULTIPLE EVENTS

Because cell growth and division are tightly regulated, a single genetic change is unlikely to send a cell into a pattern of uncontrolled proliferation that leads to cancer. The ability of a cell to grow and divide depends on numerous factors, including the balance between growth-promoting signals and apoptotic signals, the state of the telomeres, contacts with neighboring cells, and the delivery of oxygen and nutrients to support expansion. In general, *several regulatory pathways must be overridden by separate genetic events.* This is known as the multiple-hit hypothesis for carcinogenesis.

In this scenario, a tumor gradually evolves as a harmless cell becomes a small clump of cells that continue to multiply until they become a mass of cells (a tumor) and may be able to spread to other tissues as a malignant tumor. Of the approximately 140 genes that are known to be associated with carcinogenesis, a typical cancer contains mutations ("hits") in only two to eight. Each of these genetic changes contributes only a small growth advantage, but over time, following the principle of natural selection, the mass of altered cells can become large. This explains why most cancers develop later in life: it usually takes years for the mutations to accumulate and for the cancerous cells to overcome growth-limiting checkpoints.

Aside from the critical mutations that drive cancer development, additional mutations occur that may have no bearing on a tumor's growth. A current challenge in cancer research is to distinguish the "driver" mutations from the "passenger" mutations. There is no defined sequence of events for transforming a normal cell to a tumor, so cancer is not a single disease. The situation is made more difficult by the heterogeneity of tumors: No two individuals harbor the same mutations, and even within a single tumor, there may be considerable genetic variation among its billions of cells. The goal of the Cancer Genome Atlas project (https://cancergenome.nih.gov) is to collect and share genetic information for different types of cancers, which could shed more light on the early development of cancer. Most cancer treatments address the late stages of the disease, when tumors are already too large or too dispersed to be efficiently eliminated.

We have already seen that oncogenic mutations in growth signaling pathways can produce constitutively active receptors and kinases so that cells grow and divide even in the absence of a growth signal (see Box 10.B). Inactivating genetic events also contribute to cancer. For example, the childhood cancer known as retinoblastoma, which is characterized by retinal tumors, is linked to mutations in a gene known as a **tumor suppressor gene**. Children who inherit a defective copy of the tumor suppressor gene are at higher risk of developing cancer.

DNA REPAIR PATHWAYS ARE CLOSELY LINKED TO CANCER

Genetic changes—whether they activate a growth-promoting oncogene or inactivate a tumor suppressor gene—can take the form of point mutations, large or small deletions, chromosomal rearrangements, or inappropriate methylation that leads to gene silencing (Section 21.1). In many tumors, the state of the DNA goes from bad to worse, reflecting failure in the mechanisms that detect damaged DNA and promote its restoration. These steps require a host of proteins, including kinases and other intracellular signaling components that respond rapidly (usually within minutes) to DNA damage.

One key player in the damage response pathway is the protein kinase known as ATM, which is defective in ataxia telangiectasia. This disease is characterized by neurodegeneration, premature aging, and a propensity to develop cancer. ATM is a large protein (350,000 D) that includes a DNA-binding domain. It is activated in response to DNA damage, such as double-strand breaks. Among the substrates for ATM's kinase activity are proteins involved in initiating cell division, the protein BRCA1 (whose gene is commonly mutated in breast cancer), and the tumor suppressor known as p53.

Breast cancer strikes about one in nine women in developed countries. About 10% of breast cancers have a familial form, and about half of these exhibit mutations in the *BRCA1* or *BRCA2* genes. A woman bearing one of these mutated genes has a 70% chance of developing breast cancer. BRCA1 and BRCA2 are large multidomain proteins that function in part as scaffolding proteins to link proteins that detect DNA damage to proteins that can repair the damage or halt the cell cycle. For example, BRCA2 binds to Rad51, a protein necessary for the recombination repair pathway. The loss of BRCA1 or BRCA2 increases the likelihood that a cell would attempt to divide without first repairing damaged DNA.

The tumor suppressor gene p53 is found to be mutated in at least half of all human tumors. The level of p53 in the cell is controlled by its rate of degradation (it is ubiquitinated and targeted to a proteasome for destruction; Section 12.1). The concentration of p53 increases when its degradation is slowed. This can occur by a decrease in the rate at which ubiquitin is attached to it or when it is phosphorylated by the action of a kinase such as ATM. Thus, *DNA damage, which activates ATM, leads to an increase in the cellular concentration of p53* (**Fig. 20.27**).

Other modifications, such as acetylation and glycosylation, also increase the activity of p53, which responds not just to DNA damage but also to other forms of cellular stress such as low oxygen and elevated temperatures. Covalent modification of p53 triggers a conformational

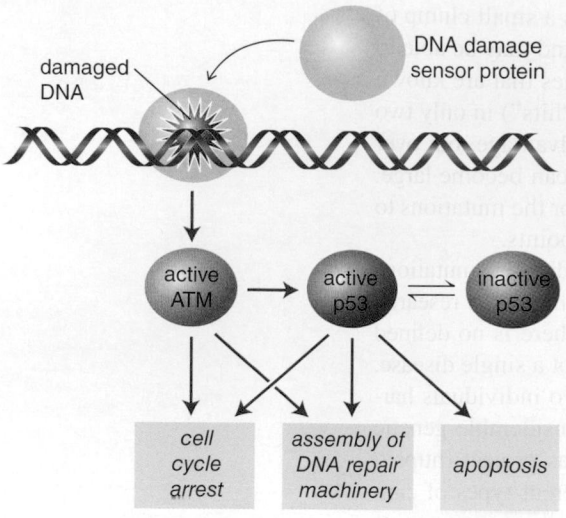

FIGURE 20.27

Overview of p53 function. DNA damage (via ATM) and other factors regulate the level of p53 activity, which in turn affects the cell's ability to complete the cell cycle, repair damaged DNA, or undergo apoptosis.

FIGURE 20.28

p53–DNA complex. This model shows a p53 dimer (gray) bound to DNA (blue). The six residues that are most commonly mutated in p53 from cancer cells are shown in red. These residues either interact directly with the DNA or are involved in stabilizing p53 structure. [*Structure (pdb 2GEQ) determined by W. C. Ho, M. X. Fitzgerald, and R. Marmorstein.*]

change that allows the protein to bind to specific DNA sequences in order to promote the transcription of several dozen different genes. p53 binds as a tetramer (a dimer of dimers) to DNA, encircling it (**Fig. 20.28**).

p53 stimulates production of a protein that blocks the cell's progress toward cell division. This regulatory mechanism buys time for the cell to repair DNA using enzymes whose synthesis is also stimulated by p53. Moreover, activated p53 turns on the gene for a subunit of ribonucleotide reductase (see Section 18.5), which promotes synthesis of the deoxynucleotides required for DNA repair.

Some of p53's other target genes encode proteins that carry out apoptosis, a multistep process in which the cell's contents are apportioned into membrane-bounded vesicles that are subsequently engulfed by macrophages. For multicellular organisms, death by apoptosis is often a better option than allowing a malfunctioning cell to persist.

It is possible that p53's ultimate effects are dose-dependent, with lower doses (reflecting mild DNA damage) pausing the cell cycle and higher doses (reflecting severe, irreparable DNA damage) leading to cell death. *The position of p53 at the interface of pathways related to DNA repair, cell cycle control, and apoptosis indicates why the loss of the p53 gene is so strongly associated with the development of cancer.*

 CONCEPT CHECK

1. Explain how mutations, environmental factors, and random events can contribute to the development of cancer.

2. Summarize the multiple hit hypothesis for carcinogenesis.

3. Describe how defects in the genes for BRCA1 or p53 can lead to cancer.

KEY CONCEPTS
- Eukaryotic DNA is wrapped around histones to form nucleosomes, which can form higher-order structures.
- Both histones and DNA may be chemically modified

DNA PACKAGING

During replication and repair, DNA molecules in bacteria and eukaryotes are relatively extended and accessible to the proteins that carry out these processes. But for much of the time, chromosomes are kept in a more compact arrangement that allows large amounts of genetic information to be packed into a small space and kept safe.

DNA IS NEGATIVELY SUPERCOILED

In order to facilitate processes such as replication and transcription, which require opening up the double helix, the DNA molecules in cells are slightly underwound. In other words, the two strands of the helix make fewer than the expected number of helical turns around each other. However, in order to maintain a conformation close to the stable B form, the DNA molecule twists up on itself, much like the cord of an old-fashioned telephone. This phenomenon, termed **supercoiling,** is readily apparent in small circular DNA molecules (**Fig. 20.29**).

The geometry of a DNA molecule can be described by the branch of mathematics known as topology. For example, consider a strip of paper. When the strip is looped once, it is said to have one writhe. If the ends of the paper are gently pulled in opposite directions, the strip is deformed into a shape called a twist.

A writhe ⇌ A twist

Twisting the strip further introduces more writhes; twisting it in the opposite direction removes writhes, or introduces negative writhes. The same topological terms apply to DNA: *Each writhe, or supercoil, in DNA is the result of overtwisting or undertwisting the DNA helix.* The twisted molecule, like the piece of paper, prefers to writhe, since this is more energetically favorable.

To demonstrate supercoiling for yourself, cut a flat rubber band to obtain a linear piece a few inches long. Hold the ends apart, twist one of them, and then bring the ends closer together. You will see the twists collapse into writhes (supercoils). The same thing happens if you let the rubber band relax and then twist it in the opposite direction. Note that the twisted rubber band (representing a double-stranded DNA molecule) adopts a more compact shape. Naturally occurring DNA molecules are negatively supercoiled, which allows them to take up less space. It also means that if the DNA were stretched out (that is, its writhes converted to twists), the two strands would unwind slightly.

TOPOISOMERASES ALTER DNA SUPERCOILING

Normal replication and transcription require that cells actively maintain the supercoiling state of DNA by adding or removing supercoils. Of course, cells cannot grab the ends of a long DNA molecule in order to twist or untwist it. Instead, topological changes occur when a topoisomerase cuts one or both strands of the supercoiled DNA, alters the structure, and then reconnects the broken strands. Type I topoisomerases cut one DNA strand, and the type II enzymes cut both strands of DNA and require ATP.

Type I topoisomerases occur in all cells and alter supercoiling by altering the DNA's helical twisting. Type IA enzymes nick the DNA (cut the backbone of one strand), pass the intact strand through the break, then seal the broken strand in order to unwind the DNA by one turn (**Fig. 20.30**). Type IB enzymes also cut one strand but hold the DNA on one

Supercoiled DNA molecules. The circular DNA molecules are slightly underwound, so they coil up on themselves, forming supercoils. [*Dr. Gopal Murti/Photo Researchers Inc.*]

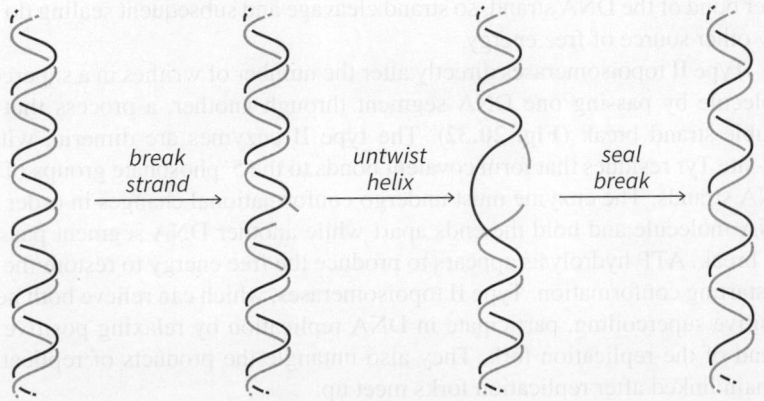

break strand → untwist helix → seal break

FIGURE 20.30

Action of a type I topoisomerase. In this scheme, the supercoiling of a DNA segment is decreased when a DNA strand is broken, the helix untwisted by passing the other strand through the break, and then sealed.

 Count the number of helical turns in the structure shown here.

FIGURE 20.31

Human topoisomerase I. The protein (blue) surrounds a DNA molecule (shown end-on in this model). [*Structure (pdb 1A36) determined by L. Stewart, M. R. Redinbo, X. Qiu, J. J. Champoux, and W. G. J. Hol.*]

side of the nick while allowing the DNA on the other side of the nick to rotate one or more turns before the broken strand is sealed. In both cases, unwinding is driven by the strain in the supercoiled DNA, so a type I topoisomerase can "relax" both negatively and positively supercoiled DNA.

Human topoisomerase I, a type IB enzyme, has four domains that encircle the DNA (**Fig. 20.31**). The surface of the protein that contacts the DNA is rich in positively charged groups that can interact with the backbone phosphate groups of about 10 base pairs. When a type I topoisomerase cleaves one strand of the DNA, an active-site tyrosine residue forms a covalent bond with the backbone phosphate at one side of the nick:

$$CH_2$$

Tyr

$$O^-$$

$$O - P - O - CH_2 \quad O \quad \text{Base}$$

$$O^-$$

H H

H H

$$O - P - O \cdots$$

$$O^-$$

DNA

Formation of this diester linkage conserves the free energy of the broken phosphodiester bond of the DNA strand, so strand cleavage and subsequent sealing do not require any other source of free energy.

Type II topoisomerases directly alter the number of writhes in a supercoiled DNA molecule by passing one DNA segment through another, a process that requires a double-strand break (**Fig. 20.32**). The type II enzymes are dimeric, with two active-site Tyr residues that form covalent bonds to the 5′ phosphate groups of the broken DNA strands. The enzyme must undergo conformational changes in order to cleave a DNA molecule and hold the ends apart while another DNA segment passes through the break. ATP hydrolysis appears to produce the free energy to restore the enzyme to its starting conformation. Type II topoisomerases, which can relieve both negative and positive supercoiling, participate in DNA replication by relaxing positive supercoils ahead of the replication fork. They also untangle the products of replication, which remain linked after replication forks meet up:

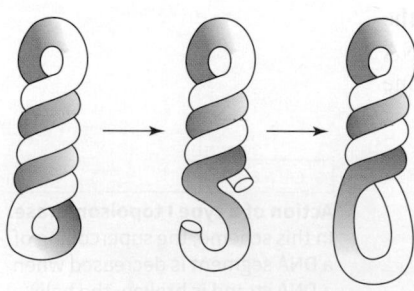

FIGURE 20.32

Action of a type II topoisomerase. The twisted worm shape represents a supercoiled double-stranded DNA molecule. A type II topoisomerase breaks both strands of the DNA, passes another segment of DNA through the break, then seals the break. In this diagram, the degree of supercoiling decreases.

Bacteria contain a type II topoisomerase called DNA gyrase, which can introduce additional negative supercoils into DNA, so its net effect is to further underwind the DNA helix. A number of antibiotics inhibit DNA gyrase without affecting eukaryotic type II topoisomerases. For example, ciprofloxacin acts on DNA gyrase to enhance the rate of DNA cleavage or reduce the rate of sealing broken DNA. The result is a large number of DNA breaks that interfere with the transcription and replication required for normal cell growth and division, and the bacteria die.

Ciprofloxacin

EUKARYOTIC DNA IS PACKAGED IN NUCLEOSOMES

Eukaryotic cells lack DNA gyrase but maintain negative supercoiling by wrapping DNA in **nucleosomes.** These complexes of DNA and protein are the basic unit of eukaryotic DNA packaging. *The core of a nucleosome consists of eight **histone** proteins: two each of the histones known as H2A, H2B, H3, and H4. Approximately 146 base pairs of DNA make about 1.65 turns around the histone octamer* (**Fig. 20.33**). A complete nucleosome contains the core structure plus histone H1, a small protein that appears to bind outside the core. Neighboring nucleosomes are separated by short stretches of DNA of 20 to 40 bp. Winding DNA in nucleosomes helps protect the DNA from chemical damage and generates the negative supercoils needed to unwind DNA to initiate replication.

The histone proteins interact with DNA in a sequence-independent manner, primarily via hydrogen bonding and ionic interactions with the sugar–phosphate backbone. Although prokaryotes lack histones, other DNA-binding proteins may help package DNA in bacterial cells. The histones are among the most highly conserved proteins known, in keeping with their essential function in packaging the genetic material in all eukaryotic cells. Each histone pairs with another, and the set of eight forms a compact structure. However, the tails of the

(a)   (b)

FIGURE 20.33

Structure of the nucleosome core. (a) Top view. (b) Side view (space-filling model). The DNA (dark blue) winds around the outside of the histone octamer. [*Structure (pdb 1AOI) determined by K. Luger, A. W. Maeder, R. K. Richmond, D. F. Sargent, and T. J. Richmond.*]

FIGURE 20.34

A eukaryotic nucleus. Heterochromatin is the darkly staining material (red in this colorized electron micrograph); euchromatin stains more lightly (yellow). [*Gopal Murti/Science Photo Library/Photo Researchers.*]

B-DNA, diameter 20 Å   Nucleosome   30-nm fiber   1 μm   0.4 μm

FIGURE 20.35

Levels of chromatin structure. The DNA helix (blue) is wrapped around a histone octamer (orange) to form a nucleosome; nucleosomes aggregate to form the 30-nm fiber; this packs into loops in the fully condensed chromosome. The approximate diameter of each structure is indicated.

histones, which are flexible and charged, extend outward from the core of the nucleosome (see Fig. 20.33). These histone tails are subject to covalent modification, which helps control gene expression (Section 21.1).

During DNA replication, nucleosomes are disassembled as the DNA spools through the replication machinery. A protein complex called the replication-coupling assembly factor helps reassemble nucleosomes on newly replicated DNA, using the displaced histones as well as newly synthesized histones imported from the cytoplasm to the nucleus.

The nucleosomes themselves form higher-order structures that are not entirely understood, although chromosomes seem to be organized in long loops. Some of the cell's DNA appears to be highly condensed at the edges of the nucleus. This DNA is known as **heterochromatin,** which is transcriptionally silent. **Euchromatin** is less condensed and appears to be transcribed at a higher rate. The two forms of **chromatin** are distinguishable by electron microscopy (**Fig. 20.34**).

When the cell is ready to divide, the DNA is further condensed. Nucleosomes compact the DNA only by a factor of about 30 to 40, but the chain of nucleosomes itself can coil into a solenoid (coil) with a diameter of about 30 nm (**Fig. 20.35**). The DNA in the 30-nm fiber is presumably well protected from nuclease attack and is inaccessible to the proteins that carry out replication and transcription. During cell division, chromosomes condense even further so that each one has an average length of about 10 μm and a diameter of 1 μm. This makes sense for a dividing cell, since fully extended DNA molecules would become hopelessly tangled rather than segregating neatly to form two equivalent sets of chromosomes.

 CONCEPT CHECK

1. Describe the relationship between writhing (supercoiling) and helical twisting.

2. Explain why negative supercoiling of DNA assists processes such as replication.

3. Explain how topoisomerases alter DNA supercoiling by transiently cutting one or both strands of DNA.

4. Explain why negative supercoiling of DNA assists processes such as replication.

5. Describe the action of type I and type II topoisomerases and their source of free energy.

6. Explain the role of histones in packaging DNA.

7. Compare the structure and activity of heterochromatin and euchromatin.

8. How could covalent modification of histones or DNA alter gene expression?

SUMMARY

20.1 The DNA Replication Machinery

- DNA replication requires a host of enzymes and other proteins located in a stationary factory. Helicases separate the two DNA strands at a replication fork, and SSB binds to the exposed single strands.

- DNA polymerase can only extend a preexisting chain and therefore requires an RNA primer synthesized by a primase. Two polymerases operate side-by-side to replicate DNA, so the leading strand of DNA is synthesized continuously while the lagging strand is synthesized discontinuously as a series of Okazaki fragments. The RNA primers of the Okazaki fragments are removed, the gaps are filled in by DNA polymerase, and the nicks are sealed by DNA ligase.

- DNA polymerase and other polymerases catalyze a reaction in which the 3′ OH group of the growing chain nucleophilically attacks the phosphate group of an incoming nucleotide that base pairs with the template strand. A sliding clamp promotes the processive activity of DNA polymerase.

- Many DNA polymerases contain a second active site that hydrolytically excises a mismatched nucleotide.

20.2 Telomeres

- In eukaryotes, the extreme 3′ end of a DNA strand cannot be replicated, so the enzyme telomerase adds repeating sequences to the 3′ end to create a structure known as a telomere.

20.3 DNA Damage and Repair

- Normal replication errors, spontaneous deamination, radiation, and chemical damage can cause mutations in DNA.

- Mechanisms for repairing damaged DNA include direct repair, mismatch repair, base excision repair, nucleotide excision repair, end-joining, and recombination.

20.4 Clinical Connection: Cancer as a Genetic Disease

- Cancer is uncontrolled cell division triggered by several genetic changes. These mutations occur as a result of inheritance, environmental factors, or random chance.

- Cellular pathways that detect and repair damaged DNA or prevent the cell from dividing are commonly disrupted in cancer.

20.5 DNA Packaging

- DNA unwinding is facilitated by the negative supercoiling (underwinding) of DNA molecules. Enzymes called topoisomerases can add or remove supercoils by temporarily introducing breaks in one or both DNA strands.

- Eukaryotic DNA is wound around a core of eight histone proteins to form a nucleosome, which represents the first level of DNA compaction in the nucleus.

KEY TERMS

replication	Okazaki fragment	mutation	constitutive
semiconservative replication	processivity	point mutation	apoptosis
DNA polymerase	proofreading	transition mutation	carcinogenesis
replication fork	nick	transversion mutation	tumor suppressor gene
factory model of replication	nick translation	abasic site	supercoiling
helicase	DNA ligase	mutagen	nucleosome
topoisomerase	replisome	carcinogen	histone
primer	centromere	mismatch repair	heterochromatin
primase	telomere	base excision repair	euchromatin
leading strand	telomerase	nucleotide excision repair	chromatin
lagging strand	reverse transcriptase	nonhomologous end-joining	
discontinuous synthesis	cDNA	recombination	

BIOINFORMATICS

Brief Bioinformatics Exercises

20.1 DNA Replication and the KEGG Database

20.2 Viewing and Analyzing Uracil–DNA Glycosylase in DNA Repair

PROBLEMS

20.1 The DNA Replication Machinery

1. Semiconservative replication is shown in Figure 20.1. Draw a diagram that illustrates the composition of DNA daughter molecules for the second, third, and fourth generations.

2. DNA replication of the circular chromosome in E. coli begins when a protein called DnaA binds to the replication origin on the DNA. Is the replication origin likely to be richer in G:C or A:T base pairs?

3. DNA replication of the circular chromosome in *E. coli* begins when a protein called DnaA binds to the replication origin on the DNA.

(a) DnaA initiates replication only when the DNA that constitutes the DNA replication origin is negatively supercoiled. Why?

(b) Is the replication origin likely be richer in G:C or A:T base pairs?

4. DNA helicases can be considered to be molecular motors that convert the chemical energy of NTP hydrolysis into mechanical energy for separating DNA strands. The bacteriophage T7 genome encodes a protein that assembles into a hexameric ring with helicase activity.

(a) In order to unwind DNA, the helicase requires two single-stranded DNA tails at one end of a double-stranded DNA segment. However, the helicase appears to bind to only one of the single strands, apparently by encircling it. Is this consistent with its ability to unwind a double-stranded DNA helix?

(b) Kinetic measurements indicate that the T7 helicase moves along the DNA at a rate of 132 bases per second. The protein hydrolyzes 49 dTTP per second. What is the relationship between dTTP hydrolysis and DNA unwinding?

(c) What does the structure of T7 helicase suggest about its processivity?

5. At eukaryotic origins of replication, helicase cannot be activated until the polymerase is also positioned on the DNA. Explain what would happen if the helicase became active in the absence of DNA polymerase.

6. Temperature-sensitive mutations allow a protein to function at a low temperature (the permissive temperature) but not a high (nonpermissive) temperature. Temperature-sensitive mutations in some replication proteins result in the immediate cessation of bacterial growth; for other mutant proteins, growth comes to a halt more gradually when the bacteria are exposed to a nonpermissive temperature. What happens to DNA replication and bacterial growth when the temperature is suddenly increased and the temperature-sensitive mutation is in **(a)** helicase or **(b)** DnaA (see Problem 3)?

7. Based on your knowledge of replication proteins, compare DNA polymerase and single-strand binding protein (SSB) with respect to **(a)** affinity for DNA and **(b)** cellular concentration.

8. Explain why treatment of a preparation of *E. coli* single strand binding protein (SSB) with acetic anhydride produces a chemically modified protein that is unable to bind to DNA. [*Hint*: Acetic anhydride reacts with primary amino groups.]

9. In the chain-termination method of DNA sequencing (described in Section 3.4), the Klenow fragment of DNA polymerase I (see Fig. 20.7) is used to synthesize a complementary strand using single-stranded DNA as a template. Along with a primer and the four dNTP substrates, a small amount of 2′, 3′-dideoxynucleoside triphosphate (ddNTP) is added to the reaction mixture. When the ddNTP is incorporated into the growing nucleotide chain, polymerization stops. Explain.

2′,3′-Dideoxynucleoside
triphosphate (ddNTP)

10. You have discovered a drug that inhibits the activity of the enzyme inorganic pyrophosphatase. What effect would this drug have on DNA synthesis? (See Fig. 20.5.)

11. Which eukaryotic DNA polymerase would you expect to have greater processivity: polymerase α, δ, and ε? Explain.

12. An *in vitro* system was developed in which simian virus 40 (SV40) can be replicated using purified mammalian proteins. Make a list of the proteins required to replicate SV40 DNA in a cell-free system.

13. Describe the ways in which a cell minimizes the incorporation of mispaired nucleotides during DNA replication.

14. DNA polymerases include two Mg^{2+} ions in the active site.

(a) Describe how Mg^{2+} could enhance the nucleophilicity of the 3′OH group that attacks the α-phosphate of an incoming nucleotide.

(b) The polymerization mechanism includes the formation of a pentacovalent phosphorus. Sketch this structure. How could an Mg^{2+} ion contribute to transition state stabilization during DNA polymerization?

15. A reaction mixture contains purified DNA polymerase, the four dNTPs, and one of the DNA molecules whose structures are represented below. Which reaction mixture produces PP_i?

(a) |||||||||||||||||||||||||

(b)

(c)

(d)

(e)

(f)

16. The "flap" endonuclease recognizes the junction between RNA and DNA near the 5′ end of an Okazaki fragment. What feature(s) of this structure might the enzyme recognize?

17. Why would it not make sense for the cell to wait to combine Okazaki fragments into one continuous lagging strand until the entire DNA molecule had been replicated?

18. Explain how DNase (an endonuclease that cleaves the backbone of one strand of a DNA molecule), *E. coli* DNA polymerase I (which includes 5′ → 3′ exonuclease activity), and DNA ligase could be used in the laboratory to incorporate radioactive nucleotides into a DNA molecule.

19. The mechanism of *E. coli* DNA ligase involves the transfer of the adenylyl (AMP) group of NAD^+ to the ε-amino group of the side chain of an essential lysine residue on the enzyme. The adenylyl group is subsequently transferred to the 5′ phosphate group of the nick. The first step is shown. Draw the complete mechanism of *E. coli* DNA ligase.

Enzyme—$(CH_2)_4$—NH_2

Nicotinamide—Ribose—O—P—O—P—O—Ribose—Adenine

20. Why would ligase enzymes that use NAD^+ as a cofactor be attractive drug targets for treating diseases caused by bacteria?

20.2 Telomeres

21. Give the name of the enzyme that catalyzes each of the following reactions:

(a) makes a DNA strand from a DNA template;

(b) makes a DNA strand from an RNA template;

(c) makes an RNA strand from a DNA template.

22. Human cells infected by HIV increase the expression of SAMDH1, a phosphohydrolase that removes triphosphate groups from dNTP substrates. Explain how SAMDH1 inhibits HIV replication.

23. In some species, G-rich telomeric DNA folds up on itself to form a four-stranded structure. In this structure, four guanine residues assume a hydrogen-bonded planar arrangement with an overall geometry that can be represented as

This is called a G quartet, and it may play a role as a negative regulator of telomerase activity.

(a) Draw the complete structure of this G quartet, including the hydrogen bonds between the purine bases.

(b) Show schematically how a single strand of four repeating TTAGGG sequences can fold to generate a structure with three stacked G quartets linked by TTA loops.

24. Why might a drug that induces the formation of the G quartet (see Problem 23) be effective as an antitumor agent?

25. In an experiment, the AAUCCC segment of the RNA template of telomerase is mutated. Will this mutation change the telomeric sequence? Explain.

26. What cellular changes occur when the RNA template on the telomerase is mutated as described in Problem 25?

20.3 DNA Damage and Repair

27. Classify the following base changes as transitions or transversions:
(a) C → A; **(b)** A → C; **(c)** G → T; **(d)** C → T; **(e)** T → G; **(f)** A → G.

28. The free deoxyribonucleotides dATP and dGTP are about 13,000 times more susceptible to oxidative damage than when they are incorporated into DNA.

(a) Explain why this minimizes the incidence of mutations.
(b) In eukaryotic cells, a specific triphosphatase hydrolyzes oxo-dGTP to its monophosphate form. Does this enzyme activity inhibit or promote DNA mutations?
(c) Explain how small molecules that inhibit the activity of the triphosphatase are able to prevent the growth of highly metabolically active cancer cells.

29. Draw the structure of an oxoguanine:adenine base pair. (*Hint:* The oxoguanine base pivots around the glycosidic bond in order to form two hydrogen bonds with adenine.)

30. Compounds such as ethidium bromide and acridine orange can be used to stain DNA bands in a gel. These compounds are known as intercalating agents, and they interact with the DNA by slipping in between stacked base pairs. This interaction increases the fluorescence of the intercalating agent and allows for visualization of DNA bands following electrophoresis. Care must be taken when working with intercalating agents because they are powerful mutagens. Explain why.

Ethidium bromide

Acridine orange

31. Oxidative deamination of nitrogenous bases can occur upon exposure to nitrous acid (HNO_2). Deamination of cytosine produces uracil, as described in the text, and deamination of adenine produces hypoxanthine.

(a) Draw the structure of hypoxanthine.
(b) Hypoxanthine can base-pair with cytosine. Draw the structure of this base pair.
(c) What is the consequence to the DNA if this deamination is not repaired?

32. Oxidative deamination of guanine produces xanthine.

(a) Draw the structure of xanthine.
(b) Xanthine base-pairs with cytosine. Would this cause a mutation? Explain.

33. How do the mutagens ethidium bromide (Problem 30) and nitrous acid (Problem 31) differ in the way they damage DNA?

34. The compound 5-bromouracil, a thymine analog, can be incorporated into DNA in the place of thymine. 5-Bromouracil readily converts to an enol tautomer, which can base pair with guanine. (The keto and enol tautomers freely interconvert through the movement of a hydrogen between an adjacent nitrogen and oxygen.) Draw the structure of the base pair formed by the enol form of 5-bromouracil and guanine. What kind of mutation can 5-bromouracil induce?

5-Bromouracil

35. Experiments with cells are performed in a tissue culture hood, which filters the air to minimize the chance of contaminating the cells with airborne bacteria. When the experiments are completed, the cells are removed from the hood, and an ultraviolet light is switched on until the hood is used again. What is the rationale for this procedure?

36. The fact that DNA has evolved to contain the bases A, C, G, and T makes the DNA molecule easy to repair. For example, deamination of adenine produces hypoxanthine, deamination of guanine produces xanthine, and deamination of cytosine produces uracil. Why are these deaminations repaired quickly?

37. Studies of bacteria and other organisms indicate that mutations occur more frequently at certain positions. These "hotspots" are due to the presence of 5-methylcytosine, which may undergo oxidative deamination.

(a) Draw the structure of deaminated 5-methlycytosine.
(b) By what other name is the base known?
(c) What kind of mutation results from 5-methylcytosine deamination?
(d) Why is the altered base difficult for the cell to repair?

38. The adenine analog 2-aminopurine (shown below) is a potent mutagen in bacteria. The 2-aminopurine substitutes for adenine during

DNA replication and gives rise to mutations because it pairs with cytosine instead of thymine. Structural studies show that the "neutral wobble" base pair (so called because it is the dominant structure at neutral pH) is in equilibrium with a "protonated Watson–Crick" structure, which forms at lower pH.

(a) Two hydrogen bonds form between cytosine and 2-aminopurine in the neutral wobble pair. Draw the structure of this base pair.

(b) At lower pH, either the cytosine or the 2-aminopurine can become protonated. The two protonated forms are in equilibrium in the protonated Watson–Crick base pair structure; the proton essentially "shuttles" from one base to the other in the pair, and the hydrogen bond is maintained. Draw the two possible structures for the base pair, one with a protonated 2-aminopurine, and one with a protonated cytosine.

Cytosine   2-Aminopurine

Protonated   Protonated
cytosine   2-aminopurine

39. Chemical methylating agents can react with guanine residues in DNA to produce O^6-methylguanine, which causes mutations because it can pair with thymine. The structure of O^6-methylguanine is shown in the text. This type of chemical damage is difficult to repair because the O^6-methylguanine:T base pair is structurally similar to a normal G:C base pair and the mismatch repair system has difficulty recognizing it. The resulting mutations can generate oncogenes. The O^6-methylguanine can form a "wobble" pair with thymine (similar to the 2-aminopurine:C base pair in Problem 38). The methylated guanine can also form a base pair with protonated cytosine at lower pH values.

(a) At physiological pH, the O^6-methylguanine:T wobble base pair predominates. Two hydrogen bonds form between the two bases. Draw the structure of this base pair. Does a mutation result?

(b) At lower pH values, cytosine becomes protonated (see Problem 38b). Draw the structure of the O^6-methyguanine:protonated cytosine base pair (three hydrogen bonds form between the two bases). Does a mutation result?

(c) Why might O^6-methylguanine be referred to as a "suicide substrate" for the methyltransferase that removes the methyl group?

40. The endonuclease of the mismatch repair system in bacteria distinguishes between the parent and newly synthesized daughter DNA (which contains the incorrect base) by recognizing that the daughter DNA is not methylated. The eukaryotic mismatch repair system does not use methylation in this way. What other structural feature might eukaryotes use to distinguish between parent and daughter DNA?

41. What is the most common DNA lesion in individuals with the disease xeroderma pigmentosum?

42. Thymine dimers can be restored to their original form by DNA photolyases that cleave the dimer. The photolyases are so named because they are activated by absorption of light. Why is this a good biochemical strategy?

43. A strain of mutant bacterial cells lacks the enzyme uracil-DNA glycosylase. What is the consequence for the organism?

44. (a) In many cases, a point mutation in DNA has no effect on the encoded amino acid sequence. Explain.

(b) In other cases, a point mutation might have an effect on the encoded amino acid sequence but does not lead to deleterious consequences. Explain why.

45. The incomplete removal of an RNA primer leaves ribonucleotides incorporated into DNA. List the enzyme activities that would be required to repair DNA containing a stretch of several ribonucleotides.

46. During replication in *E. coli*, certain types of DNA damage, such as thymine dimers, can be bypassed by DNA polymerase V. This polymerase tends to incorporate guanine residues opposite the damaged thymine residues and has a higher overall error rate than other polymerases. Thymine dimers can be bypassed by other polymerases, such as DNA polymerase III (which carries out most DNA replication in *E. coli*). DNA polymerase III incorporates adenine residues opposite the damaged thymine residues, but much more slowly than DNA polymerase V. Polymerase III is a highly processive enzyme, whereas polymerase V adds only 6–8 nucleotides before dissociating from the DNA. Explain how DNA polymerases III and V together carry out the efficient replication of UV-damaged DNA with minimal errors.

47. Eukaryotic cells contain a number of DNA polymerases. Several of these enzymes were tested for their ability to cleave nucleotides from the 3′ end of a DNA chain (3′ → 5′ exonuclease activity). The enzymes were also tested for the accuracy of DNA polymerization, expressed as the rate of base substitution. The results are summarized in the table.

Polymerase	3′ → 5′ exonuclease activity	Base substitution rate ($\times 10^{-5}$)
α	No	16
β	No	67
δ	Yes	1
ε	Yes	1
η	No	3500

(a) Is there a correlation between the presence of 3′ → 5′ exonuclease activity and the error rate (base substitutions) during DNA polymerization?

(b) Express the error rate of each polymerase in terms of how often a wrong base is incorporated.

(c) Polymerization errors can result from the ability to insert an incorrect (mispaired) base or from the inability to efficiently insert the correct (template-matched) base. To tell which mechanism accounts for the high error rate of polymerase η, the catalytic efficiency (k_{cat}/K_M) of the polymerization reaction was measured for matched and unmatched bases. The results were compared to the catalytic efficiency of another polymerase, HIV reverse transcriptase (HIV RT). The data are summarized in the table below. Compare the efficiency of DNA polymerase η and HIV RT in incorporating correct bases and incorrect bases. What do these results reveal about the cause of errors during polymerization by polymerase η?

(d) The overexpression of genes that code for DNA polymerases similar to polymerase η has been observed in bacteria. What effect would this have on the mutation rate in these bacteria?

Polymerase	Template base	Incoming base	k_{cat}/K_M ($\mu M \cdot min^{-1} \cdot 10^3$)
Polymerase η	T	A	420
	T	G	22
	T	C	1.6
	G	C	760
	G	G	8.7
HIV RT	T	A	800
	T	G	0.07

20.4 Clinical Connection: Cancer as a Genetic Disease

48. Mutations in the genes that code for repair enzymes can lead to the transformation of a normal cell into a cancerous cell. Explain why.

49. Some types of cancer are clearly linked to specific DNA repair pathways. For example, individuals with mutations in the human homolog of the bacterial MutS gene are predisposed to develop colon cancer. Explain why.

50. Although DNA damage is a cause of cancer, some cancer chemotherapies are based on drugs that damage DNA. For example, the drug carboplatin introduces a platinum ion that cross-links adjacent guanine residues in a DNA strand. Explain why this leads to death of the cancer cell.

51. The product of the retinoblastoma gene, the Rb protein, binds to and inhibits the activity of a transcription factor that induces the expression of genes required for the cell to synthesize DNA. Explain how a mutation in the retinoblastoma gene promotes cancer.

52. Explain why inheriting a defective gene for the retinoblastoma protein, BRCA1, or p53 does not guarantee that an individual will develop cancer.

53. The human immune system can recognize markers on the surface of cancerous cells so that these cells can be selectively killed. What feature of a transformed cell would change its appearance?

54. One of the proteins that senses damaged DNA may structurally resemble PCNA, the eukaryotic sliding-clamp protein that enhances the processivity of DNA polymerase (see Fig. 20.8). Explain why such a protein is suited for monitoring the chemical integrity of DNA.

55. The Ras signaling pathway (Section 10.3) activates a transcription factor called Myc, which turns on a gene for a protein that blocks the activity of the protein that adds ubiquitin to p53. Is this mechanism consistent with the ability of excessive Ras signaling to promote cell growth?

56. In the 1920s, Otto Warburg noted that cancer cells primarily use glycolysis rather than aerobic respiration to produce ATP (Section 19.4). In 2006, researchers discovered that p53 stimulates the expression of cytochrome *c* oxidase. How does the connection between p53 and cytochrome *c* oxidase explain Warburg's observation?

57. Why do unicellular organisms have no need for apoptosis?

20.5 DNA Packaging

58. **(a)** Why do type II topoisomerase enzymes require ATP whereas type I topoisomerases do not?
(b) Explain why a type I topoisomerase sometimes participates in nucleotide excision repair.

59. What is the role of DNA gyrase during bacterial DNA replication?

60. The percentages of arginine and lysine residues in the histones of calf thymus DNA are shown in the table below. Why do histones have a large number of Lys and Arg residues?

	% Arg	% Lys
H1	1	29
H2A	9	11
H2B	6	16
H3	13	10
H4	14	11

61. Why is 0.5 M NaCl effective at dissociating histones from DNA in a sample of chromatin?

62. The histone protein H4 from cows and the H4 from peas, representing groups that diverged from one another over a billion years ago, differ by only two amino acid residues out of 102 total residues. The changes are conservative—Val 60 in calf thymus H4 is replaced by an Ile in peas; Lys 77 in the calf H4 is replaced by an Arg in peas. Propose a hypothesis that explains why the H4 histone protein is so highly evolutionarily conserved.

63. A variety of compounds inhibit topoisomerases. Novobiocin is an antibiotic and, like ciprofloxacin, it inhibits DNA gyrase. Doxorubicin and etoposide are anticancer drugs that inhibit eukaryotic topoisomerases. What properties distinguish the antibiotics from the anticancer drugs?

64. The anticancer drug doxorubicin (see Problem 63) is more correctly termed as topoisomerase II poison because it blocks catalysis after the DNA is cleaved and before it is re-ligated. In contrast, the anticancer drug aclarubicin is termed a "catalytic inhibitor" because it prevents the binding of topoisomerase II to the DNA. What distinguishes an inhibitor from a poison?

65. Deoxyribonuclease (DNase) processively nicks the phosphodiester backbone of double-stranded DNA to yield small oligonucleotide fragments and can be used therapeutically to treat patients with cystic fibrosis (CF). The enzyme is inhaled into the lungs, where it hydrolyzes the DNA contained in the sputum to decrease its viscosity and improve lung function. Genetic engineering techniques have been used to synthesize hyperactive DNase variants that produce nicks in the DNA more rapidly than the wild-type enzyme. A more efficient DNase can be used to achieve the same results at a lower enzyme concentration.

(a) The engineered hyperactive DNase variants are listed in the table. (A note on nomenclature: Q9R means that the Gln at position 9 in the wild-type DNase I has been changed to an Arg.) What structural feature do all of the DNase variants have in common? Why might these changes improve the catalytic efficiency of DNase?

(b) The enzymatic activities of the DNase variants were tested using a DNA hyperchromicity assay in which the absorbance of a solution of intact DNA is first measured at 260 nm. Then the enzyme is added, and the solution is monitored for an increase in absorbance that occurs when the concentration of single-stranded DNA in the solution increases. Why is the hyperchromicity assay a useful tool to assess the activity of the DNase variants?

(c) The DNA hyperchromicity assay was used to measure the K_M and V_{max} values for each variant. The results are shown in the table. How do the amino acid changes affect the activity of the enzyme variants as compared to the wild-type? Which variant(s) is(are) most efficient?

DNase variant	Charge (relative to wild-type)	K_M (µg/mL DNA)	V_{max} (A_{260} units · min^{-1} · mg^{-1})
Wild-type		1.0	1.0
N74K	+1	0.77	3.6
E13R/N74K	+2	0.20	5.3
E13R/N74K/T205K	+3	0.18	7.7
E13R/T14K/N74K/ T205K	+4	0.37	3.5
Q9R/E13R/H44K/ N74K/T205K	+5	0.11	2.4
Q9R/E13R/T14K/ H44R/N74K/T205K	+6	0.20	2.5

(d) The DNase variants were evaluated for their ability to cut or nick DNA. A *cut* refers to the hydrolysis of phosphodiester bonds on both strands, whereas a *nick* is the hydrolysis of just one strand. This was assessed by using a circular plasmid substrate, which is most stable in its supercoiled form. If one strand is nicked, the plasmid forms a relaxed circle, but if the backbones of both strands are cut, the circle linearizes, as shown below.

Supercoiled, relaxed circular, and linear DNA can be detected by their different rates of migration during electrophoresis through agarose gels. In a series of experiments, the plasmid was incubated with DNase and then the products were analyzed by agarose gel electrophoresis. The results are shown above right. Describe the results for each lane. Compare the selected variants to the wild-type DNase with regard to their ability to cut or nick the DNA. (C, control; WT, wild-type)

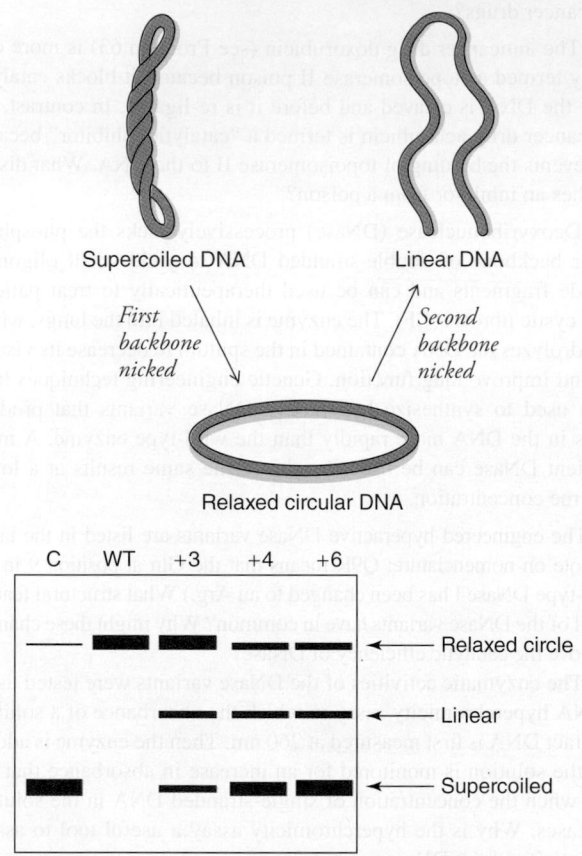

Supercoiled DNA   Linear DNA

First backbone nicked   *Second backbone nicked*

Relaxed circular DNA

C   WT   +3   +4   +6

← Relaxed circle

← Linear

← Supercoiled

(e) The variants were tested for their ability to act on DNA at high and low substrate concentration. High-molecular-weight (MW) DNA

is present in the lung secretions of CF patients in fairly high concentrations. DNase might also be effective as a therapeutic agent to treat the autoimmune disease systemic lupus erythematosus, in which DNA is present in the serum at low concentrations. What variant would be a good drug candidate to treat a CF patient? A lupus patient? Explain.

Nicking activity (compared to wild type)

	Low DNA concentration		High DNA concentration	
	Low MW	High MW	Low MW	High MW
Wild-type	1	1	1	1
N74K	26		10.4	
E13R/N74K	211	31	24.3	13
E13R/N74K/ T205K	7		1.3	
E13R/T14K/ N74K/T205K	7		0.7	

66. During sperm development, about 95% of the cell's DNA is associated with small proteins known as protamines rather than histones. Protamine–DNA complexes pack more compactly than histone–DNA complexes.

(a) Explain the advantage of replacing histones with protamines during sperm development.
(b) What type of genes would you expect to find in the remaining nucleosomes? (*Hint:* These genes are not transcribed until after fertilization.)

67. A sample of chromatin is briefly treated with micrococcal nuclease, which catalyzes the hydrolysis of double-stranded DNA. **(a)** Analysis by agarose gel electrophoresis indicates that the products are double-stranded DNA segments about 200 bp long. Explain these results. **(b)** If the digestion is allowed to proceed for an extended period of time, 146-bp DNA bands are observed on the agarose gel. Explain.

68. *In vivo*, DNA undergoes transient local melting, producing small "bubbles" of single-stranded DNA. Does this DNA "breathing" activity explain why the rate of C → T mutations is lower when the DNA is part of a nucleosome?

SELECTED READINGS

Bauer, N. C., Corbett, A. H., and Doetsch, P. W., The current state of eukaryotic DNA base damage and repair, *Nuc. Acids Res.* **43**, 10083–10101 (2015). [Describes several types of DNA damage, direct repair, base excision repair, and nucleotide excision repair.]

Deweese, J. E., Osheroff, M. A., and Osheroff, N., DNA topology and topoisomerases, *Biochem. Mol. Biol. Educ.* **37**, 2–10 (2009). [Includes discussion of topoisomerase action and inhibitors.]

Doublié, S. and Zahn, K. E., Structural insights into eukaryotic DNA replication, *Front. Microbiol.* doi: 10.3389/fmicb.2014.00444 (2014). [Summarizes structural and functional information for the three main eukaryotic DNA polymerases.]

Dunn, B., Solving an age-old problem, *Nature* **483**, S2–S6 (2012). [Discusses major advances in research on the causes, genetic basis, treatment, and prevention of cancer.]

Luger, K., Dechassa, M. L., and Tremethick, D. J., New insights into nucleosome and chromatin structure: an ordered state or a

disordered affair? *Nat. Rev. Mol. Cell Biol.* **13**, 436–447 (2012). [Discusses the functional implications of irregular and dynamic nucleosome structures.]

Miracco, E. J., Jiang, J., Cash, D. D., and Feigon, J., Progress in structural studies of telomerase, *Curr. Opin. Struct. Biol.* **24**, 115–124 (2014). [Summarizes research on telomerase structure from a variety of organisms.]

O'Donnell, M., Langston, L., and Stillman, B., Principles and concepts of DNA replication in bacteria, archaea, and eukarya. *Cold Spring Harb. Perspect. Biol.* doi: 10.1101/cshperspect.a010108 (2013). [Reviews the fundamental events of replication, including the process of initiating replication.]

Tomasetti, C. and Vogelstein, B., Variation in cancer risk among tissues can be explained by the number of stem cell divisions, *Science* **347**, 78–81 (2015). [Proposes that random mutations are more significant causes of cancer than hereditary or environmental factors.]

TRANSCRIPTION AND RNA

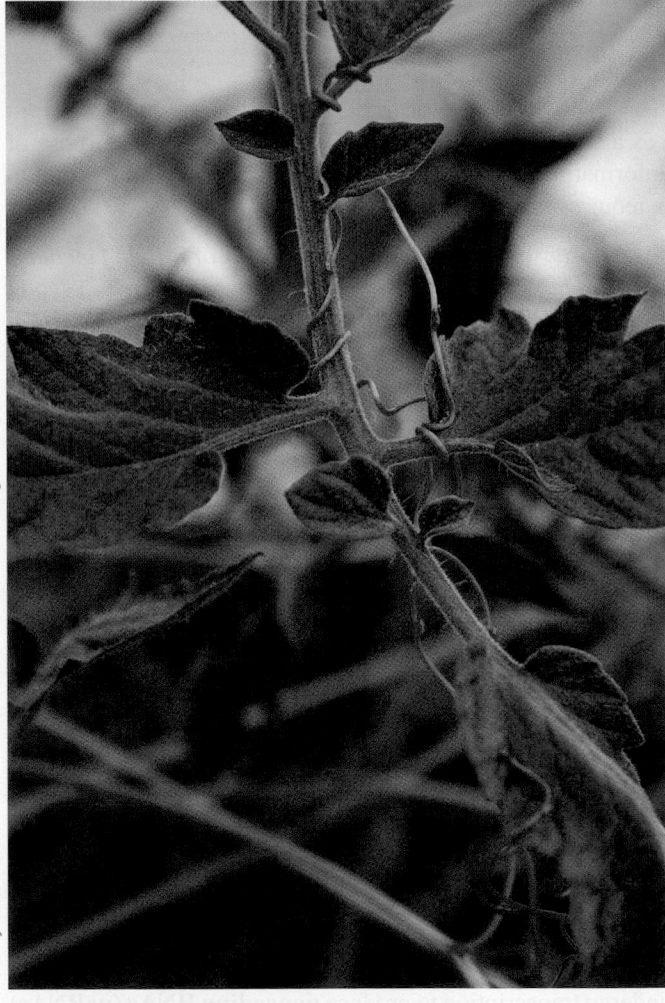

Quincy Russell, Mona Lisa Production/Science Source Images

The parasitic dodder vine can sniff out a suitable host, such as a tomato plant, and tap into its vascular system. In addition, the vine shares some of its RNA with its host and imports the host's RNA into its own system. Because RNA contains biological information, the RNA exchange may be a way for the parasite to synchronize its life cycle with the tomato plant's in order to reproduce before its support system dies.

DO YOU REMEMBER?

- DNA and RNA are polymers of nucleotides, each of which consists of a purine or pyrimidine base, deoxyribose or ribose, and phosphate (Section 3.2).
- The biological information encoded by a sequence of DNA is transcribed to RNA and then translated into the amino acid sequence of a protein (Section 3.3).
- Genes can be identified by their nucleotide sequences (Section 3.4).
- DNA replication is carried out by stationary protein complexes (Section 20.1).

Transcription is the fundamental mechanism by which a gene is expressed. It is the conversion of stored genetic information (DNA) to a more active form (RNA). The information contained in the sequence of deoxynucleotides in DNA is preserved in the sequence of ribonucleotides in the RNA transcript. Like DNA replication, transcription is characterized by template-directed nucleotide polymerization that requires a certain degree of accuracy. However, unlike DNA synthesis, RNA synthesis takes place selectively, on a gene-by-gene basis. Numerous protein cofactors interact with each other, with RNA polymerase, and with the DNA template to regulate where and when transcription takes place, to accurately transcribe the DNA, and to convert the initial RNA product to its mature, functional form.

21.1 TRANSCRIPTION INITIATION

Transcription, unlike replication, must be highly selective, since only a small portion of the genome is involved. Most bacteria contain a few thousand genes, eukaryotic cells up to around 30,000, and these genes are usually separated by stretches of DNA that are not transcribed. Not surprisingly, the process of identifying a gene and transcribing it into RNA is complex.

WHAT IS A GENE?

We can consider a **gene** as a segment of DNA that is transcribed for the purpose of expressing the encoded genetic information or transforming it to a form that is more useful to the cell. This definition of a gene requires some qualification:

1. *For protein-coding genes, the RNA transcript (called **messenger RNA or mRNA**) includes all the information specifying the sequence of amino acids in a polypeptide.* But keep in mind that not all RNA molecules are translated into protein. **Ribosomal RNA (rRNA), transfer RNA (tRNA),** and other types of RNA molecules carry out their functions without undergoing translation.

2. *Most RNA transcripts correspond to a single functional unit*, for example, one polypeptide. But some RNAs, particularly in prokaryotes, code for multiple proteins and result from the transcription of an **operon,** a set of contiguous genes whose products have related metabolic functions. In a few rare cases, a single mRNA carries information for two proteins in overlapping sequences of nucleotides.

3. *RNA transcripts typically undergo **processing**—the addition, removal, and modification of nucleotides—before becoming fully functional.* Because of variations in mRNA processing and post-translational modification, *several different forms of a protein may be derived from a single gene.*

4. Finally, *the proper transcription of a gene may depend on DNA sequences that are not transcribed but help position the **RNA polymerase** at the transcription start site or are involved in the regulation of gene expression.*

Most of this chapter focuses on the transcription of eukaryotic protein-coding genes. As introduced in Section 3.4, the human genome includes an estimated 21,000 such genes, which have an average length of about 27,000 bp. The actual fraction of protein-coding sequences is quite small, however (about 1.4% of the genome), because a typical gene includes sequences that are removed from the RNA before it is translated into protein. Nevertheless, an estimated 80% of the human genome may undergo transcription to produce **noncoding RNAs (ncRNAs)** of various sizes. Some of these are listed in **Table 21.1.**

TABLE 21.1 Some Noncoding RNAs

TYPE	TYPICAL SIZE (NUCLEOTIDES)	FUNCTION
Ribosomal RNA (rRNA)	120–4718	Translation (ribosome structure and catalytic activity)
Transfer RNA (tRNA)	54–100	Delivery of amino acids to ribosome during translation
Small interfering RNA (siRNA)	20–25	Sequence-specific inactivation of mRNA
Micro RNA (miRNA)	20–25	Sequence-specific inactivation of mRNA
Large intervening noncoding RNA (lincRNA)	Up to 17,200	Transcriptional control
Small nuclear RNA (snRNA)	60–300	RNA splicing
Small nucleolar RNA (snoRNA)	70–100	Sequence-specific methylation of rRNA

FIGURE 21.1

Histone modifications. Partial sequences of the four histones are shown using one-letter amino acid codes. The added groups are represented by colored symbols (ac = acetyl, me = methyl, ph = phosphoryl, pr = propionyl, rib = ADP-ribose, and ub = ubiquitin). Cit represents citrulline (deiminated arginine). Serine and threonine residues may also be modified by the addition of *N*-acetylglucosamine groups. Note that some residues can be modified in multiple ways. This diagram is a composite; not all modifications occur in all organisms. [*After a diagram by Ali Shilatifard, St. Louis University School of Medicine.*]

? **Which histone residues undergo acetylation? Methylation? Phosphorylation?**

The functions of some types of ncRNA molecules have been elucidated and are described later in this chapter, but others remain mysterious. It is possible that they represent transcriptional "noise," or random synthesis, which is consistent with the observation that they are degraded soon after their synthesis. However, evidence that some of these sequences are conserved among species indicates that they may play a role in large-scale regulation of transcriptional activity.

DNA PACKAGING AFFECTS TRANSCRIPTION

Recall from Section 20.5 that in eukaryotes, DNA is packaged in nucleosomes, which may form structures that further compact the DNA. Highly condensed chromatin tends to be transcriptionally inactive. The degree of DNA packaging is controlled in part by the covalent modification of histone proteins that form the core of each nucleosome (Fig. 20.34). Various groups, including methyl, acetyl, and phosphoryl groups, may be added to histone residues, particularly in the N-terminal regions that interact closely with the DNA in each nucleosome (**Fig. 21.1**).

The addition or removal of the various histone-modifying groups can potentially introduce considerable variation in the fine structure of chromatin, offering a mechanism for promoting or preventing gene expression. For example, a lysine residue is positively charged and interacts strongly with the negatively charged DNA backbone, which helps stabilize nucleosome structure. Acetylation of the lysine side chain neutralizes it and weakens its interaction with the DNA, destabilizing the nucleosome and making it more accessible to the transcription machinery.

Acetyl-Lys

Some histone modifications associated with "silent" and "active" chromatin are shown in **Table 21.2**.

Many histone modifications are interdependent. For example, in histone H3, methylation of Lys 9 inhibits phosphorylation of Ser 10, but phosphorylation of Ser 10 promotes acetylation of Lys 14. These relationships, along with the huge number of possible combinations of covalently attached groups (as high as 10^{11} different arrangements, in theory) suggest the existence of a **histone code** that marks the transcription-readiness of different regions of a

TABLE 21.2 Transcriptional Activity and Histone Modification

TRANSCRIPTIONALLY SILENT CHROMATIN	
H3K9me2	Lys 9 of histone H3 is dimethylated
H3K27me3	Lys 27 of histone H3 is trimethylated
TRANSCRIPTIONALLY ACTIVE CHROMATIN	
H3K4me3	Lys 4 of histone H3 is trimethylated
H3K9ac	Lys 9 of histone H3 is acetylated
H4K16ac	Lys 16 of histone H4 is acetylated

chromosome. Phosphorylation of tyrosine 57 in histone H2A is known to be required for efficient transcription, but the effects of many of the histone modifications are not yet understood.

Some of the acetyltransferases, methyltransferases, and kinases that alter histones may act on many residues in different histone proteins; other enzymes that act as "writers" are highly specific for one residue in one histone. Because the modifying groups are often derived from common metabolites, such as acetyl-CoA, a cell's metabolic activities may influence the histone marks and thereby help control gene transcription.

In addition to altering nucleosome structure, the histone code can be interpreted by proteins that fine-tune the rate or timing of transcription. The proteins that function as "readers" of the histone code typically have domains that recognize specific chemical groups. For example, a tudor domain binds to dimethylarginine residues; a chromodomain binds to methylated lysine, and a bromodomain binds to acetylated lysine residues (**Fig. 21.2**).

Enzymes such as deacetylases, demethylases, and phosphatases—known as "erasers"—can remove the groups that modify histones. The time course for altering the histone code is highly variable, with some modifications occurring in a highly dynamic fashion and others remaining unchanged for the lifetime of the cell.

It is unlikely that histone modifications completely obliterate nucleosome structure (unwrapping all the DNA would be energetically costly). Instead, the chromatin is rearranged so that critical DNA sequences are exposed rather than wrapped tightly around histones in the nucleosome core. The work of remodeling chromatin is carried out by a protein complex that neatly grasps the nucleosome (**Fig. 21.3**). The protein uses the free energy of ATP hydrolysis to loosen a segment of DNA from the histone core. The loop of unattached DNA travels

FIGURE 21.2

A bromodomain. The protein surface is shown in this model. A single bromodomain (~110 amino acids) includes a cavity for an acetyl-Lys group (ball-and-stick model). [*Courtesy Ming-Ming Zhou, Mt. Sinai School of Medicine. From* Science 285, 1201 (1999).]

FIGURE 21.3

Model of a nucleosome bound to a chromatin-remodeling complex. The X-ray structure of the nucleosome (see Fig. 20.29) was modeled into the structure of the yeast RSC complex (gray) determined by cryoelectron microscopy. RSC stands for Remodels the Structure of Chromatin. [*Courtesy of Andres Leschziner, Harvard University.*]

FIGURE 21.4

Nucleosome sliding. Chromatin-remodeling complexes may act by loosening a portion of DNA (blue) wound around a histone octamer (orange). The nucleosome appears to slide along the DNA, exposing different segments of DNA.

around the nucleosome as a wave so that a segment of DNA that was part of the nucleosome core is now available to interact with the transcription machinery (**Fig. 21.4**). In eukaryotes, transcription starts in nucleosome-free stretches of DNA.

DNA ALSO UNDERGOES COVALENT MODIFICATION

In many organisms, including plants and animals, DNA methyltransferases add methyl groups to cytosine residues. The methyl group projects out into the major groove of DNA and can potentially alter interactions with DNA-binding proteins.

5-Methylcytosine residue

In mammals, the methyltransferases target C residues next to G residues, so that about 80% of CG sequences (formally represented as CpG) are methylated. CpG sequences occur much less frequently than statistically predicted, but clusters of CpG (called **CpG islands**) are often located near the starting points of genes. Interestingly, these CpG islands are usually unmethylated. This suggests that *methylation may be part of the mechanism for marking or "silencing" DNA that contains no genes.*

After DNA is replicated, methyltransferases modify the new strand, using the methylation pattern of the parent DNA strand as a guide. In this way, the daughter cells produced by cell division continue the gene-expression program of the parent cell.

Variations in DNA methylation may be responsible for **imprinting,** in which the level of expression of a gene depends on its parental origin. Recall that an individual receives one copy of a gene from each parent. Normally, both copies (alleles) are expressed, but a gene that has been methylated may not be expressed. The gene behaves as if it has been "imprinted" and knows its parentage. Imprinting explains why some traits are transmitted in a maternal or paternal fashion, contrary to the Mendelian laws of inheritance. Methylation and other DNA modifications are said to be **epigenetic** (from the Greek *epi*, "above") because they represent heritable information that goes beyond the sequence of nucleotides in the DNA. Stable histone modifications can also carry epigenetic information.

TRANSCRIPTION BEGINS AT PROMOTERS

Prokaryotes typically have compact genomes without much nontranscribed DNA, whereas in eukaryotes, protein-coding genes may be separated by large tracts of DNA (see Fig. 3.12). But in both types of organisms, the efficient expression of genetic information usually involves the initiation of RNA synthesis at a particular site, known as a **promoter,** near

Transcription
start site
+1

−35 region   −10 region

5′⋯NNNTTGACANnnnnnnnnnnnnnnnnnnnnnnnTATAATNnnnnnnnnnnn⋯3′

FIGURE 21.5

The *E. coli* promoter. The first nucleotide to be transcribed is position +1. The two consensus promoter sequences, centered around positions −10 and −35, are shaded. N represents any nucleotide.

? **Approximately how many turns of the DNA helix separate the −10 region from the −35 region and the transcription start site?**

the protein-coding sequence. The DNA sequence at the promoter is recognized by specific proteins, which either are part of the RNA polymerase protein or subsequently recruit the appropriate RNA polymerase to the DNA to begin RNA synthesis.

In bacteria such as *E. coli,* the promoter comprises a sequence of about 40 bases on the 5′ side of the transcription start site. By convention, such sequences are written for the coding, or nontemplate, DNA strand (see Section 3.3) so that the DNA sequence has the same sense and 5′ → 3′ directionality as the transcribed RNA. The *E. coli* promoter includes two conserved segments, whose **consensus sequences** (sequences indicating the nucleotides found most frequently at each position) are centered at positions −35 and −10 relative to the transcription start site (position +1; **Fig. 21.5**).

E. coli RNA polymerase is a five-subunit enzyme with a mass of about 450 kD and a subunit composition of $\alpha_2\beta\beta'\omega\sigma$. The σ subunit, or σ factor, recognizes the promoter, thereby precisely positioning the RNA polymerase enzyme to begin transcribing. Although bacterial cells contain only one core RNA polymerase ($\alpha_2\beta\beta'\omega$), they contain multiple σ factors, each specific for a different promoter sequence. Since different genes may have similar promoters, *bacterial cells can regulate patterns of gene expression through the use of different σ factors.* Once transcription is under way, the σ factor is jettisoned, and the remaining subunits of RNA polymerase extend the transcript.

In eukaryotes, some promoters for protein-coding genes include a **TATA box,** an element that resembles the prokaryotic promoter. Most eukaryotic genes lack this sequence but may instead exhibit a variety of other conserved promoter elements located both upstream (on the 5′ side) and downstream (on the 3′ side) of the transcription initiation site (**Fig. 21.6**). A given gene may contain several—although not all—of these elements, which may act synergistically. For example, the DPE (downstream promoter element) is always accompanied by an Inr (initiator) element.

The majority of mammalian genes lack well-defined promoters with a precise transcription start site, so transcription initiates over a stretch of 100–150 nucleotides. This DNA typically contains unmethylated CpG islands or other epigenetic marks, but little is known about how transcription factors and RNA polymerase interact with this DNA. Eukaryotic RNA polymerase does not include a subunit corresponding to the prokaryotic σ factor. Instead, RNA polymerase locates a transcription start site through a series of complex protein–protein and protein–DNA interactions. Some evidence suggests that the human genome contains as many as 500,000 potential transcription start sites (many more than the number of genes). Even

FIGURE 21.6

Sequences of some eukaryotic promoter elements. The consensus sequence for each element in humans is given. Some positions can accommodate two or three different nucleotides. N represents any nucleotide.

? Compare these sequences to the *E. coli* promoter (see **Fig. 21.5**).

more surprising is the observation that transcription at many of these sites can proceed in both directions, that is, using each DNA strand as a template. The purpose of all this transcriptional activity is not understood.

TRANSCRIPTION FACTORS RECOGNIZE EUKARYOTIC PROMOTERS

In eukaryotes, the initiation of transcription typically requires a set of five highly conserved proteins known as **general transcription factors.** They are abbreviated as TFIIB, TFIID, TFIIE, TFIIF, and TFIIH (the II indicates that these transcription factors are specific for RNA polymerase II, the enzyme that transcribes protein-coding genes). Some of these general transcription factors interact specifically with some of the promoter elements shown in Figure 21.6. For example, TFIIB binds to the BRE sequences, and the TATA-binding protein (TBP), a subunit of TFIID, binds to the TATA box. In fact, TBP also binds to promoter regions that lack an obvious TATA box.

The various transcription factors play numerous roles in preparing the DNA for transcription and recruiting RNA polymerase. TBP, a saddle-shaped protein about $32 \times 45 \times 60$ Å, consists of two structurally similar domains that sit astride the DNA at an angle (**Fig. 21.7**). This protein–DNA interaction introduces two sharp kinks into the DNA. The kinks are caused by the insertion of two phenylalanine side chains like a wedge between pairs of DNA nucleotides.

Once TBP is in place at the promoter, the conformationally altered DNA may serve as a stage for the assembly of a preinitiation complex containing additional transcription factors and RNA polymerase (**Fig. 21.8**). TFIIB, for example, helps position the DNA near the polymerase active site. TFIIE joins the complex and recruits TFIIH, a helicase that unwinds the DNA in an ATP-dependent manner. The result is an open structure called a transcription bubble that is stabilized in part by the binding of TFIIF, which interacts with the nontemplate DNA strand.

FIGURE 21.7

Structure of TBP bound to DNA. The TBP polypeptide (green) forms a pseudosymmetrical structure that straddles a segment of DNA (shown in blue and viewed end-on). The insertion of TBP Phe residues (orange) bends the DNA in two places. [*Structure (pdb 1YTB) determined by Y. Kim, J. H. Geiger, S. Hahn, and P. B. Sigler.*]

Transcription bubble

The structural changes in the DNA may extend beyond the immediate site of the transcription bubble. For instance, the TFIID component known as TAF1 has histone acetyltransferase activity, which may allow it to alter nucleosome packing by neutralizing lysine side chains. TAF1 may also diminish histone H1 cross-linking of neighboring nucleosomes by helping to link the small protein ubiquitin to H1, thereby marking it for proteolytic destruction by a proteasome (see Section 12.1).

In TAF1, two bromodomains are arranged side by side, which might allow the protein to bind cooperatively to a multiply acetylated histone protein. The histone acetyltransferase activity of TAF1 itself may lead to localized hyperacetylation, a positive feedback mechanism for promoting gene transcription.

→ TFIID(TBP) → TFIIA → TFIIB → Pol II/TFIIF → TFIIE → TFIIH

FIGURE 21.8

Assembly of the human preinitiation complex. This model is based on cryoelectron microscopy images, with each component in a different color. Pol II (the large gray shape) is RNA polymerase II. [*Courtesy of Eva Nogales, Lawrence Berkeley National Laboratory.*]

ENHANCERS AND SILENCERS ACT AT A DISTANCE FROM THE PROMOTER

Additional sets of protein–protein and protein–DNA interactions may participate in the highly regulated expression of many eukaryotic genes. Whereas the rate of prokaryotic gene transcription varies about 1000-fold between the most and least frequently expressed genes, the gene transcription rate in eukaryotes may vary by as much as 10^9-fold. Some of this fine control is due to **enhancers,** DNA sequences that range from 50 to 1500 bp and are located up to 120 kb upstream or downstream of the transcription start site. A single gene may have more than one enhancer functionally associated with it, and hundreds of thousands of these elements may be scattered around the human genome.

The proteins that bind to enhancers are commonly called **activators,** although they are also known simply as transcription factors (hence the term *general transcription factor* for the ones that position RNA polymerase at the transcription start site). By facilitating (or inhibiting) transcription, these DNA-binding proteins can shape an organism's gene expression patterns in response to internal or external signals. We have already seen how some transcription factors are linked to signal transduction pathways (Chapter 10). For example, steroid hormones directly activate DNA-binding proteins, and Ras signaling indirectly activates several transcription factors. These proteins interact with DNA in a variety of ways (**Box 21.A**).

When an activator binds to the enhancer, a protein complex known as Mediator links the enhancer-bound activator to the transcription machinery poised at the transcription start site. Note that this interaction requires that the DNA form a loop in order to connect the enhancer and promoter region (**Fig. 21.9**). The packaging of DNA in nucleosomes may facilitate this long-range interaction by minimizing the length of the intervening DNA loop. In addition

Box 21.A   DNA-Binding Proteins

The proteins that directly participate in or regulate processes such as DNA replication, repair, and transcription must interact intimately with the DNA double helix. In fact, many of the proteins that promote or suppress transcription recognize and bind to particular sequences in the DNA. However, there do not seem to be any strict rules by which certain amino acid side chains pair with certain nucleotide bases. In general, interactions are based on van der Waals interactions and hydrogen bonds, often with intervening water molecules.

An examination of the structures of a large variety of protein–DNA complexes in prokaryotic and eukaryotic cells reveals that the DNA-binding proteins fall into a limited number of classes, depending on the structural motif that contacts the DNA. Many of these motifs are likely the result of convergent evolution and may therefore represent the most stable and evolutionarily versatile ways for proteins to interact with DNA.

By far the most prevalent mode of protein–DNA interaction involves a protein α helix that binds in the major groove of DNA. This DNA-binding motif may take the form of a helix–turn–helix (HTH) structure in which two perpendicular α helices are connected by a small loop of at least four residues. The HTH motifs are colored red in the structure at right, and the DNA is blue. In most cases, the side chains of one helix insert into the major groove and directly contact the exposed edges of bases in the DNA. Residues in the other helix and the turn may interact with the DNA backbone.

In prokaryotic and eukaryotic proteins, the HTH helices are usually part of a larger bundle of several α helices, which form a stable domain with a hydrophobic core. Prokaryotic transcription factors tend to be homodimeric proteins (as in the bacteriophage λ repressor) that recognize palindromic DNA sequences. In contrast, eukaryotic transcription factors more commonly are heterodimeric

A portion of the bacteriophage λ repressor bound to DNA. [*Structure (pdb 1LMB) determined by L. J. Beamer and C. O. Pabo.*]

or contain multiple domains that recognize a nonsymmetrical series of binding sites. For this reason, eukaryotic DNA-binding proteins are able to interact with a wider variety of target DNA sequences.

Many eukaryotic transcription factors include a DNA-binding motif in which one zinc ion (sometimes two) is tetrahedrally coordinated by cysteine or histidine side chains. The metal ion stabilizes a small protein domain (which is sometimes involved in protein–protein rather than protein–DNA interactions). In most cases, the DNA-binding motif, known as a zinc finger (see Fig. 4.16), consists of two antiparallel β strands followed by an α helix. In the structure shown below, one of the three zinc fingers is colored red. The Zn^{2+} ions are represented by purple spheres. As in the HTH proteins, the helix of each zinc finger motif inserts into the major groove of DNA, where it interacts with a three-base-pair sequence.

Zinc finger domains from the mouse transcription factor Zif268. [*Structure (pdb 1AAY) determined by M. Elrod-Erickson, M. A. Rould, and C. O. Pabo.*]

Some homodimeric DNA-binding proteins in eukaryotes include a leucine zipper motif that mediates protein dimerization. Each subunit has an α helix about 60 residues long, which forms a coiled coil with its counterpart in the other subunit (see Section 5.3). Leucine residues, appearing at every seventh position, or about every two turns of the α helix, mediate hydrophobic contacts between the two helices (they do not actually interdigitate, as the term *zipper* might suggest). The DNA-binding portions of a leucine zipper protein are extensions of the dimerization helices that bind in the major groove.

A portion of the yeast transcription factor GCN4. [*Structure (pdb 1DGC) determined by W. Keller, P. Konig, and T. J. Richmond.*]

In a few proteins, a β sheet interacts with the DNA (TBP is one such protein; see Fig. 21.7). In a few cases, two antiparallel β strands constitute the DNA-binding segment, fitting into the major groove so that protein side chains can form hydrogen bonds with the functional groups on the edges of the DNA bases.

The DNA-binding proteins described here interact with a limited portion of the DNA (typically just a few base pairs), marking the positions of regulatory DNA sequences and making additional protein–protein contacts that control processes such as gene transcription. Proteins that carry out catalytic functions (polymerases, for example) interact with the DNA much more extensively—but in a sequence-independent manner—and they tend to envelop the entire DNA helix.

? Explain why sequence-specific DNA-binding proteins usually interact with the major groove rather than with the minor groove.

FIGURE 21.9

Overview of enhancer function. An activator protein binds to a gene's enhancer sequence. Mediator, binding to the activator as well as to the general transcription factors and RNA polymerase at the gene's transcription start site, activates the polymerase to promote gene expression.

170Å

400Å

TFIIH

TFIIK
TRIMER

DNA
(DOWNSTREAM)

POL II
(RPB4/7-SIDE VIEW)

MEDIATOR
HEAD

MEDIATOR
MIDDLE

MEDIATOR
TAIL

FIGURE 21.10

Mediator bound to RNA polymerase. This model, which is based on data from X-ray crystallography, cryoelectron microscopy, and cross-linking studies, shows the yeast Mediator complex (yellow, blue, and tan) interacting with RNA polymerase (gray). Additional conformational changes in the complex likely occur before transcription initiation. [*Courtesy of Philip J. Robinson, Michael J. Trnka, David A. Bushnell, Ralph E. Davis, Pierre-Jean Mattei, Alma L. Burlingame, Roger D. Kornberg.*]

? **Is there room for Mediator to bind the preinitiation complex shown in Fig. 21.8?**

to enhancers, *a gene may have associated* **silencer** *sequences that bind proteins known as* **repressors**. Mediator may also relay silencer–repressor signals to the transcription machinery in order to repress gene transcription. The simple scheme shown in Figure 21.9 does not convey the complexity of many activator and repressor pathways, which may involve competition for binding sites among the many protein factors.

The Mediator complex, which contains 25 polypeptides in yeast and at least 31 in mammals, is a three-lobed structure (**Fig. 21.10**). Mediator interacts with RNA polymerase as well as with the general transcription factors. As many as 60 proteins may congregate at a transcription initiation site.

The multiplicity of activators and repressors, along with variability in Mediator subunits, could constitute a sophisticated system for fine-tuning transcription, which could help account for the different patterns of gene expression that distinguish the 200 or so cell types found in the human body. Although gene expression can be regulated at multiple points—transcription, RNA processing, protein synthesis, or protein processing—control of transcription is the largest source of variation in the amounts of proteins a cell produces. Furthermore, many genetic features that distinguish individuals result from variations in gene regulatory sequences rather than differences in the protein-coding segments of genes.

PROKARYOTIC OPERONS ALLOW COORDINATED GENE EXPRESSION

Although prokaryotic cells regulate transcription initiation using less complicated mechanisms than eukaryotes use, functionally related genes may be organized in operons to ensure their coordinated expression in response to some metabolic signal. About 13% of prokaryotic genes are found in operons, including the three genes of the well-studied *E. coli lac* operon, whose protein products are involved in lactose metabolism.

When bacterial cells grown in the absence of the disaccharide lactose are transferred to a medium containing lactose, the synthesis of two proteins required for lactose catabolism is quickly enhanced. These proteins are lactose permease, a transporter that allows lactose to enter the cells, and β-galactosidase, an enzyme that catalyzes the hydrolysis of lactose to its component monosaccharides, which can then be oxidized to produce ATP:

Lactose

H_2O ↘ β-galactosidase

Galactose + Glucose

β-Galactosidase and lactose permease, along with a third enzyme (thiogalactoside transacetylase, whose physiological function is not clear), are encoded by the three-gene *lac* operon. The three genes (designated Z, Y, and A) are transcribed as a single unit from a promoter, P. Near the start of the β-galactosidase gene is a regulatory site called the operator, O, which binds a protein called the *lac* repressor. The repressor itself is a product of the I gene, which lies just upstream of the *lac* operon:

In the absence of lactose, the Z, Y, and A genes are not expressed because the *lac* repressor binds to the operator. The repressor is a tetrameric protein that functions as a dimer of dimers so that it can bind simultaneously to two segments of operator DNA (**Fig. 21.11**). The repressor does not interfere with RNA polymerase binding, but it prevents the polymerase from initiating transcription at the promoter.

When the cell is exposed to lactose, a lactose isomer called allolactose acts as an inducer of the *lac* operon (allolactose is generated from lactose by trace amounts of β-galactosidase present in the bacterial cell).

FIGURE 21.11

***lac* repressor bound to DNA.** In this model, DNA segments are blue, the monomeric units of the repressor are light green, light blue, light pink, and dark pink (each amino acid is represented by a sphere). The two operator sequences to which the repressor binds are separated by a stretch of DNA of 93 bp, which probably forms a loop. [*Structure (pdb 1LBG) determined by M. Lewis, G. Chang, N. C. Horton, M. A. Kercher, H. C. Pace, and P. Lu.*]

Allolactose

The inducer binds to the *lac* repressor, triggering a conformational change that causes it to release its grip on the operator sequences. As a result, the promoter is freed for transcription, and the production of β-galactosidase and lactose permease can increase 1000-fold within a few minutes. This simple regulatory system ensures that the proteins required for metabolizing lactose are synthesized only when lactose is available as a metabolic fuel.

 CONCEPT CHECK

1. Discuss reasons why it is difficult to define a gene.

2. Explain why histone acetylation and chromatin remodeling are prerequisites for eukaryotic gene transcription.

3. Compare transcription initiation in prokaryotes and eukaryotes. Which prokaryotic elements also appear in eukaryotes?

4. Summarize the roles of enhancers, silencers, activators, repressors, and mediator in regulating gene transcription.

5. Explain how hormone signaling could affect the activity of activators and repressors.

6. Explain how a mutation in a DNA segment that does not code for a protein could affect production of that protein.

7. Describe how genetic engineers could use elements of the *lac* operon to design a system for turning the expression of particular genes on or off.

21.2 RNA POLYMERASE

KEY CONCEPTS

- Transcription in eukaryotes may require alteration of histones and chromatin structure.
- During transcription, the DNA strands separate, and RNA polymerase constructs an RNA molecule that forms a short hybrid helix with the template strand.
- RNA polymerase is a processive enzyme that proofreads its work.
- The shift from initiation to elongation involves structural changes in RNA polymerase, including phosphorylation of its C-terminal domain.
- The mechanisms for transcription termination differ in prokaryotes and eukaryotes.

Like DNA replication, *RNA transcription is carried out by immobile protein complexes that reel in the DNA.* These transcription factories in eukaryotic nuclei can be visualized by immunofluorescence microscopy and are distinct from the replication factories where DNA is synthesized (**Fig. 21.12**). If the RNA polymerase were free to track along the length of a DNA molecule, rotating around the helical template, the newly synthesized RNA strand would become tangled with the DNA. In fact, except for a short 8- to 9-bp hybrid DNA–RNA helix at the polymerase active site, the newly synthesized RNA is released as a single-stranded molecule.

Bacterial cells contain just one type of RNA polymerase, but eukaryotic cells contain three (plus additional polymerases for chloroplasts and mitochondria). Eukaryotic RNA polymerase I is responsible for transcribing rRNA genes, which are present in multiple copies. RNA polymerase III mainly synthesizes tRNA molecules and other small RNAs. Protein-coding genes are transcribed by RNA polymerase II, which is the main focus of this section.

Once the transcription start site has been selected, the DNA opens into a transcription bubble, and RNA polymerase can begin its work. Most RNA polymerases can melt apart about 15 bp of DNA on their own. Only RNA polymerase II uses a separate helicase (TFIIH), and in this case, the initial transcription bubble is much larger, perhaps reflecting the lack of defined promoter sequences to precisely locate the transcription start site. But once RNA polymerase II enters the elongation phase, its transcription bubble shrinks to the standard 15 bp size.

FIGURE 21.12

Spatial separation of transcription and replication. In these fluorescence microscopy images of mouse cells, sites of DNA replication are green (a) and sites of RNA transcription are red (b). The merged images are shown in (c). A single nucleus may contain 2000 to 3000 transcription sites or "factories." [*Courtesy Ronald Berezney. From Wei, X. et al.,* Science *281, 1502–1505 (1998).*]

RNA POLYMERASES HAVE A COMMON STRUCTURE AND MECHANISM

Mammalian RNA polymerase has a mass of over 500 kD and at first glance has the same fingers, thumb, and palm domains as DNA polymerase (**Fig. 21.13**). The highly

FIGURE 21.13

Structure of mammalian RNA polymerase II. The model shows the 12 protein subunits in different colors. A magenta sphere represents an Mg^{2+} ion in the active site. An arrow indicates where the highly flexible C-terminal domain of the largest subunit would be located. [*Structure (pdb 5FLM) determined by C. Bernecky, F. Herzog, W. Baumeister, J. M. Plitzko, and P. Cramer.*]

? Compare this structure to that of DNA polymerase (Fig. 20.7). Locate the long green "bridge helix" to the right of the active site.

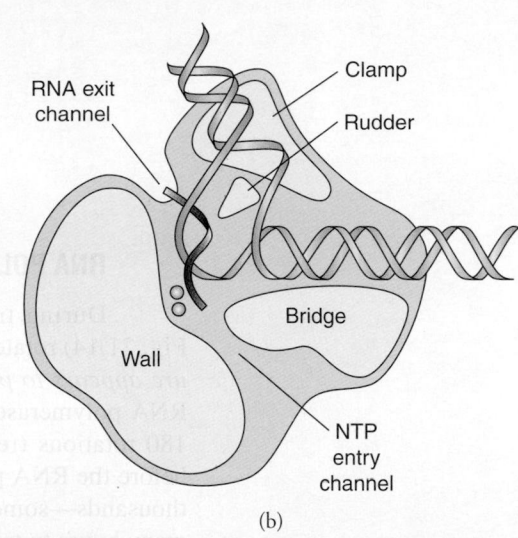

(a)   (b)

FIGURE 21.14

RNA polymerase with DNA and RNA. (a) Surface view of human RNA polymerase II, looking from the bottom of the protein as depicted in Figure 21.13. The template DNA strand is blue, the nontemplate (coding) strand is purple, and a six-base segment of RNA is red. The magenta Mg^{2+} ions mark the location of the active site. [*Structure (pdb 5IYD) determined by Y. He, C. Yan, J. Fang, C. Inouye, R. Tjian, I. Ivanov, and E. Nogales.*] (b) Schematic diagram of RNA polymerase showing the approximate positions of internal structures that facilitate the polymerization reaction. DNA enters from the right, and the strands separate before reannealing on the left. Nucleotide substrates reach the active site through an opening at the bottom of the enzyme. [*Based on a drawing by Roger Kornberg.*]

conserved sequences of eukaryotic RNA polymerases indicate that they have virtually identical structures. The core structure of RNA polymerase and its catalytic mechanism are also very similar between eukaryotes and prokaryotes. The differences are mainly at the enzyme surface, where the proteins interact with transcription factors and other regulatory proteins.

The active site of RNA polymerase is located at the bottom of a positively charged cleft between the two largest subunits. The DNA to be transcribed enters the active-site cleft of RNA polymerase, and the two DNA strands separate to form the transcription bubble. The nontemplate (coding) strand lies outside of the active-site cavity, but the template strand threads through the polymerase, making an abrupt right-angle turn where it encounters a wall of protein. Here, the template base points away from the standard B-DNA conformation and toward the floor of the active site. This geometry allows the deoxynucleotide residue to base pair with an incoming ribonucleoside triphosphate, which enters the active site through a channel in the floor (**Fig. 21.14**).

Note that the incoming nucleotide has only a 25% chance of pairing correctly (since there are four possible ribonucleotide substrates: ATP, CTP, GTP, and UTP). When the correct nucleotide forms hydrogen bonds with the template base, a loop of protein called the trigger closes over it. This movement brings a histidine side chain close enough to donate a proton to the second phosphate of the NTP substrate, a step necessary to incorporate a monophosphate residue into the growing RNA chain. A mispaired nucleotide or a dNTP does not allow the trigger loop to move close enough for catalysis to occur; in this way, selection of the correct nucleotide is linked to the polymerization reaction, with an error rate of only about 1 in 10,000.

Catalysis requires two metal ions (Mg^{2+}) coordinated by negatively charged side chains. Like DNA polymerase, *RNA polymerase catalyzes nucleophilic attack of the 3′ OH group of the growing polynucleotide chain on the 5′ phosphate of an incoming nucleotide* (see Fig. 20.5). The RNA molecule is therefore extended in the 5′ → 3′ direction. No primer is needed, so the RNA chain begins with the joining of two ribonucleotides. As the RNA strand is synthesized, it forms a hybrid double helix with the DNA template strand for eight or nine base pairs. The conformation of the hybrid is intermediate to the A form (as in double-stranded RNA) and the B form (as in double-stranded DNA):

A-DNA   DNA–RNA Hybrid   B-DNA

RNA POLYMERASE IS A PROCESSIVE ENZYME

During transcription, a portion of RNA polymerase known as the clamp (see Fig. 21.14) rotates by about 30° to close down snugly over the DNA template. *Clamp closure appears to promote the high processivity of RNA polymerase.* In experiments where RNA polymerase was immobilized and a magnetic bead was attached to the DNA, up to 180 rotations (representing thousands of base pairs at 10.4 bp per turn) were observed before the RNA polymerase slipped. This processivity is essential, since genes are usually thousands—sometimes millions—of nucleotides long, and the largest ones may require many hours to transcribe.

With each reaction cycle, the bridge helix located near the active site (see Figs. 21.13 and 21.14) appears to oscillate between a straight and bent conformation. This alternating movement seems to act as a ratchet to aid translocation of the template so that the next nucleotide can be added to the growing RNA chain. Throughout transcription, the sizes of the transcription bubble and the DNA–RNA hybrid helix remain constant. A protein loop known as the rudder (see Fig. 21.14b) may help separate the RNA and DNA strands so that a single RNA strand is extruded from the enzyme as the template and nontemplate DNA strands reanneal to restore the double-stranded DNA.

Like DNA polymerase, *RNA polymerase carries out proofreading.* If a deoxynucleotide or a mispaired ribonucleotide is mistakenly incorporated into RNA, the DNA–RNA hybrid helix becomes distorted. This causes polymerization to cease, and the newly synthesized RNA "backs out" of the active site through the channel by which ribonucleotides enter (**Fig. 21.15**). Eukaryotic transcription factor TFIIS binds to the RNA polymerase and stimulates it to act as an endonuclease to trim away the RNA containing the error. Transcription may resume if the 3′ end of the truncated transcript is then repositioned at the active site. Proofreading activity lowers the error rate to about 1 in 1,000,000.

TRANSCRIPTION ELONGATION REQUIRES A CONFORMATIONAL CHANGE IN RNA POLYMERASE

One of the puzzles of RNA polymerase action is that the enzyme appears to initiate RNA synthesis repeatedly, producing and releasing many short transcripts (up to about 12 nucleotides) before committing to elongating a transcript. This suggests that *the transcription machinery must undergo a transition from initiation mode to elongation mode.* Several structural changes must occur at this point. While the first few ribonucleotides are polymerized, the transcription machinery remains firmly associated with the promoter. As a result, the template DNA enters the RNA polymerase active site but has nowhere to go after being transcribed. In prokaryotes, the buildup of strain—termed DNA scrunching—may provide the driving force for the polymerase to eventually escape the promoter and discard its σ factor. In eukaryotes, TFIIB occupies part of the RNA polymerase active site and must be displaced to accommodate an RNA longer than a few residues. Moreover, the exit channel for the RNA is initially partially blocked. The shift to an elongation conformation relieves these constraints and allows the polymerase to advance beyond the promoter.

In eukaryotes, the shift in RNA polymerase involves the C-terminal domain of the largest subunit of RNA polymerase (a structure that is disordered and therefore not visible in the models shown in Figs. 21.13 and 21.14). The C-terminal domain

FIGURE 21.15

Schematic view of backtracking RNA in RNA polymerase. If polymerization stops due to a polymerization error, the 3′ end of the RNA transcript may back up into the channel for incoming nucleotides. The enzyme can then cleave off the 3′ end of the RNA and resume transcription.

? Identify the oldest and newest segments of RNA.

INITIATION   ELONGATION

phosphorylation

Promoter region

FIGURE 21.16

The transition from transcription initiation to elongation. During initiation, the nonphosphorylated C-terminal domain (CTD) of RNA polymerase (RNAP) serves as a binding site for a Mediator complex (M). When the C-terminal domain undergoes phosphorylation, RNA polymerase switches to an elongation mode, dissociating from the Mediator complex and the general transcription factors (GTFs) that remain at the promoter. Other proteins may bind to the phosphorylated CTD during elongation.

of mammalian RNA polymerase contains 52 seven–amino acid pseudorepeats with the consensus sequence

$$\underset{1 \quad 2 \quad 3 \quad 4 \quad 5 \quad 6 \quad 7}{\text{Tyr–Ser–Pro–Thr–Ser–Pro–Ser}}$$

Serine residues 2 and 5 (and possibly 7) of each heptad can potentially be phosphorylated. During the initiation phase of transcription, the C-terminal domain of RNA polymerase is not phosphorylated, but elongating RNA polymerase bears numerous phosphate groups. The Ser 5 phosphorylation that triggers the switch from an initiating to an elongating RNA polymerase is carried out by the kinase activity of TFIIH. Other kinases continue the phosphorylation process, mainly targeting Ser 2.

When RNA polymerase becomes phosphorylated at its C-terminal domain, it can no longer bind a Mediator complex; this would allow the polymerase to abandon transcription-initiating factors and advance along the template DNA. In fact, when RNA polymerase "clears" the promoter, it leaves behind some general transcription factors, including TFIID (**Fig. 21.16**). These proteins, along with the Mediator complex, can reinitiate transcription by recruiting another RNA polymerase to the promoter. Consequently, *the first RNA polymerase to transcribe a gene acts as a "pioneer" polymerase that helps pave the way for additional rounds of transcription.* Histone acetyltransferases associated with the pioneer RNA polymerase may alter the nucleosomes of a gene undergoing transcription.

During transcription elongation, other proteins bind to the phosphorylated C-terminal domain of RNA polymerase II, taking the place of the jettisoned initiation factors. Although RNA polymerase by itself can transcribe a DNA sequence *in vitro*, the presence of these additional factors accelerates transcription. The phosphorylated domain of an elongating RNA polymerase also serves as a docking site for proteins that begin processing the nascent (newly made) RNA transcript. The rate of RNA synthesis seems to vary erratically from about 500 to 5000 nucleotides per minute, depending on the DNA sequence being transcribed. In prokaryotes, RNA polymerase may pause as often as once every 100 bp, which may be a way to permit protein synthesis to keep pace with transcription.

TRANSCRIPTION IS TERMINATED IN SEVERAL WAYS

In both prokaryotes and eukaryotes, transcription termination involves cessation of RNA polymerization, release of the complete RNA transcript, and dissociation of the polymerase from the DNA template. In prokaryotes such as *E. coli,* termination occurs mainly by one of two mechanisms. In about half of *E. coli* genes, the 3′ end includes palindromic sequences followed by a stretch of T residues. The RNA corresponding to the gene can form a stem–loop or hairpin structure followed by a stretch of U residues. The other half of *E. coli* genes lack a hairpin sequence, and their termination depends on the action of a protein such as Rho, a hexameric helicase that may act by prying the nascent RNA away from the DNA and pushing the polymerase off the template. Both types of termination mechanisms can be explained in terms of destabilizing the DNA–RNA hybrid helix that forms in the transcription bubble during elongation. Hairpin formation or the ATP-dependent action of Rho exerts a force that causes the RNA polymerase to advance, extending the leading end of the transcription bubble without extending the RNA (**Fig. 21.17**). The hybrid helix, now shortened and consisting of relatively weak U:A base pairs, is easily disrupted, freeing the RNA transcript.

(a)

(b)

FIGURE 21.17

Mechanisms for transcription termination in prokaryotes. (a) Formation of an RNA hairpin shortens the length of the DNA–RNA hybrid helix, which is rich in easily separated A:U base pairs. (b) The movement of Rho along the RNA transcript pushes the polymerase forward, leaving a short hybrid helix from which the RNA more easily dissociates.

In eukaryotic protein-coding genes, termination is somewhat imprecise. RNA polymerase may slow or pause periodically while splicing (see below) occurs, and transcription may not terminate until this task is complete. The polymerase also pauses following a polyadenylation signal, where the RNA is cleaved. While paused, the RNA polymerase may undergo a conformational change that allows regulatory proteins to bind to its phosphorylated C-terminal domain and trigger termination of elongation. Alternatively, after the mRNA has been cut, an exonuclease may "eat away" the tail of the RNA still being synthesized, until it catches up with the polymerase and nudges it away from the DNA template, acting similarly to the Rho protein (Fig. 21.17b). In either scenario, the termination point is imprecise, but it doesn't really matter because the coding portion of the mRNA has already been synthesized.

The production of an RNA molecule from a DNA template, a major step in the central dogma of molecular biology, makes a lot of sense. The RNA molecule may undergo extensive modification before it is used, and it is relatively quickly degraded, but the original information in the DNA remains intact and can be transcribed over and over again.

 CONCEPT CHECK

1. Describe the overall structure of RNA polymerase and the arrangement of DNA and RNA within the protein.

2. Compare RNA polymerase and DNA polymerase in terms of mechanism, accuracy, and processivity.

3. Explain what would happen if the DNA–RNA hybrid helix were too long or too short.

4. Describe the role of the C-terminal domain during transcription initiation, elongation, and termination.

5. Describe the events that accompany the switch from transcription initiation to elongation.

6. Explain what the "pioneer" polymerase accomplishes.

RNA PROCESSING

In a prokaryotic cell, an mRNA transcript is typically translated immediately after it is synthesized. In a eukaryotic cell, however, transcription occurs in the nucleus (where the DNA is located), but translation takes place in the cytosol (where ribosomes are located). The separation of these processes gives eukaryotic cells two advantages: (1) They can modify the mRNA to produce a greater variety of gene products, and (2) the extra steps of RNA processing and transport provide additional opportunities for regulating gene expression.

In this section we examine some major types of RNA processing. Keep in mind that RNA is probably never found alone in the cell but rather interacts with a variety of proteins that covalently modify the transcript, splice out unneeded sequences, export the RNA from the nucleus, deliver it to a ribosome (if it is an mRNA), and eventually degrade it when it is no longer useful to the cell.

KEY CONCEPTS
- A 5′ cap and 3′ poly(A) tail are added to a eukaryotic mRNA during transcription.
- An RNA-containing spliceosome catalyzes the removal of introns and the joining of exons to generate different protein products from the same gene.
- The degradation of mRNA, such as through RNA interference, regulates gene expression following transcription.
- rRNA and tRNA transcripts are modified to produce functional molecules.

EUKARYOTIC mRNAs RECEIVE A 5′ CAP AND A 3′ POLY(A) TAIL

mRNA processing begins well before transcription is complete, as soon as the transcript begins to emerge from RNA polymerase. Many of the various enzymes required for capping the 5′ end of the mRNA, for extending the 3′ end, and for splicing are recruited to the phosphorylated domain of RNA polymerase, *so processing is closely linked to transcription*. In fact, the presence of processing enzymes may actually promote transcriptional elongation.

At least three enzyme activities modify the 5′ end of the emerging mRNA to produce a structure called a **cap** *that protects the polynucleotide from 5′ exonucleases.* First, a triphosphatase removes the terminal phosphate from the 5′ triphosphate end of the mRNA. Next, a guanylyltransferase transfers a GMP unit from GTP to the remaining 5′ diphosphate. These two reactions, which are carried out by a bifunctional enzyme in mammals, create a 5′–5′ triphosphate linkage between two nucleotides. Finally, methyltransferases add a methyl group to the guanine and to the 2′ OH group of ribose residues (**Fig. 21.18**).

The 3′ end of an mRNA is also modified. Processing begins following the synthesis of the polyadenylation sequence AAUAAA, which is a signal for a protein complex to cleave the transcript and extend it by adding adenosine residues. In fact, the RNA cleavage reaction, which occurs while the RNA polymerase is still operating, triggers transcription termination.

The enzyme poly(A) polymerase generates a 3′ **poly(A) tail** (also called a polyadenylate tail) of about 200 A residues. The enzyme resembles other polymerases in structure and catalytic mechanism, but it does not need a template to direct the addition of nucleotides.

Multiple copies of a binding protein associate with the mRNA tail. The poly(A)-binding protein consists of four copies of an RNA-binding domain (called an RBD, or RRM for RNA recognition motif) plus a C-terminal domain that mediates protein–protein contacts. A portion of the poly(A)-binding protein bound to RNA is shown in **Figure 21.19**. Each domain of about 80 amino acids can interact with two to six RNA nucleotides. Hence an mRNA's poly(A) tail can carry a large contingent of binding proteins, which help protect the 3′ end of the transcript from exonucleases. They may also provide a "handle" for proteins that deliver mRNA to ribosomes. Other RNA-binding proteins with different types of nucleotide-binding domains participate in RNA processing and other events.

SPLICING REMOVES INTRONS FROM EUKARYOTIC RNA

Genes were once believed to be continuous stretches of DNA, but experimental work in the 1970s showed that hybridized DNA and mRNA

FIGURE 21.18

An mRNA 5′ cap.

? How many phosphoanhydride bonds are cleaved during cap construction?

(a)

(b)

molecules included large loops of unpaired DNA (**Fig. 21.20**). As the gene is being transcribed, portions of the sequence called **introns** (intervening sequences) are cut out, and the remaining portions (expressed sequences, or **exons**) are joined together.

These mRNA **splicing** reactions do not occur in prokaryotes, whose protein-coding genes are continuous, but are the rule in complex eukaryotes. In simple organisms such as yeast, only a few genes contain introns, but in humans, almost all genes contain at least one intron. A typical gene consists of eight exons with an average length of 145 bp that are separated by introns with an average length of 3365 bp.

Like capping, splicing commences before RNA polymerase has finished transcribing a gene, and some components of the splicing machinery assemble on the phosphorylated C-terminal domain of RNA polymerase. Most mRNA splicing is carried out by a **spliceosome,** a complex of five small RNA molecules (called **snRNAs,** for **small nuclear RNAs**) and over a hundred proteins (**Fig. 21.21**).

The spliceosome recognizes conserved sequences at the 5′ intron/exon junction and at a conserved A residue within the intron, called the branch point (**Fig. 21.22**). Recognition depends on base pairing between the conserved mRNA sequences and snRNA sequences. However, the sequence conservation is relatively weak, and introns can be enormous, ranging from less than a hundred nucleotides to a record of 2.4 million nucleotides. These factors make it difficult to identify introns and exons in genomic DNA sequences.

Splicing is a two-step transesterification process (**Fig. 21.23**). Each step requires an attacking nucleophile (a ribose OH group) and a leaving group (a phosphoryl group). Two catalytically essential Mg^{2+} ions enhance the nucleophilicity of the attacking hydroxyl group and stabilize the phosphate leaving group. Because there is no net change in the number of phosphodiester bonds, the cutting-and-pasting process, which is catalyzed by the

(a)

(b)

FIGURE 21.20

Hybridization of ovalbumin DNA and mRNA. The template DNA strand for the chicken ovalbumin gene was allowed to hybridize with the corresponding mRNA. Complementary sequences, representing exons, have annealed, while single-stranded DNA, coding for introns that have been spliced out of the mRNA, forms loops. (a) Electron micrograph. (b) Interpretive drawing. The mRNA is shown as a dashed line, the introns of the single-stranded DNA (blue line) are labeled I–VII, and the exons are labeled 1–7. [*Courtesy Pierre Chambon and Fabienne Perrin.*]

snRNAs, needs no external source of free energy. However, proteins are essential for the overall reaction, which includes ATP hydrolysis–driven conformational changes to position the segments of RNA being processed.

Some types of introns, particularly in prokaryotic and protozoan rRNA genes, undergo self-splicing; that is, they catalyze their own transesterification reactions without the aid of proteins. These rRNA molecules were the first RNA enzymes (**ribozymes**) to be described, in 1982. One hypothesis for the evolutionary origin of splicing suggests that introns, and the splicing machinery itself, are the result of RNA molecules that spliced themselves *into* mRNA molecules, which were converted to DNA by the action of a reverse transcriptase (see Box 20.A) and then incorporated into the genome through recombination.

Introns typically comprise over 90% of a gene's total length, which means that a lot of RNA must be transcribed and then discarded. Moreover, a cell must spend energy to synthesize the RNA and proteins that make up the spliceosome and that destroy intronic RNA and incorrectly spliced transcripts. Finally, the complexity of the splicing process creates many opportunities for things to go wrong: A majority of mutations linked to inherited diseases involve defective splicing. So just what is the advantage of arranging a gene as a set of exons separated by introns?

One answer to this question is that splicing allows cells to increase variation in gene expression through alternative splicing. At least 95% of human protein-coding genes exhibit splice variants. Variation may result from selecting alternative sequences to serve as 5′ or 3′ splice sites, from skipping an exon, or from retaining an intron. Thus, certain exons present in the gene may or may not be included in the mature RNA transcript (**Fig. 21.24**). The signals that govern exon selection and splice sites probably involve RNA-binding proteins that recognize sequences or secondary structures within introns as well as exons. As a result of alternative splicing, *a given gene can generate more than one protein product*, and gene expression can be finely tailored to suit the needs of different types of cells. The evolutionary advantage of this regulatory flexibility clearly outweighs the cost of making the machinery that cuts and pastes RNA sequences. Alternative splicing also explains why humans are vastly more complex than organisms such as roundworms, which contain a comparable number of genes (see Table 3.4).

FIGURE 21.21

The yeast spliceosome. This model contains 37 of the >100 proteins (gray), three of the five snRNAs (green), and a portion of an intron (orange) [*Structure (3JB9) determined by C. Yan, J. Hang, R. Wan, M. Huang, C. Wong, and Y. Shi.*]

mRNA TURNOVER AND RNA INTERFERENCE LIMIT GENE EXPRESSION

Although mRNA accounts for only about 5% of cellular RNA (rRNA accounts for about 80% and tRNA for about 15%), it continuously undergoes synthesis and degradation. The life span of a given mRNA molecule is another regulated aspect of gene expression: mRNA molecules decay at different rates. In mammalian cells, mRNA life spans range from less than an hour to about 24 hours.

FIGURE 21.22

Consensus sequence at eukaryotic mRNA splice sites. Nucleotides shown in bold are invariant.

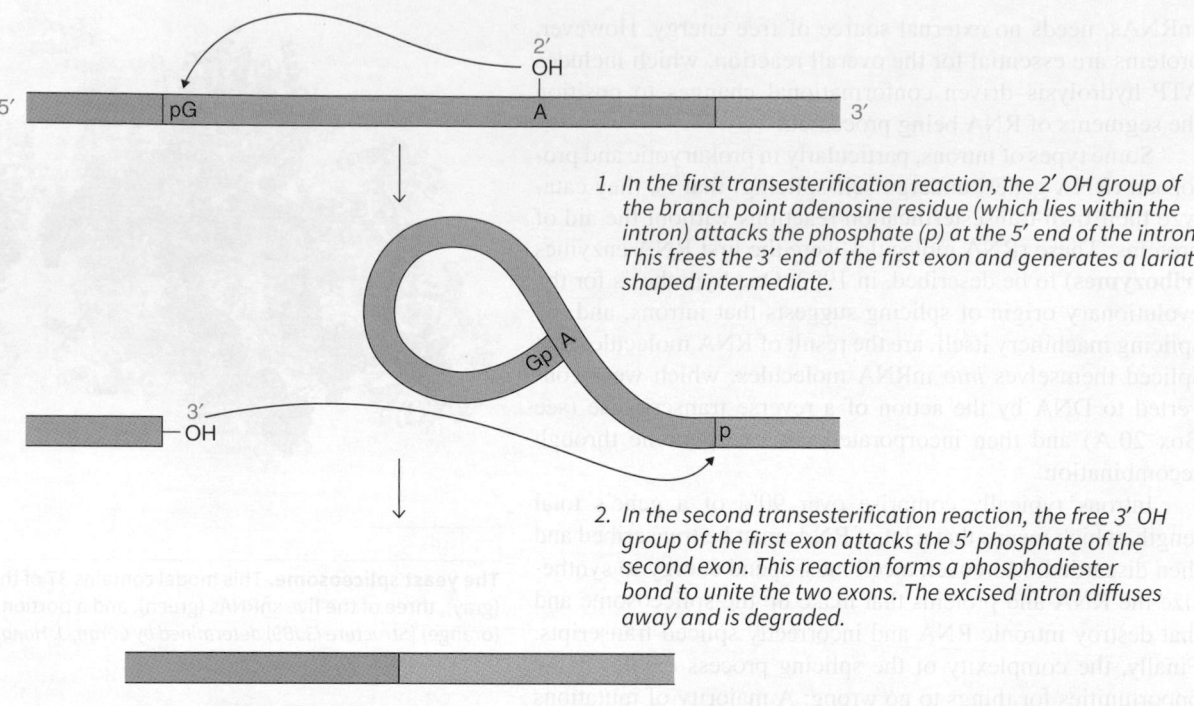

1. In the first transesterification reaction, the 2' OH group of the branch point adenosine residue (which lies within the intron) attacks the phosphate (p) at the 5' end of the intron. This frees the 3' end of the first exon and generates a lariat-shaped intermediate.

2. In the second transesterification reaction, the free 3' OH group of the first exon attacks the 5' phosphate of the second exon. This reaction forms a phosphodiester bond to unite the two exons. The excised intron diffuses away and is degraded.

FIGURE 21.23

mRNA splicing.

The rate of mRNA turnover depends in part on how rapidly its poly(A) tail is shortened by the activity of deadenylating exonucleases. Poly(A) tail shortening is followed by decapping, which allows exonucleases access to the 5' end of the transcript and eventually leads to destruction of the entire message (**Fig. 21.25**). *In vivo,* the RNA cap and tail are close together because a protein involved in translation binds to both ends of the mRNA, effectively circularizing it. RNA-binding regulatory proteins are almost certainly involved in monitoring RNA integrity. For example, transcripts with a premature stop codon are preferentially degraded, thereby avoiding the waste of synthesizing a nonfunctional truncated polypeptide.

Sequence-specific degradation of certain RNAs, a phenomenon called **RNA interference (RNAi),** provides another mechanism for regulating gene expression after transcription has occurred. RNA interference was discovered by researchers who were attempting to boost gene expression in various types of cells by introducing extra copies of genetic information in the form of RNA. They observed that instead of increasing gene expression, the RNA—particularly if it was double-stranded—actually blocked production of the gene's product. This interference or gene-silencing effect results from the ability of the introduced RNA to target a complementary cellular mRNA for destruction. Endogenously produced RNAs, known as **small interfering RNAs (siRNAs)** and **micro RNAs (miRNAs),** appear to mediate RNA interference in virtually all types of eukaryotic cells, including human cells.

FIGURE 21.24

Alternative splicing. The rat gene for the muscle protein α-tropomyosin (top) encodes 12 exons. The mature mRNA transcripts in different tissues consist of different combinations of exons (some exons are found in all transcripts), reflecting alternative splicing pathways. [*After Breitbart, R. E., Andreadis, A., and Nadal-Ginard, B.,* Annu. Rev. Biochem. *56, 481 (1987).*]

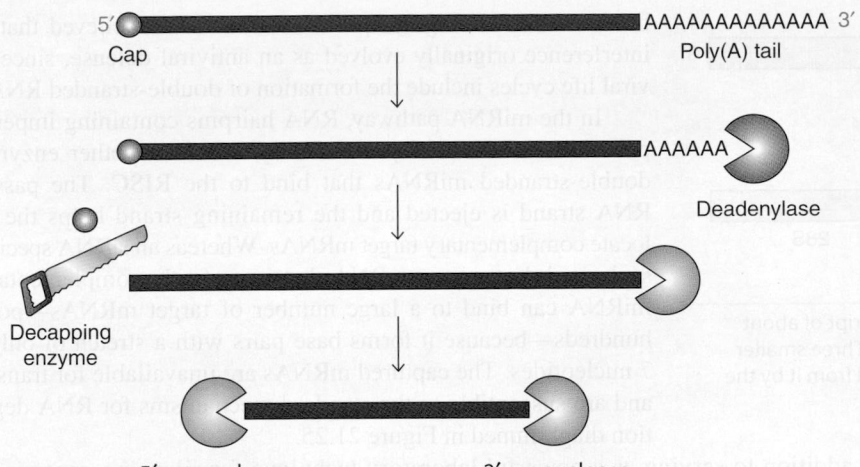

FIGURE 21.25

mRNA decay. A mature mRNA bears a 5′ cap and a 3′ poly(A) tail. After a deadenylase has shortened the tail, a decapping enzyme removes the methylguanosine cap at the 5′ end. The mRNA can then be degraded by exonucleases from both ends.

For siRNA, the RNA interference pathway begins with the production of double-stranded RNA, which may result when a single polynucleotide strand folds back on itself in a hairpin. A ribonuclease called Dicer cleaves the double-stranded RNA to generate segments of 20 to 25 nucleotides with a two-nucleotide overhang at each 3′ end (**Fig. 21.26**). These siRNAs bind to a multiprotein complex called the RNA-induced silencing complex (RISC), where one strand of the RNA (the "passenger" strand) is separated from the other by a helicase and/or degraded by a nuclease. The remaining strand serves as a guide for the RISC to identify and bind to a complementary mRNA molecule. The "Slicer" activity of the RISC, a protein known as Argonaute, then cleaves the mRNA, rendering it unfit for translation.

Like RNA splicing, RNA interference at first appears wasteful, but it provides cells with a mechanism for eliminating mRNAs—which could otherwise be translated into protein many

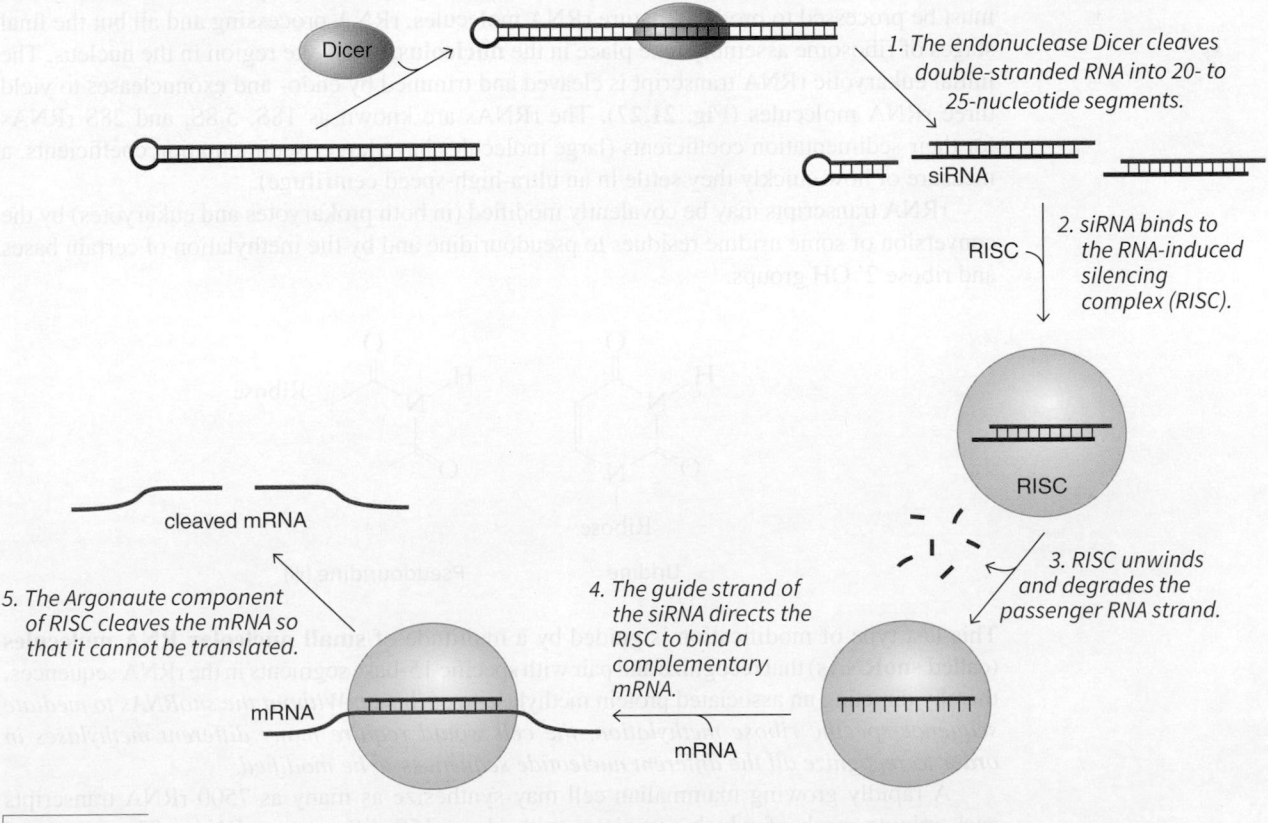

1. The endonuclease Dicer cleaves double-stranded RNA into 20- to 25-nucleotide segments.

2. siRNA binds to the RNA-induced silencing complex (RISC).

3. RISC unwinds and degrades the passenger RNA strand.

4. The guide strand of the siRNA directs the RISC to bind a complementary mRNA.

5. The Argonaute component of RISC cleaves the mRNA so that it cannot be translated.

FIGURE 21.26

RNA interference. The steps involving siRNA are shown. The miRNA pathway for inactivating mRNA is similar.

FIGURE 21.27

Eukaryotic rRNA processing. The initial transcript of about 13.7 kb has a sedimentation coefficient of 45S. Three smaller rRNA molecules (18S, 5.8S, and 28S) are derived from it by the action of nucleases.

times over—in a highly specific manner. It is believed that RNA interference originally evolved as an antiviral defense, since many viral life cycles include the formation of double-stranded RNAs.

In the miRNA pathway, RNA hairpins containing imperfectly paired nucleotides are processed by Dicer and other enzymes to double-stranded miRNAs that bind to the RISC. The passenger RNA strand is ejected and the remaining strand helps the RISC locate complementary target mRNAs. Whereas an siRNA specifically seeks and destroys an mRNA that is perfectly complementary, an miRNA can bind to a large number of target mRNAs—possibly hundreds—because it forms base pairs with a stretch of only 6 or 7 nucleotides. The captured mRNAs are unavailable for translation and are susceptible to the standard mechanisms for RNA degradation diagrammed in Figure 21.25.

In addition to serving as a powerful laboratory technique for silencing genes in order to explore their functions, the RNA interference system is being exploited for practical purposes. Several clinical trials are under way to test whether siRNAs can turn off the expression of viral genes in order to block viral replication. Other diseases in which gene silencing would be desirable, such as cancer, are amenable to RNAi therapy, provided that the siRNA can be delivered selectively to cancerous cells. In general, introducing exogenous RNAs into cells is challenging, since nucleic acids don't easily cross cell membranes, and the presence of extracellular RNA may trigger the body's innate RNA-degrading antiviral defenses. Apples and potatoes have been engineered to use RNAi to prevent the synthesis of the oxidative enzymes that cause browning. It is hoped that these crops will be more acceptable to consumers who avoid traditional genetically modified foods that contain foreign genes (see Box 3.A).

rRNA AND tRNA PROCESSING INCLUDES THE ADDITION, DELETION, AND MODIFICATION OF NUCLEOTIDES

rRNA transcripts, which are generated mainly by RNA polymerase I in eukaryotes, must be processed to produce mature rRNA molecules. rRNA processing and all but the final stages of ribosome assembly take place in the **nucleolus,** a discrete region in the nucleus. The initial eukaryotic rRNA transcript is cleaved and trimmed by endo- and exonucleases to yield three rRNA molecules (**Fig. 21.27**). The rRNAs are known as 18S, 5.8S, and 28S rRNAs for their sedimentation coefficients (large molecules have larger sedimentation coefficients, a measure of how quickly they settle in an ultra-high-speed centrifuge).

rRNA transcripts may be covalently modified (in both prokaryotes and eukaryotes) by the conversion of some uridine residues to pseudouridine and by the methylation of certain bases and ribose 2′ OH groups.

Uridine

Pseudouridine (ψ)

This last type of modification is guided by a multitude of **small nucleolar RNA molecules** (called **snoRNAs**) that recognize and pair with specific 15-base segments in the rRNA sequences, thereby directing an associated protein methylase to each site. *Without the snoRNAs to mediate sequence-specific ribose methylation, the cell would require many different methylases in order to recognize all the different nucleotide sequences to be modified.*

A rapidly growing mammalian cell may synthesize as many as 7500 rRNA transcripts each minute, each of which associates with about 150 different snoRNAs. The processed

FIGURE 21.28

Some modified nucleotides in tRNA molecules. The parent nucleotide is in black; the modification is shown in red.

rRNAs eventually combine with some 80 different ribosomal proteins to generate fully functional ribosomes, a task that requires careful coordination between RNA synthesis and ribosomal protein synthesis.

tRNA molecules, produced by the action of RNA polymerase III in eukaryotes, undergo nucleolytic processing and covalent modification. The initial tRNA transcripts are trimmed by ribonuclease P (see below).

Some tRNA transcripts undergo splicing to remove introns. In some bacteria, newly made tRNAs end with a 3′ CCA sequence, which serves as the attachment point for an amino acid that will be used for protein synthesis. In most organisms, however, the three nucleotides are added to the 3′ end of the immature molecule by the action of a nucleotidyl transferase.

Up to 25% of the nucleotides in tRNA molecules are covalently modified. The alterations range from simple additions of methyl groups to complex restructuring of the base. Some of the 100 or so known nucleotide modifications are shown in **Figure 21.28**. These are yet more examples of how cells alter genetic information as it is transcribed from relatively inert DNA to highly variable and much more dynamic RNA molecules.

RNAs HAVE EXTENSIVE SECONDARY STRUCTURE

Some of the modified nucleotides common in tRNAs also occur in other types of RNA, including mRNA. For example, N^6-methyladenosine occurs at thousands of highly conserved sites in mammalian mRNAs, which suggests that it has some functional significance. This modified nucleotide seems to function like the epigenetic marks in DNA, as it has reader, writer, and eraser enzymes associated with it. N^6-Methyladenosine residues seem to favor mRNA degradation, while 5-hydroxymethylcytosine residues enhance mRNA translation. It is likely that RNA modifications provide binding sites for proteins or influence RNA secondary structure.

Unlike DNA, whose conformational flexibility is considerably constrained by its double-stranded nature, single-stranded RNA molecules can adopt highly convoluted shapes through base pairing between different segments. In addition to the standard (Watson–Crick) types of base pairs, RNA accommodates nonstandard base pairs as well as hydrogen-bonding interactions among three bases (**Fig. 21.29**).

Base stacking stabilizes the RNA tertiary structure, achieving the same sort of balance between rigidity and flexibility exhibited by protein enzymes. *A folded RNA molecule can then bind substrates, orient them, and stabilize the transition state of a chemical reaction.*

FIGURE 21.29

Some nonstandard base pairs. R represents the ribose–phosphate backbone.

All cells contain two essential ribozymes: the tRNA-processing RNase P and the ribosomal RNA that catalyzes peptide bond formation during protein synthesis. There are at least six other naturally occurring types of catalytic RNAs (such as those involved in splicing) and many more synthetic ribozymes. Self-splicing introns are not true catalysts, since they cannot participate in more than one reaction cycle, but the RNA component of RNase P, with its numerous base-paired stems and compact protein-like structure, is a true catalyst (**Fig. 21.30**). At one time it was believed that the enzyme's RNA molecule merely helped align the tRNA substrate for the protein to cleave, but the bacterial RNase P RNA is able to cleave its substrate in the absence of the RNase P protein.

The existence of RNA enzymes such as RNase P lends support to the theory of an early **RNA world** when RNA functioned as a repository of biological information (like modern DNA) as well as a catalyst (like modern proteins). Experiments with synthetic RNAs *in vitro* have demonstrated that RNA can catalyze a wide variety of chemical reactions, including the biologically relevant synthesis of glycosidic bonds (the type of bond that links the base and ribose in a nucleoside) and RNA-template–directed RNA synthesis. Apparently, most ribozymes that originated in an early RNA world were later supplanted by protein catalysts, leaving only a few examples of RNA's catalytic abilities.

FIGURE 21.30

RNase P. In this model of the *Thermotoga maritima* enzyme, the 347-nucleotide RNA is gold and the product tRNA is red. The small protein component (117 amino acids) is green. [*Structure (pdb 3Q1Q) determined by N. J. Reiter, A. Osterman, A. Torres-Larios, K. K. Swinger, T. Pan, and A. Mondragon.*]

 CONCEPT CHECK

1. Describe the structure and function of the mRNA 5′ cap and 3′ tail.

2. Draw diagrams to recount the events of RNA splicing.

3. Explain why splicing does not require free energy input.

4. What are the advantages and disadvantages of splicing protein-coding genes?

5. Summarize the steps of RNA interference.

6. Compare RNA polymerase, poly(A) polymerase, and the tRNA CCA-adding enzyme with respect to substrates, products, and requirement for a template.

7. Explain why the products of transcription exhibit much more variability than the genes that encode them.

SUMMARY

21.1 Initiation Transcription

• Transcription is the process of converting a segment of DNA into RNA. An RNA transcript may represent a protein-coding gene or it may participate in protein synthesis or other activities, including RNA processing.

• Gene expression may be regulated by altering histones through acetylation, phosphorylation, and methylation, by methylating DNA, and by rearranging nucleosomes.

• Transcription begins at a DNA sequence known as a promoter. A gene to be transcribed must be recognized by a regulatory factor such as the σ factor in prokaryotes.

• In eukaryotes, a set of general transcription factors interact with DNA at the promoter to form a complex that recruits RNA polymerase and may further alter chromatin structure.

• Regulatory DNA sequences may affect transcription through binding proteins that interact with RNA polymerase via the Mediator complex.

• The bacterial *lac* operon illustrates the regulation of transcription by a repressor protein.

21.2 RNA Polymerase

• Eukaryotic RNA polymerase II transcribes protein-coding genes. It requires no primer and polymerizes ribonucleotides to generate

an RNA chain that forms a short double helix with the template DNA.

• The polymerase acts processively along the DNA template but reverses to allow the excision of a mispaired nucleotide.

• The elongation phase of transcription in eukaryotes is triggered by phosphorylation of the C-terminal domain of RNA polymerase II.

• Transcription termination in prokaryotes involves destabilization of the DNA–RNA hybrid helix. In eukaryotes, transcription termination is linked to polymerase pausing and RNA cleavage.

21.3 RNA Processing

• mRNA transcripts undergo processing that includes the addition of a 5′ cap structure and a 3′ poly(A) tail. mRNA splicing, carried out by RNA–protein complexes called spliceosomes, joins exons and eliminates introns.

• RNA interference is a pathway for inactivating mRNAs according to their ability to pair with a complementary siRNA or miRNA.

• rRNA and tRNA transcripts are processed by nucleases and enzymes that modify particular bases.

• The chemical and structural variability of RNA molecules makes it possible for some to function as enzymes.

KEY TERMS

transcription	histone code	activator	small nuclear RNA (snRNA)
gene	CpG island	silencer	ribozyme
mRNA	imprinting	repressor	RNAi
rRNA	epigenetics	cap	siRNA
tRNA	promoter	poly(A) tail	miRNA
operon	consensus sequence	intron	nucleolus
RNA processing	TATA box	exon	small nucleolar RNA (snoRNA)
RNA polymerase	general transcription factor	splicing	RNA world
ncRNA	enhancer	spliceosome	

BIOINFORMATICS

Brief Bioinformatics Exercises

21.1 Viewing and Analyzing RNA Polymerase

21.2 RNA Polymerase, Transcription, and the KEGG Database

21.3 Viewing and Analyzing the *lac* Repressor

PROBLEMS

21.1 Initiation Transcription

1. Why does the genome contain so many more genes for rRNA than mRNA?

2. Why is it effective for a bacterial cell to organize genes for related functions as an operon? How do eukaryotes achieve the same benefits?

3. Proteins can interact with DNA through relatively weak forces, such as hydrogen bonds and van der Waals interactions, as well as through stronger electrostatic interactions such as ion pairs. Which types of interactions predominate for sequence-specific DNA-binding proteins and for sequence-independent binding proteins?

4. Certain proteins that stimulate expression of a gene bind to DNA in a sequence-specific manner and also induce conformational

changes in the DNA. Describe the purpose of these two modes of interaction with the DNA.

5. (a) The sense (coding strand) of the *E. coli* promoter for the *rrnA1* gene is shown below. The transcription initiation site is shown by +1. Identify the −35 and −10 regions for this gene.
(b) Predict the effect of a mutation in one of the bases in either the −35 or the −10 region of the promoter.
(c) Which region contains an AT-rich region and why is this region composed of A:T and not G:C base pairs?

AAAATAAATGCTTGACTCTGTAGCG-

+1
GGAAGGCGTATTATCCAACACCC

6. Sp1 is a sequence-specific human DNA-binding protein that binds to a region on the DNA called the GC box, a promoter element with the sequence GGGCGG. Binding of Sp1 to the GC box enhances RNA polymerase II activity 50- to 100-fold. How would you use affinity chromatography (see Section 4.6) to purify Sp1?

7. Identify possible eukaryotic promoter elements in the sequence of the mouse β globin gene shown below. The first nucleotide to be transcribed is indicated by +1.

$$+1$$
GAGCATATAAGGTGAGGTAGGATCAGTTGCTCCTCACATTT

8. (a) Draw the structures of the amino acid side chains that correspond to the following histone modifications: (i) acetylation of lysine; (ii) phosphorylation of serine; (iii) phosphorylation of histidine. **(b)** How do these modifications (i) change the character of their respective side chains? (ii) decrease the binding affinity between the histones and the DNA?

9. A specific type of histone methyltransferase (HMT) catalyzes the methylation of a single lysine or a single arginine in a histone protein (usually H3 or H4). Draw the structures of methylated lysine and methylated arginine residues.

10. The reversal of histone arginine methylation converts the methylated arginine to citrulline in a reaction that consumes H_2O. Draw the resulting amino acid residue. What is the other product of the reaction?

11. DNA methylation requires the methyl group donor S-adenosylmethionine, which is produced by the condensation of methionine with ATP. The sulfonium ion's methyl group is used in methyl-group transfer reactions.

(a) The demethylated S-adenosylmethionine is then hydrolyzed to produce adenosine and a nonstandard amino acid. Draw the structure of this amino acid. How does the cell convert this compound back to methionine to regenerate S-adenosylmethionine?

(b) The proper regulation of gene expression requires methylation as well as demethylation of cytosine residues in DNA. If a demethylase carries out a hydrolytic reaction to restore cytosine residues, what is the other reaction product?

S-Adenosylmethionine

12. Mice that are homozygous for a mutation that renders the DNA methyltransferase enzyme nonfunctional usually die *in utero*. Why does this mutation have such serious consequences?

13. In addition to the modifications listed in Table 21.2, histones may also be modified by the attachment of a single ubiquitin molecule to a lysine side chain. As described in Section 12.1, proteins destined for the proteasome are also tagged with ubiquitin. How does ubiquitin tagging compare with the modification of a histone with ubiquitin? Are ubiquitinated histones marked for proteolytic destruction by the proteasome?

14. Enzymes that catalyze histone acetylation (histone acetyltransferases, or HAT) are closely associated with transcription factors, which are proteins that promote transcription. Why is this a good biochemical strategy?

15. Explain why epigenetic marks such as methylation of cytosine residues are almost completely erased during the early stages of embryogenesis.

16. 5-Methylcytosine residues can be converted to 5-formylcytosine and 5-carboxylcytosine.

(a) Draw the structures of these modified bases.

(b) Demethylation of 5-methylcytosine residues in DNA is indirect; instead of removing the methyl group, the base is oxidized and then the entire formylcytosine is removed and replaced by the base excision repair pathway (see Fig. 20.18). List the enzymes that are involved in this process.

17. The enzyme EZH2 is a histone lysine methyltransferase that is upregulated in various types of cancers. How does this upregulation affect lysine residues 4 and 27 of histone 3? How does this affect the transcription of genes associated with this histone?

18. Three human TBP-associated factors (TAFs) contain protein domains that are homologous to those found in histones H2B, H3, and H4. Why is this finding not surprising?

19. The T7 bacteriophage RNA polymerase recognizes specific promoter sequences and melts open the DNA to form a transcription bubble without the need for transcription factors.

(a) Dissociation constants (K_d) were measured for the interaction between the polymerase and DNA segments containing the promoter sequences. In some cases, the DNA contained a bulge, caused by a mismatch of one, four, or eight bases, to mimic the intermediates in the formation of a transcription bubble. To which DNA segment does the polymerase bind most tightly? Explain in terms of the DNA structure.

DNA promoter segment	K_d (nM)
Fully base paired	315
One-base bulge	0.52
Four-base bulge	0.0025
Eight-base bulge	0.0013

(b) Use the data in the table to calculate the $\Delta G°'$ for the binding of T7 RNA polymerase to fully base-paired DNA and to DNA with an eight-base bulge. [*Note:* The K_d is the inverse of K_{eq}.] Assume a temperature of 25°C.

(c) What do these results reveal about the thermodynamics of melting open a DNA helix for transcription? What is the approximate free energy cost of forming a transcription bubble equivalent to eight base pairs?

20. In bacteria, the core RNA polymerase binds to DNA with a dissociation constant of 5×10^{-12} M. The polymerase in complex with its σ factor has a dissociation constant of 10^{-7} M. Explain.

21. One of the genes expressed by the *lac* operon is *lacY*, which encodes a lactose permease transporter that allows lactose to enter the cell. Why does the expression of this gene assist in the expression of the operon?

22. The genes of the *lac* operon are not expressed when the *lac* repressor binds to the operator. But removal of the *lac* repressor is not sufficient to allow gene expression—a protein called catabolite activator protein (CAP) is also required to assist RNA polymerase and facilitate transcription. CAP can interact with the *lac* promoter only when it binds to its ligand cAMP. In *E. coli*, the intracellular concentration of cAMP falls when glucose is present. Describe the activity of the *lac* operon in each of the following scenarios:

(a) Both lactose and glucose are present.

(b) Glucose is present but lactose is absent.

(c) Both glucose and lactose are absent.

(d) Lactose is present and glucose is absent.

23. Researchers have isolated bacterial cells with mutations in various segments of the *lac* operon. What is the effect on gene expression if a mutation in the operator occurs so that the repressor cannot bind? What happens when lactose is added to the growth medium of these mutants?

24. A bacterial strain expresses a mutant *lac* repressor protein that retains its ability to bind to the operator but cannot bind lactose. What is the effect on gene expression in these mutants? What happens when lactose is added to the growth medium?

25. The compound phenyl-β-D-galactose (phenyl-Gal) is not an inducer of the *lac* operon because it is unable to bind to the repressor. However, it can serve as a substrate for β-galactosidase, which cleaves phenyl-Gal to phenol and galactose. How can the addition of phenyl-Gal to growth medium distinguish between wild-type bacterial cells and cells that have a mutation in the *lacI* gene?

26. In bacterial cells, the genes that code for the enzymes of the tryptophan biosynthetic pathway are organized in an operon as shown below. Another gene encodes a repressor protein that binds tryptophan. How does the repressor protein control the expression of the genes in the *trp* operon?

P	trpL	trpE	trpD	trpC	trpB	trpA

21.2 RNA Polymerase

27. The promoters for genes transcribed by eukaryotic RNA polymerase I exhibit little sequence variation, yet the promoters for genes transcribed by eukaryotic RNA polymerase II are highly variable. Explain.

28. RNA synthesis is much less accurate than DNA synthesis. Why does this not harm the cell?

29. Radioactively labeled γ-[^{32}P]GTP is added to a bacterial culture undergoing transcription. Is the resulting RNA labeled? If so, where?

30. (a) Explain why the adenosine derivative cordycepin inhibits RNA synthesis.

Cordycepin

(b) How does this provide evidence to support the hypothesis that transcription occurs in the $5' \rightarrow 3'$ direction, not the $3' \rightarrow 5'$ direction?

31. The antibiotic rifampicin binds to the β subunit of bacterial RNA polymerase. In the presence of rifampicin, cultured bacterial cells are capable of synthesizing only short RNA oligomers.
(a) At what point in the transcription process does rifampicin exert its inhibitory effect?
(b) Why is rifampicin used to treat bacterial infections?

32. The activity of RNA polymerase II is inhibited by the mushroom toxin α-amanitin ($K_d = 10^{-8}$ M). In contrast, RNA polymerase III is only moderately inhibited by the toxin ($K_d = 10^{-6}$ M), and RNA polymerase I is not affected at all. What would be the effect of adding 10 nM α-amanitin to cells in culture?

33. The three different eukaryotic RNA polymerases were discovered in the 1970s by researchers who loaded cell extracts onto a DEAE ion-exchange column (see Section 4.6) and then eluted the proteins with a salt gradient. Collected fractions were assayed for RNA polymerase activity in the presence and in the absence of Mg^{2+} ions and in the presence of the mushroom toxin α-amanitin. In addition to the difference in α-amanitin sensitivity described in Problem 32, the investigators noted that RNA polymerase I was fully active in the presence of 5 mM Mg^{2+} ions whereas polymerases II and III were only 50% active. How did these results support the conclusions that the three peaks constituted three different forms of RNA polymerase?

34. The C-terminal domain of RNA polymerase II projects away from the globular portion of the protein. Why?

35. Experiments were carried out to truncate RNA polymerase II so that its C-terminal domain (CTD) was missing. What are the consequences for the cells?

36. The DNA sequence of a hypothetical *E. coli* terminator is shown below. N stands for any of the four nucleotides.

$5' \cdots$ N N A A G C G C C G N N N N C C G G C G C T T T T T T N N N $\cdots 3'$
$3' \cdots$ N N T T C G C G G C N N N N G G C C G C G A A A A A A N N N $\cdots 5'$

(a) Write the sequence of the mRNA transcript that is formed if the bottom strand is the noncoding (template) strand.
(b) Draw the RNA hairpin structure that would form in this RNA transcript.

37. In an experiment, inosine triphosphate (ITP) is added to bacterial cells in culture. ITP is used by the cells in place of GTP. Inosine (I) forms base pairs with cytidine (C), and the I:C base pairs form two hydrogen bonds.
(a) Write the sequence of the mRNA transcript that is formed from the gene shown in Problem 36 if the bottom strand is the noncoding (template) strand.
(b) Draw the RNA hairpin structure that would form in this RNA transcript. Compare the stability of this RNA hairpin with the RNA hairpin you drew in Problem 36b. How is termination of transcription affected by the ITP substitution?

38. Formation of an RNA hairpin cannot be the sole factor in the termination of transcription in prokaryotes. Why?

39. The addition of β,γ-imido nucleoside triphosphates to cells in culture has been shown to inhibit Rho-dependent termination. Explain why.

β,γ-Imido nucleoside triphosphate

40. In bacteria, the organization of functionally related genes in an operon allows the simultaneous regulation of expression of those genes. If the operon consists of genes encoding the enzymes for a biosynthetic pathway, then the pathway activity as a whole can be feedback-inhibited when the concentration of the pathway's final product accumulates.
(a) In one mode of feedback regulation, a repressor protein binds to a site in the operon (called the operator) to decrease the rate of transcription only when the repressor has bound a molecule representing the operon's ultimate metabolic product. Draw a diagram showing how such a regulatory system would work.

(b) Feedback regulation of gene expression can also occur after RNA synthesis has begun. In this case, the presence of the operon's ultimate product causes transcription to terminate prematurely or leads to an mRNA that cannot be translated. Draw a diagram illustrating this control mechanism. Assume that the feedback mechanism includes a protein to which the product binds.

(c) How would the feedback inhibition system in part (b) differ if no protein were involved?

41. In some bacteria, several genes required for the biosynthesis of the redox cofactor flavin adenine dinucleotide (FAD; Fig. 3.2c) are arranged in an operon. Comparisons of the sequences of this operon in different species reveal a conserved sequence in the untranslated region at the 5' end of the operon's mRNA. The tertiary structure of an RNA molecule typically includes regions of base pairing and unpaired loops (stem–loop structures). By examining an RNA sequence and noting which positions are most conserved, it is possible to predict the stem–loop structure of the RNA. A portion of a conserved mRNA sequence called RFN, which regulates the expression of the FAD-synthesizing operon, is shown here.

$$\cdots\text{G A U U C A G U U U A A G C U G A A G C}\cdots$$

(a) Draw the stem–loop structure for this RNA segment.

(b) In order to function as an FAD sensor, the RFN element (which consists of about 165 nucleotides) must alter its conformation when FAD binds. How could researchers assess RNA conformational changes?

(c) FAD can be considered as a derivative of flavin mononucleotide (FMN; a coenzyme that resembles FAD but lacks its AMP moiety), which in turn is derived from riboflavin (Fig. 3.2c). The ability of FAD, FMN, and riboflavin to bind to the RFN element was measured as a dissociation constant, K_d; results are shown in the table. Which compound would be the most effective regulator of FAD biosynthesis in the cell? What portion of the FAD molecule is likely to be important for interacting with the mRNA?

Compound	K_d (nM)
FAD	300
FMN	5
Riboflavin	3000

42. A number of human neurological diseases result from the presence of trinucleotide repeats in certain protein-coding genes. The severity of each disease is correlated with the number of repeats, which may increase due to the slippage of DNA polymerase during replication.

(a) The most common repeated triplet is CAG, which is almost always located within an open reading frame. What amino acid is encoded by this triplet (see Table 3.3), and how would the repeats affect the protein?

(b) To test the effect of CAG repeats on transcription, researchers used a yeast expression system with genes engineered to contain CAG repeats. In addition to the expected transcripts corresponding to the known lengths of the genes, RNA molecules up to three times longer were obtained. Based on your knowledge of RNA synthesis and processing, what factors could account for longer-than-expected transcripts of a given gene?

(c) Unexpectedly long transcripts could result from slippage of RNA polymerase II during transcription of the CAG repeats. In this scenario, the polymerase temporarily ceases polymerization, slides backward along the DNA template, then resumes transcription, in effect retranscribing the same sequence. Slippage may be triggered by the formation of secondary structure in the DNA template strand. Draw a diagram showing how a DNA strand containing CAG repeats could form a secondary structure that might prevent the advance of RNA polymerase.

43. In *E. coli*, replication is several times faster than transcription. Occasionally, the replication fork catches up to an RNA polymerase that is moving in the same direction as replication fork movement. When this occurs, transcription stops and the RNA polymerase is displaced from the template DNA. DNA polymerase can use the existing RNA transcript as a primer to continue replication.

(a) Draw a diagram of this process, showing how such collisions would produce a discontinuous leading strand.

(b) Explain why most *E. coli* genes are oriented such that replication and transcription proceed in the same direction.

21.3 RNA Processing

44. The bacterial enzyme polynucleotide phosphorylase (PNPase) is a 3'→5' exoribonuclease that degrades mRNA.

(a) The enzyme catalyzes a phosphorolysis reaction, as does glycogen phosphorylase (see Section 13.3), rather than hydrolysis. Write an equation for the mRNA phosphorolysis reaction.

(b) *In vitro*, PNPase also catalyzes the reverse of the phosphorolysis reaction. What does this reaction accomplish and how does it differ from the reaction carried out by RNA polymerase?

(c) PNPase includes a binding site for long polyribonucleotides, which may promote the enzyme's processivity. Why would this be an advantage for the primary activity of PNPase *in vivo*?

45. In *E. coli*, mRNA degradation is carried out by an endonuclease, but the mRNA must first be modified by a 5' pyrophosphohydrolase. What reaction does this enzyme catalyze?

46. Complete the following table about nucleic acid polymerases.

Polymerase	Type of template	Substrates	Reaction product
DNA polymerase			
Human telomerase			
RNA polymerase			
Poly(A) polymerase			
Bacterial CCA-adding enzyme			

47. Tell whether the following elements are found on DNA or RNA: **(a)** cap; **(b)** CpG islands; **(c)** –35 region; **(d)** –10 region; **(e)** poly(A) tail; **(f)** TATA box; **(g)** enhancers; **(h)** 3' CCA sequence; **(i)** 5' and 3' splice sites; **(j)** promoter; **(k)** branch point.

48. Why are only mRNAs capped and polyadenylated? Why do these post-transcriptional modifications not take place on rRNA or tRNA?

49. Explain why capping the 5' end of an mRNA molecule makes it resistant to 5'→3' exonucleases. Why is it necessary for capping to occur before the mRNA has been completely synthesized?

50. Some short regulatory bacterial RNAs are capped at the 5' end by the structure shown below.

(a) What metabolite is the source of the capping group?

(b) How might the cap structure link the activity of the attached RNA to the cell's overall metabolic state?

(c) Would the cap help protect the RNA from degradation by 5' endonucleases?

51. The poly(A) polymerase that modifies the 3′ end of mRNA molecules differs from other polymerases.

(a) The active sites of DNA and RNA polymerases are large enough to accommodate a double-stranded polynucleotide, but the active site of poly(A) polymerase is much narrower. Explain.

(b) Explain how the substrate specificity of poly(A) polymerase differs from that of a conventional RNA polymerase.

52. A poly(A)-binding protein (PABP) has an affinity for RNA molecules with poly(A) tails. What is the effect of adding PABP to a cell-free system containing mRNA and RNases?

53. The only mRNA transcripts that lack poly(A) tails are those encoding histones. Why do these mRNA transcripts not require poly(A) tails?

54. ATP can be labeled with ^{32}P at any one of its three phosphate groups, designated α, β, and γ (see Fig. 12.12). A eukaryotic cell carrying out transcription and RNA processing is incubated with labeled ATP. Where will the radioactive isotope appear in RNA if the ATP is labeled with ^{32}P at the **(a)** α position, **(b)** β position, or **(c)** γ position?

55. Genetic engineers must modify eukaryotic genes so that they can be expressed in bacterial host cells. Explain why the DNA from a eukaryotic gene cannot be placed directly into the bacteria but is first transcribed to mRNA and then reverse-transcribed back to cDNA.

56. Introns in eukaryotic protein-coding genes may be quite large, but almost none are smaller than about 65 bp. What are some reasons for this minimum intron size?

57. Biochemistry textbooks published a few decades ago often used the phrase "one gene, one protein." Why is this phrase no longer accurate?

58. Introns are removed co-transcriptionally rather than post-transcriptionally. Why is this a good cellular strategy?

59. A portion of the gene for the β chain of hemoglobin is shown below. The sequence on the left includes the 5′ splice site and the sequence on the right includes the 3′ splice site for the intron between exons 1 and 2. Identify the 5′ splice site and the 3′ splice site.

··· CCCTGGGCAGGTTGGTA ····

···· TTTCCCACCCTTAGGCTGCT ···

60. The hemoglobin β chain gene contains three exons, so two introns must be removed from the primary mRNA transcript (see Problem 59). What types of mutations would result in splicing errors? How would the mutations affect the protein translated from the improperly spliced mRNA?

61. The ribozyme known as RNase P processes certain immature tRNA and rRNA molecules. Comparisons of RNase P RNAs from different species reveal conserved features that appear to be involved in endonuclease activity. For example, an unpaired uridine at position 69 is universally conserved. This residue does not pair with another nucleotide but forms a bulge in the RNA secondary structure. To test whether the identity or the geometry of U69 is critical for RNase P activity, several mutants were constructed and their endonuclease activity studied. The results for each mutant are given as a rate constant (k) for catalysis and a dissociation constant (K_d) for substrate binding.

RNase P RNA	k (min^{-1})	K_d (nM)
Wild-type U69	0.26	1.7
U69 → G69	0.0034	73
U69 → C69	0.0056	3
U69 deletion	0.0056	7
U69 + U70	0.0054	181

(a) What is the effect of mutating U69 to a G or a C residue? Do these data reveal whether the U69 bulge is more important for substrate binding or catalysis?

(b) What is the effect of increasing the size of the bulge by adding a second U residue (the U69 + U70 mutant), or removing the bulge by deleting the U69 residue?

62. Some tRNA molecules include sulfur-containing nucleotides. Draw the structures of 4-thiouridine and 2-thiocytidine.

63. RNA interference was investigated as a method to silence the gene for vascular endothelial growth factor (VEGF), a protein required for angiogenesis (development of blood vessels) in most cancers. The addition of siRNA specific for the VEGF gene almost completely eliminated the secretion of VEGF from prostate cancer cells in culture. A portion of the gene sequence (bases 189–207) is shown here. Design an siRNA targeted to this region of the gene.

5′ ··· GGAGTACCCTGATGAGATC ··· 3′

64. Explain why the vast majority of nucleic acids with catalytic activity are RNA rather than DNA.

65. Based on the information in this chapter, give at least three reasons why a silent mutation in a gene (that is, a mutation that does not alter the amino acid sequence of the encoded protein) could decrease the amount of protein expressed.

SELECTED READINGS

Gilmour, D. S. and Fan, R., Derailing the locomotive: transcription termination, *J. Biol. Chem.* **283**, 661–664 (2008). [Outlines the mechanisms for transcription termination in prokaryotes and eukaryotes.]

Liu, X., Bushnell, D. A., and Kornberg, R. D., RNA polymerase II transcription: structure and mechanism, *Biochim. Biophys. Acta* **1829**, 2–8 (2013). [Summarizes the essential components of the transcription initiation and elongation complexes.]

Murakami, K. S., Structural biology of bacterial RNA polymerase, *Biomolecules* **5**, 848–864 (2015). [Summarizes what is known about the structure and function of the bacterial enzyme.]

Nilsen, T. W. and Graveley, B. R., Expansion of the eukaryotic proteome by alternative splicing, *Nature* **463**, 457–463 (2010). [Reviews the advantages of alternative splicing and how it might be regulated.]

Roy, A. L. and Singer, D. S., Core promoters in transcription: old problems, new insights, *Trends Biochem. Sci.* **40**, 165–171 (2015). [Points out that a majority of mammalian genes do not have recognizable promoter sequences.]

Sashital, D. and Doudna, J. A., Structural insights into RNA interference, *Curr. Opin. Struct. Biol.* **20**, 90–97 (2010). [Includes information on the structure of RISC.]

Sharp, P. A., The discovery of split genes and RNA splicing, *Trends Biochem. Sci.* **30**, 279–281 (2005). [A brief summary of splicing by one of its discoverers.]

Washburn, R. S. and Gottesman, M. E., Regulation of transcription elongation and termination. *Biomolecules* **5**, 1063–1078 (2015). [Focuses on the transcription in *E. coli*.]

CHAPTER 22

SYNTHESIS OF PROTEINS

In many insects that undergo metamorphosis, the larval stages are characterized by the accumulation of large amounts of storage proteins. During the pupal stage, these molecules are broken down to supply the amino acids needed to synthesize proteins for the adult body.

TessarTheTegu/Shutterstock

DO YOU REMEMBER?

- DNA and RNA are polymers of nucleotides, each of which consists of a purine or pyrimidine base, deoxyribose or ribose, and phosphate (Section 3.2).
- The biological information encoded by a sequence of DNA is transcribed to RNA and then translated into the amino acid sequence of a protein (Section 3.3).
- Amino acids are linked by peptide bonds to form a polypeptide (Section 4.1).
- Protein folding and protein stabilization depend on noncovalent forces (Section 4.3).
- rRNA and tRNA transcripts are modified to produce functional molecules (Section 21.3).

In the decade that followed Watson and Crick's 1953 elucidation of DNA structure, nearly all the components required for expressing genetic information—that is, making a protein—were identified, including mRNA, tRNA, and ribosomes. We can examine how these molecules participate in protein synthesis by considering one step at a time, beginning with the attachment of a specific amino acid to the appropriate tRNA molecule. We can then look at how the tRNAs align with an mRNA sequences in a ribosome so that peptide bonds can link the amino acids in the order specified by the mRNA. We will also look at some of the steps required to convert a newly made polypeptide to a fully functional protein.

22.1 tRNA AND THE GENETIC CODE

KEY CONCEPTS

- tRNAs are compact L-shaped molecules with an anticodon at one end and an aminoacyl group at the other end.
- An aminoacyl–tRNA synthetase uses ATP to activate an amino acid and transfer it to the appropriate tRNA.
- Because of wobble pairing, a tRNA anticodon may pair with more than one mRNA codon.

In protein synthesis, the final step of the central dogma of molecular biology, a sequence of nucleotides (the first language) is translated to a sequence of amino acids (the second language). Soon after Crick's work on the structure of DNA, he hypothesized that **translation** required "adaptor" molecules (subsequently identified as tRNA) that carried an amino acid and recognized genetic information in the form of nucleotides. The correspondence between DNA sequences and protein sequences was indisputable, but it required some biochemical detective work to discover the nature of the genetic code. Ultimately, *the genetic code was shown to be based on three-nucleotide* **codons** *that are read in a sequential and nonoverlapping manner.*

THE GENETIC CODE IS REDUNDANT

A triplet code is a mathematical necessity, since the number of possible combinations of three nucleotides of four different kinds (4^3, or 64) is more than enough to specify the 20 amino acids found in polypeptides (a doublet code, with 4^2, or 16, possibilities, would be inadequate). Genetic experiments with mutant bacteriophages demonstrated that triplet codons are read sequentially. For example, a mutation resulting from the deletion of a nucleotide within a gene can be corrected by a second mutation that inserts another nucleotide into the gene. The second mutation can restore gene function because it maintains the proper **reading frame** for translation. Since a given nucleotide sequence in an mRNA molecule can potentially have three different reading frames (**Fig. 22.1**), the selection of the proper one depends on the precise identification of a translation start site.

The genetic code, shown in **Table 22.1**, is said to be redundant because *several mRNA codons may correspond to the same amino acid*. In fact, most amino acids are specified by two or more codons (arginine, leucine, and serine each have six codons). Only methionine and tryptophan have only one codon each (they are also among the amino acids that occur least frequently in polypeptides; see Fig. 4.3). The methionine codon also functions as a translation initiation point. Three codons, known as stop or nonsense codons, signal translation termination. In Table 22.1, codons are shaded according to the overall hydrophobic, polar, or ionic character of the corresponding amino acid (using the scheme introduced in Fig. 4.2). Codons for chemically similar amino acids appear to cluster; for example, U at the second codon position invariably specifies a hydrophobic amino acid. This apparently nonrandom pattern of codon–amino acid correspondence suggests that the genetic code might have evolved from a simpler system involving only two nucleotides and a handful of amino acids.

The genetic code is essentially universal (there are only a few minor variations in mitochondria and some unicellular eukaryotes). The common genetic code makes genetic engineering possible: a bacterium decodes a human gene in the same way a human cell does. The universal code also allows scientists to deduce evolutionary relationships based on DNA sequence differences (Section 1.4). This would not be possible if each organism had its own way of interpreting genetic information.

tRNAs HAVE A COMMON STRUCTURE

Each tRNA interacts specifically with one codon via its **anticodon** sequence. A bacterial cell typically contains 30 to 40 different tRNAs, and a mammalian cell as many as 150 (this is an obvious example of redundancy in biological systems, since only 20 different

A C C A U C U C G A G A G U			
Thr	Ile	Ser	Arg

A C C A U C U C G A G A G U			
Pro	Ser	Arg	Glu

A C C A U C U C G A G A G U			
His	Leu	Glu	Ser

FIGURE 22.1

Reading frames. Even with a nonoverlapping genetic code based on nucleotide triplets, a given nucleotide sequence has three possible reading frames. An mRNA molecule can therefore potentially specify three different amino acid sequences.

TABLE 22.1 The Standard Genetic Code

FIRST POSITION (5′ END)	SECOND POSITION				THIRD POSITION (3′ END)
	U	**C**	**A**	**G**	
U	UUU Phe	UCU Ser	UAU Tyr	UGU Cys	U
	UUC Phe	UCC Ser	UAC Tyr	UGC Cys	C
	UUA Leu	UCA Ser	UAA Stop	UGA Stop	A
	UUG Leu	UCG Ser	UAG Stop	UGG Trp	G
C	CUU Leu	CCU Pro	CAU His	CGU Arg	U
	CUC Leu	CCC Pro	CAC His	CGC Arg	C
	CUA Leu	CCA Pro	CAA Gln	CGA Arg	A
	CUG Leu	CCG Pro	CAG Gln	CGG Arg	G
A	AUU Ile	ACU Thr	AAU Asn	AGU Ser	U
	AUC Ile	ACC Thr	AAC Asn	AGC Ser	C
	AUA Ile	ACA Thr	AAA Lys	AGA Arg	A
	AUG Met	ACG Thr	AAG Lys	AGG Arg	G
G	GUU Val	GCU Ala	GAU Asp	GGU Gly	U
	GUC Val	GCC Ala	GAC Asp	GGC Gly	C
	GUA Val	GCA Ala	GAA Glu	GGA Gly	A
	GUG Val	GCG Ala	GAG Glu	GGG Gly	G

amino acids are routinely incorporated into polypeptides). tRNAs that bear the same amino acid but have different anticodons are called **isoacceptor tRNAs.** The structures of all tRNA molecules are similar—even those that carry different amino acids.

Each tRNA molecule contains about 76 nucleotides (the range is 54 to 100), of which up to one-quarter are post-transcriptionally modified (the structures of some of these modified nucleotides are shown in Fig. 21.28). Many of the tRNA bases pair intramolecularly, generating the short stems and loops of what is commonly called a cloverleaf secondary structure (**Fig. 22.2a**). A segment at the 5′ end of the tRNA pairs with bases near the 3′ end to form the acceptor stem (an amino acid attaches to the 3′ end). Several other base-paired stems end in small loops. The D loop often contains the modified base dihydrouridine (abbreviated D), and the TψC loop usually contains the indicated sequence (ψ is the symbol for the nucleotide pseudouridine; see Section 21.3). The variable loop, as its name implies, ranges from 3 to 21 nucleotides in different tRNAs. The anticodon loop includes the three nucleotides that pair with an mRNA codon.

The various elements of tRNA secondary structure fold into a compact L shape that is stabilized by extensive stacking interactions and nonstandard base pairs (**Fig. 22.2b**). Virtually all the bases are buried in the interior of the tRNA molecule, except for the anticodon triplet and the CCA sequence at the 3′ end. The narrow elongated structure of tRNA molecules allows them to align side-by-side so that they can interact with adjacent mRNA codons during translation. However, the tRNA anticodon is located a considerable distance (about 75 Å) from the 3′ aminoacyl group, whose identity is specified by that anticodon.

tRNA AMINOACYLATION CONSUMES ATP

Aminoacylation, the attachment of an amino acid to a tRNA, is catalyzed by an aminoacyl–tRNA synthetase (AARS). *To ensure accurate translation, the synthetase must attach the appropriate amino acid to the tRNA bearing the corresponding anticodon.* As expected, most AARSs

FIGURE 22.2

Structure of yeast tRNA^Phe. (a) Secondary structure. The 76 nucleotides of this tRNA molecule, which can carry a phenylalanine residue at its 3′ end, form four base-paired stems arranged in a cloverleaf pattern. Invariant bases are shown in boldface. ψ is pseudouridine and Y is a guanosine derivative. Some C and G residues in this structure are methylated. (b) Tertiary structure, with the various structures colored as in part (a). The long arm of the L consists primarily of the anticodon loop and D loop, and the short arm is primarily made up of the TψC loop and acceptor stem. The anticodon and acceptor ends of the molecule are separated by about 75 Å. [*Structure (pdb 4TRA) determined by E. Westhof, P. Dumas, and D. Moras.*]

? Why is it important that the bases in the anticodon loop point outward?

interact with the tRNA anticodon as well as the aminoacylation site at the other end of the tRNA molecule.

An AARS catalyzes the formation of an ester bond between an amino acid and an OH group of the ribose at the 3′ end of a tRNA to yield an aminoacyl–tRNA:

Aminoacyl–tRNA

The tRNA molecule is then said to be "charged" with an amino acid. The aminoacylation reaction has two steps and requires the free energy of ATP (**Fig. 22.3**). The overall reaction is

$$\text{amino acid} + \text{tRNA} + \text{ATP} \rightarrow \text{aminoacyl–tRNA} + \text{AMP} + \text{PP}_i$$

Most cells contain 20 different AARS enzymes, corresponding to the 20 standard amino acids (isoacceptor tRNAs are recognized by the same AARS). Although all AARSs catalyze the same reaction, they do not exhibit a conserved size or quaternary structure. Nevertheless, the enzymes fall into two groups based on several shared structural and functional features (**Table 22.2**). For example, the class I enzymes attach an amino acid to the 2′ OH group of the tRNA ribose, whereas the class II enzymes attach an amino acid to the 3′ OH group (this distinction is ultimately of no consequence, as the 2′-aminoacyl group shifts to the 3′ position before it takes part in protein synthesis).

1. The amino acid reacts with ATP to form an aminoacyl–adenylate (aminoacyl–AMP). The subsequent hydrolysis of the PP$_i$ product makes this step irreversible in vivo.

2. The amino acid, which has been "activated" by its adenylation, reacts with tRNA to form an aminoacyl–tRNA and AMP.

FIGURE 22.3

The aminoacyl–tRNA synthetase reaction.

? How many "high-energy" phosphoanhydride bonds break in this process? How many "high-energy" acyl-phosphate bonds form?

TABLE 22.2 Classes of Aminoacyl–tRNA Synthetases		
AMINO ACIDS		
Class I	Arg	Leu
	Cys	Met
	Gln	Trp
	Glu	Tyr
	Ile	Val
Class II	Ala	Lys
	Asn	Pro
	Asp	Phe
	Gly	Ser
	His	Thr

Some bacteria appear to lack the full complement of 20 AARSs. The enzymes most commonly missing are GlnRS and AsnRS (which aminoacylate tRNA^{Gln} and tRNA^{Asn}). In these organisms, Gln–tRNA^{Gln} and Asn–tRNA^{Asn} are synthesized indirectly. First, GluRS and AspRS with relatively low tRNA specificity charge tRNA^{Gln} and tRNA^{Asn} with their corresponding acids (glutamate and aspartate). Next, an amidotransferase converts Glu–tRNA^{Gln} and Asp–tRNA^{Asn} to Gln–tRNA^{Gln} and Asn–tRNA^{Asn} using glutamine as an amino-group donor. In some microorganisms, this is the only pathway for producing asparagine.

The structure of a complex of *E. coli* GlnRS and its cognate (matching) tRNA (tRNA^{Gln}) shows the extensive interaction between the protein and the concave face of the tRNA molecule (the inside of the L; **Fig. 22.4**). AARSs are modular proteins with a catalytic domain, where amino acid activation and transfer to a tRNA occur, as well as a domain that binds the tRNA anticodon or, in some cases, another part of the tRNA such as the variable loop. Most AARSs can activate an amino acid in the absence of a tRNA molecule, but GlnRS, GluRS, and ArgRS require a cognate tRNA molecule for aminoacyl–AMP formation. This suggests that the anticodon-recognition site and the aminoacylation active site somehow communicate with each other, which might help guarantee the attachment of the correct amino acid to the tRNA.

SOME SYNTHETASES HAVE PROOFREADING ACTIVITY

Studies of individual AARS enzymes indicate that the amino acid binding site may be tailored precisely to the geometry and electrostatic properties of a particular amino acid, making it less likely that one of the other 19 amino acids would be activated or transferred to a tRNA molecule. For example, TyrRS (the enzyme responsible for synthesizing Tyr–tRNA^{Tyr}) can distinguish between tyrosine and phenylalanine, which have similar shapes, by their ability to form hydrogen bonds with the protein.

In some cases, *the specificity of tRNA aminoacylation may be enhanced through proofreading by the AARS*. For example, IleRS almost always produces Ile–tRNA^{Ile} and only rarely transfers valine (about once every 50,000 reactions) even though valine differs from isoleucine only by a single methylene group and should easily fit into the IleRS active site. The high fidelity of IleRS requires two active sites that participate in a "double-sieve" mechanism to prevent the synthesis of mischarged tRNA^{Ile}.

The first active site activates isoleucine and presumably other amino acids that are chemically similar to and smaller than isoleucine (such as valine, alanine, and glycine) but excludes larger amino acids (such as phenylalanine and tyrosine). The second active site, which hydrolyzes aminoacylated tRNA^{Ile}, admits only aminoacyl groups that are smaller than isoleucine. Thus, the activating and editing active sites together ensure that IleRS produces only Ile–tRNA^{Ile}. The two active sites are on separate domains of the synthetase, so a newly aminoacylated tRNA must visit the proofreading hydrolytic active site before dissociating from the enzyme.

TRNA ANTICODONS PAIR WITH MRNA CODONS

During translation, tRNA molecules align with mRNA codons, base pairing in an antiparallel fashion, for example

tRNA anticodon $\quad 3' -A-A-G- 5'$

mRNA codon $\quad\quad 5' -U-U-C- 3'$

At first glance, this sort of specific pairing would require the presence of 61 different tRNA molecules, one to recognize each of the "sense" codons listed in Table 22.1. In fact, *many isoacceptor tRNAs can bind to more than one of the codons that specify their amino acid*. For example, yeast tRNA^{Ala} has the anticodon sequence 3′–CGI–5′ (I represents the purine nucleotide inosine, a deaminated form of adenosine) and can pair with the alanine codons GCU, GCC, and GCA.

tRNA anticodon $\quad -C-G-I- \quad\quad -C-G-I- \quad\quad -C-G-I-$

mRNA codon $\quad\quad\; -G-C-U- \quad\quad -G-C-C- \quad\quad -G-C-A-$

As Francis Crick originally proposed in the **wobble hypothesis,** the third codon position and the 5′ anticodon position experience some flexibility, or wobble,

FIGURE 22.4

Structure of GlnRS with tRNA^{Gln}. In this complex, the synthetase is green and the cognate tRNA is red. Both the 3′ (acceptor) end of the tRNA (top right) and the anticodon loop (lower left) are buried in the protein. ATP at the active site is shown in yellow. [*Structure of the E. coli complex (pdb 1QRT) determined by J. G. Arnez and T. A. Steitz.*]

in the geometry of their hydrogen bonding. The base pairs permitted by wobbling are given in **Table 22.3**. The wobble hypothesis explains why many bacterial cells can bind all 61 codons with a set of less than 40 tRNAs (the reasons why mammalian cells contain over 150 tRNAs are not clear). Variations in tRNA anticodon sequences allow nonstandard amino acids to be occasionally incorporated into polypeptides at positions corresponding to stop codons (**Box 22.A**).

TABLE 22.3 Allowed Wobble Pairs at the Third Codon–Anticodon Position

5' ANTICODON BASE	3' CODON BASE
C	G
A	U
U	A, G
G	U, C
I	U, C, A

 CONCEPT CHECK

1. Summarize the features of the genetic code.

2. Explain why a universal genetic code is useful for genetic engineers.

3. Draw a simple diagram of a tRNA molecule and label its parts.

4. Describe the free energy cost of aminoacyl–tRNA synthesis.

5. Explain why accurate aminoacylation is essential for accurate translation.

6. Explain why cells don't require 61 different codons.

7. Identify some codons whose meanings would change following a single-nucleotide substitution. Identify some codons whose meanings would not change.

A CLOSER LOOK

Box 22.A The Genetic Code Expanded

In addition to the 20 standard amino acids listed in Figure 4.2, some amino acid variants can be incorporated into proteins during translation (keep in mind that a mature protein may contain a number of modified amino acids, but these changes almost always take place *after* the protein has been synthesized). Addition of a nonstandard amino acid during protein synthesis requires a dedicated tRNA and a stop codon that can be reinterpreted. The expanded genetic code includes two naturally occurring amino acids, selenocysteine and pyrrolysine, plus a number of amino acids produced in the laboratory.

Selenocysteine occurs in a few proteins in both prokaryotes and eukaryotes, which explains why selenium is an essential trace element. Humans may produce as many as two dozen selenoproteins.

$$
\begin{array}{c}
| \\
NH \\
| \\
CH-CH_2-Se-H \\
| \\
C{=}O \\
|
\end{array}
$$

Selenocysteine (Sec) residue

Selenocysteine (Sec), which resembles cysteine, is generated from serine that has been attached to tRNA^{Sec} by the action of SerRS. A separate enzyme then converts Ser–tRNA^{Sec} to Sec–tRNA^{Sec}. This charged tRNA has an ACU anticodon (reading in the $3' \rightarrow 5'$ direction), which recognizes a UGA codon. Normally, UGA functions as a stop codon, but a hairpin secondary structure in the selenoprotein's mRNA provides the contextual signal for selenocysteine to be delivered to the ribosome at that point.

A few prokaryotic species incorporate pyrrolysine (Pyl) into certain proteins. Synthesis of these proteins requires a 21st type of AARS that directly charges the tRNA^{Pyl} with pyrrolysine.

$$
\begin{array}{c}
| \\
NH \\
| \\
CH-CH_2-CH_2-CH_2-CH_2-NH-C \\
| \\
C{=}O \\
|
\end{array}
$$

Pyrrolysine (Pyl) residue

The Pyl–tRNA^{Pyl} recognizes the stop codon UAG, which is reinterpreted as a Pyl codon with the help of a protein that recognizes secondary structure in the mRNA bound to the ribosome.

In the laboratory, proteins containing unnatural amino acids can be synthesized in bacterial, yeast, and mammalian cells. These experimental systems rely on a pair of genetically engineered components: a tRNA that can "read" a stop codon and an AARS that can attach the unnatural amino acid to the tRNA. When the cell translates an mRNA containing the stop codon, the novel amino acid is incorporated at that codon. Dozens of amino acid derivatives with fluoride, reactive acetyl and amino groups, fluorescent tags, and other modifications have been introduced into specific proteins using this technology. Because the novel amino acids are genetically encoded, they appear only at the expected positions in the translated protein—a more reliable outcome than chemically modifying a protein in a test tube.

? **What is the disadvantage for a cell to have a variant tRNA that can insert an amino acid at a stop codon?**

RIBOSOME STRUCTURE

In order to synthesize a protein, genetic information (in the form of mRNA) and amino acids (attached to tRNA) must get together so that the amino acids can be covalently linked in the specified order. This is the job of the **ribosome,** and it is a huge job: A mammalian cell may contain as many as 10^8 protein molecules. An analysis of the human proteome, a set of about 20,000 proteins encoded by the human genome, by mass spectrometry (Section 4.6) reveals that 10,000–12,000 proteins make up the core proteome in different types of human cells, with about 2350 proteins accounting for 75% of the total protein mass in every cell. These abundant "housekeeping" proteins include histones, ribosomal proteins, metabolic enzymes, and cytoskeletal proteins.

THE RIBOSOME IS MOSTLY RNA

The ribosome is a large complex containing both RNA and protein. At one time, ribosomal RNA (rRNA) was believed to serve as a structural scaffolding for ribosomal proteins, which presumably carried out protein synthesis, but it is now clear that rRNA itself is central to ribosomal function.

A bacterial cell may contain 20,000 ribosomes and a yeast cell, about 200,000. This accounts for the observation that at least 80% of a cell's RNA is located in ribosomes (tRNA comprises about 15% of cellular RNA; mRNA accounts for only a few percent of the total). A ribosome consists of a large and a small subunit containing rRNA molecules, all of which are described in terms of their sedimentation coefficients, S. Thus, the 70S bacterial ribosome has a large (50S) and a small (30S) subunit (the sedimentation coefficient indicates how quickly a particle settles during ultracentrifugation; it is related to the particle's mass). The 80S eukaryotic ribosome is made up of a 60S large subunit and a 40S small subunit. The compositions of prokaryotic and eukaryotic ribosomes are listed in **Table 22.4**. Regardless of its source, about two-thirds of the mass of a ribosome is due to the rRNA; the remainder is due to dozens of different proteins (over 80 in eukaryotes). The core of the ribosome is probably the most highly conserved structure across all forms of life.

The structures of intact ribosomes from both prokaryotes and eukaryotes have been elucidated by X-ray crystallography—a monumental undertaking, given the ribosome's large size (about 2500 kD in bacteria and about 4300 kD in eukaryotes). The small ribosomal subunit from the heat-tolerant bacterium *Thermus thermophilus* is shown in **Figure 22.5**. The overall shape of the subunit is defined by the 16S rRNA (1542 nucleotides in *E. coli*), which has numerous base-paired stems and loops that fold into several domains. This multidomain structure appears to confer some conformational flexibility on the 30S subunit—a requirement for protein synthesis. Twenty-one small polypeptides dot the surface of the structure.

(a)

(b)

FIGURE 22.5

Structure of the 30S ribosomal subunit from *Thermus thermophilus.* (a) The 30S subunit with the rRNA in gray and the proteins in purple. (b) Structure of the 16S rRNA alone. Note how the overall shape of the 30S subunit reflects the structure of the rRNA. [*Structure (pdb 1J5E) determined by B. T. Wimberly, D. E. Brodersen, W. M. Clemons, Jr., R. J. Morgan-Warren, A. P. Carter, C. Vonrhein, T. Hartsch, and V. Ramakrishnan.*]

TABLE 22.4 Ribosome Components

	RNA	POLYPEPTIDES
***E. coli* ribosome (70S)**		
Small subunit (30S)	16S	21
Large subunit (50S)	23S, 5S	31
Mammalian ribosome (80S)		
Small subunit (40S)	18S	33
Large subunit (60S)	28S, 5.8S, 5S	49

Compared to the 30S subunit, the prokaryotic 50S subunit is solid and immobile. Its 23S rRNA (2904 nucleotides in *E. coli*) and 5S rRNA (120 nucleotides) fold into a single mass (**Fig. 22.6**). As in the small subunit, the ribosomal proteins associate with the surface of the rRNA, but the surfaces of the large and small subunits that make contact in the intact 70S ribosome are largely devoid of protein. *This highly conserved rRNA-rich subunit interface is the site where mRNA and tRNA bind during protein synthesis.*

Eukaryotic ribosomes are about 40–50% larger than those from bacteria and contain many additional proteins and more extensive rRNA. The RNA sequences that have no counterparts in bacterial ribosomes are known as expansion segments; these structures, along with the unique eukaryotic protein components, surround a core structure that is shared with the simpler bacterial ribosome (**Fig. 22.7**).

THREE tRNAs BIND TO THE RIBOSOME

Up to three tRNA molecules may bind to the ribosome at a given time (**Fig. 22.8**). The binding sites are known as the **A site** (for *aminoacyl*), which accommodates an incoming aminoacyl–tRNA; the **P site** (for *peptidyl*), which binds the tRNA with the growing polypeptide chain; and the **E site** (for *exit*), which transiently binds a deacylated tRNA after peptide bond formation. The anticodon ends of the tRNAs extend into the 30S subunit to pair with mRNA codons, while their aminoacyl ends extend into the 50S subunit, which catalyzes peptide bond formation.

In bacteria, the two ribosomal subunits and the various tRNAs are held in place mainly by RNA–RNA contacts, with a number of stabilizing Mg^{2+} ions. In eukaryotic ribosomes, numerous proteins form intersubunit bridges. In both cases, the mRNA, which threads through the 30S subunit, makes a sharp bend between the codons in the A site and P site, where an Mg^{2+} ion interacts with mRNA backbone phosphate groups (see Fig. 22.8). The kink allows two tRNAs to fit side-by-side while interacting with consecutive mRNA codons. It may also help the ribosome maintain the reading frame by preventing it from slipping along the mRNA.

In a prokaryotic cell, DNA, mRNA, and ribosomes are in the same cellular compartment. But in eukaryotic cells, mRNA is synthesized and processed inside the nucleus and then exported across the double nuclear membrane via the nuclear pore, a large basket-shaped contraption that regulates molecular traffic into and out of the nucleus. Ribosomes in the cytoplasm then translate the mRNA into protein. Cryoelectron tomography, a method for reconstructing three-dimensional cellular structures by analyzing frozen slices by electron

FIGURE 22.6

Structure of the 50S ribosomal subunit from *Haloarcula marismortui*. The 50S subunit is shown with rRNA in gray and proteins in green. Most of the ribosomal proteins are not visible in this view. The protein-free central area forms the interface with the 30S subunit. [*Structure (pdb 1JJ2) determined by D. J. Klein, T. M. Schmeing, P. B. Moore, and T. A. Steitz.*]

FIGURE 22.7

Eukaryotic ribosomal subunits. The solvent-exposed surfaces of the 40S subunit (left) and 60S subunit (right) are shown with the conserved ribosomal core structures in gray. Proteins that are unique to eukaryotes are shown in transparent yellow, and rRNA expansion segments are red. Newly synthesized proteins emerge from the ribosome through the exit tunnel. [*From M. Yusupov,* Science *334, 1524–1529 (2011). Reprinted with permission from AAAS. Courtesy Marat Yusupov.*]

? Which areas of the ribosome appear to be most conserved between bacteria and eukaryotes? Why?

FIGURE 22.8

Model of the complete bacterial ribosome. The large subunit is shown in shades of gold (rRNA) and brown (proteins), and the small subunit in shades of blue (rRNA) and purple (proteins). The three tRNAs are colored magenta (A site), green (P site), and yellow (E site). An mRNA molecule is shown in dark gray. Note that the anticodon ends of the tRNAs contact the mRNA in the small subunit, while their aminoacyl ends are buried in the large subunit, where peptide bond formation occurs. [*From M. Schmeing,* Nature *461, 1234–1242 (2009). Reprinted by permission from Macmillan Publishers, Ltd. Photo Courtesy of M. Schmeing, McGill University.*]

microscopy, has been used to visualize ribosomes and other structures (**Fig. 22.9**). The ribosomes appear to cluster near the nuclear envelope in a network of actin filaments and microtubules (see Section 5.3).

 CONCEPT CHECK

1. Describe the overall structure of a ribosome and its three tRNA binding sites.
2. Summarize the structural importance of ribosomal RNA and ribosomal proteins.
3. Compare bacterial and eukaryotic ribosomes.

►NE ■Nuclear density ■ER ■Actin & Intermediate filaments ■Microtubules ■Large & ■Small ribosomal subunits ■NPC

FIGURE 22.9

Cryoelectron tomography of a human cell. The image in part (a), from a cancerous HeLa cell, has been color-coded in part (b). Gold = chromatin, light purple = endoplasmic reticulum, red = actin filaments, green = microtubules, blue = large ribosomal subunits, yellow = small ribosomal subunits, and dark purple = nuclear pore complexes. The arrowhead marks the nuclear envelope. [*Courtesy of Wolfgang Baumeister, Max Planck Institute of Biochemistry*]

TRANSLATION

Like DNA replication and RNA transcription, protein synthesis can be divided into separate phases for initiation, elongation, and termination. These stages require an assortment of accessory proteins that bind to tRNA and to the ribosome in order to enhance the speed and accuracy of translation.

INITIATION REQUIRES AN INITIATOR tRNA

In both prokaryotes and eukaryotes, protein synthesis begins at an mRNA codon that specifies methionine (AUG). In bacterial mRNAs, this initiation codon lies about 10 bases downstream of a conserved mRNA sequence called a Shine–Dalgarno sequence (**Fig. 22.10**). This sequence base pairs with a complementary sequence at the 3′ end of the 16S rRNA, thereby positioning the initiation codon in the ribosome. Eukaryotic mRNAs lack a Shine–Dalgarno sequence that can pair with the 18S rRNA. Instead, translation usually begins at the first AUG codon of an mRNA molecule.

The initiation codon is recognized by an initiator tRNA that has been charged with methionine. This tRNA does not recognize other Met codons that occur elsewhere in the coding sequence of the mRNA. In bacteria, the methionine attached to the initiator tRNA is modified by the transfer of a formyl group from tetrahydrofolate (see Section 18.2). The resulting aminoacyl group is designated fMet, and the initiator tRNA is known as $tRNA_f^{Met}$:

$$
\begin{array}{c}
CH_3 \\
| \\
S \\
| \\
CH_2 \\
| \\
CH_2 \quad\;\; O \\
O \quad\quad\quad\;\; | \quad\quad\; \| \\
\| \quad\quad\quad\quad\quad\quad\quad\quad \\
HC-NH-CH-C-O-tRNA_f^{Met}
\end{array}
$$

N-Formylmethionine–$tRNA_f^{Met}$
(fMet–$tRNA_f^{Met}$)

Because the amino group of fMet is derivatized, it cannot form a peptide bond. Consequently, fMet can be incorporated only at the N-terminus of a polypeptide. Later, the formyl group or the entire fMet residue may be removed. In eukaryotic and archaebacterial cells, the initiator tRNA, designated $tRNA_i^{Met}$, is charged with methionine but is not formylated.

Initiation in *E. coli* requires three **initiation factors (IFs)** called IF-1, IF-2, and IF-3. IF-3 binds to the small ribosomal subunit to promote the dissociation of the large and small subunits. fMet–$tRNA_f^{Met}$ binds to the 30S subunit with the assistance of IF-2, a GTP-binding protein. IF-1 sterically blocks the A site of the small subunit, thereby forcing the initiator tRNA into the P site. An mRNA molecule may bind to the 30S subunit either before or after the initiator tRNA has bound, indicating that a codon–anticodon interaction is not essential for initiating protein synthesis.

KEY CONCEPTS

- Translation begins with the binding of a methionine-bearing initiator tRNA to the P site of a ribosome.
- For each new aminoacyl-tRNA, the ribosome senses correct codon–anticodon pairing.
- Ribosomal RNA promotes peptide bond formation by a transpeptidation reaction.
- G proteins such as EF-Tu and EF-G use the free energy of GTP to increase the efficiency of translation.
- Translation terminates when a release factor rather than an aminoacyl–tRNA binds to a stop codon positioned in the A site.
- Multiple ribosomes can simultaneously translate an mRNA.

mRNA
```
· · · A A A A C C        C U A U U U A A U G G C A A C A · · ·
              A G G A G
              U C C U C
          3′  A U      C A C U A G · · ·
```
rRNA

FIGURE 22.10

Alignment of a Shine–Dalgarno sequence with 16S rRNA. A region near the 3′ end of the 16S rRNA molecule (shaded purple) is complementary to the Shine–Dalgarno sequence (blue) in an mRNA, about 10 nucleotides upstream of the initiation codon (green). This mRNA–rRNA interaction helps position the bacterial mRNA at the start of translation. Note that the Shine–Dalgarno sequence shown is a consensus sequence that varies slightly from gene to gene.

FIGURE 22.11

Summary of translation initiation in *E. coli*. Similar events occur during translation initiation in eukaryotes, when a 40S and a 60S subunit associate following the binding of Met–tRNA$_i^{Met}$ to an initiation codon.

1. mRNA and fMet–tRNA$_f^{Met}$ in complex with IF-2–GTP bind to the small (30S) ribosomal subunit.

2. Association of the large (50S) subunit with the 30S subunit triggers IF-2 to hydrolyze its bound GTP.

The tRNA bearing the initial fMet is positioned in the P site of the ribosome.

After the 30S–mRNA–fMet–tRNA$_f^{Met}$ complex has assembled, the 50S subunit associates with it to form the 70S ribosome. This change causes IF-2 to hydrolyze its bound GTP to GDP + P_i and dissociate from the ribosome. The ribosome is now poised—with fMet–tRNA$_f^{Met}$ at the P site—to bind a second aminoacyl–tRNA in order to form the first peptide bond (**Fig. 22.11**).

In eukaryotes, translation initiation requires at least 12 distinct initiation factors. Among these are proteins that recognize the 5′ cap and poly(A) tail of the mRNA (see Section 21.3) and interact so that the mRNA actually forms a circle. Initiation may also require the helicase activity of the ribosome to remove secondary structure in the mRNA that would impede translation. The 40S subunit scans the mRNA in an ATP-dependent manner until it encounters the first AUG codon, which is typically 50 to 70 nucleotides downstream of the 5′ cap (**Fig. 22.12**). The initiation factor eIF2 (the *e* signifies *eukaryotic*) hydrolyzes its bound GTP and dissociates, and the 60S subunit then joins the 40S subunit to form the intact 80S ribosome. IF-2 and eIF2 operate much like the heterotrimeric **G proteins** that participate in intracellular signal transduction pathways (see Section 10.2). In each case, *GTP hydrolysis induces conformational changes that trigger additional steps of the reaction sequence.*

FIGURE 22.12

Circularization of eukaryotic mRNA at translation initiation. A number of initiation factors form a complex that links the 5′ cap and 3′ poly(A) tail of the mRNA. The small (40S) ribosomal subunit binds to the mRNA and locates the AUG start codon that is complementary to the anticodon of the initiator tRNA.

THE APPROPRIATE tRNAs ARE DELIVERED TO THE RIBOSOME DURING ELONGATION

All tRNAs have the same size and shape so that they can fit into small slots in the ribosome. In each reaction cycle of the elongation phase of protein synthesis, an aminoacyl–tRNA enters the A site of the ribosome (the initiator tRNA is the only one that enters the P site without first binding to the A site). After peptide bond formation, the tRNA moves to the P site, and then to the E site. As Figure 22.8 shows, there isn't much room to spare. In addition, all tRNAs must be able to bind interchangeably with protein cofactors.

Aminoacyl–tRNAs are delivered to the ribosome in a complex with a GTP-binding **elongation factor (EF)** known as EF-Tu in *E. coli*. EF-Tu is one of the most abundant *E. coli* proteins (about 100,000 copies per cell, enough to bind all the aminoacyl–tRNA molecules). An aminoacyl–tRNA can bind on its own to a ribosome *in vitro*, but EF-Tu increases the rate *in vivo*.

Because EF-Tu interacts with all 20 types of aminoacyl–tRNAs (representing more than 20 different tRNA molecules), it must recognize common elements of tRNA structure, primarily the acceptor stem and one side of the TψC loop (**Fig. 22.13**). A highly conserved protein pocket accommodates the aminoacyl group. Despite the differing chemical properties of their amino acids, *all aminoacyl–tRNAs bind to EF-Tu with approximately the same affinity* (uncharged tRNAs bind only weakly to EF-Tu). Apparently, the protein interacts with aminoacyl–tRNAs in a combinatorial fashion, offsetting less-than-optimal binding of an aminoacyl group with tighter binding of the acceptor stem and vice versa. This allows EF-Tu to deliver and surrender all 20 aminoacyl–tRNAs to a ribosome with the same efficiency.

The incoming aminoacyl–tRNA is selected on the basis of its ability to recognize a complementary mRNA codon in the A site. Due to competition among all the aminoacyl–tRNA molecules in the cell, this is the rate-limiting step of protein synthesis. Before the 50S subunit catalyzes peptide bond formation, the ribosome must verify that the correct aminoacyl–tRNA is in place. When tRNA binds to the A site of the 30S subunit, two highly conserved residues (A1492 and A1493) of the 16S rRNA "flip out" of an rRNA loop in order to form hydrogen bonds with various parts of the mRNA codon as it pairs with the tRNA anticodon. *These interactions physically link the two rRNA bases with the first two base pairs of the codon and anti-codon so that they can sense a correct match between the mRNA and tRNA* (**Fig. 22.14**). Incorrect base pairing at the first or second codon position would prevent this three-way mRNA–tRNA–rRNA interaction. As expected from the wobble hypothesis (Section 22.1), the A1492/A1493 sensor does not monitor nonstandard base pairing at the third codon position.

As the rRNA nucleotides shift to confirm a correct codon–anticodon match, the conformation of the ribosome changes in such a way that the G protein EF-Tu is induced to hydrolyze its bound GTP. As a result of this reaction, EF-Tu dissociates from the ribosome, leaving behind the tRNA with its aminoacyl group to be incorporated into the growing polypeptide chain.

FIGURE 22.13

Structure of an EF-Tu–tRNA complex. The protein (blue) interacts with the acceptor end and TψC loop of an aminoacyl–tRNA (red). [*Structure (pdb 1TTT) determined by P. Nissen, M. Kjeldgaard, S. Tharp, G. Polekhina, L. Reshetnikova, B. F. C. Clark, and J. Nyborg.*]

(a)

(b)

FIGURE 22.14

The ribosomal sensor for proper codon–anticodon pairing. These images show the A site of the 30S subunit in the (a) absence and (b) presence of mRNA and tRNA analogs. The rRNA is gray, with the "sensor" bases in red. The mRNA analog, representing the A-site codon, is purple, and the tRNA analog (labeled ASL) is gold. A ribosomal protein (S12) and two Mg^{2+} ions (magenta) are also visible. Note how rRNA bases A1492 and A1493 flip out to sense the codon–anticodon interaction. [*Courtesy Venki Ramakrishnan. From* Science 292, 897–902 (2001). Reproduced with permission of AAAS.]

Correct
aminoacyl–tRNA

A site

1. EF-Tu–GTP delivers an
   aminoacyl–tRNA to the
   A site of the ribosome.

2. If the tRNA anticodon
   matches the mRNA
   codon, EF-Tu
   hydrolyzes its GTP
   and dissociates from
   the ribosome, leaving
   the aminoacyl–tRNA in
   the A site.

3. If the tRNA anticodon
   and mRNA codon are
   mismatched, the
   aminoacyl–tRNA
   dissociates before
   EF-Tu hydrolyzes GTP.

Incorrect
aminoacyl–tRNA

Ribosome with a
mismatched tRNA

Ribosome ready
for transpeptidation

FIGURE 22.15

Function of EF-Tu in translation elongation in *E. coli*.

However, if the tRNA anticodon is not properly paired with the A-site codon, the 30S conformational change and GTP hydrolysis by EF-Tu do not occur. Instead, the aminoacyl–tRNA, along with EF-Tu–GTP, dissociates from the ribosome. Because a peptide bond cannot form until after EF-Tu hydrolyzes GTP, *EF-Tu ensures that polymerization does not occur unless the correct aminoacyl–tRNA is positioned in the A site.* The energetic cost of proofreading at the decoding stage of translation is the free energy of GTP hydrolysis (catalyzed by EF-Tu). The function of EF-Tu is summarized in **Figure 22.15**. In eukaryotes, elongation factor eEF1α performs the same service as the prokaryotic EF-Tu. The functional correspondence between some prokaryotic and eukaryotic translation cofactors is given in **Table 22.5**.

The ribosome itself performs a bit of proofreading. The departure of EF-Tu–GDP leaves behind the aminoacyl–tRNA, whose acceptor end can now slip all the way into the A site of the 50S ribosomal subunit. The 30S subunit closes in around the tRNA, but at this point, the only interactions that hold the aminoacyl–tRNA in place are codon–anticodon contacts. If there is still a slight mismatch, such as a G:U base pair at the first or second position, not detectable by the A1492/A1493 sensor, the strain of not being able to form a perfect Watson–Crick pair will be felt by the ribosome and possibly the tRNA itself, and the tRNA will slip out of the A site. In this way, the ribosome verifies correct codon–anticodon pairing twice for each aminoacyl–tRNA: when EF-Tu first delivers it to the ribosome and after EF-Tu departs. Ribosomal proofreading helps limit the error rate of translation to about 10^{-4} (one mistake for every 10^4 codons).

TABLE 22.5 Prokaryotic and Eukaryotic Translation Factors

PROKARYOTIC PROTEIN	EUKARYOTIC PROTEIN	FUNCTION
IF-2	eIF2	Delivers initiator tRNA to P site of ribosome
EF-Tu	eEF1α	Delivers aminoacyl–tRNA to A site of ribosome during elongation
EF-G	eEF2	Binds to A site to promote translocation following peptide bond formation
RF-1, RF-2	eRF1	Binds to A site at a stop codon and induces peptide transfer to water

FIGURE 22.16

The peptidyl transferase reaction. Note that the nucleophilic attack of the aminoacyl group on the peptidyl group produces a free tRNA in the P site and a peptidyl–tRNA in the A site.

? Compare this reaction to the condensation reaction of two amino acids shown in Section 4.1.

THE PEPTIDYL TRANSFERASE ACTIVE SITE CATALYZES PEPTIDE BOND FORMATION

When the ribosomal A site contains an aminoacyl–tRNA and the P site contains a peptidyl–tRNA (or, prior to formation of the first peptide bond, an initiator tRNA), the peptidyl transferase activity of the large subunit catalyzes a **transpeptidation** reaction in which the free amino group of the aminoacyl–tRNA in the A site attacks the ester bond that links the peptidyl group to the tRNA in the P site (**Fig. 22.16**). This reaction lengthens the peptidyl group by one amino acid at its C-terminal end. Thus, *a polypeptide grows in the N → C direction*. No external source of free energy is required for transpeptidation because the free energy of the broken ester bond of the peptidyl–tRNA is comparable to the free energy of the newly formed peptide bond. (Recall, though, that ATP was consumed in charging the tRNA with an amino acid.)

The peptidyl transferase active site lies in a highly conserved region of the bacterial 50S subunit, and the newly formed peptide bond is about 18 Å away from the nearest protein. Thus, the ribosome is a ribozyme (an RNA catalyst). How does rRNA catalyze peptide bond formation? Two highly conserved rRNA nucleotides, G2447 and A2451 in *E. coli*, do not function as acid–base catalysts, as was initially proposed. Rather, these residues help position the substrates for reaction, an example of induced fit (Section 6.3). Binding of a tRNA at the A site triggers a conformational change that exposes the ester bond of the peptidyl–tRNA in the P site. At other times, the ester bond must be protected so that it does not react with water, a reaction that would prematurely terminate protein synthesis. Proximity and orientation effects in the ribosome increase the rate of peptide bond formation about 10^7-fold above the uncatalyzed rate. Some antibiotics exert their effects by binding to the peptidyl transferase active site to directly block protein synthesis (**Box 22.B**).

During transpeptidation, the peptidyl group is transferred to the tRNA in the A site, and the P-site tRNA becomes deacylated. The new peptidyl–tRNA then moves into the P site, and the deacylated tRNA moves into the E site. The mRNA, which is still base paired with the peptidyl–tRNA anticodon, advances through the ribosome by one codon. Experiments designed to assess the force exerted by the ribosome demonstrate that formation of a peptide bond causes the ribosome to loosen its grip on the mRNA, although it is not clear how events at the peptidyl transferase site in the large subunit are communicated to the mRNA decoding site in the small subunit. *The movement of tRNA and mRNA, which allows the next codon to be translated, is known as* **translocation**. This dynamic process requires the G protein called elongation factor G (EF-G) in *E. coli*.

Box 22.B Antibiotic Inhibitors of Protein Synthesis

Antibiotics interfere with a variety of cellular processes, including cell-wall synthesis, DNA replication, and RNA transcription. Some of the most effective antibiotics, including many in clinical use, target protein synthesis. Because bacterial and eukaryotic ribosomes and translation factors differ, these antibiotics can kill bacteria without harming their mammalian hosts.

Puromycin, for example, resembles the 3' end of Tyr–tRNA and competes with aminoacyl–tRNAs for binding to the ribosomal A site. Transpeptidation generates a puromycin–peptidyl group that cannot be further elongated because the puromycin "amino acid" group is linked by an amide bond rather than an ester bond to its "tRNA" group. As a result, peptide synthesis comes to a halt.

Puromycin

Tyrosyl–tRNA

The antibiotic chloramphenicol interacts with the active-site nucleotides, including the catalytically essential A2451, to prevent transpeptidation.

Chloramphenicol

Other antibiotics with more complicated structures interfere with protein synthesis through different mechanisms. For example, erythromycin physically blocks the tunnel that conveys the nascent polypeptide away from the active site. Six to eight peptide bonds form before the constriction of the exit tunnel blocks further chain elongation.

Streptomycin kills cells by binding tightly to the backbone of the 16S rRNA and stabilizing an error-prone conformation of the ribosome. In the presence of the antibiotic, the ribosome's affinity for aminoacyl–tRNAs increases, which increases the likelihood of codon–anticodon mispairing and therefore increases the error rate of translation. Presumably, the resulting burden of inaccurately synthesized proteins kills the cell.

The drugs described here, like all antibiotics, lose their effectiveness when their target organisms become resistant to them. For example, mutations in ribosomal components can prevent antibiotic binding. Alternatively, an antibiotic-susceptible organism may acquire a gene, often present on an extrachromosomal plasmid, whose product inactivates the antibiotic. Acquisition of an acetyltransferse gene leads to the addition of an acetyl group to chloramphenicol, which prevents its binding to the ribosome. Acquisition of a gene for an ABC transporter (Section 9.3) can hasten a drug's export from the cell, rendering it useless.

? Explain why some of the side effects of antibiotic use in humans can be traced to impairment of mitochondrial function.

EF-G bears a striking resemblance to the EF-Tu–tRNA complex (**Fig. 22.17**), and the ribosomal binding sites for the two proteins overlap. Structural studies show that the EF-G–GTP complex physically displaces the peptidyl–tRNA in the A site, causing it to translocate to the P site. This movement would also bump the deacylated tRNA from the P site to the E site. EF-G binding to the ribosome stimulates its GTPase activity. After EF-G hydrolyzes its bound GTP, it dissociates from the ribosome, leaving a vacant A site available for the arrival of another aminoacyl–tRNA and another round of transpeptidation.

The GTPase activity of G proteins such as EF-Tu and EF-G allows the ribosome to cycle efficiently through all the steps of translation elongation. Because GTP hydrolysis is irreversible, the elongation reactions—aminoacyl–tRNA binding, transpeptidation, and translocation—proceed unidirectionally. The *E. coli* ribosomal elongation cycle is shown in **Figure 22.18**. Eukaryotic cells contain elongation factors that function similarly to EF-Tu

FIGURE 22.17

Structure of EF-G from *T. thermophilus*. [Structure (pdb 2BV3) determined by S. Hansson, R. Singh, A. T. Gudkov, A. Liljas, and D. T. Logan.]

? Compare the size and shape of this complex to the size and shape of the complex containing EF-Tu and an aminoacyl–tRNA (see Fig. 22.13).

FIGURE 22.18

The *E. coli* ribosomal elongation cycle.

delivery of
aminoacyl–tRNA

P site   A site

mRNA

transpeptidation

EF-Tu

GTP   GDP

EF-Tu

GDP

EF-G–GDP

GTP

translocation

GTP

EF-G–GTP

and EF-G (Table 22.5). These G proteins are continually recycled during protein synthesis. In some cases, accessory proteins help replace bound GDP with GTP to prepare the G protein for another reaction cycle.

RELEASE FACTORS MEDIATE TRANSLATION TERMINATION

As the peptidyl group is lengthened by the transpeptidation reaction, it exits the ribosome through a tunnel in the center of the large subunit. The tunnel, about 100 Å long and 15 Å in diameter in the bacterial ribosome, shelters a polypeptide chain of up to 30 residues. The tunnel is defined by ribosomal proteins as well as the 23S rRNA. A variety of groups—including rRNA bases, backbone phosphate groups, and protein side chains—form a mostly hydrophilic surface for the tunnel. There are no large hydrophobic patches that could potentially impede the exit of a newly synthesized peptide chain.

Translation ceases when the ribosome encounters a stop codon (see Table 22.1). With a stop codon in the A position, the ribosome cannot bind an aminoacyl–tRNA but instead binds a protein known as a **release factor (RF)**. In bacteria such as *E. coli*, RF-1 recognizes stop codons UAA and UAG, and RF-2 recognizes UAA and UGA. In eukaryotes, one protein—called eRF1—recognizes all three stop codons.

The release factor must specifically recognize the mRNA stop codon; it does this via an "anticodon" sequence of three amino acids, such as Pro–Val–Thr in RF-1 and Ser–Pro–Phe in RF-2, that interact with the first and second bases of the stop codon. At the same time, a loop of the release factor with the conserved sequence Gly–Gly–Gln projects into the peptidyl transferase site of the 50S subunit (**Fig. 22.19**). The amide group of the Gln residue promotes transfer of the peptidyl group from the P-site tRNA to water, apparently by stabilizing the transition state of this hydrolysis reaction. The product of the reaction is an untethered polypeptide that can exit the ribosome. At one time, release factors were believed to act by mimicking tRNA molecules, as EF-G does. However, it now appears that release factors undergo conformational changes upon binding to the ribosome, so they don't operate straightforwardly as tRNA surrogates.

In *E. coli*, an additional RF (RF-3), which binds GTP, promotes the binding of RF-1 or RF-2 to the ribosome. In eukaryotes, eRF3 performs this role. Hydrolysis of the GTP bound to RF-3 (or eRF3) allows the release factors to

FIGURE 22.19

Structure of RF-1. The RF-1 protein is shown as a purple ribbon with its anticodon-binding Pro–Val–Thr (PVT) sequence in red and its peptidyl transferase–binding Gly–Gly–Gln (GGQ) sequence in orange. This image shows the structure of RF-1 as it exists when bound to a ribosome. [*Structure of the* T. thermophilus *ribosome with RF-1 (pdb 3D5A) determined by M. Laurberg, H. Asahara, A. Korostelev, J. Zhu, S. Trakhanov, and H. F. Noller.*]

1. A release factor (RF-1 or RF-2) recognizes a stop codon in the A site.

2. RF-1/RF-2 causes the ribosome to transfer the peptidyl group to water to release the polypeptide chain.

3. GTP hydrolysis by RF-3 allows the release factors to dissociate. Additional steps are required to prepare the ribosome for another round of translation.

RF

mRNA

Stop codon

Polypeptide

GTP
GDP

RF-3–GTP   RF-3–GDP

FIGURE 22.20

Translation termination in *E. coli*.

dissociate (**Fig. 22.20**). This leaves the ribosome with a bound mRNA, an empty A site, and a deacylated tRNA in the P site.

In bacteria, preparing the ribosome for another round of translation is the responsibility of a **ribosome recycling factor (RRF)** that works in concert with EF-G. RRF apparently slips into the ribosomal A site. EF-G binding then translocates RRF to the P site, thereby displacing the deacylated tRNA. Following GTP hydrolysis, EF-G and RRF dissociate from the ribosome, leaving it ready for a new round of translation initiation. In eukaryotes, a protein with ATPase activity binds to the ribosome containing eRF3 and helps dissociate the large and small subunits. Initiation factors appear to bind during this process, so translation termination and reinitiation are tightly linked.

TRANSLATION IS EFFICIENT *in vivo*

A ribosome can extend a polypeptide chain by approximately 20 amino acids every second in bacteria and by about 4 amino acids every second in eukaryotes. At these rates, most protein chains can be synthesized in under a minute. As we have seen, various G proteins trigger conformational changes that keep the ribosome operating efficiently through many elongation cycles. Cells also maximize the rate of protein synthesis by forming **polysomes.** These structures include a single mRNA molecule simultaneously being translated by multiple ribosomes (**Fig. 22.21**). As soon as the first ribosome has cleared the initiation codon, a second ribosome can assemble and begin translating the mRNA. The circularization of eukaryotic mRNAs (see Fig. 22.12) may promote repeated rounds of translation. Because the stop codon at the 3′ end of the coding sequence may be relatively close to the start codon at the 5′ end, the ribosomal components released at termination can be easily recycled for reinitiation.

(a)

(b)

FIGURE 22.21

A polysome. (a) In this electron micrograph, a single mRNA strand encoding silkworm fibroin is studded with ribosomes. Arrows indicate growing fibroin polypeptide chains. [*Courtesy Oscar L. Miller, Jr. and Steven L. McKnight, University of Virginia.*] **(b)** In this cryoelectron microscopy–based reconstruction of a polysome, ribosomes are blue and gold, and the emerging polypeptides are red and green. [*Courtesy Wolfgang Baumeister and Julip Ortiz, from Cell 136, 261–271 (2009).*]

? Identify the polysomes in Fig. 22.9.

CONCEPT CHECK

1. Draw a diagram of a gene with and without introns and label the promoter, transcription termination site, start codon, and stop codon.

2. Make a list of all the substances that must interact with a ribosome in order to produce a polypeptide.

3. Compare the way a ribosome recognizes a start codon and a stop codon.

4. Describe the functions of bacterial IF-2, EF-Tu, EF-G, RF-1, and RRF.

5. How does the ribosome prevent polymerization of the wrong amino acid?

6. Compare translation initiation, elongation, and termination in bacteria and eukaryotes.

7. Summarize the different types of RNA–RNA interactions that occur during translation.

8. Explain how polysomes increase the rate of protein synthesis.

22.4 POST-TRANSLATIONAL EVENTS

The polypeptide released from a ribosome is not yet fully functional. For example, it must fold to its native conformation; it may need to be transported to another location inside or outside the cell; and it may undergo **post-translational modification,** or **processing.**

CHAPERONES PROMOTE PROTEIN FOLDING

Studies of protein folding *in vitro* have revealed numerous insights into the pathways by which proteins (usually relatively small ones that have been chemically denatured) assume a compact globular shape with a hydrophobic core and a hydrophilic surface (see Section 4.3). Protein folding *in vivo* is only partly understood. For one thing, a protein can begin to fold as soon as its N-terminus emerges from the ribosome, even before it has been fully synthesized. In addition, a polypeptide must fold in an environment crowded with other proteins with which it might interact unfavorably. Finally, for proteins with quaternary structure, individual polypeptide chains must assemble with the proper stoichiometry and orientation. All of these processes may be facilitated in a cell by proteins known as **molecular chaperones.**

To prevent improper associations within or between polypeptide chains, chaperones bind to exposed hydrophobic patches on the protein surface (recall that hydrophobic groups tend to aggregate, which could lead to nonnative protein structure or protein aggregation and precipitation). *Many chaperones are ATPases that use the free energy of ATP hydrolysis to drive conformational changes that allow them to bind and release a polypeptide substrate while it assumes its native shape.* Chaperones were originally identified as heat-shock proteins (Hsp) because their synthesis is induced by high temperatures—conditions under which proteins tend to denature (unfold) and aggregate.

The first chaperone a bacterial protein meets, called trigger factor, is poised just outside the ribosome's polypeptide exit tunnel, bound to a ribosomal protein (**Fig. 22.22**). When trigger factor binds to the ribosome, it opens up to expose a hydrophobic patch facing the exit tunnel. Hydrophobic segments of the emerging polypeptide bind to this patch and are thereby prevented from sticking to each other or to other cellular components. Trigger factor may dissociate from the ribosome but remain associated with the nascent (newly formed) polypeptide until another chaperone takes over. Eukaryotes lack trigger factor, although they have other small heat-shock proteins that function in the same manner to protect newly made proteins. These chaperones are extremely abundant in cells, so there is at least one per ribosome.

KEY CONCEPTS

- Molecular chaperones promote protein folding, starting when the polypeptide emerges from the ribosome.
- Membrane and secreted proteins are delivered to a membrane translocon by a signal recognition particle.
- Post-translational protein modifications include proteolysis, glycosylation, and other covalent changes.

FIGURE 22.22

Trigger factor bound to a ribosome. The ribosome-binding portion of trigger factor is shown with its domains in different colors. It binds to the 50S ribosomal subunit where a polypeptide (magenta helix) emerges. PT represents the peptidyl transferase active site. [*Courtesy Nenad Ban, Eidgenossische Technische Hochschule Honggerberg, Zurich.*]

FIGURE 22.23

The GroEL/GroES chaperonin complex.
The two seven-subunit GroEL rings, viewed from the side, are colored red and yellow. A seven-subunit GroES complex (blue) caps the so-called *cis* GroEL ring. [*Structure (pdb 1AON) determined by Z. Xu, A. L. Horwich, and P. B. Sigler.*]

Trigger factor may hand off a new polypeptide to another chaperone, such as DnaK in *E. coli* (it was named when it was believed to participate in DNA synthesis). DnaK and other heat-shock proteins in prokaryotes and eukaryotes interact with newly synthesized polypeptides and with existing cellular proteins and therefore can prevent as well as reverse improper folding. These chaperones, in a complex with ATP, bind to a short extended polypeptide segment with exposed hydrophobic groups (they do not recognize folded proteins, whose hydrophobic groups are sequestered in the interior). ATP hydrolysis causes the chaperone to release the polypeptide. As the polypeptide folds, the heat-shock protein may repeatedly bind and release it.

Ultimately, protein folding may be completed by multisubunit chaperones, called chaperonins, which form cagelike structures that physically sequester a folding polypeptide. The best-known chaperonin complex is the GroEL/GroES complex of *E. coli*. Fourteen GroEL subunits form two rings of seven subunits, with each ring enclosing a 45-Å-diameter chamber that is large enough to accommodate a folding polypeptide. Seven GroES subunits form a domelike cap for one GroEL chamber (**Fig. 22.23**). The GroEL ring nearest the cap is called the *cis* ring, and the other is the *trans* ring.

Each GroEL subunit has an ATPase active site. All seven subunits of the ring act in concert, hydrolyzing their bound ATP and undergoing conformational changes. *The two GroEL rings of the chaperonin complex act in a reciprocating fashion to promote the folding of two polypeptide chains in a safe environment* (**Fig. 22.24**). Note that 7 ATP are consumed for each 10-second protein-folding opportunity. If the released substrate has not yet achieved its native conformation, it may rebind to the chaperonin complex. Only about 10% of bacterial proteins seem to require the GroEL/GroES chaperonin complex, and most of these range in size from 10 to 55 kD (a protein larger than about 70 kD probably could not fit inside the protein-folding chamber). Immunocytological studies indicate that some proteins never stray far from a chaperonin complex, perhaps because they tend to unfold and must periodically restore their native structures.

In eukaryotes, a chaperonin complex known as TRiC functions analogously to bacterial GroEL, but it has eight subunits in each of its two rings, and its fingerlike projections take the place of the GroES cap.

1. A GroEL ring binds 7 ATP and an unfolded polypeptide, which associates with hydrophobic patches on the GroEL subunits.

2. A GroES cap binds, triggering a conformational change that retracts the hydrophobic patches, thereby releasing the polypeptide into the GroEL chamber, where it can fold.

3. Within about 10 seconds, the cis GroEL ring hydrolyzes its 7 ATP.

4. A second polypeptide substrate and 7 ATP bind to the trans GroEL ring.

5. The cis ring releases its GroES cap, the 7 ADP, and the folded substrate polypeptide.

6. The trans GroEL ring can now bind a GroES cap, and steps 3–5 repeat.

FIGURE 22.24

The chaperonin reaction cycle.

THE SIGNAL RECOGNITION PARTICLE TARGETS SOME PROTEINS FOR MEMBRANE TRANSLOCATION

For a cytosolic protein, the journey from a ribosome to the protein's final cellular destination is straightforward (in fact, the journey may be short, since some mRNAs are directed to specific cytosolic locations before translation commences). In contrast, an integral membrane protein or a protein that is to be secreted from the cell follows a different route since it must pass partly or completely through a membrane. These proteins account for about one-third of a cell's total proteins.

In both prokaryotic and eukaryotic cells, *most membrane and secretory proteins are synthesized by ribosomes that dock with the plasma membrane (in prokaryotes) or endoplasmic reticulum (in eukaryotes).* The protein that emerges from the ribosome is inserted into or through the membrane co-translationally. The membrane translocation system is fundamentally similar in all cells and requires a ribonucleoprotein known as the **signal recognition particle (SRP).**

As in other ribonucleoproteins, including the ribosome and the spliceosome (see Section 21.3), the RNA component of the SRP is highly conserved and is essential for SRP function. The *E. coli* SRP consists of a single multidomain protein and a 4.5S RNA. The mammalian SRP contains a larger RNA and six different proteins, but its core is virtually identical to the bacterial SRP. The RNA component of the SRP includes several nonstandard base pairs, including G:A, G:G, and A:C, and interacts with several protein backbone carbonyl groups (most RNA–protein interactions involve protein side chains rather than the backbone).

How does the SRP recognize membrane and secretory proteins? Such proteins typically have an N-terminal **signal peptide** consisting of an α-helical stretch of 6 to 15 hydrophobic amino acids preceded by at least one positively charged residue. For example, human proinsulin (the polypeptide precursor of the hormone insulin) has the following signal sequence:

$$\text{M A L W M R L L P L L A L L A L W G P D P A A A F V N} \cdots$$

The hydrophobic segment and a flanking arginine residue are shaded in green and pink.

The signal peptide binds to the SRP in a pocket formed mainly by a methionine-rich protein domain. The flexible side chains of the hydrophobic Met residues allow the pocket to accommodate helical signal peptides of variable sizes and shapes. In addition to the Met residues, the SRP binding pocket contains a segment of RNA, whose negatively charged backbone interacts electrostatically with the positively charged N-terminus and basic residue of the signal peptide (**Fig. 22.25**).

Electron microscopic studies indicate that the SRP binds to the ribosome at the polypeptide exit tunnel. The SRP recognizes the signal peptide as it emerges, most likely competing with trigger factor or its eukaryotic counterpart. SRP binding to the signal peptide halts translation elongation, possibly via conformational changes in the ribosome resulting from SRP RNA–rRNA interactions. The stalled ribosome–SRP complex then docks at a receptor on the membrane in a process that requires GTP hydrolysis by the SRP. When translation resumes, the growing polypeptide is translocated through the membrane via a structure known as a **translocon.**

The translocon proteins, called SecY in prokaryotes and Sec61 in eukaryotes, form a transmembrane channel with a constricted pore that limits the diffusion of other substances across the membrane (**Fig. 22.26**). Insertion of the signal peptide itself nudges aside two Sec helices in order to open up the pore wide enough to allow a polypeptide segment to pass through. In addition to allowing a secreted protein to pass entirely across a membrane in this manner, the translocon has a lateral opening that allows one or more hydrophobic protein segments to move sideways into the lipid bilayer, which explains how transmembrane proteins (Section 8.3) become incorporated into the membrane.

The driving force for co-translational protein translocation is provided by the ribosome: *the process of polypeptide synthesis simply pushes the chain through the channel.* In situations where a completed polypeptide is inserted across or into a membrane post-translationally (after it has left the ribosome), the transport system—which may involve the Sec translocon or other machinery—relies on some sort of motor protein

signal sequence binding groove

FIGURE 22.25

The SRP signal peptide binding domain. This model shows the molecular surface of a portion of the *E. coli* signal recognition particle. The protein is magenta with hydrophobic residues in yellow. Adjacent RNA phosphate groups are red, and the rest of the RNA is dark blue. A signal peptide binds in the SRP groove, making hydrophobic as well as electrostatic contacts with the protein and RNA. [*From R. Batey, Science 287, 1232–1239 (2000). Reprinted with permission from AAAS. Courtesy Robert Batey.*]

tRNA

NC

Active PCC

FIGURE 22.26

Ribosome bound to Sec61. In this cryoEM-based image, a yeast ribosome (partially cut away) is bound to Sec 61 (pink), which is positioned at the end of the polypeptide exit tunnel. Portions of the nascent polypeptide (NC, green) are visible in the exit tunnel. [*From R. Beckmann, Science 326, 1369–1373 (2009). Reprinted with permission from AAAS. Courtesy Roland Beckmann.*]

1. *The SRP binds to the signal peptide when it emerges from the ribosome. Polypeptide elongation halts.*

2. *The SRP delivers the ribosome to the endoplasmic reticulum membrane. Translation resumes, and the polypeptide crosses the membrane via a translocon.*

3. *Signal peptidase removes the signal peptide from the growing polypeptide.*

4. *The rest of the polypeptide is translocated into the ER lumen. From here it is transported by vesicles to the extracellular space.*

FIGURE 22.27

Membrane translocation of a eukaryotic secretory protein.

that uses the free energy of the ATP hydrolysis reaction.

When the signal peptide emerges on the far side of the membrane, it may be cleaved off by an integral membrane protein known as a signal peptidase. This enzyme recognizes extended polypeptide segments such as those flanking the hydrophobic segment of a signal peptide, but it does not recognize α-helical structures, which are common in mature membrane proteins. The steps of translocating a eukaryotic secretory protein are summarized in **Figure 22.27**.

After translocation in eukaryotic cells, *chaperones and other proteins in the endoplasmic reticulum may help the polypeptide fold into its native conformation, form disulfide bonds, and assemble with other protein subunits.* Extracellular proteins are then transported from the endoplasmic reticulum through the Golgi apparatus and to the plasma membrane via vesicles. The proteins may undergo processing (described next) en route to their final destination.

MANY PROTEINS UNDERGO COVALENT MODIFICATION

Proteolysis is part of the maturation pathway of many proteins. For example, after it has entered the endoplasmic reticulum lumen and had its signal peptide removed and its cysteine side chains cross-linked as disulfides, the insulin precursor undergoes proteolytic processing. The prohormone is cleaved at two sites to generate the mature hormone (**Fig. 22.28**). Over 200 other types of post-translational modification have been documented.

Many extracellular eukaryotic proteins are **glycosylated** at asparagine, serine, or threonine side chains to generate glycoproteins (see Section 11.3). The short sugar chains (oligosaccharides) attached to glycoproteins may protect the proteins from degradation or mediate molecular recognition events. Methyl, acetyl, and propionyl groups may be added to various side chains. N-terminal groups are frequently modified by acetylation (up to 80% of human proteins), and C-terminal groups by amidation. Fatty acyl chains and other lipid groups are added to proteins to anchor them to membranes (Section 8.3). The addition and removal of phosphoryl groups is a powerful mechanism for allosterically regulating cellular signaling components (Section 10.2) and metabolic pathways (Section 19.2).

Modification of a protein by covalently attaching another protein occurs during protein degradation, when ubiquitin is covalently linked to a target protein (Section 12.1). A related protein, called SUMO (for *s*mall *u*biquitin-like *mo*difier), is also covalently attached to a target protein's lysine side chains, but rather than mark the protein for degradation, as ubiquitin does, SUMO is involved in various other processes, including protein transport into the nucleus.

All the post-translational modifications mentioned above are catalyzed by specific enzymes, which act more or less reliably depending on the nature of the modification and the

FIGURE 22.28

Conversion of proinsulin to insulin. The prohormone, with three disulfide bonds, is proteolyzed at two bonds (indicated by arrows) to eliminate the C chain. The mature insulin hormone consists of the disulfide-linked A and B chains.

cellular context. One consequence of post-translational processing therefore is that proteins may exhibit a great deal of variation beyond the sequence of amino acids that is specified by the genetic code.

 CONCEPT CHECK

1. Compare protein folding as catalyzed by heat-shock proteins and by the chaperonin complex.

2. How does the SRP recognize a membrane or secretory protein?

3. What are the functions of the translocon and the signal peptidase?

4. Describe the types of reactions that a polypeptide may undergo following its synthesis.

SUMMARY

22.1 tRNA and the Genetic Code

• The sequence of nucleotides in DNA is related to the sequence of amino acids in a protein by a triplet-based genetic code that must be translated by tRNA adaptors.

• tRNA molecules have similar L-shaped structures with a three-base anticodon at one end and an attachment site for a specific amino acid at the other end.

• Attachment of an amino acid to a tRNA is catalyzed by an aminoacyl–tRNA synthetase in a reaction that requires ATP. Various proofreading mechanisms ensure that the correct amino acid becomes linked to the tRNA.

22.2 Ribosome Structure

• The ribosome, the site of protein synthesis, consists of two subunits containing both rRNA and protein. The ribosome includes a binding site for mRNA and three binding sites (called the A, P, and E sites) for tRNA.

22.3 Translation

• Translation of mRNA requires an initiator tRNA bearing methionine (formyl-methionine in bacteria). Proteins known as initiation factors facilitate the separation of the ribosomal subunits and their reassembly with the initiator tRNA and an mRNA to be translated.

• During the elongation phase of protein synthesis, an elongation factor (EF-Tu in *E. coli*) interacts with aminoacyl–tRNAs and delivers them to the A site of the ribosome. Correct pairing between the mRNA codon and the tRNA anticodon allows the EF-Tu to hydrolyze its bound GTP and dissociate from the ribosome.

• Transpeptidation, or formation of a peptide bond, is catalyzed by rRNA in the large ribosomal subunit. The growing polypeptide chain becomes attached to the tRNA in the A site, which then moves to the P site. This movement is assisted by a GTP-binding protein elongation factor (EF-G in *E. coli*).

• Translation terminates when a release factor recognizes an mRNA stop codon in the A site of the ribosome. Additional factors prepare the ribosome for another round of translation.

22.4 Post-Translational Events

• Chaperone proteins bind newly synthesized polypeptides to facilitate their folding. Large chaperonin complexes form a barrel-shaped structure that encloses a folding protein.

• Proteins to be secreted must pass through a membrane. An RNA–protein complex called a signal recognition particle directs polypeptides bearing an N-terminal signal sequence to a membrane for translocation.

• Additional modifications to newly synthesized proteins include proteolytic processing and the attachment of carbohydrate, lipid, or other groups.

KEY TERMS

translation	ribosome	elongation factor (EF)	post-translational modification
codon	A site	transpeptidation	molecular chaperone
reading frame	P site	translocation	signal recognition particle (SRP)
anticodon	E site	release factor (RF)	signal peptide
isoacceptor tRNA	initiation factor (IF)	ribosome recycling factor (RRF)	translocon
wobble hypothesis	G protein	polysome	glycosylation

BIOINFORMATICS

Brief Bioinformatics Exercises

22.1 Viewing and Analyzing Transfer RNA

22.2 Ribosomes, Protein Processing, and the KEGG Database

PROBLEMS

22.1 tRNA and the Genetic Code

1. How many combinations of four nucleotides are possible with a hypothetical quadruplet code?

2. Synthetic biologists at the Scripps Institute in La Jolla, CA, expanded the genetic repertoire by adding two new bases into living bacterial cells. The two bases are named d5SICS and dNaM, and they pair with one another. How many different amino acids could theoretically be encoded by a synthetic nucleic acid containing six different nucleotides?

3. Cells have a mechanism for tagging and destroying proteins containing a C-terminal poly(Lys) sequence. What is the source of these proteins and why is destroying them helpful for the cell?

4. A portion of DNA from a phage genome is shown.
(a) How many reading frames are possible? Specify the amino acid sequences for all the possible reading frames.
(b) Which frames represent open reading frames?

5′-CGATGAGCCTTTCAGCACCGC -
3′-GCTACTCGGAAAGTCGTGGCG-

TTAGTGAGGTTGCGCGCCACG
AATCACTCCAACGCGCGGTGC

5. Explain why a degenerate genetic code helps protect an organism from the effects of mutations.

6. Marshall Nirenberg and his colleagues deciphered the genetic code in the early 1960s. Their experimental strategy involved constructing RNA templates by using various ribonucleotides and polynucleotide phosphorylase, an enzyme that links available nucleotides together in random order. What protein sequence was obtained when the following templates were added to a cell-free translation system?
(a) poly(U); **(b)** poly(C); **(c)** poly(A).

7. The experimental strategy described in Problem 6 was used to synthesize poly(UA). Because the polymerization of nucleotides by polynucleotide phosphorylase is random, all possible codons containing U and A could occur in the RNA template.
(a) What amino acids would be incorporated into a polypeptide synthesized by a cell-free translation system using this template?
(b) What amino acids would be incorporated into a protein when poly(UC) is used as a template?
(c) Repeat the exercise for poly(UG).
(d) How do these results show that the genetic code is redundant?

8. The elucidation of the genetic code was completed using H. Gobind Khorana's method of synthesizing polynucleotides with precise rather than random sequences.
(a) What polypeptides are synthesized from a nonrandom poly-(UAUC) template?
(b) Explain why a single RNA template can yield more than one polypeptide.

9. What polypeptide is synthesized from a nonrandom poly(AUAG) template? Compare your results with your solution to Problem 8.

10. Organisms differ in their codon usage. For example, in yeast, only 25 of the 61 amino acid codons are used with high frequency. The cells contain high concentrations of the tRNAs that pair best with those codons, that is, form standard Watson–Crick base pairs. Explain how a point mutation in a gene, which does not change the identity of the encoded amino acid, could decrease the rate of synthesis of the protein corresponding to that gene.

11. (a) Would you expect that the highly expressed genes in a yeast cell would have sequences corresponding to the cell's set of 25 preferred codons (see Problem 10)? Would you expect this to be the case for genes that are expressed only occasionally?
(b) The genomes of many bacterial species appear to contain genes acquired from other species, including mammals. Even when a gene's function cannot be identified, the gene's nonbacterial origin can be recognized. Explain.

12. Name the amino acids that are attached to the 3′ end of the *E. coli* tRNA molecules with the anticodon sequences **(a)** GUG, **(b)** GUU, or **(c)** CGU.

13. The 5′ nucleotide of a tRNA anticodon is often a nonstandard nucleotide such as a methylated guanosine. Why doesn't this interfere with the ribosome's ability to read the genetic code?

14. Draw the "wobble" base pair that forms between inosine and adenosine.

Inosine

15. A new tRNA discovered in *E. coli* contains a uridine modified to form uridine-5′-oxyacetic acid (cmo^5U). The modified uridine can base pair with G, A, and U. What mRNA codons are recognized by $tRNA^{Leu}_{cmo^5 UAG}$?

16. Some RNA transcripts are substrates for an adenosine deaminase. This "editing enzyme" converts adenosine residues to inosine residues, which can base pair with guanosine residues. Explain how the action of the deaminase could potentially increase the number of gene products obtained from a given gene.

17. In some types of cells, an aminoacyl–tRNA synthetase appears to play a secondary role in promoting certain events in RNA splicing (see Section 21.3).
(a) What structural feature of the synthetase is most likely involved in both the aminoacylation and splicing reactions?
(b) The aminoacyl–tRNA synthetases are among the oldest proteins. Why is this relevant to the presence of multiple activities in a single synthetase molecule?

18. Why doesn't GlyRS need a proofreading domain?

19. IleRS uses a double-sieve mechanism to accurately produce Ile–$tRNA^{Ile}$ and prevent the synthesis of Val–$tRNA^{Ile}$. Which other pairs of amino acids differ in structure by a single carbon and might have AARSs that use a similar double-sieve proofreading mechanism?

20. An examination of AlaRS (the enzyme that attaches alanine to $tRNA^{Ala}$) suggests that the enzyme's aminoacylation active site cannot discriminate between Ala, Gly, and Ser.
(a) List all the products of the AlaRS reaction.
(b) A double-sieve mechanism like the one in IleRS cannot entirely solve the problem of misacylated $tRNA^{Ala}$. Explain.
(c) Many organisms express a protein called AlaXp, which is a soluble analog of the AlaRS editing domain that is able to hydrolyze Ser–$tRNA^{Ala}$ but not Gly–$tRNA^{Ala}$. What is the purpose of AlaXp?

21. A tRNA molecule cannot be aminoacylated unless it bears a 3′ CCA sequence. Many tRNA precursors are synthesized without this sequence, so a CCA-adding enzyme must append the three nucleotides to the 3′ end of the immature tRNA molecule.
(a) The CCA-adding enzyme does not require a polynucleotide template. What does this imply about the mechanism for adding the CCA sequence?

(b) What can you conclude about the substrate specificity of the CCA-adding enzyme?

(c) Most CCA-adding enzymes consist of a single polymerase domain, but in one species of bacteria, the enzyme has two polymerase domains. Explain how this CCA-adding enzyme operates.

22. Predict the effect on a protein's structure and function for all possible nucleotide substitutions at the first position of a Lys codon in the gene encoding the protein.

23. Protein engineers who study the effects of nonstandard amino acids on protein structure and function can use a cell-based system for synthesizing proteins containing nonstandard amino acids. In theory, a polypeptide containing the amino acid norleucine could be produced by cells growing in media containing high concentrations of norleucine and lacking norleucine's standard counterpart, leucine.

$$
\begin{array}{c}
COO^- \\
| \\
H - C - CH_2 - CH_2 - CH_2 - CH_3 \\
| \\
NH_3^+
\end{array}
$$
Norleucine

Experimental results have shown that peptides containing norleucine in place of leucine were not produced unless the cells contained a mutant LeuRS. Explain these results.

22.2 Ribosome Structure

24. The sequences for all the ribosomal proteins in *E. coli* have been elucidated and have been found to contain large amounts of lysine and arginine residues. Why is this finding not surprising? What kinds of interactions are likely to form between the ribosomal proteins and the ribosomal RNA?

25. In eukaryotes, the primary rRNA transcript is a 45S rRNA that includes the sequences of the 18S, 5.8S, and 28S rRNAs separated by short spacers (see Fig. 21.27). What is the advantage of this operon-like arrangement of rRNA genes?

26. The sequence of ribosomal RNA is highly conserved, even though there are many rRNA genes in the genome. How does this observation argue for a functional (not just structural) role for rRNA?

27. Ribosomal inactivating proteins (RIPs) are RNA *N*-glycosidases found in plants. They catalyze the hydrolysis of specific adenine residues in RNA. RIPs are highly toxic but might be useful as antitumor drugs because ribosome synthesis is upregulated in transformed cells. Give a general explanation that describes how RIPs inactivate ribosomes.

28. What is the effect of adding EDTA, a chelating agent specific for divalent cations, to a bacterial cell extract carrying out protein synthesis?

29. In an experiment, >95% of the proteins are extracted from the 50S ribosomal subunit. Only the 23S rRNA and some protein fragments remain. Peptidyl transferase activity is unaffected. What can you conclude from these results? Propose a role for the protein fragments left behind. Why might it have been difficult to remove them?

30. Ribosomal proteins can be separated by two-dimensional electrophoresis (a technique that separates proteins based on both charge and size). One of the proteins in the large ribosomal subunit is sometimes acetylated at its N-terminus. Explain why two-dimensional electrophoresis of ribosomal proteins yields two spots corresponding to this protein.

31. Like their protein counterparts, RNA molecules fold into a variety of structural motifs. Ribosomal proteins contain a so-called RNA-recognition motif. Rho factor and the poly(A) binding protein contain this same motif. Why is this observation not surprising?

32. Propose at least one hypothesis to explain why eukaryotic ribosomes are more complex than prokaryotic ribosomes.

22.3 Translation

33. In this chapter, we have frequently compared eukaryotic translation to *bacterial* translation rather than to *prokaryotic* translation in general. This distinction is intentional, because the protein-synthesizing machinery in Archaea is more similar to the eukaryotic system than to the system in Bacteria. Is this consistent with the evolutionary scheme outlined in Figure 1.15?

34. Complete the following table about eukaryotic replication, transcription, and translation.

Process	Replication	Transcription	Translation
Substrates			
Product			
Template or guide			
Primer			
Enzyme			
Cellular location			

35. The direction of protein synthesis was determined by carrying out an experiment in a cell-free system in which the mRNA consisted of a polymer of A residues with C at the 3′ end, as shown. What polypeptide was synthesized? What would the result be if the mRNA were read in the 3′ → 5′ direction? How does this directionality allow prokaryotes to begin translation before transcription is complete?

5′-AAAA · · · AAAC-3′

36. Mycobacteriophage genes occasionally begin with GUG or even more rarely, UUG (rather than AUG). Which amino acids correspond to these codons?

37. The translation initiation sequence for the ribosomal protein L10 is shown. Draw a diagram that shows how the Shine–Dalgarno sequence aligns with the appropriate sequence on the 16S rRNA. Identify the initiation codon.

5′-CUACCAGGAGCAAAGCUAAUGGCUUUA-3′

38. An *E. coli* phage replicase gene has the mRNA initiation sequence shown below. Indicate the correct start codon and the Shine–Dalgarno sequence upstream of it.

5′-UAACUAGGAUGAAAUGCAUGUCUAAG · · ·

39. Why does eukaryotic translation initiation require proteins that recognize the 5′ cap and the poly(A) tail on the RNA?

40. S1, a protein in the small ribosomal subunit, has a high affinity for single-stranded RNA and has been shown to be important in initiation. What role might S1 play during initiation?

41. What happens when colicin E3, which cleaves on the 5′ side of A1493 in the 16S rRNA, is added to a bacterial culture?

42. Explain why modification or mutagenesis of prokaryotic 16S rRNA at position 1492 or 1493 increases the error rate of translation.

43. The pairing of an aminoacyl–tRNA with EF-Tu offers an opportunity for proofreading during translation. EF-Tu binds all 20 aminoacyl–tRNAs with approximately equal affinity so that it can deliver them and surrender them to the ribosome with the same efficiency. Based on the experimentally determined binding constants for EF-Tu and correctly charged and mischarged aminoacyl–tRNAs (shown here), explain how the tRNA–EF-Tu recognition system could prevent the incorporation of the wrong amino acid in a protein.

Aminoacyl–tRNA	Dissociation constant (nM)
Ala-tRNA^{Ala}	6.2
Gln-tRNA^{Ala}	0.05
Gln-tRNA^{Gln}	4.4
Ala-tRNA^{Gln}	260

44. The affinity of a ribosome for a tRNA in the P site is about 50 times higher than for a tRNA in the A site. Explain why this promotes translational accuracy.

45. The bacterial elongation factors EF-Tu and EF-G are essential for translation *in vivo*, but bacterial ribosomes can translate mRNA into protein *in vitro* in the absence of EF-Tu and EF-G. Why are these factors not required *in vitro*? How does their absence affect the accuracy of translation?

46. Predict the effect on protein synthesis if EF-Tu were able to recognize and form a complex with fMet–tRNA$_f^{Met}$.

47. Identify the peptide encoded by the DNA sequence shown below (the top strand is the coding strand):

GCGATATGTGTGAACATGTGAACTCTTAGAC
CGCTATACACACTTGTACACTTGAGAATCTG

48. The sequence of a portion of the coding strand of a gene is shown, along with the sequence of a mutant form of the gene.

wild-type   AC ATG CGA TCG CAA ATC CTA
mutant   AC ATG CAG ATC GCA AAT CCT

(a) Give the polypeptide sequence that corresponds to the wild-type gene.
(b) How does the mutant gene differ from the wild-type gene, and how does the mutation affect the encoded polypeptide?

49. The gene for CFTR, the protein that is mutated in cystic fibrosis, contains 250,000 nucleotides. After splicing, the mature mRNA contains 6129 nucleotides. How many nucleotides are required to encode the 1480 residues of the protein? What is the function of the "extra" mRNA nucleotides?

50. Calculate the approximate energetic cost (in kJ) for the ribosomal synthesis of one mole of a 25-residue polypeptide. Why is the actual energetic cost *in vivo* probably higher than this value?

51. In prokaryotes, translation can begin even before an mRNA transcript has been completely synthesized. Why is co-transcriptional translation not possible in eukaryotes?

52. All cells contain an enzyme that hydrolyzes peptidyl–tRNA molecules that are not bound to a ribosome. Cells that are deficient in peptidyl–tRNA hydrolase grow very slowly. What is the function of this enzyme, which cleaves the peptidyl group from the tRNA? What does this tell you about the ability of ribosomes to carry out protein synthesis?

53. In an experimental system, the rate of the peptidyl transferase reaction varies with the identity of the amino acid residue attached to the tRNA in the P site, as shown in the table.

Peptidyl group	Rate of peptidyl transfer (s^{-1})
Ala	57
Asp	8
Lys	100
Phe	16
Pro	0.14

(a) Explain why transpeptidation involving Pro is much slower than for other amino acids.
(b) What can you conclude about the electrostatic environment of the peptidyl transferase active site?
(c) For the nonpolar amino acids, what factor seems to facilitate transpeptidation?

54. The peptidyl transferase center of the ribosome actually catalyzes two reactions involving the ester bond linking a polypeptide to the tRNA: aminolysis and hydrolysis. Explain.

55. The rate of the peptidyl transferase reaction increases as the pH increases from 6 to 8. Use your knowledge of the peptidyl transferase reaction mechanism to explain this result.

56. The events of transpeptidation resemble peptide bond hydrolysis in reverse (see Fig. 6.10).
(a) Draw the "tetrahedral intermediate" of the transpeptidation reaction.
(b) Researchers hypothesized that residue A2451 protonated at position N1 stabilizes the reaction intermediate. Draw this protonated adenine and explain how it might stabilize the tetrahedral intermediate. Would this catalytic mechanism be enhanced as the pH increases? [*Note*: This hypothesis turned out to be incorrect, as described in the text.]

57. A "nonsense suppressor" mutation results from a mutation in a tRNA anticodon sequence so that the tRNA can pair with a stop codon (also called a nonsense codon).
(a) What would be the effect of a nonsense suppressor mutation on protein synthesis in the cell?
(b) Would all of the cell's proteins be affected by the mutation?
(c) Could the aminoacyl–tRNA synthetase that aminoacylates the nonmutated tRNA play a role in minimizing the effects of a nonsense suppressor mutation?

58. Many naturally occurring antibiotics target bacterial ribosomes in order to inhibit protein synthesis. In theory, any of the events of translation—including ribosome assembly, tRNA binding, codon–anticodon pairing, and peptide bond formation—are potential targets for antibiotics. The synthetic antibiotics known as oxazolidinones inhibit protein synthesis. Experiments show that these compounds do not directly inhibit peptide bond formation. They do inhibit translation initiation, but only at very high concentrations. This suggests that the oxazolidinones act by interfering with another aspect of translation. To test different possibilities, researchers used an *E. coli lacZ* expression system (*lacZ* codes for the enzyme β-galactosidase; see Section 21.1) and measured β-galactosidase enzymatic activity in the presence and absence of an oxazolidinone.
(a) First, the researchers engineered a stop codon into the N-terminal region of the polypeptide sequence and then measured β-galactosidase activity. What is the expected effect of the stop codon? Why was the stop codon introduced near the N-terminus of the protein rather than the C-terminus?
(b) The researchers observed that the level of β-galactosidase expressed from *lacZ* containing a stop codon was extremely low but increased about eightfold in the presence of the oxazolidinone. What does this suggest about the action of the oxazolidinone?
(c) Next, the researchers used a *lacZ* gene containing either a one-nucleotide insertion or a one-nucleotide deletion. How would these mutations affect the measured β-galactosidase activity?
(d) When the insertion and deletion mutations were tested in the presence of the antibiotic, β-galactosidase activity levels increased 15–25 times relative to levels in the absence of the antibiotic. What does this suggest about the action of the oxazolidinone?
(e) Next, the researchers measured β-galactosidase activity using *lacZ* genes in which the codon for an active-site Glu residue was mutated. Which codons specify Glu (see Table 22.1)? Describe how a single-base substitution in a Glu codon can give rise to codons specific for Ala, Gln, Gly, and Val.

(f) All of the genes with mutations at the Glu codon yielded very low levels of β-galactosidase activity, and the oxazolidinone had no effect. What does this tell about the action of the oxazolidinone?

(g) Based on your understanding of oxazolidinone action, explain why the compound inhibits bacterial growth.

(h) Oxazolidinones bind to a single site on the ribosome that is in the 50S subunit near the peptidyl transferase active site. Is this information consistent with the experimental results described above?

22.4 Post-Translational Events

59. In 1957, Christian Anfinsen carried out a denaturation experiment *in vitro* with ribonuclease, a pancreatic enzyme consisting of a single 124-amino-acid chain cross-linked by four disulfide bonds (see Problem 4.41). Urea (a denaturing agent) and 2-mercaptoethanol (a reducing agent) were added to a solution of purified ribonuclease, resulting in protein unfolding with a concomitant loss of biological activity. When urea and 2-mercaptoethanol were removed, the ribonuclease spontaneously folded back to its native conformation and regained full enzymatic activity. Why could proper protein folding occur in this experiment in the absence of molecular chaperones?

60. In another set of experiments, the lysine side chains on the surface of ribonuclease (see Problem 59) were covalently attached to an eight-residue chain of polyalanine. The presence of these polyalanine chains did not affect the ability of the ribonuclease to fold properly. What do these experiments reveal about the driving force for protein folding?

61. Chaperones are located not just in the cytosol but in the mitochondria as well. What is the role of mitochondrial chaperones?

62. The chaperone Hsp90 interacts with the tyrosine kinase domain of a growth factor receptor that has lost its ligand-binding domain but retains its tyrosine kinase domain (see Box 10.B). The antibiotic geldanamycin inhibits Hsp90 function. What is the effect of adding geldanamycin to cells expressing the abnormal growth factor receptor?

63. Multidomain proteins tend to fold better inside cagelike chaperonin structures (such as GroEL/GroES in *E. coli*) than with cytosolic chaperones. Explain why.

64. In immature red blood cells, globin synthesis is carefully regulated. The genes for α and β globin (see Section 5.1) are on separate chromosomes, and there are two α globin genes for every β globin gene. If too many β chains are produced, they form a functionally useless tetrameric hemoglobin. Excess α chains tend to precipitate and damage red blood cells.

(a) Explain why it is advantageous for the cell to synthesize a slight excess of α chains.

(b) Red blood cells express a protein that appears to stabilize the α chains and prevents their precipitation. Why is the stabilizing protein necessary?

(c) The disease β thalassemia results from a defect in a β globin gene. Heterozygotes (who have one normal and one abnormal gene) may develop a mild anemia, but homozygotes (who have two defective β globin genes) exhibit severe anemia. Explain why an extra copy of an α globin gene in an individual lacking one β globin gene would result in more severe anemia.

(d) Explain why a mutation in an α globin gene would reduce the severity of anemia in β thalassemia.

(e) Immature red blood cells produce a kinase that phosphorylates the ribosomal initiation factor eIF2. The phosphorylated eIF2 is unable to exchange bound GDP for GTP. How does this affect the rate of protein synthesis in the cell?

(f) In the presence of heme, the kinase is inactive. How does this mechanism regulate hemoglobin synthesis?

65. The N-terminal sequence of the secreted protein bovine proalbumin is shown below. Identify the essential features of the signal peptide in this protein.

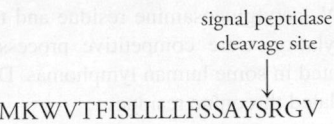

MKWVTFISLLLLLFSSAYSRGV

66. The mammalian signal recognition particle (SRP) consists of one molecule of RNA and six proteins. In addition to interacting with the signal peptide, what might be the role of the RNA in the SRP?

67. One of the six proteins in the mammalian signal recognition particle (SRP) contains a cleft lined with hydrophobic amino acids. Propose a role for this protein in the SRP.

68. Why does the mammalian signal recognition particle (SRP) bind to the nascent polypeptide as it emerges from the ribosome? Why doesn't the SRP wait until translation is complete before escorting the new polypeptide to the ER membrane?

69. A eukaryotic "cell-free" translation system contains all the components required for protein synthesis—ribosomes; tRNAs; aminoacyl–tRNA synthetases; initiation, elongation, and termination factors; amino acids; GTP; and Mg^{2+} ions. An exogenous mRNA added to this mixture can direct protein synthesis *in vitro*. When an mRNA encoding a secretory protein is added to this cell-free system, along with the SRP, the entire protein is synthesized. When microsomes (sealed vesicles derived from ER membranes) are subsequently added, the protein is not translocated into the microsomal lumen and the signal sequence is not removed. What does this observation reveal about the role of SRP in the synthesis of secretory proteins?

70. The elongation factor EF-Tu performs proofreading by monitoring codon–anticodon base pairing during translation. When a match is confirmed, a conformational change occurs and EF-Tu hydrolyzes its bound GTP (see Fig. 22.15). Could the SRP use a similar mechanism to perform a proofreading function?

71. Polyglutamine diseases are neurodegenerative disorders caused by mutations in the DNA that produce triplet CAG repeats. Proteins translated from the mutated genes contain long stretches of glutamine residues that interfere with protein folding. Polyglutamine proteins tend to aggregate and form cytosolic and nuclear inclusion bodies. The result is a loss of neuron function, although the mechanism is unknown. Recent studies show that post-translational modification of the polyglutamine proteins may play a role in the progression of the disease. Interestingly, some post-translational modifications are neurotoxic while others are protective.

(a) Protein kinase B (see Section 10.2) phosphorylates a polyglutamine protein on an essential serine residue, resulting in decreased toxicity. Draw the structure of a phosphorylated serine residue.

(b) Proteins are marked for degradation by the attachment of the protein ubiquitin to a lysine side chain on the condemned protein (see Section 12.1). An isopeptide bond forms between the lysine side chain and the carboxyl terminal group of the ubiquitin. Ubiquitination enhances the degradation of the inclusion body protein. Draw the structure of the linkage between ubiquitin and a polyglutamine protein.

(c) Polyglutamine proteins interact with a histone acetyltransferase (see Section 21.1), sequestering the enzyme in inclusion bodies and hastening its degradation. What are the cellular consequences of this interaction?

72. The protein c-Myc is a leucine zipper protein that regulates gene expression in cell proliferation and differentiation. Its activity was known to be regulated by phosphorylation of a specific threonine residue, but later studies showed that this same threonine could be modified by an *N*-acetylglucosamine residue and that phosphorylation and glycosylation were competitive processes. The specific threonine is mutated in some human lymphomas. Draw the structure of the *O*-glycosylated threonine residue.

73. In the *N*-myristoylation process, myristic acid (14:0) is attached to an N-terminal glycine residue of a protein during translation. Draw the structure of a myristoylated N-terminal glycine residue.

74. In the palmitoylation process, palmitate (16:0) is attached to the side chain of an internal cysteine residue of a protein. Draw the structure of a palmitoylated cysteine residue. What proteins involved in cell signaling pathways include this modification, and what is the role of the palmitate?

SELECTED READINGS

Moore, P. B., The ribosome returned, *J. Biol.* **8,** 8 (2009). [Reviews many aspects of translation, including the accuracy of decoding.]

Neumann, H., Rewiring translation—genetic code expansion and its applications, *FEBS Lett.* **586,** 2057–2064 (2012). [Describes how researchers introduce unusual amino acids into proteins, and what questions can be answered through these techniques.]

Nirenberg, M., Historical review: deciphering the genetic code—a personal account, *Trends Biochem. Sci.* **29,** 46–54 (2004). [Describes the experimental approaches used to elucidate the genetic code.]

Ramakrishnan, V., The ribosome emerges from a black box, *Cell* **159,** 979–984 (2014). [A short review of the ribosome's structure and operation, by a leading ribosome researcher.]

Saibil, H., Chaperone machines for protein folding, unfolding, and disaggregation, *Nat. Rev. Mol. Cell Biol.* **14,** 630–642 (2013). [Reviews the structures and mechanisms of a variety of molecular chaperones.]

Schmeing, T. M. and Ramakrishnan, V., What recent ribosome structures have revealed about the mechanism of translation, *Nature* **461,** 1234–1242 (2009). [Discusses initiation, elongation, release, and recycling, including the proteins that interact with the ribosome at each step.]

Numbers and Greek letters are alphabetized as if they were spelled out.

A-DNA. A conformation of DNA in which the double helix is wider than the standard B-DNA helix and in which base pairs are inclined to the helix axis.

A site. The ribosomal binding site that accommodates an aminoacyl–tRNA.

Abasic site. The deoxyribose residue remaining after the removal of a base from a DNA strand.

ABC transporter. A member of a family of structurally similar transmembrane proteins that use the free energy of ATP to drive conformational changes that move substances across the membrane.

Acid. A substance that can donate a proton.

Acid–base catalysis. A catalytic mechanism in which partial proton transfer from an acid or partial proton abstraction by a base lowers the free energy of a reaction's transition state.

Acid catalysis. A catalytic mechanism in which partial proton transfer from an acid lowers the free energy of a reaction's transition state.

Acid dissociation constant (K_a). The dissociation constant for an acid in water.

Acidic solution. A solution whose pH is less than 7.0 ($[H^+] > 10^{-7}$ M).

Acidosis. A pathological condition in which the pH of the blood drops below its normal value of 7.4.

Actin filament. A 70-Å-diameter cytoskeletal element composed of polymerized actin subunits. Also called a microfilament.

Action potential. The momentary reversal of membrane potential that occurs during transmission of a nerve impulse.

Activation energy (free energy of activation, $\Delta G^{\ddagger}$). The free energy of the transition state minus the free energies of the reactants in a chemical reaction.

Activator. A protein that binds at or near a gene so as to promote its transcription.

Active site. The region of an enzyme in which catalysis takes place.

Active transport. The transmembrane movement of a substance from low to high concentrations by a protein that couples this endergonic transport to an exergonic process such as ATP hydrolysis. See also secondary active transport.

Acyl group. A portion of a molecule with the formula —COR, where R is an alkyl group.

Adipose tissue. Tissue consisting of cells that are specialized for the storage of triacylglycerols. See also brown adipose tissue.

Aerobic. Occurring in or requiring oxygen.

Affinity chromatography. A procedure in which a molecule is isolated by its ability to bind specifically to a second immobilized molecule.

Agonist. A substance that binds to a receptor so as to evoke a cellular response.

Aldose. A sugar whose carbonyl group is an aldehyde.

Alkalosis. A pathological condition in which the pH of the blood rises above its normal value of 7.4.

Allele. An alternate form of a gene; a diploid organism may contain two alleles for each gene.

Allosteric protein. A protein in which the binding of ligand at one site affects the binding of other ligands at other sites. Some enzymes are allosteric proteins. See also cooperative binding.

Allosteric regulation. Binding of an activator or inhibitor to one subunit of a multisubunit enzyme, which increases or decreases the catalytic activity of all the subunits. See also positive effector and negative effector.

α-amino acid. See amino acid.

α anomer. A sugar in which the OH substituent of the anomeric carbon is on the opposite side of the ring from the CH_2OH group of the chiral center that designates the D or L configuration.

α carbon. See Cα.

α helix. A regular secondary structure of polypeptides, with 3.6 residues per right-handed turn and hydrogen bonds between each backbone C=O group and the backbone N—H group that is four residues further.

Amino acid (α-amino acid). A compound consisting of a carbon atom to which are attached a primary amino group, a carboxylate group, a side chain (R group), and an H atom.

Amphiphilic (amphipathic). Having both polar and nonpolar regions and therefore being both hydrophilic and hydrophobic.

Amyloid deposit. An accumulation of certain types of insoluble protein aggregates in tissues (e.g., in the brain in Alzheimer's disease).

Anabolism. The reactions by which biomolecules are synthesized from simpler components.

Anaerobic. Occurring independently of oxygen.

Anaplerotic reaction. A reaction that replenishes the intermediates of a metabolic pathway.

Anemia. A condition caused by the insufficient production of or loss of red blood cells.

Anneal. To allow base pairing between complementary single polynucleotide strands so that double-stranded segments form.

Anomers. Sugars that differ only in the configuration around the carbonyl carbon that becomes chiral when the sugar cyclizes.

Antagonist. A substance that binds to a receptor but does not elicit a cellular response.

Antenna pigment. A molecule that transfers its absorbed energy to other pigment molecules and eventually to a photosynthetic reaction center.

Anticodon. The sequence of three nucleotides in a tRNA that recognizes an mRNA codon through complementary base pairing.

Antiparallel. Running in opposite directions.

Antiparallel β sheet. See β sheet.

Antiport. Transport that involves the simultaneous transmembrane movement of two molecules in opposite directions.

Antisense strand. See noncoding strand.

Apoptosis. Programmed cell death that results from extracellular or intracellular signals and involves the activation of enzymes that selectively degrade cellular structures.

Aquaporin. A membrane protein that facilitates the transmembrane movement of water molecules.

Archaea. One of the two major groups of prokaryotes.

Atherosclerosis. A disease characterized by the formation of cholesterol-containing fibrous plaques in the walls of blood vessels.

ATPase. An enzyme that catalyzes the hydrolysis of ATP to ADP + P_i.

Autoactivation. A process by which the product of an activation reaction also acts as a catalyst for the same reaction, so that it appears that the compound catalyzes its own activation.

Autophosphorylation. The phosphorylation of a kinase by another molecule of the same kinase.

Axon. The extended portion of a neuron that conducts an action potential from the cell body to a synapse with a target cell.

B-DNA. The standard conformation of double-helical DNA.

Backbone. The atoms that form the repeating linkages between successive residues of a polymeric molecule, exclusive of the side chains.

Bacteria. One of the two major groups of prokaryotes.

Bacteriophage. A virus specific for bacteria. Also known as a phage.

Base. (1) A substance that can accept a proton. (2) A purine or pyrimidine component of a nucleoside, nucleotide, or nucleic acid.

Base catalysis. A catalytic mechanism in which partial proton abstraction by a base lowers the free energy of a reaction's transition state.

Base excision repair. A DNA repair pathway in which a damaged base is removed by a glycosylase so that the resulting abasic site can be repaired.

Base pair. The specific hydrogen-bonded association between nucleic acid bases. The standard base pairs are A:T and G:C. See also bp.

Basic solution. A solution whose pH is greater than 7.0 ($[H^+] < 10^{-7}$ M).

β anomer. A sugar in which the OH substituent of the anomeric carbon is on the same side of the ring as the CH_2OH of the chiral center that designates the D or L configuration.

β barrel. A protein structure consisting of a β sheet rolled into a cylinder.

β oxidation. A series of enzyme-catalyzed reactions in which fatty acids are progressively degraded by the removal of two-carbon units as acetyl-CoA.

β sheet. A regular secondary structure in which extended polypeptide chains form interstrand hydrogen bonds. In parallel β sheets, the polypeptide chains all run in the same direction; in antiparallel β sheets, neighboring chains run in opposite directions.

Bilayer. An ordered, two-layered arrangement of amphiphilic molecules in which polar segments are oriented toward the two solvent-exposed surfaces and the nonpolar segments associate in the center.

Bile acid. A cholesterol derivative that acts as a detergent to solubilize lipids for digestion and absorption.

Bimolecular reaction. A reaction involving two molecules, which may be identical or different.

Binding change mechanism. The mechanism whereby the subunits of ATP synthase adopt three successive conformations to convert ADP + P_i to ATP as driven by the dissipation of the transmembrane proton gradient.

Biofilm. A complex of bacterial cells and a protective extracellular matrix containing polysaccharides.

Bioinformatics. The use of computers in collecting, storing, accessing, and analyzing biological data, such as molecular sequences and structures.

Bisubstrate reaction. An enzyme-catalyzed reaction involving two substrates.

Blunt ends. The fully base-paired ends of a DNA fragment that are generated by a restriction endonuclease that cuts both strands at the same point.

Bohr effect. The decrease in O_2 binding affinity of hemoglobin in response to a decrease in pH.

bp. Base pair, the unit of length used for DNA molecules.

Brown adipose tissue. A type of adipose tissue in which fatty acid oxidation is uncoupled from ATP production so that the free energy of the fatty acids is released as heat.

Buffer. A solution of a weak acid and its conjugate base, which resists changes in pH upon the addition of acid or base.

C-terminus. The end of a polypeptide that has a free carboxylate group.

Cachexia. The severe loss of body mass that occurs in cancer and some other chronic diseases.

Cα. The alpha carbon, the carbon of an amino acid whose substituents are an amino group, a carboxylate group, an H atom, and a variable R group.

Calvin cycle. The sequence of photosynthetic reactions in which ribulose-5-phosphate is carboxylated, converted to three-carbon carbohydrate precursors, and regenerated.

cAMP. Cyclic AMP, an intracellular second messenger.

Cap. A 7-methylguanosine residue that is post-transcriptionally added to the 5′ end of a eukaryotic mRNA.

Carbanion. A compound that bears a negative charge on a carbon atom.

Carbohydrate. A compound with the formula $(CH_2O)_n$, where $n \geq 3$. Also called a saccharide.

Carbon fixation. The incorporation of CO_2 into biologically useful organic molecules.

Carcinogen. An agent that causes a mutation in DNA that leads to cancer.

Carcinogenesis. The process of developing cancer.

Catabolism. The degradative metabolic reactions in which nutrients and cell constituents are broken down for energy and raw materials.

Catalyst. A substance that promotes a chemical reaction without undergoing permanent change. A catalyst increases the rate at which a reaction approaches equilibrium but does not affect the free energy change of the reaction.

Catalytic constant (k_{cat}). The ratio of the maximal velocity (V_{max}) of an enzyme-catalyzed reaction to the enzyme concentration. Also called a turnover number.

Catalytic perfection. A state achieved by an enzyme that operates at the diffusion-controlled limit.

Catalytic triad. The hydrogen-bonded Ser, His, and Asp residues that participate in catalysis in serine proteases.

cDNA. See complementary DNA.

Cellular respiration. The metabolic phenomenon whereby organic molecules are oxidized, with the electrons eventually transferred to molecular oxygen.

Central dogma of molecular biology. The idea that genetic information in the form of DNA is rewritten, or transcribed, to RNA, which then is translated to direct protein synthesis. Information passes from DNA to RNA to protein.

Centromere. The region of a replicated eukaryotic chromosome where the two identical DNA molecules are attached and where the spindle fibers attach during cell division.

C4 pathway. A photosynthetic process used in some plants to concentrate CO_2 by incorporating it into oxaloacetate (a C_4 compound).

Channeling. The transfer of an intermediate product from one enzyme active site to another in such a way that the intermediate remains in contact with the protein.

Chaperone. See molecular chaperone.

Chemical labeling. A technique for identifying functional groups in a macromolecule by treating the molecule with a reagent that reacts with those groups.

Chemiosmotic theory. The postulate that the free energy of electron transport is conserved in the formation of a transmembrane proton gradient that can be subsequently used to drive ATP synthesis.

Chemoautotroph. An organism that obtains its building materials and free energy from inorganic compounds.

Chirality. The asymmetry or "handedness" of a molecule such that it cannot be superimposed on its mirror image.

Chloroplast. The plant organelle in which photosynthesis takes place.

Chromatin. The complex of DNA and protein that comprises the eukaryotic chromosomes.

Chromatography. A technique for separating the components of a mixture of molecules based on their partition between a mobile solvent phase and a porous matrix (stationary phase), often performed in a column.

Chromosome. The complex of protein and a single DNA molecule that comprises some or all of an organism's genome.

Citric acid cycle. A set of eight enzymatic reactions, arranged in a cycle, in which energy in the form of ATP, NADH, and QH_2 is recovered from the oxidation of the acetyl group of acetyl-CoA to CO_2.

Clinical trial. A three-phase series of tests of a drug's safety and effectiveness in human subjects.

Clone. An organism or collection of identical cells derived from a single parental cell.

Coagulation. The process of forming a blood clot.

Coding strand. The DNA strand that has the same sequence (except for the replacement of U with T) as the transcribed RNA; it is the nontemplate strand. Also called the sense strand.

Codon. The sequence of three nucleotides in DNA or RNA that specifies a single amino acid.

Coenzyme. A small organic molecule that is required for the catalytic activity of an enzyme. A coenzyme may be tightly associated with the enzyme as a prosthetic group.

Cofactor. A small organic molecule (coenzyme) or metal ion that is required for the catalytic activity of an enzyme.

Coiled coil. An arrangement of polypeptide chains in which two α helices wind around each other.

Competitive inhibition. A form of enzyme inhibition in which a substance competes with the substrate for binding to the enzyme active site and thereby appears to increase K_M.

Complement. (1) A molecule that pairs in a reciprocal fashion with another. (2) A set of circulating proteins that sequentially activate each other and lead to the formation of a pore in a microbial cell membrane.

Complementary DNA (cDNA). A segment of DNA synthesized from an RNA template.

Condensation reaction. The formation of a covalent bond between two molecules, during which the elements of water are lost.

Conformation. The three-dimensional shape of a molecule that it attained through rotation of its bonds.

Conjugate base. The compound that forms when an acid donates a proton.

Consensus sequence. A DNA or RNA sequence showing the nucleotides most commonly found at each position.

Conservative substitution. A change of an amino acid residue in a protein to one with similar properties (e.g., Leu to Ile or Asp to Glu).

Constitutive. Being expressed at a continuous, steady rate rather than induced.

Convergent evolution. The independent development of similar characteristics in unrelated species.

Cooperative binding. A situation in which the binding of a ligand at one site on a macromolecule affects the affinity of other sites for the same ligand. See also allosteric protein.

Cori cycle. A metabolic pathway in which lactate produced by glycolysis in the muscles is transported via the bloodstream to the liver, where it is used for gluconeogenesis. The resulting glucose returns to the muscles.

Covalent catalysis. A catalytic mechanism in which the transient formation of a covalent bond between the catalyst and a reactant lowers the free energy of a reaction's transition state.

CpG island. A cluster of CG sequences that often marks the beginning of a gene in a mammalian genome.

CRISPR. Clustered regularly interspersed short palindromic repeats, short DNA segments involved in bacterial defense against bacteriophages.

CRISPR-Cas9 system. A gene-editing tool that uses the bacterial endonuclease Cas9, which is directed to cut a specific DNA segment that is complementary to a CRISPR-like guide RNA. The CRISPR-Cas9 system can be designed to inactivate a target gene or to replace it with an altered version of the gene.

Cristae. The invaginations of the inner mitochondrial membrane.

Cross-talk. The interactions of different signal transduction pathways through activation of the same signaling components.

Cryoelectron microscopy. A variation of electron crystallography in which electron diffraction data from a molecular structure are collected at very low temperatures.

Cyclic electron flow. The light-driven circulation of electrons between Photosystem I and cytochrome b_6f, which leads to the production of ATP but not NADPH.

Cytochrome. A protein that carries electrons via a prosthetic Fe-containing heme group.

Cytokinesis. The splitting of the cell into two following mitosis.

Cytoskeleton. The network of intracellular fibers that gives a cell its shape and structural rigidity.

D sugar. A monosaccharide isomer in which the asymmetric carbon farthest from the carbonyl group has the same spatial arrangement as the chiral carbon of D-glyceraldehyde.

Dark reactions. The photosynthetic reactions in which NADPH and ATP produced by the light reactions are used to incorporate CO_2 into carbohydrates.

Deamination. The hydrolytic or oxidative removal of an amino group.

ΔG. See free energy.

$\Delta G^{\ddagger}$. See activation energy.

$\Delta G^{\circ\prime}$. See standard free energy change.

$\Delta G_{reaction}$. The difference in free energy between the reactants and products of a chemical reaction; $\Delta G_{reaction} = \Delta G_{products} - \Delta G_{reactants}$.

$\Delta\psi$. See membrane potential.

Denaturation. The loss of ordered structure in a polymer, such as the disruption of native conformation in an unfolded polypeptide or the unstacking of bases and separation of strands in a nucleic acid.

Denitrification. The conversion of nitrate (NO_3^-) to nitrogen (N_2).

Deoxyhemoglobin. Hemoglobin that does not contain bound oxygen or is not in the oxygen-binding conformation.

Deoxynucleotide. A nucleotide in which the pentose is 2′-deoxyribose.

Deoxyribonucleic acid. See DNA.

Desensitization. A cell's adaptation to long-term stimulation through a reduced response to the stimulus.

Diabetes mellitus. A disease caused by a deficiency of insulin or the inability to respond to insulin, and characterized by elevated levels of glucose in the blood.

Diazotroph. A bacterium that carries out nitrogen fixation, the conversion of N_2 to NH_3.

Dielectric constant. A measure of the ability of a substance to interfere with electrostatic interactions; a solvent with a high dielectric constant is able to dissolve salts by shielding the attractive electrostatic forces that would otherwise bring the ions together.

Diffraction pattern. The record of the radiation scattered from an object, for example, in X-ray crystallography.

Diffusion-controlled limit. The theoretical maximum rate of an enzymatic reaction in solution, about 10^8 to 10^9 $M^{-1} \cdot s^{-1}$.

Dimer. An assembly consisting of two monomeric units.

Diploid. Having two equivalent sets of chromosomes.

Dipole–dipole interaction. A type of van der Waals interaction between two strongly polar groups.

Disaccharide. A carbohydrate consisting of two monosaccharides.

Discontinuous synthesis. A mechanism whereby the lagging strand of DNA is synthesized as a series of fragments that are later joined.

Dissociation constant (K_d). The ratio of the products of the concentrations of the dissociated species to those of their parent compounds at equilibrium.

Disulfide bond. A covalent —S—S— linkage, often between two Cys residues in a protein.

Divergent evolution. The accumulation of changes in species that share an ancestor.

DNA (Deoxyribonucleic acid). A polymer of deoxynucleotides whose sequence of bases encodes genetic information in all living cells.

DNA chip. See microarray.

DNA fingerprinting. A technique for distinguishing individuals on the basis of DNA polymorphisms, such as the number of short tandem repeats.

DNA ligase. An enzyme that catalyzes the formation of a phosphodiester bond to join two DNA segments.

DNA polymerase. See polymerase.

Domain. A stretch of polypeptide residues that fold into a globular unit with a hydrophobic core.

$\mathcal{E}$. See reduction potential.

$\mathcal{E}°'$. See standard reduction potential.

E site. The ribosomal binding site that accommodates a deacylated tRNA before it dissociates from the ribosome.

Edman degradation. A procedure for the stepwise removal and identification of the N-terminal residues of a polypeptide.

EF. See elongation factor.

Ehlers–Danlos syndrome. A genetic disease characterized by elastic skin and joint hyperextensibility, caused by mutations in genes for collagen or collagen-processing proteins.

EI complex. The noncovalent complex that forms between an enzyme and a reversible inhibitor.

Eicosanoids. Compounds derived from the C_{20} fatty acid arachidonic acid, which act in or near the cells that produce them and mediate pain, fever, and other physiological responses.

Electron crystallography. A technique for determining molecular structure by analyzing the pattern of diffraction of a beam from an electron microscope. See also cryoelectron microscopy.

Electron tomography. A technique for reconstructing three-dimensional structures by analyzing electron micrographs of consecutive tissue slices.

Electron transport chain. The series of small molecules and protein prosthetic groups that transfer electrons from reduced cofactors such as NADH to O_2 during cellular respiration.

Electronegativity. A measure of an atom's affinity for electrons.

Electrophile. A compound containing an electron-poor center. An electrophile (electron-lover) reacts readily with a nucleophile (nucleus-lover).

Electrophoresis. A procedure in which macromolecules are separated on the basis of charge or size by their differential migration through a gel-like matrix under the influence of an applied electric field. In polyacrylamide gel electrophoresis (PAGE), the matrix is cross-linked polyacrylamide. In SDS-PAGE, the detergent sodium dodecyl sulfate is used to denature proteins.

Electrostatic catalysis. A catalytic mechanism in which sequestering the reacting groups away from the aqueous solvent lowers the free energy of a reaction's transition state.

Elongation factor (EF). A protein that interacts with tRNA and/or the ribosome during polypeptide synthesis.

Enantiomers. Stereoisomers that are nonsuperimposable mirror images of one another.

Endergonic reaction. A reaction that has an overall positive free energy change (a nonspontaneous process).

Endocytosis. The inward folding and budding of the plasma membrane to form a new intracellular vesicle. See also receptor-mediated endocytosis and pinocytosis.

Endonuclease. An enzyme that catalyzes the hydrolysis of the phosphodiester bonds between two nucleotide residues within a polynucleotide strand.

Endopeptidase. An enzyme that catalyzes the hydrolysis of a peptide bond within a polypeptide chain.

Endothermic reaction. A reaction that absorbs heat from the surroundings so that its change in enthalpy (ΔH) is greater than zero.

Enhancer. A eukaryotic DNA sequence located some distance from the transcription start site, where an activator of transcription may bind.

Enthalpy (H). A thermodynamic quantity that is taken to be equivalent to the heat content of a biochemical system.

Entropy (S). A measure of the degree of randomness or disorder of a system.

Enzyme. A biological catalyst. Most enzymes are proteins; a few are RNA.

Epigenetics. The inheritance of patterns of gene expression mediated by chromosomal modifications that do not alter the DNA sequence.

Epimers. Sugars that differ only by the configuration at one C atom (excluding the anomeric carbon).

Equilibrium constant (K_{eq}). The ratio, at equilibrium, of the product of the concentrations of reaction products to that of the reactants.

ES complex. The noncovalent complex that forms between an enzyme and its substrate in the first step of an enzyme-catalyzed reaction.

Essential compound. An amino acid, fatty acid, or other compound that an animal cannot synthesize and must therefore obtain in its diet.

Euchromatin. The transcriptionally active, relatively uncondensed chromatin in a eukaryotic cell.

Eukarya. See eukaryote.

Eukaryote. An organism consisting of a cell (or cells) whose genetic material is contained in a membrane-bounded nucleus.

Evolution. Change over time, often driven by the process of natural selection.

Exciton transfer. A mode of decay of an energetically excited molecule, in which electronic energy is transferred to a nearby unexcited molecule.

Exergonic reaction. A reaction that has an overall negative free energy change (a spontaneous process).

Exocytosis. The fusion of an intracellular vesicle with the plasma membrane in order to release the contents of the vesicle outside the cell.

Exon. A portion of a gene that appears in both the primary and mature mRNA transcripts.

Exonuclease. An enzyme that catalyzes the hydrolytic excision of a nucleotide residue from the end of a polynucleotide strand.

Exopeptidase. An enzyme that catalyzes the hydrolytic excision of an amino acid residue from one end of a polypeptide chain.

Exosome. A vesicle released from a cell, possibly as a form of intercellular communication. Also known as a microvesicle.

Exothermic reaction. A reaction that releases heat to the surrounding so that its change in enthalpy (ΔH) is less than zero.

Extracellular matrix. The extracellular proteins and polysaccharides that fill the space between cells and form connective tissue in animals.

Extrinsic protein. See peripheral membrane protein.

$\mathcal{F}$. See Faraday constant.

F-actin. The polymerized form of the protein actin. See also G-actin.

Factory model of replication. A model for DNA replication in which DNA polymerase and associated proteins remain stationary while the DNA template is spooled through them.

Faraday constant ($\mathcal{F}$). The charge of one mole of electrons, equal to 96,485 coulombs·mol^{-1} or 96,485 J·V^{-1}·mol^{-1}.

Fatty acid. A carboxylic acid with a long-chain hydrocarbon side group.

Feed-forward activation. The activation of a later step in a reaction sequence by the product of an earlier step.

Feedback inhibitor. A substance that inhibits the activity of an enzyme that catalyzes an early step of the substance's synthesis.

Fermentation. An anaerobic catabolic process.

Fibrous protein. A protein characterized by a stiff, elongated conformation, that tends to form fibers.

First-order reaction. A reaction whose rate is proportional to the concentration of a single reactant.

Fischer projection. A graphical convention for specifying molecular configuration in which horizontal lines represent bonds that extend above the plane of the paper and vertical bonds extend below the plane of the paper.

5′ end. The terminus of a polynucleotide whose C5′ is not esterified to another nucleotide residue.

Flip-flop. See transverse diffusion.

Flippase. See translocase.

Fluid mosaic model. A model of biological membranes in which integral membrane proteins float and diffuse laterally in a fluid lipid layer.

Fluorescence. A mode of decay of an excited molecule, in which electronic energy is emitted in the form of a photon.

Flux. The rate of flow of metabolites through a metabolic pathway.

454 sequencing. See pyrosequencing.

Fractional saturation (Y). The fraction of a protein's ligand-binding sites that are occupied by ligand.

Free energy (G). A thermodynamic quantity whose change indicates the spontaneity of a process. For spontaneous processes, $\Delta G < 0$, whereas for a process at equilibrium, $\Delta G = 0$.

Free energy of activation. See activation energy ($\Delta G^{\ddagger}$).

Free radical. A molecule with an unpaired electron.

Futile cycle. Two opposing metabolic reactions that function together to provide a control point for regulating metabolic flux.

G. See free energy.

G-actin. The monomeric form of the protein actin. See also F-actin.

G protein. A guanine nucleotide–binding and –hydrolyzing protein, involved in a process such as signal transduction or protein synthesis, that is inactive when it binds GDP and active when it binds GTP.

G protein–coupled receptor (GPCR). A transmembrane protein that binds an extracellular ligand and transmits the signal to the cell interior by interacting with an intracellular G protein.

Gas constant (R). A thermodynamic constant equivalent to $8.3145\ J \cdot K^{-1} \cdot mol^{-1}$.

Gated channel. A transmembrane channel that opens and closes in response to a signal such as changing voltage, ligand binding, or mechanical stress.

Gel filtration chromatography. See size-exclusion chromatography.

Gene. A unique sequence of nucleotides that encodes a polypeptide or RNA; it may include nontranscribed and nontranslated sequences that have regulatory functions.

Gene expression. The transformation by transcription and translation of the information contained in a gene to a functional RNA or protein product.

Gene therapy. The manipulation of genetic information in the cells of an individual in order to produce a therapeutic effect.

General transcription factor. One of a set of eukaryotic proteins that are typically required for the synthesis of mRNAs. See also transcription factor.

Genetic code. The correspondence between the sequence of nucleotides in a nucleic acid and the sequence of amino acids in a polypeptide; a series of three nucleotides (a codon) specifies an amino acid.

Genome. The complete set of genetic instructions in an organism.

Genome map. A reconstruction of an organism's genome, based on DNA sequences and physical DNA features.

Genome-wide association study (GWAS). An attempt to correlate genetic variations with a trait such as a particular disease.

Genomics. The study of the size, organization, and gene content of organisms' genomes.

Globin. The polypeptide component of myoglobin and hemoglobin.

Globular protein. A water-soluble protein characterized by a compact, highly folded structure.

Glucogenic amino acid. An amino acid whose degradation yields a gluconeogenic precursor. See also ketogenic amino acid.

Gluconeogenesis. The synthesis of glucose from noncarbohydrate precursors.

Glucose–alanine cycle. A metabolic pathway in which pyruvate produced by glycolysis in the muscles is converted to alanine and transported to the liver, where it is converted back to pyruvate for gluconeogenesis. The resulting glucose returns to the muscles.

Glycan. See polysaccharide.

Glycerophospholipid. An amphipathic lipid in which two hydrocarbon chains and a polar phosphate derivative are attached to a glycerol backbone.

Glycogen storage disease. An inherited defect in an enzyme or transporter that affects the formation, structure, or degradation of glycogen.

Glycogenolysis. The enzymatic degradation of glycogen to glucose-1-phosphate.

Glycolipid. A lipid to which carbohydrate is covalently attached.

Glycolysis. The 10-reaction pathway by which glucose is broken down to 2 pyruvate with the concomitant production of 2 ATP and the reduction of 2 NAD^+ to 2 NADH.

Glycomics. The systematic study of the structures and functions of carbohydrates, including large glycans and the small oligosaccharides of glycoproteins.

Glycoprotein. A protein to which carbohydrate is covalently attached.

Glycosaminoglycan. An unbranched polysaccharide consisting of alternating residues of an amino sugar and a sugar acid.

Glycosidase. An enzyme that catalyzes the hydrolysis of glycosidic bonds.

Glycoside. A molecule containing a saccharide linked to another molecule by a glycosidic bond to the anomeric carbon.

Glycosidic bond. The covalent linkage between two monosaccharide units in a polysaccharide, or the linkage between the anomeric carbon of a saccharide and an alcohol or amine.

Glycosylation. The attachment of carbohydrate chains to a protein through N- or O-glycosidic linkages.

Glycosyltransferase. An enzyme that catalyzes the addition of a monosaccharide residue to a polysaccharide.

Glyoxylate pathway. A variation of the citric acid cycle in plants that allows acetyl-CoA to be converted quantitatively to gluconeogenic precursors.

Glyoxysome. A membrane-bounded plant organelle in which the reactions of the glyoxylate pathway take place.

Gout. An inflammatory disease, usually caused by impaired uric acid excretion and characterized by painful deposition of uric acid in the joints.

GPCR. See G protein–coupled receptor.

GWAS. See genome-wide association study.

H. See enthalpy.

Half-reaction. The single oxidation or reduction process, involving the reduced and oxidized forms of a substance, that must be combined with another half-reaction to form a complete oxidation–reduction reaction.

Haploid. Having one set of chromosomes.

Haworth projection. A drawing of a sugar ring in which ring bonds that project in front of the plane of the paper are represented by heavy lines and ring bonds that project behind the plane of the paper are represented by light lines.

Helicase. An enzyme that unwinds DNA.

Heme. A protein prosthetic group that binds O_2 (in myoglobin and hemoglobin) or undergoes redox reactions (in cytochromes).

Henderson–Hasselbalch equation. The mathematical expression of the relationship between the pH of a solution of a weak acid and its pK: pH = pK + log ([A^-]/[HA]).

Hetero-. Different. In a heteropolymer, the subunits are not all identical.

Heterochromatin. Highly condensed, non-expressed eukaryotic DNA.

Heterotroph. An organism that obtains its building materials and free energy from organic compounds produced by other organisms.

Hexose. A six-carbon sugar.

High-performance liquid chromatography (HPLC). An automated chromatographic procedure for fractionating molecules using high pressure and computer-controlled solvent delivery.

Highly repetitive DNA. Clusters of short DNA sequences that are repeated side-by-side and are present at millions of copies in the human genome.

Histone code. The correlation between patterns of covalent modification of histone proteins and the transcriptional activity of the associated DNA.

Histones. Highly conserved basic proteins that form a core to which DNA is bound in a nucleosome.

Homeostasis. The maintenance of constant internal conditions.

Homo-. The same. In a homopolymer, all the subunits are identical.

Homologous genes. Genes that are related by evolution from a common ancestor.

Homologous proteins. Proteins that are related by evolution from a common ancestor.

Horizontal gene transfer. The transfer of genetic material between species.

Hormone. A substance that is secreted by one tissue and induces a physiological response in other tissues.

Hormone response element. A DNA sequence to which an intracellular hormone–receptor complex binds so as to regulate gene expression.

HPLC. See high-performance liquid chromatography.

Hydration. The molecular state of being surrounded by and interacting with solvent water molecules; that is, solvated by water.

Hydrogen bond. A partly electrostatic, partly covalent interaction between a donor group such as O—H or N—H and an electronegative acceptor atom such as O or N.

Hydrolysis. The cleavage of a covalent bond accomplished by adding the elements of water; the reverse of a condensation.

Hydronium ion. A proton associated with a water molecule, H_3O^+.

Hydrophilic. Having high enough polarity to readily interact with water molecules. Hydrophilic substances tend to dissolve in water.

Hydrophobic. Having insufficient polarity to readily interact with water molecules. Hydrophobic substances tend to be insoluble in water.

Hydrophobic effect. The tendency of water to minimize its contacts with nonpolar substances, thereby inducing the substances to aggregate.

Hypercholesterolemia. Elevated levels of cholesterol in the blood.

Hyperglycemia. Elevated levels of glucose in the blood.

IF. See initiation factor.

Illumina sequencing. A procedure for determining the sequence of nucleotides in DNA by detecting the presence of a fluorescent marker attached to one of the four nucleotides added to a growing DNA strand.

Imine. A molecule with the formula $>C=NH$.

Imino group. A portion of a molecule with the formula $>C=NH$.

Imprinting. A heritable variation in the level of expression of a gene according to its parental origin.

In vitro. In the laboratory (literally, in glass).

In vivo. In a living organism.

Induced fit. An interaction between a protein and its ligand that induces a conformational change in the protein to enhance the protein's interaction with the ligand.

Inhibition constant (K_I). The dissociation constant for the complex between an enzyme and a reversible inhibitor.

Initiation factor (IF). A protein that interacts with mRNA and/or the ribosome and that is required to initiate translation.

Insulin resistance. The inability of cells to respond to insulin.

Integral membrane protein. A membrane protein that is embedded in the lipid bilayer. Also called an intrinsic protein.

Intermediate. See metabolite.

Intermediate filament. A 100-Å-diameter cytoskeletal element consisting of coiled-coil polypeptide chains.

Intermembrane space. The compartment between the inner and outer mitochondrial membranes, which is equivalent to the cytosol in ionic composition.

Intrinsic protein. See integral membrane protein.

Intrinsically unstructured protein. A protein whose tertiary structure includes highly flexible extended segments that can adopt different conformations.

Intron. A portion of a gene that is transcribed but excised by splicing prior to translation.

Invariant residue. A residue in a protein that is the same in all evolutionarily related proteins.

Ion exchange chromatography. A fractionation procedure in which charged molecules are selectively retained by a matrix bearing oppositely charged groups.

Ion pair. An electrostatic interaction between two ionic groups of opposite charge.

Ionic interaction. An electrostatic interaction between two groups that is stronger than a hydrogen bond but weaker than a covalent bond.

Ionization constant of water (K_w). A quantity that relates the concentrations of H^+ and OH^- in pure water: $K_w = [H^+][OH^-] = 10^{-14}$.

Irregular secondary structure. A segment of a polymer in which each residue has a different backbone conformation; the opposite of regular secondary structure.

Irreversible inhibitor. A molecule that binds to and permanently inactivates an enzyme.

Isoacceptor tRNA. A tRNA that carries the same amino acid as another tRNA but has a different codon.

Isoelectric point (pI). The pH at which a molecule has no net charge.

Isoprenoid. A lipid constructed from five-carbon units with an isoprene skeleton. Also called a terpenoid.

Isozymes. Different proteins that catalyze the same reaction.

K. See dissociation constant.

k. See rate constant.

K_a. See acid dissociation constant.

kb. Kilobase pairs; 1000 base pairs.

k_{cat}. See catalytic constant.

k_{cat}/K_M. The apparent second-order rate constant for an enzyme-catalyzed reaction; it indicates the enzyme's overall catalytic efficiency.

K_d. See dissociation constant.

K_{eq}. See equilibrium constant.

Ketogenesis. The synthesis of ketone bodies from acetyl-CoA.

Ketogenic amino acid. An amino acid whose degradation yields compounds that can be converted to fatty acids or ketone bodies but not to glucose. See also glucogenic amino acid.

Ketone bodies. Compounds (acetoacetate and 3-hydroxybutyrate) that are produced from acetyl-CoA by the liver and used as metabolic fuels in other tissues when glucose is unavailable.

Ketose. A sugar whose carbonyl group is a ketone.

K_I. See inhibition constant.

Kinase. An enzyme that transfers a phosphoryl group between ATP and another molecule.

Kinetics. The study of chemical reaction rates.

K_M. See Michaelis constant.

K_w. See ionization constant of water.

Kwashiorkor. A form of severe malnutrition resulting from inadequate protein intake; marked by abdominal swelling and reddish hair. See also marasmus.

L sugar. A monosaccharide isomer in which the asymmetric carbon farthest from the carbonyl group has the same spatial arrangement as the chiral carbon of L-glyceraldehyde.

Lagging strand. The DNA strand that is synthesized as a series of discontinuous fragments that are later joined.

Lateral diffusion. The movement of a membrane component within one leaflet of a bilayer.

Le Châtelier's principle. The observation that a change in concentration, temperature, volume, or pressure in a system at equilibrium causes the equilibrium to shift in order to counteract the change.

Leading strand. The DNA strand that is synthesized continuously during DNA replication.

Ligand. (1) A small molecule that binds to a larger molecule. (2) A molecule or ion bound to a metal ion.

Ligase. See DNA ligase.

Light-harvesting complex. A pigment-containing protein that collects light energy in order to transfer it to a photosynthetic reaction center.

Light reactions. The photosynthetic reactions in which light energy is absorbed and used to generate NADPH and ATP.

Lineweaver–Burk plot. A rearrangement of the Michaelis–Menten equation that permits the determination of K_M and V_{max} from a linear plot.

Lipid. Any member of a broad class of macromolecules that are largely or wholly hydrophobic and therefore tend to be insoluble in water but soluble in organic solvents.

Lipid bilayer. See bilayer.

Lipid-linked protein. A protein that is anchored to a biological membrane via a covalently attached lipid.

Lipolysis. The degradation of a triacylglycerol so as to release fatty acids.

Lipoprotein. A globular particle, containing lipids and proteins, that transports lipids between tissues via the bloodstream.

Lock-and-key model. An early model of enzyme action, in which the substrate fit the enzyme like a key in a lock.

Locus. The chromosomal location of a gene or other DNA feature.

London dispersion forces. The weak van der Waals interactions between nonpolar groups as a result of fluctuations in their electron distributions that create a temporary separation of charge (polarity).

Low-barrier hydrogen bond. A short, strong hydrogen bond in which the proton is equally shared by the donor and acceptor atoms.

Lysosome. A membrane-bounded organelle in a eukaryotic cell that contains a battery of hydrolytic enzymes and that functions to digest ingested material and to recycle cell components.

Major groove. The wider of the two grooves on a DNA double helix.

Marasmus. Body wasting due to inadequate intake of all types of foods. See also kwashiorkor.

Mass action ratio. The ratio of the product of the concentrations of reaction products to that of the reactants.

Mass spectrometry. A technique for identifying molecules by measuring the mass-to-charge ratios of gas-phase ions, such as peptide fragments.

Matrix. See mitochondrial matrix.

Melting temperature (T_m). The midpoint temperature of the melting curve for the thermal denaturation of a macromolecule. For a lipid, the temperature of transition from an ordered crystalline state to a more fluid state.

Membrane potential ($\Delta\psi$). The difference in electrical charge across a membrane.

Messenger RNA (mRNA). A ribonucleic acid whose sequence is complementary to that of a protein-coding gene in DNA.

Metabolic acidosis. A low blood pH caused by the overproduction or retention of hydrogen ions.

Metabolic alkalosis. A high blood pH caused by the excessive loss of hydrogen ions.

Metabolic fuel. A molecule that can be oxidized to provide free energy for an organism.

Metabolic pathway. A series of enzyme-catalyzed reactions by which one substance is transformed into another.

Metabolic syndrome. A set of symptoms related to obesity, including inulin resistance, atherosclerosis, and hypertension.

Metabolically irreversible reaction. A reaction whose value of ΔG is large and negative so that the reaction cannot proceed in reverse.

Metabolism. The total of all degradative and biosynthetic cellular reactions.

Metabolite. A reactant, intermediate, or product of a metabolic reaction.

Metabolome. The complete set of metabolites produced by a cell or tissue.

Metabolomics. The study of all the metabolites produced by a cell or tissue.

Metal ion catalysis. A catalytic mechanism that requires the presence of a metal ion to lower the free energy of a reaction's transition state.

Micelle. A globular aggregate of amphiphilic molecules in aqueous solution that are oriented such that polar segments form the surface of the aggregate and the nonpolar segments form a core that is out of contact with the solvent.

Michaelis constant (K_M). For an enzyme that follows the Michaelis–Menten model, $K_M = (k_{-1} + k_2)/k_1$; K_M is equal to the substrate concentration at which the reaction velocity is half-maximal.

Michaelis–Menten equation. A mathematical expression that describes the activity of an enzyme in terms of the substrate concentration ([S]), the enzyme's maximal velocity (V_{max}), and its Michaelis constant (K_M): $v_0 = V_{max}[S]/(K_M + [S])$.

Micro RNA (miRNA). A 20- to 25-nucleotide double-stranded RNA that binds to and inactivates a number of complementary mRNA molecules in RNA interference.

Microarray. A collection of DNA sequences that hybridize with RNA molecules and that

can therefore be used to identify active genes. Also called a DNA chip.

Microbiome. The collection of microorganisms that live in or on the human body.

Microenvironment. A group's immediate neighbors, whose chemical and physical properties may affect the group.

Microfilament. See actin filament.

Microtubule. A 240-Å-diameter cytoskeletal element consisting of a hollow tube of polymerized tubulin subunits.

Microvesicle. See exosome.

Minor groove. The narrower of the two grooves on a DNA double helix.

(–) end. The end of a polymeric filament where growth is slower. See also (+) end.

miRNA. See micro RNA.

Mismatch repair. A DNA repair pathway that removes and replaces mispaired nucleotides on a newly synthesized DNA strand.

Mitochondrial matrix. The gel-like solution of enzymes, substrates, cofactors, and ions in the interior of the mitochondrion.

Mitochondrion (*pl.* mitochondria). The double-membrane-enveloped eukaryotic organelle in which aerobic metabolic reactions occur, including those of the citric acid cycle, fatty acid oxidation, and oxidative phosphorylation.

Mixed inhibition. A form of enzyme inhibition in which an inhibitor binds to the enzyme such that it causes the apparent V_{max} to decrease and the apparent K_M to increase or decrease.

Mobilization. The process in which polysaccharides, triacylglycerols, and proteins are degraded to make metabolic fuels available.

Moderately repetitive DNA. Sequences of DNA that are present at hundreds of thousands of copies in the human genome.

Molecular chaperone. A protein that binds to unfolded or misfolded proteins in order to promote their normal folding.

Monogenic disease. A disease linked to a defect in a single gene.

Monomer. A structural unit from which a polymer is built up.

Monosaccharide. A carbohydrate consisting of a single sugar molecule.

Motor protein. An intracellular protein that couples the free energy of ATP hydrolysis to molecular movement relative to another protein that often acts as a track for the linear movement of the motor protein.

mRNA. See messenger RNA.

Multienzyme complex. A group of noncovalently associated enzymes that catalyze two or more sequential steps in a metabolic pathway.

Multifunctional enzyme. A protein that carries out more than one chemical reaction.

Mutagen. An agent that induces a mutation in an organism.

Mutation. A heritable alteration in an organism's genetic material.

Myelin sheath. The multilayer coating of sphingomyelin-rich membranes that insulates a mammalian neuron.

***N*-linked oligosaccharide.** An oligosaccharide linked to the amide group of a protein Asn residue.

***N*-terminus.** The end of a polypeptide that has a free amino group.

Native structure. The fully folded conformation of a macromolecule.

Natural selection. The evolutionary process by which the continued existence of a replicating entity depends on its ability to survive and reproduce under the existing conditions.

ncRNA. See noncoding RNA.

Near-equilibrium reaction. A reaction whose ΔG value is close to zero, so that it can operate in either direction depending on the substrate and product concentrations.

Negative effector. A substance that diminishes an enzyme's activity through allosteric inhibition.

Nernst equation. An expression of the relationship between the actual ($\mathcal{E}$) and standard reduction potential ($\mathcal{E}^{\circ\prime}$) of a substance A: $\mathcal{E} = \mathcal{E}^{\circ\prime} - (RT/n\mathcal{F}) \ln([A_{reduced}]/[A_{oxidized}])$.

Neurotransmitter. A substance released by a nerve cell to alter the activity of a target cell.

Neutral solution. A solution whose pH is equal to 7.0 ($[H^+] = 10^{-7}$ M).

Nick. A single-strand break in a double-stranded nucleic acid.

Nick translation. The progressive movement of a single-strand break (nick) in DNA through the actions of an exonuclease that removes residues followed by a polymerase that replaces them.

Nitrification. The conversion of ammonia (NH_3) to nitrate (NO_3^-).

Nitrogen cycle. A set of reactions, including nitrogen fixation, nitrification, and denitrification, for the interconversion of different forms of nitrogen.

Nitrogen fixation. The process by which atmospheric N_2 is converted to a biologically useful form such as NH_3.

NMR spectroscopy. See nuclear magnetic resonance spectroscopy.

Noncoding RNA (ncRNA). An RNA molecule that is not translated into protein.

Noncoding strand. The DNA strand that has a sequence complementary (except for the replacement of U with T) to the transcribed RNA; it is the template strand. Also called the antisense strand.

Noncompetitive inhibition. A form of enzyme inhibition in which an inhibitor binds to an enzyme such that the apparent V_{max} decreases but K_M is not affected.

Noncyclic electron flow. The light-driven linear path of electrons from water through Photosystems II and I, which leads to the production of O_2, NADPH, and ATP.

Nonessential amino acid. An amino acid that an organism can synthesize from common intermediates.

Nonhomologous end-joining. A ligation process that repairs a double-stranded break in DNA.

Nonreducing sugar. A saccharide with an anomeric carbon that has formed a glycosidic bond and cannot therefore act as a reducing agent.

Nonspontaneous process. A thermodynamic process that has a net increase in free energy ($\Delta G > 0$) and can occur only with the input of free energy from outside the system. See also endergonic reaction.

Nuclear magnetic resonance (NMR) spectroscopy. A spectroscopic method in which the signals emitted by atomic nuclei in a magnetic field can be used to determine the three-dimensional molecular structure of a protein or nucleic acid.

Nuclease. An enzyme that degrades nucleic acids.

Nucleic acid. A polymer of nucleotide residues. The major nucleic acids are deoxyribonucleic acid (DNA) and ribonucleic acid (RNA). Also known as a polynucleotide.

Nucleolus. The region of the eukaryotic nucleus where rRNA is processed and ribosomes are assembled.

Nucleophile. A compound containing an electron-rich group. A nucleophile (nucleus-lover) reacts with an electrophile (electron-lover).

Nucleoside. A compound consisting of a nitrogenous base linked to a five-carbon sugar (ribose or deoxyribose).

Nucleosome. The disk-shaped complex of a histone octamer and DNA that represents the fundamental unit of DNA organization in eukaryotes.

Nucleotide. A compound consisting of a nucleoside esterified to one or more phosphate groups. Nucleotides are the monomeric units of nucleic acids.

Nucleotide excision repair. A DNA repair pathway in which a damaged single-stranded segment of DNA is removed and replaced with normal DNA.

***O*-linked oligosaccharide.** An oligosaccharide linked to the hydroxyl group of a protein Ser or Thr side chain.

Okazaki fragments. The short segments of DNA formed in the discontinuous lagging-strand synthesis of DNA.

Oligonucleotide. A polynucleotide consisting of a few nucleotide residues.

Oligopeptide. A polypeptide consisting of a few amino acid residues.

Oligosaccharide. A polymeric carbohydrate containing a few monosaccharide residues. In glycoproteins, the groups are known as *N*-linked and *O*-linked oligosaccharides.

Omega-3 fatty acid. A fatty acid with a double bond starting at the third carbon from the methyl (omega) end of the molecule.

Oncogene. A mutant gene that interferes with the normal regulation of cell growth and contributes to cancer.

Open reading frame (ORF). A portion of the genome that potentially codes for a protein.

Operon. A prokaryotic genetic unit that consists of several genes with related functions that are transcribed as a single mRNA molecule.

Ordered mechanism. A multisubstrate reaction with a compulsory order of substrate binding to the enzyme.

ORF. See open reading frame.

Orientation effects. See Proximity and orientation effects.

Orphan gene. A gene that appears to have no counterpart in the genome of another species.

Osmosis. The movement of solvent from a region of low solute concentration to a region of high solute concentration.

Osteogenesis imperfecta. A disease caused by mutations in collagen genes and characterized by bone fragility and deformation.

Oxidant. See oxidizing agent.

Oxidation. A reaction in which a substance loses electrons.

Oxidative phosphorylation. The process by which the free energy obtained from the oxidation of metabolic fuels is used to generate ATP from ADP + P_i.

Oxidizing agent. A substance that can accept electrons, thereby becoming reduced. Also called an oxidant.

Oxyanion hole. A cavity in the active site of a serine protease that accommodates the reactants during the transition state and thereby lowers its energy.

Oxyhemoglobin. Hemoglobin that contains bound oxygen or is in the oxygen-binding conformation.

P site. The ribosomal binding site that accommodates a peptidyl–tRNA.

Palindrome. A segment of DNA that has the same sequence on each strand when read in the $5' \rightarrow 3'$ direction.

Parallel β sheet. See β sheet.

Partial oxygen pressure (pO_2). The concentration of gaseous O_2 in units of torr.

Passive transport. The thermodynamically spontaneous protein-mediated transmembrane movement of a substance from high to low concentration.

Pasteur effect. The greatly increased sugar consumption of yeast grown under anaerobic conditions compared to that of yeast grown under aerobic conditions.

PCR. See polymerase chain reaction.

Pentose. A five-carbon sugar.

Pentose phosphate pathway. A pathway for glucose degradation that yields ribose-5-phosphate and NADPH.

Peptide. A short polypeptide.

Peptide bond. An amide linkage between the α-amino group of one amino acid and the α-carboxylate group of another. Peptide bonds link the amino acid residues in a polypeptide.

Peptidoglycan. The cross-linked polysaccharides and polypeptides that form bacterial cell walls.

Peripheral membrane protein. A protein that is weakly associated with the surface of a biological membrane. Also called an extrinsic protein.

Peroxisome. A eukaryotic organelle with specialized oxidative functions, including fatty acid degradation.

p_{50}. The ligand concentration (or pressure for a gaseous ligand) at which a binding protein such as hemoglobin is half-saturated with ligand.

pH. A quantity used to express the acidity of a solution, equivalent to $-\log[H^+]$.

Phage. See bacteriophage.

Pharmacokinetics. The behavior of a drug in the body, including its metabolism and excretion.

Phosphatase. An enzyme that hydrolyzes phosphoryl ester groups.

Phosphodiester bond. The linkage in which a phosphate group is esterified to two alcohol groups (e.g., two ribose units that join the adjacent nucleotide residues in a polynucleotide).

Phosphoinositide signaling system. A signal transduction pathway in which hormone binding to a cell-surface receptor induces phospholipase C to catalyze the hydrolysis of phosphatidylinositol bisphosphate to yield the second messengers inositol trisphosphate and diacylglycerol.

Phospholipase. An enzyme that hydrolyzes one or more bonds in a glycerophospholipid.

Phosphorolysis. The cleavage of a chemical bond by the substitution of a phosphate group rather than water.

Photoautotroph. An organism that obtains its building materials from inorganic compounds and its free energy from sunlight.

Photon. A packet of light energy.

Photooxidation. A mode of decay of an excited molecule, in which oxidation occurs through the transfer of an electron to an acceptor molecule.

Photophosphorylation. The synthesis of ATP from ADP + P_i coupled to the dissipation of a proton gradient that has been generated through light-driven electron transport.

Photoreceptor. A light-absorbing molecule, or pigment.

Photorespiration. The consumption of O_2 and evolution of CO_2 by plants (a dissipation of the products of photosynthesis), a consequence of the competition between O_2 and CO_2 for ribulose bisphosphate carboxylase.

Photosynthesis. The light-driven incorporation of CO_2 into organic compounds.

pI. See isoelectric point.

P_i. Inorganic phosphate or a phosphoryl group: HPO_3^- or PO_3^{2-}.

Pigment. See photoreceptor.

Ping pong mechanism. An enzymatic reaction in which one or more products are released before all the substrates have bound to the enzyme.

Pinocytosis. Endocytosis of small amounts of extracellular fluid and solutes.

pK. A quantity used to express the tendency for an acid to donate a proton (dissociate); equal to $-\log K$, where K is the dissociation constant.

Planck's law. An expression for the energy (E) of a photon: $E = hc/\lambda$, where c is the speed of light, λ is its wavelength, and h is Planck's constant (6.626×10^{-34} J·s).

Plasmid. A small circular DNA molecule that autonomously replicates and may be used as a vector for recombinant DNA.

(+) end. The end of a polymeric filament where growth is faster. See also (−) end.

P:O ratio. The ratio of the number of molecules of ATP synthesized from ADP + P_i to the number of atoms of oxygen reduced.

Point mutation. The substitution of one base for another in DNA, arising from mispairing during DNA replication or from chemical alterations of existing bases.

Polarity. Having an uneven distribution of charge.

Poly(A) tail. The sequence of adenylate residues that is post-transcriptionally added to the 3′ end of eukaryotic mRNAs.

Polygenic disease. A disease linked to variations in more than one gene.

Polymer. A molecule consisting of numerous smaller units that are linked together in an organized manner.

Polymerase. An enzyme that catalyzes the addition of nucleotide residues to a polynucleotide. DNA is synthesized by DNA polymerase, and RNA is synthesized by RNA polymerase.

Polymerase chain reaction (PCR). A procedure for amplifying a segment of DNA by repeated rounds of replication centered between primers that hybridize with the two ends of the DNA segment of interest.

Polynucleotide. See nucleic acid.

Polypeptide. A polymer consisting of amino acid residues linked in linear fashion by peptide bonds.

Polyprotein. A polypeptide that undergoes proteolysis after its synthesis to yield several separate protein molecules.

Polyprotic acid. A substance that has more than one acidic proton and therefore has multiple ionization states.

Polysaccharide. A polymeric carbohydrate containing multiple monosaccharide residues. Also called a glycan.

Polysome. An mRNA transcript bearing multiple ribosomes in the process of translating the mRNA.

Porin. A β barrel protein in the outer membrane of bacteria, mitochondria, or chloroplasts that forms a weakly solute-selective pore.

Positive effector. A substance that boosts an enzyme's activity through allosteric activation.

Post-translational processing. The removal or derivatization of amino acid residues following their incorporation into a polypeptide.

pO_2. See partial oxygen pressure.

PP_i. A pyrophosphoryl group: $H_3P_2O_6$, $H_2P_2O_6^-$, $HP_2O_6^{2-}$, or $P_2O_6^{3-}$.

Primary structure. The sequence of residues in a polymer.

Primase. The enzyme that synthesizes a segment of RNA to be extended by DNA polymerase during DNA replication.

Primer. An oligonucleotide that base pairs with a template polynucleotide strand and is extended through template-directed polymerization.

Prion. An infectious protein that causes its cellular counterparts to misfold and aggregate, thereby leading to the development of a disease such as transmissible spongiform encephalopathy.

Probe. A labeled single-stranded DNA or RNA segment that can hybridize with a DNA or RNA of interest in a screening procedure.

Processing. See RNA processing and post-translational modification.

Processivity. A property of a motor protein or other enzyme that undergoes many reaction cycles before dissociating from its track or substrate.

Product inhibition. A form of enzyme inhibition in which the reaction product acts as a competitive inhibitor.

Prokaryote. A unicellular organism that lacks a membrane-bounded nucleus. All bacteria and archaea are prokaryotes.

Promoter. The DNA sequence at which RNA polymerase binds to initiate transcription.

Proofreading. An additional catalytic activity of an enzyme, which acts to correct errors made by the primary enzymatic activity.

Prosthetic group. An organic group (such as a coenzyme) that is permanently associated with a protein.

Protease. An enzyme that catalyzes the hydrolysis of peptide bonds.

Protease inhibitor. An agent, often a protein, that reacts incompletely with a protease so as to inhibit further proteolytic activity.

Proteasome. A multiprotein complex with a hollow cylindrical core in which cellular proteins are degraded to peptides in an ATP-dependent process.

Protein. A macromolecule that consists of one or more polypeptide chains.

Proteoglycan. An extracellular aggregate of protein and glycosaminoglycans.

Proteome. The complete set of proteins synthesized by a cell.

Proteomics. The study of all the proteins synthesized by a cell.

Protofilament. One of the 13 linear polymers of tubulin subunits that forms a microtubule.

Proton jumping. The rapid movement of a proton among hydrogen-bonded water molecules.

Proton wire. A series of hydrogen-bonded water molecules and protein groups that can relay protons from one site to another.

Protonmotive force. The free energy of the electrochemical proton gradient that forms during electron transport.

Proximity and orientation effects. A catalytic mechanism in which reacting groups are brought close together in an enzyme active site to accelerate the reaction.

Purine. A derivative of the compound purine, such as the nucleotide base adenine or guanine.

Pyrimidine. A derivative of the compound pyrimidine, such as the nucleotide base cytosine, uracil, or thymine.

Pyrosequencing. A procedure for determining the sequence of nucleotides in DNA by detecting a flash of light generated by the addition of a new nucleotide to a growing DNA strand. Also known as 454 sequencing.

Q cycle. The cyclic flow of electrons involving a semiquinone intermediate in Complex III of mitochondrial electron transport and in photosynthetic electron transport.

qPCR. See real-time PCR.

Quantum yield. The ratio of carbon atoms fixed or oxygen molecules produced to the number of photons absorbed by the photosynthetic machinery.

Quaternary structure. The spatial arrangement of a macromolecule's individual subunits.

Quorum sensing. The ability of cells to monitor population density by detecting the concentrations of extracellular substances.

R. See gas constant.

R group. A symbol for a variable portion of a molecule, such as the side chain of an amino acid.

R state. One of two conformations of an allosteric protein; the other is the T state.

Raft. An area of a lipid bilayer with a distinct lipid composition and near-crystalline consistency.

Random mechanism. A multisubstrate reaction without a compulsory order of substrate binding to the enzyme.

Rate constant (k). The proportionality constant between the velocity of a chemical reaction and the concentration(s) of the reactant(s).

Rate-determining reaction. The slowest step in a multistep sequence, such as a metabolic pathway, whose rate determines the rate of the entire sequence.

Rate equation. A mathematical expression for the time-dependent progress of a reaction as a function of reactant concentration.

Rational drug design. The synthesis of more effective drugs based on detailed knowledge of the target molecule's structure and function.

Reactant. One of the starting materials for a chemical reaction.

Reaction center. A chlorophyll-containing protein where photooxidation takes place.

Reaction coordinate. A line representing the progress of a reaction, part of a graphical presentation of free energy changes during a reaction.

Reaction specificity. The ability of an enzyme to discriminate between possible substrates and to catalyze a single type of chemical reaction.

Reading frame. The grouping of nucleotides in sets of three whose sequence corresponds to a polypeptide sequence.

Real-time PCR. A variation of the polymerase chain reaction in which the level of gene expression can be measured. Also known as quantitative PCR (qPCR).

Receptor. A binding protein that is specific for its ligand and elicits a discrete biochemical effect when its ligand is bound.

Receptor-mediated endocytosis. Endocytosis of an extracellular component as a result of its specific binding to a cell surface receptor.

Receptor tyrosine kinase. A cell-surface receptor whose intracellular domains become active as Tyr-specific kinases as a result of extracellular ligand binding.

Recombinant DNA. A DNA molecule containing DNA segments from different sources.

Recombination. The exchange of polynucleotide strands between separate DNA segments; recombination is one mechanism for repairing damaged DNA by allowing a homologous segment to serve as a template for replacement of the damaged bases.

Redox center. A group that can undergo an oxidation–reduction reaction.

Redox reaction. A chemical reaction in which one substance is reduced and another substance is oxidized.

Reducing agent. A substance that can donate electrons, thereby becoming oxidized. Also called a reductant.

Reducing sugar. A saccharide with an anomeric carbon that has not formed a glycosidic bond and can therefore act as a reducing agent.

Reductant. See reducing agent.

Reduction. A reaction in which a substance gains electrons.

Reduction potential (ℰ). A measure of the tendency of a substance to gain electrons.

Regular secondary structure. A segment of a polymer in which the backbone adopts a regularly repeating conformation; the opposite of irregular secondary structure.

Release factor (RF). A protein that recognizes a stop codon and causes a ribosome to terminate polypeptide synthesis.

Renaturation. The refolding of a denatured macromolecule so as to regain its native conformation.

Replication. The process of making an identical copy of a DNA molecule. During DNA replication, the parental polynucleotide strands separate so that each can direct the synthesis of a complementary daughter strand, resulting in two complete DNA double helices.

Replication fork. The point in a replicating DNA molecule where the two parental strands separate in order to serve as templates for the synthesis of new strands.

Replisome. The complex of proteins that carries out the various activities of DNA replication.

Repressor. A protein that binds at or near a gene so as to prevent its transcription.

Residue. A term for a monomeric unit after it has been incorporated into a polymer.

Resonance stabilization. The effect of delocalization of electrons in a molecule that cannot be depicted by a single structural diagram.

Respiration. See cellular respiration.

Respiratory acidosis. A low blood pH caused by insufficient elimination of CO_2 (carbonic acid) by the lungs.

Respiratory alkalosis. A high blood pH caused by the excessive loss of CO_2 (carbonic acid) from the lungs.

Restriction digest. The generation of a set of DNA fragments by the action of a restriction endonuclease.

Restriction endonuclease. A bacterial enzyme that cleaves a specific DNA sequence.

Reverse transcriptase. A DNA polymerase that uses RNA as its template.

RF. See release factor.

Ribonucleic acid. See RNA.

Ribosomal RNA (rRNA). The RNA molecules that provide structural support for the ribosome and catalyze peptide bond formation.

Ribosome. The RNA-and-protein particle that synthesizes polypeptides under the direction of mRNA.

Ribosome recycling factor (RRF). A protein that binds to a ribosome after protein synthesis to prepare it for another round of translation.

Ribozyme. An RNA molecule that has catalytic activity.

RNA (Ribonucleic acid). A polymer of ribonucleotides, such as messenger RNA (mRNA), transfer RNA (tRNA), and ribosomal RNA (rRNA).

RNA interference (RNAi). A phenomenon in which short RNA segments direct the degradation of complementary mRNA, thereby inhibiting gene expression.

RNA polymerase. See polymerase.

RNA processing. The addition, removal, or modification of nucleotides in an RNA molecule that is necessary to produce a fully functional RNA.

RNA world. A hypothetical time before the evolution of DNA or protein, when RNA stored genetic information and functioned as a catalyst.

RRF. See ribosome recycling factor.

rRNA. See ribosomal RNA.

S. See entropy.

Saccharide. See carbohydrate.

Salvage pathway. A pathway that reincorporates an intermediate of nucleotide degradation into a new nucleotide, thereby minimizing the need for the nucleotide biosynthetic pathways.

Saturated fatty acid. A fatty acid that does not contain any double bonds in its hydrocarbon chain.

Saturation. The state in which all of a macromolecule's ligand-binding sites are occupied by ligand.

Schiff base. An imine that forms between an amine and an aldehyde or ketone.

Scissile bond. The bond that is to be cleaved during a proteolytic reaction.

SDS-PAGE. A form of polyacrylamide gel electrophoresis in which denatured polypeptides are separated by size in the presence of the detergent sodium dodecyl sulfate. See electrophoresis.

Second messenger. An intracellular ion or molecule that acts as a signal for an extracellular event such as ligand binding to a cell-surface receptor.

Second-order reaction. A reaction whose rate is proportional to the square of the concentration of one reactant or to the product of the concentrations of two reactants.

Secondary active transport. Transmembrane transport of one substance that is driven by the free energy of an existing gradient of a second substance.

Secondary structure. The local spatial arrangement of a polymer's backbone atoms without regard to the conformations of its substituent side chains.

Selection. A technique for distinguishing cells that contain a particular feature, such as resistance to an antibiotic.

Semiconservative replication. The mechanism of DNA duplication in which each new molecule contains one strand from the parent molecule and one newly synthesized strand.

Sense strand. See coding strand.

Serine protease. A peptide-hydrolyzing enzyme that has a reactive Ser residue in its active site.

Set-point. A body weight that is maintained through regulation of fuel metabolism and that resists change when an individual attempts to alter fuel consumption or expenditure.

Signal peptide. A short sequence in a membrane or secretory protein that binds to the signal recognition particle in order to direct the translocation of the protein across a membrane.

Signal recognition particle (SRP). A complex of protein and RNA that recognizes membrane and secretory proteins and mediates their binding to a membrane for translocation.

Signal transduction. The process by which an extracellular signal is transmitted to the cell interior by binding to a cell-surface receptor such that binding triggers a series of intracellular events.

Silencer. A DNA sequence some distance from the transcription start site, where a repressor of transcription may bind.

Single-nucleotide polymorphism (SNP). A nucleotide sequence variation in the genomes of two individuals from the same species.

siRNA. See small interfering RNA.

Site-directed mutagenesis. A technique in which a cloned gene is mutated in a specific manner.

Size-exclusion chromatography. A procedure in which macromolecules are separated on the basis of their size and shape. Also called gel filtration chromatography.

Small interfering RNA (siRNA). A 20- to 25-nucleotide double-stranded RNA that targets for destruction a fully complementary mRNA molecule in RNA interference.

Small nuclear RNA (snRNA). Highly conserved RNAs that participate in eukaryotic mRNA splicing.

Small nucleolar RNA (snoRNA). RNA molecules that direct the sequence-specific methylation of eukaryotic rRNA transcripts.

snoRNA. See small nucleolar RNA.

SNP. See single-nucleotide polymorphism.

snRNA. See small nuclear RNA.

Solute. The substance that is dissolved in water or another solvent to make a solution.

Solvation. The state of being surrounded by solvent molecules.

Specificity pocket. A cavity on the surface of a serine protease, whose chemical characteristics determine the identity of the substrate residue on the N-terminal side of the bond to be cleaved.

Sphingolipid. An amphipathic lipid containing an acyl group, a palmitate derivative, and a polar head group attached to a serine backbone. In sphingomyelins, the head group is a phosphate derivative.

Sphingomyelin. See sphingolipid.

Spliceosome. A complex of protein and snRNA that carries out the splicing of immature mRNA molecules.

Splicing. The process by which introns are removed and exons are joined to produce a mature RNA transcript.

Spontaneous process. A thermodynamic process that has a net decrease in free energy ($\Delta G < 0$) and occurs without the input of free energy from outside the system. See also exergonic reaction.

SRP. See signal recognition particle.

Stacking interactions. The stabilizing van der Waals interactions between successive (stacked) bases in a polynucleotide.

Standard conditions. A set of conditions including a temperature of 25°C, a pressure of 1 atm, and reactant concentrations of 1 M. Biochemical standard conditions include a pH of 7.0 and a water concentration of 55.5 M.

Standard free energy change ($\Delta G°'$). The force that drives reactants to reach their equilibrium values when the system is in its biochemical standard state.

Standard reduction potential ($\mathcal{E}°'$). A measure of the tendency of a substance to gain electrons (to be reduced) under standard conditions.

Steady state. A set of conditions under which the formation and degradation of individual components are balanced such that the system does not change over time.

Stereocilia. Microfilament-stiffened cell processes on the surface of cells in the inner ear, which are deflected in response to sound waves.

Sticky ends. Single-stranded extensions of DNA that are complementary, often because they have been generated by the action of the same restriction endonuclease.

Stroma. The gel-like solution of enzymes and small molecules in the interior of the chloroplast; the site of carbohydrate synthesis.

Substrate. A reactant in an enzymatic reaction.

Substrate-level phosphorylation. The transfer of a phosphoryl group to ADP that is directly coupled to another chemical reaction.

Subunit. One of several polypeptide chains that make up a protein.

Sugar–phosphate backbone. The chain of (deoxy)ribose groups linked by phosphodiester bonds in a polynucleotide chain.

Suicide substrate. A molecule that chemically inactivates an enzyme only after undergoing part of the normal catalytic reaction.

Supercoiling. A topological state of DNA in which the helix is underwound or overwound so that the molecule tends to writhe or coil up on itself.

Symport. Transport that involves the simultaneous transmembrane movement of two molecules in the same direction.

Synaptic vesicle. A vesicle loaded with neurotransmitters to be released from the end of an axon.

T state. One of two conformations of an allosteric protein; the other is the R state.

TATA box. A eukaryotic promoter element with an AT-rich sequence located upstream from the transcription start site.

Tautomer. One of a set of isomers that differ only in the positions of their hydrogen atoms.

Telomerase. An enzyme that uses an RNA template to polymerize deoxynucleotides and thereby extend the 3′-ending strand of a eukaryotic chromosome.

Telomere. The end of a linear eukaryotic chromosome, which consists of tandem repeats of a short G-rich sequence on the 3′-ending strand and its complementary sequence on the 5′-ending strand.

Terpenoid. See isoprenoid.

Tertiary structure. The entire three-dimensional structure of a single-chain polymer, including the conformations of its side chains.

Tetramer. An assembly consisting of four monomeric units.

Tetrose. A four-carbon sugar.

Thalassemia. A hereditary disease caused by insufficient synthesis of hemoglobin, which results in anemia.

Thermogenesis. The process of generating heat by muscular contraction or by metabolic reactions.

Thick filament. A muscle cell structural element that is composed of several hundred myosin molecules.

Thin filament. A muscle cell structural element that consists primarily of an actin filament.

Thioester. A compound containing an ester linkage to a sulfur rather than an oxygen atom.

3′ end. The terminus of a polynucleotide whose C3′ is not esterified to another nucleotide residue.

Thylakoid. The membranous structure in the interior of a chloroplast that is the site of the light reactions of photosynthesis.

T_m. See melting temperature.

Topoisomerase. An enzyme that alters DNA supercoiling by breaking and resealing one or both strands.

Trace element. An element that is present in small quantities in a living organism.

Transamination. The transfer of an amino group from an amino acid to an α-keto acid to yield a new α-keto acid and a new amino acid.

Transcription. The process by which RNA is synthesized using a DNA template, thereby transferring genetic information from the DNA to the RNA.

Transcription factor. A protein that promotes the transcription of a gene by binding to DNA sequences at or near the gene or by interacting with other proteins that do so. See also general transcription factor.

Transcriptome. The set of all the RNA molecules produced by a cell.

Transcriptomics. The study of the genes that are transcribed in a certain cell type or at a certain time.

Transfer RNA (tRNA). The small L-shaped RNAs that deliver specific amino acids to ribosomes according to the sequence of a bound mRNA.

Transgenic organism. An organism that stably expresses a foreign gene.

Transition mutation. A point mutation in which one purine (or pyrimidine) is replaced by another purine (or pyrimidine).

Transition state. The point of highest free energy, or the structure that corresponds to that point, in the reaction coordinate diagram of a chemical reaction.

Transition state analog. A stable substance that geometrically and electronically resembles the transition state of a reaction and that therefore may inhibit an enzyme that catalyzes the reaction.

Translation. The process of transforming the information contained in the nucleotide sequence of an RNA to the corresponding amino acid sequence of a polypeptide as specified by the genetic code.

Translocase. An enzyme that catalyzes the movement of a lipid from one bilayer leaflet to another. Also called a flippase.

Translocation. The movement of tRNA and mRNA, relative to the ribosome, that occurs following formation of a peptide bond and that allows the next mRNA codon to be translated.

Translocon. The complex of membrane proteins that mediates the transmembrane movement of a polypeptide.

Transmissible spongiform encephalopathy (TSE). A fatal neurodegenerative disease caused by infection with a prion.

Transpeptidation. The ribosomal process in which the peptidyl group attached to a tRNA is transferred to the aminoacyl group of another tRNA, forming a new peptide bond and lengthening the polypeptide by one residue at its C-terminus.

Transposable element. A segment of DNA, sometimes including genes, that can move (be copied) from one position to another in a genome.

Transverse diffusion. The movement of a membrane component from one leaflet of a bilayer to the other. Also called flip-flop.

Transversion mutation. A point mutation in which a purine is replaced by a pyrimidine or vice versa.

Treadmilling. The addition of monomeric units to one end of a polymer and their removal from the opposite end such that the length of the polymer remains unchanged.

Triacylglycerol. A lipid in which three fatty acids are esterified to a glycerol backbone. Also called a triglyceride.

Triglyceride. See triacylglycerol.

Trimer. An assembly consisting of three monomeric units.

Triose. A three-carbon sugar.

Triple helix. The right-handed helical structure formed by three left-handed helical polypeptide chains in collagen.

Trisaccharide. A carbohydrate consisting of three monosaccharides.

tRNA. See transfer RNA.

TSE. See transmissible spongiform encephalopathy.

Tumor suppressor gene. A gene whose loss or mutation may lead to cancer.

Turnover number. See catalytic constant.

Uncompetitive inhibition. A form of enzyme inhibition in which an inhibitor binds to an enzyme–substrate complex such that the apparent V_{max} and K_M are both decreased to the same extent.

Uncoupler. A substance that allows the proton gradient across a membrane to dissipate without ATP synthesis so that electron transport proceeds without oxidative phosphorylation.

Unimolecular reaction. A reaction involving one molecule.

Uniport. Transport that involves transmembrane movement of a single molecule.

Unsaturated fatty acid. A fatty acid that contains at least one double bond in its hydrocarbon chain.

Urea cycle. A cyclic metabolic pathway in which amino groups are converted to urea for disposal.

Usher syndrome. A genetic disease characterized by profound deafness and retinitis pigmentosa that leads to blindness, caused in some cases by a defective myosin protein.

v. Velocity (rate) of a reaction.

van der Waals interaction. A weak non-covalent association between molecules that arises from the attractive forces between polar groups (dipole–dipole interactions) or between nonpolar groups whose fluctuating electron distribution gives rise to temporary dipoles (London dispersion forces).

van der Waals radius. The distance from an atom's nucleus to its effective electronic surface.

Variable residue. A position in a polypeptide that is occupied by different residues in evolutionarily related proteins; its substitution has little or no effect on protein function.

Vector. A DNA molecule, such as a plasmid, that can accommodate a segment of foreign DNA.

Vesicle. A fluid-filled sac enclosed by a lipid-bilayer membrane.

Vitamin. A metabolically required substance that cannot be synthesized by an animal and must therefore be obtained from the diet.

V_{max}. Maximal velocity of an enzymatic reaction.

v_0. Initial velocity of an enzymatic reaction.

Warburg effect. The increased rate of glycolysis observed in cancerous tissues.

Wobble hypothesis. An explanation for the nonstandard base pairing between tRNA and mRNA at the third codon position, which allows a tRNA to recognize more than one codon.

X-Ray crystallography. A method for determining three-dimensional molecular structures from the diffraction pattern produced by exposing a crystal of a molecule to a beam of X-rays.

Y. See fractional saturation.

Z. The net charge of an ion.

Z-scheme. A Z-shaped diagram indicating the electron carriers and their reduction potentials in the photosynthetic electron transport system of plants and cyanobacteria.

Zinc finger. A protein structural motif consisting of 20–60 residues, including Cys and His residues to which one or two Zn^{2+} ions are tetrahedrally coordinated.

Zymogen. The inactive precursor (proenzyme) of a proteolytic enzyme.

Chapter 1

1. Since ΔH is negative, heat is released during the reaction.

2. Because entropy decreases, ΔS is negative.

3. $\Delta G = \Delta H - T\Delta S$

$= -20,000 \text{ J} \cdot \text{mol}^{-1} - 298 \text{ K}(-50 \text{ J} \cdot \text{K}^{-1} \cdot \text{mol}^{-1})$

$= -20,000 + 14,900 \text{ J} \cdot \text{mol}^{-1}$

$= -5100 \text{ J} \cdot \text{mol}^{-1} \text{ or } -5.1 \text{ kJ} \cdot \text{mol}^{-1}$

Because ΔG is less than 0, the reaction is spontaneous.

4. ΔG would be zero when the ΔH and $T\Delta S$ terms are equal:

$\Delta H = T\Delta S$

$T = \Delta H/\Delta S$

$= (15,000 \text{ J} \cdot \text{mol}^{-1})/(75 \text{ J} \cdot \text{K}^{-1} \cdot \text{mol}^{-1})$

$= 200 \text{ K or } 73°\text{C}$

Chapter 2

1. HCl dissociates completely, so the added $[H^+]$ is equal to $[HCl]$:

$[H^+] = \dfrac{(0.05 \text{ L})(0.025 \text{ M})}{0.3 \text{ L}} = 0.0042 \text{ M}$

$pH = -\log[H^+] = -\log 0.0042 = 2.4$

2. NaOH dissociates completely, so the added $[OH^-]$ is equal to $[NaOH]$:

$[OH^-] = \dfrac{(0.25 \text{ L})(0.015 \text{ M})}{0.75 \text{ L}} = 0.005 \text{ M}$

$K_w = 10^{-14} = [H^+][OH^-]$

$[H^+] = \dfrac{10^{-14}}{0.005} = 2 \times 10^{-12} \text{ M}$

$pH = -\log[H^+] = 11.7$

3. The final volume is $1 \text{ L} + 50 \text{ mL} + 50 \text{ mL} = 1.1 \text{ L}$

$[\text{acetic acid}] = [HA] = \dfrac{(0.05 \text{ L})(0.02 \text{ M})}{1.1 \text{ L}} = 9.1 \times 10^{-4} \text{ M}$

$[\text{acetate}] = [A^-] = \dfrac{(0.05 \text{ L})(0.03 \text{ M})}{1.1 \text{ L}} = 13.6 \times 10^{-4} \text{ M}$

$pH = pK + \log \dfrac{[A^-]}{[HA]}$

$= 4.76 + \log \dfrac{13.6 \times 10^{-4}}{9.1 \times 10^{-4}}$

$= 4.76 + 0.18 = 4.94$

4. The final volume is $500 \text{ mL} + 10 \text{ mL} + 20 \text{ mL} = 0.53 \text{ L}$

$[\text{boric acid}] = [HA] = \dfrac{(0.01 \text{ L})(0.05 \text{ M})}{0.53 \text{ L}} = 9.4 \times 10^{-4} \text{ M}$

$[\text{borate}] = [A^-] = \dfrac{(0.02 \text{ L})(0.02 \text{ M})}{0.53 \text{ L}} = 7.5 \times 10^{-4} \text{ M}$

$pH = pK + \log \dfrac{[A^-]}{[HA]}$

$= 9.24 + \log \dfrac{7.5 \times 10^{-4}}{9.4 \times 10^{-4}}$

$= 9.24 - 0.10 = 9.14$

5. $\qquad pH = pK + \log \dfrac{[A^-]}{[HA]}$

$\log \dfrac{[A^-]}{[HA]} = pH - pK = 5.0 - 4.76 = 0.24$

$\dfrac{[A^-]}{[HA]} = 1.74 \text{ or } [A^-] = 1.74[HA]$

$[A^-] = 1.74[HA] = 1.74(0.025 \text{ M} - [A^-])$

$[A^-] = 0.0435 - 1.74[A^-]$

$2.74[A^-] = 0.0435 \text{ M}$

$[A^-] = 0.016 \text{ M or } 16 \text{ mM}$

6. $\qquad pH = pK + \log \dfrac{[A^-]}{[HA]}$

$\log \dfrac{[A^-]}{[HA]} = pH - pK = 3.0 - 2.15 = 0.85$

$\dfrac{[A^-]}{[HA]} = 7.08 \text{ or } [HA] = \dfrac{[A^-]}{7.08}$

$[HA] = \dfrac{[A^-]}{7.08} = \dfrac{(0.025 \text{ M} - [HA])}{7.08}$

$7.08[HA] = 0.025 \text{ M} - [HA]$

$8.08[HA] = 0.025 \text{ M}$

$[HA] = 0.0031 \text{ M or } 3.1 \text{ mM}$

7. (a) At pH 5.5, pH $< pK_2$, so the $H_2PO_4^-$ form predominates.

   (b) At pH 8.5, pH $> pK_2$, so the HPO_4^{2-} form predominates.

8. The pK of the ammonium ion is 9.25, so at pH 9.25, $[NH_3] = [NH_4^+]$.

9. Since the dicarboxylic acid has a pK of 4.2 and the monocarboxylic acid has a pK of 5.64, the monocarboxylate form predominates between pH 4.2 and pH 5.64 (above pH 5.64, the dicarboxylate form predominates).

10. $\qquad pH = pK + \log \dfrac{[A^-]}{[HA]}$

$\log \dfrac{[A^-]}{[HA]} = pH - pK$

$\dfrac{[A^-]}{[HA]} = 10^{(pH-pK)}$

$\dfrac{[A^-]}{[HA]} = 10^{(9.4-9.24)} = 10^{0.16} = 1.44$

Since the starting solution contains $(0.5 \text{ L})(0.05 \text{ mol} \cdot \text{L}^{-1}) = 0.025$ mole of boric acid (HA), the amount of borate (A$^-$) needed is $1.44(0.025) = 0.036$ moles. The stock sodium borate is 5.0 M, so the volume of sodium borate to be added is

$$\frac{0.036 \text{ mol}}{5.0 \text{ mol} \cdot \text{L}^{-1}} = 0.0072 \text{ L or } 7.2 \text{ mL}$$

11.
$$pH = pK + \log \frac{[A^-]}{[HA]}$$

$$\log \frac{[A^-]}{[HA]} = pH - pK$$

$$\frac{[A^-]}{[HA]} = 10^{(pH - pK)}$$

$$\frac{[A^-]}{[HA]} = 10^{(6.5 - 7.0)} = 10^{-0.5} = 0.316$$

Since the starting solution contains $(0.5 \text{ L})(0.01 \text{ mol} \cdot \text{L}^{-1}) = 0.005$ mole of imidazole (A$^-$), the amount of imidazolium chloride (HA) needed is $0.005 \text{ mol}/0.316 = 0.016$ moles. The stock imidazolium chloride is 1 M, so the volume of imidazolium chloride to be added is

$$\frac{0.016 \text{ mol}}{1.0 \text{ mol} \cdot \text{L}^{-1}} = 0.016 \text{ L or } 16 \text{ mL}$$

Chapter 4

1. At pH 8.0, groups with pK values less than 8.0 are mostly deprotonated, and groups with pK values greater than 8.0 are mostly protonated. The tripeptide has a net charge of 0.

Group	Charge
N-terminus	+1
Asp	−1
His	0
Lys	+1
C-terminus	−1
Net charge	**0**

2. At pH 7.0, groups with pK values less than 7.0 are mostly deprotonated, and groups with pK values greater than 7.0 are mostly protonated. The tripeptide has a net charge of −3.

Group	Charge
N-terminus	+1
3 Glu	−3
C-terminus	−1
Net charge	**−3**

3. At pH 6.0, groups with pK values less than 6.0 are mostly deprotonated, and groups with pK values greater than 6.0 are mostly protonated. The dipeptide has a net charge of −1.

Group	Charge
N-terminus	+1
Asp	−1
Tyr	0
C-terminus	−1
Net charge	**−1**

4. In order for glutamate to have no net charge, its α-carboxyl group must be unprotonated (negatively charged), its side chain must be protonated (neutral), and its α-amino group must be protonated (positively charged). Because deprotonation of the side

chain or the α-amino group would change the amino acid's net charge, the pK values of these groups (4.0 and 9.0) should be used to calculate the pI:

$$pI = \tfrac{1}{2}(4.0 + 9.0) = 6.5$$

5. In order for alanine to have no net charge, its α-carboxyl group ($pK \approx 3.5$) must be unprotonated (negatively charged) and its α-amino group ($pK \approx 9.0$) must be protonated (positively charged):

$$pI = \tfrac{1}{2}(3.5 + 9.0) = 6.25$$

Chapter 5

1. $Y = \dfrac{pO_2}{K + pO_2}$

$$= \frac{6.8}{2.8 + 6.8} = 0.71$$

2. Y increases.

$$Y = \frac{pO_2}{K + pO_2}$$

$$= \frac{6.8}{1.4 + 6.8} = 0.83$$

3.
$$Y = 0.60 = \frac{pO_2}{2.8 + pO_2}$$

$$Y = 0.60(2.8 + pO_2) = pO_2$$

$$1.68 + 0.6\, pO_2 = pO_2$$

$$1.68 = 0.4\, pO_2$$

$$pO_2 = 4.2 \text{ torr}$$

Chapter 7

1. Since $v = k[X][Y]$, $k = v/([X][Y])$

$$k = (8 \text{ μM} \cdot \text{s}^{-1})/[(8 \text{ μM})(8 \text{ μM})]$$

$$= 0.125 \text{ μM}^{-1} \cdot \text{s}^{-1}$$

2. $v = k[X][Y]$

$$= (0.125 \text{ μM}^{-1} \cdot \text{s}^{-1})(20 \text{ μM})(10 \text{ μM})$$

$$= 25 \text{ μM} \cdot \text{s}^{-1}$$

3. Since $v = k[X][Y]$, $[X][Y] = [X]^2 = v/k$

$$[X]^2 = (12.5 \text{ mM} \cdot \text{s}^{-1})/(0.5 \text{ mM}^{-1} \cdot \text{s}^{-1}) = 25 \text{ mM}^2$$

$$[X] = [Y] = 5 \text{ mM}$$

4. $v_0 = \dfrac{(5 \text{ nM} \cdot \text{s}^{-1})(10 \text{ mM})}{(1 \text{ mM}) + (10 \text{ mM})}$

$$= \frac{50}{11} \text{ nM} \cdot \text{s}^{-1} = 4.5 \text{ nM} \cdot \text{s}^{-1}$$

5. $v_0 = \dfrac{V_{max}[S]}{K_M + [S]}$

$$K_M = \frac{V_{max}[S]}{v_0} - [S]$$

$$K_M = \frac{(9.5 \text{ μM} \cdot \text{s}^{-1})(1 \text{ μM})}{5 \text{ μM} \cdot \text{s}^{-1}} - 1 \text{ μM}$$

$$= 1.9 \text{ μM} - 1 \text{ μM} = 0.9 \text{ μM}$$

6.

$$v_0 = \frac{V_{max}[S]}{K_M + [S]}$$

$$[S] = \frac{V_{max}[S]}{v_0} - K_M$$

$$V_{max}[S] = v_0 K_M + v_0[S]$$

$$V_{max}[S] - v_0[S] = v_0 K_M$$

$$[S](V_{max} - v_0) = v_0 K_M$$

$$[S] = \frac{v_0 K_M}{V_{max} - v_0}$$

$$= \frac{(3.5\ \mu M \cdot s^{-1})(0.9\ \mu M)}{9.5\ \mu M \cdot s^{-1} - 3.5\ \mu M \cdot s^{-1}}$$

$$= 3.15\ \mu M/6 = 0.525\ \mu M$$

7. Calculate the reciprocals of [S] and v_0 and construct a plot of $1/v_0$ versus $1/[S]$. The intercept on the $1/[S]$ axis is $-0.1\ mM^{-1}$, which is equal to $-1/K_M$. Therefore, $K_M = 10\ mM$. The intercept on the $1/v_0$ axis is $0.05\ mM^{-1} \cdot s$, which is equal to $1/V_{max}$. Therefore, $V_{max} = 20\ mM \cdot s^{-1}$.

8. Since $K_M^{app} = 2\ K_M$, $\alpha = 2$.

$$K_I = \frac{[I]}{\alpha - 1} = \frac{5\ \mu M}{2 - 1} = \frac{5\ \mu M}{1} = 5\ \mu M$$

9. First, calculate a using Equation 7-29:

$$\alpha = 1 + \frac{[I]}{K_I} = 1 + \frac{6\ \mu M}{2\ \mu M} = 1 + 3 = 4$$

The apparent K_M is $\alpha\ K_M = (4)(8\ \mu M) = 32\ \mu M$.

10. For inhibitor A, $\alpha = 2$ and $K_I = \dfrac{[I]}{\alpha - 1} = \dfrac{2\ \mu M}{2 - 1} = 2\ \mu M$

For inhibitor B, $\alpha = 4$ and $K_I = \dfrac{[I]}{\alpha - 1} = \dfrac{9\ \mu M}{4 - 1} = 3\ \mu M$

The ratio is $3\ \mu M/2\ \mu M$ or 1.5.

Chapter 9

1.

$$\Delta \psi = 0.058 \log \frac{[Na^+]_{in}}{[Na^+]_{out}}$$

$$\log[Na^+]_{in} = \frac{\Delta \psi}{0.058} + \log[Na^+]_{out}$$

$$\log[Na^+]_{in} = \frac{-0.05}{0.058} + \log(0.160)$$

$$\log[Na^+]_{in} = -0.86 - 0.796$$

$$\log[Na^+]_{in} = -1.66$$

$$[Na^+]_{in} = 0.022\ M = 22\ mM$$

2.

$$\Delta \psi = 0.058 \log \frac{[Na^+]_{in}}{[Na^+]_{out}}$$

$$= 0.058 \log \frac{(0.010)}{(0.100)}$$

$$= 0.058 \log (0.10)$$

$$= -0.058\ V = -58\ mV$$

3.

$$\Delta G = RT\ (-4.6)$$

$$= (8.3145\ J \cdot K^{-1} \cdot mol^{-1})(298\ K)(-4.6)$$

$$= -11{,}397\ J \cdot mol^{-1} = -11.4\ kJ \cdot mol^{-1}$$

4.

$$\Delta G = RT \ln \frac{[glucose]_{in}}{[glucose]_{out}}$$

$$= RT \ln \frac{(0.0005)}{(0.005)}$$

$$= (8.3145\ J \cdot K^{-1} \cdot mol^{-1})(293\ K)(-2.3)$$

$$= -5610\ J \cdot mol^{-1} = -5.61\ kJ \cdot mol^{-1}$$

5.

$$\Delta G = RT \ln \frac{[glucose]_{in}}{[glucose]_{out}}$$

$$= RT \ln \frac{(0.008)}{(0.0008)}$$

$$= (8.3145\ J \cdot K^{-1} \cdot mol^{-1})(293\ K)(2.3)$$

$$= +5610\ J \cdot mol^{-1} = +5.61\ kJ \cdot mol^{-1}$$

6.

$$\Delta G = RT \ln \frac{[X]_{in}}{[X]_{out}} + ZF\Delta\psi$$

$$= (8.3145\ J \cdot K^{-1} \cdot mol^{-1})(293\ K) \ln \frac{(0.050)}{(0.010)}$$

$$+ (1)(96{,}485\ J \cdot V^{-1} \cdot mol^{-1})(-0.05\ V)$$

$$= 3920\ J \cdot mol^{-1} - 4820\ J \cdot mol^{-1}$$

$$= -900\ J \cdot mol^{-1} = -0.9\ kJ \cdot mol^{-1}$$

This process is spontaneous.

7.

$$\Delta G = RT \ln \frac{[X]_{in}}{[X]_{out}} + ZF\Delta\psi$$

$$= (8.3145\ J \cdot K^{-1} \cdot mol^{-1})(293\ K) \ln \frac{(0.025)}{(0.100)}$$

$$+ (1)(96{,}485\ J \cdot V^{-1} \cdot mol^{-1})(0.05\ V)$$

$$= -3380\ J \cdot mol^{-1} + 4820\ J \cdot mol^{-1}$$

$$= 1440\ J \cdot mol^{-1} = 1.44\ kJ \cdot mol^{-1}$$

This process is not spontaneous.

Chapter 10

1. Because 45% of the receptors are occupied, $[R \cdot L] = 12.6\ \mu M$ and $[R] = 15.4\ \mu M$.

$$K_d = \frac{[R][L]}{[R \cdot L]}$$

$$= \frac{(15.4 \times 10^{-6})(10 \times 10^{-6})}{(12.6 \times 10^{-6})}$$

$$= 12.2 \times 10^{-6}\ M = 12.2\ \mu M$$

2.

$$K_d = \frac{[R][L]}{[R \cdot L]}$$

$$[R \cdot L] = \frac{[R][L]}{K_d}$$

$$= \frac{(0.008)(0.020)}{(0.005)}$$

$$= 0.032\ M = 32\ mM$$

3. Let $[R \cdot L] = x$ and $[R] = 20\ mM - x$.

$$K_d = \frac{[R][L]}{[R \cdot L]}$$

$$[R \cdot L] = x = \frac{[R][L]}{K_d}$$

$$x = \frac{(0.02 - x)(0.005)}{(0.01)}$$

$$x = \frac{0.0001 - 0.005x}{0.01}$$

$$0.01x = 0.0001 - 0.005x$$

$$0.015x = 0.0001$$

$$x = 0.0067 \text{ M} = 6.7 \text{ mM}$$

The percentage of receptors occupied by ligand is 6.7 mM/20 mM or 33%.

Chapter 12

1. $\Delta G^{\circ\prime} = -RT \ln K_{eq}$

$$= -(8.3145 \text{ J} \cdot \text{K}^{-1} \cdot \text{mol}^{-1})(298 \text{ K}) \ln (0.55)$$

$$= 1480.5 \text{ J} \cdot \text{mol}^{-1} = 1.48 \text{ kJ} \cdot \text{mol}^{-1}$$

2. $\Delta G^{\circ\prime} = -RT \ln K_{eq}$

$$= -(8.3145 \text{ J} \cdot \text{K}^{-1} \cdot \text{mol}^{-1})(310 \text{ K}) \ln (0.55)$$

$$= 1541 \text{ J} \cdot \text{mol}^{-1} = 1.54 \text{ kJ} \cdot \text{mol}^{-1}$$

Increasing the temperature makes the change in free energy slightly more positive.

3. Since $\Delta G^{\circ\prime} = -RT \ln K_{eq}$,

$$\ln K_{eq} = -\Delta G^{\circ\prime}/RT$$

$$K_{eq} = e^{-\Delta G^{\circ\prime}/RT}$$

$$= e^{-(-5,000 \text{ J} \cdot \text{mol}^{-1})/(8.3145 \text{ J} \cdot \text{K}^{-1} \cdot \text{mol}^{-1})(310 \text{ K})}$$

$$= e^{1.94} = 7$$

4. $\Delta G = \Delta G^{\circ\prime} + RT \ln \dfrac{[\text{glucose-6-phosphate}]}{[\text{glucose-1-phosphate}]}$

$$= -7100 \text{ J} \cdot \text{mol}^{-1}$$

$$+ (8.3145 \text{ J} \cdot \text{K}^{-1} \cdot \text{mol}^{-1})(310 \text{ K})\ln(0.020/0.004)$$

$$= -7100 \text{ J} \cdot \text{mol}^{-1} + 4150 \text{ J} \cdot \text{mol}^{-1}$$

$$= -2950 \text{ J} \cdot \text{mol}^{-1} = -2.9 \text{ kJ} \cdot \text{mol}^{-1}$$

Because $\Delta G < 0$, the reaction is spontaneous.

5. $\quad K_{eq} = 17.6 = \dfrac{[\text{glucose-6-phosphate}]}{[\text{glucose-1-phosphate}]}$

[From sample calculation 12.2]

[glucose-1-phosphate] = [glucose-6-phosphate]/17.6

$$= 0.0045/17.6$$

$$= 2.5 \text{ mM}$$

$$= 0.0025 \text{ M}$$

6. $\quad \Delta G = \Delta G^{\circ\prime} + RT \ln \dfrac{[\text{glucose-6-phosphate}]}{[\text{glucose-1-phosphate}]}$

$$\ln \frac{[\text{glucose-6-phosphate}]}{[\text{glucose-1-phosphate}]} = \frac{\Delta G - \Delta G^{\circ\prime}}{RT}$$

$$\ln \frac{[\text{glucose-6-phosphate}]}{[\text{glucose-1-phosphate}]} = \frac{-4000 - (-7100 \text{ J} \cdot \text{mol}^{-1})}{(8.3145 \text{ J} \cdot \text{K}^{-1} \cdot \text{mol}^{-1})(310 \text{ K})}$$

$$\ln \frac{[\text{glucose-6-phosphate}]}{[\text{glucose-1-phosphate}]} = \frac{3100}{2600}$$

$$\ln \frac{[\text{glucose-6-phosphate}]}{[\text{glucose-1-phosphate}]} = 1.2$$

$$\frac{[\text{glucose-6-phosphate}]}{[\text{glucose-1-phosphate}]} = 3.3$$

Chapter 15

1. Use Equation 15-1.

$$\mathcal{E} = \mathcal{E}^{\circ\prime} - \frac{RT}{n\mathcal{F}} \ln \frac{[\text{A}_{reduced}]}{[\text{A}_{oxidized}]}$$

$$= 0.031 \text{ V} - \frac{(8.3145 \text{ J} \cdot \text{K}^{-1} \cdot \text{mol}^{-1})(310 \text{ K})}{(2)(96,485 \text{ J} \cdot \text{V}^{-1} \cdot \text{mol}^{-1})} \ln \frac{(5 \times 10^{-4})}{(4 \times 10^{-4})}$$

$$= 0.031 \text{ V} - 0.003 \text{ V} = 0.028 \text{ V}$$

2. Rearrange Equation 15-2:

$$\mathcal{E} = \mathcal{E}^{\circ\prime} - \frac{0.026 \text{ V}}{n} \ln \frac{[\text{A}_{reduced}]}{[\text{A}_{oxidized}]}$$

$$\mathcal{E}^{\circ\prime} = \mathcal{E} + \frac{0.026 \text{ V}}{n} \ln \frac{[\text{A}_{reduced}]}{[\text{A}_{oxidized}]}$$

$$= 0.2 \text{ V} + \frac{0.026 \text{ V}}{2} \ln \frac{(2.5 \times 10^{-6})}{(1 \times 10^{-4})}$$

$$= 0.2 \text{ V} + (-0.048 \text{ V}) = 0.152 \text{ V}$$

3. The relevant half-reactions are

malate $\longrightarrow$ oxaloacetate $+ 2 \text{ H}^+ + 2 \ e^-$	$\mathcal{E}^{\circ\prime} = +0.166 \text{ V}$
ubiquinone $+ 2 \text{ H}^+ + 2 \ e^- \longrightarrow$ ubiquinol	$\mathcal{E}^{\circ\prime} = +0.045 \text{ V}$

net: malate + ubiquinone $\longrightarrow$ $\qquad\qquad \Delta\mathcal{E}^{\circ\prime} = 0.211 \text{ V}$
oxaloacetate + ubiquinol

$$\Delta G^{\circ\prime} = -n\mathcal{F}\Delta\mathcal{E}^{\circ\prime}$$

$$= -(2)(96,485 \text{ J} \cdot \text{V}^{-1} \cdot \text{mol}^{-1})(0.211 \text{ V})$$

$$= -40,700 \text{ J} \cdot \text{mol}^{-1} = -40.7 \text{ kJ} \cdot \text{mol}^{-1}$$

4. The relevant half-reactions are

acetaldehyde $+ 2 \text{ H}^+ + 2 \ e^- \longrightarrow$ ethanol	$\mathcal{E}^{\circ\prime} = -0.197 \text{ V}$
NADH $\longrightarrow$ NAD$^+ + \text{H}^+ + 2 \ e^-$	$\mathcal{E}^{\circ\prime} = +0.315 \text{ V}$

net: acetaldehyde + NADH $\longrightarrow$ $\qquad\qquad \Delta\mathcal{E}^{\circ\prime} = 0.118 \text{ V}$
ethanol + NAD$^+$

$$\Delta G^{\circ\prime} = -n\mathcal{F}\Delta\mathcal{E}^{\circ\prime}$$

$$= -(2)(96,485 \text{ J} \cdot \text{V}^{-1} \cdot \text{mol}^{-1})(0.118 \text{ V})$$

$$= -22,800 \text{ J} \cdot \text{mol}^{-1} = -22.8 \text{ kJ} \cdot \text{mol}^{-1}$$

5. For the reverse reaction,

cytochrome $c_1(\text{Fe}^{3+}) + e^- \longrightarrow$ cytochrome $c_1(\text{Fe}^{2+})$ $\mathcal{E}^{\circ\prime} = 0.22 \text{ V}$
cytochrome c (Fe^{2+}) $\longrightarrow$ cytochrome c (Fe^{3+}) $+ e^-$ $\mathcal{E}^{\circ\prime} = -0.235 \text{ V}$

$$\Delta\mathcal{E}^{\circ\prime} = 0.22 \text{ V} + (-0.235 \text{ V}) = -0.015 \text{ V}$$

$$\Delta G^{\circ\prime} = -n\mathcal{F}\Delta\mathcal{E}^{\circ\prime}$$

$$= -(1)(96,485 \text{ J} \cdot \text{V}^{-1} \cdot \text{mol}^{-1})(-0.015 \text{ V})$$

$$= 1400 \text{ J} \cdot \text{mol}^{-1} = 1.4 \text{ kJ} \cdot \text{mol}^{-1}$$

6. Use the equation derived in Sample Calculation 15-3:

$$\Delta G = 2.303 \, RT \, (\text{pH}_{in} - \text{pH}_{out}) + Z\mathcal{F}\Delta\psi$$

$$\Delta G = 2.303 \, (8.3145 \text{ J} \cdot \text{K}^{-1} \cdot \text{mol}^{-1})(310 \text{ K})(7.5 - 7.25)$$

$$+ (1)(96,485 \text{ J} \cdot \text{V}^{-1} \cdot \text{mol}^{-1})(0.170 \text{ V})$$

$$= 1780 \text{ J} \cdot \text{mol}^{-1} + 16,400 \text{ J} \cdot \text{mol}^{-1}$$

$$= 18,200 \text{ J} \cdot \text{mol}^{-1} = 18.2 \text{ kJ} \cdot \text{mol}^{-1}$$

7. Use the equation derived in Sample Calculation 15-3:

$$\Delta G = 2.303\,RT\,(\text{pH}_{in} - \text{pH}_{out}) + Z\mathcal{F}\Delta\psi$$

$$
\begin{aligned}
(\text{pH}_{in} - \text{pH}_{out}) &= \frac{\Delta G - Z\mathcal{F}\Delta\psi}{2.303\,RT} \\[6pt]
&= \frac{30{,}500\ \text{J}\cdot\text{mol}^{-1} - (1)(96{,}485\ \text{J}\cdot\text{V}^{-1}\cdot\text{mol}^{-1})(0.170\ \text{V})}{2.303\ (8.3145\ \text{J}\cdot\text{K}^{-1}\cdot\text{mol}^{-1})(298\ \text{K})} \\[6pt]
&= \frac{30{,}500\ \text{J}\cdot\text{mol}^{-1} - 16{,}400\ \text{J}\cdot\text{mol}^{-1}}{5700\ \text{J}\cdot\text{mol}^{-1}} \\[6pt]
&= 2.5
\end{aligned}
$$

The pH of the matrix must be 2.5 units greater than the pH of the cytosol.

Chapter 16

1. $E = (3.6 \times 10^{-19}\ \text{J})(6.022 \times 10^{23}\ \text{mol}^{-1})$

$\quad = 217{,}000\ \text{J}\cdot\text{mol}^{-1} = 217\ \text{kJ}\cdot\text{mol}^{-1}$

2. $E = \dfrac{hc}{\lambda} \times N$

$\quad \lambda = \dfrac{hcN}{E}$

$$
\quad = \frac{(6.626 \times 10^{-34}\ \text{J}\cdot\text{s})(2.998 \times 10^{8}\ \text{m}\cdot\text{s}^{-1})(6.022 \times 10^{23}\ \text{mol}^{-1})}{300 \times 10^{3}\ \text{J}\cdot\text{mol}^{-1}}
$$

$\quad = 398 \times 10^{-9}\ \text{m} = 398\ \text{nm}$

Chapter 1

1. (a) ketone; **(b)** ether; **(c)** phosphoric acid ester; **(d)** thiol.

3.

Vitamin C

Vitamin K₃

Coenzyme Q

5. Amino acids, monosaccharides, nucleotides, and lipids are the four types of biological small molecules. Amino acids, monosaccharides, and nucleotides can form polymers of proteins, polysaccharides, and nucleic acids, respectively.

7. Gln has an amido group and Cys has a sulfhydryl group.

9. (a) Fructose has the same molecular formula, $C_6H_{12}O_6$, as glucose. **(b)** Fructose is a ketone, whereas glucose is an aldehyde.

11. Nucleotides consist of a five-carbon sugar, a nitrogenous ring, and one or more phosphoryl groups linked covalently together.

13. It is a lipid (it is actually lecithin). It is mostly C and H, with relatively little O and only one N and one P. It has too little O to be a carbohydrate, too little N to be a protein, and too little P to be a nucleic acid.

15. A diet high in protein results in a high urea concentration, since urea is the body's method of ridding itself of extra nitrogen. Nitrogen is found in proteins but is not found in significant amounts in lipids or carbohydrates. A low-protein diet provides the patient with just enough protein for tissue repair and growth. In the absence of excess protein consumption, urea production decreases, and this puts less strain on the patient's weakened kidneys.

17. You could add the compound that contains the most nitrogen, compound B, which is melamine. [Melamine is a substance that in the past was added to some pet foods and milk products from China so that they would appear to contain more protein. Melamine is toxic to pets and children.] Compound C is an amino acid, so it would already be present in protein-containing food.

19. DNA forms a more regular structure because DNA consists of only four different nucleotides, whereas proteins are made up of as many as 20 different amino acids. In addition, the 20 amino acids have much more individual variation in their structures than do the four nucleotides. Both of these factors result in a more regular structure for DNA. The cellular role of DNA relies on the *sequence* of the nucleotides that make up the DNA, not on the overall shape of the DNA molecule itself. On the other hand, proteins fold into unique shapes, as illustrated by endothelin in Figure 1.4. The ability of proteins to fold into a wide variety of shapes means that proteins can also serve a wide variety of biochemical roles in the cell. According to Table 1.2, the major roles of proteins in the cell are to carry out metabolic reactions and to support cellular structures.

21. Pancreatic amylase is unable to digest the glycosidic bonds that link the glucose residues in cellulose. Figure 1.6 shows the structural differences between starch and cellulose. Pancreatic amylase binds to starch prior to catalyzing the hydrolysis of the glycosidic bond; thus the enzyme and the starch must have shapes that are complementary. The enzyme would be unable to bind to the cellulose, whose structure is much different from that of starch.

23. (a)

25. A positive entropy change indicates that the system has become more disordered; a negative entropy change indicates that the system has become more ordered. **(a)** negative; **(b)** positive; **(c)** positive; **(d)** positive; **(e)** negative.

27. The dissolution of urea in water is an endothermic process and has a positive ΔH value. In order for the process to be spontaneous, the process must also have a positive ΔS value in order for the free energy change of the process to be negative. Solutions have a higher degree of entropy than the solvent and solute alone.

29. The dissolution of calcium chloride in water is a highly exothermic process, as indicated by the negative value of ΔH. This means that when calcium chloride dissolves in water, the system loses heat to the surroundings and the surroundings become warm. The plastic bag holding the calcium chloride solution becomes warm and can be used as a hot pack by the camper at cold temperatures.

31. First, calculate ΔH and ΔS, as described in Sample Calculation 1.1:

$$\Delta H = H_B - H_A$$
$$\Delta H = 50 \text{ kJ} \cdot \text{mol}^{-1} - 43 \text{ kJ} \cdot \text{mol}^{-1}$$
$$\Delta H = 7 \text{ kJ} \cdot \text{mol}^{-1}$$
$$\Delta S = S_B - S_A$$
$$\Delta S = 37 \text{ J} \cdot \text{K}^{-1} \cdot \text{mol}^{-1} - 11 \text{ J} \cdot \text{K}^{-1} \cdot \text{mol}^{-1}$$
$$\Delta S = 26 \text{ J} \cdot \text{K}^{-1} \cdot \text{mol}^{-1}$$

(a) $\Delta G = (7000 \text{ J} \cdot \text{mol}^{-1}) - (4 + 273 \text{ K})(26 \text{ J} \cdot \text{K}^{-1} \cdot \text{mol}^{-1})$
$\Delta G = 200 \text{ J} \cdot \text{mol}^{-1}$

The reaction is not favorable at 4°C.

(b) $\Delta G = (7000 \text{ J} \cdot \text{mol}^{-1}) - (37 + 273 \text{ K})(26 \text{ J} \cdot \text{K}^{-1} \cdot \text{mol}^{-1})$
$\Delta G = -1060 \text{ J} \cdot \text{mol}^{-1}$

The reaction is favorable at 37°C.

33. Process **(a)** is always spontaneous; processes **(b)** and **(c)** are likely to be spontaneous, depending on the temperature, and process **(d)** is never spontaneous.

35. (a) Entropy decreases when the antibody–protein complex binds because the value of ΔS is negative.

(b) $\Delta G = \Delta H - T\Delta S$
$\Delta G = -87,900 \text{ J} \cdot \text{mol}^{-1} - (298 \text{ K})(-118 \text{ J} \cdot \text{K}^{-1} \cdot \text{mol}^{-1})$
$\Delta G = -52.7 \text{ kJ} \cdot \text{mol}^{-1}$

The negative value of ΔG indicates that the complex forms spontaneously.

(c) The second antibody binds to cytochrome *c* more readily than the first because the change in free energy of binding is a more negative value. [From Raman, C. S., Allen, M. J., and Nall, B. T. *Biochemistry* **34,** 5831–5838 (1995).]

37. (a) The reaction is not favorable because the ΔG value for the reaction is positive, indicating an endergonic process.

(b)

GAP + P_i + NAD$^+$ → 1,3BPG + NADH	$\Delta G = +6.7$ kJ · mol^{-1}
1,3BPG + ADP → 3PG + ATP	$\Delta G = -18.8$ kJ · mol^{-1}

GAP + P_i + NAD$^+$ + ADP → 3PG + NADH + ATP $\Delta G = -12.1$ kJ · mol^{-1}

The coupled reaction is spontaneous because the ΔG value is negative.

39. B (most oxidized), C, A (most reduced)

41. (a) oxidized; (b) oxidized; (c) reduced; (d) oxidized.

43. (a) Palmitate's carbon atoms, which have the formula —CH$_2$—, are more reduced than CO$_2$, so their reoxidation to CO$_2$ releases free energy.
(b) Because the —CH$_2$— groups of palmitate are more reduced than those of glucose (—HCOH—), their conversion to the fully oxidized CO$_2$ would be even more thermodynamically favorable (have a larger negative value of ΔG) than the conversion of glucose carbons to CO$_2$. Therefore, palmitate carbons provide more free energy than glucose carbons.

45. The first biological molecules would have had to polymerize, and they would have had to find some way to make copies of themselves.

47. It is difficult to envision how a single engulfment event could have given rise to a stable and heritable association of the eukaryotic host and the bacterial dependent within a single generation. It is much more likely that natural selection gradually promoted the interdependence of the cells. Over many generations, genetic information supporting the association would have become widespread.

49. (a) H15 and H7 are closely related, as are H4 and H14.
(b) H4 and H14 are most closely related to H3.

Chapter 2

1. Because the partial negative charges are arranged symmetrically (and the shape of the molecule is linear), the molecule as a whole is not polar.

$$\overset{\delta^-}{O}=\overset{\delta^+}{C}=\overset{\delta^-}{O}$$

3. Water has the higher boiling point because, although each molecule has the same geometry and can form hydrogen bonds with its neighbors, the hydrogen bonds formed between water molecules are stronger than those formed between H$_2$S molecules. The electronegativity difference between H and O is greater than that between H and S and results in greater differences in the partial charges on the atoms in the water molecule.

5. The arrows point toward hydrogen acceptors and away from hydrogen donors:

Aspartame

Uric acid

Sulfanilamide

7. From the highest melting point to the lowest melting point: (c), (b), (e), (a), (d). Compound (c) (urea, melting point 133°C) has three functional groups that can serve as hydrogen donors and/or acceptors. Compound (b) (acetamide, melting point 80.16°C) has one less —NH group than Compound (c) and therefore forms fewer hydrogen bonds. Compound (e) (propionaldehyde, melting point −80°C) has one functional group that can serve as a hydrogen bond acceptor, but it has no donors, so dipole–dipole forces are the strongest intermolecular forces in a sample of this compound. Compound (a) (methyl ethyl ether, melting point −113°C) also has one functional group that can serve as a hydrogen bond acceptor (it has no donors), but it has hydrocarbon portions that interact with one another via London dispersion forces. Compound (d) (pentane, melting point −139.67°C) is nonpolar and experiences only London dispersion forces, so it has the lowest melting point in the group.

Compound (d) is completely nonpolar and cannot form hydrogen bonds, so it will most likely be insoluble in water. Compounds (e), (b), and (a) can form hydrogen bonds with water, but these molecules also have hydrocarbon portions that will not interact with water favorably, so they will be only slightly soluble in water. Compound (c) forms many hydrogen bonds with water and is likely to be very soluble in water.

9. (a) H < C < S < N < O < F
(b) The greater an atom's electronegativity, the more polar its bond with H and the greater its ability to act as a hydrogen bond acceptor. Thus, N, O, and F, which have relatively high electronegativities, can act as hydrogen bond acceptors, whereas C and S, whose electronegativities are only slightly greater than hydrogen's, cannot.

11. (a) hydrogen bonding; (b) ionic interactions; (c) van der Waals forces (London dispersion forces).

13. Methanol, which has the highest dielectric constant, would be the best solvent for the cationic NH$_4^+$. The polarity of the alcohols, which all contain a primary —OH group, varies with the size of the hydrocarbon portion. 1-Butanol, with the largest hydrophobic group, is the least polar and therefore has the lowest dielectric constant.

15. Aquatic organisms that live in the pond are able to survive the winter. Since the water at the bottom of the pond remains in the liquid form instead of freezing, the organisms are able to move around. The ice on top of the pond also serves as an insulating layer from the cold winter air.

17. Structure A depicts a polar compound, while structure B depicts an ionic compound similar to a salt like sodium chloride. This is more consistent with glycine's physical properties as a white crystalline solid with a high melting point. While structure A could be water soluble because of its ability to form hydrogen bonds, the high solubility of glycine in water is more consistent with an ionic compound whose positively and negatively charged groups are hydrated in aqueous solution by water molecules.

19. (a) Surface tension is defined as the force that must be applied to surface molecules in a liquid so that they may experience the same forces as the molecules in the interior of the liquid. Water's surface tension is greater than ethanol's because the strength and number of water's intermolecular forces (hydrogen bonds) are both greater. Ethanol's —OH group also forms hydrogen bonds, but the hydrocarbon portion of the molecule cannot interact favorably with water, and weaker London dispersion forces form instead.
(b) The kinetic energy of the water molecules increases when temperature increases. Intermolecular forces are weaker in strength as a consequence of the increased molecular motion. Because surface tension increases when the strength of intermolecular forces increases as described in part (a), surface tension decreases when temperature increases.

21. Compounds (a) and (e) are amphiphilic, compound (b) is nonpolar, and compounds (c) and (d) are polar.

23. (a) In the nonpolar solvent, AOT's polar head group faces the interior of the micelle, and its nonpolar tails face the solvent.

(b) The protein, which contains numerous polar groups, interacts with the polar AOT groups in the micelle interior.

25. (a) The nonpolar core of the lipid bilayer helps prevent the passage of water since the polar water molecules cannot easily penetrate the hydrophobic core of the bilayer. **(b)** Most human cells are surrounded by a fluid containing about 150 mM Na^+ and slightly less Cl^- (see Fig. 2.12). A solution containing 150 mM NaCl mimics the extracellular fluid and therefore helps maintain the isolated cells in near-normal conditions. If the cells were placed in pure water, water would tend to enter the cells by osmosis; this might cause the cells to burst.

27. Vesicles consist of a lipid bilayer that closes up to enclose an aqueous compartment. The polar drug readily dissolves in this aqueous compartment. Delivery to the cell is accomplished when the vesicle membrane fuses with the cell membrane, releasing the drug into the cytosol.

29. The amount of Na^+ (atomic weight 23 g · mol^{-1}) lost in 18 minutes, assuming a fluid loss rate of 2 L per hour and a sweat Na^+ concentration of 50 mM (Box 2.B), is

$$0.30\,h \times \frac{2\,L}{h} \times \frac{0.05\,mol}{L} \times \frac{23\,g}{mol} \times \frac{1000\,mg\,Na^+}{g\,Na^+} \times \frac{1\,oz\,chips}{200\,mg\,Na^+}$$

$$= 3.45\ oz\ chips$$

It would take 3.45 ounces of potato chips (about a handful) to replace the lost sodium ions.

31. Since the molecular mass of H_2O is 18.0 g · mol^{-1}, a given volume (for example, 1 L or 1000 g) has a molar concentration of 1000 g · L^{-1} ÷ 18.0 g · mol^{-1} = 55.5 M. By definition, a liter of water at pH 7.0 has a hydrogen ion concentration of 1.0×10^{-7} M. Therefore the ratio of $[H_2O]$ to $[H^+]$ is 55.5 M/(1.0×10^{-7} M) = 5.55×10^8.

33. In aqueous solution, where virtually all biochemical reactions take place, an extremely strong acid such as HCl dissociates completely, so that all its protons are donated to water: $HCl + H_2O \rightarrow H_3O^+ + Cl^-$. This leaves H_3O^+ as the only acidic species remaining.

35.

	Acid, base, or neutral?	pH	$[H^+]$ (M)	$[OH^-]$ (M)
Blood	base	7.42	3.8×10^{-8}	2.6×10^{-7}
Saliva	neutral	7.00	1.0×10^{-7}	1.0×10^{-7}
Urine	acid	6.80	1.6×10^{-7}	6.3×10^{-8}
Gastric juice	acid	2.10	7.9×10^{-3}	1.3×10^{-12}

37. Conjugate base:

(a) $C_2O_4^{2-}$ **(b)** SO_3^{2-} **(c)** HPO_4^{2-} **(d)** CO_3^{2-} **(e)** AsO_4^{3-} **(f)** PO_4^{3-} **(g)** O_2^{2-}

Conjugate acid:

(a) $H_2C_2O_4$ **(b)** H_2SO_3 **(c)** H_3PO_4 **(d)** H_2CO_3 **(e)** $H_2AsO_4^-$ **(f)** $H_2PO_4^-$ **(g)** H_2O_2

39. Convert all the data to either K_a or pK values to evaluate (pK = $-\log K_a$). The greater the K_a value, the stronger the acid—that is, the greater the tendency for the proton to be donated. (The lower the pK value, the stronger the acid.) From strongest to weakest acid: E, D, B, A, C. Note that the stronger the acid, the weaker its conjugate base. For example, citric acid is a stronger acid than citrate, and succinic acid is a stronger acid than succinate.

	Acid	K_a	pK
A	citrate	1.74×10^{-5}	4.76
B	succinic acid	6.17×10^{-5}	4.21
C	succinate	2.29×10^{-6}	5.64
D	formic acid	1.78×10^{-4}	3.75
E	citric acid	7.41×10^{-4}	3.13

41. The carbonate ions accept protons from water and form hydroxide ions (as shown in the equation below), resulting in basic urine.

$$CO_3^{2-}(aq) + H_2O(l) \rightarrow HCO_3^-(aq) + OH^-(aq)$$

43. (a) The final concentration of HCl is (0.0015 L)(2 mol/L) ÷ 2 L = 0.0015 M. Since HCl is a strong acid and dissociates completely, the added $[H^+]$ is equal to [HCl]. (The existing hydrogen ion concentration in the water itself, 1.0×10^{-7} M, can be ignored because it is much smaller than the hydrogen ion concentration contributed by the hydrochloric acid.)

$$pH = -\log[H^+]$$
$$pH = -\log(0.0015)$$
$$pH = 2.8$$

(b) The final concentration of NaOH is (0.0015 L)(2 mol/L) ÷ 2 L = 0.0015 M. Since NaOH dissociates completely, the added $[OH^-]$ is equal to the [NaOH]. (The existing hydroxide ion concentration in the water itself, 1.0×10^{-7} M, can be ignored because it is much smaller than the hydroxide ion concentration contributed by the NaOH.)

$$K_w = 1.0 \times 10^{-14} = [H^+][OH^-]$$
$$[H^+] = \frac{1.0 \times 10^{-14}}{[OH^-]}$$
$$[H^+] = \frac{1.0 \times 10^{-14}}{(0.0015\ M)}$$
$$[H^+] = 6.6 \times 10^{-12}\ M$$
$$pH = -\log[H^+]$$
$$pH = -\log(6.6 \times 10^{-12})$$
$$pH = 11.2$$

45. Use the pK value in Table 2.4 and the Henderson–Hasselbalch equation to calculate the ratio of the concentrations of imidazole (A^-) and the imidazolium ion (HA):

$$pH = pK + \log\frac{[A^-]}{[HA]}$$
$$\log\frac{[A^-]}{[HA]} = pH - pK$$
$$\frac{[A^-]}{[HA]} = 10^{(pH-pK)}$$
$$\frac{[A^-]}{[HA]} = 10^{(7.8-7.0)}$$
$$\frac{[A^-]}{[HA]} = \frac{6.3}{1}$$

47.
$$pH = pK + \log\frac{[A^-]}{[HA]}$$

$$\log\frac{[A^-]}{[HA]} = pH - pK = 5.0 - 4.76 = 0.24$$

$$\frac{[A^-]}{[HA]} = 1.74 \text{ or } [A^-] = 1.74\,[HA]$$

$$[A^-] = 1.74[HA] = 1.74(0.05\text{ M} - [A^-])$$

$$[A^-] = 0.087 - 1.74[A^-]$$

$$2.74[A^-] = 0.087\text{ M}$$

$$[A^-] = 0.032\text{ M or 32 mM}$$

49. Adding NaOH to the acetic acid will convert some of the acetic acid (HA) to acetate (A⁻):

$$NaOH + CH_3COOH \rightarrow Na^+ + CH_3COO^- + H_2O$$

For every mole of NaOH added, one mole of CH_3COOH will be consumed, and one mole of CH_3COO^- will be generated. If x is the number of moles of NaOH added, then x will also be the number of moles of A⁻ generated.

Calculate the initial amount of acetic acid:

The initial amount of acetic acid is 0.10 mol (see Solution 48), so the final amount of acetic acid will be 0.10 mol – x.

$$\frac{[A^-]}{[HA]} = 1.74 = \frac{x}{0.10\text{ mol} - x}$$

$$x = 1.74(0.10\text{ mol} - x) = 0.174\text{ mol} - 1.74\,x$$

$$2.74x = 0.174\text{ mol}$$

$$x = 0.174\text{ mol}/2.74 = 0.0635\text{ mol}$$

Calculate the mass of NaOH to add:

$$0.0635\text{ mol} \times 40.0\text{ g} \cdot \text{mol}^{-1} = 2.54\text{ g}$$

51. The pK of the fluorinated compound would be lower (it is 9.0); that is, the compound becomes less basic and more acidic. This occurs because the F atom, which is highly electronegative, pulls on the nitrogen's electrons, loosening its hold on the proton.

53.
$$pH\,2 \quad {}^+H_3N-CH_2-COOH$$
$$pH\,7 \quad {}^+H_3N-CH_2-COO^-$$
$$pH\,10 \quad H_2N-CH_2-COO^-$$

The carboxylic acid group has a pK of 2.35, and the amino group has a pK of 9.78. The Henderson–Hasselbalch equation can be used to calculate the exact percentage of protonated/unprotonated forms of each functional group, but that really isn't necessary. Instead, the pK values for each group should be compared to the pH. At pH = 2, the pH is below both pK values, so both functional groups are mostly protonated. At pH = 7, the pH is well above the pK for the carboxylic acid group but below the pK for the amino group. Therefore the carboxylic acid group is unprotonated and the amino group is protonated. At pH = 10, the pH is above the pK values of both functional groups. Thus, both groups are mostly unprotonated.

55. (a)
$$HO-(H_2C)_2-H\overset{+}{N}N-(CH_2)_2-SO_3^- \rightleftharpoons$$
Weak acid (HA)

$$HO-(H_2C)_2-NN-(CH_2)_2-SO_3^- + H^+$$
Conjugate base (A)

(b) The pK for HEPES is 7.55; therefore, its effective buffering range is 6.55–8.55.

(c) $1.0\text{ L} \times \dfrac{0.10\text{ mole}}{L} \times \dfrac{260.3\text{ g}}{\text{mol}} = 26\text{ g}$

Weigh 26 g of the HEPES salt and add to a beaker. Dissolve in slightly less than 1.0 liter of water (leave "room" for the HCl solution that will be added in the next step).

(d) At the final pH,

$$\frac{[A^-]}{[HA]} = 10^{(pH-pK)} = 10^{(8.0-7.55)} = 10^{0.45} = 2.82$$

For each mole of HCl added, x, one mole of HEPES salt (A⁻) will be converted to a mole of HEPES acid (HA). The starting amount of A⁻ is $(1.0\text{ L})(0.10\text{ mol} \cdot \text{L}^{-1}) = 0.10$ moles. After the HCl is added, the amount of A⁻ will be 0.10 moles – x, and the amount of HA will be x. Consequently,

$$\frac{[A^-]}{[HA]} = 2.82 = \frac{0.10\text{ mole} - x}{x}$$

$$2.82x = 0.10\text{ mol} - x$$

$$3.82x = 0.10\text{ mol}$$

$$x = 0.10\text{ mol}/3.82 = 0.0262\text{ mol}$$

Calculate how much 6.0 M HCl to add:

$$\frac{0.0262\text{ mol}}{6.0\text{ mol} \cdot \text{L}^{-1}} = 0.0044\text{ L, or 4.4 mL}$$

To make the buffer, dissolve 26 g of HEPES salt [see part c] in less than 1.0 L. Add 4.4 mL of 6.0 M HCl, then add water to bring the final volume to 1.0 L.

57. (a) The three ionizable protons of phosphoric acid have pK values of 2.15, 6.82, and 12.38 (Table 2.4). The pK values are the midpoints of the titration curve:

(b)
$$H_3PO_4 \rightarrow H^+ + H_2PO_4^-$$
$$H_2PO_4^- \rightarrow H^+ + HPO_4^{2-}$$
$$HPO_4^{2-} \rightarrow H^+ + PO_4^{3-}$$

(c) The dissociation of the second proton has a pK of 6.82, which is closest to the pH of blood. Therefore the weak acid present in blood is $H_2PO_4^-$ and the weak base is HPO_4^{2-}.

(d) The dissociation of the third proton has a pK of 12.38. Therefore, a buffer solution at pH 11 would consist of the weak acid HPO_4^{2-} and its conjugate base, PO_4^{3-} (supplied as the sodium salts Na_2HPO_4 and Na_3PO_4).

59. (a)

$$HOH_2C-\underset{\underset{CH_2OH}{|}}{\overset{\overset{CH_2OH}{|}}{C}}-NH_3^+ \rightleftharpoons HOH_2C-\underset{\underset{CH_2OH}{|}}{\overset{\overset{CH_2OH}{|}}{C}}-NH_2 + H^+$$

Weak acid (HA)   Conjugate base (A$^-$)

(b) The pK of Tris is 8.30; therefore, its effective buffering range is 7.30–9.30.

(c) Rearranging the Henderson–Hasselbalch equation gives

$$\frac{[A^-]}{[HA]} = 10^{(pH-pK)} = 10^{(8.2-8.3)} = 10^{-0.1} = 0.79$$

Since $[A^-] + [HA] = 0.10$ M, $[A^-] = 0.10$ M $- [HA]$,

and $\dfrac{(0.10\,\text{M} - [HA])}{[HA]} = 0.79$

$0.79\,[HA] = 0.10\,\text{M} - [HA]$

$1.79\,[HA] = 0.10\,\text{M}$

$$[HA] = \frac{0.10\,\text{M}}{1.79} = 0.056\,\text{M} = 56\,\text{mM}$$

$[A^-] + [HA] = 0.10$ M $= 100$ mM, so $[A^-] = 44$ mM

(d) When HCl is added, an equivalent amount of Tris base (A$^-$) is converted to Tris acid (HA).

Let $x =$ moles of H^+ added $= (0.0015\,\text{L})(3.0\,\text{mol} \cdot \text{L}^{-1}) = 0.0045$ moles $= 4.5$ mmol.

The final amount of A$^-$ is 44 mmol $-$ 4.5 mmol $=$ 39.5 mmol.

The final amount of HA is 56 mmol $+$ 4.5 mmol $=$ 60.5 mmol.

Use the Henderson–Hasselbalch equation to calculate the new pH:

$$pH = pK + \log\frac{[A^-]}{[HA]}$$

$$pH = 8.3 + \log\frac{39.5\,\text{mmol} \div 1001.5\,\text{mL}}{60.5\,\text{mmol} \div 1001.5\,\text{mL}}$$

$$pH = 8.3 + (-0.2)$$

$$pH = 8.1$$

The buffer is effective: The pH decreases about 0.1 unit (from 8.2 to 8.1) with the addition of the strong acid. In comparison, the addition of the same amount of acid to water, which is not buffered, results in a pH change from approximately 7.0 to 2.3 (see Problem 43a).

(e) When NaOH is added, an equivalent amount of Tris acid (HA) is converted to Tris base (A$^-$). Let $x =$ moles of OH^- added $= (0.0015\,\text{L})(3.0\,\text{mol} \cdot \text{L}^{-1}) = 0.0045$ moles $= 4.5$ mmol.

The final amount of A$^-$ is 44 mmol $+$ 4.5 mmol $=$ 48.5 mmol.

The final amount of HA is 56 mmol $-$ 4.5 mmol $=$ 51.5 mmol.

Use the Henderson–Hasselbalch equation to calculate the new pH:

$$pH = pK + \log\frac{[A^-]}{[HA]}$$

$$pH = 8.3 + \log\frac{48.5\,\text{mmol} \div 1001.5\,\text{mL}}{51.5\,\text{mmol} \div 1001.5\,\text{mL}}$$

$$pH = 8.3 + (-0.026)$$

$$pH = 8.27$$

The buffer is effective: The pH increases only 0.07 unit (from 8.2 to 8.27) with the addition of the strong base. In comparison, the addition of the same amount of base to water, which is not buffered, results in a pH change from approximately 7.0 to 11.6 (see Problem 43b).

61. $H^+(aq) + HCO_3^-(aq) \rightleftharpoons H_2CO_3(aq) \rightleftharpoons H_2O(l) + CO_2(aq)$

Failure to eliminate CO_2 in the lungs would cause a buildup of $CO_2(aq)$. This would shift the equilibrium of the above equations to the left. The increase in $CO_2(aq)$ would lead to the increased production of carbonic acid, which would in turn dissociate to form additional hydrogen ions, causing acidosis.

63. During hyperventilation, too much CO_2 (which is equivalent to H^+ in the form of carbonic acid) is given off, resulting in respiratory alkalosis. By repeatedly inhaling the expired air, the individual can recover some of this CO_2 and restore acid–base balance.

65. Ammonia and ammonium ions are in equilibrium, as represented by the following equation:

$$NH_4^+ \rightleftharpoons H^+ + NH_3$$

Carbonic acid and bicarbonate ions are in equilibrium, as represented by the following equation:

$$H_2CO_3 \rightleftharpoons H^+ + HCO_3^-$$

Phosphate ions are in equilibrium, according to the following equation:

$$H_2PO_4^- \rightleftharpoons H^+ + HPO_4^{2-}$$

In metabolic acidosis, the concentration of protons increases, so the equilibrium shifts to form $H_2PO_4^-$, carbonic acid, and ammonium ions. In order to bring the pH back to normal, the kidney excretes $H_2PO_4^-$ and ammonium ions, and bicarbonate ions are reabsorbed. The result is a decrease in the concentration of protons and an increase in blood pH.

67. The concentrations of both Na^+ and Cl^- are greater outside the cell than inside (see Fig. 2.12). Therefore the movement of these ions into the cell is thermodynamically favorable. Na^+ movement into the cell drives the exit of H^+ via an exchange protein in the plasma membrane (the favorable movement of Na^+ into the cell "pays for" the unfavorable movement of H^+ out of the cell). Similarly, the movement of Cl^- into the cell drives the movement of HCO_3^- out of the cell through another exchange protein.

69. Because it is small and nonpolar (see Solution 24), CO_2 can quickly diffuse across cell membranes to exit the tissues and enter red blood cells.

Chapter 3

1. The heat treatment destroys the polysaccharide capsule of the wild-type *Pneumococcus*, but the DNA survives the heat treatment. The DNA then "invades" the mutant *Pneumococcus* and supplies the genes encoding the enzymes needed for capsule synthesis that the mutant lacks. The mutant is now able to synthesize a capsule and has the capacity to cause disease, which results in the death of the mice and the appearance of encapsulated *Pneumococcus* in the mouse tissue.

3. Some of the labeled "parent" DNA appears in the progeny, but none of the labeled protein appears in the progeny. This indicates that the bacteriophage DNA is involved in the production of progeny bacteriophages, but bacteriophage protein is not required.

5.

3-Methylguanine

7. The base 5-chlorouracil is a substitute for thymine (5-methyluracil).

9. Thymine (5-methyl uracil) contains a methyl group attached to C5 of the pyrimidine ring of uracil.

11. The compound is a thymidine analog. [From Maity, J. and Stromberg, R., *Molecules* **18**, 12740–12750 (2013).]

5-Bromo-2′-deoxyuridine

13. (a)

Flavin adenine dinucleotide (FAD)

(b) A diphosphate bridge links the ribose groups in each dinucleotide. This linkage is a variation of the monophosphate bridge (phosphodiester linkage) in DNA and RNA.

(c) The adenosine group in CoA bears a phosphoryl group on C3′.

15. It is a G:C base pair.

17. The statement is false because the greater stability of GC-rich DNA is due to the stronger stacking interactions involving G:C base pairs and does not depend on the number of hydrogen bonds in the base pairs.

19. The sugar–phosphate backbone is on the outside of the molecule. The polar sugar groups can form hydrogen bonds with the surrounding water molecules. The negatively charged phosphate groups interact favorably with positively charged ions. The nonpolar nitrogenous bases are found on the inside of the molecule and interact favorably via stacking interactions. In this way, contact with the aqueous solution is minimized, as described by the hydrophobic effect.

21. (a) Using Chargaff's rules (see Solution 20), the number of C residues must also be 22,356. Subtracting (2 × 22,356) from 98,534 yields 53,822 (A + T) residues. Dividing this number by 2 yields 26,911 residues each of A and T. **(b)** GenBank reports only the sequence of a single strand of DNA, since the sequence of the complementary strand can easily be deduced using Watson–Crick base-pairing.

23. (a) Proteins are more likely to bind to the major groove, which can easily accommodate proteins. The larger surface area of the major groove allows multiple favorable interactions between the DNA and protein. **(b)** The positively charged side chains of the Lys and Arg residues form ion pairs with the negatively charged phosphate groups on the DNA backbone. These are strong interactions, so the histones have a high affinity for DNA.

25. (a) The DNA contains 50% G + C, so its melting point would be approximately 90°C. **(b)** The DNA would need to be cooled gradually to 65–70°C (20–25°C below its melting temperature).

27. The positively charged sodium ions can form ion pairs with the negatively charged phosphate groups on the DNA backbone and "shield" the negative charges from one another. This increases the overall stability of DNA and makes it more difficult to melt.

29. Heating denatures the target DNA (separates its two strands) so that a single-stranded probe can more easily form sequence-specific hydrogen bonds with it.

31. (a) The DNA isolated after one generation is a homogeneous sample of DNA with a density intermediate between DNA containing all ^{14}N and all ^{15}N. **(b)** The DNA isolated after the second generation is heterogeneous. Half of the DNA has the same density as the first generation; half of the DNA consists of all ^{14}N DNA and has a lower density.

= ¹⁵N
= ¹⁴N

First-generation daughter molecules

+

Second-generation daughter molecules

+   +   +

(c) A hypothetical scheme for conservative DNA replication is shown below. The DNA isolated after both the first and the second generations is heterogeneous. Half of the DNA has the same high density as the original DNA (all ¹⁵N); half of the DNA consists of all ¹⁴N DNA and has a lower density.

= ¹⁵N
= ¹⁴N

First-generation daughter molecules

+

Second-generation daughter molecules

+   +   +

33. (a) The cell-free system produces polypeptides consisting of alternating Val and Cys residues. Some of the polypeptides begin with Val; others begin with Cys, depending on which reading frame is used. **(b)** The production of this peptide does not allow one to unambiguously assign the GUG and UGU codons. Additional experimental data are required to make this assignment.

35. (a) 3′-TATGCTACGACTTAAACTCA-5′
(b) 5′-AUACGAUGCUGAAUUUGAGU-3′

37. There are 216 codons possible: 6³ = 216. [From Malyshev, D. A., Dhami, K., Lavergne, T., Chen, T., Dai, N., Foster, J. M., Corrêa Jr., I. R., and Romesberg, F. E., *Nature* **509**, 385–388 (2014).]

39. The same segment of DNA can encode two different proteins if each strand is a coding strand.

41. (a) Asparagine has two codons, AAU and AAC (see Table 3.3). An A→G mutation at the second position could generate a codon for serine (AGU or AGC). **(b)** The CGA codon codes for the amino acid arginine; the mutation (C→T in the DNA) converts the codon to a Stop codon. When the mRNA for the gene is translated, protein synthesis terminates prematurely and the truncated protein is nonfunctional.

43. (a) The siRNA is an "antisense" mRNA, so its sequence should be complementary to that of the mRNA. The solution below shows an siRNA that corresponds to the 5′ end of the mRNA:

mRNA 5′→3′ AUG GGC UCC AUC GGU GCA GCA AGC AUG GAA
siRNA 3′→5′ UAC CCG AGG UAG CCA CGU CGU UCG UAC CUU

(b) When the antisense RNA binds to the mRNA, the mRNA can no longer serve as a template for protein synthesis. (In the cell, the binding of antisense RNA to mRNA involves an enzyme that degrades the mRNA into smaller fragments.) **(c)** The siRNA molecules must be able to cross the cell membrane to enter the cell and find the target mRNA.

45. Prokaryotes tend to have smaller genomes than eukaryotes, so evolution has shaped the prokaryote genome to pack in genes more efficiently. Eukaryotes, with larger genomes and more noncoding DNA, have more space to arrange genes.

47. The 35 million differences out of 3.0 billion total nucleotides represent approximately 1%, or a bit less than the original claim. (This number reflects single-base differences and does not account for insertions and deletions of multiple bases.)

49. (a) The first reading frame is the longest ORF.

First reading frame:

TAT GGG ATG GCT GAG TAC AGC ACG TTG AAT GAG
Tyr - Gly - Met - Ala - Glu - Tyr - Ser - Thr - Leu - Asn - Glu-

GCG ATG GCC GCT GGT GAT G
Ala - Met - Ala - Ala - Gly - Asp-

Second reading frame:

T ATG GGA TGG CTG AGT ACA GCA CGT TGA ATG AGG
-Met - Gly - Trp - Leu - Ser - Thr - Ala - Arg - Stop - Met - Arg-

CGA TGG CCG CTG GTG ATG
Arg - Trp - Pro - Leu - Val - Met

Third reading frame:

TA TGG GAT GGC TGA GTA CAG CAC GTT GAA TGA GGC
---Trp - Asp - Gly - Stop - Val - Gln - His - Val - Glu - Stop - Gly-

GAT GGC CGC TGG TGA TG
Asp - Gly - Arg - Trp - Stop

(b) Assuming the reading frame has been correctly identified, the most likely start site is the first Met residue in the first ORF.

51. Because ~1.4% of the human genome consists of protein-coding genes, about 1.4% of the 10 million SNPs, or 140,000, are likely to affect these genes. Since the total number of genes is only ~21,000, in theory, every gene would differ.

53. (a) The strongest associations are located between positions 67,370,000 and 67,470,000. **(b)** Gene B contains SNPs associated with the disease whereas Genes A and C do not. [From Duerr, R. H., et al., *Science* **314**, 1461–1463 (2006).]

55. MspI, AsuI, EcoRI, PstI, SauI, and NotI generate sticky ends. AluI and EcoRV generate blunt ends.

57. AluI cleaves the sequence at two locations and EcoRI and NotI cleave the sequence at one location each. The other enzymes listed in Table 3.5 do not cleave this segment of the plasmid DNA.

59. (a) A HaeII digest of the linear 2743-bp pGEM-3Z plasmid (see Figure 3.16) would produce four bands, as shown below (1–323 produces a 323-bp band, 324–693 produces a 370-bp band, 694–2564 produces an 1871-bp band, and 2565–2743 produces a 179-bp band). **(b)** A digest of the circular plasmid produces the 370-bp band and the 1871-bp band but also produces a 502-bp band (2565–323) in place of the 323-bp band and the 179-bp band. [From Promega.]

61. A restriction endonuclease is often used to prepare fragments of DNA for insertion into a cloning vector. Since the cloned DNA contains the recognition site (whose sequence is known), this sequence can be used as a starting point to sequence the unknown DNA segment.

63. Primers with high GC content have high T_m values. If the annealing temperature is much lower than the melting temperature, improper base pairing may occur (see Figure 3.8) and the desired gene fragment may not be amplified.

65. You can use a DNA polymerase that is not heat-stable. You would have to cool the reaction mixture to a temperature at which the polymerase works best and you would have to add the enzyme at each reaction cycle because it would be destroyed every time the temperature was raised to melt the double-stranded DNA.

67. To design an oligonucleotide probe for a gene, the researcher must apply the genetic code in reverse, that is, select codons that correspond to the amino acids in the protein. Most amino acids can be encoded by more than one codon, so the researcher would have to choose one of them and hope that it matched the DNA well enough for the probe to successfully hybridize with the DNA (see Solution 66). Met and Trp, however, are encoded by only one codon each, so by using these codons as part of the probe, the researcher can be assured of a perfect match with the DNA, at least for these three nucleotides.

69. If you were to use plasmids, you would need at least 150,000 clones to accommodate the 3×10^6-kb genome in fragments of 20 kb each (see Table 3.6). This is an almost unmanageable number. However, if you were to use yeast artificial chromosomes, you would need only a minimum of 3000 clones to accommodate the 3×10^6-kb genome in fragments of 1000 kb.

Chapter 4

1. (a) His, Phe, Pro, Tyr, Trp; **(b)** His, Phe, Tyr, Trp; **(c)** His, Cys; **(d)** Gly; **(e)** Arg, Lys; **(f)** Asp, Glu; **(g)** Cys, Met.

3. From least soluble to most soluble: Phe, Ile, Trp, Ser, His, Asp. You can use Table 4.3 as a guide, but you should also be able to do this type of problem without using the table.

5. Threonine has two chiral carbons; therefore four stereoisomers are possible.

7.

9.

Aspartame (Asp–Phe–OMe)

11. Histones contain an abundance of the positively charged amino acids lysine and arginine. The positive charges of these amino acid side chains form ion pairs with the negatively charged phosphate groups on the backbone of the DNA molecule and minimize charge–charge repulsion of the negatively charged phosphate groups.

13. (a) The three amino acids are Ser, Tyr, and Gly. **(b)** Cyclization of the polypeptide backbone occurs between the carbonyl carbon of Ser and the amide nitrogen of Gly. **(c)** Oxidation results in a double bond in the Tyr side chain between Cα and Cβ (the second carbon of the side chain):

15.

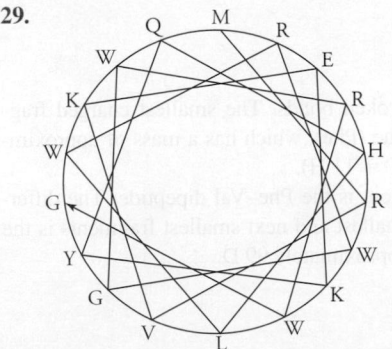

17. **(a)** The net charge is 0, since the Pro side chains are neutral, the Tyr side chains are protonated and neutral, and there is no free N-terminus or C-terminus in the cyclic molecule. **(b)** If the molecule were linear, it would contain a free amino group (+1 charge) and a free carboxylate group (–1 charge), but its net charge would still be 0.

19. There are six possible sequences: HPR, HRP, PHR, PRH, RHP, and RPH.

21. In a free amino acid, the charged amino and carboxylate groups, which are separated only by the α-carbon, electronically influence each other. When the amino acid forms a peptide bond, one of these groups is neutralized, thereby altering the electronic properties of the remaining group.

23. **(a)** tertiary; **(b)** quaternary; **(c)** primary; **(d)** secondary.

25. Both the DNA helix and the α helix turn in the right-handed direction. Both helices have tightly packed interiors; in DNA the interior of the helix is occupied by nitrogenous bases, and in the α helix, the atoms of the polypeptide backbone contact one another. In the α helix, the side chains extend outward from the helix; no such structure exists in the DNA helix.

27. The amino group of Pro is linked to its side chain (see Fig. 4.2), which limits the conformational flexibility of a peptide bond involving the amino group. The geometry of this peptide bond is incompatible with the bond angles required for a polypeptide to form an α helix.

29.

The polar amino acid residues are shown in red and the nonpolar residues in blue. The polar residues are mainly on one side of the helix while the nonpolar residues are on the other side. Quite a few of the polar side chains are positively charged. [This is an example of an amphipathic helix. From Martoglio, B., Graf, R., and Dobberstein, B., *EMBO J.* **16**, 6636–6645 (1997).]

31. The highlighted region, consisting of mainly nonpolar residues, is about the right length to span the membrane (see Problem 30).

PPEEETGERVQLAHHFSEPEITLIIFGVMAGVIGTILLI-SYGIRRLIKKSPSDVKPLPSPDTD

33. Triose phosphate isomerase is an example of an α/β protein.

35. Because Gly lacks a side chain, it often occurs in a protein conformation where no other residue can fit. Gly can take the place of a larger residue more easily than a larger residue, such as Val, can take its place.

37. There are many possible answers for this question. An example is shown for each:

(a) A Lys-Glu ion pair

(b) Hydrogen bonding between Tyr and Ser

(c) van der Waals forces between Leu and Val

39. A polypeptide synthesized in a living cell has a sequence that has been optimized by natural selection so that it folds properly (with hydrophobic residues on the inside and polar residues on the outside). The random sequence of the synthetic peptide cannot direct a coherent folding process, so hydrophobic side chains on different molecules aggregate, causing the polypeptide to precipitate from solution.

41. Anfinsen's ribonuclease experiment demonstrated that a protein's primary structure dictates its three-dimensional structure. Although some proteins, like ribonuclease, can renature spontaneously *in vitro,* most proteins require the assistance of molecular chaperones to fold properly *in vivo.*

43. When the temperature increases, the vibrational and rotational energy of the atoms making up the protein molecules also increases, which increases the chance that the proteins will denature. Increasing the synthesis of chaperones under these conditions allows the cell to renature, or refold, proteins denatured by heat.

45. Proline does not fit well into the structure of the α helix, because of both its geometry (see Problem 29) and the absence of a peptide —NH to contribute to hydrogen bonding (see Problem 37). This amino acid substitution would produce a protein with decreased stability, which would affect the ability of red blood cells to deform their shape in order to squeeze through capillaries. The cells would become damaged and would be removed from circulation, causing anemia. [From Johnson, C. P., Gaetani, M., Ortiz, V., Bhasin, N., Harper, S., Gallagher, P. G., Speicher, D. W., and Discher, D. E., *Blood* **109**, 3538–3543 (2007).]

47. If the proteins were homodimers, they would be more likely to have two identical sites to interact with their palindromic recognition sites in the DNA. Heterodimeric proteins would likely lack the necessary symmetry. (In fact, these enzymes are homodimeric.)

49. The positively charged Arg residues and the negatively charged Asp residues are likely to be found on the surface of the monomer. These residues likely form ion pairs that stabilize the dimeric form. When these residues were mutated to residues with neutral side chains, the ion pairs could not form between the dimers and the equilibrium shifted in favor of the monomers. [From Huang, Y., Misquitta, S., Blond, S. Y., Adams, E., and Colman, R. F., *J. Biol. Chem.* **283**, 32800–32888 (2008).]

51. The extra copy of chromosome 21 increases the amount of the precursor protein and therefore contributes to a higher concentration of the amyloid-β fragment. As a result, amyloid fibers begin forming in the brains of these individuals at a younger age.

53. (a) The extensive α-helical secondary structure in myoglobin makes it unlikely to easily adopt the all-β conformation necessary for amyloid formation. **(b)** This result suggests that any polypeptide—even one whose native conformation is all α-helical—can assume the β conformation if conditions permit.

55. The mutation disrupts the secondary structure of the protein, which in turn alters the tertiary structure of each monomer, disrupting the hydrophobic contacts and hydrogen bonds that hold the tetramer together. [From Yang, M., Lei, M., and Huo, S. *Protein Sci.* **12**, 1222–1231 (2003).]

57. (a) The only ionizable groups in the dipeptide are the C-terminus (p$K \approx 3.5$) and the N-terminus (p$K \approx 9.0$). pI = ½ (3.5 + 9.0) = 6.25. **(b)** In order for the tripeptide to have no net charge, its C-terminal group must be unprotonated (negatively charged), its Tyr side chain must be protonated (neutral), and its N-terminal group must be protonated (positively charged). Because deprotonation of the side chain (p$K \approx 10.5$) or deprotonation of the N-terminal group (p$K \approx 9.0$) would change the net charge, the pK values of these groups (~10.5 and ~9.0) should be used to calculate the pI: pI = ½ (10.5 + 9.0) = 9.8.

59. (a) The protein will likely have a net charge of 0 or greater. The pK value of the C-terminal carboxylate group is 3.5 (Table 4.1), so at pH values below 3.5, the group will be predominantly neutral. All other ionizable groups in proteins have higher pK values and therefore will also be predominantly neutral or positively charged below pH 3.5. **(b)** The protein will likely be negatively charged. The pK value of the N-terminal amino group is 9.0 (Table 4.1), so at pH values above 9.0, the group will be predominantly neutral. All other ionizable groups (except Arg) in proteins have lower pK values and therefore will also be predominantly neutral or negatively charged below pH 9.0.

61. (a)

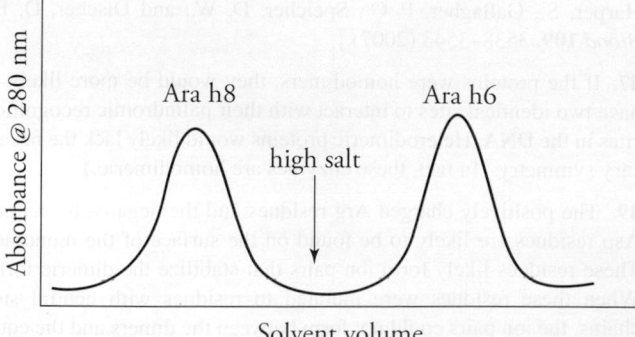

(b) A simplified version of the anion exchange chromatogram is shown.

Ara h8

Ara h6

high salt

Absorbance @ 280 nm

Solvent volume

(c) Treatment of the proteins with a reducing agent followed by iodoacetic acid converts five neutral disulfide bridges in Ara h6 to ten

negatively charged side chains. The treatment produces a modified Ara h6 protein with a more acidic pI. The two proteins were successfully separated because the Ara h8 protein is less attracted to the positively charged anion exchange beads and elutes sooner. The Ara h6 binds more strongly to the anion exchange beads and elutes only after salt is added to the elution buffer. [From Riecken, S., Lindner, B., Petersen, A., Jappe, U., and Becker, W.-M., *Biol. Chem.* **389**, 415–423 (2008).]

63. The protein elutes from the size-exclusion column as an intact dimer with a molecular weight of 160 kD. The protein consists of two identical monomers with a molecular weight of 80 kD. The monomers are held together by hydrophobic interactions, which are disrupted by the detergent when the protein is subjected to SDS-PAGE.

65. Edman degradation of a polypeptide with a disulfide cross-link would not work properly when the first Cys became exposed at the N-terminus of the polypeptide (the Cys would not be released in the next reaction since it would still be covalently linked to a Cys residue farther along the polypeptide chain). Reduction before sequencing breaks the disulfide bond, and alkylation of the two free Cys groups prevents re-formation of the bond.

67. Sequencing this particular peptide using chemical or enzymatic cleavage methods would be difficult. There is only one N-terminal Met, so CNBr would not cleave the peptide. Cleavage with trypsin would not occur because the peptide lacks Lys and Arg residues. Cleavage with elastase would yield single amino acids, so locating areas of overlap would be difficult. Cleavage with chymotrypsin would produce three fragments, but the lack of a second cleavage agent would make it difficult to put the peptide fragments in the correct order. [From Burstein, I. and Schechter, Y., *Biochem.* **17**, 2392–2400 (1978).]

69. Leu and Ile are isomers and have the same mass; therefore, mass spectrometry cannot distinguish them.

71.

Phe   Val   Ala   Asp

(a) Dashed lines indicate broken bonds. The smallest charged fragment is the N-terminal residue (Phe), which has a mass of approximately 149 D (9 C + 1 N + 1 O + 11 H). **(b)** The next smallest fragment is the Phe–Val dipeptide. The difference in mass between the smallest and next smallest fragments is the mass of the Val residue, or approximately 99 D.

Chapter 5

1. Globin lacks an oxygen-binding group and therefore cannot bind O_2. Heme alone is easily oxidized and therefore cannot bind O_2. The bound heme gives a protein such as myoglobin the ability to bind O_2. In turn, the protein helps prevent oxidation of the heme Fe atom.

3. Myoglobin facilitates O_2 diffusion in the cell by acting as a "molecular bucket brigade," accepting O_2 delivered to the cell by hemoglobin and then transferring it to mitochondrial proteins. Myoglobin is 50% saturated with O_2 when the intracellular pO_2 is equal to its p_{50} value, so it functions most effectively under these conditions. At an intracellular pO_2 greater than p_{50}, O_2 remains bound to myoglobin and is not transferred; at pO_2 less than p_{50}, myoglobin doesn't bind sufficient O_2. In either case, the transfer of O_2 from hemoglobin to mitochondrial proteins is compromised.

5. Equation 5.4 can be rearranged to solve for pO_2 at $Y = 0.30$ (30% saturated) and $Y = 0.85$ (85% saturated), letting $K = p_{50} = 2.8$ torr:

(a)
$$Y = \frac{pO_2}{p_{50} + pO_2}$$

$$pO_2 = \frac{p_{50} \times Y}{1 - Y}$$

$$pO_2 = \frac{2.8 \text{ torr} \times 0.30}{(1 - 0.30)}$$

$$pO_2 = 1.2 \text{ torr}$$

(b)
$$Y = \frac{pO_2}{p_{50} + pO_2}$$

$$pO_2 = \frac{p_{50} \times Y}{1 - Y}$$

$$pO_2 = \frac{2.8 \text{ torr} \times 0.80}{(1 - 0.80)}$$

$$pO_2 = 11.2 \text{ torr}$$

7. (a) The interior of the roast is purple because it does not have access to O_2 and thus the myoglobin in the meat is primarily in the deoxygenated form, which is purple. When sliced, the myoglobin is exposed to O_2 and becomes oxygenated, turning red. (b) When meat is cooked, the globin chains are denatured and the iron is oxidized to Fe^{3+}, converting the myoglobin to met-myoglobin, which is brown. (c) In a vacuum package, the cells rapidly use up any available O_2 in metabolism, which results in mostly deoxygenated myoglobin, which is purplish. Meat packaged in oxygen-permeable packaging continues to be metabolically active so that most of the myoglobin is oxygenated and red, which is more appealing to consumers. [Source: "Color Changes in Cooked Beef," James Claus, National Cattlemen's Association.]

9. (a) The p_{50} values from the oxygen binding curve are 0.6, 1.0, and 1.4 torr for the tuna, bonito, and mackerel, respectively. (b) The tuna has the lowest p_{50} value and the highest O_2 binding affinity; the mackerel has the highest p_{50} value and the lowest O_2 affinity. Interestingly, the p_{50} values are the same when adjusted for the body temperatures of the fish habitats, which average 25°C, 20°C, and 13°C for the tuna, bonito, and mackerel, respectively. [From Marcinek, D. J., Bonaventura, J., Wittenberg, J. B., and Block, B. A., *Am. J. Physiol. Regul. Integr. Comp. Physiol.* **280,** R1123–R1133 (2001).]

11. In the arteries, nearly all the hemoglobin is oxygenated and therefore takes on the color of the Fe(II), in which the sixth coordination site is occupied by O_2. Blood that has passed through the capillaries and given up some of its oxygen contains a mixture of oxy- and deoxyhemoglobin. Deoxyhemoglobin, in which the Fe(II) has only five ligands, imparts a bluish tinge to venous blood.

13. Use Equation 5.4 to calculate the fractional saturation (Y) for hyperbolic binding, letting $K = p_{50} = 26$ torr:

$$Y = \frac{pO_2}{p_{50} + pO_2}$$

At 30 torr, $Y = \dfrac{30 \text{ torr}}{26 \text{ torr} + 30 \text{ torr}} = 0.54$

At 100 torr, $Y = \dfrac{100 \text{ torr}}{26 \text{ torr} + 100 \text{ torr}} = 0.79$

Therefore, if hemoglobin exhibited hyperbolic oxygen-binding behavior, it would be only 79% saturated in the lungs (where $pO_2 \approx 100$ torr)

and would exhibit a loss of saturation of only 25% (79% – 54%) in the tissues (where $pO_2 \approx 30$ torr). Hemoglobin's sigmoidal binding behavior allows it to bind more O_2 in the lungs so that it can deliver relatively more O_2 to the tissues (for an overall change in saturation of about 40%; see Fig. 5.7).

15.
$$Y = \frac{(pO_2)^n}{(p_{50})^n + (pO_2)^n}$$

$$Y = \frac{(25 \text{ torr})^3}{(40 \text{ torr})^3 + (25 \text{ torr})^3}$$

$$Y = 0.20$$

$$Y = \frac{(pO_2)^n}{(p_{50})^n + (pO_2)^n}$$

$$Y = \frac{(120 \text{ torr})^3}{(40 \text{ torr})^3 + (120 \text{ torr})^3}$$

$$Y = 0.96$$

17. Because CO binds to Hb 250 times more strongly than O_2 (see Box 5.A), the Hb·CO will remain in the liganded state throughout the course of the experiment. The use of oxygenated hemoglobin might prove to be problematic because O_2 can dissociate from the hemoglobin during the experiment. [From Berenbrink, M., *Resp. Physiol. Neurobi.*, **154,** 165–184 (2006).]

19. (a) His 146, which is protonated, and Asp 94 form an ion pair in the deoxy form of hemoglobin. The protons are produced by cellular respiration. As described by the Bohr effect, an increase in hydrogen ion concentration favors the deoxy form of hemoglobin so that O_2 can be delivered to the tissues. (b) The presence of the negatively charged Asp 94 increases the pK value of the imidazole ring of His and promotes the formation of the ion pair between the two side chains. The increased pK value means that the imidazole ring's affinity for protons is increased. [From Berenbrink, M., *Resp. Physiol. Neurobi.* **154,** 165–184 (2006).]

21. Negatively charged Glu side chains on the surface of the oxygenated lamprey hemoglobin monomer resist association due to the charge–charge repulsion. But when the pH decreases, excess protons bind to the Glu side chains, neutralizing them. The monomers associate to form the deoxygenated tetramer. In this manner, O_2 is delivered to lamprey tissue when the pH decreases in metabolically active tissue. [From Qiu, Y., Maillett, D. H., Knapp, J., Olson, J. S., and Riggs, A. F., *J. Biol. Chem.* **275,** 13517–13528 (2000).]

23. (a) Curve A represents fetal hemoglobin, which has a higher O_2 affinity than adult hemoglobin (curve B). Curve A shows greater fractional saturation than curve B at any given pO_2. (b) In the γ chain, the positively charged His H21 of the β chain is replaced by the neutral Ser, resulting in a central cavity with fewer positive charges. The allosteric effector BPG cannot bind to Hb F as effectively as it binds to Hb A because the negatively charged BPG forms fewer ion pairs in the central cavity of Hb F. BPG plays a role in decreasing O_2 binding affinity, so the inability of BPG to bind to Hb F means that fetal hemoglobin has a higher O_2 binding affinity. Oxygen transfer from the maternal circulation to the fetal circulation can be accomplished more efficiently if the fetal hemoglobin has a higher O_2 affinity than maternal hemoglobin.

25. (a) The p_{50} for hemoglobin in these animals is lower (by ~10 torr) than in their sea-level counterparts. The animals indigenous to high altitudes require a hemoglobin with a higher O_2 affinity (low p_{50} corresponds to high affinity, since this means that a lower pO_2 is required to half-saturate the hemoglobin with oxygen). These animals have hemoglobin molecules with higher O_2-binding affinities that are able to bind O_2 effectively even at the low pO_2 at high altitudes.

(b) Fetal hemoglobin has a higher O_2 affinity than normal adult hemoglobin because amino acid changes in the central cavity of hemoglobin have replaced positively charged amino acid side chains with neutral amino acid side chains (see Solution 23). The neutral side chains cannot form ion pairs with the negatively charged BPG, so BPG (which binds only to the T form) does not bind, the R (high affinity) form of hemoglobin is favored, and the p_{50} decreases. The persistence of fetal hemoglobin in these animals endows the animals with yet another high-affinity O_2 binding strategy so that the hemoglobin can bind O_2, even at the low pO_2 typical of high altitudes.

27. (a) Hb A has a sigmoidal curve, which means that the binding and release of O_2 from normal hemoglobin is cooperative. The hyperbolic binding curve of Hb Great Lakes indicates that there is little cooperativity in O_2 binding and release. **(b)** Hb Great Lakes has a higher affinity for O_2. More than 60% of the mutant hemoglobin has bound O_2. Hb A is about 30% oxygenated. **(c)** Both hemoglobins are essentially 100% oxygenated and therefore have equal affinities. **(d)** Normal hemoglobin is more efficient at O_2 delivery. It delivers about 70% of its bound O_2 (since it is 100% oxygenated at 75 torr and 30% oxygenated at 20 torr). Hb Great Lakes is less efficient since less than 40% of its bound oxygen is delivered to the tissues. [From Rahbar, S., Winkler, K., Louis, J., Rea, C., Blume, K., and Beutler, E., *Blood* **58**, 813–817 (1981).]

29. (a) The oxygen affinity of Hb Providence is greater. The substitution of the neutral Asn for the positively charged Lys results in decreased binding of BPG in the central cavity of hemoglobin, since Lys forms an ion pair with the negatively charged BPG. BPG binds to the T form but not the R form of hemoglobin. Therefore, decreased BPG binding means the R form is favored and the oxygen affinity of the mutant is increased.

(b)

$$-HN-\underset{\underset{\underset{NH_2}{|}}{\overset{\overset{CH_2}{|}}{\underset{\underset{C=O}{|}}{CH}}}{}-C- \quad \overset{NH_4^+}{\underset{H_2O}{\rightleftharpoons}} \quad -HN-\underset{\underset{\underset{O^-}{|}}{\overset{\overset{CH_2}{|}}{\underset{\underset{C=O}{|}}{CH}}}{}-C-$$

(c) The oxygen affinity of Hb Providence Asp is even greater than that of Hb Providence Asn. The presence of the negatively charged Asp repels the negatively charged phosphate groups of BPG, resulting in an even greater decrease in affinity for BPG. Since BPG binds only to deoxyhemoglobin, the inability of BPG to bind to Hb Providence Asp results in a stabilization of the oxygenated form of hemoglobin and an increase in its oxygen affinity. [From Bonaventura, J., Bonaventura, C., Sullivan, B., Ferruzzi, G., McCurdy, P. R., Fox, J., and Moo-Penn, W. F., *J. Biol. Chem.* **251**, 7563–7571 (1976).]

31. A decrease in pH diminishes hemoglobin's affinity for O_2 (the Bohr effect), thereby favoring deoxyhemoglobin. Since only deoxyhemoglobin S polymerizes, sickling of cells is most likely to occur when the parasite-induced drop in pH promotes the formation of deoxyhemoglobin.

33. Hb S migrates more slowly because the Glu → Val mutation makes the protein less negatively charged. The Ala → Asp mutation inHb Ohio makes the protein more negatively charged, so it migrates faster.

35. The substitution at the C-terminus could affect the position of His F8 in such a way that O_2 either binds more readily or dissociates less readily. The substitution is quite near His 146, which binds protons (see Problem 19), and may decrease this side chain's proton-binding affinity. Since His 146 is located in the central cavity, BPG would bind less readily, which favors the oxygenated or R form of hemoglobin.

37. Both types of molecules are proteins and consist of polymers of amino acids. Both contain elements of secondary structure. But globular proteins are water-soluble and nearly spherical in shape. Examples include proteins such as hemoglobin and myoglobin as well as enzymes. Their cellular role involves participating in the chemical reactions of the cell in some way. In contrast, fibrous proteins tend to be water-insoluble and have an elongated shape. Their cellular role is structural—as elements of the cytoskeleton of the cell and the connective tissue matrix between cells.

39. (a) Keratin and collagen; **(b)** myosin and kinesin; **(c)** actin and tubulin; **(d)** actin, myosin, tubulin, and kinesin.

41. Because phalloidin binds to F-actin but not to G-actin, the addition of phalloidin fixes actin in the filamentous form. This impairs cell motility because cell movement requires both actin polymerization at the leading edge of the cell and depolymerization at the trailing edge. In the presence of phalloidin, depolymerization does not occur and cell movement is not possible.

43. During rapid microtubule growth, β-tubulin subunits containing GTP accumulate at the (+) end because GTP hydrolysis occurs following subunit incorporation into the microtubule. In a slowly growing microtubule, the (+) end will contain relatively more GTP that has already been hydrolyzed to GDP. A protein that preferentially binds to (+) ends that contain GTP rather than GDP could thereby distinguish fast- and slow-growing microtubules.

45. Polymers composed of β-tubulin molecules allowed to polymerize in the presence of a nonhydrolyzable analog of GTP are more stable. When the β-tubulin subunits are exposed to GTP in solution, the GTP binds to the β-tubulin and then is hydrolyzed to GDP, which remains bound to the β-tubulin. Additional αβ heterodimers are then added. The microtubule ends with GDP bound to the β-tubulin are less stable than those bound to GTP because protofilaments with GDP bound are curved rather than straight and tend to fray. If a nonhydrolyzable analog is bound, it will resemble GTP and the protofilament will be straight rather than curved. It is less likely to fray and the resulting protofilament is more stable as a result.

47. Microtubules form the mitotic spindle during cell division. Because cancer cells divide rapidly and hence undergo mitosis at a rate more rapid than in most other body cells, drugs that target tubulin and interfere with the formation of the mitotic spindle in some way will slow the growth of cancerous tumors.

49. Colchicine, which promotes microtubule depolymerization, inhibits the mobility of the neutrophils because cell mobility results from polymerization and depolymerization of microtubules.

51. (a) As shown in Figure 5.24, microtubules link replicated chromosomes to two points at opposite sides of the cell. Vinblastine's ability to stabilize the microtubules at the (+) end while destabilizing the (−) end disrupts this linkage. Mitosis slows down or completely halts as a result. [From Panda, D., Jordan, M. A., Chu, K. C., and Wilson, L., *J. Biol. Chem.* **271**, 29807–29812 (1996).]
(b) Mitotic spindles are composed of microtubules and are required for cell division. The formation of the mitotic spindle is a highly dynamic process. The presence of a drug that affects the formation of the mitotic spindle would therefore have a much greater effect on a population of rapidly dividing cells.

53. A fibrous protein such as keratin does not have a discrete globular core. Most of the residues in its coiled-coil structure are exposed to the solvent. The exception is the strip of nonpolar side chains at the interface of the two coils.

55. The difference in sulfur content can be explained by the amount of disulfide bonding in each type of keratin. Hard keratins are less pliable than soft keratins and are stabilized by a greater number of disulfide bonds than in the soft keratins.

57.

59. Consumption of animal protein high in collagen is not required for human collagen synthesis, and this diet is unlikely to cause wrinkles to disappear. A diet that includes high-quality protein and vitamin C–containing fruits and vegetables is sufficient to promote synthesis of collagen.

61. (a) Minoxidil inhibits the lysyl hydroxylase enzyme, and as a result, fewer Lys residues are hydroxylated. **(b)** Since minoxidil inhibits lysyl hydroxylase, newly synthesized collagen chains would be underhydroxylated. Lysines lacking hydroxyl groups decrease the stability of the collagen and increase the likelihood that the collagen will be degraded once it is secreted from the fibroblast. This would be effective in reducing collagen concentrations in patients with fibrosis. **(c)** A similar explanation indicates that long-term minoxidil use in patients without fibrosis has the potential to compromise collagen synthesis in skin fibroblasts. The under-hydroxylated collagen synthesized in the presence of minoxidil will be less stable, which might affect skin structure. The medical literature reports only scalp irritation, dryness, scaling, itching, and redness as a side effect in some men who received topical minoxidil treatments for nearly two years, however. [From Murad, S., Walker, L. C., Tajima, S., and Pinnell, Sr. R., *Arch. Biochem. Biophys.* **308**, 42–47 (1994) and Price, V. H., *N. Eng. J. Med.* **341**, 964–973 (1999).]

63. (a) (Pro–Pro–Gly)$_{10}$ has a melting temperature of 41°C, while (Pro–Hyp–Gly)$_{10}$ has a melting temperature of 60°C. (Pro–Hyp–Gly)$_{10}$ and (Pro–Pro–Gly)$_{10}$ both have an imino acid content of 67%, but (Pro–Hyp–Gly)$_{10}$ contains hydroxyproline, whereas (Pro–Pro–Gly)$_{10}$ does not. Hydroxyproline therefore has a stabilizing effect relative to proline. **(b)** (Pro–Pro–Gly)$_{10}$ and (Gly–Pro–Thr(Gal))$_{10}$ have the same melting point, indicating that they have equal stabilities. This is interesting because (Pro–Pro–Gly)$_{10}$ has an imino acid content

of 67%, whereas (Gly–Pro–Thr(Gal))$_{10}$ has an imino acid content of only 33%. The glycosylated threonine must have a stabilizing effect similar to that of proline. It is possible that the galactose, which contains many hydroxyl groups, provides additional sites for hydrogen bonding and would thus contribute to the stability of the triple helix. **(c)** The inclusion of (Gly–Pro–Thr)$_{10}$ is important because the results show that this molecule doesn't form a triple helix. This molecule is included as a control to show that the increased stability of the (Gly–Pro–Thr(Gal))$_{10}$ is due to the galactose, not to the threonine residue itself. [From Bann, J. G., Peyton, D. H., and Bächinger, H. P., *FEBS Lett.* **473**, 237–240 (2000).]

65. For many proteins, bacterial expression systems offer a convenient source of protein for structural studies. However, even if collagen genes were successfully introduced into bacterial cells, the cells would not be able to produce mature collagen molecules because collagen is processed after it is synthesized. Bacteria are unable to undertake some of the processing steps, such as cleavage by extracellular proteases and covalent modification of Pro and Lys residues.

67. Because collagen has such an unusual amino acid composition (nearly two-thirds of the protein's residues are Gly, Pro, or Pro derivatives), it contains relatively fewer of the other amino acids and is therefore not as good a source of amino acids as proteins containing a greater variety of amino acids. In particular, gelatin lacks tryptophan and contains only small amounts of methionine.

69. The collagenase in the ointment hydrolyzes collagen, a major structural component of the dead skin in the wound, allowing the healing process to take place with the synthesis of new skin collagen. The ointment must be applied carefully only to the dead skin of the wound, avoiding the surrounding tissue. [Source: Santyl.]

71. (a) The collagenase hydrolyzes the peptide bonds of collagen, thereby breaking down the meat to make it easier to chew and digest. **(b)** The collagenase in the fresh fruit will degrade the gelatin (collagen) so that it cannot gel. Cooked fruit contains no collagenase activity because the heat of cooking destroys the collagenase.

73. (a) The open reading frame is shown below. Collagen has the structure (Gly–X–Y)$_n$, where X is often proline and Y is often hydroxyproline. The codon for Gly is GGX, and the codon for Pro is CCX (where X represents any nucleotide). In the mutant polypeptide, a Gly residue has been replaced by Ala.

Normal α1(II) collagen gene:

T | AAC | GGC | GAG | AAG | GGA | GAA | GTT | GGA | CCT | CCT
– Asn – Gly – Glu – Lys – Gly – Glu – Val – Gly – Pro – Pro–

Mutant α1(II) collagen gene:

T | AAC | GGC | GAG | AAG | GCA | GAA | GTT | GGA | CCT | CCT
– Asn – Gly – Glu – Lys – Ala – Glu – Val – Gly – Pro – Pro–

(b) The T_m value for the mutant collagen would be lower than the T_m for normal collagen. The substitution of an Ala for a Gly would disrupt the triple helix in this region of the molecule. In order for the three chains to pack together to form the triple helix, every third residue must be a Gly. There are three amino acids per turn; therefore, the side chain of the Gly ends up on the interior of the triple helix. There is not sufficient room to accommodate a larger side chain of an amino acid such as Ala.
(c) The patient's bones could not form properly due to the defects in the structure of collagen, and the patient died as a result. [From Freisinger, P., Ala-Kokkon, L., LeGuelle, D., Franc, S., Bouvier, R., Ritvaniemi, P., Prockop, D. R., and Bonaventure, J., *J. Biol. Chem.* **269**, 13663–13699 (1994).]

75.

ATP binding to the trailing head induces a conformational change that causes it to release actin.

Hydrolysis of ATP to ADP + P_i triggers a conformational change that swings the trailing head forward. This also increases the affinity of the head for actin.

The new leading head binds to the actin filament. This causes the release of the ADP from the new trailing head.

P_i dissociates from the leading head, preparing the myosin for another reaction cycle.

[From Walker, M. L., Burgess, S. A., Sellers, J. R., Wang, F., Hammer, J. A., III, Trinick, J., and Knight, P. J., *Nature* **405,** 804–807 (2000).]

77. Myosin's two heads act independently, so only one binds to the actin filament at a time. Consequently, when myosin advances from one actin subunit to another, the protein dissociates completely from its track, much like a one-legged hop (see Solution 75). In contrast, kinesin's two heads work together, so one head remains bound to the microtubule track when the other is released. This mechanism is similar to a two-legged walk.

79. Normal bone development involves the formation of bone tissue in response to stresses placed on the bone. When muscle activity is impaired, as in muscular dystrophy, the forces that shape bone development are also abnormal, leading to abnormal bone growth.

Chapter 6

1. A globular protein can bind substrates in a sheltered active site and can support an arrangement of functional groups that facilitates the reaction and stabilizes the transition state. Most fibrous proteins are rigid and extended and therefore cannot surround the substrate to sequester it or promote its chemical transformation.

3. The rate enhancement is calculated as the ratio of the catalyzed rate to the uncatalyzed rate:

$$\frac{61 \text{ s}^{-1}}{1.3 \times 10^{-10} \text{ s}^{-1}} = 4.7 \times 10^{11}$$

[From Bryant, R. A. R., and Hansen, D. E., *J. Am. Chem. Soc.,* **118,** 5498–5499 (1996).]

5. For adenosine deaminase:

$$\frac{370 \text{ s}^{-1}}{1.8 \times 10^{-10} \text{ s}^{-1}} = 2.1 \times 10^{12}$$

For triose phosphate isomerase:

$$\frac{4300 \text{ s}^{-1}}{4.3 \times 10^{-6} \text{ s}^{-1}} = 1.0 \times 10^{9}$$

The rate of the uncatalyzed reaction is slower for the adenosine deaminase reaction than for the triose phosphate isomerase reaction. But adenosine deaminase is able to catalyze its reaction so that it occurs more quickly than the reaction catalyzed by triose phosphate isomerase. Therefore, the rate enhancement for the adenosine deaminase reaction is greater.

7. **(a)** Bonds hydrolyzed in the presence of a peptidase enzyme are indicated by the arrows below. **(b)** A peptidase belongs to the hydrolase class of enzymes.

9. **(a)** oxidoreductase, **(b)** isomerase, **(c)** lyase, **(d)** hydrolase.

11. (a)

Succinate → (succinate dehydrogenase) → Fumarate

(b)

Malate → (malate dehydrogenase) → Oxaloacetate

(c) Both succinate dehydrogenase and malate dehydrogenase are oxidoreductases.

13.

Aspartate + α–Ketoglutarate

↕ aspartate aminotransferase

Oxaloacetate + Glutamate

15. (a) Reaction 4, **(b)** Reaction 1, **(c)** Reaction 3, **(d)** Reaction 2.

17. The common names are **(a)** hexokinase and **(b)** nitrite reductase.

19. At first, the temperature increases the reaction rate because heat increases the proportion of reacting groups that can achieve the transition state in a given time. However, when the temperature rises above a certain point, the heat causes the enzyme, which is a protein, to denature. Because most proteins are only marginally stable (see Section 4.3), denaturation occurs rapidly, accounting for the steep drop in enzymatic activity.

21.

23. Chymotrypsin can degrade neighboring molecules by catalyzing the hydrolysis of peptide bonds on the carboxyl side of Phe, Tyr, and Trp. If the chymotrypsin is stored in a solution of weak acid, His 57 would be protonated and would be unable to accept a proton from Ser 195 to begin hydrolysis.

25. (a) Gly, Ala, and Val have side chains that lack the functional groups required for acid–base or covalent catalysis. **(b)** Mutating one of these residues may alter the conformation at the active site enough to disrupt the arrangement of other groups that are involved in catalysis.

27. (a) In order for any molecule to act as an enzyme, it must be able to recognize and bind a substrate specifically, it must have the appropriate functional groups to carry out a chemical reaction, and it must be able to position those groups for reaction. **(b)** Functional groups on the nitrogen bases can participate in chemical reactions in much the same way as amino acid side chains on proteins. For example, the amino groups on adenine, guanine, and cytosine bases could act as nucleophiles and could also act as proton donors. **(c)** DNA, as a double-stranded molecule, has limited conformational freedom. RNA, which is single-stranded, is able to assume a greater range of conformations. This flexibility allows it to bind to substrates and carry out chemical transformations.

29. His 57 abstracts a proton from Ser 195, thus rendering the serine oxygen a better nucleophile. When Ser 195 is modified by formation of a covalent bond with DIP, the proton is no longer available and Ser 195 is unable to function as a nucleophile. Modification of His 57 with TPCK alkylates the nitrogen of the imidazole ring so that it can no longer abstract protons.

31. (a)

His
Lys   Ser

(b) The experiment indicates that a Lys residue may be part of the active site. It is also possible that a modified Lys elsewhere on the enzyme affects catalytic activity by altering the conformation of the enzyme.

33. The enzyme would be inactive because Cys often functions as a nucleophile and would be unable to do so if a carboxymethyl group were covalently attached to the sulfur.

35. (a)

(b) Since benzoate resembles the substrate, it is likely that benzoate binds to the active site of the enzyme. Under these conditions, FDNP does not have access to the active site and will be unable to react with the tyrosine side chain (which is also assumed to be part of the enzyme's active site because of its unusual reactivity). [From Nishino, T., Massey, V., and Williams, C. H., *J. Biol. Chem.* **255,** 3610–3616 (1979).]

37. The His residue may accept a proton from the sulfhydryl group of Cys, forming the —S⁻ nucleophile. The Ser residue may play a role in properly orienting the Cys and His residues for this proton transfer. [From Sanchez-Pulido, L. and Ponting, C. P., *Bioinformatics* **32,** 1441–1445 (2016).]

39. (a)

His 124
Asp 70
RNA Substrate

(b) His 124 most likely has a pK of less than 5.0 because the His shown in the relay is unprotonated. It must be unprotonated in order to accept a proton from the water molecule. If His 124 had a higher pK value, it would be more likely to be partially protonated at physiological pH and less able to accept a proton.

(c) Alanine has an aliphatic side chain and is unable to accept a proton in the relay, as shown in part a. [From Oda, Y., Yoshida, M., and Kanaya, S., *J. Biol. Chem.* **268,** 88–92 (1993).]

41. (a) RNase A proceeds via acid–base catalysis. His 12 acts as a base and abstracts a proton from the 2′ OH group on the RNA, allowing the oxygen to nucleophilically attack the phosphorus atom. His 119, acting as an acid, protonates the leaving group and promotes the breaking of the bond. **(b)** His 12 has a pK value of 5.4 and His 119 has a pK value of 6.4. His 12 is unprotonated and has the lower pK, while His 119 is protonated and has the higher pK. **(c)** The pH optimum is 6 because at this pH His 12 is unprotonated (the pH is greater than the pK) and His 119 is protonated (the pH is less than the pK). At a pH less than 6, His 12 would be protonated and would not serve as a nucleophile to abstract the 2′-hydroxyl hydrogen. At pH greater than 6, His 119 would be unprotonated and would not be able to donate a hydrogen to the scissile bond. **(d)** Ribonuclease does not hydrolyze DNA because DNA contains deoxyribose and is missing the 2′ hydroxyl group, which serves as the attacking nucleophile for the phosphorus (once the His 12 has removed the hydrogen).

43. (a)

Asp 32
C=O
O⁻
...
OH
R_C
N
C
R_N
O
Asp 215

Asp 32
C=O
OH
...
R_C
N
OH
R_N
C
O⁻
C=O
Asp 215

Asp 32
C=O
O⁻
...
OH
OH
R_N
C
O
C=O   R_CNH₂
Asp 215

(b) Renin uses an acid–base catalytic strategy. **(c)** The p*K* value of Asp 32 is lower than that of Asp 215 since the Asp 32 is unprotonated and the Asp 215 residue is unprotonated.

45.

—(CH₂)₂—C—OH
O
OH
H
N
R_C
R_N
C
O⁻
Glu 224
Zn²⁺
His His Glu

—(CH₂)₂—C—O⁻
O
R_N
C
OH
H₂N
R_C
Glu 224
Zn²⁺
His His Glu

[From Li, L., Binz, T., Niemann, H., Singh, B. R., *Biochemistry* **39**, 2399–2405 (2000).]

47. (a) The deamidation reaction for asparagine is shown. The deamidation reaction for glutamine is similar.

~NH—CH—C~ + H₂O ⟶ ~NH—CH—C~ + NH₄⁺
O
CH₂
C=O
NH₂
Asparagine

O
CH₂
C=O
O⁻
Aspartate

(b)

CH₂
CH₂
C=O
NH₂
H—A
→(1)→
CH₂
CH₂
C=O
NH₃⁺
O
H H
A⁻
→(2)→
CH₂
CH₂
HO—C—O⁻
NH₃⁺
→(3)→

CH₂
CH₂
C=O
O⁻
←(4)←
NH₄⁺
NH₃

CH₂
CH₂
C=O
O—H

(c) Ser and Thr residues could stabilize the transition state. They could also serve as bases (if unprotonated) and accept a proton from water to form a hydroxide ion that would act as the attacking nucleophile. Ser and Thr (in the unprotonated form) could also act as attacking nucleophiles themselves.

(d) The mechanism for the deamidation of an amino-terminal Gln residue is shown. Amino terminal Asn residues are not deamidated because a four-membered ring, which is unstable, would result. [From Wright, H. T., *Crit. Rev. Biochem. Mol. Biol.* **26**, 1–52 (1991).]

NH₂—CH—C—NH— ⟶ NH—CH—C—NH—
O
CH₂
CH₂
C=O
NH₂
H—A
O
CH₂
CH₂
+ NH₃ + HA

(e) Water is a substrate in the reaction. The Asn and Gln residues on the surface of the protein have much greater access to water molecules than interior Asn and Gln residues. [From Wright, H. T., *Crit. Rev. Biochem. Mol. Biol.* **26**, 1–52 (1991).]

49. The ability of an enzyme to accelerate a reaction depends on the free energy difference between the enzyme-bound substrate and the enzyme-bound transition state. As long as this free energy difference is less than the free energy difference between the unbound substrate and the uncatalyzed transition state, the enzyme-mediated reaction proceeds more quickly.

51. (a) In a serine protease, there is no need to exclude water from the active site, since it is a reactant for the hydrolysis reaction catalyzed by the enzyme.

(b) Yes. The smaller substrate (xylose) is unable to trigger the same conformational changes elicited by glucose binding, so the enzyme is not able to fully prevent the ATP hydrolysis reaction.

53. Like serine proteases, a tetrahedral intermediate forms. The intermediate might be stabilized by an "oxyanion hole" similar to that of the serine proteases.

55. The enzyme likely proceeds via a tetrahedral intermediate (shown) in the first step when the urea reacts with water and releases the first molecule of ammonia. [From Kosikowska, P. and Berliki, L., *Expert Opin. Ther. Patents* **21**, 945–957 (2011).]

O⁻
P
H₂N   NH₃⁺
O⁻

57. A mutation can increase or decrease an enzyme's catalytic activity, depending on how it affects the structure and activity of groups in the active site.

59. (a) Trypsin cleaves peptide bonds on the carboxyl side of Lys and Arg residues, which are positively charged at physiological pH. These residues fit into the specificity pocket and interact electrostatically with Asp 189.

(b) A mutant trypsin with a positively charged Lys residue in its specificity pocket would no longer prefer basic side chains because the like charges would repel one another. The mutant trypsin might instead prefer to cleave peptide bonds on the carboxyl side of negatively charged residues such as Glu and Asp, whose side chains could interact electrostatically with the positively charged Lys residue.

(c) If the substrate specificity pocket does not include a positively charged Lys residue, then there would be no reason to expect the mutant enzyme to prefer substrates with acidic side chains. Instead, the mutant enzyme would be more likely to prefer substrates with nonpolar side chains such as Leu or Ile. [From Graf, L., Craik, C. S., Patthy, A., Roczniak, S., Fletterick, R. J., and Rutter, W. J. *Biochemistry* **26**, 2616–2623 (1987).]

61. During chymotrypsin activation, chymotrypsin cleaves other chymotrypsin molecules at a Leu, a Tyr, and an Asn residue. Only one of these (Tyr) fits the standard description of chymotrypsin's specificity. Clearly, chymotrypsin has wider substrate specificity, probably determined in part by the identities of residues near the scissile bond.

63. No, the compound shown in Problem 10 would not be hydrolyzed by chymotrypsin. The side chain on the carboxyl side of the amide bond is an arginine side chain, which would not fit into chymotrypsin's specificity pocket.

65. (a) Persistent activation of trypsinogen to trypsin also results in the activation of chymotrypsinogen to chymotrypsin (see Solution 62) and causes proteolytic destruction of the pancreatic tissue. **(b)** Since trypsin is at the "top of the cascade," it makes sense to inactivate it by using a trypsin inhibitor. [From Hirota, M., Ohmuraya, M., and Baba, H., *Postgrad. Med. J.* **82**, 775–778 (2006).]

67. A protease with extremely narrow substrate specificity (that is, a protease with a single target) would pose no threat to nearby proteins because these proteins would not be recognized as substrates for hydrolysis.

69. The function of factor IXa is to activate factor X in order to promote thrombin activation and fibrin formation. Factor VIIa can also activate factor X, so the physiological effect is similar.

71. Because factor IX is not the only trigger for blood coagulation, partial restoration of factor IX function is enough to supplement factor VIIa–tissue factor. Even if initiation of coagulation is slower than normal, once thrombin has been generated, fibrin production is guaranteed.

73. (a)

$$
\begin{array}{c}
O \\
\| \\
-HN-CH-C- \\
| \\
CH_2 \quad \text{γ-Carboxy-Glu} \\
| \\
CH \\
-OOC \qquad COO^-
\end{array}
$$

(b) The γ-carboxyglutamate side chain has an overall charge of –2 and more effectively binds Ca^{2+} ions than a glutamate side chain with only a single negative charge.

75. The thrombin variant can promote coagulation but cannot be inhibited normally, so it would lead to an increased risk of clotting.

77. The leech, a "sanguivorous" organism, secretes an anticoagulant in its saliva in order to prevent blood clotting while feeding upon the host. It is therefore a good source of anticoagulants.

Chapter 7

1. The hyperbolic shape of the velocity versus substrate curve suggests that the enzyme and substrate physically combine so that the enzyme becomes saturated at high concentrations of substrate. The lock-and-key model describes the interaction between an enzyme and its substrate in terms of a highly specific physical association between the enzyme (lock) and the substrate (key).

3. $v = -\dfrac{d[S]}{dt}$

$v = -\dfrac{0.05\ M}{440\ y \times 365\ d\cdot y^{-1} \times 24\ h\cdot d^{-1} \times 3600\ s\cdot h^{-1}}$

$v = -3.6 \times 10^{-12}\ M\cdot s^{-1}$

5. $v = -\dfrac{d[S]}{dt}$

$v = -\dfrac{0.05\ M}{6 \times 10^6\ y \times 365\ d\cdot y^{-1} \times 24\ h\cdot d^{-1} \times 3600\ s\cdot h^{-1}}$

$v = -2.6 \times 10^{-16}\ M\cdot s^{-1}$

7. (a) $v = \dfrac{d[P]}{dt} = \dfrac{50 \times 10^{-6}\ M}{50\ d \times 24\ h\cdot d^{-1} \times 3600\ s\cdot h^{-1}}$

$v = 11.6 \times 10^{-12}\ M\cdot s^{-1}$

[From Bryant, R. A. R. and Hansen, D. E., *J. Am. Chem. Soc.* **118**, 5498–5499 (1996).]

9. (a) $v = -\dfrac{d[S]}{dt}$

$v = -\dfrac{0.045\ M}{60\ s}$

$v = -7.5 \times 10^{-4}\ M\cdot s^{-1}$

(b) $v = -\dfrac{d[S]}{dt} = \dfrac{2\ d[P]}{dt}$

$v = \dfrac{7.5 \times 10^{-4}\ M\ \text{maltose}}{s} \times \dfrac{2\ \text{mol glucose}}{1\ \text{mol maltose}}$

$v = 1.5 \times 10^{-3}\ M\cdot s^{-1}$

11. $v = k[\text{sucrose}]$
$v = k[\text{trehalose}]$

The reactions are both first-order overall.

13. Using Equation 7.2 and the k value provided in Problem 18:

$$v = k[\text{trehalose}]$$

$[\text{trehalose}] = \dfrac{v}{k}$

$[\text{trehalose}] = \dfrac{3.0 \times 10^{-3}\ M\cdot s^{-1}}{2.6 \times 10^3\ s^{-1}}$

$[\text{trehalose}] = 1.15 \times 10^{-6}\ M = 1.9\ \mu M$

15. (a) The reaction is a second-order reaction because the units of the rate constant k are $M^{-1}\cdot s^{-1}$.

(b) Convert the partial pressure of CO_2 into units of molar concentration by using the ideal gas law:

$$PV = nRT$$

$$\frac{n}{V} = \frac{P}{RT}$$

$$\frac{n}{V} = \frac{45 \text{ torr} \times \dfrac{1 \text{ atm}}{760 \text{ torr}}}{\dfrac{0.0821 \text{ L} \cdot \text{atm}}{\text{K} \cdot \text{mol}} \times 310 \text{ K}}$$

$$\frac{n}{V} = 0.0023 \text{ M}$$

Next, substitute values for concentration and k into the rate equation:

$$\text{rate} = k[RNH_2][CO_2]$$
$$= 4950 \text{ M}^{-1} \cdot \text{s}^{-1} \times 0.5 \times 10^{-3} \text{ M} \times 0.0023 \text{ M}$$
$$= 5.7 \times 10^{-3} \text{ M} \cdot \text{s}^{-1}$$

(c) The rate constant k would increase with increasing pH because an amino group is more likely to be unprotonated (and thus able to react) at a higher pH.

(d) $\text{rate} = k[RNH_2][CO_2]$

$$[CO_2] = \frac{\text{rate}}{k[RNH_2]}$$

$$[CO_2] = \frac{0.045 \text{ M} \cdot \text{s}^{-1}}{4950 \text{ M}^{-1} \cdot \text{s}^{-1} \times 0.5 \times 10^{-3} \text{ M}}$$

$$[CO_2] = 0.018 \text{ M}$$

$$P = \frac{nRT}{V}$$

$$P = \frac{0.018 \text{ mol}}{\text{L}} \times \frac{0.0821 \text{ L} \cdot \text{atm} \times \dfrac{760 \text{ torr}}{1 \text{ atm}}}{\text{K} \cdot \text{mol}} \times 310 \text{ K}$$

$$P = 348 \text{ torr}$$

[From Gros, G., Forster, E., and Lin, L., *J. Biol. Chem.* **251**, 4398–4407 (1976).]

17. See Figure 7.4 for the answers to parts a–d.

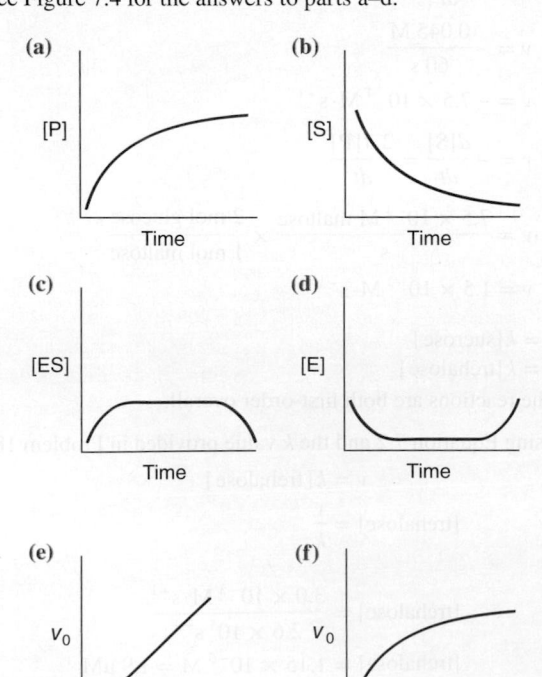

(a) [P] vs Time

(b) [S] vs Time

(c) [ES] vs Time

(d) [E] vs Time

(e) v_0 vs [E]

(f) v_0 vs [S]

19. The apparent K_M would be greater than the true K_M because the experimental substrate concentration would be less than expected if some of the substrate has precipitated out of solution during the reaction.

21. (a) Velocity measurements can be made using any convenient units. K_M is by definition a substrate concentration, so its value does not reflect how the velocity is measured.
(b) It's not necessary to know the enzyme concentration in order to determine K_M or V_{max}; only [S] and v_0 must be known so that a Lineweaver–Burk plot can be constructed. The value of $[E]_T$ is required to calculate k_{cat} since $k_{cat} = V_{max}/[E]_T$, according to Equation 7-23.

23. Rearrange Equation 7.21 to solve for K_M, then substitute the values for v_0, substrate concentration and V_{max}:

$$v_0 = \frac{V_{max}[S]}{K_M + [S]}$$

$$K_M = \frac{[S](V_{max} - v_0)}{v_0}$$

$$K_M = \frac{(10 \text{ mM})(0.36 - 0.23 \text{ U} \cdot \text{min}^{-1} \cdot \text{mL}^{-1})}{0.23 \text{ U} \cdot \text{min}^{-1} \cdot \text{mL}^{-1}}$$

$$K_M = 5.6 \text{ mM}$$

[From Doğru, Y. Z. and Erat, M., *Food Res. Int.* **49**, 411–415 (2012).]

25. Rearrange Equation 7.21 to solve for substrate concentration, then substitute the values for v_0, K_M and V_{max}:

$$v_0 = \frac{V_{max}[S]}{K_M + [S]}$$

$$[S] = \frac{v_0 K_M}{V_{max} - v_0}$$

$$[S] = \frac{(0.3 \text{ μmol} \cdot \text{min}^{-1} \cdot \text{mL}^{-1})(0.6 \text{ mM})}{(1.1 - 0.3 \text{ μmol} \cdot \text{min}^{-1} \cdot \text{mL}^{-1})}$$

$$[S] = 0.2 \text{ mM}$$

[From Botman, D., Tigchelaar, W., and Van Noorden, C. J. F., *J. Histochem. Cytochem.* **62**, 813–826 (2014).]

27. The K_M of enzyme 1 is about 2 mM and the K_M for enzyme 2 is about 5 mM. The V_{max} for enzyme 1 is about 3.5 mM $\cdot$ s^{-1}; the V_{max} for enzyme 2 is about 5.5 mM $\cdot$ s^{-1}. **(a)** The [S] is about half as large as the K_M when [S] = 1 mM for enzyme 1. For enzyme 2, [S] is about one-fifth of the K_M. **(b)** The [S] is five times greater than the K_M for enzyme 1. For enzyme 2, the [S] is about two times greater than the K_M.

29.
$$v_0 = \frac{V_{max} \times 5K_M}{5K_M + K_M}$$

$$v_0 = \frac{V_{max} \times 5K_M}{6K_M}$$

$$v_0 = 0.83 V_{max}$$

When [S] = $5K_M$, the velocity is 83% of the maximum velocity.

$$v_0 = \frac{V_{max} \times 20K_M}{20K_M + K_M}$$

$$v_0 = \frac{V_{max} \times 20K_M}{21K_M}$$

$$v_0 = 0.95 V_{max}$$

When [S] = $20K_M$, the velocity is 95% of the maximum velocity. Therefore, a fourfold increase in substrate concentration causes a smaller proportional increase in velocity (from 83% to 95%). Estimating V_{max} from a plot of v_0 versus [S] is difficult because the substrate concentration must be quite high in order to achieve a velocity close to 100%. (And it is possible that these points on the hyperbolic curve cannot be experimentally measured because high concentrations of the substrate may not be soluble in the reaction medium.) It is better to obtain V_{max} by fitting experimental data to the equation of a hyperbola or, alternatively, from a Lineweaver–Burk plot.

31. (a) N-Acetyltyrosine ethyl ester, with its lower K_M value, has a higher affinity for chymotrypsin. The aromatic tyrosine residue more easily fits into the nonpolar "pocket" on the enzyme (see Fig. 6.17) than does the smaller aliphatic valine residue. **(b)** The value of V_{max} is not related to the value of K_M, so no conclusion can be drawn.

33. (a)
$$k_{cat} = \frac{V_{max}}{[E]_T}$$
$$k_{cat} = \frac{0.36 \text{ U·min}^{-1}\text{·mL}^{-1}}{2.0 \times 10^{-2} \text{ U·mL}^{-1}}$$
$$k_{cat} = 18 \text{ min}^{-1} = 0.30 \text{ s}^{-1}$$

(b)
$$k_{cat} = \frac{V_{max}}{[E]_T}$$
$$k_{cat} = \frac{0.39 \text{ U·min}^{-1}\text{·mL}^{-1}}{2.0 \times 10^{-2} \text{ U·mL}^{-1}}$$
$$k_{cat} = 19.5 \text{ min}^{-1} = 0.33 \text{ s}^{-1}$$

35. (a) The maximum rate at which two molecules can collide with one another in solution is 10^8 to 10^9 $M^{-1}\text{·s}^{-1}$. Enzymes with k_{cat}/K_M values in this range can be considered to be diffusion-controlled, which means the reaction is catalyzed as rapidly as the two reactants can encounter each other in solution. Thus, enzymes B and C are diffusion-controlled but enzyme A is not.

Enzyme	k_{cat}/K_M ($M^{-1}\text{·s}^{-1}$)
A	1.7×10^7
B	2×10^9
C	4.2×10^8

(b) Enzyme B has the greatest catalytic efficiency, so Product Q will be more abundant.

37. The V_{max} can be calculated by taking the reciprocal of the y intercept:
$$V_{max} = \frac{1}{y \text{ int}}$$
$$V_{max} = \frac{1}{4.41 \times 10^{-4} \text{ μM}^{-1}\text{·h}}$$
$$V_{max} = 2.27 \times 10^3 \text{ μM·h}^{-1}$$

The K_M can be determined by first calculating the x intercept and then taking its reciprocal:
$$x \text{ int} = -\frac{b}{m}$$
$$x \text{ int} = -\frac{4.41 \times 10^{-4} \text{ μM}^{-1}\text{·h}}{0.26 \text{ μM}^{-1}\text{·h·μM}}$$
$$x \text{ int} = -1.70 \times 10^{-3} \text{ μM}^{-1}$$
$$K_M = -\frac{1}{x \text{ int}}$$
$$K_M = -\frac{1}{-1.70 \times 10^{-3} \text{ μM}^{-1}}$$
$$K_M = 590 \text{ μM}$$

39. (a) The wild-type enzyme has the lowest K_M, indicating that it has the highest affinity for aspartate. The aspartate substrate is not able to bind to the mutant enzymes as tightly. There is an 18-fold increase in K_M when Arg 386 is mutated to lysine and an 80-fold

increase in K_M when Arg 292 is mutated to lysine (which is similar to the double mutant). The Arg 292 must play a critical role in binding the aspartate substrate to the enzyme.
(b) Although both Arg and Lys have positive charges that would attract the carboxylate group on the substrate, there must be other interactions that are important in addition to ionic interactions. For example, the ability of Arg and Lys side chains to form hydrogen bonds differs. Also, steric considerations may play a role, since the Lys side chain has a different shape than the Arg side chain.
(c) Substitution of Arg with Lys greatly decreases the catalytic efficiency of the mutant enzyme. Changing only one arginine decreases the catalytic efficiency by four orders of magnitude, but the catalytic efficiency decreases by six orders of magnitude when both arginines are replaced.

Enzyme	k_{cat}/K_M ($M^{-1}\text{·s}^{-1}$)
Wild-type AspAT (Arg 292, Arg 386)	1.3×10^5
Mutant AspAT (Lys 292, Arg 386)	1.4×10^1
Mutant AspAT (Arg 292, Lys 386)	1.3×10^2
Mutant AspAT (Lys 292, Lys 386)	1.8×10^{-1}

(d) Changing the residue from Arg to Lys may cause a three-dimensional conformational change that would interfere with the ability of the catalytic apparatus to function properly. If proton donation is important to the mechanism, this would be affected as well, since the lysine side chain donates a proton more readily than the resonance-stabilized guanidino group in the side chain of arginine. [From Vacca, R. A., Giannattasio, S., Graber, R., Sandmeier, E., Marra, E., and Christen, P., J. Biol. Chem. **272**, 21932–21937 (1997).]

41. (a) If an irreversible inhibitor is present, the enzyme's activity would be exactly 100 times lower when the sample is diluted 100-fold. Dilution would not change the degree of inhibition.
(b) If a reversible inhibitor is present, dilution would lower the concentrations of both the enzyme and the inhibitor enough that some inhibitor would dissociate from the enzyme. The enzyme's activity would therefore not be exactly 100 times less than the diluted sample; it would be slightly greater because the proportion of uninhibited enzyme would be greater at the lower concentration.

43. (a) A. competitive; **B.** uncompetitive; **C.** noncompetitive; **D.** mixed; **E.** mixed.
(b) A. K_M increases and V_{max} stays the same; **B.** K_M and V_{max} both decrease; **C.** K_M stays the same and V_{max} decreases; **D.** K_M increases and V_{max} decreases; **E.** K_M and V_{max} both decrease.

45. Indole is a competitive inhibitor of chymotrypsin because its structure resembles the side chain of tryptophan, which fits into the specificity pocket of chymotrypsin. Thus, indole and tryptophan side chains compete with each other for binding to the active site. Indole does not fit into the specificity pocket of either trypsin or elastase and therefore is not an effective competitive inhibitor of either of these enzymes.

47. Since competitive inhibitors compete with substrate for binding to the active site of the enzyme, it is possible (especially if the concentration of the substrate increases in vivo) that this inhibition can be overcome. There might also be cellular complications if these competitive inhibitors, which resemble substrates, accumulate. Noncompetitive and uncompetitive inhibitors do not resemble the substrate, and inhibition cannot be overcome by increasing the concentration of substrate. If these inhibitors bind to their targets very tightly, small amounts can be used to inhibit the enzyme effectively. [From Westley, A. M. and Westley, J., J. Biol. Chem. **271**, 5347–5352 (1996).]

49. (a) NADPH is structurally similar to NADP$^+$ and is likely to be a competitive inhibitor.

(b) The V_{max} is the same in the presence and absence of the inhibitor since inhibition can be overcome at high substrate concentrations. The K_M increases because a higher concentration of substrate is needed to achieve half-maximal activity in the presence of an inhibitor.

(c) The K_M is 400 times greater for NAD$^+$, indicating that the enzyme prefers NADP$^+$ as a cofactor. The differences in V_{max} are not as great. [From Hansen, T., Schicting, B., and Schonheit, P., *FEMS Microbiol. Lett.* **216**, 249–253 (2002).]

51. The compound is a transition state analog (it mimics the planar transition state of the reaction) and therefore acts as a competitive inhibitor.

53. The structure of coformycin structurally resembles the proposed transition state for adenosine deaminase and this supports the proposed structure. However, 1,6-dihydroinosine has a K_I of 1.5×10^{-13} M whereas coformycin's K_I is about 0.25 μM; thus 1,6-dihydroinosine more closely resembles the transition state than does coformycin.

55. $\quad \alpha = \dfrac{K_M^{app}}{K_M}$

$\alpha = \dfrac{40\ \mu M}{10\ \mu M} = 4$

$\alpha = 1 + \dfrac{[I]}{K_I}$

$4 = 1 + \dfrac{30\ \mu M}{K_I}$

$K_I = 10\ \mu M$

[From Gross, R. W. and Sobel, B. E., *J. Biol. Chem.* **258**, 5221–5226 (1983).]

57. The K_M is calculated from the x intercept, the V_{max} from the y intercept (see Solution 37). In the presence of the inhibitor, the V_{max} and the K_M values decrease to a similar extent, as shown in the table. Dodecyl gallate is an uncompetitive inhibitor and does not compete with the substrate for binding to the enzyme. [From Kubo, I., Chen, Q.-X., and Nihei, K.-I., *Food Chem.* **81**, 241–247 (2003).]

	Without inhibitor	With inhibitor
x intercept (mM^{-1})	−0.99	−2.70
K_M (mM)	1.01	0.37
y intercept (min·OD^{-1})	1.51	4.27
V_{max} (OD·min^{-1})	0.66	0.23

59. (a) The Lineweaver–Burk plot is shown below. The K_M is calculated from the x intercept, the V_{max} from the y intercept (see Solution 37). Troglitazone is a competitive inhibitor and competes with progesterone for binding to the active site of the hydroxylase. The K_M value increases over twofold and the V_{max} values are essentially the same.

◆ No inhibitor
■ With inhibitor

	Without inhibitor	With inhibitor
x intercept (μM^{-1})	−0.62	−0.27
K_M (μM)	1.6	3.7
y intercept (pmol^{-1}·min)	86	80
V_{max} (pmol·min^{-1})	0.012	0.012

(b) $\quad \alpha = \dfrac{K_M^{app}}{K_M}$

$\alpha = \dfrac{3.7\ \mu M}{1.6\ \mu M} = 2.3$

$\alpha = 1 + \dfrac{[I]}{K_I}$

$K_I = \dfrac{[I]}{\alpha - 1}$

$K_I = \dfrac{10\ \mu M}{2.3 - 1}$

$K_I = 7.6\ \mu M$

[From Arlt, W., Auchus, J., and Miller, W. L., *J. Biol. Chem.* **276**, 16767–16771 (2001).]

61. To calculate K_I, rearrange Equation 7.28 and solve for α:

$$\alpha = \dfrac{\dfrac{V_{max}[S]}{v_0} - [S]}{K_M}$$

Substitute the values for V_{max} and K_M (in the absence of vanadate). The substrate concentration is 6.67 μM. The values for α are presented in the table:

[Vanadate] (μM)	v_0 (nM·s^{-1})	α
0.0	5.70	1.0
0.2	3.83	1.7
0.4	3.07	2.1
0.7	2.35	2.9
1.0	2.04	3.3
2.0	1.18	6.0
4.0	0.71	10.1

A plot of α versus inhibitor concentration, shown below, gives a slope of 2.25 μM^{-1}. The value of K_I is the reciprocal of the slope, or 0.44 μM.

[From Huyer, G., Liu, S., Kelly, J., Moffat, J., Payette, P., Kennedy, B., Tsaprailis, G., Gresser, M., and Ramachandran, C., *J. Biol. Chem.* **272**, 843–851 (1997).]

63. (a) The Lineweaver–Burk plot is shown below. This is an example of mixed inhibition.

◆ Without PA
■ With PA

(b) The K_M increases by 50% in the presence of PA, and there is a nearly fourfold decrease in V_{max} as shown in the table.

	Without inhibitor	With inhibitor
x intercept, $(mL \cdot mg^{-1})$	–9.60	–6.57
K_M, $(mg \cdot mL^{-1})$	0.104	0.152
y intercept, $(min \cdot mL \cdot nmol^{-1})$	4.03	15.64
V_{max}, $(nmol \cdot mL^{-1} \cdot min^{-1})$	0.248	0.064

[From Kishikawa, K., Chalfant, C. E., Perry, D. K., Bielawska, A., and Hannun, Y. A., *J. Biol. Chem.* **274**, 21335–21341 (1999).]

65. (a) The combination that gives the most effective inhibition is CTP + UTP. More than 90% of the ATCase activity is inhibited when CTP and UTP are both present. In contrast, CTP alone inhibits only ~60% of ATCase activity, and UTP alone inhibits only 5% of ATCase activity.
(b) UTP and CTP are both pyrimidines. Regulation of the ATCase in this manner makes metabolic sense because CTP is synthesized from UTP and thus is produced in larger quantities first. The CTP produced can then inhibit the ATCase enzyme, but not completely. Only when UTP levels rise do UTP and CTP act synergistically to completely inhibit ATCase.
(c) The graph below shows the effect of CTP and UTP combined on ATCase activity. The K_M value is increased, indicating that a higher concentration of aspartate is required to achieve half- maximal activity. Although there isn't enough information to make a definite conclusion, the V_{max} value is probably unchanged, as the V_{max} values are unchanged in the presence of ATP alone and CTP alone. [From Wild, J. R., Loughrey-Chen, S. J., and Corder, *T. S., Proc. Natl. Acad. Sci. USA*, **86**, 46–50 (1989).]

◆ Control ■ With CTP
▲ With ATP ● With CTP and UTP

67. (a) The M_1 isoform of pyruvate kinase is not an allosteric enzyme, as shown by the hyperbolic shape of the curve. F16BP does not influence the activity of the enzyme. **(b)** Substitution of Ala for the Arg at an intersubunit contact converts the pyruvate kinase from a nonallosteric form to an allosteric form, as shown by the sigmoidal curve. F16BP is an allosteric activator of the mutant enzyme, as shown by the shift of the curve to the left in the presence of F16BP. It's possible that the substitution of the neutral Ala for the positively charged Arg produces intersubunit contacts, perhaps in the form of ion pairs, that allow the subunits to behave in a cooperative manner. [From Ikeda, Y., Tanaka, T., and Noguchi, T., *J. Biol. Chem.* **272**, 20495–20501 (1997).]

69. With a mass of 357 $g \cdot mol^{-1}$, only one hydrogen bond donor group, and only six hydrogen bond acceptor groups, the drug obeys the rules.

71. The liver is damaged by acetamidoquinone because the cytochrome P450 enzymes that convert the acetaminophen to its more toxic derivative are located in the liver.

73. (a) Hypertension (high blood pressure) can be treated with enalapril (or a similar drug) which blocks angiotensin II production and thereby keeps blood vessels in a more relaxed (lower-pressure) state.

(b)

Chapter 8

1. (a) $H_3C-(CH_2)_{14}-COO^-$
Palmitate (16:0)

(b) $H_3C-(CH_2)_5-CH=CH-(CH_2)_7-COO^-$
Palmitoleate (16:1n-7)

(c) $H_3C-(CH_2)_4-(CH=CH-CH_2)_3-(CH_2)_3-COO^-$
γ-Linolenate (18:3n-6)

(d) $H_3C-(CH_2)_7-CH=CH-(CH_2)_{13}-COO^-$
Nervonate (24:1n-9)

3.
$H_3C-(CH_2)_4-CH=CH-CH_2-CH=CH-(CH_2)_4-CH=CH-(CH_2)_3-COO^-$
Sciadonate (all-*cis*-$\Delta^{5,11,14}$-eicosatrienoate)

[From Sayanova, O., Haslam, R., Venegas Caleron, M., and Napier, J. A., *Plant Physiol.* **144**, 455–467 (2007).]

5. (a) $H_3C-(CH_2)_{13}-CH=CH-(CH_2)_2-CH=CH-(CH_2)_3-COO^-$
cis, cis-$\Delta^{5,9}$-Tetracosodienoate

(b) $H_3C-(CH_2)_4-(CH=CH-CH_2)_2-(CH_2)_3-CH=CH$
$-(CH_2)_2-CH=CH-(CH_2)_3-COO^-$
all *cis*-$\Delta^{5,9,15,18}$-Tetracosotetraenoate

7. $H_3C-(CH_2)_7-C=C-(CH_2)_7-COOH$ (with H's shown above and below the double bond)

Elaidic acid (*trans*-Δ^9-octadecenoic acid)

$$H_3C-(CH_2)_7-\overset{\displaystyle H}{\underset{\displaystyle H}{C}}=C-(CH_2)_7-COOH$$

Elaidic acid

The melting point of elaidic acid is higher than the melting point of oleic acid, since the trans double bond of elaidic acid gives it an elongated shape, whereas the cis double bond oleic acid gives it a bent shape.

9. (a) Because it has a head group with just one anionic charge, SQDG is likely to substitute for phosphatidylglycerol (net charge −1) rather than phosphatidylethanolamine (with a positive and a negative charge). **(b)** When phosphorus is limited, the organism produces relatively more of the sulfur-containing SQDG to substitute for phospholipids in the cell membrane.

11. (a)
$$H_3C-(CH_2)_{14}-\overset{O}{\overset{\|}{C}}-O-\underset{\underset{CH_2-OH}{|}}{\overset{\overset{CH_2-OH}{|}}{C}}-H \quad + \quad 2\ \ ^-O-\overset{O}{\overset{\|}{C}}-(CH_2)_{14}-CH_3$$

(b) The monoacylglycerol and fatty acid products are both amphipathic molecules, with a polar head and one nonpolar tail. These molecules form micelles as a consequence of the hydrophobic effect (see Section 2.2).

13. The polar head group is indicated in blue; the two nonpolar tails in red:

$$\underset{\underset{\underset{\underset{\text{(inositol ring)}}{O=P-O^-}}{O}}{H_2C-CH-CH_2}}{\overset{\overset{\overset{R_1\quad R_2}{O=C\quad C=O}}{O\quad O}}{}}$$

15.
DPPC
$$O=P-O-(CH_2)_2-N^+(CH_3)_3$$
$$H_2C-CH-CH_2$$
$$O=C\quad C=O$$
$$(CH_2)_{14}\ (CH_2)_{14}$$
$$CH_3\quad CH_3$$

17.
$$\text{Glucose}-O-CH_2$$
$$HO-CH-CH$$
$$CH\quad NH$$
$$CH\quad C=O$$
$$(H_2C)_{12}\ (CH_2)_{26}$$
$$CH_3\quad CH_2$$
$$O-C-(CH_2)_6(CH_2CH=CH)_2(CH_2)_4CH_3$$
$$\overset{\|}{O}$$

19. All except phosphatidylcholine have hydrogen-bonding head groups.

21. The polar head group consists of the phosphate derivative (shown in red in Section 8.1) and the glycerol backbone (shown in black). The two nonpolar tails are shown in blue.

23. (a)
$$O=\overset{\overset{O^-}{|}}{\underset{\underset{O}{|}}{P}}-O-CH_2-\underset{\underset{NH_3^+}{|}}{CH}-COO^-$$
$$H_2C-\underset{\underset{OH}{|}}{CH}-\underset{\underset{\underset{\underset{R_2}{C=O}}{|}}{O}}{CH_2} \quad + \quad R_1-\overset{O}{\overset{\|}{C}}-O^- \quad + \quad H^+$$

(b)
$$\underset{\underset{\underset{\underset{R_1\ R_2}{O=C\ C=O}}{O\ O}}{}}{H_2C-CH-CH_2} \quad + \quad O=\overset{\overset{O^-}{|}}{\underset{\underset{O^-}{|}}{P}}-O-(CH_2)_2-\overset{\overset{CH_3}{|}}{\underset{\underset{CH_3}{|}}{N^+}}-CH_3 \quad + \quad H^+$$

(c)
$$\underset{\underset{\underset{\underset{R_1\ R_2}{O=C\ C=O}}{O\ O}}{}}{H_2C-CH-CH_2} \quad + \quad HO-CH_2-\underset{\underset{OH}{|}}{CH}-CH_2-OH \quad + \quad H^+$$
with phosphate group $O=P-O^-$ shown above.

25. The spicy ingredient in the food is a powder made from peppers that contains the hydrophobic compound capsaicin. Yogurt containing whole milk also contains hydrophobic ingredients that can cleanse the palate of the irritating capsaicin. Water is polar, so it does not dissolve the capsaicin and cannot cleanse the palate.

27. Vitamin A, and the compound from which it is derived, β-carotene, are lipid-soluble molecules. The vegetables in a typical salad do not contain large amounts of lipid. The addition of the lipid-rich avocado provides a means to solubilize the β-carotene and thus increase its absorption. [From Unlu, N. Z., Bohn, T., Clinton, S. K., and Schwartz, S. J., *J. Nutr.* **135,** 431–436 (2005).]

29. Oxidation of retinal (an aldehyde) yields retinoic acid (a carboxylic acid).

31. (a) The traditional definition of a vitamin (a substance that an organism requires but cannot synthesize) implies that vitamins must be obtained in the diet. Because vitamin D_3 can be produced from cholesterol, it is not strictly a vitamin. However, its production does require ultraviolet light, which an individual must obtain through exposure to the sun and which may not always be available.
(b) This is because, vitamin D synthesis requires exposure to sunshine, which is less abundant at higher latitudes.

33. (b) is polar, **(d)** is nonpolar, and **(a), (c),** and **(e)** are amphipathic.

35. (a) A hydrocarbon chain is attached to the glycerol backbone at position 1 by a vinyl ether linkage. In a glycerophospholipid, an acyl group is attached by an ester linkage. **(b)** The presence of this plasmalogen would not have a great effect since it has the same head group and same overall shape as phosphatidylcholine.

37. Lipids that form bilayers are amphipathic, whereas triacylglycerols are nonpolar. Amphipathic molecules orient themselves so that their polar head groups face the aqueous medium on the inside and outside of the cell. Also, triacylglycerols, which lack a large head group, are cone-shaped rather than cylindrical and thus would not fit well in a bilayer structure.

39. Phospholipase A_1 catalyzes the hydrolysis of one of the acyl chains on a phospholipid. The resulting product, a lysophospholipid, has a cone-shaped structure. Lysophospholipids do not assemble to form bilayers as the cylindrical-shaped glycerophospholipids do. The conversion of a significant portion of glycerophospholipids to lysophospholipids results in the destruction of the red blood cell membrane.

41. The melting point of elaidic acid is higher than the melting point of oleic acid, since the *trans* double bond in elaidic acid gives it an elongated shape, whereas the *cis* double bond in oleic acid gives it a bent shape.

43. In general, animal triacylglycerols contain longer and/or more saturated acyl chains than plant triacylglycerols, since these chains have higher melting points and are more likely to be in the crystalline phase at room temperature. The plant triacylglycerols contain shorter and/or less saturated acyl chains in order to remain fluid at room temperature.

45. Unsaturated fatty acids closer to the hoof have a lower melting point due to the presence of double bonds that prevent them from packing together tightly. These lipids therefore help membranes remain fluid even at low temperatures as the reindeer walks through the snow. Saturated fatty acids, which pack together more efficiently and have higher melting points, would decrease membrane fluidity at low temperatures. By increasing the proportion of unsaturated fatty acids from the top of the leg (warm) to the hoof (cold), the reindeer can maintain constant membrane fluidity.

47. The cyclopropane ring in lactobacillic acid produces a bend in the aliphatic chain and thus its melting point is closer to the melting point of oleate, which also has a bend due to the double bond. The presence of bends decreases the opportunity for van der Waals forces to act among neighboring molecules. Less heat is required to disrupt the intermolecular forces, resulting in a melting point that is lower than that of a saturated fatty acid with a similar number of carbons. Therefore stearate has the highest melting point (69.6°C), followed by lactobacillic acid (28°C) and oleate (13.4°C).

49. Cholesterol's planar ring system interferes with the movement of acyl chains and thus tends to decrease membrane fluidity. At the same time, cholesterol prevents close packing of the acyl chains, which tends to prevent their crystallization. The net result is that cholesterol helps the membrane resist melting at high temperatures and resist crystallization at low temperatures. Therefore, in a membrane containing cholesterol, the shift from the crystalline form to the fluid form is more gradual than it would be if cholesterol were absent.

51. The cold temperatures might induce the activation of desaturase enzymes in the plant that convert the 18:0, 18:1, and 18:2 fatty acids to 18:3 fatty acids. Of these four fatty acids, 18:3 is the most unsaturated and has the lowest melting point. Membranes composed of phospholipids containing unsaturated fatty acids maintain their fluidity in cold temperatures. [From Shi, Y., An, L., Zhang, M., Huang, C., Zhang, H., and Xu, S., *Protoplasma* **232**, 173–181 (2008).]

53. The flippase enzyme has a strong preference for PS, which is consistent with the observation that PS is located exclusively in the cytosolic leaflet of the membrane. Translocation of PE also explains the predominance of PE in the cytosolic leaflet. PC and SM, both choline-containing lipids, are not translocated and remain in the extracellular leaflet of the membrane. The flippase has a preference for amino-containing phospholipid head groups and eschews choline-containing phospholipids. The flippase requires ATP, probably because translocation occurs against a concentration gradient, and magnesium ions. A cysteine amino acid side chain must be essential for translocation activity since chemical modification of this side chain abolishes translocation activity. [From Daleke, D. L. and Heustis, W. H., *Biochemistry* **24**, 5406–5416 (1985).]

55. (a) Detergents are required to solubilize a transmembrane protein because the protein domains that interact with the nonpolar acyl chains are highly hydrophobic and would not form favorable interactions with water. **(b)** A schematic diagram of the detergent SDS interacting with a transmembrane protein is shown below. The polar head group of SDS is represented by a circle and the nonpolar tails as wavy lines. The nonpolar tails of SDS interact with the nonpolar

regions of the protein, effectively masking these regions from the polar solvent. The polar head groups of SDS interact favorably with water. The presence of the detergent effectively solubilizes the transmembrane protein so that it can be purified.

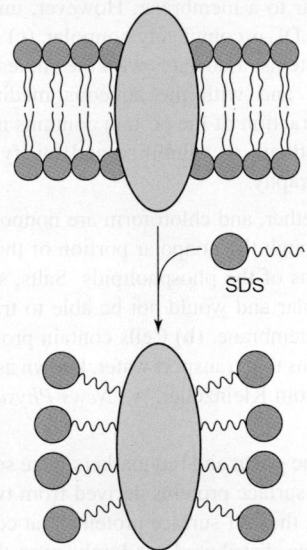

57. (a)

```
      O
      ‖
—HN—CH—C—
      |
      CH₂
      |
      O
      |
    O=C
      |
    (CH₂)₇
      |
     CH
     ‖
     CH
      |
   CH₂)₅
      |
     CH₃
```

(b)

```
      O
      ‖
—HN—CH—C—
      |
      CH₂
      |
      O
      |
    O=C
      |
    (CH₂)₆
      |
     CH₃
```

59. (a) A fully hydrogen-bonded β barrel can form only if the number of strands is even. A β sheet with an odd number of strands could not close up on itself to form a barrel. **(b)** The strands are antiparallel because adjacent strands can be easily linked by loops on the solvent-exposed portions of the protein. **(c)** A β barrel could contain some parallel β strands, but these could not be consecutive. A β barrel with consecutive parallel strands could occur only if the strands were linked by additional membrane-spanning segments (such as a transmembrane α helix or a structure that passed through the center of the barrel).

61. Melittin is more conformationally restricted in the membrane when it is associated with glycerophospholipids containing more saturated fatty acids (like oleate, 18:1) and less restricted when associated with lipids containing more unsaturated fatty acids (like arachidonate, 20:4). The presence of double bonds in the acyl chains produces "bends" that prevent the acyl chains from packing closely together. The unsaturated fatty acids thus provide a less conformationally restricted environment for the melittin peptide. In contrast, when there are fewer double bonds, such as in oleate, the fatty acyl chains pack more closely together and interact via van der Waals forces. This environment restricts the conformation of melittin. [From Raghuraman, H. and Chattopadhyay, A., *Biochim. Biophys. Acta* **1665**, 29–39 (2004).]

63. (a) Cholesterol and cholesteryl esters are highly nonpolar and must be solubilized via their association with protein and phospholipids in order to be transported through the aqueous bloodstream. **(b)** The LDL particle is similar to a membrane in that the surrounding layer consists of phospholipids (however, this is a single layer, rather than a bilayer). Cholesterol is associated with this layer of phospholipids in a manner similar to a membrane. However, unlike a membrane, the interior of the LDL is completely nonpolar. **(c)** Apolipoprotein B, like membrane proteins, associates with the cholesteryl esters via its nonpolar domains and with the aqueous medium via its polar domains. A large fraction of the protein contains nonpolar domains, which makes it difficult to solubilize and purify using traditional column chromatography.

65. (a) Alcohols, ether, and chloroform are nonpolar molecules and can easily pass through the nonpolar portion of the lipid bilayer, the aliphatic acyl chains of the phospholipids. Salts, sugars, and amino acids are highly polar and would not be able to traverse the nonpolar portion of the membrane. **(b)** Cells contain proteins that serve as transporters. Proteins that transport water, known as aquaporins, have been identified. [From Kleinzeller, A., *News Physiol. Sci.* **12**, 49–54 (1997).]

67. After fusion, the green and red markers were segregated because they represent cell-surface proteins derived from two different kinds of cells. Over time, the cell-surface proteins that could diffuse in the lipid bilayer became distributed randomly over the surface of the hybrid cell, so the green and red markers were intermingled. At 15°C, the lipid bilayer was in a gel-like rather than a fluid state, which prevented membrane protein diffusion. Edidin's experiment supports the fluid mosaic model by demonstrating the ability of proteins to diffuse through a fluid membrane.

Chapter 9

1. (a) Use Equation 9.2 and substitute the values for $[Na^+]_{in}$ and $[Na^+]_{out}$:

$$\Delta\psi = 0.058 \text{ V} \log \frac{[Na^+]_{in}}{[Na^+]_{out}}$$

$$\Delta\psi = 0.058 \text{ V} \log \frac{15 \times 10^{-3} \text{ M}}{150 \times 10^{-3} \text{ M}}$$

$$\Delta\psi = -0.058 \text{ V} = -58 \text{ mV}$$

(b)
$$\Delta\psi = 0.058 \text{ V} \log \frac{[Na^+]_{in}}{[Na^+]_{out}}$$

$$\Delta\psi = 0.058 \text{ V} \log \frac{56 \times 10^{-3} \text{ M}}{35 \times 10^{-3} \text{ M}}$$

$$\Delta\psi = 0.012 \text{ V} = 12 \text{ mV}$$

3. Use Equation 9.2 and substitute the values for $[Na^+]_{in}$ and $[Na^+]_{out}$:

$$\Delta\psi = 0.058 \text{ V} \log \frac{[Na^+]_{in}}{[Na^+]_{out}}$$

$$+0.050 \text{ V} = 0.058 \text{ V} \log \frac{[Na^+]_{in}}{[Na^+]_{out}}$$

$$0.86 = \log \frac{[Na^+]_{in}}{[Na^+]_{out}}$$

$$10^{0.86} = \frac{[Na^+]_{in}}{[Na^+]_{out}}$$

$$\frac{7.3}{1} = \frac{[Na^+]_{in}}{[Na^+]_{out}}$$

When a nerve cell is depolarized, sodium ions enter the cell; therefore the $[Na^+]_{in}/[Na^+]_{out}$ ratio is more than 100-fold greater in the depolarized cell than in the resting cell.

5. Use Equation 9.4 and let $Z = 1$ for Na^+ and 2 for Ca^{2+} and $T = 293$ K:

$$\Delta G = RT \ln \frac{[Na^+]_{in}}{[Na^+]_{out}} + Z\mathcal{F}\Delta\psi$$

$$\Delta G = (8.3145 \times 10^{-3} \text{ kJ} \cdot \text{K}^{-1} \cdot \text{mol}^{-1})(293 \text{ K}) \ln \frac{10 \times 10^{-3} \text{ M}}{450 \times 10^{-3} \text{ M}}$$
$$+ (1)(96,485 \times 10^{-3} \text{ kJ} \cdot \text{V}^{-1} \cdot \text{mol}^{-1})(-0.070 \text{ V})$$

$$\Delta G = -9.27 \text{ kJ} \cdot \text{mol}^{-1} - 6.75 \text{ kJ} \cdot \text{mol}^{-1}$$

$$\Delta G = -16.0 \text{ kJ} \cdot \text{mol}^{-1}$$

$$\Delta G = RT \ln \frac{[Ca^{2+}]_{in}}{[Ca^{2+}]_{out}} + Z\mathcal{F}\Delta\psi$$

$$\Delta G = (8.3145 \times 10^{-3} \text{ kJ} \cdot \text{K}^{-1} \cdot \text{mol}^{-1})(293 \text{ K}) \ln \frac{0.1 \times 10^{-6} \text{ M}}{4 \times 10^{-3} \text{ M}}$$
$$+ (2)(96,485 \times 10^{-3} \text{ kJ} \cdot \text{V}^{-1} \cdot \text{mol}^{-1})(-0.070 \text{ V})$$

$$\Delta G = -25.8 \text{ kJ} \cdot \text{mol}^{-1} - 13.5 \text{ kJ} \cdot \text{mol}^{-1}$$

$$\Delta G = -39.3 \text{ kJ} \cdot \text{mol}^{-1}$$

Since the concentrations of both ions are greater outside the cell than inside and the cell potential is negative, the passive movement of the ions will be from outside the cell to inside. In order to maintain the ion concentrations given in the problem, energy-consuming active transport processes are required.

7. Use Equation 9.4 and let $Z = 2$ and $T = 310$ K:

(a) $\Delta G = RT \ln \dfrac{[Ca^{2+}]_{in}}{[Ca^{2+}]_{out}} + Z\mathcal{F}\Delta\psi$

$$\Delta G = (8.3145 \times 10^{-3} \text{ kJ} \cdot \text{K}^{-1} \cdot \text{mol}^{-1})(310 \text{ K}) \ln \frac{0.1 \times 10^{-6} \text{ M}}{1 \times 10^{-3} \text{ M}}$$
$$+ (2)(96,485 \times 10^{-3} \text{ kJ} \cdot \text{V}^{-1} \cdot \text{mol}^{-1})(-0.05 \text{ V})$$

$$\Delta G = -23.7 \text{ kJ} \cdot \text{mol}^{-1} - 9.6 \text{ kJ} \cdot \text{mol}^{-1}$$

$$\Delta G = -33.3 \text{ kJ} \cdot \text{mol}^{-1}$$

The negative value of ΔG indicates a thermodynamically favorable process.

(b) $\Delta G = RT \ln \dfrac{[Ca^{2+}]_{in}}{[Ca^{2+}]_{out}} + Z\mathcal{F}\Delta\psi$

$$\Delta G = (8.3145 \times 10^{-3} \text{ kJ} \cdot \text{K}^{-1} \cdot \text{mol}^{-1})(310 \text{ K}) \ln \frac{0.1 \times 10^{-6} \text{ M}}{1 \times 10^{-3} \text{ M}}$$
$$+ (2)(96,485 \times 10^{-3} \text{ kJ} \cdot \text{V}^{-1} \cdot \text{mol}^{-1})(+0.05 \text{ V})$$

$$\Delta G = -23.7 \text{ kJ} \cdot \text{mol}^{-1} + 9.6 \text{ kJ} \cdot \text{mol}^{-1}$$

$$\Delta G = -14.1 \text{ kJ} \cdot \text{mol}^{-1}$$

The negative value of ΔG indicates a thermodynamically favorable process, but not as favorable as in part a.

9. Use Equation 9.4 and let $Z = 1$ and $T = 293$ K:

$$\Delta G = RT \ln \frac{[Na^+]_{in}}{[Na^+]_{out}} + Z\mathcal{F}\Delta\psi$$

$$\Delta G = (8.3145 \times 10^{-3} \text{ kJ} \cdot \text{K}^{-1} \cdot \text{mol}^{-1})(293 \text{ K}) \ln \frac{30 \times 10^{-3} \text{ M}}{150 \times 10^{-3} \text{ M}}$$
$$+ (1)(96,485 \times 10^{-3} \text{ kJ} \cdot \text{V}^{-1} \cdot \text{mol}^{-1})(+0.05 \text{ V})$$

$$\Delta G = -3.9 \text{ kJ} \cdot \text{mol}^{-1} + 4.8 \text{ kJ} \cdot \text{mol}^{-1}$$

$$\Delta G = 0.88 \text{ kJ} \cdot \text{mol}^{-1}$$

This process is not spontaneous.

11. (a) Since all the terms on the right side of Equation 9.1 are constant, except for T, the following proportion for the two temperatures (310 K and 313 K) applies:

$$\frac{-70 \text{ mV}}{310 \text{ K}} = \frac{\Delta\psi}{313 \text{ K}}$$

$$\Delta\psi = -70.7 \text{ mV}$$

The difference in membrane potential at the higher temperature would not significantly affect the neuron's activity. **(b)** It is more likely that an increased temperature would increase the fluidity of cell membranes. This in turn might alter the activity of membrane proteins, including ion channels and pumps, which would have a more dramatic effect on membrane potential than temperature alone.

13. Use Equation 9.3 and substitute the values for extracellular and cytosolic glucose:

(a) $\Delta G = RT \ln \dfrac{[\text{glucose}]_{in}}{[\text{glucose}]_{out}}$

$\Delta G = (8.3145 \times 10^{-3} \text{ kJ} \cdot \text{K}^{-1} \cdot \text{mol}^{-1})(293 \text{ K}) \ln \dfrac{0.5 \times 10^{-3} \text{ M}}{5 \times 10^{-3} \text{ M}}$

$\Delta G = -5.6 \text{ kJ} \cdot \text{mol}^{-1}$

(b) $\Delta G = RT \ln \dfrac{[\text{glucose}]_{in}}{[\text{glucose}]_{out}}$

$\Delta G = (8.3145 \times 10^{-3} \text{ kJ} \cdot \text{K}^{-1} \cdot \text{mol}^{-1})(293 \text{ K}) \ln \dfrac{5 \times 10^{-3} \text{ M}}{0.5 \times 10^{-3} \text{ M}}$

$\Delta G = 5.6 \text{ kJ} \cdot \text{mol}^{-1}$

15. Use Equation 9.4 and let $Z = -1$ and $T = 310$ K:

$\Delta G = RT \ln \dfrac{[\text{Glutamate}]_{in}}{[\text{Glutamate}]_{out}} + Z\mathcal{F}\Delta\psi$

$\Delta G = (8.3145 \times 10^{-3} \text{ kJ} \cdot \text{K}^{-1} \cdot \text{mol}^{-1})(310 \text{ K}) \ln \dfrac{10 \times 10^{-3} \text{ M}}{0.1 \times 10^{-3} \text{ M}}$

$\quad + (-1)(96{,}485 \times 10^{-3} \text{ kJ} \cdot \text{V}^{-1} \cdot \text{mol}^{-1})(-0.070 \text{ V})$

$\Delta G = 11.9 \text{ kJ} \cdot \text{mol}^{-1} + 6.8 \text{ kJ} \cdot \text{mol}^{-1}$

$\Delta G = 18.7 \text{ kJ} \cdot \text{mol}^{-1}$

17. (a) The permeability coefficients are positively correlated with membrane solubility. The greater the permeability coefficient, the more easily the solute passes through the membrane.
(b) Glucose has a slightly larger permeability coefficient than mannitol and therefore moves across the synthetic bilayer more easily.
(c) Both solutes have higher permeability coefficients for the red blood cell membrane, indicating that transport is occurring via a protein transporter rather than diffusion through the membrane. The transporter binds glucose specifically and transports it rapidly across the membrane whereas it is less specific for mannitol and transports it less effectively.

19. (a) Phosphate ions are negatively charged, and lysine side chains most likely carry a full positive charge at physiological pH. It is possible that an ion pair forms between the phosphate and the lysine side chains and that the lysine side chains serve to funnel the phosphate ions through the porin. **(b)** If the hypothesis described in part a is correct, the replacement of lysines with the negatively charged glutamates would abolish phosphate transport by the porin, due to charge–charge repulsion. Possibly the mutated porin might even transport positively charged ions instead of phosphate. [From Sukhan, A. and Hancock, R. E. W., *J. Biol. Chem.* **271,** 21239–21242 (1996).]

21. (a) Acetylcholine binding triggers the opening of the channel, an example of a ligand-gated transport protein. **(b)** Na^+ ions flow into the muscle cell, where their concentration is low. **(c)** The influx of positive charges causes the membrane potential to increase.

23. The transfer to pure water increases the influx of water by osmosis, and the cell begins to swell. Swelling, which puts pressure on the cell membrane, causes mechanosensitive channels to open. As soon as the cell's contents flow out, the pressure is relieved and the cell can return to its normal size. Without these channels, the cell would swell and burst.

25. The hydroxyl and amido groups act as proton donors to coordinate the negatively charged chloride ion. Cations could not interact with the protons and so would be excluded.

27. (a) A transport protein, like an enzyme, carries out a chemical reaction (in this case, the transmembrane movement of glucose) but is not permanently altered in the process. Because the transport protein binds glucose, its rate does not increase in direct proportion to increasing glucose concentration, and it becomes saturated at high glucose concentrations. **(b)** The transport protein has a maximum rate at which it can operate (corresponding to V_{max}, the upper limit of the curve). It also binds glucose with a characteristic affinity (corresponding to K_M, the glucose concentration at half-maximal velocity). The estimated V_{max} for this transporter is about 0.8×10^6 mM $\cdot$ cm $\cdot$ s^{-1} and the K_M is about 0.5 mM.

29. Intracellular exposure of the glucose transporter to trypsin indicates that there is at least one cytosolic domain of the transport protein that is essential for glucose transport. Hydrolysis of one or more peptide bonds in this domain(s) abolishes glucose transport. But extracellular exposure of the ghost transporter to trypsin has no effect, so there is no trypsin-sensitive extracellular domain that is essential for transport. This experiment also shows that the glucose transporter is asymmetrically arranged in the erythrocyte membrane.

31. As the glutamate (charge –1) enters the cell, 4 positive charges also enter (3 Na^+, 1 H^+) for a total of 3 positive charges. Since 1 K^+ exits the cell at the same time, a total of 2 positive charges are added to the cell for each glutamate transported inside.

33. (a) CO_2 produced by respiring tissues enters the red blood cell and combines with water to form carbonic acid, which then dissociates to form H^+ and HCO_3^- ions. The HCO_3^- ions are transported out of the cell by Band 3 in exchange for Cl^- ions, which enter the cell. The HCO_3^- ions travel through the circulation to the lungs, where they recombine with H^+ ions to form carbonic acid, which subsequently dissociates to form water and CO_2. The CO_2 is then exhaled in the lungs.
(b) When red blood cells pass through tissues, HCO_3^- ions are transported out of the cell by Band 3. However, in the lungs, HCO_3^- ions are transported into red blood cells by Band 3 (in exchange for Cl^- ions, which exit the cell). The intracellular HCO_3^- ions combine with H^+ ions that dissociate from hemoglobin (promoting conversion of

hemoglobin from the deoxy T form to the oxy R form), forming carbonic acid. The carbonic acid dissociates into water and CO_2, which diffuses out of the cell and is exhaled.

35. (a) The maximal velocity is estimated to be between 6 and 7 pmol choline $\cdot$ mg^{-1} $\cdot$ mL^{-1}. The K_M is the substrate concentration at half maximal velocity and is about 40 mM. (b) The choline transporter responds to increased choline concentrations by increasing its rate of transport so that efficient uptake occurs over a range of choline concentrations near the K_M. At low concentrations of choline (10 μM), the transporter operates at 20% of its maximal velocity, whereas at high concentrations of choline (80 μM), the transporter operates at nearly 100% of its maximal velocity. The K_M value of 40 μM is between the low and high physiological concentrations of choline. (c) It is possible that the choline transporter cotransports hydrogen ions and choline. A hydrogen ion might be exported when choline is imported. This is an example of antiport transport. (d) TEA is structurally similar to choline and acts as a competitive inhibitor. TEA might bind to the choline transporter, preventing choline from binding. In this manner, TEA is brought into the cell and choline transport is inhibited. [From Sinclair, C. J., Chi, K. D., Subramanian, V., Ward, K. L., and Green, R. M., *J. Lipid Res.* **41**, 1841–1847 (2000).]

37.

Aspartyl phosphate

39. The proline transporter can bind L-hydroxyproline but not D-proline, indicating that the transporter is stereoselective. The transporter does not co-transport proline with H$^+$ ions. Because the rate of transport was not affected by Na$^+$ or K$^+$ it is unlikely that a secondary active transport mechanism is used. The proline transporter does not have a binding site for ouabain as the Na,K-ATPase does (see Problem 38). The proline transporter likely uses an active transport mechanism that requires energy in the form of ATP, since intracellular ATP depletion decreases the rate of transport. [From L'Hostis, C. L., Geindre, M., and Deshusses, J., *Biochem. J.* **291**, 297–301 (1993).]

41. (a) Glucose uptake increases as sodium concentration increases in pericytes. In endothelial cells, glucose uptake is constant regardless of sodium ion concentration. (b) The shape of the curve for the pericytes indicates that a protein transporter is involved. Glucose uptake initially increases as sodium ion concentration increases, then reaches a plateau at high sodium ion concentration, indicating that the transporter is saturated and is operating at its maximal capacity. (c) It is likely that the pericytes use secondary active transport to import glucose. Sodium ions and glucose molecules enter the cell in symport. The sodium ions are then ejected from the cell by the Na,K-ATPase transporter.

43. The ABC transporters bind ATP, then undergo a conformational change as the ATP is hydrolyzed and P$_i$ is released, leaving ADP. Vanadate, a phosphate analog, might serve as a competitive inhibitor by binding to the phosphate portion of the ATP binding site. With ATP unable to bind, the necessary conformational change cannot occur and the transporter is inhibited.

45. Both transporters are examples of secondary active transport. The H$^+$/Na$^+$ exchanger uses the free energy of the Na$^+$ gradient (established by the Na,K-ATPase) to remove H$^+$ from the cell as Na$^+$

enters. Similarly, a preexisting Cl$^-$ gradient (see Fig. 2.12) allows the cell to export HCO$_3^-$ as Cl$^-$ enters.

47. Di- and tripeptides enter the cell in symport with H$^+$ ions. The H$^+$ ions leave in exchange for Na$^+$ via the antiport protein. The Na$^+$ ions are ejected via the Na,K-ATPase. This is an example of secondary active transport in which the expenditure of ATP by the Na,K-ATPase pump is the driving force for peptide entry into the cell.

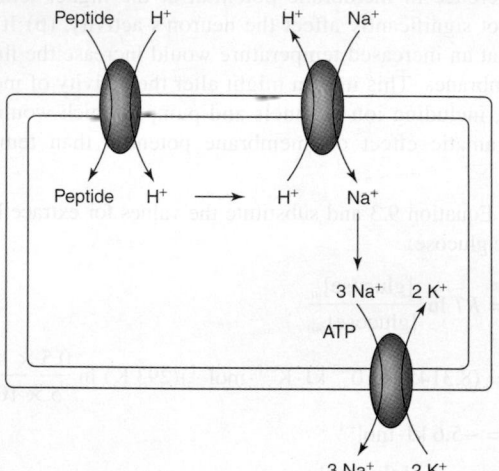

49. Acetylcholinesterase inhibitors prevent the enzyme from breaking down acetylcholine. This increases the concentration of acetylcholine in the synaptic cleft and increases the chances that acetylcholine will bind to a dwindling number of receptors in the postsynaptic cell. [From Thanvi, B. R., and Lo, T. C. N., *Postgrad. Med. J.*, **80**, 690–700 (2004).]

51.

53. Serotonin recycling depends on a transporter that uses the free energy of the Na$^+$ gradient to move serotonin back into the cell. The Na$^+$ gradient is established through the action of the Na,K-ATPase.

55. The tetanus toxin cleaves the SNAREs, which are required for the fusion of synaptic vesicles with the neuronal plasma membrane. This prevents the release of acetylcholine, interrupting communication between nerves and muscles and causing paralysis.

57. By adding a phosphate group, the kinase increases the size and negative charge of the lipid head group, which then occupies a larger volume and more strongly repels neighboring negatively charged lipid head groups. The phosphatidylinositol would become more cone-shaped, thereby increasing bilayer curvature, which is a necessary step in the formation of a new vesicle by budding.

59. Two lipid bilayers separate the damaged organelle from the rest of the cell.

61. Receptor-mediated endocytosis reaches a maximum when all receptor sites are occupied, whereas the rate of pinocytosis is proportional to amount of substance internalized. Receptor-mediated endocytosis is a far more efficient way to deliver substances to the interior of the cell.

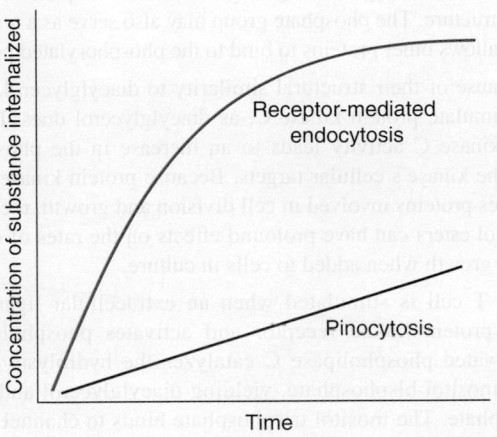

Chapter 10

1. Signal molecules that are lipids (cortisol and thromboxane) or very small (nitric oxide) can diffuse through lipid bilayers and do not need a receptor on the cell surface.

3. (a) Because 50% of the receptors are occupied, $[R \cdot L] = [R] = 20$ mM. Use Equation 10.1 to calculate K_d:

$$K_d = \frac{[R][L]}{[R \cdot L]}$$

$$K_d = \frac{(20 \times 10^{-3}\,\text{M})(8 \times 10^{-3}\,\text{M})}{(20 \times 10^{-3}\,\text{M})}$$

$$K_d = 8 \times 10^{-3}\,\text{M} = 8\ \text{mM}$$

(b) The dissociation constant K_d is equal to the ligand concentration when half of the receptor molecules have bound ligand.

5. Let $[R \cdot L] = x$ and $[R] = 0.020 - x$

$$K_d = \frac{[R][L]}{[R \cdot L]}$$

$$[R \cdot L] = x = \frac{[R][L]}{K_d}$$

(a)

$$x = \frac{(0.020 - x)(5 \times 10^{-3}\,\text{M})}{2.5 \times 10^{-3}\,\text{M}}$$

$$x = (0.020 - x)2$$

$$x = 0.040 - 2x$$

$$3x = 0.040$$

$$x = 0.0133$$

$$x = 13.3\ \text{mM} = [R \cdot L]$$

The percentage of receptors occupied by ligand is 13.3 mM/20 mM or 66.5%.

(b)

$$x = \frac{(0.020 - x)(5 \times 10^{-3}\,\text{M})}{0.6 \times 10^{-3}\,\text{M}}$$

$$x = \frac{(0.020 - x)5}{0.6}$$

$$0.6x = 0.1 - 5x$$

$$5.6x = 0.1$$

$$x = 0.018 = 18\ \text{mM} = [R \cdot L]$$

The percentage of receptors occupied by ligand is 18 mM/20 mM or 90%. A higher percentage of the receptors are occupied by a ligand with a smaller K_d.

7. The experimental results would produce a curve that is similar in shape but shifted to the right (similar to the plot shown in Figure 7.9). As an antagonist that competes with adenosine for binding the same receptors, caffeine acts like a competitive inhibitor.

9. (a)

$$\frac{[R \cdot L]}{[R]_T} = \frac{[L]}{K_d + [L]}$$

$$\frac{[R \cdot L]}{[R]_T} = \frac{\dfrac{K_d}{3}}{K_d + \dfrac{K_d}{3}}$$

$$\frac{[R \cdot L]}{[R]_T} = \frac{0.33 K_d}{K_d + 0.33 K_d}$$

$$\frac{[R \cdot L]}{[R]_T} = \frac{0.33 K_d}{1.33 K_d}$$

$$\frac{[R \cdot L]}{[R]_T} = 0.25$$

(b)

$$\frac{[R \cdot L]}{[R]_T} = \frac{[L]}{K_d + [L]}$$

$$\frac{[R \cdot L]}{[R]_T} = \frac{K_d}{K_d + K_d}$$

$$\frac{[R \cdot L]}{[R]_T} = \frac{K_d}{2 K_d}$$

$$\frac{[R \cdot L]}{[R]_T} = 0.50$$

(c)

$$\frac{[R \cdot L]}{[R]_T} = \frac{[L]}{K_d + [L]}$$

$$\frac{[R \cdot L]}{[R]_T} = \frac{3 K_d}{K_d + 3 K_d}$$

$$\frac{[R \cdot L]}{[R]_T} = \frac{3 K_d}{4 K_d}$$

$$\frac{[R \cdot L]}{[R]_T} = 0.75$$

11. If the number of receptors decreases to 200, a greater proportion of these receptors must have bound ligand, since occupation of 150 receptors is required for a maximal response (see Problem 10). A nearly 17-fold increase in the concentration of ligand is required in order to achieve a maximal response.

$$\frac{[R \cdot L]}{[R]_T} = \frac{[L]}{[L] + K_d}$$

$$\frac{150}{200} = \frac{[L]}{[L] + 1.0 \times 10^{-10}\,\text{M}}$$

$$0.75([L] + 1.0 \times 10^{-10}\,\text{M}) = [L]$$

$$0.75[L] + 7.5 \times 10^{-11}\,\text{M} = [L]$$

$$7.5 \times 10^{-11}\,\text{M} = 0.25[L]$$

$$[L] = 3.0 \times 10^{-10}\,\text{M}$$

13. The K_d is obtained from the x-intercept in the double-reciprocal plot.

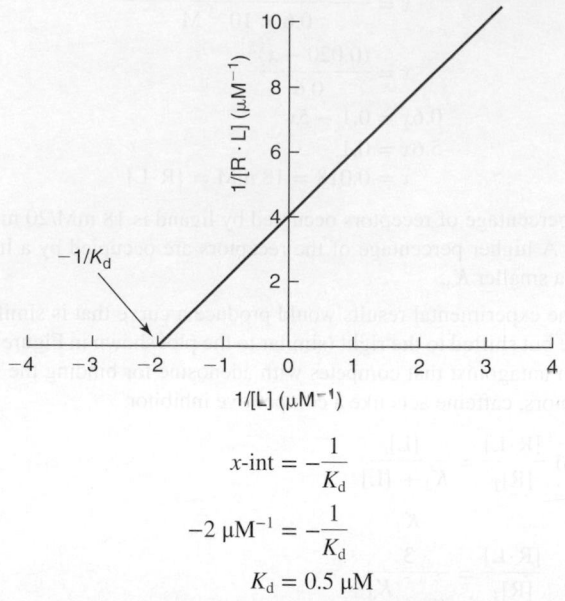

$$x\text{-int} = -\frac{1}{K_d}$$

$$-2\ \mu M^{-1} = -\frac{1}{K_d}$$

$$K_d = 0.5\ \mu M$$

15. Cell-surface receptors are difficult to purify because they are usually integral membrane proteins and require the addition of detergents to dissociate them from the membrane. The receptor proteins constitute a very small proportion of all of the proteins in the cell; this makes it difficult for the experimenter to isolate the receptor protein from other cellular proteins.

17. The different types of G protein–linked receptors are found in different types of cells. The cellular response elicited when a ligand binds to a receptor depends on how that particular cell integrates and processes the signal. Different cells have different intracellular components, which results in different responses to what appears to be the same signal.

19. If receptors are removed from the cell surface, the ligand cannot bind and an intracellular response cannot occur.

21. (a)

$$\begin{array}{c} \text{O} \\ \text{H} \quad \parallel \\ -\text{N}-\text{CH}-\text{C}- \\ | \\ \text{CH}_2 \\ | \\ \text{S} \\ | \\ \text{C}-(\text{CH}_2)_{14}-\text{CH}_3 \\ \parallel \\ \text{O} \end{array}$$

(b) The GPCR is a lipid-linked protein. The palmitoyl group interacts with the fatty acyl chains of membrane phospholipids, anchoring the receptor to the membrane. **(c)** If the Cys residue is mutated to a Gly, the palmitoyl chain cannot be attached. This might interfere with the localization of the receptor to the membrane and result in the loss of its function.

23. (a) The synthesis of the GPCR may be impaired in such a way that it is not targeted to the cell membrane; the receptor may be present on the cell membrane but a mutation alters the binding site so that the ligand fails to bind; the mutation may render the GPCR unable to interact with a G protein, or the GPCR may be more sensitive to phosphorylation by the GPCR kinase, allowing arrestin to bind more easily. **(b)** The GCPR may adopt an active conformation in the absence of agonist; the mutation may increase the synthesis rate so there are more GCPR molecules on the cell surface; an allosteric modulator may bind to the GPCR and increase its sensitivity to agonists. [From Vassart, G. and Costagliola, S., *Nat. Rev. Endocrinol.* **7**, 362–372 (2011).]

25. Both hormones lack tyrosine's carboxylate group and have hydroxyl groups attached to the ring and to the β carbon. In epinephrine, the amino nitrogen bears a methyl group.

27. The inhibition of the intrinsic GTPase activity results in a continuously active G protein. This increases the activity of adenylate cyclase, which results in an increase in the concentration of intracellular cAMP. In intestinal cells, the increased cAMP concentration leads to the loss of water and electrolytes from the cells and results in diarrhea that can be fatal.

29. The ADP-ribosylated G protein cannot respond to ligand binding to its associated GPCR and therefore cannot interact with adenylate cyclase. As a result, the enzyme remains active and cellular cAMP levels remain high, bringing about the symptoms associated with whooping cough.

31. The phosphate group is large and bulky and may cause a conformational change that alters the activity of the protein. The negatively charged phosphate group can interact with other amino acid side chains in the protein, forming either hydrogen bonds or ion pairs that change protein structure. The phosphate group may also serve as a recognition site that allows other proteins to bind to the phosphorylated protein.

33. Because of their structural similarity to diacylglycerol, phorbol esters stimulate protein kinase C, as diacylglycerol does. Increased protein kinase C activity leads to an increase in the phosphorylation of the kinase's cellular targets. Because protein kinase C phosphorylates proteins involved in cell division and growth, the addition of phorbol esters can have profound effects on the rates of cell division and growth when added to cells in culture.

35. The T cell is stimulated when an extracellular ligand binds to a G protein–linked receptor and activates phospholipase C. The activated phospholipase C catalyzes the hydrolysis of phosphatidylinositol bisphosphate, yielding diacylglycerol and inositol trisphosphate. The inositol trisphosphate binds to channel proteins in the endoplasmic reticulum and allows calcium ions to flow into the cytosol. Calcium ions then bind to calmodulin, causing a conformational change that allows it to bind and activate calcineurin. The activated calcineurin then activates NFAT as described in the problem.

37. (a) Overexpression of PTEN in mammalian cells promotes apoptosis. PTEN removes a phosphate group from inositol trisphosphate, and when this occurs, inositol trisphosphate is no longer able to activate protein kinase B. In the absence of protein kinase B, cells are not stimulated to grow and proliferate and instead undergo apoptosis.
(b) Yes, mutations in which the gene coding for PTEN is either nonfunctional or absent are commonly found in cancer cells. In the absence of PTEN, inositol trisphosphate is not dephosphorylated, and protein kinase B is highly active. As described in Problem 43, activation of protein kinase B promotes cell growth, so a highly active protein kinase B could account for the uncontrolled cell growth observed in cancer cells.

39. (a) Acetylcholine in the synaptic cleft is rapidly hydrolyzed by acetylcholinesterase (see Section 9.4). **(b)** Various calcium pumps, using the energy of ATP hydrolysis, pump calcium ions out of the cytosol. The Ca^{2+} concentration in the cell drops, the ion channels close, and Ca^{2+} ions dissociate from calmodulin, changing its conformation and rendering it unable to bind to NO synthase. Without bound calmodulin, the NO synthase is inactive. **(c)** The signaling molecule NO rapidly decomposes. In its absence, guanylate cyclase is not activated. Any cGMP still present is hydrolyzed to GMP by cGMP phosphodiesterase. In the absence of cGMP, protein kinase G reassociates with its regulatory subunits and is inactivated. Any phosphorylated proteins are acted upon by phosphatase enzymes, which remove the phosphate groups by hydrolysis.

41. NO synthase is activated when calcium ions bind to calmodulin (see Solution 38b). Clotrimazole is a calmodulin antagonist and prevents calmodulin from binding to NO synthase. In the absence of active NO synthase, cyclic GMP is not produced and protein kinase G is not activated. Cellular targets of protein kinase G are not phosphorylated in the presence of clotrimazole.

43. Inhibition of cGMP phosphodiesterase, the enzyme that hydrolyzes cGMP, increases the intracellular cGMP concentration. The activity of protein kinase G is increased as a result, as is the phosphorylation of the enzyme's targets, which are proteins involved in smooth muscle cell relaxation. The overall result is increased blood flow.

45. By inactivating a signaling molecule required for cell proliferation, LF can block the production of white blood cells. This impairs the ability of the immune system to fight off the infection, so the bacteria are able to multiply in the body.

47. PTEN is a phosphatase (see Problem 37); therefore if H_2O_2 inactivates cellular phosphatases (see Solution 46), it likely inhibits PTEN as well. [From Gough, D. R. and Cotter, T. G., *Cell Death and Disease*, **2**, e213 (2011).]

49. In the presence of Ras with this particular mutation, the signaling pathway would be constitutively active. Ras is active when bound to GTP and is inactivated when GTP is hydrolyzed to GDP. If hydrolysis of GTP cannot occur, Ras cannot be inactivated and the signaling pathway cannot be turned off. The cell is stimulated to grow and proliferate even in the absence of growth-signaling ligand.

51. Overexpression of IRS-1 would stimulate the activity of the signaling pathway and would lead to enhanced activation of protein kinases B and C. The activation of protein kinase B would result in the activation of glycogen synthase, so the cultured cells would be expected to show an increase in glycogen synthesis. Enhanced activation of protein kinase C would stimulate translocation of glucose transporters to the plasma membrane; thus, an increase in glucose import would be observed in the cultured cells.

53. A ligand could bind to a G protein–coupled receptor and subsequently activate phospholipase C. This enzyme catalyzes the cleavage of phosphatidylinositol bisphosphate to diacylglycerol and inositol trisphosphate. Diacylglycerol activates protein kinase C, which then activates the MAP kinase cascade, as shown in Problem 52. This is an example of cross-talk, in which signaling pathways share intracellular components.

55. In order to become activated, two inactive PKR proteins must come close enough to phosphorylate each other (autophosphorylation). A long RNA molecule can bind two PKR proteins simultaneously, holding them in close proximity so that they can activate each other. Short RNA molecules prevent PKR activation because when a short RNA molecule occupies the PKR RNA-binding site, the PKR cannot bind to another RNA where it might encounter a second PKR and get phosphorylated. [From Nallagatla, S. R., Toroney, R., and Bevilacqua, P. C., *Curr. Opin. Struct. Biol.* **21**, 119–127 (2011).]

57. Substances cannot enter the nucleus unless they possess a nuclear localization signal, a sequence that interacts with the nuclear pore and allows entry into the nucleus. The nuclear localization signal on the progesterone receptor must be exposed, even when ligand is not bound. But the nuclear localization signal on the glucocorticoid receptor must be masked. When ligand binds, a conformational change occurs that unmasks the nuclear localization signal, and the complex can pass through the nuclear pore and enter the nucleus.

59. Arachidonic acid is the substrate for the production of prostaglandins, many of which have inflammatory properties. Stimulating the release of arachidonic acid from the membrane by C1P increases the concentration of substrate available for prostaglandin synthesis. One of the enzymes that catalyzes the first step in the

production of prostaglandins is COX-2, which is stimulated by S1P. Both C1P and S1P can potentially increase production of prostaglandins, which accounts for their inflammatory properties, as shown in the diagram.

61. S1P might use a variety of mechanisms to activate Ras, either through receptor tyrosine kinases or activation of protein kinase C. Ras then activates the MAP kinase pathway (see Problem 52), which leads to the phosphorylation of transcription factors that promote the expression of proteins involved in the cell cycle, ultimately leading to cell survival.

63. (a)

(b) Without knowing the mechanism of the enzyme, it is not possible to say for certain why acetylating the serine inhibits cyclooxygenase activity. But it's possible that the acetylation alters the structure of the active site such that the arachidonate substrate is unable to bind. It's also possible that serine participates in catalysis, possibly as a nucleophile, as in chymotrypsin. An acetylated serine would be unable to function as a nucleophile, which would explain why the modified enzyme is catalytically inactive. **(c)** Because a covalent bond forms between the acetyl group and the Ser side chain, aspirin is an irreversible inhibitor (see Section 7.3).

65. Phospholipase A_2 catalyzes the release of arachidonate from membrane phospholipids. Blocking this reaction would prevent the COX-catalyzed conversion of arachidonate to proinflammatory prostaglandins.

67.

Chapter 11

1. (a) aldose; (b) ketose; (c) aldose

3. Sugar (c) has the most stereoisomers. The sugar has 4 chiral carbons, so there are $2^4 = 16$ stereoisomers. The other sugars (a) and (d) shown in Problem 1 have $2^3 = 8$ stereoisomers and (b) has only $2^1 = 2$ stereoisomers.

5. (a) L; (b) D; (c) D; (d) L. Sugars (b) and c are epimers.

7. Fructose and galactose are isomers of glucose.

9. (a)

CH₂OH
|
C=O
|
HO—C—H
|
HO—C—H
|
H—C—OH
|
CH₂OH

(b)

CH₂OH
|
C=O
|
H—C—OH
|
H—C—OH
|
HO—C—H
|
CH₂OH

(c) Tagatose is less effectively absorbed in the small intestine because the transport proteins in the epithelial cells lining the small intestine do not bind and transport tagatose as efficiently as they do sugars that are more naturally and commonly present in the diet.

11. The β anomer of galactose is more stable because most of the bulky hydroxyl substituent are in the equatorial position.

Beta-Galactose

13.

α-D-Ribose

β-D-Ribose

15. (a) A five-membered ring results.

α-D-Fructose

β-D-Fructose

(b) A six-membered ring results.

α-D-Fructose

β-D-Fructose

17. All the sugar molecules will be converted to product because the α and β anomers are in equilibrium. Depletion of molecules in the β form will cause more of the α anomers to convert to β anomers, which will then be converted to product.

19. Glucose-6-phosphate is more likely than glucose to remain in the cell because the phosphorylated glucose is negatively charged and cannot easily cross the nonpolar lipid bilayer by passive diffusion. Glucose-6-phosphate cannot exit the cell via transport proteins either, as these transporters are specific for glucose, not glucose-6-phosphate.

21. (a)

COO⁻
|
H—C—OH
|
HO—C—H
|
H—C—OH
|
H—C—OH
|
CH₂OH

Gluconate

(b)

(c)

CH₂OH
|
H—C—OH
|
HO—C—H
|
H—C—OH
|
H—C—OH
|
CH₂OH

Sorbitol

23. (a)

α-D-Glucose → 1,2,3,4,6-Penta-O-methyl-D-glucose

(b)

1,2,3,4,6-Penta-O-methyl-D-glucose → 2,3,4,6-Tetra-O-methyl-D-glucose

25.

COO⁻
|
C=O
|
H—C—H
|
H—C—OH
|
H—C—OH
|
CH₂OH

KDG

27. [From Vivoli, M., Isupov, M. N., Nicholas, R., Hill, A., Scott, A. E., Kosma, P., Prior, J. L., and Harmer, N. J., *Chem. Biol.* **22**, 1622–1632 (2015).]

CH₂OH
|
CH₂
|
...
|
C=O
|
CH₃

2-O-acetyl-6-deoxy-α-D-mannoheptopyranose

29. Individuals whose ancestors domesticated cows and goats relied on the milk produced from these animals as a food source and became lactose tolerant as a result of natural selection. Those individuals who maintained the ability to produce lactase were more likely to survive to adulthood to produce offspring with the same metabolic capability.

31. Trehalose is not a reducing sugar because the anomeric carbons of both glucose residues are involved in the glycosidic bond.

33. The numerous hydroxyl groups on the trehalose molecule enable it to form a network of hydrogen bonds with numerous water molecules. In this manner, cellular structures are stabilized in a dehydrated form.

35.

37. In order for sorbitol to be catabolized to yield energy, it would need to enter the same catabolic pathway as glucose. The enzymes that catalyze glucose catabolism are specific for glucose and do not bind to sorbitol, so the sugar alcohol is not metabolized and passes through the body undigested. In this manner, sorbitol contributes no calories to the food containing it, but it is nearly identical to its parent monosaccharide in taste and sweetness.

39. Humans do not have one of the enzymes needed to hydrolyze the glycosidic bonds in order to convert the trisaccharide raffinose to its constituent monosaccharides. The glycosidic linkages in raffinose are an $\alpha(1 \rightarrow 6)$ linkage between galactose and glucose and an $\alpha(1 \rightarrow 2)$ glycosidic linkage between glucose and fructose. Humans have the enzymes to digest the second glycosidic bond but not the first.

41. The fungus must have contained an α-galactosidase enzyme capable of digesting the $\alpha(1 \rightarrow 6)$ linkage and a sucrase-like enzyme that could hydrolyze the $(1 \rightarrow 2)$ glycosidic bond formed between the α anomer of glucose and the β anomer of fructose. [From Feng, S., Saw, C. L., Lee, Y. K., and Huang, D., *J. Agric. Food Chem.* **56**, 10078–10084 (2008).]

43. There is one reducing end.

45. Potato starch consists of a mixture of amylose and amylopectin. The iodine can fit into the center of the helical structure of the amylose and produce a blue color. Apples, on the other hand, contain mostly cellulose. Cellulose consists of sheets of extended $\beta(1 \rightarrow 4)$ glycan chains and does not form a helical structure. Consequently, iodine does not form a complex with the cellulose and a blue color is not produced.

47. Pectin is a highly hydrated polysaccharide, so it thickens the fruit preparation and helps turn it into a gel.

49. The cellulose-based plant cell wall is strong and rigid, but it must be remodeled as the plant cell grows. The cell uses cellulase to weaken the cell wall so that the cell can expand.

51. Four. Both starch and glycogen contain glucose residues linked by $\alpha(1 \rightarrow 4)$ glycosidic bonds and $\alpha(1 \rightarrow 6)$ glycosidic bonds at branch points. Sucrose consists of an α-glucose linked by a $(1 \rightarrow 2)$ glycosidic bond to β-fructose. Lactose consists of a β-galactose linked by a $\beta(1 \rightarrow 4)$ bond to glucose.

53.

[From Jackson, M. and Brennan, P., *J. Biol. Chem.* **284**, 1949–1953 (2009).]

55.

57. The residues of the disaccharide are glucuronate linked by a $\beta(1 \rightarrow 3)$ glycosidic bond to N-acetylgalactosamine-4-sulfate. Disaccharides are linked to each other by $\beta(1 \rightarrow 4)$ bonds.

59.

61. (a) The monosaccharide is N-acetylglucosamine. (b) The amide bond forms between Ala and the carboxylate group on the C4 substituent in the disaccharide. The arrow indicates the linkage site.

63. (a) Sialic acid is negatively charged. The presence of sialic acid on their surfaces weakens the attachment of the tumor cells to each other and may promote the detachment process. (b) A drug could act as an inhibitor of one of the enzymes in the biochemical pathway for synthesizing sialic acid. Alternatively, foreign sialic acid precursors could be administered that would be taken up by the tumor cells and used for sialic acid synthesis. The use of a foreign precursor would result in the synthesis of a sialic acid derivative that is more immunogenic. [From Fuster, M. M. and Esko, J. D., *Nat. Rev. Cancer* **5**, 526–542 (2005).]

Chapter 12

1. (a) chemoautotroph; (b) photoautotroph; (c) chemoautotroph; (d) heterotroph; (e) heterotroph; (f) chemoautotroph; (g) photoautotroph.

3. The pH of the stomach is ~2. At this pH, the salivary amylase is denatured and can no longer catalyze the hydrolysis of glycosidic bonds in dietary carbohydrates.

5. Maltase is required to hydrolyze the $\alpha(1 \rightarrow 4)$ glycosidic bonds in maltotriose and maltose (see Solution 4). Isomaltase is needed to hydrolyze the $\alpha(1 \rightarrow 6)$ glycosidic bonds in the limit dextrins because α-amylase only catalyzes the hydrolysis of $\alpha(1 \rightarrow 4)$ glycosidic bonds and cannot accommodate branch points (see Problem 4). These enzymes are required to completely hydrolyze starch to its component monosaccharides, since only monosaccharides can be absorbed.

7. Sugar alcohols are not present in abundance naturally, which explains the absence of transporters for these molecules. Passive diffusion is less effective than passive transport (as shown in Problem 9.17(b)).

9. Because the products of nucleic acid digestion are relatively large, charged nucleotides, a transport protein is required to facilitate their movement across the cell membrane. The transport protein most likely uses the free energy of a Na^+ gradient (active transport), as is the case for intestinal monosaccharide and amino acid transporters.

11. The low pH denatures the protein, unfolding it so peptide bonds are more accessible to proteolytic digestion by stomach enzymes.

13. The pH optimum for pepsin is 2, which is the pH of the stomach. The pH optimum for trypsin and chymotrypsin is 7–8, as the small intestine is slightly basic (see Table 2.3). Each enzyme functions optimally in the conditions of its environment.

15. Amino acids enter the cells lining the small intestine via secondary active transport. This system is similar to the process for glucose absorption shown in Figure 9.18.

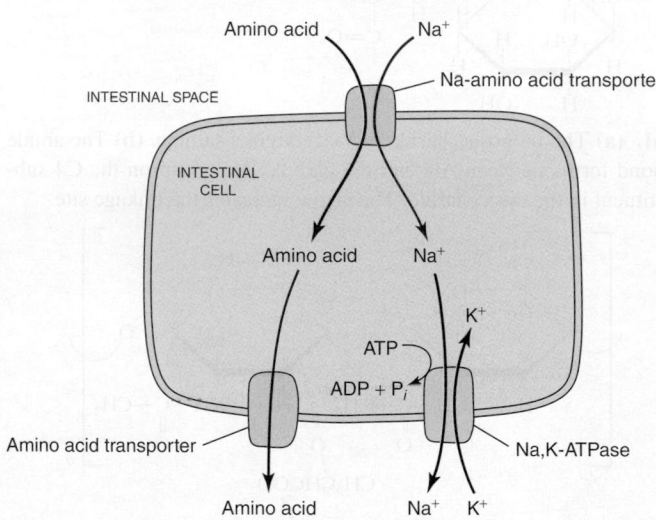

17. (a)

Triacylglycerol + H_2O →(Gastric lipase) Diacylglycerol + Fatty acid (R_3—C—O$^-$)

b. Both diacylglycerol and fatty acids are amphipathic molecules—they have both hydrophilic and hydrophobic domains. These molecules can form micelles that emulsify the dietary triacylglycerols, which are nonpolar and are unable to form micelles.

19.

Cholesteryl ester → Cholesteryl esterase → Fatty acid + Cholesterol

21. (a) The polar glycogen molecule is fully hydrated, so its weight reflects a large number of closely associated water molecules. Fat is stored in anhydrous form. Therefore, a given weight of fat stores more free energy than the same weight of glycogen. **(b)** Because it must be hydrated, a glycogen molecule occupies a large effective volume of the cytoplasm, which it shares with other glycogen molecules, enzymes, organelles, and so on. Because hydrophobic fat molecules are sequestered from the bulk of the cytoplasm, they do not have the same potential for interfering with other cellular constituents, so their collective volume is virtually unlimited.

23. The phosphorylated glucose molecule is not recognized by the glucose transporter. Removal of the phosphate group allows the glucose to more easily leave the cell.

25.

	Acetyl-CoA	GAP	Pyruvate
Glycolysis		✔	✔
Citric acid cycle	✔		
Fatty acid metabolism	✔		
TAG synthesis		✔	
Photosynthesis		✔	
Transamination			✔

27. (a) NAD^+; **(b)** NADPH; **(c)** NADH; **(d)** $NADP^+$.

29. Individuals with gastrointestinal disorders might have a gastrointestinal tract that is not colonized by the appropriate vitamin B_{12}–synthesizing bacteria. A deficiency in haptocorrin or intrinsic factor would be manifested as a vitamin B_{12} deficiency, since these proteins are essential for absorption of the vitamin. Vegetarians and vegans who consume no animal products would also be at risk for a deficiency of vitamin B_{12}.

31. (a)

γ-Carboxyglutamate

(b) The additional carboxylate group on the glutamate residue confers a –2 charge on the side chain, generating a high-affinity binding site for the Ca^{2+} ions essential for blood clotting.

33. (a) vitamin C; **(b)** biotin; **(c)** pyridoxine; **(d)** pantothenic acid.

35. (a) The prosthetic group is related to NAD^+, as it contains a nicotinamide–ribose–phosphate group in which the amide group is a thioamide. **(b)** A histidine and a lysine side chain hold the prosthetic group in place.

His
Lys
phosphate
nicotinamide
ribose

37. (a) Because K_{eq} is the ratio of the product concentration to the reactant concentration at equilibrium, the reaction with the larger K_{eq} will have a higher concentration of product. Therefore, the concentration of B in Tube 1 will be greater than the concentration of D in Tube 2.

(b) Use Equation 12.2. For the A $\rightleftharpoons$ B reaction,

$$\Delta G^{\circ\prime} = -RT \ln K_{eq}$$
$$\Delta G^{\circ\prime} = -(8.3145 \text{ J} \cdot \text{K}^{-1} \cdot \text{mol}^{-1})(310 \text{ K}) \ln 10$$
$$\Delta G^{\circ\prime} = -5900 \text{ J} \cdot \text{mol}^{-1} = -5.9 \text{ kJ} \cdot \text{mol}^{-1}$$

For the C $\rightleftharpoons$ D reaction,

$$\Delta G^{\circ\prime} = -(8.3145 \text{ J} \cdot \text{K}^{-1} \cdot \text{mol}^{-1})(310 \text{ K}) \ln 0.1$$
$$\Delta G^{\circ\prime} = 5900 \text{ J} \cdot \text{mol}^{-1} = 5.9 \text{ kJ} \cdot \text{mol}^{-1}$$

39. (a) Since $K_{eq} = 1$, $\ln K_{eq} = 0$ and $\Delta G^{\circ\prime}$ is also equal to zero (Equation 12.2). **(b)** Since $K_{eq} = 1$, the concentrations of reactants and products must be equal at equilibrium. If the reaction started with 1 mM F, the equilibrium concentrations will be 0.5 mM E and 0.5 mM F. **(c)** Since $K_{eq} = 1$, the concentrations of reactants and products must be the same at equilibrium. Therefore the reaction will proceed in the forward direction with 1.5 mM E being converted to 1.5 mM F, resulting in equal concentrations (3.5 mM of each) at equilibrium.

41. Rearrange Equation 12.3 to solve for the [glyceraldehyde-3-phosphate]/[dihydroxyacetone phosphate] ratio:

$$\Delta G = \Delta G^{\circ\prime} + RT \ln \frac{[\text{glyceraldehyde-3-phosphate}]}{[\text{dihydroxyacetone phosphate}]}$$

$$e^{(\Delta G - \Delta G^{\circ\prime})/RT} = \frac{[\text{glyceraldehyde-3-phosphate}]}{[\text{dihydroxyacetone phosphate}]}$$

$$e^{(2000 \text{ J} \cdot \text{mol}^{-1} - 7560 \text{ J} \cdot \text{mol}^{-1})/8.3145 \text{ J} \cdot \text{K}^{-1} \cdot \text{mol}^{-1}(310\text{K})}$$

$$= \frac{[\text{glyceraldehyde-3-phosphate}]}{[\text{dihydroxyacetone phosphate}]}$$

$$\frac{[\text{glyceraldehyde-3-phosphate}]}{[\text{dihydroxyacetone phosphate}]} = e^{-2.16} = 0.12$$

43. Use Equation 12.3 and the $\Delta G^{\circ\prime}$ for the reaction calculated in Solution 37(a):

$$\Delta G = \Delta G^{\circ\prime} + RT \ln \frac{[\text{D}]}{[\text{C}]}$$
$$\Delta G = 5900 \text{ J} \cdot \text{mol}^{-1} + (8.3145 \text{ J} \cdot \text{K}^{-1} \cdot \text{mol}^{-1})(310 \text{ K})$$
$$\ln\left(\frac{0.2 \times 10^{-3} \text{ M}}{0.8 \times 10^{-3} \text{ M}}\right)$$
$$\Delta G = 2326 \text{ J} \cdot \text{mol}^{-1} = 2.3 \text{ kJ} \cdot \text{mol}^{-1}$$

The reaction will proceed in the reverse direction as written, with D converted to C until the ratio of [C]/[D] = 1/10.

45. (a)

Glucose + P_i → Glucose-6-phosphate + H_2O $\Delta G^{\circ\prime} = +13.8 \text{ kJ mol}^{-1}$

Glucose-1-phosphate + H_2O → Glucose + P_i $\Delta G^{\circ\prime} = -20.9 \text{ kJ mol}^{-1}$

Glucose-1-phosphate → Glucose-6-phosphate $\Delta G^{\circ\prime} = -7.1 \text{ kJ mol}^{-1}$

The value of $\Delta G^{\circ\prime}$ obtained above is the same as the value presented in Sample Calculation 12.2. Free energy is independent of path, so the values of $\Delta G^{\circ\prime}$ are the same whether the reaction is visualized as occurring in one step or two steps. $\Delta G^{\circ\prime}$ is negative, which indicates that the reaction is spontaneous under standard conditions.

(b)

$$\Delta G = \Delta G^{\circ\prime} + RT \ln \frac{[\text{G6P}]}{[\text{G1P}]}$$
$$\Delta G = -7100 \text{ J} \cdot \text{mol}^{-1} + (8.3145 \text{ J} \cdot \text{K}^{-1} \cdot \text{mol}^{-1})(310 \text{ K})$$
$$\ln\left(\frac{10 \times 10^{-3} \text{ M}}{0.5 \times 10^{-3} \text{ M}}\right)$$
$$\Delta G = -7100 \text{ J} \cdot \text{mol}^{-1} + 7700 \text{ J} \cdot \text{mol}^{-1} = 0.6 \text{ kJ} \cdot \text{mol}^{-1}$$

The value of ΔG is positive and therefore the reaction is not spontaneous under these conditions.

47. (a) The equilibrium constant can be derived by rearranging Equation 12.2:

$$K_{eq} = e^{-\Delta G^{\circ\prime}/RT}$$
$$K_{eq} = e^{-7900 \text{ J} \cdot \text{mol}^{-1}/(8.3145 \text{ J} \cdot \text{K}^{-1} \cdot \text{mol}^{-1})(298 \text{ K})}$$
$$K_{eq} = e^{-3.19} = 0.041$$

(b) Use Equation 12.3:

$$\Delta G = \Delta G^{\circ\prime} + RT \ln \frac{[\text{DHAP}]}{[\text{GAP}]}$$
$$\Delta G = 7900 \text{ J} \cdot \text{mol}^{-1} + (8.3145 \text{ J} \cdot \text{K}^{-1} \cdot \text{mol}^{-1})(310 \text{ K}) \ln \frac{0.6 \times 10^{-3} \text{ M}}{0.2 \times 10^{-3} \text{ M}}$$
$$\Delta G = 7900 \text{ J} \cdot \text{mol}^{-1} + 2900 \text{ J} \cdot \text{mol}^{-1} = 11 \text{ kJ} \cdot \text{mol}^{-1}$$

(c) The reaction is not spontaneous as written. The reverse reaction, in which $\Delta G = -11 \text{ kJ} \cdot \text{mol}^{-1}$, would be spontaneous.

49. Convert Calories to joules (see Solution 48): A 760-Calorie drink contains 760,000 calories, which is equivalent to 3.18×10^6 J or 3180 kJ. The hot chocolate drink contains 3180 kJ/30.5 kJ $\cdot$ mol^{-1} or the equivalent of about 104 moles of ATP.

51. Creatine, which undergoes phosphorylation to serve as a reservoir to phosphorylate ADP, becomes depleted after a few seconds of muscle activity. Therefore, it would not be expected to play a role in maintaining ATP concentrations for longer periods.

53. The hydrolysis of pyrophosphate releases 19.2 kJ $\cdot$ mol^{-1} (see Table 12.4). These two reactions are coupled, and the overall ΔG value for the coupled reactions is negative; thus UDP–glucose formation occurs spontaneously.

55. (a)   ADP + P_i → ATP + H_2O $\Delta G^{\circ\prime} = +30.5 \text{ kJ} \cdot \text{mol}^{-1}$

$$\frac{2200 \text{ Cal}}{\text{d}} \times \frac{1 \text{ mol ATP}}{30.5 \text{ kJ}} \times \frac{4.184 \text{ kJ}}{1 \text{ Cal}} \times 0.33 = 100 \text{ mol ATP} \cdot \text{d}^{-1}$$

(b) $100 \text{ mol ATP} \times \frac{505 \text{ g}}{\text{mol}} \times \frac{1 \text{ lb}}{2200 \text{ g}} = 23 \text{ lb}$

(c) ATP does not accumulate but instead is constantly recycled. As ATP is used, its hydrolysis products, ADP and P_i, serve as reactants for ATP synthesis in the process of oxidative phosphorylation.

57. (a) Reactions involving phosphoenolpyruvate, 1,3-bisphosphoglycerate, and phosphocreatine could drive the synthesis of ATP because transfer of a phosphoryl group from one of these compounds to ADP occurs with a greater change in free energy than the transfer of free phosphate to ADP. **(b)** Reactions involving the conversion of phosphate to pyrophosphate and the phosphorylation of glucose (to glucose-1-phosphate and glucose-6-phosphate) and glycerol (to glycerol-3-phosphate) would require the hydrolysis of ATP to drive the reaction, because transfer of a phosphoryl group from ATP occurs with a greater change in free energy than the transfer of free phosphate to one of these compounds.

59. (a) Use Equation 12.2 to solve for $\Delta G^{\circ\prime}$:

$$\Delta G^{\circ\prime} = -RT \ln K_{eq}$$
$$\Delta G^{\circ\prime} = -(8.3145 \text{ J} \cdot \text{K}^{-1} \cdot \text{mol}^{-1})(298 \text{ K}) \ln 0.4$$
$$\Delta G^{\circ\prime} = 2200 \text{ J} \cdot \text{mol}^{-1} = 2.2 \text{ kJ} \cdot \text{mol}^{-1}$$

The reaction will proceed in the opposite direction as written standard conditions.

(b) Use Equation 12.3 to solve for ΔG:

$$\Delta G = \Delta G^{\circ\prime} + RT \ln \frac{[\text{F6P}]}{[\text{G6P}]}$$

$$\Delta G = 2200 \text{ J}\cdot\text{mol}^{-1} + (8.3145 \text{ J}\cdot\text{K}^{-1}\cdot\text{mol}^{-1})(310 \text{ K}) \ln \frac{0.3 \times 10^{-3} \text{ M}}{1 \times 10^{-3} \text{ M}}$$

$$\Delta G = 2200 \text{ J}\cdot\text{mol}^{-1} - 3100 \text{ J}\cdot\text{mol}^{-1} = -0.9 \text{ kJ}\cdot\text{mol}^{-1}$$

Under these conditions, the reaction will proceed as written.

61. (a) The equilibrium constant can be determined by rearranging Equation 12.2 (see Sample Calculation 12.2). The $\Delta G^{\circ\prime}$ value for the reaction given is obtained by reversing the $\Delta G^{\circ\prime}$ for the hydrolysis of glucose-6-phosphate (G6P) found in Table 12.4:

$$K_{eq} = e^{-\Delta G^{\circ\prime}/RT}$$

$$K_{eq} = e^{-13{,}800 \text{ J}\cdot\text{mol}^{-1}/(8.3145 \text{ J}\cdot\text{K}^{-1}\cdot\text{mol}^{-1})(298 \text{ K})}$$

$$K_{eq} = e^{-5.57} = 0.0038$$

(b) Use the equilibrium constant expression and the K_{eq} calculated in part a to solve for the equilibrium concentration of G6P:

$$K_{eq} = \frac{[\text{G6P}]}{[\text{glucose}][\text{P}_i]}$$

$$0.0038 = \frac{[\text{G6P}]}{(5 \times 10^{-3} \text{ M})(5 \times 10^{-3} \text{ M})}$$

$$[\text{G6P}] = 9.5 \times 10^{-8} \text{ M}$$

Under the given conditions, the reaction would produce only 9.5×10^{-8} M glucose-6-phosphate from 5 mM glucose and thus is not a feasible route to the production of this compound for the glycolytic pathway.

(c)
$$K_{eq} = \frac{[\text{G6P}]}{[\text{glucose}][\text{P}_i]}$$

$$0.0038 = \frac{250 \times 10^{-6} \text{ M}}{[\text{glucose}](5 \times 10^{-3} \text{ M})}$$

$$[\text{glucose}] = 13 \text{ M}$$

Driving the reaction to the right using this method is not feasible because it is impossible to achieve a concentration of 13 M glucose inside the cell.

(d) The equilibrium constant can be determined by rearranging Equation 12.2 (see Sample Calculation 12.2). The $\Delta G^{\circ\prime}$ value for the reaction given can be found in the text.

$$K_{eq} = e^{-\Delta G^{\circ\prime}/RT}$$

$$K_{eq} = e^{-(-16{,}700 \text{ J}\cdot\text{mol}^{-1})/(8.3145 \text{ J}\cdot\text{K}^{-1}\cdot\text{mol}^{-1})(298 \text{ K})}$$

$$K_{eq} = e^{6.74} = 850$$

(e) Use the equilibrium constant expression and the K_{eq} calculated in part a to solve for the equilibrium concentration of G6P:

$$K_{eq} = \frac{[\text{G6P}][\text{ADP}]}{[\text{glucose}][\text{ATP}]}$$

$$850 = \frac{(250 \times 10^{-6} \text{ M})(1.25 \times 10^{-3} \text{ M})}{[\text{glucose}](5 \times 10^{-3} \text{ M})}$$

$$[\text{glucose}] = 7.4 \times 10^{-8} \text{ M}$$

(f) The reaction can be accomplished at a much lower glucose concentration (7.4×10^{-8} M) when the phosphorylation of glucose is coupled to ATP hydrolysis than if glucose is phosphorylated directly (13 M, see Solution (c)). This is possible because the second reaction couples the phosphorylation of glucose with the exergonic hydrolysis of ATP. Thus an unfavorable reaction is converted to a favorable reaction.

63.

I. $\dfrac{\begin{array}{ll} \text{GAP} \rightleftharpoons \text{1,3BPG} & \Delta G^{\circ\prime} = 6.7 \text{ kJ}\cdot\text{mol}^{-1} \\ \text{1,3BPG} + \text{H}_2\text{O} \rightleftharpoons \text{3PG} + \text{P}_i & \Delta G^{\circ\prime} = -49.3 \text{ kJ}\cdot\text{mol}^{-1} \end{array}}{\begin{array}{ll} \text{GAP} + \text{H}_2\text{O} \rightleftharpoons \text{3PG} + \text{P}_i & \Delta G^{\circ\prime} = -42.6 \text{ kJ}\cdot\text{mol}^{-1} \end{array}}$

II. $\dfrac{\begin{array}{ll} \text{GAP} \rightleftharpoons \text{1,3BPG} & \Delta G^{\circ\prime} = 6.7 \text{ kJ}\cdot\text{mol}^{-1} \\ \text{1,3BPG} + \text{ADP} \rightleftharpoons \text{3PG} + \text{ATP} & \Delta G^{\circ\prime} = -18.8 \text{ kJ}\cdot\text{mol}^{-1} \end{array}}{\begin{array}{ll} \text{GAP} + \text{ADP} \rightleftharpoons \text{3PG} + \text{ATP} & \Delta G^{\circ\prime} = -12.1 \text{ kJ}\cdot\text{mol}^{-1} \end{array}}$

The second scenario is more likely. The first coupled reaction is more exergonic, but the second coupled reaction "captures" some of this free energy in the form of ATP, which the cell can use.

65. (a) The value of C is determined from the slope of the plot. The slope is equal to 1.8×10^{-6} mM^{-1}.

(b)
$$C = \frac{[\text{PP}_i]}{K_{eq}[\text{ATP}]}$$

$$K_{eq} = \frac{[\text{PP}_i]}{C[\text{ATP}]}$$

$$K_{eq} = \frac{(1.0 \times 10^{-3})}{(1.8 \times 10^{-3})(14 \times 10^{-6})}$$

$$K_{eq} = 4.0 \times 10^{4}$$

(c) Use Equation 12.2 to solve for $\Delta G^{\circ\prime}$:

$$\Delta G^{\circ\prime} = -RT \ln K_{eq}$$

$$\Delta G^{\circ\prime} = -(8.3145 \text{ J}\cdot\text{K}^{-1}\cdot\text{mol}^{-1})$$
$$(298 \text{ K}) \ln 4.0 \times 10^{4}$$

$$\Delta G^{\circ\prime} = -26{,}000 \text{ J}\cdot\text{mol}^{-1} = -26 \text{ kJ}\cdot\text{mol}^{-1}$$

(d)

$\dfrac{\begin{array}{ll} \text{Nick} \rightleftharpoons \text{phosphodiester bond} & \Delta G^{\circ\prime} = ? \\ \text{ATP} \rightleftharpoons \text{AMP} + \text{PP}_i & \Delta G^{\circ\prime} = -48.5 \text{ kJ}\cdot\text{mol}^{-1} \end{array}}{\begin{array}{ll} \text{ATP} + \text{nick} \rightleftharpoons \text{AMP} + \text{PP}_i + \\ \quad \text{phosphodiester bond} & \Delta G^{\circ\prime} = -26 \text{ kJ}\cdot\text{mol}^{-1} \end{array}}$

The $\Delta G^{\circ\prime}$ for the formation of the phosphodiester bond is $22.5 \text{ kJ}\cdot\text{mol}^{-1}$.

(e) The $\Delta G^{\circ\prime}$ value for the hydrolysis of a phosphodiester bond in DNA is $-22.5 \text{ kJ}\cdot\text{mol}^{-1}$, whereas the $\Delta G^{\circ\prime}$ value for the hydrolysis of a typical phosphomonoester bond is $-13.8 \text{ kJ}\cdot\text{mol}^{-1}$. Therefore, the phosphodiester bond in DNA is less stable than the typical phosphomonoester bond. [From Dickson, K., Burns, C. M., and Richardson, J., *J. Biol. Chem.* **275**, 15828–15831 (2000).]

Chapter 13

1. (a) Reactions 1, 3, 7, and 10; **(b)** Reactions 2, 5, and 8; **(c)** Reaction 6; **(d)** Reaction 9; **(e)** Reaction 4.

3. (a)

$$\Delta G^\circ = -RT \ln \frac{[\text{glucose-6-phosphate}][\text{ADP}]}{[\text{glucose}][\text{ATP}]}$$

$$-16.7 \text{ kJ} \cdot \text{mol}^{-1} = -(8.3145 \times 10^{-3} \text{ kJ} \cdot \text{K}^{-1} \cdot \text{mol}^{-1})$$

$$(298 \text{ K}) \ln \frac{[\text{glucose-6-phosphate}][\text{ADP}]}{[\text{glucose}][\text{ATP}]}$$

$$16.7 \text{ kJ} \cdot \text{mol}^{-1} = 2.48 \text{ kJ} \cdot \text{mol}^{-1}$$

$$\ln \frac{[\text{glucose-6-phosphate}][\text{ADP}]}{[\text{glucose}][\text{ATP}]}$$

$$6.73 = \ln \frac{[\text{glucose-6-phosphate}][\text{ADP}]}{[\text{glucose}][\text{ATP}]}$$

$$e^{6.73} = \frac{[\text{glucose-6-phosphate}][\text{ADP}]}{[\text{glucose}][\text{ATP}]}$$

$$840 = \frac{[\text{glucose-6-phosphate}][\text{ADP}]}{[\text{glucose}][\text{ATP}]}$$

$$840 = \frac{[\text{glucose-6-phosphate}](1)}{[\text{glucose}](10)}$$

$$8.4 \times 10^3 = \frac{[\text{glucose-6-phosphate}]}{[\text{glucose}]}$$

(b)

$$\Delta G^\circ = -RT \ln \frac{[\text{glucose}][\text{ATP}]}{[\text{glucose-6-phosphate}][\text{ADP}]}$$

$$16.7 \text{ kJ} \cdot \text{mol}^{-1} = -(8.3145 \times 10^{-3} \text{ kJ} \cdot \text{K}^{-1} \cdot \text{mol}^{-1})$$

$$(298 \text{ K}) \ln \frac{[\text{glucose}][\text{ATP}]}{[\text{glucose-6-phosphate}][\text{ADP}]}$$

$$-16.7 \text{ kJ} \cdot \text{mol}^{-1} = 2.45 \text{ kJ} \cdot \text{mol}^{-1} \ln \frac{[\text{glucose}][\text{ATP}]}{[\text{glucose-6-phosphate}][\text{ADP}]}$$

$$-6.73 = \ln \frac{[\text{glucose}][\text{ATP}]}{[\text{glucose-6-phosphate}][\text{ADP}]}$$

$$e^{-6.73} = \frac{[\text{glucose}][\text{ATP}]}{[\text{glucose-6-phosphate}][\text{ADP}]}$$

$$1.2 \times 10^{-3} = \frac{[\text{glucose}][\text{ATP}]}{[\text{glucose-6-phosphate}][\text{ADP}]}$$

$$1.2 \times 10^{-3} = \frac{[\text{glucose}](10)}{[\text{glucose-6-phosphate}](1)}$$

$$1.2 \times 10^{-4} = \frac{[\text{glucose}]}{[\text{glucose-6-phosphate}]}$$

In order to reverse the reaction, the ratio of glucose-6-phosphate to glucose would have to be $8.3 \times 10^3 : 1$.

5. The low K_M means that the enzyme will be saturated with glucose and will therefore operate at maximum velocity. Even if the concentration of glucose were to fluctuate slightly, the brain's ability to catabolize glucose would not be affected.

7. (a)

Fluorodeoxyglucose

(b) Phosphorylation traps FDG in the cell because the negatively charged phosphate group prevents the sugar from either passing through the membrane and/or from binding to glucose transporters. **(c)** Phosphorylated FDG cannot be isomerized by phosphoglucose isomerase. The first ring-opening step can occur, but the conversion of the aldose to the ketose requires the presence of an oxygen atom

at the C2 position. Therefore any FDG that enters the cell becomes phosphorylated and is trapped inside the cell in that form. **(d)** Although FDG is used to detect tumors because they take up glucose rapidly, other cells in the body (such as the brain) that take up glucose rapidly will also take up FDG.

9. (a) R; **(b)** T; **(c)** R.

11. The negatively charged fructose-1,6-bisphosphate is repelled by the negatively charged Glu 161 in the active site in the low-affinity form of the enzyme, whereas the substrate is attracted to the positively charged Arg 162 positioned in the active site in the high-affinity form.

13. (a) When F26BP binds to the allosteric site on PFK, one of its phosphate groups may interact with the serine in some manner, perhaps by hydrogen bonding. But if the serine is replaced with the negatively charged aspartate, the negative charges on F26BP are repelled by it and F26BP cannot bind to the enzyme. **(b)** Apparently PFK cannot be fully active in the absence of the F26BP allosteric activator, which plays an important role in stimulating glycolysis. In its absence, the glycolytic pathway is less active, which accounts for the observed decrease in glucose consumption and ethanol production. [From Heinisch, J. J., Boles, E., and Timpel, C., *J. Biol. Chem.* **271**, 15928–15933 (1996).]

15. In the presence of iodoacetate, fructose-1,6-bisphosphate accumulates, which suggests that iodoacetate inactivates the enzyme that uses fructose 1,6-bisphosphate as a substrate. Because iodoacetate reacts with Cys residues, the inactivation of the enzyme by the reagent suggests that a Cys residue is in the active site. Later studies showed that Cys was not part of the active site; instead, the reaction of the Cys with iodoacetate probably caused a conformational change that rendered the enzyme inactive.

17. Rearrange Equation 12.3 to solve for the [GAP]/[DHAP] ratio:

$$\Delta G = \Delta G^{\circ\prime} + RT \ln \frac{[\text{GAP}]}{[\text{DHAP}]}$$

$$e^{(\Delta G - \Delta G^{\circ\prime})/RT} = \frac{[\text{GAP}]}{[\text{DHAP}]}$$

$$e^{(4400 \text{ J} \cdot \text{mol}^{-1} - 7900 \text{ J} \cdot \text{mol}^{-1})/(8.3145 \text{ J} \cdot \text{K}^{-1} \cdot \text{mol}^{-1})(310 \text{ K})} = \frac{[\text{GAP}]}{[\text{DHAP}]}$$

$$\frac{[\text{GAP}]}{[\text{DHAP}]} = e^{-1.36} = \frac{0.26}{1}$$

The ratio of [GAP] to [DHAP] is 0.26:1, which seems to indicate that the formation of DHAP, not the formation of GAP, is favored. However, GAP, the product of the triose phosphate isomerase reaction, is the substrate for the glyceraldehyde-3-phosphate dehydrogenase reaction that occurs next in the pathway. The continuous removal of the product GAP by the action of the dehydrogenase shifts the equilibrium toward formation of GAP from DHAP.

19. Arsenate is a metabolic poison, and the cells eventually die. In the presence of arsenate, 1,3-bisphosphoglycerate is not formed; instead, this step is essentially skipped. Two moles of ATP per mole of glucose are normally generated at this step. If ATP is not generated at this step, the net ATP yield for the glycolytic pathway is zero, and the cells die because they are unable to meet their energy requirements.

21. (a) In the GAPDH reaction presented in the text, NAD$^+$ is reversibly reduced to NADH and H$^+$ and GAP is phosphorylated to 13BPG, coupling oxidation and phosphorylation. A subsequent step generates ATP when 13BPG donates a phosphate to ADP. *T. tenax* expresses this same enzyme, with the exception that NADP$^+$ is reduced to NADPH. In contrast, the second enzyme catalyzes the irreversible production of 3PG in a reaction that involves only oxidation and not phosphorylation. A subsequent step will not produce ATP.

(b) Estimated K_M values and their effects on enzyme activity are shown in the table. Inhibitors increase the K_M and activators decrease the K_M. Interestingly, in the presence of the activators, substrate binding to the enzyme becomes less cooperative (as illustrated by the hyperbolic curves) while substrate binding becomes more cooperative in the presence of the inhibitor (as illustrated by the sigmoidal curve).

(c) NADP+ may serve as an inhibitor to ensure that both forms of GAPDH are not simultaneously active. AMP and G1P may indicate that cellular energy in the form of ATP is needed and stimulate the activity of the enzyme. [From Brunner, N. A., Brinkmann, H., Siebers, B., and Hensel, R., *J. Biol. Chem.* **273,** 6149–6156 (1998).]

	K_M (mM)	Activator or inhibitor?
Control	3.0	
NADP+	5.0	Inhibitor
AMP	1.5	Activator
G1P	0.5	Activator

23. Phosphate levels increase because phosphate is a reactant for GAPDH. ATP levels decrease, since ATP is a product of the phosphoglycerate kinase reaction, whose substrate 1,3-bisphosphoglycerate is generated by the GAPDH reaction. Levels of 2,3-bisphosphoglycerate also decrease when levels of 1,3-bisphosphoglycerate decrease.

25. (a) The glucose–lactate pathway releases 196 kJ·mol⁻¹ of free energy, enough theoretically to drive the synthesis of $(196/30.5) \times 0.33$, or about 2 ATP.

(b) The complete oxidation of glucose releases 2850 kJ·mol⁻¹ of free energy, enough for $(2850/30.5) \times 0.33$, or about 31 ATP.

(c) Yes, this explains the Pasteur effect. If yeast metabolize glucose in the presence of oxygen, more ATP can be produced from glucose, so less glucose will be catabolized to meet the yeast cell's energy needs.

27. (a) If PFK is the only enzyme affected by aluminum treatment, then the production of pyruvate from perfused livers of control and aluminum-treated rats should be the same (see Box 13.A). The PFK reaction is bypassed, so inhibition of this enzyme should have no effect.

(b) If the same experiment were carried out using glucose as an energy source, pyruvate production would be much lower in the livers from the aluminum-treated rats than in livers from the control rats. This result is due to the inhibition of the PFK enzyme, which is essential in the production of pyruvate from glucose. [From Xu, Z.-X., Fox, L., Melethil, S., Winberg, L., and Badr, M., J., *Pharmacol. Exp. Therapeutics* **254,** 301–305 (1990).]

29. (a) Methanol reacts with alcohol dehydrogenase (as ethanol does) to produce formaldehyde (Box 13.B).

$$H_3C-OH \xrightleftharpoons[\text{alcohol dehydrogenase}]{\text{NAD}^+ \quad \text{NADH, H}^+} H-\overset{\displaystyle O}{\overset{\|}{C}}-H$$

Methanol   Formaldehyde

(b) Administering ethanol is a good antidote because ethanol will compete with methanol for binding to alcohol dehydrogenase and will produce the less harmful acetaldehyde. This allows time for methanol to be eliminated from the system. [From Cooper, J. A. and Kini, M., *Biochemical Pharmacology* **11,** 405–416 (1962).]

31. If hexokinase cannot be inhibited, all glucose that enters the yeast cell will be phosphorylated at the expense of ATP to produce glucose-6-phosphate. Glucose-6-phosphate is isomerized to fructose-6-phosphate, which is then phosphorylated to fructose-1,6-bisphosphate, again at the expense of ATP. If

concentrations of glucose are high, and if there is no mechanism to inhibit hexokinase, then cellular ATP will be depleted in these early steps. The early reactions of glycolysis proceed at a rate greater than the rate of the later ATP-generating steps of glycolysis. ATP is therefore used faster than it is regenerated, and the yeast mutants die as a result. [From Teusink, B., Walsh, M. C., van Dam, K., and Westerhoff, H. V., *Trends Biochem. Sci.* **23,** 162–169 (1998).]

33. (a) When glucose is converted to pyruvate, NAD+ is reduced to NADH in the glyceraldehyde-3-phosphate dehydrogenase reaction. Without a mechanism for regenerating NAD+, glycolysis could not continue. **(b)** If the trypanosome reduced pyruvate to lactate, this step would regenerate NAD+ and no other pathway for regenerating NAD+ would be needed. **(c)** A drug that interfered with one of the enzymes in the trypanosome's NADH-oxidizing pathway would likely have no effect on the host since mammals lack the enzymes that would be inactivated by the drug.

35. Pyruvate accumulates when pyruvate carboxylase is deficient because in the absence of the enzyme, pyruvate cannot be converted to oxaloacetate. Pyruvate can be converted to lactate by the reversible lactate dehydrogenase reaction, so elevated pyruvate leads to elevated levels of lactate. If both lactate and pyruvate are elevated to the same extent, the ratio in the patient will be the same as in a normal infant.

37. Insulin, the hormone of the fed state, might be expected to suppress the transcription of the gluconeogenic enzymes pyruvate carboxylase, PEPCK, fructose-1,6-bisphosphatase, and glucose-6-phosphatase. [In fact, insulin has been shown to suppress the transcription of PEPCK and glucose-6-phosphatase.]

39. (a) The phosphatase enzyme is active under fasting conditions. The phosphatase removes the phosphate group from F26BP, forming fructose-6-phosphate. Thus, F26BP is not present to stimulate glycolysis (or to inhibit gluconeogenesis); therefore gluconeogenesis is active. **(b)** The hormone of the fasted state is glucagon. **(c)** When glucagon binds to its receptors, cellular cAMP levels rise as described in Section 10.2. This activates protein kinase A, which phosphorylates the bifunctional enzyme, resulting in the activation of the phosphatase activity and the inhibition of the kinase activity.

41. (a) Increasing the activity of the kinase that produces F26BP would increase the concentration of this metabolite, which stimulates PFK and inhibits fructose-1,6-bisphosphatase. This would have the effect of stimulating glycolysis and inhibiting gluconeogenesis. Stimulation of PFK would increase the concentration of fructose-1,6-bisphosphate, which activates pyruvate kinase via feed-forward activation. Additional stimulation of pyruvate kinase by brazilin would lead to an increase in flux through the glycolytic pathway. **(b)** If brazilin acts on the liver to increase glycolysis and decrease gluconeogenesis, this could help alleviate the high blood glucose concentrations that occur in diabetes. An active gluconeogenic pathway in the liver would result in efflux of glucose from the liver, which is not desirable in the diabetic patient. [From You, E.-J., Khill, L.-Y., Kwak, W.-J., Won, H.-S., Chae, S.-H., Lee, B.-H., and Moon, C.-K., *J. Ethnopharmacol.* **102,** 53–57 (2005).]

43. (a) A diagram of this pathway, referred to as the Cori cycle, is shown in Figure 19.3. Lactate is released from the muscle as a result of anaerobic glycolytic activity. It travels via the bloodstream to the liver, where it is taken up, converted to pyruvate, and then transformed back to glucose via gluconeogenesis. The cost of running this cycle is 4 ATP, since 2 ATP are generated in glycolysis and the cost of gluconeogenesis is 6 ATP.

(b) The transformations described comprise the glucose–alanine cycle, which is shown in Figure 19.4. Prolonged operation of this cycle results in muscle breakdown, since muscle protein is the source of the amino acids required in the transamination of pyruvate to alanine.

45. Amylase hydrolyzes starch into oligosaccharides (see Section 12.1), which are subsequently hydrolyzed to monosaccharides. Although amylases are naturally present in bread flour, adding additional enzyme speeds up the starch degradation process to provide monosaccharides for glycolysis, which ultimately results in the conversion of pyruvate to ethanol, yielding CO_2 that causes the bread to rise.

47. (a) See Sample Calculation 12.2. K_{eq} can be calculated by rearranging Equation 12.2, then the equilibrium constant expression can be used to solve for the $[P_i]/[G1P]$ ratio. The values for glycogen ($n-1$ residues) and glycogen (n residues) are not substantially different from one another and can be set to 1.

$$K_{eq} = e^{-\Delta G^{\circ\prime}/RT}$$

$$K_{eq} = e^{-(3100\ \mathrm{J \cdot mol^{-1}})/(8.3145\ \mathrm{J \cdot K^{-1} \cdot mol^{-1}})(298\ \mathrm{K})}$$

$$K_{eq} = e^{-1.25}$$

$$K_{eq} = \frac{0.29}{1} = \frac{[\text{glycogen } (n-1 \text{ residues})][\text{G1P}]}{[\text{glycogen } (n \text{ residues})][P_i]}$$

$$\frac{0.29}{1} = \frac{1[\text{G1P}]}{1[P_i]}$$

$$\frac{3.5}{1} = \frac{[P_i]}{[\text{G1P}]}$$

(b) Use Equation 12.3; substitute in values of 1 for glycogen ($n-1$ residues) and glycogen (n residues), and 50/1 for the $[P_i]/[G1P]$ ratio given in the problem:

$$\Delta G = \Delta G^{\circ\prime} + RT \ln \frac{[\text{glycogen } (n-1)\text{residues}][\text{G1P}]}{[\text{glycogen } (n \text{ residues})][P_i]}$$

$$= 3100\ \mathrm{J \cdot mol^{-1}} + (8.3145\ \mathrm{J \cdot K^{-1} \cdot mol^{-1}} \times 310\ \mathrm{K})\ln \frac{(1)(1)}{(1)(50)}$$

$$= 3100\ \mathrm{J \cdot mol^{-1}} + (-10{,}100\ \mathrm{J \cdot mol^{-1}}) = -7.0\ \mathrm{kJ \cdot mol^{-1}}$$

(c) Production of glucose-1-phosphate requires only an isomerization reaction catalyzed by phosphoglucomutase to convert it to glucose-6-phosphate, which can enter glycolysis. This skips the hexokinase step and saves a molecule of ATP. Hydrolysis, which produces glucose, would require expenditure of an ATP to phosphorylate glucose to glucose-6-phosphate.

49. An increase in the activity of glycogen phosphorylase in the fat body increases degradation of glycogen to glucose. Because fructose-2,6-bisphosphate concentrations are low, glycolysis will not be stimulated (F26BP is a potent activator of the glycolytic enzyme PFK). Instead, glucose can be used to synthesize trehalose, which leaves the fat body and enters the hemolymph. In this way, the fat body produces sugars for use by other tissues in the fasting insect. [From Meyer-Fernandes, J. R., Clark, C. P., Gondim, K. C., and Wells, M. A., *Insect Biochem. Mol. Biol.* **31**, 165–170 (2001).]

51. (a) The first committed step of the pentose phosphate pathway is the first reaction, which is catalyzed by glucose-6-phosphate dehydrogenase and is irreversible. Once glucose-6-phosphate has passed this point, it has no other fate than conversion to a pentose phosphate.
(b) The hexokinase reaction does not commit glucose to the glycolytic pathway, since the product of the reaction, glucose-6-phosphate, can also enter the pentose phosphate pathway.

53.

55. G16BP inhibits hexokinase but stimulates PFK and pyruvate kinase. This means that glycolysis will be active, but only if the substrate is glucose-6-phosphate, since glucose cannot be phosphorylated in the absence of hexokinase activity. The pentose phosphate pathway is inactive, since 6-phosphogluconate dehydrogenase is inhibited. Phosphoglucomutase is activated, which converts glucose-1-phosphate (the product of glycogenolysis) to glucose-6-phosphate. Thus, in the presence of G16BP, glycogenolysis is active and produces substrate for glycolysis but not the pentose phosphate pathway. This is a more efficient process than using glucose taken up from the blood, which would need to be phosphorylated at the expense of ATP. [From Beitner, R., *Trends Biol. Sci.* **4**, 228–230 (1979).]

57. The low PFK activity means that glycolysis cannot generate a sufficient amount of the ATP required by myosin for muscle contraction (see Section 5.4), and the patient experiences cramping as a result. The effort of exercise may also damage muscle cells, releasing myoglobin into the blood, which subsequently appears in the urine.

59. (a) In cells with a pyruvate kinase deficiency, the $[ADP]/[ATP]$ ratio increases and the $[NAD^+]/[NADH]$ ratio decreases. Pyruvate kinase catalyzes the second ATP-generating step in glycolysis; in the absence of this reaction, ATP levels decrease and the $[ADP]/[ATP]$ ratio increases. The substrate PEP accumulates in a pyruvate kinase deficiency, stimulating PFK and increasing the concentration of F16BP. This ultimately leads to an increase in glyceraldehyde-3-phosphate, which reacts with NAD^+ to form NADH. In the absence of pyruvate kinase, pyruvate is not produced, so lactate cannot subsequently reoxidize NADH to NAD^+; therefore the $[NAD^+]/[NADH]$ ratio decreases.
(b) Cells swell and lyse in the absence of ATP because the Na,K-ATPase pump (see Fig. 9.15) is unable to function. Consequently, Na^+ ions enter the cell, along with water, which eventually causes the cells to swell and lyse, resulting in anemia.

61. (a) McArdle's disease is also known as Type V glycogen storage disease, as it results from a deficiency of muscle (and not liver) glycogen phosphorylase. Therefore symptoms are confined to the muscle and glycogen metabolism in the liver is normal.
(b) Blood glucose concentrations are regulated by pancreatic hormones acting on the liver to stimulate glycogen synthesis or degradation, whatever is appropriate. Since the patient's liver enzymes appear to function normally, the blood glucose concentration is properly regulated and the patient is neither hypo- nor hyperglycemic.

63. (a) Glucose-6-phosphatase catalyzes the last reaction in gluconeogenesis (and glycogenolysis) in the liver. Glucose-6-phosphate is converted to glucose, and the glucose transporters export glucose to the circulation, where it is available to other body tissues that do

not carry out gluconeogenesis and that do not store glycogen. In the absence of this enzyme, glucose-6-phosphate cannot be converted to glucose and instead accumulates in the liver and is converted to glucose-1-phosphate, which is used for glycogen synthesis. Glycogen synthesis is therefore elevated in the livers of patients with this disease. The accumulation of glycogen enlarges the liver and causes the abdomen to protrude. **(b)** Although the kidney stores a limited amount of glycogen, it does participate in both gluconeogenesis and glycogenolysis, so the lack of glucose-6-phosphatase causes the kidney to enlarge for the same reasons as described in part a. **(c)** Patients with von Gierke's disease cannot convert glucose-6-phosphate to glucose in the absence of glucose-6-phosphatase, thus these patients are hypoglycemic.

65. (a) Inhibition of glycogen phosphorylase blocks glycogenolysis, and inhibition of the bisphosphatase blocks gluconeogenesis. As a result, the liver is unable to produce glucose and the blood glucose level is low.
(b) The inhibition of fructose-1,6-bisphosphatase prevents gluconeogenesis even when substrates such as glycerol and dihydroxyacetone phosphate are supplied. Galactose relieves the hypoglycemia because it can be phosphorylated, converted to glucose-6-phosphate, and dephosphorylated to yield glucose that enters the circulation.

67. (a) Since serum withdrawal decreases G6PD activity, serum withdrawal should decrease the $NADPH/NADP^+$ ratio. NADPH is a product of the G6PD reaction and $NADP^+$ is a reactant, so decreased enzyme activity increases the oxidized form of the coenzyme and decreases the reduced form. **(b)** In the presence of DHEA, the $NADPH/NADP^+$ ratio should decrease, since DHEA inhibits G6PD activity, for the reasons outlined in part a. **(c)** In the absence of any other factor, adding H_2O_2 should not affect the ratio, since G6PD can increase its activity to produce more NADPH to react with the H_2O_2. **(d)** In the absence of the serum containing growth factors, G6PD cannot handle the increased load of H_2O_2, and the ratio would decrease. [From Tian, W.-N., Braunstein, L. D., Pang, J., Stuhlmeier, K. M., Xi, Q.-C., Tian, X., and Stanton, R., *J. Biol. Chem.*, **273**, 10609–10617 (1998).]

Chapter 14

1. In mammalian cells, pyruvate can be converted to lactate by lactate dehydrogenase. Pyruvate can also be transformed into oxaloacetate; this reaction is catalyzed by pyruvate carboxylase. Pyruvate can be converted to acetyl-CoA by the pyruvate dehydrogenase complex. Pyruvate can be converted to alanine by transamination.

3. The purpose of steps 4 and 5 is to regenerate the enzyme. In step 3, the product acetyl-CoA is released, but the lipoamide prosthetic group of E2 is reduced. In step 4, the E3 reoxidizes the lipoamide group by accepting the protons and electrons from the reduced lipoamide. In step 5, the enzyme is reoxidized by NAD^+. The product NADH then diffuses away.

5. Arsenite reacts with the reduced lipoamide group on E2 of the pyruvate dehydrogenase complex to form a compound with the structure shown in the figure. The enzyme cannot be regenerated and can no longer catalyze the conversion of acetyl-CoA to pyruvate. The α-ketoglutarate dehydrogenase complex has a lipoamide group on its E2 subunits and will be inhibited as well. The entire citric acid cycle cannot function, glucose cannot be oxidized aerobically, and respiration comes to a halt, which explains why these compounds are so toxic.

7. In both cases, the activity of the pyruvate dehydrogenase complex decreases, as both NADH and acetyl-CoA are products of the reaction. Rising concentrations of NADH and acetyl-CoA decrease

pyruvate dehydrogenase activity by competing with NAD^+ and CoASH for binding sites on the enzyme.

9. Ca^{2+} inhibits PDH kinase and activates PDH phosphatase. In this way, the pyruvate dehydrogenase complex is active and can funnel substrates from glycolysis into the citric acid cycle to provide ATP needed by myosin (see Section 5.4) for muscle contraction.

11. Citrate synthase uses a base catalyst strategy. In the rate-limiting step of the reaction, an unprotonated Asp residue acts as a base by accepting a proton from the acetyl group of the acetyl-CoA to form an enolate anion.

13. In the citrate synthase mechanism, a proton is removed from the acetyl group of acetyl-CoA. The histidine then forms a hydrogen bond to stabilize the resulting enediolate intermediate. The alanine side chain cannot form this hydrogen bond; thus, the reaction cannot continue. [From Pereira, D. S., Donald, L. J., Hosfield, D. J., and Duckworth, H. W., *J. Biol. Chem.* **269**, 412–417 (1994).]

15.

$$CoA-S-C\begin{smallmatrix}OH\\\\CH_2\end{smallmatrix}$$

17. The yeast mutant would be unable to carry out the citric acid cycle and so would be unable to carry out aerobic oxidation of fuel molecules in order to generate ATP. The yeast mutant would obtain all of its energy from the anaerobic oxidation of glucose, even in the presence of oxygen.

19. The substrate is isocitrate, a compound with three carboxylate groups that are negatively charged at physiological pH. Four of the five conserved amino acid residues have positively charged side chains that could form ion pairs with isocitrate. The enzyme–substrate complex should not be too stable, however; it is the role of the enzyme to stabilize the transition state. So it is possible that the positively charged side chains in the binding pocket stabilize the transition state. This stabilization must be important in the conversion of substrate to product, which explains why these residues are highly conserved throughout evolution. [From Hurley, J. H., Thorsness, P. E., Ramalingam, V., Helmers, N. H., Koshland, D. E., and Stroud, R. M., *Proc. Natl. Acad. Sci. USA* **86**, 8635–8639 (1989).]

21. TPP attacks the carbonyl carbon of the succinyl phosphonate, but the subsequent step (the departure of the leaving group) cannot occur because the C—P bond cannot be broken. A covalent bond forms between the enzyme and the inhibitor (although the reaction is reversible, the reverse reaction occurs slowly) and the substrate cannot bind. [From Bunik, V. I., Denton, T. T., Xu, H., Thompson, C. M., Cooper, A. J. L., and Gibson, G., *Biochemistry* **44**, 10552–10561 (2005).]

23. Succinate accumulates because it cannot be converted to fumarate. Succinyl-CoA also accumulates because the succinyl-CoA synthetase reaction is reversible. However, the succinyl-CoA ties up some of the cell's CoA supply, so the α-ketoglutarate dehydrogenase reaction, which requires CoA, slows. As a result, α-ketoglutarate accumulates.

25. Rearrange Equation 12.3 to solve for the [malate]/[fumarate] ratio:

$$\Delta G = \Delta G^{\circ\prime} + RT \ln \frac{[\text{malate}]}{[\text{fumarate}]}$$

$$e^{(\Delta G - \Delta G^{\circ\prime})/RT} = \frac{[\text{malate}]}{[\text{fumarate}]}$$

$$e^{(0\,\text{J}\cdot\text{mol}^{-1} + 3400\,\text{J}\cdot\text{mol}^{-1})/(8.3145\,\text{J}\cdot\text{K}^{-1}\cdot\text{mol}^{-1})(310\,\text{K})} = \frac{[\text{malate}]}{[\text{fumarate}]}$$

$$\frac{[\text{malate}]}{[\text{fumarate}]} = e^{1.32} = \frac{3.7}{1}$$

The ratio of malate to fumarate is 3.7 to 1, indicating that the reaction proceeds in the direction of formation of malate. This is not a control point for the citric acid cycle because the ΔG is close to zero, indicating it is a near-equilibrium reaction.

27. (a) The isotopic label on C4 of oxaloacetate is released as $^{14}CO_2$ in the α-ketoglutarate dehydrogenase reaction. (b) The isotopic label on C1 of acetyl-CoA is scrambled at the succinyl-CoA synthetase step. Because succinate is symmetrical, C1 and C4 are chemically equivalent, so in a population of molecules, both C1 and C4 would appear to be labeled (half the label would appear at C1 and half at C4). Consequently, one round of the citric acid cycle would yield oxaloacetate with half the labeled carbon at C1 and half at C4. Both of these labeled carbons would be lost as $^{14}CO_2$ in a second round of the citric acid cycle.

29. Acetyl-CoA binds to pyruvate carboxylase and acts as an activator. A thioester functional group must be required for binding, since S-acetonyl-CoA, which lacks a thioester functional group, cannot bind to this site. [From Rubenstein, P. and Dryer, R., *J. Biol. Chem.* **255,** 7858–7862 (1980).]

31. During the resting state, citric acid cycle activity is low as a result of the low activities of the two enzymes described. Any available citrate will inhibit citrate synthase, while the low concentrations of Ca^{2+} ions ensure that the activity of isocitrate dehydrogenase is also low. Upon beginning exercise, the Ca^{2+} concentration increases in muscle cells, which increases the activity of isocitrate dehydrogenase. This depletes the cellular concentration of citrate, depriving citrate synthase of its inhibitor, so the activity of citrate synthase also increases. Thus, when the cell transitions from resting to exercise mode, the activities of the enzymes involved in the citric acid cycle increase in order to meet the increased demand for ATP in the working muscle.

33. Citrate accumulates when aconitase is inhibited (see Solution 35). Aconitase is also inhibited when fluoroacetate is added because citrate synthase reacts with oxaloacetate and fluoroacetyl-CoA to produce fluorocitrate (see figure), which inhibits aconitase. This same result is observed when fluorocitrate is added directly. Therefore all three mechanisms either directly or indirectly inhibit aconitase and thus have similar effects on cells in culture.

$$FH_2C-COO^- \longrightarrow FH_2C-\overset{\displaystyle O}{\overset{\displaystyle \|}{C}}-S-CoA$$

Fluoroacetate   Fluoroacetyl-CoA

$$\xrightarrow[\text{Oxaloacetate}]{\text{citrate synthase}} \begin{array}{c} F-CH-COO^- \\ | \\ HO-C-COO^- \\ | \\ CH_2-COO^- \end{array}$$

Fluorocitrate

35. (a) When yeast use glucose as a carbon source, they obtain ATP from the anaerobic oxidation of glucose by glycolysis, without the citric acid cycle. But when the food source is shifted from glucose to acetate, yeast must obtain all their energy from the citric acid cycle and oxidative phosphorylation. Acetate is converted to acetyl-CoA, which enters the citric acid cycle. In the absence of glycolysis, flux through the citric acid cycle increases. Because isocitrate dehydrogenase is one of the main regulatory enzymes of the cycle, increasing its expression increases the rate of its reaction and increases the flux through the citric acid cycle. In this manner, yeast can effectively use acetate as a food source. (b) For the wild-type yeast, the [NAD$^+$]/[NADH] ratio increases slightly as the cells switch from using glycolysis to using the citric acid cycle to generate ATP, as described in part a. Flux through the citric acid cycle increases, but perhaps the isocitrate dehydrogenase cannot keep up with the demand, so the ratio increases. There is a more dramatic change in the ratio for the mutant because the isocitrate dehydrogenase enzyme is nonfunctional, so NAD$^+$ (a reactant) accumulates and the ratio increases. These cells will eventually die if not given a carbon source other than acetate. [From Minard, K. I. and McAlister-Henn, L., *Arch. Biochem. Biophys.* **483,** 136–143 (2009).]

37. Because of the enzyme deficiency, the citric acid cycle cannot be completed; in the absence of a functional citric acid cycle, glucose must be oxidized anaerobically. Lactate is a product of anaerobic oxidation of glucose; thus lactate levels are elevated in the patients. The pyruvate → lactate transformation is reversible, so if lactate levels rise due to increased glycolytic activity, pyruvate levels rise as well. However, lactate levels rise more (because there are other options for pyruvate as described in Solution 1), so the [lactate]/[pyruvate] ratio is increased in the patient. [From Bonnefont, J.-P., et al., *J. Pediatrics,* **121,** 255–258 (1992).]

39. (a)

malate + NAD$^+$ → oxaloacetate + NADH + H$^+$ $\Delta G^{\circ\prime} = 29.7\ \text{kJ}\cdot\text{mol}^{-1}$
oxaloacetate + acetyl-CoA → citrate + CoA $\Delta G^{\circ\prime} = -31.5\ \text{kJ}\cdot\text{mol}^{-1}$

malate + NAD$^+$ + acetyl-CoA → $\Delta G^{\circ\prime} = -1.8\ \text{kJ}\cdot\text{mol}^{-1}$
citrate + NADH + H$^+$ + CoA

(b) The equilibrium constants for both reactions can be determined by rearranging Equation 12.2 (see Sample Calculation 12.2):

For the coupled reaction:

$$K_{eq} = e^{-\Delta G^{\circ}/RT}$$

$$K_{eq} = e^{-(-1800\,\text{J}\cdot\text{mol}^{-1})/(8.3145\,\text{J}\cdot\text{K}^{-1}\cdot\text{mol}^{-1})(298\,\text{K})}$$

$$K_{eq} = e^{0.73} = 2.1$$

For the uncoupled reaction:

$$K_{eq} = e^{-\Delta G^{\circ}/RT}$$

$$K_{eq} = e^{-29,700\,\text{J}\cdot\text{mol}^{-1}/(8.3145\,\text{J}\cdot\text{K}^{-1}\cdot\text{mol}^{-1})(298\,\text{K})}$$

$$K_{eq} = e^{-12} = 6.2 \times 10^{-6}$$

The equilibrium constant for the coupled reaction is 2.1/$(6.2 \times 10^{-6}) = 3.4 \times 10^5$ times greater than the equilibrium constant for the uncoupled reaction.

41. (a) Substrate availability: Acetyl-CoA and oxaloacetate levels regulate citrate synthase activity. (b) Product inhibition: Citrate inhibits citrate synthase; NADH inhibits isocitrate dehydrogenase and α-ketoglutarate dehydrogenase; and succinyl-CoA inhibits α-ketoglutarate dehydrogenase. (c) Feedback inhibition: NADH and succinyl-CoA inhibit citrate synthase.

43. The alternate pathway bypasses the succinyl-CoA synthetase reaction of the standard citric acid cycle, a step that is accompanied by the phosphorylation of a nucleoside diphosphate. The alternative pathway therefore generates one less nucleoside triphosphate than the standard citric acid cycle. There is no difference in the number of reduced cofactors generated.

45. (a) The citric acid cycle is a multistep catalyst. Degrading an amino acid to a citric acid cycle intermediate boosts the catalytic activity of the cycle but does not alter the stoichiometry of the overall reaction (acetyl-CoA → 2 CO_2). **(b)** Pyruvate derived from the degradation of an amino acid can be converted to acetyl-CoA by the pyruvate dehydrogenase complex; these amino acid carbons can then be completely oxidized by the citric acid cycle.

47. Any metabolite that can be converted to oxaloacetate can enter gluconeogenesis and serve as a precursor for glucose. Biological molecules that are degraded to acetyl-CoA cannot be used as glucose precursors because acetyl-CoA enters the citric acid cycle and its two carbons are oxidized to carbon dioxide. Thus, glyceraldehyde-3-phosphate, tryptophan, phenylalanine, and pentadecanoate can serve as gluconeogenic substrates because at least one of their breakdown products can be converted to oxaloacetate. Palmitate and leucine are not glucogenic because their breakdown products are acetyl-CoA or one of its derivatives. Acetyl-CoA cannot be converted to pyruvate in mammals.

49. Alanine can be converted to pyruvate via a reversible transamination reaction. In gluconeogenesis, pyruvate is converted to oxaloacetate via the pyruvate carboxylase reaction, then oxaloacetate is converted to phosphoenolpyruvate, and so on to produce glucose. If pyruvate carboxylase is deficient, alanine is converted to pyruvate, but the gluconeogenic pathway can go no further.

51. Pyruvate carboxylase requires biotin as a cofactor (see Table 12.2). If the pyruvate carboxylase defect decreases the enzyme's affinity for biotin, administering large doses of the vitamin might help treat the disease. Biotin treatment is ineffective for more severe forms of the disease in which the mutation does not occur in the biotin binding region or in which the enzyme is expressed at extremely low levels or not at all.

53. Succinyl-CoA is converted to oxaloacetate in the citric acid cycle. Oxaloacetate can be converted to glucose by gluconeogenesis.

55. As the citric acid cycle oxidizes acetyl carbons to CO_2, NAD⁺ and ubiquinone collect the electrons and become reduced. Oxygen is required as the final acceptor in the electron transport chain that reoxidizes NADH and QH_2.

57. Ethanol is converted to acetaldehyde and then to acetate, and acetate is converted to acetyl-CoA. Acetyl-CoA then enters the glyoxylate pathway. The first step is the synthesis of citrate from acetyl-CoA and oxaloacetate. Citrate is isomerized to isocitrate. Isocitrate lyase then splits isocitrate into succinate and glyoxylate. Succinate leaves the glyoxysome and enters the mitochondrion, where it participates in the citric acid cycle. The glyoxylate fuses with a second molecule of acetyl-CoA to yield malate. Malate leaves the glyoxysome and enters the cytosol, where it is converted to oxaloacetate via the malate dehydrogenase reaction. Oxaloacetate can then enter gluconeogenesis to form glucose.

59. The mutant cannot convert isocitrate to α-ketoglutarate, so the citric acid cycle is blocked at this point. Adding glutamate to the culture medium bypasses the block, because glutamate can be converted to α-ketoglutarate and the citric acid cycle can continue.

61. (a) Hydroxycitrate differs by the addition of a hydroxyl group. **(b)** Based on its structural similarity to the substrate citrate, hydroxycitrate is most likely to act as a competitive inhibitor. **(c)** Acetyl-CoA synthesized in the mitochondria from pyruvate, the product of carbohydrate catabolism, is made available in the cytosol by the action of ATP-citrate lyase (see Fig. 14.18). Inhibiting this enzyme might decrease the amount of acetyl-CoA available for fatty

acid synthesis. **(d)** Since cholesterol synthesis also begins with acetyl-CoA, hydroxycitrate might reduce the production of cholesterol and the structurally related steroid hormones. (Although hydroxycitrate inhibits ATP-citrate lyase *in vitro*, controlled clinical trials have shown that the compound has no significant effect on weight loss in humans.)

63. (a) The acetyl-CoA reacts with glyoxylate to produce malate, which can then be converted to oxaloacetate and used to synthesize glucose. Acetyl-CoA can also condense with oxaloacetate to form citrate, which can be converted to isocitrate and then to α-ketoglutarate. Glutamate dehydrogenase converts α-ketoglutarate to glutamate by reductive amination. Alternatively, α-ketoglutarate can be acted upon by alanine aminotransferase to produce glutamate and pyruvate. **(b)** Aspartate undergoes transamination to form oxaloacetate, which can condense with acetyl-CoA to form citrate. Citrate is then converted to isocitrate, which is converted to α-ketoglutarate. α-Ketoglutarate can then be converted to glutamate.

65. (a) Isocitrate lyase and malate synthase are found in the glyoxysome, so the pathogen is using the glyoxylate pathway. The glyoxylate pathway is linked to the citric acid cycle, so citrate synthase and malate dehydrogenase are upregulated as well. The glyoxylate pathway allows the pathogen to use two-carbon sources to synthesize malate, and from there, glucose and larger biological macromolecules. The phagosome is not likely to be a rich source of glucose and amino acids for the pathogen to use for biosynthesis, so the glyoxylate pathway is essential to its survival. **(b)** The host does not have glyoxylate pathway enzymes, so inhibitors of isocitrate lyase and malate synthase would be good targets, assuming that there is a way to deliver these inhibitors to the phagosome. [From Lorenz, M. C. and Fink, G. R., *Euk. Cell* **1,** 657–662 (2002).]

67. (a)

PEP
CO_2 ⟍ PEP carboxylase
OAA
NADH ⟍
NAD⁺ ⟋ malate dehydrogenase
malate
H_2O ⟍ fumarase
fumarate
QH_2 ⟍
Q ⟋ succinate dehydrogenase
succinate

(b) If the bacteria were allowed to grow aerobically, the carbon atoms of fuel molecules would be completely oxidized to carbon dioxide. The citric acid cycle would function in a forward, catabolic direction. Depriving the bacteria of oxygen ensures that the final three steps of the citric acid cycle will function in a reverse, anabolic direction to produce the desired product, succinate. [From Agarwal, L., Isar, J., Meghwanshi, G. K., and Saxena, R. K., *Enzyme Microb. Tech.* **40,** 629–636 (2007).]

69. Glutamate is transaminated to α-ketoglutarate, then isocitrate dehydrogenase, working in reverse, adds CO_2 to α-ketoglutarate to generate isocitrate, which is in equilibrium with citrate. ATP-citrate lyase can then break down the citrate into oxaloacetate and the acetyl-CoA needed for fatty acid synthesis.

Chapter 15

1. (a) Reverse the NADH half-reaction and the sign of its $\mathcal{E}^{\circ\prime}$ to indicate oxidation; then combine the half-reactions and their $\mathcal{E}^{\circ\prime}$ values.

$$\text{pyruvate} + 2\,H^+ + 2\,e^- \rightarrow \text{lactate} \qquad \mathcal{E}^{\circ\prime} = -0.185\ \text{V}$$
$$\text{NADH} \rightarrow \text{NAD}^+ + H^+ + 2\,e^- \qquad \mathcal{E}^{\circ\prime} = 0.315\ \text{V}$$

$$\text{NADH} + \text{pyruvate} + H^+ \rightarrow \text{NAD}^+ + \text{lactate} \quad \Delta\mathcal{E}^{\circ\prime} = 0.130\ \text{V}$$

Use Equation 15.4 to calculate $\Delta G^{\circ\prime}$ for this reaction:

$$\Delta G^{\circ\prime} = -n\mathcal{F}\Delta\mathcal{E}^{\circ\prime}$$
$$\Delta G^{\circ\prime} = -(2)(96{,}485\ \text{J}\cdot\text{V}^{-1}\cdot\text{mol}^{-1})(0.130\ \text{V})$$
$$\Delta G^{\circ\prime} = -25{,}100\ \text{J}\cdot\text{mol}^{-1} = -25.1\ \text{kJ}\cdot\text{mol}^{-1}$$

The reduction of pyruvate by NADH (Section 13.1) is spontaneous under standard conditions.

(b) Reverse the cytochrome a_3 half-reaction and the sign of its $\mathcal{E}^{\circ\prime}$ value to indicate oxidation; multiply the coefficients by 2 so that the number of electrons transferred will be equal; then combine the half-reactions and their $\mathcal{E}^{\circ\prime}$ values.

$$2\,\text{cyt}\ a_3\ (\text{Fe}^{2+}) \rightarrow 2\,\text{cyt}\ a_3\ (\text{Fe}^{3+}) + 2\,e^- \qquad \mathcal{E}^{\circ\prime} = -0.385\ \text{V}$$
$$\tfrac{1}{2}O_2 + 2\,H^+ + 2\,e^- \rightarrow H_2O \qquad \mathcal{E}^{\circ\prime} = 0.815\ \text{V}$$

$$2\,\text{cyt}\ a_3\ (\text{Fe}^{2+}) + \tfrac{1}{2}O_2 + 2\,H^+ \rightarrow 2\,\text{cyt}\ a_3\ (\text{Fe}^{3+}) + H_2O$$
$$\Delta\mathcal{E}^{\circ\prime} = 0.430\ \text{V}$$

Use Equation 15.4 to calculate $\Delta G^{\circ\prime}$ for this reaction:

$$\Delta G^{\circ\prime} = -n\mathcal{F}\Delta\mathcal{E}^{\circ\prime}$$
$$\Delta G^{\circ\prime} = -(2)(96{,}485\ \text{J}\cdot\text{V}^{-1}\cdot\text{mol}^{-1})(0.430\ \text{V})$$
$$\Delta G^{\circ\prime} = -83{,}000\ \text{J}\cdot\text{mol}^{-1} = -83\ \text{kJ}\cdot\text{mol}^{-1}$$

The reaction is spontaneous under standard conditions.

3. (a) Fumarate is the oxidized compound; succinate is the reduced compound; $\mathcal{E}^{\circ\prime} = 0.031\ \text{V}$ (Table 15.1). Use Equation 15.1:

$$\mathcal{E} = \mathcal{E}^{\circ\prime} - \frac{RT}{n\mathcal{F}} \ln \frac{[\text{succinate}]}{[\text{fumarate}]}$$

$$\mathcal{E} = 0.031\ \text{V} - \frac{(8.3145\ \text{J}\cdot\text{K}^{-1}\cdot\text{mol}^{-1})(310\ \text{K})}{(2)(96{,}485\ \text{J}\cdot\text{V}^{-1}\cdot\text{mol}^{-1})} \ln \frac{(90 \times 10^{-6}\ \text{M})}{(60 \times 10^{-6}\ \text{M})}$$

$$\mathcal{E} = 0.031\ \text{V} - (0.0134\ \text{V} \times 0.405) = 0.025\ \text{V}$$

(b) Ubiquinone (Q) is the oxidized compound; ubiquinol (QH_2) is the reduced compound; $\mathcal{E}^{\circ\prime} = 0.045\ \text{V}$ (Table 15.1). Use Equation 15.1:

$$\mathcal{E} = \mathcal{E}^{\circ\prime} - \frac{RT}{n\mathcal{F}} \ln \frac{[QH_2]}{[Q]}$$

$$\mathcal{E} = 0.045\ \text{V} - \frac{(8.3145\ \text{J}\cdot\text{K}^{-1}\cdot\text{mol}^{-1})(310\ \text{K})}{(2)(96{,}485\ \text{J}\cdot\text{V}^{-1}\cdot\text{mol}^{-1})} \ln \frac{(6 \times 10^{-6}\ \text{M})}{(30 \times 10^{-6}\ \text{M})}$$

$$\mathcal{E} = 0.045\ \text{V} - (0.0134\ \text{V})(-1.61) = 0.066\ \text{V}$$

5. Reverse the malate half-reaction and the sign of its $\mathcal{E}^{\circ\prime}$ to indicate oxidation; then combine the half-reactions and their $\mathcal{E}^{\circ\prime}$ values. (Note that ubiquinone is Q and ubiquinol is QH_2.)

$$Q + 2\,H^+ + 2\,e^- \rightarrow QH_2 \qquad \mathcal{E}^{\circ\prime} = 0.045\ \text{V}$$
$$\text{malate} \rightarrow \text{oxaloacetate} + 2H^+ + 2\,e^- \qquad \mathcal{E}^{\circ\prime} = 0.166\ \text{V}$$

$$Q + \text{malate} \rightarrow QH_2 + \text{oxaloacetate} \qquad \Delta\mathcal{E}^{\circ\prime} = 0.211\ \text{V}$$

Use Equation 15.4 to calculate $\Delta G^{\circ\prime}$ for this reaction:

$$\Delta G^{\circ\prime} = -n\mathcal{F}\Delta\mathcal{E}^{\circ\prime}$$
$$\Delta G^{\circ\prime} = -(2)(96{,}485\ \text{J}\cdot\text{V}^{-1}\cdot\text{mol}^{-1})(0.211\ \text{V})$$
$$\Delta G^{\circ\prime} = -40{,}700\ \text{J}\cdot\text{mol}^{-1} = -40.7\ \text{kJ}\cdot\text{mol}^{-1}$$

The reduction of oxidation of malate by ubiquinone is spontaneous under standard conditions.

7. The relevant reactions and their $\mathcal{E}^{\circ\prime}$ values are obtained from Table 15.1 or given in the problem:

$$\text{cyt}\ c\ (\text{Fe}^{2+}) \rightarrow \text{cyt}\ c\ (\text{Fe}^{3+}) + e^- \qquad \mathcal{E}^{\circ\prime} = -0.235\ \text{V}$$
$$\text{Cu}_A\ (\text{Cu}^{2+}) + e^- \rightarrow \text{Cu}_A\ (\text{Cu}^+) \qquad \mathcal{E}^{\circ\prime} = 0.245\ \text{V}$$

$$\text{cyt}\ c\ (\text{Fe}^{2+}) + \text{Cu}_A\ (\text{Cu}^{2+}) \rightarrow \text{cyt}\ c\ (\text{Fe}^{3+}) + \text{Cu}_A\ (\text{Cu}^+)$$
$$\Delta\mathcal{E}^{\circ\prime} = 0.010\ \text{V}$$

Use Equation 15.4 to calculate $\Delta G^{\circ\prime}$ for this reaction:

$$\Delta G^{\circ\prime} = -n\mathcal{F}\Delta\mathcal{E}^{\circ\prime}$$
$$\Delta G^{\circ\prime} = -(1)(96{,}485\ \text{J}\cdot\text{V}^{-1}\cdot\text{mol}^{-1})(0.010\ \text{V})$$
$$\Delta G^{\circ\prime} = -965\ \text{J}\cdot\text{mol}^{-1} = -0.96\ \text{kJ}\cdot\text{mol}^{-1}$$

9. Consult Table 15.1 for the relevant half-reactions. Reverse the NADH half-reaction and the sign of its $\mathcal{E}^{\circ\prime}$ value to indicate oxidation; then combine the half-reactions and their $\mathcal{E}^{\circ\prime}$ values.

$$\text{acetoacetate} + 2\,H^+ + 2\,e^- \rightarrow \text{3-hydroxybutyrate} \quad \mathcal{E}^{\circ\prime} = -0.346\ \text{V}$$
$$\text{NADH} \rightarrow \text{NAD}^+ + H^+ + 2\,e^- \qquad \mathcal{E}^{\circ\prime} = 0.315\ \text{V}$$

$$\text{acetoacetate} + \text{NADH} + H^+ \rightarrow \text{NAD}^+ + \text{3-hydroxybutyrate}$$
$$\Delta\mathcal{E}^{\circ\prime} = -0.031\ \text{V}$$

Use Equation 15.4 to calculate $\Delta G^{\circ\prime}$ for this reaction:

$$\Delta G^{\circ\prime} = -n\mathcal{F}\Delta\mathcal{E}^{\circ\prime}$$
$$\Delta G^{\circ\prime} = -(2)(96{,}485\ \text{J}\cdot\text{V}^{-1}\cdot\text{mol}^{-1})(-0.031\ \text{V})$$
$$\Delta G^{\circ\prime} = 6000\ \text{J}\cdot\text{mol}^{-1} = 6.0\ \text{kJ}\cdot\text{mol}^{-1}$$

Consult Table 15.1 for the relevant half-reactions involving acetoacetate and $FADH_2$. Reverse the $FADH_2$ half-reaction and the sign of its $\mathcal{E}^{\circ\prime}$ value to indicate oxidation; then combine the half-reactions and their $\mathcal{E}^{\circ\prime}$ values.

$$\text{acetoacetate} + 2\,H^+ + 2\,e^- \rightarrow \text{3-hydroxybutyrate} \quad \mathcal{E}^{\circ\prime} = -0.346\ \text{V}$$
$$FADH_2 \rightarrow FAD + 2\,H^+ + 2\,e^- \qquad \mathcal{E}^{\circ\prime} = 0.0\ \text{V}$$

$$\text{acetoacetate} + FADH_2 \rightarrow FAD + \text{3-hydroxybutyrate}$$
$$\Delta\mathcal{E}^{\circ\prime} = -0.346\ \text{V}$$

Use Equation 15.4 to calculate $\Delta G^{\circ\prime}$ for this reaction:

$$\Delta G^{\circ\prime} = -n\mathcal{F}\Delta\mathcal{E}^{\circ\prime}$$
$$\Delta G^{\circ\prime} = -(2)(96{,}485\ \text{J}\cdot\text{V}^{-1}\cdot\text{mol}^{-1})(-0.346\ \text{V})$$
$$\Delta G^{\circ\prime} = 66{,}800\ \text{J}\cdot\text{mol}^{-1} = 66.8\ \text{kJ}\cdot\text{mol}^{-1}$$

The reduction of acetoacetate by NADH is more favorable than reduction by $FADH_2$, as shown by the $\Delta G^{\circ\prime}$ values calculated above. Although neither reaction is spontaneous under standard conditions, reduction by $FADH_2$ is far more unfavorable. The reduction of acetoacetate by NADH is, in fact, a favorable process under cellular conditions, which differ from standard conditions.

11. Consult Table 15.1 for the relevant half-reactions involving succinate and NAD^+. Reverse the succinate half-reaction and the sign of its $\mathcal{E}^{\circ\prime}$ value to indicate oxidation; then combine the half-reactions and their $\mathcal{E}^{\circ\prime}$ values.

$$\text{succinate} \rightarrow \text{fumarate} + 2\,H^+ + 2\,e^- \qquad \mathcal{E}^{\circ\prime} = -0.031\ \text{V}$$
$$\text{NAD}^+ + H^+ + 2\,e^- \rightarrow \text{NADH} \qquad \mathcal{E}^{\circ\prime} = -0.315\ \text{V}$$

$$\text{succinate} + \text{NAD}^+ + H^+ \rightarrow \text{NADH} + \text{fumarate} \quad \Delta\mathcal{E}^{\circ\prime} = -0.346\ \text{V}$$

Use Equation 15.4 to calculate $\Delta G^{\circ\prime}$ for this reaction:

$$\Delta G^{\circ\prime} = -n\mathcal{F}\Delta\mathcal{E}^{\circ\prime}$$
$$\Delta G^{\circ\prime} = -(2)(96{,}485\ \text{J}\cdot\text{V}^{-1}\cdot\text{mol}^{-1})(-0.346\ \text{V})$$
$$\Delta G^{\circ\prime} = 66{,}800\ \text{J}\cdot\text{mol}^{-1} = 66.8\ \text{kJ}\cdot\text{mol}^{-1}$$

Consult Table 15.1 for the relevant half-reactions involving succinate and FAD. Reverse the succinate half-reaction and the sign of its $\mathcal{E}^{\circ\prime}$

value to indicate oxidation; then combine the half-reactions and their $\mathcal{E}^{\circ\prime}$ values.

$$\text{succinate} \rightarrow \text{fumarate} + 2\,H^+ + 2\,e^- \qquad\qquad \mathcal{E}^{\circ\prime} = -0.031\,V$$
$$FAD + 2\,H^+ + 2\,e^- \rightarrow FADH_2 \qquad\qquad \mathcal{E}^{\circ\prime} = 0.0\,V$$

$$\overline{\text{succinate} + FAD \rightarrow \text{fumarate} + FADH_2 \qquad \Delta\mathcal{E}^{\circ\prime} = -0.031\,V}$$

Use Equation 15.4 to calculate $\Delta G^{\circ\prime}$ for this reaction:

$$\Delta G^{\circ\prime} = -n\mathcal{F}\Delta\mathcal{E}^{\circ\prime}$$
$$\Delta G^{\circ\prime} = -(2)(96{,}485\,J\cdot V^{-1}\cdot mol^{-1})(-0.031\,V)$$
$$\Delta G^{\circ\prime} = 6000\,J\cdot mol^{-1} = 6.0\,kJ\cdot mol^{-1}$$

The oxidation of succinate by FAD is more favorable than oxidation by NAD^+, as shown by the $\Delta G^{\circ\prime}$ values calculated above. Although neither reaction is spontaneous under standard conditions, oxidation by NAD^+ is far more unfavorable. The oxidation of succinate by FAD is, in fact, a favorable process under cellular conditions (citric acid cycle, Section 14.2), which differ from standard conditions.

13. (a) Reverse the half-reaction for the iron–sulfur protein to indicate that it is being oxidized. Add the two half-reactions to obtain the $\Delta\mathcal{E}^{\circ\prime}$ for the reaction.

$$FeS\,(red) \rightarrow FeS\,(ox) + e^- \qquad\qquad \mathcal{E}^{\circ\prime} = -0.280\,V$$
$$\text{cyt}\ c_1\,(Fe^{3+}) + e^- \rightarrow \text{cyt}\ c_1\,(Fe^{2+}) \qquad\qquad \mathcal{E}^{\circ\prime} = 0.215\,V$$

$$\overline{FeS\,(red) + \text{cyt}\ c_1\,(Fe^{3+}) \rightarrow FeS\,(ox) + \text{cyt}\ c_1\,(Fe^{2+})}$$
$$\Delta\mathcal{E}^{\circ\prime} = -0.065\,V$$

Use Equation 15.4 to calculate $\Delta G^{\circ\prime}$ for this reaction:

$$\Delta G^{\circ\prime} = -n\mathcal{F}\Delta\mathcal{E}^{\circ\prime}$$
$$\Delta G^{\circ\prime} = -(1)(96{,}485\,J\cdot V^{-1}\cdot mol^{-1})(-0.065\,V)$$
$$\Delta G^{\circ\prime} = 6300\,J\cdot mol^{-1} = 6.3\,kJ\cdot mol^{-1}$$

(b) The positive $\Delta G^{\circ\prime}$ indicates that the electron transfer is unfavorable under standard conditions. However, cellular conditions are not necessarily standard conditions, and the ΔG for this reaction is likely to be negative. Also, since this reaction occurs as part of the electron transport chain, the electrons gained by cytochrome c_1 will be passed along to Complex IV, in effect coupling the two reactions, which would also tend to make the process more favorable than the $\Delta G^{\circ\prime}$ indicates.

(c) Solve Equation 15.4 for $\Delta\mathcal{E}$:

$$\Delta G = -n\mathcal{F}\Delta\mathcal{E}$$
$$\mathcal{E} = -\frac{\Delta G}{n\mathcal{F}}$$
$$\mathcal{E} = -\frac{-6.0 \times 10^3\,J\cdot mol^{-1}}{(1)(96{,}485\,J\cdot V^{-1}\cdot mol^{-1})}$$
$$\mathcal{E} = 0.062\,V$$

15. (a) Consult Table 15.1 for the relevant half-reactions involving NADH and coenzyme Q. Reverse the NADH half-reaction and the sign of its $\mathcal{E}^{\circ\prime}$ value to indicate oxidation; then combine the half-reactions and their $\mathcal{E}^{\circ\prime}$ values.

$$Q + 2\,H^+ + 2\,e^- \rightarrow QH_2 \qquad\qquad \mathcal{E}^{\circ\prime} = 0.045\,V$$
$$NADH \rightarrow NAD^+ + H^+ + 2\,e^- \qquad\qquad \mathcal{E}^{\circ\prime} = 0.315\,V$$

$$\overline{NADH + Q + H^+ \rightarrow NAD^+ + QH_2 \qquad \Delta\mathcal{E}^{\circ\prime} = 0.360\,V}$$

Use Equation 15.4 to calculate $\Delta G^{\circ\prime}$ for this reaction:

$$\Delta G^{\circ\prime} = -n\mathcal{F}\Delta\mathcal{E}^{\circ\prime}$$
$$\Delta G^{\circ\prime} = -2(96{,}485\,J\cdot V^{-1}\cdot mol^{-1})(0.360\,V)$$
$$\Delta G^{\circ\prime} = -69{,}500\,J\cdot mol^{-1} = -69.5\,kJ\cdot mol^{-1}$$

The phosphorylation of ADP to ATP requires $+30.5\,kJ\cdot mol^{-1}$ of free energy. Assuming that this reaction is 35% efficient (see Solution 14), 0.8 mol ATP can be synthesized under standard conditions.

$$\frac{69.5\,kJ\cdot mol^{-1}}{30.5\,kJ\cdot mol^{-1}}\,(0.35) = 0.80\,mol$$

(b) Consult Table 15.1 for the relevant half-reactions involving ubiquinol and cytochrome c. Reverse the ubiquinol half-reaction and the sign of its $\mathcal{E}^{\circ\prime}$ value to indicate oxidation; multiply the coefficients in the cytochrome c equation by 2 so that the number of electrons transferred is equal; then combine the half-reactions and their $\mathcal{E}^{\circ\prime}$ values.

$$QH_2 \rightarrow Q + 2\,H^+ + 2\,e^- \qquad\qquad \mathcal{E}^{\circ\prime} = -0.045\,V$$
$$2\ \text{cyt}\ c\,(Fe^{3+}) + 2\,e^- \rightarrow 2\ \text{cyt}\ c\,(Fe^{2+}) \qquad\qquad \mathcal{E}^{\circ\prime} = 0.235\,V$$

$$\overline{QH_2 + 2\ \text{cyt}\ c\,(Fe^{3+}) \rightarrow Q + 2\,H^+ + 2\ \text{cyt}\ c\,(Fe^{2+})}$$
$$\Delta\mathcal{E}^{\circ\prime} = 0.190\,V$$

Use Equation 15.4 to calculate $\Delta G^{\circ\prime}$ for this reaction:

$$\Delta G^{\circ\prime} = -n\mathcal{F}\Delta\mathcal{E}^{\circ\prime}$$
$$\Delta G^{\circ\prime} = -2(96{,}485\,J\cdot V^{-1}\cdot mol^{-1})(0.190\,V)$$
$$\Delta G^{\circ\prime} = -36{,}700\,J\cdot mol^{-1} = -36.7\,kJ\cdot mol^{-1}$$

The phosphorylation of ADP to ATP requires $+30.5\,kJ\cdot mol^{-1}$ of free energy. Assuming that this reaction is 35% efficient, 0.42 mol ATP can be synthesized.

$$\frac{36.7\,kJ\cdot mol^{-1}}{30.5\,kJ\cdot mol^{-1}}\,(0.35) = 0.42\,mol$$

(c) Consult Table 15.1 for the relevant half-reactions involving cytochrome c and oxygen. Reverse the cytochrome c half-reaction and the sign of its $\mathcal{E}^{\circ\prime}$ value to indicate oxidation; multiply the coefficients by 2 so that the number of electrons transferred is equal; then combine the half-reactions and their $\mathcal{E}^{\circ\prime}$ values.

$$2\ \text{cyt}\ c\,(Fe^{2+}) \rightarrow 2\ \text{cyt}\ c\,(Fe^{3+}) + 2e^- \qquad\qquad \mathcal{E}^{\circ\prime} = -0.235\,V$$
$$\tfrac{1}{2}O_2 + 2\,H^+ + 2\,e^- \rightarrow H_2O \qquad\qquad \mathcal{E}^{\circ\prime} = 0.815\,V$$

$$\overline{2\ \text{cyt}\ c\,(Fe^{2+}) + \tfrac{1}{2}O_2 + 2\,H^+ \rightarrow 2\ \text{cyt}\ c\,(Fe^{3+}) + H_2O}$$
$$\Delta\mathcal{E}^{\circ\prime} = 0.580\,V$$

Use Equation 15.4 to calculate $\Delta G^{\circ\prime}$ for this reaction:

$$\Delta G^{\circ\prime} = -n\mathcal{F}\Delta\mathcal{E}^{\circ\prime}$$
$$\Delta G^{\circ\prime} = -2(96{,}485\,J\cdot V^{-1}\cdot mol^{-1})(0.580\,V)$$
$$\Delta G^{\circ\prime} = -112{,}000\,J\cdot mol^{-1} = -112\,kJ\cdot mol^{-1}$$

The phosphorylation of ADP to ATP requires $+30.5\,kJ\cdot mol^{-1}$ of free energy. Assuming that this reaction is 35% efficient, 1.3 mol ATP can be synthesized.

$$\frac{112\,kJ\cdot mol^{-1}}{30.5\,kJ\cdot mol^{-1}}\,(0.35) = 1.3\,mol$$

17. (a) Since all of these inhibitors interfere with electron transfer somewhere in the electron transport chain, oxygen consumption decreases when any of the inhibitors are added to a suspension of respiring mitochondria. Adding any of these inhibitors prevents electrons from being transferred to the oxygen, the final electron acceptor.

(b) In rotenone- or amytal-blocked mitochondria, NADH and Complex I redox centers are reduced while components from ubiquinone on are oxidized. In antimycin A–blocked mitochondria, NADH, Complex I redox centers, ubiquinol, and Complex III redox centers are reduced while cytochrome c and Complex IV redox centers are oxidized. In cyanide-blocked mitochondria, all of the electron transport components are reduced and only oxygen remains oxidized.

19. (a) Adding tetramethyl-*p*-phenylenediamine to rotenone-blocked and antimycin A–blocked mitochondria effectively bypasses the block as the compound donates its electrons to Complex IV and electron transport resumes. Adding tetramethyl-*p*-phenylenediamine is not an effective bypass for cyanide-blocked mitochondria because cyanide inhibits electron transport in Complex IV. **(b)** Similarly, ascorbate, which donates its electrons to cytochrome *c* and then to Complex IV, can act as an effective bypass for antimycin A–blocked mitochondria but not cyanide-blocked mitochondria.

21. Cyanide binds to the Fe^{2+} in the Fe–Cu center of cytochrome a_3 (see Problem 17). When the iron in hemoglobin is oxidized from Fe^{2+} to Fe^{3+}, cytochrome a_3 can donate an electron to reduce the hemoglobin to Fe^{2+}. This oxidizes the iron in cytochrome a_3 to Fe^{3+}. Cyanide does not bind to Fe^{3+}, so it is released and Complex IV can again function normally. The cyanide binds to the Fe^{2+} in hemoglobin, where it does not interfere with respiration (although it does interfere with oxygen delivery).

$$HbO_2 \ (Fe^{2+})$$
$$\downarrow$$
$$HbO_2 \ (Fe^{3+})$$
$$\downarrow \quad \text{cytochrome } a_3 \ (Fe^{2+}\text{—Cu)—CN}^-$$
$$\downarrow \quad \text{cytochrome } a_3 \ (Fe^{3+}\text{—Cu)}$$
$$HbO_2 \ (Fe^{2+})$$

23. All these enzymes catalyze reactions in which electrons are transferred from reduced substances, such as NADH, to ubiquinone. The flavin group, whose reduction potential is lower than that of ubiquinone and higher than that of NADH (Table 15.1), is ideally suited to shuttle electrons between the reduced NADH and the oxidized ubiquinone.

25. Like the lipids that compose the membrane, coenzyme Q is amphiphilic, with a hydrophilic head and a hydrophobic tail. The hydrophobic tail of coenzyme Q forms favorable van der Waals interactions with the hydrophobic acyl chains of the phospholipids that compose the cell membrane. Coenzyme Q literally dissolves in the membrane, which facilitates rapid diffusion.

27. Cytochrome *c* is a water-soluble, peripheral membrane protein and is easily dissociated from the membrane by adding salt solutions that interfere with the ionic interactions that tether it to the inner mitochondrial membrane. Cytochrome c_1 is an integral membrane protein and is largely water-insoluble due to the nonpolar amino acids that interact with the acyl chains of the membrane lipids. Detergents are required to dissociate cytochrome c_1 from the membrane because amphiphilic detergents can disrupt the membrane and coat membrane proteins, acting as substitute lipids in the solubilization process.

29. The dead algae are a source of food for aerobic microorganisms lower in the water column. As the growth of these organisms increases, the rates of respiration and O_2 consumption increase to the point where the concentration of O_2 in the water becomes too low to sustain larger aerobic organisms.

31. In the absence of myoglobin, the mice developed several compensatory mechanisms to ensure adequate oxygen delivery to tissues. The symptoms described all involve increasing the amount of available hemoglobin. In this manner, hemoglobin takes over some of the functions usually performed by myoglobin.

33. Myoglobin functions in muscle cells to facilitate oxygen diffusion throughout the cell (see Problem 31) and possibly assumes this same role in tumor cells. The resulting increase in oxygen concentration may allow the tumor cells to oxidize a higher percentage of the available glucose aerobically and thus obtain

more ATP per molecule of glucose than if glucose were oxidized anaerobically. [From Kristiansen, G., et al., *J. Biol. Chem.*, **286**, 43417–43428 (2011).]

35. As a result of the exercise program, the number of mitochondria in the muscle cells of the participants increased, as indicated by an increase in DNA content. The increase in Complex II activity is of similar magnitude, which might be the result of the increased number of mitochondria. However, the total electron transport chain activity is twofold greater after the exercise intervention, indicating that mitochondrial function increased as well. These results suggest that even though oxidative damage to mitochondria increases with age (see Box 15.A), exercise can help maintain or increase mitochondrial function. [From Menshikova, E. V., Ritov, V. B., Fairfull, L., Ferrell, R. E., Kelley, D. E., and Goodpaster, B. H., *J. Gerontol. A-Biol.* **61**, 534–540 (2006)].

37. (a) The free energy change for generating the electrical imbalance is calculated using Equation 15.6:

$$\Delta G = Z \mathcal{F} \Delta \psi$$
$$\Delta G = (1)(96,485 \ \text{J} \cdot \text{V}^{-1} \cdot \text{mol}^{-1})(0.081 \ \text{V})$$
$$\Delta G = 7800 \ \text{J} \cdot \text{mol}^{-1} = 7.8 \ \text{kJ} \cdot \text{mol}^{-1}$$

(b) The free energy change for generating the electrical imbalance is calculated using Equation 15.6:

$$\Delta G = Z \mathcal{F} \Delta \psi$$
$$\Delta G = (1)(96,485 \ \text{J} \cdot \text{V}^{-1} \cdot \text{mol}^{-1})(0.150 \ \text{V})$$
$$\Delta G = 14,500 \ \text{J} \cdot \text{mol}^{-1} = 14.5 \ \text{kJ} \cdot \text{mol}^{-1}$$

39. Use the rearranged form of Equation 15.7 shown in Sample Calculation 15.3 and solve for $(pH_{in} - pH_{out})$; then enter the values given in the problem:

$$\Delta G = 2.303 \, RT(pH_{in} - pH_{out}) + Z\mathcal{F}\Delta\psi$$
$$pH_{in} - pH_{out} = \frac{\Delta G - Z\mathcal{F}\Delta\psi}{2.303RT}$$
$$pH_{in} - pH_{out} = \frac{(29,200 \ \text{J} \cdot \text{mol}^{-1}) - (1)(96,485 \ \text{J} \cdot \text{V}^{-1} \cdot \text{mol}^{-1})(0.170 \ \text{V})}{2.303(8.3145 \ \text{J} \cdot \text{K}^{-1} \cdot \text{mol}^{-1})(310 \ \text{K})}$$
$$pH_{in} - pH_{out} = \frac{12800 \ \text{J} \cdot \text{mol}^{-1}}{5900 \ \text{J} \cdot \text{mol}^{-1}} = 2.2$$

41. (a) The impermeability of the membrane to ions is an important feature of chemiosmosis, because the free movement of ions other than H^+ would dissipate the electrical component of the proton gradient and thereby compromise the ability of the gradient to supply free energy for ATP synthesis.
(b) If the membrane were permeable to other ions, the proton gradient would be a simple chemical gradient rather than an electrochemical gradient. However, it would still be able to serve as a source of free energy for ATP synthesis.

43. A phosphate ion enters the mitochondrial matrix in symport with a proton. This proton moves in a direction opposite to the protons translocated by Complexes I, III, and IV. Thus, the difference in pH between the mitochondrial matrix and the intermembrane space becomes smaller.

45. A total of 32 ATP are obtained from the exergonic oxidation of glucose under aerobic conditions (see Fig. 14.13), whereas only 2 ATP are produced when glucose is oxidized in the absence of oxygen to lactate or ethanol (Section 13.1). Organisms that can oxidize glucose in the presence of oxygen have an advantage over anaerobic organisms because these organisms can extract more energy per glucose. This may have been important in evolution.

47. **(a)** Aerobic respiration yields 32 ATP (see Problem 45), each of which requires 30.5 kJ·mol^{-1} to synthesize:

32 ATP × 30.5 kJ·mol^{-1} ATP × 100/2850 kJ·mol^{-1} = 34%

Lactate fermentation yields 2 ATP:

2 ATP × 30.5 kJ·mol^{-1} ATP × 100/196 kJ·mol^{-1} = 31%

Alcoholic fermentation yields 2 ATP:

2 ATP × 30.5 kJ·mol^{-1} ATP × 100/235 kJ·mol^{-1} = 26%

(b) Organisms that can oxidize glucose in the presence of oxygen have an advantage over anaerobic organisms because these organisms can extract more energy per glucose. This may have been important in evolution.

49. Stimulation of the glycolytic enzymes hexokinase and phosphofructokinase increases glycolytic flux and allows the yeast cells to obtain ATP more quickly under anaerobic conditions. [From Beitner, R., Cohen, T. J., Nordenberg, J., and Haberman, S., *Biochim. Biophys. Acta* **586,** 266–277 (1979).]

51. If the translocase is unable to function, ATP cannot exit the mitochondrial matrix and ADP cannot enter. Without ADP, the substrate for ATP synthase, ATP synthesis will not occur. Since electron transport and oxidative phosphorylation are coupled, a decrease in the rate of oxidative phosphorylation decreases the rate of electron transport.

53. The donation of a pair of electrons to cytochrome *c* and then to Complex IV results in the synthesis of about 1.3 ATP per atom of oxygen (½ O$_2$; see Solution 15c). Therefore, the P:O ratio of this compound is 1.3.

55. Dinitrophenol (DNP) uncouples electron transport from oxidative phosphorylation by dissipating the proton gradient. Electron transport still occurs, but the energy released by electron transport is dissipated as heat instead of being harnessed to synthesize ATP. One might think that DNP would be an effective diet aid because the sources of the electrons for the electron transport chain are dietary carbohydrates and fatty acids. If the energy of these compounds is dissipated as heat instead of used to synthesize ATP (which would then be used for, among other processes, the synthesis of fatty acids in adipocytes), weight gain from the ingestion of food could theoretically be prevented.

57. FCCP is a lipid-soluble compound that can pass through both the outer and inner mitochondrial membranes and into the matrix. FCCP has an acidic hydrogen on the central nitrogen that dissociates in the basic environment of the matrix. The FCCP passes back through the inner membrane to the intermembrane space, where it is reprotonated and the entire cycle begins again. In this manner, the FCCP shuttles protons into the mitochondrial matrix and dissipates the proton gradient.

59. **(a)** The P:O ratio is decreased. In this patient, electron transport and oxidative phosphorylation are uncoupled. NADH is oxidized in the electron transport chain, oxygen is reduced to water, but ATP will not be synthesized even if ADP is available. This decreases the P:O ratio because oxygen consumption occurs to a greater extent than ADP phosphorylation.

(b) If the energy released in electron transport is not used to synthesize ATP, it is released as heat, which accounts for the patient's elevated body temperature. If sufficient ATP is not synthesized to meet energy needs, the rate of electron transport increases, which leads to an increased consumption of O$_2$ and an increase in the concentration of reduced coenzymes that enter electron transport, thus increasing metabolic rate.

(c) The patient will not be able to carry out strenuous exercise under aerobic conditions because her muscle cells are incapable of generating enough ATP to power the muscle.

61. For yeast, the ratio is 10 H$^+$ per 3 ATP, or 3.33. For the chloroplast, the ratio is 14 H$^+$ per 3 ATP, or 4.67.

63. One complete revolution of the *c* ring generates 3 ATP.

$$\frac{6000 \text{ revolutions}}{\text{min}} \times \frac{3 \text{ ATP}}{\text{revolution}} \times \frac{1 \text{ min}}{60 \text{ s}} = 300 \text{ ATP·s}^{-1}$$

65. An increase in the amount of glucose led to an increase in glycolytic flux and an increase in the concentration of NADH, which was subsequently reoxidized during electron transport. Because this did not occur with the *pdc*–mutants, pyruvate decarboxylase must play a role in reoxidizing the NADH produced during glycolysis. In the absence of the enzyme, glycolytic flux could not increase due to the deficiency of NAD$^+$. The NAD$^+$ deficiency also prevented the decarboxylation of pyruvate by pyruvate dehydrogenase.

67. **(a)** When UCP1 is stimulated in normal mice, oxidative phosphorylation is uncoupled from electron transport. This means that the ATP yield per substrate molecule oxidized decreases, since ATP must then be produced via substrate level phosphorylation. The cell's energy needs are not met, which increases the rate of glycolysis and the citric acid cycle, and eventually electron transport, in a vain attempt to synthesize more ATP. Since oxygen is the final electron acceptor in electron transport, oxygen consumption increases in order to keep up with the increased rate of electron transport. In knockout mice, since there is no UCP1, oxidative phosphorylation is not uncoupled. Therefore, ATP can be synthesized via oxidative phosphorylation. Since the energy needs of the cell are being met, the rate of electron transport and oxygen consumption do not increase.

(b) In the normal mice, the cold temperature stimulated the uncoupling protein. As a result, the energy of electron transport was dissipated as heat rather than used to synthesize ATP. This helped the mice maintain normal body temperature. But the UCP1-knockout mice lacked the uncoupling protein and were unable to uncouple oxidation from phosphorylation. Thus, they could not generate "extra" heat and their body temperatures decreased as a result. [From Enerbäck, S., Jacobsson, A., Simpson, E. M., Guerra, C., Yamashita, H., Harper, M.-E., and Kozak, L. P., *Nature* **387,** 90–94 (1997).]

69. The synthesis of the uncoupling protein increases with decreasing temperature, presumably by an increase in transcription of the mRNA that codes for the uncoupling protein (although a decrease in the rate of mRNA degradation would produce the same results). The increased amount of mRNA likely results in an increase in concentration of the uncoupling protein so that the mitochondrial proton gradient would be dissipated. Thus, at low temperatures, the potato could generate heat rather than ATP. [From Laloi, M., Klein, M., Riesmeier, J. W., Müller-Röber, B., Fleury, C., Bouillaud, F., and Ricquier, D., *Nature* **389,** 135–136 (1997).]

71. **(a)** Substrate-level phosphorylation is catalyzed by phosphoglycerate kinase and pyruvate kinase in glycolysis, and by succinyl-CoA synthetase in the citric acid cycle.

(b) C$_6$H$_{12}$O$_6$ + 6 O$_2$ → 6 CO$_2$ + 6 H$_2$O.

Chapter 16

1. **(a)** proton translocation C, M; **(b)** photophosphorylation C; **(c)** photooxidation C; **(d)** quinones C, M; **(e)** oxygen reduction M; **(f)** water oxidation C; **(g)** electron transport C, M; **(h)** oxidative phosphorylation M; **(i)** carbon fixation C; **(j)** NADH oxidation M; **(k)** Mn cofactor C; **(l)** heme groups C, M; **(m)** binding change mechanism C, M; **(n)** iron–sulfur clusters C, M; **(o)** NADP$^+$ reduction C.

3.

5. (a) Use Planck's law (Equation 16.1) to determine the energy of a single photon (see Sample Calculation 16.1), then multiply by Avogadro's number (N) to calculate the energy of a mole of photons:

$$E = \frac{hc}{\lambda}$$

$$E = \frac{(6.626 \times 10^{-34}\,\text{J}\cdot\text{s})(2.998 \times 10^{8}\,\text{m}\cdot\text{s}^{-1})}{680 \times 10^{-9}\,\text{m}}$$

$$E = 2.9 \times 10^{-19}\,\text{J}$$

$$E = (2.9 \times 10^{-19}\,\text{J}\cdot\text{photon}^{-1}) \times (6.022 \times 10^{23}\,\text{photons}\cdot\text{mol}^{-1})$$

$$E = 1.8 \times 10^{5}\,\text{J}\cdot\text{mol}^{-1} = 180\,\text{kJ}\cdot\text{mol}^{-1}$$

(b) $E = \dfrac{hc}{\lambda}$

$$E = \frac{(6.626 \times 10^{-34}\,\text{J}\cdot\text{s})(2.998 \times 10^{8}\,\text{m}\cdot\text{s}^{-1})}{450 \times 10^{-9}\,\text{m}}$$

$$E = 4.4 \times 10^{-19}\,\text{J}$$

$$E = (4.4 \times 10^{-19}\,\text{J}\cdot\text{photon}^{-1}) \times (6.022 \times 10^{23}\,\text{photons}\cdot\text{mol}^{-1})$$

$$E = 2.7 \times 10^{5}\,\text{J}\cdot\text{mol}^{-1} = 270\,\text{kJ}\cdot\text{mol}^{-1}$$

(c) $E = \dfrac{hc}{\lambda}$

$$E = \frac{(6.626 \times 10^{-34}\,\text{J}\cdot\text{s})(2.998 \times 10^{8}\,\text{m}\cdot\text{s}^{-1})}{600 \times 10^{-9}\,\text{m}}$$

$$E = 3.3 \times 10^{-19}\,\text{J}$$

$$E = (3.3 \times 10^{-19}\,\text{J}\cdot\text{photon}^{-1}) \times (6.022 \times 10^{23}\,\text{photons}\cdot\text{mol}^{-1})$$

$$E = 2.0 \times 10^{5}\,\text{J}\cdot\text{mol}^{-1} = 200\,\text{kJ}\cdot\text{mol}^{-1}$$

Wavelength and energy are inversely related; photons with a shorter wavelength have greater energy.

7. If the synthesis of each ATP requires 30.5 kJ·mol^{-1}, then

(a) About 5.9 mol ATP (180/30.5) could be synthesized.
(b) About 8.8 mol ATP (270/30.5) could be synthesized.
(c) About 6.5 mol ATP (200/30.5) could be synthesized.

9. Photosynthetic bacteria can absorb light in the infrared and ultraviolet regions of the electromagnetic spectrum.

11. (a) The sequences would provide only limited information, since even highly homologous proteins such as myoglobin and hemoglobin (which have very similar structures) have limited sequence identity (see Section 5.1). The three-dimensional structures of the proteins would be more likely to indicate whether they function similarly. **(b)** Low-resolution models that included the heme groups would be more useful, since the positions of the heme groups within the proteins might indicate whether they function similarly in accepting and donating electrons to their redox partners. It is possible that the heme groups could have similar orientations even if the apoproteins did not resemble each other in overall tertiary structure.

13. Photooxidation would not be a good protective mechanism since it might interfere with the normal redox balance among the electron-carrying groups in the thylakoid membrane. Releasing the energy by exciton transfer or fluorescence (emitting light of a longer wavelength) could potentially funnel light energy back to the overactive photosystems. Dissipation of excess energy as heat would be the safest mechanism, since the photosystems do not have any way to harvest thermal energy to drive chemical reactions.

15. Since the energy of one mole of photons (see Solution 6) is

$$E = \frac{hc}{\lambda} \times N$$

$$\lambda = \frac{hcN}{E}$$

$$= \frac{(6.626 \times 10^{-34}\,\text{J}\cdot\text{s})(2.998 \times 10^{8}\,\text{m}\cdot\text{s}^{-1})(6.022 \times 10^{23}\,\text{photons}\cdot\text{mol}^{-1})}{188{,}000\,\text{J}\cdot\text{mol}^{-1}}$$

$$= 636 \times 10^{-9}\,\text{m} = 636\,\text{nm}$$

17. If the photosystems were in close proximity, then Photosystem II might not undergo photooxidation and instead act as a light-harvesting complex for Photosystem I. Because the reaction center of Photosystem II absorbs light with a peak wavelength of 680 nm, it could pass energy to the reaction center of Photosystem I, which absorbs lower-energy (longer-wavelength) light at its P700 group.

19. Myxothiazol inhibits electron transfer in the cytochrome b_6f complex.

21. The alanine residue at position 251 is essential for photosynthesis and photoautotrophic growth and may be part of the binding site for herbicides. When alanine is mutated to cysteine, which resembles alanine except that a sulfhydryl group has replaced a hydrogen, the mutated protein is similar to the wild-type. When alanine is changed to an amino acid that is similar in size but more polar (such as Gly, Pro, or Ser), photosynthesis is impaired. Replacement with a larger, nonpolar amino acid results in a mutant that is further impaired, and replacement with a larger polar amino acid results in an organism that is not photosynthetically competent. Therefore, the residue at position 251 must be relatively small and nonpolar. [From Lardans, A., et al., *J. Biol. Chem.* **273**, 11082–11091 (1998).]

23. Use Equation 15.7, as applied in Sample Calculation 15.3. The matrix and the stroma are both *in*.

$$\Delta G = 2.303\,RT(\text{pH}_{in} - \text{pH}_{out}) + Z\mathcal{F}\Delta\psi$$

$$\Delta G = 2.303\,(8.3145\,\text{J}\cdot\text{K}^{-1}\cdot\text{mol}^{-1})(298\,\text{K})(0.75)$$

$$+ (1)(96{,}485\,\text{J}\cdot\text{V}^{-1}\cdot\text{mol}^{-1})(0.20\,\text{V})$$

$$\Delta G = 4300\,\text{J}\cdot\text{mol}^{-1} + 19{,}300\,\text{J}\cdot\text{mol}^{-1} = 23.6\,\text{kJ}\cdot\text{mol}^{-1}$$

For both organelles, the translocation is endergonic. In the chloroplast, most of this energy is due to the concentration (pH) difference on the two sides of the membrane (see Solution 25); in the mitochondrion, most of the energy is due to the membrane potential ($\Delta\psi$).

25. Use Equation 16.1 and multiply by 2 to calculate the energy of the photons:

$$E = \frac{hc}{\lambda} \times 2$$

$$E = \frac{(6.626 \times 10^{-34}\,\text{J}\cdot\text{s})(2.998 \times 10^{8}\,\text{m}\cdot\text{s}^{-1})(2)}{700 \times 10^{-9}\,\text{m}}$$

$$E = 5.7 \times 10^{-19}\,\text{J}$$

The energy of two photons is nearly 1.6 fold greater than the energy change for the oxidation of one molecule of water by NADP^{+} so, in theory, the two photons supply sufficient energy to drive the oxidation.

27. Some of the label appears in molecular oxygen released by the water-splitting reaction of Photosystem II. Some of the label appears in other compounds, including intermediates of the Calvin cycle, since H_2O participates in the rubisco reaction (Fig. 16.23).

29. Complex III of the mitochondrial electron transport chain is analogous to the cytochrome $b_6 f$ complex. In the presence of antimycin A, electron flow in the cytochrome $b_6 f$ complex would be inhibited. Electrons would not reach Photosystem I, and NADPH would not be produced. Proton translocation from the stroma to the thylakoid lumen also would not occur, so CF_1 would not be stimulated and ATP synthesis would not occur.

31. Because oligomycin inhibits the F_0 subunit of mitochondrial ATP synthase, protons from the intermembrane space cannot translocate to the mitochondrial matrix. F_1 is not stimulated, and ATP synthesis does not occur. The cytosolic ratio of ATP/ADP decreases because less ATP is exported to the cytosol from the matrix when ATP synthesis is inhibited. Because oligomycin does not inhibit CF_0, ATP synthesis is not inhibited, so the ATP/ADP ratio in the chloroplast is not affected. In some cases the decreased cytosolic ATP/ADP ratio might stimulate ATP synthesis in the chloroplast, resulting in an increased chloroplastic ATP/ADP ratio. [From Gardeström, P. and Lernmark, U., *J. Bioenerg. Biomembr.* **27,** 415–421 (1995).]

33. The net equation would be

$$CO_2 + 2 H_2S \rightarrow (CH_2O) + S_2 + H_2O$$

35. When rubisco reacts with oxygen in a process called photorespiration, the products are 3-phosphoglycerate and 2-phosphoglycolate. The latter product is further metabolized by pathways that consume ATP and NADPH and thus waste some of the energy captured by photosynthesis. If rubisco could bind only CO_2, the enzyme would produce two molecules of 3-phosphoglycerate, which is a precursor to many biological macromolecules. In a plant with a rubisco that binds only CO_2, photorespiration would not occur, and more of the energy obtained in photosynthesis would be available for biosynthetic reactions.

37. Rearrange Equation 12.3 to solve for the [products]/[reactants] ratio:

$$\Delta G = \Delta G^{\circ\prime} + RT \ln \frac{[products]}{[reactants]}$$

$$e^{(\Delta G - \Delta G^{\circ\prime})/RT} = \frac{[products]}{[reactants]}$$

$$e^{(-41,000 \text{ J}\cdot\text{mol}^{-1} + 35,100 \text{ J}\cdot\text{mol}^{-1})/(8.3145 \text{ J}\cdot\text{K}^{-1}\cdot\text{mol}^{-1})(298 \text{ K})} = \frac{[products]}{[reactants]}$$

$$\frac{[products]}{[reactants]} = e^{-2.38} = \frac{0.092}{1}$$

The ratio of products to reactants is 0.092 to 1.

39. (a) The compound resembles the transition state of the rubisco carboxylase reaction (see Fig. 16.23) and therefore inhibits the enzyme by binding in the active site.
(b) At night, the compound inhibits rubisco in order to prevent the Calvin cycle from consuming NADPH and ATP. During the day, when the light reactions are supplying NADPH and ATP, the inhibitor is broken down to reactivate rubisco so that it can fix carbon.

41. ATP and NADPH are produced by the light reactions during the day. When photosynthesis is occurring, synthesis of starch (an ATP-requiring process) occurs, and glycolysis, an opposing process, is inhibited. At night, when the light reactions do not occur and ATP and NADPH concentrations are low, the inhibition of phosphofructokinase is relieved. Starch is broken down, and the resulting glucose-1-phosphate enters glycolysis to produce ATP to meet the

energy needs of the plant. [From Cséke, C., Nishizawa, A. N., and Buchanan, B. B., *Plant Physiol.* **70,** 658–661 (1982).]

43.

The activity of Photosystem I generates reduced ferredoxin, which is a substrate for the reductase. The product of the reductase reaction, reduced thioredoxin, can then reduce the disulfides of the Calvin cycle enzymes. Conformational changes in the enzymes upon exposure of free sulfhydryl groups could increase their activity.

45. Rearrange Equation 12.3 to solve for the [products]/[reactants] ratio:

$$\Delta G = \Delta G^{\circ\prime} + RT \ln \frac{[products]}{[reactants]}$$

$$e^{(\Delta G - \Delta G^{\circ\prime})/RT} = \frac{[products]}{[reactants]}$$

$$e^{(-29,700 \text{ J}\cdot\text{mol}^{-1} + 14,200 \text{ J}\cdot\text{mol}^{-1})/(8.3145 \text{ J}\cdot\text{K}^{-1}\cdot\text{mol}^{-1})(298 \text{ K})} = \frac{[products]}{[reactants]}$$

$$\frac{[products]}{[reactants]} = e^{-6.26} = \frac{0.0019}{1}$$

The enzyme is likely to be regulated because it catalyzes an irreversible step of the Calvin cycle (as evidenced by the large negative value of ΔG).

47. PEPC supplies oxaloacetate for the citric acid cycle, which can generate ATP for the biosynthetic processes occurring in the germinating seedling. The PEPC enzyme also plays a role in providing oxaloacetate for the glyoxylate cycle (see Box 14.B), which provides a route for the synthesis of glucose from fatty acids, a pathway that is lacking in animals.

49. The Bt toxin is not toxic to humans because the stomach is acidic (see Section 12.1). The toxin forms a pore only under high pH conditions.

51. Glyphosate is not toxic to humans because humans do not contain the enzyme EPSPS. This pathway was lost during evolution and the result is that phenylalanine is an essential amino acid that must be obtained in the diet (see Table 12.1).

Chapter 17

1. The lipoproteins increase in density as the percentage of protein content increases and the percentage of lipid content decreases. Thus, chylomicrons have the lowest density and HDL have the highest density.

3. Cholesteryl esters and triacylglycerols are nonpolar; cholesterol and phospholipids are amphipathic.
The nonpolar cholesteryl esters and triacylglycerols are in the core of the lipoprotein particle. The hydroxyl group of the cholesterol

molecule is on the surface of the lipoprotein particle while the remaining portion of the molecule is oriented toward the nonpolar core. The phospholipid head group is on the surface of the lipoprotein while the fatty acyl chains are oriented toward the nonpolar core.

5. (a) HDL remove excess cholesterol from tissues and transport it back to the liver. This prevents the accumulation of cholesterol in vessel walls that leads to atherosclerosis.

(b) HDL level alone does not indicate the risk of developing atherosclerosis, since the level of LDL, the activity of the LDL receptor, and other factors such as smoking or vessel wall injuries resulting from infection can all influence the likelihood of developing the disease.

7. (a) The products are one palmitoyl methyl ester, two oleoyl methyl esters, and one glycerol:

$$H_3C-(CH_2)_{14}-\overset{\overset{\displaystyle O}{\|}}{C}-O-CH_3$$

$$H_3C-(CH_2)_7-CH=CH-(CH_2)_7-\overset{\overset{\displaystyle O}{\|}}{C}-O-CH_3$$

$$\begin{array}{ccc} H_2C-CH-CH_2 \\ | \quad | \quad | \\ HO \quad OH \quad OH \end{array}$$

(b) Water would participate in hydrolysis reactions, yielding glycerol plus three fatty acids.

9.

11. In the absence of cytosolic carnitine acyltransferase activity, fats can't be transported into the mitochondrial matrix for β oxidation and therefore cannot be degraded as a source of energy. The body instead relies on glucose oxidation for energy. The increased consumption of glucose leads to hypoglycemia. The disease can be treated with a diet high in protein and carbohydrates and low in fat and by avoiding fasting.

13. If medium-chain fatty acids are metabolized normally in the absence of carnitine, then fatty acids of this size must be able to cross the inner mitochondrial membrane and enter the mitochondrial matrix without the aid of carnitine. Carnitine is required for transport of fatty acids longer than 10 carbons.

15. The patient should avoid fasting and sustained exercise, since both of these conditions result in the release of free fatty acids into the bloodstream for uptake by the tissues. The patient should also consume a diet that is low in fat and high in carbohydrates.

17. Oxidation of the methylene groups is accomplished by the action of a dehydrogenase, which removes electrons. The oxygen atom that becomes part of the carbonyl group is derived from a water molecule added by enoyl-CoA hydratase.

19. The products are three molecules of propionyl-CoA, three molecules of acetyl-CoA, and one molecule of 2-methylpropionyl-CoA.

21. (a) Oleate is an 18-carbon fatty acid. If it were completely saturated, 120 ATP would be generated via β oxidation (see Problem 23b). The double bond at the 9,10 position must be converted from the *cis* to the *trans* isomer. β Oxidation continues after this isomerization step, but with the loss of 1 QH$_2$. Thus 1.5 ATP should be subtracted from the total to yield 118.5 ATP. The pathway shows a total of

120.5 ATP, but 2 ATP are subtracted to account for the ATP spent in fatty acid activation:

$$H_3C-(CH_2)_7-CH=CH-(CH_2)_7-\overset{\overset{O}{\|}}{C}-SCoA$$

Oleoyl-CoA

3 cycles of β oxidation → 3 (1 QH$_2$, 1 NADH, 1 acetyl-CoA) =
3(1.5 + 2.5 + 10) = 42 ATP

$$H_3C-(CH_2)_7-\underset{H}{\overset{}{C}}=\underset{H}{\overset{}{C}}-CH_2-\overset{\overset{O}{\|}}{C}-SCoA$$

enoyl-CoA isomerase ↓

$$H_3C-(CH_2)_8-\underset{H}{\overset{}{C}}=\overset{H}{\overset{}{C}}-\overset{\overset{O}{\|}}{C}-SCoA$$

*1 cycle of β oxidation
(acyl-CoA dehydrogenase
reaction is bypassed)* → 1 NADH, 1 acetyl-CoA =
2.5 + 10 = 12.5 ATP

$$H_3C-(CH_2)_8-\overset{\overset{O}{\|}}{C}-SCoA$$

4 cycles of β oxidation → 4 (1 QH$_2$, 1 NADH, 1 acetyl-CoA) =
4(1.5 + 2.5 + 10) = 56 ATP

Acetyl-CoA (10 ATP)

(b) Linoleate is an 18-carbon fatty acid and would yield 120 ATP via β oxidation if it were saturated. The double bond at the 9,10 position must be converted from the *cis* to the *trans* isomer, resulting in the loss of 1 QH$_2$ (1.5 ATP). In addition, an NADPH-dependent reduction of the 12,13 double bond costs 2.5 ATP. This reduces the ATP total by 4, so oxidation of linoleate yields 116 ATP. (Subtract 2 ATP for activation from the total shown in the figure.)

$$H_3C-(CH_2)_4-CH=CH-CH_2-CH=CH-(CH_2)_7-\overset{\overset{O}{\|}}{C}-SCoA$$

Linoleyl-CoA

3 cycles of β oxidation → 3 (1 QH$_2$, 1 NADH, 1 acetyl-CoA) =
3(1.5 + 2.5 + 10) = 42 ATP

$$H_3C-(CH_2)_4-CH=CH-CH_2-\underset{H}{\overset{}{C}}=\underset{H}{\overset{}{C}}-CH_2-\overset{\overset{O}{\|}}{C}-CoA$$

enoyl-CoA isomerase ↓

$$H_3C-(CH_2)_4-CH=CH-CH_2-CH_2-\underset{H}{\overset{}{C}}=\overset{H}{\overset{}{C}}-\overset{\overset{O}{\|}}{C}-CoA$$

*1 cycle of β oxidation
(acyl-CoA dehydrogenase
reaction is bypassed)* → 1 NADH, 1 acetyl-CoA =
2.5 + 10 = 12.5 ATP

$$H_3C-(CH_2)_4-CH=CH-CH_2-CH_2-\overset{\overset{O}{\|}}{C}-S-CoA$$

acyl-CoA dehydrogenase → 1 QH$_2$ = 1.5 ATP

$$H_3C-(CH_2)_4-CH=CH-\underset{H}{\overset{}{C}}=\overset{H}{\overset{}{C}}-\overset{\overset{O}{\|}}{C}-S-CoA$$

2,4-dienoyl-CoA reductase ⟨ NADPH / NADP$^+$ *Costs 2.5 ATP*

$$H_3C-(CH_2)_4-CH_2-CH=CH-CH_2-\overset{\overset{O}{\|}}{C}-S-CoA$$

enoyl-CoA isomerase ↓

$$H_3C-(CH_2)_4-CH_2-CH_2-\underset{H}{\overset{}{C}}=\underset{H}{\overset{}{C}}-\overset{\overset{O}{\|}}{C}-S-CoA$$

*1 cycle of β oxidation
(acyl CoA dehydrogenase
reaction is bypassed)* → 1 NADH, 1 acetyl-CoA =
(2.5 + 10) = 12.5 ATP

3 cycles of β oxidation → 3 (1 QH$_2$, 1 NADH, 1 acetyl-CoA) =
3(1.5 + 2.5 + 10) = 42 ATP

Acetyl-CoA (10 ATP)

23. Adding up the ATP as shown in the figure below gives a total of 155 ATP. Subtracting the 2 ATP required for activation brings the total to 153 ATP.

$$H_3C-(CH_2)_{20}-CH_2-CH_2-\overset{\overset{O}{\|}}{C}-SCoA$$

acyl-CoA oxidase ⟨ FAD / FADH$_2$ ⟩ ⟨ H$_2$O$_2$ / O$_2$ ⟩ → 2 H$_2$O + O$_2$

$$H_3C-(CH_2)_{20}-\underset{H}{\overset{}{C}}=\overset{H}{\overset{}{C}}-\overset{\overset{O}{\|}}{C}-SCoA$$

enoyl-CoA

*6 cycles of β oxidation
(acyl-CoA dehydrogenase
reaction is bypassed)* → 6 NADH, 6 acetyl-CoA =
6(2.5) + 6(10) ATP = 75 ATP

$$H_3C-(CH_2)_8-CH_2-CH_2-\overset{\overset{O}{\|}}{C}-SCoA$$

-- PEROXISOME
                                                   MITOCHONDRION

5 cycles of β oxidation → 5 QH$_2$, 5 NADH, 5 acetyl-CoA =
5(1.5) + 5(2.5) + 5(10) ATP = 70 ATP

Acetyl-CoA (10 ATP)

25. A patient with pernicious anemia would have an excess of methylmalonate in the blood or urine because the methylmalonyl-CoA mutase reaction (Figure 17.7), which requires the cobalamin cofactor, would be unable to take place, leading to accumulation of reactants.

27. (a) The free energy cost of synthesizing ATP from ADP + P$_i$ is 30.5 kJ · mol^{-1}. Glucose oxidation could theoretically yield 2850/30.5, or 93.4 ATP. The ATP yield per carbon atom is 93.4/6 = 15.6. Palmitate oxidation could theoretically yield 9781/30.5, or 320.7 ATP. The ATP yield per carbon atom is 320.7/16 = 20.4 ATP.
(b) *In vivo*, the catabolism of glucose leads to the production of 32 ATP (see Solution 15.45), which is 32/6, or 5.3 ATP per carbon atom. The catabolism of palmitate leads to the production of 106 ATP (see Solution 20a), which is 106/16, or 6.6 ATP per carbon atom.
(c) In theory as well as *in vivo*, fatty acid oxidation yields more ATP per carbon atom than glucose oxidation (this is primarily because the carbons of carbohydrates are already partially oxidized, whereas fatty acid carbons are usually fully reduced). Both pathways are equally efficient from a thermodynamic point of view, recovering slightly more than 30% of the free energy available (for glucose 5.3/15.6, and for palmitate, 6.6/20.4).

29. Acetyl-CoA carboxylase generates malonyl-CoA, the substrate for fatty acid synthase. Mice without acetyl-CoA carboxylase are unable to synthesize fatty acids and hence store less fat in their bodies. Malonyl-CoA also inhibits carnitine acyltransferase. In the absence of acetyl-CoA carboxylase to produce this inhibitor, transport of fatty acids into mitochondria cannot be regulated, and mitochondrial β oxidation proceeds continuously.

31. As shown in the mechanism for malonyl-CoA formation (Solution 30), ATP is required to generate a carboxyphosphate intermediate. The result is that malonyl-CoA contains some of the free energy of ATP. This free energy is released when the acetyl group condenses with malonyl-ACP and CO_2 is released. The release of CO_2 is entropically favored and drives the reaction to completion.

33. Epinephrine signaling via an adrenergic receptor activates a G protein, which activates adenylate cyclase to produce cyclic AMP to activate cellular kinases, including protein kinase A (PKA). PKA phosphorylates its substrates, including acetyl-CoA carboxylase, thereby inactivating the enzyme. As a result of lower acetyl-CoA carboxylase activity, the rate of fatty acid synthesis drops. Epinephrine signaling also leads to phosphorylation and activation of glycogen phosphorylase (see Section 13.3), which mobilizes glucose from glycogen. These responses are consistent: When the cell needs to mobilize metabolic fuel (for example, by glycogenolysis), storage of fuel (for example, by fatty acid synthesis) is inhibited.

35. The citrate transporter (see Fig. 17.11), which helps move acetyl groups from the mitochondrial matrix to the cytosol, transports citrate to the cytosol. There, ATP-citrate lyase converts citrate to acetyl-CoA and oxaloacetate. Acetyl-CoA is then available for fatty acid synthesis in the cytosol. Since all of the acetyl units that will be used to synthesize fatty acids must be transported out of the mitochondrial matrix into the cytosol using this system, citrate levels are high when fatty acid synthesis is occurring at a high rate.

37. The synthesis of palmitate from acetyl-CoA costs 42 ATP. Seven rounds of the synthase reaction are required. ATP is required to convert each of 7 acetyl-CoA to malonyl-CoA for a total of 7 ATP. Two NADPH are required for seven rounds of synthesis, which is equivalent to $2 \times 7 \times 2.5 = 35$ ATP.

39. Mammalian fatty acid synthase differs structurally from bacterial fatty acid synthase; thus, triclosan can inhibit the bacterial enzyme but not the mammalian enzyme. Mammalian fatty acid synthase is a multifunctional enzyme made up of two identical polypeptides (see Fig. 17.12). In bacteria, the enzymes of the fatty acid synthetic pathway are separate proteins. Triclosan actually inhibits the bacterial enoyl-ACP reductase. The enzymes of the mammalian multifunctional enzyme must be arranged in such a way as to preclude the binding of triclosan to the active site of the enoyl-ACP reductase.

41. In mammals, fatty acid desaturation requires O_2 for dehydrogenation. Because bacteria use a different enzyme to introduce a double bond, inhibition of that enzyme could block bacterial growth without harming the mammalian host.

43. (a) Approximately 10,000 (104) cells will be resistant.
(b) The probabilities are multiplied:

$$\frac{1}{10^8} \times \frac{1}{10^8} \times \frac{1}{10^8} = \frac{1}{10^{24}}$$

In theory, it is highly unlikely that even one cell could survive the combination of three drugs.

45. DHA is an *n*-3 fatty acid and is essential for growth. A well-nourished mother will produce breast milk containing DHA, but a bottle-fed infant whose sole source of food is infant formula will not obtain this essential fatty acid unless it is added to the formula.

47. (a) Liver fatty acid synthase activity increases with consumption of a high-carbohydrate diet. Glucose in excess of what is required to meet immediate energy needs is oxidized to pyruvate by glycolysis, then converted to acetyl-CoA by pyruvate dehydrogenase. Excess acetyl-CoA is used to synthesize fatty acids, which are ultimately used to synthesize triacylglycerols for storage in adipose tissue.
(b) Liver fatty acid synthase activity is decreased with consumption of a high-fat diet. Endogenous fatty acid synthesis is not required if fatty acids are obtained from the diet.
(c) Mammary gland fatty acid synthase activity increases in mid to late pregnancy to provide fatty acids for triacylglycerols in breast milk for the neonate.

49. Fatty acid degradation produces acetyl-CoA, but in the absence of glucose, acetyl-CoA cannot enter the citric acid cycle due to insufficient oxaloacetate, a substrate for the first step. When carbohydrate concentrations drop, any available oxaloacetate is diverted to gluconeogenesis to produce the glucose necessary for red blood cell and brain function. If acetyl-CoA cannot enter the citric acid cycle (and acetyl-CoA cannot be converted to pyruvate in mammals), it is converted to ketone bodies.

51. Oxidation of fatty acids yields acetyl-CoA, which enters the citric acid cycle by reacting with oxaloacetate. Oxaloacetate is produced by the action of pyruvate carboxylase on pyruvate produced by glycolysis. Without glucose metabolism to generate pyruvate, the citric acid cycle cannot proceed. Thus, the statement "fats burn in the flame of carbohydrates" refers to the fact that normal carbohydrate metabolism (leading to the production of oxaloacetate) is absolutely required for normal fatty acid metabolism, which provides acetyl-CoA for the citric acid cycle.

53. A patient with a pyruvate carboxylase deficiency has elevated levels of pyruvate (see Solution 14.58) as well as lactate because the cells rely on anaerobic metabolism in the absence of proper citric acid cycle function. NADH is required to transform pyruvate to lactate (see Section 13.1), decreasing the cellular [NADH]/[NAD^+] ratio. NADH is required to convert acetoacetate to β-hydroxybutyrate (Reaction 4 in Figure 17.16), so the equilibrium of this reaction favors acetoacetate when cellular NADH concentrations are diminished. [From Saudubray, J. M., Marsac, C., Charpentier, C., Cathelineau, L., Besson Leaud, M., and Leroux, J. P., *Acta Pædiat. Scand.* **65**, 717–724 (1976).]

55. All cells can obtain glycerol-3-phosphate from glucose, because glucose enters glycolysis to form dihydroxyacetone phosphate, which is subsequently transformed to glycerol-3-phosphate via the glycerol-3-phosphate dehydrogenase reaction. Cells capable of carrying out gluconeogenesis, such as liver cells, can transform pyruvate to dihydroxyacetone phosphate. (Interestingly, adipocytes do not carry out gluconeogenesis but do express PEPCK and are thus able to convert pyruvate to glycerol-3-phosphate.)

57. The experimental results indicate that HMG-CoA reductase is regulated by phosphorylation. In normal cells, delivery of LDL particles results in an increase in cellular cholesterol, which stimulates a kinase that phosphorylates HMG-CoA reductase on the essential serine residue. The phosphorylated form of the enzyme is less active. In the mutant cells, the alanine cannot be phosphorylated, so the enzyme activity is unaffected by the presence of increasing cellular cholesterol.

59. Three phosphoanhydride bonds must be hydrolyzed in order to provide sufficient free energy to synthesize phosphatidylcholine from choline and diacylglycerol. Choline is activated by conversion to phosphocholine, with a phosphate group donated by ATP. The phosphocholine is subsequently converted to CDP–choline, a reaction in which a second phosphoanhydride bond is hydrolyzed. Pyrophosphate is the leaving group in this reaction, and the third phosphoanhydride bond is hydrolyzed when the pyrophosphate is hydrolyzed to orthophosphate.

61. The first enzyme in the pathway shown in Figure 17.19 is choline kinase, which catalyzes the ATP-dependent phosphorylation of choline to produce phosphocholine, which is needed for phosphatidylcholine synthesis. Increasing the activity of the enzyme that catalyzes the first step of the pathway allows the cancer cell to synthesize the membrane lipids required for cellular growth and proliferation.

63. (a) The enzyme transfers a methyl group from a donor molecule three times in order to convert phosphoethanolamine to phosphocholine.

$$^{2-}O_3P-O-CH_2-CH_2-NH_3^+ \longrightarrow {}^{2-}O_3P-O-CH_2-CH_2-N(CH_3)_3^+$$

Phosphoethanolamine   Phosphocholine

(b) The methyltransferase would be a suitable drug target only if the enzyme were unique to the parasite and was not expressed in humans. (In fact, this is the case.)

65. Ergosterol has two additional double bonds (one in a ring and one in the extended chain) and one additional methyl group.

67. (a) Insufficient apolipoprotein B-100 affects the production of LDL. Without enough of the protein portion of the lipoprotein, these individuals are less able to export triacylglycerols and other lipids from the liver in the form of LDL, and the lipids remain in the liver.
(b) LDL particles are a major vehicle for cholesterol transport from the liver to other tissues, so low levels of LDL would cause hypocholesterolemia.

Chapter 18

1. The ATP-induced conformational change must decrease $\mathcal{E}°'$ from -0.29 V to about -0.40 V. The decrease in reduction potential allows the protein to donate electrons to N_2, since electrons flow spontaneously from a substance with a lower reduction potential to a substance with a higher reduction potential. Without the conformational change, nitrogenase could not reduce N_2.

3. Leghemoglobin, like myoglobin, is an O_2-binding protein. Its presence decreases the concentration of free O_2 that would otherwise inactivate the bacterial nitrogenase.

5. (a)

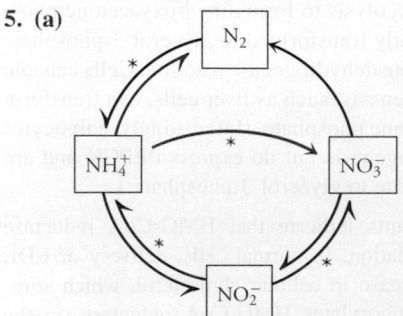

(b) Plants need a source of nitrogen in the form of ammonia or nitrate, so the organisms that carry out the processes marked with an asterisk in part a can potentially support plant growth.

7. (a) The low K_M of glutamine synthetase for ammonium ions ensures that the first method (6a) will be used when the concentration of available ammonia is low.
(b) The second method (6b) will be used when the concentration of ammonia is high.
(c) Incorporation of one mole of ammonia into an amino acid using the first method (6a) costs one mole of ATP, whereas ATP is not

required when the prokaryotic cell uses the second method (6b). The prokaryotic cell is at a disadvantage when the ammonia concentration is low because energy must be expended in order to synthesize amino acids under these conditions.

9. (a)

Glutamine synthetase   Glutamine synthetase
(more active)   (less active)

$$CH_2 \quad + \quad ATP \quad \longrightarrow \quad CH_2 \quad + \quad PP_i$$

(with OH below left CH₂; right side showing phosphate linked to adenosine)

(b) α-Ketoglutarate would inhibit adenylylation, preserving the enzyme in its unmodified, active form. Transamination reactions involving α-ketoglutarate produce glutamate, a reactant for the glutamine synthetase reaction.

11. (a)

Asp $\xrightarrow[\text{α-Ketoglutarate}]{\text{Glu}}$ Oxaloacetate

(b)

Ala $\xrightarrow[\text{α-Ketoglutarate}]{\text{Glu}}$ Pyruvate

(c)

Glu $\xrightarrow[\text{Oxaloacetate}]{\text{Asp}}$ α-Ketoglutarate

The products are all intermediates of the citric acid cycle (oxaloacetate and α-ketoglutarate) or closely associated with the citric acid cycle (pyruvate).

13. (a) leucine; **(b)** methionine; **(c)** tyrosine; **(d)** valine.

15. A malnourished alcoholic may not obtain sufficient vitamin B_6, the vitamin from which the PLP cofactor for the transaminase reaction is derived. In the absence of PLP, the transaminase is inactive. The ALT test may show low activity levels for serum transaminases, leading to the erroneous conclusion that the patient's liver is healthy. In order to obtain a more accurate result, the clinician running the ALT assay should add vitamin B_6 to the patient's serum sample.

17. (a)

Aspartate $\xrightarrow[\text{AMP}+PP_i]{NH_4^+ \quad ATP}$ Asparagine

(b) The reaction shows NH_4^+ rather than glutamate as the nitrogen donor. ATP is hydrolyzed to yield ATP and PP_i. This is similar to the glutamine synthetase reaction with the exception that ATP is hydrolyzed to ADP and P_i.

19. Cells that undergo rapid cell division (and thus need to replicate their DNA) require a rapid rate of nucleotide synthesis. Examples include skin cells, cells lining the gastrointestinal tract, and bone marrow cells. Tumor cells also have a high rate of cell division.

21.

Pyruvate $\xrightarrow{\textit{transamination}}$ Ala

3-Phosphoglycerate $\longrightarrow$ Ser $\longrightarrow$ Gly

Cys

α-Ketoglutarate $\xrightarrow{\textit{transamination}}$ Glu $\xrightarrow{\text{Gln synthetase}}$ Gln

Oxaloacetate $\xrightarrow{\textit{transamination}}$

Asp

Asn synthetase

Glu $\longrightarrow$ Arg

Asn

Pro

23. Cysteine is the source of taurine. The cysteine sulfhydryl group is oxidized to a sulfonic acid, and the amino acid is decarboxylated.

25. If an essential amino acid is absent from the diet, then the rate of protein synthesis drops significantly, since most proteins contain an assortment of amino acids, including the deficient one. The other amino acids that would normally be used for protein synthesis are therefore broken down and their nitrogen excreted. The decrease in protein synthesis, coupled with the normal turnover of body proteins, leads to the excretion of nitrogen in excess of intake.

27. **(a)** Tyrosine is not an essential amino acid because it can be synthesized from phenylalanine. As long as the diet contains sufficient amounts of phenylalanine, tyrosine is not essential.

(b) Phenylalanine hydroxylase catalyzes the transformation of phenylalanine to tyrosine, so if this enzyme were not functional, tyrosine would become essential and would need to be obtained in the diet.

29. **(a)** Lysine might regulate its own synthesis by acting as an allosteric inhibitor of dihydropicolinate synthase. This enzyme catalyzes the first unique reaction in a series of eight reactions that lead to lysine.

(b) Using the same reasoning, methionine might act as a feedback inhibitor for acyltransferase. Methionine might also regulate the activity of homoserine dehydrogenase, another major branch point of the pathway, even though this enzyme is also involved in threonine biosynthesis.

(c) Similarly, threonine might act as a feedback inhibitor for homoserine kinase and might also inhibit homoserine dehydrogenase.

31. **(a)** Arginine residues are converted to citrulline residues by a process of deamination (water is a reactant and ammonia is a product). Note that free citrulline produced by the urea cycle (Section 18.4) or in the generation of nitric oxide is not incorporated into polypeptides by ribosomes since there is no codon for this nonstandard amino acid.

(b) The nonstandard amino acid citrulline is not normally incorporated into polypeptides, so its presence appears foreign to the immune system, increasing the risk of triggering an autoimmune response.

33. Pyruvate can be transaminated to alanine, carboxylated to oxaloacetate, or oxidized to acetyl-CoA to enter the citric acid cycle. α-Ketoglutarate, succinyl-CoA, fumarate, and oxaloacetate are all citric acid cycle intermediates; they can also all enter gluconeogenesis. Acetyl-CoA can enter the citric acid cycle, be converted to acetoacetate, or be used for fatty acid synthesis. Acetoacetate is a ketone body and can be converted to acetyl-CoA for the citric acid cycle or fatty acid synthesis.

35. Threonine catabolism yields glycine and acetyl-CoA. The acetyl-CoA is a substrate for the citric acid cycle, which ultimately provides ATP for the rapidly dividing cell. Glycine is a source of one-carbon groups, which become incorporated into methylenetetrahydrofolate via the glycine cleavage system. THF delivers one-carbon groups for the synthesis of purine nucleotides and for the methylation of dUMP to produce dTMP; nucleotides are needed in large amounts in rapidly dividing cells. [From Wang, J., Alexander, P., Wu, L., Hammer, R., Cleaver, O., and McKnight, S. L., *Science* **325**, 435–439 (2009).]

37. The fate of propionyl-CoA produced by degradation of isoleucine is identical to that of propionyl-CoA produced in the oxidation of odd-chain fatty acids (see Fig. 17.7). Propionyl-CoA is converted to (*S*)-methylmalonyl-CoA by propionyl-CoA carboxylase. A racemase converts the (*S*)-methylmalonyl-CoA to the (*R*) form. A mutase converts the (*R*)-methylmalonyl-CoA to succinyl-CoA, which enters the citric acid cycle.

39. **(a)** Acetyl-CoA can enter the citric acid cycle if sufficient oxaloacetate is available; if not, excess acetyl-CoA is converted to ketone bodies. Leucine differs from isoleucine in that leucine is exclusively ketogenic, generating a ketone body and a ketone body precursor upon degradation. Isoleucine produces propionyl-CoA along with acetyl-CoA; the former can be converted to succinyl-CoA (see Solution 37) and then to glucose. Thus, isoleucine is glucogenic as well as ketogenic.

(b) Persons deficient in HMG-CoA lyase are unable to degrade leucine and must restrict this amino acid in their diets. A low-fat diet is also recommended because this same enzyme is involved in the production of ketone bodies (see Reaction 3 in Figure 17.16). A diet high in fat would generate a high concentration of acetyl-CoA, which could not be converted to ketone bodies in the absence of this enzyme.

41. **(a)** Insulin inhibits the enzyme, whereas glucagon stimulates the enzyme.

(b) In the presence of phenylalanine, the activity of the enzyme increases dramatically, more so in the presence of glucagon. Phenylalanine acts as an allosteric activator of phenylalanine hydroxylase and plays a role in converting the enzyme from the inactive dimeric form to the active tetrameric form.

(c) The incorporation of phosphate into the active form of phenylalanine hydroxylase indicates that the enzyme is regulated by phosphorylation as well as allosteric control. Glucagon signaling must lead to phosphorylation of the enzyme.

(d) Phenylalanine hydroxylase is most active when the glucagon concentration is high, corresponding to the fasting state. Under these circumstances, phenylalanine can be degraded to produce acetoacetate (a ketone body) and fumarate (which can be converted to glucose); both compounds provide necessary resources in the fasting state.

43. Persons with NKH lack a functioning glycine cleavage system. This is the major route for the disposal of glycine, and in its absence, glycine accumulates in body fluids. The presence of excessive glycine, a neurotransmitter, in the cerebrospinal fluid explains the effects on the nervous system.

45. The glutamate dehydrogenase reaction converts α-ketoglutarate to glutamate. In the presence of excess ammonia, α-ketoglutarate in the brain could be depleted, diminishing flux through the citric acid cycle.

47. Arginine allosterically stimulates *N*-acetylglutamate synthase. Arginine is the substrate for the urea cycle enzyme arginase, which catalyzes the hydrolysis of arginine, producing urea and regenerating ornithine so that the urea cycle can continue. Allosteric stimulation of carbamoyl phosphate synthetase by *N*-acetylglutamate provides carbamoyl phosphate that will enter the urea cycle, and the presence of arginine under these conditions ensures that sufficient ornithine will be available to react with the carbamoyl phosphate.

49. One nitrogen atom is derived from ammonia that is incorporated into carbamoyl phosphate for entrance into the urea cycle. The other nitrogen atom comes from aspartate, which serves as a substrate in the argininosuccinase reaction. Ultimately, both nitrogen atoms that appear in urea originated from excess dietary protein.

51. Glutamate dehydrogenase, glutamine synthetase, and carbamoyl phosphate synthetase can potentially "mop up" NH_4^+.

53. (a) A urea cycle enzyme deficiency decreases the rate at which nitrogen can be eliminated as urea. Since the sources of nitrogen for urea synthesis include free ammonia, low urea cycle activity may lead to high levels of ammonia in the body.
(b) A low-protein diet might reduce the amount of nitrogen to be excreted.

55. Adding arginine, the product of the argininosuccinase reaction, would increase flux through the urea cycle.

57. (a) The condensation of ammonia and α-ketoglutarate produces the amino acid glutamate.
(b) The reverse reaction, in which glutamate is deaminated to produce α-ketoglutarate, replenishes the citric acid cycle intermediate.

59. The amino acids released by protein degradation are used as metabolic fuels. The amino groups are removed and the nitrogen atoms are eventually excreted in the form of urea. The carbon skeletons can be completely catabolized to CO_2 by muscle cells to produce ATP via the citric acid cycle and oxidative phosphorylation, or they can be partially broken down and transported to the liver to be used in gluconeogenesis, which indirectly supplies the muscle cells with glucose.

61. (a) *H. pylori* urease converts urea to NH_3 and CO_2. The ammonia has a pK of 9.25, so it combines with protons to produce NH_4^+. The resulting decrease in hydrogen ion concentration helps the bacteria maintain an intracellular pH higher than the environmental pH.
(b) Urease on the cell surface increases the pH of the fluid surrounding the cell, creating a more hospitable microenvironment for bacterial growth.

63. In an autoimmune disease, the body's own white blood cells become activated to mount an immune response that leads to pain, inflammation, and tissue damage. The activity of the white blood cells, which proliferate rapidly, can be diminished by methotrexate, as occurs in rapidly dividing cancer cells.

65. This is an example of divergent evolution, as the three types of enzymes likely diverged from a common ancestor and have retained their basic structure and mechanism (see Section 6.4).

67. (a) A 10-fold increase in the K_M for APRT results in decreased affinity of the enzyme for one of its substrates. The result is the accumulation of adenine, which is oxidized to dihydroxyadenine and forms kidney stones. **(b)** One way to treat this condition is to administer a compound that would bind to xanthine dehydrogenase and prevent adenine from binding. This would prevent the conversion of adenine to dihydroxyadenine.

69. Phosphoribosyl pyrophosphate (PRPP) is a reactant in the salvage reactions involving IMP and GMP, so if these reactions cannot occur, the PRPP that would normally be used in the salvage reactions accumulates. PRPP stimulates amidophosphoribosyl transferase by feed-forward activation (see Solution 69), which accelerates the synthesis of purine nucleotides and increases the concentration of their degradation product, uric acid.

71. 5-Fluorouracil structurally resembles uracil and competitively inhibits thymidylate synthase. When the enzyme is inhibited, dUMP is not converted to dTMP and there is no subsequent phosphorylation to dTTP, so the concentration of dTTP in the cells declines. In the presence of the inhibitor, dUMP accumulates and is converted to dUTP. The concentration of dUTP in cells is normally low because thymidylate synthase acts quickly on dUMP to convert it to dTMP. Cancer cells die in the presence of 5-fluorouracil because the cells lack the dTTP required for DNA synthesis. Normal cells are not affected because DNA synthesis is not as rapid as in cancer cells.

Chapter 19

1. The two main metabolites at the "crossroads" are pyruvate and acetyl-CoA. Pyruvate is the main product of glycolysis. It can be converted to acetyl-CoA by pyruvate dehydrogenase. Pyruvate is produced from a transamination reaction involving alanine. Pyruvate can be carboxylated to oxaloacetate for gluconeogenesis. Acetyl-CoA is a product of fatty acid degradation and one of the reactants in the citric acid cycle. Acetyl-CoA is a product of the degradation of ketogenic amino acids. Acetyl-CoA can be used to synthesize fatty acids and ketone bodies.

3. The Na,K-ATPase pump requires ATP to expel Na^+ ions while importing K^+ ions, both against their concentration gradients (see Fig. 9.15). Inhibition by oubain reveals that the brain devotes half of its ATP production solely to power this pump. Because the brain does not store much glycogen, it must obtain glucose from the circulation. Glucose is oxidized aerobically in order to maximize ATP production.

5. (a) The reaction is probably a near-equilibrium reaction because the reactants and products have the same total number of phosphoanhydride bonds.
(b) In highly active muscle, ATP is rapidly converted to ADP. Adenylate kinase catalyzes the conversion of two ADP to ATP and AMP as a way to generate additional ATP to power the actin–myosin contractile mechanism (see Section 5.4).

7. Glucose enters glycolysis, which produces the ATP required for malonyl-CoA biosynthesis (see Section 17.3). The end product of glycolysis is pyruvate, which is converted to the starting material for fatty acid synthesis—acetyl-CoA—via the pyruvate dehydrogenase reaction (see Section 14.1). Glucose can also serve as the source of the NADPH required for fatty acid biosynthesis, via the pentose phosphate pathway (see Section 13.4).

9. The lactate dehydrogenase reaction, which reduces pyruvate to lactate, occurs with concomitant oxidation of NADH to NAD^+; the latter serves as a reactant in the glyceraldehyde-3-phosphate dehydrogenase reaction in glycolysis. If pyruvate were the end product of glycolysis, all of the cellular NAD^+ would become reduced to NADH and the glycolytic pathway would grind to a halt for lack of NAD^+.

11. Plasma alanine would be elevated as well. The increased plasma concentration of pyruvate would accelerate the operation of the glucose–alanine cycle. Pyruvate is taken up by the muscle and transaminated to alanine, which is then released to the circulation to be taken up by the liver. Since plasma pyruvate levels are elevated, alanine levels would also be elevated.

13. Increased levels of citrulline indicate that the cytosolic argininosuccinate synthetase reaction in the urea cycle (see Section 18.4) is not occurring normally. This reaction requires aspartate as a reactant in addition to citrulline. The accumulation of citrulline may occur because there is a shortage of aspartate. A deficiency of pyruvate carboxylase results in a decreased concentration of oxaloacetate that could otherwise be transaminated to aspartate. Hyperammonemia is the result of urea cycle impairment, since ammonia is not being converted to urea for excretion by the kidneys. [From Coude, F. X., Ogier, H., Marsac, C., Munnich, A., Charpentier, C., and Saudubray, J. M., *Pediatrics* **68,** 914 (1981).]

15. Fatty acid oxidation must be a major source of metabolic free energy during metamorphosis.

17. (a) *Phosphorolytic cleavage*

glycogen$_{(n\ residues)}$ + P$_i$ → glycogen$_{(n-1\ residues)}$ + glucose-1-phosphate

Hydrolytic cleavage

glycogen$_{(n\ residues)}$ + H$_2$O → glycogen$_{(n-1\ residues)}$ + glucose

(b) Phosphorolytic cleavage yields glucose-1-phosphate, which is negatively charged due to its phosphate group and cannot exit the cell via the glucose transporter. In addition, glucose-1-phosphate can be isomerized to glucose-6-phosphate (and can enter glycolysis) without the expenditure of ATP. Hydrolytic cleavage yields neutral glucose, which can leave the cell via the glucose transporter. Converting free glucose to glucose-6-phosphate so that it can enter glycolysis requires the expenditure of ATP in the hexokinase reaction.

19. Because glucokinase is not saturated at physiological glucose concentrations, it can respond to changes in glucose availability with an increase or decrease in reaction velocity. Consequently, the entry of glucose into glycolysis and subsequent metabolic pathways depends on the glucose concentration. Because hexokinase is saturated at physiological glucose concentrations, its rate does not change with changes in glucose concentration.

21. Insulin stimulates uptake of glucose via GLUT4 receptors in adipocytes. Glucose is converted to glycerol-3-phosphate via glycolysis and the glycerol-3-phosphate dehydrogenase reaction. The glycerol-3-phosphate is used as the backbone of the triacylglycerol molecule.

23. Phosphorylated GSK3 is inactive (see Solution 22) and unable to phosphorylate and inactivate glycogen synthase. In the presence of adiponectin, GSK3 phosphorylation is blocked so GSK3 remains active. GSK3 is thus able to phosphorylate and inactivate glycogen synthase. As a result, glycogen synthesis is inhibited and glycogen degradation can proceed.

25. (a) Stimulation of the rate of phosphate removal from glucose-6-phosphate explains why the glucose concentration increased and the glucose-6-phosphate concentration decreased. The phosphoenolpyruvate concentration increased because flux from dihydroxyacetone phosphate to phosphoenolpyruvate increased (the glycolytic reactions are all near equilibrium).

(b) Removing the phosphate from glucose-6-phosphate is the final step of gluconeogenesis (and glycogenolysis), allowing glucose to leave the liver. Increasing the rate of this step enhances the overall rate of gluconeogenesis. At the same time, glycolysis is inhibited because the very first step of glycolysis is the phosphorylation of glucose to glucose-6-phosphate. If phosphate groups are continually removed, glucose-6-phosphate cannot enter glycolysis and thus glycolysis is effectively inhibited. [From Ichai, C., Guignot, L., El-Mir, M. Y., Nogueira, V., Guigas, B., Chauvin, C., Fontaine, E., Mithieux, G., and Leverve, X. M., *J. Biol. Chem.* **276,** 28126–28133 (2001).]

27. Phosphorylase kinase is regulated by phosphorylation, which causes a conformational change that activates the enzyme. The enzyme is phosphorylated by protein kinase A (Figure 19.11), so the activity of the enzyme therefore depends somewhat on the G protein–coupled receptor pathway activated by either glucagon or epinephrine. But phosphorylase kinase is not fully active until calmodulin (a calcium-binding protein that is part of its structure) binds calcium and undergoes its own conformational change. The intracellular concentration of calcium increases when the phosphoinositide signaling system is activated (see Section 10.2), so the activation of phosphorylase kinase depends on this signaling pathway as well.

29. (a) AMPK increases the expression of GLUT4, which increases ATP production via glucose catabolism.

(b) AMPK decreases expression of the gluconeogenic enzyme glucose-6-phosphatase, since gluconeogenesis consumes cellular ATP.

31. Insulin activation of cAMP phosphodiesterase destroys cAMP so that it can no longer activate protein kinase A, the enzyme that activates phosphorylase kinase (as a step toward activating glycogen phosphorylase) and inactivates glycogen synthase. The result is that insulin promotes glycogen synthesis.

33. These two enzymes are part of the gluconeogenic pathway. Their concentrations increase when dietary fuels are not available so that the liver can supply other tissues with newly synthesized glucose.

35. If 3000 g of fat are utilized at a rate of 75 g/day, the fast can last for 40 days before death occurs.

37. After a few days of a diet low in carbohydrate, glycogen stores are depleted and the liver converts fatty acids to ketone bodies to be used as fuel for muscle and other tissues. Acetone is produced from the nonenzymatic decarboxylation of the ketone body acetoacetate. The relatively nonpolar acetone passes from blood capillaries into the lung alveoli, and its smell can be detected in exhaled breath.

39. (a) Leptin stimulates glucose uptake by skeletal muscle.

(b) Glycogenolysis is inhibited, probably by direct inhibition of glycogen phosphorylase.

(c) Leptin increases the activity of cAMP phosphodiesterase; the result is that cellular cAMP concentration decreases. In this way, leptin acts as a glucagon antagonist in the same manner that insulin does; glucagon's signal transduction pathway leads to an increase in cAMP concentration.

41. Phosphorylation and inactivation of GSK3 indirectly results in the activation of glycogen synthase, as described in Solution 22. Promoting the storage of glucose as glycogen in muscle and liver reverses the symptoms of insulin resistance.

43. (a) The cytochrome *c* content is high because of the large number of mitochondria that allow brown adipose tissue to oxidize metabolic fuels aerobically, funneling the reduced coenzymes through the electron transport chain. Uncoupling oxidative phosphorylation permits the energy of electron transport to be dissipated as heat (see Box 15.B).

(b) When subjects were exposed to cold, uptake of the labeled glucose into brown adipose tissue increased to provide reduced coenzymes for electron transport and thermogenesis.

45. Endogenous fatty acid synthesis is required when dietary fatty acid intake is insufficient. Stimulation of acetyl-CoA carboxylase ensures that the body will have enough fatty acids in the absence of dietary lipids (although essential fatty acids will still be lacking under these circumstances). During starvation (and untreated diabetes, which is similar to the starved state), body tissues do not have the resources to synthesize fatty acids, so acetyl-CoA carboxylase is inhibited. In the starved state, fatty acids are mobilized to provide fuel to body tissues.

47. (a) Acetate is converted to acetyl-CoA, which is converted to malonyl-CoA by acetyl-CoA carboxylase. Fatty acid synthase normally acts on malonyl-CoA for fatty acid synthesis, but in the presence of the inhibitor, fatty acid synthesis does not occur. Therefore, radioactive malonate accumulates, and virtually no label appears in fatty acids.

(b) Compound D causes hypothalamic levels of NPY to decrease. This decreases appetite, which leads to weight loss. In this manner, the inhibitor fools the body into thinking that it is in the fed state, when in fact it is being starved.

(c) Acetyl-CoA carboxylase (ACC) catalyzes the reaction that converts acetyl-CoA to malonyl-CoA. In the presence of the ACC inhibitor, malonyl-CoA cannot be produced and hepatic malonyl-CoA levels would decrease. Subsequent administration of Compound D would have no effect on the ACC inhibitor–treated animals, because Compound D's target is fatty acid synthase, which is downstream of ACC. If malonyl-CoA inhibits feeding, the ACC inhibitor–treated and Compound D–treated mice would have lower levels of malonyl-CoA, and feeding would not be inhibited, that is, the mice would eat normally. These experiments were carried out and this is in fact what occurred.

(d) When malonyl-CoA accumulates, long-chain fatty acyl-CoA molecules also accumulate in the cytosol. Malonyl-CoA inhibits carnitine acyltransferase, which prevents translocation of fatty acyl-CoAs into the mitochondrial matrix for β oxidation. It is possible that long-chain fatty acyl-CoAs serve as signaling molecules to begin the pathway that suppresses appetite. [From Loftus, T. M., Jaworksy, D. E., Grehywot, G. L., Townsend, C. A., Ronnett, G. V., Lane, M. D., and Kuhajda, F. P., *Science* **288**, 2379–2381 (2000).]

49. (a) In the absence of the phosphatase, the intracellular domain of the insulin receptor remains phosphorylated longer. The IRS-1 substrate may also retain its phosphate group, enabling it to stimulate the intracellular processes that allow glucose to be brought into cells via the GLUT4 transporter. In the absence of phosphatase activity to turn off the signaling pathway, the action of insulin is potentiated, and a normal physiological effect is observed with a lower concentration of the hormone.

(b) Injecting the mice with insulin would produce greater phosphorylation of the insulin receptor and possibly IRS-1, since there is no phosphatase to remove the phosphate groups.

(c) Compounds that inhibit the activity of the PTP-1B phosphatase might be effective in treating diabetes. A concern would be the specificity of these compounds, since unpredictable side effects would result if the drug inhibits other cellular phosphatases or if PTP-1B has other cellular targets unrelated to insulin signaling.

51. The bypass surgery eliminates a portion of the stomach and the small intestine, which are normally sources of hormones that regulate fuel metabolism and insulin sensitivity. The surgery apparently alters hormonal signaling enough to restore insulin sensitivity.

53. Metformin activates AMPK and, in so doing, promotes the phosphorylation of acetyl-CoA carboxylase, the enzyme that catalyzes the first committed step of lipid synthesis (see Section 17.2). Phosphorylation inactivates acetyl-CoA carboxylase, which decreases the concentration of malonyl-CoA and inhibits lipid synthesis. The decrease in malonyl-CoA relieves the inhibition of fatty acid transport and allows β oxidation to occur. The result is a decrease in the concentration of the plasma triacylglycerols that predispose the patient to atherosclerosis. Phosphorylation of protein kinase B by AMPK

would increase translocation of GLUT4 transporters to the plasma membrane. In this manner, AMPK is acting as insulin does, promoting the uptake of glucose into muscle and adipose tissue. Stimulation of AMPK by metformin benefits the patient with metabolic syndrome, who is resistant to insulin, because AMPK triggers the same responses as insulin signaling. [From Hardie, D. G., *FEBS Lett.* **582**, 81–89 (2008).]

55. The glycolytic intermediate 3-phosphoglycerate is used to synthesize serine, which can then be used to synthesize glycine and cysteine. Alanine can be synthesized from pyruvate in a one-step transamination reaction.

57. The transfer of a phosphate group from phosphoenolpyruvate to the active site histidine residue in the mutase produces pyruvate even in the absence of pyruvate kinase activity. The subsequent spontaneous hydrolysis of the phospho-His allows the mutase to remain active. This alternative conversion of phosphoenolpyruvate to pyruvate occurs without concomitant ATP production. This strategy allows the cancer cell to use glycolysis as a source of biosynthetic intermediates and not as an ATP-generating pathway. [From Vander Heiden, M. G., Locasale, J. W., Swanson, K. D., Sharf, H., Heffron, G. J., Amador-Noguez, D., Cristofk, H. R., Wagner, G., Rabinowitz, J. D., Asara, J. M., and Cantley, L. C., *Science* **329**, 1492–1499 (2010).]

59. (a) The negatively charged phosphate group makes glucose unable to fit in the glucose transporter.

(b) The lack of a hydroxyl group at C2 prevents the sugar from being isomerized to fructose-6-phosphate and continuing through the steps of glycolysis.

(c) The brain maintains a high metabolic rate and relies on glucose to supply its energy needs, so a significant portion of the FDG tracer accumulates in the brain. Since FDG is injected into the bloodstream, some of it is filtered out by the kidneys and is stored in the bladder until it can be eliminated. (For these reasons, PET scans using FDG are not suitable for monitoring brain or bladder cancers.)

61. AMPK stimulates pathways involved in ATP production, so the rates of glycolysis and fatty acid oxidation increase. Cancer cells normally rely on anaerobic metabolism to meet their energy needs, so the stimulation of AMPK has the effect of increasing aerobic metabolism, which increases the production of reactive oxygen species.

63. (a) The fumarase-deficient FH63 cells are similar to the control cells. Knocking out LDH increases apoptosis, but cells that lack both fumarase and LDH show the highest apoptosis (lowest survival).

(b) The FH63 cells from the renal cancer patients use an "aerobic glycolysis" survival strategy. In the absence of fumarase, the citric acid cycle cannot be completed to regenerate oxaloacetate, so the cell cannot oxidize glucose aerobically. Instead, glucose is oxidized to pyruvate and then reduced to lactate to generate NAD^+ to ensure the continued operation of glycolysis. Many types of tumor cells use aerobic glycolysis as a survival strategy, even in the presence of oxygen and a functioning citric acid cycle, but these renal cancer cells must use glycolysis due to the lack of a functioning citric acid cycle and an inability to generate ATP via oxidative phosphorylation. Cells lacking the LDH gene rely on the citric acid cycle exclusively, and it is interesting that this survival strategy is not as effective as the aerobic glycolysis used by the renal cancer cells. The survival of the renal cancer cells decreases when the LDH gene is knocked out because pyruvate cannot be converted to lactate with concomitant regeneration of NAD^+ in these cells. [From Xie, H., Valera, V. A., Merino, M. J., Amato, A. M., Signoretti, S., Linehan, W. M., Sukhatme, V. P., and Seth, P., *Mol. Cancer Ther.* **8**, 626–635 (2009).]

65. **(a)** The structure of 2-hydroxyglutarate is shown.
(b) Normal isocitrate dehydrogenase catalyzes an oxidative decarboxylation reaction to yield α-ketoglutarate (2-oxoglutarate), while the mutant enzyme catalyzes a decarboxylation reaction to yield 2-hydroxyglutarate.
(c) Production of this compound in the glioma cells depletes the cell's reduced cofactors, since a reduced coenzyme is required as a proton and an electron donor to form the hydroxyl group at the C2 position. [From Reitman, Z. and Yan, H., *J. Natl. Cancer Inst.* **102**, 932–941 (2010).]

$$
\begin{array}{c}
\text{COO}^- \\
| \\
\text{H}-\text{C}-\text{OH} \\
| \\
\text{CH}_2 \\
| \\
\text{CH}_2 \\
| \\
\text{COO}^-
\end{array}
$$

2-Hydroxyglutarate

Chapter 20

1. Parental ^{15}N-labeled DNA strands are shown in black, and newly synthesized ^{14}N DNA strands are shown in gray. The original ^{15}N-labeled parental DNA strands persist throughout succeeding generations, but their proportion of the total DNA decreases as new DNA is synthesized:

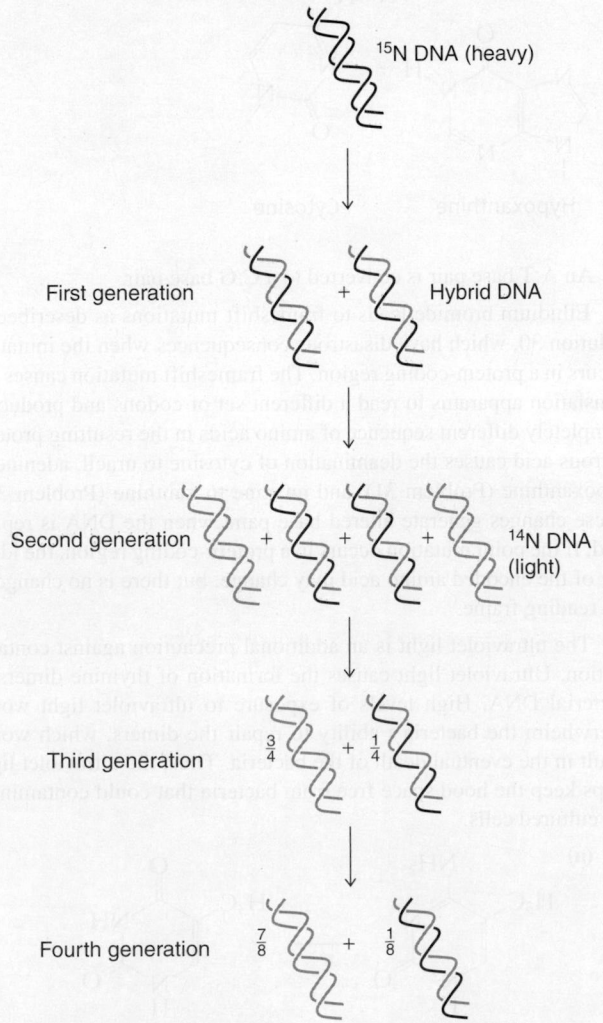

3. **(a)** Negatively supercoiled DNA is more easily unwound, and thus the separation of the DNA strands occurs more easily, which facilitates the replication process.
(b) The origin is more likely to be richer in A:T base pairs, since these experience fewer stacking interactions and are more easily separated, which would allow easier access for the replication proteins.

5. Helicase unwinds the parental DNA, exposing extensive segments of single-stranded DNA. Without a polymerase to convert these template strands into double-stranded DNA, the single strands would be susceptible to endonucleases, especially if the supply of replication protein A were limited.

7. **(a)** SSB, which coats single-stranded DNA exposed at the replication fork, has a relatively low affinity for DNA because it is displaced as the polymerase proceeds. The polymerase, along with accessory proteins such as the clamp, has a relatively high affinity for the DNA.
(b) The cell most likely contains large amounts of SSB in order to coat all the single-stranded DNA and prevent it from forming secondary structures. Although the cell requires multiple copies of SSB for each replication fork, in theory, only two DNA polymerase enzymes are required (one for the leading strand and one for the lagging strand).

9. The ddNTP lacks the 3′ hydroxyl group that serves as the attacking nucleophile for the incoming dNTP.

11. DNA polymerase α is least processive because it synthesizes only a short DNA segment before polymerase δ or ε takes over. DNA polymerase ε is most processive because it synthesizes the leading strand continuously. DNA polymerase δ has intermediate processivity because it synthesizes Okazaki fragments (~200 bp long).

13. First, the cell contains roughly equal concentrations of the four deoxynucleotide substrates for DNA synthesis; this minimizes the chance for an overabundant dNTP to take the place of another or for the wrong dNTP to take the place of a scarce dNTP. Second, DNA polymerase requires accurate pairing between the template base and the incoming base. Third, the 3′→5′ exonuclease proofreads the newly formed base pair. Fourth, the removal of the RNA primer and some of the adjacent DNA helps minimize errors introduced by primase and by the DNA polymerase at the 5′ end of a new DNA segment. Finally, DNA repair mechanisms can excise mispaired or damaged nucleotides.

15. PP_i is the product of the polymerization reaction catalyzed by DNA polymerase. This reaction also requires a template DNA strand and a primer with a free 3′ end. **(a)** There is no primer strand, so no PP_i is produced. **(b)** There is no primer strand, so no PP_i is produced. **(c)** PP_i is produced. **(d)** No PP_i is produced because there is no 3′ end that can be extended. **(e)** PP_i is produced. **(f)** PP_i is produced.

17. The DNA molecule (chromosome) is much longer than an Okazaki fragment and must be condensed and packaged in some way to fit inside the nucleus (in a eukaryote) or cell (in a prokaryote). If the cell waited until the entire DNA molecule had been replicated, the newly synthesized lagging strand, in the form of many Okazaki fragments, might already be packaged and inaccessible to the endonuclease, polymerase, and ligase enzymes necessary to produce a continuous lagging strand.

19.

$$E-Lys-NH_2 + N-R-O-P-O-P-O-R-A$$

$$\xrightarrow[1]{\quad NAD^+ \quad \rightarrow NMN^+ \quad}$$

$$E-Lys-NH_2-P-O-R-A$$

2

3 → AMP

21. (a) DNA polymerase; **(b)** reverse transcriptase or telomerase; **(c)** primase or RNA polymerase.

23. (a)

(b)

25. The resulting telomeres will have a sequence complementary to the mutated sequence of the telomerase-associated RNA template. This experiment was important because it established the mechanism of the enzyme and verified the role of the RNA template in extending chromosome length.

27. (a) transversion; **(b)** transversion; **(c)** transversion; **(d)** transition; **(e)** transversion; **(f)** transition.

29.

8-Oxoguanine   Adenine

31. (a)

Hypoxanthine

(b)

Hypoxanthine   Cytosine

(c) An A:T base pair is converted to a C:G base pair.

33. Ethidium bromide leads to frameshift mutations as described in Solution 30, which have disastrous consequences when the mutation occurs in a protein-coding region. The frameshift mutation causes the translation apparatus to read a different set of codons and produce a completely different sequence of amino acids in the resulting protein. Nitrous acid causes the deamination of cytosine to uracil, adenine to hypoxanthine (Problem 31), and guanine to xanthine (Problem 32). These changes generate altered base pairs when the DNA is replicated. If the point mutation occurs in a protein-coding region, the identity of the encoded amino acid may change, but there is no change in the reading frame.

35. The ultraviolet light is an additional precaution against contamination. Ultraviolet light causes the formation of thymine dimers in bacterial DNA. High levels of exposure to ultraviolet light would overwhelm the bacteria's ability to repair the dimers, which would result in the eventual death of the bacteria. Thus, the ultraviolet light helps keep the hood space free from bacteria that could contaminate the cultured cells.

37. (a)

5-Methylcytosine   Thymine

(b) Thymine. **(c)** A C:G to T:A transition occurs. **(d)** The cell has difficulty repairing the deaminated 5-methylcytosine since the resulting base is indistinguishable from thymine that occurs normally.

39. (a)

O^6-Methylguanine   Thymine

The methylation of guanine causes a G:C to A:T transition.

(b)

O^6-Methylguanine

No, since G normally base pairs with C, a mutation does not result. A mutation occurs only when O^6-methylguanine pairs with T. [From Leonard, G.A., Thomson, J., Watson, W.P., and Brown, T., *Proc. Natl. Acad. Sci. USA* **87,** 9573–9576 (1990).]

(c) During the course of the reaction, a methyl group from the O^6-methylguanine is transferred to a cysteine residue in the enzyme's active site, inactivating the enzyme. Normally, enzymes are regenerated after a cycle of catalysis.

41. Most likely the thymine–thymine dimer, since this lesion forms upon exposure of the DNA to ultraviolet light.

43. The mutant bacteria are unable to repair deaminated cytosine (uracil). In these cells, the rate of change of G:C base pairs to A:T base pairs is much greater than normal.

45. First, an endoribonuclease cleaves the phosphodiester bond on the 5′ side of the ribonucleotides. Next, an endonuclease such as RNase H removes the ribonucleotides. After a DNA polymerase replaces the excised ribonucleotides with deoxyribonucleotides, DNA ligase seals the nick.

47. (a) The polymerases that lack 3′ → 5′ exonuclease activity have higher error rates than polymerases that contain the exonuclease activity.
(b) α, once every 6250 bases; β, once every 1493 bases; δ, once every 100,000 bases; ε, once every 100,000 bases; η, once every 29 bases.
(c) Polymerase η and HIV RT incorporate the correct base with approximately the same efficiency (for example, 420 and 760 μM · min^{-1} × 10^3 for polymerase η and 800 μM · min^{-1} × 10^3 for HIV RT). However, polymerase η incorporates mispaired bases much more efficiently than HIV RT does (22, 1.6, and 8.7 μM · min^{-1} × 10^3 for polymerase η vs. 0.07 μM · min^{-1} × 10^3 for HIV RT). These results indicate that the high error rate of polymerase η results from its ability to incorporate the wrong base rather than its inability to incorporate the correct base.
(d) Overexpression of an error-prone DNA polymerase similar to polymerase η would increase the mutation rate. [From Matsuda, T.,

Bebenek, K., Masutani, C., Hanaoka, F., and Kunkel, T.A., *Nature* **404,** 1011–1013 (2000).]

49. MutS participates in mismatch repair by binding to the mismatched base pair and inducing a conformational change that causes the DNA to bend. This recruits an endonuclease that nicks the DNA and a helicase that peels away the strand containing the incorrect base. In the absence of functional MutS, mismatched base pairs are not detected and repaired, leading to mutations in genes involved in regulating cell growth in the colon and predisposing the individual to colon cancer.

51. Normally, the Rb protein acts as a tumor suppressor by preventing the cell from synthesizing DNA, a prerequisite for cell division. A mutation in the Rb protein may yield a protein that is unable to bind to and inhibit its target transcription factor. The transcription factor is thus free to induce the expression of genes required for DNA synthesis. This may result in the loss of control of the cell cycle, converting a non-cancerous cell to a cancerous cell. [From Giacinti, C. and Giordano, A., *Oncogene* **25,** 5220–5227 (2006).]

53. The same genetic changes that alter the growth behavior of a cancerous cell may also affect the expression of many proteins, resulting in the display of abnormal proteins on the cell surface.

55. According to the pathway described in the problem, overactivation of Ras would lead to a decrease in ubiquitinated p53. This would decrease the rate of p53 degradation, leaving more available to halt the cell cycle. This would actually counteract the growth-promoting activity of the Ras pathway.

57. A damaged unicellular organism does not harm others of its kind; it can go on to proliferate or die according to natural selection. In a multicellular organism, a cell that fails to respond to normal growth control signals could threaten the well-being of the whole organism. Thus apoptosis is necessary only in multicellular organisms.

59. DNA gyrase is a type II topoisomerase in *E. coli*. It can introduce negative supercoils into the DNA ahead of the replication fork. In the absence of DNA gyrase, strand separation would cause overwinding of the DNA ahead of the replication fork, generating positive supercoils that would hinder DNA unwinding.

61. A solution of 0.5 M NaCl effectively dissociates the positively charged histone proteins from the negatively charged DNA by disrupting the ionic bonds that hold these two chromatin components together. The charged sodium and chloride ions can form substitute ionic interactions with the histone protein and DNA.

63. Novobiocin and ciprofloxacin are useful as antibiotics because they inhibit prokaryotic DNA gyrase but not eukaryotic topoisomerases. They can kill disease-causing prokaryotes without harming host eukaryotic cells. Doxorubicin and etoposide inhibit eukaryotic topoisomerases and can be used as anticancer drugs. Although these drugs inhibit topoisomerases from both cancer cells and normal cells, cancer cells have a higher rate of DNA replication and are more susceptible to the effects of the inhibitors than are normal cells.

65. (a) In all of the variants, a neutral or negatively charged amino acid is replaced by a positively charged amino acid (Lys or Arg). It's reasonable to hypothesize that an enzyme with an increased number of positive charges might bind more tightly to the negatively charged DNA backbone.
(b) Intact, double-stranded DNA has a lower absorbance at 260 nm than does single-stranded DNA (see Fig. 3.7). An increase in absorbance at 260 nm over time is a useful measurement of the catalytic activity of DNase, since the products of the reaction are short, single-stranded oligonucleotides.
(c) All of the variants have lower K_M values and higher V_{max} values than the wild-type DNase, indicating that the variants catalyze the reaction more efficiently than the wild-type enzyme. The lower K_M values indicate tighter binding, possibly due to the formation of ion

pairs between the variant enzymes and the DNA. The replacement of three amino acids yields a variant enzyme in which the K_M and V_{max} values are optimized.

(d) The plasmid DNA normally exists in a supercoiled circle, as shown in the control lane. The wild-type DNase nicks the DNA on one strand to convert the plasmid to a relaxed circular DNA. The E13R/N74K/T205K (+3) mutant cleaves supercoiled DNA better than the E13R/T14K/N74K/T205K (+4) mutant. Both mutants produce linear DNA, whereas the wild-type DNase does not. This indicates that the variants can cut both strands, whereas the wild-type enzyme cuts only one strand.

(e) All of the mutants have a greater nicking activity than the wild-type for low- and high-molecular-weight DNA at low concentrations. The +2 mutant is especially active under these conditions and would be a good choice to use for a lupus patient. The +1 and +2 mutants can degrade low-molecular-weight DNA at a high DNA concentration better than the wild-type enzyme, but the +3 and +4 mutants are not as active. Not all of the data are given, so it is difficult to draw conclusions about the ability of the mutants to degrade high-molecular-weight DNA, but the +2 mutant is clearly more effective than the wild-type DNase and would be a good choice to use for a CF patient. [From Pan, C. Q., and Lazarus, R. A., *J. Biol. Chem.* **273**, 11701–11708 (1998).]

67. (a) About 146 bp of DNA are wrapped around a histone octamer to form a nucleosome. The nucleosomes are separated by about 20–40 bp of DNA. The "linker" DNA is more susceptible to hydrolysis, so brief treatment with the nuclease yields DNA fragments about 200 bp long.

(b) If the nuclease treatment is prolonged, the entire stretch of "linker" DNA is digested, leaving only the 146 bp segment of DNA that is wrapped around the histone octamer and thus protected from digestion.

Chapter 21

1. Messenger RNA is translated into protein, and a single mRNA can be used to translate many proteins. In this way, the mRNA is "amplified." Ribosomal RNA performs a structural role and is not amplified, so many more rRNA genes are needed to express sufficient rRNA to meet the needs of the cell.

3. Sequence-specific interactions require contact with the bases of DNA, which can participate in hydrogen bonding and van der Waals interactions with protein groups. Electrostatic interactions involve the ionic phosphate groups of the DNA backbone and are therefore sequence independent.

5. (a) The promoter region is shaded.

(b) Mutations causing the promoter sequence to more closely resemble the consensus sequence will increase the rate of transcription, whereas mutations that cause the promoter sequence to deviate from the consensus sequence will have the opposite effect.

(c) The –10 region is AT-rich. A:T base pairs have weaker stacking interactions (see Section 3.2) and are easier to melt apart than G:C base pairs, which have stronger stacking interactions. This facilitates DNA unwinding to expose the template for transcription.

AAAATAAATGCTTGACTCTGTAGCG-

+1

GGAAGGCGTATTATCCAACACCC

7. See Figure 21.6. A TATA box and an Inr site are identified.

+1

GAGCATATAAGGTGAGGTAGGATCAGTTGCTCCTCACATTT

TATA box Inr

9.

11. (a) The nonstandard amino acid is homocysteine, which can accept a methyl group donated by methyl-tetrahydrofolate to regenerate methionine (see Section 18.2).

(b) The other product is methanol, CH_3OH.

13. No, histones modified with ubiquitin are not marked for proteolytic destruction by the proteasome because the amino acid side chains of the histone proteins have only one ubiquitin attached. In order to enter the proteasome for degradation, a chain of at least four ubiquitin molecules is required (see Section 12.1).

15. The epigenetic marks that are present on the maternal and paternal DNA reflect the activities of their cells of origin (egg and sperm). The marks must be erased because the cells of the early embryo must be undifferentiated, that is, able to develop into any type of cell in the future adult.

17. Upregulation of the methylase increases the extent of methylation of Lys 4 (K4) and Lys 27 (K27) in histone 3. Since H3K4me3-associated genes normally tend to be in transcriptionally active chromatin (see Table 21.2), these genes are hyperactivated in cancer cells. The opposite occurs with H3K27me3-associated genes, which are normally transcriptionally inactive and are hypersilenced in cancer cells. [From Stark, G. R., Wang, Y., and Lu, T., *Cell Res.*, **21**, 375–380 (2011).]

19. (a) The polymerase binds most tightly to the DNA segment with the largest bulge. This DNA mimics the transcription bubble, in which the DNA strands are separated.

(b) Since K_d is a dissociation constant, the apparent equilibrium constant for the binding is $1/K_d$. The equilibrium constant can be determined by using Equation 12.2.

For fully based-paired DNA:

$$\Delta G^{\circ\prime} = -RT \ln(1/K_d)$$
$$\Delta G^{\circ\prime} = -(8.3145 \text{ J} \cdot \text{K}^{-1} \cdot \text{mol}^{-1})(298 \text{ K}) \ln(1/315 \times 10^{-9})$$
$$\Delta G^{\circ\prime} = -37,000 \text{ J} \cdot \text{mol}^{-1} = -37 \text{ kJ} \cdot \text{mol}^{-1}$$

For the eight-base bulge,

$$\Delta G^{\circ\prime} = -RT \ln(1/K_d)$$
$$\Delta G^{\circ\prime} = -(8.3145 \text{ J} \cdot \text{K}^{-1} \cdot \text{mol}^{-1})(298 \text{ K}) \ln(1/0.0013 \times 10^{-9})$$
$$\Delta G^{\circ\prime} = -68,000 \text{ J} \cdot \text{mol}^{-1} = -68 \text{ kJ} \cdot \text{mol}^{-1}$$

Polymerase binding to the eight-base bulge DNA (a mimic of melted DNA) is more favorable than binding to fully base-paired DNA.

(c) Melting open a DNA helix is thermodynamically unfavorable. Some of the favorable free energy of binding the polymerase to the DNA is spent in forming the transcription bubble. When the transcription bubble is preformed (for example, in the DNA with an eight-base bulge), this energy is not spent and is reflected in the apparent energy of polymerase binding. The difference in $\Delta G^{\circ\prime}$ values for polymerase binding to double-stranded DNA and to the eight base bulge is $-68 - (-37) = -31$ kJ $\cdot$ mol^{-1}. This value estimates the free energy cost ($+31$ kJ $\cdot$ mol^{-1}) of melting open eight base pairs of DNA. [From Bandwar, R. P. and Patel, S. S., *J. Mol. Biol.* **324**, 63–72 (2002).]

21. The lactose permease allows lactose to enter the cell, which increases the intracellular lactose concentration. Lactose can then serve as a substrate to form allolactose, which binds to the repressor protein to remove it from the operator. The presence of additional lactose assists in the full expression of the operon.

23. If the repressor cannot bind to the operator, the genes of the *lac* operon are constitutively expressed; that is, the genes are expressed irrespective of whether lactose is present or absent in the growth medium. Adding lactose has no effect on gene expression.

25. Wild-type cells cannot grow in the presence of phenyl-Gal. The wild-type cells produce a small amount of β-galactosidase in the absence of *lac* operon expression, but not enough to be able to cleave phenyl-Gal to phenol and galactose. The *lacI* mutants, however, will thrive in this growth medium. The mutation in the *lacI* gene results in the expression of a nonfunctional repressor (or perhaps no repressor); in any case, the *lac* operon is constitutively expressed and β-galactosidase is produced in sufficient amounts to act on phenyl-Gal to release galactose. The use of this growth medium permits selection of repressor mutants, since the mutants survive while the wild-type cells do not.

27. RNA polymerase I transcribes rRNA genes. Although there are multiple copies of these genes, there are only a few types of rRNA molecules and the cell requires them in equal amounts. Therefore, rRNA gene transcription does not need to be selective, and one promoter suffices. In contrast, the protein-coding genes transcribed by RNA polymerase II must be expressed at different times and at different levels. The regulation of transcription of these genes is more elaborate, requiring different promoter sequences that can potentially interact with different sets of transcription factors.

29. The 5′ end of any prokaryotic RNAs ending in G will be labeled. The 5′ ends of RNA transcripts have 5′ triphosphate groups containing the labeled γ-phosphate. The phosphodiester bonds in the RNA will not be labeled because these phosphate groups come from the α-phosphates of the nucleoside triphosphates (the γ-phosphates are released as pyrophosphate).

*pppG + pppN $\rightleftharpoons$ *pppGpN + PP$_i$

31. (a) Rifampicin inhibits the transition from RNA chain initiation to elongation. Normally, RNA polymerase initiates RNA synthesis repeatedly, releasing many short transcripts before committing to elongation. In the presence of rifampicin, the RNA polymerase cannot convert from initiation mode to elongation mode and remains bound to its promoter. Synthesis of longer RNA transcripts is not possible in the presence of the drug.
(b) The prokaryotic RNA polymerase is a five-subunit enzyme with a subunit composition of $\alpha_2\beta\beta'\omega\sigma$ and is the only RNA polymerase in prokaryotic cells. In contrast, eukaryotic cells have three different RNA polymerases that differ in structure from the prokaryotic polymerase. Thus, rifampicin can be effectively used as an antibiotic to treat bacterial diseases in eukaryotic organisms

because the drug is able to block transcription in the disease-causing prokaryotic cells without inhibiting transcription in the eukaryotic organism.

33. Differential binding to the ion exchange column indicates that the RNA polymerases have different affinities for DEAE, indicating that the amino acid sequences of the polymerases are different. The different sensitivities to Mg^{2+} ions and to the mushroom toxin also indicate differences in the mechanisms of the three enzymes, even though all three enzymes catalyze RNA polymerization.

35. The C-terminal domain (CTD) is phosphorylated when RNA polymerase transitions from initiation mode to elongation mode. If the CTD is missing, it cannot be phosphorylated, and elongation cannot occur. Transcripts that do manage to clear the promoter in the absence of a phosphorylated domain will not be properly processed, since the phosphorylated CTD serves as a docking site for enzymes involved in co-transcriptional modification.

37. (a) 5′ · · · NNAAICICCINNNNCCIICICUUUUUUNNN · · · 3′

(b)

```
        N     N
         N   N
          N   C
       N
         I --- C
         C --- I
         C --- I
         I --- C
         C --- I
         I --- C
         A --- U
         A --- U
  5′ · · · NN     UUUUNNN · · · 3′
```

The RNA terminator hairpin is much less stable when I is substituted for G because I:C base pairs have only two hydrogen bonds whereas G:C base pairs have three. The stability of the RNA hairpin is important in termination. When ITP is substituted for GTP in culture, the hairpin might not form; as a consequence, RNA transcripts will not terminate properly.

39. The β,γ-imido nucleoside triphosphate can be used as a substrate by RNA polymerase, since it is the phosphoanhydride bond closest to the ribose that is cleaved when the nucleotide is incorporated into the growing RNA transcript. Elongation is therefore not affected by the presence of the modified nucleotide. Rho-dependent termination is affected because Rho is a helicase that uses the energy of ATP hydrolysis to pry apart the DNA–RNA hybrid. The β,γ-imido nucleoside triphosphate cannot be hydrolyzed to its corresponding diphosphate and inorganic phosphate, so Rho-dependent termination cannot occur.

41. (a)

```
         U  A
        U     A
        U     G
        G --- C
        A --- U
        C --- G
        U --- A
 · · · GAU --- AGC · · ·
```

(b) Researchers could probe the conformation of an RNA molecule in the presence and absence of FAD by monitoring a conformation-dependent property of the RNA molecule, such as its susceptibility to an endonuclease. **(c)** The FMN component of FAD is the most effective, since it has the lowest dissociation constant. The phosphate group is important for RNA binding, since riboflavin, which lacks a phosphate group, binds much less tightly. [From Winkler, W. C., Cohen-Chalamish, S., and Breaker, R. B., *Proc. Nat. Acad. Sci. USA* **99**, 15908–15913 (2002).]

43. (a)

(b) Head-on collisions between DNA polymerase and RNA polymerase would halt both replication and transcription since advancement of the replication fork would be blocked by the RNA polymerase moving toward it. To resolve the stand-off, both enzymes would need to dissociate from the DNA, and replication would need to be reinitiated, which would be time-consuming. Cells in which transcription and replication occur in the same direction can undertake both processes more efficiently.

45. Bacterial mRNAs have a 5′ triphosphate group. The pyrophosphohydrolase removes two of the phosphoryl groups as pyrophosphate (PP_i), leaving a 5′ monophosphate (this apparently makes the mRNA a better substrate for the endonuclease).

47. (a) RNA; **(b)** DNA; **(c)** DNA; **(d)** DNA; **(e)** RNA; **(f)** DNA; **(g)** DNA; **(h)** RNA; **(i)** RNA; **(j)** DNA; **(k)** RNA.

49. The capped mRNA has a 5′–5′ triphosphate linkage, which is not recognized by exonucleases (which normally cleave 5′–3′ phosphodiester bonds). Capping must occur as soon as the 5′ end of the mRNA emerges from the RNA polymerase so that the message will not be degraded by exonucleases.

51. (a) The active site of poly(A) polymerase is narrower because it does not need to accommodate a template strand.
(b) Poly(A) polymerase binds only ATP, whereas other RNA polymerases bind ATP, CTP, GTP, and UTP.

53. Histone synthesis is required for only a very short period of time during the cell cycle when DNA is being replicated; thus stabilization of the mRNA coding for these proteins is not required.

55. Bacterial genes contain no introns, so the cells, which lack splicing machinery, cannot express eukaryotic genes containing introns and exons. Mature eukaryotic mRNA, which has already been spliced, contains only exons, which, after they have been converted back to DNA, can be transcribed and translated by the bacteria.

57. The mRNA from a gene may be alternatively spliced to yield several different types of proteins. This increases the diversity of the proteins produced by the cell without a correspondingly large number of genes.

59.

5′ splice site
↓
· · · CCCTGGGCAGGTTGGTA · · ·

3′ splice site
↓
· · · TTTCCCACCCTTAGGCTGCT · · ·

61. (a) Mutating U69 to G or C dramatically decreases activity as shown by the nearly 50-fold decrease in the value of k. Mutating the U residue to the similar-sized C residue has little effect on binding, but a mutation to the larger G residue has a much greater effect on binding as evidenced by the larger K_d for the U69G mutant. In general, catalysis seems to be affected more than substrate binding, indicating the importance of the bulge to catalysis.
(b) Deleting the U69 residue, which deletes the bulge, dramatically decreases activity but has little effect on substrate binding. Increasing the size of the bulge dramatically decreases both the activity and substrate binding affinity. These results confirm that the presence of the bulge is essential for catalysis and that the geometry of the bulge is important for substrate binding.

63.   5′ · · · G G A G U A C C C U G A U G A G A U C · · · 3′
3′ · · · C C U C A U G G G A C U A C U C U A G · · · 5′

[From Takei, Y., Kadomatsu, K., Yuzawa, Y., et al., *Cancer Res.* **64**, 3365–3370 (2004).]

65. The mutation could alter the promoter sequence to decrease the frequency of transcription; the mutation could increase the binding affinity for a repressor protein (or decrease the binding affinity for an activator); the mutation could introduce secondary structure into the mRNA that would hinder RNA polymerase or promote premature transcription termination; the mutation could slow the rate of mRNA splicing; the mutation could render the mRNA a better target for RNA interference.

Chapter 22

1. A hypothetical quadruplet code would have 4^4, or 256, possible combinations.

3. The poly(Lys) peptide results from translation of the poly(A) tail of a mRNA whose stop codon is missing as the result of a mutation or faulty transcription (AAA is a codon for lysine). With the addition of the poly(Lys) segment, the protein is likely to be nonfunctional, so it is best to destroy it and reuse its amino acids.

5. Because the genetic code is redundant, a mutation that alters a codon may not alter the amino acid encoded by that codon. This is particularly true for mutations at the third codon position. Changes at the first or second position almost always change the encoded amino acid, but the new amino acid may be chemically similar to the old one (for example, Val → Ala).

7. (a) The possible codons are UUU, UUA, UAA, UAU, AAA, AAU, AUU, and AUA. UUU codes for Phe, UUA for Leu, UAA is a stop codon, UAU codes for Tyr, AAA for Lys, AAU for Asn, and both AUU and AUA for Ile. Therefore, the amino acids in the polypeptide would be Phe, Leu, Tyr, Lys, Asn, and Ile. **(b)** The possible codons are UUU, UUC, UCC, UCU, CCC, CCU, CUU, and CUC. UUU and UUC code for Phe, UCC and UCU code for Ser, CCC and CCU for Pro, and CUU and CUC for Leu. Therefore, the polypeptide would contain Phe, Ser, Pro, and Leu. **(c)** The possible codons are UUU, UUG, UGG, UGU, GGG, GGU, GUU, and GUG. UUU codes for Phe, UUG for Leu, UGG for Trp, UGU for Cys, GGG and GGU for Gly, and GUU and GUG for Val. Therefore, the polypeptide would contain Phe, Leu, Trp, Cys, Gly, and Val. **(d)** These results show that the genetic code is redundant because all three templates result in the incorporation of Leu.

9. The result is an Ile–Asp–Arg tripeptide. Longer polypeptides such as the ones described in Solution 8 are not produced because UAG is a stop codon.

11. (a) For maximum efficiency in translation, one would expect that the highly expressed genes would contain the preferred codons, so that the cell would have high levels of the appropriate tRNAs to deliver the amino acids to make those proteins. For genes that are expressed more rarely, the delay that would result from using rare codons or rare tRNAs would probably not have a large impact on the cell. **(b)** Because different organisms have different codon-usage patterns, the gene's sequence may reveal whether it contains codons preferred by bacteria or codons preferred by mammals.

13. The 5′ nucleotide is at the wobble position, which can participate in non–Watson–Crick base pairings with the 3′ nucleotide of an mRNA codon. Because the first two codon positions are more important for specifying an amino acid (see Table 22.1), wobble at the third position may not affect translation.

15. CUG, CUA, and CUU. [From Sørensen, M. A., et al., *J. Mol. Biol.* **354**, 16–24 (2005).]

17. (a) The RNA-binding site of the synthetase is necessary for binding tRNA for aminoacylation and for binding RNA during splicing.
(b) Because of their long evolutionary history, aminoacyl–tRNA synthetases have had many opportunities to diversify in structure and in function in order to assume additional roles in the cell.

19. Gly and Ala; Val and Leu; Ser and Thr; Asn and Gln; Asp and Glu.

21. (a) The enzyme itself must act as a template to direct the addition of two C residues followed by one A residue.
(b) The enzyme must recognize only ATP and CTP as substrates, excluding GTP, UTP, and all dNTPs.
(c) In the two-domain enzyme, one polymerase adds the two C residues, and the other domain then adds the terminal A residue.

23. In order for a cell to incorporate a nonstandard amino acid into a polypeptide, the amino acid must first be attached to a tRNA corresponding to one of the 20 standard amino acids. The aminoacyl–tRNA can then bind to the ribosome and its amino acid can be incorporated into the growing polypeptide at positions corresponding to the codon for the standard amino acid. The failure of cells to synthesize norleucine-containing peptides most likely reflects the inability of LeuRS to efficiently attach norleucine to tRNA^{Leu}. A mutant LeuRS, which presumably lacks the proofreading activity of the wild-type LeuRS, was able to produce norleucine–tRNA^{Leu}, and the cells' ribosomes used this aminoacylated tRNA to translate Leu codons.

25. The assembly of functional ribosomes requires equal amounts of the rRNA molecules. Therefore, it is advantageous for the cell to synthesize the rRNAs all at once.

27. Ribosomal inactivating proteins catalyze the removal of adenine residues from ribosomal RNA. This is analogous to removing a side chain from an amino acid residue in a protein. Like proteins, ribosomal RNAs have specific residues that are essential to their function; removing these residues causes loss of activity.

29. The peptidyl transferase activity lies entirely within the 23S rRNA; that is, 23S rRNA is a ribozyme. The proteins might be necessary to assist the 23S rRNA in forming the necessary three-dimensional structure required for catalytic activity, just as the proper conformation is required for protein enzymes. Extremely strong intermolecular interactions between the proteins and the rRNA confirm the importance of the proteins and explain why the extraction process failed to remove them.

31. All three types of proteins have an RNA recognition motif because all of them bind to RNA. The ribosomal proteins bind to rRNA to form the ribosome. Rho factor is a bacterial transcriptional terminator that acts as a helicase to pry the nascent RNA away from its DNA template (see Section 21.2). The eukaryotic poly(A) binding protein binds to the poly(A) tail at the 3′ end of mRNA (see Fig. 21.19).

33. Molecular sequence data indicate that Bacteria split from a group that later diverged to form the Archaea and Eukarya. One

would therefore expect Archaea and Eukarya to be more similar in some ways than Archaea and Bacteria or Eukarya and Bacteria.

35. The polypeptide would contain all lysine residue with asparagine at the C-terminus. If the mRNA were read from 3′ → 5′, the polypeptide would consist of an N-terminal glutamine followed by a series of lysine residues. Transcription and translation both take place in the 5′ → 3′ direction; this allows bacterial cells to begin translating nascent mRNA before transcription is complete. If translation took place in the 3′ → 5′ direction, the ribosome would have to wait for mRNA synthesis to be completed before translation could begin.

37. The sequence on the 16S rRNA that aligns with the Shine–Dalgarno sequence is shown. The initiation codon is highlighted in gray.

$$5' \cdots \text{CUACCAGGAGCAAAGCUAAUGGCUUUA} \cdots 3'$$
$$3' \cdots \text{UCCUC} \cdots 5'$$

39. The presence of a 5′ cap and a 3′ poly(A) tail indicates that the RNA is messenger RNA that was transcribed by RNA polymerase II (see Section 21.2). These modifications help the ribosome to distinguish mRNA, which is a template for translation, from other types of RNA, which are not.

41. Colicin E3 is lethal to the cells because it prevents accurate and efficient translation. Cleavage of the 16S rRNA at A1493 destroys the part of the 30S ribosomal subunit that verifies codon–anticodon pairing (see Fig. 22.14). As a result, the ribosome is less able to incorporate the correct aminoacyl group into a growing polypeptide. In addition, EF-Tu hydrolysis of GTP is slow, because EF-Tu does not receive a signal from the ribosome that an mRNA–tRNA match has occurred, so the speed of translation decreases.

43. The correctly charged tRNAs (Ala–tRNA^{Ala} and Gln–tRNA^{Gln}) bind to EF-Tu with approximately the same affinity, so they are delivered to the ribosomal A site with the same efficiency. The mischarged Ala–tRNA^{Gln} binds to EF-Tu more loosely, indicating that it may dissociate from EF-Tu before it reaches the ribosome. The mischarged Gln–tRNA^{Ala} binds to EF-Tu much more tightly, indicating that EF-Tu may not be able to dissociate from it at the ribosome. These results suggest that either a higher or a lower binding affinity could affect the ability of EF-Tu to carry out its function, which would decrease the rate at which mischarged aminoacyl–tRNAs bind to the ribosomal A site during translation.

45. In a living cell, EF-Tu and EF-G enhance the rate of protein synthesis by rendering various steps of translation irreversible. They also promote the accuracy of protein synthesis through proofreading. In the absence of the elongation factors, translation would be too slow and too inaccurate to support life. These constraints do not apply to an *in vitro* translation system, which can proceed in the absence of EF-Tu and EF-G. However, the resulting protein is likely to contain more misincorporated amino acids than a protein synthesized in a cell.

47. The mRNA has the sequence

GCGAUAUGUGUGAACAUGUGAACUCUUAGAC

The start codon and stop codon are highlighted. The encoded protein has the sequence Met–Cys–Glu–His–Val–Asn–Ser.

49. To encode 1480 amino acids, 1480 codons or 4440 nucleotides (1480 × 3) are needed, plus a stop codon. The 1686 additional mRNA nucleotides (6129 − 4443) include segments at the 3′ and 5′ ends where translation factors and the ribosome bind.

51. In prokaryotes, both mRNA and protein synthesis take place in the cytosol, so a ribosome can assemble on the 5′ end of an mRNA even while RNA polymerase is synthesizing the 3′ end of the transcript. In eukaryotes, RNA is produced in the nucleus, but ribosomes are located in the cytosol. Because transcription and translation occur in separate compartments, they cannot occur simultaneously. A eukaryotic mRNA must be transported from the nucleus to the cytosol before it can be translated.

53. (a) The ribosome positions the peptidyl group for reaction with the incoming aminoacyl group, so a peptidyl group with a constrained geometry, like Pro, is unable to react optimally.
(b) Because Lys (with a positively charged side chain) reacts much faster than Asp (with a negatively charged side chain), the active site must be more accommodating of cationic groups than anionic groups.
(c) Transpeptidation of Ala is faster than for Phe, so for nonpolar amino acids, small size is more favorable. [From Wohlgemuth, I., Brenner, S., Beringer, M., and Rodnina, M. V., *J. Biol. Chem.* **283**, 32229–32235 (2008).]

55. Transpeptidation involves the nucleophilic attack of the amino group of the aminoacyl–tRNA on the carbonyl carbon of the peptidyl–tRNA (see Fig. 22.16). The higher the pH, the more nucleophilic the amino group (the less likely it is to be protonated).

57. (a) The mutation would allow the aminoacylated tRNA rather than a release factor to enter the ribosome and pair with a stop codon. The result would be incorporation of an amino acid into a polypeptide rather than translation termination, so the ribosome would continue to read mRNA codons and produce elongated polypeptides. The inability of the mutated tRNA to recognize its amino acid–specifying codons would have a minor impact on protein synthesis, since the cell likely contains other isoacceptor tRNAs that can recognize the same codons.
(b) Not all proteins would be affected. Only proteins whose genes include the stop codon that is read by the mutated tRNA would be affected. Proteins whose genes include one of the other two stop codons would be synthesized normally.
(c) Aminoacyl–tRNA synthetases usually recognize both the anticodon and acceptor ends of their tRNA substrates. A mutation in the tRNA anticodon, such as a nonsense suppressor mutation, might interfere with tRNA recognition, so the mutated tRNA molecule might not undergo aminoacylation. This would minimize the ability of the mutated tRNA to insert the amino acid at a position corresponding to a stop codon.

59. Anfinsen's ribonuclease experiment demonstrated that a protein's primary structure dictates its three-dimensional structure. Molecular chaperones assist in the protein-folding process but do not contribute any additional information regarding the tertiary structure of the protein. The purified ribonuclease was able to refold without the assistance of chaperones because other cellular components were absent. Chaperones are required *in vivo* because they prevent the interaction and aggregation of proteins and other cellular components.

61. Mitochondria contain ribosomes that synthesize proteins encoded by mitochondrial DNA. Like cytosolic proteins, mitochondrial proteins require the assistance of chaperones for proper folding. Other proteins are synthesized in the cytosol and are transported partially unfolded through pores in the mitochondrial membranes; these proteins also require the assistance of chaperones to fold properly once they reach their destination.

63. The different domains in a multidomain protein associate with one another via van der Waals forces, since the domain interfaces eventually end up in the interior of the protein. A cagelike chaperonin structure allows these proteins to fold in a protected environment where the hydrophobic regions of the protein are not exposed to other intracellular proteins with which they could potentially aggregate.

65. The basic residue is highlighted in gray; the hydrophobic core is underlined.

MKWVT<u>FISLLLLLFS</u>SAYSRGV

67. The hydrophobic cleft of this particular protein might allow it to recognize the hydrophobic core of the signal sequence. Other proteins in the SRP might be involved in pausing translation and docking the ribosome with the endoplasmic reticulum.

69. In the cell-free system, the SRP can bind to the exit tunnel of the ribosome, but translation is not arrested when no membrane is present. This indicates that the SRP must interact with both the nascent polypeptide and the ER membrane in order to pause translation. When microsomal membranes are subsequently added, the protein is not translocated, indicating that translocation must occur co-translationally, not post-translationally. Proteins that are not translocated retain their signal sequences because they do not have access to the signal peptidase, which is located in the microsomal lumen.

71. (a) **(b)**

Polyglutamine protein

(c) The acetyltransferase acetylates lysine residues in histones, neutralizing the positive charge of the lysine side chain and weakening its interaction with the DNA, so that transcriptional activity is increased. If the acetyltransferase is inactive, the DNA will be less transcriptionally active and certain genes will not be expressed. The loss of transcriptional activity could contribute to the progression of the polyglutamine disease. [From Pennuto, M., Palazzolo, I., and Poletti, A., *Hum. Mol. Gen.* **18**, R40–R47 (2009).]

73.

$$H_3C-(H_2C)_{12}-\overset{\overset{\displaystyle O}{\|}}{C}-\overset{\overset{\displaystyle H}{|}}{N}-\overset{\overset{\displaystyle |}{CH}}{\underset{\underset{\displaystyle H}{|}}{}}-\overset{\overset{\displaystyle O}{\|}}{C}-$$

Page references followed by T indicate tables. Page references followed by F indicate figures.

Common Functional Groups and Linkages in Biochemistry

Compound Name	Structure[a]	Functional Group
Amine[b]	RNH_2 or RNH_3^+ R_2NH or $R_2NH_2^+$ R_3N or R_3NH^+	$-N<$ or $^+\!N-$ (amino group)
Alcohol	ROH	$-OH$ (hydroxyl group)
Thiol	RSH	$-SH$ (sulfhydryl group)
Ether	ROR	$-O-$ (ether linkage)
Aldehyde	$\overset{\displaystyle O}{\overset{\|}{R-C-H}}$	$\overset{\displaystyle O}{\overset{\|}{-C-}}$ (carbonyl group), $\overset{\displaystyle O}{\overset{\|}{R-C-}}$ (acyl group)
Ketone	$\overset{\displaystyle O}{\overset{\|}{R-C-R}}$	$\overset{\displaystyle O}{\overset{\|}{-C-}}$ (carbonyl group), $\overset{\displaystyle O}{\overset{\|}{R-C-}}$ (acyl group)
Carboxylic acid[b] (Carboxylate)	$\overset{\displaystyle O}{\overset{\|}{R-C-OH}}$ or $\overset{\displaystyle O}{\overset{\|}{R-C-O^-}}$	$\overset{\displaystyle O}{\overset{\|}{-C-OH}}$ (carboxyl group) or $\overset{\displaystyle O}{\overset{\|}{-C-O^-}}$ (carboxylate group)
Ester	$\overset{\displaystyle O}{\overset{\|}{R-C-OR}}$	$\overset{\displaystyle O}{\overset{\|}{-C-O-}}$ (ester linkage)
Amide	$\overset{\displaystyle O}{\overset{\|}{R-C-NH_2}}$ $\overset{\displaystyle O}{\overset{\|}{R-C-NHR}}$ $\overset{\displaystyle O}{\overset{\|}{R-C-NR_2}}$	$\overset{\displaystyle O}{\overset{\|}{-C-N<}}$ (amido group)
Imine[b]	$R=NH$ or $R=NH_2^+$ $R=NR$ or $R=NHR^+$	$>C=N-$ or $>C=N<^H$ (imino group)
Phosphoric acid ester[b]	$\overset{\displaystyle O}{\overset{\|}{R-O-P-OH}}$ $\underset{\displaystyle OH}{}$ or $\overset{\displaystyle O}{\overset{\|}{R-O-P-O^-}}$ $\underset{\displaystyle O^-}{}$	$\overset{\displaystyle O}{\overset{\|}{-O-P-O-}}$ (phosphoester linkage) $\underset{\displaystyle OH}{}$ $\overset{\displaystyle O}{\overset{\|}{-P-OH}}$ or $\overset{\displaystyle O}{\overset{\|}{-P-O^-}}$ (phosphoryl group, P_i) $\underset{\displaystyle OH}{}$ $\quad$ $\underset{\displaystyle O^-}{}$
Diphosphoric acid ester[b]	$\overset{\displaystyle O \quad O}{\overset{\| \quad \|}{R-O-P-O-P-OH}}$ $\underset{\displaystyle OH \quad OH}{}$ or $\overset{\displaystyle O \quad O}{\overset{\| \quad \|}{R-O-P-O-P-O^-}}$ $\underset{\displaystyle O^- \quad O^-}{}$	$\overset{\displaystyle O \quad O}{\overset{\| \quad \|}{-O-P-O-P-O-}}$ (phosphoanhydride linkage) $\underset{\displaystyle OH \quad OH}{}$ $\overset{\displaystyle O \quad O}{\overset{\| \quad \|}{-P-O-P-OH}}$ or $\overset{\displaystyle O \quad O}{\overset{\| \quad \|}{-P-O-P-O^-}}$ $\underset{\displaystyle OH \quad OH}{}$ $\quad$ $\underset{\displaystyle O^- \quad O^-}{}$ (diphosphoryl group, pyrophosphoryl group, PP_i)

[a]R represents any carbon-containing group. In a molecule with more than one R group, the groups may be the same or different.

[b]Under physiological conditions, these groups are ionized and hence bear a positive or negative charge.

AMINO ACID STRUCTURES AND ABBREVIATIONS

Hydrophobic amino acids

Alanine (Ala, A)

Valine (Val, V)

Phenylalanine (Phe, F)

Tryptophan (Trp, W)

Leucine (Leu, L)

Isoleucine (Ile, I)

Methionine (Met, M)

Proline (Pro, P)

Polar amino acids

Serine (Ser, S)

Threonine (Thr, T)

Tyrosine (Tyr, Y)

Cysteine (Cys, C)

Asparagine (Asn, N)

Glutamine (Gln, Q)

Histidine (His, H)

Glycine (Gly, G)

Charged amino acids

Aspartate (Asp, D)

Glutamate (Glu, E)

Lysine (Lys, K)

Arginine (Arg, R)